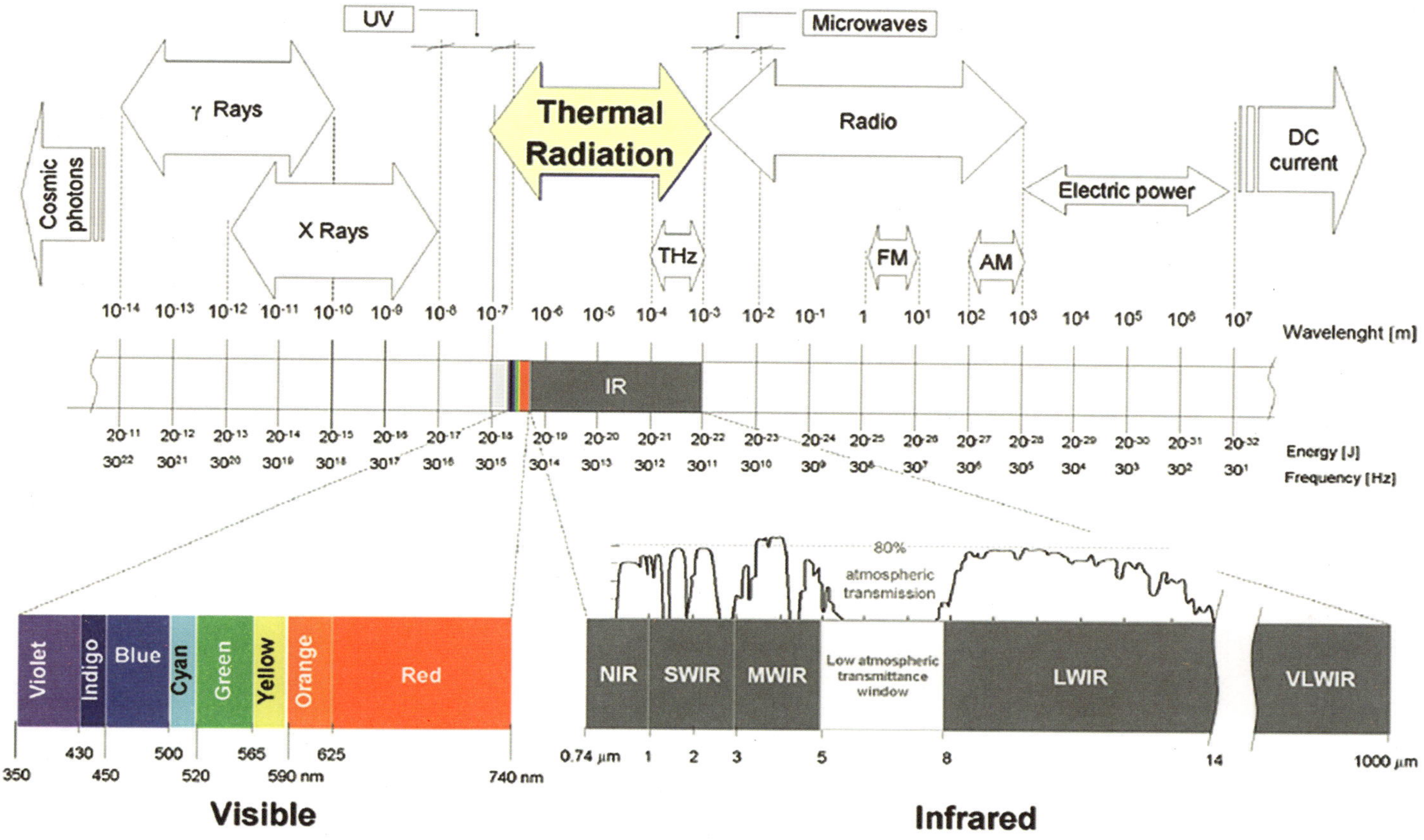
UV
Microwaves
γ Rays
Thermal Radiation
Radio
DC current
Cosmic photons
X Rays
Electric power
THz
FM
AM
Wavelenght [m]
IR
Energy [J]
Frequency [Hz]
80%
atmospheric transmission
Violet
Indigo
Blue
Cyan
Green
Yellow
Orange
Red
350
430
450
500
520
565
590 nm
625
740 nm
NIR
SWIR
MWIR
Low atmospheric transmittance window
LWIR
VLWIR
0.74 μm
1
2
3
5
8
14
1000 μm
Visible
Infrared

Third Edition

Fundamentals of MICROFABRICATION AND NANOTECHNOLOGY

VOLUME I

Solid-State Physics, Fluidics, and Analytical Techniques in Micro- and Nanotechnology

Third Edition

Fundamentals of MICROFABRICATION AND NANOTECHNOLOGY

VOLUME I

Solid-State Physics, Fluidics, and Analytical Techniques in Micro- and Nanotechnology

Marc J. Madou

CRC Press
Taylor & Francis Group
Boca Raton London New York

CRC Press is an imprint of the
Taylor & Francis Group, an **informa** business

CRC Press
Taylor & Francis Group
6000 Broken Sound Parkway NW, Suite 300
Boca Raton, FL 33487-2742

Printed and bound in India by Replika Press Pvt. Ltd.

International Standard Book Number: 978-1-4200-5511-5 (Hardback)

Visit the Taylor & Francis Web site at
http://www.taylorandfrancis.com

and the CRC Press Web site at
http://www.crcpress.com

I dedicate this third edition of Fundamentals of Microfabrication to my family in the US and in Belgium and to all MEMS and NEMS colleagues in labs in the US, Canada, India, Korea, Mexico, Malaysia, Switzerland, Sweden and Denmark that I have the pleasure to work with. The opportunity to carry out international research in MEMS and NEMS and writing a textbook about it has been rewarding in terms of research productivity but perhaps even more in cultural enrichment. Scientists have always been at the frontier of globalization because science is the biggest gift one country can give to another and perhaps the best road to a more peaceful world.

Contents

Roadmap

In *Solid-State Physics, Fluidics, and Analytical Techniques in Micro- and Nanotechnology* we lay the foundations for a qualitative and quantitative theoretical understanding of micro- and nanoelectromechanical systems, i.e., MEMS and NEMS. In integrated circuits (ICs), MEMS, and NEMS, silicon (Si) is still the substrate and construction material of choice. A historical note about the ascent of silicon, MEMS and NEMS is the topic of Chapter 1. The necessary solid-state physics background to understanding the electronic, mechanical, and optical properties of solids relied on in ICs, MEMS and NEMS is covered in Chapters 2–5. Many important semiconductor devices are based on crystalline materials because of their reproducible and predictable electrical properties. We cover crystallography in Chapter 2. The ultimate theory in modern physics today to predict physical, mechanical, chemical, and electrical properties of atoms, molecules, and solids is quantum mechanics. Quantum mechanics and the band theory of solids are presented in Chapter 3. The relevance of quantum mechanics in the context of ICs and NEMS cannot be underestimated, and the profound implications of quantum physics for nanoelectronics and NEMS are a recurring topic throughout this book. Given the importance of single-crystal Si (SCS) for IC, MEMS, and NEMS applications, we analyze silicon crystallography and band structure in more detail in Chapter 4. This chapter also elucidates all the single-crystal Si properties that conspired to make Si so important in electronic, optical, and mechanical devices that one might rightly call the second half of the 20th century the Silicon Age. Photonics, treated in Chapter 5, involves the use of radiant energy and uses photons the same way that electronic applications use electrons. We review the distinctive optical properties of bulk 3D metals, insulators, and semiconductors and summarize effects of electron and photon confinement in lower-dimensional structures. We show how evanescent fields on metal surfaces enable the guiding of light below the diffraction limit in plasmonics. Plasmonics is of growing importance for use in submicron lithography, near-field optical microscopy, enhancement of light/matter interaction in sensors, high-density data storage, and highly integrated optic chips. We also delve into the fascinating new topic of metamaterials, man-made structures with a negative refractive index, and explain how this could make for perfect lenses and could change the photonic field forever. In Chapter 6 we introduce fluidics, compare various fluidic propulsion mechanisms, and discuss the influence of miniaturization on fluid behavior. Given the high level of interest, fluidics for miniaturized analytical equipment is covered in this chapter as well. Chapter 7 combines a treatise on electrochemical and optical analytical processes whose implementation is often attempted in miniaturized components and systems.

Note to the Reader: *Solid-State Physics, Fluidics, and Analytical Techniques in Micro- and Nanotechnology* was originally composed as part of a larger book that has since been broken up into three separate volumes. *Solid-State Physics, Fluidics, and Analytical Techniques in Micro- and Nanotechnology* represents the first volume in this set. The other two volumes include *Manufacturing Techniques for Microfabrication and Nanotechnology* and *From MEMS to Bio-NEMS: Manufacturing Techniques and Applications*. Cross-references to these books appear throughout the text and will be referred to as Volume II and Volume III, respectively. The interested reader is encouraged to consult these volumes as necessary.

Author

Dr. Madou is the Chancellor's Professor in Mechanical and Aerospace Engineering (MEA) at the University of California, Irvine. He is also associated with UC Irvine's Department of Biomedical Engineering and the Department of Chemical Engineering and Materials Science. He is a Distinguished Honorary Professor at the Indian Institute of Technology, Kanpur, India and a World Class University Scholar (WCU) at UNIST in South Korea.

Dr. Madou was Vice President of Advanced Technology at Nanogen in San Diego, California. He specializes in the application of miniaturization technology to chemical and biological problems (bio-MEMS). He is the author of several books in this burgeoning field he helped pioneer both in academia and in industry. He founded several micromachining companies.

Many of his students became well known in their own right in academia and through successful MEMS start-ups. Dr. Madou was the founder of the SRI International's Microsensor Department, founder and president of Teknekron Sensor Development Corporation (TSDC), Visiting Miller Professor at UC Berkeley, and Endowed Chair at the Ohio State University (Professor in Chemistry and Materials Science and Engineering).

Some of Dr. Madou's recent research work involves artificial muscle for responsive drug delivery, carbon-MEMS (C-MEMS), a CD-based fluidic platform, solid-state pH electrodes, and integrating fluidics with DNA arrays, as well as label-free assays for the molecular diagnostics platform of the future.

To find out more about those recent research projects, visit http://www.biomems.net.

Acknowledgments

I thank all of the readers of the first and second editions of Fundamentals of Microfabrication as they made it worthwhile for me to finish this completely revised and very much expanded third edition. As in previous editions I had plenty of eager reviewers in my students and colleagues from all around the world. Students were especially helpful with the question and answer books that come with the three volumes that make up this third edition. I have acknowledged reviewers at the end of each chapter and students that worked on questions and answers are listed in the questions sections. The idea of treating MEMS and NEMS processes as some of a myriad of advanced manufacturing approaches came about while working on a WTEC report on International Assessment Of Research And Development In Micromanufacturing (http://www.wtec.org/micromfg/report/Micro-report.pdf). For that report we travelled around the US and abroad to visit the leading manufacturers of advanced technology products and quickly learned that innovation and advanced manufacturing are very much interlinked because new product demands stimulate the invention of new materials and processes. The loss of manufacturing in a country goes well beyond the loss of only one class of products. If a technical community is dissociated from manufacturing experience , such as making larger flat-panel displays or the latest mobile phones, such communities cannot invent and eventually can no longer teach engineering effectively. An equally sobering realization is that a country might still invent new technologies paid for by government grants, say in nanofabrication, but not be able to manufacture the products that incorporate them. It is naïve to believe that one can still design new products when disconnected from advanced manufacturing: for a good design one needs to know the latest manufacturing processes and newest materials. It is my sincerest hope that this third edition motivates some of the brightest students to start designing and making things again rather than joining financial institutions that produce nothing for society at large but rather break things.

Introduction

MEMS and NEMS Foundations

Miniaturization science is the science of making very small things. In top-down micro- and nanomachining, one builds down from large chunks of material; in bottom-up nanochemistry, one builds up from smaller building blocks. Both require a profound understanding of the intended application, different manufacturing options, materials properties, and scaling laws. The resulting three-dimensional structures, ranging in size from subcentimeters to subnanometers, include electronics, photonics, sensors, actuators, micro- and nanocomponents, and micro- and nanosystems.

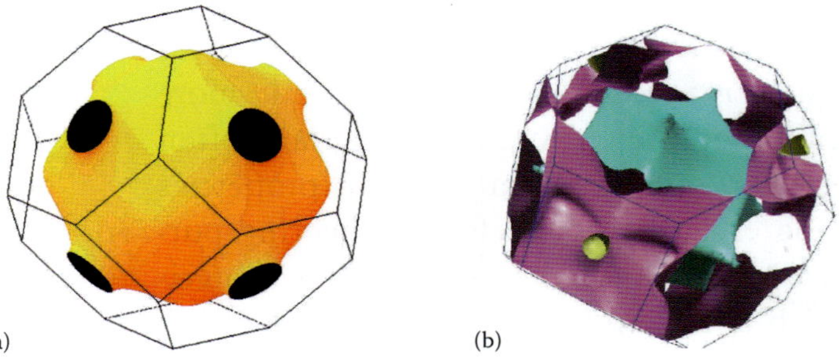

(a) (b)

(a) Copper Fermi surface—FCC sixth band with eight short necks touching the eight hexagonal zone faces. (Fermi surface database at http://www.phys.ufl.edu/fermisurface.) (b) Platinum Fermi surface—FCC fourth, fifth, and sixth bands. (Fermi surface database at http://www.phys.ufl.edu/fermisurface.)

No one behind, no one ahead. The path the ancients cleared has closed. And the other path, everyone's path, easy and wide, goes nowhere. I am alone and find my way.

Dharmakirti
(7th century India)

Introduction

Introduction

In Volume I, we lay the foundations for a qualitative and quantitative understanding of micro- and nanoelectromechanical systems, i.e., MEMS and NEMS. In integrated circuits (ICs), MEMS, and NEMS, silicon (Si) is still the substrate and building material of choice. A historical note about the history of the ascent of silicon, MEMS, and NEMS is the topic of Chapter 1.

The necessary solid-state physics background of electronic, mechanical, and optical properties of solids relied on in MEMS and NEMS is covered in Chapters 2–5. Solid-state physics is the study of solids. A major part of solid-state physics is focused on crystals because the periodicity of atoms in a crystal facilitates mathematical modeling, but more importantly because crystalline materials often have electrical, optical, or mechanical properties that can be easier exploited for engineering purposes. In Chapter 2, we detail crystalline materials in which atoms are arranged in a pattern that repeats periodically in three dimensions. The materials covered here prepare the reader for Chapter 3, which explains the band theory of solids based on quantum mechanics. The relevance of quantum mechanics in the context of ICs and NEMS cannot be underestimated, and the profound implications of quantum physics for nanoelectronics and NEMS is a recurring topic throughout this book. This is followed in Chapter 4 by a description of the single-crystal Si band structure, the growth of single crystals of Si, its doping, and oxidation. In this chapter, we also review the single-crystal Si properties that conspired to make Si so important in electronic, optical, and mechanical devices so that one might rightly call the second half of the 20th century the Silicon Age. Although the emphasis in this book is on nonelectronic applications of miniaturized devices, we briefly introduce different types of diodes and two types of transistors (bipolar and MOSFET). In Chapter 5, we introduce photonics. We compare electron and photon propagation in materials and contrast electron and photonic confinement structures and the associated evanescent wave phenomena. We also delve into the fascinating new topic of metamaterials, artificially engineered materials possessing properties (e.g., optical, electrical) that are not encountered in naturally occurring ones. An introduction to diode lasers, quantum well lasers, and quantum cascade lasers concludes the photonics section.

Fluidics and electrochemical and optical analytical techniques are important current applications of MEMS and NEMS. In Chapter 6 we introduce fluidics, compare various fluidic propulsion mechanisms, and discuss the influence of miniaturization on fluid behavior. Given the current academic and industrial interest, fluidics in miniaturized analytical equipment is detailed separately at the end of this chapter. Chapter 7 combines a treatise on

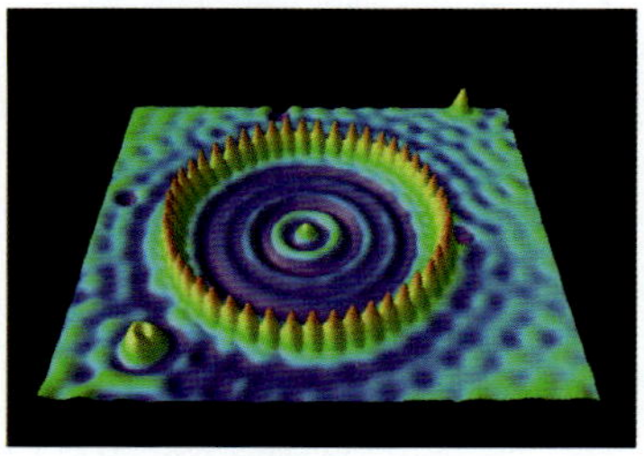

STM image showing standing waves in a 2D electron gas trapped in a "quantum corral" made by positioning Fe atoms on a Cu (111) surface.[1]

electrochemical and optical analytical techniques. Using sensor examples, we introduce some of the most important concepts in electrochemistry, i.e., the electrical double layer, potentiometry, voltammetry, two- and three-electrode systems, Marcus' theory of electron transfer, reaction rate- and diffusion rate-controlled electrochemical reactions, and ultramicroelectrodes. Many researchers use MEMS and NEMS to miniaturize optical components or whole instruments for absorption, luminescence, or phosphorescence spectroscopy. Optical spectroscopy is concerned with the production, measurement, and interpretation of electromagnetic spectra arising from either emission or absorption of radiant energy by matter. The sensitivity of these optical sensing techniques and the analysis of how amenable they are to miniaturization (scaling laws) are also compared herein. Chapter 7 ends with a comparison of the merits and problems associated with electrochemical and optical measuring techniques.

Reference

1. Crommie, M. F., C. P. Lutz, and D. M. Eigler. 1993. Confinement of electrons to quantum corrals on a metal surface. *Science* 262:218–20.

1

Historical Note: The Ascent of Silicon, MEMS, and NEMS

Silicon Valley is the only place on Earth not trying to figure out how to become Silicon Valley (Robert Metcalfe, father of the ethernet). (From PG&E and USGS.)

Outline

Silicon in Integrated Circuits

In 1879, William Crookes recounted his experiments on passing electric discharges through an evacuated glass tube for the Royal Society, thus describing the first cathode ray tube (CRT). Four years later, Thomas Alva Edison and Francis Upton discovered the "Edison effect." They introduced a metal plate into an incandescent electric light bulb (invented by Edison in 1879) in an attempt to keep the bulb from turning black (Figure 1.1). It did not work, but they discovered that there was a current between the lighted filament and the separate metal plate when the plate was positively charged but not when it was negatively charged. This led Edison and Upton to stumble on the basic principle of the operation of the vacuum tube (rectification!).

The first diode tube we owe to John Fleming, who, in 1904, filed a patent for a "valve" vacuum tube, also called a *Fleming valve* or *Fleming diode.*

FIGURE 1.1 Edison bulb used to demonstrate the "Edison effect."

FIGURE 1.3 Lee De Forest in 1906 with the Audion, the first triode.

Early researchers thought of electricity as a kind of fluid, leading to the inherited jargon such as current, flow, and valve. Fleming recognized the importance of Edison and Upton's discovery and demonstrated it could be used for the rectification of alternating currents. Interestingly, Fleming, at first, tried to get reliable rectification from single-crystal rectifiers used in crystal set radios (Figure 1.2), but could never get them to work well enough, so he switched to tubes!

J.J. Thomson, in 1887, convincingly showed that an electrical current was really an electron flow, and Fleming could explain the rectification in his diode tube as electrons boiling of the heated filament and flowing to the metal plate (thermionic emission). Because the plate was not hot enough to emit electrons, no current could go in the opposite direction. Thus, the Edison effect always produced direct current only. In 1906, an American scientist, Lee De Forest (Figure 1.3), invented the vacuum tube amplifier or triode based on the two-element vacuum tube invented by Fleming. De Forest, reportedly a tireless self-promoter, added an electrode—the *grid*—to the Fleming diode, and inserted it between the anode and the cathode. With this grid the diode became an active device, i.e., it could be used for the amplification of signals (say, for example, in radios) and as a switch (for computers). Hence, the amplifying vacuum tube, the ancestor of the transistor, was born. A gate in a dam controls huge amounts of flowing water with relatively small movements. Similarly, a small signal applied to the grid controls the much larger signal between anode and cathode.

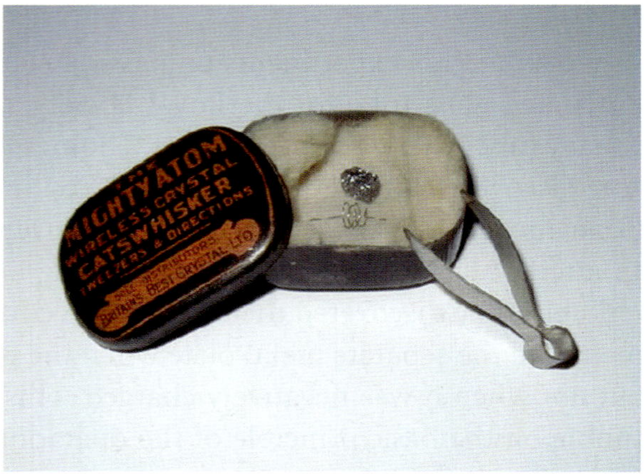

FIGURE 1.2 Early crystal set radio with a galena (lead sulphide) and the "cat's whisker" (the small coil of wire) that was used to make contact with the crystal.

Vacuum tubes—miniature particle accelerators—dominated the radio and television industries until the 1960s, and were the genesis of today's huge electronics industry. However, tubes were fragile, large, very power hungry, and costly to manufacture. The industry needed something better. That today's world is largely electronic—e.g., automobiles, home appliances, even books, writing tablets, and tally sheets—is because of solid-state electronics,* not vacuum tubes.

It is true, albeit unfortunate, that World War II and the subsequent Cold War era is what spurred research and development in solid-state electronic devices. Human foibles led to faster development of

* Based on or consisting chiefly or exclusively of semiconducting materials, components, and related devices.

RAdio Detecting And Ranging or RADAR, SOund Navigation And Ranging or SONAR, and many other technological innovations. As we all know, the list of innovations made to feed human aggression did not abate. Alan Turing led the team in England that in 1943 built the Colossus coding and deciphering machine. The Colossus was a special-purpose computer used to break the German code ULTRA, encrypted using ENIGMA machines. Breaking the German code was one of the keys to the success of the D-Day invasion. The Harvard Mark I and later II, III, and IV were general-purpose electromechanical calculators (sponsored by the U.S. Navy) to compute artillery and navigation tables—the same purpose intended, 100 years earlier, by Babbage for his analytical engine (Figure 1.4).

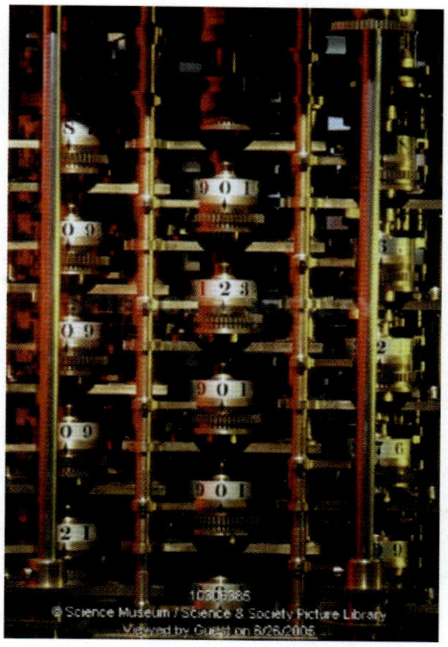

FIGURE 1.4 Charles Babbage (1791–1871) first conceived the idea of an advanced calculating machine to calculate and print mathematical tables in 1812. It was a decimal digital machine, with the value of a number being represented by the positions of toothed wheels marked with decimal numbers.

John Mauchly and Presper Eckert started work on the first electronic computer, the ENIAC (Electronic Numerical Integrator and Computer), at the University of Pennsylvania in 1943. The ENIAC, having been a secret during the war, was unveiled in Philadelphia in 1946. This computer featured 17,468 vacuum tubes used as switches and consumed 174 kW of power, enough to light 10 homes! Several tubes would fail every day until the engineers decided to never turn off the machine. This increased the average time until a tube would fail to 2 days. ENIAC was designed to calculate munition trajectory tables for the U.S. Army. It was U-shaped, 25 m long, 2.5 m high, 1 m wide, and weighed more than 30 tons (see Figure 1.5a). Programming was done by plugging cables and setting switches. By the mid-1970s, identical ENIAC functions could be achieved by a 1.5- × 1.5-cm silicon die, and the original Pentium processor, if fabricated using ENIAC technology, would cover more than 10 square miles.

One of ENIAC's heirs was a computer called the UNIVAC (Universal Automatic Computer), considered by most historians to be the first commercially successful digital computer (Figure 1.5b). First constructed by the Eckert-Mauchly Computer Corporation (EMCC), it was taken over by Sperry-Rand. At 14.5 ft. long, 7.5 ft. high, and 9 ft. wide, the UNIVAC, priced at $1 million, was physically

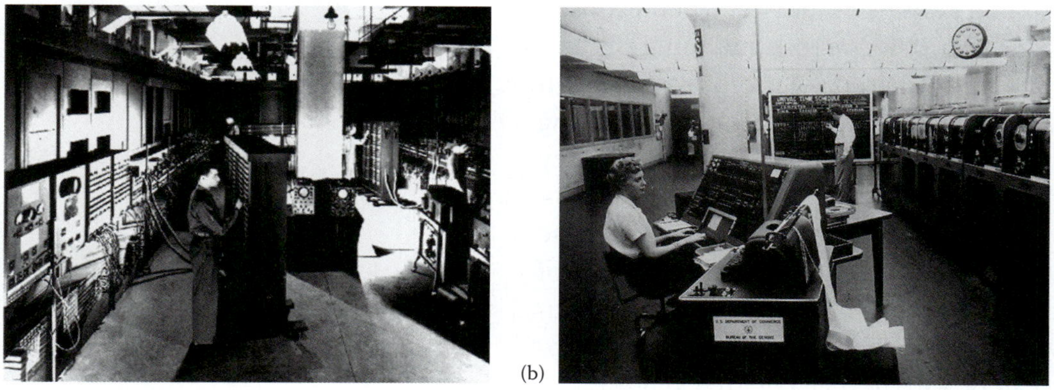

(a) (b)

FIGURE 1.5 Electronic Numerical Integrator and Computer (ENIAC), the world's first large-scale, general-purpose electronic computer (a), and the UNIVAC, the first commercial computer (b). First-generation computers based on vacuum tubes.

FIGURE 1.6 Von Neumann in his living room. (Photograph by Alan Richards hanging in Fuld Hall, Institute for Advanced Study, Princeton, NJ. Courtesy of the Archives of the Institute for Advanced Study.)

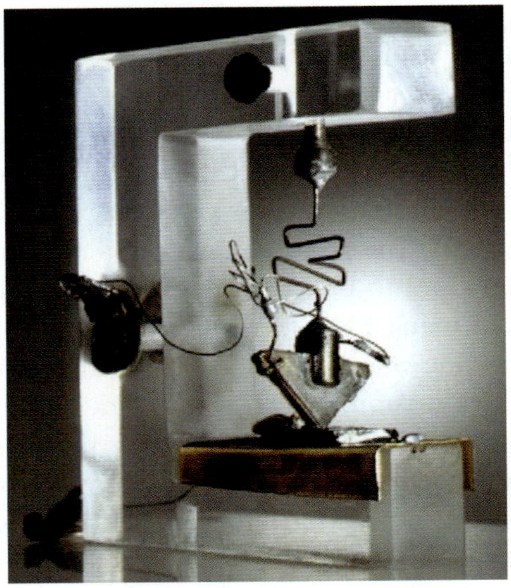

FIGURE 1.7 The first point-contact germanium bipolar transistor. Notice the paper clip! Roughly 50 years later, electronics accounted for 10% ($4 trillion) of the world's aggregate GDP.

smaller than ENIAC but more powerful. ENIAC and UNIVAC constitute first-generation computers based on vacuum tubes.

It was the concept of the stored program, invented by John von Neumann in 1945 (Figure 1.6); the magnetic core memory, invented by An Wang at Harvard and used in grids by J.W. Forrester and colleagues at MIT for random access memory (RAM); and William Shockley's transistor, based on transistors for switches instead of tubes, that would make a second generation of computers possible, thus starting the computer revolution.

The year 1940 gave rise to an important milestone in solid-state electronics history with the invention of a silicon-based solid-state p-n junction diode by Russell Ohl at Bell Labs.[1] This device, when exposed to light, produced a 0.5 V across the junction and represented the first Si-based solar cell. Bell Labs, in 1945, established a group charged with developing an alternative to the vacuum tube. Led by Shockley (1910–1989) and including John Bardeen (1908–1991) and Walter Brattain (1902–1987), in 1947 the group made an odd-looking device consisting of semiconducting germanium (Ge), gold strips, insulators, and wires, which they called a transistor (subject of U.S. Patent #2,524,035 [1950][2]) (Figure 1.7; notice the paper clip!). For this invention Shockley, Bardeen, and Brattain were awarded the 1956 Nobel Prize for Physics (Bardeen went on to claim a second Nobel Prize for Physics in 1972 for superconductivity). Bardeen called Ohl's junction diode fundamental to the invention of the transistor. Brattain was the unassuming experimentalist, Bardeen the theorist, and Shockley* the inventor and leader (Figure 1.8).

This trio thus succeeded in creating an amplifying circuit using a point-contact bipolar transistor that *trans-ferred* resistance (hence *transistor*). Two

FIGURE 1.8 Shockley (seated), Bardeen, and Brattain.

* Unfortunately, Shockley became associated with racist ideas and briefly pursued a U.S. Senate seat (as a Republican).

wires made contact with the germanium crystal near the junction between the p- and n-zones just like the "cat whiskers"* in a crystal radio set. A few months later, Shockley devised the junction transistor, a true solid-state device that did not need the whiskers of the point-contact transistor (see also Figure 1.2). Junction transistors were much easier to manufacture than point-contact transistors, and by the mid-1950s the former had replaced the latter in telephone systems. G. Teal and J.B. Little, also from Bell Labs, were able to grow large single crystals of germanium by 1951, which led to the start of commercial production of germanium transistors in the same year. Christmas 1954 saw the first transistor radio (the Regency TR-1) built by Industrial Development Engineering Associates, which sold for \$49.95 (Figure 1.9). This radio featured four germanium transistors from Texas Instruments. Although germanium was used in early transistors, by the late 1960s silicon, because of its many advantages, had taken over.

Silicon has a wider bandgap (1.1 eV for Si vs. 0.66 eV for Ge), allowing for higher operating temperatures (125–175°C vs. 85°C), a higher intrinsic resistivity (2.3×10^5 Ωcm vs. 47 Ωcm), and a better native oxide (SiO_2 vs. GeO_2 [water soluble!]).

FIGURE 1.9 Movie producer mogul Michael Todd (husband of Elizabeth Taylor in the mid-1950s) placed Regency TR-1s in gift books to commemorate his movie *Around the World in 80 Days*. The one pictured was for Shirley MacLaine.

The latter results in a higher-quality insulator that protects and "passivates" underlying circuitry, helps in patterning, and is useful as a mask for dopants. Finally, silicon is cheaper and much more abundant (sand!) than germanium. Second-generation computers relied on transistors instead of vacuum tubes for switches (logic gates). In recent years, germanium is making a comeback based mostly on its higher carrier mobility (three times higher than silicon-based ones), of great interest for faster circuitry, and because Ge has a lattice constant similar to GaAs, making it easier to integrate GaAs optical components with Ge-CMOS circuits.

Transistors perform functions similar to vacuum tubes, but they are much smaller, cheaper, less power hungry, and more reliable. Michael Riordan and Lillian Hoddeson's *Crystal Fire* gives, in the author's opinion, one of the best popular accounts of the invention of the transistor.[3]

The honeymoon with the transistor was quickly over; by the second half of the 1950s, new circuits on the drawing board were so big and complex that it was virtually impossible to wire that many different parts together reliably. A circuit with 100,000 components could easily require 1 million, mostly manual, soldering operations that were time consuming, expensive, and inherently unreliable. The answer was the "monolithic" idea, in which a single bloc of semiconductor is used for all the components and interconnects, invented by two engineers working at competing companies: Jack Kilby at Texas Instruments (Figure 1.10) and Robert Noyce (Figure 1.11) at Fairchild Semiconductor.

In 1958, Jack Kilby at Texas Instruments formed a complete circuit on a piece of germanium, landing U.S. Patent #3,138,743 (1959). His circuit was a simple IC oscillator with three types of components: transistors, resistors, and capacitors (Figure 1.12). Kilby got his well-deserved Nobel Prize for this work only in 2000. Technological progress and engineering feats are not often awarded a Nobel Prize, and if awarded at all they are often belated or controversial (see Kary Mullis, Nobel Laureate Chemistry 1993 for the invention of PCR).

Robert Noyce—Mr. Intel (Integrated Electronics)—then at Fairchild, introduced, with Jean Horni, planar technology, wiring individual devices together

* A cat whisker is a piece (often springy) of pointed metal wire.

FIGURE 1.10 Jack Kilby with notebook. (TI downloadable pictures.)

on a silicon wafer surface. Noyce's "planar" manufacturing, in which all the transistors, capacitors, and resistors are formed together on a silicon chip with the metal wiring embedded on the silicon, is still used today. By 1961, Fairchild and Texas Instruments had devised methods whereby large

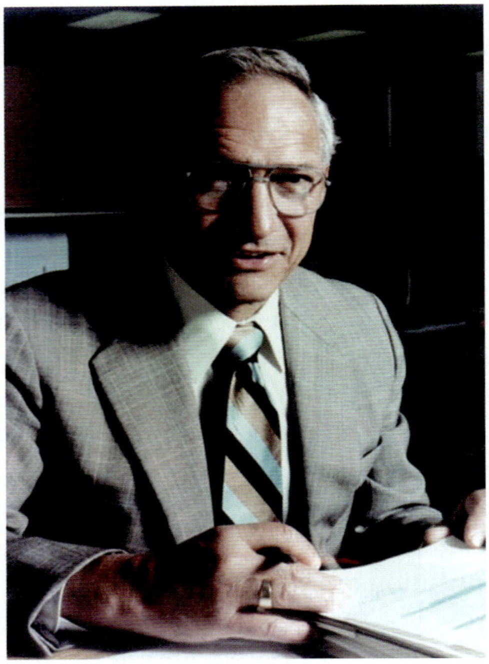

FIGURE 1.11 Robert Noyce in 1990.

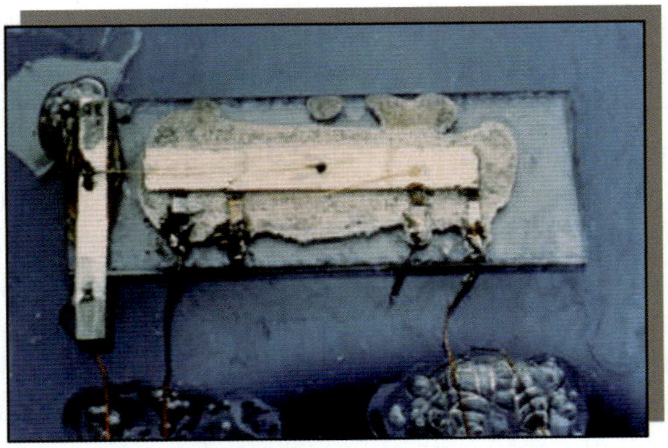

FIGURE 1.12 The first integrated circuit (germanium) in 1958 by Jack S. Kilby at Texas Instruments contained five components of three types: transistors, resistors, and capacitors.

numbers of transistors were produced on a thin slice of Si—and IC production on an industrial scale took off. The transistors on ICs were not the bipolar type but rather field effect transistor devices. The concept of a field effect transistor (FET) was first proposed and patented in the 1930s; however, it was the bipolar transistor that made it first to commercial products. Shockley resurrected the idea of the FET in the early 1950s, but it took until 1962 before a working FET was fabricated. These new FETs proved to be more compatible with both IC and Si technology.

Integrated circuits made not only third-generation computers possible but also cameras, clocks, PDAs, RF-IDs, etc. The National Academy of Sciences declared ICs the progenitor of the "Second Industrial Revolution," and Jerry Sanders, founder of Advanced Microdevices, Inc., called ICs the crude oil of the 1980s. A very well-written popular account of the invention of the IC is T.R. Reid's *The Chip: How Two Americans Invented the Microchip and Launched a Revolution.*[4]

Robert Noyce, Gordon Moore, and Andrew Grove left Fairchild to start Intel in 1968 with the aim of developing random access memory (RAM) chips. The question these inventors wanted answered was this: since transistors, capacitors, and resistors can be put on a chip, would it be possible to put a computer's central processor unit (CPU) on one? The answer came swiftly; by 1969 Ted Hoff had designed the Intel 4004, the first general-purpose 4-bit microprocessor. The Intel 4004 microprocessor was a 3-chip set with a 2-kbit read-only memory (ROM) IC, a 320-bit RAM

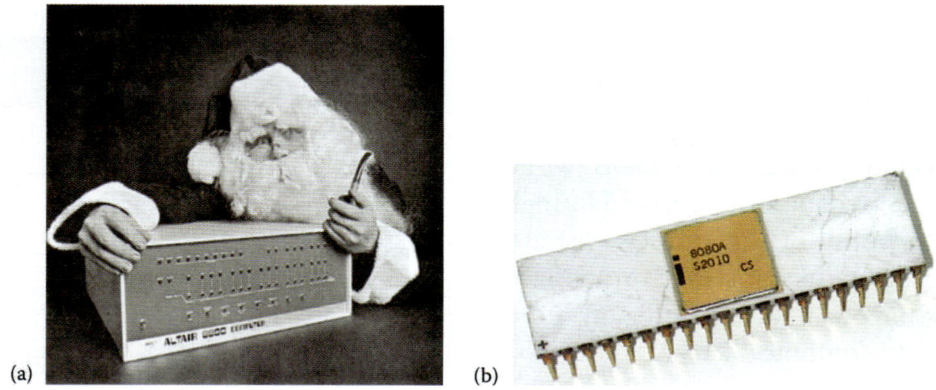

FIGURE 1.13 (a) The Altair 8800 computer and (b) the Intel 8080 microprocessor.

IC, and a 4-bit processor, each housed in a 16-pin dual in-line package (DIP). The processor, made in a 10-μm silicon gate pMOS process, contained 2,250 transistors and could execute 60,000 operations per second on a die size of 13.5 mm^2. It came on the market in 1971, giving rise to the fourth generation of computers based on microprocessors and the first personal computer (PC). The era of a computer in every home—a favorite topic among science fiction writers—had arrived!

The first desktop-size PC appeared in 1975, offered by Micro Instrumentation Telemetry Systems (MITS) as a mail-order computer kit. The computer, the Altair 8800, named after a planet on a *Star Trek* episode, retailed for $397. It had an Intel 8080 microprocessor, 256 bytes of memory (not 256K), no keyboard, no display, and no auxiliary storage device, but its success was phenomenal, and the demand for the microcomputer kit was overwhelming (Figure 1.13). Scores of small entrepreneurial companies responded to this demand by producing computers for the new market. In 1976, Bill Gates, Paul Allen, and Monte Davidoff wrote their first software program for the Altair—a BASIC (Beginners All-purpose Symbolic Instruction Code) interpreter (a high-level language translator that converts individual high-level computer language program instructions [source code] into machine instructions).

The first major electronics firm to manufacture and sell personal computers, Tandy Corporation (Radio Shack), introduced its computer model (TRS-80) in 1977. It quickly dominated the field because of the combination of two attractive features: a keyboard and a cathode ray display terminal. It was also popular because it could be programmed, and the user was able to store information by means of a cassette tape. In 1976, Steve Wozniak, who could not afford an Altair, built his own computer using a cheaper microprocessor and adding several memory chips. As a circuit board alone, it could do more than the Altair. Wozniak and Steve Jobs called it Apple I, and Jobs took on the task of marketing it while Wozniak continued with improvements (Figure 1.14). By 1977, Wozniak had built the Apple II and quit his day job. The Apple II had 16–64K RAM and secondary memory storage in the shape of a cassette tape or a 5.25-in. floppy disk drive and cost $1,300. At that time, Wozniak and Jobs formed Apple Computer,

FIGURE 1.14 (a) Jobs and Wozniak with the board for Apple I and (b) the Apple II.

Inc. When it went public in 1980, its stock value was $117 million; three years later it was worth $985 million.

Vacuum tubes coexisted with their progeny, the transistor, and even with ICs for a short while. Although solid-state technology overwhelmingly dominates today's world of electronics, vacuum tubes are still holding out in some areas. You might, for example, still have a CRT (cathode ray tube) as your television or computer screen. Tubes also remain in two small but vibrant areas for entirely different reasons. The first involves microwave technology, which still relies on vacuum tubes for their power-handling capability at high frequencies. The other—the creation and reproduction of music—is a more complicated story. Tubes distort signals differently than transistors when overdriven, and this distortion is regarded as being more "pleasant" by much of the music community.

Extrapolating back to 1961, Gordon Moore in 1965 (Figure 1.15), while at Fairchild, predicted that transistors would continue to shrink, doubling in density on an IC every 18–24 months, while the price would continue to come down—this prediction we know today as Moore's Law. History has proven Moore right as evidenced by past and projected feature sizes of ICs in the International Technology Roadmap for Semiconductors (ITRS) shown in Appendix 1A and on the Internet at http://public.itrs.net (updated in 2007).[5] In this International Technology Roadmap for Semiconductors, technology modes have been defined. These modes are the feature sizes that have to be in volume manufacturing at a fixed date (year of production). The feature size is defined as half-pitch, i.e., half of a dense pair of lines and spaces (see figure in Appendix 1A).

FIGURE 1.15 Gordon Moore, cofounder of Intel.

TABLE 1.1 Integration Scale and Circuit Density

IC Evolution	Acronym	Number of Logic Gates	Year of Introduction
Zero-scale integration	ZSI	1	1950
Small-scale integration	SSI	2–30	1965
Medium-scale integration	MSI	$30–10^3$	1970
Large-scale integration	LSI	$10^3–10^5$	1980
Very large-scale integration	VLSI	$10^5–10^7$	1985
Ultra-large-scale integration	ULSI	$10^7–10^9$	1990
Giga-scale integration	GSI	$10^9–10^{11}$	2005
Tera-scale integration	TSI	$10^{11}–10^{13}$	2020

In Table 1.1 the increasing numbers of devices integrated on an IC are tabulated. As we will learn in Chapters 1 and 2 on lithography in Volume II, new lithography techniques, novel device structures, and the use of new materials drive Moore's Law.

The state of the art in ICs today is a 2-GB DRAM* with 60-nm features (Samsung). Intel introduced the Core 2 Quad "Kentsfield" chip in January 2007, a chip featuring a 65-nanometer technology mode.

The 32-nm node should be achieved in 2009. The first Moore's Law is the good news, but there is a second Moore's Law that is a bit problematic; this second law states that the cost of building a chip factory doubles with every other chip generation, i.e., every 36 months. Today's technology involves Si wafers with a 12-in. diameter and factories that cost $3–4 billion to construct. With this type of start-up costs, few countries can afford to enter the IC market, and the search is on for alternative, less-expensive

* Dynamic random access memory. A type of memory component used to store information in a computer system. "Dynamic" means the DRAMs need a constant "refresh" (pulse of current through all the memory cells) to keep the stored information.

bottom-up NEMS techniques (below). From Appendix 1B, the IC market for 2003 was $166 billion worldwide, with 2004 projected at $241 billion (data by the World Semiconductor Trade Statistics [WSTS]; http://www.wsts.org).[6] Also from Appendix 1B we learn that the IC business is feeding a trillion-dollar electronic equipment business. It is worth pointing out that China is expected to control 5% of the IC market by 2010.

MEMS

Single-crystal silicon is not only an excellent electronic material but it also exhibits superior mechanical properties; the latter gave birth to the **m**icro**e**lectro**m**echanical **s**ystems (MEMS) field on the coattails of the IC industry. Originally MEMS constituted mostly mechanical types of devices based on single-crystal silicon with at least one or more of their dimensions in the micrometer range. As MEMS applications broadened, in Europe the acronym MST for *microsystem technology* became more popular. In Japan one refers to *micromachining*, and *microengineering* is popular in the United Kingdom. Development of single-crystal silicon mechanical MEMS involves the fabrication of micromechanical parts, e.g., a thin membrane in the case of a pressure sensor or a cantilever beam for an accelerometer. These micromechanical parts are fabricated by selectively etching areas of a Si substrate away to leave behind the desired geometries. The terms *MEMS* and *micromachining* came into use in 1982 to name these fabrication processes. Around the same time references to "bulk" micromachining techniques also appeared. Richard Feynman's December 26, 1959 presentation "There's Plenty of Room at the Bottom" is considered by many to be the starting bongo for MEMS (Figure 1.16) (http://www.its.caltech.edu/~feynman/plenty.html),[7] but in a practical sense, it was the invention of the transistor and the processes developed to fabricate transistors, six years earlier, that enabled MEMS.

An early milestone for the use of single-crystal silicon in MEMS was the 1956 discovery of porous Si by Uhlir.[8] His discovery eventually led to all types of interesting new, single-crystal Si-based devices, from reference electrodes for electrochemical sensors, biosensors, quantum structures, and permeable membranes to photonic crystals and photoluminescent and electroluminescent devices.

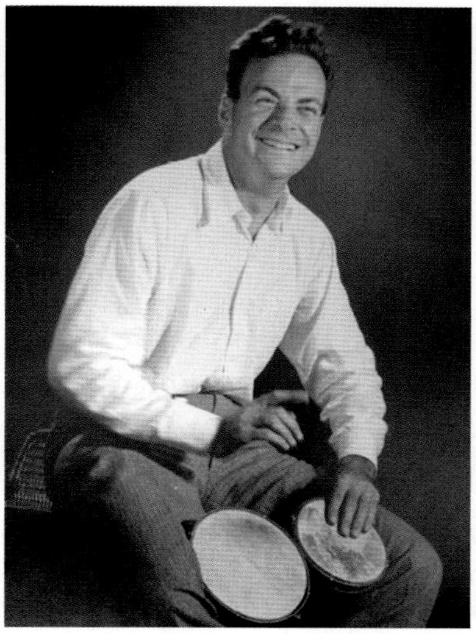

FIGURE 1.16 Richard Feynman on the bongo drums.

The first impetus for the use of single-crystal silicon as a micromechanical element in MEMS can be traced to the discovery of its large piezoresistance. Piezoresistance is the change in the resistivity of certain materials as a result of an applied mechanical strain. Charles Smith, of the Case Institute of Technology (now part of the Case Western Reserve University), during a sabbatical leave at Bell Labs in 1953, studied the piezoresistivity of semiconductors and published the first paper on the piezoresistive effect in Si and Ge in 1954.[9] The piezoresistive coefficients Smith measured demonstrated that the gauge factor* of Si and Ge strain gauges (see Figure 1.17) was 10–20 times larger than the gauge factor of metal film strain gauges, and, therefore, semiconductor gauges were expected to be much more sensitive.

Motivated by these results, companies such as Kulite and Honeywell started developing Si strain gauges commercially from 1958 onward. Pfann and colleagues, in 1961, proposed a dopant diffusion technique for the fabrication of silicon piezoresistive sensors for the measurement of stress, strain, and pressure.[10] Based on this idea, Kulite integrated

* A strain gauge is a device used to measure deformation (strain) of an object. The gauge factor of a strain gauge relates strain to change in electrical resistance.

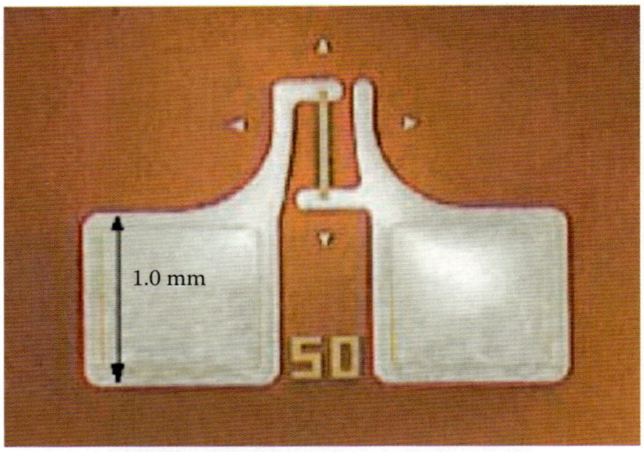

FIGURE 1.17 The Si single-crystal gauge element can be seen as the vertical bar centered between the two solder pads. The single-crystal silicon strain gauge offers sensitivities 20–50 times greater than metal foil gauges. A microphotograph of the LN-100. (BF Goodrich Advanced Micro Machines.)

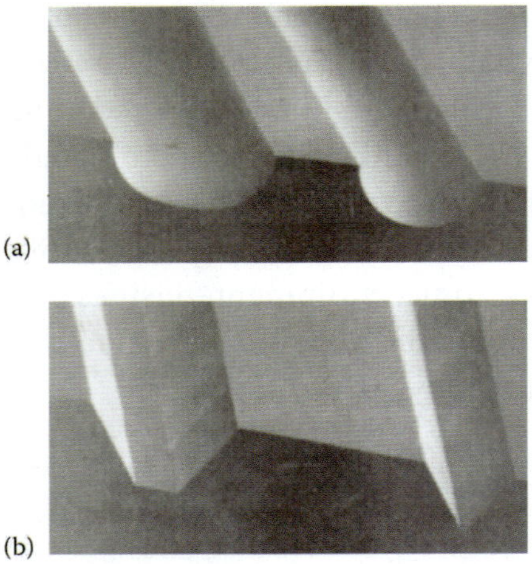

FIGURE 1.19 Isotropic and anisotropic etching profiles in single-crystal Si. Isotropic etching of grooves in (100) Si (a) and anisotropic etching of grooves in (100) Si (b) using rectangular mask openings. Features are in the 100-μm range.

Si strain gauges on a thin Si substrate with diffused resistors in 1961. As early as 1962, Tufte and coworkers at Honeywell, using a combination of wet isotropic etching, dry etching (using a plasma instead of a solution), and oxidation, made the first thin Si piezoresistive diaphragms for pressure sensors.[11] Isotropic etching of Si had been developed earlier for transistor fabrication. In the mid-1960s, Bell Labs started work on single-crystal silicon etchants with directional preferences, i.e., anisotropic etchants, such as mixtures of, at first, KOH, water, and alcohol and later KOH and water.[12] Both chemical and electrochemical anisotropic etching methods were pursued. The aspect ratio (height-to-width ratio) of features in MEMS is typically much higher than in ICs (Figure 1.18). The first high-aspect-ratio cuts in

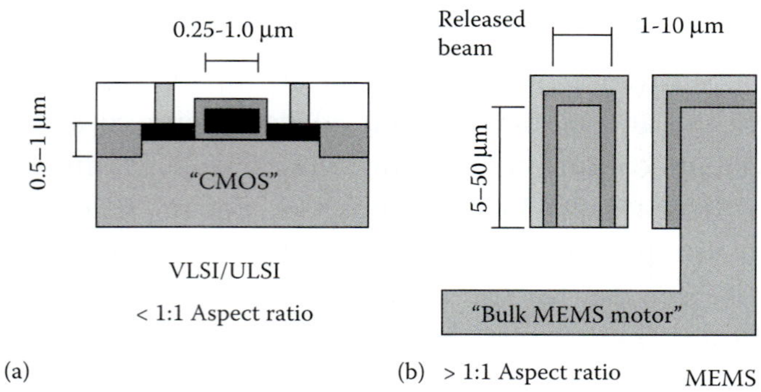

FIGURE 1.18 Aspect ratio (height-to-width ratio) typical in (a) fabrication of integrated circuits and (b) microfabricated component.

silicon were used in the fabrication of dielectrically isolated structures in ICs such as those for beam leads. In the mid-1970s, a surge of activity in anisotropic etching was associated with the work on V-groove and U-groove transistors. Isotropic and anisotropic etching profiles are compared in Figure 1.19. Figure 1.19a shows the isotropic etching of grooves in (100) Si, and Figure 1.19b shows the anisotropic etching of grooves in (100) Si. In both cases, rectangular mask openings are used.

Most single-crystal silicon MEMS devices feature bonding of one Si wafer to another or to a differing substrate, say a glass pedestal, and some MEMS involve cavity-sealing techniques, perhaps for a vacuum reference or to accommodate a deflecting cantilever beam. The most prominent techniques developed to achieve these features are field-assisted bonding, invented by Wallis and Pomerantz in 1969,[13] and Si fusion bonding (SFB) by Shimbo in 1986.[14] Field-assisted thermal bonding, as shown in Figure 1.20, also known as *anodic* bonding, *electrostatic* bonding, or the *Mallory* process, is commonly used for joining glass to silicon at high temperatures (e.g., 400°C) and high voltages (e.g., 600 V). The ability to bond two Si wafers directly, at high temperatures (>800°C) in an oxidizing environment, without intermediate layers or applying an electric field, simplified the fabrication of many devices in silicon fusion bonding (SFB).

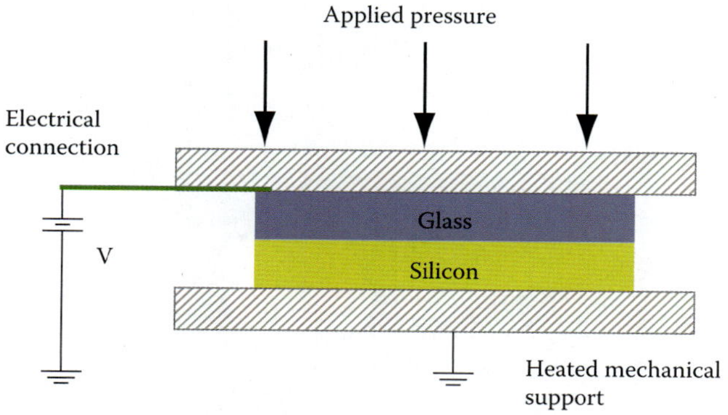

FIGURE 1.20 During anodic bonding, the negative potential applied to the borosilicate glass plate, which has been heated to 500°C, allows the migration of positive ions (mostly Na^+) away from the wafer's interface, creating a strong electric field between the glass and the Si wafer.

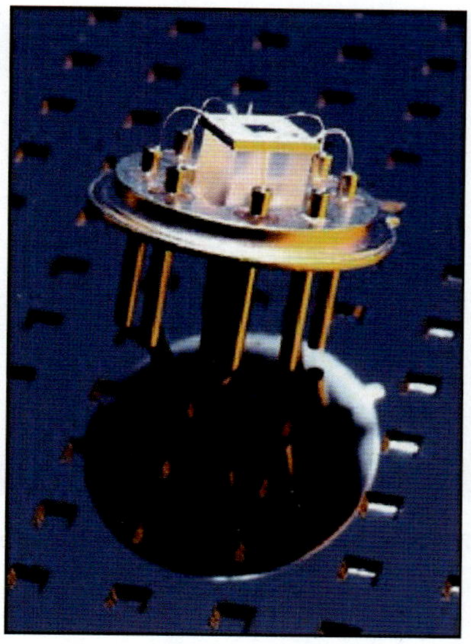

FIGURE 1.21 Piezoresistive pressure sensor featuring a Si/glass bond achieved by anodic bonding. The Bosch engine control manifold absolute pressure (MAP) sensor is used in automobile fuel injection systems. By measuring the manifold pressure, the amount of fuel injected into the engine cylinders can be calculated. Micromachined silicon piezoresistive pressure sensors are bonded at the wafer level to a glass wafer using anodic bonding before dicing. The glass pedestal that is created by this process provides stress isolation for the silicon sensor from package-induced thermal stresses. (Photo courtesy of Robert Bosch GmbH, Germany.)

The first Si accelerometer was demonstrated in 1970 at Kulite. In 1972, Sensym became the first company to make stand-alone Si sensor products. By 1974, National Semiconductor Corporation, in California, carried an extensive line of Si pressure transducers as part of the first complete silicon pressure transducer catalog.[15] Other early commercial suppliers of micromachined pressure sensor products were Foxboro/ICT, Endevco, Kulite, and Honeywell's Microswitch. To achieve better sensitivity and stability than possible with piezoresistive pressure sensors, capacitive pressure sensors were first developed and demonstrated by Dr. James Angell at Stanford University around 1977.[16] In Figure 1.21 we show a typical piezoresistive single-crystal silicon pressure sensor, with the silicon sensor anodically bonded to a glass substrate.

In many cases, it is desirable to stop the etching process when a certain cavity depth or a certain membrane thickness is reached. High-resolution silicon micromachining relies on the availability of effective etch-stop layers rather than the use of a stopwatch to control the etch depth. It was the discovery/development of impurity-based etch stops in silicon that allowed micromachining to become a high-yield commercial production process. The most widely used etch-stop technique is based on the fact that anisotropic etchants do not attack heavily boron-doped (p++) silicon layers. Selective p++ doping is typically implemented using gaseous or solid boron diffusion sources with a mask (such as silicon dioxide). The boron etch-stop effect was first noticed by Greenwood in 1969,[17] and Bohg in 1971[18] found that an impurity concentration of about $7 \times 10^{19}/cm^3$ resulted in the anisotropic etch rate of Si decreasing sharply.

Innovative, micromachined structures, different from the now mundane pressure sensors, accelerometers, and strain gauges, began to be explored by the mid- to late 1970s. Texas Instruments produced a thermal print head in 1977,[19] and IBM produced ink-jet nozzle arrays the same year.[20] In 1980, Hewlett Packard made thermally isolated diode detectors,[21] and fiberoptic alignment structures were manufactured at Western Electric. Chemists worldwide took notice when Terry, Jerman, and Angell, from Stanford University, integrated a gas chromatograph on a Si wafer in 1979 as shown in Figure 1.22.[22,23] This first analytical chemistry application would eventually lead to the concept of total analytical systems on a chip, or μ-TAS. An important milestone in the MEMS world was the founding of NovaSensor,

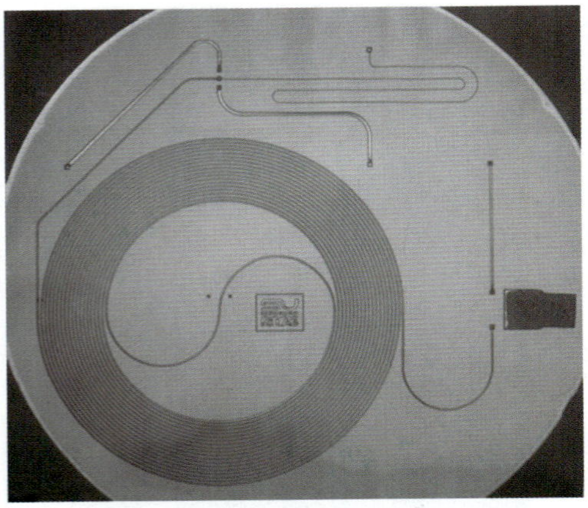

FIGURE 1.22 Gas chromatograph on a Si wafer. (Courtesy Hall Jerman.)

FIGURE 1.24 Integrated photonic mirror array from Transparent Networks. MEMS-VLSI integration achieved through wafer bonding. There are 1200 3D mirrors on the chip; each is 1 × 1 mm, with a ±10° tilt in two axes. (Courtesy of Janusz Bryzek.)

in 1985, by Kurt Petersen, Janusz Bryzek, and Joe Mallon (Figure 1.23). This was the first company totally dedicated to the design, manufacture, and marketing of MEMS sensors.

Kurt Petersen developed the first torsional, scanning micromirror in 1980 at IBM.[24] A more recent version of a movable mirror array is shown in Figure 1.24. Mirror arrays of this type led to the infamous stock market optical MEMS bubble of 2000, one of the bigger disappointments befalling the MEMS community.

The first disposable blood pressure transducer became available in 1982 from Foxboro/ICT, Honeywell for $40. Active on-chip signal conditioning also came of age around 1982. European and Japanese companies followed the U.S. lead more than a decade later; for example, Druck Ltd., in the United Kingdom, started exploiting Greenwood's micromachined pressure sensor in the mid-1980s. Petersen's 1982 paper extolling the excellent mechanical properties of single-crystalline silicon helped galvanize academia's involvement in Si micromachining in a major way.[25]

In MEMS the need sometimes arises to build structures on both sides of a Si wafer; in this case, a double-sided alignment system is required. These systems started proliferating after the EV Group (formally known as Electronic Visions) created the world's first double-side mask aligner with bottom side microscope in 1985 (http://www.evgroup.com). In the mid-1990s, new high-density plasma etching equipment became available, enabling directional deep dry reactive ion etching (DRIE) of silicon. Dry plasma etching was now as fast as wet anisotropic etching, and as a consequence the MEMS field underwent a growth spurt.

U.S. government agencies started large MEMS programs beginning in 1993. Older MEMS researchers remember the idealistic and inspired leadership of Dr. Kaigham (Ken) Gabriel (Figure 1.25) at Defense Advanced Research Projects Agency (DARPA). Gabriel got many important new MEMS products launched.

When the first polysilicon MEMS devices, made in a process called surface micromachining pioneered at University of California, Berkeley by Muller

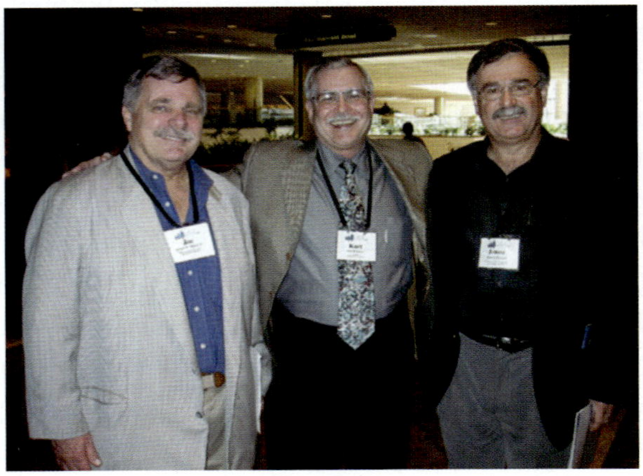

FIGURE 1.23 NovaSensor founders in 2003 at the Boston Transducer Meeting. Left to right: Joe Mallon, Kurt Petersen, and Janusz Bryzek.

FIGURE 1.25 Dr. Kaigham Gabriel: Early champion of MEMS work at DARPA.

and Howe, appeared in 1983,[26] bulk single-crystal Si micromachining started to get some stiff competition. This was exacerbated when, from the mid-1990s onward, MEMS applications became biomedical and biotechnology oriented. The latter applications may involve inexpensive disposables or implants, and Si is not a preferred inexpensive substrate nor is it biocompatible. Glass and polymers became very important substrates in microfluidics, and many researchers started using a flexible rubber (polydimethylsiloxane or PDMS) as a building material in a process called soft lithography. The latter manufacturing method, which dramatically shortened the time between novel fluidic designs and their testing, was invented in the late 1990s by Harvard's Whitesides.[27]

In the early years of MEMS, it was often projected that the overall MEMS market would grow larger than that for ICs. This notion was based on the expectation of many more applications for the former than the latter. This market projection has not been fulfilled. Including nonsilicon devices such as read-write heads, one can claim a MEMS market of about 10% of the total IC market today. From Appendix 1B we learn that Si sensors and actuators amount to only 4% of IC sales.

MEMS never really constituted a paradigm shift away from the IC concept but rather a broadening of it: incorporating more diverse materials, higher aspect ratio structures, and a wider variety of uses in smaller and more fragmented applications. In almost all respects MEMS remained IC's poor cousin: using second-hand IC equipment, with less than 5% of IC sales, and no Nobel laureates to trumpet breakthrough new concepts. Today, the Si MEMS market prospects are looking much better as MEMS are finally penetrating mass consumer products from projectors, to game controllers, to portable computers to cameras, mobile phones, and iPods with MEMS digital micromirror devices (DMDs), oscillators, accelerometers, gyros, etc. Even MEMS foundries are now thriving inside and outside the United States. This new generation of MEMS products fits high-throughput production lines on large Si substrates and is succeeding in the marketplace. The IC world has started to absorb the Si-MEMS world.

Over the years, the many MEMS applications did lead to a plethora of MEMS acronyms, some of them perhaps coined by assistant professors trying to get tenure faster. Here are some attempts at 15 minutes of fame:

- BioMEMS = MEMS applied to the medical and biotechnology field
- Optical MEMS = mechanical objects + optical sources/detectors
- Power-MEMS
- C-MEMS (carbon MEMS for this author but ceramic MEMS for others)
- HI-MEMS = hybrid insect-microelectromechanical systems
- RF-MEMS = radiofrequency MEMS
- Cif-MEMS = CMOS IC Foundry MEMS
- COTS MEMS = commercial off-the-shelf MEMS
- MOEMS = microoptical electromechanical systems
- P-MEMS = polymer MEMS
- CEMS = cellular engineering microsystems
- HARMEMS = high-aspect-ratio MEMS

We do expect that there are many more MEMS applications yet to be realized and that MEMS will facilitate the handshake between the macro and nano world in nanoelectromechanical systems (NEMS).

In Table 1.2 we sketch our attempt at a Si/MEMS history line. Many more MEMS milestones than listed in the preceding text are captured here.

Table 1.2 MEMS History

Year	Fact
1824	Berzelius discovers Si
1910	First patent on the MOS transistor concept
1927	Field effect transistor patented (Lilienfield)
1939	First pn junction transistor (J. Bardeen, W.H. Brattain, W. Shockley)
1947 (23 December) 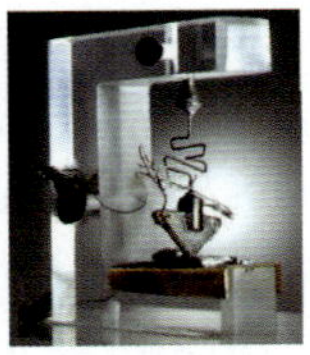	Invention of the transistor made from germanium at Bell Telephone Laboratories
1954	Evidence of piezoresistive effect in Si and Ge by Smith[9]
1956	An early milestone for the use of single-crystal silicon in MEMS was the discovery of porous Si by Uhlir[8]
1958  1960s IC (1st Monolithic BJT IC, 4BJT)	Jack Kilby of Texas Instruments invents the IC, using GE devices. A patent was issued to Kilby in 1959. A few months later, Robert Noyce of Fairchild Semiconductor announced the development of a planar Si IC
1958	Silicon strain gauges commercially available
1958	First IC (oscillator)
1959	R. Feynman famous talk: "There's Plenty of Room at the Bottom"[7]
1961	Fabrication of the first piezoresistive sensor, pressure (Kulite)
1967	Anisotropic deep silicon etching (H.A. Waggener[12])
1967 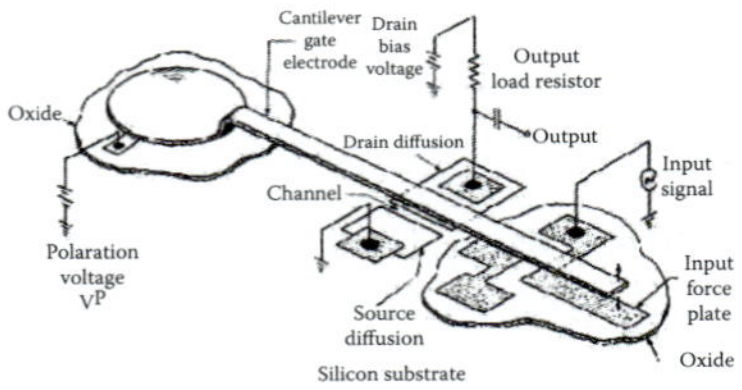	First surface micromachining process (H. Nathanson[28]): resonant gate before it was called MEMS
1969–1970	Anodic bonding of glass to Si[13]
1972	National Semiconductor: commercialize a Si MEMS pressure sensor
1975	Gas chromatograph on a Si wafer by S.C. Terry, J.H. Jerman, and J.B. Angell at Stanford University[23]
1977	First capacitive pressure sensor (Stanford)[16]
1977 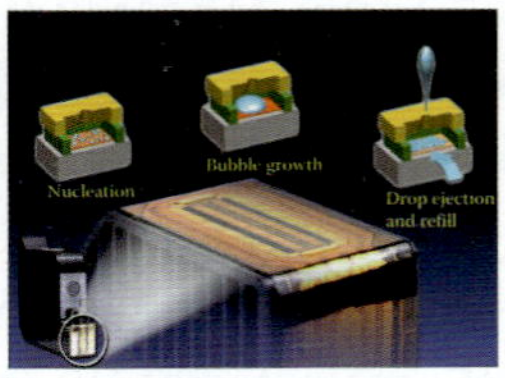	IBM–HP: micromachined ink-jet nozzle[24]
1980	K.E. Petersen, silicon torsional scanning mirror[24]

Table 1.2 MEMS History (Continued)

Year	Fact
1982	Review paper "Silicon as a mechanical material" published by K.E. Petersen[25]
1982	Disposable blood pressure transducer (Foxboro/ICT, Honeywell, $40)
1982	The use of x-ray lithography in combination with electroplating and molding (or LIGA), introduced by Ehrfeld and his colleagues[29]
1983	Integrated pressure sensor (Honeywell)
1983	"First" polysilicon MEMS device (Howe, Muller UCB[26]); see also Nathanson in 1967[28]
1986	Silicon to silicon wafer bonding (M. Shimbo[14])
1987	Texas Instrument's Larry Hornbeck invents the digital micromirror devices (DMDs)
1988	Rotary electrostatic side drive motors (Fan, Tai, Muller[30]) Electrostatic micromotor (UC-Berkeley BSAC)
1988	First MEMS conference (first transducers conference held in 1987)
1989 	Lateral comb drive (Tang, Nguyen, Howe[31])
1990	The concept of miniaturized total chemical analysis system or μ-TAS is introduced by Manz et al.[32] This may be seen as the beginning of BIOMEMS
1992	Grating light modulator (Solgaard, Sandejas, Bloom)
1992	First MUMPS process (MCNC) (with support of DARPA). Now owned by MEMSCAP
1992	First MEMS CAD tools: MIT, S.D. Senturia, MEMCAD 1.0 Michigan, Selden Crary, CAEMEMS 1.0
1992	Single-crystal reactive etching and metallization (SCREAM) developed at Process (Cornell)
1993	Analog devices: commercialize multiaxis accelerometer integrating electronics (ADXL50)
1995	Intellisense Inc. introduces MEMS CAD IntelliSuite. MEMCAD 2.0 is launched, and ISE introduces SOLIDIS and ICMAT
1996 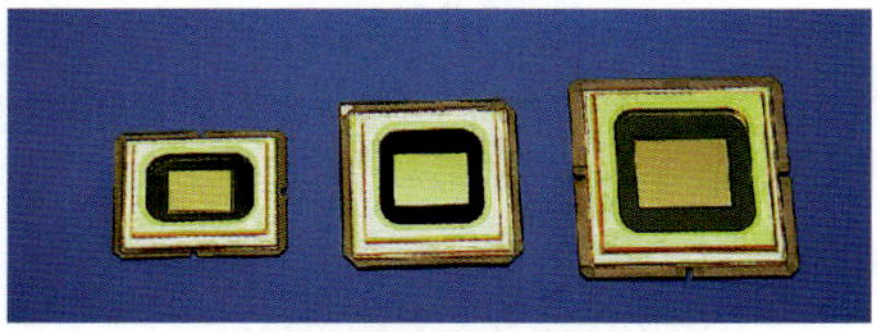 TI's VGA (640 × 480), the SVGA (800 × 600), and the XGA (1024 × 768)	The first digital mirror device (DMD)–based products (Texas Instruments) appear on the market
1996	DRIE (Bosch Process)
1997	Printing meets lithography when George M. Whitesides et al. at Harvard discover soft lithography[27]
1998	First PCR-microchips
1998	Sandia's ultraplanar multilevel technology SUMMiT-IV and -V technologies. Four- and five-level poly-Si processes

(continued)

Table 1.2 MEMS History (Continued)

Year	Fact
1999  Affymetrix genechip	DNA microarray techniques
1999	Electrokinetic platforms (Caliper, Aclara, and Agilent)
2000	Nortel buys Xros for $3.25 billion
2002  Lucent 3D optical switch	The telecom recession puts many things on standby
2004	MEMS rebuilds. First application of accelerometer in consumer electronics (CE) to hard drive protection in notebooks. IBM puts dual-axis accelerometer in the notebook (now Lenovo)
2006	Sony (PS3) and Nintendo (Wii) introduce motion-based game controllers
2007	Apple announces the iPhone with motion-based features

NEMS

The criteria we use in this book for classifying something as a nanoelectromechanical system (NEMS) are not only that the miniaturized structures have at least one dimension that is smaller than 100 nanometers, but also that they are crafted with a novel technique (so beer making is out) or have been intentionally designed with a specific nanofeature in mind (so medieval church stained glass is out). This definition fits well within the one adopted by the National Nanotechnology Institute (NNI; http://www.nano.gov/html/facts/whatisnano.html):

1. Nanotechnology involves R&D at the 1- to 100-nm range.
2. Nanotechnology creates and uses structures that have novel size-based properties.
3. Nanotechnology takes advantage of the ability to control or manipulate at the atomic scale.

Paul Davis (http://cosmos.asu.edu), a theoretical physicist and well-known science popularizer, said, "The nineteenth century was known as the machine age, the twentieth century will go down in history as the information age and I believe the twenty-first century will be the quantum age." We believe that the current nanotechnology revolution, underpinned by quantum mechanics, is already leading the way toward that reality.

The manufacture of devices with dimensions between 1 and 100 nanometers is either based on top-down manufacturing methods (starting from bigger building blocks, say a whole Si wafer, and chiseling them into smaller and smaller pieces by cutting, etching, and slicing), or it is based on bottom-up manufacturing methods (in which small particles such as atoms, molecules, and atom clusters are added for the construction of bigger functional constructs). The top-down approach to nanotechnology we call *nanofabrication* or *nanomachining*, an extension of the MEMS approach. The bottom-up approach we like to refer to as *nanochemistry*. An example of this second approach is the self-assembly of a monolayer (SAM) from individual molecules on a gold surface. Bottom-up methods are nature's way of growing materials and organisms, and in biomimetics one studies how nature, through eons of time, developed manufacturing methods, materials, structures, and intelligence and tries to mimic or replicate what nature does in the laboratory to produce MEMS or NEMS structures.

A history-line with the most important NEMS milestones on it is difficult to put together as so many authors of such charts automatically include themselves or their institution on it first (one author of an early MEMS/NEMS timeline puts himself on it three times and his institution four times!). It sometimes seems that science and engineering are starting to resemble FOX News more by the day. What follows are some milestones toward nanotechnology that many scientists/engineers might agree to.

Norio Taniguchi introduced the term *nanotechnology* in 1974, in the context of traditional machining with tolerances below 1 micron. The 1959 Feynman lecture "There's Plenty of Room at the Bottom," which helped launch the MEMS field (see above), was geared more toward NEMS than MEMS (http://www.its.caltech.edu/~feynman/plenty.html).[7] Feynman proclaimed that he knew of no principles of physics that would prevent the direct manipulating of individual atoms. In his top-down *gedanken* experiment, he envisioned a series of machines each an exact duplicate, only smaller and smaller, with the smallest in the series being able to manipulate individual atoms (see Figure 1.26).

In 1981, Gerd Binning and Heinrich Rohrer of IBM Zurich invented the scanning tunneling microscope (STM), enabling scientists to see and move individual atoms. Such a microscope, shown in Figure 1.27, measures the amount of electrical current flowing between a scanning tip and the conductive surface that is being measured. This unexpectedly simple instrument allowed for the imaging of micro- and nanostructures, catapulted nanotechnology onto the world stage, and got its inventors the 1986 Nobel Prize (http://www.zurich.ibm.com/imagegallery/st/nobelprizes). Just as 350 years before the microscope changed the way we viewed the world, the STM impacts our current view of biology, chemistry, and physics.

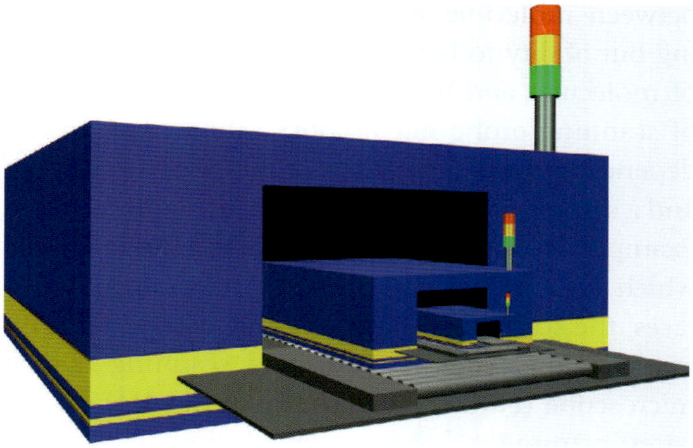

FIGURE 1.26 Master and slave hands on a set of Feynman machines.

In fast succession, a series of similar instruments, all called scanning proximal probes, followed the introduction of the STM. For example, Binnig, Quate (Stanford), and Gerber (IBM) developed the atomic force microscope (AFM) in 1986. An AFM, in contact mode, measures the repulsive force interaction between the electron clouds on the probe tip atoms and those on the sample—making it possible to image both insulating and conducting surfaces. This results in the visualization of the interactions

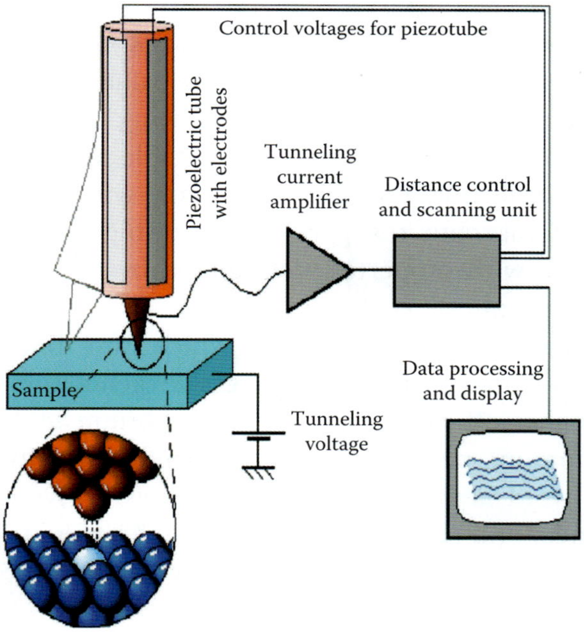

FIGURE 1.27 Scanning tunneling microscope (STM). Operational principle of an STM. (Courtesy Michael Schmidt, TU Wien.)

between molecules at the nanoscale, thus increasing our ability to better understand the mechanism of molecular and biological processes. Other forms of scanning probe microscopes—those that do not depend on tunneling or forces between a probe tip and a sample surface—have also been demonstrated. Examples include the scanning thermal microscope, which responds to local thermal properties of surfaces, the scanning capacitance microscope for dopant profiling, and the near-field scanning optical microscope (NSOM, also known as SNOM). In the latter instrument,[33] the wavelength limitation of the usual far-field optics of a light microscope is avoided by mounting the light detector (say an optical fiber) on an AFM tip, at a distance from the sample that is a fraction of the wavelength used; this way it is possible to increase the resolution of a light microscope considerably. These new tools were an important catalyst behind the surge in nanotechnology activities worldwide, and this illustrates that progress in science is inextricably linked to the development of new measurement tools.

The discovery, in the early 1980s at Bell Labs by David L. Allara (now at Pennsylvania State University) and Ralph G. Nuzzo (now at University of Illinois, Urbana-Champaign) of the self-assembly of disulfide and, soon thereafter, of alkanethiol monolayers (SAMs) on metal surfaces coincided with the maturation of STM technology.[34,35] SAMs, especially on Au, turned out to be a valuable type of sample for STM investigation, showing these films to spontaneously assemble into stable and highly organized molecular layers, bonding with the sulfur atoms onto the gold and resulting in a new surface with properties determined by the alkane head group.

Like SAMs, dendrimers, which are branching polymers sprouting successive generation of branches off like a tree, are an important tool for bottom-up nanotechnologists. Dendrimers (from the word *dendron*, Greek for tree) were invented, named, and patented by Dr. Donald Tomalia (now CTO at Dentritic NanoTechnologies, Inc.) in 1980 while at Dow Chemical.[36]

In 1970, Arthur Ashkin was the first to report on the detection of optical scattering and gradient forces on micron-sized particles.[37] In 1986, Ashkin

FIGURE 1.28 Steven Chu (Stanford University), recipient of the 1997 Nobel Prize in Physics for his work on cooling and trapping of atoms.

and colleagues reported the first observation of what is now commonly referred to as an optical trap, i.e., a tightly focused beam of light capable of holding microscopic particles stable in three dimensions.[38] One of the authors of this seminal 1986 paper, Steven Chu (Figure 1.28), would go on to make use of optical tweezing techniques in his work on cooling and trapping of atoms. Where Ashkin was able to trap larger particles (10 to 10,000 nanometers in diameter), Chu extended these techniques to the trapping of individual atoms (0.1 nanometer in diameter). This research earned him the 1997 Nobel Prize in Physics (with Claude-Cohen Tannoudji and William D. Phillips). In another heralded experiment, Steven Chu was also the one who demonstrated that by attaching polystyrene beads to the ends of DNA one can pull on the beads with laser tweezers to stretch the DNA molecule (http://www.stanford.edu/group/chugroup/steve_personal.html).

In 1985, Robert F. Curl Jr., Harold W. Kroto, and the late Richard E. Smalley serendipitously (while investigating the outer atmosphere of stars) discovered a new form of carbon: buckminsterfullerene, also known as buckyball or C60, shown in Figure 1.29.[39] They were awarded the Nobel Prize in 1996.

Perhaps a more important discovery, because of its generality and broader applicability, is the one by NEC's Sumio Iijima, who, in 1991, discovered

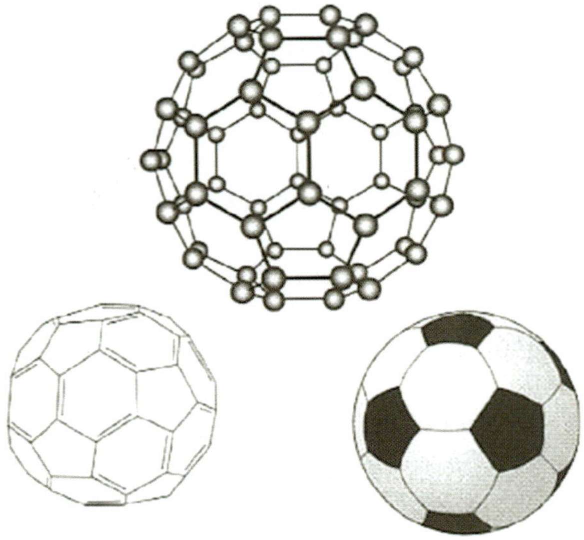

FIGURE 1.29 Buckminsterfullerene C60 or buckyball with 60 atoms of carbon; each is bound to three other carbons in an alternating arrangement of pentagons and hexagons.

carbon nanotubes, with an electrical conductivity that is up to six orders of magnitude higher than that of copper (http://www.nec.co.jp/rd/Eng/innovative/E1/myself.html). Like buckyballs, cylindrical nanotubes each constitute a lattice of carbon atoms, and each atom is again covalently bonded to three other carbons.

Carbon nanotubes exist as single-walled (SWNT) and multiwalled (MWNT); the ones depicted in Figure 1.30a are multiwall nanotubes. Unique among the elements, carbon can bond to itself to form extremely strong two-dimensional sheets, as it does in graphite, as well as buckyballs and nanotubes.

Cees Dekker demonstrated the first carbon nanotube transistor in 1998 at the Delft University of Technology[40] (see Figure 1.30b). In this device, a semiconducting carbon nanotube of only about 1 nm in diameter bridges two closely separated metal electrodes (400 nm apart) atop a silicon surface coated with silicon dioxide. Applying an electric field to the silicon (via a gate electrode) turns on and off the flow of current across the nanotube by controlling the movement of charge carriers in it. By carefully controlling the formation of metal gate electrodes, Dekker's group (http://www.ceesdekker.net) was able to create transistors with an output signal 10 times stronger than the input. At around the same time, the first nanotransistor was built at Lucent Technologies (1997). The MOS semiconductor transistor was 60 nm wide, including the source, drain, and gate; its thickness was only 1.2 nm. Other companies have since built smaller nanotransistors.

At one point, in the late 1990s, it was—with just a bit of exaggeration—hard to find a proposal to a government agency that did not involve carbon nanotubes. But perhaps more real progress was mapped in the meantime in the area of nanocrystals or quantum dots (QD).

Quantum dots possess atom-like energy states. The behavior of such small particles was beginning to be understood with work by the Russians Ekimov and Efros from 1980 to 1982.[41,42] They recognized that nanometer-sized particles of CdSe, with their very high surface-to-volume ratio, could

(a)

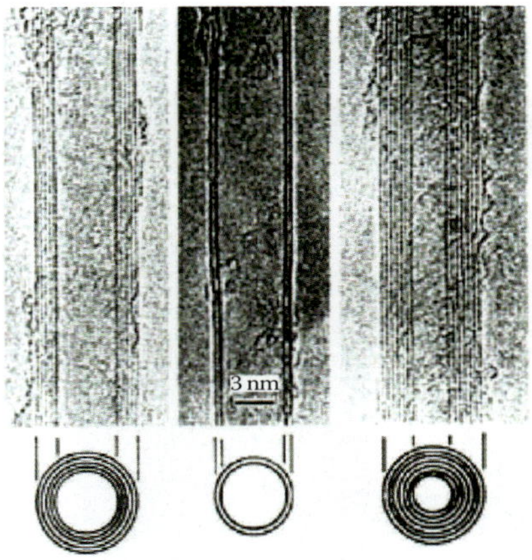

(b)

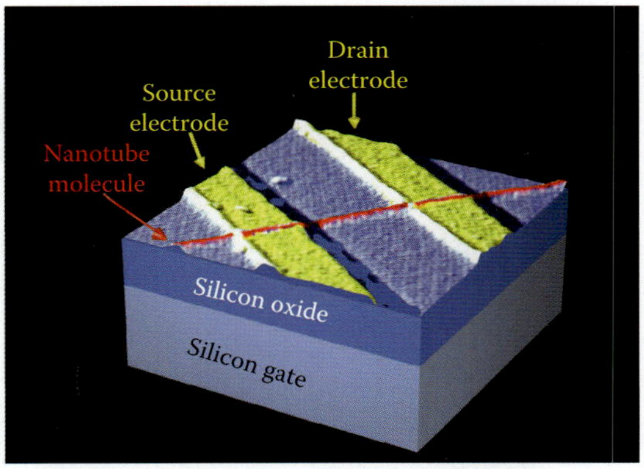

FIGURE 1.30 (a) Multiwall nanotubes with Russian inset doll structure; several inner shells are shown a typical radius of the outermost shell >10 nm. (From S. Iijima, *Nature* 354, 54, 1991.) (b) First nanotube transistor. This three-terminal device consists of an individual semiconducting nanotube on two metal nanoelectrodes with the substrate as a gate electrode.

trap electrons and that these trapped electrons might affect the crystal's response to electromagnetic fields, that is, absorption, reflection, refraction, and emission of light. Louis E. Brus, a physical chemist at Bell Laboratories at the time, and now at Columbia University, put this to practice when he learned to grow CdSe nanocrystals in a controlled manner.[43,44] Murray, Norris, and Bawendi synthesized the first high-quality quantum dots in 1993.[45] Crystallites from ~12 to ~115 Å in diameter with consistent crystal structure, surface derivatization, and a high degree of monodispersity were prepared in a single reaction based on the pyrolysis of organometallic reagents by injection into a hot coordinating solvent. The confinement of the wave-functions in a nanocrystal or quantum dot lead to a blue energy shift, and by varying the particle size one can produce any color in the visible spectrum, from deep (almost infra-) reds to screaming (almost ultra-) violet as illustrated in Figure 1.31. Today, quantum dots form an important alternative to organic dye molecules. Unlike fluorescent dyes, which tend to decompose and lose their ability to fluoresce, quantum dots maintain their integrity, withstanding many more cycles of excitation and light emission (they do not bleach as easily!). Combining a number of quantum dots in a bead conjugated to a biomolecule is used as a spectroscopic signature—like a barcode on a commercial product—for tagging those biomolecules.

Carbon nanotubes are only one type of nanowire. In terms of investigating and exploiting quantum confinement effects, semiconductor wires, with diameters in the 10s of nanometers, often single crystalline, represent the smallest dimension for efficient transport of electrons and excitons and are the logical interconnects and critical devices for nanoelectronics and nanooptoelectronics of the future. Over the past decade, there has been major progress in chemical synthesis technologies for growing these nanoscale semiconductor wires. As originally proposed by R.S. Wagner and W.C. Ellis from Bell Labs for the Au-catalyzed Si whisker growth, a vapor-liquid-solid mechanism is still mostly used.[46] But the field got a shot in the arm (a rebirth so to speak) with efforts by Charles Lieber (Harvard), Peidong Yang (http://www.cchem.berkeley.edu/pdygrp/main.html), James Heath (http://www.its.caltech.edu/~heathgrp), and Hongkun Park (http://www.people.fas.harvard.edu/~hpark). Lieber's group at Harvard (http://cmliris.harvard.edu) reported arranging indium phosphide semiconducting nanowires into a simple configuration that resembled the lines in a tick-tack-toe board. The team used electron beam lithography to place electrical contacts at the ends of the nanowires to show that the array was electronically active. The tiny arrangement was not a circuit yet, but it was the first step, showing that separate nanowires could communicate with one another.

Molecules are 30,000 times smaller than a transistor (180 nm on a side), so obviously it is of some use to investigate whether molecules can act as switches. Mark Ratner and Ari Aviram had suggested this as far back as 1974.[47] The suggestion remained a pipe dream until the advent of scanning probe microscopes in the 1980s, which gave researchers finally

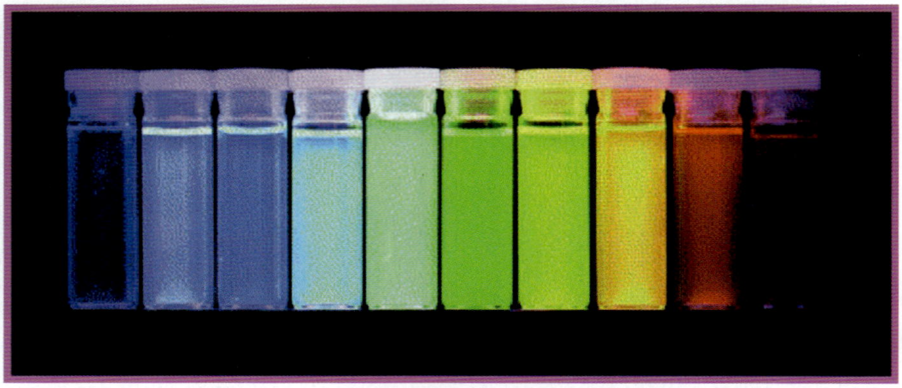

FIGURE 1.31 Different-sized quantum dots in response to near-UV light. Also composition of core affects wavelength. Red: bigger dots! Blue: smaller dots!

the tools to probe and move individual molecules around. This led to a large number of studies in the late 1990s that demonstrated that individual molecules can conduct electricity just like metal wires, and turning individual molecules into switches came not far behind. In 1997, groups led by Robert Metzger (http://bama.ua.edu/~rmmgroup) of the University of Alabama, Tuscaloosa, and Mark Reed of Yale University (http://www.eng.yale.edu/reedlab) created molecular diodes. In July 1999, another group headed by James Heath and Fraser Stoddart of the University of California, Los Angeles (UCLA) (http://stoddart.chem.ucla.edu) also created a rudimentary molecular switch, a molecular structure that carries current but, when hit with the right voltage, alters its molecular shape and stops conducting. Heath's team placed molecules called rotaxanes, which function as molecular switches, at each junction of a circuit. By controlling the input voltages the scientists showed that they could make 16-bit memory circuits work. The field of moletronics was born.

As shown in Figure 1.32, rotaxanes are "mechanically linked" molecules that consist of a dumbbell-shaped molecule, with a cyclic molecule linked around it between the two ends. The two ends of the dumbbell molecule are very big and prevent the cyclic molecule from slipping off the end. A number of factors (e.g., charge, light, pH) can influence the position of the cyclic molecule on the dumbbell.

The use of x-ray lithography in combination with electroplating and molding (or LIGA), introduced by Ehrfeld and his colleagues in 1982,[29] demonstrated to the world that lithography may be merged with more traditional manufacturing processes to make master molds of unprecedented aspect ratios and tolerances to replicate microstructures in ceramics, plastics, and metals. The hard x-rays used enable nano-sized patterns to be printed.

At around 1997, Whitesides et al. introduced soft lithography, including the use of pattern transfer of self-assembled monolayers (SAMs) by elastomeric stamping.[27] This technique formed a bridge between top-down and bottom-up machining; a master mold is made based on "traditional" lithography, and the stamp generated from this master is inked with SAMs to print (stamp) substrates with nano-sized patterns.

Imposing boundaries on photons, by making them move in a material with a periodic dielectric constant in one, two, or three directions, leads to photonic crystals. Photonic crystals were first studied by Lord Rayleigh in 1887, in connection with the peculiar reflective properties of a crystalline mineral with periodic "twinning" planes.[48] He identified a narrow bandgap prohibiting light propagation through the planes. This bandgap was angle-dependent because of the differing periodicities experienced by light propagating at non-normal incidences, producing a reflected color that varies sharply with angle. A similar effect is seen in nature, such as in butterfly wings (Figure 1.33) and abalone shells.

A one-dimensional periodic structure, such as a multilayer film (a Bragg mirror), is the simplest type of photonic crystal, and Lord Rayleigh showed that any such one-dimensional system has a bandgap. The possibility of two- and three-dimensionally periodic crystals with corresponding two- and three-dimensional bandgaps was suggested 100

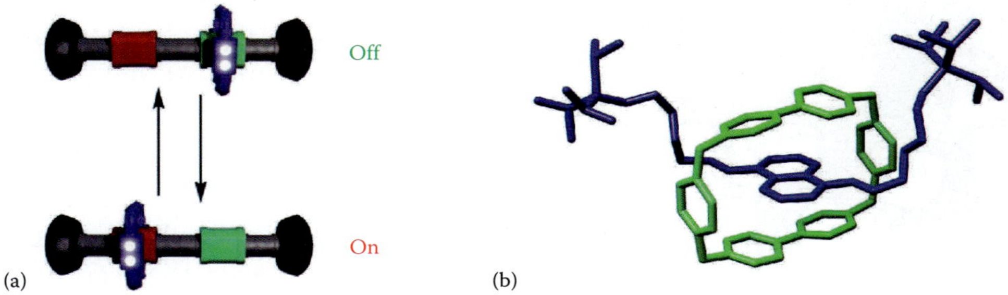

FIGURE 1.32 (a) A diagram of rotaxane. The usefulness of rotaxanes is because there are a number of positions along the dumbbell molecule that the cyclic molecule can attach to temporarily. The dumbbell can be thought of as a train track, with the positions on the dumbbell molecule as stations and the cyclic molecule as the train. (b) Crystal structure of rotaxane with a cyclobis(paraquat-p-phenylene) macrocycle.

FIGURE 1.33 A full-grown *Morpho rhetenor* butterfly, a native to South America. (From University of Southhampton. Color by nanostructure instead of dyes.)

years after Rayleigh by Eli Yablonovitch[49] (http://www.ee.ucla.edu/labs/photon/homepage.html) and Sajeev John[50] (http://www.physics.utoronto.ca/~john) in 1987. Yablonovitch (in 1991)[51] demonstrated the first microwave photonic bandgap (PBG) structure experimentally with 1-mm holes drilled in a dielectric material as illustrated in Figure 1.34 and known today as *yablonovite*. Since then, several research groups verified this prediction, which has ignited a worldwide rush to build tiny "chips" that control light beams instead of electron streams. In photonic crystals the repeat unit in the lattice is of the same size as the incoming wavelength, so homogeneous (effective) media theory cannot be applied. Photonic crystals feature lattice spacings ranging from the macroscopic (say 1 mm, for operating in the microwave domain-like yablonovite) to the 100s of nanometer range (to operate in the visible range). We cover photonic crystals here under NEMS, although only photonic crystals for the visible range qualify as nanotechnology. The potential applications of photonics are limitless, not only as a tool for controlling quantum optical systems but also in more efficient miniature lasers for displays and telecommunications, in solar cells, LEDs, optical fibers, nanoscopic lasers, ultrawhite pigments, radiofrequency antennas and reflectors, photonic integrated circuits, etc.

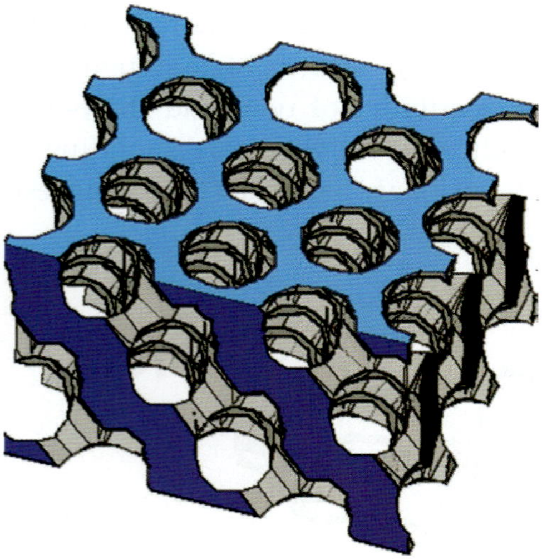

FIGURE 1.34 Holes drilled in dielectric: known now as *yablonovite*, after Yablonivitch (http://www.ee.ucla.edu/~pbmuri). The holes are 1 mm in size, and this photonic crystal is meant to operate in the microwave range.

In 1967 Victor Veselago, a Russian physicist, predicted that composite metamaterials might be engineered with negative magnetic permeability and negative permittivity.[52] Metamaterials are artificially engineered materials possessing properties that are not encountered in nature. Whereas photonic materials do exist in nature, metamaterials do not; moreover, in the case of metamaterials, the building blocks are small compared with the incoming wavelength so that effective media theory can be applied. In conventional materials the plane of the electrical field, the plane of the magnetic field, and the direction in which light travels are all oriented at right angles to each other and obey the right-hand rule. In Veselago's imaginary metamaterial, the above quantities obey a left-hand rule (as if they were reflected in a mirror). These materials would interact with their environment in exactly the opposite way from natural materials (see negative refractive index water in Figure 1.35). One intriguing prediction was that the left-hand rule would allow for a flat superlens to focus light to a point and that could image with a resolution far beyond the diffraction limit associated with far-field illumination. Veselago's prediction that such perfect lenses could be made from metamaterials lay dormant until 1996–2000, when the remarkable John Pendry, a physicist at Imperial College in London, showed that certain metals could be engineered to respond to electric fields as though the field parameters were negative.[53–58] In 2001, researchers at Imperial College and Marconi Caswell Ltd. (London) announced a magnetic resonance

FIGURE 1.35 Refraction illustrated (a) empty glass: no refraction; (b) typical refraction with pencil in water with $n = 1.3$; (c) what would happen if the refractive index were negative with $n = -1.3$ (see metamaterials). (From Gennady Shvets, The University of Texas at Austin; http://www.ph.utexas.edu/~shvetsgr/lens.html.)

imaging system using a magnetic metamaterial based on Pendry's design.[59] Physicist Richard Shelby's group at the University of California, San Diego demonstrated a left-handed composite metamaterial that exhibited a negative index of refraction for microwaves.[60] The simple arrangement consisted of a planar pattern of copper split-ring resonators (SRRs) and wires on a thin fiberglass circuit board. Operating in the microwave range these metal patterns are large (5 mm repeat unit) but progress toward metamaterials operating in the visible was very swift. By 2005, Zhang's group at University of California, Berkeley made a 35-nm thick Ag superlens and imaged objects as small as 40 nm with 365 nm light, clearly breaking the diffraction limit of far-field imaging[61] (find Zhang's group at http://xlab.me.berkeley.edu). By 2007, a left-handed material operating in the visible range (780 nm) was demonstrated[62] by Soukoulis (http://cmpweb.ameslab.gov/personnel/soukoulis) at the U.S. Department of Energy's Ames Laboratory on the Iowa State University campus and Wegener's group from the University of Karlsruhe (http://www.aph.uni-karlsruhe.de/wegener), Germany.

As in the case of photonic crystals, only the metamaterials operating in the visible qualify as nanotechnology, but for comprehensiveness sake we cover all of them together here.

In 2000, IBM scientists placed a magnetic cobalt atom inside an elliptical coral of atoms. They observed the Kondo effect, i.e., electrons near the atom align with the atom's magnetic moment, effectively canceling it out. When the atom was placed at one focus of the elliptical coral, a second Kondo effect was observed at the other focus, even though no atom was there (see Figure 1.36). Hence some of the properties (info) carried by an atom are transferred to the other focus (www.research.ibm.com). This quantum mirage effect "reflects" information using the wave nature of the electrons rather than transmission of info using electrons in a wire. It has the potential to be able to transfer data within future nanoscale electronic circuits where wires would not work. This would allow miniaturization of circuits well below what is envisioned today.

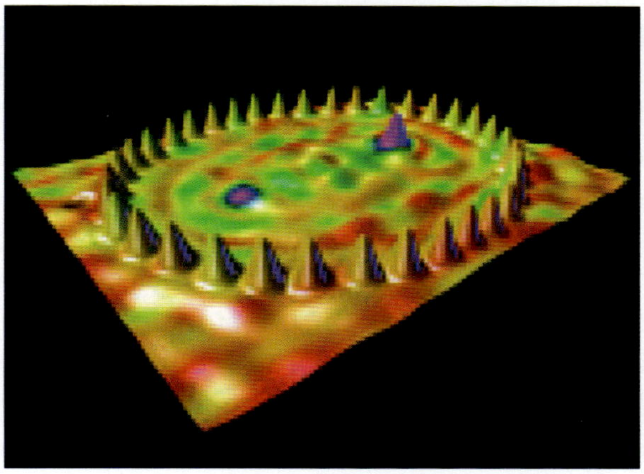

FIGURE 1.36 Quantum mirage phenomenon (http://domino.research.ibm.com/comm/pr.nsf/pages/rsc.quantummirage.html).

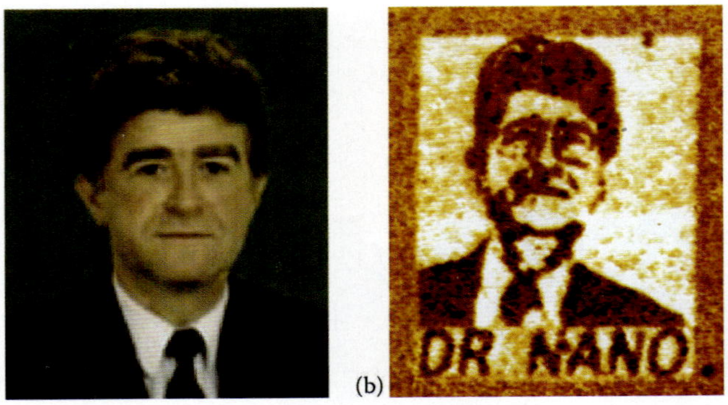

FIGURE 1.37 NSF's Dr. Mike Roco, a photo (a) and a nanograph (b). The nanograph of Dr. M. Roco was recorded at Oak Ridge National Laboratory using piezoresponse force microscopy, one of the members of the family of techniques known as scanning probe microscopy, which can image and manipulate materials on the nanoscale. Each picture element is approximately 50 nanometers in diameter; the distance from chin to eyebrow is approximately 2.5 microns. (Courtesy Dr. Roco.)

In 1999 President Clinton announced the National Nanotechnology Initiative (NNI); this first formal government program for nanotechnology accelerated the pace of nano research (the program had been around unofficially since 1996). In December 2003, George W. Bush signed the 21st Century Nanotechnology Research and Development Act. In this government NEMS program, Dr. Mike Roco (Figure 1.37) played a similar catalyzing role that Dr. Ken Gabriel played earlier in the MEMS field (see above).

Two other important nanotech promoters are Ray Kurzweil (http://www.kurzweilai.net/index.html) and K. Eric Drexel (http://www.foresight.org). Whereas Feynman continues to receive almost universal praise for his inspiring 1959 speculative talk, K. Eric Drexel, who in 1981 described mechanochemistry in his speculative paper "Molecular Engineering: An Approach to the Development of General Capabilities for Molecular Manipulation," continues to receive mostly harsh criticism—sometimes bordering on derision. In this paper and in two books,[63,64] Drexel builds nanotechnology, bottom-up, atom by atom, rather than whittling down materials as Feynman had suggested. Drexel also makes more of the fact that nature and molecular biology are proof of concept for this type of molecular technology. Drexel's early warnings about "gray goo," his emphasis on assemblers—small machines that would guide chemical bonding operations by manipulating reactive molecules—and building nanotechnology in a dry environment probably explain most of the hostility toward his work (even by those who do not even make an attempt to understand it; see for example Atkinson[65] in Nanocosm—we might as well listen to Newt Gingrich talk about nanotechnology). Drexel, unfortunately, has been associated too much with the nano pop culture. *Nano!* by Ed Regisis is an engaging and entertaining book that describes some of the researchers involved in nanotechnology; he is uncharacteristically positive about Drexler.[66]

Market projections on NEMS are today even wilder than the early ones on MEMS. Overall, though, this author believes that it is in nanotechnology, especially when considering bottom-up manufacturing, that a paradigm shift away from IC-type manufacturing is taking shape, and that it is nanotechnology that holds the potential of having a much larger impact on society than IC technology ever did. Indeed, the nanoscale is unique because at this length scale important material properties, such as electronic conductivity, hardness, or melting point, start depending on the size of the chunk of material in a way they do not at any other scale. Moreover, in biotechnology, molecular engineering already has made major progress, and the confluence of miniaturization science and molecular engineering is perhaps the most powerful new avenue for progress of humankind in general. In this regard, NEMS must be seen as a support technology to extract yet more benefits from the ongoing biotechnology revolution.

In Table 1.3 we show a milestone chart in nanotechnology. It can be argued that molecular scientists and genetic engineers were practicing nanotechnology long before the name became popular with electrical and mechanical engineers. Molecular biology or "wet nanotechnology" has been called "nanotechnology that works." But adding breakthroughs pertaining to molecular biology to this table would make it much too long. For the same reason we also did not list any IC-related milestones.

Table 1.3 A Milestone Chart in Nanotechnology

3.5 billion years ago	The first living cells emerge. Cells house nanoscale biomachines that perform such tasks as manipulating genetic material and supplying energy
400 BC	Democritus coins the word "atom," which means "not cleavable" in Greek
1857	Michael Faraday introduces "colloidal gold" to the Royal Society
1887	Photonic crystals are studied by Lord Rayleigh, in connection with the peculiar reflective properties of a crystalline mineral with periodic "twinning" planes[48]
1905	Albert Einstein publishes a paper that estimates the diameter of a sugar molecule as 1 nm. Jean-Baptist Perrin confirmed these results experimentally and was awarded the 1926 Nobel Prize for this work
1931	Max Knoll and Ernst Ruska develop the electron microscope, which enables nanometer imaging
1932	Langmuir establishes the existence of monolayers (Nobel Prize in 1932)
1959	Richard Feynman gives his famed talk "There's Plenty of Room at the Bottom," on the prospects for miniaturization[7]
1967  Victor Veselago	Victor Veselago, a Russian physicist, predicted that composite metamaterials might be engineered with negative magnetic permeability and negative permittivity[52]
1968	Alfred Y. Cho and John Arthur of Bell Laboratories and their colleagues invent molecular-beam epitaxy, a technique that can deposit single atomic layers on a surface
Early 1970s	Groups at Bell Laboratories and IBM fabricate the first two-dimensional quantum wells
1974	Norio Taniguchi conceives the word *nanotechnology* to signify machining with tolerances of less than a micron
1974	Mark Ratner and Ari Aviram suggest using molecules as switches[47]
1980	The behavior of quantum dots began to be understood with work by the Russians Ekimov and Efros in 1980–1982.[41,42] Louis E. Brus learned to grow CdSe nanocrystals in a controlled manner[43,44]
1980	Dendrimers (from the word *dendron*, Greek for tree) were invented, named, and patented by Dr. Donald Tomalia[36]
1981 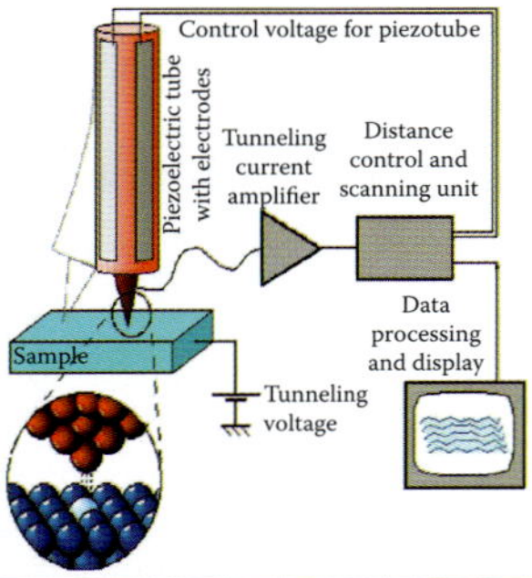	Gerd Binnig and Heinrich Rohrer create the scanning tunneling microscope, which can image individual atoms
Early 1980s	The discovery, in the early 1980s, by David L. Allara and Ralph G. Nuzzo of the self-assembly of disulfide and, soon thereafter, of alkanethiol monolayers (SAMs) on metal surfaces
1982	The use of x-ray lithography in combination with electroplating and molding (or LIGA) is introduced by Ehrfeld and his colleagues[29]
1984	Pohl develops near-field scanning optical microscope (NSOM, also known as SNOM)[33]
1985 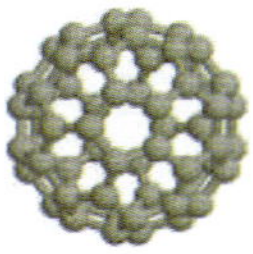	Robert F. Curl, Jr., Harold W. Kroto, and Richard E. Smalley discover buckminsterfullerenes, also known as buckyballs, which measure about a nanometer in diameter[39]

(*continued*)

Table 1.3 A Milestone Chart in Nanotechnology (Continued)

1986	Ashkin and colleagues reported the first observation of what is now commonly referred to as an optical trap, i.e., a tightly focused beam of light capable of holding microscopic particles stable in three dimensions[38]
1986	K. Eric Drexel publishes *Engines of Creation*, a futuristic book that popularizes nanotechnology
1987	The possibility of two- and three-dimensionally periodic crystals with corresponding two- and three-dimensional bandgaps was suggested 100 years after Rayleigh, by Eli Yablonovitch[49] and Sajeev John[50]
1989 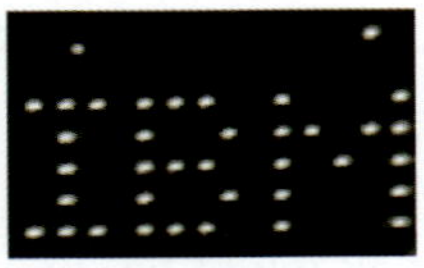	Donald M. Eigler of IBM writes the letters of his company's name using 35 individual xenon atoms on a nickel surface (in high vacuum and at liquid helium temperatures)
1991	Yablonovitch[52] demonstrates the first microwave photonic bandgap (PBG) structure experimentally (holes drilled in dielectric), known now as *yablonovite*
1991	Sumio Iijima of NEC in Tsukuba, Japan, discovers carbon nanotubes. The first single-walled nanotubes (SWNT) were produced in 1993
1993	The first high-quality quantum dots are synthesized by Murray, Norris, and Bawendi[45,67]
1993	Warren Robinett of the University of North Carolina and R. Stanley Williams of UCLA devise a virtual reality system connected to an STM that lets users see and touch atoms
1997	The first complete metal oxide semiconductor transistor (60 nm wide) is invented by Lucent Technologies. The key breakthrough was the 1.2-nm-thick gate oxide
1997	Whitesides et al.[27] introduced soft lithography, including the use of pattern transfer of self-assembled monolayers (SAMs) by elastomeric stamping
1998	Cees Dekker's group at the Delft University of Technology in the Netherlands creates a transistor from a carbon nanotube[40]
1999 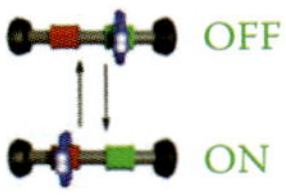	James Heath and Fraser Stoddart of UCLA (http://stoddart.chem.ucla.edu/) create rudimentary molecular switches, with molecules called rotaxanes, which function as molecular switches
1999	James M. Tour, now at Rice University, and Mark A. Reed of Yale University demonstrate that single molecules can act as molecular switches[68]
1999	The Clinton administration announces the National Nanotechnology Initiative, which provides a big boost in funding and gives the field greater visibility
1999 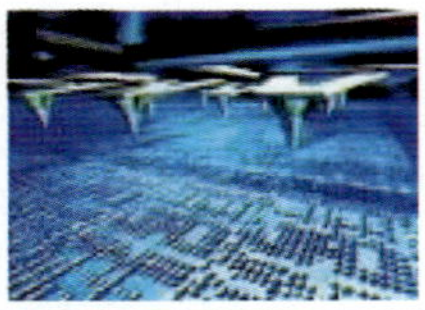	Thermomechanical memory device, unofficially known as "Millipede," first demonstrated at IBM Zurich
2000	Eigler and other IBM scientists devise a quantum mirage—placing a magnetic atom at the focus of an elliptical ring of atoms creates a mirage atom at the other focus—transmitting info without wires
2000	John Pendry, a physicist at Imperial College in London, showed that certain metals could be engineered to respond to electric fields as though the field parameters were negative[55]
2001	Researchers at Imperial College and Marconi Caswell (London) announced a magnetic resonance imaging system using a magnetic metamaterial based on Pendry's design[59]
2004	Physicists at the University of Manchester make graphene sheets[69]
2005	Zhang et al. demonstrate the near-field superlens—imaging objects in the tens of nanometer range with 365 nm light[61]
2007	The first left-handed material in the visible range[62]

Acknowledgments

Special thanks to Xavier Casadevall i Solvas and Drs. Sylvia Daunert and Benjamin Park.

Appendix 1A: International Technology Roadmap for Semiconductors (ITRS)

The complete 2003 ITRS and past editions of the ITRS editions are available for viewing and printing as an electronic document at http://public.itrs.net. The International Technology Roadmap for Semiconductors (ITRS) predicts the main trends in the semiconductor industry spanning across 15 years into the future. The participation of experts from Europe, Japan, Korea, and Taiwan as well as the United States ensures that the ITRS is a valid source of guidance for the semiconductor industry as it strives to extend the historical advancement of semiconductor technology and the worldwide integrated circuit (IC) market. The 2003 ITRS edition, used as the source for the tables below, extends to the year 2018. The 2003 ITRS does not predict a further acceleration in the timing of introduction of new technologies as the industry struggles through the worst recession of its history during the past couple of years. As projected, though, the half-pitch of 90 nm (hp90 nm) for DRAMs was introduced in 2004 (Intel's Prescott Pentium IV). Traditionally, the ITRS has focused on the continued scaling of CMOS (complementary metal-oxide-silicon) technology. By 2001, the horizon of the Roadmap started challenging the most optimistic projections for continued scaling of CMOS (e.g., MOSFET channel lengths below 9 nm). By that time it also became difficult for most people in the semiconductor industry to imagine how one could continue to afford the historic trends of increase in process equipment and factory costs for another 15 years! Thus, the ITRS started addressing post-CMOS devices. The Roadmap became necessarily more diverse for these devices, ranging from more familiar nonplanar CMOS devices to exotic new devices such as spintronics. Whether extensions of CMOS or radical new approaches, post-CMOS technologies must further reduce the cost per function and increase the performance of integrated circuits. Thus, new technologies may involve not only new devices but also new manufacturing paradigms.

The ITRS technology nodes in the table below are defined as the minimum metal pitch used on any product, for example, either DRAM half-pitch or Metal 1 (M1) half-pitch in Logic/MPU (see also figure below the tables). In 2003, DRAMs continue to have the smallest metal half-pitch; thus, it continues to represent the technology node. Commercially used numbers for the technology generations typically differ from the ITRS technology node numbers. However, the most reliable technology standard in the semiconductor industry is provided by the above definition, which is quite clear in that the patterning and processing (e.g., etching) capabilities of the technology are represented as the pitch of the minimum metal lines. The above definition is maintained not only for the 2003 version but also as a continuation from previous ITRS editions. Therefore, the official 2003 ITRS metal hpXX node indicator has been added to differentiate the ITRS definition from commercial technology generation numbers. Interim shrink-level node trend numbers are calculated and included for convenience of monitoring the internode progress of the industry.

Near-Term Years

Year of Production	2003	2004	2005	2006	2007	2008	2009
Technology node		hp90			hp65		
DRAM half-pitch (nm)	100	90	80	70	65	57	50
MPU/ASIC MI half-pitch (nm)	120	107	95	85	75	67	60
MPU/ASIC Poly Si half-pitch (nm)	107	90	80	70	65	57	50
MPU printed gate length (nm)	65	53	45	40	35	32	28
MPU physical gate length (nm)	45	37	32	28	25	22	20

Long-Term Years

Year of Production	2010	2012	2013	2015	2016	2018
Technology node	hp45		hp32		hp22	
DRAM half-pitch (nm)	45	35	32	25	22	18
MPU/ASIC MI half-pitch (nm)	54	42	38	30	27	21
MPU/ASIC Poly Si half-pitch (nm)	45	35	32	25	22	18
MPU printed gate length (nm)	25	20	18	14	13	10
MPU physical gate length (nm)	18	14	13	10	9	7

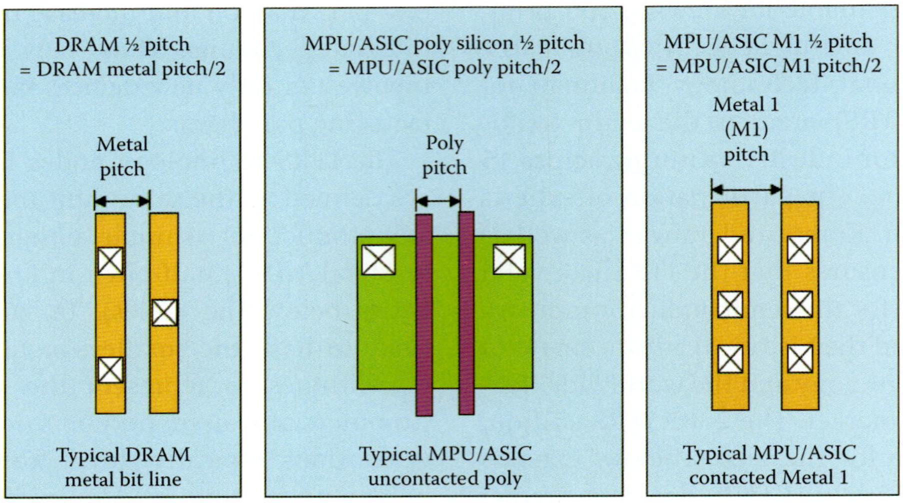

Appendix 1B: Worldwide IC and Electronic Equipment Sales

	Amounts in US $M				Year on Year Growth in %			
	2003	2004	2005	2006	2003	2004	2005	2006
Americas	**32,330.7**	**39,514.2**	**41,734.6**	**40,089.1**	3.4	22.2	5.6	−3.9
Europe	**32,310.0**	**40,537.5**	**43,693.5**	**43,082.1**	16.3	22.5	7.8	−1.4
Japan	**38,942.2**	**47,822.9**	**51,066.8**	**50,306.9**	27.7	22.8	6.8	−1.5
Asia Pacific	**62,842.6**	**85,756.0**	**95,253.0**	**96,546.2**	22.8	36.5	11.1	1.4
Total world*	**166,425.5**	**213,630.6**	**231,748.0**	**230,024.4**	18.3	28.4	8.5	−0.7
Discrete semiconductors	**13,347.0**	**16,043.4**	**17,036.5**	**16,689.1**	8.1	20.2	6.2	−2.0
Optoelectronics	**9,544.7**	**13,100.8**	**14,851.7**	**15,281.2**	40.6	37.3	13.4	2.9
Sensors and actuators	**3,569.2**	**4,827.6**	**5,739.0**	**6,262.1**	†	35.3	18.9	9.1
Integrated circuits	**139,964.7**	**179,658.8**	**194,120.8**	**191,792.0**	16.1	28.4	8.0	−1.2
Bipolar	216.8	239.7	200.8	150.6	−4.2	10.6	−16.3	−25.0
Analog	26,793.9	33,652.2	36,971.8	36,952.4	12.0	25.6	9.9	−0.1
Micro	43,526.1	52,412.0	57,218.6	57,564.8	14.3	20.4	9.2	0.6
Logic	36,921.9	46,421.3	50,631.8	49,571.9	18.1	25.7	9.1	−2.1
Memory	32,506.0	46,933.6	49,097.9	47,552.3	20.2	44.4	4.6	−3.1
Total products*	**166,425.5**	**213,630.6**	**231,748.0**	**230,024.4**	18.3	28.4	8.5	−0.7

* All numbers are displayed as rounded to one decimal place, but totals are calculated to three decimal places precision.

† WSTS included actuators in this category from 2003. Before only sensors were reported. Therefore, a growth rate is not meaningful to show.

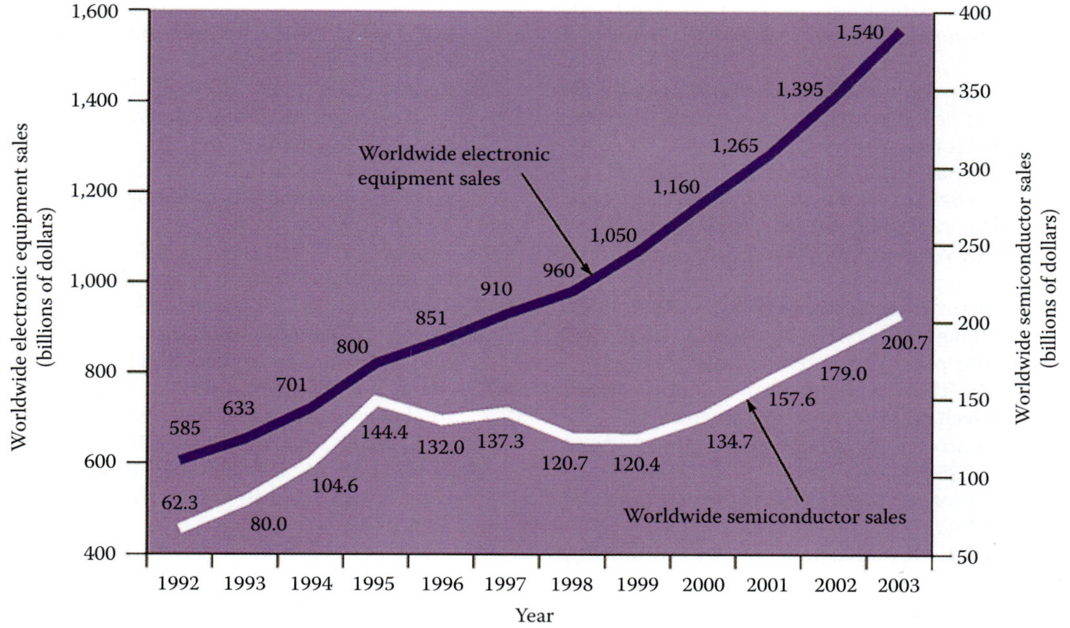

Percent semiconductor	10.7	12.6	14.9	18.1	15.5	15.1	12.6	11.5	11.6	12.5	12.8	13.0

Questions

1.1: Why is silicon so important to MEMS and NEMS?

1.2: Compare the pros and cons of transistors and vacuum tubes.

1.3: Why was the Si-MEMS market at one point in time expected to be much larger than the IC market?

1.4: Can you list some of the current technological and economic barriers that restrict the wider commercialization of Si-MEMS?

1.5: (a) State Moore's first law (we are talking about Moore, Intel's cofounder). (b) What is Moore's second law?

1.6: Why did surface micromachining catch on so fast with the IC industry?

1.7: Why are MEMS market forecasts so difficult to prepare? How would you go about making a better MEMS market forecast?

1.8: How does radar work? How is it useful?

1.9: What is the biggest advantage Ge has over Si IC circuits?

1.10: What is a strain gauge and what is its gauge factor?

1.11: What is the definition of nanotechnology?

1.12: List at least five commercial products that incorporate nanotechnology.

1.13: What year was the word "nanotechnology" first used?

1.14: What was Feynman's role in catalyzing the genesis of MEMS and NEMS?

1.15: Why was the honeymoon with the transistor over so quickly? What technology took over very fast?

1.16: What does ITRS stand for? What does it mean?

1.17: List a number of nanostructures that have been fabricated with bottom-up methodologies.

1.18: What is a photonic crystal?

1.19: What is a metamaterial?

1.20: What are the important differences between typical devices made in the IC industry and MEMS?

References

We took advantage of Google and Wikipedia on numerous occasions and also relied on the following references:

1. Ohl, R. S. 1946. Light-sensitive electric device, Bell Labs. US Patent 2402662.
2. Bardeen, J., and W. Brattain. 1950. Three electrode circuit element utilizing semiconductive materials, Bell Labs. US Patent 2524035.

3. Riordan, M., and L. Hoddeson. 1998. *Crystal fire: the birth of the information age.* New York: W. W. Norton & Company.
4. Reid, T. R. 2001. *The chip: how two Americans invented the microchip and launched a revolution.* New York: Random House Publishing House.
5. ITRS. 2007. International Technology Roadmap for Semiconductors. *Presentations from the 2007 ITRS Conference,* December 5, 2007, Makuhari Messe, Japan. http://www.itrs.net/Links/2007Winter/2007_Winter_Presentations/Presentations.html.
6. World Semiconductor Trade Statistics (WSTS). 2004. WSTS Semiconductor Market Forecast Spring 2004. Press release. http://www.wsts.org/plain/content/view/full/869.
7. Feynmann, R. 1959. There is plenty of room at the bottom. http://www.its.caltech.edu/~feynman/plenty.html.
8. Uhlir, A. 1956. Electrolytic shaping of germanium and silicon. *Bell Syst Tech J* 35:333–47.
9. Smith, C. S. 1954. Piezoresistance effect in germanium and silicon. *Phys Rev* 94:42–49.
10. Pfann, W. G. 1961. Improvement of semiconducting devices by elastic strain. *Solid State Electron* 3:261–67.
11. Tufte, O. N., P. W. Chapman, and D. Long. 1962. Silicon diffused-element piezoresistive diaphragms. *J Appl Phys* 33:3322–27.
12. Waggener, H. A., R. C. Kragness, and A. L. Taylor. 1967. *International Electron Devices Meeting, IEDM '67 68.* Washington, DC, IEEE.
13. Wallis, P. R., and D. I. Pomeranz. 1969. Field assisted glass-metal sealing. *J Appl Phys* 40:3946–49.
14. Shimbo, M., K. Furukawa, K. Fukuda, and K. Tanzawa. 1986. Silicon-to-silicon direct bonding method. *J Appl Phys* 60:2987–89.
15. National Semiconductor. 1974. *Transducers, Pressure, and Temperature* (catalog). Sunnyvale, CA: Author.
16. Angell, J. B., S. C. Tery, and P. W. Barth. 1983. Silicon micromechanical devices. *Scientific American* 248:44–55.
17. Greenwood, J. C. 1969. Ethylene diamine-cathechol-water mixture shows preferential etching of a p-n junction. *J Electrochem Soc* 116:1325–26.
18. Bohg, A. 1971. Ethylene diamine-pyrocatechol-water mixture shows etching anomaly in boron-doped silicon. *J Electrochem Soc* 118:401–02.
19. Texas Instruments. 1977. Texas Instruments Thermal Character Print Head. Austin, TX: Author.
20. Bassous, E., H. H. Taub, and L. Kuhn. 1977. Ink jet printing nozzle arrays etched in silicon. *Appl Phys Lett* 31:135–37.
21. O'Neil, P. 1980. A monolithic thermal converter. *Hewlett-Packard J,* 12.
22. Terry, S. C., J. H. Jerman, and J. B. Angell. 1979. A gas chromatograph air analyzer fabricated on a silicon wafer. *IEEE Trans Electron Devices* 26:1880–86.
23. Terry, S. C. 1975. *A gas chromatography system fabricated on a silicon wafer using integrated circuit technology.* PhD diss., Stanford University.
24. Petersen, K. E. 1980. Silicon torsional scanning mirror. *IBM J Res Dev* 24:631–37.
25. Petersen, K. E. 1982. Silicon as a mechanical material. *Proc IEEE* 70:420–57.
26. Howe, R. T., and R. S. Muller. 1983. Polycrystalline silicon micromechanical beams. *J Electrochem Soc* 130:1420–23.
27. Xia, Y., and G. M. Whitesides. 1998. Soft lithography. *Angew Chem Int Ed Engl* 37:551–75.
28. Nathanson, H. C., W. E. Newell, R. A. Wickstrom, and J. R. Davis. 1967. The resonant gate transistor. *IEEE Trans Electron Devices* ED-14:117–33.
29. Becker, E. W., W. Ehrfeld, D. Munchmeyer, H. Betz, A. Heuberger, S. Pongratz, W. Glashauser, H. J. Michel, and V. R. Siemens. 1982. Production of separation nozzle systems for uranium enrichment by a combination of x-ray lithography and galvanoplastics. *Naturwissenschaften* 69:520–23.
30. Fan, L. S., Y. C. Tai, and R. S. Muler. 1989. IC-processed electrostatic micro-motors. *Sensors and Actuators* A20:41–48.
31. Tang, W. C., T. C. Nguyen, and R. T. Howe. 1989. Laterally driven polysilicon resonant microstructures. *IEEE Micro Electro Mechanical Systems* 20:25–32.
32. Manz, A., N. Graber, and H. M. Widmer. 1990. Miniaturized total chemical analysis systems: A novel concept for chemical sensing. *Sens Actuators* B1:244–248.
33. Pohl, D. W., W. Denk, and M. Lanz. 1984. Optical stethoscopy: image recording with resolution l/20. *Appl Phys Lett* 44:651–53.
34. Allara, D. L., and R. G. Nuzzo. 1985. Spontaneously organized molecular assemblies; II. Quantitative infrared spectroscopic determination of equilibrium structures of solution adsorbed n-alkanoic acids on an oxidized aluminum surface. *Langmuir* 1:52–66.
35. Allara, D. L., and R. G. Nuzzo. 1985. Spontaneously organized molecular assemblies; I. Formation, dynamics and physical properties of n-alkanoic acids adsorbed from solution on an oxidized aluminum surface. *Langmuir* 1:45–52.
36. Tomalia, D. A., H. Baker, J. R. Dewald, M. Hall, G. Kallos, S. Martin, J. Roeck, J. Ryder, and P. Smith. 1985. A new class of polymers: starburst-dendritic macromolecules. *Polym J* 17:117–32.
37. Ashkin, A. 1970. Acceleration and trapping of particles by radiation pressure. *Phys Rev Lett* 24:156–59.
38. Ashkin, A., J. M. Dziedzic, J. E. Bjorkholm, and S. Chu. 1986. Observation of a single-beam gradient force optical trap for dielectric particles. *Optics Letters* 11:288–90.
39. Kroto, H. W., J. R. Heath, S. C. O'Brien, R. F. Curl, and R. E. Smalley. 1985. C60: buckminsterfullerene. *Nature* 318:162–63.
40. Tans, S. J., A. R. M. Verschueren, and C. Dekker. 1998. Room-temperature transistor based on a single carbon nanotube. *Nature* 393:49–52.
41. Ekimov, A. I., A. L. Efros, and A. A. Onushchenko. 1985. Quantum size effect in semiconductor microcrystals. *Solid State Commun* 56:921–24.
42. Efros, A. L., and A. L. Efros. 1982. Interband absorption of light in a semiconductor sphere. *Sov Phys Semicond* 16:772–74.
43. Brus, L. E. 1984. On the development of bulk optical properties in small semiconducting crystallites. *J Luminescence* 31/32: 381.
44. Brus, L. E. 1984. Electron-electron and electron-hole interactions in small semiconductor crystallites: the size dependence of the lowest excited electronic state. *J Chem Phys* 80:4403–07.
45. Murray, C. B., D. J. Norris, and M. G. Bawendi. 1993. Synthesis and characterization of nearly monodisperse CdE (E=S, Se, Te) semiconductor nanocrystallites. *J Am Chem Soc* 115:8706–15.
46. Wagner, R. S., and W. C. Ellis. 1964. Vapour-liquid-solid mechanism of single crystal growth. *Appl Phys Lett* 4:89–90.

47. Aviram, A., and M. A. Ratner. 1974. Molecular rectifiers. *Chem Phys Lett* 29:277.
48. Rayleigh, J. W. S. 1888. On the remarkable phenomenon of crystalline reflexion described by Prof. Stokes. *Phil Mag* 26:256–65.
49. Yablonovitch, E. 1987. Inhibited spontaneous emission in solid state physics and electronics. *Phys Rev Lett* 58:2059.
50. John, S. 1987. Strong localization of photons in certain disordered dielectric superlattices. *Phys Rev Lett* 58:2486.
51. Yablonovitch, E., T. J. Gmitter, and K. M. Leung. 1991. Photonic band structure: the face-centered cubic case employing nonspherical atoms. *Phys Rev Lett* 67: 2295–98.
52. Veselago, V. G. 1968. The electrodynamics of substances with simultaneously negative values of e and m. *Sov Phys Usp* 10:509–14.
53. Pendry, J. B. 1999. Photonic gap materials. *Curr Sci* 76:1311.
54. Pendry, J. B. 2000. Negative refraction makes a perfect lens. *Phys Rev Lett* 85:3966.
55. Pendry, J. B. 2004. Manipulating the near field with metamaterials. *Optics Photonics News* 15:33–37.
56. Pendry, J. B. 2007. Metamaterials and the control of electromagnetic fields. *Proceedings of the Ninth Rochester Conference on Coherence and Quantum Optics*. Washington, DC: Optical Society of America.
57. Pendry, J. B., A. J. Holden, D. J. Robbins, and W. J. Stewart. 1998. Low frequency plasmons in thin wire structures. *J Phys Cond Matter* 10:4785.
58. Ramakrishna, S. A., J. B. Pendry, M. C. K. Wiltshire, and W. J. Stewart. 2003. Imaging the near field. *J Mod Optics* 50:1419–30.
59. Wiltshire, M. C. K., J. B. Pendry, I. R. Young, D. J. Larkman, D. J. Gilderdale, and J. V. Hajnal. 2001. Microstructured magnetic material for RF flux guides in magnetic resonance imaging (MRI). *Science* 291:848–51.
60. Shelby, R. A., D. R. Smith, and S. Schultz. 2001. Experimental verification of a negative refractive index. *Science* 292:77–79.
61. Fang, N., H. Lee, and X. Zhang. 2005. Sub-diffraction-limited optical imaging with a silver superlens. *Science* 308:534–37.
62. Dolling, G., M. Wegener, C. M. Soukoulis, and S. Linden. 2007. Negative-index metamaterials at 780 nm wavelength. *Opt Lett* 32:53–55.
63. Drexel, K. E. 1987. *Engines of creation*. New York: Anchor Books.
64. Drexel, K. E. 1992. *Nanosystems, molecular machinery, manufacturing, and computation.* New York: John Wiley & Sons, Inc.
65. Atkinson, W. I. 2003. *Nanocosm: nanotechnology and the big changes coming from the inconceivable small.* New York: AMACOM.
66. Regis, E. 1995. *Nano: remaking the world atom by atom.* Boston: Little, Brown and Company.
67. Murray, C. B., C. R. Kagan, and M. G. Bawendi. 1995. Self-organization of CdSe nanocrystallites into three-dimensional quantum dot superlattices. *Science* 270:1335–38.
68. Reed, M. A., C. Zhou, C. J. Muller, T. P. Burgin, and J. M. Tour. 1997. Conductance of a molecular junction. *Science* 278:252–54.
69. Geim, A. K., and K. S. Novoselov. 2007. The rise of graphene. *Nature Materials* 6:183–191.

2

Crystallography

The designer Tokujin Yoshioka makes his *Venus – Natural Crystal Chair* by submerging a block of polyester fibers in the shape of a straight-backed dining chair in a vat of water and then adding a mineral to crystallize it. (Courtesy of Mr. Tokujin Yoshioka.)

Outline

Introduction

Crystallography is the science of analyzing the crystalline structure of materials. The spatial arrangement of atoms within a material plays a most important role in determining the precise properties of that material. Based on the degree of order, materials are classified as amorphous, with no recognizable long-range order; polycrystalline, with randomly ordered domains (10 Å to a few μm); and single crystalline, where the entire solid is made up of repeating units in an orderly array. This classification is illustrated in Figure 2.1. Amorphous solids (e.g., glasses and plastics) are homogeneous and isotropic because there is no long-range order or periodicity in the internal arrangement.

Many engineering materials are aggregates of small crystals of varying sizes and shapes. The size of the single-crystal grains may be as small as a few nanometers but could also be large enough to be seen by the naked eye. Regions between grains are called grain boundaries. These polycrystalline materials have properties determined by both the chemical nature of the individual crystals and their aggregate properties, such as size and shape distribution, and in the orientation relationships between them. In the case of thin polycrystalline films, material properties might deviate significantly from bulk crystalline behavior, as we discover in

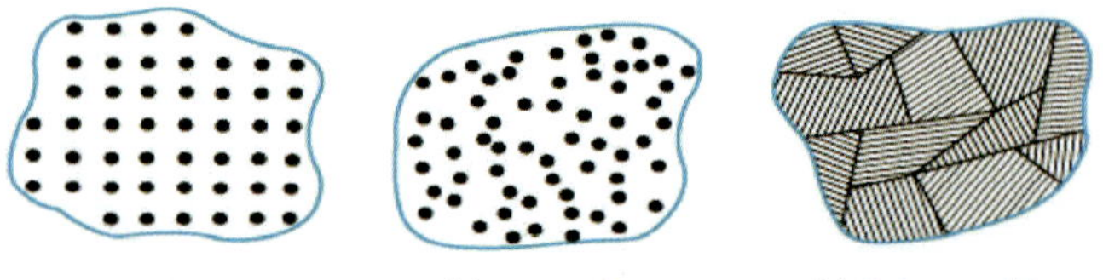

FIGURE 2.1 Classification of materials as crystalline (a), amorphous (b), and polycrystalline (c). (Drawing by Mr. Chengwu Deng.)

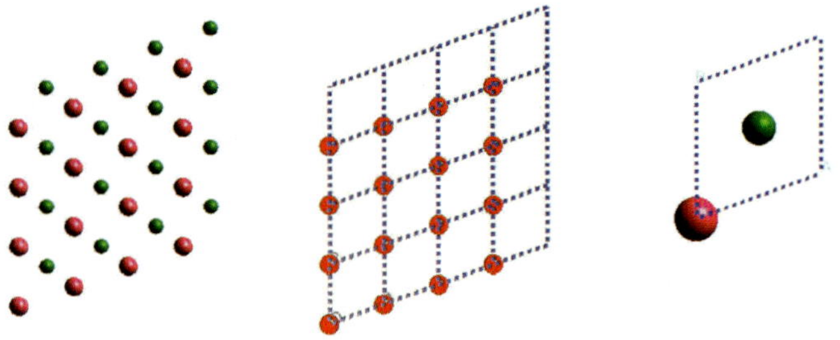

FIGURE 2.2 Any crystal lattice can be simplified to a three-dimensional array of periodically located points in space. Such a periodic array, specifying how the repeated units of a crystal are arranged, is called a *Bravais lattice*. A real crystal is made up of a basis and a lattice. (Drawing by Mr. Chengwu Deng.)

Volume II, Chapter 7, where we deal with thin film properties and surface micromachining. In the case of nanoparticles, deviation from expected bulk theory is even more pronounced (see Chapter 3 on quantum mechanics and the band theory of solids in the current volume). The crystal structure of a nanoparticle is not necessarily the same as that of the bulk material. Nanoparticles of ruthenium (2–3 nm in diameter), for example, have body-centered cubic (lattice point at each corner plus one at the center; see below) and face-centered cubic structures (lattice points at each corner as well as in the centers of each face; see below) not found in bulk ruthenium.

Most important, semiconductor devices are based on crystalline materials because of their reproducible and predictable electrical properties. Crystals are anisotropic—their properties vary with crystal orientation. In this chapter we explain the importance of the symmetry of point groups and space groups in determining, respectively, bulk physical properties and microscopic properties of crystalline solids, properties relied on for building miniaturized electronics, sensors, and actuators. We also launch the concept of reciprocal space (also called Fourier space, **k**-space, or momentum space), clarify the conditions for x-ray diffraction in terms of such a reciprocal space, and offer an introduction to Brillouin zones. All these elements are needed for our introduction to the band theory of solids in Chapter 3. We finish Chapter 2 with a description of crystal defects.

Bravais Lattice, Unit Cells, and the Basis

Under special conditions almost every solid can be made into a crystal (helium is the only substance that does not form a solid). Atoms organize themselves into crystals because energy can be minimized that way. Any crystal lattice can be simplified to a three-dimensional (3D) array of periodically located points in space as shown in Figure 2.2 in the case of a two-dimensional (2D) crystal. Such a periodic array, specifying how the repeated units of a crystal are arranged, is called a *Bravais lattice*. Bravais, in 1848, demonstrated that there are only 14 ways of arranging points symmetrically in space that do not lead to voids in a crystal (in 2D there are only five such lattices). All crystalline materials, including nanomaterials, assume one of the 14 Bravais lattices. A real crystal can be described in terms of a Bravais lattice, with one specific atom (or ion) or a group of atoms (a molecule), called a *basis*, attached to each lattice point (Figure 2.2). The basis superposed on the Bravais lattice renders the complete crystal structure. The 3D Bravais lattice can be mathematically defined by three noncoplanar basis vectors, $\mathbf{a}_1$, $\mathbf{a}_2$, and $\mathbf{a}_3$, which are the three independent shortest vectors connecting lattice points. These vectors form a parallelepiped called a *primitive cell*, i.e., a cell that can reproduce the entire crystal lattice by translation alone. Such a primitive cell is a minimum volume cell with a density of only one lattice point per cell—there are lattice points at each of the eight corners of the parallelepiped, but each corner point is shared among the eight cells that come together there (1/8 of a point at each corner). The lattice translational vector **r** is given by:

$$\mathbf{r} = n_1\mathbf{a}_1 + n_2\mathbf{a}_2 + n_3\mathbf{a}_3 \quad (2.1)$$

where n_1, n_2, and n_3 are integers. A displacement of any lattice point by **r** will result in a new position in the lattice that has the same positional appearance as the original position. A lattice translation vector, **r**, as described in Equation 2.1, connects two points in the lattice that exhibit identical point symmetry.

Nonprimitive unit cells or simple unit cells are also called *conventional* unit cells or *crystallographic* unit cells. They are not necessarily unique and need not be the smallest cell possible. Primitive cells are chosen with the shortest possible vectors, whereas unit cells are chosen for the highest symmetry and may contain more than one lattice point per cell. The unit cell in a lattice, like a primitive cell, is representative of the entire lattice. The simplest unit cell belongs to a cubic lattice, which is further divided into simple cubic (SC), face-centered cubic (FCC), and body-centered cubic (BCC) as illustrated in Figure 2.3.

An FCC lattice has the closest atomic packing, then BCC, and then SC. For a simple cubic crystal (SC) unit cell, as shown in Figure 2.3, $\mathbf{a}_1 = \mathbf{a}_2 = \mathbf{a}_3$, and the axes angles are $\alpha = \beta = \gamma = 90°$. The dimension a $(= \mathbf{a}_1 = \mathbf{a}_2 = \mathbf{a}_3)$ is known as the *lattice constant.* For SC the conventional unit cell coincides with the primitive cell. This is not true for FCC and BCC as we shall see in Figure 2.5 below.

The 14 possible Bravais lattices can be subdivided into 7 different "crystal classes" based on the choice of conventional unit cells. These 7 crystal classes are cubic, tetragonal, trigonal, hexagonal, monoclinic, orthorhombic, and triclinic. Each of these systems is characterized by a set of symmetry elements, and the more symmetry elements a crystal exhibits, obviously, the higher its symmetry. A cubic crystal has the highest possible symmetry and a triclinic crystal the lowest. The 14 Bravais lattices categorized according to the 7 crystal systems are shown in Figure 2.4.

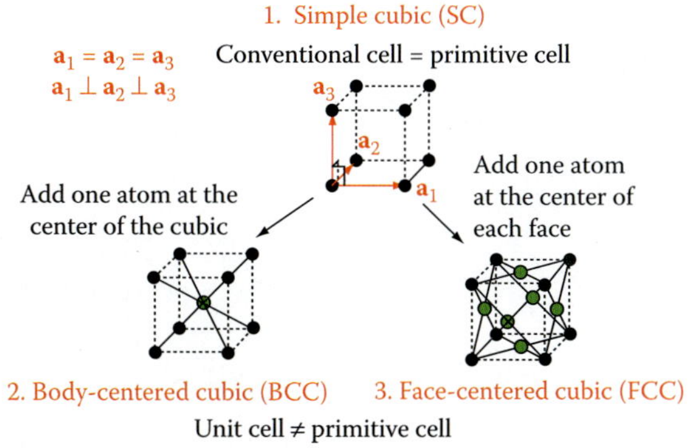

FIGURE 2.3 The simplest unit cell belongs to a cubic lattice, which is further divided into: simple cubic (SC), face-centered cubic (FCC), and body-centered cubic (BCC).

A Wigner-Seitz cell is a primitive cell with the full symmetry of the Bravais lattice. It is an important construct for the understanding of Brillouin zones, the boundaries of which satisfy the Laue conditions for diffraction (see below). To appreciate how Wigner-Seitz cells are constructed, we illustrate some simple examples for the case of two different types of 2D lattices in Figure 2.5. Lines are drawn passing through the middle points of dotted lines connecting nearest neighbors. In 3D, the Wigner-Seitz cells

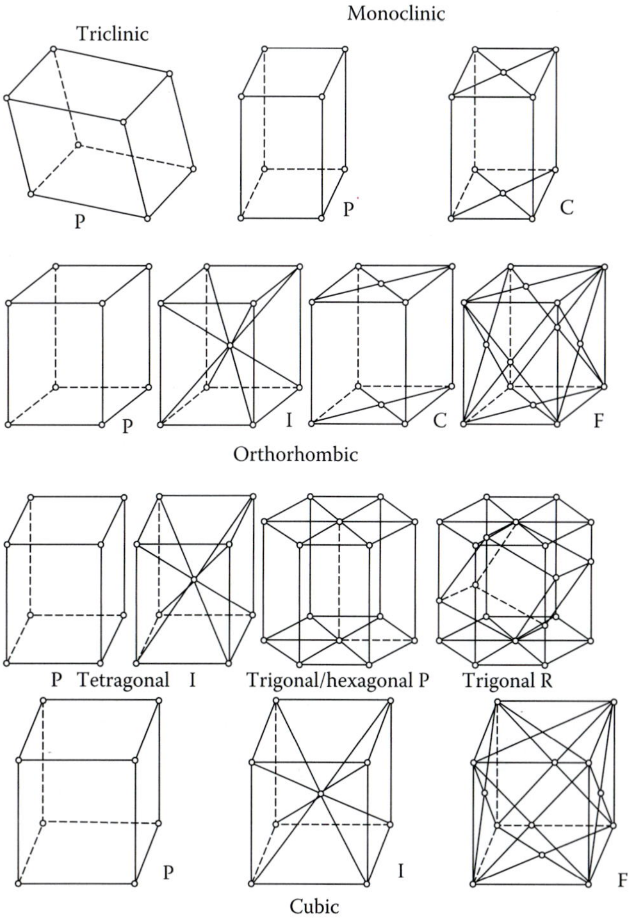

FIGURE 2.4 Conventional unit cells for the 14 Bravais lattices arranged according to the 7 crystal systems. *P* means lattice points on corners only, *C* means lattice points on corners as well as centered on faces, *F* means lattice points on corners as well as in the centers of all faces, and lattice points on corners as well as in the center of the unit cell body are indicated by *I*.

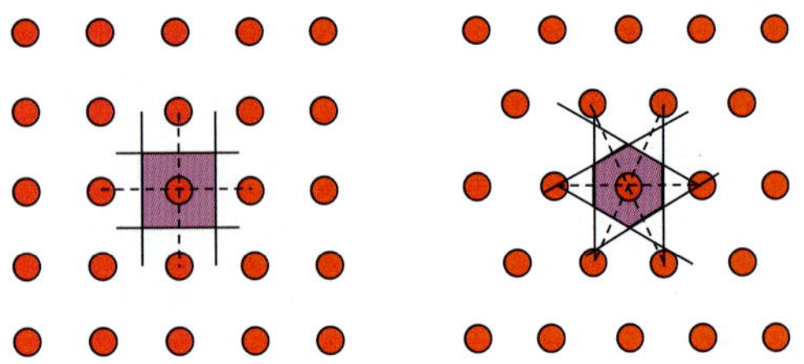

FIGURE 2.5 Wigner-Seitz primitive cells for two types of simple 2D lattices.

are polyhedra constructed about each atom by drawing planes that are the perpendicular bisectors of the lines between nearest neighbors. The Wigner-Seitz cell about a lattice point is the region of space that is closer to that point than any other lattice point. Wigner-Seitz cells for FCC and BCC Bravais lattices are shown in Figure 2.6. In the same figure we also show conventional unit and primitive unit cells for these lattices.

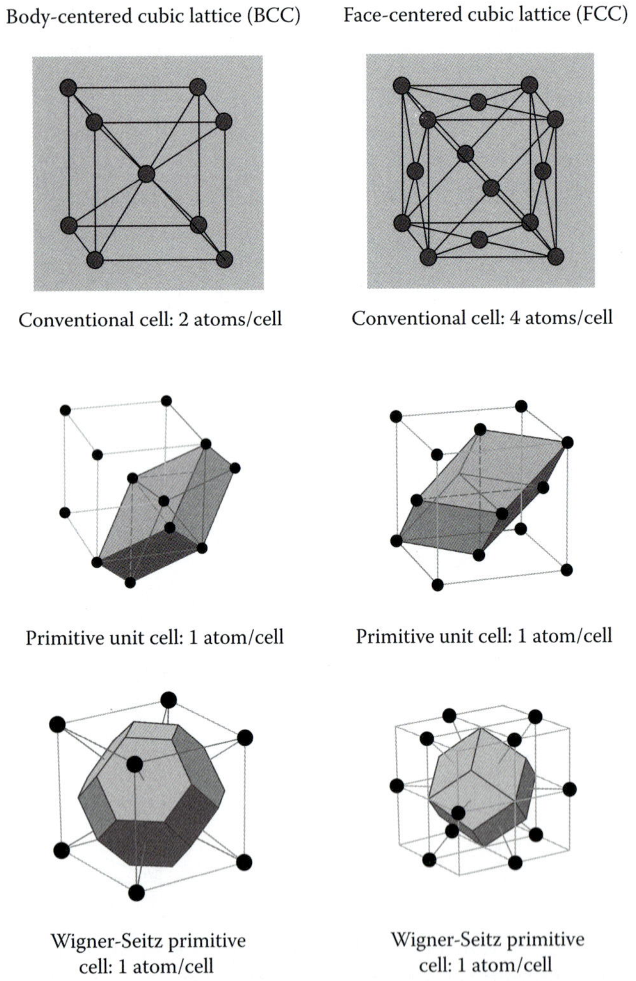

FIGURE 2.6 Conventional unit cells, primitive unit cells, and Wigner-Seitz primitive cells for BCC and FCC lattices. The BCC Wigner-Seitz unit cell is a truncated octahedron. The FCC Wigner-Seitz primitive unit cell is a rhombic dodecahedron.

Point Groups and Space Groups

A lattice translation as described by Equation 2.1 is a type of symmetry operation where a displacement of a crystal parallel to itself carries the crystal structure into itself (Figure 2.7).

Rotation and reflection or a combination of rotation and reflection—a so-called compound symmetry operation—about various points are other symmetry operations that may "carry the crystal into itself" (see Figure 2.8). The point around which the symmetry operation is carried out may be a lattice point or a special point within the elementary parallelepiped. There are five types of rotation axes possible, i.e., one- (360°), two- (180°), three- (120°), four- (90°), and sixfold (60°) rotation. One sees from Figure 2.8 why fivefold rotational symmetry does not occur in nature; it just cannot be stacked without leaving holes. This explains, for example, why we do not see ice crystals with a pentagon shape (Figure 2.9). Mirror reflection takes place about a plane through a lattice or special point. An inversion operation is an example of a compound symmetry operation and is achieved by rotation of π, followed by a reflection in a plane normal to the rotation axis; the effect is also illustrated in Figure 2.8. The collection of point symmetry elements possessed by a crystal is called a *point group* and is defined as the collection of symmetry operations which, when applied about a point, leave the lattice invariant. There are 32 crystallographic point groups in all.

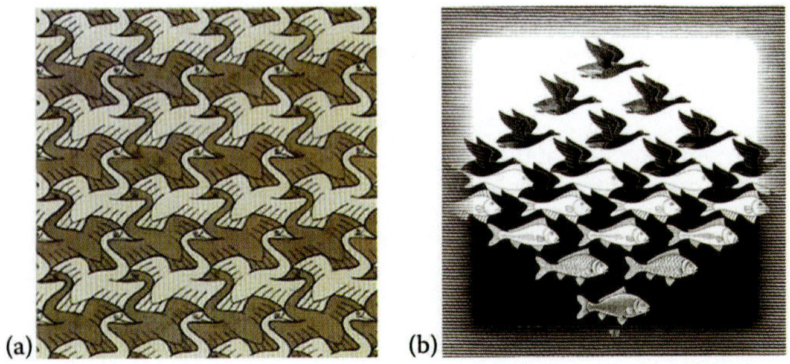

FIGURE 2.7 The drawing on the left (a) is crystal-like and can be carried into itself by a translation that is not possible in the figure on the right (b). The latter is missing a translation vector and is not crystal-like.

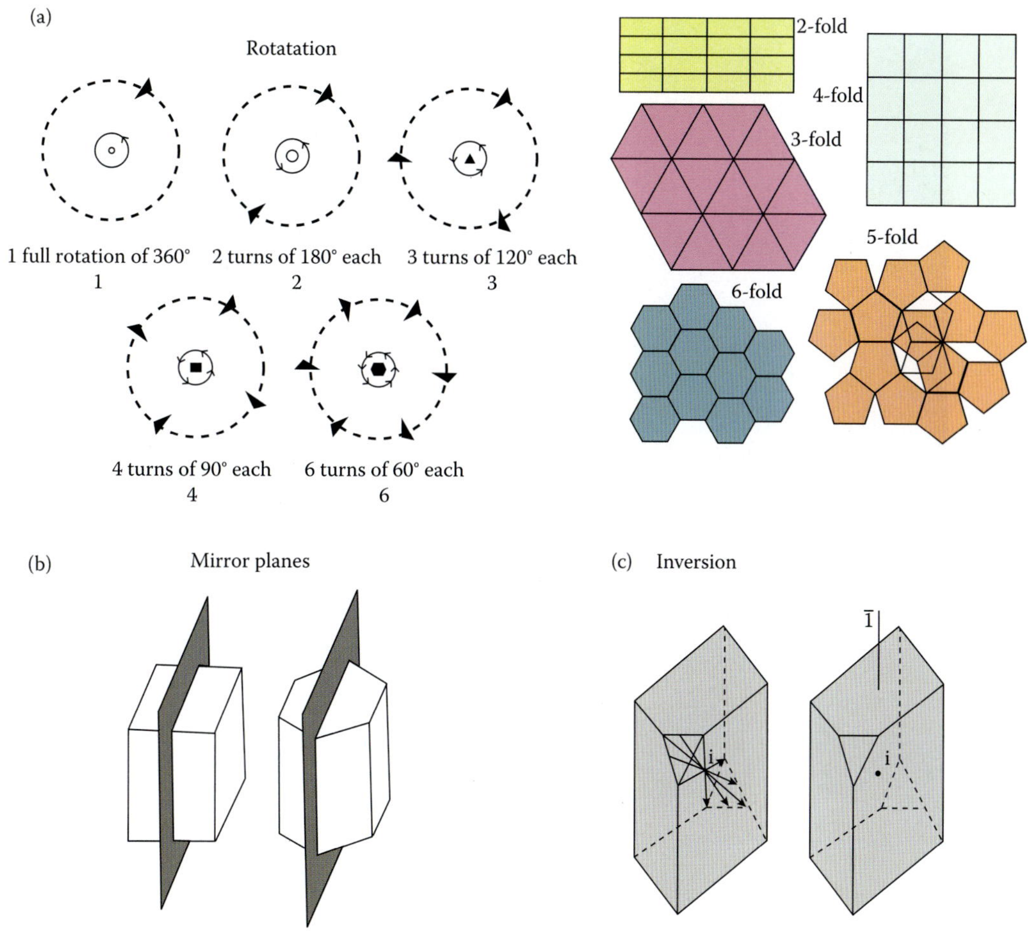

FIGURE 2.8 Point symmetry operations: (a) rotation, (b) reflection, and (c) a compound symmetry operation: inversion. The latter is made up of a rotation of π followed by reflection in a plane normal to the rotation axis. This is also called inversion through a point (i). The symbol for the inversion axis is $\bar{1}$.

The importance of the 32 point group symmetries and corresponding crystal classes is revealed by the important physical properties of crystalline solids they control, including electrical conductivity, thermal expansion, birefringence, piezoresistance, susceptibility, elastic stiffness coefficient, etc. Some properties that depend on the direction along which they are measured relative to the crystal axes are listed in Table 2.1. This orientation dependence of physical and chemical properties is called *anisotropy*. Anisotropy also explains why crystals do not grow into spheres but as polyhedra and why certain crystal directions etch faster than others. The reason for this anisotropy is the regular stacking of atoms in a crystal; as one passes along a given direction, one encounters atoms or groups of atoms at different intervals and from different angles than if one travels through the crystal from another direction. Single molecules in a liquid or a gas can also be anisotropic, but because they are free to move, liquids and gases are isotropic. In a crystal the anisotropy of atoms and groups of atoms is locked into the crystal structure. As a first example we consider a physical quantity such as current density **J** (column 4), and its cause the electrical field **E** (column 5). These quantities are linked, to a first approximation, in a linear relationship described by a tensor equation:

$$\mathbf{J} = \sigma \mathbf{E} \tag{2.2}$$

One remembers that a tensor field represents a single physical quantity that is associated with certain places in three-dimensional space and instants of time. The crystalline property, in column 1, is a tensor field of the rank listed in column 2. Also recall that a scalar (e.g., mass, temperature, charge) is a tensor of zeroth rank, and a vector (e.g., position, velocity, flow of heat) is a tensor of first rank. Other quantities such as stress inside a solid or fluid may be

FIGURE 2.9 Ice crystals. No pentagons are found in ice crystal stacking.

characterized by tensors of order two or higher. The physical properties listed in Table 2.1 are exploited to build electronics, sensors, and actuators in MEMS and NEMS.

We elaborate a bit further here on the anisotropy of the electrical conductivity σ in a single crystal, where a causal or forcing term, the electrical field **E**, causes a current density **J**. Because both are vectors, this case is referred to as a *vector-vector* effect. In general, the current vector may not have the same direction as the electric vector. Assuming a linear relationship between electrical field (cause) and

Table 2.1 Linear Physical Properties of Solids

Property and Symbol	Rank of Tensor	Number of Independent Components	Dependent Physical Quantity	Causal or Forcing Term
Pyroelectric coefficient p	1	3	Electric polarization dP	Temperature change dT
Conductivity σ	2	6	Current density **J**	Field **E**
Resistivity ρ	2	6	Field **E**	Current density **J**
Susceptibility χ	2	6	Electric polarization P	Electric field **E**
Thermal expansion α	2	6	Strain ε	Temperature change dT
Piezoelectric coefficient d	3	18	Electric polarization P_s	Stress S
Elastic stiffness constants χ	4	21	Stress S	Strain ε
Elastic compliance σ	4	21	Stress ε	Stress S
Piezoresistance π	4	21	Resistivity change ρ	Stress S

Unfortunately, symbols customarily used for these properties do sometimes overlap.

current (effect), we can describe the components of the current relative to an arbitrarily chosen Cartesian coordinate system as:

$$
\begin{aligned}
J_x &= \sigma_{xx}E_x + \sigma_{xy}E_y + \sigma_{xz}E_z \\
J_y &= \sigma_{yx}E_x + \sigma_{yy}E_y + \sigma_{yz}E_z \\
J_z &= \sigma_{zx}E_x + \sigma_{zy}E_y + \sigma_{zz}E_z
\end{aligned} \tag{2.3}
$$

The quantities σ_{ik} are components of a 3×3 "conductivity tensor." The resistivity tensor $\rho(= 1/\sigma)$ tensor, like the conductivity tensor, is a second rank tensor described by:

$$\mathbf{E} = \rho\mathbf{J} \tag{2.4}$$

Based on the basic symmetry of the equations of motion, Onsager demonstrated that the tensor is symmetric, i.e., $\sigma_{ik} = \sigma_{ki}$, so that the nine coefficients are always found to reduce to six. Taking advantage of this symmetry argument and multiplying the expressions in Equation 2.3 by E_x, E_y, and E_z, respectively, one obtains on adding:

$$
\begin{aligned}
J_xE_x + J_yE_y + J_zE_z &= \sigma_{xx}E^2{}_x + \sigma_{yy}E^2{}_y + \sigma_{zz}E^2{}_z \\
&+ 2\sigma_{xy}E_xE_y + 2\sigma_{yz}E_yE_z + 2\sigma_{zx}E_zE_x
\end{aligned} \tag{2.5}
$$

To make the mixed terms on the right side of Equation 2.5 disappear, one chooses a new coordinate system with the coordinates along the principal axes of the quadratic surface represented by this right-hand side (rhs) term, and in this new coordinate system one obtains:

$$J_x = \sigma_1E_x;\ J_y = \sigma_2E_y;\ J_z = \sigma_3E_z \tag{2.6}$$

where σ_1, σ_2, and σ_3 are the principal conductivities. The current and field vectors only have the same direction when the applied field falls along any one of the principal axes of the crystal. From Equation 2.6, no matter how low the symmetry of a crystal, it can always be characterized by three conductivities (σ_1, σ_2, and σ_3) or three specific resistivities (ρ_1, ρ_2, and ρ_3). In cubic crystals the three quantities are equal, and the specific resistivity does not vary with direction. In hexagonal, trigonal, and tetragonal crystals, two of the three principal conductivities (or resistivities) are the same. In such a case, the resistivity only depends on the angle θ between the direction in which ρ is measured and the hexagonal, trigonal, or tetragonal axis. One then finds:

$$\rho(\phi) = \rho_{per}\sin^2\phi + \rho_{par}\cos^2\phi \tag{2.7}$$

where the subscripts stand for perpendicular and parallel to the axis.

Another vector-vector effect example in Table 2.1 is the one involving thermal conductivity, where a thermal current vector is caused by a thermal gradient. Scalar-tensor effects lead to similar relations as vector-vector effects. For example, the deformation tensor of a solid resulting from a temperature change (scalar) involves three principal expansion coefficients, α_1, α_2, and α_3. The latter will again all be equal in the case of a cubic crystal, and the angular dependence of α for hexagonal, trigonal, and tetragonal crystals is given by an expression analogous to Equation 2.7.

A simple example of a scalar-vector effect from Table 2.1, illustrating the importance of crystal symmetry or lack thereof, involves pyroelectricity (see first row in Table 2.1). Pyroelectricity is the ability of a material to spontaneously polarize and produce a voltage as a result of changes in temperature. It must be a change in temperature: incident light may heat a pyroelectric crystal, thus changing its dipole moment and causing current to flow, but because pyroelectrics respond to the rate of change of temperature only, the light or heat source must be pulsed or modulated! A pyroelectric is a ferroic material, i.e., a class of smart multifunctional materials having both sensing and actuating functions. *Ferroic materials* is a simplified term to represent ferroelastic, ferromagnetic, and ferroelectric materials. In pyroelectricity, the opposite faces of certain crystals [e.g., tourmaline (Na, Ca)(Li, Mg, Al)(Al, Fe, Mn)$_6$(BO_3)$_3$(Si_6O_{18})(OH)$_4$ the "Ceylon magnet," ZnO, $BaTiO_3$, and $PbTiO_3$] become electrically charged as a result of a change in temperature. This is illustrated for tetragonal $BaTiO_3$ in Figure 2.10. In Table 2.1 we consider the linear relationship between the electric polarization **P** (a vector) and a temperature change (a scalar). Electric polarization of materials is covered from a theoretical point of view in Chapter 5. For practical applications of pyroelectricity in actuator construction refer to

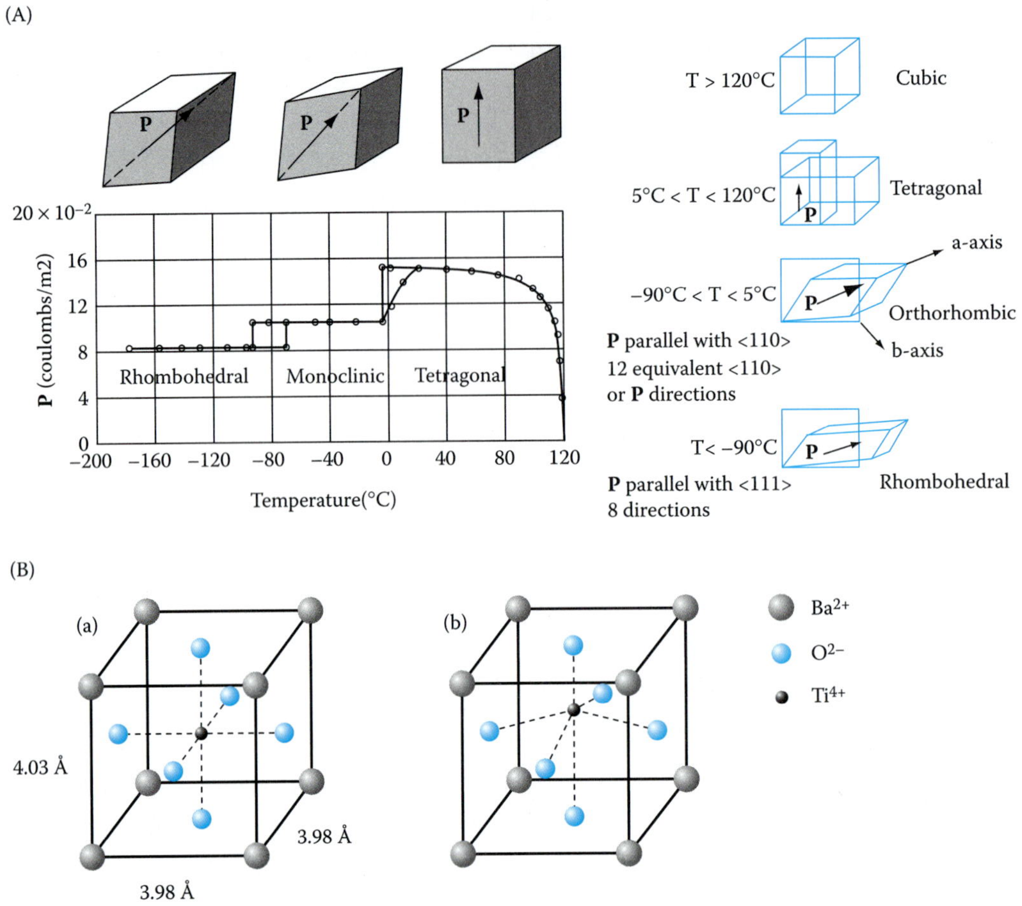

FIGURE 2.10 (A) In the pyroelectric crystal $BaTiO_3$, **P** changes with temperature only when the material is in its tetragonal state. Pyroelectricity only occurs in a crystal lacking an inversion center. This is clear from (B) (a). In cubic $BaTiO_3$ the oxygen ions are at face centers; Ba^{2+} ions are at cube corners; and Ti^{4+} is at cube center. (B) (b) In tetragonal $BaTiO_3$, the Ti^{4+} is off-center, and the unit cell has a net polarization. (Drawing by Mr. Chengwu Deng.)

Volume III, Chapter 8. The crystalline property, the pyroelectric coefficient **p**, is a tensor of rank 1 (a vector in other words). One can predict readily that pyroelectricity only occurs in crystals that lack an inversion center or a center of symmetry, i.e., in noncentrosymmetric crystals with one or more polar axes. Indeed, one could not have a crystal with one face positively charged and one negatively charged—i.e., with a polar axis—as a result of a uniform change in temperature if these crystal faces were equivalent.

More complicated cases involve vector-tensor effects and tensor-tensor effects. Piezoelectricity is an example of a vector-tensor effect; an electric field (vector) causes a mechanical deformation (tensor). Elastic deformation under the influence of a stress tensor is an example of a tensor-tensor effect. These effects require many more constants than the simple examples presented above. In discussing actuators in Volume III, Chapter 8, we will find out that an increase in symmetry introduces major simplifications in the coefficient matrices and that to describe the tensors correctly the point-group symmetry of the crystal must be known.

The combined effects of rotation or reflections from the point groups with translation from the Bravais lattice results in two additional symmetries: screw axes and glide planes. A screw axis combines rotation and translation, and a glide plane combines reflection with translation (Figure 2.11). Considering the various combinations involving the 32 point groups, screw axes, and glide planes, as well as the different Bravais lattices, a total of 230 different possible "space groups" results. In protein crystals there are only 65 space groups because all natural products are chiral so that inversion and mirror symmetry

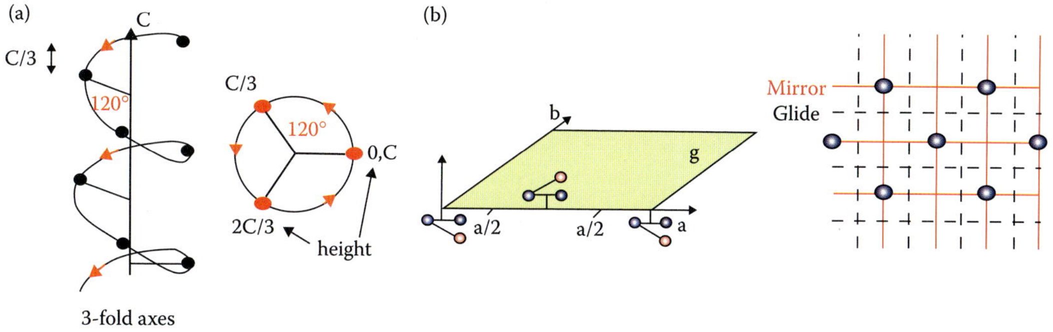

FIGURE 2.11 Example of a screw axis and a glide plane. (a) *N*-fold screw axes *C*: a combination of a rotation of 360°/*n* around *C* and a translation by an integer of *C*/*n*. (b) Glide plane: a translation parallel to the glide plane *g* by *a*/2.

operations are not allowed. A space group is a group that includes both the point symmetry elements and the translations of a crystal. These space groups are most important when studying the microscopic properties of solids. The space groups for most inorganic substances are known and can be found in tables in the Inorganic Crystal Structure Database (ICSD). These tables make it possible to calculate the exact distances and angles between different atoms in a crystal. The external shape of a crystal is referred to as the *habit*. Not all crystals have well-defined external faces. Natural faces always have low indices, i.e., their orientation can be described by Miller indices that are small integers as introduced next. The faces that we see are the lowest energy faces as the surface energy is minimized during growth.

Miller Indices

To identify a plane or a direction in a crystal, a set of integers *h*, *k*, and *l*, called the *Miller indices*, are widely used. To determine the Miller indices of a plane, one takes the intercept of that plane with the axes and expresses these intercepts as multiples of the base vectors $\mathbf{a}_1$, $\mathbf{a}_2$, and $\mathbf{a}_3$. The reciprocal of these three integers is taken, and, to obtain whole numbers, the three reciprocals are multiplied by the smallest common denominator. The resulting set of numbers is written down as (*hkl*). By taking the reciprocal of the intercepts, infinities (∞) are avoided in plane identification. Parentheses or braces are used to specify planes.

The rules for determining the Miller indices of a direction or an orientation in a crystal are as follows: translate the orientation to the origin of the unit cell, and take the normalized coordinates of its other vertex. For example, the body diagonal in a cubic lattice as shown in the right-most panel in Figure 2.12 is 1**a**, 1**a**, and 1**a** or the [111] direction. The Miller indices for a direction are thus established using the same procedure for finding the components of a vector. Brackets or carets specify directions.

Directions [100], [010], and [001] are all crystallographically equivalent and are jointly referred to as the family, form, or group of <100> directions. A form, group, or family of faces that bear like relationships to the crystallographic axes—e.g., the planes (001), (100), (010), $(00\bar{1})$, $(\bar{1}00)$, and $(0\bar{1}0)$—are all equivalent, and they are marked as {100} planes (see Figure 2.13). A summary of the typical representation for Miller indices is shown in Table 2.2. The orientation of a plane is defined by the direction

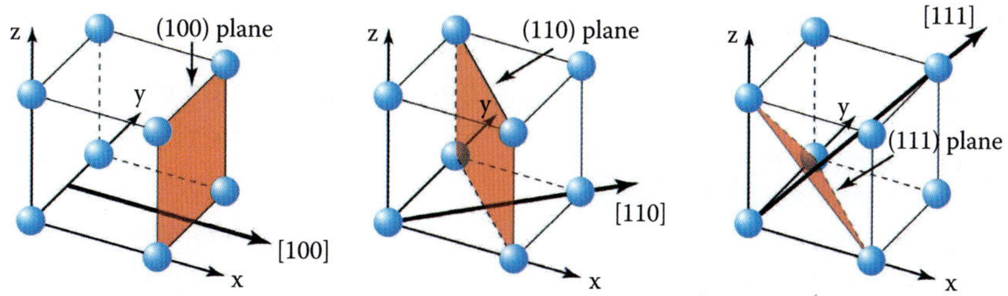

FIGURE 2.12 Miller indices for planes and directions in an SC cubic crystal. Shaded planes are from left to right (100), (110), and (111). (Drawing by Mr. Chengwu Deng.)

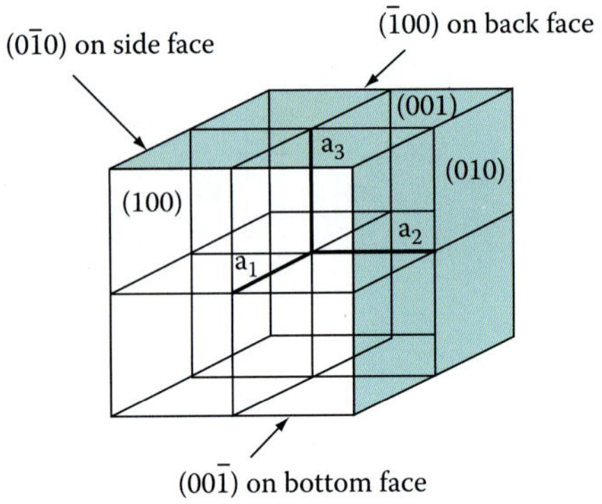

FIGURE 2.13 Miller indices for the planes of the {100} family of planes.

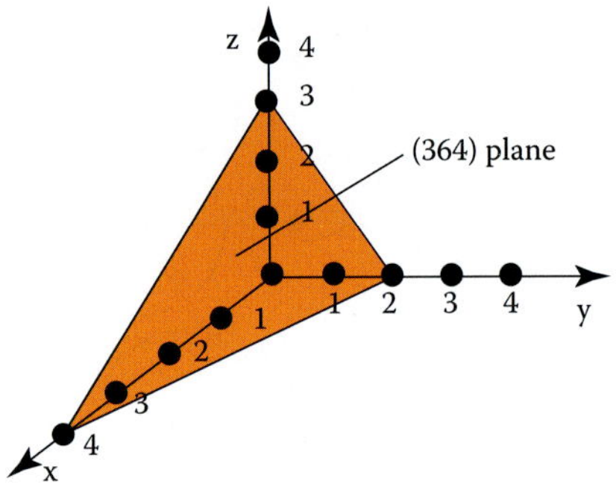

FIGURE 2.14 The (364) plane in a SC cubic lattice.

of a normal to the plane or the vector product ($\mathbf{A} \times \mathbf{B} = \mathbf{C}$). For a cubic crystal (such as silicon or gallium arsenide), the plane (*hkl*) is perpendicular to the direction [*hkl*]. In other words, the indices of a plane are the same numbers used to specify the normal to the plane. Using a simple cubic lattice as an example, you can check that crystal planes with the smallest Miller indices, such as {100}, {110}, {111}, have the largest density of atoms. Usually crystals are cleaved along these planes and are grown in directions perpendicular to them.

When one comes across more complicated planes than the ones considered above, the mathematical vector algebra approach to calculate the Miller indices becomes useful. For examples, consider the plane in Figure 2.14, defined by three points P1, P2, and P3, where P1: (400), P2: (020), and P3: (003).

Step 1. Define the following vectors:

$$\mathbf{r} = x\mathbf{a}_1 + y\mathbf{a}_2 + z\mathbf{a}_3$$

$$\mathbf{r}_1 = 4\mathbf{a}_1 + 0\mathbf{a}_2 + 0\mathbf{a}_3$$

$$\mathbf{r}_2 = 0\mathbf{a}_1 + 2\mathbf{a}_2 + 0\mathbf{a}_3 \qquad (2.8)$$

$$\mathbf{r}_3 = 0\mathbf{a}_1 + 0\mathbf{a}_2 + 3\mathbf{a}_3$$

Table 2.2 Miller Indices Symbols

Notation	Interpretation
(*hkl*)	Crystal plane
{*hkl*}	Equivalent planes
[*hkl*]	Crystal direction
<*hkl*>	Equivalent directions

and find the differences:

$$\mathbf{r} - \mathbf{r}_1 = (x - 4)\,\mathbf{a}_1 + (y - 0)\,\mathbf{a}_2 + (z - 0)\,\mathbf{a}_3$$

$$\mathbf{r}_2 - \mathbf{r}_1 = (0 - 4)\,\mathbf{a}_1 + (2 - 0)\,\mathbf{a}_2 + (0 - 0)\,\mathbf{a}_3 \qquad (2.9)$$

$$\mathbf{r}_3 - \mathbf{r}_1 = (0 - 4)\,\mathbf{a}_1 + (0 - 0)\,\mathbf{a}_2 + (3 - 0)\,\mathbf{a}_3$$

Step 2. Calculate the scalar triple product of these three vectors [$\mathbf{A} \cdot (\mathbf{B} \times \mathbf{C})$], which in this case is a plane and its volume is zero because the vectors are coplanar [$\mathbf{A} \cdot (\mathbf{B} \times \mathbf{C}) = 0$]:

$$(\mathbf{r} - \mathbf{r}_1) \cdot [(\mathbf{r}_2 - \mathbf{r}_1) \times (\mathbf{r}_3 - \mathbf{r}_1)] = 0 \qquad (2.10)$$

For vectors **A**, **B**, **C** with coordinates (A_1, A_2, A_3), (B_1, B_2, B_3), and (C_1, C_2, C_3), the requirement in Equation 2.10 is equivalent to:

$$\mathbf{A} \cdot \mathbf{B} \times \mathbf{C} = \begin{vmatrix} A_1 & A_2 & A_3 \\ B_1 & B_2 & B_3 \\ C_1 & C_2 & C_3 \end{vmatrix} = A_1(B_2C_3 - B_3C_2)$$

$$+A_2(B_3C_1 - B_1C_3) + A_3(B_1C_2 - B_2C_1) \qquad (2.11)$$

For our example this leads to:

$$\begin{vmatrix} x - 4 & y & z \\ -4 & 2 & 0 \\ -4 & 0 & 3 \end{vmatrix} = (x - 4)6 + 12y + 8z = 0$$

or $3x + 6y + 4z = 12$

Once we have the equation for the plane, we easily find the Miller indices.

Step 3. Determine the Miller indices.

1. The intercepts with the axes: $x = 4$ (for $y = z = 0$), $y = 2$ (for $x = z = 0$), and $z = 3$ (for $x = y = 0$)
2. The reciprocals 1/4, 1/2, and 1/3, or
3. The Miller indices for the plane are (364)

Adjacent planes (hkl) in a simple cubic crystal are spaced a distance d_{hkl} from each other, with d_{hkl} given by:

$$d_{hkl} = \frac{a}{\sqrt{h^2 + k^2 + l^2}} \tag{2.12}$$

where a is the lattice constant. Equation 2.12 provides the magnitude of d_{hkl} and follows from simple analytic geometry. To generalize this expression, notice that for a plane (hkl) with $hx + ky + lz = a$, the distance from any point (x_1, y_1, z_1) to this plane is:

$$d_{hkl} = \left| \frac{hx_1 + ky_1 + lz_1 - a}{\left(h^2 + k^2 + l^2\right)^{\frac{1}{2}}} \right| \tag{2.13}$$

Hence when that point is at origin (0, 0, 0) we find Equation 2.12 back.

- Example 2.1: With a = 5 Å, we find d = a = 5 Å for (100) planes and d = a/$\sqrt{2}$ = 3.535 Å for (110) planes.

Because a, b = |a||b| cos θ the angle between plane (h_1, k_1, l_1) and plane (h_2, k_2, l_2) can be calculated as:

$$\cos\theta = \frac{(h_1h_2 + k_1k_2 + l_1l_2)}{\sqrt{h_1^2 + k_1^2 + l_1^2}\sqrt{h_2^2 + k_2^2 + l_2^2}} \tag{2.14}$$

- Example 2.2: The angle between *a* (100) and *a* (111) plane is cos θ = 1/$\sqrt{1}\sqrt{3}$ = 0.58 or θ = 54.74°.

X-Ray Analysis

Introduction

X-ray analysis reveals the symmetries of crystals (lattice type), distances between atomic planes (lattice parameter), the positions of atoms in crystals, the types of atoms from the intensities of diffracted x-rays, and the degree of crystallinity (ordering). To perform x-ray crystallography, it is necessary to grow crystals with edges of around 0.1–0.3 mm. This is usually not a problem for inorganic materials, but in the case of organic materials, such as proteins (for example, see the crystal structure of the GFP protein in Figure 7.108) and nucleic acids (see x-ray diffraction image in Figure 2.17), it often is a challenge: imagine trying to crystallize a molecule with 10,000 atoms! A crystallographer must combine ingenuity and patience to trick these molecules into crystallizing. In this section we learn about diffraction and the all-important Bragg and Laue x-ray diffraction laws and how the latter are used to deduce the 3D structure of molecules.

Fourier Transforms

Diffraction forms the basis for x-ray crystallography. The first step toward interpreting diffraction patterns was a mathematical trick discovered by the French mathematician Joseph Fourier, who in 1807 introduced Fourier transforms for solving heat conduction problems. The result of a Fourier transform is that periodic functions in the time domain, e.g., light waves, can be completely characterized by information in the frequency domain, i.e., by frequencies and amplitudes of sine, cosine functions. Fourier analysis provides us with the tools to express most functions as a superposition of sine and cosine waves of varying frequency. For periodic signals a discrete sum of sines/cosines of different frequencies is multiplied by a different weighting coefficient in a so-called Fourier series (FS). For nonperiodic functions, one needs a continuous set of frequencies so the integral of sines/cosines is multiplied by a weighting function in a so-called Fourier transform (FT) (Figure 2.15). One important property of Fourier transforms is that they can be inverted. If you apply a Fourier transform to some function, you can take the result and run it through an inverse Fourier transform to get the original function back. The inverse Fourier transform is essentially just another Fourier transform. Fourier and inverse Fourier transforms, which take the signal back and forth between time and frequency domains, are:

$$\text{From time to frequency: } X(f) = \frac{1}{\sqrt{2\pi}} \int_{-\infty}^{\infty} x(t)\, e^{-2\pi ift} dt \tag{2.15}$$

$$\text{From frequency to time: } x(t) = \frac{1}{\sqrt{2\pi}} \int_{-\infty}^{\infty} X(f)\, e^{2\pi ift} df \tag{2.16}$$

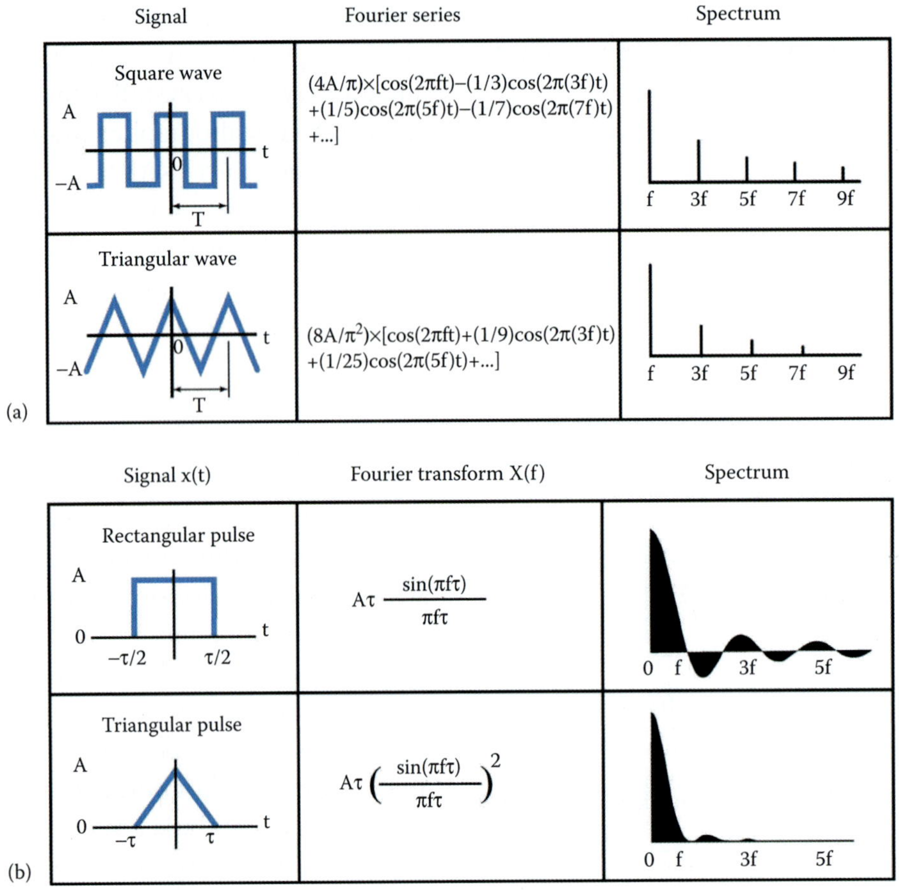

FIGURE 2.15 (a) For periodic signals a discrete sum of sines/cosines of different frequencies is multiplied by a different weighing coefficient in a so-called Fourier series (FS). (b) For nonperiodic functions, one needs a continuous set of frequencies so the integral of sines/cosines is multiplied by a weighting function in a so-called Fourier transform (FT).

In Chapter 5 we will see that the formation of an image, according to Abbe's theory, is a two-stage, double-diffraction process: an image is the diffraction pattern of the diffraction pattern of an object. In x-ray diffraction there is no lens to focus the x-rays, so we have to use a computer to reassemble the image: the x-ray diffraction patterns from a crystal are related to the object diffracting the waves through a Fourier transform.

Fourier transforms are actually even more general than revealed here: a FT allows for a description given in one particular "space" to be transformed to a description in the reciprocal of that space, and time-frequency transformation is just one example. It is interesting to note that strong criticism by peers blocked publication of Fourier's work until 1822 (*Theorie Analytique De La Chaleur*). Today, Fourier analysis is used in GSM (global system for mobile communications)/cellular phones, most DSP (digital signal processing)-based applications, music, audio, accelerator control, image processing, x-ray spectrometry, chemical analysis (FT spectrometry), radar design, PET scanners, CAT scans and MRI, speech analysis (e.g., voice-activated "devices," biometry), and even stock market analysis.

X-Ray Diffraction

Introduction

Wilhelm Conrad Roentgen discovered x-rays in 1895 and received the Nobel Prize in Physics in 1901 for his discovery. X-rays are scattered by the electrons in atoms because electromagnetic radiation (including x-rays) interacts with matter through its fluctuating electric field, which accelerates charged particles. You can think of electrons oscillating in position and, through their accelerations, re-emitting electromagnetic radiation. The scattered radiation interferes both constructively and destructively, producing a diffraction pattern that can be recorded on a photographic plate. For x-rays, electrons, and neutrons incident on a single crystal, diffraction occurs because of interference between waves scattered elastically from the atoms in the crystal. Intensity of scattered radiation is proportional to the square of the charge/mass ratio, and the proton is about 2000 times as massive as the electron. Because electrons have a much higher charge-to-mass ratio than atomic nuclei or even protons, they are much more efficient in this process. With x-rays the interaction is with the electron mantle of the atoms. In the case of electron beams, say in an electron microscope, scattering is from both the electron mantle and the atom nuclei, and neutrons interact with the nucleus only.

The final result of a crystallographic experiment is not a picture of the atoms, but a map of the distribution of electrons in the molecule, i.e., an electron density map (Figure 2.16). Because the electrons are mostly tightly localized around the nuclei, the electron density map gives us a pretty good picture of the molecule. As we do not have a lens we do not get the electron density map directly; the x-ray diffraction patterns from a crystal are related to the object diffracting the waves through a Fourier transform.

If one thinks of electron density as a mathematical function:

$$\rho(x, y, z) \qquad (2.17)$$

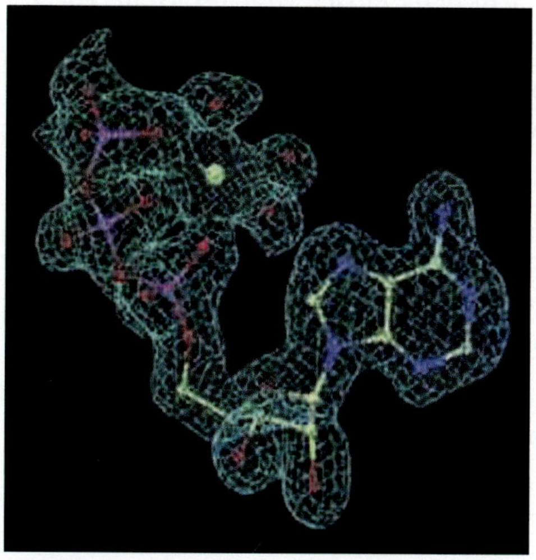

FIGURE 2.16 Electron density map of adenosine triphosphate (ATP).

with x, y, z indices for real space, then the diffraction pattern is the Fourier transform of that electron density function and given as:

$$F(hkl) = T[\rho(x, y, z)] \qquad (2.18)$$

where $F(hkl)$ is the structure factor (a scattered wave, therefore a complex number with amplitude and phase) with hkl indices in reciprocal space, and T is the forward Fourier transform of $\rho(x, y, z)$. As we saw in Equation 2.16, the reverse relationship holds also, namely:

$$\rho(x, y, z) = T^{-1}\{T[\rho(x, y, z)]\} \qquad (2.19)$$

where T^{-1} is the inverse Fourier transform. This expression tells us that the inverse Fourier transform of the Fourier transform of an object is the original object. The latter is a rewording of Abbe's treatment of image generation as presented in Chapter 5. The intensity in an x-ray diffraction photograph is the square of the amplitude of the diffracted waves, $|F(X)|^2$, or the recorded diffraction pattern of an object is the square of the Fourier transform of that object. In a diffraction pattern, each point arises from the interference of rays scattered from all irradiated portions of the object. To determine the image, one must measure or calculate the structure factor $F(X)$ at many or all points of the diffraction pattern. Each $F(X)$ is described by an amplitude and a phase, but in recording the intensity of the diffracted x-rays, the phase information is lost. This is referred to as the *phase* problem. X-ray phases could be obtained directly if it were possible to rediffract (focus) the scattered rays with an x-ray lens to form an image; unfortunately, an x-ray lens does not exist. With x-rays we can thus detect diffraction from molecules, but we have to use a computer to reassemble an image as shown in Figure 2.16. The process is summarized in Figure 2.17. We will now learn how the x-ray diffraction pattern, $F(X)$, comes about and how the expected intensities in the diffraction pattern are calculated to solve Equation 2.19, i.e., how to reconstitute the image that led to the measured x-ray diffraction pattern.

Bragg's Law

In 1913 W.H. and W.L. Bragg (a father and son team) proposed that the condition for constructive

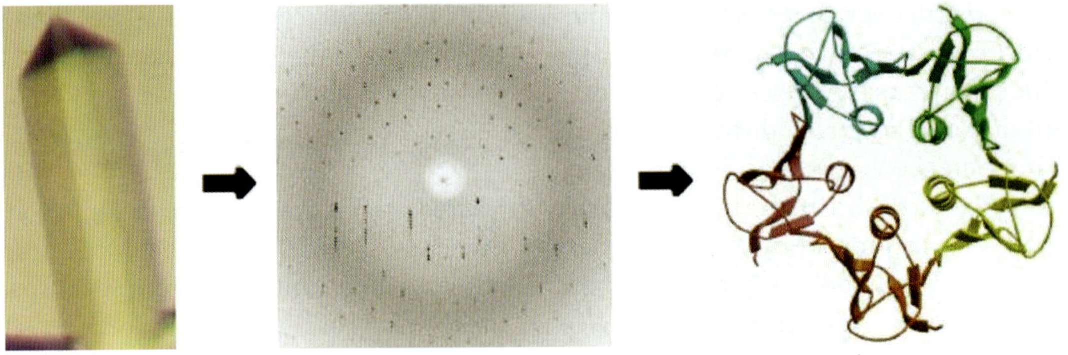

FIGURE 2.17 With x-rays, we can detect diffraction from molecules, but we have to use a computer to reassemble the electron density/molecular structure image.

specular reflection of x-rays from a set of crystal planes separated by a distance d_{hkl} could be represented as:

$$2d_{hkl} \sin \theta = n\lambda \tag{2.20}$$

This expression basically tells us that constructive interference of waves reflected by successive crystal planes occurs whenever the path difference ($2d_{hkl}\sin \theta$) is an integral multiple (n) of the wavelength λ. Also, for each (hkl) family of planes, x-rays will only diffract at one angle θ. The integer n is known as the order of the corresponding reflection. Because Bragg reflection can only occur for $\lambda \leq 2d$, one needs x-rays with wavelengths in the Ångstrom range to resolve crystal planes. The Bragg equation is easily derived from an inspection of Figure 2.18. Bragg's law is a result of the periodicity of the lattice with the atoms in the crystal basis controlling the relative intensity of the various orders (n) of diffraction from a set of parallel (hkl) planes. This basic equation is the starting point for understanding crystal diffraction of x-rays, electrons, neutrons, and any other particles that have a de Broglie wavelength

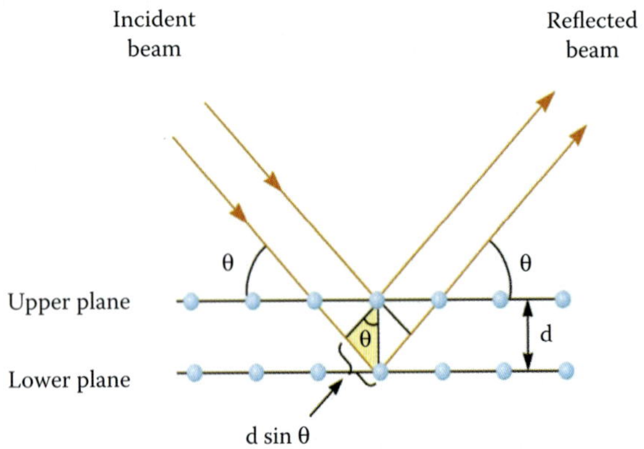

FIGURE 2.18 Schematic used to derive the Bragg equation.

(Chapter 3) less than the interatomic spacing. Although the reflection from each plane is specular, only for certain values of θ will the reflections from all planes add up in phase to give a strong reflected beam. Each plane reflects only 10^{-3} to 10^{-5} of the incident radiation, i.e., it is not a perfect reflector. Hence, 10^3 to 10^5 planes contribute to the formation of the Bragg-reflected beam in a perfect crystal.

The composition of the basis determines the relative intensity of the various orders of diffraction.

Laue Equations

In 1912 von Laue predicted that diffraction patterns of x-rays on crystals would be entirely analogous to the diffraction of light by an optical grating.* In the von Laue approach there is no ad hoc assumption of specular reflection, as in the case of Bragg's law. Instead, this more general approach considers a crystal as composed of sets of ions or atoms at the sites of a Bravais lattice that reradiate the incoming x-rays. For both optical gratings and crystals only the repeat distances of the periodic structure and the wavelength of the radiation determine the diffraction angles.

Let us inquire first into the interference conditions for waves originating from different but identical atoms in a single row—the one-dimensional diffraction case. The scattering atoms in a line form secondary, coherent x-ray sources (scattering from two atoms is shown in Figure 2.19).

* We will encounter the diffraction of light by an optical grating again in Volume II, Chapter 1 on photolithography, where we discuss patterning of a photoresist with UV light, using a mask with a grating structure on it.

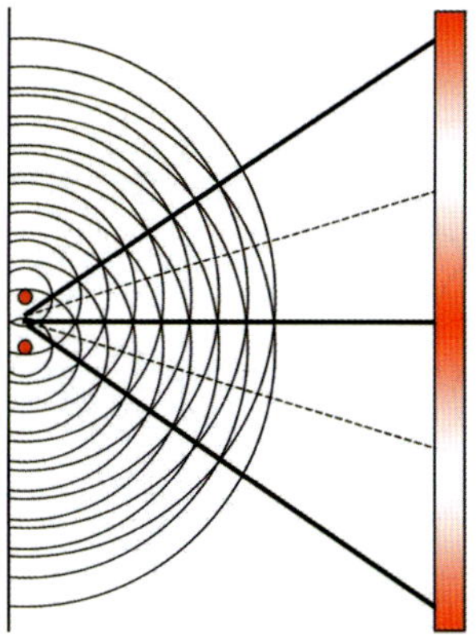

FIGURE 2.19 Two scattering atoms act as coherent secondary sources.

Constructive interference will occur in a direction such that contributions from each lattice point differ in phase by 2π. This is illustrated for the scattering of an incident x-ray beam by a row of identical atoms with lattice spacing $\mathbf{a}_1$ in Figure 2.20. The direction of the incident beam is indicated by wave vector $\mathbf{k}_0$ or the angle α_0, and the scattered beam is specified by the direction of $\mathbf{k}$ or the angle α. Because we assume elastic scattering, the two wave vectors $\mathbf{k}_0$ and $\mathbf{k}$ have the same magnitude, i.e., $2\pi/\lambda$ but with differing direction. A plane wave* $e^{ik.r}$ is constant in a plane perpendicular to $\mathbf{k}$ and is periodic parallel to it, with a wavelength $\lambda = 2\pi/\mathbf{k}$ (see Appendix 2A). The path difference $A_1B - A_2C$ in Figure 2.20 must equal $e\lambda$ with e = 0, 1, 2, 3, ... For a fixed incident x-ray with wavelength λ and direction $\mathbf{k}$, and an integer value of e, there is only one possible scattering angle α defining a cone of rays drawn about a line through the lattice points (see Figure 2.20). Because crystals are periodic in three directions, the Laue equations in 3D are then:

$$\begin{aligned} \mathbf{a}_1(\cos\alpha - \cos\alpha_0) &= e\lambda \\ \mathbf{a}_2(\cos\beta - \cos\beta_0) &= f\lambda \\ \mathbf{a}_3(\cos\gamma - \cos\gamma_0) &= g\lambda \end{aligned} \quad (2.21)$$

For constructive interference from a three-dimensional lattice to occur, the three equations above must all be satisfied simultaneously, i.e., six angles α, β, γ, a, α_0, β_0, and γ_0; three lattice lengths $\mathbf{a}_1$, $\mathbf{a}_2$, and $\mathbf{a}_3$; and three integers (e, f, and g) are fixed. Multiplying both sides of Equation 2.21 with $2\pi/\lambda$ and rewriting the expression in vector notation we obtain:

$$\begin{aligned} \mathbf{a}_1 \cdot (\mathbf{k} - \mathbf{k}_0) &= 2\pi e \\ \mathbf{a}_2 \cdot (\mathbf{k} - \mathbf{k}_0) &= 2\pi f \\ \mathbf{a}_3 \cdot (\mathbf{k} - \mathbf{k}_0) &= 2\pi g \end{aligned} \quad (2.22)$$

with $\mathbf{a}_1$, $\mathbf{a}_2$, and $\mathbf{a}_3$ being the primitive vectors of the crystal lattice. When two of the conditions in Equation 2.22 are met, one entire plane array scatters in phase. This is depicted in Figure 2.21, where two cones of allowed diffracted rays are depicted. The two conditions are met simultaneously only in two directions along which the cones intersect. To satisfy all three Laue equations simultaneously, the diffracted beam can only have one allowed direction because three cones can mutually intersect along only one line.

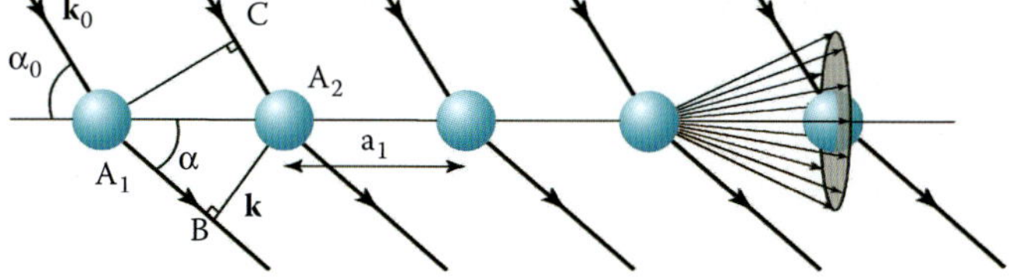

FIGURE 2.20 Scattering of an incident x-ray beam (incident direction is $\mathbf{k}_0$) by a row of identical atoms with lattice spacing $\mathbf{a}_1$. The scattered beam is specified by the direction $\mathbf{k}$. The path difference $A_1B - A_2C$ must equal $e\lambda$, with e = 0,1,2,3,.... (Drawing by Mr. Chengwu Deng.)

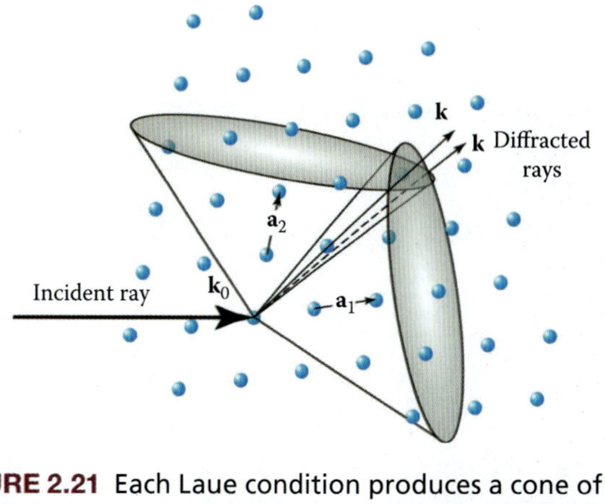

FIGURE 2.21 Each Laue condition produces a cone of allowed rays. In a plane array the entire plane scatters in phase in two directions. These two directions are along the intersection of the two cones. (Drawing by Mr. Chengwu Deng.)

* The complex exponential representation for periodic functions is convenient for adding waves, taking derivatives of wave functions, and so on. It is equivalent to a linear combination of a sine and cosine function, because $e^{i\theta} = \cos\theta + i \sin\theta$ (Euler) (see also Appendix 2A).

FIGURE 2.22 Max von Laue (1897–1960).

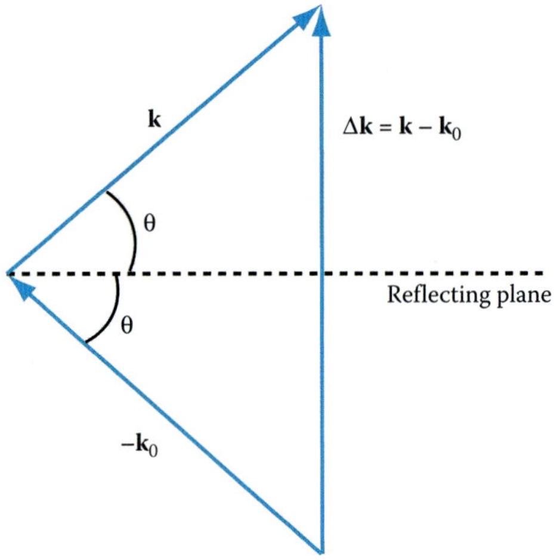

FIGURE 2.24 In case of mirror-like Bragg reflection, the vector **Δk**, the summation of the unit vectors representing incoming ($\mathbf{k}_0$) and reflected rays (**k**), is normal to the plane that intersects the 2θ angle between them. (Drawing by Mr. Chengwu Deng.)

If we further define a vector $\Delta\mathbf{k} = \mathbf{k} - \mathbf{k}_0$, Equation 2.22 simplifies to:

$$\begin{aligned} \mathbf{a}_1 \cdot \Delta\mathbf{k} &= 2\pi e \\ \mathbf{a}_2 \cdot \Delta\mathbf{k} &= 2\pi f \\ \mathbf{a}_3 \cdot \Delta\mathbf{k} &= 2\pi g \end{aligned} \tag{2.23}$$

Dealing with 12 variables for each reflection simultaneously [six angles (α, β, γ, a, α_0, β_0, and γ_0), three lattice lengths ($\mathbf{a}_1$, $\mathbf{a}_2$ and $\mathbf{a}_3$), and three integers (e, f, and g)] is a handful; this is the main reason why the Laue equations are rarely referred to directly, and a simpler representation is used instead. The reflecting conditions can indeed be described more simply by the Bragg equation. Historically, von Laue (Figure 2.22) developed his equations first; it was one year after his work that the father and son team William Henry and William Lawrence Bragg (Figure 2.23) introduced the simpler Bragg's law. Max von Laue and the Braggs received the Nobel Prize in Physics in 1914 and 1915, respectively.

Further below we will learn that constructive interference of diffracted x-rays will occur provided that the change in wave vector, $\Delta\mathbf{k} = \mathbf{k} - \mathbf{k}_0$, is a vector of the reciprocal lattice.

Bragg's law is equivalent to the Laue equations in one dimension as can be appreciated from an inspection of Figures 2.24 and 2.25, where we use a two-dimensional crystal for simplicity. Suppose that vector $\Delta\mathbf{k}$ in Figure 2.24 satisfies the Laue condition; because incident and scattered waves have the same magnitude (elastic scattering), it follows that incoming ($\mathbf{k}_0$) and reflected rays (**k**) make the same angle θ with the plane perpendicular to $\Delta\mathbf{k}$.

FIGURE 2.23 Father and son Bragg: Sir William Henry and William Lawrence Bragg.

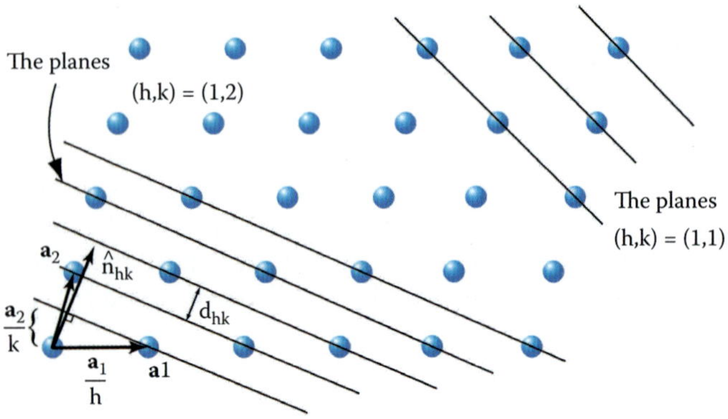

FIGURE 2.25 Connecting Bragg's law with Laue equations and Miller indices. (Drawing by Mr. Chengwu Deng.)

The magnitude of vector $\Delta\mathbf{k}$, from Figure 2.24, is then given as:

$$|\Delta\mathbf{k}| = 2\mathbf{k}\sin\theta \qquad (2.24)$$

We now derive the relation between the reflecting planes to which $\Delta\mathbf{k}$ is normal and the lattice planes with a spacing d_{hkl} (see Figure 2.25 and Bragg's law in Equation 2.20). The normal unit vector $\hat{\mathbf{n}}_{hk}$ and the interplanar spacing d_{hk} in Figure 2.25 characterize the crystal planes (*hk*). From Equation 2.23 we deduce that the direction cosines of $\Delta\mathbf{k}$, with respect to the crystallographic axes, are proportional to e/a_1, f/a_2, and g/a_3 or:

$$e/a_1{:}f/a_2{:}g/a_3 \qquad (2.25)$$

From the definition of the Miller indices, an (*hkl*) plane intersects the crystallographic axes at the points a_1/h, a_2/k, and a_3/l, and the unit vector $\hat{\mathbf{n}}_{hkl}$, normal to the (*hkl*) plane, has direction cosines proportional to:

$$h/a_1,\ k/a_2,\ \text{and}\ l/a_3 \qquad (2.26)$$

Comparing Equations 2.25 and 2.26 we see that $\Delta\mathbf{k}$ and the unit normal vector $\hat{\mathbf{n}}_{hkl}$ have the same directions; all that is required is that e = nh, f = nk, and g = nl, where *n* is a constant. The factor *n* is the largest common factor of the integers *e*, *f*, and *g* and is itself an integer. From the above, Laue's equations can also be interpreted as reflection from the *h,k,l* planes. From Figure 2.25 it can be seen that the spacing between the (*hk*) planes, and by extension between (*hkl*) planes, is given as*:

$$d_{hkl} = \frac{\hat{\mathbf{n}}_{hkl}\cdot\mathbf{a}_1}{h} = \frac{\hat{\mathbf{n}}_{hkl}\cdot\mathbf{a}_2}{k} = \frac{\hat{\mathbf{n}}_{hkl}\cdot\mathbf{a}_3}{l} \qquad (2.27)$$

Because $\Delta\mathbf{k}$ is in the direction of the normal $\hat{\mathbf{n}}_{hkl}$ and comes with a magnitude given by Equation 2.24, we obtain Bragg's law from the Laue's equations as:

$$\mathbf{a}_1\cdot\Delta\mathbf{k} = \mathbf{a}_1\cdot\hat{\mathbf{n}}_{hkl}2k\sin\theta = 2\pi e \qquad (2.28)$$

or:

$$hd_{hkl}\frac{4\pi}{\lambda}\sin\theta = 2\pi e \qquad (2.29)$$

* Notice that in the case of a cubic crystal, Equation 2.27 can be simplified to give the distance between planes as in Equation 2.12.

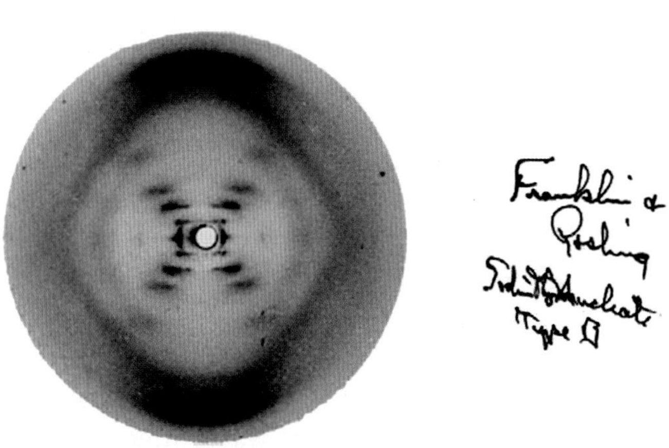

FIGURE 2.26 Sodium deoxyribose nucleate from calf thymus. (Structure B, Photo 51, taken by Rosalind E. Franklin and R.G. Gosling.) Linus Pauling's annotations are to the right of the photo (May 2, 1952).

and with e = nh:

$$2d_{hkl}\sin\theta = n\lambda \qquad (2.30)$$

In the Bragg equation we treat x-ray diffraction from a crystal as a reflection from reciprocal lattice planes rather than scattering from atoms. This construction has fewer variables than the Laue equations because reflections are wholly represented in two dimensions only.

In Figure 2.26 we reproduce what is possibly the most famous x-ray diffraction photograph. It was this photograph—photo 51 of DNA taken by Rosalind Franklin and R.G. Gosling—that convinced Watson and Crick that the DNA molecule was helical. The discovery of the double helix followed soon after, as well as an enduring controversy that Franklin probably did not get the credit she deserved in the elucidation of the DNA structure.

X-Ray Intensity and Structure Factor F(hkl)

So far we have considered only the condition for diffraction from simple lattices for which only corner points of the unit cell are occupied. The intensity of a beam diffracted from an actual crystal will depend on the grouping of atoms in the unit cell and on the scattering power of these atoms. In this section we discuss the intensity and phase of the diffracted rays and the structure factor, *F*(*hkl*). If we treat the incident x-ray waves as plane waves and the atoms as ideal point scatterers, the scattered waves are spherical waves (Figure 2.19) close to the source, i.e., near-field or Fresnel diffraction patterns form at finite distances from the crystal and Fraunhofer diffraction patterns at infinity (far-field). In a typical x-ray diffraction experiment, a Fraunhofer diffraction pattern is registered at about 50 to 150 mm behind the specimen. This distance is relatively large compared to the size of the diffracting crystal unit cell (~1–50 nm in dimension) and the wavelength of the incident radiation (typically 1.54178 Å for Cu Kα and 0.71073 Mo Kα). Thus, x-ray diffraction methods provide a direct way to display the decomposition of x-rays in component waves (frequencies), and for this reason x-ray diffraction may be called *spatial frequency spectrum* analysis or *harmonic* analysis. Near-field and far-field optics are compared in more detail in Chapter 5.

The number of scattered x-ray photons picked up by the detector results in an intensity proportional to the square of the amplitude (peak height) of the diffracted waves. The scattering intensity depends on the number and distribution of electrons associated with the scattering atoms in the basis, i.e., the structure factor *F*(*hkl*). The phase of the diffracted rays is the relative time of arrival of the scattered radiation at a particular point in space, say at the emulsion of a photographic film. The phase information is lost when the x-ray diffraction pattern is recorded on the film—one could say that the film integrates intensity over time—in other words, we cannot measure x-ray phases directly. When using a lens such as in a microscope, light first strikes the imaged object and is diffracted in various directions. The lens then collects the diffracted rays and reassembles them to form an image. The *phase problem* is a major concern in structure analysis as we need both intensity and phase to feed into the Fourier transform. Today several techniques exist to regenerate the *lost phase* information of x-ray diffraction, but the topic falls outside the scope of this book. We need to extract the structure factor *F*(*hkl*) from the intensities and phase of the diffraction spots and then do an inverse Fourier transform (T^{-1}) to obtain the crystal structure/electron density function, $\rho(x, y, z)$ (see Figure 2.17).

We now mathematically derive the intensity profile of x-rays scattered from a crystal. The result, as Laue predicted, is the same as for visible light diffracted from an optical grating. When an incident x-ray beam travels inside a crystal, we assume that the beam is not much influenced by the presence of the crystal; in other words, the refractive index for x-rays is close to unity, and there is not much loss of energy from the beam through scattering, i.e., elastic scattering dominates!

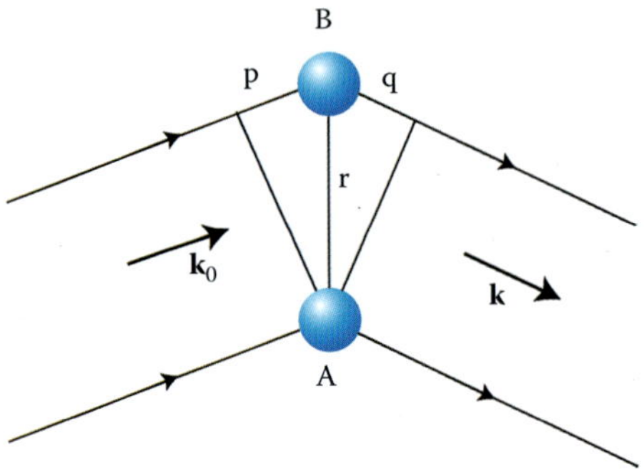

FIGURE 2.27 Scattering of x-rays from two nearby atoms A and B with identical scattering density. (Drawing by Mr. Chengwu Deng.)

With reference to Figure 2.27, we assume two parallel plane x-ray waves of wavelength λ and frequency ν (hence velocity $c = \lambda v$), scattered elastically from two nearby atoms A and B of identical scattering density. The wave vector for the incoming wave is $\mathbf{k}_0$ and that of the diffracted beam is $\mathbf{k}$. Because we assume elastic scattering:

$$|\mathbf{k}_0| = |\mathbf{k}| = k = 2\pi/\lambda \tag{2.31}$$

Scattering atom A is at the origin, and scattering atom B is at a distance $\mathbf{r}$ away from the origin. The path difference, or phase factor, between the waves can be calculated from Figure 2.27 as:

$$p + q = \mathbf{r}\cdot(\mathbf{k} - \mathbf{k}_0) = \mathbf{r}\cdot\Delta\mathbf{k} \tag{2.32}$$

The equations for the wave amplitudes are (for those readers less familiar with wave equations, consult Appendix 2A for more details):

$$\psi_A(x,t) = \frac{Af}{l_A} e^{i(\mathbf{k}_0 \cdot \mathbf{l}_A - \omega \cdot t)} \tag{2.33}$$

and:

$$\psi_B(x,t) = \frac{Af}{l_B} e^{i(\mathbf{k}_0 \cdot \mathbf{l}_B - \omega t + \mathbf{r} \cdot \Delta k)} \tag{2.34}$$

where *A* is the amplitude and *f* is the atomic scattering factor defined as the ratio of the amplitude of an electromagnetic wave scattered by atom A and that of a wave scattered by a free electron. The position of the detector with respect to atom A is given by l_A, and the position of the detector relative to atom B is l_B. The product $\mathbf{r} \cdot \Delta\mathbf{k}$ is the phase factor we calculated in Equation 2.32. Generalizing, an x-ray wave scattered from the j_{th} atom in a crystal is:

$$\Psi_j(x,t) = \frac{Af_j}{l_j} e^{i(\mathbf{k}_0 \cdot \mathbf{l}_j - \omega t + \mathbf{r}_j \cdot \Delta\mathbf{k})} \qquad (2.35)$$

where $\mathbf{r}_j$ is the position of scatterer j relative to scatterer A, and l_j is the position of the detector with respect to scatterer j. The total scattered wave amplitude at the detector is the sum of all the contributing atoms:

$$\Psi(x,t) = \sum_{\text{all atoms}} \frac{Af_j}{l_j} e^{i(\mathbf{k}_0 \cdot \mathbf{l}_j - \omega t)} e^{i(\mathbf{r}_j \cdot \Delta\mathbf{k})}$$
$$\cong \frac{A}{L} e^{i(\mathbf{k}_0 \cdot \mathbf{L} - \omega t)} \sum_{\text{all atoms}} f_j e^{i(\mathbf{r}_j \cdot \Delta\mathbf{k})} \qquad (2.36)$$

For a small sample, the distances l_j are all about the same (the crystal is small compared to the distance between it and the detector), so in Equation 2.36 we can replace l_j with L.

Constructive and destructive interference between the scattered waves that reach the detector is the result of the sum of all scatterers and the scattering vector, $\Delta\mathbf{k}$, and determines where the detector should be put. The term $F(hkl)$:

$$F(hkl) = \sum_{\text{all atoms}} f_j e^{i(r_j \cdot \Delta k)} \qquad (2.37)$$

from the expression 2.36 for the wave amplitude, is the geometrical structure factor. It is defined, in analogy with the atomic scattering factor, as the ratio of the amplitude of the wave scattered by all atoms in a unit cell and that scattered by a free electron for the same incident beam. It incorporates the scattering of all atoms of the unit cell and sums up the extent to which interference of the waves scattered from atoms within the basis diminishes the intensity of the diffraction peaks. So even if the Laue condition is met, the structure factor may be zero and no diffraction will be observed. In the case that there is only one type of atom in the basis, the atomic scattering factor f_j disappears ($f_j = 1$)—identical atoms have identical scattering factors.

Assume now a crystal with base vectors $\mathbf{a}_1$, $\mathbf{a}_2$, and $\mathbf{a}_3$ and a total number of atoms along each axis of M, N, and P, respectively, and also accept that there is only a single atom at each lattice point (i.e., $f_j = 1$), then the amplitude of the total wave will be proportional to:

$$\Psi(x,t) \propto F(h,k,l) = \sum_{\text{all atoms}} e^{i(r_j \cdot \Delta k)}$$
$$= \sum_{m=0}^{M-1} \sum_{n=0}^{N-1} \sum_{p=0}^{P-1} e^{i[(ma_1 + na_2 + pa_3) \cdot \Delta k]}$$

after rearrangement we obtain:

$$\Psi(x,t) \propto F(h,k,l) = \sum_{m=0}^{M-1} e^{ima_1 \cdot \Delta k} \sum_{n=0}^{N-1} e^{ina_2 \cdot \Delta k} \sum_{p=0}^{P-1} e^{ipa_3 \cdot \Delta k} \qquad (2.38)$$

The intensity of the scattered wave is the square of the wave amplitude, and taking the value of one of the sums in Equation 2.38 for a crystal of dimension $M\mathbf{a}_1$ in the direction of $\mathbf{a}_1$, we obtain:

$$\sum_{m=0}^{M-1} e^{ima_1 \cdot \Delta k} = \frac{1 - e^{iM(a_1 \cdot \Delta k)}}{1 - e^{i(a_1 \cdot \Delta k)}}$$

$$\text{and } I \propto |\Psi(x,t)|^2 \propto \frac{[1 - e^{-iM(a_1 \cdot \Delta k)}][1 - e^{iM(a_1 \cdot \Delta k)}]}{[1 - e^{-i(a_1 \cdot \Delta k)}][1 - e^{i(a_1 \cdot \Delta k)}]}$$

which simplifies to

$$I \propto \frac{2 - e^{-iM(a_1 \cdot \Delta k)} - e^{iM(a_1 \cdot \Delta k)}]}{[2 - e^{-i(a_1 \cdot \Delta k)} - e^{i(a_1 \cdot \Delta k)}]} = \frac{2 - 2\cos(M\mathbf{a}_1 \cdot \Delta\mathbf{k})}{2 - 2\cos(\mathbf{a}_1 \cdot \Delta\mathbf{k})}$$

$$I \propto \frac{1 - \cos(M\mathbf{a}_1 \cdot \Delta\mathbf{k})}{1 - \cos(\mathbf{a}_1 \cdot \Delta\mathbf{k})} \quad \text{and with the identity}$$

$$\cos 2x = \cos^2 x - \sin^2 x = 1 - 2\sin^2 x$$

$$I \propto \frac{\sin^2\left(\frac{1}{2} M\mathbf{a}_1 \cdot \Delta\mathbf{k}\right)}{\sin^2\left(\frac{1}{2}\mathbf{a}_1 \cdot \Delta\mathbf{k}\right)} \qquad (2.39)$$

This is the same result as the light intensity profile expected from an M-slit diffraction grating (see Chapter 5 on photonics). If M is large (~10^8 for a macroscopic crystal), it has very narrow, intense peaks. Between the peaks the intensity is essentially zero. In Figure 2.28 we have plotted $y = \sin^2 Mx/\sin^2 x$.

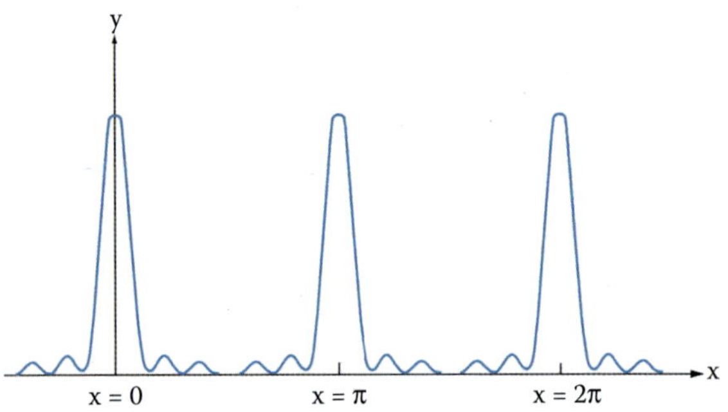

FIGURE 2.28 Graphical presentation of y = sin²Mx/sin²x. The width of the peaks and the prominence of the ripples are inversely proportional to *M*.

This function is virtually zero except at the points where x = nπ (*n* is an integer including zero), where it rises to the maximum value of M^2. The width of the peaks and the prominence of the ripples are inversely proportional to *M*.

Remember that there are three sums in Equation 2.38. For simplicity we only evaluated one sum to calculate the intensity in Equation 2.39. The total intensity equals:

$$I \propto \frac{\sin^2\left(\frac{1}{2}M\mathbf{a}_1\cdot\Delta\mathbf{k}\right)}{\sin^2\left(\frac{1}{2}\mathbf{a}_1\cdot\Delta\mathbf{k}\right)} \times \frac{\sin^2\left(\frac{1}{2}N\mathbf{a}_2\cdot\Delta\mathbf{k}\right)}{\sin^2\left(\frac{1}{2}\mathbf{a}_2\cdot\Delta\mathbf{k}\right)} \times \frac{\sin^2\left(\frac{1}{2}P\mathbf{a}_3\cdot\Delta\mathbf{k}\right)}{\sin^2\left(\frac{1}{2}\mathbf{a}_3\cdot\Delta\mathbf{k}\right)} \tag{2.40}$$

so that the diffracted intensity will equal zero unless all three quotients in Equation 2.40 take on their maximum values at the same time. This means that the three arguments of the sine terms in the denominators must be simultaneously equal to integer multiples of 2π, or the peaks occur only when:

$$\mathbf{a}_1\cdot\Delta\mathbf{k} = 2\pi e$$

$$\mathbf{a}_2\cdot\Delta\mathbf{k} = 2\pi f$$

$$\mathbf{a}_3\cdot\Delta\mathbf{k} = 2\pi g$$

These are, of course, the familiar Laue equations. As we will see below, to solve for the wave vector **Δk** it is very convenient to introduce the concept of a reciprocal lattice so that the set of all wave vectors **Δk** yields plane waves with the periodicity of a given Bravais lattice.

From the above, the intensity of the x-ray diffraction is:

$$I \propto |F(X)|^2 = \left|\sum_{\text{all atoms}} f_j e^{i(\mathbf{r}_j\cdot\Delta\mathbf{k})}\right|^2 \tag{2.41}$$

So what we measure from the x-ray film is intensity *I*, and the dependence of that intensity on the atomic position follows from the Fourier expansion of the electron density function in Equation 2.17:

$$\rho(x,y,z) = \sum_h\sum_k\sum_l F(hkl)e^{i(\mathbf{r}_j\cdot\Delta\mathbf{k})} \tag{2.42}$$

where *F*(*hkl*) is the coefficient to be determined, and *h*, *k*, and *l* are integers over which the series is summed. Because of the 3D periodicity, a triple summation is required:

$$F(hkl)=\int_0^1\int_0^1\int_0^1 \rho(x,y,z)e^{i(\mathbf{r}_j\cdot\Delta\mathbf{k})}dxdydz \tag{2.43}$$

If we knew all the *F*(*hkl*) values, we could calculate the electron density, and vice versa. Unfortunately, as remarked above, Equation 2.42 requires values of *F*(*hkl*), but the measured intensities only give us $|F(hkl)|^2$. The apparent impasse is known as the *phase problem* and arises because we need to know both the amplitude and the phase of the diffracted waves to compute the inverse Fourier transform.

Reciprocal Space, Fourier Space, k-Space, or Momentum Space

Introduction

The crystal lattice discussed above is a lattice in real space. Practice has shown the usefulness of defining a lattice in reciprocal space for the interpretation of diffraction patterns of single crystals, the simple reason being that a diffraction pattern reveals the reciprocal lattice. We did already consider some reciprocal features when we introduced the Miller indices, which are derived as the reciprocal of the unit cell intercepts. Because the lattice distances between planes are proportional to the reciprocal of the distances in the real crystal (i.e., they are proportional to $1/d_{hkl}$), the array is called a reciprocal lattice. We will learn in this section that the condition for nonzero intensity in scattered x-rays is that the scattering vector $\Delta\mathbf{k} = \mathbf{k} - \mathbf{k}_0$ is a translation vector, $\mathbf{G}_{hkl}$, of the reciprocal lattice, i.e., $\Delta\mathbf{k} = \mathbf{G}_{hkl}$. One can describe a reciprocal lattice the same way we describe a real one, but one must keep in mind that in one case the points are the position of real objects (atoms or the base), whereas in the other case they mark the positions of abstract points—magnitude and direction of momentum. Reciprocal space, also called Fourier space, **k**-space, or momentum space, plays a fundamental role in most analytical studies of periodic structures. After the introduction of the reciprocal lattice we will discuss the conditions for x-ray diffraction in terms of such a lattice and end with an introduction to Brillouin zones. The concepts introduced here are not only needed to understand x-ray diffraction better, they also are required preparation for Chapter 3, where we deal with band theory of solids.

The Reciprocal Lattice

Graphical Presentation of the Reciprocal Lattice

Before introducing a reciprocal lattice mathematically, let us learn how to draw one as we demonstrate in Figure 2.29. In this example we start with the two-dimensional unit cell of a monoclinic crystal with $\mathbf{a}_1$ and $\mathbf{a}_2$ in the plane of the paper. The figure also shows the edges of four (*hkl*) planes: (100), (110), (120), and (010). These planes all are perpendicular to the face of the paper. To construct the reciprocal lattice we draw from a common origin a normal to each plane. Next we place a point on the normal to each plane (*hkl*) at a distance $2\pi/d_{hkl}$ from the origin. Each of the points thus obtained maintains the important features of the stack of parallel planes it represents. The direction of the point from the origin preserves the orientation of the planes, and the distance of the point from the origin preserves the interplanar distance. A doubling of the periodicity in real space will produce twice as many diffraction spots in reciprocal space. It is convenient to let the reciprocal lattice vector, $\mathbf{G}_{hkl}$, be 2π times the reciprocal of the interplanar distance, d_{hkl}. This convention converts the units from periods per unit length to radians per unit length. This simplifies comparison of different periodic phenomena, e.g., a crystal lattice and light that interacts with it. For instance, $\Delta\mathbf{k}$, the wave vector, has an absolute value of $2\pi/\lambda$. If we choose this scaling factor, we are able to compare the two values directly, and all it really does in our drawing is to expand the size of the reciprocal lattice.

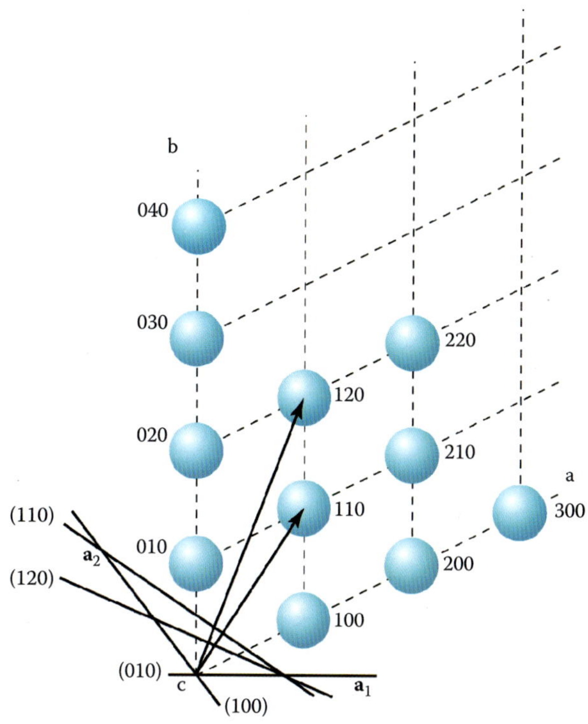

FIGURE 2.29 Graphical construction of the reciprocal lattice. (Drawing by Mr. Chengwu Deng.)

Definition of the Reciprocal Lattice

Now that we can draw a reciprocal lattice in 2D, let us define what a reciprocal lattice is. Imagine a set of points **R** constituting a Bravais lattice and a plane wave, $e^{i\mathbf{k}\cdot\mathbf{r}}$, interacting with that lattice. For most values of **k**, a plane wave will not have the periodicity of the Bravais lattice of points **R**. Only a very special set of the wave vectors, $\mathbf{G}_{hkl} = h\cdot\mathbf{a}_1^* + k\cdot\mathbf{a}_2^* + l\cdot\mathbf{a}_3^*$, of the reciprocal lattice satisfies that condition. Mathematically, $\mathbf{G}_{hkl}$ belongs to the reciprocal lattice of points **R** of the Bravais lattice, if the relation:

$$e^{i\mathbf{G}_{hkl}\cdot(\mathbf{r}+\mathbf{R})} = e^{i\mathbf{G}_{hkl}\cdot\mathbf{r}} \tag{2.44}$$

holds for any **r**, and for all **R** in the Bravais lattice. From Equation 2.44 it follows that we can describe the reciprocal lattice as the set of wave vectors $\mathbf{G}_{hkl}$ satisfying:

$$e^{i\mathbf{G}_{hkl}\cdot\mathbf{R}} = 1 \quad (2.45)$$

for all **R** in the Bravais lattice. Because lattice scatterers are displaced from one another by the lattice vectors **R**, the condition that all scattered rays interfere constructively is that the Laue equations hold simultaneously for all values of Bravais lattice vectors **R**, or:

$$\mathbf{R}\cdot\Delta\mathbf{k} = 2\pi m \quad (2.46)$$

with integer *m*. Because exp[i2π(integer)] = 1, this can be written in the equivalent form:

$$e^{i\mathbf{R}\cdot\Delta\mathbf{k}} = 1 \quad (2.47)$$

We see from Equation 2.47 that if $\Delta\mathbf{k}$ is equal to any reciprocal lattice vector $\mathbf{G}_{hkl}$, then the Laue equations for wave diffraction are satisfied. The diffraction condition is thus simply:

$$\Delta\mathbf{k} = \mathbf{G}_{hkl} \quad (2.48)$$

or from Equation 2.23 (with e = h, f = k and g = l):

$$\mathbf{G}_{hkl}\cdot\mathbf{a}_1 = 2\pi h$$
$$\mathbf{G}_{hkl}\cdot\mathbf{a}_2 = 2\pi k \quad (2.49)$$
$$\mathbf{G}_{hkl}\cdot\mathbf{a}_3 = 2\pi l$$

The expression $\Delta\mathbf{k} = \mathbf{G}_{hkl}$ can be represented in the form of a vector triangle as illustrated in Figure 2.30. Because **k** and $\mathbf{k}_0$ are of equal length, i.e. 2π/λ, the triangle *O*, *O*′, *O*″ has two equal sides. The angle between **k** and $\mathbf{k}_0$ is 2θ, and the *hkl* plane dissects it.

In the two-dimensional reciprocal lattice shown at the bottom of Figure 2.31, the $\mathbf{G}_{hkl}$ vectors give the outline of the unit cell. Vector $\mathbf{a}_1^* = \mathbf{G}_{100}$ and $\mathbf{a}_3^* = \mathbf{G}_{001}$ and:

$$|a_1^*| = \frac{2\pi}{d_{100}} \text{ and } |a_3^*| = \frac{2\pi}{d_{001}} \quad (2.50)$$

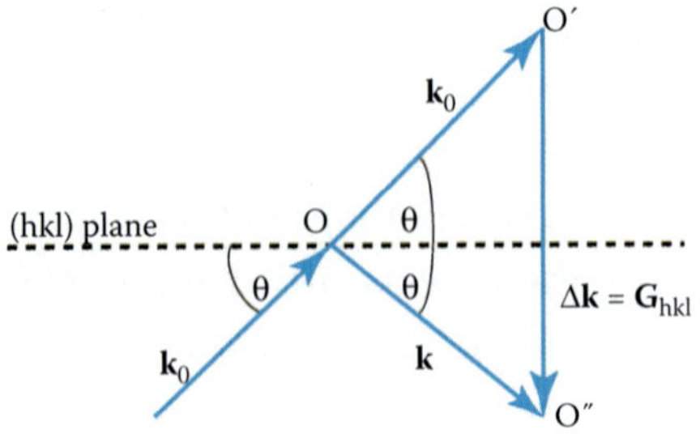

FIGURE 2.30 Vector triangle representation of $\Delta\mathbf{k} = \mathbf{G}_{hkl}$. (Drawing by Mr. Chengwu Deng.)

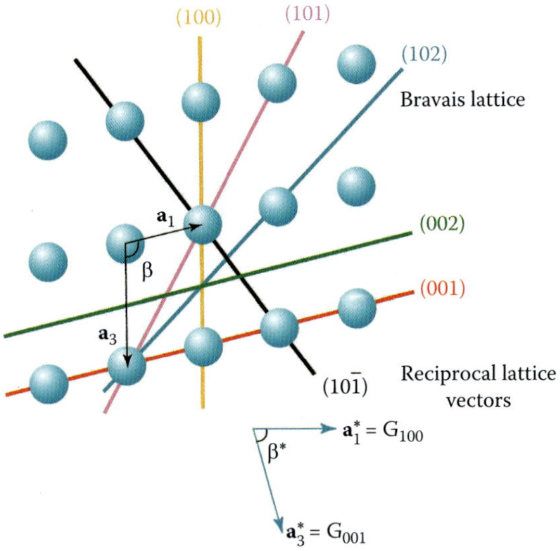

FIGURE 2.31 Reciprocal lattice vectors $a_1^* = G_{100}$ and $a_3^* = G_{001}$ in a monoclinic unit cell and their relation to the Bravais lattice. (Drawing by Mr. Chengwu Deng.)

The angle β* between the reciprocal lattice vectors in Figure 2.31 is the complement of β in the Bravais lattice.

The components of any vector referred to the reciprocal lattice represent the Miller indices of a plane whose normal is along that vector, whereas the spacing of the plane is given by the inverse of the magnitude of that vector multiplied by 2π. For example, the reciprocal lattice vector $\mathbf{u}^*$ = [123] is normal to the planes with Miller indices (123) and has an interplanar spacing $2\pi/|\mathbf{u}^*|$.

Explicit Algorithm for Constructing the Reciprocal Lattice

We can write out the change in wave vector, $\mathbf{G}_{hkl} = \Delta\mathbf{k} = \mathbf{k} - \mathbf{k}_0$, in the following expression:

$$\mathbf{G}_{hkl} = \Delta\mathbf{k} = e\mathbf{a}_1^* + f\mathbf{a}_2^* + g\mathbf{a}_3^* \quad (2.51)$$

where *e*, *f*, and *g* are the integers from the Laue equations (Equation 2.23) and $\mathbf{a}_1^*$, $\mathbf{a}_2^*$ and $\mathbf{a}_3^*$ are basis vectors of the reciprocal lattice to be determined. Equation 2.51 is a solution of the Laue Equation 2.23 if all of the following relations are satisfied:

$$\mathbf{a}_1^*\cdot\mathbf{a}_1 = \mathbf{a}_2^*\cdot\mathbf{a}_2 = \mathbf{a}_3^*\cdot\mathbf{a}_3 = 2\pi$$
$$\mathbf{a}_1^*\cdot\mathbf{a}_2 = \mathbf{a}_2^*\cdot\mathbf{a}_1 = \mathbf{a}_3^*\cdot\mathbf{a}_2$$
$$= \mathbf{a}_1^*\cdot\mathbf{a}_3 = \mathbf{a}_2^*\cdot\mathbf{a}_3 = \mathbf{a}_3^*\cdot\mathbf{a}_1 = 0 \quad (2.52)$$

From Equation 2.52, $\mathbf{a}_1^*$ is perpendicular to primitive lattice vectors $\mathbf{a}_2$ and $\mathbf{a}_3$ of the direct lattice. A vector

TABLE 2.3 Bravais Lattice after Fourier Transform

Real Space	Fourier Space
Normals to the planes (vectors)	Points
Spacing between planes	2π/distance planes
λ is distance, wavelength	$2\pi/\lambda$ is momentum or wave number
First Brillouin zone (see below)	Wigner-Seitz cell

that is perpendicular to $\mathbf{a}_2$ and $\mathbf{a}_3$ is given by the vector product $\mathbf{a}_2 \times \mathbf{a}_3$. To construct $\mathbf{a}_1^*$ of the reciprocal lattice completely, we must normalize $\mathbf{a}_2 \times \mathbf{a}_3$ to satisfy the expression $\mathbf{a}_1^* \cdot \mathbf{a}_1 = 2\pi$. All the equations in 2.52 are thus satisfied when choosing:

$$\mathbf{a}_1^* \equiv \frac{2\pi \mathbf{a}_2 \times \mathbf{a}_3}{\mathbf{a}_1 \cdot \mathbf{a}_2 \times \mathbf{a}_3}$$
$$\mathbf{a}_2^* \equiv \frac{2\pi \mathbf{a}_3 \times \mathbf{a}_1}{\mathbf{a}_1 \cdot \mathbf{a}_2 \times \mathbf{a}_3}$$
$$\mathbf{a}_3^* \equiv \frac{2\pi \mathbf{a}_1 \times \mathbf{a}_2}{\mathbf{a}_1 \cdot \mathbf{a}_2 \times \mathbf{a}_3} \quad (2.53)$$

The denominators have been written the same way because of the property of the triple scalar product: $\mathbf{a}_1 \cdot \mathbf{a}_2 \times \mathbf{a}_3 = \mathbf{a}_2 \cdot \mathbf{a}_3 \times \mathbf{a}_1 = \mathbf{a}_3 \cdot \mathbf{a}_1 \times \mathbf{a}_2$. The magnitude of this triple product is the volume of the primitive cell. In Table 2.3 we summarize the properties of a Bravais lattice before and after a Fourier transform.

The Ewald Construction

The Ewald construction is a way to visualize the Laue condition that the change in wave vector, $\Delta\mathbf{k}$, must be a vector of the reciprocal lattice. We reconsider for a moment the vector representation of the von Laue condition $\Delta\mathbf{k} = \mathbf{G}_{hkl}$ represented in the form of a vector triangle in Figure 2.30. Because $\mathbf{k}$ and $\mathbf{k}_0$ are of equal length ($= 2\pi/\lambda$) the triangle O, O', O'' has two equal sides, and we can draw a sphere with $|\mathbf{k}| = |\mathbf{k}_0| = 2\pi/\lambda$ as illustrated in Figure 2.32. The angle between $\mathbf{k}$ and $\mathbf{k}_0$ is 2θ, and the *hkl* plane dissects it. Diffraction of the incoming beam represented by the vector $\mathbf{k}_0$ giving the vector $\mathbf{k}$ may be thought of as reflection from the dotted line in this diagram (as in Bragg's Law). We will now see that with $\Delta\mathbf{k} = \mathbf{G}_{hkl}$ the diffraction vector corresponds to the distance between planes in reciprocal space. We thus superimpose an imaginary sphere of x-ray radiation on the reciprocal lattice as illustrated

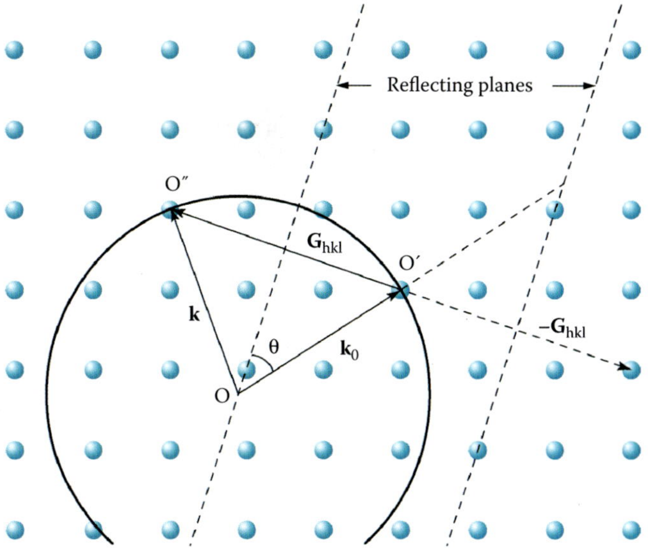

FIGURE 2.32 The reciprocal lattice and the geometry of diffraction clarified by the Ewald sphere. The sphere with center *O* intersects the reciprocal lattice center *O′*. (Drawing by Mr. Chengwu Deng.)

in Figure 2.32. Somewhat confusingly, one might consider two origins: *O*, which is the center of the sphere and may be considered as the position of the crystal, is the source of the secondary beam $\mathbf{k}$, and O' is the origin of reciprocal space, the origin of the diffraction vector $\Delta\mathbf{k} = \mathbf{G}_{hkl}$, and the center of the reciprocal lattice. As the crystal rotates, the reciprocal lattice rotates in exactly the same way. Any points in reciprocal space that intersect the surface of the sphere reveals where diffraction peaks will be observed if the structure factor is nonzero. In other words, reflection is only observed if the sphere intersects a point where $\Delta\mathbf{k} = \mathbf{G}_{hkl}$. The diffraction angle, θ, is then half the angle between the incident and diffracted wave vectors.

It is useful to think of the crystal at the center of the Ewald sphere (O) being linked to the center (origin) of the reciprocal lattice (O') by something like a bicycle chain—the two "objects" rotate exactly in step with each other.

The Ewald construction also makes for a good link with the Brillouin zones discussed in the next section. Notice in Figure 2.32 that both $\mathbf{G}_{hkl}$ and $-\mathbf{G}_{hkl}$ are vectors in reciprocal space. It can be seen that a reflecting plane bisects the vector $\mathbf{G}_{hkl}$ at $\mathbf{G}_{hkl}/2$ and another reflecting plane cuts $-\mathbf{G}_{hkl}$ at $-\mathbf{G}_{hkl}/2$. The incident wave vector $\mathbf{k}_0$ starting from the Ewald circle's center O must terminate at the $-\mathbf{G}_{hkl}/2$ reflecting plane for

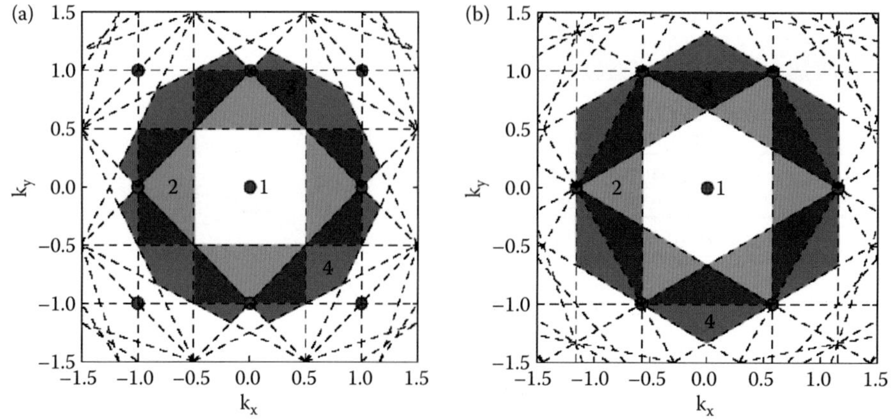

FIGURE 2.33 The Brillouin zones for (a) a square 2D lattice and (b) a triangular 2D lattice. The solid circles are the lattice points, and the dashed lines are the Bragg lines. The first four Brillouin zones are marked with different gray scales.

diffraction to occur. This observation gives rise to the idea of using the reciprocal lattice to construct Brillouin zones, the boundaries of which satisfy the Laue conditions for diffraction. Bragg planes bisect the lines joining the origin to points of the reciprocal lattice, and one can define the first Brillouin zone as the set of points that can be reached from the origin without crossing any other Bragg planes. Recall that this was also the way we learned how to construct Wigner-Seitz cells. So the first Brillouin zone is the Wigner-Seitz cell of the reciprocal lattice.

Brillouin Zones

Leon Brillouin (1889–1969) was a French-American physicist who was a professor at the Sorbonne, Collège de France, Wisconsin, Columbia, and Harvard. He also worked briefly at IBM. In the previous section

Lattice real space
Lattice **k**-space
BCC Wigner-Seitz cell
BCC BZ (FCC lattice in **k**-space)
FCC Wigner-Seitz cell
FCC BZ (BCC lattice in **k**-space)

FIGURE 2.34 The Wigner-Seitz cell of BCC lattice in real space transforms to a Brillouin zone in an FCC lattice in reciprocal space, whereas the Wigner-Seitz cell of an FCC lattice transforms to a Brillouin zone of a BCC lattice in reciprocal space.

we used the Ewald sphere (Figure 2.32) as a way to introduce the important concept of Brillouin zones. A Brillouin zone is defined as a Wigner-Seitz cell in the reciprocal lattice and gives a geometric interpretation of the diffraction condition. The Wigner-Seitz cell of the reciprocal lattice is the set of points laying closer to $\Delta\mathbf{k} = 0$ than to any other reciprocal lattice point. The Brillouin construction exhibits all wave vectors **k** that can be Bragg reflected by the crystal. The constructions divide the Fourier space into zones, out of which the first Brillouin zone is of greatest importance.

We have just seen that x-ray waves traveling in a crystal lattice undergo Bragg reflection at certain wave vectors. The first Brillouin zone is the set of points in **k**-space that can be reached from the origin without crossing any Bragg reflection plane. The second Brillouin zone is the set of points that can be reached from the first zone by crossing only one Bragg plane. The *n*th Brillouin zone can be defined as the set of points that can be reached from the origin by crossing $n - 1$ Bragg planes, but no fewer. Each Brillouin zone is a primitive cell of the reciprocal lattice. In Figure 2.33 we illustrate the first four Brillouin zones in square and triangular two-dimensional Bravais lattices. For examples of 3D Wigner-Seitz cells, corresponding to Brillouin zones, see Figures 2.6 and 2.34.

Nothing Is Perfect

Even a crystal having a mass of less than 0.1 g is likely to have more than 10^{23} ions, and it is unrealistic to expect that the growth of such crystal from solution can lead to zero impurities or imperfections. Successive purification steps may remove impurities, but even in the purest crystal, thermodynamics predicts the existence of some structural intrinsic imperfections. There are at least four types of defects one distinguishes: 1) point defects where the irregularity in structure extends over only a few atoms in size (0D), 2) linear defects where the irregularity extends across a single row of atoms (1D), 3) planar defects with an irregularity extending across a plane of atoms (2D), and 4) volumetric defects with the irregularity taking place over 3D clusters of atoms. In addition, defects can be categorized as either intrinsic, where defects are induced because of physical laws, and extrinsic, where defects are present because of the environment and/or processing conditions.

A physical law imposing the presence of intrinsic defects in a crystal is the minimization of the Gibbs free energy (G). The Gibbs free energy is higher in a crystal without vacancies than one with vacancies (G = H - TS = minimum). This is because defects are energetically unfavorable but are entropically favorable. Formation of vacancies (*n*) does indeed increase the enthalphy *H* of the crystal because of the energy required to break bonds ($\Delta H = n\Delta H_f$), but vacancies also increase *S* of the crystal as a result of an increase in configurational entropy. The configurational entropy is given as:

$$S = k\ln W \tag{2.54}$$

where *W* is the number of microstates. If the number of atoms in the crystal is *N* and the number of vacancies is *n*, then the total number of sites is $n + N$, and the number of all possible microstates *W* may be calculated from:

$$W = \frac{(N+n)!}{n!N!} \tag{2.55}$$

and the increase in entropy ΔS is then given by:

$$\Delta S = k\ln W = k\ln\frac{(N+n)!}{n!N!} \tag{2.56}$$

and the total free energy change as:

$$\Delta G = n\Delta H_f - T\Delta S \tag{2.57}$$

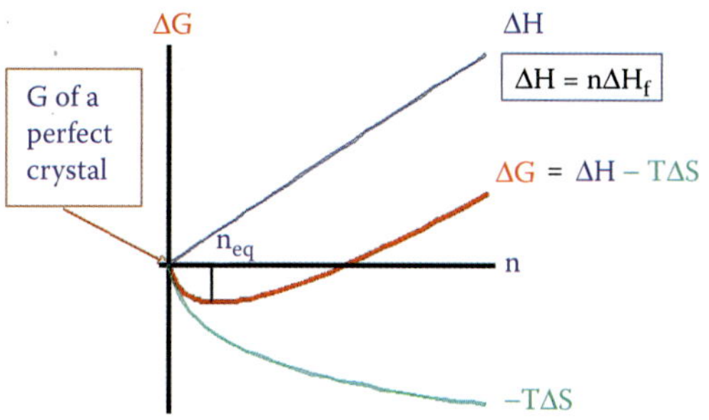

FIGURE 2.35 Change in Gibbs free energy *G* of a crystal as a result of the number of vacancies *n*.

This expression is plotted in Figure 2.35. It can be seen that for a crystal in equilibrium, vacancies are required to be present at any temperature above 0 K. The equilibrium concentration n_{eq} is calculated from:

$$\left|\frac{\partial \Delta G}{\partial n}\right|_{n=n_{eq}} = 0 \tag{2.58}$$

or we obtain:

$$\frac{n_{eq}}{N} = \exp\left(-\frac{\Delta H_f}{kT}\right) \tag{2.59}$$

so the n_{eq} should increase with increasing temperature. For Al, $\Delta H_f = 0.70$ eV/vacancy and for Ni, $\Delta H_f = 1.74$ eV/vacancy, leading to the values for n_{eq}/N at three different temperatures as shown in Table 2.4.

Defects, even in very small concentrations, can have a dramatic impact on the properties of a material. Actually, without defects solid-state electronic devices could not exist, metals would be much stronger, ceramics would be much tougher, and some crystals would have no color. Vacancies even make a small contribution to the thermal expansion of a crystal. Some commonly observed defects are summarized in Table 2.5; they are categorized here as point defects (0D), line defects (1D), and plane defects (2D).

The simplest sorts of defects are those based on points (0D). In Figure 2.36 we review point defects, including vacancies, interstitial atoms, small and

TABLE 2.4 n_{eq}/N Values for Al and Ni at Three Different Temperatures

n_{eq}/N	0 K	300 K	900 K
Al	0	$1.45\ 10^{-12}$	$1.12\ 10^{-4}$
Ni	0	$5.59\ 10^{-30}$	$1.78\ 10^{-10}$

Table 2.5 Common Defects in Crystals

Type of Imperfection	Description
Point defects:	
Interstitial	Extra atom in an interstitial site
Schottky defect	Atom missing from correct site
Frenkel defect	Atom displaced to interstitial site creating nearby vacancy
Line defects:	
Edge dislocations	Row of atoms marking edge of a crystallographic plane extending only part way in crystal
Screw dislocations	Row of atoms about which a normal crystallographic plane appears to be spiral
Plane defects:	
Lineage boundary	Boundary between two adjacent perfect regions in the same crystal that are slightly tilted with respect to each other
Grain boundary	Boundary between two crystals in a polycrystalline solid
Stacking fault	Boundary between two parts of a closest packing having alternate stacking sequences

large substitutional atoms, and the intrinsic Frenkel and Schottky defects.

A missing atom is a vacancy; a dissimilar atom in a nonlattice spot is an interstitial; and a different atom in a lattice position is substitutional. Vacancies are required in ionic solids, just like they are in metals, but the vacancies must be formed in such a way that the solid remains charge neutral. The two main ways to create point defects in ionic solids without causing a charge imbalance are through correlated vacancies (Schottky defects, for the German who described them in 1930) and through correlated vacancy/interstitial pairs (Frenkel defects, for the Russian who first described them in 1924). Impurities, such as dopants in single-crystal Si (see Chapter 4), are atom(s) of a type that do not belong in the perfect crystal structure. An impurity atom in a pure crystal will generally raise the enthalpy (H) somewhat and increase the entropy ($S = k\ln W$) a lot. In other words, there will always be some impurities present for the same reason as vacancies form in a pure crystal. When doping a crystal we introduce impurities on purpose (see Chapter 4), perhaps turning the material into an extrinsic semiconductor. Because of defects, metal oxides may also act as semiconductors. Some nonstoichiometric solids are engineered to be n-type or p-type semiconductors. Nickel oxide (NiO) gains oxygen on heating in air, producing Ni^{3+} sites acting as electron traps, resulting in p-type semiconductor behavior. On the other hand, ZnO loses oxygen on heating, and the excess Zn metal atoms in the sample are ready to donate electrons, leading to n-type semiconductor behavior.

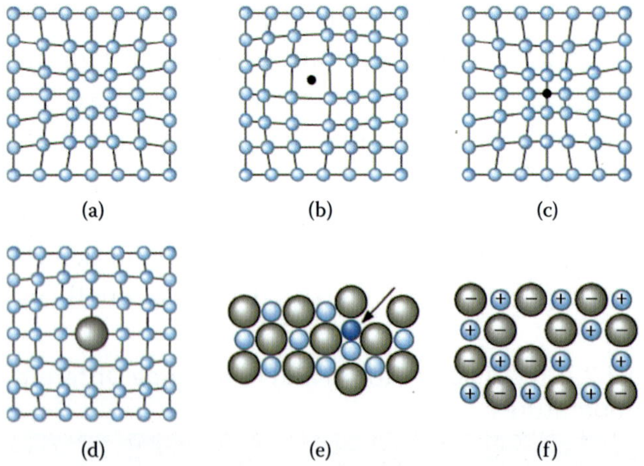

FIGURE 2.36 Point defects: vacancy (a), interstitial atom (b), small (c) and large (d) substitutional atom, Frenkel (e), and Schottky defect (f).

Color centers are imperfections in crystals that cause color. The simplest color center is found in sodium chloride, normally a colorless crystal. When sodium chloride is bombarded with high-energy radiation, a Cl^- can be ejected, creating a vacancy. Momentarily the crystal is no longer electrically neutral, and to regain stability, it grabs an available electron and sticks it in the vacancy previously occupied by the ejected Cl^-. With the electron replacing the ejected Cl^-, there are now equal numbers of positive and negative charges in the crystal, and the electron is held firmly in its site by the surrounding positively charged Na^+ ions. This process turns the colorless salt crystal into an orange/brown. These electrons are color centers, often referred to as F-centers, from the German word *Farben*, meaning color. The color center can also exist in an excited state, and the energy needed to reach that excited state is equal to the energy of a visible photon. The color center absorbs a "violet" photon, causing a jump to the excited state, and the crystal appears with the color orange/brown (the complement of violet). Analogous color

(a) (b) (c)

FIGURE 2.37 Color centers in some well-known minerals: (a) the Dresden green diamond, (b) smoky quartz, and (c) amethyst or violet quartz.

centers occur in several minerals. For example, a diamond with C vacancies (missing carbon atoms) absorbs light, and these centers lead to a green color (Figure 2.37). In some cases, impurities are involved in forming the color centers. Replacement of Al^{3+} for Si^{4+} in quartz gives rise to the color of smoky quartz (Figure 2.37). An iron impurity is responsible for the violet color in amethyst through the creation of a color center. The color center, not the iron impurity, is responsible for absorbing the "yellow" photon that makes amethyst violet (Figure 2.37).

Line defects are 1D imperfections in a crystal structure in which a row of atoms has a local structure that differs from the surrounding crystal. These defects are extrinsic because their presence is not required by thermodynamics. They are created by material processing conditions and by mechanical forces acting on the material. In a typical material, about 5 of every 100 million atoms (0.000005%) belong to a line defect. Line defects or dislocations have a dramatic impact on the mechanical properties of metals and some ceramics. Two Hungarians, Egon Orowan and Michael Polanyi, and an Englishman, G.I. Taylor, discovered dislocations in 1934. As we will see, dislocations really govern the mechanical properties of solids, and their discovery was a very important milestone in the material science field. There are two pure types of dislocations, edge dislocations and screw dislocations, but sometimes a mixed type is displayed. An edge dislocation is the simplest type of dislocation and can be viewed as an extra half-plane of atoms inserted into the crystal. This plane terminates somewhere inside the crystal. The boundary of the half-plane is the dislocation, shown in Figure 2.38A. A screw dislocation, shown in Figure 2.38B, is a dislocation produced by skewing a crystal so that one atomic plane produces a spiral ramp about the dislocation. A mixed dislocation is a dislocation that contains some edge components and some screws components as illustrated in Figure 2.38C. Line dislocations cannot terminate inside an otherwise perfect crystal but do end at a crystal surface, an internal surface, or interface (e.g., a grain boundary) or they form dislocation loops. Line defects are mostly caused by the misalignment of ions or the presence of vacancies along a line. When lines of ions are missing in an otherwise perfect array of ions, an edge dislocation appears.

Dislocations are visible in transmission electron microscopy (TEM) where diffraction images of dislocations appear as dark lines. When a crystal surface is etched the rate of material removal at the location of a dislocation is faster, and this results in an etch pit. Symmetrical strain fields of edge dislocations produce a conical pit, whereas for a screw dislocation a spiral etch pit results.

A Burgers vector **b** is a measure of the magnitude and direction of a dislocation. Imagine going around a dislocation line, and exactly going back as many atoms in each direction as you have gone forward, you will not come back to the same atom where you have started. The Burgers vector points from the start atom to the end atom of your journey. This "journey" is called Burgers circuit in dislocation theory. A Burgers circuit is thus a clockwise trace around the core of a dislocation, going from lattice point to lattice point, and it must go an equal number of steps left and right and an equal number

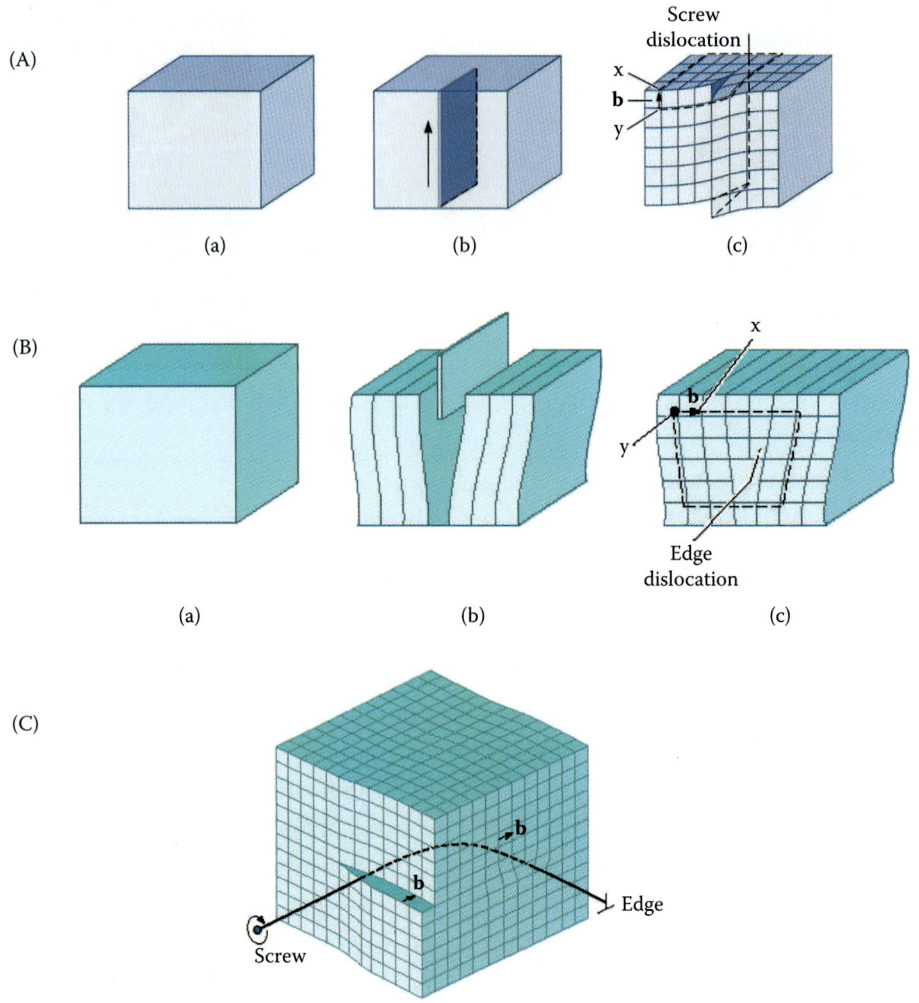

FIGURE 2.38 (A) Edge dislocation: The perfect crystal in (a) is cut, and an extra plane of atoms is inserted (b). The bottom edge of the extra plane is an edge dislocation (c). A Burgers vector **b** is required to close a loop of equal atom spacings around the edge dislocation. (B) Screw dislocation: The perfect crystal (a) is cut and sheared over one atom spacing (b and c). The line along which shearing occurs is a screw dislocation. (C) A mixed dislocation: The screw dislocation at the front face of the crystal gradually changes to an edge dislocation at the side of the crystal.

of steps up and down. Referring to Figure 2.38A, the extra half-plane of lattice points in the edge dislocation causes the Burgers circuit to be open. The vector that points from the end of the Burgers circuit to its beginning is shown here as the Burgers vector, **b**. The vector always points from one lattice point to another; it always has the same length and direction for a given dislocation, regardless where the circuit starts. For an edge dislocation, **b** is always perpendicular to the dislocation line. If **b** is parallel to the dislocation line, the dislocation is a screw dislocation. If **b** is neither perpendicular nor parallel to the line, a mixed dislocation is involved.

Plastic deformation of a material refers to irreversible deformation or change in shape that remains even when the force or stress causing it is removed, whereas elastic deformation is deformation that is fully recovered when the stress causing it is removed. Before the discovery of dislocations, materials scientists faced a big theoretical problem; they calculated that plastic deformation of a perfect crystal should require stresses 100 to 1000 times higher than those observed in tensile tests! So the problem was to explain why metals yielded so easily to plastic deformation. How they got these very large theoretical numbers can be understood from Figure 2.39. Planes of atoms in a crystal slip with respect to each other, in contrast to flow in a fluid, where the solid remains crystalline. The theoretical maximal shear stress or yield strength (failure), τ_{max}, needed

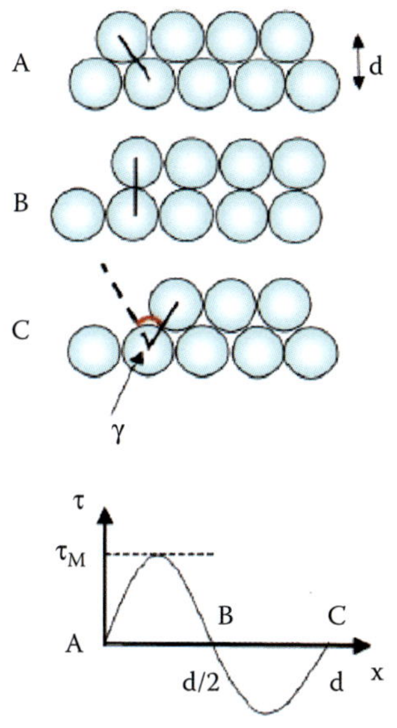

FIGURE 2.39 Calculation of the theoretical shear stress in a crystal.

to produce slip in an ideal crystal is calculated from the type and strength of the bonds involved, the spacing of the crystal, *d*, and the crystal symmetry. In yield, atoms slide tangentially from one equilibrium position to another. Thus, the shear stress, τ, is a periodic function:

$$\tau = \tau_{max} \sin\frac{2\pi x}{d} \qquad (2.60)$$

where *x* is the direction of the shear. With *x* small this may be rewritten as:

$$\tau = \tau_{max}\frac{2\pi x}{d}$$
$$x = \gamma d = \frac{\tau}{\mu}d \therefore \tau = \tau_{max}\frac{2\pi}{d}\frac{\tau}{\mu}d \qquad (2.61)$$
$$\tau_{max} = \frac{\mu}{2\pi}$$

where we have used γ = τ/μ (Hooke's law) as the shear strain, with μ as the shear modulus (see Figure 2.40).

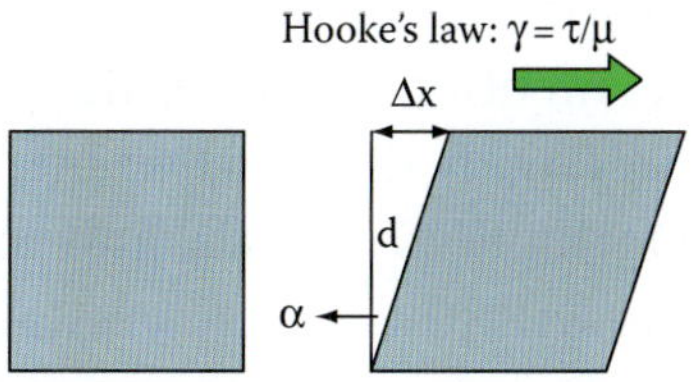

FIGURE 2.40 Shear stress, τ; shear strain, γ; and shear modulus, μ. The shear stress τ produces a displacement Δx of the upper plane as indicated; the shear strain, γ, with Δx/d = tan α is defined γ = τ/μ.

TABLE 2.6 Theoretical Yield Strengths (Shear and Tensile) of Some Important Materials

Material	Shear Strength (GPa)	Tensile Strength (GPa)
Metallic Cu	1.2	3.9
NaCl	2.84	0.43
Quartz	4.4	16
Diamond	121	205

Calculated values for μ are in the range of 1–150 GPa (see also Table 2.6).

But when the experiment is carried out, permanent deformation takes place by a stress as low as 0.5 MPa! The reason is that real materials have lots of dislocations, from $10^2/cm^2$ in the best Ge and Si crystals to $10^{12}/cm^2$ in heavily deformed metals. Therefore, the strength of the material depends on the force required to make dislocations move, not the bonding energy between all atoms in two planes as calculated above. Dislocations can move if the atoms from surrounding planes break their bonds and rebond with the atoms at the terminating edge as shown in Figure 2.41. Instead of all the atoms in

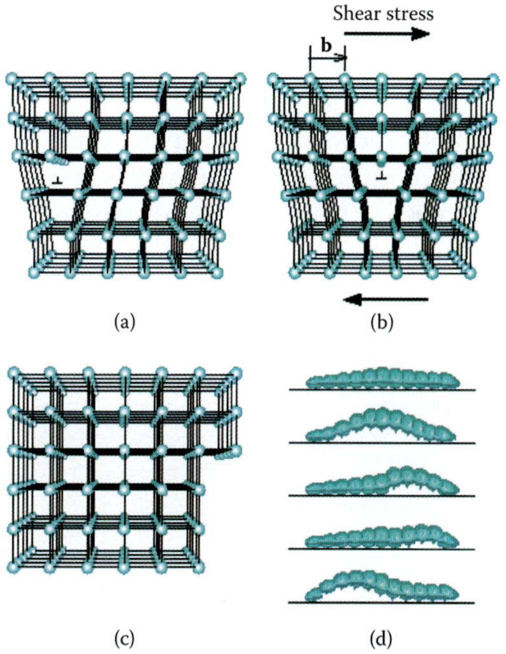

FIGURE 2.41 Dislocation movement: When a shear stress is applied to the dislocation in (a), the atoms are displaced, causing the dislocation to move one Burgers vector **b** in the slip direction (b). Continued movement of the dislocation eventually creates a step (c), and the crystal is deformed. (d) The caterpillar does not move its complete body at a single time, but it moves one segment at a time as it pulls itself forward.

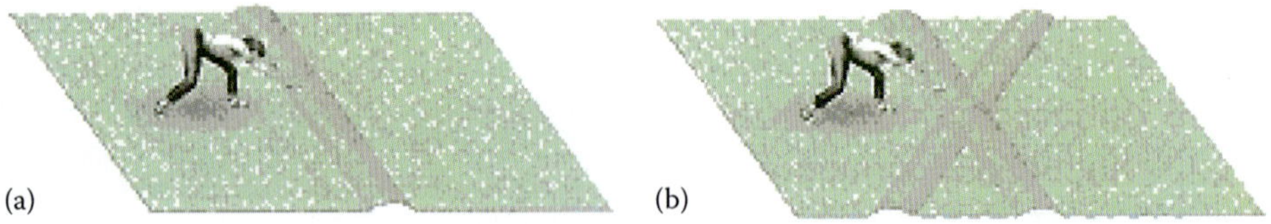

FIGURE 2.42 Moving a carpet over the floor to illustrate the effect of a line dislocation in a crystal: (a) dislocation; (b) work hardening.

a plane breaking at the same time, atoms are gliding gently in the direction of applied stress normal to dislocation lines. Thus, the direction of movement, i.e., the deformation, is the same as the Burgers vector, **b**. As a consequence **b** is also called the slip vector.

The interatomic forces in a crystal offer little resistance to the gliding motion of dislocations. An analogy is that of moving a carpet across the floor. This is difficult because of the friction developed from the contact of the whole surface of the carpet with the floor. But with a wrinkle in the carpet, as shown in Figure 2.42, the carpet can now be moved by pushing the wrinkle across the floor. The work involved is much less because only the friction between a small section of carpet and the floor has to be overcome. A similar phenomenon occurs when one plane of atoms moves past another by means of a dislocation defect.

Figure 2.43 shows how an applied shear stress, τ, exerts a force on a dislocation and is resisted by a frictional force, **F**, per unit length. The work done by the shear stress (W_τ) equals the work done by the frictional force (W_F), or:

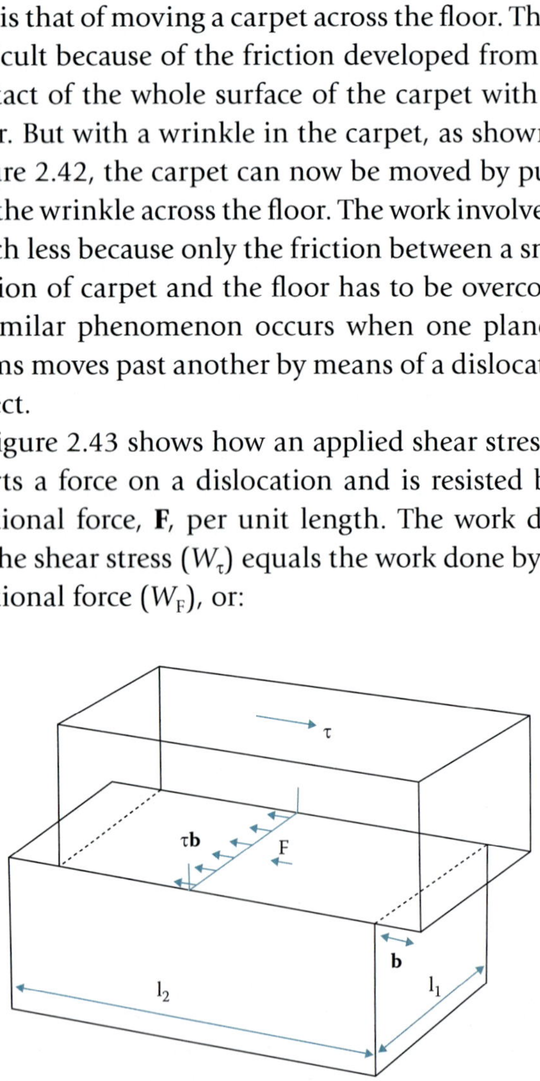

FIGURE 2.43 An applied shear stress, τ, exerts a force on a dislocation and is resisted by a frictional force, **F**, per unit length. The slip vector or Burgers vector is **b**. (Drawing by Mr. Chengwu Deng.)

$$W_\tau = (\tau\, l_1 l_2) \times \mathbf{b} \quad \text{and} \quad W_F = (F\, l_1) \times l_2 \tag{2.62}$$

With $W_\tau = W_F$, the lattice friction stress for dislocation motion is:

$$\tau = \frac{\mathbf{F}}{\mathbf{b}} \tag{2.63}$$

where **F** is the force per unit length of the dislocation and **b** is its Burgers vector or slip vector. Thus, the applied stress produces a force per unit length everywhere along the dislocation line equal to τ**b** and perpendicular to the line element. It can be shown that:

$$\tau = \frac{\mathbf{F}}{\mathbf{b}} = \mu e^{\frac{2\pi \mathbf{a}}{\mathbf{b}}} \tag{2.64}$$

This lattice friction stress is much less than the theoretical shear strength. Dislocation motion, from Equation 2.64, is most likely on closed packed planes (large **a**, interplanar spacing) in closed packed directions (small **b**, in-plane atomic spacing). It is thus easiest to create dislocations in the closest packed crystals, and they are typically very soft and ductile.

When dislocations move it is said that slip occurred, and the lattice planes that slipped are called—imaginatively—slip planes as elucidated in Figure 2.44 for an edge dislocation and a screw dislocation. As mentioned above, the preferred slip planes are those with the greatest interplanar distance, e.g., (111) in FCC crystals, and the slip directions are those with the lowest resistance, i.e., the closest packed direction. Slip lines are the intersection of a slip plane with a free surface.

The motion of dislocations can be blocked by another dislocation, a grain boundary, or a point

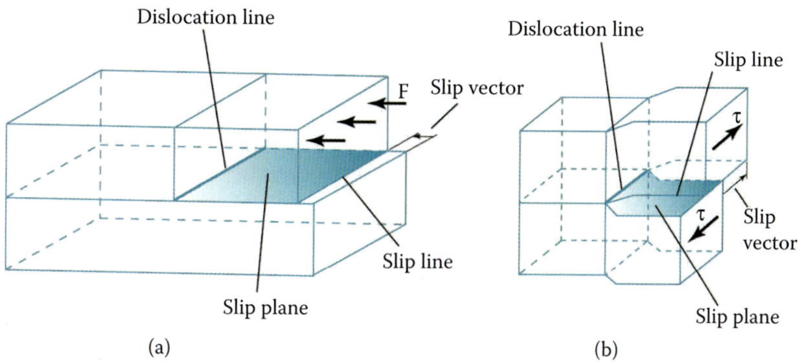

FIGURE 2.44 Schematic of slip line, slip plane, and slip (Burgers) vector for (a) an edge dislocation and (b) for a screw dislocation. (Drawing by Mr. Chengwu Deng.)

defect. Two dislocations may repel or attract each other, depending on their directions. Atoms near the core of a dislocation have a higher energy because of distortion. To minimize this energy, dislocations tend to shorten as if the line had a line tension, *T*, i.e., strain energy per unit length. The line tension is given as:

$$T \approx \frac{1}{2}\mu \mathbf{b}^2 \tag{2.65}$$

Dislocations might get pinned by interstitials and bow with a radius *R* when subjected to a shear stress like the string shown in Figure 2.45:

$$\tau \mathbf{b} L = 2T \sin\frac{\theta}{2} \tag{2.66}$$

With Equation 2.65 this leads to:

$$R = \frac{\mu \mathbf{b}}{2\tau} \tag{2.67}$$

Plastic deformation (or yielding) occurs by sliding (or slip) of parallel lattice planes past each other. Pure metals have low resistance to dislocation motion, thus they exhibit low yield strength. Adding impurities in *solution strengthening* may increase the yield strength. A well-known example is alloying Zn and Cu to form brass with a strength increase of up to 10 times over pure Cu. The bigger Zn atoms make the slip plane "rougher," thus increasing the resistance to motion. In general, impurities diffuse to dislocations and form "clouds" that pin dislocations, increasing the elastic limit. Small particles, precipitates, can also promote strengthening by impeding dislocation motion. Precipitates cause bowing of dislocations as illustrated in Figure 2.45. A critical condition is reached when the dislocation takes on a semicircular configuration between two particles separated by a distance *L*. Beyond this point a dislocation may loop and escape between the finely dispersed particles in the metal. In the semicircular situation $\tau \mathbf{b} L = 2T$ or $\tau = 2T/\mathbf{b}L$, and with Equation 2.67 $\tau = \mu\mathbf{b}/L$; thus, making *L* smaller will disadvantage dislocation looping. Finally, metals can be hardened by work hardening. Dislocations move when metals are subjected to "cold work," and their density can increase up to 10^{12} dislocations/cm^2 because of the formation of new dislocations and dislocation multiplication (see below). The consequent increasing overlap between the strain fields of adjacent dislocations gradually increases the resistance to further dislocation motion. This causes a hardening of the metal as the deformation progresses (Figure 2.42b). This effect is known as strain hardening. The effect of strain hardening can be removed by appropriate heat treatment (annealing), which promotes the recovery and subsequent recrystallization of the material. Annealing decreases the dislocation density to around 10^6 dislocations/cm^2.

Some dislocations form during the process of crystallization, but more are created during plastic deformation. Frank and Read elucidated one possible mechanism by which dislocations multiply;

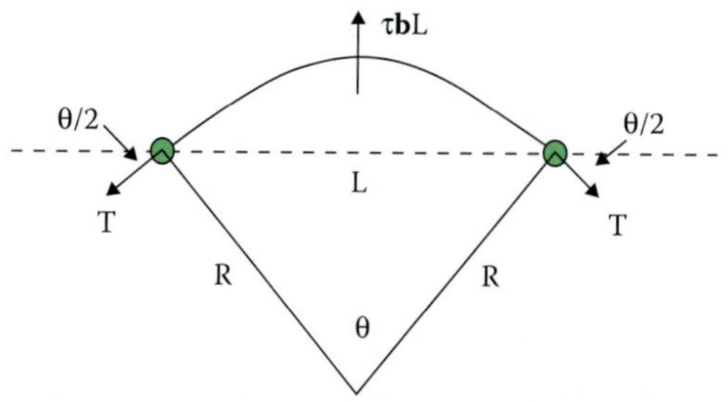

FIGURE 2.45 A pinned dislocation bows under a shear stress.

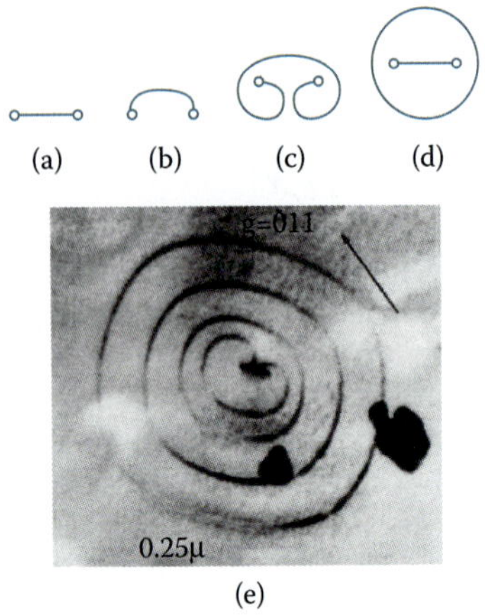

FIGURE 2.46 A Frank-Read source can generate dislocations. (a) A dislocation is pinned at its ends by lattice defects. (b) As the dislocation continues to move, the dislocation bows, eventually bending back on itself. (c) Finally the dislocation loop forms, and (d) a new dislocation is created. (e) Electron micrograph of a Frank-Read source (×330,000).

they found that a pinned dislocation (as shown in Figure 2.45), under an applied stress, produces additional dislocations. This mechanism is at least partly responsible for strain hardening. The Frank-Read dislocation multiplication mechanism is illustrated in Figure 2.46. First, a dislocation is pinned at its ends by lattice defects, and as the dislocation continues to move, it bows and eventually bends back on itself. Then the dislocation loop forms, and a new dislocation is created. Figure 2.46 also shows an electron micrograph of a Frank-Read source.

Schmid's Law, illustrated in Figure 2.47, gives the relationship between shear stress, the applied stress, and the orientation of the slip system. Slip on a given

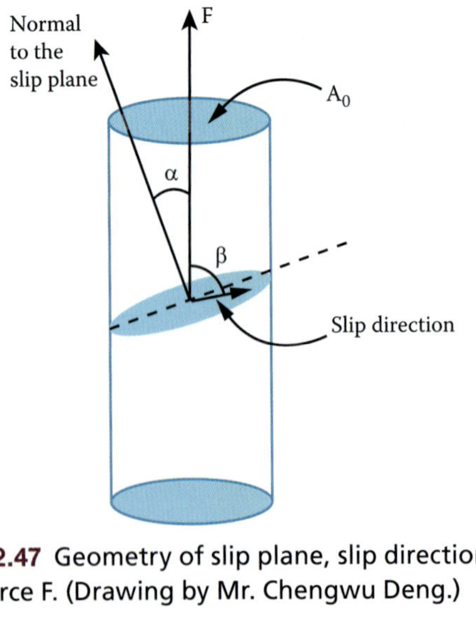

FIGURE 2.47 Geometry of slip plane, slip direction, and tensile force F. (Drawing by Mr. Chengwu Deng.)

slip system begins when the shear stress resolved on that system reaches a critical value. Consider a cylindrical crystal of cross-section A_0 under the influence of a tensile force *F*. Let the normal to the active slip plane make an angle α with *F*, and let the angle between the slip direction and *F* be β. The resolved shearing force, i.e., the force acting per unit area of the slip plane in the slip direction, is then given by:

$$\tau = \frac{F}{A}\cos\alpha\cos\beta \tag{2.68}$$

Because the area of the slip plane is *A*/cosα, the tensile stress per unit area normal to the slip plane is:

$$\sigma = \frac{F}{A}\cos^2\alpha \tag{2.69}$$

When increasing the force *F*, the rate of plastic flow increases very rapidly when the resolved shear stress τ reaches a critical value τ_c. In general τ_c decreases with increasing temperature and increases as a result of alloying or cold working (see above).

Besides point and line defects there are also surface defects to reckon with in single crystals. Surface defects include grain boundaries, phase boundaries, and free surfaces. The crystal structure is disturbed at grain boundaries, and different crystal orientations are present in different grains (Figure 2.48).

Dislocations are blocked by grain boundaries, so slip is blocked. The smaller the grain size, the larger the surface of the grain boundaries and the larger the elastic limit. The latter is expressed in the empirical Hall-Petch equation for the maximum elastic yield strength (stress at which the material permanently deforms) of a polycrystalline material:

$$\sigma_Y = \sigma_0 + \frac{K}{\sqrt{d}} \tag{2.70}$$

where *d* is the average diameter of the grains in micrometers, with σ_0 the frictionless stress (N/m^2) that opposes dislocation, and *K* a constant. The strength of a material thus depends on grain size. In a small grain, a dislocation gets to the boundary and stops, i.e., slip stops. In a large grain, the dislocation can travel farther. So small grain size equates to more strength. For example, the elastic limit of copper doubles when the grain size falls from 100 μm

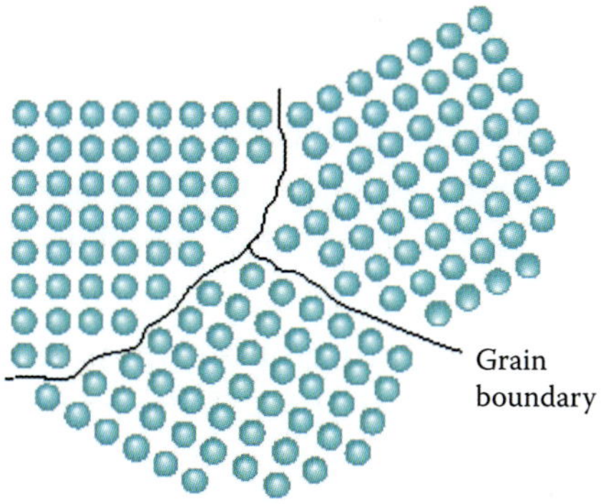

(a)

(b)

FIGURE 2.48 Crystal structure is disturbed at grain boundaries. Schematic representation of grain boundaries (a) and microscope picture (b). (From Askeland D. R., and P. P. Phule, *The science and engineering of materials*, Brooks/Cole, Pacific Grove, CA, 2003.)

to 25 μm. Instead of yield strength one might also plot the hardness (H) of the material as a function of grain size, and a similar relationship is obtained (see Volume II, Figure 7.54): smaller grain size corresponds to a harder material. Indentation (hardness) testing is very common for bulk materials where the direct relationship between bulk hardness and yield strength is well known. As we will learn in Volume II, Chapter 7, in the section on thin film properties, the Hall-Petch relationship has been well established for grain sizes in the millimeter through submicrometer regimes but is less well-known in the nano regime. Based on Equation 2.70 one expects that nano-sized grains would produce materials with yet greater mechanical integrity, but in reality this is not the case! There is a reverse Hall-Petch effect, i.e., the strength of materials, from a small grain size on, starts to decrease with decreasing grain size. Plastic deformation occurs at lower and lower stresses as the grains shrink. In other words, an optimal grain size (d_c) exists as suggested by the plot in Volume II, Figure 7.54. The classic Hall-Petch relationship is based on the idea that grain boundaries act as obstacles to dislocation movement, and because dislocations are carriers of plastic deformation, this manifests itself macroscopically as an increase in material strength. The Hall-Petch behavior breaks down when the smallest dislocation loop no longer fits inside a grain. Lattice dislocation cannot be the way very small-grained materials deform. The deformation mechanism for materials with very small grains (<20 nm) is indeed different, and it has been suggested that plastic deformation in this case is no longer dominated by dislocation motion; it is believed that individual atoms migrate, diffuse, and slide along grain boundaries and through triple junctions (Y-shaped grain boundary intersections).[1]

In Volume III, Chapter 7, on scaling, we learn that the surface-to-volume ratio (S/V) of particles scales as $1/r$, where r is the characteristic dimension of the particles. The smaller a particle, the more of its atoms find themselves at its surface. A bulk solid material will typically have less than 1% of its atoms on its surface but 10-nm particles have about 15% of surface atoms (Figure 2.49). The high S/V ratio of nanoparticles makes them

Full-Shell Clusters		Total Number of Atoms	Surface Atoms (%)
1 Shell		13	92
2 Shells		55	76
3 Shells	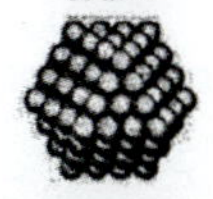	147	63
4 Shells		309	52
5 Shells	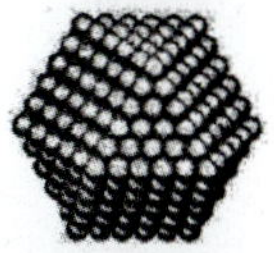	561	45
7 Shells		1415	35

FIGURE 2.49 Nanoparticles, clusters of atoms in shells.

more reactive as catalysts in chemical reactions and lowers their melting temperature, T_m. There is an inverse linear relationship between melting temperature and the surface-to-volume ratio. This makes sense because atoms on a surface are more easily accessed and rearranged than atoms in the bulk. As a consequence, the melting temperature of particles is always lower than the bulk. Grains in a solid are analogous to particles, and the melting temperature of solids decreases with grain size. Although the temperature dependence of σ_0 and K in Equation 2.70 can be neglected for conventional grain sizes, for grains smaller than 20 nm this assumption breaks down. In nanomaterials the melting temperature decreases because of the smaller grain size.

If we could grow a crystal without dislocations, it should approach the theoretical shear stress τ calculated in Equation 2.61. Sometimes fine metallic whiskers can be grown virtually free of dislocations, and these are very strong indeed.

In Figure 2.50 we show a picture of a zinc whisker. Metal whiskers are a crystalline metallurgical phenomenon whereby metal grows tiny, filiform hairs. The effect is primarily seen with elemental metals but is also observed with alloys. The mechanism behind metal whisker growth is not well understood but seems to be encouraged by compressive mechanical stresses, including residual stresses caused by electroplating, mechanically induced stresses, stresses induced by diffusion of different metals, and thermally induced stresses. Metal whiskers differ from metallic dendrites in several respects; dendrites are fern-shaped and grow across the surface of the metal, whereas metal whiskers are hair-like and project at a right angle to the surface. Although the precise mechanism for whisker formation remains unknown, it is known that whisker formation does not require either dissolution of the metal or the presence of an electromagnetic field. Whiskers, like carbon nanotubes, approach ideal lattices.

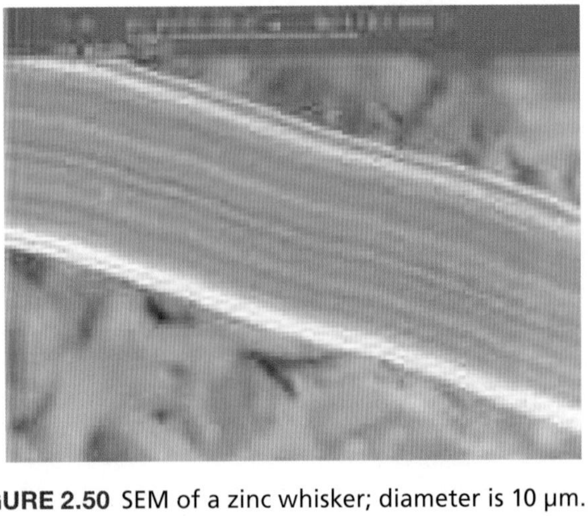

FIGURE 2.50 SEM of a zinc whisker; diameter is 10 μm.

As we are moving to smaller and smaller functional devices in MEMS and NEMS, surfaces and crystal imperfections contribute more and more importantly to overall performance and behavior. Many of the observations described here will have to be revisited when exploring nanomaterials and nanodevices. These effects will be treated in Volume II, Chapter 7 on thin films and in Volume III, Chapter 3 on nanotechnology.

Acknowledgments

Special thanks to Xavier Casadevall I Solvas, Robin Gorkin, Chengwu Deng, Kartikeya Malladi, Leyla Esfandiari, Fatima Alim, and Drs. Sean Parkin, Han Xu, and Guangyao Jia.

Appendix 2A: Plane Wave Equations

We define a plane wave,* $e^{i\mathbf{k}.\mathbf{r}}$, as a wave that is constant in a plane perpendicular to **k** (in rad/m) and is periodically parallel to it, with a wavelength $\lambda = 2\pi/\mathbf{k}$, where **k** is the wave number (because it measures the number of wavelengths in a complete cycle) with a value of $2\pi/\lambda$ and where $\omega = 2\pi\nu = c\mathbf{k}$ (free space) is the period of the wave (see also Table 2A.1). A bit simpler: a plane wave is a continuous wave (CW) whose amplitude and phase are constant in the directions transverse to the propagation direction.

* The complex exponential representation for periodic functions is convenient for adding waves, taking derivatives of wave functions, and so on. It is equivalent to a linear combination of a sine and cosine function because $e^{i\theta} = \cos\theta + i \sin\theta$.

Table 2A.1 Plane Wave Parameters

Parameter	Symbol/Value
Amplitude	E_m, B_m
Phase	$\phi = kx - \omega t$
Velocity	$v = \frac{\omega}{\mathbf{k}}$
Wavelength	λ
Period	T
Wave vector	$\mathbf{k} = \frac{2\pi}{\lambda}$
Angular frequency	$\omega = \frac{2\pi}{T}$
Wave number	$\bar{\mathbf{k}} = \frac{1}{\lambda}$
Cyclic frequency	$v = \frac{1}{T}$

The amplitude of a wave propagating in the x-direction is mathematically introduced as

$$\psi = \psi_0 \sin(kx + \omega t + \delta)$$
$$\omega = 2\pi f, \text{ the cyclic frequency}$$
$$k = \frac{2\pi}{\lambda}, \text{ the wave vector}$$
$$\delta, \text{ the phase at } t = 0 \text{ and } x = 0$$

Calculations are greatly simplified by using complex numbers:

$$\psi = \psi_0 \text{Im } e^{i(kx+\omega t+\delta)}$$
$$i = \sqrt{-1}$$
$$|e^{ix}| = 1$$
$$\text{Im }(e^{ix}) = \sin x$$
$$\text{Re }(e^{ix}) = \cos x$$
$$\sin x = \frac{e^{ix} + e^{-ix}}{2i}$$
$$\cos x = \frac{e^{ix} + e^{-ix}}{2}$$

Questions

2.1: Scientists are considering using nanoparticles of magnetic materials such as iron-platinum (Fe-Pt) as a medium for ultrahigh density data storage. Arrays of such particles potentially can lead to storage of trillions of bits of data per square inch—a capacity that is 10 to 100 times higher than any current storage devices such as computer hard disks. If these scientists consider iron (Fe) particles that are 3 nm in diameter, what is the number of atoms in one such particle?

2.2: Assuming that silica (SiO_2) has 100% covalent bonding, describe how oxygen and silicon atoms in silica (SiO_2) are joined.

2.3: What is the difference between lattice and basis and between unit cell and primitive cell?

2.4: What are the net numbers of Na^+ and Cl^- ions in the NaCl unit cell represented below? The crystal is an example of which type of cubic lattice? Identify the atom positions of the Na and Cl atoms in the NaCl structure.

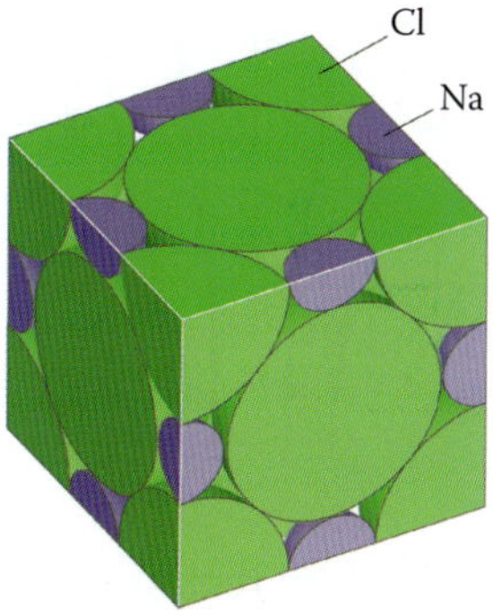

2.5: Consider the plane defined by the three points, P1(2,4,–3), P2(–1,2,1), and P3(3,0,–2). Calculate the points where this plane intersects with the axes and derive the Miller indices associated with this plane.

2.6: Calculate (a) the angle between [111] and the direction normal to (111) plane in a simple cubic crystal and (b) the angle between the [121] direction and the direction normal to (113) plane in a simple cubic crystal.

2.7: A signal $x(t)$ has a Fourier transform $X(f)$. Express the Fourier transforms of x_1 in terms of $X(f)$, the function $x_1(t) = x(3 - t)$.

2.8: In an x-ray set-up with a crystal-to-detector distance of 100 mm (D), you find that the highest resolution reflection is at $x = 20$ mm, $y = 15$ mm, relative to the direct beam position. The wavelength $\lambda = 1.54$ Å. What is the Bragg angle of this reflection? What is the d-spacing of the crystal? If the detector is circular and has a radius of 100 mm, and you would like to collect data so that the highest resolution reflection is at the detector edge, would you move the detector closer to or further away from the crystal?

2.9: Describe the difference between a reciprocal lattice and a real one.

2.10: Edge dislocations may be used to getter impurities in semiconductors. In an edge dislocation, there are always two regions, a compressive region, where the layer is inserted, and a tensile region (see figure below). Which of these two regions will best accommodate (via elastic interactions) *substitutional* atoms whose radius is larger than that of the host atoms?

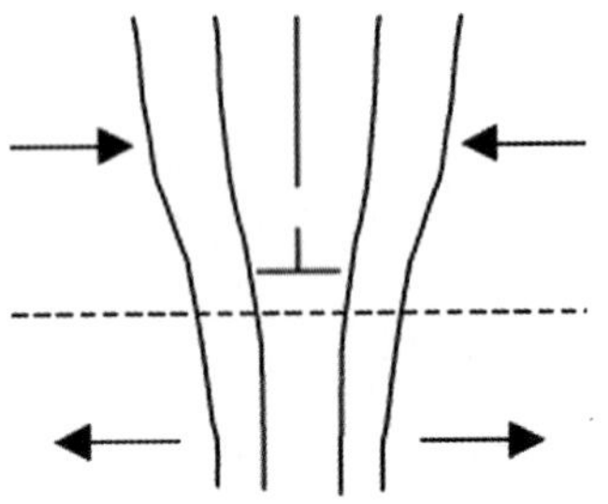

2.11: Show that the reciprocal lattice of a face-centered cubic (FCC) lattice is a body-centered cubic (BCC) lattice or, conversely, that the reciprocal lattice of a BCC lattice is an FCC lattice.

2.12: An electron moves with speed u = 0.7c. Calculate its total energy and its kinetic energy in eV.

2.13: Calculate the number of atoms in a 100 μm long Ag line (1 μm wide and 1 μm high). If using STM we put one atom down per second, how long will it take to finish this Ag line?

2.14: What is the Miller index for the plane shown below?

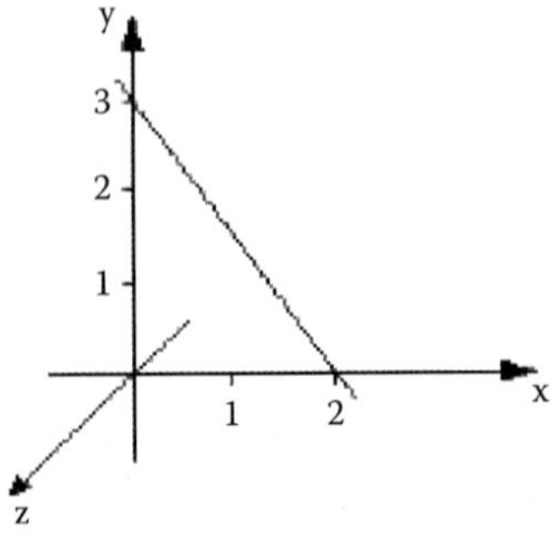

2.15: What is the number of nearest neighbors for the following crystal lattices:
(a) simple cubic (SC)
(b) face-centered cubic (FCC)
(c) body-centered cubic (BCC)

2.16: Calculate the angle θ of reflection for an x-ray experiment with λ = 1.54 Å, for a cubic crystal with a lattice parameter of **a** = 5 Å.

2.17: X-rays with wavelength 1.54 Å are "reflected" from the (110) planes of a cubic crystal with unit cell **a** = 6 Å. Calculate the Bragg angle, θ, for all orders of reflection, *n*.

2.18: What is the closest packed crystal? Simple cubic (SC), body-centered cubic (BCC), or face-centered cubic (FCC)?

2.19: Is five-fold symmetry ever found in crystal lattices? Why or why not?

2.20: Calculate the number of atoms in 100 g of silver.

2.21: The derivation of Bragg's law results in nλ = 2dsinθ. What does *n* represent and why is it usually omitted? Can you give an example to show why n is not needed?

2.22: Calculate the packing factor in a FCC lattice.

2.23: What is the rule for determining the *slip direction* in a close-packed material?

2.24: What causes work hardening?

2.25: How can we determine the direction cosine between two vectors?

2.26: How can we identify the direction of slip in a crystal (Burgers vector)?

2.27: Why do very thin metal wires/whiskers exhibit very high strengths?

2.28: Why does the strength of glass fibers increase as the diameter goes down?

2.29: Find the angle between the planes 110 and 100 in a simple cubic crystal.

2.30: Could you use x-ray diffraction to determine the coefficient of thermal expansion (i.e., change in length of a material due to change in temperature)?

2.31: Diamond and graphite are examples of which type of crystalline solids: molecular, covalent network, ionic, or metallic?

2.32: What value does the atomic scattering factor f approach as 2θ approaches 0?

2.33: How does light affect the color of a crystal?

Further Reading

Vectors and Tensors

Danielson D. A. 1997. *Vectors and tensors in engineering and physics.* New York: Addison-Wesley Publishing Company, Inc.

Crystallography

Books reviewed or listed in *Crystallography News*. See http://img.cryst.bbk.ac.uk/bca/cnews/books/index.html.

The International Union of Crystallography (IUcr) produces a number of books on crystallography, in association with Oxford University Press and Springer. See http://www.iucr.org/iucr-top/publ/index.html.

Hannay, N. B. 1967. *Solid-state chemistry.* Englewood Cliffs, NJ: Prentice-Hall, Inc.

Ladd, M. F. C., and R. A. Palmer. 2003. *Structure determination by X-ray crystallography.* 4th ed. New York: Springer.

Sands, D. S. 1969: *Introduction to crystallography.* New York: W.A. Benjamin, Inc.

Solid-State Physics

Brown, F. C. 1967. *The physics of solids.* New York: W.A. Benjamin, Inc.

Dekker, A. J. 1970. *Solid state physics.* New York: Prentice-Hall, Inc.

Kittel, C. 2005. *Introduction to solid state physics.* 8th ed. New York: John Wiley & Sons, Inc.

"Phase Problem"

Eisenberg, D., and D. Crothers. 1979. *Physical chemistry: with applications to the life sciences.* San Francisco: Benjamin-Cummings/Addison Wesley Longman.

Glusker, J. P., and K. N. Trueblood. 1972. *Crystal structure analysis: a primer.* New York: Oxford University Press.

Holmes K. C., and D. M. Blow. 1965. *Use of X-ray diffraction in the study of protein and nucleic acid structure.* New York: Krieger Pub Co/Interscience.

Wilson, H. R. 1966. *Diffraction of X-rays by proteins, nucleic acids, and viruses.* New York: St. Martin's Press.

Dislocations

Hull, D. 1965. *Introduction to dislocations.* Oxford, UK: Pergamon Press.

Weertman, J., and J. R. Weertman. 1964. *Elementary dislocation theory.* New York: MacMillan.

Reference

1. Schiotz, J., and K. W. Jacobson. 2003. A maximum in the strength of nanocrystalline copper. *Science* 301:1357–59.

3

Quantum Mechanics and the Band Theory of Solids

Some of the actors in this chapter: (a) Fermi, (b) Schrödinger, (c) de Broglie, (d) Feynman, (e) Born, (f) Bohr, (g) Rutherford, (h) Dirac; and below is Einstein.

The only reason for time is so that everything doesn't happen at once.

Albert Einstein

Nothing is real.

John Lennon, 1940–1980

Outline

Introduction

Classical Theory Starts Faltering

Quantum Mechanics to the Rescue

Beyond Schrödinger's Equation

Introduction

As we came to understand from Chapter 2, when atoms are brought together in crystals, their outermost electrons are influenced by a periodic potential. Therefore, the possible electron energies form bands of allowed values separated by bands of forbidden values. This band structure is of fundamental importance in explaining the properties of metals, semiconductors, and insulators. The most accepted and most accurate theory in modern physics today, to predict physical, mechanical, chemical, and electrical properties of atoms, molecules, and solids—including the band theory of solids—is quantum mechanics.

At the end of the nineteenth century, physicists thought they had a good grasp of the physical world with mechanics (e.g., Newton, Hamilton, and Lagrange), statistical mechanics (mostly Boltzmann and Gibbs), hydrodynamics (Stokes), and electrodynamics (Maxwell). There had been a growing unease, though, about the incapability of classical theories to explain a wide variety of experiments, such as the magnitude of the electrical and thermal conductivity of metals and the heat capacity of metals and insulators. Confidence was further challenged by a series of new discoveries: radioactivity (1896), the electron (1897), the quantum (1900), the photoelectric effect (1887), and x-rays (1895). The result was the development of a set of new theories, all explaining extreme aspects of nature better than classical theories. Although the classical theories worked well at everyday velocities and scales, at the extremes, new theories were needed. For very fast phenomena, special relativity worked better; for very small particles, quantum mechanics; and for very large phenomena, general relativity was demonstrated to be superior.

Before introducing quantum mechanics, we start out by detailing the areas where classical physics started faltering at the end of the nineteenth century. We specifically describe the shortfalls of the classical theories in explaining electrical and thermal conductivity of metals, heat capacity, the observation of a positive Hall effect, the ultraviolet catastrophe, and the photoelectric effect. All the above phenomena follow logically from quantum theory.

The relevance of quantum mechanics in the context of ICs and NEMS cannot be underestimated, and the profound implications of quantum physics for nanoelectronics and NEMS are a recurring topic throughout this book. In the IC and NEMS world, we are moving fast into the realm of quantum mechanics. Moore's law might remain valid until about 2020, but by then the scale of electronic components will be at the molecular/atomic level, and hence can no longer be described by classical mechanics. Quantum computing and nanotechnology, including nanotubes, nanowires, biological nanostructures, and quantum dots, all require some grounding in quantum mechanics to be understood at all. Quantum mechanics must now become a familiar tool not only to physicists but also to materials scientists, biologists, and electrical, mechanical, and bioengineers.

Classical Theory Starts Faltering

Introduction

In this section, we cover examples where microscopic classical models started failing at explaining a wide variety of experimental results obtained in the late nineteenth century. The examples include early models devised to explain electrical conductivity in metals, heat capacity of metals and insulators, the temperature dependence of electrical conductivity and heat capacity, thermal conductivity, the Hall effect, blackbody radiation, and the photoelectric effect. The failing of each model ultimately originated in the mistaken assumption that lattice vibrations (phonons), electrons, and photons could all take on continuous energy values, an error that quantum mechanics corrects.

Electronic Conductivity

DC Electrical Conductivity

Paul Drude (1863–1906) (Figure 3.1) was intrigued with the huge resistivity range in materials—from 10^{18} $\Omega\cdot$cm for fused quartz to 10^{-6} $\Omega\cdot$cm for silver (Figure 3.2)—and developed one of the first models to explain electrical conductivity in metals.

Modern condensed matter physics really started with the discovery of the electron by J.J. Thompson

FIGURE 3.1 Paul Drude (1863–1906).

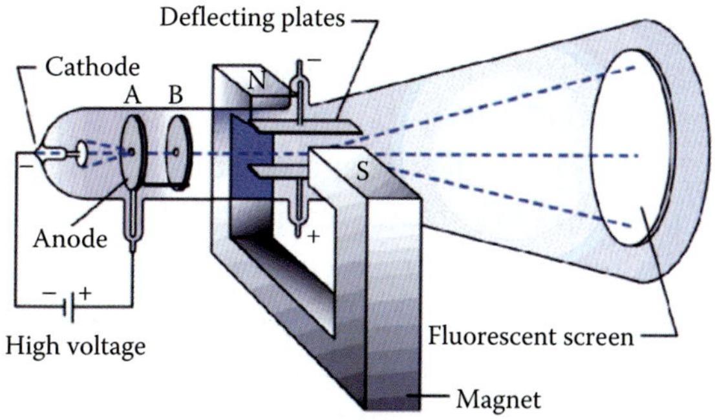

FIGURE 3.3 J.J. Thomson demonstrated that an electron behaves as a particle using an apparatus as shown here (A and B are anodes). The electron stream is deflected by electric and magnetic fields.

in 1897. In a series of experiments performed before the year 1900, Thompson demonstrated that electrons behave as particles of mass m that carry a fixed amount of negative electrical charge, e. These particles move in trajectories governed by the laws of electricity, magnetism, and classical mechanics, and Thompson determined the ratio e/m directly using an instrument as shown in Figure 3.3. Thompson received the Nobel Prize for his work on the electron in 1906. It was in the famous oil-drop experiment of Robert Millikan (1868–1953), carried out in 1909, that the size of the charge on an electron was finally established. Millikan also determined that there was a smallest "unit" charge, or that charge is "quantized."

Drude relied on the newly introduced concept of electrons to postulate a theory of metallic conductivity. His work was before the development of quantum mechanics, so he relied on classical physics models only. He introduced the idea of a free electron gas (FEG), in which he assumed that an electron gas surrounds the positive ion cores of the metal and that electron-ion interactions are negligible. Applying the kinetic theory of gases to the free electron gas, he had electrons only moving in straight lines and colliding only with ion cores, thus neglecting electron-electron interactions. Drude also envisioned collisions in which electrons instantaneously lose all the energy they previously gained from an electric field. He approximated the mean free path of the electrons, λ, with the interionic core spacing in the solid lattice, a. It should be noted that in reality electron densities (10^{22}–10^{23} cm^{-3}) are thousands of times greater

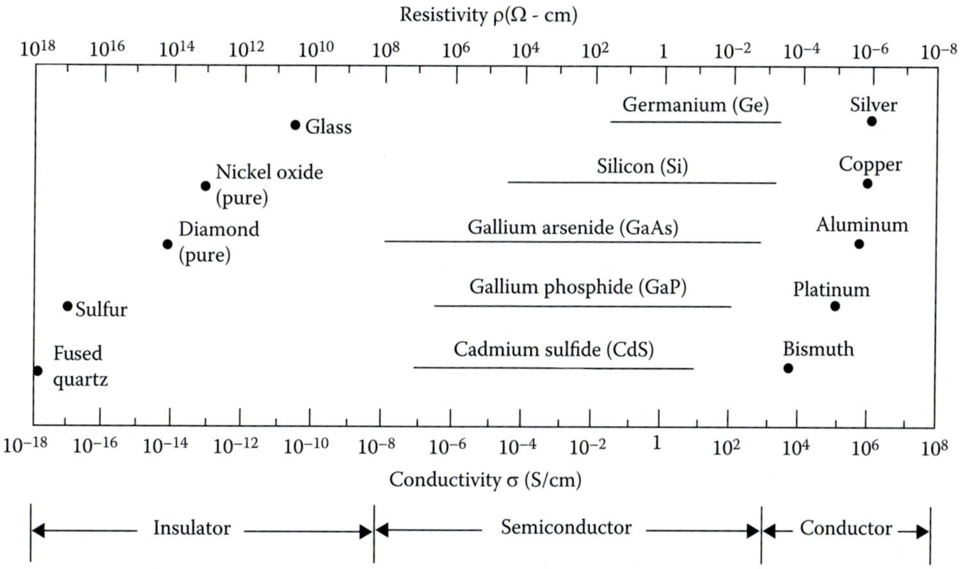

FIGURE 3.2 Conductivity/resistivity range for common materials.

than those of a gas at normal conditions, and the assumption that there are no strong electron-electron and electron-ion interactions is wrong. In spite of this, the model successfully explains the form of Ohm's law. However, it does fail to explain many other important aspects of conductors, such as the exact magnitude of electrical conductivity and heat capacity and their temperature dependence, or the relationship between electrical and thermal conduction.

In what follows we apply macroscopic boundary conditions to a piece of metal. In the nanoworld, as we will learn, things are different: with a macroscale conductor wire, the mean free path of the electrons, λ, is very small compared with the wire length, and the motion of electrons is diffusive. But, with a nanowire, the wire is short compared with λ and the electron motion may turn ballistic.

We start with the experimental observation that the current in a conductor is proportional to the applied voltage ($V \propto I$), i.e., Ohm's law. If an electrical field **E** (V/m) is applied over a conductor of length L, a charge, Q, flows and an electrical current, $I = dQ/dt$, results [SI unit: 1 ampere (A) = 1 coulomb per second (C/s)]. The current density is given as $J = I/A$ or dI/dA, with A the cross-sectional area of the current pathway or also:

$$I = \int J \, dA \tag{3.1}$$

where **J** is given by (see Chapter 2):

$$J = \sigma E \tag{2.2}$$

or also (see Chapter 2):

$$E = \rho J \tag{2.4}$$

σ is the electrical conductivity, and ρ is the electrical resistivity ($\sigma = 1/\rho$). The SI unit for resistivity is in ohm·meter (Ω·m) and ohm = volt/ampere (V/A); the SI unit for conductivity is in siemens per meter (S/m^{-1}) and siemens = ampere/volt = ohm^{-1}. The conductivity, σ, and resistivity, ρ, are intrinsic characteristics of a material and are independent of sample geometry but are linked to the crystal structure as we may glean from Equations 2.3–2.7 in Chapter 2. From Equation 2.2, σ measures the current density for a given electric field. For now, we assume

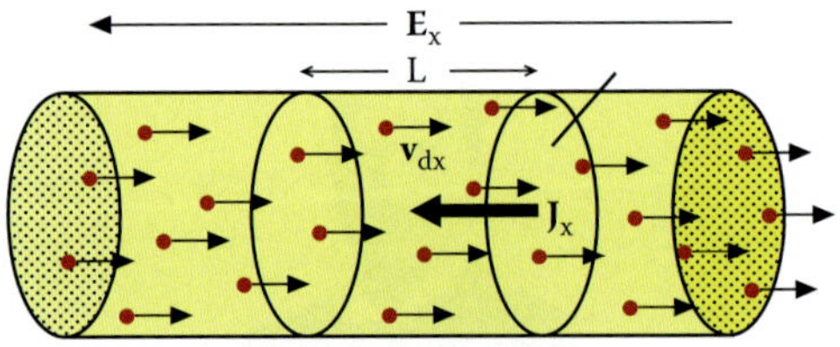

FIGURE 3.4 Electron conduction in a section of metal wire in the presence of an electrical field. The average distance traveled by an electron is the mean free path λ, which Drude assumed to be equal to the lattice constant, *a*.

an isotropic solid, and because $J = I/A$ and $E = V/L$, Equation 2.2 can be rewritten as $J = I/A = \sigma V/L$ or:

$$V = I(L/\sigma A) = I(\rho L/A) = IR \tag{3.2}$$

i.e., the familiar Ohm's law, with the resistance, R, linked to the geometry of the sample (L and A). The SI unit for resistance is ohm (Ω).

Drude's microscopic interpretation of Ohm's law can be understood from an inspection of Figure 3.4, where we consider the conduction of electrons in a metal wire on application of an electrical field in the x-direction, E_x. There are n electrons per unit volume (electron density), and they all move in the direction of the current, I, with a drift velocity in the x-direction, v_{dx}. The drift velocity, v_{dx}, is the net motion of electrons opposite to the electrical field (Figure 3.5). The number of electrons crossing area A in a time dt is $nAv_{dx}dt$. In the time segment dt, the electrons have traveled a distance L along the wire. The current density in the x-direction, J_x, is then $\frac{I}{A}$ or $\frac{\Delta Q}{A\Delta t} = \frac{(ne)(AL)}{AL/v_{dx}}$, and with Q the charge we calculate:

$$J_x = env_{dx} \tag{3.3}$$

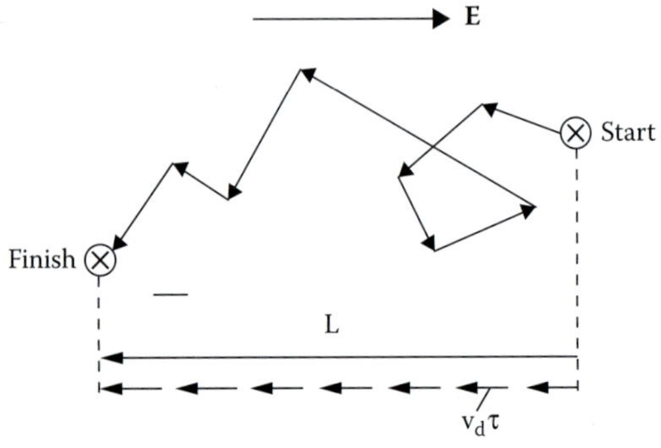

FIGURE 3.5 Drift of an electron opposite the electrical field. During an average time τ, the electron travels a mean free path $\lambda = v_{dx}\tau$. In a time dt, the electrons have traveled a distance L along the wire.

where e is the charge of the electron. By convention, a positive charge moves with the electrical field, and Equation 3.3, more correctly, should read $\mathbf{J}x = env_{dx}$, with $\mathbf{J}_x$ and $\mathbf{v}_{dx}$ vectors with a direction and magnitude. For simplicity, we often leave the vector notation out, and introduce it only when we want to draw special attention to it. Drude treated the free electron gas as a gas of molecules where distribution of velocities follows the Maxwell-Boltzmann distribution (see below). In such a gas, at T = 0, all the free electrons in a conductor have zero kinetic energy. When heating the conductor, the lattice ions acquire an average kinetic energy of $\frac{3}{2} k_B T$, where k_B is the Boltzmann constant. This average energy is imparted to the electron gas by collisions between electrons and lattice ions. The latter is a consequence of the equipartition theorem. At ordinary temperatures (~300 K) the mean kinetic energy of the electrons based on this model is about 0.04 eV. From this mean kinetic energy, we approximate the thermal velocity, v_{th}, of the electrons in a metal from the root mean square (rms speed) or $v_{rms} = \sqrt{\frac{3k_BT}{m_e}}$ (see Equation 3.20 below), which is slightly larger than the average v_{avg} or mean speed $\bar{v} = \sqrt{\frac{8k_BT}{\pi m_e}}$ (see Equation 3.21 below), as $v_{th} = \bar{v} \approx v_{rms} = 1.57 \times 10^5$ m/s or almost 1000 km/s (this is about 1% of the speed of light!). However, when calculating $\mathbf{v}_{dx}$ in Equation 3.3 for a current of 1 A, one obtains velocities of the order of 10^{-4} m/s or only 0.1 mm/s! The reason for the huge difference between drift and thermal velocity is that electrons travel at fast thermal velocities for a short, average time, τ, and then "scatter" because of collisions with atoms, grain boundaries, impurities, or material surfaces (especially in very thin films or small particles). The drift velocity is also not how fast "electricity travels," for electric fields travel essentially at the speed of light. To reconcile this discrepancy, think of electrons in a wire as a pipe full of water; when a little water enters one end of the pipe, almost immediately some water flows out at the other end.

The very small drift velocity caused by the electric field has essentially no effect on the very large mean speed of the electrons: in other words, v_{th} does not depend on E. Drude assumed that all of the electrons' forward velocity is reduced to zero after each collision and must then be accelerated again by the electrical field.

The forces exerted on a colliding electron in an electric field, **E**, are given by Newton's second law, according to which forces give rise to a change in the momentum of particles $[(\mathbf{F} = d(m_e\mathbf{v}_{dx})/dt = d\mathbf{p}_x/dt]$ or:

$$\frac{d\mathbf{p}_x}{dt} = -e\mathbf{E}_x + \mathbf{F}_{coll} = -e\mathbf{E}_x - \frac{\mathbf{p}_x}{\tau} = -e\mathbf{E}_x - \frac{m_e\mathbf{v}_{dx}}{\tau} \quad (3.4)$$

where $-e\mathbf{E}_x$ is the force on the electron garnered from the electrical field, m_e the mass of an electron in vacuum, $\mathbf{F}_{coll}$ the collision force, and $\boldsymbol{p}_x$ the electron's momentum in the x-direction ($=m_e\mathbf{v}_{dx}$). Because

$$-e\mathbf{E}_x - m_e\mathbf{v}_{dx}/\tau = 0 \quad (3.5)$$

the result is a constant average velocity:

$$\mathbf{v}_{dx} = -\frac{e\mathbf{E}_x\tau}{m_e} \quad (3.6)$$

where we introduced a minus sign because, as remarked above, the drift velocity is opposite the electrical field **E**. Substituting this result in Equation 3.3 and generalizing for three directions, we obtain the famous Drude result (ignoring vector notation):

$$\sigma = \frac{ne^2\tau}{m_e} \text{ and } \rho = \frac{m_e}{ne^2\tau} \quad (3.7)$$

Scatter time τ is also known as the relaxation time, the collision time, or the mean free time a randomly picked electron travels before the next collision. Scatter time τ decreases with increasing temperature T, i.e., more scattering at higher temperatures leads to higher resistivity in a metal. During an average time τ, electrons travel a mean free path λ, i.e., $\lambda = v_{th}\tau$, and in terms of the mean free path and the mean speed the resistivity is given as:

$$\rho = \frac{m_e v_{th}}{ne^2\lambda} \quad (3.8)$$

According to Ohm's law, the resistivity is independent of the field, and in Equation 3.8 only λ and v_{th} could possibly be dependent on the field. But we just saw that the very small drift velocity caused by the electric field has essentially no effect on the very large mean speed of the electrons or v_{th} does not depend on **E**. Drude assumed the electron mean free path, λ, to be equal to the lattice constant, *a*, which is of the order of 1 nm, and of course also independent of the electric field **E**; neither λ nor v_{th} depends on E in accordance with Ohm's law. In the absence of an electric field, the electrons perform a random, thermal motion, and there is no conduction, but when an electric field is applied, electrons move into a direction opposite to the field, thus generating a current. The value of the resistance is finite because electrons collide with the lattice ions, and they are stopped frequently in their tracks before being accelerated again. This simple model does explain correctly the form of Ohm's law.

From Equation 3.7 there are two contributions to the conductivity. One is the number of charges (*ne*), and the second is how easily those charges can be accelerated ($e\tau/m_e$). To really understand the differences between the numbers of "free charges" in different materials requires the quantum description introduced below. However, let us accept for now that in a material such as silver, the number of electrons that are free to move equals $5.8 \times 10^{28}/m^3$ and that in zinc it is even larger, $1.3 \times 10^{29}/m^3$. From these numbers one would expect zinc to be the better conductor! However, as we shall see, this is not the end of the story. The second contribution to the conductivity in Equation 3.7 is the charge carrier's drift mobility, μ_e (in the case of electrons), which, in cm^2/Vs, is given as:

$$\mu_e = \frac{e\tau}{m_e} \tag{3.9}$$

Therefore, the conductivity is the product of the number of free charges and the mobility of those charge carriers:

$$\sigma = \mu_e ne \tag{3.10}$$

The drift mobility, μ_e, may also be defined as the drift velocity per unit applied electric field, or:

$$\mu_e = \frac{v_d}{E} \tag{3.11}$$

The drift mobility, from Equation 3.9, is linked in turn to the mean scattering time between collisions, τ. Coming back to the comparison of the conductivity of silver and zinc, typical mobilities in silver are much higher than those in zinc so that the electrical conductivity is higher, even though zinc has more electrons. This makes silver the better conductor after all. Once we have introduced quantum mechanics, it will become clear that the mean scattering time, τ, has nothing to do with the stagnant lattice ions in the crystal but is determined by lattice vibrations, imperfections, and impurities instead. Consequentially, the mobility of a metal can be reduced by introducing defects (e.g., kinking a wire) or by increasing the number of lattice vibrations, i.e., phonons (by raising the temperature).

The Maxwell-Boltzmann Distribution

The next question is, how does resistivity of a metal depend on temperature? To answer this we calculate, à la Drude, how the drift velocity, v_d, depends on temperature. With electrons behaving like an ideal gas, the distribution of electron speeds is described by a Maxwell-Boltzmann (MB) distribution. In a MB distribution, the probability of finding particles in a particular energy state varies exponentially as the negative of the energy divided by k_BT or:

$$f_{MB}(E) = Ae^{-\frac{E}{k_BT}} \tag{3.12}$$

where *A* is a normalization constant, and $e^{-\frac{E}{k_BT}}$ is called the Boltzmann factor. Based on Equation 3.12, at a given temperature, particles are distributed among all available levels, and the ground state always contains the bulk of the particles, whereas other levels will contain exponentially less. The Maxwell-Boltzmann distribution is illustrated in Figure 3.6.

The number of particles per unit volume, i.e., density *n(E)*, that have energies between *E* and *E* + *dE* is *n(E)dE*, where $n(E) = G(E)f_{MB}(E)$. The function *G(E)dE* is the total number of possible (allowed-to-occupy) energy states per unit volume with energies between *E* and *E* + *dE*. It is also referred to as the density of state function, often abbreviated as DOS. The other function, $f_{MB}(E)$, is the Maxwell-Boltzmann distribution, representing the probability function

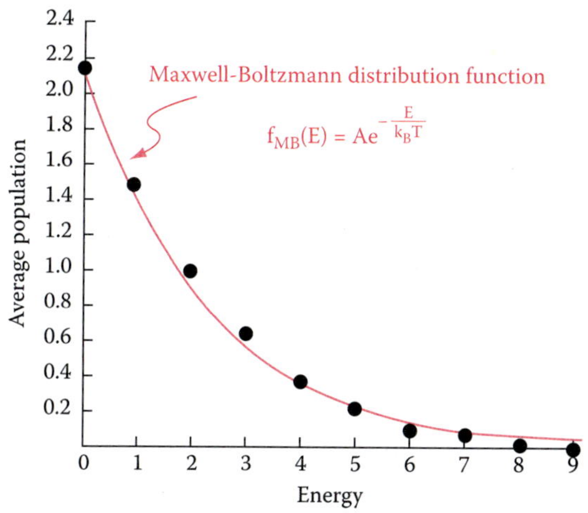

FIGURE 3.6 The Maxwell-Boltzmann distribution function: $f_{MB}(E) = Ae^{-\frac{E}{k_BT}}$.

of occupancy of a state with energy *E*. Because $E = mv^2/2$, we may, using Equation 3.12, rewrite *n(E)dE* as:

$$n(E)dE = G(E)Ae^{-\frac{mv^2}{2k_BT}}dE \quad (3.13)$$

To find the number of gas molecules with a speed in the range *dv*, irrespective of the direction of the velocity, we need the speed distribution function, *n(v)*. The number of allowed states (velocities) between *v* and *v* + *dv*, i.e., the number of states between a sphere of radius *v* and a sphere of radius *v* + *dv*, as illustrated in Figure 3.7, is:

$$4\pi v^2 dv = G(E)dE \quad (3.14)$$

For every value of *v* there is a value of *E* in *G(E)*, and substituting Equation 3.14 in Equation 3.13 we obtain:

$$n(E)dE = n(v)dv = A4\pi v^2 e^{-\frac{mv^2}{2k_BT}}dv \quad (3.15)$$

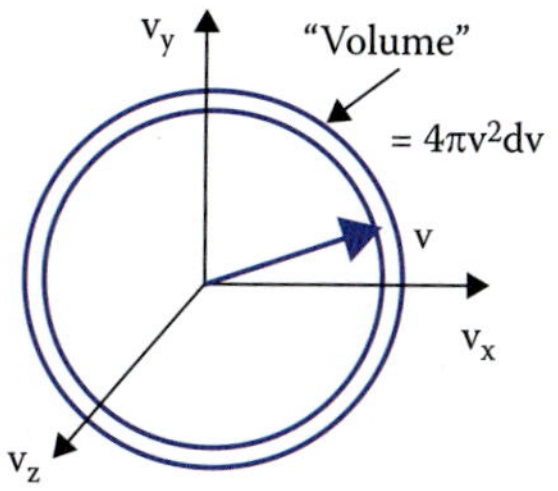

FIGURE 3.7 The velocity is a vector in three-dimensional space, and one needs to take into account that, with the increase of the magnitude of *v*, the space accessible to a particle increases.

To calculate the value of *A*, we remember that we can derive the number of molecules per unit volume or volume density of all particles (all velocities/energies) as:

$$\frac{N}{V} = \int_0^\infty n(v)dv$$

or

$$A = \frac{N}{V}\left(\frac{m}{2\pi kT}\right)^{\frac{3}{2}} \quad (3.16)$$

$$\left(\text{since} \int_0^\infty v^2 e^{-av^2} dv = \frac{1}{4a}\sqrt{\frac{\pi}{a}}\right)$$

From Equations 3.15 and 3.16, the Maxwell-Boltzmann speed distribution is derived as:

$$n(v)dv = \frac{4\pi N}{V}\left(\frac{m}{2\pi k_B T}\right)^{\frac{3}{2}} v^2 e^{-\frac{1}{2}\frac{mv^2}{k_BT}}dv \quad (3.17)$$

This expression gives us the speed distribution of *n* particles as a function of their mass and the temperature: it represents the probability that a particle has a speed in the range *v* to *v* + *dv* as illustrated in Figure 3.8. As the temperature increases, the curve

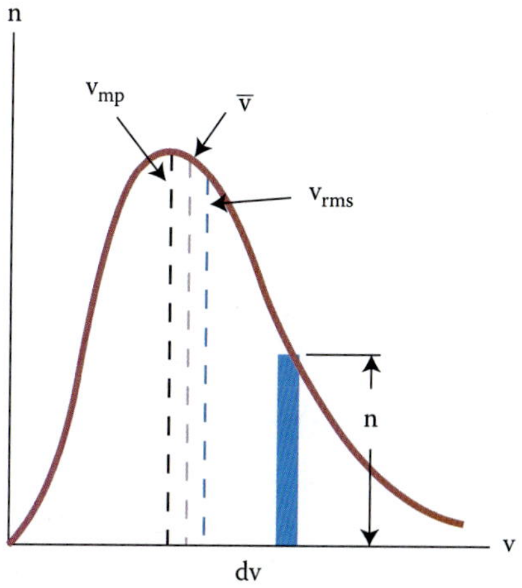

FIGURE 3.8 The Maxwell-Boltzmann distribution of speeds of particles as a function of the temperature and mass of the particles. The most probable speed v_{mp} is the speed at which the distribution curve reaches a peak, $\bar{v}$ is the average speed, and the root mean square speed is v_{rms}.

broadens and extends to higher speeds. Using $dE = mvdv$ ($E = mv^2/2$), the Maxwell-Boltzmann distribution of kinetic energies is:

$$n(E)dE = \frac{2N}{V\sqrt{\pi}}\left(\frac{1}{k_BT}\right)^{\frac{3}{2}}\sqrt{E}\, e^{-\frac{E}{k_BT}}dE \quad (3.18)$$

Comparing Equation 3.18 with Equation 3.13 yields the 3D density of states function for gas molecules:

$$G(E) = \frac{2N}{V\sqrt{\pi}}\left(\frac{1}{kT_B}\right)^{\frac{3}{2}}\sqrt{E} \quad (3.19)$$

This function increases smoothly with the square root of the energy.

The most probable speed $v_{mp} = \sqrt{\frac{2k_BT}{m}}$ is the speed at which the speed distribution (Equation 3.17) reaches a peak; this can be calculated from $\frac{dn(v)}{dv}$, setting it to zero, and solving for v. Although Equation 3.17 gives the distribution of speeds or, in other words, the fraction of molecules having a particular speed, we are often more interested in quantities such as the root mean square speed or the average speed of the particles rather than the actual distribution. The root mean square speed is given as:

$$v_{rms} = \sqrt{\frac{\int_0^\infty v^2 n(v)dv}{N/V}} = \sqrt{\frac{3k_BT}{m}} \quad (3.20)$$

and the average or mean speed $\bar{v}$ is given as:

$$\bar{v} = \sqrt{\frac{\int_0^\infty vn(v)dv}{N/V}} = \sqrt{\frac{8k_BT}{\pi m}} \quad (3.21)$$

In the case in which the Maxwell-Boltzmann distribution is applied to electrons rather than gas molecules, we must replace m with m_e, the mass of an electron. The Maxwell-Boltzmann distribution, shown in Figure 3.8, is a classical distribution of particles; in quantum statistics, other particle distribution functions are introduced. In the Maxwell-Boltzmann distribution function, it is assumed that all the particles are distinguishable; particles are physically identical but distinguishable in position and trajectory. Particles are considered distinguishable if the distance separating them is large compared with their de Broglie wavelength (see below). Another way of saying this is that the average distance between particles must be large compared with the quantum uncertainty. For an ideal gas, this criterion is certainly fulfilled, but it is not true for electrons and, as we remarked before, this is ultimately the reason why the Drude model fails to explain conductivity correctly: the Maxwell-Boltzmann distribution is not valid for a collection of electrons.

Drude Fails

Above we saw that during an average time τ, electrons travel a mean free path λ, i.e., $\lambda = v_{th}\tau$, and because Drude assumed the electron mean free path, λ, to be equal to the lattice constant, a, which is of the order of 1 nm, this yields a typical value for τ of about 10^{-14} s (because we calculated $v_{th} = \bar{v} \approx v_{rms} = 1.57 \times 10^5$ m/s). Based on Equations 3.8 and 3.21, where we replaced m with m_e and used $\bar{v}$ for calculating v_{th} (Equation 3.21), one derives for the resistivity:

$$\rho = \frac{1}{\sigma} = \frac{m_e v_{th}}{ne^2\lambda} = \frac{m_e}{ne^2a}\sqrt{\frac{8k_BT}{\pi m_e}} \quad (3.22)$$

Using the lattice constant, a, for the mean free path and the Maxwell-Boltzmann equation at T = 300 K to calculate v_{th}, values for the resistivity of a metal are obtained that are six times too large. In addition, from Equation 3.22 the temperature dependence of resistivity is determined by v_{th}, which in this model is proportional to $\sqrt{T}$. In practice, the temperature dependence of the resistivity is represented by the empirical relationship:

$$\rho = \rho_0 + \alpha T \quad (3.23)$$

where ρ_0 is the resistivity at a reference temperature, usually room temperature, and α is the temperature coefficient. An expanded version of Equation 3.23 is known as the Matthiessen rule. This rule is just an approximation (a rule of thumb that works pretty well), not a true physical law. According to Matthiessen rule, the resistivity can be expressed as a sum of

TABLE 3.1 Resistivity Values of Common Metals

Material	ρ_0 (μΩ·cm)	α (μΩ·cm/K)*	Resistivity at 100°C (μΩ·cm)
Aluminum	2.284	0.00390	3.0640
Copper, annealed	1.7241	0.00393	2.5101
Copper, hard-drawn	1.771	0.00382	2.5350
Brass	7.000	0.00200	7.4000
Gold	2.440	0.00340	3.1200
Iron	9.710	0.00651	11.0120
Lead	20.6480	0.00336	21.3200
Nickel	11.000	0.00600	12.2000
Silver	1.590	0.00380	2.3500
Steel	10.400	0.00500	11.4000
Nichrome	100.000	0.00040	100.0800
Platinum	10.000	0.00300	10.6000
Tungsten	5.600	0.00450	6.5000

*Determined at 25°C.

terms resulting from (nearly) independent contributions, for example, $\rho = \rho_0 + \rho_T T + \rho_I C + \rho_{other}$. Here, ρ_0 is the extrapolation of the resistivity to 0 K, $\rho_T T$ is the roughly linear and independent contribution caused by temperature, $\rho_I C$ is the roughly linear contribution caused by solid solution impurities, and ρ_{other} represents the contributions from other scattering centers, such as dislocations and precipitates. Typical values of ρ_0 and α are listed in Table 3.1, along with the calculated resistivity at 100°C.

Experimentally determined resistivity versus temperature plots for a metal, an insulator, and a superconductor are shown in Figure 3.9. Superconductivity is the flow of electric current without resistance. It has been observed in certain metals, alloys, and ceramics at temperatures near absolute zero, and in some cases at temperatures hundreds of degrees above absolute zero. At low temperatures, all materials, other than superconductors, are insulators or metals. For pure metals, the resistivity increases rapidly with increasing temperature, whereas for insulators the resistivity decreases rapidly with increasing temperature [this was first observed by Michael Faraday (1791–1867) in 1833]. Semiconductors have resistivities intermediate between metals and insulators at room temperature. Instead of a square root dependence of temperature, the plot in Figure 3.9 reveals that, for a metal, there is proportionality with T at higher temperatures, and, at low temperatures, the resistivity is proportional to T^5! The latter is known as the Bloch-Gruneisen T^5 law.

From the above observations it is clear that some of the Drude assumptions are very wrong. Moreover, Drude's model cannot explain why one material acts as an insulator and another as a metal. This was all very confusing around Drude's time. Because solids contain a number of atoms and electrons with a similar density, why the large conductivity

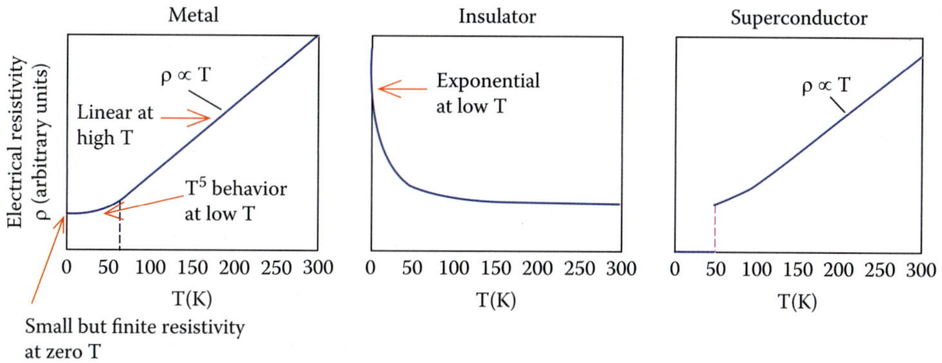

FIGURE 3.9 The three states of solid-state matter as defined by their electrical resistivity in the low temperature limit. The resistivity at low temperatures is finite for a metal, very large for an insulator, and zero for a superconductor.

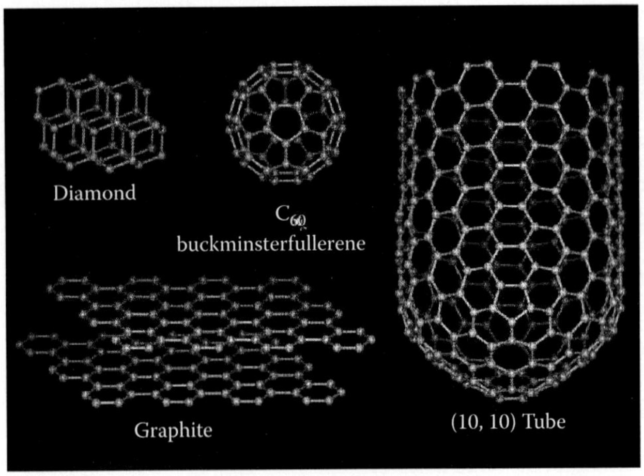

FIGURE 3.10 Carbon may act as an insulator (diamond), as a semimetal (graphite), and as a superconductor (buckminsterfullerene and carbon nanotube).

differences shown in Figure 3.2? More intriguing yet, in the case of carbon, the same material may act as a good insulator (diamond), as a semimetal (graphite), and even as a superconductor (buckminsterfullerenes and carbon nanotubes)* (see Figure 3.10). The answer, we will learn, is that electrons are *fermions*, and only electrons with energy on the order of a few k_BT contribute to the conduction process, at room temperature. A *fermion* is a particle, such as an electron, proton, or neutron, having half-integral spin and obeying statistical rules requiring that not more than one in a set of identical particles may occupy a particular quantum state. For fermions, the Boltzmann distribution must be replaced by a Fermi-Dirac distribution. Because of the wave nature of electrons and the exclusion principle (to be discussed below), the energy distribution of electrons in a metal does not even resemble a Maxwell-Boltzmann distribution. Moreover, the collision between ions and electrons cannot be pictured as that of two hard objects. Instead it involves the scattering of electron waves by lattice ions.

Drude AC Electrical Conductivity and Dielectric Functions

Drude AC Electrical Conductivity In the previous sections we considered the dc conductivity of metals; we now consider their ac electrical conductivity. We assume that the wavelength of the electromagnetic (EM) field is large compared with the electronic mean free path λ, so that electrons "see" a homogeneous field when moving between collisions. In Equation 3.4, we used Newton's second law to launch the change of momentum of a free electron in the presence of an applied dc electrical field. Here we use the same expression to derive the change of momentum of electrons in a time-harmonic electric field, given by:

$$\mathbf{E}(\omega,t) = \mathbf{E}(\omega)e^{-iwt} \tag{3.24}$$

The equation for the momentum per electron is:

$$\frac{d\mathbf{p}(\omega,t)}{dt} = -e\mathbf{E}(\omega,t) - \frac{\mathbf{p}(\omega,t)}{\tau} \tag{3.25}$$

where τ is the relaxation time or the mean free time a randomly picked electron travels before the next collision in a metal. Because $\frac{d\mathbf{p}(\omega,t)}{dt} = -i\omega\mathbf{p}(\omega,t)$, this expression may be rewritten as:

$$\begin{aligned}\mathbf{p}(\omega,t) &= -\left[\frac{e\mathbf{E}(\omega,t)}{\frac{1}{\tau} - i\omega}\right] \text{ or also}\\ &= -\left[\frac{e\mathbf{E}(\omega,t)}{\frac{1}{\tau^2} + \omega^2}\right]\left(\frac{1}{\tau} + i\omega\right)\\ &= -\left[\frac{e\mathbf{E}(\omega)}{\frac{1}{\tau^2} + \omega^2}\right]\left(\frac{1}{\tau} + i\omega\right)\left(\cos\omega t - i\sin\omega t\right)\\ &= -\left[\frac{e\mathbf{E}(\omega)}{\frac{1}{\tau^2} - \omega^2}\right]\left(\frac{1}{\tau}\cos\omega t + \omega\sin\omega t\right)\\ &\quad + i\left(\omega\cos\omega t - \frac{1}{\tau}\sin\omega t\right)\end{aligned} \tag{3.26}$$

We analyze this expression now for two different and important situations: 1) with ω << 1/τ, where the electrons can follow the changing electrical field, and 2) with ω >> 1/τ, i.e., at very high frequencies

* The latter were not yet discovered at the time; if they had been, this would have led to even more consternation. How can the same material have all these different resistivities?

(optical frequencies we will see), where the electrons cannot follow the fast changing electrical field anymore.

1. With $\omega \ll 1/\tau$:

$$p(\omega,t) = -\left[\frac{eE(\omega)}{\frac{1}{\tau^2}}\right]\left(\frac{1}{\tau}\cos\omega t - i\frac{1}{\tau}\sin\omega t\right)$$

$$= -\left[\frac{eE(\omega)}{\frac{1}{\tau}}\right]e^{-iwt} = -e\tau E(\omega,t) \qquad (3.27)$$

At these frequencies, **p**(ω,*t*) is in phase with **E**(ω,*t*).

2. With $\omega \gg 1/\tau$:

$$p(\omega,t) = -\left[\frac{eE(\omega)}{\omega^2}\right](\omega\sin\omega t + i\omega\cos\omega t)$$

$$= -\left[\frac{eE(\omega)i\omega}{\omega^2}\right](\cos\omega t - i\sin\omega t)$$

$$= -\left[\frac{eE(\omega)i\omega}{\omega^2}\right]e^{-iwt} = -\left[\frac{eE(\omega,t)i}{\omega}\right]$$

$$= \frac{eE(\omega,t)}{i\omega} \qquad (3.28)$$

At very high frequencies, where the electrons are too slow to follow the very fast changing electrical field vector, **p**(ω,*t*) is out of phase with **E**(ω,*t*); moreover, **p**(ω,*t*) tends to zero. The transition between these two behaviors occurs when the ac frequency exceeds the collision frequency τ. With a typical value for τ of metal of about 10^{-14} s, that transition frequency is at roughly 10^{14} Hz ($\tau \sim 10^{-14}$ s, $v\lambda = c$ or $v = \frac{3.10^{10}\,\text{cm/s}}{5000\cdot10^{-8}\,\text{cm}} \approx 10^{14}$ Hz), corresponding to optical frequencies (light waves). The equations derived here apply for the ac behavior of metals as well their interaction with light (metal optics).

Relying on Equation 3.3, rendered in vector format as $\mathbf{J}(\omega) = -en\mathbf{v}_{dx}$, and given that $\mathbf{p}(t) = m_e\mathbf{v}(t)$, we can also write:

$$J(\omega) = -\frac{enp(\omega)}{m_e} = \frac{\left[\frac{e^2n}{m_e}\right]E(\omega)}{\frac{1}{\tau} - i\omega} = \sigma(\omega)E(\omega) \qquad (3.29)$$

so that the Drude complex AC conductivity is calculated as:

$$\sigma(\omega) = \frac{\sigma_0}{1 - i\omega\tau} \qquad (3.30)$$

which simplifies to the DC conductivity, i.e., σ_0 ($= \frac{ne^2\tau}{m_e}$, Equation 3.7), in the case of very low ac frequencies ($\omega \lll 1/\tau$).

The real and imaginary parts of the complex conductivity are:

$$\sigma' = \frac{\sigma_0}{1+\omega^2\tau^2}(\text{Re}) \qquad (3.31)$$

and

$$\sigma'' = \frac{\sigma_0\omega\tau}{1+\omega^2\tau^2}(\text{Im}) \qquad (3.32)$$

The real (σ') and imaginary (σ'') parts of the complex conductivity $\sigma(\omega)$ are plotted versus $\omega\tau$ in Figure 3.11. The maximum in σ'' is called the Drude peak and is characteristic for each metal.

We consider now what happens with the current and the conductivity as a function of frequency. At very low ac frequencies ($\omega \ll 1/\tau$), electrons have many collisions before the direction of the wave changes; this situation corresponds to the Ohm's law

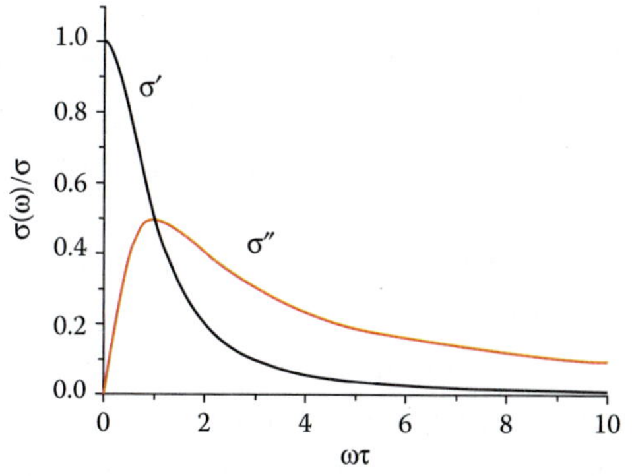

FIGURE 3.11 Frequency dependency of the real (σ') and imaginary (σ'') in parts of the conductivity. The maximum in σ'' at $\sigma''/\sigma_0 = 0.5$ is the Drude peak at $\omega\tau = 1$ and is a characteristic of a metal. The DC conductivity σ_0 is reached at $\omega\tau = 0$ ($\sigma''/\sigma_0 = 1$).

regime—**J** follows **E**—and σ is real. At very high frequencies (ω >> 1/τ), electrons might have only one collision or less when the direction of the ac field is changed. In this case **J** is imaginary and out of phase with **E**, and σ is also imaginary. As we noted above, the transition from one regime to the other occurs at optical frequencies, and qualitatively we can say that with ωτ << 1, electrons are in phase and reirradiate, i.e., they are reflected and the metal appears shiny. With ωτ >> 1, electrons are out of phase as they are too slow, there is less interaction, and we have transmission. More specifically, free electrons do not influence E-fields in metals with frequencies greater than that of visible light, or the electron gas is transparent in the UV range! No energy is absorbed from the field in this range, and no joule heating occurs.

Next, we will investigate how the permittivity or dielectric constant of a medium changes as a function of frequency of the applied electrical field (again from ac fields to electromagnetic radiation).

Dielectric Functions and an Introduction to Metal Optics The permittivity or dielectric constant of a medium describes how an electric field both affects and is affected by that medium and can be looked at as the quality of a material that allows it to store electrical charge. The dielectric constant ε or permittivity (*As/Vm*) is a materials-dependent "constant," and its frequency dependence defines the so-called dielectric function ε(ω). In the most general terms, dielectric functions are dependent on both frequency (ω) and wave-vector **k**: ε(ω,**k**). To understand the dielectric behavior and optical properties of metals better, we need to explain how plasma oscillations come about and analyze how the dielectric "constant" ε of a metal changes with frequency. A plasma, in general, is a medium with equal amounts of positive and negative charges, of which at least one charge type is mobile. In case of a plasma in a metal, the mobile charges are the free electrons, and these charges are balanced by the positive, immobile, metal ion cores. Drude's free electron gas (FEG), with an ac field applied, can exhibit collective longitudinal oscillations of the plasma because the displacement of all the electrons against the ion bodies of the material induces a dipole moment

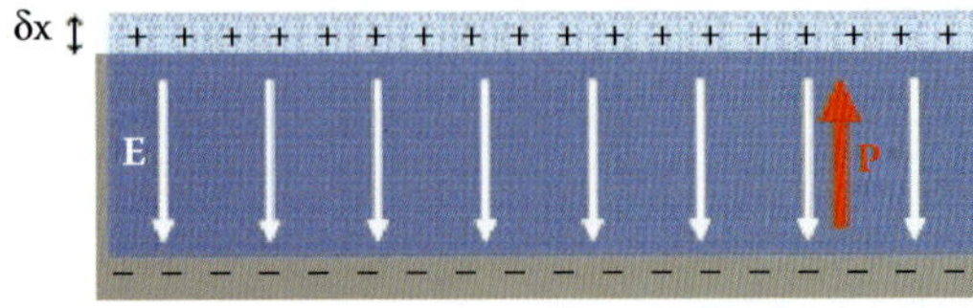

FIGURE 3.12 Drude bulk plasma oscillation in a metal film. **E** is the electrical field, and **P** is the polarization. The entire free electron gas is displaced over a small distance, δx.

and an electric field opposing that displacement, as shown in Figure 3.12. In longitudinal oscillations, the displacement is in the same direction as the wave motion, as in sound waves. To appreciate better how oscillations of free electrons are induced by a time harmonic electrical field $\mathbf{E}(\omega,t) = \mathbf{E}(\omega)e^{-i\omega t}$, we rewrite Equation 3.25 for one dimension with $\mathbf{p}_x(\omega,t) = m_e\mathbf{v}_x(\omega,t) = m_e d_x/dt$ as:

$$m_e \frac{d^2x}{dt^2} + m_e\gamma\frac{dx}{dt} = -eE(t) = -eE(\omega)e^{-i\omega t} \quad (3.33)$$

In the Drude model, electrons are free, and a damping factor γ comes about only because of electron scattering. The damping factor is related to the scattering constant as 1/γ = τ (as we saw, typically ~10^{-14} s in a metal). The solution of Equation 3.33—the displacement *x* of the entire free electron gas—is then given as:

$$x(t) = \frac{eE(\omega)}{m_e(\omega^2 + i\gamma\omega)} \quad (3.34)$$

Polarization comes about as a result of small displacement of charges in an electrical field. The macroscopic polarization density **P** (C/m²), illustrated in Figure 3.12, is a vector field that represents the density of permanent or induced electric dipole moments and is given by the product of the displacement *x*(*t*) and the electron density ne(**P** = – nex) or:

$$\mathbf{P} = -\frac{ne^2\mathbf{E}(t)}{m_e(\omega^2 + i\gamma\omega)} \quad (3.35)$$

The polarization **P** is also related to the electrical field as:

$$\mathbf{P}(\omega,t) = \chi_e\varepsilon_0\mathbf{E}(\omega,t) \quad (3.36)$$

where χ_e is the electric susceptibility tensor of the material, a proportionality constant relating the electrical field **E** to the induced dielectric polarization density **P**, and ε_0 is the permittivity of free space with a value of 8.854 pF/m. In the case of a linear homogeneous and isotropic material, χ_e becomes a scalar constant. The polarization **P** is further linked to the electric displacement field **D**, expressed in coulombs per square meter (C/m²), and to the dielectric constant $\varepsilon(\omega,\mathbf{k})$ or permittivity (*As*/*Vm*), a materials-dependent constant, as:

$$\mathbf{D} = \varepsilon_0\mathbf{E} + \mathbf{P} = (1+\chi_e)\varepsilon_0\mathbf{E} = \varepsilon(\omega,\mathbf{k})\mathbf{E} \tag{3.37}$$

In the expression $\mathbf{D} = \varepsilon_0\mathbf{E} + \mathbf{P}$ we have separated the electric displacement **D** in its materials (**P**) and vacuum parts (ε_0**E**). The permittivity of a material is usually given as a relative permittivity $\varepsilon_r(\omega)$ (also called the dielectric constant). The permittivity $\varepsilon(\omega,\mathbf{k})$ of a medium is an intensive parameter and is calculated by multiplying the relative permittivity by ε_0, or $\varepsilon(\omega) = \varepsilon_0\varepsilon_r(\omega)$. The permittivity of air is $\varepsilon_{air} = 8.876$ pF/m, so the relative permittivity of air is $\varepsilon_{r,air} = 1.0005$, and for vacuum it is 1 by definition. From Equation 3.37, it then also follows that $\varepsilon_r(\omega) = 1 + \chi_e$. The electric displacement field, **D**, is in turn related to the electric field **E** (in units of V/m) by the constitutive relation:

$$\mathbf{D} = \varepsilon(\omega,\mathbf{k})\ \mathbf{E} \tag{3.38}$$

Using Equation 3.36 for the polarization **P**, we can rewrite Equation 3.37 as:

$$\mathbf{D} = \varepsilon(\omega,\mathbf{k})\mathbf{E} = \varepsilon_0\mathbf{E} + \mathbf{P} = \varepsilon_0\mathbf{E} - \frac{ne^2\mathbf{E}}{m_e(\omega^2 + i\gamma\omega)} \tag{3.39}$$

From Equation 3.39 we arrive at the following expression for the complex dielectric function:

$$\varepsilon(\omega,\mathbf{k}) = \varepsilon_0\left[1 - \frac{ne^2}{m_e\varepsilon_0(\omega^2 + i\gamma\omega)}\right] = \varepsilon_0\left[1 - \frac{\omega_p^2}{(\omega^2 + i\gamma\omega)}\right] \tag{3.40}$$

The plasma frequency ω_p in Equation 3.40 is defined as:

$$\omega_p = \sqrt{\frac{ne^2}{m_e\varepsilon_0}} \tag{3.41}$$

where *n* is the density of electrons in the metal. To obtain $\varepsilon_r(\omega)$, the relative dielectric response of a material to electromagnetic waves at various frequencies, one divides by ε_0 because $\varepsilon(\omega,\mathbf{k})/\varepsilon_0 = \varepsilon_r(\omega,\mathbf{k})$. The corresponding real and imaginary components of the complex dielectric function are:

$$\varepsilon_r'(\omega,\mathbf{k}) = 1 - \frac{\omega_p^2\tau^2}{1+\omega^2\tau^2} \quad \text{Re} \tag{3.42}$$

and

$$\varepsilon_r''(\omega,\mathbf{k}) = \frac{\omega_p^2\tau}{\omega(1+\omega^2\tau^2)} \quad \text{Im} \tag{3.43}$$

In a metal, the damping term γ is just the electron collision rate, which is the inverse of the mean electron collision time, τ, i.e., $\gamma = \tau^{-1}$. As stated before, the collision rate can be quite rapid—tens of femtoseconds. But for optical frequencies (e.g., for $\lambda = 500$ nm, $\omega = 2\pi c/\lambda = 3.8 \times 10^{15}$ rad/s) $(\omega\tau)^2 >> 1$. Under this approximation, we find:

$$\varepsilon_r'(\omega,\mathbf{k}) \approx 1 - \frac{\omega_p^2}{\omega^2} \tag{3.44}$$

and

$$\varepsilon_r''(\omega,\mathbf{k}) \approx \frac{\omega_p^2}{\omega^3\tau} = \frac{\omega_p^2\gamma}{\omega^3} \tag{3.45}$$

This approximation may break down in the far-infrared spectral region, where damping can be significant. Note that damping (γ) is absolutely necessary to have an imaginary part of $\varepsilon_r(\omega)$. The dispersion of the real part of the dielectric function $\varepsilon_r'(\omega,\mathbf{k})$ (Equation 3.44) is plotted in Figure 3.13. The Drude model predicts a monotonous decrease of $\varepsilon_r'(\omega,\mathbf{k})$ for decreasing frequency, and experiments confirm

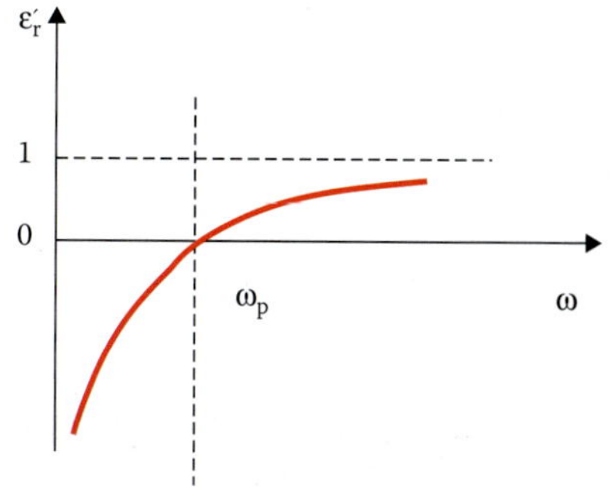

FIGURE 3.13 The dielectric function: ε_r' as a function of ω. The dielectric constant ε_r' becomes zero when $\omega = \omega_p$, and this supports free longitudinal collective modes for which all electrons oscillate in phase.

that $\varepsilon_r'(\omega,\mathbf{k})$ becomes zero when $\omega = \omega_p$. When $\varepsilon_r'(\omega)$ becomes zero at $\omega = \omega_p$, the material can support free longitudinal collective modes for which all electrons oscillate in phase. Such a collective charge oscillation is called a *plasmon*. Note that in the limit of γ going to 0, Equation 3.40 leads to the same result as shown in Equation 3.44 (no imaginary component to the dielectric constant at $\gamma = 0$). We come back to the field of plasmonics in Chapter 5 on photonics, and at that point we will not only analyze plots of $\varepsilon_r(\omega)$ for metals but also of $n(\omega)$ (refractive index), $\alpha(\omega)$ (absorption), and $R(\omega)$ (reflectance) for all types of materials.

With reference to Figure 3.12, we can derive the plasma frequency ω_p in yet another way. Let us look at the case where $\varepsilon_r'(\omega) = 0$ at $\omega = \omega_p$; because $\varepsilon_r'(\omega) = 1 + \chi_e'$, this means that at this frequency, $\chi_e' \approx -1$, and with the polarization, $\mathbf{P}(\omega,t) = \chi_e\varepsilon_0\mathbf{E}(\omega,t)$ **P** (Equation 3.36), this results in $\mathbf{P}(\omega,t) \approx -\varepsilon_0\mathbf{E}(\omega,t)$. The displacement δx of the electron gas of density n creates an electric field $\mathbf{E} = -\mathbf{P}/\varepsilon_0 = ne\delta x/\varepsilon_0$; this opposing field acts as the restoring force ($F_r = -eE$) on the electron gas. The equation of motion resulting from this restoring force F_r in such a simple electron gas oscillator is given as:

$$F_r = -eE = -\frac{ne^2\delta x}{\varepsilon_0} = m_e\frac{\partial^2(\delta x)}{\partial^2 t^2} \quad (3.46)$$

This results in oscillations at the plasma frequency of $\omega_p = \sqrt{\frac{ne^2}{m_e\varepsilon_0}}$ (Equation 3.41).

The plasma frequency ω_p comes with a free space wavelength of:

$$\lambda_p = \frac{2\pi c}{\omega_p} \quad (3.47)$$

In Table 3.2 we compare the plasma frequency and the free space wavelength for a generic metal and a generic semiconductor. Using Equation 3.41 we can now calculate ω_p for a variety of materials. For silver, for example, with a σ of 6.2×10^7 Ω-m, one obtains $\omega_p = 9.65 \times 10^{14}$ Hz (= 311 nm or 4 eV). The lower the resistivity (higher n), the higher the plasma frequency. Consequently, plasma frequencies ω_p are in the optical range (UV) for metals and in the THz to the infrared region for semiconductors and insulators. Metals reflect light in the visible region and are transparent at very high frequencies (UV and x-rays). With a free space wavelength less than λ_p, a wave propagates; when it is longer, the wave is reflected ($\omega > \omega_p$, or $\lambda < \lambda_p$).

Table 3.2 Plasma Frequency (ω_p) and Free Space Frequency (λ_p) for a Generic Semimetal and Semiconductor

Property	Metal	Semiconductor
Electron density: n, cm^{-3}	10^{22}	10^{18}
Plasma frequency: ω_p, Hz	5.7×10^{15}	5.7×10^{13}
Space frequency: λ_p, cm	3.3×10^{-5}	3.3×10^{-3}
Spectral range	UV	Infrared region

In a metal where both bound electrons $[\varepsilon_B(\omega)]$ and conduction electrons $[\varepsilon_r(\omega)]$ contribute an effective dielectric constant, $\varepsilon_{Eff}(\omega)$ must be defined as:

$$\varepsilon_{Eff}(\omega) = \varepsilon_B(\omega) + \varepsilon_r(\omega) \quad (3.48)$$

where $\varepsilon_r(\omega)$ is the complex dielectric constant at $\omega\tau >> 1$, given by (see Equations 3.44 and 3.45):

$$\varepsilon_r(\omega) = \varepsilon_r'(\omega) + i\varepsilon_r''(\omega) = 1 - \frac{\omega_p^2}{\omega^2} + i\frac{\omega_p^2}{\omega^3\tau} \quad (3.49)$$

Plasma oscillations can be excited in dielectric films as well. A dielectric material is a nonconducting substance whose bound charges are polarized under the influence of an externally applied electric field. In this case, although qualitatively the same effect as in a metal, the oscillations of the bound electrons are with respect to the immobile ions instead of with respect to the films boundary, and a restoring force related to the strength of the bond of the electrons to those ions must be introduced. In Chapter 5 on photonics, after introducing the Maxwell equations, the Drude plasmon model will be upgraded with the more complete Drude-Lorentz model that includes such a restoring force term. The restoring force pulls charges back in their equilibrium position. The Drude-Lorentz model without the restoring force reduces back to the Drude model, in which conduction electrons are not bound to atoms. There we will also calculate the effective dielectric constant $\varepsilon_{Eff}(\omega)$ (with bound and free electron contributions) of Equation 3.48.

Specific Heat Capacity of Metal and Insulators

Different materials require different amounts of heat to increase their temperature. Heat capacity per unit mass of a substance is known as specific heat, and as we will see, it is a measure of the number of degrees of freedom of the system. When using classical models, as in the case of electrical conductivity, things went wrong for calculations of the specific heat capacity of metals and insulators.

Based on the Drude assumption that electrons in a metal behave as a monatomic gas of *N* classical particles, they should be able to take up translational kinetic energy when the metal is heated. According to the equipartition principle of energy in a gas with *N* particles, the electron internal energy *U*, at temperature *T*, expressed per mole is:

$$\frac{U}{n} = \frac{3}{2}\frac{N}{n}k_BT = \frac{3}{2}N_AkT = \frac{3}{2}RT \tag{3.50}$$

with n the number of moles in the system, N_A Avogadro's number, *R* the ideal gas constant, and k_B Boltzmann's constant. Remember also from thermodynamics that the first derivatives of the fundamental thermodynamic equations, in *U* (internal energy) or *S* (entropy), correspond to the intensive parameters *T*, *P*, and μ (chemical potential), whereas second derivatives correspond to important materials properties such as the molar heat capacity, or:

$$C_v = T\left(\frac{\partial s}{\partial T}\right)_v = \frac{T}{n}\left(\frac{\partial S}{\partial T}\right)_v = \frac{1}{n}\left(\frac{\partial Q}{\partial T}\right)_v \tag{3.51}$$

This equation basically tells us that the molar heat capacity at constant volume is the quasistatic heat flux per mole required to produce a unit increase in the temperature for a system maintained at constant volume. Materials with high specific heat capacity (C_v) require more energy to reach a given temperature. In the case of a metal, one expects contributions to the heat capacity from both the lattice and the free electrons. The electronic contribution to the molar specific heat capacity at constant volume, C_v, from Equation 3.50, is:

$$C_{v,el} = \frac{\delta}{\delta T}\left(\frac{U}{n}\right)_v = \frac{3}{2}R \tag{3.52}$$

or about 12.5 J/(mol·k).

Atoms in a lattice usually have no translational energy, and as temperature increases only vibrational energy increases. In the case of a monovalent metal, the lattice specific heat contribution, $C_{v,lat}$, equals 3*R*. This follows from the fact that a classical atom oscillator has 3 degrees of freedom in vibration, or U = 3NkT. The vibration of a classical 1D harmonic oscillator is U = kT, with $\frac{1}{2}kT$ for kinetic and $\frac{1}{2}kT$ for potential energy (Figure 3.14). From the latter we expect $C_{v,lat}$ to be constant at 3*R* or 25 J/(mol·k), the so-called Dulong-Petit law. Experimentally, C_v, the sum of vibrational and electronic contributions ($C_v = C_{v,el} + C_{v,lat}$), is proportional to T^3 at low *T* and approaches a constant 3*R* at high *T* (see Figure 3.15). Diamond reaches the Dulong-Petit value 25 J/(mol·k) only at high temperature, whereas lead reaches the Dulong-Petit value at relatively low temperature.

From the above, at high temperature, electrons are expected to contribute to the lattice heat capacity for a total $C_v = C_{v,el} + C_{v,lat}$ or $\frac{9}{2}R$, which is obviously a different result from Dulong-Petit's 3*R*! The total C_v for metals is found to be only slightly higher than that for insulators, which only feature lattice contributions. The question is then: where is the electronic contribution? The absence of a measurable contribution by electrons to the C_v was historically

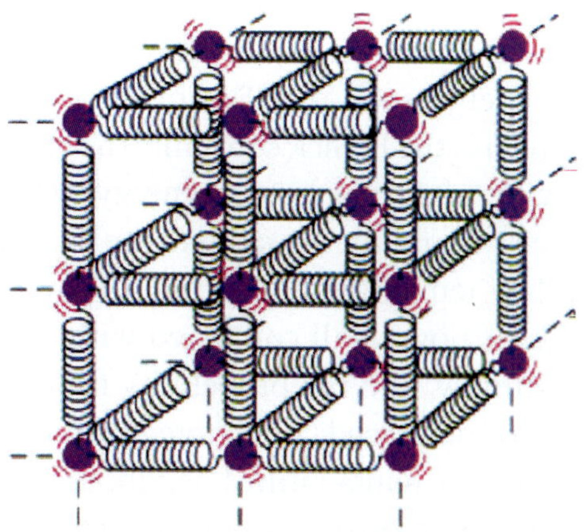

FIGURE 3.14 C_v: energy needed to raise *T* of 1 mol by 1 K at constant *V*, volume. The internal energy resides in vibrations of the solid constituents. The vibration of a classical 1D harmonic oscillator is U = kT. In a solid there are three perpendicular modes of vibration or three harmonic oscillators, and the total energy of a solid is thus U = 3NkT = 3RT.

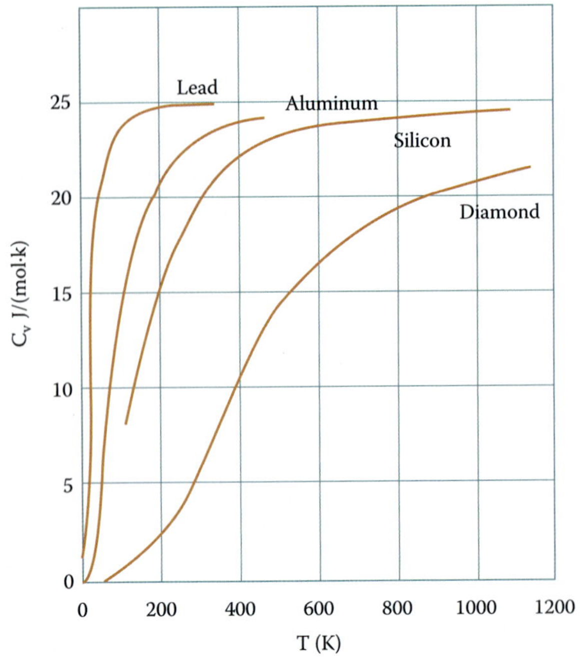

FIGURE 3.15 Dulong-Petit law with C_v = 3R or 25 J/(mol·k) at high temperatures.

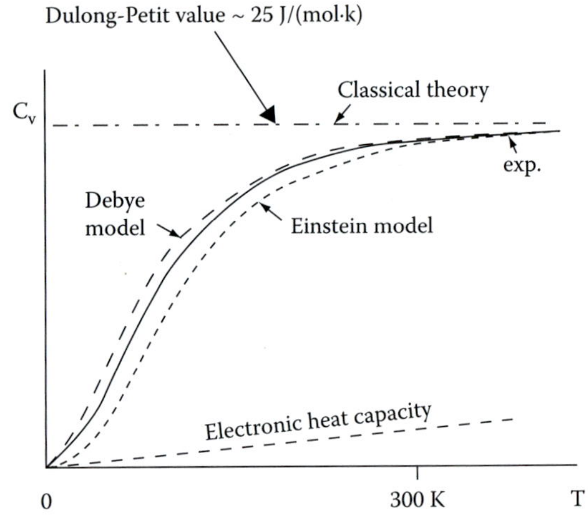

FIGURE 3.16 Schematic representation of the temperature dependence of the molar heat capacity, experimental and according to four models. For the Debye and Einstein models, see further below.

one of the major objections to the classical free electron model: because electrons are free to carry current, why would they not be free to absorb heat energy? Conduction electrons only contribute a small part of the heat capacity of metals, but, as we will see below, they are almost entirely responsible for the thermal conductivity.

In the above model, we used the principle of equipartition of the energy, which we borrowed from classical ideal gas theory. But, as Planck discovered in 1901, a vibrating system can only take up energy in quanta, the size of which is proportional to the vibration energy such that E = hν, where ν is frequency and h is the Planck constant. The chance that an atom vibrator can pick up an energy $h\nu$ is proportional to $e^{-h\nu/kT}$, and the average energy of a vibrating atom, at low temperature, may fall so low that the value of kT is now small compared with the size of the quantum $h\nu$, and the probability factor $e^{-h\nu/kT}$ is dramatically decreased. Therefore, Dulong-Petit is incorrect at low temperatures as vibrations of successively lower frequencies fail to become excited. Atoms in a crystal do not obey Maxwellian statistics at low temperatures, and one needs to invoke the Einstein or Debye law instead. The reason that electrons do not contribute to the heat capacity at room temperature is, just like in the case of electrical conductivity, that they are fermions and cannot be treated as an ordinary Maxwell-Boltzmann gas. At room temperature, only the very few electrons in the so-called Maxwell-Boltzmann tail of the Fermi distribution can contribute to the heat capacity. Thus, at ordinary temperatures the electronic heat capacity is almost negligible, and it is the atomic vibration, i.e., the contribution of phonons, that dominates. In Figure 3.16 we show the experimental molar heat capacity as a function of temperature and the expectations for those values from classical theory and from the Einstein and Debye models, which are discussed at the end of this chapter. It is seen here that the electronic contribution $C_{v,el}$ varies linearly with temperature.

Thermal Conductivity

Heat conduction is the transfer of thermal energy from a hot body to a cold body. In a solid both electrons and phonons can move, and depending on the material involved, one or the other tends to dominate. We intuitively expect electrical and thermal conductivities somehow to be linked; from experience we know that, in general, good electrical conductors are also good thermal conductors. The connection between electrical and thermal conductivities for metals was first expressed in the Wiedemann-Franz law in 1853, suggesting that electrons carry thermal energy and electrical charge. Insulators are often transparent, and conducting

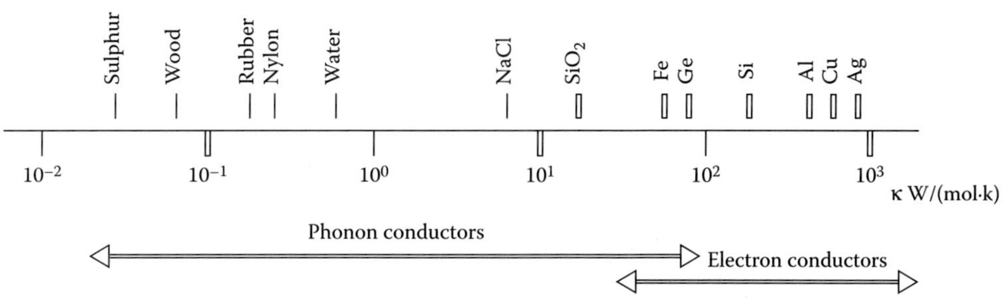

FIGURE 3.17 Phonon and electron conductors.

metals are reflecting and shiny when polished. Most metals are shiny because when light strikes a metal, the light is scattered by the moving electrons. For an exception, think about diamond, which is a transparent insulator but conducts heat better than aluminum or copper. For a glaring difference between heat and electrical conduction, consider that electrical conductivity of materials spans 25 orders of magnitude, whereas thermal conductivity only spans about four orders. In metals, heat is mostly transferred by electrons, but in electrical insulators, there are few free electrons, so the heat must be conducted in some other way, i.e., lattice vibrations or phonons. Thus, materials are divided into phonon conductors and electron conductors of heat as illustrated in Figure 3.17.

There is a major difference between heat conduction by electrons and by phonons; for phonons, the number changes with the temperature, but the energy is quantized, whereas for electrons, the number is fixed, but the energy varies. We will analyze now in more detail what that relation between electronic conductivity, σ, and thermal conductivity, κ is, and will discover that, when using classical models, the experimental results again cannot be properly explained.

Recall the electrical result **J** = σ**E** (Equation 2.2) and also $\mathbf{J} = \sigma\mathbf{E} = \sigma\frac{dV}{dx}$. The thermal equivalent for this expression is Fourier's law for heat conduction, Q, which states that:

$$Q = \kappa A \frac{T_h - T_c}{L} = \kappa A \frac{dT}{dx} \tag{3.53}$$

with T_h the hot temperature and T_c the cold temperature, κ the thermal conductivity, *L* the thermal conduit path-length, and *A* its cross-sectional area. From the first law of thermodynamics (heat conservation), we know that the rate of heat conduction must equal the rate of change of energy storage:

$$\kappa \frac{\partial^2 T}{\partial x^2} = C_{v,el} \frac{\partial T}{\partial t} \tag{3.54}$$

The specific heat capacity, in general, is made up of a phonon and an electron term ($C_v = C_{v,el} + C_{v,lat}$). In most metals, the contribution of the electrons to heat conductivity greatly exceeds that of the phonons, and the phonon term can be neglected ($C_v = C_{v,el}$). Because of electrical neutrality, equal numbers of electrons move from hot to cold as the reverse, but their thermal energies are different.

Equations 3.53 and 3.54 only apply under the conditions that:

t >> scattering mean free time of the energy carriers (τ)
L >> scattering mean free path of the energy carriers (λ)

These conditions break down for applications involving thermal transport in small length/time scales, e.g., nanoelectronics, nanostructures, NEMS, ultrafast laser processing, etc. (see below and Volume III, Chapter 7 on scaling). For now, however, we are interested in the relation between electronic conductivity, σ, and thermal conductivity, κ, for somewhat larger systems.

To arrive at the expression for thermal conductivity, κ, we inspect Figure 3.18 and consider a metal bar [area *A* is a unit area (A = 1) here] with a temperature gradient *dT*/*dx* where we are interested in a small volume of material, with a length 2λ, with λ the mean free path between collisions. The idea is that an electron must have undergone a collision in this space and hence will have the energy/temperature of this location. To calculate the energy

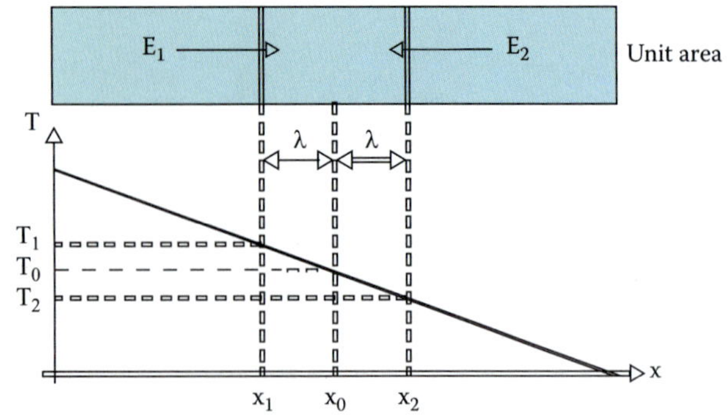

FIGURE 3.18 Fourier's Law for heat conduction: we are interested in a small volume of material, with a length 2λ, where λ is the mean free path between collisions with the lattice.

flowing per unit time per unit area from left to right (E_1), we multiply the number of electrons *N* crossing by the energy of one electron:

$$\text{energy} = \frac{1}{2}kT_1 = \frac{1}{2}k\left[T_0 + \lambda\left(-\frac{dT}{dx}\right)\right]$$

$$\text{number} = \frac{1}{6}nv_{dx} \tag{3.55}$$

$$E_1 = \frac{nv_{dx}}{2}\frac{1}{2}k\left(T_0 - \lambda\frac{dT}{dx}\right)$$

with v_{dx} the drift velocity in the *x*-direction, and dT/dx a thermal gradient in the *x*-direction. We know that for charge neutrality the same number of electrons must flow in the opposite direction to maintain charge neutrality, but their energy per electron is lower.

Based on:

$$E_2 = \frac{nv_{dx}}{2}\frac{1}{2}k\left(T_0 + \lambda\frac{dT}{dx}\right) \tag{3.56}$$

the thermal energy transferred per unit time per unit area is:

$$Q = E_1 - E_2 = -\frac{nv_{dx}}{2}k\lambda\frac{dT}{dx} \tag{3.57}$$

Comparing this result with Equation 3.53 (with A = 1), we obtain:

$$\kappa = \frac{1}{2}nv_{dx}k\lambda \tag{3.58}$$

This says that the thermal conductivity is larger if there are more electrons (*n* large), the electrons move faster (v_{dx} large), and they move more easily (large λ with fewer collisions). Because $N_Ak = R$ with *R* the ideal gas constant, $N/N_A = n$ (number of moles of electrons), and the electronic contribution to the molar specific heat capacity at constant volume, $C_{v,el}$, from Equation 3.52 is $3/2R$. The bulk heat conductivity of a solid per mole (n = 1) can then be rewritten as:

$$\kappa = \frac{1}{3}C_{v,el}v_{dx}\lambda \text{ or also } \frac{1}{3}C_{v,el}v_{dx}^2\tau \text{ sine } \lambda = v_{dx}\tau \tag{3.59}$$

As mentioned above, in most metals the conduction by electrons greatly exceeds that of the phonons; typically the phonon contribution at room temperature is only 1% of the electron contribution.

In 1853, long before Drude's time, Gustav Wiedemann and Rudolf Franz published a paper claiming that the ratio of thermal and electrical conductivities of all metals has almost the same value at a given temperature:

$$\frac{\kappa}{\sigma} = \frac{C_{v,el}mv_{dx}^2}{3ne^2} = \frac{3}{2}\left(\frac{k_B}{e}\right)^2 T \tag{3.60}$$

calculated through Drude's application of classical gas law: $C_{v,el} = \frac{3}{2}nk_B$ and $\frac{1}{2}mv_{dx}^2 = \frac{3}{2}k_BT$. Ludwig Lorenz realized in 1872 that this ratio scaled linearly with temperature, and thus a Lorenz number, *L*, was defined as:

$$\frac{\kappa}{\sigma T} \equiv L \tag{3.61}$$

which is very nearly constant for all metals (at room temperature and above). A typical value for *L*, say for Ag, is 2.31 10^{-8} WΩK^{-2} at 0°C and 2.37 10^{-8} WΩK^{-2} at 100°C (see Table 3.3).

Although Equation 3.61 is the correct relationship, we now know that the value for *L* calculated from it is wrong. However, Drude, using classical values for the electron velocity v_{dx} and heat capacity $C_{v,el}$, somehow got a number very close to the experimental value. But how lucky that Drude dude was: by a tremendous coincidence, the error in each term he made was about two orders of magnitude … in the opposite direction [the electronic $C_{v,el}$

TABLE 3.3 Some Typical Lorenz Numbers

Lorenz Number L in 10^{-8} WΩK²		
Metal	273 K	373 K
Ag	2.31	2.37
Au	2.35	2.40
Cd	2.42	2.43
Cu	2.23	2.33
Ir	2.49	2.49
Mo	2.61	2.79
Pb	2.47	2.56
Pt	2.51	2.60
Sn	2.52	2.49
W	3.04	3.20
Zn	2.31	2.33

is 100 times smaller than the classical prediction, but $(v_{dx})^2$ is 100 times larger]. So the classical Drude model gives the prediction:

$$L_{Drude} = 1.12\,10^{-8}\,W\Omega K^{-2} \qquad (3.62)$$

But in Drude's original paper, he inserted also an erroneous factor of two, as a result of a mistake in the calculation of the electrical conductivity. So he originally reported:

$$L = 2.24 \times 10^{-8}\,W\Omega K^{-2} \qquad (3.63)$$

The correct value for the Lorenz number, L, derived from quantum mechanics is $\frac{\pi^2 k_B^2}{3e^2}$ or 2.45×10^{-8} WΩK² (see Equation 3.340 below). So although Drude's predicted electronic heat capacity was far too high (by a factor of 100!), his prediction of L made the free electron gas (FEG) model seem more impressive than it really was and led to a general acceptance of the model.

Hall Coefficients

Conductivity measurements do not yield information about the sign of charge carriers; for this we need the Hall effect. The Hall effect uses current and a magnetic field to determine mobility and the sign of charge carriers. The Hall effect, discovered in 1879 by American physics graduate student (!) Edwin Hall, is simple to understand.[1] With reference to Figure 3.19, consider a fairly strong magnetic field **B** (~2000 Gauss) applied perpendicular to a thin metal film (thickness is d and width is w) carrying a current I. The magnetic Lorentz force is

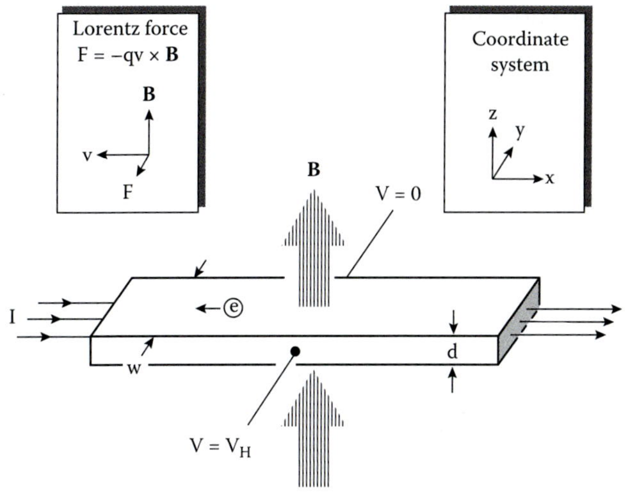

FIGURE 3.19 The Hall effect. Magnetic field **B** is applied perpendicular to a thin metal film sample carrying current I. V_H is the Hall voltage.

normal to both the direction of the electron motion and the magnetic field. The path of the charge carriers shifts as a result of this Lorentz* force, and, as a consequence, the Hall voltage, $V_H = E_y w$ (where E_y is the Hall field strength), builds up until it prevents any further transverse displacement of electrons. In other words, the Hall field increases until it is equal to and opposite of the Lorentz force. The orientation of the fields and sample is illustrated in Figure 3.19.

In mathematical terms, the above translates as follows. With both electric and magnetic fields present, a charge carrier with charge q experiences a Lorentz force in the lateral direction:

$$\mathbf{F}_L = q(\mathbf{v}_d \times \mathbf{B}_z) \qquad (3.64)$$

We do use the drift velocity v_d of the carriers because the other velocities (and the forces caused by these components) cancel to zero on average. Note that instead of the usual term "electron," the term "charge carrier" is used here because in principle an electrical current could also be carried by charged particles other than electrons, e.g., positively charged ions or holes (missing electrons). The vector product in Equation 3.64 ensures that the Lorentz force is perpendicular to $\mathbf{v}_d$ and $\mathbf{B}_z$. For the geometry assumed in Figure 3.19, the Lorentz force $\mathbf{F}_L$ has only a component in the y-direction, and we can use a

* This is the Lorentz we encounter again in Chapter 5 (Figure 5.50), not the Lorenz from the Lorenz number L.

scalar equation for the Lorentz force: $F_L = -qv_d\mathbf{B}_z$ or with Equation 3.11 (where we replaced μ_e by μ_q for generality):

$$\mathbf{F}_L = -q\mu_q E_x \mathbf{B}_z \quad (3.65)$$

As more and more carriers are deflected by the magnetic force, they accumulate on one side of the thin-film conductor, and this accumulation of charge carriers leads to the "Hall field" $\mathbf{E}_y$ that imparts a force opposite to the Lorentz force. The force from that electrical field $\mathbf{E}_y$ in the y-direction (which is of course $q \cdot \mathbf{E}_y$) must be equal to the Lorentz force with opposite signs. This way we obtain:

$$\begin{aligned} F_L &= -q\mu_q \mathbf{E}_x \mathbf{B}_z = q\mathbf{E}_y \text{ or} \\ F_L &= -\mu_q \mathbf{E}_x \mathbf{B}_z = \mathbf{E}_y \end{aligned} \quad (3.66)$$

The Hall voltage V_H now is simply the field in y-direction multiplied by the dimension w in the y-direction (=$w\mathbf{E}_y$). It is clear then that the (easily measured) Hall voltage is a direct measure of the mobility μ of the carriers involved, and that its sign or polarity will change if the sign of the charges changes.

It is customary to define a Hall coefficient, R_H, for a given material as:

$$R_H \equiv \frac{\mathbf{E}_y}{\mathbf{J}_x \mathbf{B}_z} \quad (3.67)$$

In other words, we expect that the Hall voltage, $w\mathbf{E}_y$, will be proportional to the current density $\mathbf{J}_x$ and the magnetic field strength $\mathbf{B}$, which are, after all, the main experimental parameters (besides the trivial dimensions of the specimen):

$$\mathbf{E}_y = R_H \mathbf{B}_z \mathbf{J}_x \quad (3.68)$$

Using Equations 3.10, 3.66, and 3.67, as well as $\mathbf{J}_x = \sigma\mathbf{E}_x$, we calculate R_H as:

$$R_H = -\frac{-\mu_q \mathbf{E}_x \mathbf{B}_z}{\sigma \mathbf{E}_x \mathbf{B}_z} = -\frac{\mu}{\sigma} = -\frac{\mu}{qn\mu} = -\frac{1}{qn} \quad (3.69)$$

When we calculate R_H from our measurements and assume $|q| = e$ (which Hall at the time did not know!), we can find n, the charge density. Also, the sign of V_H and thus of R_H tells us the sign of q! If R_H is negative, the predominant carriers are electrons; if positive, they are holes. If both types of carriers are present, the Hall field will change its polarity depending on the majority carrier (Figure 3.20). Also from Equation 3.69 one sees that the lower the carrier density, the higher R_H and thus the higher V_H; this is a key to some very sensitive magnetic field sensors.

For most metals the Hall constant is negative because electrons are majority carriers. But for Be and Zn, for example, there is a band overlap with dominant conduction by holes in the first band and fewer electrons in the second. As a result, R_H is positive for these metals. Of course energy bands were not heard of yet, and the results of a positive R_H were baffling at the time: how can we have $q > 0$ (even for metals!)? The Hall coefficient changes sign with the sign of the charge carrier and therefore provides an important method for investigating the electronic structure of the solid state. In particular, the positive Hall coefficients exhibited by metals such as magnesium and aluminum are a clear indication that a naive picture of a sea of conduction electrons is inappropriate because the majority carriers are clearly positively charged (and are, in fact, holes). The discrepancies between the FEG predictions and experiments nearly vanish when liquid metals are compared (see Table 3.4). This

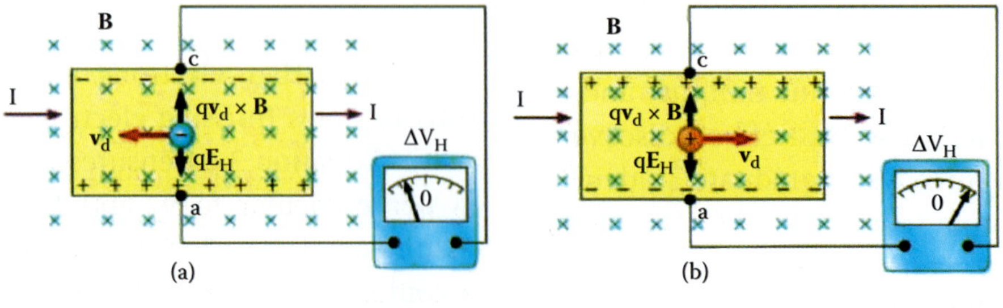

FIGURE 3.20 Electron and hole charge carriers in the Hall effect. When the charge carriers are negative, the upper edge of the conductor becomes negatively charged (a). When the charge carriers are positive, the upper edge becomes positively charged (b).

TABLE 3.4 Discrepancies between the FEG Predictions for R_H and Experiments Nearly Vanish When Liquid Metals Are Compared*

		R_H (10^{-11} m³/As)		
Metal	n_0	Solid	Liquid	FEG Value
Na	1	–25	–25.5	–25.5
Cu	1	–5.5	–8.25	–8.25
Ag	1	–9.0	–12.0	–12.0
Au	1	–7.2	–11.8	–11.8
Be	2	+24.4	–2.6	–2.53
Zn	2	+3.3	–5	–5.1
Al	2	–3.5	–3.9	–3.9

*This reveals clearly that the source of these discrepancies lies in the electron-lattice interaction. Notice positive R_H for Be and Zn.

reveals clearly that the source of these discrepancies lies in the electron-lattice interaction of a solid.

The Hall effect "oddities" will be properly explained once we have introduced quantum physics. We will also see that under special conditions of extremely low temperature, high magnetic field, and two-dimensional electronic systems (2D electron gas, in which the electrons are confined to move in planes), the Hall resistance R_H is quantized and increases as a series of steps with increasing magnetic field. These discrete energy levels are called Landau levels.

Blackbody Radiation

Planck first discovered the discontinuous behavior that characterizes the atomic world in his analysis of the light spectra emitted from blackbodies in 1900. All substances, with thermal energy, radiate EM waves, and the radiation emanating from solids consists of a continuous spectrum of wavelengths. A blackbody (also cavity radiation) is a hypothetical object that is a perfect absorber or perfect emitter of radiation (Figure 3.21).

"Blackbody" is an unfortunate name as the ideal radiating/absorbing body does not have to be black; stars and planets, to a rough approximation, are blackbodies! When a blackbody is heated, it first feels warm and then glows red or white hot, depending on the temperature. A typical spectrum of the radiated light intensity, brightness, or emittance of a blackbody is shown in Figure 3.22. Intensity is a measure of how much energy is emitted from an object per unit surface area per unit time at a given wavelength

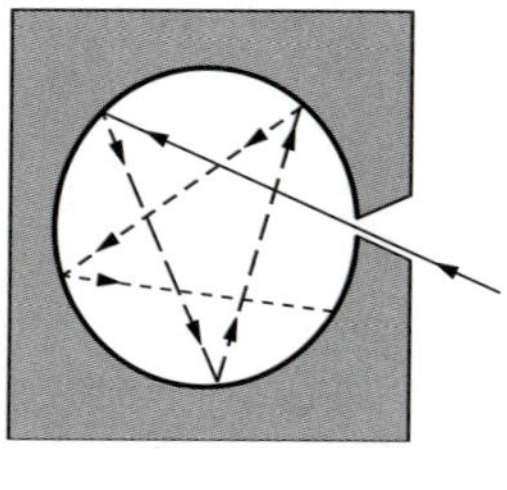

FIGURE 3.21 Ideal blackbody (also cavity radiation). "Blackbody radiation" or "cavity radiation" refers to an object or system that absorbs all radiation incident on it and reradiates energy that is characteristic of this radiating system only, not dependent on the type of radiation that is incident on it. The radiated energy can be considered to be produced by standing wave or resonant modes of the cavity, which is radiating.

and in a particular direction. A blackbody of temperature T emits a continuous spectrum peaking at λ_{max}. At very short and very long wavelengths there is little light intensity, with most energy radiated in some middle range frequencies. As the body gets hotter, the peak of the spectrum shifts toward shorter wavelengths (higher frequencies), but there is always a cutoff at very high frequencies.

It was experimentally observed that the brightness peak position shifted with temperature as:

$$T\lambda_{max} = \text{constant} = 2.898 \times 10^{-3}\ \text{m·K} \quad (3.70)$$

This is known as the Wien displacement law (see Figure 3.22). It was also known that the total energy, E, could be represented as the Stefan-Boltzmann law:

$$E = \sigma T^4 \quad (3.71)$$

where the constant $\sigma = 5.6704 \times 10^{-8}$ W/m²·K⁴. Classical interpretation predicted something

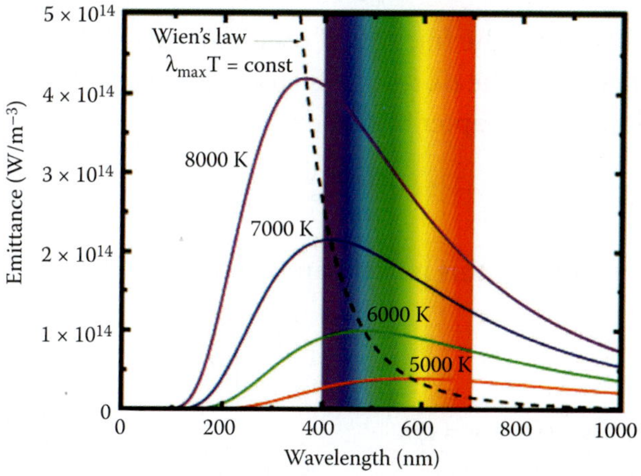

FIGURE 3.22 Blackbody radiation spectra at four different temperatures.

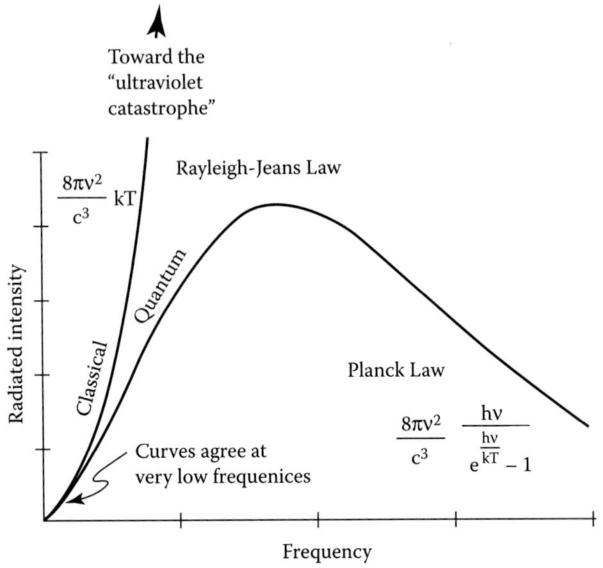

FIGURE 3.23 Planck and Rayleigh-Jeans models for blackbody radiation.

altogether different; in the classical Rayleigh-Jeans model, instead of a peak in the blackbody radiation and a falling away to zero at zero wavelength, the measurements were predicted to go off scale at the short wavelength end as shown in Figure 3.23.

Here is how the British physicists Lord Rayleigh and Jeans derived their model. They interpreted the blackbody radiation coming from a solid as electromagnetic radiation from oscillators that vibrate at every possible wavelength λ. In Figure 3.23 the radiated intensity, $E(\lambda)$, in $Js^{-1}\ m^{-3}$, is the energy distribution, i.e., the energy, E, at each λ. At equilibrium, the mean energy of all oscillators at temperature T is kT. The energy E_λ in a small interval $d\lambda$ is then given by:

$$E_\lambda\, d\lambda = kTdn \tag{3.72}$$

where dn is the fraction of oscillators at the "average energy" in the $d\lambda$ interval. From the Maxwell equations (see Chapter 5) one derives:

$$\frac{dn}{d\lambda} = \frac{8\pi}{\lambda^4} \tag{3.73}$$

Substituting this relation in Equation 3.72 leads to:

$$E_\lambda d\lambda = kT\left[\frac{8\pi}{\lambda^4}\right]d\lambda$$

or also since $\nu = c/\lambda$ (3.74)

$$E_\lambda d\lambda = kT\left[\frac{8\pi\nu^2}{c^3}\right]d\nu$$

The term in brackets $\left[\frac{8\pi\nu^2}{c^3}\right]$ is the number of modes per unit frequency per unit volume.

The amount of radiation emitted in a given frequency range should be proportional to the number of modes in that range. The best of classical physics suggested that all modes had an equal chance of being produced, and that the number of modes increased proportionally to the square of the frequency. Equation 3.74 works at long wavelengths (see Figure 3.23) but fails at short wavelengths. This failure at short wavelengths is called the ultraviolet catastrophe; as λ decreases, E_λ goes to $+\infty$. Even at very low T, the exponential curves in Figure 3.23 would have a huge value for visible light. Objects would be visible in the dark. The predicted continual increase in radiated energy with frequency (dubbed the "ultraviolet catastrophe") did not happen. Nature knew better.

The UV catastrophe attracted the attention of many physicists at the end of the nineteenth century, including Max Planck. Planck took the revolutionary step that led to quantum mechanics. He concluded that the available amount of energy could only be divided into a finite number of pieces among the atom oscillators in the cavity walls, and the energy of such a piece of radiation must be related to its frequency according to a new extremely important equation:

$$E = h\nu \tag{3.75}$$

where h was a new constant now called the Planck constant. The quantization implies that a photon of blue light of given frequency or wavelength will always have the same size quantum of energy. For example, a photon of blue light of wavelength 450 nm will always have 2.76 eV of energy. It occurs in quantized chunks of 2.76 eV, and you cannot have half a photon of blue light—it always occurs in precisely the same-sized energy chunks. Planck showed that the intensity I of radiation from a blackbody could be described by the function (now known as the Planck function):

$$I(\lambda,T) = \frac{2hc^2}{\lambda^5}\frac{1}{(e^{\frac{1}{\lambda kT}} - 1)} \tag{3.76}$$

This Planck function was initially found empirically by trial and error but later derived by assuming the Planck constant and a Boltzmann distribution. Differentiating the Planck function with respect to wavelength one derives Wien's displacement law—the wavelength of the maximum:

$$\lambda_{max} = \frac{hc}{4.965kT} \tag{3.77}$$

The constant hc/4.965kT agrees with the 2.898 × 10^{-3} m·K experimental value from Equation 3.70. Integration of the Planck function with respect to wavelength over all possible directions results in the total energy emitted per unit area per unit time from the surface of a blackbody (or absorbed per unit area per unit time by a body in a blackbody radiation field). This gives us back the Stefan-Boltzmann law (Equation 3.71):

$$E = \int_0^\infty E_\lambda(T) = \frac{2\pi^5 k^4}{15c^2h^3}T^4 = \sigma T^4 \tag{3.78}$$

which also enables us to verify that the constant σ is indeed 5.6704 × 10^{-8} W/m²·K⁴.

Planck, interestingly, never appreciated how far removed from classical physics his work really was. He spent most of his life trying to reconcile the new ideas with classical thermodynamics. The Planck constant *h* was a bit like an "uninvited guest" at a dinner table; no one was comfortable with this new guest. But today we know that discontinuities in the nanoworld are meted out in units based on this Planck constant. This constant and its particular magnitude constitute one of the great mysteries in nature. It is the underlying reason for the perceived weirdness of the nanoworld, the existence of a "least thing that can happen" quantity—a quantum. The ubiquitous occurrence of discontinuities in the nanoworld constantly upsets our common-sense understanding of the apparent continuity of the macroscopic world.

Light as a Stream of Particles

Photoelectric Effect

The photoelectric effect, discovered by chance, by Heinrich Hertz and his student Hallwachs in 1888,

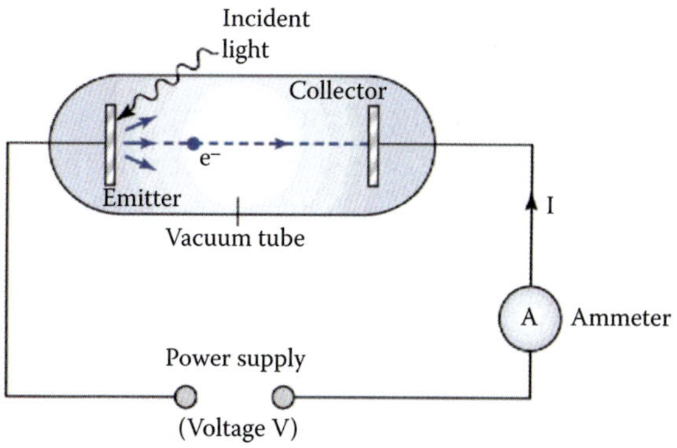

FIGURE 3.24 Experimental setup for studying the photoelectric effect.

is a process whereby light falling on negatively charged Zn, in an evacuated vessel, knocks electrons out of the surface as illustrated in Figure 3.24. The details of the photoelectric effect were in direct contradiction to the expectations of very well-developed classical physics of light waves, and the correct explanation by Einstein in 1905 marked one of the major steps toward quantum theory. Classical light wave theory predicts that the electrons in the metal will absorb radiation energy continuously. Once an electron has absorbed energy in excess of its binding energy, it will be ejected, and one would expect that a higher intensity would increase the chance that electrons are ejected.

As illustrated in Figure 3.24, to test this model we introduce a metal collector plate that collects electrons and a circuit that may be closed to measure the current *I*. In this setup, a retarding or stopping potential, V_0, may be applied to determine the kinetic energy of the electrons ($eV_0 = 1/2\ mv^2$). The classical wave model implies that the stopping voltage V_0 must be proportional to the intensity. Increasing frequency ν should not matter much, perhaps only causing a small decrease in current *I*, as a result of the rapid wave oscillations at high frequencies. With low intensity light, there should be a time delay to build up enough energy before current starts to flow.

The results were unexpected; no electrons were ejected, regardless of the intensity of the light, unless the frequency exceeded a certain threshold characteristic of the bombarded metal (red light did not cause the ejection of electrons, no matter what the intensity). The electrons were emitted immediately—no time lag. Increasing the intensity of the light increased the number of photoelectrons but not their maximum kinetic energy (a weak violet light would eject only a few electrons, but their maximum kinetic energies were greater than those for intense light of longer wavelengths). The kinetic energy of the ejected electrons varied linearly with the frequency of the incident radiation but was independent of the intensity, or:

$$eV_0(= \text{kinetic energy of the electrons}) = h\nu - \Phi \tag{3.79}$$

where Φ, the photoelectric work function, is the energy lost by a surface when an electron is freeing itself from its environment. This cannot be explained by the Maxwell equations.

If the charge of the electron is known, a plot of retarding or stopping voltage versus frequency of incident light, shown for three different metals in Figure 3.25, may yield a value for the Planck constant *h*. The electron charge was determined by Robert Millikan in 1909, and with that value and the slope of the lines in Figure 3.25, a value for *h* of 6.626×10^{-34} J·s was calculated, identical to the one derived from the hydrogen atom spectrum and blackbody radiation (see above). The intercept with the frequency axis (at kinetic energy zero) is the threshold frequency, ν_0, and the stop or retarding voltage axis intercept is the binding energy (−Φ) of the electron.

The photoelectric phenomenon could not be understood without the concept of a light particle, i.e., a quantum amount of light energy for a particular frequency. Einstein's paper explaining the photoelectric effect was one of the earliest applications of quantum theory and a major step in its establishment. The remarkable fact that the ejection energy was independent of the total energy of illumination showed that the interaction must be like that of a particle that gave all of its energy to the electron! This fit in well with Planck's hypothesis that light in the blackbody radiation experiment could exist only in discrete bundles with energy. In quantum theory, the frequency, ν, of the light determines the energy, *E*, of the photons and E = hν, where *h* is Planck's constant (h = 6.626069×10^{-34} J·s) (Figure 3.26).

This assumption explains quantitatively all the observations associated with the photoelectric effects. A photon hits an electron and is absorbed. The energy of the emitted electron is given by the energy of the photon minus the energy needed to release the electron from the surface. This explains the observance of a threshold value below which no electrons are emitted. Thus, it depends on the frequency of light falling on the surface but not on its intensity. It also explains why there is no time lag; a photon hits an electron, is absorbed by the electron, and the electron leaves. Higher intensity light contains more photons, and so it will knock out more electrons. However, if the frequency of the light is such that a single photon is not energetic enough to release an electron from the surface, then none will be ejected no matter how intense the light. Gilbert N. Lewis in 1926 called Einstein's light particles photons. Just as the word *photon* highlights the particle aspect of an electron, the word *graviton* emphasizes

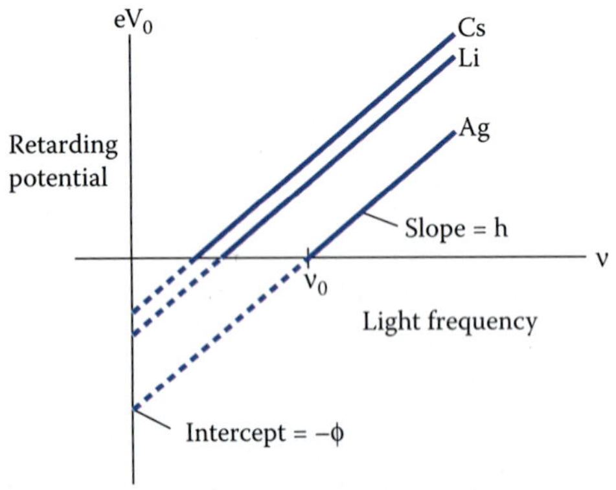

FIGURE 3.25 A plot of retarding or stopping voltage versus frequency of incident light. Slope is the Planck constant *h*. The intercept with the frequency axis (at kinetic energy zero) is the threshold frequency, ν_0, and the stop voltage axis intercept is the binding energy.

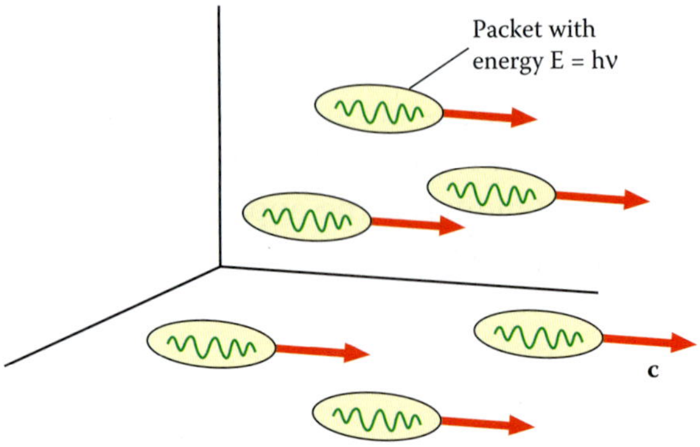

FIGURE 3.26 Light as energy packets with an energy E = hν.

the particle aspect of gravity, and *gluon* that of the strong nuclear force.

The young Einstein, in 1905, was the first scientist to interpret Planck's work as more than a mathematical trick and took the quantization of light (E = hν) for physical reality. He gave the uninvited dinner guest—the Planck constant *h*—a place at the quantum mechanics dinner table. What Einstein proposed here was much more audacious than the mathematical derivations by Planck to explain away the UV catastrophe. For a long time the science of optics had hesitated between Newton's corpuscular hypothesis and Huygens' wave theory. By the beginning of the nineteenth century the wave theory had become the dominant theory, largely because of the persistence of Augustin Fresnel (1788–1827), who described diffraction mathematically, and James Clerk Maxwell (1831–1879), who introduced the famous Maxwell equations in 1864 (Chapter 5). No wonder Einstein wrote at that time to one of his friends: "I have just published a paper about light, but I am sure nobody will understand it." Einstein reintroduced a modified form of the old corpuscular theory of light, which had been supported by Newton but which was long abandoned.

The particle nature of light was hard to swallow at the time, and it indeed still is. Remember the diffraction of x-rays on a crystal described in Chapter 2. Diffraction is something that happens with waves, not with particles. Einstein's light particles also negated the issue of "ether," a medium required for wave propagation as in the case of sound waves (sound waves cannot propagate in a vacuum); light is perfectly happy traveling in a vacuum.

How can we reconcile this duality of a photon as both wave- and particle-like in nature? We will have to learn to think of the wave as the probability of finding the particle. For example, if we know the momentum of the particle exactly, we cannot say where it is, only where it is likely to be [the Heisenberg uncertainty principle (HUP)].

Compton Scattering

The first strong support for the quantum nature of light came from monochromatic x-ray scattering on a graphite block. In 1922, Arthur Compton (Figure 3.27), at Washington University in St. Louis, saw that the wavelength of x-rays increases on scatterings off graphite, depending on the angle (Figure 3.28). This effect cannot be explained using wave theory of x-rays.

FIGURE 3.27 Arthur Harry Compton (1892–1962).

As first explained by Compton in 1923, a photon can lose part of its energy and momentum on scatterings with electrons in the graphite, and the resulting energy loss (or change in wavelength, Δλ) can be calculated from the scattering angle θ. The result is that some of the scattered radiation has a smaller frequency (longer wavelength) than the incident

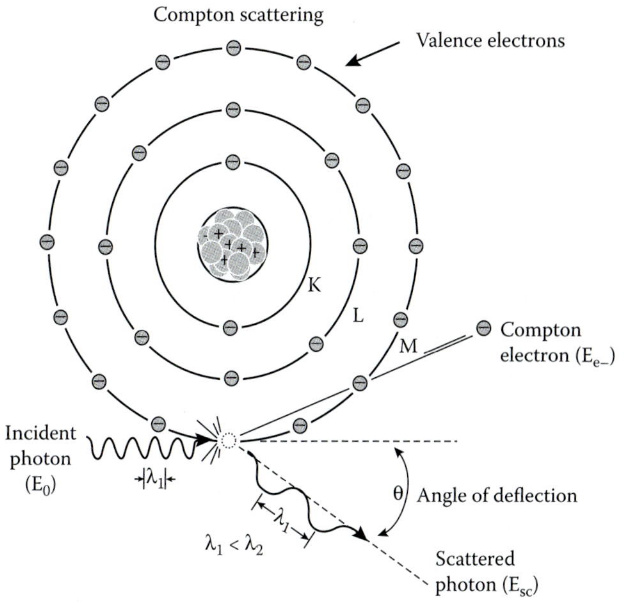

FIGURE 3.28 Compton effect. An incoming photon (E_0) can inelastically scatter from an electron and lose energy, resulting in an outgoing scattered photon (E_{sc}) with lower energy ($E_{sc} < E_0$).

radiation. This change in wavelength depends on the angle through which the radiation is scattered, and Compton concluded that an x-ray photon—or light in general—possesses momentum and thus behaves as a particle. Obviously, this was a big boost for the theory of light as composed of quanta. The three experiments that made the quantum revolution—blackbody radiation, the photoelectric effect, and the Compton effect—all indicate that light consists of particles.

Quantum Mechanics to the Rescue

Introduction

Invoking quantum mechanics we can solve the many problems with classical theories we exposed in the previous sections. Central to quantum mechanics is Schrödinger's equation, but before introducing Schrödinger's equation, we must put down some more foundations. We start with Kelvin's and Thomson's plum pudding atom model, then review Rutherford's and Bohr's improved orbital atom models, emphasize the importance of the Balmer's hydrogen emission lines in the discovery of the inner structure of an atom, and then we get baffled by de Broglie's matter waves and Heisenberg's uncertainty principle (HUP). There are four principal representations of quantum mechanics: Dirac's Hamiltonian and quantum algebra representation; the matrix representation of Born, Heisenberg, and Jordan; Schrödinger's wave equations; and Feynman's sum-over-histories approach; the latter constitutes a fundamentally new way of looking at quantum theory. With the Schrödinger formalism of quantum mechanics we tackle the band model and revisit electrical and thermal conductivity and heat capacity.

Bohr's and Rutherford's Atom

The Greek philosopher Leucippus of Miletus, who lived around 400 BC, first proposed atomic theory of matter. His disciple, Democritus of Abdera, concluded that infinite divisibility of a substance belongs only in the imaginary world of mathematics and further developed his mentor's atomic theory (Figure 3.29).

FIGURE 3.29 Democritus.

Democritus suggested that all things are "composed of minute, invisible indestructible particles of pure matter." According to the ancient Greeks, "atomos"* or atoms were all made of the same basic material, but atoms of different elements had different sizes and shapes. The sizes, shapes, and arrangements of a material's atoms determined the material's properties. For example, sour-tasting substances were believed to contain atoms with jagged edges, whereas round atoms made substances oily. It was further believed that there were four elements that all things were made of: earth, air, fire, and water. This basic theory remained unchanged until the nineteenth century, when it first became possible to test the theory with more sophisticated experiments. Lord Kelvin and J.J. Thomson developed a "raisin cake" model of the atom (also called the plum pudding model, 1897), which incorporated Thomson's negatively charged electrons as the raisins in a positively charged cake (Figure 3.30).

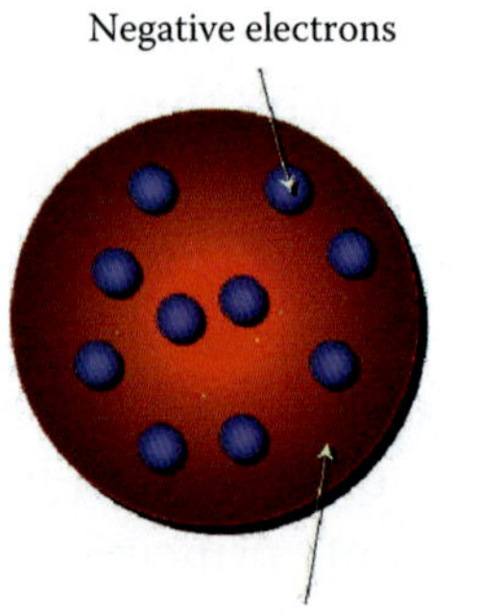

FIGURE 3.30 Kelvin's and Thomson's raisin cake.

* Atomos in Greek means "unbreakable."

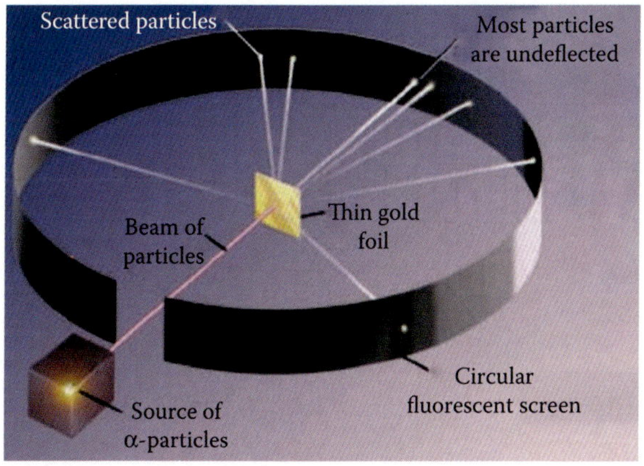

FIGURE 3.31 The Rutherford experiment. Structure of atom experiments.

To check the Kelvin and Thomson model, Geiger and Marsden, working in Rutherford's lab in 1911, directed a narrow beam of alpha particles, double-ionized helium atoms, of known energy onto a thin gold foil. A ZnS scintillation screen was used to record the striking alpha particles. Most particles went through the gold film undeflected, and some were deflected at small angles; however, they found that, once in a while, for about 1% of particles, the α-particles were scattered backward by the target (Figure 3.31).

For a moving alpha particle to be scattered through a large angle, a considerable repulsive force must be exerted. To explain the backscattering, Rutherford proposed that the positive charge in a gold atom must be concentrated in a small region. Rutherford proposed in this way the first realistic model of the atom: he concentrated 99.99% of the mass of the atom in the nucleus, which is only 10^{-15} m across (Figure 3.32). In other words, the atom (and therefore matter in general) is composed of 99.9999999999999% empty space. The proportion of nucleus to total atom size is obviously not drawn to scale in Figure 3.32.

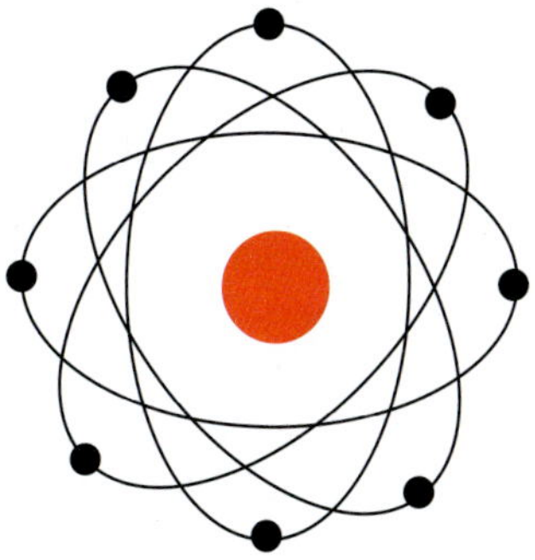

FIGURE 3.32 Rutherford's atom. Matter is mostly empty space.

The problem is that according to classical models the Rutherford model cannot lead to stable atoms. Rutherford's electrons are undergoing a centripetal acceleration and should radiate electromagnetic waves of the same frequency, so-called *bremsstrahlung* or "braking" radiation, leading to an electron "falling on a nucleus" in about 10^{-12} s! In the real world we have stable atoms, and atoms emit certain discrete characteristic frequencies of electromagnetic radiation. The Rutherford model is unable to explain these phenomena.

It was the analysis of light emitted or absorbed by atoms and molecules that led the way to better and better atom models. Cold, dilute gases absorb light at characteristic and discrete wavelengths (absorption spectra), and hot gases emit light at discrete wavelengths while continuous light spectra result when hot solids, liquids, very dense gases, or blackbodies (see Figure 3.33) emit light at all wavelengths (emission spectra).

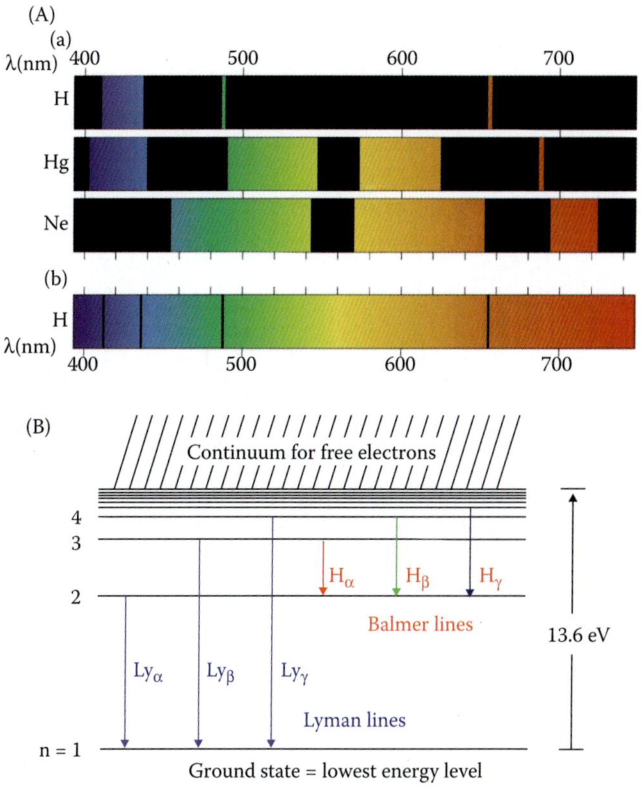

FIGURE 3.33 (Aa) Emission lines for H, Hg, and Ne; (Ab) Spectrum of sunlight with the hydrogen lines adsorbed out. (B) The hydrogen atom emission spectra with Balmer and Lyman lines for hydrogen.

When the light emitted from a hot gas is analyzed with a spectrometer, a series of discrete bright lines is observed. Each line has a different wavelength and color. This series of lines is the emission spectrum (Figure 3.33A). The absorption spectrum consists of a series of dark lines superimposed on the otherwise continuous spectrum. The dark lines of the absorption spectrum coincide with the bright lines of the emission spectrum. The continuous spectrum emitted by the sun passes through the cooler gases of the sun's atmosphere. The various absorption lines can be used to identify elements in the solar atmosphere, and this led, for example, to the discovery of helium.

Atomic hydrogen (one electron and one proton) in a gas discharge tube emits strongly at visible wavelengths H_α, H_β, and H_γ. More lines are found in the ultraviolet region, and the lines get closer and closer until a limit is reached (Figure 3.33B). These same lines are also seen in stellar spectra from absorption in the outer layers of the stellar gas.

There are four bright lines in the hydrogen emission spectrum, and in 1885, a Swiss teacher, Johann Balmer (1825–1898), guessed the following formula for the wavelength of these four lines:

$$\lambda = 364.5 \times 10^{-7} \frac{n^2}{n^2 - 4} (\mathrm{m}) \tag{3.80}$$

where n = 3, 4, 5, and 6, which are now called the *Balmer series*. Equation 3.80 may be rewritten as the so-called Balmer formula as:

$$\frac{1}{\lambda} = R_H \left(\frac{1}{n^2} - \frac{1}{k^2} \right) \tag{3.81}$$

with positive n and k integers and $n > k$ and R_H the Rydberg constant for hydrogen with a measured value of 1.096776×10^7 m^{-1}. With n = 1, we obtain the Lyman series in the UV; visible light is emitted in the Balmer series (n = 2), and with n ≥ 3 the infrared series are obtained (Paschen with n = 3, Brackett with n = 4, and Pfund with n = 5). The discrete emissions in Figure 3.33B suggest discrete energy levels, and Balmer's formula suggests that the allowed energies are given by (R_H/n^2) (in case of hydrogen) (Figure 3.34).

The "Great Dane," Niels Bohr, explained the above results for the H emission spectra by introducing four quantum postulates in a model halfway between classical physics and quantum theory. He reasoned that if electrons orbit the nucleus in circles with radii restricted to certain values, then the energy also can only take on certain discrete values, i.e., if it is quantized and there is a lowest energy orbit, and the electron is not allowed to fall to a lower energy. The allowed states are called *stationary states*. When in these permitted orbits, contrary to classical theory, the electrons do not radiate (Postulate 1). Bohr also assumed that classical mechanics applies to electrons in those stationary states (Postulate 2). He recognized that there might be a link between stable orbits and the Planck's and Einstein's quantum relation between the quantized energy of a photon and its frequency, so he proposed that radiation absorption or emission corresponds to electrons moving from one stable orbit to another, i.e., $\Delta E = h\nu$ (where h is the Planck constant = 6.6×10^{-34} J·s) (Postulate 3). In classical physics, angular momentum (L) of an object in circular motion is defined as mass (m_e) multiplied by velocity (v), multiplied by the radius (r) of the orbit, i.e., $L = m_e vr$ (or $L = pr$) (Figure 3.35). Bohr argued that allowed orbits are determined by the quantization of that angular momentum, $L = nh/2\pi$.

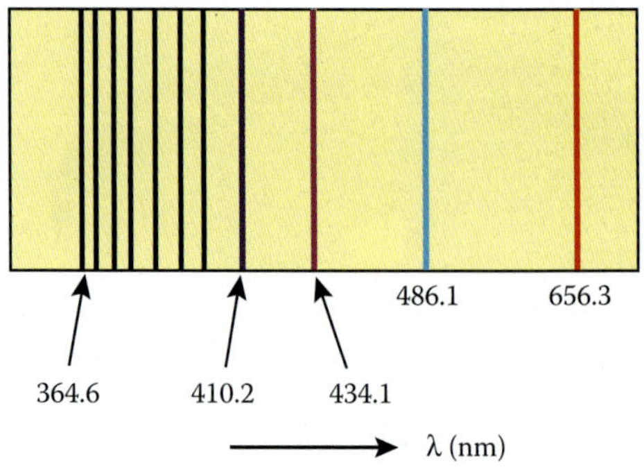

FIGURE 3.34 Some of the hydrogen emission lines.

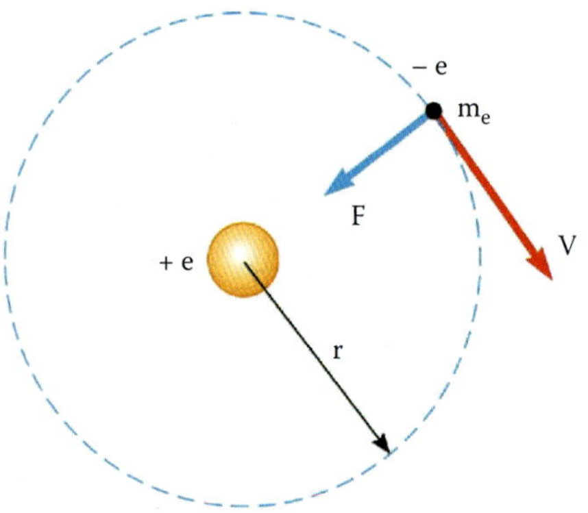

FIGURE 3.35 Momentum **p** for circular motion is $m_e vr$.

In this expression n is called the principal quantum number, and the ground state corresponds to $n = 1$. So here Bohr postulated that it was not the atom that determined the Planck constant h, but h that determined the properties of atoms (Postulate 4)!

Although this planetary kind of model has been shown to be mostly wrong, it makes for a very nice transition to full-fledged quantum mechanics. Bohr's theory, with slight modifications, is used, for example, to estimate the binding energies of dopant atoms in Chapter 4 and to explain the energy of excitons in Chapter 5.

Mathematically we can express Postulate 4 in vector notation as $|\mathbf{L}| = |\mathbf{p} \times \mathbf{r}| = m_e vr$, with v the tangential velocity and $\mathbf{p}$ the momentum of the electron (Figure 3.35). Bohr now combined his quantum model of the atom with Newton's classical model of planetary orbits to calculate the radius of the hydrogen atom. The attractive force between the electron and the proton in a hydrogen atom is caused by the Coulomb force and is given as:

$$F = \frac{e^2}{4\pi\varepsilon_0 r^2} \tag{3.82}$$

Based on Newton (Postulate 2), we may write:

$$F = \frac{m_e v^2}{r} = \frac{e^2}{4\pi\varepsilon_0 r^2} \tag{3.83}$$

Rutherford could not explain the spectral lines in and Figures 3.33 and 3.34, but he did set up the energy balance for a H atom correctly:

$$E_n \text{ (total energy)} = \text{KE (kinetic energy)} + \text{PE (potential energy)} \tag{3.84}$$

where the KE term depends on the velocity v and the PE term on the system (e.g., positional or electrostatic), or in the current case:

$$E = \frac{m_e v^2}{2} - \frac{e^2}{4\pi\varepsilon_0 r} \tag{3.85}$$

The radial acceleration is $a = v^2/r$, and with Newton's Law ($F = ma$) this yields $F = e^2/(4\pi\varepsilon_0 r^2) = mv^2/r$, which, solving for r, results in:

$$r = e^2/4\pi\varepsilon_0 m_e v^2 \tag{3.86}$$

And substituting Equation 3.86 in Equation 3.85 results in:

$$E = \frac{m_e v^2}{2} - m_e v^2 = -\frac{m_e v^2}{2} < 0 \tag{3.87}$$

meaning that, because $E < 0$, the motion of the electron is not free: it is bound by the attractive force of the nucleus as illustrated in Figure 3.36. To free the electron (and ionize the atom), the electron must receive an amount of energy, called

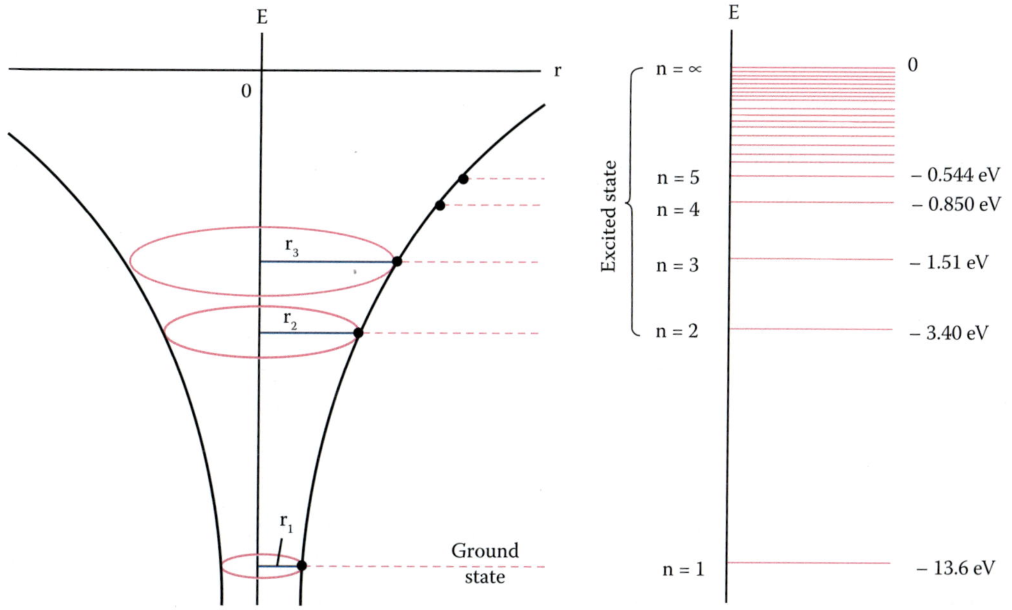

FIGURE 3.36 The electron is bound by the attractive force of the nucleus $E < 0$.

ionization energy, that will bring its total energy up to zero. Although this concept of a negative energy is somewhat counterintuitive, it does make sense; if we say that an electron has zero energy when far removed from the nucleus, then the electrons that are attached to an atom have a negative amount of energy.

The Quantum Numbers

First or Principal Quantum Number n for the Level of Energy E_n

Now we recall Bohr's fourth postulate about the quantization of angular momentum, $L_n = m_e v r_n = \frac{nh}{2\pi}$ or also $m_e v r_n = n\hbar$ with $\hbar = h/2\pi$. In the latter expression, n = 1, 2, 3, 4… and is called the principal quantum number; the ground state corresponds to n = 1. Thus, L_n is not only conserved but also constrained to discrete values by the quantum number *n*. This quantization of angular momentum is a crucial result and can be used in determining the Bohr orbit radii and Bohr energies. From the expression for *r* (Equation 3.86), we derive:

$$r_n = \frac{e^2}{4\pi\varepsilon_0 m_e (n\hbar/m_e r_n)^2} = \frac{e^2 r_n^2 m_e}{4\pi\varepsilon_0 n^2 \hbar^2} \quad (3.88)$$

Solving this expression for r_n results in:

$$r_n = \frac{n^2\hbar^2 4\pi\varepsilon_0}{m_e e^2} = n^2 a_0 \quad (3.89)$$

where a_0 is the Bohr radius, and permitted orbits have radii $r_1 = a_0$, $r_2 = 4a_0$, $r_3 = 9a_0$ … for n = 1, 2, 3, … as shown in Figure 3.37.

The Bohr radius, $a_0 = \frac{\hbar^2 4\pi\varepsilon_0}{m_e e^2}$, has a value of 0.53×10^{-10} m.

We are now also in a position to calculate Bohr orbit speeds v_n:

$$v_n = \frac{n\hbar}{m r_n} = \frac{\hbar}{n m a_0} = \frac{e^2}{4\pi n \hbar \varepsilon_0} \quad (3.90)$$

With the latter two expressions, we are able to calculate the total energy of the electron, E_n, associated

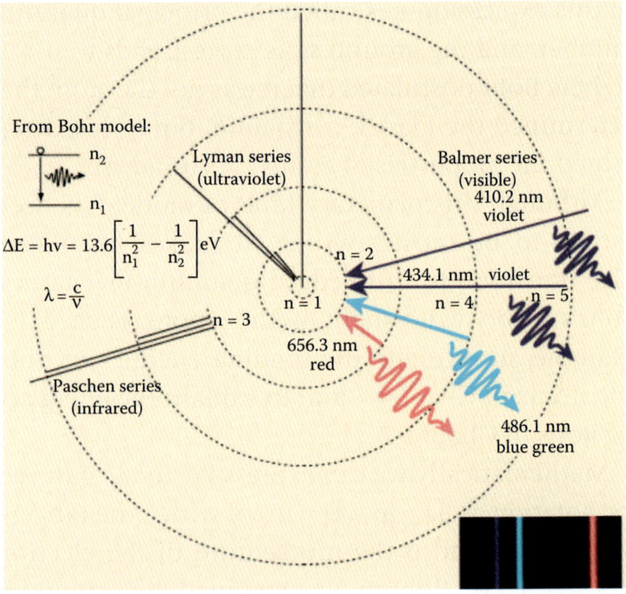

FIGURE 3.37 The Bohr atom with permitted radii. This Bohr model picture of the orbits has some usefulness for visualization, as long as it is realized that the "orbits" and the "orbit radius" just represent the most probable values of a considerable range of values.

with any integer value for *n*; by substituting the last two expressions in Equation 3.85, we obtain:

$$\begin{aligned} E_n &= \frac{m_e v^2}{2} - \frac{e^2}{4\pi\varepsilon_0 r_n} \\ &= \frac{1}{2} m_e \left[\frac{e^2}{n\hbar 4\pi\varepsilon_0}\right]^2 - \frac{e^2}{4\pi\varepsilon_0 n^2 a_0} \\ E_n &= \frac{m_e e^4}{32\hbar^2\pi^2\varepsilon_0^2 n^2} - \frac{m_e e^4}{16\hbar^2\pi^2\varepsilon_0^2 n^2} \\ &= -\frac{m_e e^4}{32\hbar^2\pi^2\varepsilon_0^2 n^2} = -\frac{E_0}{n^2} \end{aligned} \quad (3.91)$$

and $E_0 = \frac{e^2}{8\pi\varepsilon_0 a_0} = \frac{me^4}{2(\hbar 4\pi\varepsilon_0)^2}$ with a value of 13.6 eV = 21.8×10^{-19} J, sometimes called the Rydberg energy. The 13.6-eV energy value for an electron in a hydrogen atom (with n = 1) is the energy required to remove an electron from that atom (see also Figure 3.33). The possible energy levels of the hydrogen atom are labeled by the values of the quantum number *n*. The lowest energy level occurs for n = 1; this is the most negative energy level and the ground state. As *n* progressively increases, the energy increases (becomes less negative) for the excited states of the hydrogen atoms, as is clear from Figure 3.36.

Using the expression $E_n = -\frac{E_0}{n^2}$, we can now invoke Bohr's second postulate, which says that the photon energy is the difference in E_n values, and we derive Balmer's formula:

$$E_n = h\nu = \frac{hc}{\lambda} = -E_0\left[\frac{1}{k^2} - \frac{1}{n^2}\right]$$

$$\frac{1}{\lambda} = \frac{E_0}{hc}\left[\frac{1}{n^2} - \frac{1}{k^2}\right] = \frac{e^2}{8\pi\varepsilon_0 a_0 hc}\left[\frac{1}{n^2} - \frac{1}{k^2}\right] \quad (3.92)$$

$$= R_H\left[\frac{1}{n^2} - \frac{1}{k^2}\right]$$

It was considered a big success for the Bohr model that it accounted for Balmer and other series and that the calculated R_H agreed almost exactly (to within 0.1%) with the phenomenological value for the Rydberg constant for hydrogen, i.e., 1.096776×10^7 m^{-1}. The model also gives an expression for the radius of the atom that increases with n^2, the orbit speed that decreases with $1/n$, and predicts the energy levels of a hydrogen atom that increase with $-1/n^2$, and it can be extended to "hydrogen-like" atoms.

Second Quantum Number l for Orbital Angular Momentum

By 1914, Bohr had combined the quantum models of Planck and Einstein with the experimental work of Rutherford to provide a quantum model of the hydrogen atom, which fully explained the bright line spectra of hydrogen. To explain the spectra of more complicated atoms, other quantum numbers in addition to the principal quantum number *n*, defined by Bohr, had to be introduced. More detailed analysis of light spectra again led the way. Arnold Sommerfeld's (1868–1951) first major contribution was to extend Bohr's model to quantize all types of motion. He conceived elliptical atomic orbits by analogy to Johannes Kepler's elliptical planetary orbits and described mathematically orbits with different elliptical shapes but the same value of the principal quantum number *n*. This gave a number of different stationary states, some with slightly smaller and some with slightly larger energies—and hence multiple spectral lines—just as observed. The orbital angular momentum for an atomic electron can be visualized in terms of a vector model where the angular momentum vector is seen as precessing about a direction in space. Whereas the angular momentum vector has the magnitude shown in Figure 3.38, only a maximum of *l* units of $\hbar$ can be measured along a given direction, where *l* is the orbital quantum number. One of Sommerfeld's other important contributions was to link quantum theory to relativity. If one calculates the speed of an electron moving around the nucleus of an atom a value of the order of 1000 km/s is found. This is small compared with the velocity of light but large enough to have to use relativistic mechanics (see Chapter 5). Using this approach Sommerfeld was able to work out the fine structure of the hydrogen spectrum—a result that was regarded as a triumph both for relativity and quantum theory.

Third Quantum Number m, the Magnetic Number

In the 1890s, Pieter Zeeman (1865–1943) had noticed that if a magnetic field **B** was applied to a hot gas, the emission lines were further split into yet a finer structure. Sommerfeld, in 1916, showed that this was because of the direction (i.e., the orientation) of the orbiting of the electron with respect to the magnetic field. Sommerfeld was able to account for this orientation effect with the addition of a magnetic quantum number, *m*, which was an integer (Figure 3.39). Although called a "vector," the

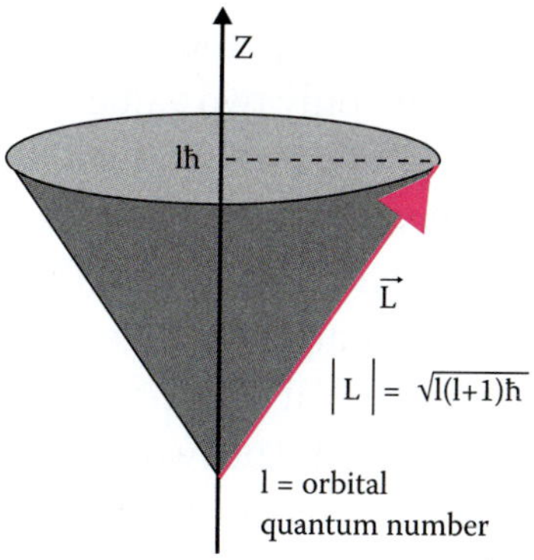

FIGURE 3.38 The Bohr-Sommerfeld atom. Sommerfeld introduced a second quantum number, *l* for elliptical orbitals. Spin-up and spin-down. Pauli's fourth quantum number: the electron spin *s* (see text later this chapter and Figure 3.39).

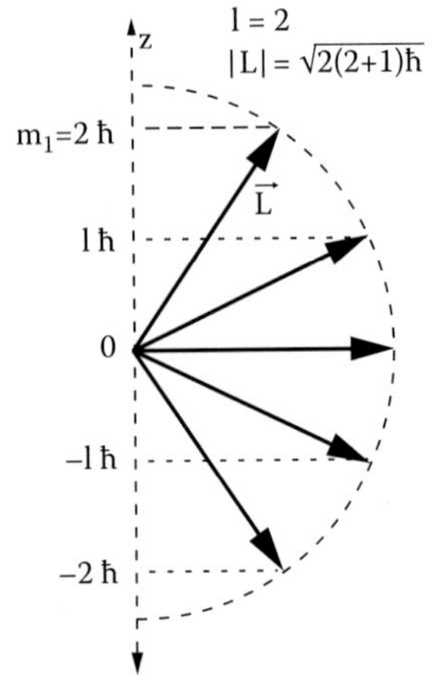

FIGURE 3.39 The quantum number m describes how the direction of the angular momentum is quantized with respect to the direction of the magnetic field.

orbital angular momentum in quantum mechanics is a special kind of vector because its projection along a direction in space is quantized to values one unit of angular momentum ($\hbar$) apart. In Figure 3.39 we show that the possible values that the "magnetic quantum number," m_l (for l = 2), can take are the values m = −2, −1, 0, +1, +2.

Fourth Quantum Number s *for Electron Spin*

More detailed measurements of the effects of a magnetic field on spectral line splitting showed yet further fine double splitting of the spectral lines. Wolfgang Pauli (1900–1958) explained this a little later, in 1921, by introducing yet another quantum number. While still a student, he proposed a fourth quantum number that he took to represent the electron spinning, not just in its orbit, but around its own axis. Because there are only two directions of spin, clockwise or spin-up and counterclockwise or spin-down, there are only two values for Pauli's electron spin quantum number. No two electrons in an atom can have identical quantum numbers. This is an example of a general principle that applies not only to electrons but also to other particles of half-integer spin (fermions). It does not apply to particles of integer spin (bosons). The two lowest-energy electron shells have an almost identical shape. Of the two, one shell is occupied or "filled" first with an electron that has an intrinsic magnetic direction that is opposite to the intrinsic magnetic field caused by the nucleus. The next shell has an electron with the opposite magnetic direction. The Dutch-American physicists Samuel Goudsmit and George Uhlenbeck discovered the intrinsic "spin" magnetism of the electron in the 1920s. It had been discovered years earlier that in a magnetic field, a beam of electrons splits into two beams (Stern-Gerlach experiment). This fourth quantum number explains this phenomenon. It is believed to be caused by some internal circulation of the electron matter, in addition to its wave flow around the equator of the shell. The wave flow around the equator of the atom also produces atomic orbital magnetic effects. Some shells have no net orbital circulation, which is explained as the result of two equal and opposite counter-rotating orbital waves. The magnetism of the nucleus itself is the result of the fundamental internal spin of the proton.

Surprisingly, it turned out that the fourth quantum number, spin, *s*, had only half of the usual quantization value of $h/2\pi$, i.e., $s = \pm 1/2\ h/2\pi$. Each quantum state, characterized by *n*, *l*, and *m*, is restricted to one electron of s = +1/2 and one electron of s = −1/2. Here classical analogies break down completely—a spin of a half implies the electron has to turn round twice to get back to where it started!

The implications of Pauli's exclusion principle are extremely profound. It is the restrictions imposed by Pauli's exclusion principle—i.e., the fact no two electrons can be in the same quantum state—that prevents all the electrons from piling up into the lowest (n = 1) energy state, and hence stops all matter from collapsing! Pauli's exclusion principle also implies that there is some sort of connectivity between the electron states in an atom: one electron must "know" which states all the other electrons are occupying to choose its own state! It is part of one of our most basic observations of nature: particles of half-integer spin must have antisymmetric wave functions, and particles of integer spin must have symmetric wave functions.

In Table 3.5 the different quantum numbers with their properties are summarized. The quantum numbers associated with the atomic electrons along with Pauli's exclusion principle provide insight into the building up of atomic structures and the periodic properties observed. For a given principal number *n*, there are $2n^2$ different possible states.

Table 3.5 Quantum Numbers and Their Properties

The principal quantum number:	
n	The principal quantum number. Quantization of angular momentum: must be a positive integer (1, 2, 3, 4 … etc.).
The angular momentum quantum number:	
l	Related to the ellipticity of the orbit: must again be an integer but for a particular orbit can be no bigger than n (l = 0, 1, 2, 3 … n − 1).
The magnetic quantum number:	
m	Quantization of the orientation of the orbit with respect to a magnetic field: can be a positive or negative integer (m = −l, −l + 1, 0, 1, 2, … l) but must be no larger than $-l \le m_l \le 1$.
The spin quantum number:	
s	The electron spin quantum number: must be +1/2 or −1/2.

n	Possible l	Possible m	Possible s	Spectroscopic Notation	Total States	Shell/Maximum Number of Electrons
1	0	0	±1/2	1s	2	K or 1st/2
2	0	0	±1/2	2s	2	L or 2nd/8
	1	−1,0,+1	±1/2	2p	6	
3	0	0	±1/2	3s	2	M or 3rd/10
	1	−1, 0, +1	±1/2	3p	6	
	2	−2, −1, 0, +1, +2	±1/2	3d	10	
4	0	0	±1/2	4s	2	N or 4th/32
	1	−1, 0, +1	±1/2	4p	6	
	2	−2, −1, 0, +1, +2	±1/2	4d	10	
	3	−3, −2, −1, 0, +1, +2, +3	±1/2	4f	14	

The Periodic Table of Dmitri Mendeleev (1834–1907)

Most of the qualities of an atom are derived from the structure of its electron cloud with the nucleus in the hinterland. This includes most chemical, material, optical, and electronic properties. Not only did the quantum model of the atom proposed by Bohr, Sommerfeld, and Pauli explain atomic spectra, it also explained the periodic table itself (inside front cover plate)! Dmitri Mendeleev (1834–1907) had devised the periodic table by grouping together elements with similar chemical properties.

The order of filling of electron energy states in an atom is dictated by energy, with the lowest available state consistent with Pauli's principle being the next to be filled. The labeling of the levels follows the scheme of the spectroscopic notation. For each value of principal quantum number, one refers to a different shell: n = 1 for the *K* shell, n = 2 for the *L* shell, n = 3 for the *M* shell, and n = 4 for the *N* shell. Different values of *l* correspond to different subshells. For example, for n = 2 (*K* shell) we have two subshells, namely, 2*s* and 2*p*. For historical reasons, the orbital quantum numbers *l* were given names associated with their appearance in spectroscopic emission and absorption patterns: l = 0 is given the letter *s* (for sharp), l = 1 is given the letter *p* (for principal), l = 2 is given the letter *d* (for diffuse), and l = 3 is given the letter *f* (for fundamental). Therefore, we can write the electronic occupation of an atom's shell as nl^e, where *n* is the principal quantum number, *l* is the appropriate letter, and *e* is the number of electrons in the orbit. So, for example, nitrogen (N) with seven electrons has the configuration: $1s^2 2s^2 2p^3$. In chemistry, this is called the Aufbau (build-up) principle, from the German word for *structure*.

As the periodic table of the elements is built up by adding the necessary electrons to match the atomic number, the electrons will take the lowest energy consistent with Pauli's exclusion principle. The maximum population of each shell is determined by the quantum numbers, and the diagram in Figure 3.40 is a convenient way to illustrate the order of filling of the electron energy states. For a single electron, the energy is determined by the principal quantum number *n*, and that quantum

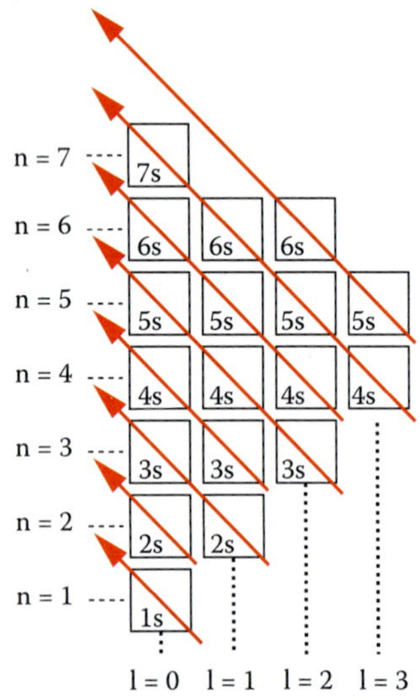

FIGURE 3.40 The periodic table of the elements is built up by adding the necessary electrons to match the atomic number.

number is used to indicate the "shell" in which the electrons reside. For a given shell in multielectron atoms, those electrons with lower orbital quantum number l will be lower in energy because of greater penetration of the shielding cloud of electrons in inner shells. These energy levels are specified by the principal and orbital quantum numbers using the spectroscopic notation. When you reach the 4*s* level, the dependence on orbital quantum number is so large that the 4*s* is lower than the 3*d*. Although there are minor exceptions, the level crossing follows the scheme indicated in the diagram, with the arrows indicating the points at which one moves to the next shell rather than proceeding to a higher orbital quantum number in the same shell.

The quantum scheme gives a firm scientific basis for Mendeleev's grouping: the chemical properties of the elements are defined by how nearly full or nearly empty a shell is. For example, full shells are associated with chemical stability (e.g., helium, neon, argon). Shells with a single electron or with one electron short of a filled shell are associated with chemical activity (e.g., sodium, potassium, chlorine, bromine). Most metals are formed from atoms with partially filled atomic orbitals, e.g., Na and Cu, which have the electronic structure:

$$\textbf{Na} \quad 1s^2 2s^2 2p^6 3s^1$$
$$\textbf{Cu} \quad 1s^2 2s^2 2p^6 3s^2 3d^{10} 4s^1$$

In the simplest picture, a metal has core electrons that are bound to the nuclei and valence electrons that can move through the metal. Insulators are formed from atoms with closed (totally filled) shells, e.g., inert gases:

$$\textbf{He} \quad 1s^2$$
$$\textbf{Ne} \quad 1s^2 2s^2 2p^6$$

or they form closed shells by covalent bonding as in the case of diamond.

With principal quantum number n = 4, the 4*f* orbitals make their capricious appearance. Because electrons in these orbitals are less strongly held by the nucleus, it is easier to excite them, and they can exhibit a myriad of distinct energy states. This gives rise to the unusual optical and magnetic properties of the rare-earth elements.

Most atoms with odd atomic numbers (1, 3, 5...) have a very slight overall atomic magnetism because of the one electron spin (and some orbital magnetism in some elements), whereas most even atomic number (2, 4, 6...) atoms have no net electron spin magnetism, and thus approximately zero resulting atomic magnetism.

Bohr's Correspondence Principle

Classical physics works for large systems; it is only at the atomic level that it fails. In 1923, Bohr proposed that any satisfactory quantum theory, then still being sought, should be in agreement with classical physics when the energy differences between quantized levels are very small or the quantum numbers are very large. This is the so-called Bohr correspondence principle. Let us consider the two main differences between the quantum theory and classical physics. The first difference is that whereas classical theory deals with continuously varying quantities, quantum theory deals with discontinuous or indivisible processes (e.g., the unit of energy packed in a quantum). The second difference is that whereas classical theory completely determines the relationship between variables at an earlier time and those at a later time, quantum laws determine only probabilities of future events in terms of given conditions in the past. The Bohr correspondence principle states that the laws of quantum physics must be so chosen that in the classical limit, where many quanta are involved (e.g., n is a large integer in E = nh), the quantum laws lead to the classical equations as an average. This requirement combined with indivisibility and incomplete determinism defines the quantum theory in an almost unique manner. In this chapter we will on a few occasions show how classical theory can be derived from an extension of quantum mechanics. An example of this principle that we may appreciate already is that of Newtonian mechanics as a special case of relativistic mechanics when the speed is much less than the speed of light ($v \ll c$) (see Chapter 5, Equation 5.180). Similarly, Schrödinger's equation agrees with Bohr's ideas in general, except that the electron wave function is not at a given radius but spread out over a range of radii.

Summary of Bohr's Model

Although Bohr's model, in which r (space), energy, and momentum are quantized, was a major step toward understanding the quantum theory of the atom, it is not in fact a correct description of the nature of electron orbits. Some of the shortcomings of the model are:

1. It fails to provide any understanding of why certain spectral lines are brighter than others.
2. There is no mechanism for the calculation of electron transition probabilities.
3. Bohr's model treats the electron as if it were a miniature planet, with definite radius and momentum. This is in direct violation of the uncertainty principle (see below), which dictates that position and momentum cannot be simultaneously determined.

The precise details of spectra and charge distribution must be left to quantum mechanical calculations. Wave functions derived from Schrödinger's equation, the model that corrects all of Bohr's errors, are determined by the values of the above reviewed quantum numbers. For each energy level E_n, there is more than one distinct state with the same energy but different quantum numbers; this is called *degeneracy*. Degeneracy has no counterpart in Bohr's model. The solution of Schrödinger's equation, introduced below, for the case of the hydrogen atom is achieved by using spherical polar coordinates and by separating the variables so that the wave function is represented by the product of three terms. The separation leads to three equations for the three spatial variables, and their solutions give rise to three quantum numbers associated with the hydrogen energy levels. The solution to these equations can exist only when a few constants, which arise in the solution, are restricted to integer values. This gives the same hydrogen atom quantum numbers: n, l, and m as introduced. The fourth quantum number is, of course, again the electron spin s = 1/2, an intrinsic property of electrons.

de Broglie Matter Waves

Matter Waves

To launch Schrödinger's equation, we revisit de Broglie's concept of "matter waves." Louis-Victor, seventh duc de Broglie (1892–1987), discovered that the secret of Planck's and Einstein's quanta could be found in a general law of nature, i.e., the dual character of waves and particles.

Einstein's special relativity* (Chapter 5) allows one to calculate the momentum, **p**, of a photon starting from:

$$E^2 = (\mathbf{p}c)^2 + (m_0c^2)^2 \tag{3.93}$$

The rest mass, m_0, is zero for a photon, so one obtains for its momentum:

$$E^2 = (\mathbf{p}c)^2 + 0 \text{ or } E = \mathbf{p}c \tag{3.94}$$

And also, from Einstein and Planck, E = hν (the photoelectric effect), or:

$$\mathbf{p} = \frac{E}{c} = \frac{h\nu}{c} = \frac{h}{\lambda} \tag{3.95}$$

de Broglie, while studying for his PhD in Paris in 1924, postulated that Equation 3.95 also applied to a moving particle such as an electron, in which case λ is the wavelength of the wave associated with the moving particle, i.e., a "matter wave." Einstein commented about this fantastic insight: "de Broglie has lifted the great veil." From Equation 3.95, **p**λ = h,

* For p = 0 Equation 3.93 leads to the familiar $E = m_0c^2$. This amounts to 511 keV for an electron and 930 GeV for a proton. Where atomic physics deals in eV to keV, nuclear physics deals in keV to MeV and particle physics involves GeV to TeV.

or momentum multiplied by wavelength, gives the Planck constant; the smaller the wavelength, the bigger the momentum of the particle! Thus, electrons, with their small mass and correspondingly small momentum, are very "wavy" particles. The de Broglie wavelength of a particle is then given as:

$$\lambda = \frac{h}{mv} \tag{3.96}$$

where v is its velocity. This simple equation proves to be one of the most useful, and famous, equations in quantum mechanics. It accounts for both waves (characterized by wavelength and frequency) and particles (characterized by position, mass, and velocity), incorporating momentum (particle aspect) and the wavelength (wave aspect).

The critical reader could very legitimately ask here: "But how can photons have a momentum if they do not have any mass?" The answer is that photons do not have a mass, but they do have energy, and as Einstein famously proved, mass and energy are the same thing. For photons, the wavelength and frequency, from Equation 3.95, are related because $\lambda v = c$ is fixed, but for matter particles λv is not a constant because they do not travel at speed *c*. For nonzero rest mass particles, such as electrons or protons, neither $\lambda v = c$ nor $E = \mathbf{p}c$ applies.

We almost have come full circle here; Einstein gave what we had come to think of as a wave (light) a particle character, and de Broglie gave what we thought of as a particle (electrons) a wave character. Radiation has wave character and particle character, and matter has particle and wave character, or at the nanoscale, nature presents itself with a wave-particle duality. This duality for material things is only significant at the very small scale. With its mass m_e of 9.11×10^{-31} kg and a velocity v of 3×10^6 m/s (i.e., c/100), an electron has a wavelength of 2.5×10^{-10} m. Larger objects, such as a car (1000 kg), moving at highway speeds (60 mph ≈ 100 km/h) have a de Broglie wavelength as well, but there is no measurement capable of resolving them; the oscillations cannot be seen. In other words, the car does not behave like a wave, with a $\lambda = \frac{h}{p} = \frac{h}{mv} = \frac{6.6 \times 10^{-34}\,\mathrm{J{\cdot}s}}{1000\ \mathrm{kg} \times 10^5\ \mathrm{m/h} \times 1\mathrm{h}/36{,}000\,\mathrm{s}} = 2.4 \times 10^{-38}\,\mathrm{m}$. A man (60 kg) running at 20 km/h (5 m/s) has a wavelength of 2.2×10^{-36} m (for some other matter wave examples see Table 3.6).

Wave Packets

Because we associate a de Broglie wave with a moving body, it is reasonable to expect that this wave moves at the same velocity as that of the particle—this turns out not to be the case. de Broglie hypothesized that the particle itself was not a wave but always had with it a pilot wave, or a wave that helps guide the particle through space and time. He also pointed out that in wave theory there is a difference between the speed of waves of some uniquely defined frequency and the speed of a localized pulse. Waves have both a phase velocity and a group velocity, and de Broglie proposed to associate the localized pulse, wave packet, or group velocity with the moving particle.

To understand *matter* waves and their link with wave packets or group velocity and phase velocity better, we reconsider here for a moment a sinusoidal wave function $\psi(x) = Ae^{ikx} = A\cos kx + iA\sin kx$

Table 3.6 de Broglie Wavelengths of Various Moving Objects

	Mass	Kinetic Energy (eV)	de Broglie Wavelength (m)
Electron	9.11×10^{-31}	1	1.23×10^{-9}
		100	1.23×10^{-10}
		10^4	1.23×10^{-11}
Neutron	1.67×10^{-27}	1.00×10^3	9.05×10^{-13}
		1.00×10^6	2.86×10^{-14}
		1.00×10^9	9.05×10^{-16}
Proton	1.67×10^{-27}	1.00	2.86×10^{-11}
		1.00×10^2	2.86×10^{-12}
		1.00×10^3	9.04×10^{-13}
Thermal neutrons (300 K)		2.50×10^{-2}	1.81×10^{-10}

that extends all the way from $x = -\infty$ to $x = +\infty$ with an amplitude A. To make this wave more localized or "lumpy," we add together two different stationary-state waves with slightly different wave vectors $\mathbf{k}$ (e.g., $\mathbf{k}_1$ and $\mathbf{k}_2$) and different amplitudes A (e.g., A_1 and A_2) and consider the result at one instance of time, e.g., at $t = 0$. For $t = 0$, the time factor of the superposed waves, $e^{-i\omega t}$ (e.g., ω_1 and ω_2), is equal for both superposed waves, i.e., $e^0 = 1$, so the wave function is represented by:

$$\begin{aligned}\psi(x) &= A_1 e^{ik_1x} + A_2 e^{ik_2x} \\ &= [A_1\cos(k_1x) + iA_1\sin(k_1x)] \\ &\quad + [A_2\cos(k_2x) + iA_2\sin(k_2x)] \\ &= [A_1\cos(k_1x) + iA_2\cos(k_2x)] \\ &\quad + i[A_1\sin(k_1x) + A_2\sin(k_2x)]\end{aligned} \tag{3.97}$$

This is represented in Figure 3.41a, where we graph the real parts of the individual waves for the case that $A_2 = -A_1$. The real part of the combined wave function is shown in Figure 3.41b. With two additional sinusoidal waves, we can choose to emphasize every other lump and suppress the in-between ones as illustrated. Going on like this, we can, in the extreme, superpose a very large number of waves with different $\mathbf{k}$ and A values to construct a wave with only one lump, as illustrated in Figure 3.41c. The latter wave pulse is called a *wave packet*. The particle represented by the lump in Figure 3.41c is more likely to be found in the region of the lump than in other regions; in other words, it has become more localized. However, the particle now does have a less definite momentum as we introduced a very large number of different waves. The above procedure of adding waves is not new to us: in Chapter 2 on Fourier transforms (Equations 2.15 and 2.16) we saw how a localized disturbance or pulse can be synthesized from a very large number of pure sinusoidal waves of different frequencies and amplitudes. Such a pulse, a wave function with a lump, is an entity with both particle and wave characters. It behaves like a particle in the sense that it is localized in space, but it also has periodicity, a property of a wave. The Fourier integral representing the synthesized pulse or wave packet is:

$$\psi(x) = \int_{-\infty}^{\infty} A(\mathbf{k})e^{ikx}d\mathbf{k} \tag{3.98}$$

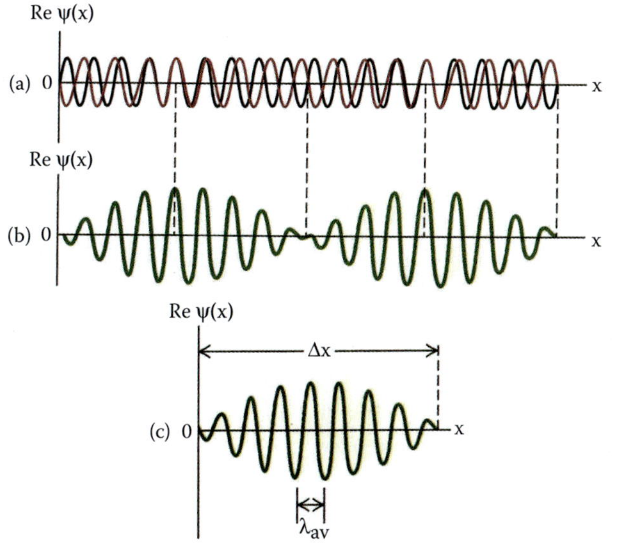

FIGURE 3.41 (a) Two sinusoidal waves with slightly different wave numbers at one instant of time ($A_2 = -A_1$) in Equation 3.97. (b) The superposition of these two waves has a wave number equal to the average of the two individual wave numbers. The amplitude varies, giving the total wave a more lumpy character. (c) Superimposing a large number of sinusoidal waves with different wave numbers and appropriate amplitudes can produce a wave pulse.

This Fourier integral superimposes wave functions with a continuous distribution of $\mathbf{k}$ values and amplitudes $A(\mathbf{k})$ that depend on $\mathbf{k}$. Below we will see how Heisenberg's uncertainty principle (HUP) follows directly from the existence of the de Broglie waves and this Fourier integral (see Figure 3.47 further below).

The synthesized pulse travels with a group velocity (V_g) that may be different from the characteristic phase velocity (V_p) of the individual waves. Let us consider the Fourier superposition of just two waves, $\Psi_1 = \sin[kx - \omega t]$ and $\Psi_2 = \sin[(k + \Delta k)x - (\omega + \Delta\omega)t]$, as represented in Figure 3.42a, where we can write:

$$\Psi_1 + \Psi_2 = \underbrace{2\cos\left[\frac{\Delta\omega}{2}t - \frac{\Delta k}{2}x\right]}_{\text{Modulation}} \times \underbrace{\sin\left[\left(k + \frac{\Delta k}{2}\right)x - \left(\omega + \frac{\Delta\omega}{2}\right)t\right]}_{\text{Sine wave}} \tag{3.99}$$

The phase velocity is directly proportional to the angular frequency, but inversely proportional to the wave number, or the phase velocity v_p is given as:

$$v_p = \frac{\omega + \frac{\Delta\omega}{2}}{\mathbf{k} + \frac{\Delta\mathbf{k}}{2}} = \frac{\omega'}{\mathbf{k}'} \quad (3.100)$$

from the sine wave in Equation 3.99. For the group velocity v_g and thus the particle velocity:

$$v_g = \frac{\frac{\Delta\omega}{2}}{\frac{\Delta k}{2}} = \frac{\Delta\omega}{\Delta k} = \frac{d\omega}{dk} \quad (3.101)$$

which is based on the modulation amplitude in Equation 3.99. This group velocity, the derivative of the angular frequency with respect to the wave vector, is the velocity at which the energy of the wave propagates. Group and phase velocities are illustrated in Figure 3.42b.

The group velocity is the derivative of the phase velocity, and it is often the case that the phase velocity is larger than the group velocity. For any wave that is not electromagnetic, the phase velocity will be larger than c, or the speed of light.

de Broglie was trying to find a velocity that was fit for all particles (not just photons), so he associated the group velocity v_g with any particle's velocity. He equated the following two expressions:

$$v_g = \frac{\partial\omega}{\partial\mathbf{k}} = \frac{\partial E}{\partial\mathbf{p}} \quad (3.102)$$

because $v_p = E/\mathbf{p}$ and $v_g = \partial E/\partial\mathbf{p}$ and (as we will see below) $E = \hbar\omega$ and $p = \hbar\mathbf{k}$, where $\hbar = h/2\pi$. In the case of light:

$$v_p = \frac{E}{\mathbf{p}} = \frac{mc^2}{mc} = c \text{ and } v_g = \frac{\partial E}{\partial p} = \frac{\partial pc}{\partial p} = c \quad (3.103)$$

We get the same result for phase and group velocity. For any other particle, from Equation 3.93 with $m_0 = (m \neq 0)$, we obtain:

$$v_g = \frac{\partial E}{\partial p} = \frac{pc^2}{\sqrt{(pc)^2 + (mc^2)^2}} = \frac{pc^2}{E} \quad (3.104)$$

with $m = m_0$, $v_g = c$ again.

de Broglie Vindicated

de Broglie's doctoral dissertation exam committee was distinctly unsure as what to do with the thesis of

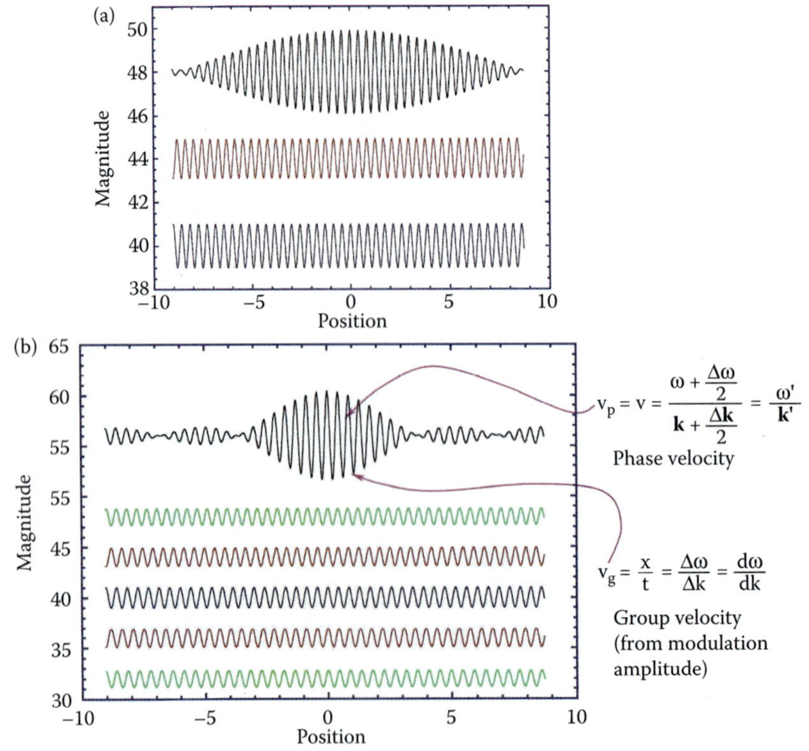

FIGURE 3.42 (a) Using Fourier's theorem to construct wave packets. (b) Phase velocity v_p from the sine wave and group velocity v_g from the modulation amplitude.

the prince. Luckily, Langevin, who was on the committee, asked Einstein's opinion on the work, and, as mentioned above, Einstein was duly impressed. Experimental evidence was soon to follow.

J.J. Thompson received the Nobel Prize in 1906 for his work demonstrating the particle nature of an electron. J.J. Thompson's son, G.P. Thompson, received the Nobel Prize in 1937 for experiments that tested Equation 3.96, demonstrating that electrons also exhibit wave-like properties. In his 1928 tests, G.P. Thompson and Reid at the University of Aberdeen observed interference patterns from electrons reflecting from a thin polycrystalline metal foil surface. Working independently, Clinton Davisson and Lester Germer, at Bell Laboratories, in 1927 found the same experimental evidence: a beam of electrons scattered from a single-crystal of nickel resulted in a diffraction pattern fitting Bragg's diffraction law (Equation 2.20 and Figure 3.43). This established the wave character of electrons, forming the basis of analytical techniques for determining the structures of molecules, solids, and surfaces, such as in low energy electron diffraction (LEED) and SEM described in Volume III, Chapter 6 on metrology. Whereas optical microscopes cannot resolve details smaller than the wavelength of visible light, with an SEM the wavelength depends on the electron's momentum in accordance with de Broglie's relation. In this case we can increase the resolution by simply accelerating the electron more.

Other particles, including protons, neutrons, and He atoms, were subsequently found to possess wave properties, including diffraction. In 1999, researchers from the University of Vienna demonstrated that the wave-particle duality even applied to molecules such as fullerene. One of the coauthors of this research, Julian Voss-Andreae, became an artist and has since created several sculptures symbolizing wave-particle duality in buckminsterfullerenes.[2]

Bohr's Model and de Broglie Matter Waves

de Broglie's idea of matter waves fits nicely with Bohr's model; the de Broglie wavelength associated with the electron is what causes the quantization that Bohr had assumed before matter waves were even known. The angular momentum of the electron is restricted to certain values because an integral number of electron wavelengths must fit into the circumference of the circular orbit. Mathematically we expressed Bohr's Postulate 4, in vector notation, as $|\mathbf{L}| = |\mathbf{p} \times \mathbf{r}| = m_e vr$, with v the tangential velocity and $\mathbf{p}$ the momentum of the electron. Bohr's model, as we saw above, assumed $\mathbf{L} = 2\pi r$, or:

$$\mathbf{L} = 2\pi r = n\lambda = \frac{nh}{\mathbf{p}} \text{ and } r = \frac{\mathbf{n}\hbar}{p}\left(\text{with } \hbar = \frac{h}{2\pi}\right) \tag{3.105}$$

or

$$L = \mathbf{p}r = \frac{nh}{2\pi} = n\hbar$$

The latter is illustrated in Figure 3.44, where we show that only an integer number of wavelengths fits into a circular orbit or, i.e., $L = nh/2\pi$ (where L

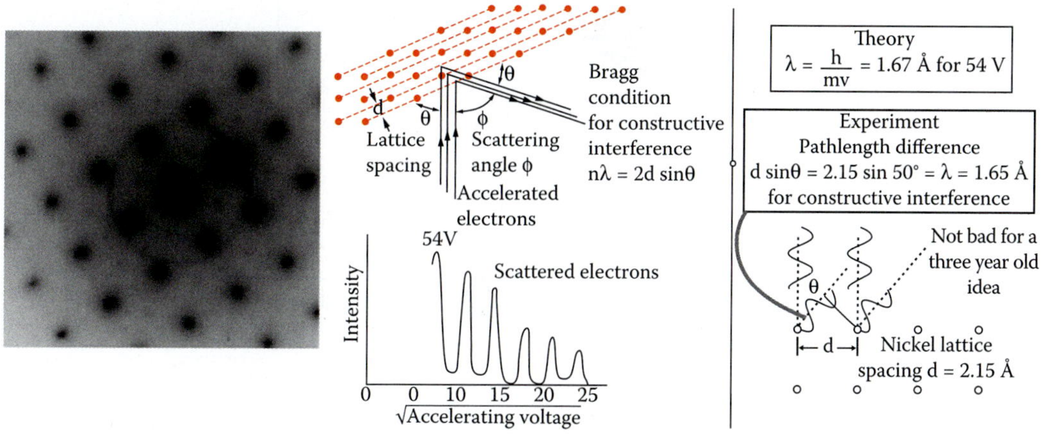

FIGURE 3.43 Davisson-Germer experiment (1927). A graph of the intensity of the scattered electron beam demonstrates that the angular position of the maxima depended on the accelerating voltage used to produce the electron beam. Using the de Broglie equation, the de Broglie wavelength of the electron could be calculated.

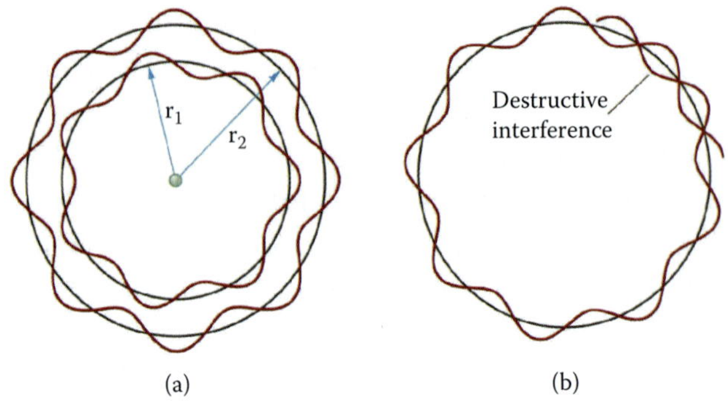

FIGURE 3.44 Bohr's atom model: only an integer number of wavelengths fits into a circular orbit, $L = nh/2\pi$ (where *L* is the length of an orbit and *h* is the Planck constant).

is the length of an orbit). Thus, a wave-mechanical picture leads naturally to the quantization of the angular momentum.

de Broglie's relations imply the quantization of orbits of electrons in atoms, $L = n\hbar$ with *n* an integer, but although this explains the root of quantization—the wave nature of matter—it does not provide a complete dynamic that would be able to describe the motion of particles that are waves at the same time. One needs a wave equation for that, and Maxwell's equations are inappropriate for this purpose because they predict that $\lambda\nu = c$, which is not valid for matter particles, where $\nu\lambda = v$. Interestingly, we will see next that Bohr's model violates Heisenberg's uncertainty principle: an electron may not move in a circle with exact radius *r*!

Heisenberg's Uncertainty Principle

The wave-particle duality introduced in the previous section forced physicists to reconsider their description of the position and momentum of very small particles, and it is at the core of Heisenberg's uncertainty principle (HUP). In Newtonian mechanics, we can always describe a particle in terms of its spatial coordinates and the three components of its velocity. However, in the nanoworld, Heisenberg's principle states that there are physical parameters in quantum physics whose values cannot be known accurately simultaneously. For example, the momentum, $\mathbf{p}_x$, and position, *x*, of an electron cannot be known simultaneously. In classical physics, knowing $\mathbf{p}_x$ would not be difficult because light is considered to have continuously varying energies and would hence cause minimal disturbance on an electron that is being observed. In quantum physics, light consists of photons that have discrete (quantized) energies, and a photon bouncing off an electron being observed gives that electron a kick, disturbing its momentum (Figure 3.45). To obtain an accurate measurement of the position of the electron, one must use a very short probing wavelength because a long wavelength would not have enough resolution to locate the electron. Thus, a short wavelength is expected to give good resolution or a small uncertainty in position. Nevertheless, a short wavelength

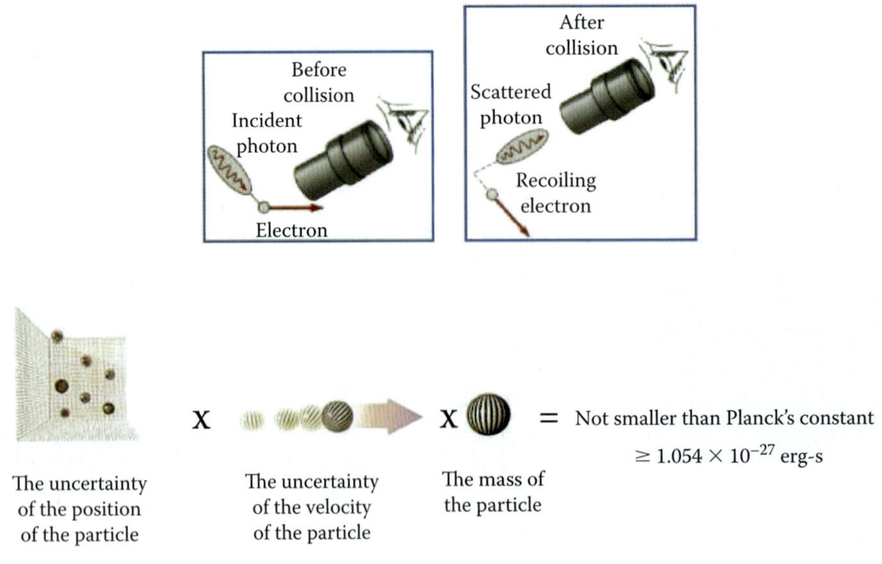

FIGURE 3.45 Looking is disturbing! Imagine trying to see an electron with a powerful optical microscope. At least one photon must scatter off the electron and enter the microscope, but in doing so it will transfer some of its momentum to the electron.

also means a large energy and hence a large momentum ($\mathbf{p}_x = E/c$), which gives the observed electron too large a kick and too large an uncertainty about its momentum $\Delta\mathbf{p}_x$ (Compton effect).

A more quantitative analysis shows that the product of the two uncertainties, Δx and $\Delta\mathbf{p}_x$, is a constant:

$$\Delta\mathbf{p}_x \Delta x \geq \frac{h}{2\pi} = \hbar \quad (3.106)$$

where h is again the Planck constant, and the constant "h-bar" has the approximate value of 10^{-34} J·s. There is an uncertainty relation as shown in Equation 3.106 for each coordinate and its corresponding momentum component. Equation 3.106 is graphically represented in Figure 3.46. Note that there is no restriction on the precision in simultaneously knowing/measuring the position along a given direction (x) and the momentum along another, perpendicular direction (y or z).

For a particle in circular motion, the equivalent Heisenberg expression to Equation 3.106 is $\Delta\mathbf{p}_r\Delta\mathbf{r} \geq \hbar$. Applying this to Bohr's model, in which an electron moves in a circle of *exact* radius r, we obtain $\Delta r = 0$ and $\Delta P_r = \infty$, or the model violates the uncertainty principle. We will see further below that despite this obvious problem, the energy level predictions of Bohr's model remain valid.

Similar ideas lead to the expression of uncertainty for other pairs of observables such as time and energy:

$$\Delta E \Delta t \geq \frac{h}{2\pi} = \hbar \quad (3.107)$$

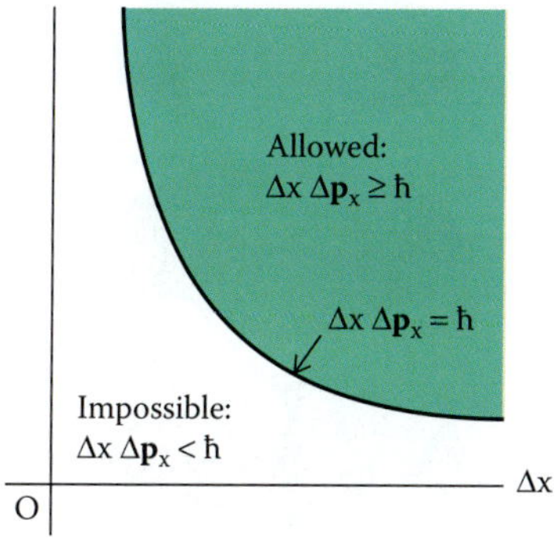

FIGURE 3.46 A graphic presentation of the Heisenberg principle from Equation 3.106 for position and momentum.

This says that if an energy state only lasts for a limited time, its energy will be uncertain. The uncertainty about the energy of a particle depends on the time interval Δt that the system remains in a given energy state. Importantly, this also means that conservation of energy can be violated if the time is short enough. From the uncertainty principles, it is possible that empty space locally does not have zero energy but may have sufficient ΔE for a very short time Δt to create particles and their antiparticles. This can be demonstrated through the Casimir effect (see below). Equation 3.107 is responsible for "lifetime broadening" of spectral lines. Short-lived excited states (small Δt) possess large uncertainty in the energy of the state (large ΔE). As a consequence, shorter laser pulses (e.g., femto- and attosecond lasers) have broader energy (therefore, wavelength) bandwidths.

The existence of a zero-point energy—vibrational energy cannot be zero even at T = 0 K (see below)—is also a consequence of Heisenberg's uncertainty principle. If the vibration would cease at T = 0 K, then the position and momentum would both be 0, violating the HUP.

One might object that perhaps light was a poor choice to measure the position and momentum of the electron and that some other method might avoid these uncertainties. No such luck: it turns out that this is the absolute best that can be achieved independently of the measuring method. Note that in the last two equations all references to light have dropped out; the result does not depend on λ, n, or c. The Heisenberg uncertainty principle follows solely from the wave-particle duality and has nothing to do with the unavoidable disturbance of the system by the measurement. Quantum mechanics tells us there are limits to measurement—not because of the limits of our instruments, but inherently.

Actually, the uncertainty principle is an inevitable consequence of de Broglie's relation and the Fourier integral representing the synthesized pulse or wave packet (Equation 3.98). In Figure 3.47 we illustrate qualitatively how $\psi(x)$ depends on $A(\mathbf{k})$. In Figure 3.47a, we show a sharp peak in $A(\mathbf{k})$ for a narrow range of wave numbers $\mathbf{k}$. The resulting real part of the wave pulse, shown in Figure 3.47b, is relatively broad: a narrow range of $\mathbf{k}$ means a narrow range of $\mathbf{p}_x = \hbar\mathbf{k}$ and thus a small $\Delta\mathbf{p}_x$, and the

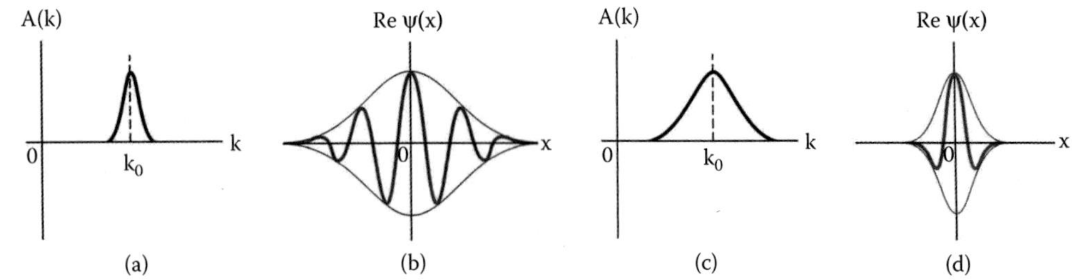

FIGURE 3.47 (a) A sharp peaked $A(\mathbf{k})$ function leads to a wave function $\psi(x)$ with a broad spatial extent (Δx) (b); (c) a broad peaked $A(\mathbf{k})$ function leads to a wave function $\psi(x)$ with a narrow spatial extent (Δx) (d).

result is a relatively large Δx. Broadening the $A(\mathbf{k})$ function in Figure 3.47c results in a more localized wave pulse with a smaller Δx, as clear from Figure 3.47d. In other words, this is Heisenberg's uncertainty principle, $\Delta \mathbf{p}_x \Delta x \geq \hbar$, in action.

Heisenberg's relations are of no practical importance in the macroworld, and in classical physics they can be ignored completely.

The Launchpad: Classical Mechanics Revisited

By 1924, the quantum concepts of Planck, Einstein, Bohr, and de Broglie were widely accepted, and between 1925 and 1927 three distinct, independent, and very different integrating theories of quantum theory were proposed: Dirac's Hamiltonian and quantum algebra representation; the matrix representation of Born, Heisenberg, and Jordan; and Schrödinger's wave equations. Much later, in the late 1940s, Feynman formulated his sum-over-histories approach.

Schrödinger, after attending a seminar on Einstein's and de Broglie's ideas that wave-like entities can behave like particles, and vice versa, thought that there must be a wave equation, $\Psi(x,t)$, to describe particles. Schrödinger almost completely dispensed with the concept of a particle and instead focused on the wave-like properties of matter. His picture of the atom has the electron standing waves vibrating in their orbitals much like the vibrations on a string—but in three dimensions instead of one. In Figure 3.48, a 2D representation of Schrödinger waves, like vibrations on a drum skin, is shown.

A concept that plays an important role in both classical and quantum theory is that of the Hamiltonian of a system. Consider an isolated system composed of one or more particles, and assume that the total energy of this system remains constant. The Hamiltonian of this system is merely its total energy:

$$H = E_n(\text{total energy}) = \text{KE (kinetic energy, depends on v)} + \text{PE (potential energy, depends on position)} \quad (3.84)$$

with the kinetic energy arising from motion and the potential energy arising from the position in a force field, *F*. The Hamiltonian in quantum mechanics is an example of an operator, a mathematical object that tells us what operation to perform on the function that follows. We rewrite Equation 3.84 here as:

$$H = E = E_k + V(x) \quad (3.108)$$

The linear momentum $\mathbf{p}$ equals mv so that $E_k = mv^2/2$ or $\mathbf{p}^2/2m$. When working with the Hamiltonian, the kinetic energy of a particle is expressed as $\mathbf{p}^2/2m$, not as $mv^2/2$. Thus, Equation 3.84 may be rewritten as:

$$H = E = \frac{\mathbf{p}^2}{2m} + V(x) \quad (3.109)$$

In case the potential is time varying, the last term in Equation 3.109 must be written out as $V(x,t)$. The total energy, E, in the absence of a potential energy

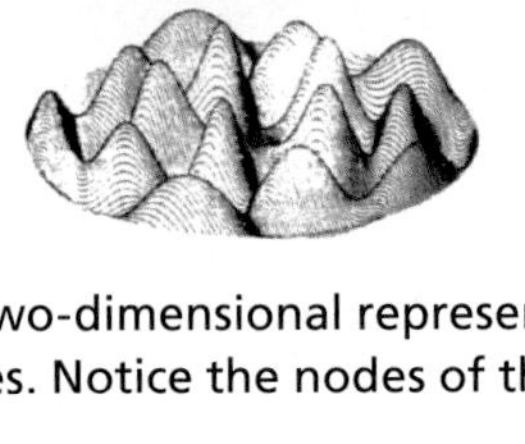

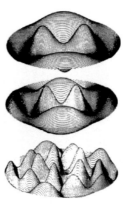

FIGURE 3.48 A two-dimensional representation of Schrödinger waves. Notice the nodes of the vibration.

equals E_k. Newton's second law relates potential energy to change in momentum. According to this law of mechanics, forces give rise to a change in the momentum of particles [(F = d(mv)/dt = dp/dt], and each force, *F*, has an associated potential energy *V*(*x*) (F = −dV/dx). The direction of the force is toward decreasing potential energy:

$$F = \frac{dp}{dt} = ma = m\frac{d^2x}{dt^2}$$
$$F = -\frac{dV(x)}{dx} \tag{3.110}$$

or:

$$F = -\frac{dV(x)}{dx} = m\frac{d^2x}{dt^2}$$

i.e., given *V*(*x*) one can solve for *x*(*t*) or *v*(*x*,*t*). We illustrate the use of the above deterministic expressions with two important examples so we might better appreciate the analogies and the differences between classical and quantum physics.

Example 3.1: Free particle [no force, V(x) = 0] moving along *x* (1D problem):

$$E = E_k = \frac{mv^2}{2} = \frac{p^2}{2m} \tag{3.111}$$

v = dx/dt, and:

$$v = \frac{dx}{dt} = \sqrt{\frac{2E_k}{m}} \tag{3.112}$$

Therefore, we can formulate the differential equation:

$$dx = \sqrt{\frac{2E_k}{m}}dt \tag{3.113}$$

Integration from x_0 to x_t and from t = 0 to t = t and using Equation 3.112 results in:

$$x_t - x_0 = \sqrt{\frac{2E_k}{m}}t = \sqrt{\frac{2p^2}{2m^2}}t = \frac{p}{m}t = vt \tag{3.114}$$

With V = 0, knowing x_0 and *p*, we can predict x_t at *t*; in other words, we can predict the trajectory at all times later.

Example 3.2: Harmonic oscillators [F = −kx with *x* the displacement]. The pendulum, which is just a mass on a spring, is an example of a harmonic oscillator, but its physical description also encompasses many objects that oscillate, from tuning forks to oscillating bridges and oscillating skyscrapers (Figure 3.49).

Let us first calculate *V*(*x*) in Equation 3.109 for the given problem. The force F = −dV(x)/dx = −kx (Hooke's law), with *k* the force or spring constant, or:

$$dV(x) = kxdx \tag{3.115}$$

Integrating, we obtain:

$$V - 0 = k\frac{1}{2}(x^2 - 0^2) \text{ since } V = 0 \text{ at } x = 0$$

or

$$V = \frac{kx^2}{2} \tag{3.116}$$

The force F = ma = −kx, and we obtain the differential equation:

$$\frac{d^2x}{dt^2} = -\frac{k}{m}x \tag{3.117}$$

with a solution:

$$x(t) = A\sin\left(\sqrt{\frac{k}{m}}t\right) \tag{3.118}$$

where *A* is the maximum displacement (x_{max} = A). For a simple harmonic oscillator ω = 2πf = 2π/T and:

$$\omega = \sqrt{\frac{k}{m}} \text{ and } T = 2\pi\sqrt{\frac{m}{k}} \tag{3.119}$$

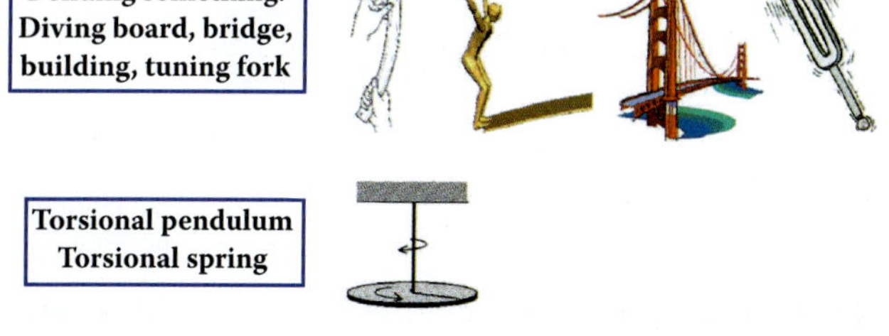

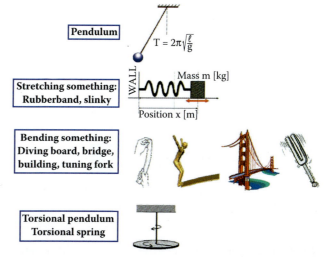

FIGURE 3.49 A wide variety of harmonic oscillator examples.

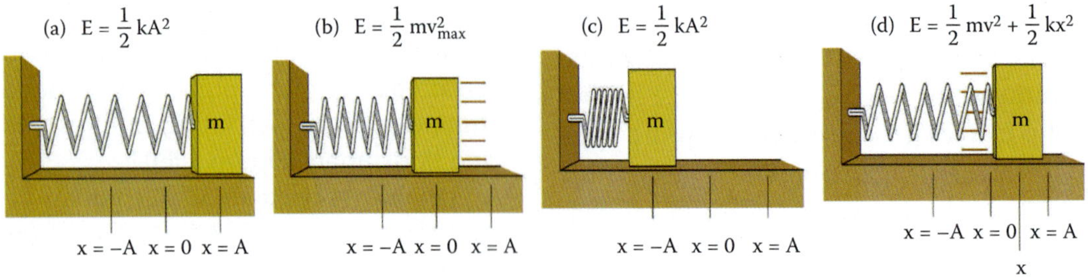

FIGURE 3.50 Spring/mass system with values for kinetic and potential energy as a function of the position of the mass on the *x*-axis. (a) x = A; (b) x = 0; (c) x = −A; and (d) x = x.

Observe that frequency only depends on characteristics of the system (*m*,*k*) and not on the amplitude *A*. The first and second derivatives of Equation 3.118 are given as:

$$v(t) = \frac{dx}{dt} = A\sqrt{\frac{k}{m}}\cos\left(\sqrt{\frac{k}{m}}t\right)$$

and

$$a(t) = \frac{d^2x}{dt^2} = -A\frac{k}{m}\sin\left(\sqrt{\frac{k}{m}}t\right) = -\frac{k}{m}x(t) \qquad (3.120)$$

with $v_{max} = A\omega$, and $a_{max} = A\omega^2$. For the momentum, p(t) = mv, we obtain:

$$p(t) = mv = m\frac{dx}{dt} = mA\sqrt{\frac{k}{m}}\cos\left(\sqrt{\frac{k}{m}}t\right) \qquad (3.121)$$

We can now solve for *E* exactly:

$$E = \frac{p^2}{2m} + \frac{kx^2}{2} \text{ or also } E = \frac{mv^2}{2} + \frac{kx^2}{2}$$

$$= \frac{\left[mA\sqrt{\frac{k}{m}}\cos\left(\sqrt{\frac{k}{m}}t\right)\right]^2}{2m} + \frac{k\left[A\sin\sqrt{\frac{k}{m}}t\right]^2}{2} \qquad (3.122)$$

Thus, the total mechanical energy of a simple oscillator is proportional to the square of the amplitude. As the amplitude (*A*) can take any value, this means that the energy (*E*) can also take any value—i.e., energy is continuous. Any energy value is allowed by simply changing the force constant *k*. At x = −A and x = A, $E = kA^2/2$, and at x = 0, $E = mv^2_{max}/2$ (see Figure 3.50).

In Figure 3.51 we show a typical nonquantized oscillator with a parabolic curve for the potential energy as a function of position $x-x_0$ of the mass. The potential sketched here is very important because it describes the potential for many systems, including vibrational and electronic states in molecules. It approximates the potential in many more systems for small departures from equilibrium. As a typical application, in Figure 3.52, we illustrate the potential energy for two hydrogen atoms approaching each other. Clearly over a considerable range of energies, a parabolic curve, typifying a simple harmonic oscillator, can represent the real situation. The horizontal lines cutting through the parabolic part of the curve in Figure 3.52 represent

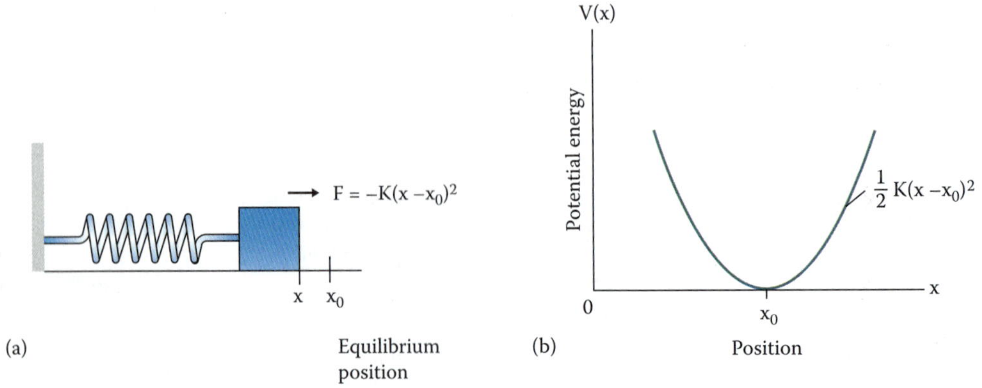

FIGURE 3.51 Typical nonquantized oscillator with a parabolic curve for the potential energy as a function of position *x* of the mass.

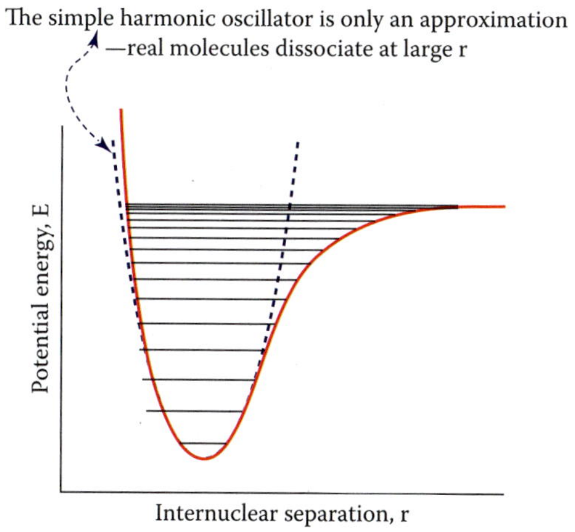

FIGURE 3.52 Application of simple oscillator approximation: two hydrogen atoms in a hydrogen molecule. For an explanation of the horizontal lines in this figure, see quantized oscillator below.

the quantization of the vibration states as introduced next.

The above examples demonstrate the essence of determinism in classical physics, i.e., given $V(x)$, one can solve for $x(t)$ or $v(x,t)$, or also at any time (t), the position $[x(t)]$ and velocity $[v(t)]$ can be determined exactly—i.e., the particle trajectory can be specified precisely. In other words, given the force, the motion can be found. In the late eighteenth century the mathematician Pierre Simon de Laplace (1749–1827) encapsulated classical determinism as follows: "…if at one time we knew the positions and motion of all the particles in the Universe, then we could calculate their behavior at any other time, in the past or the future." In classical physics, particles and trajectories are real entities, and it is assumed that the universe exists independently from the observer, that it is predictable, and that for every effect there is a cause so experiments are reproducible. Heisenberg's uncertainty principle destroyed all this. In quantum physics, measured and unmeasured particles are described differently. The measured particle has definite attributes such as position and momentum, but the unmeasured particle does not have one but all possible attribute values, as Nick Herbert describes it in his book *Quantum Reality* … somewhat like a broken television that displays all its channels at the same time.[3]

We shall see that these ideas of classical mechanics fail when we go to the atomic regime (where E and m are very small). Classical mechanics also fails when velocity is very large (as $v \to c$) because of relativistic effects.

Schrödinger's Equation

Plausibility of Schrödinger's Equation

From the evidence presented above, an atomic or subatomic particle cannot be described anymore as a simple Newtonian point, and *matter waves* must be taken into account. For matter waves, from $p = h/\lambda$ (Equation 3.95), the smaller the wavelength, the bigger the momentum, and as we calculated, electrons, for example, are very "wavy" particles. The book-keeping term for the energy of a system or the Hamiltonian, H, for a Newtonian particle with mass m is:

$$H = E = \frac{\mathbf{p}^2}{2m} + V(x) \qquad (3.109)$$

An equivalent to Newton's equation is needed to calculate how forces (described by a potential energy) affect λ or **p** in the case of a "waving" particle. Once that differential wave equation is found, we may solve it for a wave, which has an amplitude for each value of position (x) and time (t).

Erwin Schrödinger (1887–1961), in 1926, encouraged by Debye, who remarked that there should be a wave equation to describe the de Broglie waves, proposed a wave equation that can be applied to any physical system in which it is possible to describe the energy mathematically. In one dimension he postulated:

$$\frac{\partial^2\Psi(x,t)}{\partial x^2} + \frac{8\pi^2 m}{h^2}[E - V(x,t)]\Psi(x,t) = 0 \qquad (3.123)$$

with $\Psi(x,t)$ the wave function, a wave representing the spatial distribution of a "particle," and m the characteristic mass of the particle. The first term on the left is the rate of change of the wave function with distance x. The energy of the particle is E, and the potential energy function to describe the forces acting on the particle is represented by $V(x,t)$. Schrödinger's equation has the same central role in quantum mechanics that Newton's laws have in mechanics and Maxwell's equations have in electromagnetism. Solutions to Newton's equations are of the form v = f(x,t) (see Examples 3.1 and 3.2 above), whereas solutions to the wave equation (Equation 3.118) are called wave functions $\Psi(x,t)$. Schrödinger's equation is more difficult to solve than Newton's equation, but it is just as well-defined, and like Newton's equation, it describes the relation between kinetic energy, potential energy, and total energy. If one knows the forces involved, one can calculate the potential energy *V* and solve the equation to find Ψ. Solving Schrödinger's equation specifies $\Psi(x,t)$ completely, except for a constant; if $\Psi(x,t)$ is a solution, then $A\Psi(x,t)$ is a solution as well. Remember that Equation 3.123, like Newton's law, cannot be derived—it is a plausibility argument. Einstein called Ψ a *Gespensterfeld* or ghost field. Because it carries no energy, the wave function is also referred to as an empty wave. In France, the Ψ wave is sometimes called *densité de présence*, or *presence density*.

Here is how Schrödinger, using a rather sophisticated analogy with classical mechanics, came to derive Equation 3.123. He assumed a sinusoid wave of wavelength λ and frequency ν and hence a velocity v = λν. The equation of such a traveling wave, as we saw in Chapter 2, is $\Psi(x,t) = A\cos(\mathbf{k}_x x - \omega t) + B\sin(\mathbf{k}_x x - \omega t)$, where $\mathbf{k}_x$ (the wave number) = 2π/λ, and the period is T = 2π/ω = 1/ν. Rewriting this expression in terms of the complex variable form, with B = iA and using Euler's formula, one obtains as the simplest form of a wave:

$$\Psi(x,t) = Ae^{i(k_x x - \omega t)} \tag{3.124}$$

This is classical so far; indeed, we will see in the section "Solution of the Wave Equation for Free Particles", this chapter, and in Chapter 5 that Equation 3.124 is also a solution of the Maxwell equations.

Now apply Einstein's photon formula, i.e., E = hν, to the particle, with ν the frequency of the "waving" of the particle—de Broglie's idea! Because E = hν = (h/2π)(2πν) = ℏω, we find ω = E/ℏ, where we have introduced the definition of "h-bar," ℏ = h/2π. From de Broglie, p = h/λ and p = (2π/λ)(h/2π) = $\mathbf{k}_x\hbar$ and $\mathbf{k}_x = p/\hbar$. Substitute the results for ω and **k** into Equation 3.124 and obtain:

$$\Psi(x,t) = Ae^{\frac{i(px-Et)}{\hbar}} \tag{3.125}$$

Schrödinger now assumed that the total energy *E* could be expressed in terms of the kinetic energy (KE) and the potential energy (PE) of the particle:

$$H = E = E_k + V(x) = \frac{p^2}{2m} + V(x,t) \tag{3.126}$$

The derivatives of Equation 3.125 are:

$$\frac{\partial\Psi(x,t)}{\partial x} = \frac{ip}{\hbar}\Psi(x,t)$$

so that

$$\frac{\partial^2\Psi(x,t)}{\partial^2 x} = \frac{-p^2}{\hbar^2}\Psi(x,t) \tag{3.127}$$

and

$$\frac{\partial\Psi(x,t)}{\partial t} = \frac{-iE}{\hbar}\Psi(x,t)$$

Note that Ψ cannot be canceled. Making use of the correspondence of the operators on $\psi(x,t)$ we can write: $\frac{\partial^2}{\partial^2 x} = \frac{-p^2}{h^2}$ and $\frac{\partial}{\partial t} = \frac{-iE}{h}$, and substituting these into the total energy equation (Equation 3.109) one derives:

$$E = i\hbar\frac{\partial}{\partial t} = -\left(\frac{\hbar^2}{2m}\frac{\partial^2}{\partial x^2}\right) + V(x,t) \tag{3.128}$$

Applying these operators to $\Psi(x,t)$, we obtain:

$$E\Psi(x,t) = i\hbar\frac{\partial\Psi(x,t)}{\partial t} = -\left(\frac{\hbar^2}{2m}\frac{\partial^2\Psi(x,t)}{\partial x^2}\right) + V(x,t)\Psi(x,t) \tag{3.129}$$

which, the reader can easily demonstrate, is the famous Schrödinger's wave equation (Equation 3.123) for the 1D case. We may rewrite Equation 3.129 also in the form:

$$\left[-\frac{h^2}{8\pi^2 m}\frac{\partial^2}{\partial x^2} + V(x,t)\right]\Psi(x,t) = E\Psi(x,t) \tag{3.130}$$

This equation shows that *H*, the Hamiltonian operator, is given by:

$$H = -\frac{h^2}{8\pi^2 m}\frac{\partial^2}{\partial x^2} + V(x,t) \tag{3.131}$$

Thus, Equation 3.130 may simply be formulated in operator form as $H\Psi$ (operator Ψ acting on function

Ψ, an eigenfunction) = EΨ(function Ψ multiplied by a number E, an eigenvalue), where H is the 1D Hamiltonian operator and in which the energy E of the particles is called the eigenvalue, and Ψ the eigenfunction. Note again that Ψ cannot be canceled. The latter merely represents the wave associated with the particle. Expressed yet another way, kinetic and potential energies are transformed into the Hamiltonian, which acts on the wave function to generate the evolution of the wave function in time and space. Schrödinger's equation gives the quantized energies of the system and gives the form of the wave function so that other properties may be calculated.

The above does not represent a derivation of the wave equation; it is just a description of Schrödinger's thought process to make his postulate more plausible.

Wave Function Interpretation

Schrödinger did not have a clear idea of the meaning of a matter wave. However, in 1926 Max Born (1882–1970) presented a new, vivid interpretation of the particle wave function. The so-called "Copenhagen interpretation" of Schrödinger's equation is that the $\Psi(x,t)$ function is not some physical representation of a physical substance as in classical physics (e.g., the amplitude of a water wave), but rather a "probability amplitude" of the particle, which, when squared, gives the probability of finding the particle at a given place at a given time: $|\Psi(x,t)|^2dx$ = probability the particle will be found between x and $x + dx$ at time t and the wave function itself has no physical meaning. Because $\Psi(x,y,z,t)$ is complex and can be positive or negative, it cannot be the probability directly. The Born interpretation of Ψ places restrictions on the form of the wave function:

1. Ψ must be continuous (no breaks).
2. The gradient of $\Psi(d\Psi/dx)$ must be continuous (no kinks).
3. Ψ must have a single value at any point in space.
4. Ψ must be finite everywhere.
5. Ψ cannot be zero everywhere.

The Copenhagen interpretation also holds that an unmeasured particle is not real: its attributes are created or realized by the measuring act. Another way of saying this is that a wave function *collapses* up on measurement; before measurement a particle is described by a wave function described by Schrödinger's equation, but on measuring that particle's wave suddenly and discontinuously *collapses*. We will come back to this *mystic* interpretation of quantum reality when introducing *Schrödinger's cat* at the end of this chapter. Because the probability that the particle is somewhere must equal one, it holds that one can normalize this probability function as:

$$\int_{-\infty}^{+\infty} |\Psi(x,t)|^2 dx = 1 \tag{3.132}$$

Note that the probability is a real number; although Ψ is complex, $|\Psi(x,t)|^2$ is real. In this case, Ψ is said to be a normalized wave function. Electrons do not fly around the nucleus like the Earth around the sun (Rutherford, Bohr), but depending on which energy level it is in, the electron can take one of a number of stationary probability cloud configurations (Schrödinger).

The boundary conditions imposed on Ψ mean that only certain wave functions and thus only certain energies of the system are allowed. Quantization of the wave function leads to quantization of the energy.

As we will learn below, solutions of Schrödinger's equation for an atom are spherical Bessel functions. In Figure 3.53 we show, as an example of the type

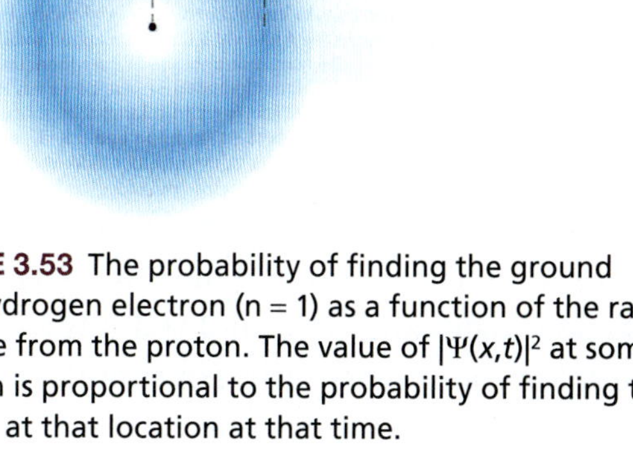

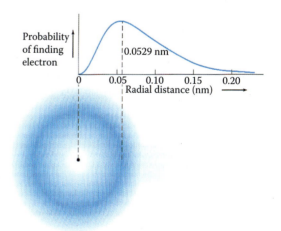

FIGURE 3.53 The probability of finding the ground state hydrogen electron (n = 1) as a function of the radial distance from the proton. The value of $|\Psi(x,t)|^2$ at some location is proportional to the probability of finding the particle at that location at that time.

of solutions obtained, the probability of finding an electron around the nucleus of a hydrogen atom. The potential that must be used in Schrödinger's equation for this case is $V(r) \propto 1/r$. Where we assume that the Coulomb force between the electron and the nucleus is the force responsible for binding the electron in the atom, this is the so-called central force or inverse square law $[F(r) \propto \frac{1}{r^2}]$.

Within Schrödinger's model the atom is regarded as a sort of vibrating balloon that extends to infinity and whose vibrations are in tune with Bohr's frequencies. The quantum numbers of Bohr and Sommerfeld are related to the number of nodes in this vibrating 3D system. This theory of matter waves also reproduces the Balmer series for the bright lines in the hydrogen atom. Schrödinger removed the mysterious discontinuous jumps between electron orbitals, replacing them with "beats" between the vibration frequencies of different quantum states. Schrödinger's elegant wave theory explains many things but lacks elsewhere, e.g., it cannot explain quantum processes such as the photoelectric effect.

The Time-Independent Schrödinger Equation (TISE) for Stationary States

The value of $|\Psi(x,t)|^2$ in Equation 3.132 at a particular point is in general a function of time. However, if the particle under consideration has a definite energy—think about an electron in a specific energy level in an atom—then the value of $|\Psi(x,t)|^2$ at each point becomes independent of time. It follows from quantum mechanics that for a particle in such a state of definite energy E, Ψ can be factored into a time-dependent component and a space-dependent component: $\Psi(x,t) = \psi(x)\psi(t)$:

$$\Psi(x,t) = \underset{\text{spatial}}{\underset{\uparrow}{\psi(x)}}\,\underset{\text{temporal}}{\underset{\uparrow}{e^{-\frac{iEt}{\hbar}}}} \tag{3.133}$$

This is the time-dependent wave function for a stationary state where we use Ψ if the wave is a function of all coordinates and time, and ψ if it is a function of the space coordinates only. For simplicity we just deal with the x-coordinate here. The probability distribution function $|\Psi|^2$ for this state is the product of Ψ and its complex conjugate Ψ^*, or:

$$\begin{aligned} |\Psi(x,t)|^2 &= \Psi^*(x,t)\Psi(x,t) \\ &= \psi^*(x)\psi(x)e^{+\frac{iEt}{\hbar}}e^{-\frac{iEt}{\hbar}} \\ &= \psi^*(x)\psi(x)e^0 \\ &= |\Psi(x)|^2 \end{aligned} \tag{3.134}$$

As $\psi(x)$ does not depend on time, we see from Equation 3.134 that the probability function does not depend on time either. As soon as one can define a state with a definite energy, one can define a stationary state.

We can now substitute the result from Equation 3.133 into the time-dependent Schrödinger's equation (Equation 3.129) to get:

$$i\hbar\psi(x)\frac{\partial\psi(t)}{\partial t} = -\left(\frac{\hbar^2}{2m}\psi(t)\frac{\partial^2\psi(x)}{\partial x^2}\right) + V(x,t)\psi(x)\psi(t) \tag{3.135}$$

Dividing both sides by $\psi(x)\psi(t)$ results in:

$$i\hbar\frac{1}{\psi(t)}\frac{\partial\psi(t)}{\partial t} = -\left(\frac{\hbar^2}{2m}\frac{1}{\psi(x)}\frac{\partial^2\psi(x)}{\partial x^2}\right) + V(x,t) \tag{3.136}$$

If the potential V is independent of t, then $V(x,t) = V(x)$, and the left side in Equation 3.136 depends on t only and the right side on x only. The only way these two sides can be equal to each other is if they are both equal to a constant, i.e., E. In the case $V(x,t)$ is independent of time, Equation 3.130 can thus be converted into a time-independent Schrödinger's equation (TISE). Hence we obtain the time-independent form of Schrödinger's equation as:

$$\left[-\frac{\hbar^2}{2m}\frac{\partial^2}{\partial x^2} + V(x)\right]\psi(x) = E\psi(x) \tag{3.137}$$

Solving this equation, say for an electron acted on by a fixed nucleus, we will see that this results in standing waves.

The more general Schrödinger equation features a time-dependent potential $V = V(x,t)$ and must be used, for example, when trying to find the wave function of an atom in an oscillating magnetic field or other time-dependent phenomena such as photon emission and absorption.

Applications of Schrödinger's Equation

Solution of the Wave Equation for Free Particles

The wave function of free particles, such as photons, phonons, plasmons, and "nearly free" particles, such as conduction electrons in metals, should all be solutions of Schrödinger's equation. In the case of freely traveling photons, the expression for planar light waves, $\Psi(x,t) = Ae^{i(k_x x-\omega t)}$ (1D case), derived from Maxwell's equations in Chapter 5, according to Bohr's correspondence principle, should also be a solution of Schrödinger's equation. To solve this first problem, we start from Equation 3.137, with V(x) = 0. This represents a free particle that experiences no force, has a definite energy *E*, and is moving in the *x*-direction with a momentum $\mathbf{p}_x$ (see Figure 3.54), say a photon:

$$-\frac{\hbar^2}{2m}\frac{d^2\psi}{dx^2} = E\psi \text{ or } \frac{d^2\psi}{dx^2} = -\frac{2mE}{\hbar^2}\psi \quad (3.138)$$

This is a second-order differential equation whose solutions are functions that, when differentiated twice, yield back the same functions multiplied by a constant. Such solutions include sines, cosines, and exponentials. Specifically, in this case, solving Equation 3.138 leads to the general solution, consisting of the superposition of two spatial wave functions of the form:

$$\psi(x) = Ae^{ik_x x} \quad (3.139)$$

The two traveling waves, one traveling in the +*x*-direction and one traveling in the –*x*-direction, can be rewritten (using Euler) as:

$$\psi(x) = A_1[\cos(\mathbf{k}_x x) + i\sin(\mathbf{k}_x x)] + A_2[\cos(-\mathbf{k}_x x) + i\sin(-\mathbf{k}_x x)] \quad (3.140)$$

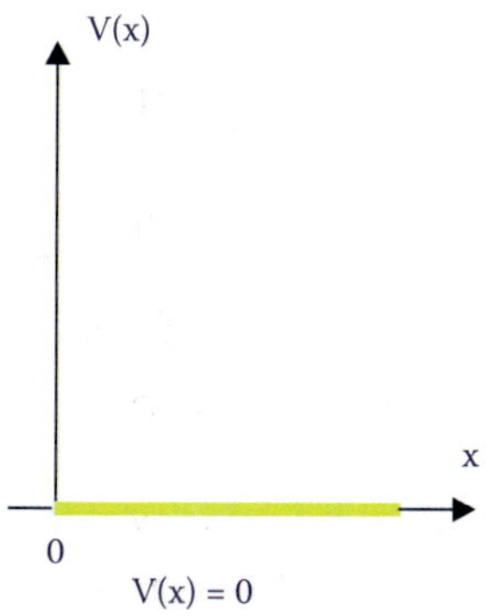

FIGURE 3.54 Photon, no boundary conditions, except for V(x) = 0.

where A_1 and A_2 are constants. We expected this result, of course, because in the preceding section we saw how Schrödinger built up his equation with the requirement that it would, at minimum, yield the same results that Maxwell's equations provide. This wave function for a free particle has a definite momentum $\mathbf{p}_x$ in the *x*-direction, or $\Delta\mathbf{p}_x = 0$. From the Heisenberg uncertainty principle (Equation 3.106), it then follows that Δx must be infinite; we have no idea where the particle is located in space. This makes sense as we are dealing with traveling waves. To confirm this further, we calculate the probability distribution function $|\Psi|^2$ for a free photon, which is the product of Ψ and its complex conjugate Ψ^*, or:

$$\begin{aligned}|\Psi(x,t)|^2 &= \Psi^*(x,t)\Psi(x,t) \\ &= (A^* e^{-ik_x x}e^{+i\omega t})(Ae^{ik_x x}e^{-i\omega t}) \\ &= A^* Ae^0 = |A|^2\end{aligned} \quad (3.141)$$

This result is independent of space and time, something we again expect for a sinusoidal wave function that extends all the way from x = –∞ to x = +∞ with an amplitude *A*. Normalization is not possible here as the wave extends to infinity; integration of Equation 3.139 over all space is infinite for any value of *A*.

By substituting $\psi(x) = Ae^{ik_x x}$ into Equation 3.138, we derive:

$$E = \mathbf{p}^2/2m \quad (3.142)$$

and also:

$$\mathbf{k}_x = \frac{\sqrt{2mE}}{\hbar} \quad (3.143)$$

With a positive $\mathbf{k}_x$, the wave function represents a free particle moving in the positive *x*-direction. If $\mathbf{k}_x$ is negative, the motion is in the negative *x*-direction. For a free particle, there is no restriction on the value of $\mathbf{k}_x$, and the associated, unquantized energy from Equation 3.143 is given as:

$$E(\mathbf{k}) = \frac{\hbar^2 k_x^2}{2m} \quad (3.144)$$

Any positive value of energy is allowed for such a free wave/particle; there is nothing that restricts the values of *E*. Equation 3.144 is the so-called dispersion relation for free particles. When we plot the energy E versus the wave vector $\mathbf{k}_x$ for free particles, we obtain the parabola shown in Figure 3.55. It is appreciated from this plot that mass, m, is inversely proportional to curvature.

Free Electrons in an Infinite Piece of Metal

For free electrons in an infinitely large 3D piece of metal, the allowed electron states are solutions of an expanded version of Schrödinger's equation in Equation 3.123, or:

$$\nabla^2\psi(\mathbf{r}) + \frac{8\pi^2 m}{h^2}[E - V(\mathbf{r})]\psi(\mathbf{r}) = 0 \quad (3.145)$$

For electrons swarming around freely in this infinite metal, the potential energy $V(r)$ is zero inside the conductor, and the solutions inside the metal are plane waves moving in the direction of $\mathbf{r}$:

$$\Psi_k(\mathbf{r}) = A\exp i\mathbf{k}\cdot\mathbf{r} \quad (3.146)$$

where $\mathbf{r}$ is any vector in real space, and $\mathbf{k}$ is any wave vector. As with a freely moving particle, normalization is impossible as the wave extends to infinity.

Plotting the energy E versus the wave number $\mathbf{k}_x$ for a free electron gas in one direction leads to the same parabolic dispersion relation shown in Figure 3.55. The density of states (DOS) function $G(E)$ is the number of possible energy states per unit volume (also degeneracy), and we will derive it further below for free electrons in a bulk piece of solid. We find that, just as in the case of the density of states of an ideal gas (Equation 3.19), it increases smoothly with the square root of the energy. Importantly, we will also find that when reducing the dimensionality of the solid, say to a plane, a line, or a dot, the density of state function changes dramatically and acquires more density of states at specific energy values; i.e., $G(E)$ is not a smooth function of E anymore.

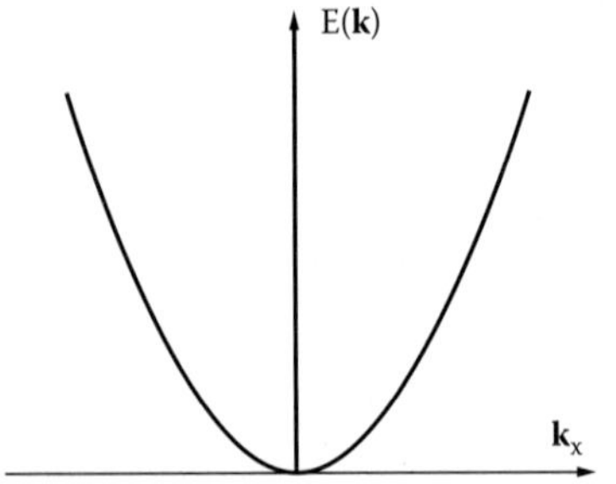

FIGURE 3.55 Plot of energy versus wave number $\mathbf{k}_x$ from Equation 3.144 for free particles. Dispersion relation $E(\mathbf{k}_x)$.

From classical theory we could not appreciate the occurrence of long electronic mean free paths; indeed, Drude used the interatomic distance a for the mean free path λ. But from experiments with very pure materials and at low temperatures it is clear the mean free path may be much longer; actually it may be as long as 10^8 or 10^9 interatomic spacings or more than 1 cm. The quantum physics answer is that the conduction electrons are not deflected by ion cores arranged in a periodic lattice because matter waves propagate freely through a periodic structure just as predicted by Equation 3.146.

Confining electrons by limiting their propagation in certain directions in a crystal introduces a varying $V(r)$ in Schrödinger's equation, and this may lead to an electronic bandgap as we will introduce below.

Particles in Infinitely Deep Potential Wells of Finite Size

The Born and von Karman's Periodic Boundary Condition According to Pauli's exclusion principle, in an atom, no two electrons can have all four quantum numbers the same. We ask ourselves now if there are similar restrictions for electrons in a larger structure. Instead of the infinitely large piece of metal, considered above, we limit the size of the metal chunk to say a 1-cm^3 piece of metal, and we find that restrictions for electron energies do indeed materialize. Discrete energy levels inevitably arise whenever a small particle such as a photon or electron is confined to a region in space. Sommerfeld, in 1928, was the first to show this. He adopted Drude's free electron or Fermi gas (FEG) and added the restriction that the electrons must behave in accordance with the rules of quantum mechanics. In his Fermi gas, electrons are free, except for their confinement within a cubic piece of crystalline conductor with a volume of $V = L^3$, and they follow Fermi-Dirac statistics instead of Maxwell-Boltzmann rules. The choice of a cube shape is a matter of mathematical convenience; a periodic boundary condition ensures that the free electron form of the wave function is NOT modified by the shape of the conductor or its boundary. This can be interpreted as follows: an electron coming to the surface is not reflected back in but reenters the metal piece from the opposite surface. This excludes the surfaces from playing any role in transport phenomena. The value L is set by the Born and von Karman's periodic boundary condition, i.e., that the wave functions must obey the following rule:

$$\psi(x + L, y + L, z + L) = \psi(x,y,z) \quad (3.147)$$

For the derivation of the possible energies, we assume N electrons (one for each metal ion) in a cube of solid conductor with sides of length L.

Outside the 3D cube of solid the potential $V \rightarrow \infty$, and the wave function ψ is zero anywhere outside the solid with $x,y,z \leq 0$ and $x,y,z \geq L$. The situation applies, for example, to totally free electrons in a metal where the ion cores do not influence their movement. Sommerfeld assumed that $V(x)$ outside the conductor equaled the work function Φ. The work function is the amount of energy required to remove an electron from the surface of a metal, i.e., the height of the wall electrons would have to scale to escape the solid, but $V = \infty$ is a good enough approximation for electrons in low-energy levels. Before we apply the Born and von Karman's periodic boundary condition to a finite-size 3D box that is infinitely deep ($V = \infty$), we consider finite-size infinitely deep 1D and 2D wells.

Infinitely Deep, Finite-Sized 1D Potential Wells—Quantum Wells We apply Equation 3.137 now to a finite-sized 1D box with infinitely high potential walls (Figure 3.56). In a 1D box, $V(x)$ in Equation 3.137—the time-independent Schrödinger equation (TISE)—is 0 everywhere inside the conductor (region II) and is infinite outside the conductor (regions I and III).

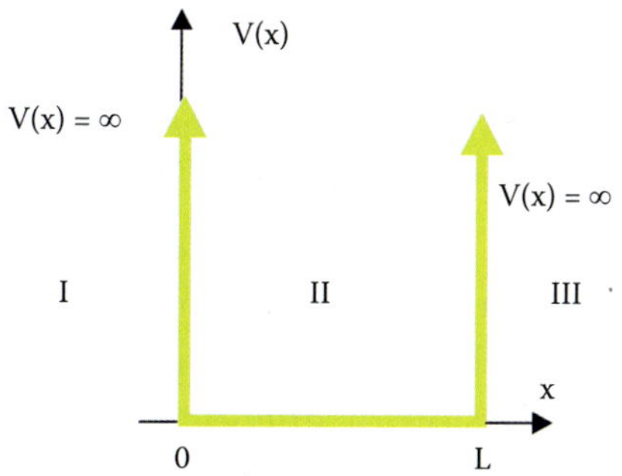

FIGURE 3.56 Electron in a box. L is not the real physical boundary of the conductor, as the surfaces of the conductor are not determining the physical properties. The value of L is set by the Born and von Karman periodic boundary condition (see Equation 3.147).

Outside the well $V \rightarrow \infty$ and $\psi = 0$ for $x \leq 0$ and $x \geq L$. For the Schrödinger equation inside the well ($0 < x < L$), we write:

$$-\frac{\hbar^2}{2m_e}\frac{d^2\psi(x)}{dx^2} = E\psi(x) \quad (3.148)$$

or:

$$\frac{d^2\psi}{dx^2} = -\mathbf{k}_x^2\,\psi \quad (3.149)$$

with:

$$\mathbf{k}_x = \sqrt{\frac{2m_eE}{\hbar^2}} \quad (3.150)$$

We are interested in the quantum mechanically allowed energy levels of the electrons in this 1D box; because $V(x) = 0$, these will be kinetic energy levels (from Equation 3.109, $E = p^2/2m_e$). From Equation 3.148 the kinetic energy is proportional to the curvature of ψ (the curvature of a function is its second derivative).

In Sommerfeld's mathematical model, for $x = 0$ and at $x = L$, the wave function must be zero. The general stationary state solution for Equation 3.149 consists of the superposition of two spatial wave functions $\psi(x)=Ae^{ik_xx}$, as shown in Equation 3.140, which we rewrite here as:

$$\psi(x) = (A_1 + A_2)\cos\mathbf{k}_x x + i(A_1 - A_2)\sin\mathbf{k}_x x \quad (3.151)$$

with one traveling in the $+x$-direction and one traveling in the $-x$-direction. The boundary conditions are set as follows: $x = 0$, $\psi(x) = A_1 + A_2 = 0$ or $A_2 = -A_1$, and we rewrite Equation 3.151 as:

$$\psi(x) = 2iA_1\sin\mathbf{k}_x x = \psi_0\sin\mathbf{k}_x x \quad (3.152)$$

From the second boundary condition that $\Psi_0\sin(\mathbf{k}_x L)$ must be 0, it follows that either $\psi_0 = 0$ or $\sin(\mathbf{k}_x L) = 0$. If $\psi_0 = 0$, the wave function is zero everywhere, or there is no probability of finding the particle in the box anywhere; therefore, this solution must be disregarded. Hence $\mathbf{k}_x L = n\pi$, where $n = 1, 2, 3,\ldots$ or:

$$\psi(x) = \psi_0\sin\left(\frac{n\pi x}{L}\right) \quad (3.153)$$

This is a standing wave that acts like a string clamped down at both ends. With n = 1 one has the basic harmonic, and with n = 2, 3,... higher harmonics result. Note that the constant, ψ_0, cannot be determined from solving Schrödinger's equation because we are dealing with a linear equation. The constant must be recovered from the normalization condition, which is required to give a probability interpretation to the wave function. The constant ψ_0 is calculated as:

$$\int_{-\infty}^{+\infty} |\Psi(x,t)|^2 dx = 1 = |\psi_0|^2 \int_0^L \sin^2\left(\frac{n\pi x}{L}\right) dx = |\psi_0|^2 \frac{L}{2} \quad (3.154)$$

or we find:

$$\Psi_n(x,t) = \sqrt{\frac{2}{L}} \sin\left(\frac{n\pi x}{L}\right) \quad (3.155)$$

with n = 1,2,3,.... These are the eigenfunctions of an electron in a box. The discrete energies—energy eigenvalues—that the electron can adopt are:

$$E_n(\mathbf{k}) = \frac{\hbar^2}{2m_e}\left(\frac{n\pi}{L}\right)^2 \quad (3.156)$$

The values of $E_n(\mathbf{k})$ for different quantum numbers *n* represent the various allowed energy levels in a 1D box, and the gap between two successive levels describes the effect of quantization (discreteness). At the lowest energy (n = 1), the *ground state*, the energy remains finite despite the fact that V = 0 inside the region. According to quantum mechanics, an electron cannot be inside the box and have zero energy. This is called the zero-point energy, an important consequence of Heisenberg's uncertainty principle (we elaborate on this point on p. 114 "Heisenberg's Uncertainty Priniciple"). If V(x) ≠ 0, everywhere in the box, all energies are shifted by V ($E_n = \frac{n^2 h^2}{8m_e L^2} + V$). For the same value of quantum number *n*, the energy is inversely proportional to the mass of the particle and to the square of the length of the box. For a heavier particle and a larger box, the energy levels become more closely spaced. Only when $m_e L^2$ is of the same order as h^2 do quantized energy levels become important in experimental measurements (with L = 1 nm, $E_1 = \frac{h^2}{8m_e L^2} = 0.36$ eV). With a 1-cm³ piece of metal (instead of 1 nm³), the energy levels become so closely spaced that they seem to be continuous ($E_1 = \frac{h^2}{8m_e L^2} = 3.6 \times 10^{-15}$ eV); in other words, the quantum mechanics formula gives the classical result for dimension such that $m_e L^2 >> h^2$. This is another nice illustration of Bohr's correspondence principle: for large dimensions, Schrödinger's equation yields the classical results back. Because the kinetic energy is proportional to the curvature of the wave function, a higher kinetic energy E_n, caused by a higher value of *n* and/or a smaller value for *L*, corresponds to a more curved wave function (shorter wavelength). On a molecular scale, because L^2 appears in the denominator of Equation 3.156, increasing *L*, the size of the box, stabilizes the particle. Chemists are familiar with this effect in the case of electrons where delocalization is a stabilizing factor. As a consequence, allyl and benzyl carbonium ions, radicals, and carbanions are relatively stable.

The wave number $\mathbf{k}_x$ given as:

$$\mathbf{k}_x = \frac{2\pi p}{h} = \frac{p}{\hbar} = \frac{\sqrt{2m_e E_n}}{\hbar} \quad (3.150)$$

is now also quantized because E_n is quantized so that $\mathbf{k}_x = n\pi/L_x$ where *n* can take on only values 1, 2, 3....

Quantization occurs because of boundary conditions and the requirement for ψ to be physically reasonable (Born interpretation). The quantum number *n* labels each allowed state (ψ_n) of the system and determines its energy (E_n). Knowing *n*, we can calculate ψ_n and E_n. The wave number is related to the momentum $\mathbf{p}_x$ of the electron, viewed as a particle, by $\mathbf{k}_x = 2\pi \mathbf{p}_x/h$, and to the wavelength (λ) of the electron, viewed as a wave, by $\mathbf{k}_x = 2\pi/\lambda$. The dispersion function $E(\mathbf{k}_x)$ for an electron in an infinite deep well of finite size is shown in Figure 3.57. The dispersion or energy spectrum $E(\mathbf{k}_x)$ is now discrete rather than continuous; the allowed wave vectors are uniformly spaced in **k**-space with a separation of π/L; and the sample size *L* determines the spacing of allowed wave vectors and single-particle energies, with a smaller box giving

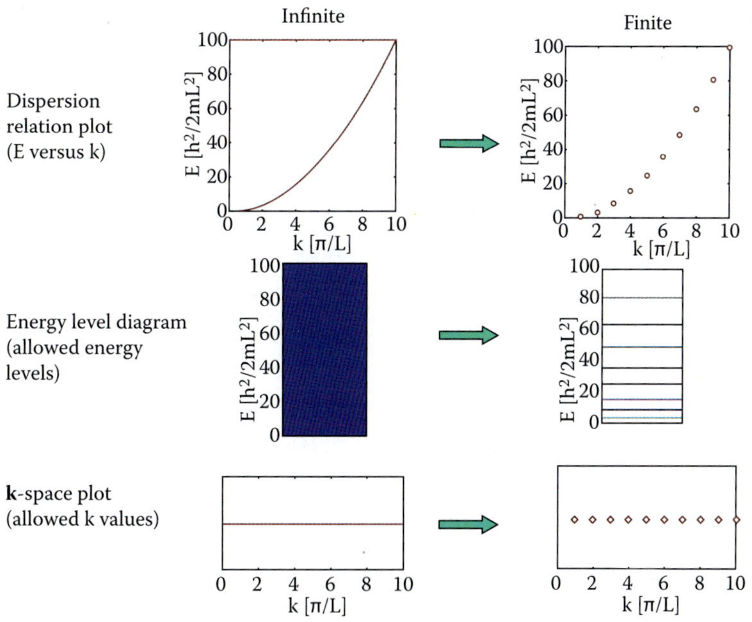

FIGURE 3.57 Dispersion $E(\mathbf{k}_x)$ function for an electron in an infinite box compared with the dispersion of the electron in a finite-size box. Finite size drastically alters allowed energy levels.

the larger spacings. For the free electron, no such values exist (see Figure 3.55).

The ground state (n = 1) of the wave function ψ_1 (Equation 3.150) has no "zero crossings." The first excited state (n = 2) has one zero crossing, and so on. Successive functions possess one more half-wave (they have a shorter wavelength). Nodes in the wave function are points at which $\psi_n = 0$ (excluding the ends, which are constrained to be zero), and the number of nodes = n − 1. The important point to notice is that the imposition of the boundary conditions has restricted the energy to discrete values. The allowed wave functions, a family of standing waves, and energies of an electron in a 1D box are summarized in Figure 3.58.

As apparent from Figure 3.58, the probability distribution of the particle in a box is not uniform; however; for a very large n, the probability is almost uniform throughout the box, as dictated again by classical physics. The probability of finding the particle at some point varies with the energy of the particle. A particle with energy E_1 is most likely to be found in the middle of the box. A particle with energy E_2 will never be found at that spot. As the energy of the particle increases, so does the number of nodes in the eigenfunction. Increasing the number of nodes (decreasing λ) corresponds to increasing kinetic energy.

Devices that come with a length L in one direction comparable with the size of the de Broglie wavelength of an electron are known as quantum wells (1D confinement). With at least one dimension between 1 and 100 nm, the excess electrons in this confined direction have no room to move in a Newtonian fashion. Instead their positions and velocities take on a probabilistic nature as their wave nature now dominates. Charge carriers are still free to move in the other two unconstrained directions though and

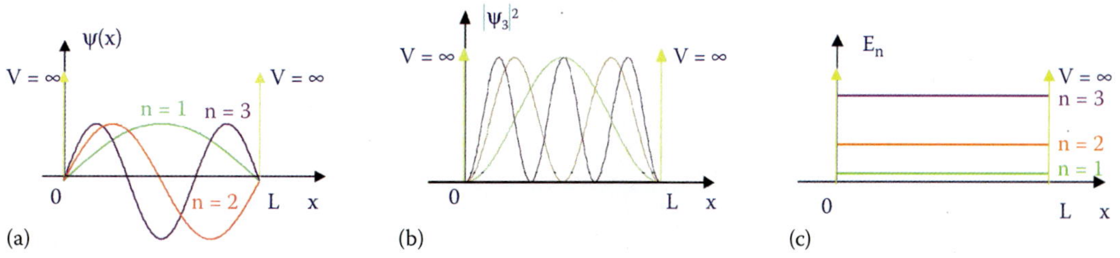

FIGURE 3.58 Particle in a box: overview. (a) Standing waves in the box. (b) The probability distribution $|\psi_3|^2$. (c) The energy is proportional to n^2.

form a 2D carrier gas in a plane perpendicular to the confinement direction (the *x-y* plane if the confinement is in the *z*-direction). Quantum wells are relatively easy and cheap to fabricate. By using two different types of semiconductor materials, one of these having a wider bandgap, quantum wells are formed on sandwiching a thin layer of a narrow bandgap material between wider bandgap semiconductor layers. This is realized, for example, by sandwiching a thin epitaxial layer of GaAs between two layers of $Al_xGa_{1-x}As$, in a so-called heterostructure as depicted in Figure 3.59. The narrower bandgap GaAs is enclosed by $Al_xGa_{1-x}As$, a material with a considerably larger bandgap to establish a potential barrier for electrons at the surface of the confined material. The motion of electrons and holes is thereby restricted in the direction perpendicular to the thin layer of GaAs (*z*-direction). In this structure, electrons are still free to move unrestricted in the *x-y* plane.

These quantum wells (QWs) were developed in the early 1970s and constituted the first lower dimensional heterostructures. The foremost advantage of such a design involves their improved optical properties. Reduced dimensionality leads to marked improved optical performances because of the change in the density of state or DOS function (see p. 161 "Fermi Surfaces, Brillouin Zones, Density of State Functions, and Conductivity as a Function of Quantum Confinement" for more details) for such a device compared with its 3D counterpart. In contrast to a bulk semiconductor, in a QW there are no allowed electron states at the very lowest energies (an electron in a box with energy = 0 does not exist; see p. 125 "Infinitely Deep, Finite-Sized 1D Potential Wells—Quantum Wells"), but there are many more available states (higher DOS) in the lowest conduction state in a QW so that many more electrons can be accommodated. Similarly, the top of the valence band has plenty more states available for holes. This means that it is possible for many more holes and electrons to combine and produce photons with identical energy for enhanced probability of stimulated emission (lasing) (see Chapter 5). Optical properties can be tuned by changing the structural parameters of a QW, principally its thickness and composition; this is known as bandgap engineering.

Today QWs form the basis of most optoelectronic devices. QWs are used, for example, in high electron mobility transistors (HEMTs) and solar cells with high efficiency. HEMTs are in use in all types of high frequency electronics such as cell phone and satellite television. The reason for the higher speed of a HEMT over a classical transistor is that the mean free path of the charge carriers in the 2D GaAs layer of a HEMT can be made larger than the gate length of the transistor, resulting in ballistic transport, i.e., charge transport without any intervening collisions with other charge carriers. Another mature application of QWs involves solid-state lasers. With the right voltages applied over a single QW, very large numbers of holes and electrons can be brought together in a tiny physical space and narrow energy range, leading to powerful surface emitting lasers as used in $5 laser pointers, compact disc players, and laser printers.

Almost 30 years after quantum wells were first developed, the 2000 Nobel Prize in Physics was awarded to Zhores Alferov and Herbert Kroemer (Figure 3.60) for their contributions in the field of semiconductor heterostructures and their high-speed and optoelectronics applications.

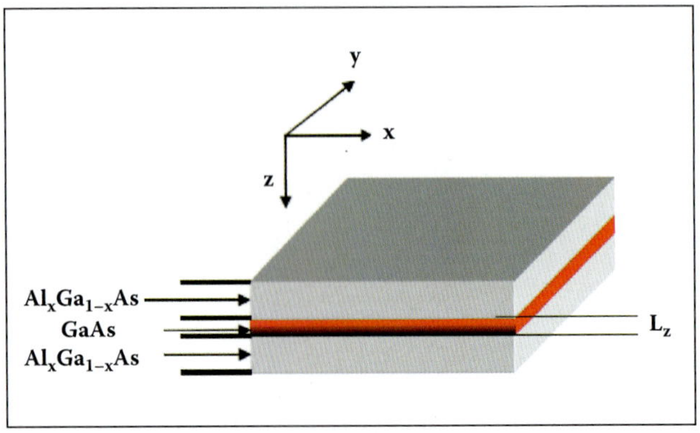

FIGURE 3.59 Quantum well with *x,y* dimensions infinite and L_z finite.[4]

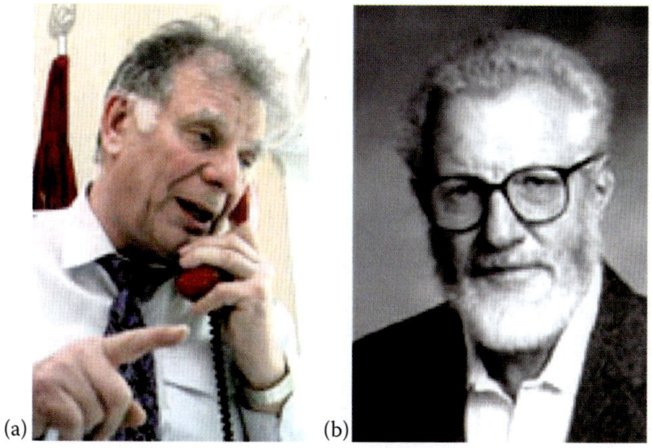

FIGURE 3.60 Zhores Alferov (a) and Herbert Kroemer (b) were awarded the Nobel Prize in Physics in 2000.

Led by the insights garnered on QWs, scientists investigated the possibility of reducing the dimensionality of heterostructures even further to create 1D (quantum wire) and 0D (quantum dot) structures (see following sections).

Infinitely Deep, Finite-Size 2D Potential Wells—Quantum Wires For a particle in a finite-sized, 2D infinitely deep potential well, we define a wave function similar to the 1D potential well, but now we obtain $\psi(x,y)$ solutions that are defined by two quantum numbers, one associated with each confined dimension. The pertinent wave functions and energies for a 2D infinitely deep potential well, as shown in Figure 3.61, are:

$$\Psi_{n_1n_2}(x,y) = \left(\frac{4}{L_1L_2}\right)^{\frac{1}{2}} \sin\left(\frac{n_1\pi x}{L_1}\right)\sin\left(\frac{n_2\pi y}{L_2}\right) \quad (3.157)$$

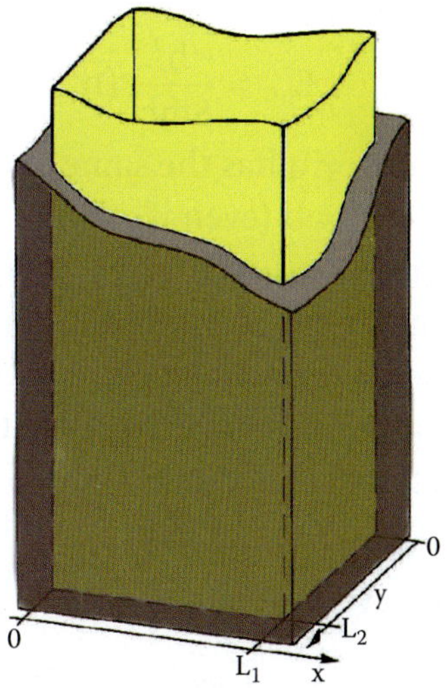

FIGURE 3.61 A 2D square well. A quantum wire.

and:

$$E_{n_1n_2}(\mathbf{k}) = E^x_{n_1} + E^y_{n_2} = \frac{h^2}{8m}\left(\frac{n_1^2}{L_1^2} + \frac{n_2^2}{L_2^2}\right) \quad (3.158)$$

with $n_1 = 1, 2, 3$ and $n_2 = 1, 2, 3$ and where L_1 is along the x-axis and L_2 along the y-axis, making the z-direction the unconstrained axis. The energy is quantized now along the x- and y-axes as a consequence of the boundary conditions, and most of the features of the 1D well are reproduced. Some of the low-energy wave functions together with their contour maps are illustrated in Figure 3.62.

One new feature in 2D potential wells, not found in 1D potential wells, is degeneracy, i.e., the occurrence of several quantum states at the same energy level. This is best understood through an inspection of Equations 3.157 and 3.158: for a square well instead of a rectangular one, $L_1 = L_2 = L$, and

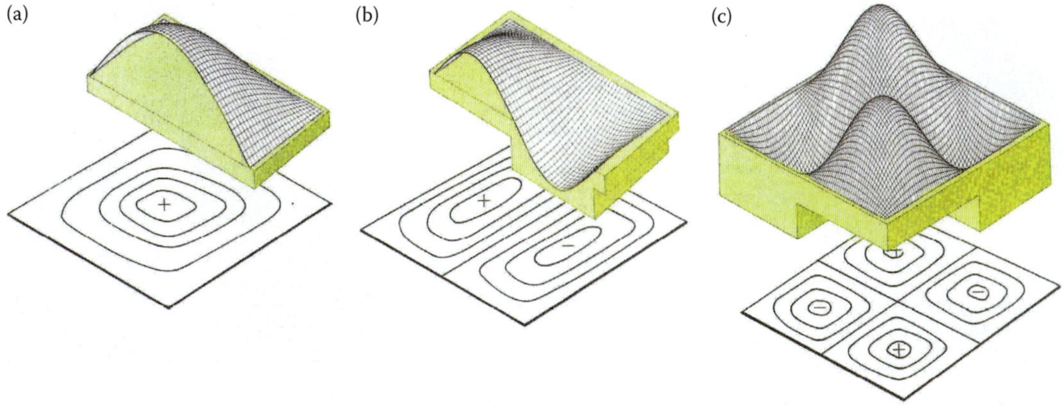

FIGURE 3.62 (a) ψ_{11}, (b) ψ_{21}, and (c) ψ_{22} for a 2D square potential well. A quantum wire.

Equation 3.158 yields now the same energy of $E_{n_1n_2}(\mathbf{k}) = E^x_{n_1} + E^y_{n_2} = \frac{h^2}{8mL^2}(n_1^2 + n_2^2)$. A state with $n_1 = b$ and $n_2 = a$ has the same energy as a state with $n_2 = a$ and $n_2 = b$ (even if $a \neq b$); these different states that correspond to a same energy are called degenerate. That these are indeed different quantum states becomes obvious when substituting the quantum numbers for such states in Equation 3.157. In the case of two degenerate states with $a = 1$ and $b = 2$, both corresponding to an energy of $\frac{5h^2}{8mL^2}$, we obtain two different wave functions:

$$\Psi_{1,2}(x,y) = \left(\frac{2}{L}\right)\sin\left(\frac{\pi x}{L}\right)\sin\left(\frac{2\pi y}{L}\right) \quad \text{and}$$

$$\Psi_{2,1}(x,y) = \left(\frac{2}{L}\right)\sin\left(\frac{2\pi x}{L}\right)\sin\left(\frac{\pi y}{L}\right).$$

Degeneracy is obviously connected to the degree of symmetry of a system; in the case we make an asymmetric box with $L_1 \neq L_2$ (rectangle instead of a square), the degeneracy disappears.

When *L* is very small along two directions (2D confinement)—of the order of the de Broglie wavelength of an electron—one obtains a quantum wire where electrons can only move freely in one direction, i.e., along the length of the quantum wire as illustrated in Figure 3.63.

Examples of 2D quantum confinement comprise nanowires and carbon nanotubes. Quantum wires represent the smallest dimension feasible for efficient transport of electrons and are thus aimed at as the ultimate interconnects in nanoelectronics and nanooptoelectronics. We mentioned above that quantum well lasers are superior over traditional-bulk solid-state lasers. Structures with yet lower dimensionality, such as nanowires and quantum dots (see next section), are even better, coming with a lower threshold current and switching on and off faster than quantum well lasers (~40 GHz and higher). Quantum wires also have been made into transistors (bipolar and FET), inverters, LEDs, and memory structures.

Compared with the fabrication of quantum wells, the realization of nanoscale quantum wires requires more difficult and precise growth control in the lateral dimension, and, as a result, quantum wire applications are only in the development stage. Quantum wire fabrication techniques include nanoscale lithography, self-organization, selective growth, and chemical and electrochemical synthesis (see Volume III, Chapter 3). In the top-down approach, taking advantage of well-developed quantum well fabrication technologies, using molecular beam epitaxy (MBE) and metalorganic vapor deposition (MOCVD), the most straightforward method to realize 1D nanostructures is etching (and regrowth) through wire-defining masks placed above a quantum well. This way, GaAs quantum wires are fabricated starting from the same thin layer of GaAs sandwiched between two layers of Al_xGa_{1-x} As we encountered in the manufacture of quantum wells. In the bottom-up approach, quantum wires are formed via direct growth in the form of semiconductor or metal nanowires and carbon nanotubes. Sumio Iijima (Figure 3.64) discovered multiwall carbon nanotubes (MWNTs) in 1991, after experimenting with

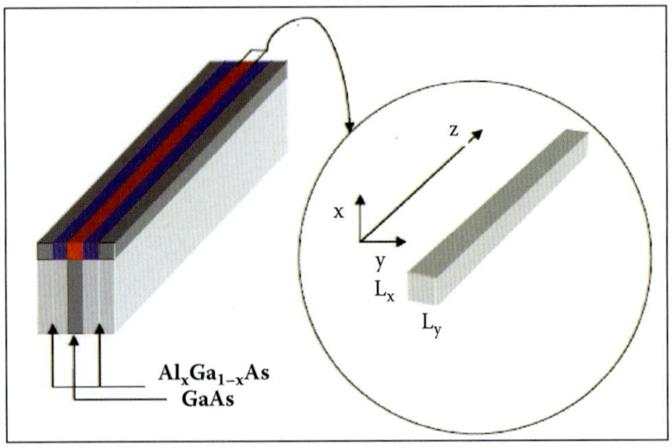

FIGURE 3.63 Quantum wire with dimensions *z* infinite and L_x, L_y finite.[4]

FIGURE 3.64 NEC's Sumio Iijima discovered multiwall carbon nanotubes (MWNTs) in 1991.

an arc-discharge technique similar to the one used by Richard Smalley and his team at Rice University in the discovery of buckminsterfullerene (C60). During arc discharge between two closely spaced graphite rods, carbon vaporizes and after condensing yields a "sooty" mass. When Iijima looked at the soot with an SEM, he noticed hollow nanotubes of carbon. One may speculate that this finding, like that of C60 "buckyballs," could have been achieved earlier if better microscopy techniques had been available to see the nano-sized products hidden in the soot.

During the past decade, there has been major progress reported in the chemical synthesis (i.e., bottom-up manufacture) of all types of nanoscale semiconductor wires. As originally proposed by R.S. Wagner and W.C. Ellis, from Bell Labs, for the Au-catalyzed Si whisker growth, a vapor-liquid-solid mechanism is still mostly used.[5] The field got a shot in the arm (a rebirth, so to speak) with efforts by Charles Lieber (Harvard) (Figure 3.65), Peidong Yang (http://www.cchem.berkeley.edu/pdygrp/main.html), James Heath (http://www.its.caltech.edu/~heathgrp), and Hongkun Park (http://www.people.fas.harvard.edu/~hpark). Lieber's group at Harvard (http://cmliris.harvard.edu) reported arranging indium phosphide semiconducting nanowires into a simple configuration that resembled the lines in a tick-tack-toe board.

The team used electron beam lithography to place electrical contacts at the ends of the nanowires to show that the array was electronically active. The tiny arrangement is not yet a circuit, but it is a first step, showing that separate nanowires can be contacted to one another.

Infinitely Deep, Finite-Sized 3D Potential Wells—Quantum Dots The solution of Schrödinger's equation for electrons in a cube of metal with side L is given as:

$$\Psi_k(\mathbf{r}) = V^{-\frac{1}{2}}\exp(i\mathbf{k}\cdot\mathbf{r}) \qquad (3.159)$$

—a 3D generalization of Equation 3.157 (with V = L³). The wave vector **k**, if you recall, points in the direction of wave propagation and has a magnitude given by $2\pi/\lambda$ for a plane wave. Generally, if a wave propagates along a displacement vector **r**, the

FIGURE 3.65 Harvard's Charles Lieber reinvigorated the nanowire field.

amount of phase accumulated by the wave is given by $\mathbf{k}\cdot\mathbf{r}$. From the generalization of Equation 3.158, electrons in a cubic box of side L ($L_1 = L_2 = L_3$) have allowed energy levels specified by three quantum numbers n_1, n_2, and n_3, or:

$$E_{n_1,n_2,n_3} = \frac{h^2}{8m}\left(\frac{n_1^2}{L_1^2} + \frac{n_2^2}{L_2^2} + \frac{n_3^2}{L_3^2}\right) = \frac{h^2}{8mL^2}(n_1^2 + n_2^2 + n_3^2) \qquad (3.160)$$

and they come with the following allowed wave vectors:

$$\mathbf{k}_x = \frac{2\pi n_x}{L}, \mathbf{k}_y = \frac{2\pi n_y}{L}, \mathbf{k}_z = \frac{2\pi n_z}{L},$$
$$n_x, n_y, n_z = 1, 2, 3, \ldots \text{(integers)} \qquad (3.161)$$

The allowed **k**-states are uniformly spaced along the axes with one state per length $2\pi/L$. This is shown for a 2D square array and a 3D cubic array in Figure 3.66. One consequence of confining a quantum particle in a cubic 3D is again "degeneracy." As we

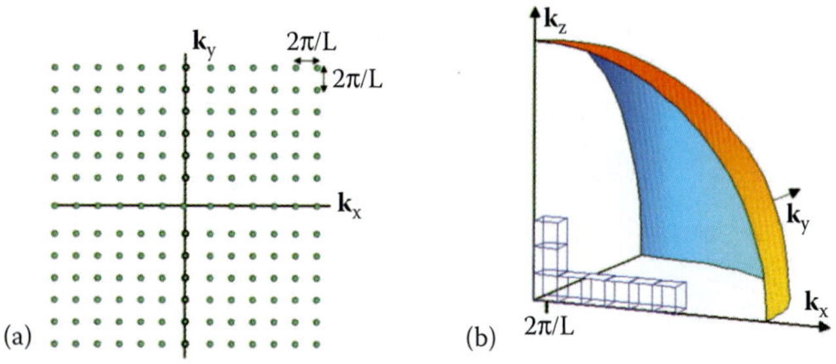

FIGURE 3.66 A 2D array of allowed **k**-state where we are looking in "**k**-space" or "reciprocal space" (so-called because **k** has units of 1/L) (a) and a cubic array of allowed states in 3D **k**-space; in three dimensions, we have states described by **k** = ($\mathbf{k}_x$; $\mathbf{k}_y$; $\mathbf{k}_z$) (b). The area per point is $(2\pi/L)^2$, and the volume per point is $(2\pi/L)^3$.

saw for the 2D case, degeneracy reflects an underlying symmetry in the $V(x,y,z)$ potential and can be removed if certain symmetry is broken (e.g., by making $L_1 \neq L_2 \neq L_3$). Each dot in Figure 3.66 represents an allowed **k**-state. An important feature here is that the larger the box, the closer the spacing in energy of single particle states, and vice versa. The **k**-states are discrete, but for a normal-sized conductor there are $\sim 10^{26}$ states, so we can treat them as a continuum.

From Figure 3.66a, each value of **k** occupies an area $A = (2\pi/L)^2$, and from Figure 3.66b each point occupies a volume $V = (2\pi/L)^3$. The number of states per unit volume of **k**-space is $1/V$ (or $L/2\pi)^3$.

When L becomes very small, of the order of the de Broglie wavelength of an electron in all three directions, the electrons lose all capacity to move (3D confinement). These 0D structures, with 3D quantum confinement, are called quantum dots or QDs, as illustrated in Figure 3.67. Quantum dots, also known as nanocrystal semiconductors, ranging from 2 to 10 nm (10–50 atoms) in diameter, are typically composed of materials from periodic groups II–VI, III–V, or IV–VI (e.g., CdS, CdSe, PbS, PbSe, PbTd, CuCl…). The trapped electrons in these dots behave as de Broglie standing waves. The confinement of the waves leads to a blue energy shift, and by varying the particle size one can produce any color in the visible spectrum, say from deep (almost infra-) reds to screaming (almost ultra-) violets as illustrated in Figure 1.31. A quantum dot (QD) is an atom-like state of matter sometimes referred to as an "artificial atom." What is so interesting about a QD is that electrons trapped in them arrange themselves

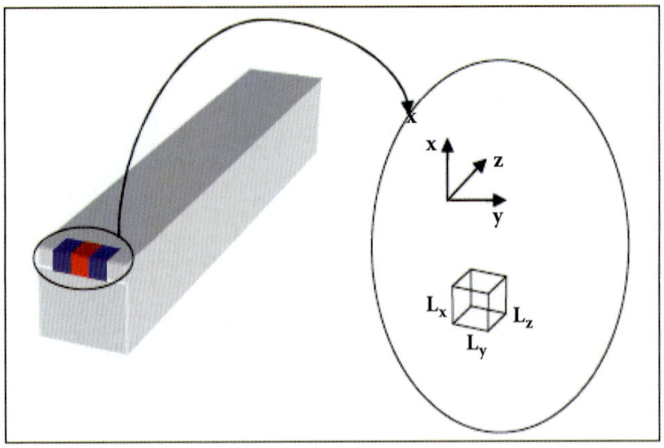

FIGURE 3.67 Quantum dot with dimensions L_x, L_y, and L_z that are finite.[4]

as if they were part of an atom, although there is no nucleus for the electrons to surround here. The type of atom the dot emulates depends on the number of atoms in the well and the geometry of the potential well $V(x)$ that surrounds them.

An important consequence of decreasing the dimensionality even further than in quantum wells and wires is that the density of state function for quantum dots features an even sharper and yet more discrete density of states. As a consequence, quantum dot lasers exhibit a yet lower threshold current than lasers based on quantum wires and quantum wells, and because of the more widely separated discrete quantum states, they are also less temperature sensitive. However, because the active lasing material volume is very small in quantum wires and dots, a large array of them has to be made to reach a high enough overall intensity. Making quantum wires and dot arrays with a very narrow size distribution to reduce inhomogeneous broadening remains a real manufacturing challenge, and as a result only quantum well lasers are commercially mature.

However, quantum dots already form an important alternative to organic dye molecules. Unlike fluorescent dyes, which tend to decompose and lose their ability to fluoresce, quantum dots maintain their integrity, withstanding many more cycles of excitation and light emission (they do not bleach as easily!). Combining a number of quantum dots in a bead conjugated to a biomolecule is used as a spectroscopic signature—like a bar code on a commercial product—for tagging those biomolecules (see Chapter 7, "Fluorophores, Quantum Dots, and Fluorescent Proteins").

In the early 1980s, Dr. Ekimov discovered quantum dots with his colleague, Dr. Efros, while working at the Ioffe Institute in St. Petersburg (then Leningrad), Russia.[6,7] This team's discovery of quantum dots occurred at nearly the same time as Dr. Louis E. Brus (Figure 3.68), a physical chemist then working at AT&T Bell Labs (he is now at Columbia), found out how to grow CdSe nanocrystals in a controlled manner.[8,9] Experimenting with CdSe nanocrystal semiconductor material, Dr. Brus and his collaborators observed solutions of astonishingly different colors made from this same substance. Their observation led to the recognition that there is a very clear transition in material behavior when

FIGURE 3.68 Columbia's Louis Brus.

a chunk of it becomes smaller than a fundamental scale, intrinsic to the substance.

Even though it was predicted in 1982 that QDs could be used as the active region of lasers, providing reduced threshold current and lower temperature dependence, it took nearly a decade to develop reliable growth techniques to produce QDs of a quality suitable for commercial applications. Major contributions to making QDs a reality were made by two Bell Labs scientists, Dr. Moungi Bawendi and Dr. Paul Alivisatos. They have since moved to MIT and University of California, Berkeley, respectively, where they continue their investigation of optical properties of quantum dots. Before 1993, QDs were prepared in aqueous solution with added stabilizing agents to avoid colloid precipitation (see colloid stability in Chapter 7, p. 518 "Intermolecular Forces"). In 1993, Bawendi and coworkers synthesized better luminescent CdSe QDs by using a high-temperature organometallic procedure instead.[10,11] Crystallites from 12 to ~115 Å in diameter with consistent crystal structure, surface derivatization, and a high degree of monodispersity were prepared in a single reaction based on the pyrolysis of organometallic reagents by injection into a hot coordinating solvent. At that time, even if one could make QDs in a narrow size range, they still came with two major problems: 1) poor fluorescence and 2) hydrophobicity, making them useless in biology. The addition of semiconductor caps, such as ZnS or CdS caps over a CdSe core, was found to dramatically increase the fluorescence quantum yield to 45% or higher.[12–14] The core determines the nanocrystal color, and the shell of the higher bandgap material (ZnS or CdS) dramatically enhances not only the brightness but also the chemical stability. A few different methods for making nanocrystals water soluble are also available today.[15] Methods for water solubilization of QDs include derivatizing the surface with mercaptoacetic acid or dithiothreitol, and Alivisatos and coworkers use a silica/siloxane coating.[16,17] The addition of these coats makes the particles water soluble, and therefore more useful in biology experiments.

The quest to find new energy solutions stimulated the development of modern quantum dot technology even further. The advantage of the surface area-to-volume ratio of nanocrystal particles for energy conversion was realized, and photoelectrochemistry research (e.g., solar energy conversion) tapped the semiconductor/liquid interface to exploit this (also visit http://www.technologyreview.com/Energy/19256).[18,19]

Quantum dots, we will learn in Chapter 7 and Volume III, Chapter 3, are manufacturable through a very wide variety of methods, including self-assembly in colloidal wet chemical synthesis, template chemistry (zeolite, alumina templates), sol-gel methods, micelle methods, organometallic synthesis, pyrolysis, lithography and etching, electrostatic confinement, scanning tunneling microscope (STM) tips, epitaxial strain, and so on. QD devices have now been demonstrated in many research laboratories, and commercial products are available on the market.

Zero Point Energy We noted before that at the lowest electron energy (n = 1), the *ground state energy*, remains finite despite the fact that V = 0 inside the solid (see Equation 3.156 and Figure 3.58c). In classical physics, both the kinetic energy (KE) and potential energy (PE) can have a zero value, but not in quantum physics! According to quantum mechanics, an electron in a box of length L must necessarily have energy. It cannot be inside the box and have zero energy. The minimum possible energy is E_1 and is called the zero point energy (ZPE). Zero point energy hypothesized by Max Planck in 1911 and developed by him and Walter Nernst between 1911 and 1916. It was first measured by Dr. Willis Lamb in 1947 as a slight upward shift of electrons in their atomic orbitals. The concept of a zero point energy has very far-reaching consequences, manifesting themselves from the quantum world to the cosmos. The existence of a zero point energy for a quantum particle is a general phenomenon and is consistent with the Heisenberg uncertainty principle; i.e., particles with zero momentum cannot be localized. From the Heisenberg uncertainty principle (Equation 3.106), it follows that for a 1D box:

$$E = \frac{p^2}{2m} \text{ and also}$$

$$\frac{(\Delta p)^2}{2m} \geq \frac{\hbar^2}{2m(\Delta x)^2} \geq \frac{\hbar^2}{2m\left(\frac{L}{2}\right)^2} = \frac{2\hbar^2}{mL^2} > 0 \quad (3.162)$$

where we used $\Delta x \leq \frac{L}{2}$, and also for E_1 (see Equation 3.156) we obtain:

$$E_1 = \frac{k_x^2 \hbar^2}{2m} = \frac{4\pi^2\hbar^2}{2mL^2} > \frac{2\hbar^2}{mL^2} > 0 \quad (3.163)$$

because $k_x = 2\pi/L$ (see Equation 3.161).

That the zero point energy is a direct consequence of the Heisenberg uncertainty principle can simply be recognized by considering an electronic vibration ceasing at T = 0. If such a thing could happen, then position and momentum both would be zero, which would violate the Heisenberg uncertainty principle.

The zero point energy can be measured through the Casimir force, a small attractive force between two close parallel, uncharged plates in a vacuum (see Volume III, Chapter 8 on actuators). This force, inversely proportional to the fourth power of the distance between the plates, comes about because, according to quantum physics, even a perfect vacuum contains a zero point energy keeping virtual particles in a continuous state of fluctuation (see Equation 3.107). Given the equivalence of mass and energy expressed by Einstein's $E = mc^2$, the vacuum energy must be able to create particles, known as virtual particles. They flash briefly into existence and expire within an interval dictated by the uncertainty principle. Casimir realized that only the virtual photons with wavelengths that fit a whole number of times into the gap between the two plates should be part of the energy calculation (i.e., no half-quantum lengths are allowed). Moving the plates together, the energy density should thus decrease as there are more allowed virtual photon states pushing against the plates from the outside than from between the two plates (see Figure 3.69).

An example of a practical application of the existence of the zero point energy for confined electrons is the quantum well laser, which is more efficient than a diode laser. In a quantum well, as we saw above, there are no allowed electron states at the very lowest energies, but there are more available states in the lowest conduction band state and at the top of the valence band so that many more holes and electrons can combine and produce photons with identical energy for enhanced probability of stimulated emission (lasing; see also Chapter 5).

Potential Wells of Finite Depth and Size, Finite Barriers, Tunneling, and Interfaces

Finite-Depth Potential Wells A potential well is a potential energy function with a minimum. An infinitely deep potential well, as considered above, is an idealization. On the atomic scale there are no infinitely high and sharp barriers, and both $\psi(x)$ and $d\psi(x)/dx$ go to zero smoothly near a boundary. If the walls of an electron *prison* are not infinitely high, as sketched in Figure 3.56, but can be scaled by the particles inside as sketched in Figure 3.70, Equation 3.137 must be rewritten as:

$$-\frac{\hbar^2}{2m}\frac{d^2\psi(x)}{dx^2} = (E - V_0)\psi(x) \quad (3.164)$$

with V_0 the potential barrier the particle must jump to get out of the well. We will now solve this equation in the three regions marked in Figure 3.70.

A quantum well with finite potential walls for trapped particles is a more realistic picture than the infinite potential wells assumed earlier and applies

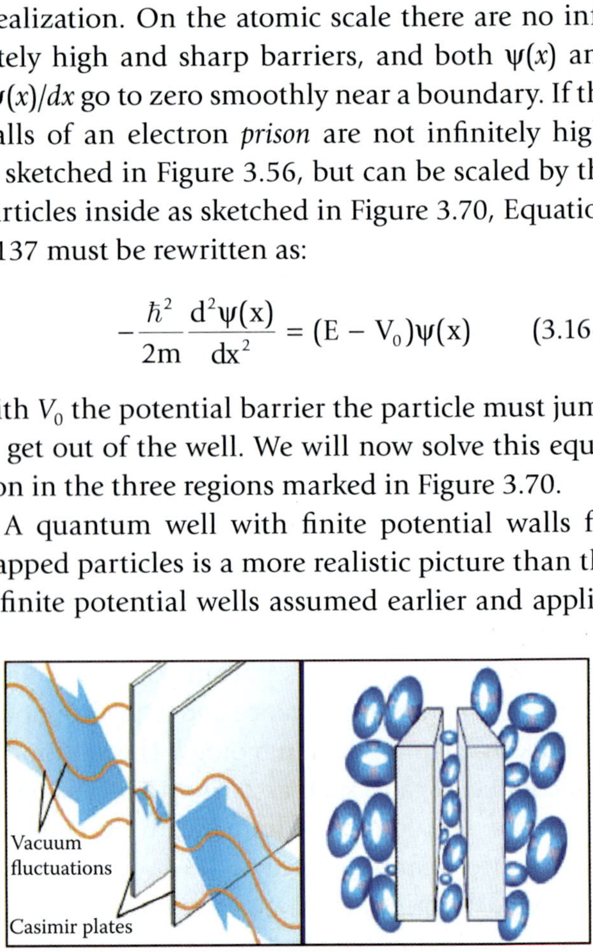

FIGURE 3.69 Vacuum fluctuations. The Casimir effect.

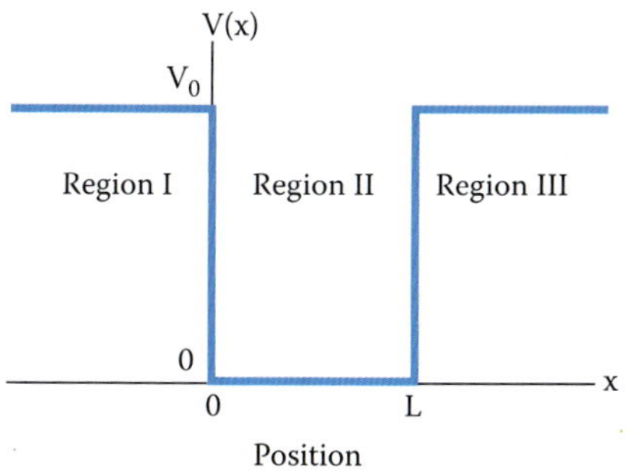

FIGURE 3.70 Quantum well with finite walls.

to most practical 3D (quantum dots), 2D (quantum wires), and 1D (quantum wells) confined structures. The representation also applies to nucleons inside the nucleus of an atom. Even the Krönig-Penney model, used to calculate the potential energy of an electron in a row of atoms in a linear solid, uses an array of periodic square wells of the type shown in Figure 3.70. In this model, each cell with $V(x) = 0$ represents an atom.

Outside the box sketched in Figure 3.70, in regions I and III, the boundary condition is that $V(x) = V_0$. These are regions that are "forbidden" to classical particles with $E < V_0$. With $E < V_0$ a classical particle cannot penetrate a barrier region: think about a particle hitting a metal foil and only penetrating the foil if its initial energy is greater than the potential energy it would possess when embedded in the foil and where otherwise it will be reflected. Defining α as:

$$\alpha^2 = \frac{2m(V_0 - E)}{\hbar^2} \tag{3.165}$$

yields:

$$\frac{d^2\psi}{dx^2} = \alpha^2\psi \tag{3.166}$$

In a region with $E < V_0$ there is an immediate effect on the waveform for the particle because $\mathbf{k}_x$ is real under these conditions, and we can write:

$$\mathbf{k}_x = \alpha = \frac{\sqrt{2m(V_0 - E)}}{\hbar} \tag{3.167}$$

We find for the general solution of Equation 3.164 in regions I and III a wave function of the form $\psi(x) = Ae^{+\alpha x} + Be^{-\alpha x}$, i.e., a mixture of an increasing and a decreasing exponential function. With a barrier that is infinitely thick we can see that the increasing exponential must be ruled out as it conflicts with the Born interpretation because it would imply an infinite amplitude. Therefore, in a barrier region the wave function must simply be the decaying exponential $Be^{-\alpha x}$, the important point being that a particle may be found inside a classically forbidden region (regions I and III).

If the barrier thickness is not infinite, then the increasing exponential component in the wave function cannot be ruled out because the wave function amplitude may not necessarily rise to infinity before the potential drops back to zero. In this case, the solutions for regions I and III are the following damped exponentials:

$$\psi_I(x) = Ae^{+\alpha x} \text{ (region I, } x < 0) \tag{3.168}$$

$$\psi_{III}(x) = Be^{-\alpha x} \text{ (region III, } x > L) \tag{3.169}$$

The values for *A* and *B* of the damped exponentials in Equations 3.168 and 3.169 are determined from the boundary conditions. These boundary conditions include the fact that wave function ψ and its derivative $d\psi/dx$ must be continuous at boundaries between regions I, II, and III (the boundary conditions require that $\psi_I = \psi_{II}$ at $x = 0$ and $\psi_{II} = \psi_{III}$ at $x = L$).

In Figure 3.71, the "leaky" waves in the forbidden regions are shown as exponential tails. The penetration depth of these waves is the distance outside the potential well over which the probability significantly decreases and is given by:

$$\delta x \approx \frac{1}{\alpha} = \frac{\hbar}{\sqrt{2m(V_0 - E)}} \tag{3.170}$$

Thus, the penetration distance that violates classical physics is proportional to Planck's constant and also depends on the value of $V_0 - E$ and on the mass of the particle. Because of this "barrier penetration," the electron density of a material extends outside the surface of the material.

In region II ($0 < x < L$) where $V(x) = 0$, the wave equation becomes:

$$\frac{d^2\psi}{dx^2} = -\mathbf{k}_x^2\psi \tag{3.171}$$

where:

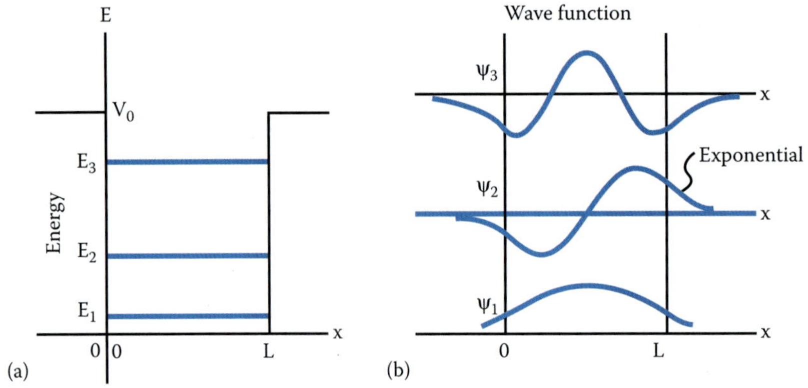

FIGURE 3.71 (a) Energies of a wave function inside the quantum well (region II). (b) The wave functions leak out of the well: see exponential tails.

$$\mathbf{k}_x = i\alpha \text{ and } \alpha = \frac{\sqrt{2mE}}{\hbar} \tag{3.172}$$

The solution here is an oscillating wave just as in the case of the well with infinite walls (Figure 3.56):

$$\psi_{II} = Ce^{i\mathbf{k}_x x} + De^{-i\mathbf{k}_x x}$$

or (3.173)

$$\psi_{II}(x) = \psi_0 \sin \mathbf{k}_x x + \psi_0' \cos \mathbf{k}_x x$$

Thus, the wave functions for a particle in a well with finite walls look very similar to the ones for the infinite square well ... except that the particle now has a finite probability of "leaking out" of the well!

Finite Height Barriers (Step Functions)

(1) *E > V₀* A potential barrier is the opposite of a potential well. It is a potential energy function with a maximum. For a barrier of finite height and thickness (Figure 3.72) and with $E > V_0$, we use again the TISE (Equation 3.137). For particles outside and above the barrier (regions I and III), V = 0 and we obtain:

$$\begin{aligned} \psi_I(x) &= Ae^{+ik_I x} + Be^{-ik_I x} (x < 0 \text{ and } V = 0) \\ \psi_{III}(x) &= Ee^{+ik_{III} x} + Fe^{-ik_{III} x} (x > L \text{ and } V = 0) \end{aligned} \tag{3.174}$$

These are oscillations with the wave vector:

$$\mathbf{k}_I = \mathbf{k}_{III} = \sqrt{\frac{2mE}{\hbar^2}} \tag{3.175}$$

We have replaced $\mathbf{k}_x$ here with the symbol $\mathbf{k}_I = \mathbf{k}_{III}$ or $\mathbf{k}_{II}$ to be more specific about the regime under consideration.

It is important to recognize here that with $E > V_0$ a particle would, in a classical picture, easily overcome the barrier, and one would expect a 100 % transmission. We will see that this is not the case when the particle is "wavy." In the barrier region (region II), where $0 < x < L$, we calculate:

$$\psi_{II}(x) = Ce^{+ik_{II} x} + De^{-ik_{II} x} \tag{3.176}$$

with the wave vector:

$$k_{II} = \frac{\sqrt{2m(E - V_0)}}{\hbar} \tag{3.177}$$

whereas in the case of $E < V_0$, $\mathbf{k}_x$ is real in a barrier region (see Equation 3.167), in the case of $E > V_0$, $\mathbf{k}_x$ (=$\mathbf{k}_{II}$) remains imaginary. If we are only considering waves moving from left to right (Figure 3.73), we can simplify the above wave functions to:

$$\text{Incident wave: } \psi_I(x) = Ae^{+ik_I x} \tag{3.178}$$

$$\text{Reflected wave: } \psi_I(x) = Be^{-ik_I x} \tag{3.179}$$

$$\text{Transmitted wave: } \psi_{III}(x) = Ee^{+ik_I x} \tag{3.180}$$

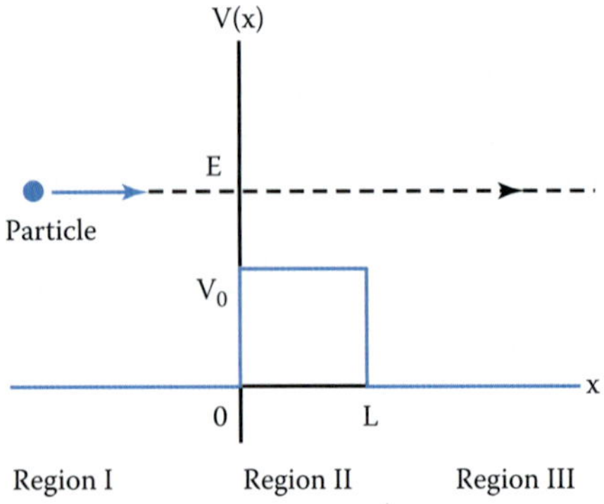

FIGURE 3.72 Finite barrier with boundary conditions.

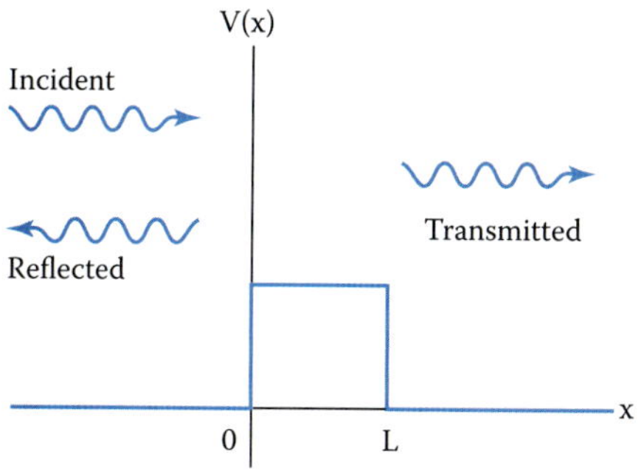

FIGURE 3.73 Wave with $E > V_0$ moving from left to right.

We define a "reflection coefficient" R as:

$$R = \frac{|B|^2}{|A|^2} \quad (3.181)$$

and likewise we can define a transmission coefficient T as:

$$T = \frac{|E|^2}{|A|^2} \quad (3.182)$$

The probability of the particles being reflected R or transmitted T is then:

$$R = \frac{|\psi_I(\text{reflected})|^2}{|\psi_I(\text{incident})|^2} = \frac{|B|^2}{|A|^2}$$

and:

$$T = \frac{|\psi_{III}(\text{transmitted})|^2}{|\psi_I(\text{incident})|^2} = \frac{|E|^2}{|A|^2} \quad (3.183)$$

The transmission probability is the probability that a particle incident on the left of the barrier emerges on the right of it.

(2) $E < V_0$ The situation where classically the particle does not have enough energy to surmount the potential barrier, $E < V_0$, is sketched in Figure 3.74.

With $E < V_0$, the particle will be reflected at x = 0 with the same kinetic energy and Equation 3.176 again applies, but the transmitted wave is now a damped exponential as we saw above. The wave function in region II becomes:

$$\psi_{II}(x) = \underset{\substack{\uparrow \\ \text{Unphysical} = 0}}{Ce^{+\alpha x}} + \underset{\substack{\uparrow \\ \text{Damped}}}{De^{-\alpha x}} \quad (3.184)$$

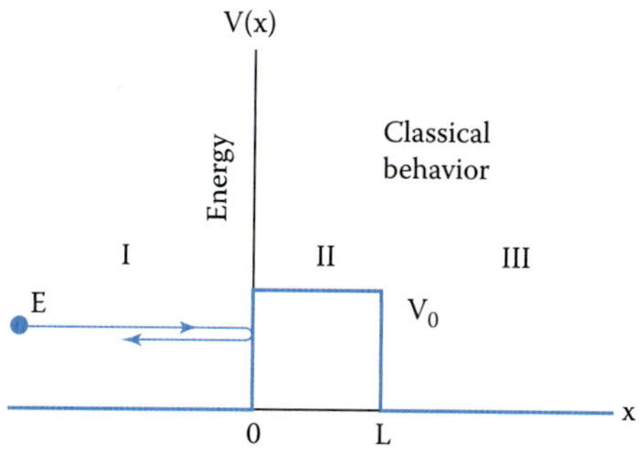

FIGURE 3.74 Wave with $E < V_0$ moving from left to right.

with $k_{II} = \alpha = \frac{\sqrt{2m(V_0 - E)}}{\hbar}$ (see Equation 3.167).

Tunneling We briefly return here to the tunneling phenomenon as described in Equation 3.167. The violation of classical physics in tunneling as described by this equation is allowed by the uncertainty principle. A particle can violate classical physics by ΔE for a short time, $\Delta t \sim \hbar/\Delta E$ (see Equation 3.107). The tunneling wave function is shown in Figure 3.75. The exponential decay of the wave function inside the barrier is given as:

$$\psi(x) = Ae^{-\alpha x} \quad (3.185)$$

with $\alpha^2 = \frac{2m(V_0 - E)}{\hbar}$ (Equation 3.165). If the barrier is narrow enough (L in Figure 3.75 is small), there will be a finite probability P of finding the particle on the other side of the barrier. The probability of an electron reaching across barrier L is:

$$p = |\psi(x)|^2 = A^2e^{-2\alpha L} \quad (3.186)$$

where A is a function of energy E and barrier height V_0. The probability of finding an electron

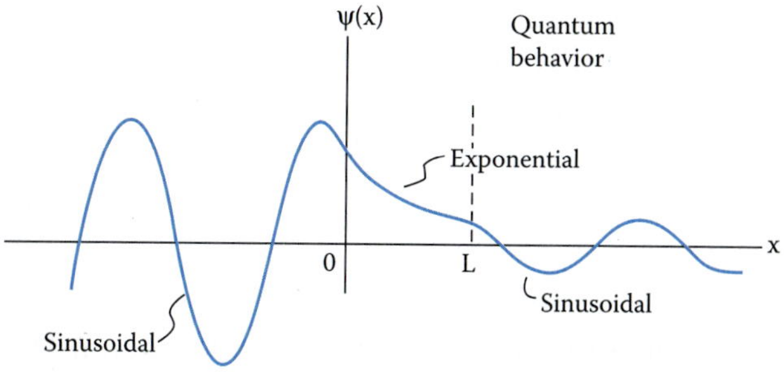

FIGURE 3.75 Tunneling wave function. The rectangular barrier stretches from x = 0 to x = L, the height of the barrier is V_0.

on the other side of a barrier of width L can be probed with a fine needle tip from a scanning tunneling microscope (STM) (see also Volume III, Chapter 6 on metrology). The tunneling current, based on Equation 3.186, picked up by the sharp needle point is given by:

$$I = f_w(E)A^2e^{-2\alpha L} \quad (3.187)$$

where $f_w(E)$ is the Fermi-Dirac function, which contains a weighted local density of electronic states in the solid surface that is being probed and states in the needle point (see also Fermi's golden rule further below). The weighted local density of electronic states is a material property of both probed surface and probe and may be obtained by measurements of the current I as a function of bias voltage V (dI/dV), which gives spatial and spectroscopic information about the quantum states of a nanostructure. From Equation 3.187, when L changes by 1 Å, the current changes by a factor of about 10! Obviously, the current is very sensitive to the gap distance. The size of the gap in practice is on the order of a couple of Angstroms! If the tip has two atoms vying for sitting at the very apex of the tungsten tip, the atom recessed by two atoms lower than the winning atom carries about 1 million times less current. That is why one wants such a fine tip.

Because particles must be either reflected (R) or transmitted (T) we have $R + T = 1$. By applying the boundary conditions $x \rightarrow \pm\infty$, $x = 0$, and $x = L$, we may also calculate the theoretical transmission or tunneling probability P from:

$$P = \frac{1}{1+G} \text{ with } G = \frac{(e^{\alpha L} - e^{-\alpha L})^2}{4\left(\frac{E}{V_0}\right)\left(1 - \frac{E}{V_0}\right)} \quad (3.188)$$

We note that P can be nonzero; i.e., particles may tunnel through a barrier even when $E < V_0$. The wave function does not fall abruptly to zero inside a region where its potential energy exceeds its total energy. This quantum mechanical result is a most remarkable feature of modern physics: there is a finite probability that the particle can penetrate the barrier and even emerge on the other side! We also notice here that even when $E > V_0$, $P < 1$. So a particle that has enough energy to overcome a barrier has a high probability to be reflected instead. This is counterintuitive as a classical particle with that energy would have a $P = 1$. We can summarize this situation as follows: quantum mechanics predicts an enhanced tunneling when $E < V_0$ and an enhanced reflection when $E > V_0$. This is equivalent to light reflecting from an interface with an abrupt change of refractive index.

Example 3.3: Assume that the work function (i.e., the energy difference between the most energetic conduction electrons and the potential barrier at the surface) of a certain metal is $\Phi = 5$ eV. Estimate the distance x outside the surface of the metal at which the electron probability density decreases to 1/1000 of that just inside the metal.

$$\frac{|\Psi(x)|^2}{|\Psi(0)|^2} = e^{-2kx} \approx \frac{1}{1000}$$

This leads to:

$$x = -\frac{1}{2k}\ln\left(\frac{1}{1000}\right) \approx 0.3 \text{ nm}$$

$$\text{Using } k = \sqrt{\frac{2m_e}{\hbar^2}(V_0 - E)} = 2\pi\sqrt{\frac{2m_e}{h^2}\Phi}$$

$$= 2\pi\sqrt{\frac{5 \text{ eV}}{1.505 \text{ eV}\cdot\text{nm}^2}} = 11.5 \text{ nm}^{-1}$$

with $m = m_e$.

In this section we considered the simplified problem of tunneling through a square barrier, but in most cases the barriers are not simply square shaped. One then needs to obtain a more general expression for the tunneling probability. These calculations are fairly involved, and we refer the reader to the specialized literature (e.g., Wolf's *Principles of Electron Tunneling Spectroscopy*[20]).

Some Tunneling History The concept of tunneling has no analogy in classical mechanics. The experimental manifestations of this effect are one of the many triumphs of the quantum theory. Based on electron tunneling, Fowler and Nordheim, in 1928, explained the main features of electron emission from cold metals by high external electric fields, which had been unexplained since its first observation by Lilienfeld in 1922.

The discovery of the Esaki tunnel diode (see Figure 4.44) was very significant in the history of tunneling as it was the first electronic device where electron tunneling was clearly manifested in a semiconductor. Esaki described the tunnel diode in his 1957 thesis and received the 1973 Nobel Prize for his invention (http://nobelprize.org/nobel_prizes/physics/laureates/1973/esaki-bio.html). The next significant event was early in the 1970s, when the first quantum wells (QWs), which were also the first low-dimensional heterostructures, were demonstrated (see the AlGaAs/GaAs/AlGaAs structure in Figure 3.59).

Esaki was not only the originator of the Esaki tunnel diode; he also invented the double-barrier resonant tunneling diodes (abbreviated DBRT diode; see Figure 5.149) in 1974. Furthermore, in 1969, Esaki* and Tsu initiated research on semiconductor superlattices based on a periodic structure of alternating layers of semiconductor materials with wide and narrow bandgaps, in other words, a series of quantum wells or multiquantum well devices (MQWs).[21] The first superlattices were fabricated using an AlGaAs/GaAs material system (Figure 5.148).

In 1981, Gerd Binnig and Heinrich Rohrer invented the scanning tunneling microscope (STM), enabling the visualization and moving of individual atoms for the first time.

Tunneling is also of great importance in the history of nuclear physics. The decay of a nucleus is the escape of particles bound inside a barrier. The phenomenon of tunneling explains α-particle decay of heavy, radioactive nuclei. Inside the nucleus, an α-particle feels the strong, short-range attractive nuclear force as well as the repulsive Coulomb force. The nuclear force dominates inside the nuclear radius where the potential can be approximated by a square well. The rate for escape can be very small; particles in the nucleus "attempt to escape" 10^{20} times per second but may succeed in escaping only once in many years! Even if the quantum state (wave function) of the nucleus is completely defined with no uncertainty, one cannot predict when a nucleus will decay (Figure 3.76). Quantum mechanics tells us only the probability per unit time that any nucleus will decay (Figure 3.76a). Electrons, we saw above, are bound with negative total energy and can never escape the nucleus (Figure 3.76b). But a nucleon, such as an α-particle, is held inside the nucleus by the strong nuclear force and can escape by tunneling through the positive Coulomb barrier, which leads to the nuclear decay processes (G. Gamov, 1928). The potential barrier at the nuclear radius is several times greater than the energy of an α-particle. Based on Equation 3.107, for a short time, a particle can "borrow" energy from the uncertainty relation, gaining enough energy to jump over the potential barrier before giving it back. When it returns to its "proper" energy state, it is just outside the barrier instead of just inside, and rushes away. The process is as if the particle tunneled through the barrier, and it is purely a quantum effect.

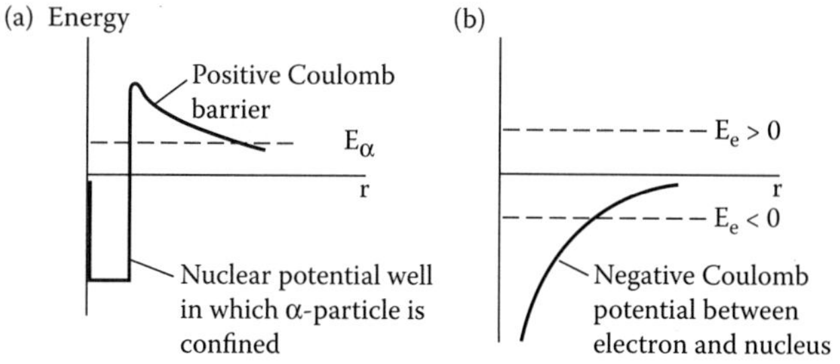

FIGURE 3.76 Nucleons escape the nucleus in radioactive decay (a); electrons cannot escape the nucleus (b).

Interfaces The theoretical treatment of particles hitting an interface is the same as the treatment of particles hitting one side of a square barrier (half square or half step function) and finds all types of applications. The results allow one to handle important problems involving the transmission and reflection of particle waves, such as encountered in vacuum/metal interfaces (work function), vacuum/semiconductor interfaces, semiconductor/metal interfaces, metal/metal interfaces (e.g., in thermocouples), and semiconductor/semiconductor interfaces (e.g., in diodes). For this analysis no new equations are needed as we can retrieve all the necessary expressions from our treatment of the full square barrier as summarized in Table 3.7.

* Esaki worked for IBM at the Thomas J. Watson Research Center, Yorktown Heights, New York, until 1992, when he returned to Japan to become president of Tsukuba University, Ibaraki.

Table 3.7 Transmitted and Reflected Particle Waves on an Interface

Energy	Equations	Wave Vectors
$E=\frac{\langle p^2\rangle}{2m}>V_0$	Incident $\psi_I(x)=Ae^{+ik_Ix}+Be^{-ik_Ix}$ Reflected Transmitted $\psi_{II}(x)=Ce^{+ik_{II}x}+De^{-ik_{II}x}$ Unphysical = 0	$k_I=\frac{\sqrt{2mE}}{\hbar}$ $k_{II}=\frac{\sqrt{2m(E-V_0)}}{\hbar}$
$E=\frac{\langle p^2\rangle}{2m}<V_0$	Incident $\psi_I(x)=Ae^{+ik_Ix}+Be^{-ik_Ix}$ Reflected Unphysical = 0 $\psi_{II}(x)=Ce^{+\alpha x}+De^{-\alpha x}$ Damped	$k_I=\frac{\sqrt{2mE}}{\hbar}$ $\alpha=\frac{\sqrt{2m(V_0-E)}}{\hbar}$

Harmonic Potential Wells

We now revisit the harmonic oscillator from Figure 3.51 and solve it in the Schrödinger way so we may learn how quantization comes about in the case of an oscillating system. Simple harmonic oscillators describe many physical situations from springs to diatomic molecules and atomic lattices. In all cases, harmonic oscillations occur because the system contains a part that experiences a restoring force (spring) proportional to the displacement from equilibrium. In Figure 3.77, depicting a harmonic potential well, we discern two regions: inside the well (region I), with $E > V(x)$, and outside of the well (region II), with $E < V(x)$.

The time-independent Schrödinger equation (TISE) for a harmonic oscillator is given as:

$$-\frac{\hbar^2}{2m}\frac{d^2\psi(x)}{dx^2}=\left(E-\frac{1}{2}kx^2\right)\psi(x) \qquad (3.189)$$

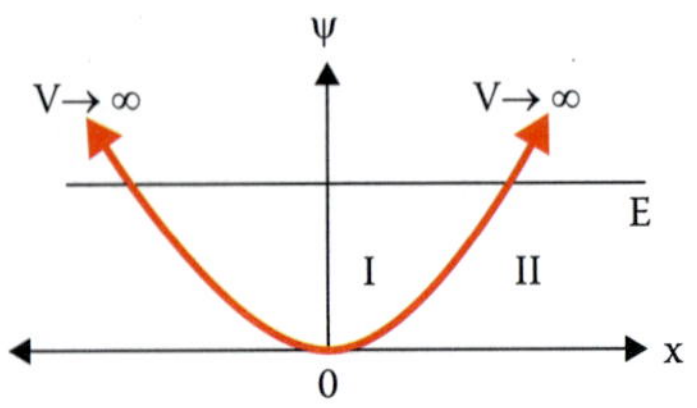

FIGURE 3.77 Harmonic potential well.

where $V(x)=\frac{1}{2}kx^2$ (in this equation k is the force constant, not the wave number **k**!) and $\omega=\left(\frac{k}{m}\right)^{\frac{1}{2}}$.

The general solutions for Equation 3.189 are bell-shaped Gaussian functions multiplied by a Hermite polynomial $H_n(y)$:

$$\psi_n(x)=N_nH_n(\alpha^{\frac{1}{2}}x)e^{-\frac{\alpha x^2}{2}}$$

$$\text{with } n=0,1,2,3\ldots \text{ and } \alpha=\left(\frac{K_m}{\hbar^2}\right)^{\frac{1}{2}}$$

$$N_n(\text{a normalization constant})=\frac{1}{(2^n n!)^{\frac{1}{2}}}\left(\frac{\alpha}{\pi}\right)^{\frac{1}{4}} \qquad (3.190)$$

Some values for $H_n(y)$ are listed in Table 3.8.

The energy is given by:

$$E_n=\left(n+\frac{1}{2}\right)\hbar\omega,\ n=0,1,2,3\ldots \qquad (3.191)$$

with n the vibrational quantum number. As shown in Figure 3.78, energies are evenly spaced and again cannot be zero in the ground state because of quantum confinement. Here we start the quantum numbers from n = 0 rather than n = 1.

The ground state (n = 0) of the wave inside the box is a simple Gaussian with no zero crossings (Figure 3.79a):

$$\psi_{n=0}(x)=\left(\frac{\alpha}{\pi}\right)^{\frac{1}{4}}e^{-\frac{\alpha x^2}{2}} \qquad (3.192)$$

with zero point energy:

$$E_0=\frac{1}{2}\hbar\omega \qquad (3.193)$$

Table 3.8 Hermite Polynomials, $H_n(y)$

n	$H_n(y)$
0	1
1	$2y$
2	$4y^2 - 2$
3	$8y^3 - 12y$
4	$16y^4 - 48y^2 + 12$
5	$32y^5 - 160y^3 + 120y$

n = vibrational quantum number.

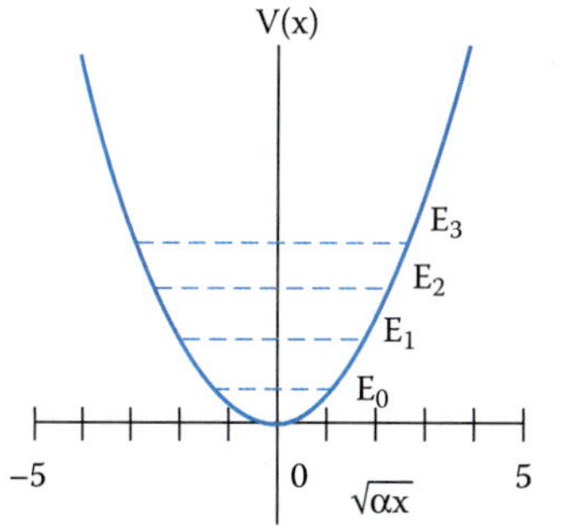

FIGURE 3.78 Energies of the quantum oscillator according to Equation 3.191, $\alpha = \left(\frac{K_m}{\hbar^2}\right)^{\frac{1}{2}}$.

The first excited state (n = 1): $\psi_{n=1}(x) = \left(\frac{\alpha}{\pi}\right)^{\frac{1}{4}} \sqrt{2\alpha} x e^{\frac{-\alpha x^2}{2}}$ has one zero crossing; the second excited state (n = 2): $\psi_{n=2}(x) = \left(\frac{\alpha}{\pi}\right)^{\frac{1}{4}} \frac{1}{\sqrt{2}}(2\alpha x^2 - 1)e^{\frac{-\alpha x^2}{2}}$ has two zero crossings; the third excited state (n = 3): $\psi_{n=3}(x) = \left(\frac{\alpha}{\pi}\right)^{\frac{1}{4}} \frac{1}{\sqrt{3}} \sqrt{\alpha}\, x(2\alpha x^2 - 3)e^{\frac{-\alpha x^2}{2}}$ has three crossings; and so on. From Figure 3.79b the lowest energy state (n = 0) of the QM oscillator has a maximum probability at the equilibrium position, whereas the classical oscillator always has its maximum probability at the extremes.

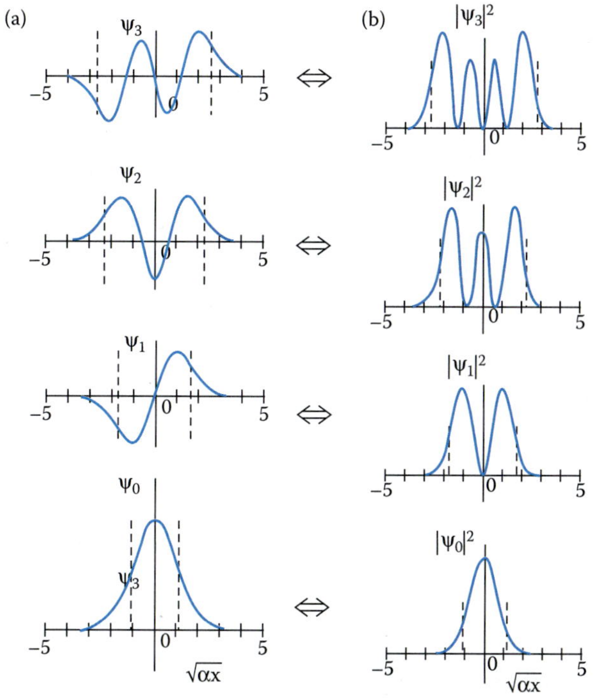

FIGURE 3.79 (a) The waves for n = 0 to n = 3. (b) Probabilities for the same waves.

As expected from Bohr's correspondence principle, the higher the quantum numbers, the better the quantized oscillator resembles the classical non-quantized oscillator from Figure 3.51. This is illustrated in Figure 3.80.

Quantized harmonic oscillators are all around us: diatomic molecules, vibrations within molecules, vibrations of atoms about equilibrium positions, oscillations of atoms or ions in crystal lattices (phonons), normal modes of electromagnetic fields in a cavity (blackbody radiation), Landau levels in the quantum Hall effect, and to a first approximation, any oscillatory behavior. We will encounter another example of the harmonic quantum oscillator solution presented here in the solution Einstein proposed in 1903 for the energies of the "atomic oscillators" in solids (see below under "Classical and Quantum Oscillators").

Example 3.4: The spacings between vibrational levels of molecules in the atmosphere, CO_2 and H_2O, are in the infrared frequency range: $\Delta E = h\nu = \hbar\omega \sim 0.01$ eV. As a consequence, Earth has an atmosphere acting as a greenhouse.

Particles in an Atom: Central Force

An electron bound to the hydrogen nucleus is an example of a central force system: the force depends on the radial distance between the electron and the nucleus only. The potential energy associated with a central or inverse square law force is very important

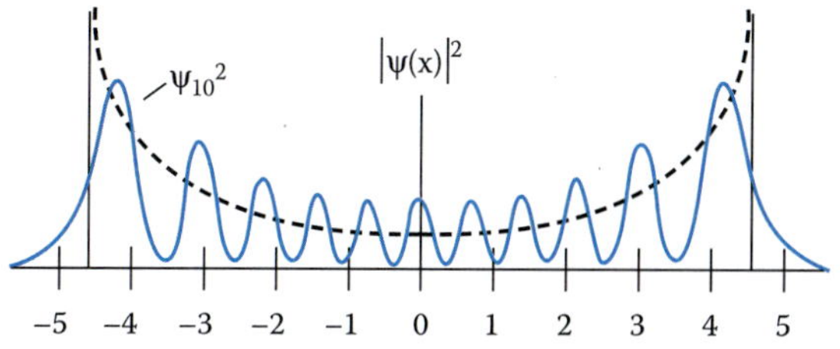

FIGURE 3.80 Quantized oscillator. Probability densities *P* for n = 10 states of a quantum mechanical harmonic oscillator. The probability densities for classical harmonic oscillators with the same energies are shown in black. In the n = 10 state, the wavelength is shortest at x = 0 and longest at x = |L|. The higher the quantum number *n*, the closer the solution resembles the classical one represented in Figure 3.51.

and was illustrated earlier for the case of a hydrogen atom in Figure 3.53. The potential energy of the electron-proton system is:

$$V(r) = -\frac{e^2}{4\pi\varepsilon_0 r} \tag{3.194}$$

The solutions of Schrödinger's equation with this potential are spherical Bessel functions. In Figure 3.53 the probability of finding the ground state for a hydrogen electron (n = 1) as a function of the radial distance from the proton was shown. The value of $|\Psi(x,t)|^2$ at some location at a given time is proportional to the probability of finding the particle at that location at that time. The wave functions for hydrogen are like vibrating strings or membranes, but the vibrations are in three dimensions and they are described by spherical Bessel functions. In Figure 3.81 we show solutions for higher quantum number cases (up to n = 4). In Figure 3.81A we provide a summary, and Figure 3.81B details some of these solutions for hydrogen. The lowest shells are spherical, 1s (n = 1, l = 0, m = 0), 2s (n = 2, l = 0, m = 0), 3s (n = 3, l = 0, m = 0), and 4s (n = 4, l = 0, m = 0), and hold two electrons each. The next shells are three 2p orbitals (n = 2, l = 1 and m = 1, 0, −1), which are dumbbell shaped and fit into the atom along perpendicular axes. A total of six electrons can fit. Level three (n = 3) has one s orbital and three p's—just like the ones at level two—and five three d's (with m = −2, −1, 0, +1, +2). The d orbitals are shaped for the most part like four-leaf clovers. At level four (n = 4) we encounter for the first time the 4f orbitals. They hold 14 electrons and look somewhat like onion blossoms. The electrons in the 4f orbitals are the furthest removed from the nucleus, are easily excited, and exhibit thousands of distinct energy states. They are associated with the many unusual optical, magnetic, and catalytic properties of the rare earth elements.

We come back to the mathematical formulation of the solutions of Schrödinger's equation for a central force when analyzing energy levels for quantum dots. These "artificial atoms" have energy quantization just as like atoms and molecules.

Summary: Most Important Periodic Potential Profiles and the Sommerfeld Model

Summarizing, the quantization for three of the most important potential profiles leads to the following mathematical solutions of Schrödinger's equation: for the central force, we obtain spherical Bessel functions; for an infinite square well potential, sines, cosines, and exponentials; and for an oscillator, Hermite polynomials. The quantized energy levels for each case are summarized in Figure 3.82 (only the 2D case is shown). Notice that the separation between consecutive energy levels increases with increasing *n* for an infinite square well potential (*b*) are evenly spaced for a harmonic well potential (*c*), and in the case of an inverse square law potential (*a*) the separation becomes closer with the larger *n*.

Sommerfeld's model assumes that the electrons in a metal experience a constant zero potential so

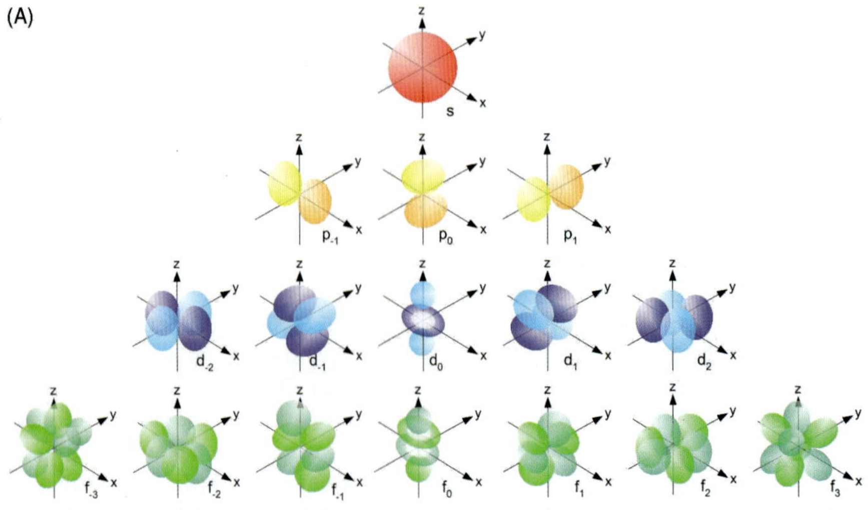

FIGURE 3.81 (A) Summary or overview of atomic orbitals up to n = 4.

(B)(a) (b) (c) (d)

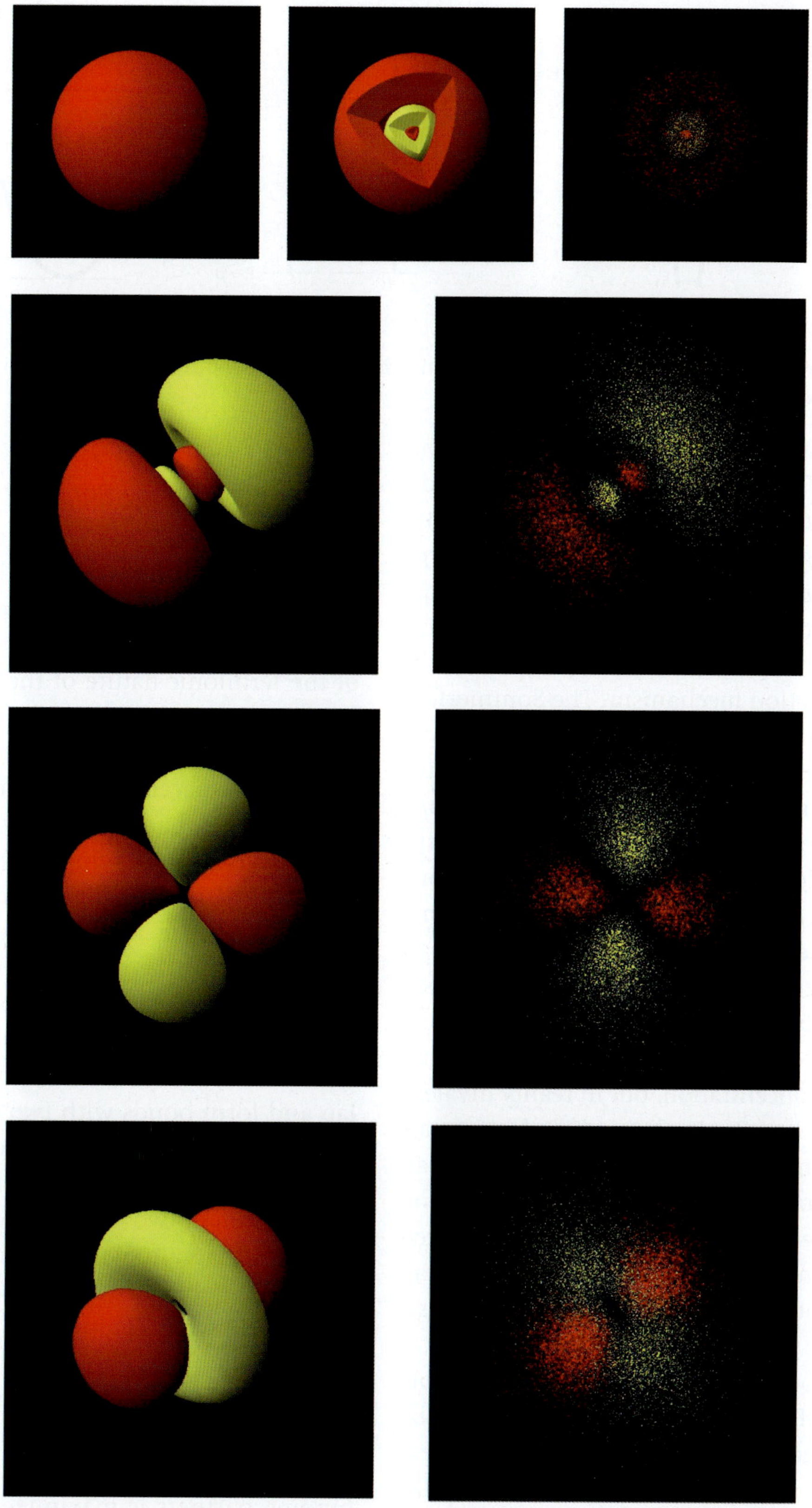

FIGURE 3.81 (*Continued*) (B) Where red = negative phase of Ψ and yellow = positive phase of Ψ. The density of dots reflects the magnitude of Ψ. (a) n = 3, l = 0, m = 0 for hydrogen. (b) n = 3, l = 1, m = −1 for hydrogen. (c) n = 3, l = 2, m = −1 for hydrogen. (d) 4 = 3, l = 3, m = 0 for hydrogen.

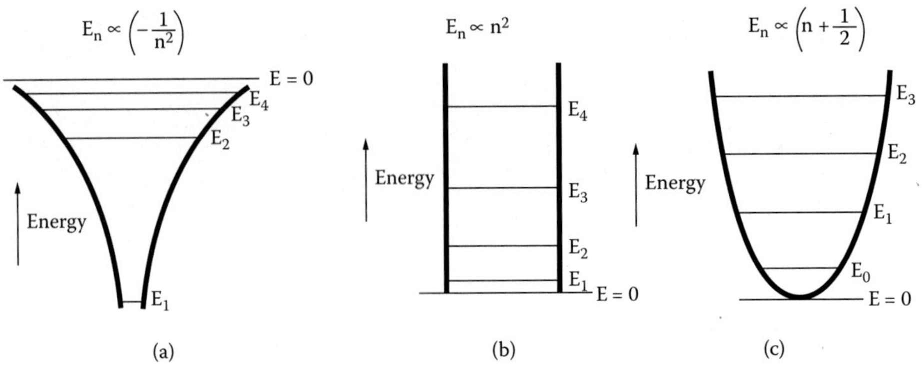

FIGURE 3.82 Quantized energy levels for a particle in an inverse square law potential (a) an infinite square well potential (b) and a harmonic well potential (c) 2D cases.

that the electrons are completely free to move about in the crystal. The important thing Sommerfeld's model does is introduce a finite surface for the metal at L and adopting a Fermi distribution for the charge carriers. What is most different from the classical theory as a result is the notion that only a few electrons, those with energies close to the Fermi level, contribute to the conduction mechanism. The Sommerfeld theory successfully explains specific heat, electrical conductivity, thermionic emission, thermal conductivity, and paramagnetism. However, the model fails to explain why some solids are good conductors, others are semiconductors, and yet others are insulators. The model also cannot account for the fact that some metals such as Be, Zn, and Cd exhibit a positive Hall constant; the free electron model always predicts a negative Hall coefficient! This model further predicts that the electrical conductivity is proportional to the electron concentration, but in reality divalent metals (Be, Cd, and Zn) and even trivalent metals (Al, In) have consistently lower conductivities than monovalent metals (Cu, Ag, and Au). Also, measurements of the Fermi surface, a concept introduced further below, indicate that it is often not spherical, contradicting Sommerfeld's model, which predicts a perfect sphere.

We prepare ourselves now to launch the more realistic Krönig-Penney model, where the potential energy of an electron in a row of atoms in a linear solid is modeled as an array of periodic square wells. We start by learning to combine atoms in simple molecules in the valence bond and molecular orbital theory; we then learn how to work with Bloch functions; and at last we are ready to introduce the band theory of solids.

Bringing Atoms Together

Molecules: Valence Bond and Molecular Orbital Theory

Molecules are formed from atoms by quantum mechanical forces. The so-called covalent bonds have no classical counterparts. They exist only because of the fermionic nature of the valence electrons. In the valence bond (VB) theory, atoms form electron-pair covalent bonds through the overlap of atomic orbitals of adjacent atoms. In the molecular orbital (MO) theory, shared, delocalized, valence electrons are viewed as occupying regions of space extending over all the binding atoms. When two atoms come together to form a molecule a valence bond is formed, and the valence electrons are in orbitals (called molecular orbitals) spread over the entire molecule, i.e., the electrons are delocalized. Half-filled atomic orbitals from the binding partners overlap and form bonds with two electrons of opposite spin occupying the molecular orbital. The latter corresponds to Pauli's exclusion principle, i.e., there are a maximum of two electrons with opposite spin per orbital. A diagram of the potential energy versus the internuclear distance for two approaching hydrogen atoms is shown in Figure 3.83a. The example involves the formation of a sigma bond (σ) from the overlap of s orbitals from the binding hydrogen atoms. At an internuclear distance, r_0, of 74 picometers, equilibrium is reached, and the attraction between the binding partners is maximized with an H-H bond strength of −436 kJ/mol. Hydrogen atoms react to form a molecule because the energy of the system is less than the sum of the individual constituents. From Figure 3.83b and c, we learn that compared

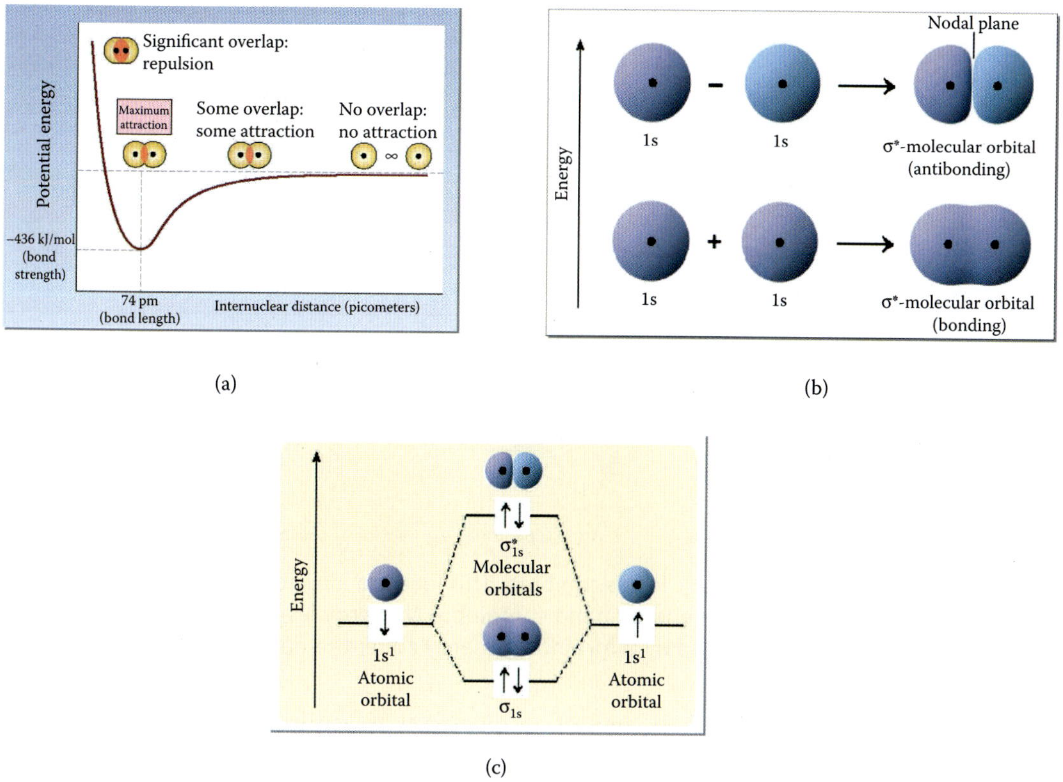

FIGURE 3.83 (a) A diagram of the potential energy versus the internuclear distance for two approaching hydrogen atoms. A sigma bond (σ) is formed from the overlap of the *s* orbitals from the binding hydrogen atoms. At an internuclear distance, r_0, of 74 picometers, equilibrium is reached, and the attraction between the binding partners is maximized with a H-H bond strength of −436 kJ/mol. (b) A bonding and antibonding molecular orbital is formed. (c) The bonding molecular orbital (σ_{1s}) is lower in energy than the parents' atomic orbitals, and the antibonding (σ^*_{1s}) is higher in energy.

with a single hydrogen atom there are twice as many theoretically permitted electron shells; the bonding molecular orbital (σ_{1s}) is lower in energy than the parents' atomic orbitals, and the antibonding ($\sigma^*{}_{1s}$) is higher in energy. The available electrons of the just created molecule are assigned, just as in the case of an atom, orbitals of successively higher energy and in the case of orbitals of equal energy—*degenerate orbitals*—these are filled one electron at a time before paring begins (Hund's rule), and only two electrons of opposite spin can occupy the same energy level (Pauli's exclusion principle). The total number of molecular orbitals equals the number of atomic orbitals contributed by the atoms constituting the molecule. When we examine a very large *N*-atom molecule, such as a long carbon chain hydrocarbon, we find a splitting of each one-atom energy level into *N* energy levels, each one corresponding to a somewhat different electron shell form.

Molecules also have excited states originating from a variety of sources. Because they are not usually spherically symmetric, molecules can rotate around several of their axes. The rotational energy is again quantized and depends on the *l* quantum number. The atoms in the molecule also vibrate compared to each other around an equilibrium distance. This can be described well by a harmonic oscillator spectrum (Equation 3.191).

To form a solid, one keeps adding atoms in three dimensions, making a very large molecule, and that is exactly what we will do next. We will see that in solids, just like in molecules, higher energy states occur where they interact to form a higher band of allowed energies.

Solids: A First Look at The Band Model of Solids

For the total number *N* of atoms in a solid (~10^{23} cm^{-3}), *N* energy levels split apart within a width ΔE at r_0 as shown in Figure 3.84a, where first two, then six, and then *N*, 3s atomic levels combine. This leads to a band of energies for each initial atomic energy level, and 2*N* electrons may occupy an energy band

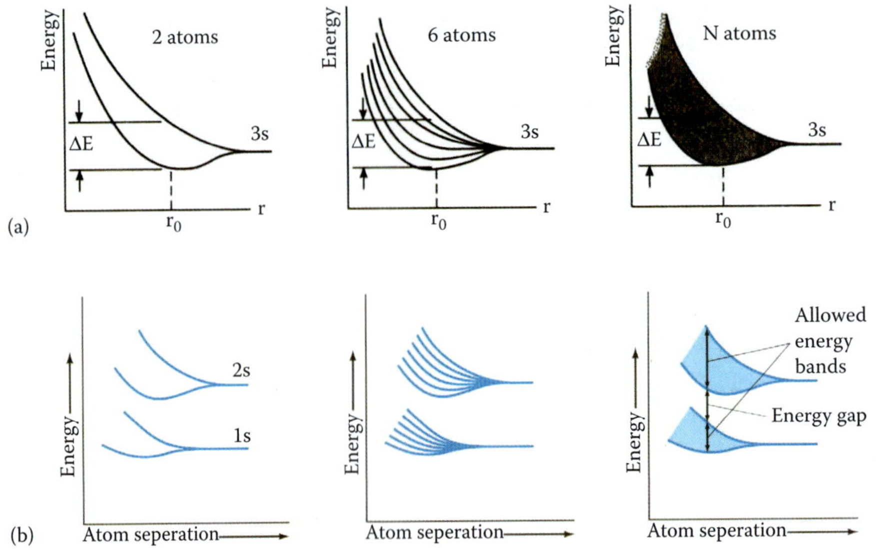

FIGURE 3.84 (a) Two, six, and then *N*, 3s atomic levels combine. This leads to a band of energies for each initial atomic energy level, and 2*N* electrons may occupy an energy band containing *N* energy levels. (b) Sets of 1s and 2s levels are combined into allowed energy bands separated by a forbidden energy zone, the bandgap, E_g.

containing *N* energy levels. In Figure 3.84b, sets of 1s and 2s levels are combined into allowed energy bands separated by a forbidden energy zone, the bandgap, E_g. A group of energy states incompletely filled at room temperature and empty at 0 K is called a conduction band. A conduction band is able to support the movement of electrons. A band of energy levels, missing some electrons at room temperature and completely filled at 0 K, is called a valence band. The valence band supports the movement of missing electrons, which are called holes. As illustrated in Figure 3.85, the relative position of conduction band, valence band, and forbidden bandgap are instructive for the classification of materials into conductors, semiconductors, and insulators. In conductors, the conduction band is partially filled, allowing electrons to move freely, or the valence band overlaps with the conduction band, again enabling free electron movement. In insulators there is a substantial forbidden energy gap between completely filled valence band and empty conduction band (~9 eV in Figure 3.85). Semiconductors are similar to insulators, but the bandgap is narrower, and electrons and/or holes are available at room temperature. Because in semiconductors the energy gap is small, thermal energy might suffice for some electrons to jump from the valence band to the conduction band; however, more often doping with impurities is needed to generate enough charge carriers at room temperature.

Altering the band structure, confining the geometry or the potential of electrons, leads to engineered states with very interesting unseen properties.

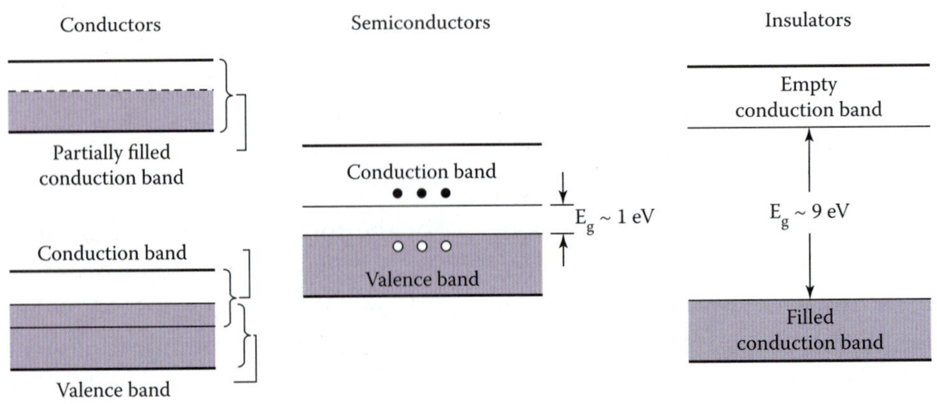

FIGURE 3.85 Classification of materials into conductors, semiconductors, and insulators based on the band model.

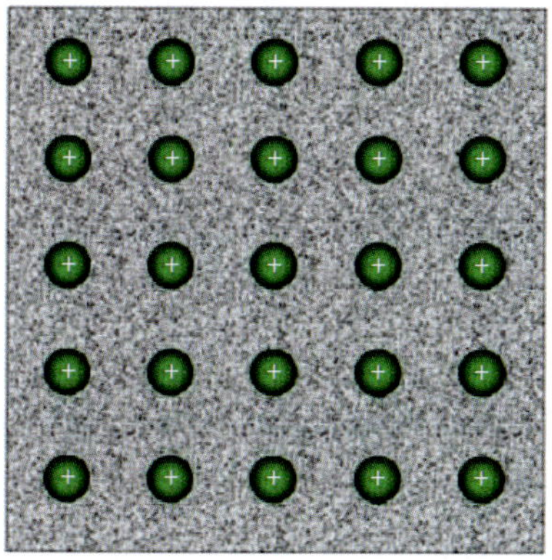

FIGURE 3.86 Positive ion cores in a metal.

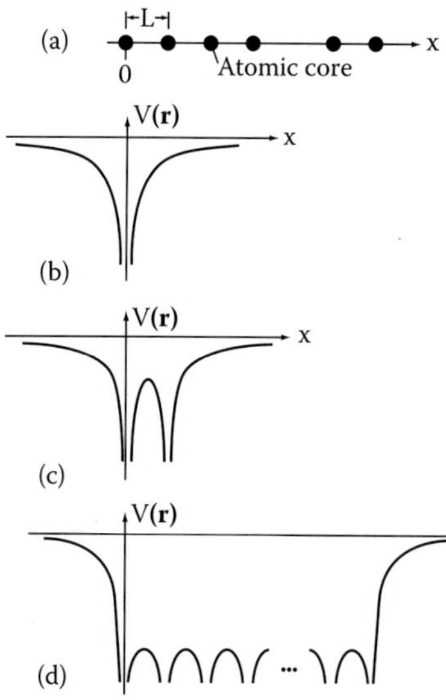

FIGURE 3.87 Bringing atoms together in a lattice. (a) Atomic cores; (b) isolated atom with binding energy as a function of *V*(**r**); (c) two atoms brought together; and (d) array of atoms with binding energy as a function of *V*(**r**).

Bloch Functions

Introduction Armed with some understanding of quantum mechanics and an introduction to the valence bond (VB) and molecular orbital (MO) theory, we are in a much better position to understand the band model of solids. The Drude free electron gas model surveyed above explained a number of important metallic properties, but it did not explain the most intriguing problems: what is it that distinguishes a metal from an insulator, and how can the same material, say carbon, form a conductor, a superconductor, a semiconductor, or an insulator? These facts remain unanswerable until one also takes into account that an electron gas moves through space occupied by a periodic array of positive-charged atomic cores (Figure 3.86). When electrons are free to roam in an infinite solid, e.g., in a metal, the electronic "orbitals" are traveling wave solutions, but when the metal is limited to a cube of size L^3 or a periodic array of positive ions is assumed, this will act to restrict the allowed energies and hence other quantities such as momentum, spin, and so on. Let us specify the force acting on the "free" electrons as $V(\mathbf{r})$.

Because the ion cores in a crystal are arranged periodically, the potential an electron feels is also periodic; with the periodicity of the underlying Bravais lattice in three dimensions, this yields the Born and von Karman's periodic boundary condition:

$$V(\mathbf{r}+\mathbf{R}) = V(\mathbf{r}) \tag{3.195}$$

for all Bravais lattice vectors **R** (see also Chapter 2). In Figure 3.87 we illustrate how the potential $V(\mathbf{r})$ for a single atom compares with that of two and with that of an array of atoms. Electrons in isolated atoms occupy discrete allowed energy levels E_1, E_2, and so on, with a potential energy of the electron at a distance r from a positively charged nucleus q given as (see Figure 3.87a):

$$V(\mathbf{r}) = \frac{-qe}{4\pi\varepsilon_0 \mathbf{r}} \tag{3.196}$$

The 1D potential energy of an electron caused by an array of nuclei of charge q separated by a distance **R** is:

$$V(\mathbf{r}) = \sum_{n} \frac{-qe}{4\pi\varepsilon_0 |\mathbf{r} - n\mathbf{R}|} \tag{3.197}$$

where n = 0, ±1, ±2, and so on. This is shown in Figure 3.87d. The periodic potential in one dimension, shown here, is a Krönig and Penney-like potential (see Figure 3.98). From this figure, $V(\mathbf{r})$ is lower in the solid than in the isolated atoms. In the lowest binding energy states, conduction electrons move in a nearly constant potential as shown in Figure 3.88.

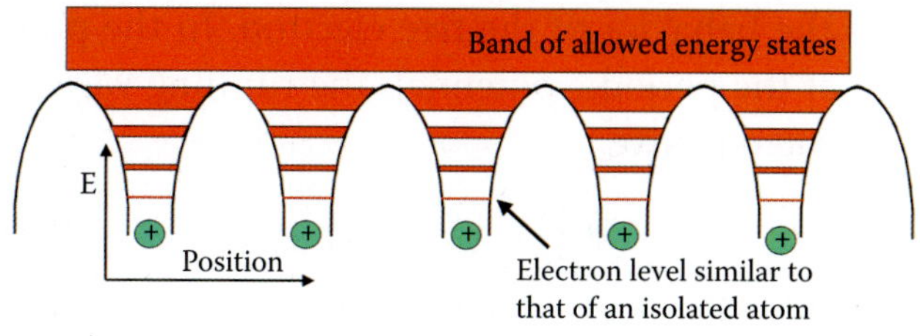

FIGURE 3.88 Band formation.

These electrons are trapped within the metal; a work function of several electron volts prevents them from escaping from the surface.

The higher energy states of tightly bound electrons are very similar to those of the isolated atoms. Lower binding electron states from the atoms become bands of allowed states in the crystal, and we will learn shortly that with partially filled bands the solid becomes a conductor.

Consider now what potentials an electron would see as it moves through a crystal lattice (limited to 1D for now). The electrostatic potential, $V(x)$, is periodic such that $V(x + L) = V(x)$. Bloch's theorem states that because the potential repeats every L length, the magnitude of the wave function (but not necessarily the phase) must also repeat every L length. This is the case because the probability of finding an electron at a given point in the crystal must be the same as found in the same location in any other unit cell. Generalizing, according to Bloch's theorem (1928), if $\psi_0(\mathbf{r})$ is a solution of Schrödinger's equation in free space, then a solution in a potential field that is periodic with period $\mathbf{R}$ is a product of $\psi_0(\mathbf{r})$ and another function $V(\mathbf{r})$, which is itself periodic with period $\mathbf{R}$. This theorem is one of the most important formal results in all of solid-state physics because it tells us the mathematical form of an electron wave function in the presence of a periodic potential energy. Thus, the wave function of the electron in a periodic potential $V(\mathbf{r})$ has the form:

$$\psi(\mathbf{r}) = \psi_0(\mathbf{r}) \cdot V(\mathbf{r}) \quad (3.198)$$

These are the so-called Bloch functions. In the Bloch approach an electron is considered to belong to the crystal as a whole rather than a particular atom. We have encountered the solution for $\psi_0(\mathbf{r})$ already. It is a running wave that can be written as a sine or cosine function or, more generally, in the form $e^{i\mathbf{k}\cdot\mathbf{r}}$, where $\mathbf{k}$ is equal to $2\pi/\lambda$, λ being the wavelength of the electron in this case. The running wave represents the behavior of the free electron. In the presence of the periodic potential, the running wave is modulated to give:

$$\psi(\mathbf{r}) = e^{i\mathbf{k}\cdot\mathbf{r}} \cdot V(\mathbf{r}) \quad (3.199)$$

where $V(\mathbf{r}+\mathbf{R}) = V(\mathbf{r})$ for all R in the Bravais lattice. The Bloch function can be interpreted (approximately) as describing the electron distribution within a single unit cell, and an overall phase variation term $e^{i\mathbf{k}\cdot\mathbf{r}}$ represents a phase difference of the wave function in adjacent unit cells. The latter can take on several values depending on the wave vector $\mathbf{k}$. For core electrons (those tightly bound to the nucleus) Equation 3.199 represents a strongly localized wave function similar to the electron orbitals around a hydrogen atom. The less strongly bound valence electrons are described by more extended wave functions that have significant amplitude between neighboring atoms. Free electrons are described by wave functions with a high energy E, such that the wave vector remains real between atoms [where $V(\mathbf{r})$ is high]. This important result shows that the wave function for the electron itself has the periodicity of the lattice. As a consequence, just like with x-rays or vibrations of atoms in crystals (phonons), only certain wavelengths (that is, energies) are permissible for electrons in crystals. Importantly, we shall see that this leads, among other things, to a natural distinction among metals, insulators, and semiconductors. From Equation 3.199 we must have standing waves in the crystal that have a period equal to a multiple of the period of the crystal's electrostatic potential (similar to a multilayer antireflection coating in optics). Indeed, when an electron travels in a solid and enters a region with a lower potential energy (e.g., closer to a positively charged atom), the kinetic energy goes up, and the wave function acquires a shorter wavelength. This behavior of electron waves entering a low potential region resembles that of light entering a high refractive index region. Just as for light, changes in wavelength give rise to reflections, in this case electron wave reflection. Inside a crystal, electrons experience a periodic potential caused by the regularly spaced atomic cores in the crystal lattice, leading to multiple electron wave reflections and electron wave interference. The multiple reflections result in eigenmodes that are affected by the exact shape of the periodic potential $V(x)$. For an infinite crystal, the eigenfunctions describe a state with a well-defined energy and a corresponding spatial distribution of the electron throughout the entire crystal. In a simple linear lattice with lattice spacing a, we have $V(\mathbf{r} + \mathbf{R}) = V(\mathbf{r})$, as shown above.

It is important to note that because the wave function repeats each unit cell, we only have to consider what happens in one unit cell to describe the entire crystal. Thus, we can restrict ourselves to values of **k** such that $-\pi/a$ to $+\pi/a$ [implying **k**a ≤ 1 or $(2\pi/\lambda)\ a \leq 1$], or we can describe the electron behavior in a solid with wave vectors that lie in the first Brillouin zone.

Bloch Function Applied to a Six-Atom Linear Lattice To get a better appreciation about Bloch functions, let us consider an array of six atoms (N = 6), as illustrated in Figure 3.89. The Bloch function for an electron in this linear solid is $\psi_n(x) = V(x)e^{ik_nx}$, with a function $V(x)$ that depends on the electronic states involved (see Figure 3.90). The Born and von Karman's periodic boundary condition (Equation 3.195) can only be satisfied if the wave vector **k** has *N* possible values from **k** = π/L to **k** = π/a, with *L* the total length of the six-atom crystal and *a* the lattice spacing. In other words, $\mathbf{k}_n = n\pi/L$, with n = 1, 2, 3… L/a, or there are N = L/a states. This is really nothing else than the electron in a box problem, except that the box is now divided in six compartments!

For N = 6, there are six different superpositions of the atomic states that form the crystal states as shown in Figure 3.91 (where we only consider the 1s combinations).

Bloch Function for Metals Let us now take an example metal and determine which states its electrons

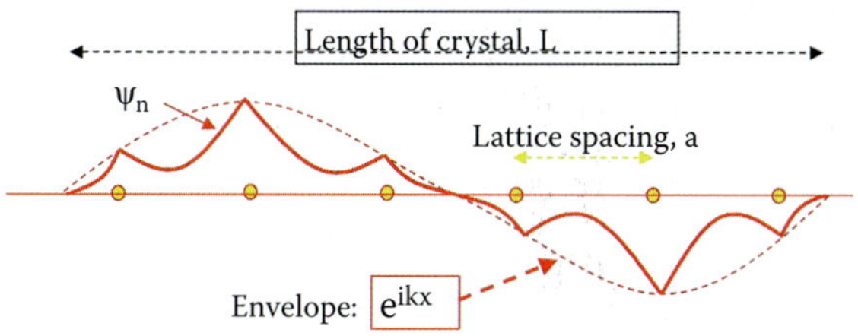

FIGURE 3.89 The states in a six-atom linear crystal (N = 6).

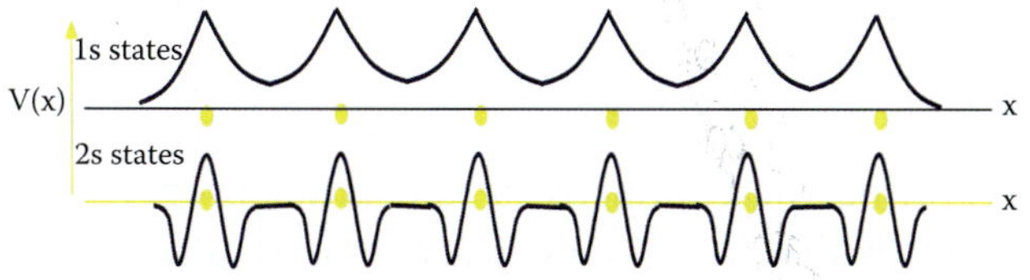

FIGURE 3.90 The periodic potential *V*(*x*) depends on the electronic states of the atoms involved.

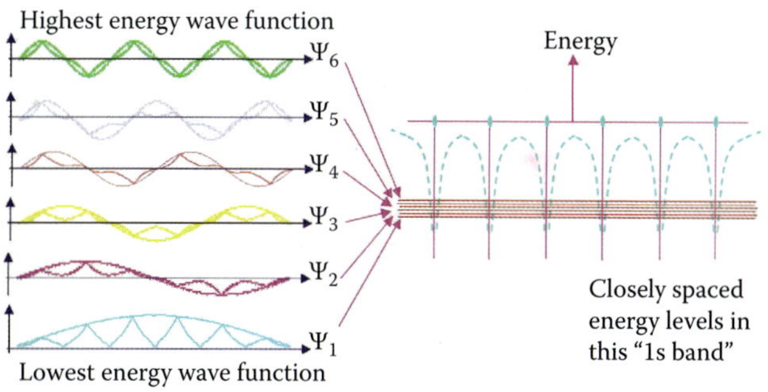

FIGURE 3.91 Six different super positions of the 1s atomic states form the crystal states.

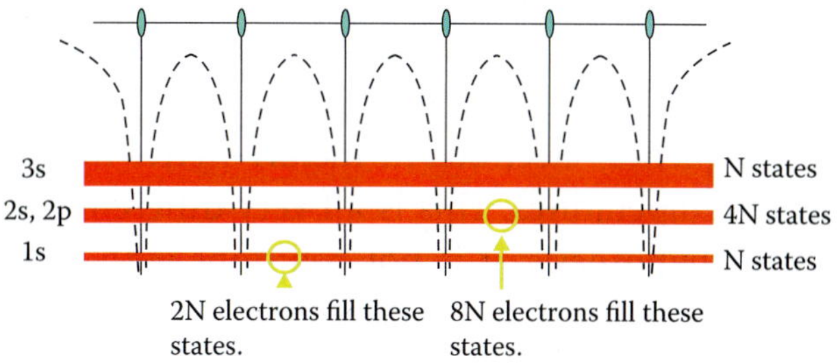

FIGURE 3.92 Filling the Bloch functions of sodium with electrons.

are occupying. In Figure 3.92 we have filled the Bloch functions of sodium with electrons according to Pauli's principle. We consider *N* sodium atoms, and because for sodium Z = 11 ($1s^22s^22p^63s^1$), this means we have 11*N* electrons to distribute. Notice that the 3s band is only half filled (*N* orbital states and *N* electrons). Electrons in this 3s band are easily promoted to higher states in the band (Figure 3.93), and this is what makes sodium a good conductor. In other words, to get a good conductor one needs a band partially filled with electrons.

The Bloch wave functions in metals are stationary waves, and in a perfectly periodic metal lattice, an electron would freely move without scattering from the atomic cores. This corresponds to a metal without any resistance at all! An electron in a periodic potential has a well-defined wave vector and momentum, and it is only when there are defects in the crystal,

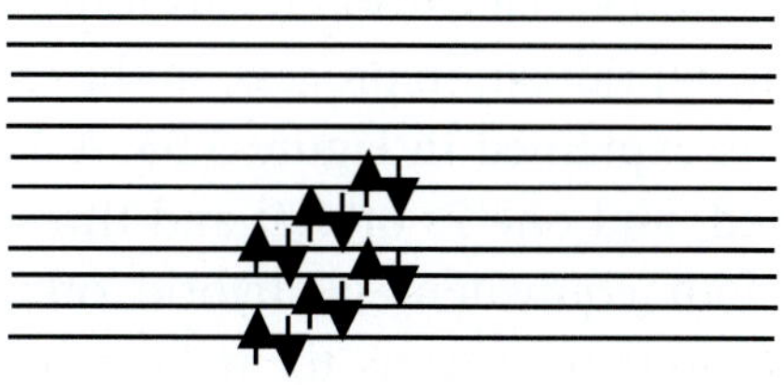

FIGURE 3.93 Half-filled 3*s* band of sodium.

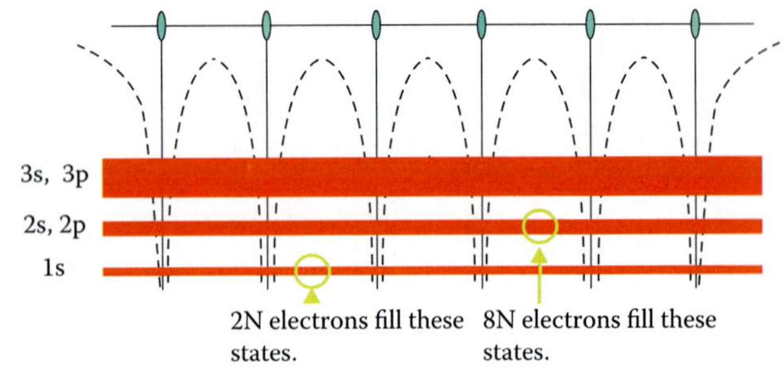

FIGURE 3.94 Filling the Bloch functions of silicon with electrons.

breaking up the lattice periodicity, that an electron may scatter to other Bloch states. Lattice vibrations also may break the perfect lattice periodicity, and as a consequence, electrons in metals scatter more at higher temperatures. This is all very different from Drude's hypothesis: it is not collisions with lattice atoms that determine the resistance of a metal but rather defects and lattice vibrations.

Bloch Functions for Semiconductors and Insulators

Next we will fill the Bloch function of a semiconductor such as Si with electrons. This is illustrated in Figure 3.94. In the case of Si, Z = 14, and the atom orbitals are $1s^2 2s^2 2p^6 3s^2 3p^2$. With a total number of N atoms, we will have 14N electrons to accommodate. At first blush it appears that, like Na, Si will also have a half-filled band because the $3s3p$ band has 4N orbital states and 4N electrons. By this analysis, Si should be a good metal, just like Na. But something unique happens for C and Si and other group IV elements.

In group IV elements, bonding orbitals are hybrid combinations of s and p states, so-called sp^3 hybrids. Hybrid atomic orbitals were introduced to reconcile the discrepancy between what atomic orbital theory predicts and what is seen experimentally. Take carbon, for example; the electron configuration is $1s^2$, $2s^2$, $2p^2$, so one expects carbon, in say methane, to form two bonds via the two unpaired p electrons; in practice we know that carbon forms four equivalent bonds. The hybridization process, in the case of C in methane, is explained in Figure 3.95. A $2s$ electron is promoted, and one $2s$ orbital and three $2p$ orbitals form four equivalent sp^3 hybrid orbitals used in binding four hydrogens to make a tetrahedral-shaped methane molecule. Similarly, in a Si lattice,

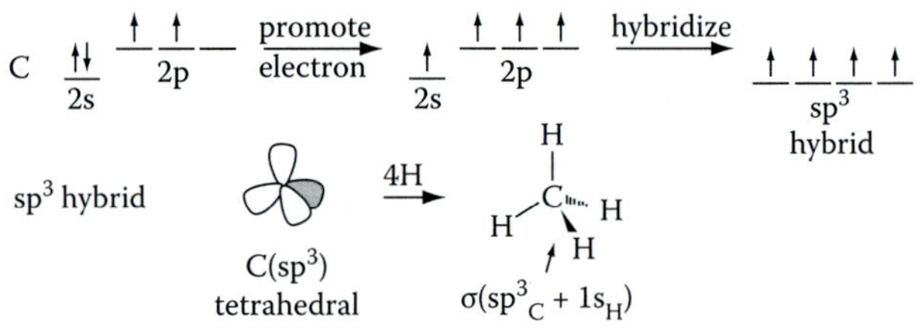

FIGURE 3.95 Hybridization of one 2s and three 2p orbitals in carbon to form four sp³ hybrid orbitals. In methane carbon, four equivalent bonds are formed with hydrogen.

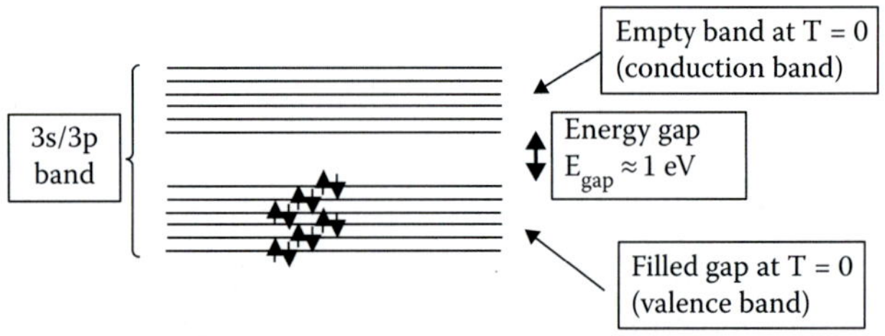

FIGURE 3.96 Two hybrid sp³ bands in Si split in an empty conduction band and a filled valence band.

four sp^3 orbitals link all the Si atoms tetrahedrally together. In the Si case, the hybridizing orbitals are one $3s$ orbital and three $3p$ orbitals, and these four equivalent bonding orbitals are completely filled in the single-crystal Si with two electrons each.

The superposition of sp^3 bands between neighboring Si atoms results in a filled bonding band, the valence band, and an empty antibonding band, the conduction band, as illustrated in Figure 3.96.

The electrons in a filled band cannot contribute to conduction because within reasonable E fields they cannot be promoted to a higher kinetic energy. Therefore, at T = 0, Si is an insulator. At higher temperatures, however, electrons are thermally promoted into the conduction band. For a Si crystal at room temperature, the amount of energy an electron must gain to overcome the bandgap is about 40 times more than the average amount of thermal energy. As a consequence, in a semiconductor the number n of free electrons increases rapidly with T (much faster than the scattering time τ decreases), whereas for a metal scattering time, τ gets shorter with increasing T (Figure 3.97).

Thus, energy bands and the gaps between them determine the conductivity and other properties of solids. Insulators have a valence band that is full and a large energy gap (a few eV). Consequently, no states of higher energy are available for electrons to go to.

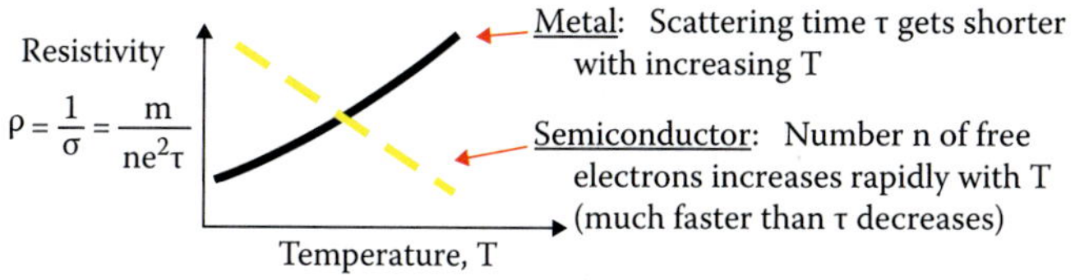

FIGURE 3.97 Resistivity as a function of temperature for a metal and a semiconductor.

Semiconductors are insulators at T = 0, but they have a small energy gap (~1 eV) between valence and conduction bands, so they conduct only at higher *T*. Metals have an upper band that is only partly full; in other words, the Fermi level lies within the valence band, and when applying an electric field, lots of states of higher energy are available for electrons to go to.

The Krönig-Penney Model

Solution of Schrödinger's Equation for a Periodic Potential

Bloch's theorem, along with the use of periodic boundary conditions, enables the calculation of the energy bands of electrons in a crystal if the potential energy function experienced by the electron is known. This was demonstrated for a simple finite square well potential model by Krönig and Penney in 1931, representing the first solution of Schrödinger's equation for a periodic potential. The Krönig-Penney model uses a simple 1D model of a crystalline solid as shown in Figure 3.98. The period of the potential is (a + b = L). The potential (*V*) is assumed equal to zero in the region of an atom, e.g., for $0 < x < a$, and to equal V_0 in the region between atoms, e.g., $-b < x < 0$. The calculations are a repeat of the calculations for a square barrier carried out above. For the zero regions, where electrons essentially act as free particles, the Schrödinger equation that applies is (Equation 3.171):

$$\frac{d^2\psi}{dx^2} = -\mathbf{k}_x^2\psi$$

with (Equation 3.172):

$$\mathbf{k}_x = i\frac{\sqrt{2mE}}{\hbar}$$

and the solution is an oscillating wave like in Equation 3.173:

$$\psi = Ae^{ik_x x} + Be^{-ik_x x}$$

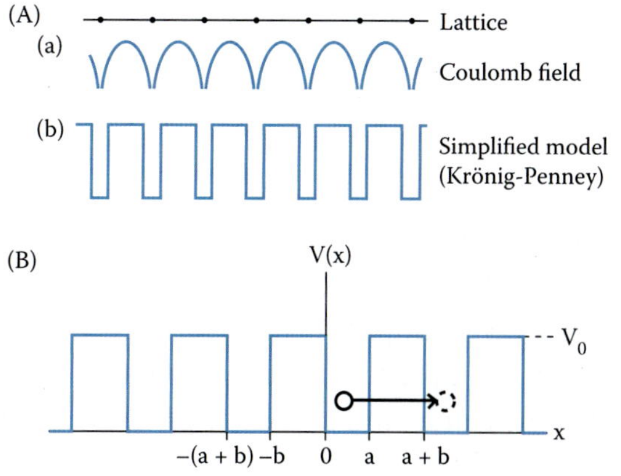

FIGURE 3.98 (A) (a) The periodic lattice potential in a real crystal (see also Figure 3.94). The bullets represent the positions of the nuclei. (b) One-dimensional periodic potential used in the Krönig-Penney model. A central question is whether an electron with energy *E* will be able to propagate from one lattice cell to another. (B) Electrons are essentially free between $0 < x < a$ (and in any similar region along the lattice) and have to tunnel through the barrier regions.

In the nonzero regions, electrons must tunnel through the rectangular barriers for which the following Schrödinger equation applies (Equation 3.166):

$$\frac{d^2\psi}{dx^2} = \alpha^2\psi$$

where α is (Equation 3.167):

$$\alpha = \frac{\sqrt{2m(V_0 - E)}}{\hbar}$$

and the solution is the summation of the damped exponentials of Equations 3.168 and 3.169:

$$\psi(x) = Ce^{+\alpha x} + De^{-\alpha x}$$

It is assumed that the energy *E* of the electron is always smaller than V_0; in other words, electrons stay in the solid.

We will not detail solving Schrödinger's equations for a linear array of atoms here. A detailed treatment can be found, for example, in A.J. Dekker[22] and better yet in S.O. Pillai.[23] Suffice it to say that one finds two solutions: one for the regions where V(x) = 0, and one for the regions where V(x) = V_0, and that these solutions come with constants *A*, *B*, *C*, and *D*. The boundary conditions are that the wave functions and their derivatives are continuous across the potential boundaries and that the solutions must be Bloch functions of the form $\psi_n(x) = V(x)e^{ik_x x}$, with $V(x+nL) = V(x)$, n = 0, +1, +2, +3, . . ., representing the symmetry of the assemblage of atoms. With application of these boundary conditions including the use of Bloch's theorem, the constants *A*, *B*, *C*, and *D* can de determined and the wave functions calculated. One can then show that, under the simplifying conditions, V_0 tends to infinity and *b* approaches zero, whereas the product $V_0 b$ remains finite, and solutions to the wave equations exist only if:

$$P\frac{\sin\beta a}{\beta a} + \cos\beta a = \cos ka \quad (3.200)$$

where *P* is $\frac{mV_0 ba}{\hbar^2}$, a measure for the potential barrier height and width, and with β equal to $\left(\frac{2mE}{\hbar^2}\right)^{\frac{1}{2}}$. In Figure 3.99 we plot $P\frac{\sin\beta a}{\beta a} + \cos\beta a$, the left side of Equation 3.200, versus β*a*, fixing *P* at a value of 3π/2, as a typical example. Because β^2 is proportional to the energy *E*, the *x*-axis in this figure is a measure of energy. Importantly, the right side in Equation 3.200, the term cosk*a*, can only have values between −1 and +1, as marked by the two dashed horizontal lines in the figure. Obviously this condition can only be satisfied with values of β*a* for which the left side lies between +1 and −1.

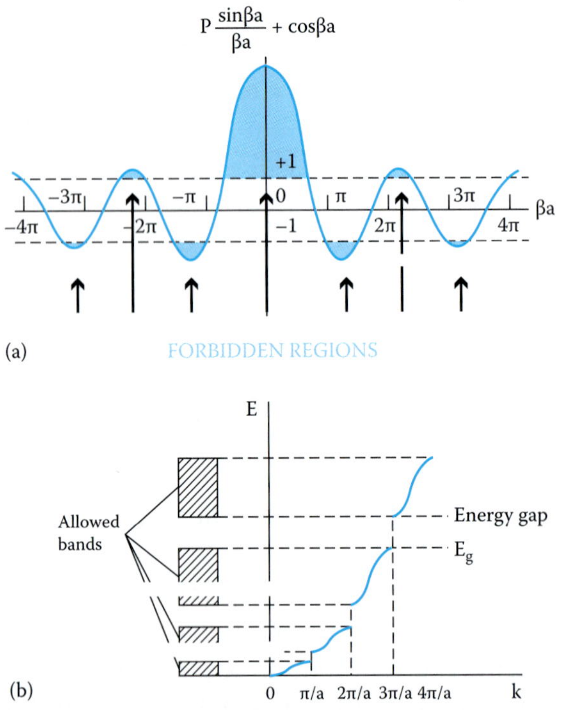

FIGURE 3.99 (a) Plot of $P\frac{\sin\beta a}{\beta a} + \cos\beta a$ versus β*a*. Arrows point to forbidden regions. (b) The energy as a function of wave number **k**. The allowed values of energy are given by those ranges of $\beta = \left(\frac{2mE}{\hbar^2}\right)^{\frac{1}{2}}$ for which the function lies between +1 and −1.

Analysis of the Solution

Some very important insights can be gained from an analysis of Equation 3.200 and an inspection of Figure 3.99:

1. Because β is related to *E*, electrons possess energies within certain bands but not outside those bands: there are allowed bands of energy and forbidden bands of energy (see arrows).
2. If V_0 increases, then *P* increases, the binding energy goes up, and the width of a particular allowed band decreases. The left side of Equation 3.200 becomes steeper, and in the limit, with *P* tending to infinity, the allowed energy bands reduce to the single energy levels we encounter in isolated atoms. In the latter situation, the equation only has solutions if sinβa = 0, in other words, if βa = ± nπ with n = 1, 2, 3,... and the energy spectrum becomes that of an electron in a constant potential box of atomic dimensions. In such cases, the energy levels are the energy levels of a potential well with L = a: $E_n = \frac{\pi^2\hbar^2}{2ma^2}n^2$ (see Equation 3.156). Each electron is confined to one atom by an infinite potential well, so electrons are completely bound to atoms.
3. In the case we decrease V_0 and reduce *P* to zero—in other words, when the binding energy goes to zero—Equation 3.200 is reduced to cos (βa) = cosk*a* (β = **k**), and the energy *E* is now given by $E = \frac{\hbar^2 k^2}{2m}$, i.e., the parabolic *E*-**k** relation for free electrons (see Figure 3.55). By varying V_0, the model thus covers the whole range from completely free electron to completely bound electron. In Figure 3.100 we illustrate the energy level structure as a function of varying degrees of binding strength.

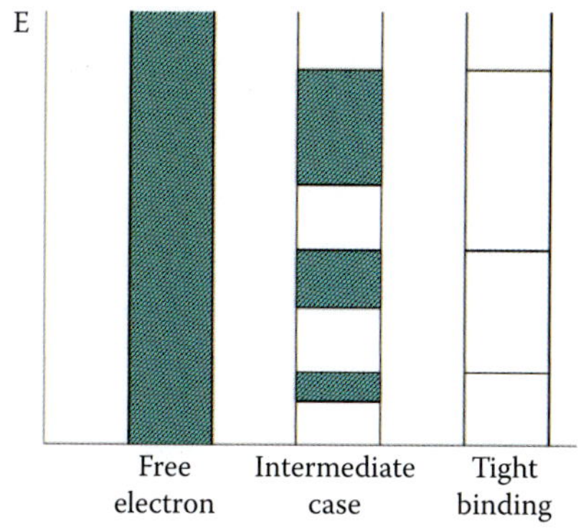

FIGURE 3.100 Energy level structure as a function of binding strength.

4. The width of allowed energy bands increases with increasing values of βa, i.e., with increasing energy, because the first term in the equation decreases on average with increasing βa.
5. From Figure 3.99a we see that at the boundary of an allowed energy band the cos (**k**a) = ±1 with **k** = nπ/a (dashed horizontal lines in the figure). Based on Equation 3.200 we can represent the energy also as a function of wave number **k**. The result is shown in Figure 3.99b and Figure 3.101 further below. This particular way of displaying the electronic levels in a periodic potential is known as the extended-zone scheme. Discontinuities in the *E*-**k** curve occur for **k** = nπ/a with n = 1, 2, 3,…. The zones in **k**-space that correspond to allowed energies for

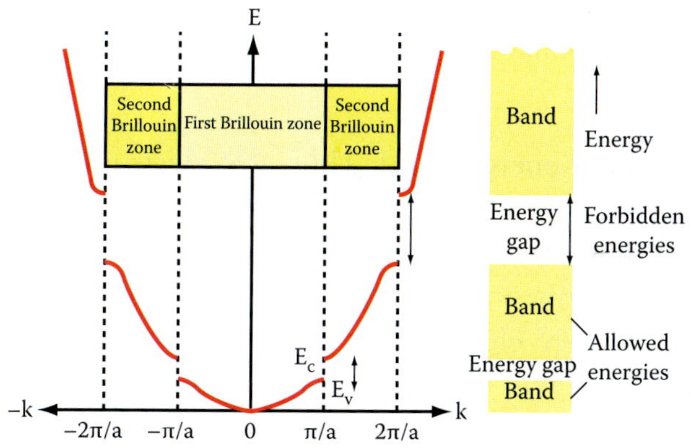

FIGURE 3.101 Energy versus wave number for motion of an electron in a one-dimensional periodic potential. The range of allowed **k** values goes from $-\pi/a$ to $+\pi/a$ corresponding to the first Brillouin zone for this system. Similarly, the second Brillouin zone consists of two parts: one extending from π/a to $2\pi/a$, and another part extending between $-\pi/a$ and $-2\pi/a$. This representation is called the extended zone scheme. Deviations from free electrons parabola are easily identified.

motion of an electron are the Brillouin zones (BZ), which we first encountered in Chapter 2.* The periodic potential $V(x)$ splits the free electron *E*-**k** curve into "energy bands" separated by gaps at each BZ boundary. Electrons can never have an energy within this energy gap. In a metal, the periodic potential $V(x)$ is very small or 0; in a semiconductor, the potential is large and consequently the energy splitting is large. The first Brillouin zone extends between **k** = −π/a and **k** = +π/a; the second is in the range of **k** from $-2\pi/a$ to $-\pi/a$ plus the range from π/a to $2\pi/a$. There can be no energy value for an electron between the bottom of the conduction band, E_c, and the top of the valence band, E_v. Therefore, the value $E_c - E_v = E_g$ is an energy gap at **k** = ±π/a. The existence of forbidden energies has very fundamental consequences. It happens for electrons in crystals with a periodically varying potential and for photons in systems with a periodically varying refractive index, in which case we call the gap a photonic bandgap (see Chapter 5). Brillouin zones are further detailed in the section below.

6. Finally, inspection of Equation 3.200 reveals that when **k** is substituted by **k** + $2\pi n/a$, where n is an integer, the right side of the equation remains the same, or **k** is not uniquely determined. In other words, in a given energy band, the energy is a periodic function of **k**. Given this periodicity, it is often convenient to introduce a "reduced wave vector" as −π/a ≤ **k** ≤ +π/a. By shifting the second Brillouin zone $2\pi/a$ left or right, one can obtain the reduced zone scheme. A representation of energy versus reduced wave vector is marked by a double pointed arrow in Figure 3.102a. Obviously one does not need all *E*-**k** curves in all BZs. All information is already contained in the first Brillouin zone in a reduced zone scheme because of the $2\pi/a$ periodicity. In Figure 3.102a, we also display the electronic

* From Chapter 2 we remember that the x-ray diffraction pattern of a crystal is a map of the reciprocal lattice. It is a Fourier transform of the lattice in real space or also the representation of the lattice in **k**-space. This is true for the x-rays considered in Chapter 2 but also for the matter waves associated with electrons, neutrons, and so on. We are presenting BZs for all dimensionalities of the electronic structure further below.

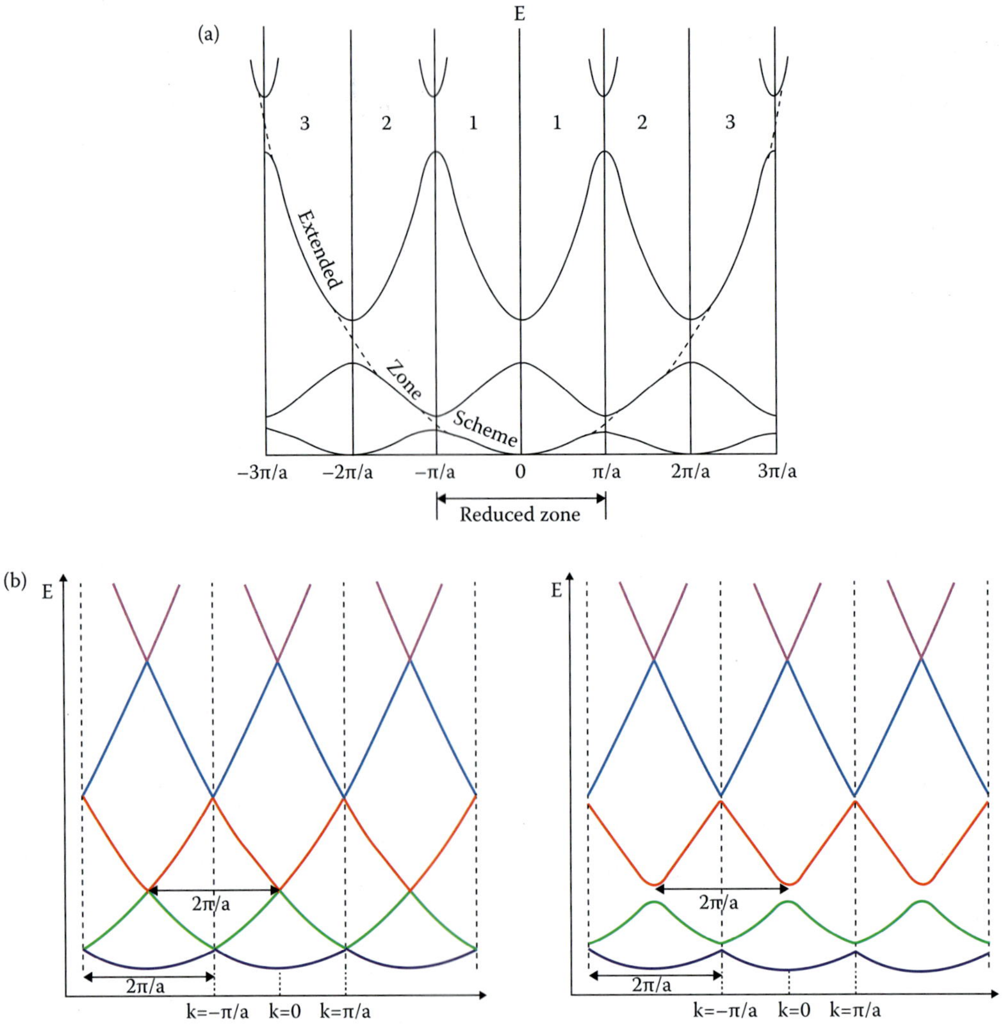

FIGURE 3.102 (a) Different representations of *E*-**k** in the presence of a periodic potential, band splitting. The extended zone scheme: plot *E*-**k** from **k** = 0 through all possible Brillouin zones (**bold curve**); the periodic or repeated zone scheme: redraw *E*-**k** in each zone and superimpose; the reduced zone scheme: all states with |**k**| > π/a are translated back into the first BZ. (b) Different *E*-**k** plots in the presence of a vanishing periodic potential: zone folding (left) and a nonvanishing periodic potential (right).

levels in a repeated or periodic zone scheme and repeat the extended zone scheme from Figure 3.101. In Figure 3.102b, we show what happens when the periodic potential becomes vanishing small; the bandgaps disappear and we get zone folding.

It is easy to show that the number of **k** values in each BZ is just *N*, the number of primitive unit cells in the sample. For this consider the finite, linear crystal with a total length of L = Na. The allowed values of the electron wave vector **k** in the first Brillouin zone for this arrangement are exactly N (k = 0, ±2π/L, ±4π/L,..., ±Nπ/L). Each primitive cell contributes exactly one independent value of **k** to each energy band. Thus, 2*N* electrons resulting from spin degeneracy can occupy each band. A monovalent element with one atom per primitive cell has only one valence electron per primitive cell and thus *N* electrons in the lowest energy band. This band will only be half-filled, and the material will be a conductor. The Fermi energy, the energy dividing the occupied and unoccupied states, for such a monovalent element will be in the middle of the valence band. If each atom contributes two valence electrons to the band (divalent element), the band, at T = 0 K, will be filled to the top. Thus, the simple rule whether an element is an insulator or a metal is given by whether the number of electrons is odd or even. This rule works surprisingly well;

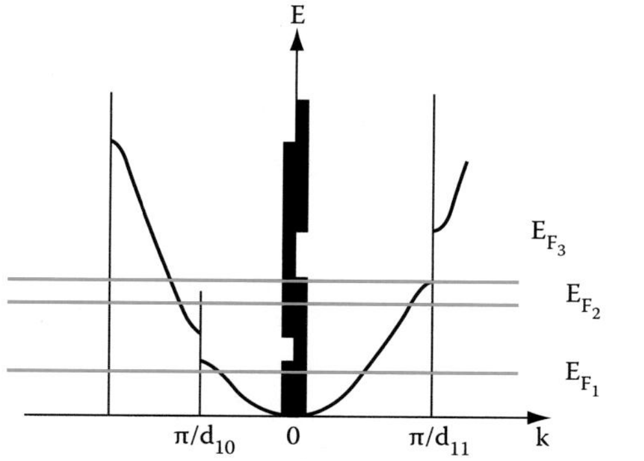

FIGURE 3.103 The presence of more than one periodicity in a crystal may cause the overlap in the integrated density of states as shown here as black rectangles along the *E*-axis. The material depicted in the *E*-**k** diagram is a semimetal conductor; see also Figure 3.104b.

the exceptions are caused by bands with more complicated structures in three dimensions than those discussed here. Indeed, the fact that a monovalent element is a conductor does not mean that a divalent element will always be an insulator. Although true in the 1D, it is not necessarily so in 2D or 3D! Bands along different directions in **k**-space can overlap, so that electrons can partially occupy both of the overlapping bands and thus form a semimetal. A semimetal is a metal with a carrier concentration several orders of magnitude smaller than the 10^{22} cm^{-3} typical for ordinary metals. Graphite and the pentavalent conducting elements, for example, are semimetals. A 2D band diagram for a semimetal is shown in Figure 3.103. In this example we show a material with three valence electrons and Fermi levels E_{F_1} to E_{F_3}. This case illustrates how bands in different directions overlap, resulting in a semimetal (this corresponds to the situation depicted in Figure 3.104b).

However, it remains true that only crystals with an even number of valence electrons in a primitive cell can be insulators. Depending on the energy band structure, for a given number of electrons, you can get a different filling, and it is the band structure that determines electronic and optical behavior. Some example band configurations leading to insulating and metallic behavior are shown in Figure 3.104.

The Effective Mass, Velocity of Charge Carriers, and Crystal Momentum

Up to this point we have implied that the mass of an electron in a solid is the same as the mass of a free electron (m_e). In reality, for some solids the measured electron mass is larger than that of the free electron, and in other cases it is smaller. The cause for this deviation is found in the interactions between the drifting electrons and the atoms in the crystal. The mass of an electron in a crystal is called the *effective mass*, m_e^*, and in general it is different from the mass of a free electron. The effective mass, m_e^*, of an electron is the mass of the free electron modified by the presence of the periodic potential of an array of positive lattice ions. Rather than moving undisturbed through the lattice, electrons are constantly jostled by atom movements (phonons). By lowering the temperature, atoms move more sluggishly, and this reduces the lattice resistance for electrons. However, temperature reduction does not reduce the

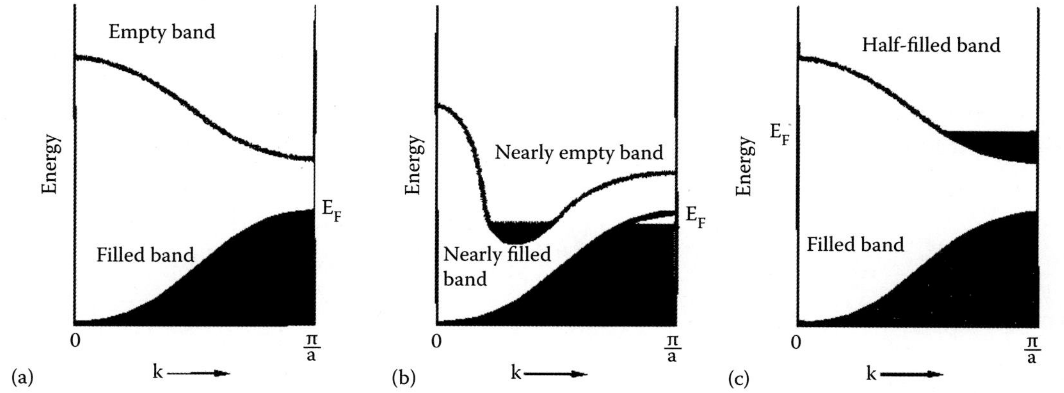

FIGURE 3.104 Occupied states and band structures leading to (a) semiconductor or insulator: two electrons per atom, *N* states per band, and two electrons/state; (b) semimetal: two electrons per atom, *N* states per band, and two electrons/state, a semimetal because bands overlap; (c) and metal: three electrons per atom (Li), *N* states per band, and two electrons/state, metal because of electron population.

influence of lattice defects; the only avenue here is to purify the crystal further, but even the purest crystal features some remaining defects. A third phenomenon controlling the speed of an electron through a crystal is the electron mobility, $\mu_e = \frac{e\tau}{m_e}$ (Equation 3.9). Applying the same voltage to equally pure samples of Si and GaAs, one finds that electrons in GaAs are accelerated much more. The electrons in Si and GaAs are of course the same, but their host lattices influence them differently. We refer to this property by saying that an electron has an effective mass in a given material; therefore, the effective mass m_e^* of an electron in gallium arsenide is less than that of an electron in silicon (making μ_e in GaAs larger; see Equation 3.9).

For a free particle such as an electron, we derived a wave solution $\psi(x) = Ae^{ik_x x}$ with energy $E = \frac{\hbar^2 \mathbf{k}_x^2}{2m_e}\left(= \frac{\mathbf{p}^2}{2m_e}\right)$ (Equation 3.144). For a free electron, the quantity $\hbar\mathbf{k}$ represents the true momentum $\mathbf{p}$ of the electron. For electrons in a crystal, we need to define a *crystal momentum* that is different from $\mathbf{p}$ because the energy for electrons in a crystal does not vary in the same fashion with $\mathbf{k}$ as it does for free electrons—in other words, the dispersion curve $E(\mathbf{k})$ is different for an electron in lattice than that of a free electron. When dealing with interactions of the electron with the lattice, we must use the conservation of crystal momentum $\hbar\mathbf{k}$ and not that of a free electron.

We will now show how the effective mass and the velocity of an electron in a lattice can both be derived from a knowledge of the energy dispersion curves $E(\mathbf{k})$. Remember that an electron state is a wave packet with a group velocity v_g given as the derivative of ω, the angular frequency of the de Broglie waves with respect to the wave number $\mathbf{k}$ or:

$$v_g = \frac{d\omega}{d\mathbf{k}} \tag{3.101}$$

From quantum theory the frequency of a wave function with energy E is $\omega = \frac{E}{\hbar}$ (because $E = \hbar\omega$), so that we may write:

$$\frac{dE}{d\mathbf{k}} = \hbar\frac{d\omega}{d\mathbf{k}} \tag{3.201}$$

Substituting this expression in Equation 3.101 results in:

$$v_g = \frac{1}{\hbar}\left[\frac{dE}{d\mathbf{k}}\right] \text{ or more generally } \mathbf{v}_g = \frac{1}{\hbar}\nabla_k E(\mathbf{k}) \tag{3.202}$$

The group velocity of a wave packet is basically the velocity at which the wave packet transports energy through the system. It also gives the average velocity of the Bloch electron. The influence the crystal is exerting on the electron motion is contained in the dispersion relation $E(\mathbf{k})$, and the velocity of an electron in a crystal then depends on the dispersion curve $E(\mathbf{k})$. Let us illustrate this point with a simple example. For a free electron, with the parabolic dispersion function $E(\mathbf{k}) = \frac{\hbar^2\mathbf{k}_x^2}{2m}$ (Equation 3.144), we derive $\frac{dE}{d\mathbf{k}} = \frac{\hbar^2\mathbf{k}}{m_e}$, and using Equation 3.202 we calculate for the free electron velocity:

$$v_g = \frac{\hbar\mathbf{k}}{m_e} = \left(\frac{\hbar\mathbf{k}}{m_e}\right)\left(\frac{2\pi\mathbf{p}}{h}\right) = \frac{\mathbf{p}}{m_e} \tag{3.203}$$

In this equation we have a linear relation of v_g with $\mathbf{k}$, which is of course expected because E is proportional to $\mathbf{k}^2$. In general, E is not proportional to $\mathbf{k}^2$, or at least only in small regions as schematically reproduced here in Figure 3.105d.

In Figure 3.105c, we show the velocity v as a function of $\mathbf{k}$. We observe that at the bottom of the energy band ($\mathbf{k} = 0$), the velocity is zero and then increases with $\mathbf{k}$ until it reaches a maximum value at $\mathbf{k} = \mathbf{k}_0$, corresponding to the inflection point in the E-$\mathbf{k}$ curve. Beyond the inflection point, the velocity starts to decrease and finally assumes the zero value at $\mathbf{k} = \pi/a$, which is at the top of the band.

An electron has a well-defined mass, and when accelerated it obeys Newtonian mechanics. To calculate the acceleration ($a = dv_g/dt$) of an electron in a crystal we derive from Equation 3.202 that:

$$a = \left(\frac{2\pi}{h}\right)\frac{d}{dt}\left(\frac{dE}{d\mathbf{k}}\right) \text{ or also } a = \left(\frac{2\pi}{h}\right)\frac{d^2E}{d\mathbf{k}^2}\frac{d\mathbf{k}}{dt} \tag{3.204}$$

The term $\frac{d^2E}{d\mathbf{k}^2}$ we can get from the E-$\mathbf{k}$ relationship, but we still need to derive $d\mathbf{k}/dt$ tocalculate a. For an

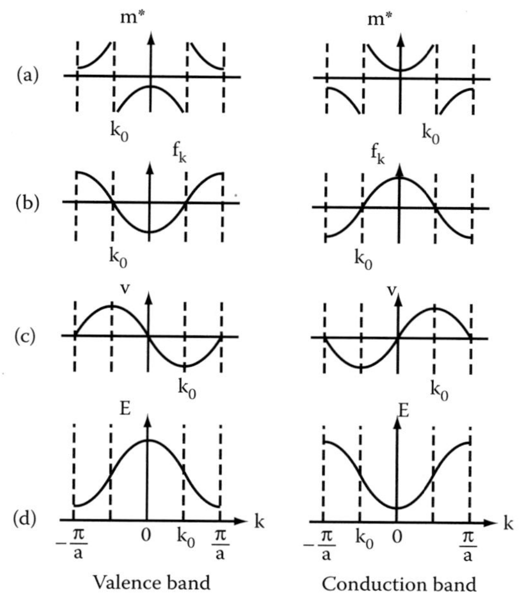

FIGURE 3.105 Effective mass m^*, f_k, velocity v, and energy E as a function of $\mathbf{k}_0$ for electrons in the conduction band (right) and holes in the valence band (left). See text for details.

electron with velocity $\mathbf{v}_g$, subjected to the influence of an external field E applied for a time dt, the distance traveled is $\mathbf{v}_g dt$, so that the work dE done by the electrical field on the electron is:

$$dE = -eE\mathbf{v}_g dt \tag{3.205}$$

Substituting the value for $\mathbf{v}_g$ from Equation 3.202 in the above equation, we get:

$$dE = -eE\left(\frac{2\pi}{h}\right)\left(\frac{dE}{dk}\right)dt \tag{3.206}$$

or:

$$\frac{\mathbf{dk}}{dt} = -\frac{2\pi eE}{h} \tag{3.207}$$

or since $\mathbf{F} = -eE$:

$$\hbar\frac{\mathbf{dk}}{dt} = \mathbf{F} \tag{3.208}$$

The last expression shows us that in a crystal $\hbar\frac{\mathbf{dk}}{dt}$ is equal to the external force on the electron, whereas in free space the force is equal to $d(mv)/dt$. The electron in a crystal is subject to both forces from the lattice and from external fields. In case there is also a magnetic field present, the force term **F** in Equation 3.208 must include the Lorentz force, $\mathbf{F}_L = -e(\mathbf{v}_g \times \mathbf{B})$ (see Equation 3.64), so that the equation of motion of an electron with group velocity $\mathbf{v}_g$ in a constant magnetic field **B** is given by:

$$\hbar\frac{\mathbf{dk}}{dt} = -e(\mathbf{v}_g \times \mathbf{B}) \tag{3.209}$$

Combining Equation 3.204 with Equation 3.207 (no magnetic field) we finally derive for a:

$$a = \left(\frac{4\pi^2}{h^2}\right)eE\left(\frac{d^2E}{\mathbf{dk}^2}\right) \tag{3.210}$$

Comparing this equation with that for a free, classical particle where we have $m_e(dv/dt) = eE$ and $a = (dv/dt) = (eE/m_e)$, we then define the effective electron mass m_e^* as:

$$m_e^* = \left(\frac{h^2}{4\pi^2}\right)\left(\frac{d^2E}{\mathbf{dk}^2}\right)^{-1} \tag{3.211}$$

so that:

$$a = \left(\frac{eE}{m_e^*}\right) \tag{3.212}$$

From Equation 3.211, the effective mass is determined by $\left(\frac{d^2E}{dk^2}\right)^{-1}$ (see Figure 3.105a). For a free electron $m_e^* = m_e$ because $E = \frac{\hbar^2\mathbf{k}^2}{2m_e}$ and $\frac{d^2E}{\mathbf{dk}^2} = \frac{\hbar^2}{m_e}$. All the equations for a free electron may be used for an electron in a crystal, provided that m_e in each case is replaced by the suitable m_e^*. For example, we may write:

$$E = \frac{\hbar^2\mathbf{k}^2}{2m_e^*} \tag{3.213}$$

Typically one considers states near the top of the valence band to be holes (particles of charge $+e$) with free electron-like dynamics but with effective mass m_h^*, and states near the bottom of the conduction band to be electrons with free electron-like dynamics but effective mass m_e^*. The effective mass is inversely proportional to the curvature of the band, and in general m^* is different in each direction of the crystal and is a tensor.

From experimental values for m_e^*, it is apparent that the effective mass need not always be larger than m_e. It can be smaller, and it may even be negative. For most metals, it is from one-half to twice m_e. For some transition metals, it is much higher than m_e, and for semiconductors it is lower. From Figure 3.105a, near $\mathbf{k} = 0$, the effective mass approaches m_e. With **k** increasing, m_e^* also increases, and it reaches a maximum value at the inflection point ($\mathbf{k}_0$) of the *E*-**k** curve. Above the inflection point ($\mathbf{k} > \mathbf{k}_0$), m_e^* becomes negative, and as **k** approaches π/a it decreases to a small negative value. Near the bottom of the band, the effective mass m_e^* has a constant positive value because the quadratic equation $E \propto k^2$ holds over a small region here (second derivative is a constant). As **k** increases, the quadratic relation no longer holds, and m_e^* changes with **k**. Beyond $\mathbf{k}_0$, the mass m^* becomes negative because the region is close to the top of the band, and a negative mass is to be expected. The way to interpret this is to consider that for $\mathbf{k} > \mathbf{k}_0$ the velocity decreases, and therefore the acceleration is negative, implying a negative mass. In this region of **k**-space, the lattice exerts such a large retarding force on the electron that it overcomes the applied force and produces a negative acceleration. In other words, in the upper half of the band the electron behaves as a positively charged particle, referred to as a hole. Because we can describe the holes in terms of an *E*-**k** diagram, we can again define an effective mass simply by putting a minus sign in front of Equation 3.210. This turns out to be positive and given by the curvature of the *E*-**k** diagram at the top of the valence band. Therefore, here we finally have our answer why even metals might produce a positive Hall effect and why the conductivity is not simply proportional to the electron density; it matters where in the band these electrons find themselves!

Neither the electron nor the hole as described by its effective mass exists outside of the material; they are more properly referred to as quasiparticles. The fictitious positively charged particle is even stranger than a conduction band electron with an effective mass. If we add a hole to a completely filled valence band, we must end up with an empty electron state. Therefore, the hole must have a charge of $+e$ (compared with an electron's $-e$). If an empty state exists at low energy, then it is energetically favorable for an electron from a higher energy state to fall into it. In terms of holes, this means it is energetically favorable for the hole to rise to the top of the valence band, i.e., the energy scales for holes are reversed from those for electrons.

In Figure 3.105b we plot the degree of freedom of an electron, f_k, as a function of **k**. Here f_k, a measure of the extent to which an electron in state **k** is free, is defined as:

$$f_k = \frac{m}{m^*} = \frac{m}{\hbar^2}\left(\frac{d^2E}{dk^2}\right) \tag{3.214}$$

If the effective mass is large, f_k is small, or the particle behaves as a heavy particle. When $f_k = 1$, the electron behaves as a free electron. In the lower half of the band, f_k is positive, and in the upper half f_k is negative.

The behavior of electrons and holes near the band edges determines most of the optical and electronic properties of a solid-state device. From the preceding, near the band edges, the electrons and the holes can be described by a simple effective mass picture, i.e., the electrons behave as if they are in free space except their masses are m_e^* and m_h^*, respectively. In Figure 3.106, we show a four-band model for a generic semiconductor with a parabolic approximation of the bands. This simplified band structure was first proposed by Kane (1957) and is valid near $\mathbf{k} = 0$. The valence band features a heavy holes (hh) band (I), a light holes (lh) band (II), and a split-off (so) band (III).

Electron transitions 1 and 2 are from the heavy holes band to the conduction band and from the light holes band to the conduction band, respectively. Split-off hole transitions are also possible.

As shown in Figure 3.106, along a given direction, the top two degenerate valence bands can be approximately fitted by two parabolic bands with different curvatures: the heavy holes band (the wider band,

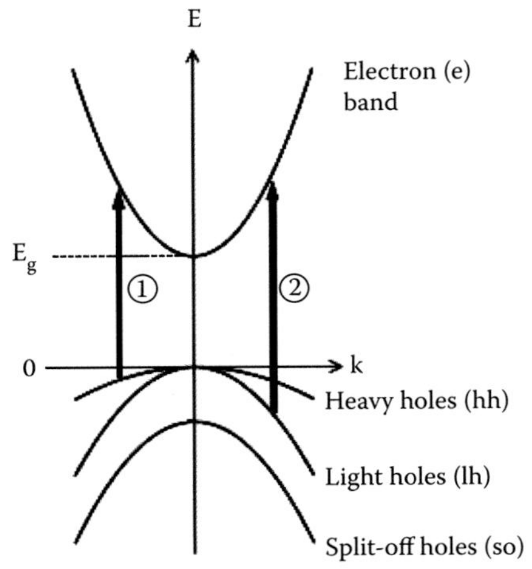

FIGURE 3.106 Four-band model. In the valence band, we have a heavy holes (hh) band (I), a light holes (lh) band (II), and a split-off (so) band (III).

with smaller $\frac{\partial^2 E(k)}{\partial k_i \partial k_j}$) and the light holes band (the narrower band, with larger $\frac{\partial^2 E(k)}{\partial k_i \partial k_j}$). Thus, the effective mass, in general, is tensorial with components:

$$\frac{1}{m^*_{ij}} \equiv \frac{1}{\hbar^2} \frac{\partial^2 E(k)}{\partial k_i \partial k_j} \tag{3.215}$$

which represents a generalization of Equation 3.211.

Example 3.5: In Si, $m_{hh} = 0.53\ m_0$ and $m_{lh} = 0.16\ m_0$; and in GaAs, $m_{hh} = 0.51\ m_0$ and $m_{lh} = 0.074\ m_0$.

Because there are multiple valence bands available for holes but there is only one conduction band for electrons, electromagnetic radiation absorption by holes is different from that of electrons. Free electrons located at wave vector **k** = 0, at the bottom of a single parabolic conduction band, may reach higher energy states only if their momentum is increased. As a consequence, in n-type semiconductors, conduction band electrons may only be excited by the simultaneous absorption of a photon and the absorption or the emission of a phonon to conserve momentum (see also direct and indirect bandgap transitions discussed further below). This three-particle process (photon-electron-phonon) is less probable than a two-particle process (electron-photon). With p-type semiconductors, one can have such direct (i.e., no phonons needed) transitions between the different degenerate valence bands (heavy and light hole bands) because they occur at the same **k**. Because no momentum or **k** change is required, holes produce stronger free carrier absorption than conduction band electrons.

Below we take a closer look at the density of states (DOS) function, which describes the number of allowed energy states that are available in a system per unit energy and per unit volume (i.e., in units of number of states/eV/cm^3). From the preceding we can appreciate that the density of states, the effective mass, and the electron mobility are all correlated. When traveling through a crystal and bumped off course by a phonon or an impurity, the new energies an electron can adopt depend on the number of available states at the bottom of the conduction band. The density of states at the bottom of the conduction band is larger for Si than it is for GaAs, and as a consequence an electron traveling through silicon has a greater chance of being knocked off course into an allowed energy state. In general, then, a small effective mass is indicative of a relatively low number of energy levels at the bottom of the conduction band. The effective mass of an electron in silicon is six times heavier than that in GaAs, and as a consequence, based on Equation 3.9, the electron mobility in GaAs is six times larger or an electron can race six times faster through a GaAs lattice compared with a Si lattice (see Table 3.9). GaAs transistors, although much more difficult to fabricate than Si transistors, are used in cases where speed is of utmost importance; this includes military applications and the latest generations of supercomputers (e.g., the Cray 3).

Table 3.9 Electron and Hole Mobilities

Material	Mobility (cm^2/V-s)
Si	$\mu_n = 1500$, $\mu_p = 460$
Ge	$\mu_n = 3900$, $\mu_p = 1900$
GaAs	$\mu_n = 8000$, $\mu_p = 380$

Particle Distributions Functions

Introduction

Particle distributions functions *f(T, E)* represent the probability that a particular state with energy *E* is occupied by a particle (say an electron) in equilibrium at a given temperature *T*. Earlier in this chapter we introduced the Maxwell-Boltzmann distribution, a classical distribution of particles, illustrated in Figure 3.6. In this distribution function, it is assumed that all the particles are distinguishable. This kind of consideration comes from the fact that all particles have characteristic wave properties according to the de Broglie hypothesis. Two particles can be considered to be distinguishable if their separation *d* is large compared with their de Broglie wavelength and are considered to be indistinguishable if their wave packets overlap significantly. In such case, a Fermi-Dirac particle distribution and (f_{FD}) must be applied. The thermal de Broglie wavelength is roughly the average de Broglie wavelength of the gas particles in an ideal gas at the specified temperature and is given by:

$$\lambda_{DB} = \sqrt{\frac{2\pi\hbar^2}{mk_BT}} \sim \frac{h}{p} \qquad (3.216)$$

We can take the average interparticle spacing in an ideal gas to be approximately $(n = V/N)^{1/3}$, where V is the volume, N is the number of particles, and n is the density of particles. When the thermal de Broglie wavelength is much smaller than the interparticle distance, the gas can be considered to be a classical or a Maxwell-Boltzmann gas, or:

$$\lambda_{DB} <<< d = n^{-\frac{1}{3}} \qquad (3.217)$$

On the other hand, when the thermal de Broglie wavelength is on the order of, or larger than, the interparticle distance, quantum effects will dominate, and the gas must be treated as a Fermi gas or a Bose gas, depending on the nature of the gas particles. Particles become indistinguishable when $d = n^{-1/3} \sim \lambda_{DB}$. The de Broglie temperature is given by:

$$T_{DB} = \frac{2\pi\hbar^2}{mk_B} n^{\frac{2}{3}} \qquad (3.218)$$

Example 3.6: For an electron gas in metals, $n = 10^{22}$ cm^{-3}, $m = m_e$, and $T_{DB} \sim 3 \times 10^4$ K, but for a gas of Rb atoms, $n = 10^{15}$ cm^{-3} and $m_{atom} = 10^5$ m_e, so we obtain a $T_{DB} \sim 5 \times 10^{-6}$ K. In other words, at room temperature electrons are indistinguishable with overlapping wave functions, whereas Rb atoms only become so at very low temperatures.

Besides the Fermi-Dirac distribution, we will also introduce the Bose-Einstein distribution function (f_{BE}) and explain under what conditions they each apply. We will see that at $T < T_{DB}$, f_{BE} and f_{FD} are strongly different from f_{MB} and at $T >> T_{DB}$: $f_{BE} \approx f_{FD} \approx f_{MB}$.

Fermi-Dirac

When assigning electrons to the energy levels one must require, as Sommerfeld did, that the allowed wave functions obey Pauli's principle (introduced in 1925), i.e., one can only put two electrons with opposite spin in each level of quantum number n. When that is done this level is filled. One then proceeds to the next higher level for the next pair of electrons. The result is obvious—all the lowest levels are filled with pairs of electrons until they reach some maximum value of energy E_F, the so-called Fermi energy. The exclusion principle applies to all "spin one-half" particles, which include electrons, protons, and neutrons. If we draw the distribution function, i.e., the probability of filling a level, $f(E,T)$, as a function of E, for two different temperatures, we find the result shown in Figure 3.107.

The probability of an energy state being occupied is called the Fermi factor, $f(E,T)$. As T approaches zero, the Fermi-Dirac distribution becomes a step function. At 0 K, $f(E, T) = 1$ until we reach the maximum level E_F, after which it falls to zero. If we increase the temperature, thermal energy can excite electrons to energy levels higher than E_F. Because the kinetic energy of the lattice ions is of the order of kT (~0.025 eV), electrons cannot gain much more than kT in collisions with lattice ions. This is determined by the Fermi-Dirac distribution, which applies for any particle that follows Pauli's exclusion principle—no more than two particles of opposite spin being allowed in a given energy level. Half-integer spin particles are called fermions, and Fermi-Dirac statistics and Pauli's exclusion principle hold. The "material" particles, such as electrons, protons, and neutrons, are all fermions, and without Pauli's exclusion principle, the plethora of chemical elements and the variety of our physical world would simply not exist. Other particles such as α-particles, deuterons, photons, and mesons do not obey Pauli's exclusion principle. Such particles are called bosons

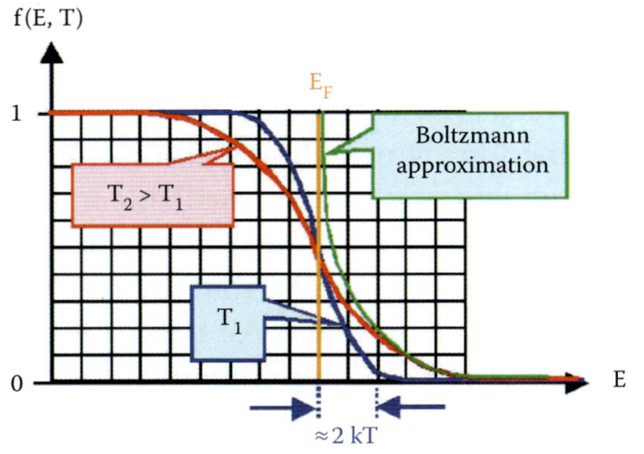

FIGURE 3.107 The Fermi-Dirac distribution function for two different temperatures. The Boltzmann approximation is indicated as well (green line).

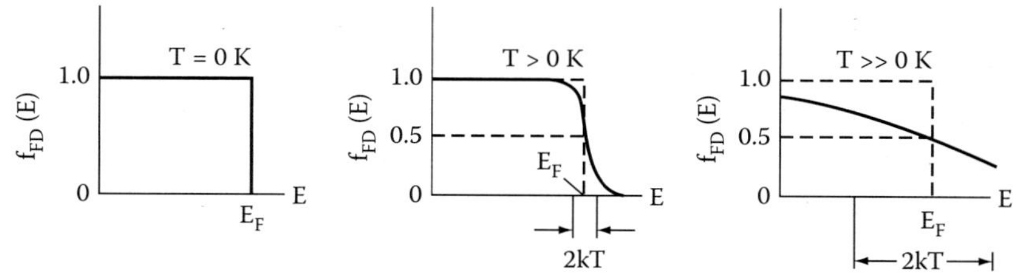

FIGURE 3.108 Evolution of Fermi distribution function as a function of temperature. At the highest temperatures the Fermi-Dirac function smears out and starts resembling a Boltzmann distribution.

and have either zero intrinsic spin or integral spin quantum numbers. Fermion and boson particle distributions are compared further below in Figure 3.108. The wave function that describes a collection of fermions must be antisymmetric with respect to the exchange of identical particles. One fermion of a system in a certain state prevents all other fermions from being in that state. The mathematical expression for the Fermi-Dirac distribution as a function of energy and temperature is:

$$f(E, T)_{FD} = \frac{1}{e^{\left[\frac{(E-\mu)}{kT}\right]}+1} \text{ or } \frac{1}{Ae^{\frac{E}{kT}}+1} \tag{3.219}$$

where μ is the chemical potential and $A = e^{-\mu}$. At T = 0, one can see that μ(T = 0) = E_F, the Fermi energy. Above T = 0 there is no abrupt energy that separates filled from unfilled levels, so the definition of the Fermi energy must be slightly modified. Note that when f(E, T) = 1/2, there is a 50-50 probability of finding the level occupied and E = E_F. Thus, at temperature *T*, the Fermi energy is defined as that energy for which the probability of being occupied is 0.5. The Fermi energy plays a very important role in the band theory of semiconductors and metals. In metals the Fermi energy is an effective cutoff level for the allowed energies of the electrons. By analogy, imagine a sea of electrons with a "depth" of E_F. At room temperature only a small fraction of the electrons will ever have an energy much above that sea level (≈$2kT_1$ at T_1). The energy E_F corresponds to the chemical potential, μ, i.e., the amount of energy needed to add an electron to the system (E_F = μ). For fermions, the chemical potential may be either positive or negative. From thermodynamics, the chemical potential, and thus the Fermi energy, is related to the Helmholtz free energy: $\mu = F(n + 1) - F(n)|_{T,V}$, where F = U − TS. For comparison, we have indicated the Maxwell-Boltzmann distribution in Figure 3.107 as well (green line).

With *T* > 0 K, the *f*(*E*, *T*) function has the general shape as sketched in Figure 3.108, but as the temperature keeps on increasing it gradually smears out, and finally at very high temperature (*T*>> 0 K) it begins to look like an ordinary Boltzmann distribution as shown in Figure 3.108 (see also Figure 3.6). At ordinary temperatures, say 1000 K, kT = 0.088 eV, whereas for gold E_F = 5.51 eV and for sodium it is 3.12 eV-typical values for metals. In other words, $E_F >> kT$, and an electron gas in a metal cannot usually be treated as an ordinary Maxwell-Boltzmann gas. With $E_F >> kT$, the electron distribution is quantum-mechanical, and the electron gas is said to be degenerate. A Fermi gas well described by a Fermi-Dirac distribution that is approximately step-like is termed degenerate.

Remember that Drude used a Maxwell-Boltzmann distribution for the free electrons in a metal. From the above we now recognize that this was the wrong particle distribution to use. The Maxwell-Boltzmann distribution, shown in Figure 3.6, is a classical distribution of particles, but with particles satisfying quantum statistics, a Fermi distribution must be used. The electronic velocity vector component distribution for a single direction v_i according to Drude (based on Equation 3.17, which gives the distribution of speeds of molecules rather than their component velocities), is given as:

$$f_{MB}(v_i) = n\left(\frac{m}{2\pi k_B T}\right)^{\frac{3}{2}} e^{-\frac{\frac{1}{2}mv_i^2}{kT}} \tag{3.220}$$

where n = N/V with *n* the electron density, *V* the volume, and *N* the sum of all electrons. The Maxwell-Boltzmann distribution assumes that the described particles are distinguishable. In other words, we can make a distinction as to which particle is in which energy state. For an ideal gas, that is certainly the case, but this is not true for electrons, and this is the reason why Drude's model ultimately failed. Sommerfeld used the same classical gas as Drude but used the quantum Fermi-Dirac distribution for the electrons' velocity vector component in one direction (v_i):

$$f_{FD}(v_i) = \frac{(m/\hbar)^3}{4\pi^3} \frac{1}{\exp\left[\left(\frac{1}{2}mv_i^2 - kT_0\right)/kT\right]+1} \quad (3.221)$$

The temperature T_0 is determined by the normalization $n = \int f(v_i)dv_i$.

We can understand now why the electron gas does not contribute to the heat capacity of metals. As a metal is heated, the only way an electron can take up energy is by moving into a somewhat higher allowed energy level. But a typical electron is buried deep inside the Fermi sea, and there are no empty levels to move to because each level above the electron is already occupied by a pair of electrons of opposing spin. Only the relatively few electrons at the top of the distribution can find empty levels to move into. These electrons find themselves in the so-called Maxwellian tail of the Fermi-Dirac function. They are the only electrons that contribute to the heat capacity, as we will calculate below. Thus, at ordinary temperatures the electronic heat capacity is almost negligible, as we can see from Figure 3.16. But why then can all the electrons in a free electron gas contribute to the electrical conductivity? We would expect that also in this case, electrons take energy from the electrical field and move into higher energy levels. What happens in this case, however, is a shift of the entire set of electron energy levels from lower to higher values. The Fermi distribution, in this case, as we will show below in Figure 3.121, is shifted bodily by the applied field. Thus, electrons can acquire an average drift velocity in the field without violating Pauli's exclusion principle.

Bose-Einstein Distribution

Bosons are a third type of particles that come with their own kind of distribution law, i.e., the Bose-Einstein distribution. Violations of the Maxwell-Boltzmann statistics are observed if the density of particles is very large (neutron stars), the particles are very light (electrons in metals, photons), or they are at very low temperatures (liquid helium). Classical and quantum mechanics particle distributions are compared in Table 3.10. From this table we recognize that with a very large *E* or small *T* all three distributions reduce to the classical Maxwell-Boltzmann form.

Whole-integer spin particles are called bosons, for which Bose-Einstein statistics are applicable. The wave function that describes a collection of bosons must be symmetric with respect to the exchange of identical particles. One boson of a system in a certain state increases the probability of finding another boson in this state. Satyendra Bose and Albert Einstein developed these statistics in the

Table 3.10 Classical and Quantum Distributions of Particles, with $\mu = E_F$ the Chemical Potential of the Particle

Distribution Name	Properties of the Particles	Examples	Distribution Function, F
Maxwell-Boltzmann	Spin does not matter. Unlimited number of particles per state. Particles are identical but distinguishable. Wave functions do not overlap.	Classical gas. Fermions and bosons at high T ($\mu - E >> kT$).	$F_{MB} = e^{\left(\frac{\mu - E}{kT}\right)}$ or $Ae^{\left(\frac{-E}{kT}\right)}$
Bose-Einstein	Boson particles are indistinguishable with integer spin (0, 1, 2...). Unlimited number of particles per state. Wave functions overlap.	Liquid ^{4}He, photons, Cooper pairs, excitons, Rb.	$F_{BE} = \frac{1}{e^{\left(\frac{E-\mu}{kT}\right)} - 1}$ or $\frac{1}{Ae^{\frac{E}{kT}} + 1}$
Fermi-Dirac	Fermion particles are identical with half-integer spins (1/2, 3/2, 5/2...). Wave functions overlap. Never more than one particle per state.	Free electrons in metals, protons, and neutrons; electrons in white dwarfs.	$F_{FD} = \frac{1}{e^{\left(\frac{E-\mu}{kT}\right)} + 1}$ or $\frac{1}{Ae^{\frac{E}{kT}} + 1}$

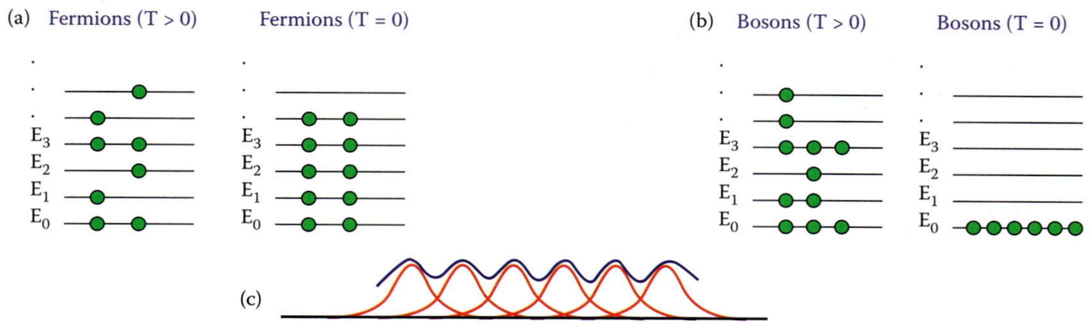

FIGURE 3.109 (a) Fermions at $T > 0$ and at $T = 0$ filling up a set of energy levels. (b) Bosons at $T > 0$ and $T = 0$. The lowest state is macroscopically populated. (c) The particles get so close together that their wave functions overlap, and the condensate is described by a single, macroscopic ψ.

1924–1925 time frame. Bose, a reader in physics at Dacca University, first derived Planck's blackbody law anew from a combination of light quanta of zero mass and statistics. By treating radiation as a quantum gas and counting particles instead of wave frequencies, Bose cut quantum theory loose from its classical antecedents. Then Einstein developed these statistics further and applied them to collections of atoms—gas or liquid—obeying the same rules, and he predicted the phenomenon of what later would be called the Bose-Einstein condensation (BEC) or *superatoms*. If one concentrates a large number of identical bosons in a small region at low temperatures, their wave functions start to overlap, and the bosons lose their individual identities and become one object. The BEC is a macroscopic matter wave, and it is formed of atoms that are delocalized and indistinguishable. In other words, order appears in momentum space (atoms adopt a collective behavior). In Figure 3.109a and b we compare how fermions and bosons fill up a set of energy levels and in c we illustrate how bosons, when they get close together, can be described by a single macroscopic wave function ψ. Einstein was himself not convinced that BEC could happen: "*Die Theorie ist schön, aber ist sie auch war*?" The Bose-Einstein distribution is given as:

$$f_{BE}(E) = \frac{1}{e^{\left(\frac{(E-\mu)}{kT}\right)} - 1} \quad \text{or} \quad \frac{1}{Ae^{\frac{E}{kT}} - 1} \qquad (3.222)$$

where $A = \exp(-\mu)$. The boson statistics would mark Einstein's last great contribution to quantum theory.

Bose-Einstein, Maxwell-Boltzmann, and Fermi-Dirac distributions are compared in Figure 3.110 for the same value of $A = 1$. It is clear that these distributions deviate the most at low temperatures, so one expects to see quantum phenomena emerge at low temperatures.

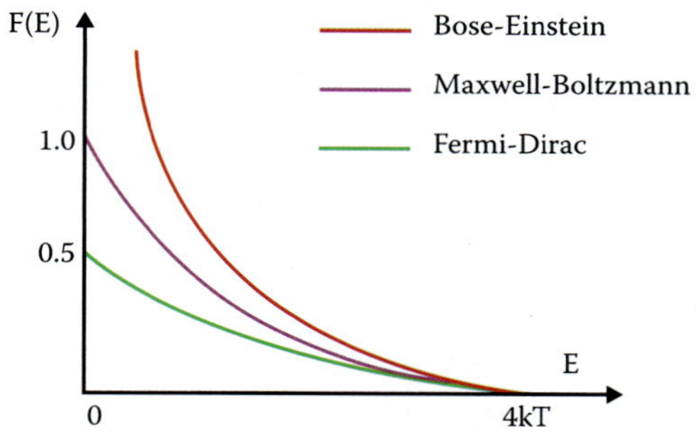

FIGURE 3.110 Bose-Einstein, Maxwell-Boltzmann, and Fermi-Dirac distributions compared.

With bosons there is no exclusion principle, and a macroscopic number of bosons occupy the lowest energy quantum state. The blackbody law, for example, is a direct result of photons, which are bosons, all trying to get into the same energy state. Among other things, bosons also explain superconductivity, lasers (detailed in Chapter 5), and superfluidity in ^{4}He. A superfluid has no viscosity; in effect it experiences no friction and is able to flow forever while at the same time the thermal conductivity becomes very large!

Fermi Surfaces, Brillouin Zones, Density of State Functions, and Conductivity as a Function of Quantum Confinement

Introduction

In this section we consider how electronic processes, like charge transport, are affected by the dimensions of the solid. For this, we need to introduce the density of state functions for electronic systems with different levels of quantum confinement. The density of state function, $G(E)$, is also called the *degeneracy* of a system and is often abbreviated as DOS. The DOS function describes the number of allowed energy states that are available in a system per unit energy and per unit volume (i.e., in units of number of states/eV/cm^3). Earlier, we showed that for a gas of molecules, the number of molecules per unit volume, i.e., density $n(E)$, that have energies between E and $E + dE$, is $n(E)dE$, where $n(E) = G(E)f_{MB}(E)$ with $f_{MB}(E)$, the Maxwell-Boltzmann distribution, representing the probability function of occupancy of a state with energy E. In the case of an electron gas, the number of electrons, $n(E)$, with energies between E and $E + dE$, is again given by the product of $G(E)f_{FD}(E)$ and a distribution function $f_{FD}(E)$, which in this case is the Fermi-Dirac distribution. Thus, combining the density of available states (DOS) with the Fermi function, one can calculate the density of filled states $n(E)$. In other words, we obtain a number for those charge carriers that can contribute to conductivity and charge transport.

Systems with 3, 2, 1, and 0 degrees of freedom for electrons to move in are referred to as bulk material, quantum wells, quantum wires, and quantum dots, respectively, as summarized in Table 3.11. In this table, λ_{DB} represents the de Broglie wavelength (see Equation 3.216). It is the size of L_1, L_2, L_3 compared with λ_{DB} that determines whether we are dealing with a bulk semiconductor (L_1, L_2, $L_3 > \lambda_{DB}$), a quantum well (L_1, $L_2 > \lambda_{DB} > L_3$), a quantum wire ($L_1 > \lambda_{DB} > L_2$, L_3), or a quantum dot ($\lambda_{DB} > L_1$, L_2, L_3). We will see that the DOS function, $G(E)$, strongly depends on the degrees of freedom of the electrons in the system.

To calculate the charge carrier transport or current density ($\mathbf{J} = -n e \mathbf{v}$), using solids of different dimensionalities, we need to know the charge carrier density n as a function of dimensionality, but also the concept of Fermi surfaces needs to be introduced, and we need to get reacquainted with the Brillouin zones of Chapter 2. A Fermi surface is a surface of constant energy E_F in **k**-space, and at absolute zero temperature, it separates occupied from unoccupied quantum states. The way Fermi surfaces fill Brillouin zones determines whether a material will be a metal, semimetal, semiconductor, or an insulator. For example, if the first Brillouin zone is completely filled and there is a large gap between it and the next Brillouin zone, we have an insulator; a smaller gap makes for a semiconductor. If the first band is not completely filled or overlaps with an empty second Brillouin zone, a conductor or a semimetal results, respectively.

Bulk Materials: L_1, L_2, $L_3 > \lambda_{DB}$

Fermi Surface for Bulk Materials A Fermi surface is a surface of constant energy E_F in **k**-space, and at absolute zero temperature, it separates occupied from unoccupied quantum states. With a large number of electrons N at $T = 0$, the electrons will occupy the lowest energy states, consistent with the exclusion

TABLE 3.11 Examples of Reduced-Dimensional Material Geometries and Definitions of Their Dimensionality and of the Associated Type of Confinement*

$L_{1,2,3} > \lambda_{DB}$	No nanostructures	No confinement	Bulk material	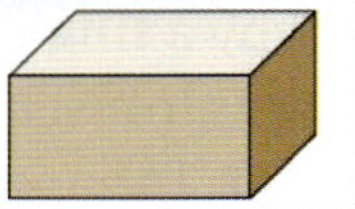
$L_{1,2} > \lambda_{DB} > L_3$	2D nanostructures	1D confinement	Wells	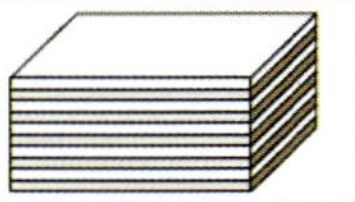
$L_1 > \lambda_{DB} > L_{2,3}$	1D nanonstructures	2D confinement	Wires	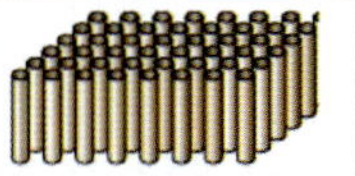
$\lambda_{DB} > L_{1,2,3}$	0D nanostructures	3D confinement	Dots	

*Here λ_{DB} represents the de Broglie wavelength.

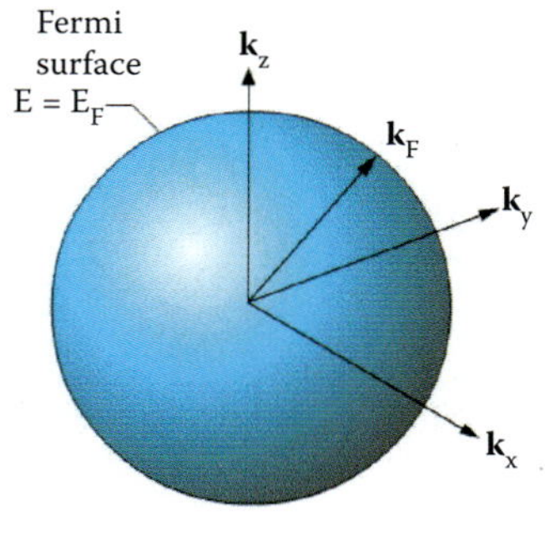

FIGURE 3.111 Free electron Fermi surface.

principle. Thus, the *N* electrons will fill up the lowest *N*/2 energy levels (two electrons per level). The energy of the last filled (or half-filled) level at T = 0 is called the Fermi energy, E_F. An unrestricted 3D Fermi electron gas is isotropic, so that a surface of constant energy *E* in **k**-space is a sphere with a surface called the Fermi surface (Figure 3.111). This surface separates the occupied from the unoccupied states at 0 K. At T = 0 K, all the free electron states are occupied up to an energy E_F, and all fall within a Fermi sphere with a Fermi wave vector, $\mathbf{k}_F$. When electrons are not free, a Fermi surface still exists, but it is distorted by the interaction with lattice ions (see, for example, Figure 3.117). The Fermi energy E_F is an effective cutoff level for the allowed energies of the electrons and corresponds to the chemical potential, μ, i.e., the amount of energy needed to add an electron to the system ($E_F = \mu$). The quantities $\mathbf{k}_F$ and E_F are called the Fermi wave vector and Fermi energy, respectively, in honor of Enrico Fermi. To calculate their exact value, we need to introduce the density of states function (see below, Equations 3.238 and 3.232, respectively).

Brillouin Zones for Bulk Materials Brillouin zones for crystals were first introduced in Chapter 2 in the context of x-ray diffraction. An x-ray diffraction pattern of a crystal represents a map of the reciprocal lattice, a Fourier transform of the lattice in real space, or also a representation of the lattice in **k**-space. Each Brillouin zone is a primitive cell of the reciprocal lattice, which corresponds to the Wigner-Seitz primitive cell of the real lattice (Figure 2.34). The boundaries of Brillouin zones satisfy the von Laue conditions for diffraction, and the Brillouin zone surface describes all **k** vectors that are constructively diffracted by the crystal. In other words, Bragg planes bound the Brillouin zones. This is true for the x-rays considered in Chapter 2 but also for the matter waves associated with electrons, neutrons, etc., discussed in this chapter. In discussing the Ewald sphere in Chapter 2 we also pointed out that the Bragg planes bisect the shortest vectors of the reciprocal lattice and that one can define the first Brillouin zone as the set of points in **k**-space that can be reached from the origin without crossing any Bragg planes. The second Brillouin zone is the set of points that can be reached from the first zone by crossing only one Bragg plane. The *n*th Brillouin zone can be defined as the set of points that can be reached from the origin by crossing *n* – 1 Bragg planes, but no fewer (see the Brillouin zone for a square and a triangular 2D lattice in Figure 2.33). To get reacquainted with the procedure of constructing Brillouin zones, we start off here with a simple square lattice of atoms with interatomic distance, *a*. Its reciprocal lattice is also a square, with reciprocal lattice base vector of length $2\pi/a$. That reciprocal lattice is shown in Figure 3.112a, where we also have drawn the three shortest vectors marked $\mathbf{G}_1$, $\mathbf{G}_2$, and $\mathbf{G}_3$. Three lines, Bragg planes, are drawn as perpendicular bisectors of these three vectors. By drawing all the lines equivalent by symmetry with those bisector lines, we obtain the regions in **k**-space that constitute the first three Brillouin zones as marked (Figure 3.112b).

We turn now to a 3D crystal where we find again that a zone structure exists that may be different for different directions in **k**-space. The coordinates $\mathbf{k}_x$, $\mathbf{k}_y$, and $\mathbf{k}_z$ specify a point in **k**-space: a value of the vector from the origin that specifies the momentum of the electron. The zones in **k**-space corresponding to allowed energies for motion of an electron in a solid are 3D Brillouin zones forming nested sets of polyhedra. Different potentials exist in different directions, so the electron wavelength and crystal momentum, $\mathbf{k} = 2\pi/\lambda$, differ with direction, and many different parabolic *E*-**k** relationships exist, depending on the crystalline momentum. For a simple example, consider a cubic crystal with a Bragg reflection (Equation 2.20 with d = a) occurring when:

$$\mathbf{k}_x = \frac{2\pi n_x}{a},\ \mathbf{k}_y = \frac{2\pi n_y}{a},\mathbf{k}_z = \frac{2\pi n_z}{a},$$
$$n_x, n_y, n_z, = \pm 1, \pm 2, \pm 3, \ldots \qquad (3.223)$$

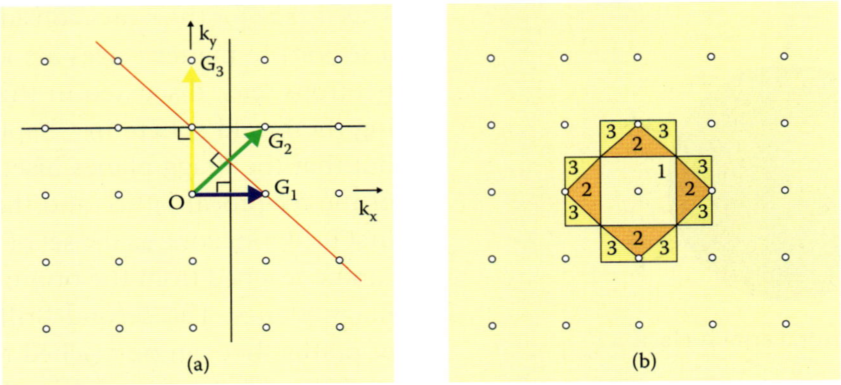

FIGURE 3.112 Illustration of the definition of the Brillouin zones for a 2D square Bravais lattice. To go from (a) to (b): the numbers in (b) denote the zone to which the region belongs. The numbers are ordered according to the length of the vector **G** used in the construction of the outer boundary of the zone (a).

These are the zone boundary values corresponding to discontinuities in the allowed energy levels. From Bragg's law, reflecting planes with the largest interplanar spacings (d = a) have the smallest values for **k**. The planes with the largest spacing are the {100} planes. Thus, the first Brillouin zone in **k**-space is bounded by the {100}* planes at $\mathbf{k}_x = \frac{\pm\pi}{a}$, $\mathbf{k}_y = \frac{\pm\pi}{a}$, $\mathbf{k}_z = \frac{\pm\pi}{a}$, which therefore is a cube with faces π/a from the origin as shown in Figure 3.113. The next set of reflecting planes are the {110} planes with spacing $d = a/2^{1/2}$. Corresponding planes in **k**-space are:

$$\pm\mathbf{k}_x \pm \mathbf{k}_y = \frac{2\pi}{a},\quad \pm\mathbf{k}_x \pm \mathbf{k}_z = \frac{2\pi}{a},\quad \pm\mathbf{k}_x \pm \mathbf{k}_z = \frac{2\pi}{a} \tag{3.224}$$

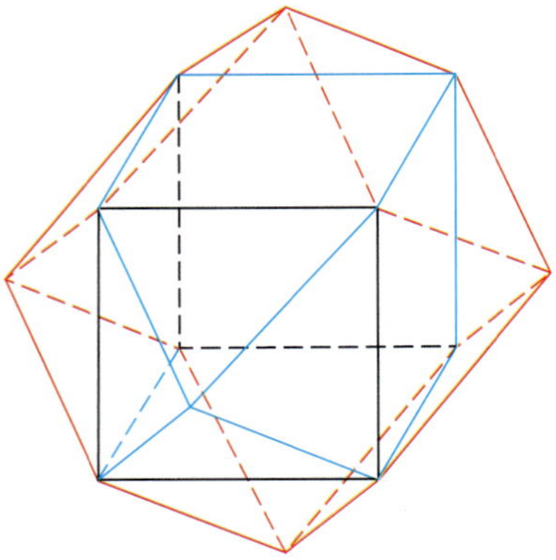

FIGURE 3.113 First two Brillouin zones for a simple cubic lattice. The first zone is the cubic volume $8\pi^3/a^3$ in **k**-space. The second zone is the **k**-space between the cube and the circumscribed dodecahedron.

These 12 planes outline the dodecahedron shown in the same figure. The second Brillouin zone is the **k**-space between the cube and the dodecahedron.

In Figure 3.114 we show what happens when one fills up the first Brillouin zone (BZ) of a face-centered cubic structure with electrons. In this case the largest spacing that satisfies the Bragg condition involves both {111} and {200} planes. These are the set of planes that would be reached first (largest a is smallest **k**). The zone boundaries then derive partly from planes derived from {200} and partly from planes derived from {111}. Envision a growing Fermi sphere inside this first Brillouin zone. Near the center of the zone the electrons are virtually free and behave like an electron gas with an energy given by Equation 3.144. At these energies, well below the zone boundaries, the surfaces of constant energies are represented as spheres within the Brillouin zone as drawn in Figure 3.114a. But when adding more and more electrons, we approach the diffraction

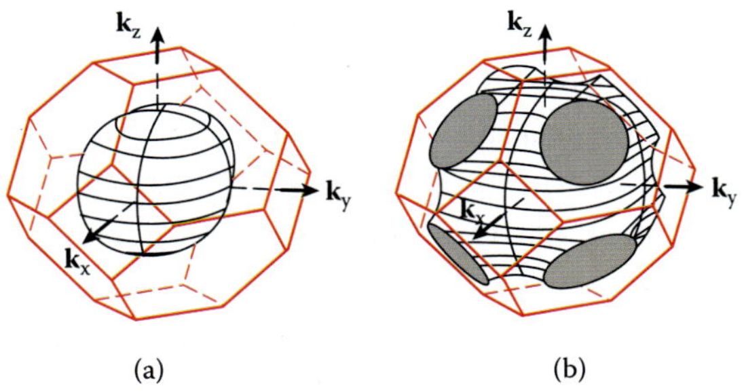

FIGURE 3.114 (a) The first Brillouin zone of a face-centered cubic (FCC) structure with surfaces of constant energy of electrons shown for nearly free electrons near the bottom of the zone and (b) electrons at the zone boundary.

* See Chapter 2 on crystallography on Miller indices.

boundaries, and the wave vectors near or at the BZ feel the periodic potential of the crystal, whereas the others do not. All wave vectors that end on a BZ will fulfill the Bragg condition and thus are diffracted. Because electrons cannot transgress this boundary, the spherical surface becomes distorted by bulging out toward the zonal plane (Figure 3.114b). These electrons no longer behave as a free electron gas. Wave vectors completely in the interior of the first BZ, or between any two BZs, will never get diffracted; they move pretty much as if the potential would be constant; i.e., they behave very close to the solutions of the free electron gas.

If the first Brillouin zone is completely filled and there is a considerable gap between it and the next Brillouin zone, we have an insulator at hand. With a smaller gap, we are dealing with a semiconductor. If the first band is not completely filled or overlaps with an empty second Brillouin zone, we are dealing with a conductor or a semimetal, respectively.

Some other jargon often associated with the difference between a metal, a semiconductor, and an insulator is the characterization of the highest occupied molecular orbital (HOMO) and the lowest unoccupied molecular orbital (LUMO). If the difference in the energy between the HOMO and the LUMO is zero, then just a little energy can promote an electron in an unoccupied level. Therefore, with an electrical potential difference, some electrons are very mobile and give rise to electrical conductivity in a metal. When the temperature increases, there are more electrons excited toward empty orbitals. However, the conductivity decreases because the vibration of the nuclei increases the collisions between the transported electrons and the nuclei, so there is a less-efficient transport or the resistance increases. When the HOMO-LUMO energy difference is nonzero, there is an electronic gap. If the bandgap is small, thermal excitations can promote electrons to unoccupied levels; consequently, those electrons can contribute to the electrical conductivity. This is the case of semiconductors, and that is why the conductivity of semiconductor increases with temperature. Insulators are characterized by a huge HOMO-LUMO energy difference, and the electrons cannot reach the unoccupied levels; there is no measurable conductivity.

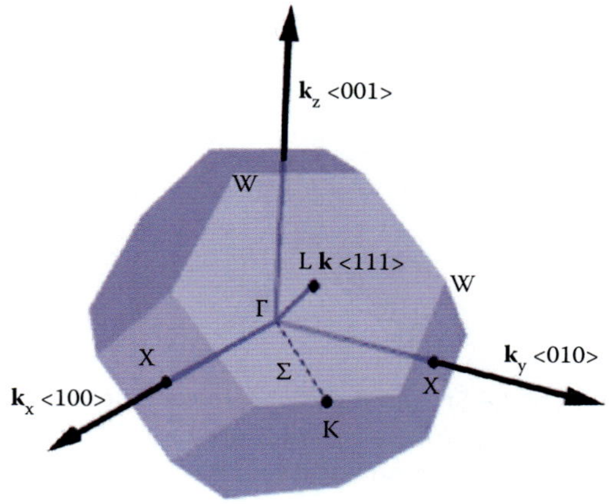

FIGURE 3.115 First Brillouin zone for Si. Points of high symmetry on the Brillouin zone have specific importance. The most important point for optoelectronic devices is the center at $\mathbf{k} = 0$, known as the gamma point Γ (000). One finds *X* at the surface boundaries (100), (001), and (010); *K* at (110); and *L* at (111). *K* is in the middle of an edge joining two hexagonal faces, *L* is at the center of a hexagonal face, and *W* is a corner point. Σ means directed from Γ to *K*.

In Figure 3.115 we show the first Brillouin zone for Si, a semiconductor. Letters are used to mark many of the high symmetry points on this first BZ boundary. The gamma point (Γ) is always the zone center, where $\mathbf{k} = 0$ ($\mathbf{k}_x = \mathbf{k}_y = \mathbf{k}_z = 0$), the *X* point is at $\mathbf{k}_x = 2\pi/a$ and $\mathbf{k}_y$ and $\mathbf{k}_z = 0$ (center of a square face), and the *L* point at $\mathbf{k}_x = \mathbf{k}_y = \mathbf{k}_z = \pi/a$ (center of a hexagonal face) (*a* is the lattice constant, i.e., cube edge). *K* is in the middle of an edge joining two hexagonal faces, *L* is at the center of a hexagonal face, and *W* is a corner point. Σ means directed from Γ to K. Most semiconductors have band edges of allowed bands at one of these points. Zone edges or surfaces are marked with symbols from the Roman alphabet, whereas the interior is marked with symbols from the Greek alphabet. Two example Greek letters not yet introduced are: Δ which means directed from Γ to *X* and Λ which means directed from Γ to L. These symbols are not marked in Figure 3.115 where we only introduced Γ and Σ but they are used in Figure 3.116. In Figure 3.116 the energy bands in 2D are plotted along the major symmetry directions in the first BZ for two of the industrially most important semiconductors, namely, Si and GaAs.

Group IV and III–V semiconductors have a band structure that appears somewhat similar because their basic character is controlled by sp^3 hybridization and tetrahedral bonding. From Figure 3.116a, in the

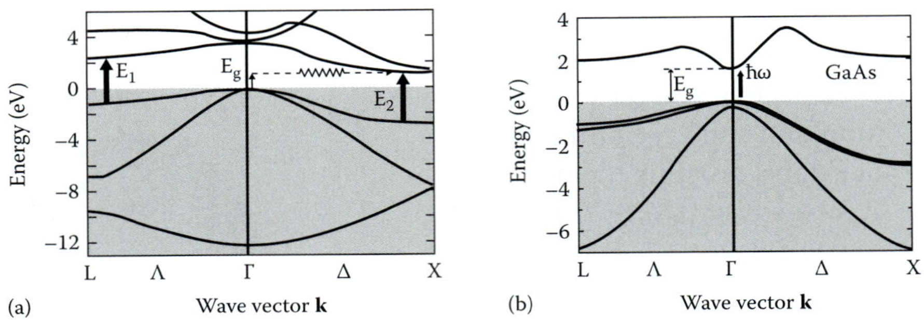

FIGURE 3.116 (a) Si is an indirect semiconductor because the maximum of the valence band (E_v at Γ) does not coincide with the minimum of the conduction band (E_c at X). For silicon, the valence band maximum (VBM) occurs at **k** = 0 (Γ point). However, the conduction band minimum (CBM) occurs to the left of *X*. This makes Si an indirect bandgap material. The shortest distance between E_c and E_v is the bandgap E_g of Si: 1.1 eV (a). GaAs is a direct bandgap semiconductor because the VBM and CBM occur at the same point in the Brillouin zone, and it features a strong absorption. For GaAs, E_g = 1.5 eV (b).

case of silicon, the valence band maximum (VBM) occurs at **k** = 0 (Γ point). However, the conduction band minimum (CBM) occurs to the left of X. Thus, the VBM and the CBM occur at different points in the Brillouin zone, and this makes Si an indirect bandgap material (with a bandgap E_g = 1.1 eV). For an electron to move from the VBM to the CBM, its **k** vector needs to be changed, making the process less likely because a phonon is required for the electron to change its momentum (three-particle process: electron, photon, and phonon). Its indirect bandgap makes Si a weak absorber of light and thus a poor optoelectronic material. The Si band diagram also has two critical points at E_1 (3.2 eV) and at E_2 (4.3 eV). Critical points (van Hove singularities) occur whenever $dE/dk = 0$, in other words, when the bands are parallel to each other. At those critical points, band transitions are more likely because no momentum change is required (parallel band effect). The Si band structure also features six equivalent conduction band minima at *X* along six equivalent <100> directions, and the valence band maxima for the heavy-hole, light-hole, and split-off bands are located exactly at the Γ point. Germanium and diamond also are indirect bandgaps. GaAs is a direct bandgap semiconductor (Figure 3.116b) because the VBM and CBM occur at the same point in the BZ, and it features a strong absorption at $\hbar\omega > E_g$ (for GaAs, E_g = 1.5 eV), making it an excellent optoelectronic material.

The Fermi surface for Cu, shown in Figure 3.117, is a sphere entirely contained within the first Brillouin zone. Even if the free electron Fermi sphere does not intersect a BZ boundary, its shape can still be affected at points close to the boundary where the energy bands begin to deviate from the free electron parabolic shape. This is the case with Cu, where in the <111> directions, contact is made with the hexagonal Brillouin zone faces. Eight short "necks" reach out to touch the eight hexagonal zone faces. The Fermi surface shown in Figure 3.117 can extend throughout many unit cells if the necks in the <111> directions are joined together with similar surfaces in adjacent cells. A section of such a continuous zone structure is shown in Figure 3.118. A magnetic field may close off these necks, a phenomenon that may be studied using cyclotron resonance. These necks are clearly evident from de Haas-van Alphen oscillations (http://www.lanl.gov/orgs/mpa/nhmfl/users/pages/deHaas.htm and http://physics.binghamton.edu/Sei_Suzuki/pdffiles/Note_dHvA.pdf) for magnetic fields in the <111> directions, which contain two periods, determined by the extremal "belly" (maximum) and "neck" (minimum) orbits. The de Haas-van Alphen

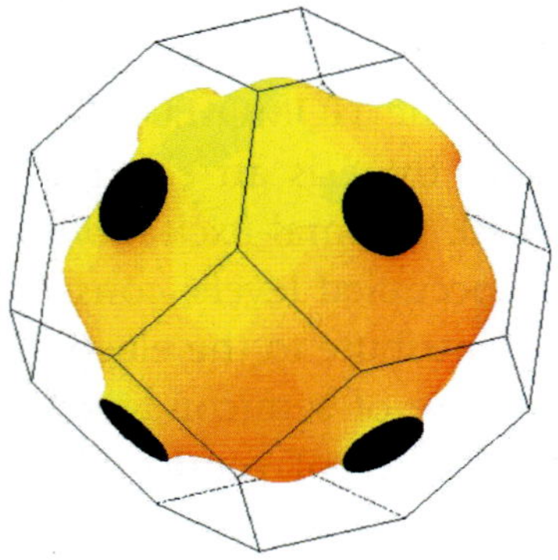

FIGURE 3.117 Brillouin zone for Cu with free electron sphere bulging out in the <111> directions to make contact with the hexagonal zone faces.

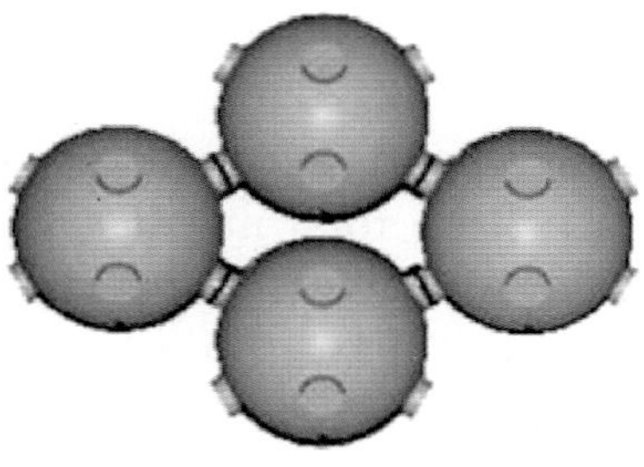

FIGURE 3.118 Continuous Fermi surfaces extending through adjacent unit cells in gold structures.

effect has its origin in the Bohr-Sommerfeld quantization of the orbits of conduction electrons under the influence of a magnetic field. A measurement of the temperature dependence of the oscillation amplitude permits a determination of the cyclotron frequency or, equivalently, the electron mass.

Density of States for Bulk Materials The DOS function $G(E)$ is a property that quantifies how closely packed energy levels are in some physical system. It is usually expressed as a function of internal energy E, $[G(E)]$, or as a function of the wave vector **k**, $[G(\mathbf{k})]$. The density of **k**-states, $G(\mathbf{k})$, in the Fermi sphere drawn in Figure 3.111 is the number of allowed states between **k** and $\mathbf{k} + d\mathbf{k}$, i.e., the number of states between a sphere of radius **k** and a sphere of radius $\mathbf{k} + d\mathbf{k}$ (analogous to the calculation of the density of states for gas molecules illustrated in Figure 3.7). In three dimensions the volume between the two shells, **V** is given by:

$$\mathbf{V}d\mathbf{k} = 4\pi\mathbf{k}^2 d\mathbf{k} \tag{3.225}$$

The density of states $G(\mathbf{k})_{3D}$ is then derived by dividing this volume by the volume of a single energy state. From Figure 3.66b we recognize that each value of **k** occupies a volume $V = (2\pi/L)^3$, so that the number of states per unit volume of **k**-space is $1/V$ or $(L/2\pi)^3$. We also introduce a factor of two here from Pauli's exclusion principle: each energy state can accommodate two electrons, or:

$$G(\mathbf{k})_{3D} d\mathbf{k} = 2\frac{4\pi\mathbf{k}^2 d\mathbf{k}}{\left(\frac{2\pi}{L}\right)^3} = \frac{L^3\mathbf{k}^2}{\pi^2} d\mathbf{k} \tag{3.226}$$

To obtain the density of states in terms of the energy $[G(E)_{3D}]$, we use the relation between E and **k**, derived earlier as $\mathbf{k} = \left(\frac{2m_e^* E}{\hbar^2}\right)^{1/2}$ (Equation 3.150). For every **k**-state there is a corresponding energy state. Differentiating the latter expression with respect to energy leads to:

$$d\mathbf{k} = \left(\frac{2m_e^* E}{\hbar^2}\right)^{-\frac{1}{2}} \frac{m_e^*}{\hbar^2} dE \tag{3.227}$$

The density of energy states $G(E)_{3D}$, the number of allowed states between E and $E + dE$, is then obtained using the chain rule and dividing by L^3 (=V) because we want an expression per unit energy and unit volume:

$$G(E)_{3D} = G(\mathbf{k})_{3D}\frac{d\mathbf{k}}{dE} = \frac{k^2}{\pi^2}\frac{d\mathbf{k}}{dE} \tag{3.228}$$

Thus, we calculate the density of states for a parabolic band in a bulk material (three degrees of freedom) as:

$$G(E)_{3D} dE = \frac{1}{2\pi^2}\left(\frac{2m_e^*}{\hbar^2}\right)^{\frac{3}{2}} E^{\frac{1}{2}} dE \tag{3.229}$$

for $E \geq 0$. There are no available states at $E = 0$, and the effective mass of the electron takes into account the effect of the periodic potential on the electron. The minimum energy of the electron is the energy at the bottom of the conduction band (CB), E_c. For the valence band (VB), similar results hold. In the case that the origin of the energies is not chosen to be the bottom of the band (i.e., $E_c \neq 0$), for the conduction band Equation 3.229 comes with an energy term $(E - E_c)^{1/2}$, and in the case of the valence band the energy term is given as $(E_v - E)^{1/2}$. Here the contributions from the light- and heavy-hole bands add to give the total DOS. Note that, just like the DOS in a 3D ideal gas (Equation 3.19), we obtain a square root dependence in energy E. This function is illustrated in Figure 3.119 (full line). Thus, in the case of free 3D motion, the electronic DOS has a smooth square root dependence on energy E with three continuously varying wave vectors $\mathbf{k}_x$, $\mathbf{k}_y$, and $\mathbf{k}_z$. This holds for materials that are large compared with the de Broglie wavelength ($L_1, L_2, L_3 > \lambda_{DB}$). An important feature of nanostructure devices is that their DOS is very different from this expression, and

G(E)dE =
$\frac{1}{V}$ × [the number of states in the energy range from E to E + dE]

G(E)f(E)dE =
$\frac{1}{V}$ × [the number of filled states in the energy range from E to E + dE]

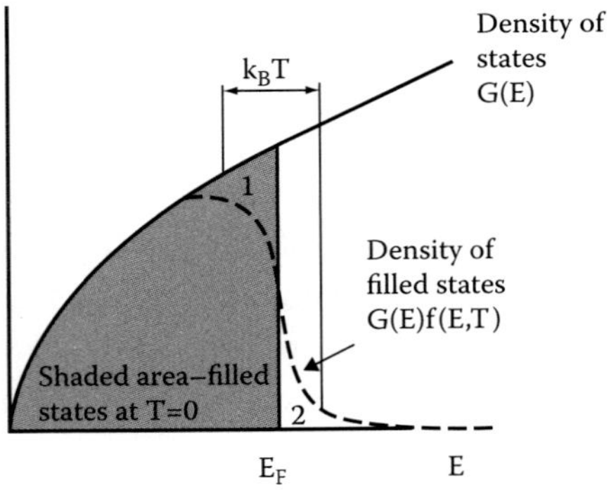

FIGURE 3.119 Density of states (DOS) (full line) and occupied states (shaded area filled at T = 0 K) around the Fermi energy [$G(E)_{3D}$]. The density of filled states is marked with a broken line. As the temperature increases above T = 0 K, electrons from region 1 are excited into region 2.

this will turn out to have important consequences for their electrical and optical properties.

Equation 3.229 gives us the number of possible electronic states of a large 3D device (L_1, L_2, L_3 > λ_{DB}). To deduce how these states are occupied with electrons, we need to multiply the density of state function [$G(E)_{3D}$] with the Fermi-Dirac distribution function, $f_{FD}(E, T)$, which gives the probability that a state of energy E is occupied by an electron. The number of electrons with energies between E and $E + dE$ is then written as:

$$n(E)_{3D}dE = G(E)f_{FD}(E)dE \quad (3.230)$$

where $G(E)_{3D}dE$ is the number of states between E and $E + dE$. This function is shown as a broken line in Figure 3.119.

The integral of $n(E)_{3D}dE$ over all energies gives the total number of electrons n_{3D} per unit volume ($=n_{3D}/V$). From Equation 3.230, with f(E) = 1 (T = 0 K), the number density of filled states n_{3D} is:

$$n_{3D} = \int_0^\infty n(E)_{3D}dE = \int_0^{E_F} G(E)_{3D}f(E)dE \text{ or}$$

$$n_{3D} = \int_0^{E_F} \frac{1}{2\pi^2}\left(\frac{2m_e^*}{\hbar^2}\right)^{\frac{3}{2}} E^{\frac{1}{2}}dE = \frac{\left(\frac{2m_e^*E_F}{\hbar^2}\right)^{\frac{3}{2}}}{3\pi^2} \quad (3.231)$$

Note that at T = 0, $n(E)_{3D}$ is zero for $E > E_F$, so we only have to integrate from E = 0 to E = E_F. In Figure 3.119 this integral corresponds to the shaded area.

For a metal with a total number n_{3D} of valence electrons per unit volume, we can now calculate the maximum energy (E_F) and the maximum **k** value ($\mathbf{k}_F$). The maximum energy (the energy at the surface of the sphere) at T = 0 K is obtained by solving for E_F:

$$E_F = \frac{\hbar^2}{2m_e^*}\left(3\pi^2 n_{3D}\right)^{\frac{2}{3}} = (0.365 \text{ eV nm}^2)(n_{3D})^{\frac{2}{3}} \quad (3.232)$$

This Fermi level is the top of the collection of electron energy levels at absolute zero temperature, and it depends on the number of electrons per unit volume (n_{3D}). Example numbers for Fermi energies for different metals can be found in Table 3.13.

Because fermions cannot exist in identical energy states (see the exclusion principle), at absolute zero, electrons pack into the lowest available energy states and build up a "Fermi sea" of electron energy states. In this state (0 K), the average energy per electron in a 3D electron gas is calculated as:

$$E_{av} = \frac{\text{Total Energy}}{\text{Number of Electrons}} = \frac{\int_0^{E_F} EG(E)_{3D}dE}{\int_0^{E_F} G(E)_{3D}dE}$$

$$= \frac{1}{n_{3D}}\int_0^{E_F} EG(E)_{3D}dE = \frac{3}{5}E_F \quad (3.233)$$

For Cu, for example, this average energy is 4 eV, huge compared with typical thermal energies of 0.025 eV (kT at 300 K). Contrast this result with that of a gas of molecules where a Boltzmann distribution leads to an energy E of zero at 0 K and of the order of kT at a temperature T!

The total number N of **k**-states within the Fermi sphere can be calculated as:

$$N = \frac{\frac{4}{3}\pi \mathbf{k}_F^3}{\left(\frac{2\pi}{L}\right)^3} = \frac{\mathbf{k}_F^3}{6\pi^2}V \quad (3.234)$$

because each **k**-state occupies $(2\pi/L)^3$ and the total volume of the Fermi sphere equals $\frac{4}{3}\pi\mathbf{k}_F^3$. The ratio of these two gives the total number of states N, each of which accommodates two electrons; in other words, the density of states n_{3D}(=N/V) at the Fermi level is:

$$n_{3D} = \frac{2N}{V} = \frac{k_F^3}{3\pi^2} = \frac{\left(\frac{2m_e^* E_F}{\hbar^2}\right)^{\frac{3}{2}}}{3\pi^2} \qquad (3.235)$$

giving us back Equation 3.231. Thus, the density of single-particle states available per unit volume per unit energy of states at the Fermi level in an unrestricted (3D) electronic device is derived as:

$$G(E)_{3D}(E = E_F) = \left.\frac{dn_{3D}}{dE}\right|_{E=E_F} = \frac{1}{2\pi^2}\left(\frac{2m_e^*}{\hbar^2}\right)^{\frac{3}{2}} E_F^{\frac{1}{2}} \qquad (3.236)$$

This is a very important quantity as the rates of many processes are proportional to $G(E)$. Increasing n_{3D}, the density of electrons in 3D increases both E_F (Equation 3.232) and $G(E)_{3D}$ (Equation 3.236). This is of course the same outcome as obtained in Equation 3.229, the only difference being that we replaced E with E_F here.

The result in Equation 3.236 can be simplified by comparing it with Equation 3.235 to obtain:

$$G(E)_{3D}(E = E_F) = \left.\frac{dn_{3D}}{dE}\right|_{E=E_F} = \frac{3n_{3D}}{2E_F} \qquad (3.237)$$

What this means is that the density of single-particle states available per unit volume and per unit energy of states at the Fermi level in 3D is 1.5 times the density of conduction electrons divided by the Fermi energy.

Example 3.7: Calculate the Fermi energy E_F of Na. Na has an atomic density of 2.53×10^{28} atoms/m³. We are assuming $m_e = m_e^*$, i.e., $h^2/2m_e = 1.505$ eV.nm².

Answer: Sodium has one valence electron (3*s*) per atom, so the electron density is $n = 2.53 \times 10^{28}/\text{m}^3$. Using Equation 3.232 we obtain:

$$E_F = \frac{1}{4\pi^2}\frac{h^2}{2m_e}\left(3\pi^2 n_e\right)^{\frac{2}{3}}$$

$$= \frac{1}{4\pi^2}\left(1.505 \text{ eV}\cdot\text{nm}^2\right)\left(3\pi^2 \times 25.3 \text{ nm}^{-3}\right)^{\frac{2}{3}} = 3.31 \text{ eV}$$

Obviously electrons in a metal have a very large kinetic energy, even at T = 0!

From Equation 3.235, $\mathbf{k}_F$ is given by:

$$\mathbf{k}_F = (3n\pi^2)^{\frac{1}{3}} \qquad (3.238)$$

which depends only on particle concentration n.

We can now also ask, what is the speed in the highest occupied state E_F? From $\mathbf{k}_F$, the Fermi wave vector, we calculate the velocity $\mathbf{v}_F$ of the electrons on the Fermi surface, i.e., the Fermi velocity (see Equation 3.203, with group velocity $\mathbf{v}_g = \mathbf{v}_F$) as:

$$\mathbf{v}_F = \frac{\mathbf{p}_F}{m_e} = \frac{\hbar \mathbf{k}_F}{m_e} = \sqrt{\frac{2E_F}{m_e^*}} \qquad (3.239)$$

The Fermi velocity is the average velocity of an electron in an atom at absolute zero. This average velocity corresponds to the average energy given above (Equation 3.233). In Equation 3.239, $\mathbf{p}_F$ is the Fermi momentum, i.e., the momentum of fermions at the Fermi surface, and with the expression for $\mathbf{k}_F$ in Equation 3.238 we obtain:

$$\mathbf{v}_F = \frac{\hbar}{m}\left(3\pi^2 n_{3D}\right)^{\frac{1}{3}} \qquad (3.240)$$

We see that in 3D, the higher the electron density, the faster the electrons are moving. In the presence of an electrical field, all the electrons in a conductor move together, so the exclusion principle does not prevent the free electrons in filled states from contributing to the conduction. This is illustrated in Figure 3.120, where we show the Fermi function in one dimension versus velocity at an ordinary temperature. Over a wide range of velocities, the Fermi function equals 1, and speeds $\mathbf{v}_F$ marked in this figure are given by Equations 3.239 and 3.240.

The dashed curve represents the Fermi function after the electric field has been applied for a period t.

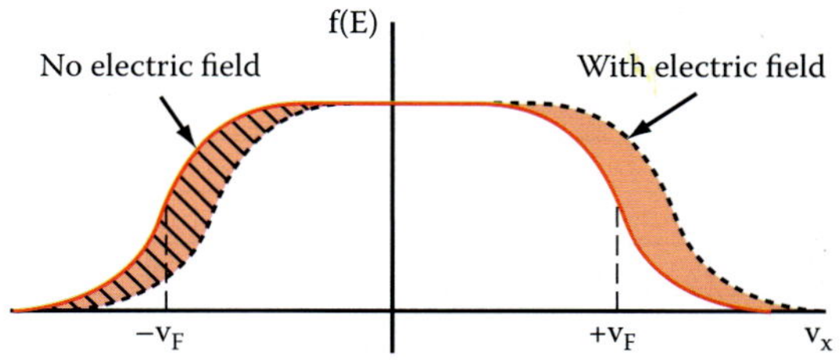

FIGURE 3.120 The Fermi function versus velocity in one dimension for a conductor with (solid) and without (broken line) electrical field in the +*x*-direction. The effect is greatly exaggerated (see text).

Table 3.12 Fermi Quantities*

Fermi Quantity	Equation 3D	Typical Value
Fermi wave vector	$\mathbf{k}_F = (3\pi^2 n)^{\frac{1}{3}}$	~10^8 cm^{-1}
Fermi energy	$E_F = \frac{\hbar^2}{2m_e^*}(3\pi^2 n)^{\frac{2}{3}}$	~1–10 eV
Fermi wavelength	$\lambda_F = 2\pi/\mathbf{k}_F$	Few nanometers for metals, several tens of nanometers for semiconductors
Fermi temperature	$T_F = E_F/k_B$	~10^4–10^5 K
Fermi momentum	$p = \hbar k_F$	
Fermi velocity	$V_F = \frac{\mathbf{p}}{m_e^*} = \frac{\hbar k_F}{m_e^*}$	~10^8 cm/s

*All values at T = 0 K.

All the electrons have been shifted to higher velocities, but the net effect is equivalent to shifting electrons near the Fermi level only.

A typical value for the Fermi velocity at 0 K is ~10^6 m/s or 1000 km/s; as pointed out before, this is a surprising result because for a classical gas at 300 K the thermal velocity $\mathbf{v}_{rms} = \sqrt{\frac{8\mathbf{k}_B T}{\pi m}} = 10^5$ m/s (see Equation 3.21) with a velocity that goes to zero at T = 0 K!

The Fermi wavelength λ_F is given as $2\pi/\mathbf{k}_F$; it represents the de Broglie wavelength associated with the Fermi wave vector $\mathbf{k}_F$. The Fermi temperature is the temperature T_F at which $k_B T_F = E_F$. Thus, it is the energy of the Fermi level of an assembly of fermions divided by Boltzmann's constant. The quantity T_F is not to be confused with the temperature of the electron gas. Below the Fermi temperature, a substance gradually expresses more and more quantum effects of cooling. For temperatures much lower than the Fermi temperature, the average energy of the phonons of the lattice will be much less than the Fermi energy, and the electron energy distribution will not differ greatly from that at T = 0. We also recognize here that photons travel much faster than electrons! Photons travel at $c = 3.0 \times 10^8$ m/s (in vacuum) versus electrons that travel at $\mathbf{v}_F$ (Fermi speed) = 1.57×10^6 m/s (copper wire).

The Fermi quantities we introduced in this section are summarized in Table 3.12, and in Table 3.13 we list calculated free electron Fermi surface parameters for some metals at room temperature.

Electronic Conductivity for Bulk Materials Drude, at the end of the nineteenth century, could not have known about the Fermi-Dirac distribution. Electrons, being fermions, follow this distribution, and at room temperature it is almost the same as at absolute zero temperature, resulting in a velocity distribution for fermions very different from the one predicted by Maxwell-Boltzmann statistics. Only electrons near the Fermi level contribute to electrical conductivity, and as a result of the wave nature of the electrons, they can pass through a perfect crystal without suffering any resistance at all. This means that the mean free path of an electron passing through a perfect crystal with all nuclei at rest is infinity rather than the interatomic spacing of the order of 1 nm assumed by Drude. In such an ideal crystal, a Bloch function ψ_k evolves into $\psi_{k+\Delta k_x}$, and

Table 3.13 Calculated Free Electron Fermi Surface Parameters for Metals at Room Temperature

Metal/Valency	Electron Concentration n (cm^{-3})	Fermi Wave Vector (cm^{-1})	Fermi Velocity (cm s^{-1})	Fermi Energy E_F (eV)	Fermi Temperature $T_F = E_F/k_B$ (in K)
Cu(1)	8.45×10^{22}	1.36×10^8	1.57×10^8	7.00	8.12×10^4
Ag(1)	5.85	1.20	1.39	5.48	6.36
Au(1)	5.90	1.20	1.39	5.51	6.39
Be(2)	24.2	1.93	2.23	14.14	16.41
Mg(2)	8.60	1.37	1.58	7.13	8.27
Zn(2)	13.10	1.57	1.82	9.39	10.90
Al(3)	18.06	1.75	2.02	11.63	13.49
Ga(3)	15.30	1.65	1.91	10.35	12.01
In(3)	11.49	1.50	1.74	8.60	9.98
Pb(4)	13.20	1.57	1.82	9.37	10.87
Sn(4)	14.48	1.62	1.88	10.03	11.64

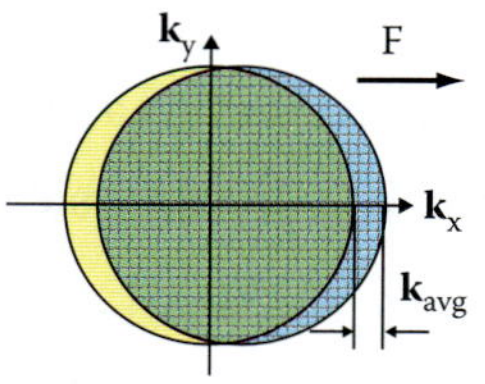

FIGURE 3.121 Displacement of the Fermi surface with an applied electrical field. Collisions with thermal vibrations and defects (not stationary ions or other electrons, as Drude envisaged) stop the Bloch oscillations and cause electrons to settle to a drift velocity. See also Figure 3.71, where the Fermi factor in one dimension for a conductor with (solid) and without (broken line) electrical field in the $+x$-direction is shown.

when this electron state reaches the Brillouin zone at $k = +\pi/a$, it re-enters the crystal at $-\pi/a$ or we get resistance-less Bloch oscillations.

In the absence of an electric field, the same number of electrons is moving in the $\pm x$-, $\pm y$-, and $\pm z$-directions, so the net current is zero. But when a field E is applied, e.g., along the x-direction, the Fermi sphere in Figure 3.121, in the absence of collisions, is displaced at a uniform rate by an amount related to the net change in momentum, $\Delta \mathbf{p}_x$, of the free electron gas (FEG) as a whole. The equation of motion (Newton's law) describes this situation as (see Equation 3.208):

$$m_e^* \frac{d\mathbf{v}_x}{dt} = \hbar\left(\frac{dk_x}{dt}\right) = -E_x e = F \quad (3.241)$$

The quantity $\hbar\mathbf{k}$ is the crystal momentum, and thus one can say that the force caused by the electric field is equal to the time derivative of the crystal momentum. Integrating the previous expression we obtain:

$$\mathbf{k}_x(t) - \mathbf{k}_x(0) = -\frac{e\mathbf{E}_x t}{\hbar} \quad (3.242)$$

The shift in Fermi sphere creates a net current flow because more electrons move in the $+x$-direction than the $-x$-direction. According to this model, the Fermi sphere moves with constant velocity in **k**-space. This means that the electron velocity increases indefinitely. This is of course not possible, and it is evident that scattering processes must limit the electron velocity and hence the finite electrical conductivity of metals. A viscous term must be introduced in the equation of motion. In a real crystal, scattering events with a scattering time τ, involving collisions of electrons near the Fermi surface (these are the only ones that can move into empty states), prevent the observation of the ideal Bloch oscillations and oppose the electrical field effect so that the Fermi sphere reaches a steady state when the new center is displaced in the x-direction by an average wave vector:

$$\mathbf{k}_{avg} = -\frac{e\mathbf{E}_x\tau}{\hbar} \quad (3.243)$$

The wave vector changes calculated from Equation 3.243 are very small changes; for example, with an electric field E of 1 V/m and with a value for the scattering time τ of ~10^{-14} s we calculate a value for $|\mathbf{k}_{avg}|$ of ~15 m^{-1}, a quantity very small compared with a BZ dimension, which we calculate as:

$$k_{BZ} \approx \frac{2\pi}{a} \approx \frac{2\pi}{3.00 \times 10^{-10} m} \approx 2 \times 10^{10}\ m^{-1}$$

The effect of the Fermi shift in Figure 3.121 are greatly exaggerated. If at t = 0 an electrical field is applied to a Fermi sphere centered at the origin of **k**-space, the sphere will move to a new position in a characteristic time between scattering events given by:

$$\tau = \frac{\lambda}{\mathbf{v}_{avg}} \quad (3.244)$$

From Equation 3.243 the average velocity of the Fermi sphere is given by:

$$\mathbf{v}_{avg} = \frac{\hbar\mathbf{k}_{avg}}{m_e^*} = -\frac{e\mathbf{E}_x\tau}{m_e^*} \quad (3.245)$$

At the steady state the current density is then given as:

$$J = -ne\mathbf{v}_{avg} = \left(\frac{ne^2\mathbf{E}_x\tau}{m_e^*}\right) = \sigma E \quad [3.3][3.6]$$

where n is the electron density and:

$$\sigma = \frac{ne^2\tau}{m_e^*} \quad (3.7)$$

These are the same expressions as derived in Drude's model, but with τ given now as $\tau = \frac{\lambda}{\mathbf{v}_{avg}}$ (Equation 3.244) and m_e^* replacing m_e. The average velocity, derived as $\bar{v} = \sqrt{\frac{8k_BT}{\pi m}}$ (Equation 3.21) from the Boltzmann distribution of speeds in Drude's model, is replaced here by $\mathbf{v}_{avg} = \frac{\hbar \mathbf{k}_{avg}}{m_e^*} = -\frac{e\mathbf{E}_x\tau}{m_e^*}$ (Equation 3.245). In Drude's model the mean free time between collisions of electrons with lattice ions, τ, is related to the average velocity in $\bar{\mathbf{v}} = \mathbf{v}_{dx} = -\frac{e\mathbf{E}_x\tau}{m_e}$ (Equation 3.6) as:

$$\tau = \frac{\lambda}{v_x} \approx \frac{a}{v_x} \tag{3.246}$$

Drude assumed the electron mean free path, λ, to be equal to the lattice constant, *a*, which is of the order of 1 nm, in which case this equation yields a typical value for τ of about 10^{-14} s. Based on Equations 3.7, 3.21, and 3.246, Drude then derived the following relationship for the resistivity:

$$\rho = \frac{1}{\sigma} = \frac{m_e \mathbf{v}_d}{ne^2\lambda} = \frac{m_e}{ne^2 a}\sqrt{\frac{8kT}{\pi m_e}} \tag{3.22}$$

Using the lattice constant, *a*, for the mean free path, this equation leads to values for the conductivity of a metal that are six times too small. Moreover, from Equation 3.22 the temperature dependence of the resistivity is determined by v_{dx}, which in this model is proportional to $\sqrt{T}$, whereas in practice the temperature dependence of the resistivity is represented by the empirical relationship:

$$\rho = \rho_0 + \alpha T \tag{3.23}$$

where ρ_0 is the resistivity at a reference temperature, usually room temperature, and α is the temperature coefficient. If the Boltzmann distribution function is applied to the electron gas, one thus immediately finds the velocity of the electron to change as $\sqrt{T}$. According to Drude's model, the mean free path is obviously temperature-independent because it is calculated from the scattering cross-section of rigid ions (with lattice constant *a*). This results in a resistivity proportional to $\sqrt{T}$, provided that the number of electrons per unit volume *n* is temperature-independent (Equation 3.22). However, people at that time had been well aware that the resistivity of typical metals increases linearly with increasing temperature well above room temperature (Equation 3.23). To be consistent with the Maxwell–Boltzmann distribution law, one then had to assume *n* to change as $\frac{1}{\sqrt{T}}$ in metals. This was not physically accepted, and the application of the Maxwell–Boltzmann distribution to the electron system was apparently the source of the problem.

In the quantum mechanical model, only a *few electrons*, all moving at the temperature-independent very high Fermi velocity $\mathbf{v}_F$ ~10^8 cm/s (Equation 3.239), carry the current, instead of *all electrons* moving at the average drift velocity $\mathbf{v}_d$ (~0.1 cm/s) in Drude's model (see p. 82 "Drude Fails"). Only electrons near the surface of the Fermi sphere find empty orbitals in which they can scatter. Electrons in the inner part of the Fermi sphere find no empty states with similar energy as those electrons near the Fermi sphere. Therefore, inner electrons cannot scatter, and those electrons do not contribute to the current transport process. Replacing v_{th} with v_F in Equation 3.8, one obtains a value for the resistivity that is 100 times larger than the experimental numbers—understandable because v_F is 16 times larger than v_{th}, and we know that the numbers obtained with v_{th} were already six times too large. The resolution lies in the calculation of the mean free path λ; in a perfectly ordered crystal $\lambda = \infty$; in a real crystal, λ is determined by scattering phenomena. We need to replace the inner atomic distance a in Equation 3.22 with the quantum mechanics value for λ, the mean free path of the conduction electrons. Experiments have shown that the electrons can move surprisingly far without any interaction; the mean free electron pass can be up to 10^8 atom distances at low temperatures. Electrons are not scattered by the regular building blocks of the lattice because of the wave character of the electrons. Scattering mechanisms instead are:

1. Lattice defects (foreign atoms, vacancies, interstitial positions, grain boundaries, dislocations, stacking disorders), and
2. Thermal vibration of the lattice (phonons).

Item 1 is more or less independent of temperature, whereas item 2 is independent of lattice defects but dependent on temperature. The mean free path now does not depend on the radius of the ions but rather on deviations of the ions from a perfectly ordered array such as seen from lattice thermal vibrations and the presence of impurities. The ion vibrations lead to an effective area *A* that results in an electron mean free path given as $\lambda = 1/n_{ion}A$. Lattice ions are basically points only, but their thermal vibration has them occupy an electron scattering area $A = \pi r^2$, where *r* is the amplitude of the thermal vibrations. The energy of thermal vibration in a simple harmonic oscillator is proportional to the square of the amplitude of the vibration (r^2). In other words, the area *A* is proportional to the energy of the vibrating lattice ions. From the equipartition theory we know that the average vibration energy is proportional to *kT*, so it follows that *A* is proportional to *T* and λ is proportional to $1/T$. Because the mean free path is inversely proportional to temperature at high temperatures, it follows that $\sigma \propto \frac{1}{T}$, in agreement with the experimental evidence (Equation 3.23). This solves the issue of the wrong temperature dependence of the resistivity. We need to also calculate a correct absolute value for the resistance using the quantum mechanical mean free path.

The quantum mechanical mean free path of the conduction electrons, say in Cu, is defined as:

$$\lambda = \mathbf{v}_F \tau \tag{3.247}$$

where $\mathbf{v}_F$ is the velocity at the Fermi surface, because all collisions involve only electrons near the Fermi surface. With Fermi velocities $\mathbf{v}_F$ of typically 10^8 cm/s (see Table 3.13) and with room-temperature resistivities for many metals of $\rho \sim 1–10\ \mu\Omega\cdot$cm, the corresponding relaxation time is $\sim 10^{-14}$ s and the resulting averaged free electron path at room temperature is about 100 Å. So it is of the order of a few 10s to 100s of interatomic distances. At low temperatures for very pure metals the mean free path can actually be made as high as a few centimeters ($\tau \approx 2 \times 10^{-9}$ s at 4 K for very pure Cu). Compared with using the lattice constant, *a*, for the mean free path, this equation leads obviously to values for the conductivity that are 100 times higher—in agreement with experimental data.

For a current **J** of 1 A/mm² in a conductor with an electron density of $n = 10^{22}$ cm^{-3}, we calculate an average speed $\mathbf{v}_{avg}$ of the Fermi sphere of $J/ne \sim 0.1$ cm/s (see Equation 3.3), which is much less than the Fermi velocity $\mathbf{v}_F \sim 10^8$ cm/s. In a conductor, charges are always moving at the Fermi velocity, but the Fermi sphere moves much slower because electrons travel at fast Fermi velocities for a short average time, τ, and then "scatter" because of collisions with atom vibrations, grain boundaries, impurities, or material surfaces (especially in very thin films).

The value for $n(E)_{3D}$ in $\mathbf{J} = -n(E)_{3D}e\mathbf{v}_{avg}$ (Equation 3.3) at $T > 0$ is obtained from Equation 3.230:

$$n(E)_{3D}dE = G(E)_{3D}f_{FD}(E)dE$$

$$= \frac{8\pi\sqrt{2}m_e^{\frac{3}{2}}}{h^3}E^{\frac{1}{2}}\frac{1}{e^{\frac{E-E_F}{kT}}+1}dE \tag{3.248}$$

For a large semiconductor, Equation 3.248 was illustrated in Figure 3.119 with a broken line. We will see below that *G*(*E*) depends strongly on dimensionality, so we also expect the current density to vary strongly with dimensionality.

Quantum Wells: $L_{1,2} > \lambda_{DB} > L_3$

Quantization As we saw earlier, devices that come with a length *L* that in one direction is comparable with the size of the electron de Broglie wavelength are known as quantum wells (1D confinement). A planar quantum well structure may be made from a thin region of a narrow gap semiconductor sandwiched between two layers of a wide bandgap semiconductor. We use for an example of such 1D confinement a quantum well made by sandwiching a layer of GaAs between two layers of $Al_xGa_{1-x}As$, as shown first in Figure 3.59 and, simplified, reproduced in Figure 3.122. Growing two different semiconductors on top of each other, as illustrated,

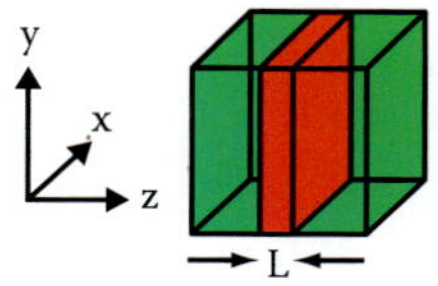

FIGURE 3.122 A quantum well can be made, for example, by sandwiching a layer of GaAs between two macroscopic layers of $Al_xGa_{1-x}As$. This creates a layer in which the electrons (and holes) behave in a 2D way (see also Figure 3.59).

forms heterojunctions. The narrower bandgap GaAs material is enclosed by a material with a considerably larger bandgap to establish a potential barrier at the surface of the confined material. Because of the potential barrier, the motion of electrons and holes is restricted in one dimension (thickness L in the z-direction in this case) and is forced to occupy discrete states of energy instead of staying arbitrarily within an energy continuum. Hence quantization of the system occurs by shrinking the thickness of the GaAs layer. A 2D electron gas in the laboratory is really a 3D electron system in which the electron motion is strongly confined in one spatial direction but in which free motion is still allowed in the other two directions. At the interface of the semiconductors GaAs and $Al_xGa_{1-x}As$ a potential well is formed. The potential well is a result of charge transfer between the two materials and their conduction band offset (E_c), confining the electrons, and the motion of electrons perpendicular to the plane of the hetero-interface (z-axis in Figure 3.122) is quantized.

The composition x of the ternary semiconductor $Al_xGa_{1-x}As$ can be varied to control the electron barrier height. A good lattice match between GaAs and $Al_xGa_{1-x}As$ over a wide range of x values minimizes lattice strain at the interfaces. When the electron and holes are confined in the same layer, one talks about a type I quantum well; with electron and holes confined in different layers, one defines a type II quantum well.

For the quantum well considered here, the $Al_xGa_{1-x}As$ layers are thick enough so that tunneling through these layers remains very limited.

For an energy E smaller than the potential barrier V, the energy of an electron in the conduction band of a quantum well is given as:

$$E_n(k_x, k_y) = E_c^0 + \underbrace{\frac{n_z^2 h^2}{8m_e^* L_z^2}}_{E_{1D}(n_z)} + \underbrace{\frac{h^2(k_x^2 + k_y^2)}{8m_e^*}}_{E_{1D}(k_x, k_y)}, \quad n = 0,1,2,... \qquad (3.249)$$

with the wave function:

$$\psi_n(x, y, z) = \psi_n(z) e^{ik_x x} e^{ik_y y} \qquad (3.250)$$

There is one quantized component in the z-direction and a "free" electron component in the x-y plane. The second term on the right side of Equation 3.249 represents the quantized energy in the z-direction (the thickness direction L of the GaAs film). This is the same expression we derived in Equation 3.156 for a finite-sized 1D box with infinitely high potential walls. For quantization to be important, the difference between the electron energy levels should be much larger than the thermal energy $k_B T$, that is, $E_n = \frac{n_z^2 h^2}{8m_e^* L_z^2} >> k_B T$, where $n_z = 1, 2, 3$ are the quantum numbers labeling.the energy levels. E_c^0 is the energy corresponding to the bottom of the conduction band. Strictly speaking, the above expressions apply only to an infinitely deep potential well. However, we can use the same equations as long as E_n is well below the bottom of the conduction band of the wide band material. Using this condition, we find, for example, that in GaAs where $m_e^*/m_e = 0.067$, the levels are quantized at room temperature when L_z = 150 Å. The third term on the right side of Equation 3.249 represents the kinetic energy of the electrons in the x-y plane where they are free to move. The $\mathbf{k}_z$-component is absent in the last term of Equation 3.249 because the motion in this direction is quantized. Equation 3.249 reveals that for each value of the quantum number n, the values of wave vector components $\mathbf{k}_x$ and $\mathbf{k}_y$ form a 2D band structure. The wave vector $\mathbf{k}_z$ in the z-direction, on the other hand, can only take on discrete values, $\mathbf{k}_z = n_z\pi/L_z$. For each value of n there is a sub-band with n the sub-band index as illustrated in Figure 3.123. In this figure we show energy levels (bottoms of

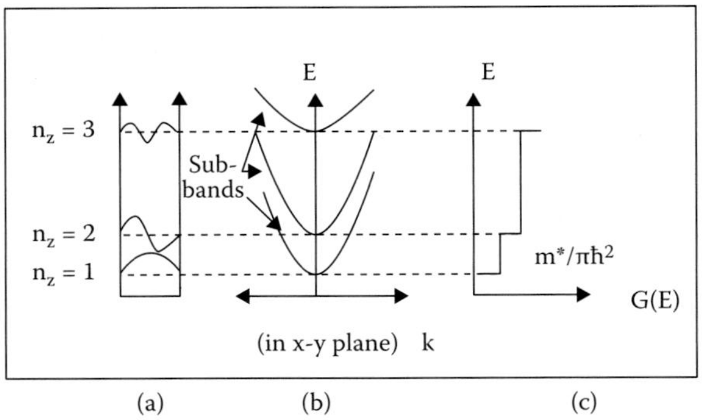

FIGURE 3.123 (a) Energy levels (bottoms of sub-bands) for a quantum well made by sandwiching a layer of GaAs between two macroscopic layers of $Al_xGa_{1-x}As$ (L_z is 150 Å). (b) Energy versus $\mathbf{k} = (\mathbf{k}_x^2 + \mathbf{k}_y^2)^{1/2}$ for 2D electron gas in GaAs quantum well. (c) Density of states for quantum well structure (Harris, 2006.[4])

sub-bands), density of states for a quantum well structure, and energy versus $\mathbf{k} = (\mathbf{k}_x^2 + \mathbf{k}_y^2)^{1/2}$ for the 2D electron gas in the GaAs quantum well.

We can carry out an analogous argument for the holes in the valence band with the difference that their quantized energy is inverted and that we need to invoke m_h^* for the effective mass of the hole. For a quantum well, the lowest-energy band-to-band (interband) transition is now different from the bandgap (E_g) transition of the bulk semiconductor. It will occur at a higher energy level (shorter wavelength) between the lowest energy state for electrons in the conduction band (n = 1) and the corresponding state for holes in the valence band. This defines the effective bandgap for a quantum well as:

$$E_g^* = E_c^0 - E_v^0 + \frac{h^2}{8L^2}\left(\frac{1}{m_e^*} + \frac{1}{m_h^*}\right) \quad (3.251)$$

The shift to higher wavelengths is referred to as a "blue shift," caused by quantization.

Fermi Surfaces and Brillouin Zone for Quantum Wells In the case of a free 3D electron gas we appreciate that the surface of the Fermi sea is a sphere of radius $\mathbf{k}_F$ connecting points of equal energy in **k**-space (see Figure 3.111). In 2D this becomes a circle connecting points of equal energy in 2D **k**-space. In Figure 3.124 we show the Fermi circles corresponding to 2D crystals with one, two, three, and four valence electrons per atom. In this figure we also show the 2D Brillouin zone (see square in dashed line), and we see how the free electron circle of a three-valent metal (red circle) cuts the Bragg planes located at π/a. The square shown in Figure 3.124 corresponds to the first Brillouin zone of a 2D square lattice (Figure 3.112b).

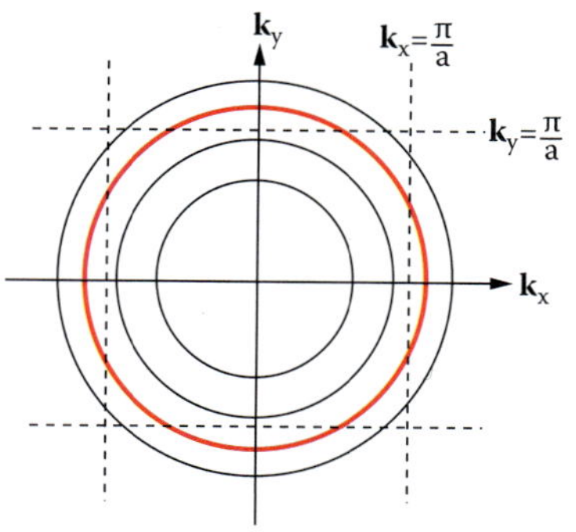

FIGURE 3.124 Free electron circles for 1, 2, 3, and 4 valence electrons. The free electron circle of a three-valent metal (red circle) cuts the Bragg planes located at π/a.

Density of States for Quantum Wells The density of states for each sub-band of a quantum well (Figure 3.123c) can be found using an approach similar to the one we used above for a 3D density of states function, that is, by counting the number of states with wave vectors **k** between **k** and $d\mathbf{k}$. In the case of 1D confinement of electrons we must find the number of **k**-states enclosed in an annulus of radius $\mathbf{k} + d\mathbf{k}$ (see Figure 3.125). Each state occupies an area A of $\left(\frac{2\pi}{L}\right)^2$. The area of the annulus, A, is given by:

$$A d\mathbf{k} = 2\pi \mathbf{k} d\mathbf{k} \quad (3.252)$$

Dividing the area of the annulus by the area occupied by a **k**-state, and remembering again to multiply by 2 for the electron spin states, we get for the number of states per unit area:

$$G(\mathbf{k})_{2D} d\mathbf{k} = 2\frac{2\pi \mathbf{k} d\mathbf{k}}{\left(\frac{2\pi}{L}\right)^2} = \frac{L^2 \mathbf{k}}{\pi} d\mathbf{k} \quad (3.253)$$

Or, in terms of energy per unit area at an energy E (dividing by L^2), the density of states for each sub-band is given as:

$$G(E)_{2D} dE = \frac{\mathbf{k} d\mathbf{k}}{\pi} = \sqrt{\frac{2m_e^* E}{\hbar^2}}\left(\frac{2m_e^* E}{\hbar^2}\right)^{-\frac{1}{2}} \frac{m_e^*}{\pi\hbar^2} dE$$
$$= \frac{m_e^*}{\pi\hbar^2} dE \quad (3.254)$$

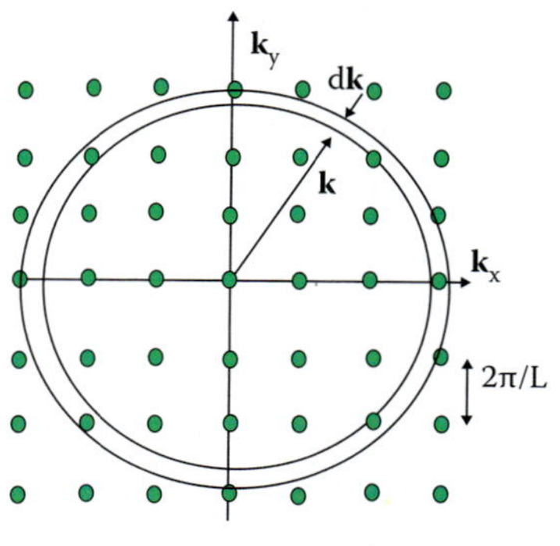

FIGURE 3.125 Density of states, $G(\mathbf{k})_{2D}$, is the number of allowed states between **k** and $\mathbf{k} + d\mathbf{k}$, i.e., the number of states between a sphere of radius **k** and a sphere of radius $\mathbf{k} + d\mathbf{k}$.

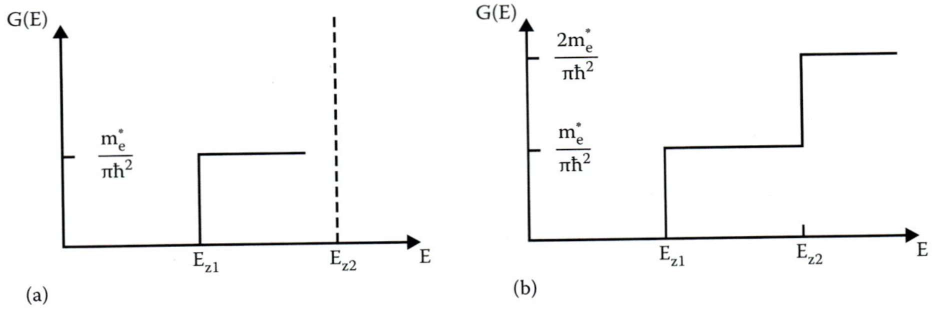

FIGURE 3.126 Filling of the first two sub-bands in a 2D structure. The first sub-band has a constant energy at $\frac{m_e^*}{\pi\hbar^2}$ (a), and the second has a constant energy at $\frac{2m_e^*}{\pi\hbar^2}$ (b).[4]

For this derivation we used Equation 3.227, which holds true for any dimensionality D of the problem. Importantly, in an ideal 2D system, the density of states is constant and does not depend on energy.

The density of states in a 2D electron gas, in which only the lowest energy sub-band (n = 1 level E_{z1}) is occupied, is illustrated in Figure 3.126a, where we assume that the confinement is in the z-direction. For all electron energies up to E_{z1} the density of states is zero because electrons cannot exist in the well at lower energies than this. Because of the freedom of motion in the plane of the hetero interface (with continuously varying wave vectors $\mathbf{k}_x$ and $\mathbf{k}_y$ in the x-y plane), the energy levels that form in the potential well are highly degenerate. If the electron density is sufficiently low, then all electrons can be accommodated in the lowest level of the well, and the freedom of motion in the transverse direction is frozen out and the electron system effectively behaves as the ideal 2D one as described by Equation 3.254 with a single step in energy only with a constant value $\frac{m_e^*}{\pi\hbar^2}$. In the regime where only the lowest sub-band is occupied, increasing energy corresponds to increasing electron motion in the x-y plane (kinetic energy).

As the electron filling of the quantum well is increased, eventually electrons begin to fill the next transverse level of the potential well (Figure 3.126b). At this point we have two so-called 2D sub-bands occupied in the quantum well. To a good approximation, electrons in the two sub-bands may be viewed as forming two independent 2D electron gas systems. Thus, the density of states is double that which we would expect in the case where just a single sub-band is occupied. For each quantum state in the quantum well, there will be a step in the density of states. The overall density of states is discontinuous, with a stepwise structure that is characteristic of quantum wells. Because of the summing over the different sub-bands, a more general description of a 2D system has Equation 3.254 modified as:

$$G(E)_{2D} = \frac{m_e^*}{\pi\hbar^2}\sum_n H(E - E_n) \qquad (3.255)$$

where $H(E - E_c)$ is the Heaviside step function.* It takes the value of zero when E is less than E_n and 1 when E is equal to or greater than E_n. E_n is the nth energy level within the quantum well.

In Figure 3.127 we summarize the characteristics of a 2D density of states function $G(E)_{2D}$. The energy spacing shown here increases with decreasing L; the thinner the 2D film, the more it approximates an ideal 2D gas with only one sub-band.

The step-like behavior of a 2D density of states function $G(E)_{2D}$ implies that the density of states in the vicinity of the bandgap is much larger than in the case of a bulk semiconductor, where the value of $G(E)_{3D}$ goes to zero (E_1 versus E_g). This makes for stronger optical transitions because a major factor

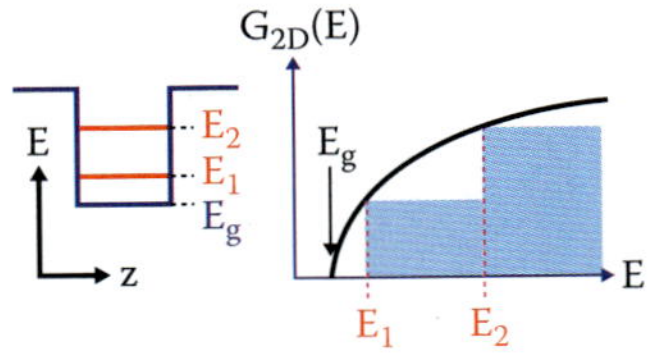

FIGURE 3.127 The density of states function for a quantum well. The solid black curve is that of a free electron. The bottom of the quantum well is at energy E_g.

* The Heaviside step function, H, also called the unit step function, is a discontinuous function whose value is zero for negative argument and one for positive argument.

in the expression for the probability of optical transitions is the density of states. The strength of an optical transition is often defined as the oscillator strength, and the oscillator strength of a quantum well in the vicinity of the bandgap is considerably enhanced compared with a bulk semiconductor. Below we will show how Fermi's golden rule for absorption describes transition rates between levels in terms of the availability of states [density of states: $G(E)$], the availability of photons (intensity E_0), and a "coupling strength" between the levels (transition matrix element $|H'_{v,c}|^2$). The enhanced oscillator strength of quantum-confined structures is put to good use in the fabrication of laser media for highly efficient and compact solid-state lasers (see Chapter 5). In Figure 3.128 we illustrate the bound-state energies for electrons in the conduction band and for holes in the valence band for a GaAs/$Al_xGa_{1-x}As$ heterojunction. If the GaAs layer is thin enough, bound states form as indicated here by dashed lines. For a perfect 2D electron gas, each bound state corresponds to a discontinuous jump in the electronic density of states function. The density of states function (DOS) can be investigated by measuring the absorption α of electromagnetic radiation associated with the excitation of an electron from the valence band to the conduction band. Because the wavelength of the radiation is long compared with the width of the well, transitions only occur between states for which the spatial variation of the wave function is similar; this leads to the selection rule $\Delta n = 0$ for the adsorption. Therefore, the allowed transitions are those indicated by the arrows in Figure 3.128a. The frequency dependence of the absorption should reflect the steps in the DOS with the steps in the absorption expected to occur at frequencies ω_n given by:

$$\hbar\omega_n = E_{Cn} - E_{Vn} \tag{3.256}$$

where $E_{Cn} - E_{Vn}$ is the energy difference between the nth bound states in the conduction and valence bands. The measured adsorption spectra for GaAs layers of thickness 140, 210, and 4000 Å are shown in Figure 3.128b, and the expected step structure is clearly visible for the two thinner layers, with arrows indicating the frequencies at which steps are expected. Peaks at energies just a little below the predicted values mark the absorption. These results from the creation of an exciton, an electron-hole bound state that is created when a photon is absorbed (see below for details). Because there is an attraction between the electron and the hole in an exciton, the photon energy required to create an exciton is lower than the predicted values that ignore such interactions. Thus, the difference in energy of the peak and the predicted absorption edge is a measure of the binding energy of the electron-hole pair. An exciton, with an energy just below the bandgap, is clearly seen in the absorption curve for the 4000-Å layers, but the step-like structure has disappeared, indicating that the DOS is a smooth curve as one expects for 3D behavior.

Electronic Conductivity of Quantum Wells The circles shown in Figure 3.124 are the Fermi circles corresponding to 2D crystals with one, two, three, and

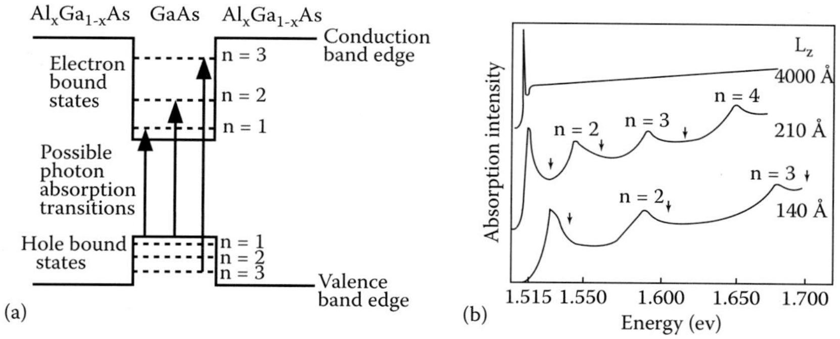

FIGURE 3.128 (a) Heterojunction of GaAs and $Al_xGa_{1-x}As$. If the GaAs layer is thin enough, bound states (dashed lines) form. When photons are absorbed, electrons are excited between these bound states. (b) The adsorption of light, measured as a function of photon energy for GaAs layers of thickness 4000, 210, and 140 Å. The arrows indicate the energies at which the onset of adsorption is expected to occur for transitions involving the nth bound state.

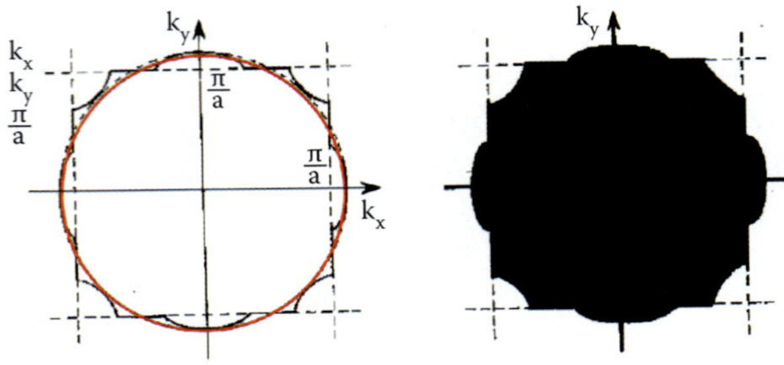

FIGURE 3.129 Deformation of the free electron circle near Bragg planes, V(x) ≠ 0.

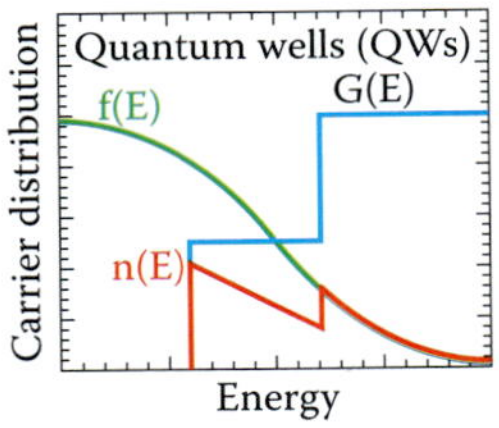

FIGURE 3.130 Quantum wells (QWs). Density of states (DOS) (blue) and occupied states (red) around the Fermi energy for a quantum well. The Fermi-Dirac function is $f(E)$, and the product of $f(E)G(E)_{2D} = n(E)_{2D}$ at $T > 0$.

four valence electrons per atom. The circles represent surfaces of constant energy for free electrons [$V_{(x)} = 0$], i.e., the Fermi surface for some particular value of the electron concentration. The total area of the filled region in **k**-space depends only on the electron concentration and is independent of any interaction of electrons with a lattice. In a solid more realistic shape of the Fermi surface depends on the lattice interaction of the electrons and will usually not be an exact circle in a lattice. There is a discontinuity introduced into a free electron Fermi circle any time it approaches a 2D Brillouin zone boundary. In Figure 3.129 we see how the shape of the Fermi circle is distorted near the surface if $V(x)$ is not zero.

Energy gaps appear at zone boundaries, and the Fermi surface intersects zone boundaries almost always perpendicularly. The crystal potential causes rounding of sharp corners on a Fermi surface. The volume enclosed by a Fermi surface only depends on electron density and not on details of the lattice interaction. In other words, the volume enclosed by a Fermi surface remains unchanged under the "deformations" just mentioned.

The number of occupied states per unit volume in the energy range E to $E + dE$ and with $f(E) = 1$ $(T = 0)$ is calculated as:

$$n_{2D} = \int_0^{\infty} n(E)_{2D} dE = \int_0^{E_F} G(E)_{2D}[f(E) = 1] dE \tag{3.257}$$

or

$$n_{2D} = \int_0^{E_F} \frac{m_e^*}{\pi\hbar^2} dE = \frac{m_e^* E_F}{\pi\hbar^2}$$

To calculate the current density for a 2D gas at a particular temperature we substitute the value for $n(E)_{2D}$ at $T > 0$ [and thus $f(T)$ can then be different from 1] in $\mathbf{J} = -n(E)_{2D} e \mathbf{v}_{avg}$ (Equation 3.3). The function $n(E)_{2D}$ for a given temperature T (>0) is shown as a red line in Figure 3.130. In the same graph we also show the Fermi-Dirac function and the DOS function (blue).

The form of the density of states function of a 2D gas can also be dramatically modified by a magnetic field, giving rise to very pronounced behavior in the conductance with the appearance of the so-called Landau levels.

Quantum Wires: $L_1 > \lambda_{DB} > L_{2,3}$

Quantization We saw earlier that when L becomes very small along two directions (2D confinement)—of the order of the de Broglie wavelength of an electron—one obtains a quantum wire where electrons can only move freely in one direction, i.e., along the length of the quantum wire as shown first in Figure 3.63 and, simplified, reproduced in Figure 3.131. These 1D electronic structures with 2D quantum confinement comprise nanowires, quantum wires, nanorods, and nanotubes.

The starting point for the fabrication of one type of quantum wire is a 2D electron gas confined in one direction as discussed above. A 2D quantum gas that is very strongly confined at some interface and where we can assume that only the lowest subband of the electron gas is occupied, so that the motion transverse to that interface is frozen out. This is the situation we encountered for a

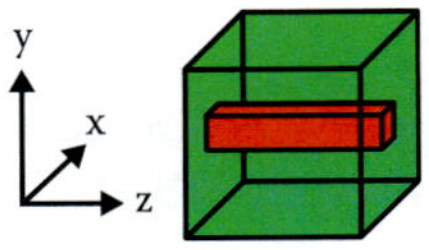

FIGURE 3.131 A rectangular quantum wire.

heterojunction with a very small L, so that Equation 3.254 is applicable $[G(E)_{2D} = \frac{m_e^*}{\pi\hbar^2}$ (no summation of steps)]. To this strong confinement, a typically weaker, lateral confinement of the electrons is added by etching nanowires in the 2D quantum well. In Figure 3.131 we drew a rectangular quantum wire with a square cross-section; in reality nanowires will typically be much wider than they are thick (the x-direction is assumed to be the film thickness direction), so the quantum confinement is most severe in the x-direction. The reason for this discrepancy is that in micro- and nanotechnology it is much easier to control a film's thickness (x-direction) than it is possible to control the lateral dimensions of a structure (y direction). Let us consider the 75-nm-wide quantum wires etched in a $GaAs/Al_xGa_{1-x}As$ heterojunction in Figure 3.132. The 2D electron gas formed at the heterointerface is confined here in a scale of just a few nanometers (in the thickness or x-direction), and so its quantized energies are large in the x direction. The lateral confinement (y direction) of electrons in the 75-nm-wide wire is much weaker than this and its quantized energies are consequently much smaller. The weaker lateral confinement in the wires gives rise to a series of relatively closely spaced energy levels (see E_Y in Figure 3.132). Therefore, transport through the wire in the z-direction will involve electrons that occupy many of these lateral sub-bands. Thus, these structures in reality are quasi-2D confined systems with free electron motion in one direction and two different types of confinement in the other two.

FIGURE 3.132 75-nm-wide quantum wires etched in a $GaAs/Al_xGa_{1-x}As$ heterojunction. The confinement of electrons in the x- and y-directions quantizes the electron energy into a set of discrete energies. The confinement in the x-direction is much stronger than in the y-direction.

For an idealized nanowire with a square cross section and the x = y dimensions in the nanoscale but continuous along the wire axis (z-direction), the energy dispersion function may be written as:

$$E = E_{n_1,n_2}(k_z) + \frac{\hbar^2 k_z^2}{2m_e^*} \tag{3.258}$$

with the wave function:

$$\psi(x, y, z) = \psi_{n_1,n_2}(x, y)e^{ik_{yz}} \tag{3.259}$$

with n_1 and n_2 the quantum numbers labeling the eigenstates in the x-y plane and $\mathbf{k}_z$ the wave vector in the z-direction, and where we assume that the dispersion relation for the electron energy is parabolic. The energy term has again two contributions: one is caused by the continuous band value $\left(\frac{\hbar^2 k_z^2}{2m_e^*}\right)$, and the other term $[E = E_{n_1,n_2}(k_z)]$ is that of the quantized values of electrons confined in two dimensions as derived in Equation 3.158 (unconstrained direction is along the z-axis). Referencing energies to the conduction band edge, the latter expression may be rewritten as:

$$E_{n_1,n_2}(\mathbf{k}_z) = E_c^0 + \frac{h^2}{8m_e^*}\left(\frac{n_1^2}{L_1^2} + \frac{n_2^2}{L_2^2}\right) + \frac{\hbar^2 k_z^2}{2m_e^*} \tag{3.260}$$

where E_c^0 is the bottom of the conduction band. The lowest sub-band is obtained for n_1 and $n_2 = 1$. The energy at the bottom of each sub-band ($\mathbf{k}_z = 0$) is simply given by:

$$E_{n_1,n_2} = E_c^0 + \frac{h^2}{8m_e^*}\left(\frac{n_1^2}{L_1^2} + \frac{n_2^2}{L_2^2}\right) \tag{3.261}$$

If we choose the cross-section of the GaAs quantum wire containing the 2D confined electron gas to be equal to 100 × 100 Å, then the lowest energies in the two lowest sub-bands are equal to 0.112 eV and 0.280 eV. This is determined from Equation 3.261 for $n_1 = 1$, $n_2 = 1$ and $n_1 = 1$, $n_2 = 2$, respectively. In the case that the nanowire is much wider (y-direction) than it is thick (x-direction), the quantum confinement is most severe in the x-direction. Because in this case $L_y >> L_x$, the n_2 levels form a staircase of small steps in the widely separated sub-bands corresponding to the various values for n_1. Thus, confinement that is different in the x and y directions splits each sub-band further up into a set of more narrowly spaced sub-bands.

In Figure 3.133 we show band structure and density of states for a quantum wire with a square cross section; similar to the situation in quantum wells, sub-bands are formed, but in quantum wires with a rectangular cross-section multiple sub-bands form at each eigenvalue $E(n_x,n_y)$, spread out as $\frac{\hbar^2 k_z^2}{2m_e^*}$.

Fermi Surface of Quantum Wires Whereas Fermi surfaces in 3D and 2D electronic structures consist of a sphere and a circle, respectively, the Fermi surface of a strictly 1D electronic structure consists of just two points at $+\mathbf{k}_F$ and $-\mathbf{k}_F$ or $\int \mathbf{dk} = 2\mathbf{k}_F$ (where $\mathbf{k}_F$ is the Fermi wave vector) or $n_{1D} = \frac{2\mathbf{k}_F}{\pi}$. This unusual Fermi surface has some pretty significant consequences. In 1981, Hiroyuki Sakaki predicted that ideal 1D electrons moving at the Fermi level in quantum wires would require very large momentum changes ($\Delta k = 2\mathbf{k}_F$) to undergo any scattering. The result is that electron scattering is strongly forbidden. This is a consequence of the fact that in one dimension, electrons can scatter only in one of two directions: forward and 180° backward. With this large reduction in scattering, electrons should achieve excellent transport properties (e.g., very high mobility).

Density of States Function for Quantum Wires For calculating the density of states for a quantum wire,

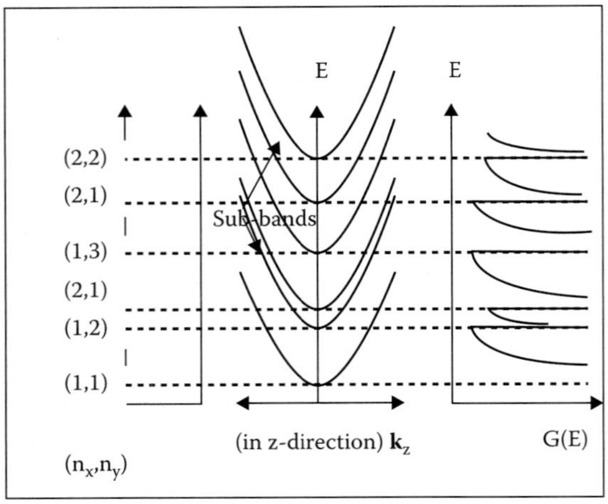

FIGURE 3.133 Band structure and density of states for a quantum wire with a square cross-section.[4]

we use the same approach we took for the 3D and 2D cases. The **k**-state has now a length l of $2\pi/L$, and we must find the number of **k**-states lying in a length of $\mathbf{k} + d\mathbf{k}$. The wire length difference, l, is given by:

$$l = 2 \tag{3.262}$$

The factor of two appears because the wave number could be either positive or negative, corresponding to the two directions along the wire.

The resulting density of states per unit length of 1D **k**-space is obtained by multiplying by 2 for spin degeneracy and dividing by $2\pi/L$:

$$G(\mathbf{k})_{1D}\,\mathbf{dk} = 2\left(\frac{2}{2\pi/L}\right)\mathbf{dk} = \frac{2L}{\pi}\mathbf{dk} \tag{3.263}$$

To obtain the density of states in terms of the energy $[G(E)_{1D}]$, we use again the relation between E and $\mathbf{k}$, derived earlier as $\mathbf{k} = \left(\frac{2m_e^*E}{\hbar^2}\right)^{\frac{1}{2}}$ (Equation 3.150) and differentiate the latter expression with respect to energy:

$$\mathbf{dk} = \left(\frac{2m_e^*E}{\hbar^2}\right)^{-\frac{1}{2}}\frac{m_e^*}{\hbar^2}dE \tag{3.227}$$

The density of energy states $G(E)_{1D}$, the number of allowed states between E and $E + dE$, per unit energy and unit length (divide by L) is then obtained using the chain rule:

$$\begin{aligned} G(E)_{1D}dE &= \frac{2dk}{\pi} = \frac{1}{\pi}\left(\frac{2m_e^*}{\hbar^2}\right)^{\frac{1}{2}}\frac{1}{E^{\frac{1}{2}}}dE \\ &= \frac{1}{h\pi}\sqrt{\frac{2m_e^*}{E}}dE \end{aligned} \tag{3.264}$$

This is in sharp contrast with the behavior of a 3D electron gas where $G(E)_{3D}$ goes to zero at low energies, and two dimensions, where $G(E)_{2D}$ steps up to a constant value at the bottom of each 2D sub-band. Remembering that the group velocity is given as:

$$\mathbf{v}_g = \frac{d\omega}{dk} \tag{3.101}$$

which we found may also be written as:

$$\mathbf{v}_g = \frac{1}{\hbar}\left[\frac{dE}{dk}\right] \text{ or more generally } \mathbf{v}_g = \frac{1}{\hbar}\nabla_k E(\mathbf{k}) \tag{3.202}$$

we can rewrite Equation 3.264 now also in terms of the group velocity as:

$$G(E)_{1D}dE = \frac{2d\mathbf{k}}{\pi} = \frac{2}{\pi}\left(\frac{dE}{d\mathbf{k}}\right)^{-1} dE = \frac{2}{\pi\hbar v_g(E)} dE \quad (3.265)$$

Thus, for a 1D system the density of states is inversely proportional to the velocity! We will use this expression in v_g when we derive the expression for the current through a nanowire. We will learn that in a nanowire the current is constant and proportional to the velocity and density of states (see further below Equation 3.269). If we add electrons to the nanowire, they initially fill only the lowest of the levels, say level (1,1) in Figure 3.133. As the energy is increased and remains less than level (1,2), the lateral motion remains frozen, and the increase in energy is transferred into motion along the length of the wire. The density of states in this level will then take the form predicted by Equation 3.265. However, in most situations the Fermi energy of the electrons in the wire is several times larger than the average spacing between the lateral energy levels in Figure 3.133, so several of these levels will be occupied, and each of these levels defines a corresponding 1D sub-band. The density of states within each sub-band is 1D, but the total density of states is obtained by summing over all sub-bands n. The summing over the individual sub-bands can be formulated mathematically as:

$$G(E)_{1D} = \frac{1}{\pi}\left(\frac{m_e^*}{\hbar^2}\right)^{\frac{1}{2}} \sum_n L\left(\frac{1}{E-E_n}\right)^{\frac{1}{2}} \quad (3.266)$$

L is the unit step function (same as H in Equation 3.255). The density of states of a quantum wire diverges as $\left(\frac{1}{E-E_n}\right)^{\frac{1}{2}}$ at each sub-band threshold and has the inverse energy dependence $E^{-1/2}$ compared with the $E^{1/2}$ dependence of a 3D electron gas (bulk semiconductor). The DOS of a quantum wire obviously has a more pronounced structure than does a 2D well, with a large number of sub-bands, each one starting as a peak. An immediate manifestation of a large $G(E)_{1D}$ at the bottom of each sub-band is again an increase of the strength of optical transitions or oscillator strength as compared with 3D and 2D electronic structures.

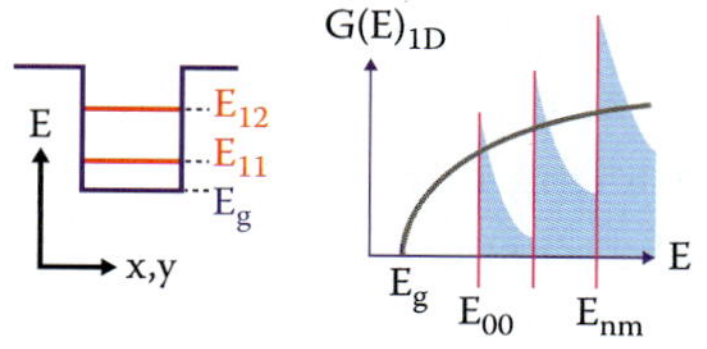

FIGURE 3.134 The density of states function for a rectangular quantum wire. The solid black curve is that of a free electron.

In Figure 3.134 we summarize the characteristics of a 1D density of state function $G(E)_{1D}$. The axis of the rectangular wire is again in the z-direction, and quantization is in the x- and y-directions. The peaks show the calculated density of states of a quantum wire over a range of energies where several different sub-bands become occupied. For comparison, the solid line shows the monotonic variation of the density of states expected for a 3D system.

Electronic Conductivity of Quantum Wires Ohm's law for macroscopic systems is given by V = IR, and in terms of conductance σ this is equivalent to σ = J/E (Equation 2.2). From the previous sections, we know that only electrons close to the Fermi level contribute to the conductance. From Bloch's theorem, we also recall that in an ideal periodic potential, electrons propagate without any scattering—and thus no resistance at all—but that electron propagation in real materials does involve scattering. The origin of such scattering can be any source of disorder that disturbs the perfect symmetry of the lattice. Examples include defects and impurities, scattering from other electrons, and lattice vibrations (phonons). Because an electric current constitutes a movement of those "bumbling" electrons under an electric field, one expects that with nanostructures featuring dimensions comparable with the fundamental size of the electron, electrical properties will be strongly influenced by quantum-mechanical transport effects.

Each of the discrete peaks in the density of states (DOS) in Figure 3.134 is caused by the filling of a new lateral sub-band. The peaks in the density of states functions at those energies where the different sub-bands begin to fill are called criticalities or Van Hove singularities. These singularities (sharp peaks) in the density of states function lead to sharp peaks in optical spectra and can also be observed directly with scanning tunneling microscopy (STM). The STM tunneling current J as a function of bias V gives spatial and spectroscopic information about the quantum states of nanostructures. At low temperatures, the derivative of the current with respect to voltage is roughly proportional to the density of states of the sample weighted by the density of states at the STM tip position in other words a convolution of the DOS of the tip and the sample. Because dJ/dV ~ DOS, the DOS is directly observed from an STM scan. This is illustrated in Figure 3.135 for a metallic and semiconducting carbon nanotube. The DOS is shown in Figure 3.135a and b, and the differential conductance is illustrated in Figure 3.135c and d. Metallic carbon tubes have a constant DOS around E = 0, whereas for semiconducting ones there is a gap at E = 0 (DOS = 0).

Singularities occur because electron transitions accumulate at frequencies for which the bands n_1,n_2 are parallel, i.e., at frequencies where the gradient is equal to zero:

$$\nabla_{\mathbf{k}}[E_{n_1}(\mathbf{k}) - E_{n_2}(\mathbf{k})] = 0 \tag{3.267}$$

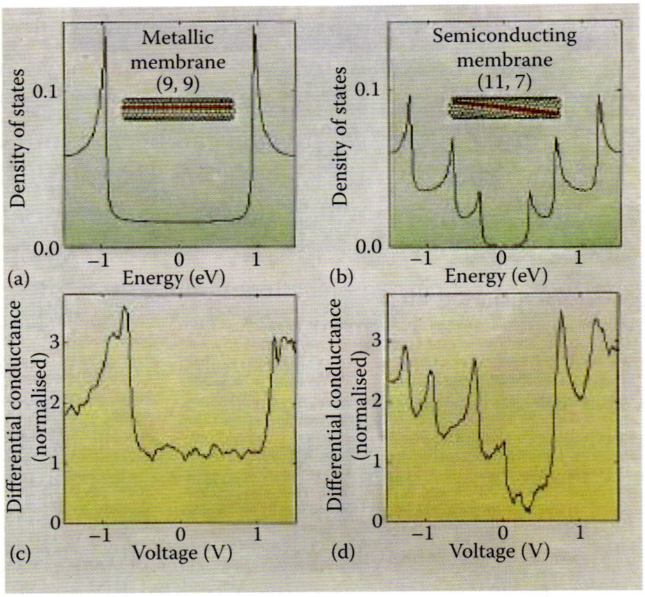

FIGURE 3.135 Van Hove singularities can be observed directly with scanning tunneling microscopy (STM). This is illustrated here for a metallic and semiconducting carbon nanotube. The DOS is shown in (a) and (b) and the differential conductance in (c) and (d). (From Dekker, C., *Physics Today*, 52, 22–28, 1999. With permission.[24])

where n_1 is an empty band and n_2 is a filled band. This can also be appreciated by considering that the density of states according to Equation 3.228 is given as $G(E) = G(k)(dE/d\mathbf{k})^{-1}$, so that singularities emerge when $d\mathbf{k}/dE = 0$ (for an example of such states, see the band diagram of Si in Figure 3.116, states at E_1 and E_2).

The number of occupied states at T = 0 [where f(E) = 1] per unit volume in the energy range E to $E + dE$ is again calculated from $n(E)_{1D}dE = G(E)_{1D}\,[f(E) = 1]dE$ and one obtains:

$$n_{1D} = \int_0^{\infty} n(E)_{1D}\,dE = \int_0^{E_F} G(E)_{1D}[f(E) = 1]dE \text{ or } n_{1D} = \frac{2\mathbf{k}_F}{\pi} \tag{3.268}$$

that $n_{1D} = \int_0^{E_F} G(E)_{1D}dE$ indeed equals $\frac{2\mathbf{k}_F}{\pi}$ is easily recognized from the fact that:

$$G(E)_{1D}dE = \frac{2d\mathbf{k}}{\pi} \tag{3.265}$$

To calculate the number of occupied states at T > 0 per unit volume in the energy range E to $E + dE$, we use again $n(E)_{1D}\,dE = G(E)_{1D}f(E)dE$, but because T is now > 0, $f(E)$ might be different from 1. That function is shown as a red line in Figure 3.136.

Consider now a quantum wire held between ideal contacts, as shown in Figure 3.137. In this very short 1D wire, the charge transport is assumed to be ballistic, i.e., electron transport without scattering, or the mean free path (λ) and device size (L) are comparable. Because the current carrying capacity of a 1D wire is finite, there must be a finite conductance even in the absence of scattering. Applying a voltage V

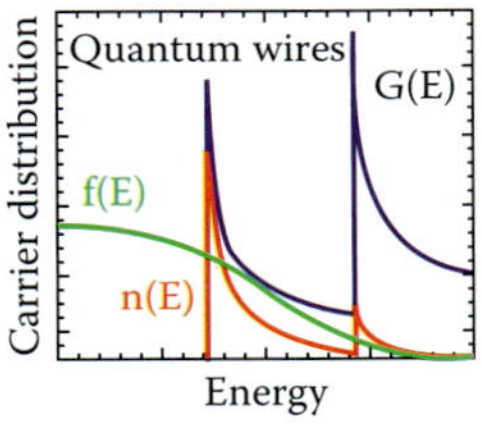

FIGURE 3.136 Quantum wires. Density of states (DOS) (blue) and occupied states (red) around the Fermi energy for a quantum wire. The Fermi-Dirac function is $f(E)$, and $n(E)$ is the product of $f(E)$ and $G(E)_{1D}$ at $T > 0$.

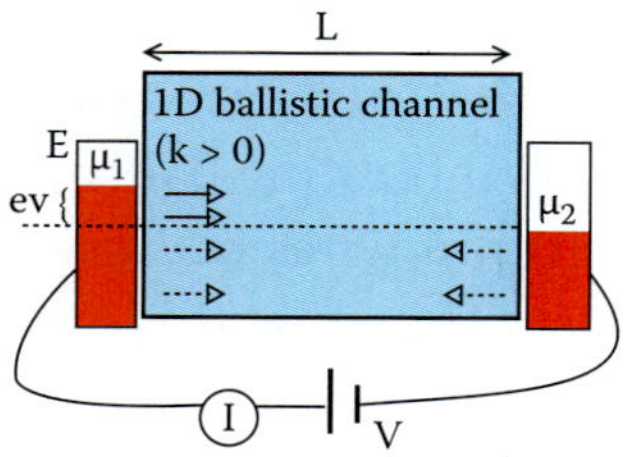

FIGURE 3.137 A quantum wire held between ideal contacts. In this very short one-dimensional wire, the charge transport is assumed to be ballistic with the mean free path (λ) and device size (*L*) comparable. Fermi level $E_F(1) = \mu_1$ and Fermi level $E_F(2) = \mu_2$ with $E_F(2) - E_F(1) = qV$, where $q = -e$ for electrons and $+e$ for holes. A net current flows between the two contacts for an applied bias *V*. A positive wave vector **k** is assumed with excess right-moving electrons constituting the current.

over the wire has an excess of electrons moving in one direction only (left to right in the current case), so we assume a positive wave vector **k** with excess right-moving electrons constituting the net current.

The chemical potential μ of the electrons in the contacts, we recall, is equivalent to the Fermi level, and here we assign the contacts Fermi level $E_F(1) = \mu_1$ and Fermi level $E_F(2) = \mu_2$ with $E_F(2) - E_F(1) = qV$, with $q = -e$ for electrons and $+e$ for holes. To calculate the current density for this 1D electron gas at T = 0, we must substitute the value for $n(E)_{1D}$ at T = 0 [where f(E) = 1] in $\mathbf{J} = -n(E)_{1D}e\mathbf{v}_{avg}$ (Equation 3.3), and because $n(E)_{1D} = \int_0^\infty n(E)_{1D}dE = \int_0^{E_F} G(E)_{1D}dE$, we can use the expression for $G(E)_{1D}$ in terms of the group velocity, $G(E)_{ID} = \frac{2}{\pi\hbar v_g(E)}$ (Equation 3.265), resulting in:

$$\mathbf{J} = -n(E)_{1D}e\mathbf{v}_{avg}(E) = -e\int_{E_F(1)}^{E_F(1)} \mathbf{v}_{avg}(E)n(E)_{1D}c$$

$$= -e\int_{E_F(1)}^{E_F(2)} \mathbf{v}_{avg}(E)G(E)_{1D}dE$$

$$\mathbf{J} = -e\int_{E_F(1)}^{E_F(2)} \mathbf{v}_{avg}(E)\frac{2}{\pi\hbar v_g(E)}dE \quad \text{and}$$

since $\mathbf{v}_{avg}(E) = \mathbf{v}_g(E)$

$$\mathbf{J} = -\frac{2e}{h}[E_F(2) - E_F(1)] = \frac{2e^2}{h}V \qquad (3.269)$$

since $E_F(2) - E_F(1) = qV$, with $q = -e$ for electrons and $+e$ for holes.

For a macroscopic metal wire, the conductance decreases continuously as the diameter of the constriction is reduced. However, as the diameter becomes of the order of the Fermi wavelength, the conductance takes on integer values of the fundamental unit of conductance σ (= J/V) = $2e^2/h$. The conductance σ (= J/V) is fixed, regardless of length *L* of the conductor, and as a consequence there is no well-defined conductivity σ. This phenomenon is called quantized conductance (QC). What is remarkable is that in the case of a quantum wire the velocity cancels with the density of states, producing a current that depends only on the voltage and fundamental constants. As a consequence, the current carried by a sub-band is independent of the sub-band shape; it need not be parabolic, and the relation holds for all cases! The amount of current is dictated only by the number of modes (also called sub-bands or channels) $M(E_F)$, which are filled between μ_1 and μ_2 with each mode contributing $2e^2/h$ or:

$$\sigma = \frac{2e^2}{h}M(E_F) \qquad (3.270)$$

The corresponding quantized resistance is:

$$R = \frac{h}{2e^2}\frac{1}{M(E_F)} \qquad (3.271)$$

As we increase the channel cross-section, the number of modes increases, so the resistance goes down. This quantized resistance *R* has a value of 12.906 kΩ and has allowed for the definition of a new practical standard for electrical resistance: the resistance unit h/e^2, roughly equal to 25.8 kΩ; the latter number is referred to as the von Klitzing constant R_K (after Klaus von Klitzing, the discoverer of quantization of the Hall effect), and since 1990, a fixed conventional value R_K –90 has been used in resistance calibrations worldwide.

Resistance of a conductor, of course, decreases as the length of the channel decreases. But at some point it reaches a minimum, given by Equation 3.271, and decreases no further. This minimum resistance comes from the contact resistance, and the smaller the device, the more important this contact resistance. One might assume that this resistance can be eliminated by improving the contacts, but this simply is not true: there is a minimum contact resistance.

Quantized conductance in a quantum wire is dramatically illustrated in Figure 3.138, where we show experiments involving quantum wires in a GaAs/$Al_xGa_{1-x}As$ heterostructure at very low temperatures (close to 0 K). A short quasi-1D "wire" is formed here between two regions of a 2D electron gas as a negative gate voltage V_g applied to the metallic gates on the surface depletes the carriers in the underlying 2D electron gas in the GaAs (Figure 3.138a). Once the narrow constriction is reduced to a width d comparable with the electron Fermi wavelength λ_F (see Figure 3.138b) 1D quantization is obtained regardless of the wire length L. In the conductance voltage plot shown in Figure 3.138c, 14 quantized conductance steps of height $2e^2/h$ are observed for a wire width d of 0.5 μm (see also Figure 3.138d). Each step corresponds to the occupancy of an additional sub-band or additional conductance channel in the quantum wire. In other words, $M(E_F) = 14$.

The conductance voltage plots in Figure 3.138d show that quantized steps are still clearly present for the 5-μm wire, indicating ballistic transport, but wash away for 10-μm wires because of crossover to diffuse transport. Thus, for this GaAs 1D system, the electron Fermi wavelength λ_F mean free path is between 5 and 10 μm. Long electron Fermi wavelengths λ_F are obviously key to integrating these 1D electronic nanostructures.

An important experimental finding is that the conductance quantization is typically destroyed in narrow constrictions longer than a few hundred nanometers; ballistic transport cannot be preserved on length scales much longer than this. The reason for this is that as the constriction length is made longer, there is a greater probability of finding some disordered region (a barrier) that can disrupt ballistic transport by causing electron scattering. When we derived the conductance quantization

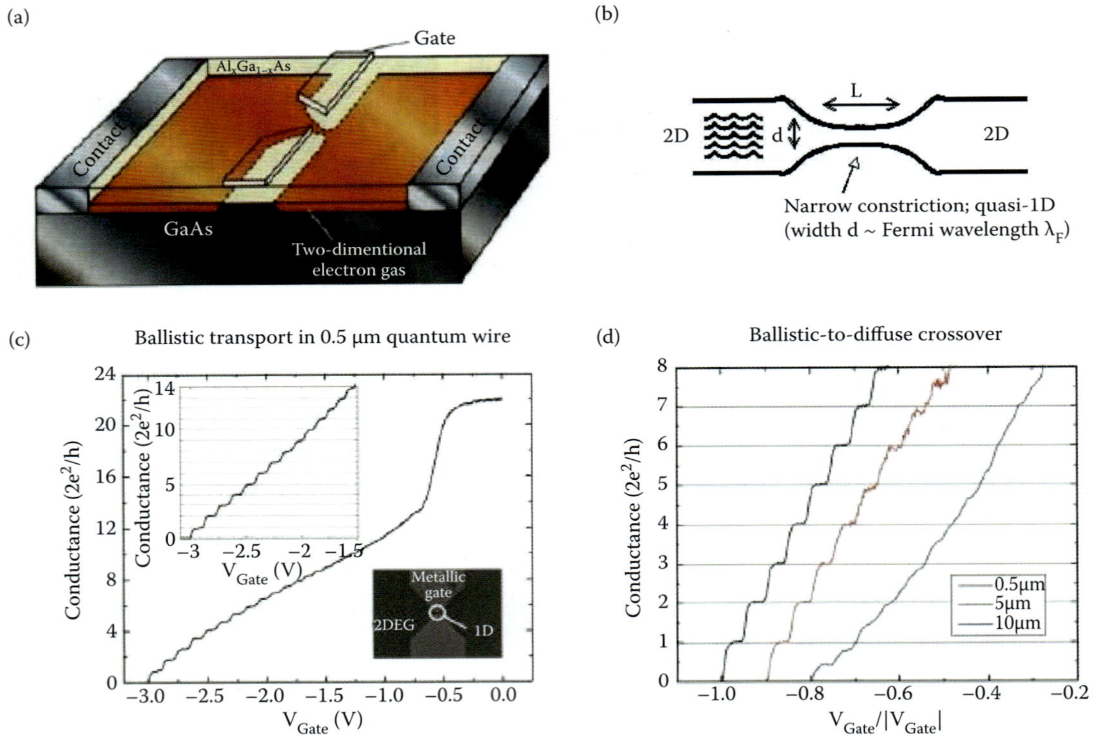

FIGURE 3.138 Conductance quantization in a short channel electrostatically defined in a GaAs/$Al_xGa_{1-x}As$ heterostructure. Measurements are carried out at very low temperatures (close to 0 K). (a) A negative gate voltage V_g applied to the metallic gates depletes the carriers in the underlying 2D electron gas. (b) Once the narrow constriction is reduced to a width d comparable to the electron Fermi wavelength λ_F, 1D quantization is obtained regardless of the wire length L. (c) In the conductance voltage plot, 14 quantized conductance steps of height $2e^2/h$ are observed for a wire width d of 0.5 μm. (d) Ballistic to diffuse transport crossover in GaAs quantum wires. The conductance voltage plots show that quantized steps are still clearly present for the 5-μm wire, indicating ballistic transport, but wash away for 10-μm wires because of crossover to diffuse transport. (From Sandia National Laboratory.)

for a ballistic system we assumed a total absence of any scattering between the different 1D sub-bands. This allowed us to make a definite association between the direction of the carriers in the wire and the reservoirs from which they originated (see Figure 3.139).

Such a distinction is no longer possible in the case where scattering occurs in the wire because electrons traveling in any particular direction may have originated from either reservoir. To compute the current that flows through such a disordered conductor, we now consider the probability that an electron incident in a given channel in one lead is transmitted into the other lead. Because of scattering in the sample an electron incident in some particular channel in the input lead will be transmitted into the different channels of the output lead with different probabilities. In the case in which a channel

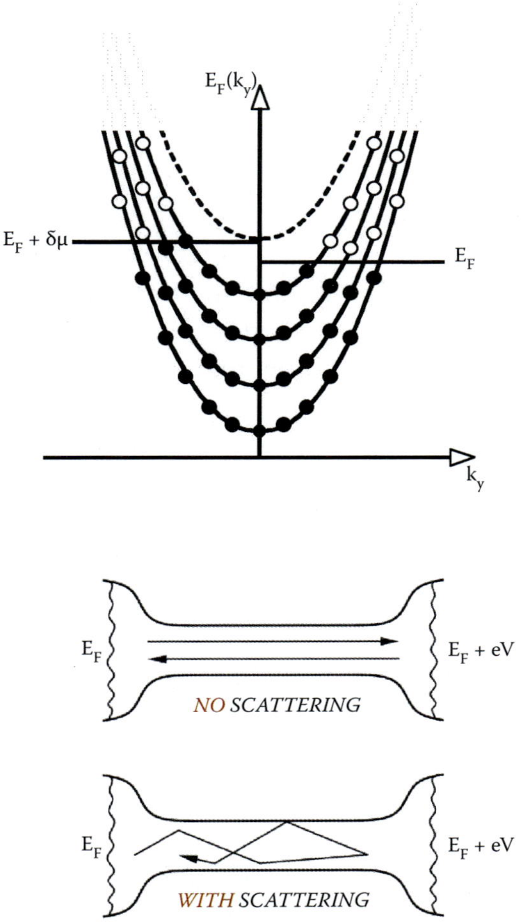

FIGURE 3.139 When there is no scattering in the wire, states are filled to higher energies in one direction than in the other.

is not perfectly conducting, e.g., there is some type of barrier in the channel, the conductance of that channel is then given by the quantum conductance times the electron transmission probability at the Fermi energy $T_n(E_F)$ through the barrier in that channel:

$$\sigma = \frac{2e^2}{h} T_n(E_F) \tag{3.272}$$

where $T_n(E_F)$ is given as:

$$T_n(E_F) = \sum_{m=1}^{N} T_{nm}(E_F) \tag{3.273}$$

Here T_{nm} is the probability that an electron incident in mode n is transmitted into mode m in the output lead and the summation runs over the number (N) of occupied modes in the leads. Therefore, the probability $T_n(E_F)$ provides a measure of the degree of disorder in the conductor. This leads us to the so-called Landauer (1927–1999) formula, wherein the total current flow through the conductor is just the sum over all the occupied channels, which leads to the following expression for the conductance:

$$\sigma = \frac{J}{V} = \frac{2e^2}{h} \sum_{n=1}^{N} T_n = \frac{2e^2}{h} T \tag{3.274}$$

where T is the total transmission coefficient of the conductor. A simple application of the Landauer formula is the problem of the ballistic quantum point contact that we considered earlier. In this case, each channel is perfectly transmitted, and the sum over the transmission probabilities in Equation 3.274 is replaced by the total number of channels $M(E_F)$; therefore, we recover the usual quantization condition given by Equation 3.270:

$$\sigma = \frac{2e^2}{h} M(E_F) \tag{3.270}$$

So we sum over the contributions of each sub-band and for perfect transmission of all the sub-bands $T(E_F) = M(E_F)$ (which was 14 in Figure 3.138). When scattering is possible, however, the transmission coefficients T_n will no longer be equal to unity, and so the conductance quantization will not be observed.

The Landauer equation can also be written in terms of transmission and reflection coefficients.

Based on Equation 3.274, the resistance of a one-channel system $[M(E_F) = 1]$ can be rewritten as:

$$\text{Resistance }(E_F) = \frac{h}{2e^2}\frac{1}{T(E_F)} = \frac{h}{2e^2}\left[1+\frac{1-T(E_F)}{T(E_F)}\right]$$
$$= \frac{h}{2e^2} + \frac{h}{2e^2}\frac{R(E_F)}{T(E_F)}$$
$$= \text{quantized contact resistance} + \text{scattering from barriers inside channel} \qquad (3.275)$$

where $R(E_F) = 1 - T(E_F)$ is the reflection coefficient, and $T(E_F) + R(E_F) = 1$. Therefore, the resistance of the quantum wire is the sum of the quantized contact resistance and a second term, as a result of scattering from barriers in the channel. It is the latter term that is zero for a perfect conductor.

At finite temperatures a thermal smearing out of the conductance staircase is observed because the Fermi-Dirac energy distribution of electrons in the left and right leads must now be taken into account. This smearing out effect is observed in Figure 3.140, where the conductance voltage plots for the same type of quantum wires of Figure 3.138 are shown at 0.3 K, 0.6 K, 1.6 K, and 4.2 K. At the highest temperatures the staircase profile is almost lost as transport becomes more and more incoherent because of dephasing, resulting from inelastic scattering, phonons, etc.

Quantum Dot: $\lambda_{DB} > L_{1,2,3}$

Quantization In quantum dots (QDs), generally consisting of 10^3 to 10^5 atoms and ranging from nanometers to tens of nanometers in size, dimensions are smaller than the de Broglie wavelength of thermal electrons in all three directions (L_1, L_2, $L_3 < \lambda_{DB}$). A quantum dot was first illustrated in Figure 3.67 and, simplified, is reproduced in Figure 3.141. Charge carriers in these "artificial atoms" are dimensionally confined in three directions (zero degrees of freedom), just like in an atom, and similarly have only discrete energy levels. The energy levels for an electron in a quantum dot are based on Equation 3.160, where we introduce the effective electron mass m_e^* and, for generality, make the dimensions L_1 # L_2 # L_3, to obtain:

$$E_{n_1,n_2,n_3} = \frac{h^2}{8m_e^*}\left(\frac{n_1^2}{L_1^2}+\frac{n_2^2}{L_2^2}+\frac{n_3^2}{L_3^2}\right) \qquad (3.276)$$

where the quantum numbers n_1, n_2, and n_3, each assuming the integral values 1, 2, 3,..., characterize quantization along the *x*-, *y*-, and *z*-axes, respectively. When considering the density of states for a zero-dimensional (0D) electron gas, no free motion is possible: there is no **k**-space to be filled with electrons, and all available states exist only at discrete energies. Because carriers cannot move there is no kinetic energy term $\left(\frac{\hbar^2 \mathbf{k}^2}{2m_e^*}\right)$ here, only the quantized energy terms survive.

In Figure 3.142 we show the band structure and the density of states for a quantum dot.

Quantum dots exhibit mesoscopic phenomena that are not observed in "simple" atoms. Mesoscale describes structures and behaviors at lengths between nanometers and 0.1 micrometers—too large for molecular or atomic characterization methods but too small to be fully characterized by their bulk behavior or to be modeled using continuum models. Phenomena at that scale, however, determine the properties of many scientifically and commercially important materials.

Fermi Surfaces and Brillouin Zone for Quantum Dots There is no Fermi surface for quantum dots

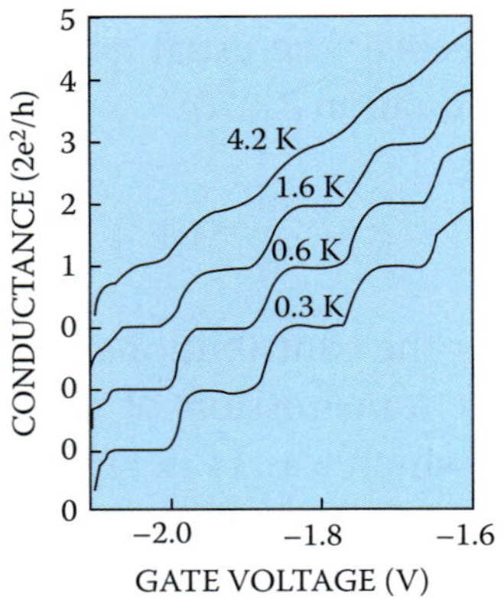

FIGURE 3.140 Conductance in number of units of $2e^2/h$ as a function of gate voltage (V) and at different temperatures. Same quantum wires as studied in Figure 3.138.

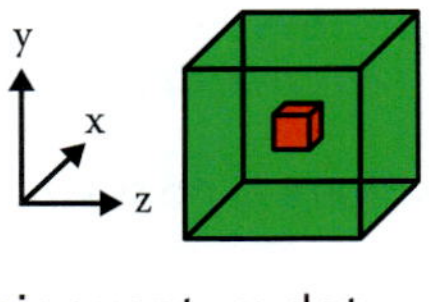

FIGURE 3.141 A cubic quantum dot.

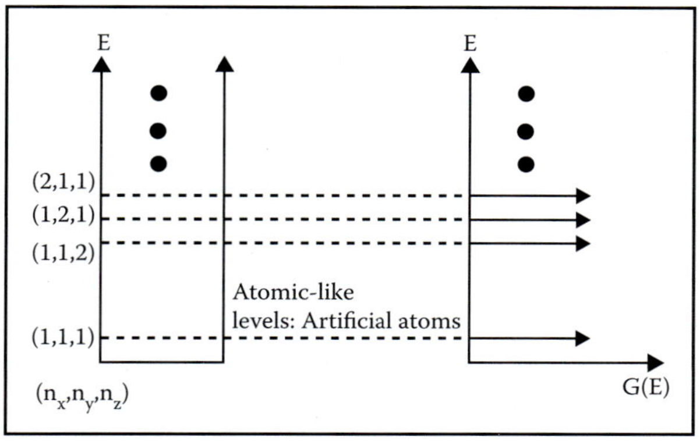

FIGURE 3.142 Energy levels and DOS of a quantum dot.[4]

because electrons are confined in all three directions (L_1, L_2, $L_3 < \lambda_{DB}$) and cannot freely move. It is important to figure out for different solids at what length scale charge carriers become frozen in fixed energy states. The Fermi wavelength is a most important parameter for this as it tells us how finely a solid needs to be structured to have an impact on its electronic properties. The size of an electron in a solid is essentially its Fermi wavelength, $\lambda_F = 2\pi/\mathbf{k}_F$, with $\mathbf{k}_F$ given by $(3n\pi^2)^{1/3}$ (Equation 3.238), where n is the electron density. In most metals the electron density is very large ($n > 10^{21}$ cm^{-3}), and the value of the Fermi wavelength is consequently of the order of a few nanometers. In semiconductors, however, the lower carrier density ($n < 10^{21}$ cm^{-3}) gives rise to larger electron wavelengths of several tens of nanometers (see also Table 3.12). Therefore, confinement and quantization phenomena are visible in semiconductors already at dimensions greater than 200 nm, whereas in metals they typically are seen at 1–2 nm. This is one important reason why quantum effects have been observed mainly in semiconductor nanostructures, which have large Fermi wavelengths, rather than in metals, which have Fermi wavelengths comparable with a few crystal lattice periods only. We can also approach the difference in quantization effects between a metal and semiconductor nanoparticle from the point of view of the separation between adjacent energy levels. The average level spacing at the Fermi level between subsequent levels in a metal can be estimated from:

$$\Delta E_{n+1-n} = \frac{1}{G(E)_{3D}(E = E_F)} = \frac{2E_F}{3n_{3D}} \quad (3.237)$$

For a spherical Au nanoparticle with a radius R = 2 nm, the average level spacing $\Delta E = E_{n+1} - E_n$ is 2 meV only. In contrast, we calculate the energy separation between adjacent levels for a 2-nm CdSe particle as 0.76 eV (see Volume III, Chapter 3). So obviously energy-level quantization is far more pronounced in semiconductor dots than in metal dots. Beyond the Fermi wavelength difference (or electron density difference) there are other reasons why semiconductors show quantum effects at a larger length scale: first, the energy level spacing—the energy difference $E_{n+1} - E_n$—increases for low-lying states in a 3D potential well; second, the effective mass for electrons is smaller in semiconductors than in metals, and according to Equation 3.235 this also decreases the effective electron density n; and, third, the appearance of the plasmon oscillations of metal particles, described in Chapter 5 on photonics, often obscures quantization effects in metals.

A Brillouin zone describes a material's periodicity and is not defined for a quantum dot.

Density of States Function for Quantum Dots

Because band structure is no longer a term that has any relevance in a quantum dot, one describes the density of states for 0D (0 degrees of freedom) with a summation of delta functions δ(E) (sharp DOS peaks):

$$G(E)_{0D} = 2\sum_{n} \delta(E - E_n) \quad (3.277)$$

where 2 is for the spin degeneracy. Such discrete atom-like densities of states are illustrated in Figure 3.143. The solid black curve is that of an electron in a large 3D semiconductor.

As we just saw above, the relative importance of the quantum effect in a given system is determined

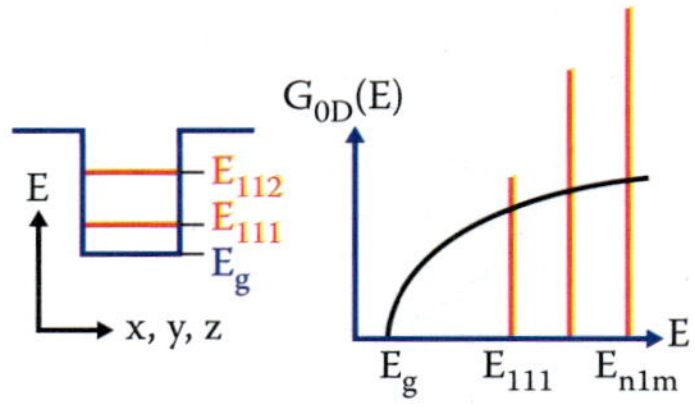

FIGURE 3.143 The density of states function for a quantum dot. The solid black curve is that of a free electron (see also Figure 3.119, full line), for example, in self-organized III-V quantum dots (e.g., InAs/GaAs).

by the energy differences between the energy levels: $\Delta E = E_{n+1} - E_n$. Quantum size effects are only observable when this energy separation is larger than the thermal energy of the carriers (kT), so that adjacent sub-bands (e.g., the E_{111} and E_{112} levels in Figure 3.143) are differently populated.

Electronic Conductivity of Quantum Dots To illustrate conduction through a quantum dot we introduce a type of quantum dot that is made by the electrostatic confinement of a 2D electron gas through a set of electrodes whose voltage can be varied at will. The specific example shown in Figure 3.144 is that of a lateral quantum dot (Figure 3.144a) made by applying a voltage on a set of gate electrodes (Figure 3.144b) to deplete the electrons in the underlying $Al_{0.3}Ga_{0.7}As$ layer and thereby defining the lateral size of a quantum dot in the 2D electron gas at the $Al_{0.3}Ga_{0.7}As$/GaAs heterojunction (Figure 3.144c). The potential well forms as a result of charge transfer between the two materials and their conduction band offset (E_c), and it confines the electrons so that the motion of electrons perpendicular to the plane of the heterointerface is quantized. The gate electrodes then further confine this 2D gas in the lateral direction to make a quantum dot (3D confinement).

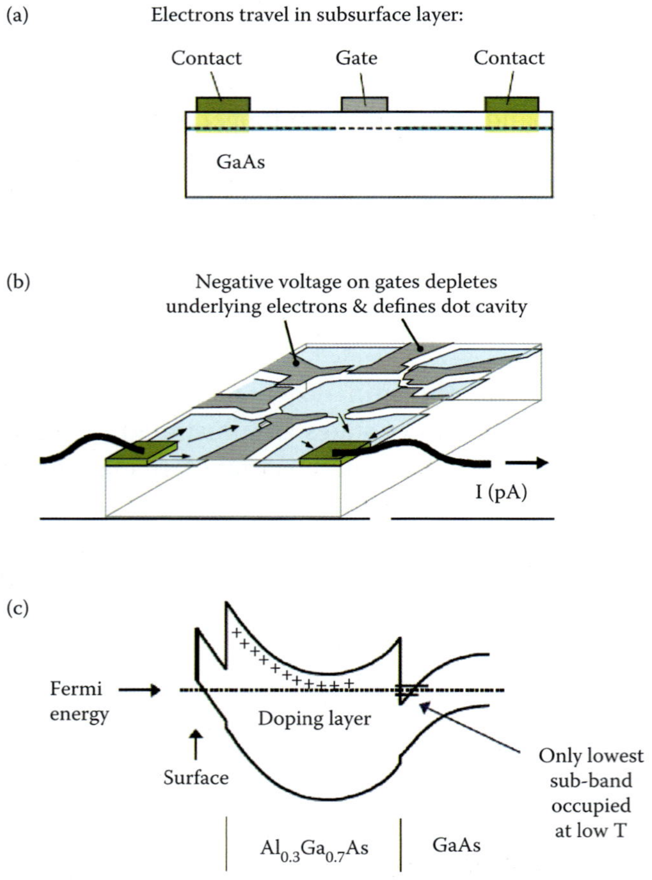

FIGURE 3.144 Electrostatically confined quantum dot in a 2D electron gas. (a) physical configuration. (b) gate electrodes and contacts. (c) energy diagram.

Any movement of electrons through a quantum dot requires the number of electrons in the dot to change one at a time. The Coulomb repulsion between electrons in the dot means that the energy of a dot containing $N + 1$ electrons is higher than one containing N electrons. This in turn requires extra energy to add an electron to the dot, and no current flows until increasing the voltage provides this energy. This is known as the Coulomb blockade.

To better visualize the electronic transport through a quantum dot, consider a classical capacitor and compare it with a quantum dot between contact pins (Figure 3.145a). In the case of the macroscopic capacitor, the charge on the plates is given as $Q = CV$ and the displacement current as $I(t) = dQ/dt$, and no charge can flow through this capacitor because the dielectric is an insulator. The electrostatic energy in the capacitor is then given as:

$$E = \int_0^{\infty} V(t)I(t)dt = \frac{Q^2}{2C} \tag{3.278}$$

In the case of a quantum particle contacted by two measuring electrodes, one can envision that the contact points form a capacitor on either side of the quantum dot; if an electron is transferred by tunneling across these capacitors (Figure 3.145a), its Coulomb energy is changed by:

$$E_c = e^2/2C \tag{3.279}$$

where it is assumed that the capacitors on either side of the quantum dot are identical. To ensure that thermal motion does not initiate a charge flow and an accompanying change in E_c, kT should be much smaller than the Coulomb blockade ($kT \ll e^2/2C$). This means that C must be less than an attofarad (10^{-18} F). The tunneling current is given as:

$$I = V_c/R_T \tag{3.280}$$

Current starts to flow at the Coulomb voltage, or:

$$V_c = e/2C \tag{3.281}$$

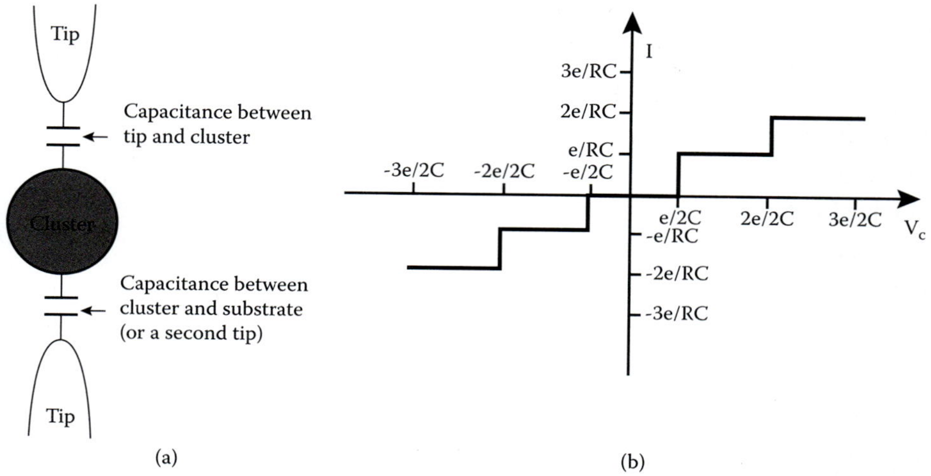

FIGURE 3.145 (a) Quantum dot probed by two pins. (b) Current-voltage plot of a quantum dot.

The electrostatic potential of the quantum dot increases significantly by introducing just one electron. Further electron transfer happens if the Coulomb energy of the quantum dot is compensated by an external voltage $V_c = \pm ne/2C$, where n is an integer. Repeated tunneling results in a "staircase" with step height in current, e/RC, as shown in Figure 3.145b.

The electron density function of a quantum dot (Equation 3.277) is illustrated in Figure 3.146: because $G(E)$ is peaked in this case (blue line spikes), so is the electron density function n (red spikes).

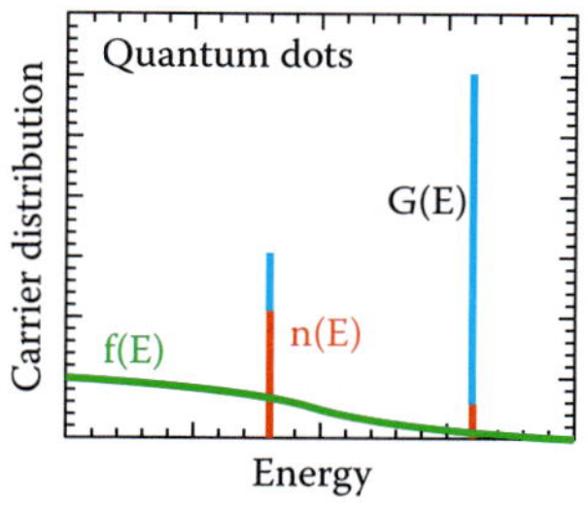

FIGURE 3.146 Density of states (DOS) (blue) and occupied states (red) around the Fermi energy for a quantum dot. The Fermi-Dirac function is *f(E)*, and the product of $f(E)G(E) = n_D$.

Generalizations and Summary

Density of States Functions From the previous sections, quantum confinement in different dimensions gives rise to very different densities of states functions compared with the situation in a bulk material, and this causes dramatic modification of the electrical and optical behavior. The higher the degree of the confinement, the fewer degrees of freedom, and the more the density of states becomes singular. Generalizing the density of states $G(E)_D$ varies with the dimension D as:

$$G(E)_D = \int_{\Sigma(E)} \frac{d\Sigma}{4\pi^3} \frac{1}{|\nabla E(\mathbf{k})|} \tag{3.282}$$

where

$\Sigma(E)$ is the equi-energetic as described for each D and

$\Delta E(\mathbf{k})$ is the gradient of E with respect to **k.** As we have seen the dependence of the density of states on the inverse of the gradient of E results in divergences in the DOS functions for lower dimensions.

For a computer program using Mathematica to calculate the densities of states in 1D, 2D, and 3D, visit Dr. Fretwell's web site at http://pyweb.swan.ac.uk/~fretwell/DOS.pdf and download densities of states.

Fermi Quantities We can also generalize Equation 3.237 for $G(E)_{3D}(E = E_F)$ the density of single-particle states available per unit volume and unit energy at the Fermi level in 3D, to:

$$G(E)_{nD}(E = E_F) = \left.\frac{dn_{nD}(E)}{dE}\right|_{E=E_F} \tag{3.283}$$

where n_{nD} is now the spatial electron density in nD dimensions: $n_{3D} = N/V$ for 3D (volume), $n_{2D} = N/A$

for 2D (area), and $n_{1D} = N/L$ (length) for 1D. From these definitions, we can find the spacing of single-particle levels in a piece of material, as well as all the other Fermi quantities described earlier. There is a qualitative change in energy dependence as the dimensions of a block of material are varied! Other things maintained equal, a higher $G(E)$ means levels are more closely spaced.

Fermi's Golden Rule in Solids

In quantum physics, Fermi's golden rule is a way to calculate the probability of a transition between two eigenstates of a quantum system, like the simple one illustrated in Figure 3.147. As sketched here, electronic levels in a solid are broadened because of atom-atom interactions. Consequently, when shining light on a semiconductor, as shown here, there is no single absorption line, but many transitions are feasible.

Fermi's golden rule (FGR), which was derived by Dirac, originates in the time-dependent perturbation theory. Time-dependent perturbation theory enables one to treat excitations that explicitly depend on time. This is of importance, for example, in discussing optical transitions where one can view the solid, with unperturbed Hamiltonian H_0, as being perturbed by a time-dependent electromagnetic field $H'(t)$ generated by an incident photon flux as:

$$\underset{\text{solid}}{H = H_0} + \underset{\text{photon}}{H'(t)} \tag{3.284}$$

Fermi's golden rule for absorption describes transition rates between energy levels in terms of the availability of states [the density of states (DOS) $G(E)$], the availability of photons [intensity ($\mathbf{E}_0^2$)], and a "coupling strength" or transition matrix element between the levels ($|\langle n|H'|n+1\rangle|^2$). The transition

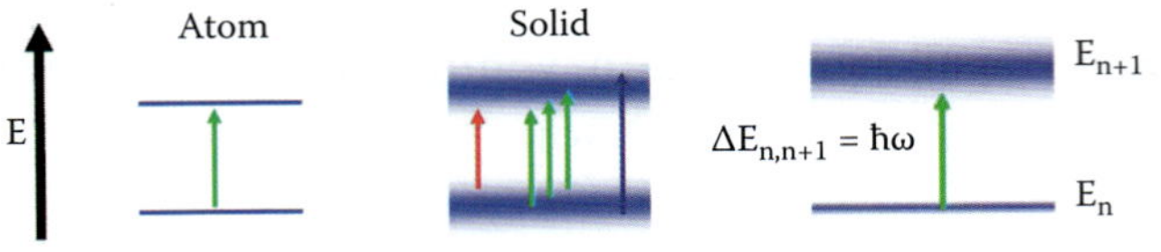

FIGURE 3.147 In solids, electronic levels broaden because of atom-atom interactions. Here we want to calculate the probability of a transition between two eigenstates E_n and E_{n+1} by using Fermi's golden rule.

rate (transitions per second) between an initial state n and a final state $n + 1$ on illumination with photons is given by the FGR as:

$$\Gamma_{n\to n+1} = \frac{2\pi}{\hbar}\mathbf{E}_0^2\left|\langle n|H'|n+1\rangle\right|^2 \times \delta(E_n - E_{n+1} - \hbar\omega) \tag{3.285}$$

where E_0 = the intensity of the photon radiation
$\hbar\omega$ = the photon energy
$\delta(E_n - E_{n+1} - \hbar\omega)$ = density of states (units C·m) at an energy $E_n - E_{n+1} - \hbar\omega$.

Note that $\langle n|H'|n+1\rangle|^2 = 0$ corresponds to a forbidden transition. Let us apply this expression to the transition of electrons from a filled valence to an empty conduction band, where $|\langle n|H'|n+1\rangle|$ becomes $|\langle v|H'|c\rangle|$. In theory, we need to calculate this matrix element for each set of initial and final states. In practice in solids, the matrix element does not vary strongly as we consider successive band states, so the total transition rate can be approximated as:

$$\begin{aligned}\Gamma &= \sum_{v,c}\Gamma_{v,c}\\ &= \mathbf{E}_0^2\sum_{v,c}\frac{2\pi}{\hbar}\left|\langle v|H'|c\rangle\right|^2\delta_{v,c}\left[E_c(\mathbf{k}) - E_v(\mathbf{k}) - \hbar\omega\right]\\ &\cong \mathbf{E}_0^2\frac{2\pi}{\hbar}\left|\langle v|H'|c\rangle\right|^2\sum_{v,c}\delta_{v,c}\left[E_c(\mathbf{k}) - E_v(\mathbf{k}) - \hbar\omega\right]\end{aligned} \tag{3.286}$$

The joint density of states in this expression is calculated from:

$$G_{v,c}(E) = \frac{2}{8\pi^3}\int d\mathbf{k}\delta_{v,c}\left[E_c(\mathbf{k}) - E_v(\mathbf{k}) - \hbar\omega\right] \tag{3.287}$$

and thus:

$$\Gamma = \mathbf{E}_0^2\frac{2\pi}{\hbar}|\langle v|H'|c\rangle|^2\, G_{v,c}\left[E_c(\mathbf{k}) - E_v(\mathbf{k}) - \hbar\omega\right] \tag{3.288}$$

Because of the form of the expression of joint density of states in Equation 3.287, optical measurements provide information about particular points in the Brillouin zone: usually high symmetry points near the band extremes. To get a better insight into this, let us perform the integral in Equation 3.287 in terms of energy dE instead of $d\mathbf{k}$. Consider a surface

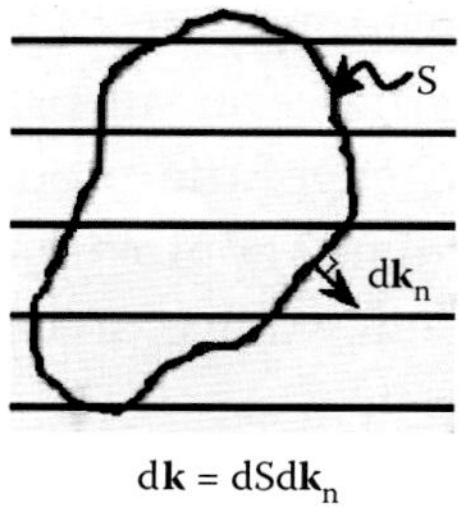

FIGURE 3.148 Energy surface *S* in **k**-space.

S in **k**-space such that $E_c(\mathbf{k}) - E_v(\mathbf{k}) = \hbar\omega$ as sketched in Figure 3.148, where we can write:

$$dE = |\nabla_{\mathbf{k}} E| d\mathbf{k}_n$$

or

$$|\nabla_{\mathbf{k}}(E_c - E_v)| d\mathbf{k}_n = d(E_c - E_v) \text{ and thus}$$

$$d\mathbf{k} = dSd\mathbf{k}_n = dS\left\{\frac{d(E_c - E_v)}{|\nabla_{\mathbf{k}}(E_c - E_v)|}\right\} \quad (3.289)$$

The joint density of states is then:

$$G_{v,c}(E) = \frac{2}{8\pi^3}\int_{k\text{-space}} \frac{dS\, d(E_c - E_v)\delta[E_c(\mathbf{k}) - E_v(\mathbf{k}) - \hbar\omega]}{|\nabla_{\mathbf{k}}(E_c - E_v)|} \quad (3.290)$$

Integrating over $d(E_c - E_v)$ results in:

$$G_{v,c}(E) = \frac{2}{8\pi^3}\int \frac{dS}{|\nabla_{\mathbf{k}}(E_c - E_v)|_{E_c - E_v = \hbar\omega}} \quad (3.291)$$

At critical points, i.e., at Van Hove singularities, where $|\nabla_{\mathbf{k}}(E_c - E_v)| \mapsto 0$, large contributions to $G_{v,c}(E)$ occur. At these points the bands are parallel to each other; this can be recognized more simply by considering that in the density of states $G(E) = G(k)(dE/d\mathbf{k})^{-1}$ singularities emerge when $d\mathbf{k}/dE = 0$ (see "Brillouin Zones for Bulk Materials"). Critical points fall then in different categories depending on whether the band separations are increasing or decreasing as we move away from the critical point. This is illustrated in a detailed Si absorption spectrum in Figure 3.149. The peaks E_1 and E_2 in Figure 3.149a correspond to the **k**-values where the conduction and valence bands are parallel to each other as shown in Figure 3.149b.

For a direct transition between parabolic bands as sketched in Figure 3.150a, the difference $E_c(\mathbf{k}) - E_v(\mathbf{k})$ is given as:

$$E_c(\mathbf{k}) - E_v(\mathbf{k}) = E_g + \frac{\hbar^2\mathbf{k}^2}{2}\left(\frac{1}{m_e^*} + \frac{1}{m_h^*}\right) = \frac{\hbar^2\mathbf{k}^2}{2m_r^*} \quad (3.292)$$

where we introduced the reduced effective mass:

$$\frac{1}{m_r^*} = \frac{1}{m_e^*} + \frac{1}{m_h^*} \quad (3.293)$$

The gradient of $E_c(\mathbf{k}) - E_v(\mathbf{k})$ is then:

$$\nabla E_c(\mathbf{k}) - E_v(\mathbf{k}) = \frac{\hbar^2\mathbf{k}}{m_r^*} \quad (3.294)$$

Substituting Equation 3.294 in Equation 3.291, we find:

$$G_{v,c}(E) = \frac{2}{8\pi^3}\int \frac{dS}{|\nabla_{\mathbf{k}}(E_c - E_v)|_{E_c - E_v = \hbar\omega}} = \frac{2}{8\pi^3}\left[\frac{4\pi k^2}{\frac{\hbar^2\mathbf{k}}{m_r^*}}\right] = \frac{m_r^*}{\pi^2\hbar^2}\mathbf{k} \quad (3.295)$$

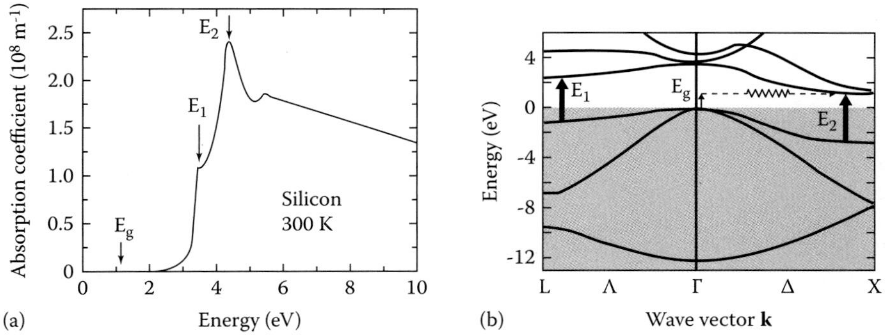

FIGURE 3.149 (a) Absorption spectrum for Si. (b) Si band structure. The indirect bandgap is at 1.1 eV, and critical points are at E_1 (3.2 eV) and E_2 (4.3 eV) semiconductors compared.

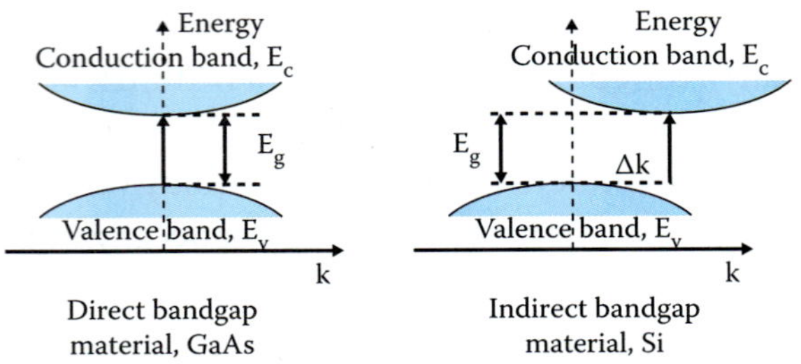

FIGURE 3.150 Direct and indirect bandgaps compared in an *E*-**k** diagram.

To evaluate **k**, note that $E_c(\mathbf{k}) - E_v(\mathbf{k}) = E_g + \frac{\hbar^2 \mathbf{k}^2}{2m_r^*}$ (Equation 3.292), so that:

$$\mathbf{k} = \left[\frac{2m_r^*}{\hbar^2}(\hbar\omega - E_g)\right]^{\frac{1}{2}} \quad (3.296)$$

And, finally, the joint density of states for a two-parabolic band system is:

$$G_{v,c}(E) = \frac{1}{2\pi^2}\left[\frac{2m_r^*}{\hbar^2}\right]^{\frac{3}{2}}(\hbar\omega - E_g)^{\frac{1}{2}} \quad (3.297)$$

Substituting this result in Equation 3.288 yields:

$$\Gamma = \mathbf{E}_0^2 \frac{2\pi}{\hbar}\left|H'_{v,c}\right|^2 \frac{1}{2\pi^2}\left[\frac{2m_r^*}{\hbar^2}\right]^{\frac{3}{2}}(\hbar\omega - E_g)^{\frac{1}{2}} \quad (3.298)$$

We can relate the transition rate Γ to the absorption coefficient α of a material. Because the absorption rate Γ of photons (with an energy $E = \hbar\omega$) by a material corresponds to an absorbed power per volume, *W*, we obtain:

$$\alpha = W = \hbar\omega\Gamma = -dI/dV \quad (3.299)$$

where *I* is the irradiance, and because $\mathbf{E}_0^2 = \frac{2I}{nc\varepsilon_0}$, we obtain:

$$\frac{dI}{dV} = -\frac{2\pi}{\hbar}\left|H'_{v,c}\right|^2 \hbar\omega \frac{2I}{2\pi^2 nc\varepsilon_0}\left[\frac{2m_r^*}{\hbar^2}\right]^{\frac{3}{2}}(\hbar\omega - E_g)^{\frac{1}{2}}$$

or

$$\alpha(\omega) = \frac{2\omega}{\pi nc\varepsilon_0}\left|H'_{v,c}\right|^2\left[\frac{2m_r^*}{\hbar^2}\right]^{\frac{3}{2}}(\hbar\omega - E_g)^{\frac{1}{2}} \quad (3.300)$$

This shows that the absorption coefficient α depends on the transition matrix and on the density of states. The transition matrix element is very difficult to calculate; however, in practice, sufficiently close to the band edge, the product of ω and the matrix element remains approximately constant over a significant frequency range, so that we may write:

$$\alpha(\omega) \propto G(E) \propto (\hbar\omega - E_g)^{\frac{1}{2}} \quad (3.301)$$

In other words, a plot of the absorption coefficient versus frequency is a good reflection of the density of states function. In the case of an indirect bandgap semiconductor, absorption is a second-order process involving both a phonon and a photon. In this case, one obtains for the absorption:

$$\alpha(\omega) \propto G(E) \propto (\hbar\omega - E_g \pm \hbar\nu_p)^2 \quad (3.302)$$

with ν_p the phonon frequency and with + for absorption and – for emission.

In Chapter 5, we will see that for many semiconductors, the effect of excitons (a bound electron-hole state) causes the density of states to be enhanced for $\hbar\omega \approx E_g$, so that the absorption coefficient in reality might have a rather different shape than the predicted dependence (see the absorption curves for various semiconductors in Figure 5.93).

Fermi's golden rule applies to all types of quantum transitions, not only simple photon absorption. The expression can be expanded to cover spontaneous emission, stimulated emission, tunneling, scattering phenomena like phonon-electron scattering, and so on. The golden rule is especially useful in the case of isolated atoms where electron wave functions have been calculated theoretically or in cases where the density of states is the most rapidly varying function, so that the shape of the absorption spectrum is dominated by the DOS. Apart from excitonic effects (see Chapter 5) and selection rules, the dimensionality of the structure has little influence on transition probability.

Heat Transport and Heat Capacity

Introduction

Using quantum mechanics, we now understand the photoelectric effect, avoided the ultraviolet catastrophe, clarified the light spectra emitted from a hot gas, and gave a firm scientific base to Dmitri

Mendeleev's periodic table. With the band model of solids, the difference between conductors, semiconductors, and insulators also became clear. Curiously, we also found that in quantum mechanics, there is a finite probability of finding a particle inside a "classically forbidden" region, i.e., inside a "barrier." We still need to demonstrate how quantum mechanics correctly explains thermal transport, heat capacity, and their temperature dependencies.

Specific Heat and Thermal Transport

Electron Gas Contribution to Specific Heat Earlier in this chapter we described how, experimentally, C_v, the sum of vibrational (phonon) and electronic contributions (electrons) to the heat capacity ($C_v = C_{v,el} + C_{v,lat}$) is proportional to T^3 at low temperatures while approaching a constant 3R at high T, and how the electronic contribution $C_{v,el}$ is found to vary linearly with temperature (see Figure 3.16). From classical theory, at high temperature, electrons are expected to contribute to the lattice heat capacity for an amount of $3/2R$, something that is not observed experimentally. The total C_v for metals is found to be only slightly higher than that for insulators, which feature lattice contributions only! We explore here how QM comes to the rescue in this case. We first consider the electronic contribution to the specific heat and then consider Einstein and Debye's calculations for the lattice contribution. Along the way we reveal details about the nature of motions of atoms in a lattice, i.e., about phonons.

In the classical model for specific heat capacity of an electron gas, we used the principle of equipartition of the energy from the classical ideal gas theory. This led to the wrong expectation of an electronic contribution to the specific heat of $3/2R$. The reason that electrons do not contribute to the heat capacity at room temperature is again, just like in Drude's conductivity case, because they are fermions and cannot be treated as an ordinary Maxwell-Boltzmann gas. At room temperature only the very few electrons in the Maxwell-Boltzmann tail of the Fermi distribution can contribute to the heat capacity. Thus, at ordinary temperatures the electronic heat capacity is almost negligible, and it is the vibrations of the atoms, i.e., the phonon contributions, that dominate (see Figure 3.16)!

Only electrons within an energy range of kT of the Fermi level can be thermally excited, and the minimum energy required for exciting all the free electrons is E_F. The fraction of electrons that can be excited at T = 300 K, say in the case of copper is given by $kT/E_F = 0.025/5 = 0.005 = 0.5\%$. Therefore, at room temperature less than 1% of the available electrons contribute to the heat capacity, whereas in the classical view all electrons are supposed to contribute. Because each electron can absorb a kinetic energy of $3/2\ kT$, the energy of a kilomole of metal with some simplifications can be written down as:

$$\begin{aligned} U &= U_0 + n\left(\frac{kT}{E_F}\right)\left(\frac{3}{2}kT\right) \\ &= U_0 + \left(\frac{nk}{E_F}\right)\left(\frac{3}{2}kT^2\right) \end{aligned} \tag{3.303}$$

Or the electronic contribution to the heat capacity is given by:

$$C_{v,el} = \left(\frac{dU}{dT}\right) = 3nk\left(\frac{kT}{E_F}\right) \tag{3.304}$$

This expression has the correct linear temperature dependence (see Figure 3.16), and in the case of copper one obtains a room temperature value for $C_{v,el}$ of $3R \times 0.005$ or $C_{v,el} = 0.015R$, whereas the lattice contribution $C_{v,lat}$ is $3R$. The value of $C_{v,el} = 0.015R$ agrees well with experimental data.

At very low temperatures, the lattice contribution falls off as T^3 and becomes small, whereas the electronic contribution turns significant. A more detailed analysis (for example, see Pillai[23]) leads to the following expression for U in a 3D structure:

$$U = U_0 + \frac{\pi^2}{6}(k_BT)^2 G(E = E_F)_{3D} \tag{3.305}$$

Or the specific heat for a 3D case is:

$$C_{v,el} = \frac{dU}{dT} = \frac{\pi^2}{3}k_B^2 T G(E = E_F)_{3D} \tag{3.306}$$

Because $G(E = E_F)_{3D}$ was derived earlier as:

$$G(E)_{3D}(E = E_F) = \left.\frac{dn_{3D}}{dE}\right|_{E=E_F} = \frac{1}{2\pi^2}\left(\frac{2m_e^*}{\hbar^2}\right)^{\frac{3}{2}} E_F^{\frac{1}{2}}$$

(Equation 3.236), and the Fermi energy was calculated as:

$$E_F = \frac{\hbar^2}{2m_e^*}(3\pi^2 n_{3D})^{\frac{2}{3}} = (0.365\ \text{eV.nm}^2)(n_{3D})^{\frac{2}{3}}$$

(see Equation 3.232), or also:

$$G(E)_{3D}(E = E_F) = \left.\frac{dn_{3D}}{dE}\right|_{E=E_F} = \frac{3n_{3D}}{2E_F} \quad (3.237)$$

Thus, for the quantum electronic heat capacity we derive:

$$C_{v,el} = \frac{dU}{dT} = \frac{\pi^2}{3} k_B^2 T G(E = E_F)_{3D}$$

$$= \frac{\pi^2}{2}\left(\frac{k_B T}{E_F}\right) n_{3D} k_B \quad (3.307)$$

again with the expected linear temperature dependence (see Figure 3.13). Different dimensionalities lead to different results as U in Equation 3.306 depends on the density of states $G(E)_{\#D}$. Classically, the heat capacity $C_{v,el}$ was given by $3/2R$ or $3/2nk_B$; from the above it follows that Fermi-Dirac statistics depress $C_{v,el}$ by a factor of $\frac{\pi^2}{3}\left(\frac{k_B T}{E_F}\right)$, or for 1 kmol of monovalent metal the heat capacity is:

$$C_{v,el} = \left(\frac{\pi^2}{2}\right)\left(\frac{kT}{E_F}\right)R = \left(\frac{\pi^2}{2}\right)\left(\frac{kT}{kT_F}\right)R$$

$$= \left(\frac{\pi^2}{2}\right)\left(\frac{T}{T_F}\right)R = AR$$

$$\text{with } A = \left(\frac{\pi^2}{2}\right)\left(\frac{T}{T_F}\right) \quad (3.308)$$

The ratio of electronic specific heat calculated classically and quantum mechanically is:

$$\frac{C_{v,el}(QM)}{C_{v,el}(\text{Classical})} = \frac{\left(\frac{\pi^2}{2}\right)\left(\frac{T}{T_F}\right)R}{\frac{3}{2}R} = 0.5710^{-4}T \quad (3.309)$$

The total specific heat C_v is the sum of vibrational and electronic contributions ($C_v = C_{v,el} + C_{v,lat}$). Equation 3.307 reveals a linear dependence of electronic specific heat with temperature, whereas the lattice specific heat varies as the cube of the absolute temperature. In other words:

$$C_v = AT + BT^3$$

electron + phonon contributions to specific heat (3.310)

with A and B constants. Because the lattice specific heat at high temperature is $3R$ (Dulong-Petit law) and is independent of temperature, Equation 3.310 may be rewritten as:

$$\frac{C_v}{T} = A + BT^2 \quad (3.311)$$

Lattice Contribution to Specific Heat

Lennard-Jones Potential In the classical model for heat capacity C_v (Dulong-Petit law), it was assumed that a solid consists of a large number of atomic oscillators executing simple harmonic oscillations about their equilibrium positions (see Figure 3.14). Atoms vibrate about their ideal lattice positions because of the thermal energy in the lattice, and the vibration amplitude is about 10% of the equilibrium atomic spacing. The equilibrium distance between atoms in a lattice can be derived from the Lennard-Jones relationship as depicted in Figure 3.151 and detailed in Chapter 7 (see Figure 7.11). According to the Lennard-Jones model, neighboring atoms are subject to two distinct forces: a repulsive force at short ranges (Pauli repulsion from Pauli's exclusion principle) and in the limit of large distances an attractive force (van der Waals force). In 1931, taking into account both repulsive and attractive interactions, John Lennard-Jones of Bristol University made a good approximation of the total potential energy

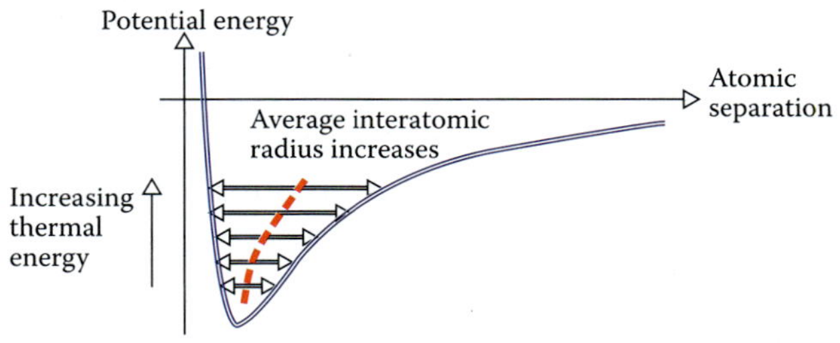

FIGURE 3.151 Lennard-Jones model approaching neutral atoms are subject to two distinct forces: a repulsive force at short ranges (Pauli repulsion from Pauli's exclusion principle) and in the limit of large distances an attractive force (van der Waals force).

of that interaction with the so-called Lennard-Jones (L-J) potential (also referred to as the L-J potential or the 6-12 potential) (see Equation 7.22). Close to equilibrium the L-J potential takes on a parabolic curve, typifying a simple harmonic oscillator.

At higher temperatures, the L-J curve is not symmetric, and the increased energy of the atoms leads to a change of average atomic spacing, which explains the thermal expansion of materials characterized by a thermal expansion coefficient α. If the curve is more symmetric, the effect is reduced and α is lower.

Classical and Quantum Oscillators The internal energy of a crystal, *U*, is caused by the vibrational energy of all the atomic oscillators. In the case of a classical harmonic oscillator, energy can take on a continuum set of values from zero to infinity. Using the classical oscillator model, Dulong and Petit derived that C_v must be 3*R*, independent of temperature because each atom oscillator in the solid has three degrees of freedom. But we know from experiments that the specific heat for all solids decreases sharply at low temperatures and approaches zero at 0 K (see Figure 3.15). Clearly, the Dulong-Petit prediction is wrong for low temperatures.

We will see next that Einstein, as usual, came up with the solution to this difficulty, and in 1903 he proposed that the energies of the "atomic oscillators" were quantized. He considered a solid of *N* atoms to behave as 3*N* independent harmonic oscillators and that, unlike electrons, the number of these oscillators increases with temperature or, conversely, phonons are eliminated with decreasing temperature and the energy of each phonon is constant: for electrons it is their energy that increases with temperature, not their number. Like Planck before him, he assumed that the atom oscillators were quantized, with discrete values given by Planck's formula $E_n = nh\nu = n\hbar\omega$. Debye's model, an improvement on Einstein's model, retained the quantization of the atom oscillator energies but regarded them as coupled, with 3*N* modes belonging to the whole system rather than the *N* single atoms. Lattice vibrations play a most important role in a material's thermal and electrical conductivities. In particular, long-wavelength phonons give rise to sound in solids—hence the name phonon (voice in Greek). Phonons are quantum mechanical particles with integer spin, i.e., bosons that are found to obey the Bose-Einstein distribution. Classical oscillators were compared with quantum mechanical oscillators earlier in this chapter. For a classical harmonic oscillator the energy values are continuous from zero to infinity. Revisiting the harmonic oscillator from Figure 3.51 and solving the time-independent Schrödinger equation (TISE) for a quantum harmonic oscillator $-\frac{\hbar^2}{2m}\frac{d^2\psi(x)}{dx^2} = (E - \frac{1}{2}kx^2)\psi(x)$ (Equation 3.189), we saw how energy quantization came about with energy steps given by $E_n = (n + \frac{1}{2})\hbar\omega$, n = 0,1,2,3, . . . (Equation 3.191). This energy quantization results in the evenly spaced energy levels shown in Figure 3.78. The horizontal lines cutting through the parabolic part of the curve represent the quantization of the vibration states. Also different from the classical oscillator is the fact that a quantum oscillator has a zero-point energy (ZPE) of $E_0 = \frac{1}{2}\hbar\omega$ (Equation 3.193). To describe phonons in solids, Debye went beyond the treatment of a set of these single oscillators and considered a network of interconnected oscillators. Phonons constitute a quantum mechanical version of the normal modes in classical mechanics, in which each part of a lattice oscillates with the same frequency.

Phonons

1D Monoatomic Lattice—No Light Interaction Possible In a crystal, the individual molecular vibrations are replaced by collective vibration modes of the whole lattice. Let us consider phonons in the simplest case of a 1D mechanical harmonic chain of length *L* with *N* identical atoms in a crystalline monoatomic 1D lattice as sketched in Figure 3.152. Here, a unit cell (length *a*, so that $N \times a = L$) contains

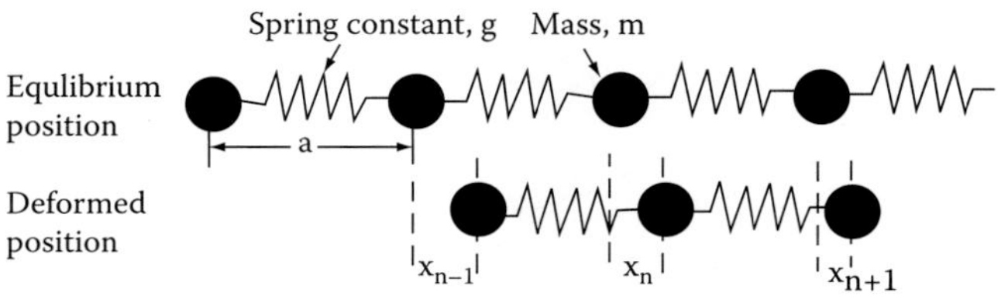

FIGURE 3.152 A 1D array of springs with all identical atoms with mass *M*. Note that *K* is a spring constant here, not a wave vector.

only one atom with mass m, and there is only one spring constant K (capitalized to distinguish it from the phonon wave vector $\mathbf{k}_{ph}$). This is the simplest mechanical model of a lattice, and we will analyze how phonons arise from it. The formalism that we develop for a 1D lattice is readily generalizable to 2D and 3D.

The atoms' equilibrium positions are na, where integer n labels atoms, but each atom executes a motion from equilibrium, which we denote as x_n. Classically the situation depicted in Figure 3.153 is described through an equation of motion where, for simplicity, we only consider nearest-neighbor interactions:

$$m\frac{d^2x_n}{dt^2} = K(x_{n+1} + x_{n-1} - 2x_n) \quad (3.312)$$

This equation describes longitudinal (L) and transverse (T) modes for small amplitude vibrations with a restoring force that depends linearly on the atom's excursion. This expression has traveling wave solutions where the displacement of an atom relative to its center position is given as:

$$x_n = x_0 \exp(-i\omega t)\exp(in\mathbf{k}_{ph}a) \quad (3.313)$$

with $\mathbf{k}_{ph}$ the phonon wave number (just like the electron case). This same result may be obtained from quantum mechanics. In the latter case $\mathbf{k}_{ph}$ takes on quantized values ($\mathbf{k}_{ph,n}$ is then the wave vector of the nth mode) because the total number of atoms N is finite (Figure 3.153). For these modes, a single quantum of energy is called a phonon, and the total number of phonon modes depends on the total number of atoms N in crystal length L. The form of the quantization depends on the choice of boundary conditions; for simplicity, we impose periodic boundary conditions, defining the $(N+1)^{th}$ atom as equivalent to the first atom. Physically, this corresponds to joining the chain at its ends (reread the section on Born and von Karman's periodic boundary condition, p. 124 "The Born and von Karman's Periodic Boundary Condition"). The resulting quantization is:

$$\mathbf{k}_{ph,n}(\min) = \frac{\pi}{Na} = \frac{\pi}{L}$$

$$\mathbf{k}_{ph,n} = \frac{\pi}{Na} = \frac{n\pi}{L} \quad (3.314)$$

$$\text{for } n = 0, = 1,2,\ldots,N$$

$$\mathbf{k}_{ph,n}(\max) = \frac{N\pi}{L} = \frac{\pi}{a}$$

with the corresponding wavelengths:

$$\lambda_{min} = 2a,\ \lambda_n = \frac{2L}{n},\ \text{and } \lambda_{max} = 2L \quad (3.315)$$

These simplest ("normal mode") cases are depicted in Figure 3.154, and any motion can be written as a superposition of the normal modes. Mode wavelengths are easy to find, but finding the energy for each these modes is a more complex issue.

In real crystals the number of atoms N is so large that the $\mathbf{k}_{ph,n}$-spacing and the corresponding frequency spacing are irrelevant so that vibration spectra look continuous.

Substituting the wave solutions for x_n, x_{n-1}, and x_{n+1} (Equation 3.313) in Equation 3.312, we obtain, after a lot of convenient cancellations, a relation of the vibration frequency as a function of the wave vector:

$$\omega_n^2 = \frac{2K}{m}(1 - \cos \mathbf{k}_{ph,n}a) \quad (3.316)$$

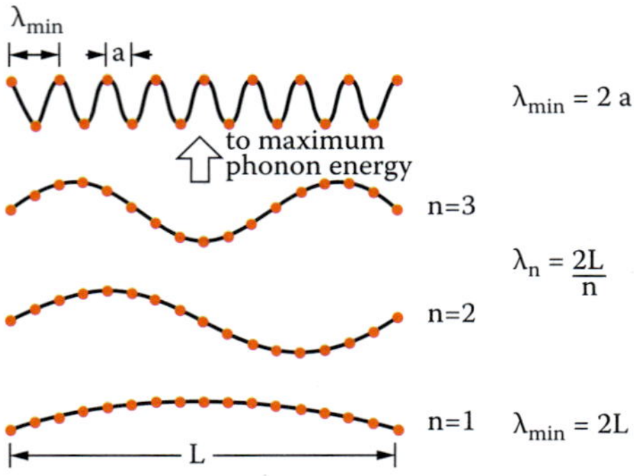

FIGURE 3.153 The simplest normal modes of lattice vibrations in a 1D crystal lattice of length L.

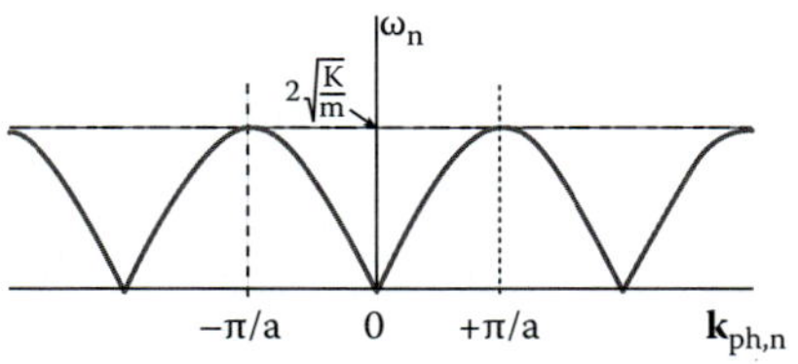

FIGURE 3.154 Phonon dispersion relation for a 1D monoatom crystal.

The vibration frequency ω_n depends only on the relative phase of the neighbors, and we now have a series of allowed modes with wave vector $\mathbf{k}_{ph,n}$ and angular frequency ω_n according to:

$$\omega_n = \sqrt{\frac{2K}{m}(1-\cos \mathbf{k}_{ph,n} a)} = 2\sqrt{\frac{K}{m}}\left|\sin\left(\frac{\mathbf{k}_{ph,n} a}{2}\right)\right| \quad (3.317)$$

This expression represents the dispersion relation for a 1D monoatomic lattice as shown in Figure 3.154. At low values of $\mathbf{k}_{ph,n}$ the oscillation frequencies are low as well, and the maximum oscillation frequency that a collective mode can have is $2\sqrt{\frac{K}{m}}$.

The maximum oscillation frequency occurs at $\mathbf{k}_{ph} = \pm\pi/a$; in other words, when adjacent atoms are 180° out of phase (the wave number π/a corresponds to a vibration wavelength of $2a$). As in the case of electron wave functions, a first Brillouin zone ranges from $-\pi/a < \mathbf{k}_{ph,n} < \pi/a$. For any wave numbers $> \pi/a$, the same physical configuration can be described by a wave number that is less than π/a: there is no new physical situation outside this zone.

In Figure 3.155 we focus for a moment on the positive $\mathbf{k}_{ph,n}$ side in the first Brillouin zone of the phonon dispersion relation. At $\mathbf{k}_{ph,n} = 0$ all atoms move in phase, and at $\mathbf{k}_{ph,n} = \pi/a$ neighboring atoms move in opposite directions. In Figure 3.155 we show both the longitudinal (LA) and transverse (TA) acoustic branches. The longitudinal branch lies above the transverse branch. For a 3D monoatomic crystal there is one longitudinal vibration mode and two transverse direction modes. In the case illustrated the two transverse waves are "degenerate," i.e., they have the same frequency. The energies involved here are typically in the range of hundreds of meV [$\omega_{max} \sim (10^{12}\text{–}10^{14})$ Hz or $\lambda \sim (5\text{–}500)$ μm and $h\nu \sim (2\text{–}200)$ meV]. These low energies are possible because the restoring force depends on an atom position difference, which can become very small for long wavelengths. With a thermal energy at room temperature of ~24 meV, many phonon modes are excited even at room temperature.

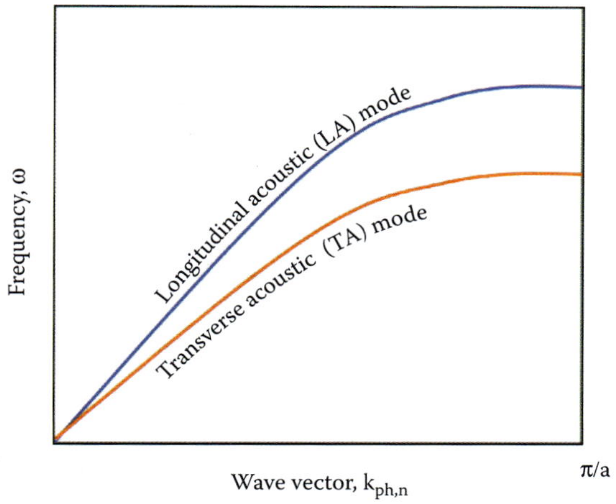

FIGURE 3.155 Positive **k** values in the first Brillouin zone for phonons.

In three dimensions, as mentioned above, there are three branches of the dispersion relation: two transverse and one longitudinal. Different crystal structures have different propagation speeds along different directions. In a monatomic lattice, all the phonon modes are called "acoustic" modes.

As illustrated in Figure 3.156, the slope of the dispersion relation is the group velocity v_g ($v_g = \partial\omega/\partial\mathbf{k}_{ph}$), whereas the phase velocity $v_{ph} = \omega/\mathbf{k}_{ph}$. At low values of $\mathbf{k}_{ph} \sim 0$ (i.e., long wavelengths), the dispersion relation is almost linear, and the speed of sound v_s through the material $v_s = v_g = v_{ph}$ is then represented as:

$$v_s = \left(\frac{d\omega}{d\mathbf{k}_{ph}}\right)_{\mathbf{k}_{ph\to 0}} \quad (3.318)$$

At $\mathbf{k}_{ph} = \pm\pi/a$, the Brillouin zone boundary (BZB), the group velocity = 0. As a result, phonons with long wavelengths can propagate for large distances so that sound propagates through solids without significant distortion. This behavior fails at short

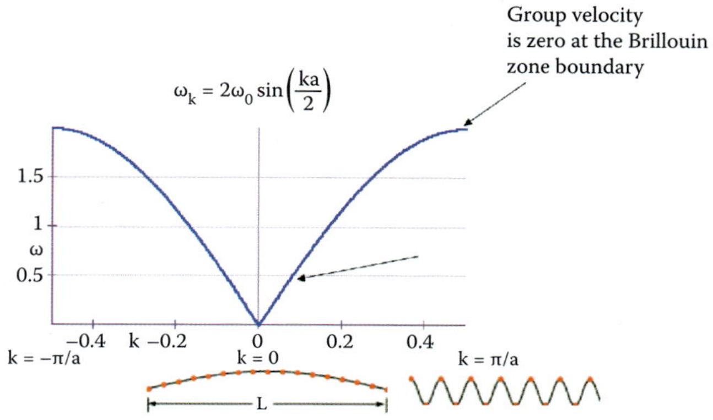

FIGURE 3.156 The slope of the dispersion relation is the group velocity v_g ($v_g = \partial\omega/\partial\mathbf{k}_{ph}$), whereas the phase velocity $v_{ph} = \omega/\mathbf{k}_{ph}$.

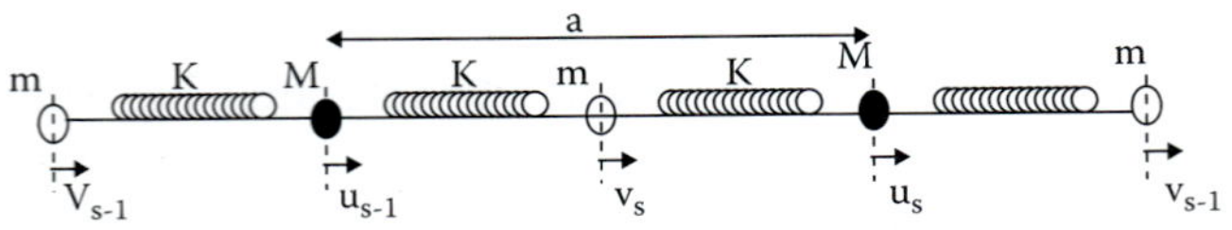

FIGURE 3.157 Vibrational modes in a crystalline diatomic 1D lattice.

wavelengths (large values of $\mathbf{k}_{ph}$) because of the microscopic details of the lattice. Sound waves in fluids only have longitudinal components, whereas sound waves in solids have longitudinal and transverse components. The reason behind this is that fluids cannot support shear stresses. Monatomic crystals only have acoustic branches. Light does not interact with the lattice modes of a monoatomic crystal; for that we need a lattice that has at least two different atoms (diatomic) so that a polarization **P** can be generated.

1D Diatomic Lattice—Light Interaction Possible
A diatomic lattice is illustrated in Figure 3.157. Here we consider two types of atoms with masses *m* (light) and *M* (heavy) characterized by displacement amplitudes u_s and v_s, respectively. In this case, *a* is the length of the unit cell, and each unit cell contains one of each kind of atom. A wave vector of π/a here implies a phase difference of 180° between the behavior inside adjacent unit cells, not between adjacent atoms as in the 1D monoatomic lattice! This internal structure of the unit cell adds an internal degree of freedom, giving rise to two phonon modes for each value of the wave vector.

Assuming again that only next neighbors interact, the equations of motion lead to an expression of the order ω^4, resulting in two independent solutions in ω^2 for the dispersion relationship:

$$mM\omega^4 - 2K(m + M)\omega^2 + 2K^2(1 - \cos \mathbf{k}_{ph}a) = 0$$

or

$$\omega^2 = \frac{K}{mM}\left\{(m + M) \pm \sqrt{(m + M)^2 - 2mM(1 - \cos \mathbf{k}_{ph}a)}\right\} \quad (3.319)$$

Each solution is twofold degenerate because the electrical field *E* can be chosen in an arbitrary direction perpendicular to the propagation vector. If the masses and spring constants are known, this equation can be solved. The dispersion relation for a diatomic linear lattice is schematically shown in Figure 3.158, where we see that in solids with more than one atom in the smallest unit cell, there is a "phonon branch" for "acoustic" phonons (red curve) and an "optical branch" phonon (blue curve). Both types of phonons come with transverse and longitudinal modes. This makes for four possible types of phonons: LA, TA, longitudinal optical (LO), and transverse optical (TO), as summarized in Figure 3.159. Figure 3.158b shows the phonon dispersion for GaAs, and all four types of phonons can be observed here.

"Acoustic phonons" have frequencies that become small at the long wavelengths and correspond to sound waves in the lattice. "Optical phonons" can only arise in crystals that have more than one atom in the smallest unit cell. They are called "optical phonons" because in ionic crystals they are excited

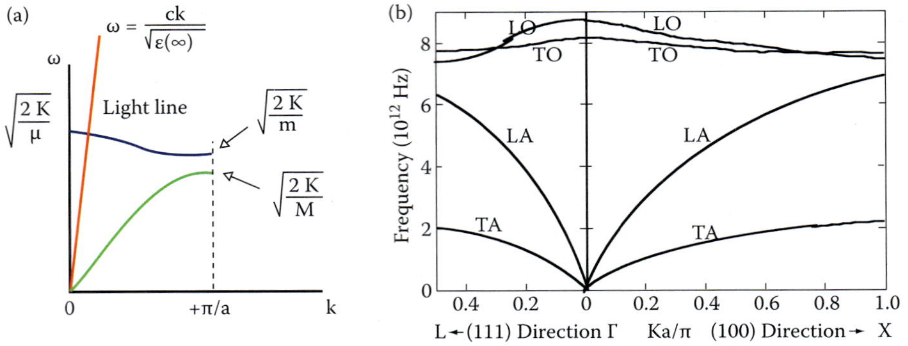

FIGURE 3.158 (a) Dispersion relationship for a diatomic linear lattice. Both optical and acoustic modes come with longitudinal and transverse waves. The straight line represents the "light line" assuming no interaction with phonons $\left(\omega = \frac{ck}{\sqrt{\varepsilon(\infty)}}\right)$ (see Chapter 5, "Summary: Generalized Dispersion Relation" for details). (b) The phonon dispersion for GaAs.

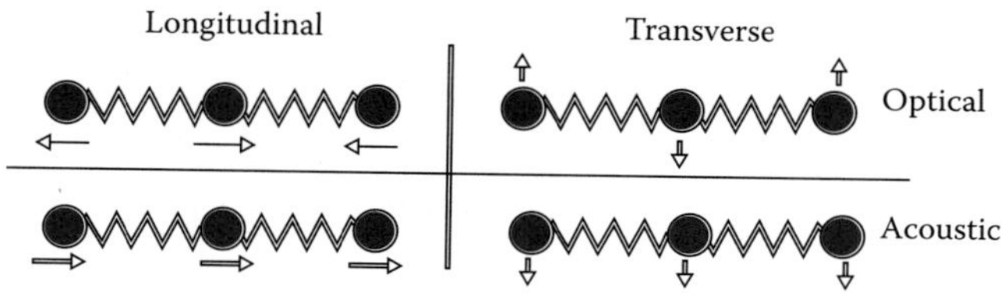

FIGURE 3.159 Four types of phonons: LA, TA, LO, and TO.

very easily by light (IR radiation). The underlying reason is that these vibrations correspond to a mode where positive and negative ions at adjacent lattice sites swing against each other, creating a time-varying electrical dipole moment **P**. Optical phonons that interact in this way with light are called infrared active. In the case of a 3D diatomic lattice there are three optical and three acoustic branches (6*N* degrees of freedom). The three optical branches can interact with light. A phonon dispersion relation can be measured experimentally by inelastic x-ray or neutron scattering.

Let us investigate the maxima and minima values in the phonon relationship for diatomic 1D crystals. For $\mathbf{k}_{ph} = \pi/a$, where neighboring cells are out of phase, the two solutions are:

$$\omega^2 = \frac{2K}{M}\text{(acoustic mode) and} \qquad \omega^2 = \frac{2K}{M}\text{(optical mode)} \tag{3.320}$$

For $\mathbf{k}_{ph}$ tending to zero ($\mathbf{k}_{ph}a \lll 1$) the acoustic branch tends to zero (see above), and the optical branch has the following (constant) solution:

$$\omega^2 \cong 2K\frac{m+M}{mM} = 2K\left(\frac{1}{m}+\frac{1}{M}\right) = \frac{2K}{\mu} \tag{3.321}$$

where μ is the reduced mass of the unit cell $\left(\mu = \frac{1}{m}+\frac{1}{M}\right)$.

Example 3.8:
For GaAs ≈ $\hbar\omega_{OP}$ ≈ 36 meV, whereas for GaN $\hbar\omega_{OP}$ ≈ 90 meV; this difference stems from the much lighter *N* atom $\left[\omega_{OP}^2 \cong 2K\frac{m+M}{mM} = 2K\left(\frac{1}{m}+\frac{1}{M}\right)\right]$.

Interaction of Radiation with Lattice Modes—Polaritons The two types of atoms (*m* and *M*) in the unit cells of diatomic lattices in general come with slightly different charges and therefore experience different forces in different directions when an electromagnetic wave is applied. This leads to a macroscopic polarization **P**. The natural frequency of the oscillation of this polarization **P**, we will see, corresponds to the optical modes we just calculated above. In Figure 3.159a the straight line represents the "light line," assuming no interaction with phonons $\left(\omega = \frac{ck}{\sqrt{\varepsilon(\infty)}}\right)$ (for a detailed explanation see section in Chapter 5, "Summary: Generalized Dispersion Relation"). In polar materials, light does interact with the optical phonon modes and ε_r, the relative dielectric constant strongly varies, and ω versus *k* is not a straight line anymore. The wavelength of light is usually much greater than the lattice spacing ($\lambda >> a$), especially for the infrared region, where we find that the frequency of the EM is resonant with the optical modes, so the incoming optical wave interacts strongly with very long wavelength phonons (compared with the lattice distance *a*).

At very long wavelengths, $\mathbf{k}_{ph}$ is close to 0 (see Figure 3.158), and we can assume that the applied field is spatially uniform $[E(t) = E_0 e^{-i\omega t}]$, which is true over many thousands of atomic spacings. The excursion of the *m* and *M* atoms driven by that uniform field is then given as:

$$u = u_0 e^{-i\omega t}\text{(for M) and } v = v_0 e^{-i\omega t}\text{(for m)} \tag{3.322}$$

Substituting these expressions in the equation of motion leads to:

$$-2Ku_0 + (2K - m\omega^2)v_0 = -qE_0 \tag{3.323}$$

For this, the solutions for amplitudes u_0 and v_0 are:

$$u_0 = \frac{+qE_0/M}{\omega_T^2 - \omega^2} \text{ and } v_0 = \frac{-qE_0/m}{\omega_T^2 - \omega^2} \tag{3.324}$$

where ω_T is the resonant frequency, also called the "restrahl" frequency, of the optical modes at $\mathbf{k} = 0$:

$$\omega_T = \sqrt{\frac{2K}{\mu}} \tag{3.325}$$

This is the so-called transverse optical (TO) frequency, and μ is the reduced mass; this frequency depends on the harmonic forces in the crystal and not the applied field. The large mass M results in a small amplitude u_0 and a large amplitude for the small mass m in v_0. The opposite signs of these amplitudes refer to the opposite motions that m and M make.

The excitation near ω_T (natural frequency of the optical $\mathbf{k} = 0$ branch) features a large amplitude, and this coupled excitation of a transverse optical phonon to electromagnetic radiation is called a polariton. In the longitudinal direction light and the lattice vibrations do not couple. There are six modes for a polariton because there are three optical phonons and three electromagnetic modes.

The polarization introduced by electromagnetic radiation in the lattice of a polar compound can then be written as:

$$\mathbf{P}_{lattice} = Ne(u_0 - v_0) \tag{3.326}$$

where e is the effective charge on the atoms in the lattice, and u_0 and v_0 are the amplitudes calculated in Equation 3.324. The total polarization of a material is given by the sum of this lattice contribution ($\mathbf{P}_{lattice}$) at low infrared frequencies, a bound electronic contribution ($\mathbf{P}_B$), and a free electron contribution ($\mathbf{P}_E$):

$$\mathbf{P} = \mathbf{P}_{lattice} + \mathbf{P}_B + \mathbf{P}_E \tag{3.327}$$

In Chapter 5 we connect the polarization $\mathbf{P}$ of a wide range of materials from insulators to metals to their susceptibility, their dielectric constant, refractive index, and absorption coefficient.

Einstein's Heat Capacity Equation Einstein, by introducing quantized noncoupled oscillators, obtained the following expression for the specific heat capacity contribution of the lattice:

$$C_v = 3R\left(\frac{h\nu}{kT}\right)^2 \left\{ \frac{\exp\left(\frac{h\nu}{kT}\right)}{\left[\exp\left(\frac{h\nu}{kT}\right) - 1\right]^2} \right\}$$

$$= 3R\left(\frac{\theta_E}{T}\right)^2 \frac{\exp\left(\frac{\theta_E}{T}\right)}{\left[\exp\left(\frac{\theta_E}{T}\right) - 1\right]^2} = 3RF_E\left(\frac{\theta_E}{T}\right) \tag{3.328}$$

with $h\nu/k = \theta_E$ the Einstein temperature and F_E the Einstein function. At high temperatures ($h\nu \ll kT$) this equation simplifies to the Dulong-Petit form, i.e., $C_v = 3R$, whereas at low temperatures ($h\nu \gg kT$) it reduces to:

$$C_v = 3R\left(\frac{h\nu}{kT}\right)^2 \exp\left(\frac{-h\nu}{kT}\right) \tag{3.329}$$

At low temperature the exponential term becomes dominant and controls the temperature variation of C_v. An exponential decay of C_v agrees with experimental data, except at very low temperatures, where C_v is proportional to T^3.

Debye's Heat Capacity Equation Debye retained the quantization of the atom oscillator energies introduced by Einstein but regarded them as coupled, where $3N$ modes now belong to the whole system rather than the N single atoms. In this model, because of the interconnections between atoms in a solid, the displacement of one or more atoms from their equilibrium positions gives rise to a set of mechanical waves propagating through the material. The amplitude of these "solid" waves is given by the displacements of the atoms from their equilibrium positions. The general Debye expression is:

$$C_v = 9R\left(\frac{T}{\theta_D}\right)^3 \int_0^{\frac{\theta_D}{T}} \frac{e^4 x^4}{(e^x - 1)^2} dx = 9RF_D\left(\frac{\theta_D}{T}\right) \tag{3.330}$$

where $x = h\nu/kT$ and $x_{max} = h\nu_D/kT$ and $k\theta_D = h\nu_D$ with θ_D the Debye temperature. At high temperatures this expression again leads to Dulong-Petit, i.e., $C_v = 3R$, and at low temperatures ($T \ll \theta_D$) we get:

$$C_v = \frac{12}{5}\pi^4 R\left(\frac{T}{\theta_D}\right)^3 \tag{3.331}$$

The model correctly predicts a T^3 temperature dependence for the specific heat of a 3D structure at very low temperatures, and for the internal energy U it predicts a T^4 dependency.

The Einstein and Debye models were compared with experiments in Figure 3.16. Because phonons follow Bose-Einstein statistics, their equilibrium distribution as a function of temperature is given as:

$$\left\langle n_{k_{ph,p}} \right\rangle = \frac{1}{\exp\left(\frac{\hbar\omega}{k_B T}\right) - 1} \quad (3.332)$$

The lattice energy is then equal to:

$$U = U_0 + \sum_p \sum_{k_{ph}} \left[\left\langle n(\omega_{k_{ph,p}}) \right\rangle + \frac{1}{2} \right] \hbar\omega_{ph,p} \quad (3.333)$$

where p stands for all polarizations of the phonons (LA, TA, LO, and TO). The second term on the right represents the energy $\hbar\omega_{k_{ph,p}}$ multiplied by the probability to have a state occupied at this energy and comes with a summation over all wave vectors k_{ph} and polarization states p. The summation over all wavevectors k_{ph} can be transformed into an integral $\int d\omega G(E) dE$. The energy density is then given by:

$$U' = \frac{U_0}{V} + \sum_p \int \left[\left\langle n(\omega_{k_{ph,p}}) \right\rangle + \frac{1}{2} \right] \hbar\omega_{k_{ph,p}} G(E) dE \quad (3.334)$$

where $G(E)$ is the density of states between ω and ω + dω. This number of vibrational states can be derived from the dispersion relation $\mathbf{k}_{ph}(\omega)$ because:

$$G(E) = \frac{\mathbf{k}^2(\omega)}{2\pi^2} \frac{d\mathbf{k}}{d\omega} \text{ in 3D} \quad (3.335)$$

Or the specific heat is:

$$C_{lattice} = \frac{dU}{dT} = \sum_p \int \frac{d\langle n \rangle}{dT} \hbar\omega_{k_{ph,p}} G(E) dE \quad (3.336)$$

Debye assumed that $\omega = v_s \mathbf{k}_{ph}$, where v_s is the sound velocity so that the Debye density of states became:

$$G(E) = \frac{\mathbf{k}^2(\omega)}{2\pi^2} \frac{d\mathbf{k}}{d\omega} = \frac{\omega^2}{2\pi^2 v_s^3} \quad (3.337)$$

Putting this expression in Equation 3.336 leads to the Debye expression as given in Equation 3.330. Because $G(E)$ depends on the dimensionality, the specific heat of the lattice will also depend on the dimensionality of the structure (3D, 2D, 1D, or 0D). In 3D materials, phonons possess three polarizations: two transverse and one longitudinal, whereas in a perfect 1D material (a nanowire), only longitudinal polarizations are possible. In real 1D materials the surface gives rise to additional surface phonon modes. The resulting change in dispersion relation modifies the group velocity and the density of states. Generalizing for different dimensions D, one derives:

$$U \propto T^{D+1} \text{ and } C_{lattice} \propto T^D \quad (3.338)$$

with D = 1, 2, 3 the dimensions of the sample.

For an insulator, heat is carried entirely by phonons, whereas in metals, both electrons and phonons may transport heat. In most metals the conduction by electrons greatly exceeds that of the phonons; typically the phonon contribution at room temperature is only 1% of the electron contribution. As a metal is heated, the only way an electron can take up energy is by moving into a somewhat higher allowed energy level. But a typical electron is buried deep inside the Fermi sea, and there are no empty levels to move to because a pair of electrons of opposing spin already occupies each level above the electron. Only the relatively few electrons at the top of the distribution can find empty levels to move into. They are the only electrons that contribute to the thermal conductivity. The thermal heat conductivity was derived earlier as:

$$\kappa = \frac{1}{3} v_{dx} C_{v,el} \lambda \quad (3.59)$$

In those thermal conductivity calculations we used classical statistics to estimate the electron mean free path, λ, and the drift velocity, v_{dx}. These terms need to be replaced now by λ_F and v_F from quantum mechanics, and using the expression for $C_{v,el}$ we launched in Equation 3.308, we derive:

$$\kappa = \frac{\pi^2}{6}\left(\frac{nk_B^2 T}{E_F}\lambda_F v_F\right) = \frac{\pi^2}{6}(nk_B^2 T \lambda_F v_F)\frac{2}{m_e^* v_F^2}$$

and with $\tau_F = \frac{\lambda_F}{v_F}$

$$\kappa = \left(\frac{\pi^2}{3}\right)\left(\frac{nk^2 T\tau_F}{m_e^*}\right) \quad (3.339)$$

$C_{v,el}$ is expressed here per unit volume rather than per kilocalorie. The expression gives the thermal conductivity in terms of the electronic properties of the material. A typical number, say for copper, is 387 $Wm^{-1}K^{-1}$. Using the quantum mechanical expression for electrical conductivity $\left[\sigma = \left(\frac{ne^2\tau_F}{m_e^*}\right) \text{from Equation 3.7}\right]$ we obtain the ratio of thermal conductivity to electrical conductivity, or the Wiedemann-Franz law, as:

$$\frac{\kappa}{\sigma} = \frac{\pi^2}{3}\left(\frac{k_B}{e}\right)^2 T \tag{3.340}$$

(compare this with the classical result in Equation 3.60). The Lorenz number, *L*, defined as $\frac{\kappa}{\sigma T} \equiv L$ (see Equation 3.61), now becomes:

$$L = \frac{\pi^2}{3}\left(\frac{k}{e}\right)^2 \tag{3.341}$$

or 2.45×10^{-8} WΩ K^{-2}. This is the correct answer; no Drude hanky-panky is required. The Lorenz number only depends on the universal constants *k* and *e* and therefore is the same for all metals.

Quantized Thermal Transport

Introduction Research in nanoscale thermal properties is recent compared with nanoscale electronic properties, which started in the 1960s. An important reason for this is the fact that measuring and modeling micro- and nanoscale thermal properties accurately is very challenging. An added difficulty is the fact that phonons are quasiparticles (eigenmodes) that can be annihilated or created at any time. Only since the mid-1980s and early 1990s have techniques become available that allow measuring temperature and heat flow at the nanoscale. Small thermocouples at the tip of atomic force microscopes in scanning thermal microscopy are successfully used to measure temperatures with <100-nm resolution. Scanning Seebeck voltage measurements have reached 2–3 nm spatial resolution with the use of sharp heated metallic tips in ultrahigh vacuum environments, and femtosecond laser technology has permitted the measurement of heat transfer and acoustic velocities in thin-film submicrometer structures. Thermoreflectance imaging using visible wavelengths is used to measure the surface temperature of IC chips with submicrometer spatial resolution and 1–100 mK temperature resolution. Very recently a phonon Hall effect under a magnetic field has also been observed. This could also provide a useful source of information about phonons and their transport in solids.

Macroscopic thermal transport theory, with Fourier's law (Equation 3.53) and the diffusion equation (Equation 3.54), breaks down for applications involving thermal transport in small length/time scales, e.g., nanoelectronics, nanostructures, NEMS, ultrafast laser materials processing, and so on. As we mentioned in the beginning of this chapter, this occurs when *t* is shorter than the scattering mean free time of the energy carriers (τ) and when *L* is shorter than the scattering mean free path of the energy carriers (λ).

Although classically temperature can in principle be defined for a single atom, the quantum mechanical definition can only be applied on the length scale of the phonon mean free path. Because most of the heat is carried by phonons with large wave vectors, the relevant mean free paths are on the order of 1–100 nm. Consequently, the notion of quantum mechanical temperature is difficult to define in nanostructures, thus limiting the applicability of Fourier's law of heat conduction and the diffusion equation because both rely on the definition of local temperature. Moreover, the diffusion equation treats electrons and phonons as classical particles without associated waves. As a consequence, when a structure's dimensions become comparable to the mean free path and wavelength of heat carriers, classical laws are no longer valid and new approaches must be taken to predict heat transfer at the nanoscale.

Research in heat conduction and phonon transport in MEMS and NEMS is a growing research field. Experimental results have shown, for example, that phonon surface and interface scattering can lower thermal conductivity of silicon thin films and nanowires in the sub-100-nm range by two orders of magnitude.[25] So whereas for bulk material interface effects only dominate at low temperatures, where the phonon mean free path is large, in nanometer-scale thin films, the room temperature thermal conductivity can be significantly modified.

It has also been shown that electrons in low-dimensional semiconductors such as quantum wells and wires have an improved thermoelectric power factor that may lead to more efficient Peltier coolers. Applications of nanoheat transport will be covered in Volume III, Chapter 8 on actuators.

Quantized Thermal Conductance Thermal conductance, like electrical conductance, becomes quantized at small length scales. The specific electronic heat for a 3D case introduced above as $C_{v,el} = \frac{\pi^2}{2}\left(\frac{k_B T}{E_F}\right) k n_{3D}$ (Equation 3.307) must be modified for different dimensionalities because U depends on the density of states $G(E)_{\#D}$ and $C_{v,el} = \frac{dU}{dT}$. The thermal electronic conductivity expression, based on Equation 3.59, with the Fermi velocity v_{dx} replacing v_d, results in:

$$\kappa_e = \frac{1}{3} v_F C_{v,el} \lambda_e \tag{3.342}$$

The latter expression has been further adapted with an effective mean free path λ_e to reflect the various possible electron scattering mechanisms of electrons in nanodevices [defect scattering, phonon scattering, and boundary scattering (grain boundary or device boundary)]. This effective mean free path λ_e and the scattering constant τ_e that goes with it are calculated according to the so-called Matthiessen rule as:

$$\begin{aligned} \frac{1}{\tau_e} &= \frac{1}{\tau_{defect}} + \frac{1}{\tau_{boundary}} + \frac{1}{\tau_{phonon}} \\ \frac{1}{\lambda_e} &= \frac{1}{\lambda_{defect}} + \frac{1}{\lambda_{boundary}} + \frac{1}{\lambda_{phonon}} \end{aligned} \tag{3.343}$$

In Figure 3.160 we show a plot of λ_e as a function of temperature. At lower temperatures defect scattering dominates, whereas at higher temperatures phonon scattering takes over. From Figure 3.160 one can also surmise that for nanodevices the device boundary ($\lambda_{boundary}$) might start taking over from defect control. For nanodevices and thin films the plateau at low temperature is decreased compared with a bulk solid. In Figure 3.161 the thermal conductivity κ_e as a function of temperature for copper and aluminum is shown; for metals, electrons dominate the thermal conductivity. Scattering of heat flow at the interfaces significantly reduces thermal conductivity. As expected from Equation 3.308, we observe a linear dependence of electronic specific heat with temperature until phonon scattering sets in.

To adapt Equation 3.59 for phonon thermal conductance κ_l, we introduce the effective mean free path λ_l for phonons and replace v_{dx} with the sound velocity in the solid at hand and use $C_{v,lat}$ for the lattice contribution to the total heat capacity C_v:

$$\kappa_l = \frac{1}{3} v_s C_{v,lat} \lambda_l \tag{3.344}$$

The effective mean free path λ_l now reflects the various possible phonon scattering mechanisms and is again calculated using Matthiessen's rule [defect and dislocation scattering, phonon-phonon scattering, and boundary scattering (grain boundary or device boundary)]. This is illustrated in Figure 3.162, where we plot λ_l as a function of T/θ_D.

Device boundary scattering effects dominate at low temperatures but are taken over by defect scattering, and finally phonon-phonon scattering causes a fast decay as T approaches θ_D.

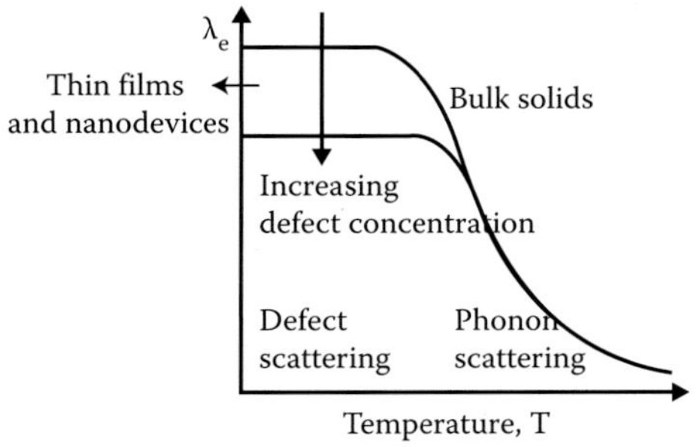

FIGURE 3.160 A plot of λ_e as a function of temperature. At lower temperatures defect scattering dominates, whereas at higher temperatures phonon scattering takes over.

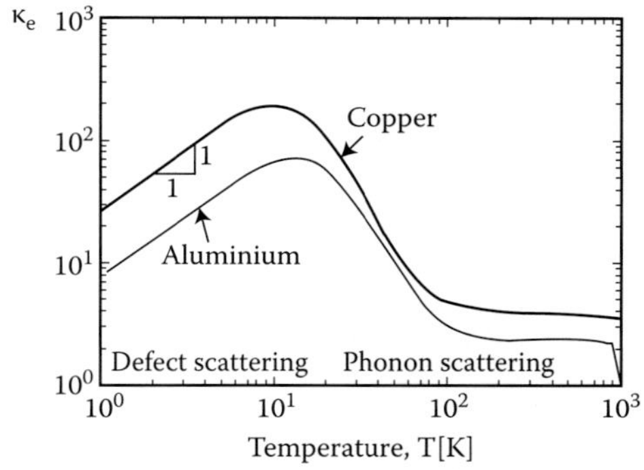

FIGURE 3.161 The thermal conductivity κ_e as a function of temperature for copper and aluminum shows a linear dependence with temperature until phonon scattering sets in.

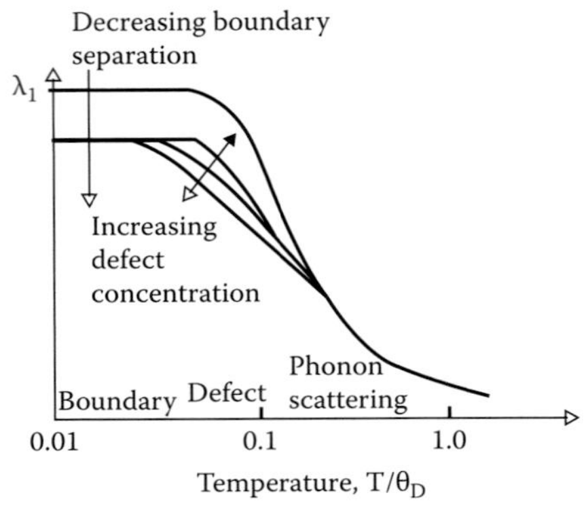

FIGURE 3.162 Plot of phonon effective mean free path λ_l as a function of T/θ_D.

Calculating the phonon contribution to the specific heat of a 3D insulating structure, we found earlier a constant value at $T >> \theta_D$, and for $T << \theta_D$ (where we replaced C_v with $C_{v,lat}$ and $C_{v,el} = 0$) we calculated: $C_v = \frac{12}{5}\pi^4 R\left(\frac{T}{\theta_D}\right)^3$ (Equation 3.331), with a T^3 temperature dependency for the specific heat and a T^4 dependency for the internal energy U. Generalizing for different dimensions D we introduced earlier, $U \propto T^{D+1}$ and $C_{v,lat} \propto T^D$ (Equation 3.338), with $D = 1$, 2, 3 for the dimensions of the sample. Substituting this expression in Equation 3.344, we obtain for the thermal conductivity κ_l as a function of T/θ_D the relationship shown in Figure 3.163. At $T >> \theta_D$, $C_{v,lat}$ is a constant, and at $T << \theta_D$, $C_{v,lat} \propto T_D$, whereas scattering at low temperatures is static (λ_{static}: boundaries and defects) and has a constant value ($\lambda_{static} = C$), and at high temperatures dynamic Umklapp or phonon-phonon scattering takes over ($\lambda_{Umklapp} = e_{\theta/T}$); the combined effect is reflected in Figure 3.163. The peak in the curve occurs when $\lambda_{static} = \lambda_{Umklapp}$.

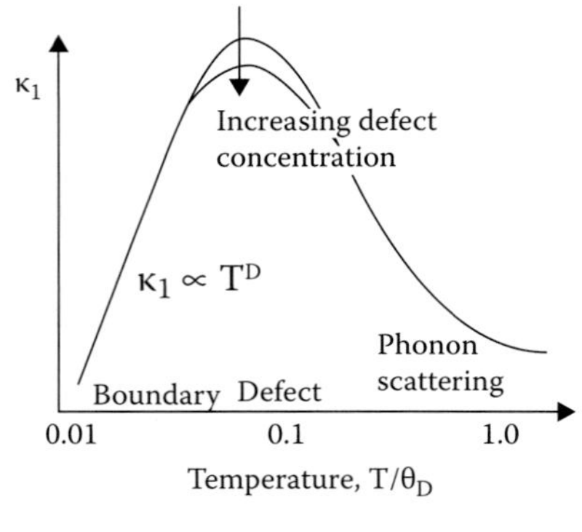

FIGURE 3.163 Phonon thermal conductivity κ_l as a function of T/θ_D.

Comparing Electrical and Thermal Conductance Quantization Quantized electrical conductance in a quantum wire (1D) was dramatically illustrated in Figure 3.138, where we illustrated experiments involving quantum wires in a $GaAs/Al_xGa_{1-x}As$ heterostructure at very low temperatures (close to 0 K). The quantum of electrical conductance we found is $2e^2/h$. In electrical experiments, the chemical potential (μ) and temperature T can be varied independently, so that at low temperatures the sharp edge of the Fermi-Dirac distribution can be swept through 1D modes. With phonons, only the temperature can be swept, and the broader Bose-Einstein distribution function smears out all features except the lowest lying modes at the lowest temperatures.

The quantum thermal conductance is proportional to T because energy is being transported as opposed to quantum electrical conductance, where charge is being transported. In ballistic thermal transport in a 1D structure, in the limit of $kT << \hbar\omega$, only the four lowest phonon modes contribute to the thermal conductance. For ideal coupling of phonon waveguide modes [transmission coefficient $T_m(\omega) = 1$] to a thermal reservoir, the following fundamental relation holds for each mode:

$$G_{th} = g_0 = \frac{\pi^2 k_B^2 T}{3h} \tag{3.345}$$

This quantum represents the maximum possible value of energy transported per phonon mode. This fundamental quantum limit of thermal conductance has been experimentally confirmed in experiments illustrated in Figure 3.164.[26] In Figure 3.164a, a log-log of $G_{th}/16\ g_0$ versus temperature in mK is presented, and as observed, at the lowest temperatures the curve saturates at a value of 1. For this measurement Schwab et al.[26] used a suspended silicon nitride island connected to the rest of the device by a set of four silicon nitride wires (60 nm thick and less than 200 nm wide). The "island" is suspended over a phonon cavity of 4 × 4 μm (Figure 3.164b). The four wires connect to two thin-film heating and sensing Cr/Au resistors (the two bright "C"-shaped objects in the center) on the suspended Si_3N_4 island. The dark regions in Figure 3.164b are empty space, i.e., the cavity underneath the island and bridges.

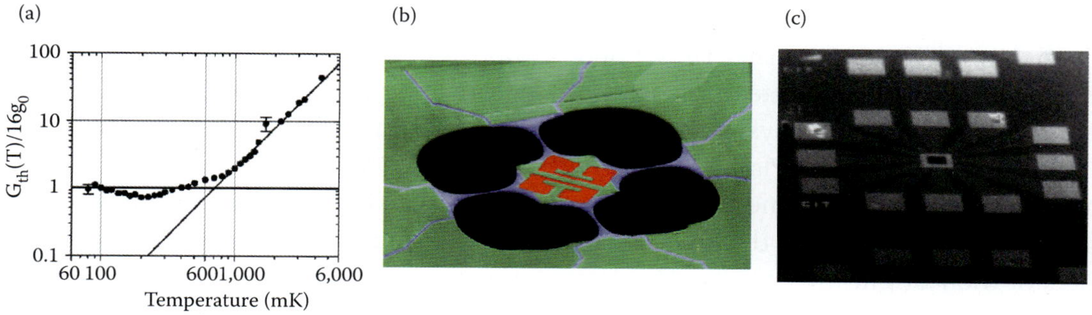

FIGURE 3.164 (a) Log-log of $G_{th}/16\ g_0$ versus temperature in mK. At the lowest temperatures the curve saturates at a value of 1. (b) Set of four 60-nm-thick and less than 200-nm-wide silicon nitride wires suspended over a phonon cavity of 4 × 4 μm. Two thin-film heating and sensing Cr/Au resistors are the bright "C"-shaped objects in the center on a suspended Si_3N_4 island. The dark regions are empty space. (c) The overall device (1.0 × 0.8 mm) has 12 wire-bond pads converging via thin-film niobium leads into the center of the device.

The overall device (1.0 × 0.8 mm) is shown in Figure 3.164c. Twelve wire-bond pads converge via thin-film niobium leads into the center of the device. These leads run atop the "phonon waveguides" to the Cr/Au resistors. Sensing is accomplished with a superconducting quantum interference device (SQUID) thermometry circuit. Because there are four modes per phonon waveguide and there are four of them, one would indeed expect that $G_{th}/16\ g_0 = 1$, as confirmed by the data in Figure 3.164a.

Summarizing quantized thermal and electronic transport:

1. Metal conductivity at room temperature is mostly independent of impurities or lattice defects. The electron concentration is determined by the nature of the metal and is comparable with the atom density. Mobility of electrons in this regime is dominated by lattice vibrations that are caused by sound or simply random thermal motion (phonons). Lattice vibrations (not atoms) scatter the electrons and determine their mean free path in an electric field, and thus for pure metals the resistivity $\rho \propto T$ (at high temperatures).
2. At very low temperatures and for a very pure metal one would expect the resistance to become zero because all lattice vibrations have died out. In reality, near zero the resistivity goes as T^5 (Bloch-Gruneisen law) because this represents the temperature dependence of electrons scattered by phonons with the number of electrons independent of temperature.
3. The film thickness and lattice defects, such as vacancies, grain boundaries, interstitials, dislocations, and impurities, ensure that the resistance remains nonzero even at the lowest temperatures. The combined effect of thermal vibrations, impurities, and defects on the resistivity is given by Matthiessen's rule, where $\rho_{total} = \rho_0 + \rho(T)$ (see expanded Equation 3.23). In this expression ρ_0 is a constant that increases with impurity content, and $\rho(T)$ is the phonon contribution to the resistivity. For pure elements the contribution of defects is on the order of 0.1% of the total, but for heavily cold worked metals it can be as high as 5%.
4. At high temperature the resistivity is limited by lattice thermal motion and at low temperature by lattice defects. The mobility in Equation 3.9 may be extracted from Hall measurements (see Equation 3.65), and when measured at low temperatures it serves as a measure of a material's purity. A typical mobility in single-crystal Si at 300 K is about 1000 cm^2/Vs. In the case of superconductivity (see Chapter 8, Volume III under "Superconductivity") the resistivity disappears above 0 K and below the critical superconducting phase change temperature T_c.

Beyond Schrödinger's Equation

The Other Models

In the quantum mechanics sections above, we relied exclusively on Schrödinger's equation. But by 1926 Schrödinger himself came to accept the serious limitations of his own theory, particularly its lack of particulate behavior. In 1925 Werner Heisenberg came up with a new approach. Instead of treating the atom as a miniature planetary system, he saw it as a virtual oscillator (remember Planck?), and he set out to find a description that connected the Bohr-Sommerfeld quantum states with the frequencies and intensities of their bright line spectra. He characterized the electron in terms of its linear momentum, **p**, and its displacement from equilibrium, **q**, and attempted to calculate its energy in a quantum state *n*. Within a few months he had succeeded—with some surprising consequences. In classical theory, the simple act of multiplication is commutative—it does not matter which way round the multiplication is done, so: 2 × 3 = 6 and 3 × 2 = 6. But in Heisenberg's mathematics, multiplication did not follow this rule: **q** × **p** gave a different answer than **p** × **q**. It was soon appreciated that this noncommutability was not really surprising: Heisenberg's new formalism turned out to be a form of matrix algebra developed almost a century before by William Hamilton. In matrix algebra the following matrices:

$$\begin{vmatrix} 3 & 5 & 4 \\ 1 & 1 & 1 \\ 2 & 3 & 5 \end{vmatrix} \times \begin{vmatrix} 1 & 3 & 5 \\ 2 & 5 & 1 \\ 4 & 3 & 2 \end{vmatrix} = \begin{vmatrix} 29 & 46 & 28 \\ 7 & 11 & 8 \\ 28 & 36 & 23 \end{vmatrix}$$

$$\mathrm{p} \times \mathrm{q} \tag{3.346}$$

and

$$\begin{vmatrix} 1 & 3 & 5 \\ 2 & 5 & 1 \\ 4 & 3 & 2 \end{vmatrix} \times \begin{vmatrix} 3 & 5 & 4 \\ 1 & 1 & 1 \\ 2 & 3 & 5 \end{vmatrix} = \begin{vmatrix} 16 & 23 & 32 \\ 13 & 18 & 18 \\ 19 & 29 & 29 \end{vmatrix}$$

$$\mathrm{q} \times \mathrm{p}$$

do not lead to the same result. Heisenberg's matrix mechanics worked perfectly if one assumed the noncommutability could be expressed as:

$$\mathbf{pq} - \mathbf{qp} = \frac{\hbar}{\mathrm{i}} \tag{3.347}$$

where *i* is the square root of −1, an imaginary number. For solid bodies on a large scale, on the other hand, $\mathbf{pq} - \mathbf{qp} = 0$, as if we were dealing with ordinary numbers. But his theory was purely mathematical—there was no longer a physical or "common-sense" picture with which to visualize the atom. The vital noncommutability was a real problem: what did it mean physically? The closest that Heisenberg could come to the physical significance of noncommutability is that the order in which measurements are made during an experiment is critical to the answer obtained. He showed that a consequence of the noncommutability of pairs of quantities such as momentum, **p**, and position, **q**, of a particle is that **p** cannot be known precisely if **q** is known, and **q** cannot be known precisely if **p** is known. Moreover, if the uncertainty in **p** is $\Delta\mathbf{p}$ and the uncertainty in **q** is $\Delta\mathbf{q}$, then $\Delta\mathbf{p} \times \Delta\mathbf{q}$ must always be equal to or greater than $h/2\pi$. This does not come from our inability to perform a careful experiment—it is absolutely fundamental to the quantum world. In other words, we obtain again the Heisenberg uncertainty principle.

In the meantime, Paul Dirac, working alone in Cambridge, had developed a third quantum theory. It is Dirac's theory that is the most comprehensive and far reaching. Dirac was able to demonstrate that both Schrödinger's and Heisenberg's theories, as well as classical physics, were special cases of his own quantum algebra or quantum mechanics. He was able to incorporate Maxwell's equations of electromagnetic radiation into his theory (the first example of a quantum field theory), as well as relativistic effects (i.e., close to the speed of light). In so doing he predicted *ab initio* that the electron should have a spin of 1/2. Dirac's theory removed the paradox of particle-wave duality: it showed that if a particle was probed in a way that was meant to demonstrate its particle-like properties, it would appear to be a particle, and if it was probed in a way that was meant to demonstrate its wave-like properties, it would appear to be a wave. It seems that it is our own inability to conjure up an appropriate or adequate mental picture of photons, atoms, electrons, and other quantum particles that is at the heart of the particle-wave duality paradox.

With his general relativity theory, Einstein reconciled Newton's laws with special relativity. But, as discussed in Chapter 5, the general relativity theory is inherently classical, and this bothered many physicists as it makes for a difficult marriage between quantum physics and general relativity. Physicists first tried to merge special relativity with quantum concepts by focusing on how electromagnetic waves interact with matter, which led to quantum electrodynamics (QED). QED is the most precise theory of natural phenomena ever advanced. A major contributor to this fourth model of quantum mechanics was Richard Feynman, who in 1965 received the Nobel Prize for his work in QED. Feynman, one of the greatest theoretical physicists since Einstein, challenged the basic assumption of probability waves associated with electrons. He instead proposed that when an electron moves from here to there it explores all possible paths, and he showed he could assign a number to each of these paths in such a way that their combined average yields exactly the same result for the probability calculated by using the wave function approach. This is known as the "sum-over-paths" or the "sum-over-histories" approach to quantum mechanics. The mathematical formulation of this has an infinite amount of integrals. In practice a large number will suffice. An excellent resource to get better acquainted with the Feynman approach to quantum mechanics can be found in Taylor et al.[27] and in Feynman's own *QED—The Strange Theory of Light and Electrons*.[28]

By the 1970s there were quantum mechanical descriptions for three of the four known forces: strong, weak, and electromagnetic. Only gravity so far has eluded becoming part of a unified field description. String theory is one of most promising candidates for a unified theory of everything.

Antimatter

Dirac's theory alluded to in the preceding section also predicted the existence of antimatter. His equations have solutions that suggest that electron states with negative energy exist—but if this were the case, why did not all the electrons fall into these negative states? He reasoned that all of these states must already be full—what we consider to be empty space is a sea of negative energy electrons. If we give such an electron enough energy, it should break free of the sea and appear as a normal electron, leaving behind a "hole" in the negative energy sea. To get from $-mc^2$ (negative energy electron) to $+mc^2$ (normal electron), one needs $2mc^2$ (Figure 3.165).

Such a hole should appear as a positively charged electron, i.e., a positron, which will recombine with a normal electron and vanish in a burst of energy equal to $2mc^2$. Carl Anderson (1905–1991), an American physicist, discovered positrons in 1932 in cosmic ray showers and received the Nobel Prize in 1936 for his discovery (Figure 3.166). The discipline of particle physics was born, and soon antiparticles of most of the known real particles were found: the antiproton in 1955, the antineutron in 1960, the antideuteron in 1965, and even antihydrogen atoms were created at the FermiLab in 1995. To create an antiparticle all that is needed is sufficient energy—$2mc^2$—to release a real particle from the negative energy sea, leaving the equivalent "hole"—the antiparticle—behind. A particle can annihilate with its antiparticle to form gamma rays. Conversely, gamma rays with sufficiently high energy can turn into a particle-antiparticle pair.

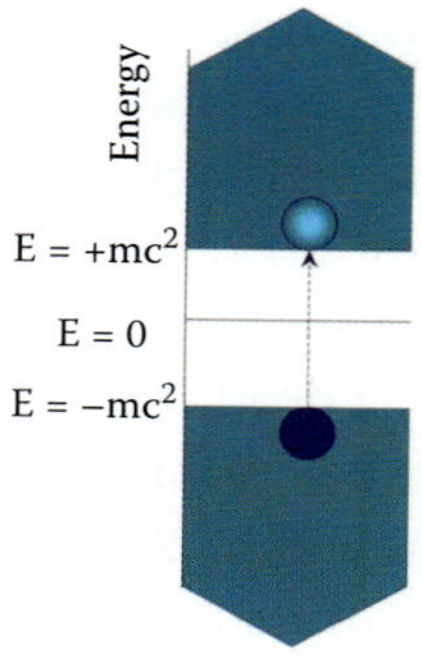

FIGURE 3.165 Positron formation out of a sea of negative energy electrons.

Quantum Reality—Schrödinger's Cat

To understand how dramatic a change of view of reality quantum physics implies, it is good to consider Schrödinger's legendary cat story. A cat is placed in a box, together with a radioactive atom. If the radioactive atom decays and the Geiger counter detects an alpha particle, the hammer hits a flask of prussic acid (HCN), killing the cat. The paradox

FIGURE 3.166 Carl Anderson (1905–1991).

lies in the clever coupling of quantum and classical domains. Before the observer opens the box, the cat's fate is tied to the wave function of the atom, which is itself a superposition of decayed and undecayed states. Thus, said Schrödinger, the cat must itself be in a superposition of dead and alive states before the observer opens the box, "observes" the cat, and "collapses" its wave function. Schrödinger devised the cat experiment to illustrate just how radically the quantum realm differs from the macroscopic, everyday world that we inhabit (Figure 3.167).

He himself had shown that a particle such as an electron exists in a number of possible states, the probability of each of which is incorporated into an equation known as the wave function. In the case of an atom of radioactive material, for example, the atom has a certain probability of decaying over a given period. Based on our "classical" intuition, we would assume that there are only two possibilities: either the atom has decayed, or it has not. According to quantum physics, however, the atom inhabits both states simultaneously. It is only when an observer tries to determine the state of the atom by measuring it that the wave function "collapses," and the atom assumes just one of its possible states: decayed or undecayed.

During the 1980s, the late theorist John Bell suggested a more palatable version of Schrödinger's experiment, one in which the decay of the atom causes a bottle of milk to spill onto the floor; thus, the superposed cat is hungry or full rather than alive or dead. But either version seems weirdly nonsensical: the outcome seems logical from a quantum physics viewpoint, but common sense tells us that a cat cannot be alive and dead (or hungry and full) at the same time.

The paradox of Schrödinger's cat has provoked a great deal of debate among theoretical physicists and philosophers. Although some thinkers have argued that the cat does exist in two superposed states, most contend that superposition only occurs when a quantum system is isolated from the rest of its environment. Various explanations have been advanced to account for this paradox—including the idea that the cat, or simply the animal's physical environment, can act as an observer. The question is, at what point, or scale, do the probabilistic rules of the quantum realm give way to the deterministic laws that govern the macroscopic world? This question has been brought into vivid relief by the work done by a National Institute of Standards and Technology (NIST) team in 1996, which included Christopher Monroe, Dawn Meekhof, Brian King, and Dave Wineland.[29] This group confined a charged beryllium atom in a tiny electromagnetic cage and then cooled it with a laser to its lowest energy state. In this state, the position of the atom and its "spin" (a quantum property that is only metaphorically analogous to spin in the ordinary sense) could be ascertained to within a very high degree of accuracy, limited by Heisenberg's uncertainty principle. This team of physicists managed to create a "Schrödinger cat-like state of matter" in a single atom. By supercooling a beryllium atom with a laser, then prodding it with a rapid sequence of laser pulses, the physicists managed to get the atom to oscillate in such a way that it exists "in the bizarre state of being in two well-separated positions at once."

FIGURE 3.167 Dead and alive.

Mathematically, the property of superposition of states is directly related to the special form of linear differential equation, which governs the behavior of a single particle in quantum theory. When this particle interacts with other particles (as in a macroscopic object) or the detecting device or its environment, the equation becomes nonlinear, implying no more superposition and hence the "collapse" of the particle to a definite state.

Bell's Theorem

Bell's theorem says that reality must be nonlocal; the properties of the measured atom are influenced by events in whose details large portions of the universe participate instantly. In local reality, influences cannot travel faster than light. Bell's theorem says that in any reality of this sort, information does not get around fast enough to explain the quantum facts: reality must be nonlocal. The theorem proves that any model of reality, whether ordinary or contextual, must be connected by influences that do not respect the optical speed limit.

In the summer of 1982, Alain Aspect and his team at the University of Paris-South carried out an experiment to establish the reality below the unreal world of the quantum, and their results dealt a fatal blow to Einstein's "hidden variables" concept. The hidden variable idea was espoused by some physicists like Einstein, who argued that the statistical nature of quantum mechanics suggests that quantum mechanics is "incomplete." There must be some other variables at play. This group of scientists maintained that physical theories must be deterministic to be complete. Bell's theorem proves that local hidden variables are impossible. The experiment concerned the behavior of photons flying off in opposite directions from a source and a detector capable of measuring the polarization of these photons. Because the two photons are emitted together, their polarizations are correlated. According to quantum theory, this polarization does not exist until it is measured, but according to the neorealists, each photon has a "real" polarization from the moment it is created. The results demonstrated that the correlation between the photons is the kind predicted by quantum theory, and, again as predicted by quantum theory, the measurement that is made on one photon has an instantaneous effect on the nature of the other photon. Some interaction links the two photons, although these photons speed away at the speed of light, and relativity theory tells us that no signal can travel faster than light.

Acknowledgments

Special thanks to Nahui Kim (UCI), Arvind Ajoy (IIT Madras), G.S. Jayadeva (IIT Madras), and Dr. Jia Guanyao (UC Irvine).

Questions

3.1: What is the physical meaning of a wave function? In what limit and in what sense does a wave function approach the classical limit?

3.2: What are allowed and forbidden regions in classical mechanics? What is the general behavior of a wave function in quantum mechanical allowed and forbidden regions? What is a turning point?

3.3: Why is the conductivity much larger in quantum theory than in classical theory?

3.4: What is the physical meaning of distribution functions [Maxwell-Boltzmann (classical), Fermi-Dirac, and Bose-Einstein]? How does one calculate the density of states (DOS) in a quantum distribution? How does one relate the Fermi energy to the density of particles?

3.5: A 1-W laser ($\lambda = 530$ nm) shines on a nickel surface that has a reflectivity of 96%. (a) How many photons per second does the laser emit? (b) Do you expect to observe the photoelectric effect? If so, what is the maximum energy of the photoelectrons? (The work function of nickel is 1.9 eV, Planck constant $h = 6.6 \times 10^{-34}$ Js.)

3.6: What is the maximum wavelength of light that can excite hydrogen atoms in the ground state ($n = 1$)? What is the color of hydrogen gas? ($R_H = 1.097 \times 10^7 m^{-1}$)

3.7: (a) What are the de Broglie frequencies and wavelengths of an electron and a proton accelerated to 100 eV? (b) In order to confine an electron to a nucleus, its de Broglie wavelength must be less than 10^{-14} m. What is its kinetic energy in that case? (Planck constant $h = 6.6 \times 10^{-34}$ Js, electron mass 9.1×10^{-31} kg, proton mass 1.7×10^{-27} kg)

3.8: An electron with total energy 2 eV impinges on a finite square potential barrier of height 1 eV. What is the velocity of the electron away from and in the barrier region? (electron mass 9.1×10^{-31} kg)

3.9: A particle is described by the wave function

$$\psi(x) = A\cos\left(\frac{2\pi x}{L}\right) \text{ for } -L/4 \le x \le L/4$$

and $\psi(x) = 0$ otherwise

(a) Determine the normalization constant A.
(b) What is the probability that the particle will be found between $x = 0$ and $x = L/8$?

3.10: Consider a particle in an infinitely deep one-dimensional potential well, where $V = 0$ for $|x| < L$. The wave function is of the form $\psi = A \sin kx + B \cos kx$.

(a) Apply the boundary conditions on the wave function to deduce that $k = n\pi/(2L)$, $n = 1,2,3,\ldots$, with $A = 0$ for odd n's, $B = 0$ for even n's.

(b) Consider the same problem with a shift of origin, so that the well now runs from $x = 0$ to $x = 2L$. What is the form of the wave function in this case?

(c) The probability density for finding the particle at a given value of x is $\propto |\psi|^2$. Use this fact to normalize the wave function; i.e., find the constant of proportionality A.

(d) The time-independent Schrödinger (TISE) equation may be written as $(p^2/2m + V)\psi = E\psi$, where $p = -i\hbar\partial/\partial x$ in 1D. Show that a particle in state n has an energy $E_n = \hbar^2\pi^2 n^2/(8mL^2)$.

3.11: What is the de Broglie wavelength of a small car moving at highway speeds (60 mph ≈ 100 km/h)? Can the wave nature of the car be detected?

3.12: The wave function e^{-ikx} is the solution of the time-independent Schrödinger (TISE) equation for a free particle [$V(x) = 0$]. What is the momentum and energy of this particle?

3.13: Through how many volts does a proton have to be accelerated for if it is to reach a million miles per hour? $\Delta V = ?$ The charge of a proton = 1.6×10^{-19} C and its mass $m = 1.67 \times 10^{-27}$ kg; $v_i = 0$ m/s; $v_f = 1 \times 10^6$ mph $= 4.46 \times 10^5$ m/s.

3.14: Do the electrons in atoms and molecules obey Newton's classical laws of motion?

3.15: Suppose we have a solution, $\psi(x)$, of the time-independent Schrödinger equation, with energy, E. How do we construct a solution of the time-dependent Schrödinger equation?

3.16: The wave function e^{-ikx} is the solution of the time-independent Schrödinger equation for a free particle ($V(x) = 0$). What is the momentum and energy of this particle and in which direction does it move?

3.17: What is the boundary condition on the wave function at a point where the potential makes a jump (a) but remains finite and (b) goes to infinity?

3.18: Why is the infinite potential well (particle in a box) an important problem in quantum mechanics?

3.19: What is degeneracy in quantum mechanics?

3.20: What is the degeneracy of the first excited states of the three-dimensional cubic infinite potential well?

3.21: Why is the problem of the harmonic oscillator important in quantum mechanics?

3.22: What are transmission and reflection coefficients of particles and how are they related?

3.23: What are the important steps in solving the Schrödinger equation for a particle penetrating a potential barrier?

3.24: What is tunnelling?

3.25: What is the constraint that leads to the quantization of the principal quantum number, n?

3.26: What is spin in quantum mechanics?

3.27: What is the Pauli exclusion principle?

3.28: *Photoelectric effect*: Electromagnetic radiation with a frequency of $1.25 \cdot 10^{15}$ Hz strikes a piece of pure silver. Calculate the maximum speed of the electrons that are emitted. The work function of silver is 4.7 eV. What will be the change in speed be if the intensity of radiation is doubled?
Thanks to Arvind Ajoy, Indian Institute of Technology Madras.

3.29: *Energy levels in a 1D quantum well*: Assume an infinite electronic 1D quantum well of width $L_1 = 100$ nm. Calculate the ground state energy and the energy gap between the first excited state and the ground state. Repeat the calculations for $L_2 = 500$ nm. Comment on the result.
Thanks to Arvind Ajoy, Indian Institute of Technology Madras.

3.30: *Degeneracy in a 2D quantum well:* Consider a symmetric 2D quantum well, such that $L_x = L_y = 100$ nm. Calculate the energy levels of the first two states $\{n_x, n_y\}$. Compare the results with calculations of an asymmetric well,

$L_x = 100$ nm, $L_y = 110$ nm. Thanks to Arvind Ajoy, Indian Institute of Technology Madras.

3.31: *Ionization energy of dopants using the Bohr model*: Consider an *n*-type silicon crystal, doped with phosphorus atoms. Each dopant atom replaces a silicon atom from its lattice site. It hence donates one extra electron to the crystal lattice. Calculate the ionization energy of this extra electron, assuming the other 4 valence electrons, the inner electrons, and the nucleus of the *P* atom as a core of net charge +1. Assume the effective mass of electrons in silicon as $m_e^* = 0.2\ m_0$ and a relative permittivity $\varepsilon_r = 12$.

Thanks to Arvind Ajoy, Indian Institute of Technology Madras

3.32: (a) Plot current density versus voltage for a silver metal wire of length 0.01 m for a voltage varying from 0.1 to 1 V. The resistivity of silver is $\rho = 1.59\ \mu\Omega$-cm at T = 300 K.

(b) Repeat part (a) by assuming Drude's model to find the resistivity. Given a thermal velocity of the electrons of $v_{th} =$ 0 $1.57 \cdot 10^5$ m/s, what is the percentage of error in current density at 1 V when compared with part (a)?

(c) Repeat part (a) for silicon assuming $\rho = 3\ \Omega$-m (the resistivity of the semiconductor varies between 0.1 and 60 3 Ω-m depending on the doping level). Comment on current densities of metal and semiconductor.

Thanks to G.S. Jayadeva, Indian Institute of Technology Madras

3.33: (a) Find the energy levels for an electron in a quantum dot made from CdSe with an effective mass $m_e^* = 0.13\ m_0$ and having a de Broglie wavelength $1.23 \cdot 10^{-10}$ m, for $n_1 = 1$, $n_2 = 1, 2$, and $n_3 = 1, 2, 3$. Assume $L_1 = L_2 = L_3 = 1 \cdot 10^{-10}$ m.

(b) Do quantum effects occur in this quantum dot?

(c) What is the kinetic energy of carriers in the quantum dot?

Thanks to G.S. Jayadeva, Indian Institute of Technology Madras

3.34: Consider a 1D bound particle. Show that:

$$\frac{d}{dt}\int_{-\infty}^{\infty} \psi^*(x,t)\psi(x,t)dx = 0$$

Thanks to G.S. Jayadeva, Indian Institute of Technology Madras

3.35: An electron is confined in the ground state in a 1D box of width 10^{-11} m. Its energy is 36 eV. Calculate:

(a) The energy of the electron in the first excited state and the associated wave number and plot the wave function.

(b) The average force on the walls of the box when the electron is in the ground state.

Thanks to G.S. Jayadeva, Indian Institute of Technology Madras

3.36: What is the low temperature limit of the Fermi-Dirac distribution?

3.37: What is the value of the Fermi-Dirac distribution at $E = E_F$?

3.38: How does the number of particle-hole pairs increase with temperature?

3.39: What physical picture was the Drude theory of electrical conduction based on and what correct and incorrect results did it lead to?

3.40: Obtain the plot for Fermi-Dirac function when $\mu = E_F = 0.9$ eV.

Thanks G.S. Jayadeva, Indian Institue of Technology Madras

3.41: What distinguishes fermions and bosons?

Further Reading

Andrade e Silva, J., and G. Lochak. 1969. *Quanta*. London: World University Library.

Ashcroft, N. W., and N. D. Mermin. 1976. *Solid state physics*. Philadelphia: Saunders College.

Atkins, P. W., and J. de Paula. 2001. *Atkins' physical chemistry*, 7th ed. Oxford, UK: Oxford University Press.

Gribbin, J. 1984. *In search of Schrodinger's cat*. New York: Bantam Books.

Griffiths, D. J. 2005. *Introduction to quantum mechanics*, 2nd ed. Upper Saddle River, NJ: Pearson Prentice Hall.

Han, M. Y. 1990. *The secret life of quanta*. Blue Ridge Summit, PA: TAB Books.

Hayward, D. O. 2002. *Quantum mechanics for chemists* (RSC Tutorial Chemistry Texts 14). Royal Society of Chemistry.

Hey, T., and P. Walters. 2003. *The new quantum universe*. New York: Cambridge University Press.

Kittel, C. 2005. *Introduction to solid state physics*, 8th ed. New York: John Wiley.

Marshall, I., and D. Zohar. 1997. *Who's afraid of Schrodinger's cat?* New York: Quill William Morrow.

McCarthy, W. 2003. *Hacking matter*. New York: Basic Books.

Moore, W. J. 1967. *Seven solid states*. New York: W. A. Benjamin, Inc.

Sakurai, J. J. 1985. *Modern quantum mechanics*. Redwood City, CA: Addison-Wesley Publishing Company, Inc.

Straus, H. L. 1968. *Quantum mechanics: an introduction*. Englewood Cliffs, NJ: Prentice-Hall.

Turton, R. 1995. *The quantum dot: a journey into the future of microelectronics*. 1995. New York: Oxford University Press.

Young, H. D., and R. A. Freeman, contributing author A. L. Ford. 2004. *Sears and Zemansky's university physics*, vol. 3, 11th ed. San Francisco: Pearson Addison Wesley.

References

1. Hall, E. H. 1879. On a new action of the magnet on electric currents. *Am J Math* 2:287–92.
2. Arndt, M., O. Nairz, J. Voss-Andreae, C. Keller, G. V. D. Zouw, and A. Zeilinger. 1999. Wave-particle duality of C60. *Nature* 401:680–82.
3. Herbert, N. 1987. *Quantum reality: beyond the new physics*. New York: Anchor Books, a Division of Random House, Inc.
4. Harris, R. A. 2006. *Influence of the shape and size of a quantum structure on its energy levels*. PhD diss., University of the Free State, Bloemfontein, South Africa.
5. Wagner, R. S., and W. C. Ells. 1964. Vapour-liquid-solid mechanism of single crystal growth. *Appl Phys Lett* 4:89–90.
6. Ekimov, A. I., A. L. Efros, and A. A. Onushchenko. 1985. Quantum size effect in semiconductor microcrystals. *Solid State Commun* 56:921–24.
7. Efros, A. L., and A. L. Efros. 1982. Interband absorption of light in a semiconductor sphere. *Sov Phys Semicond* 16:772–74.
8. Brus, L. E. 1984. On the development of bulk optical properties in small semiconducting crystallites. *J Luminescence* 31-32:381.
9. Brus, L. E. 1984. Electron-electron and electron-hole interactions in small semiconductor crystallites: the size dependence of the lowest excited electronic state. *J Chem Phys* 80:4403–07.
10. Murray, C. B., C. R. Kagan, and M. G. Bawendi. 1995. Self-organization of CdSe nanocrystallites into three-dimensional qantum dot superlattices. *Science* 270:1335–38.
11. Murray, C. B., D. J. Norris, and M. G. Bawendi. 1993. Synthesis and characterization of nearly monodisperse CdE (E=S, Se, Te) semiconductor nanocrystallites. *J Am Chem Soc* 115:8706–15.
12. Dabboussi, B. O., J. Rodriguez-Viejo, F. V. Mikulec, J. R. Heine, H. Mattoussi, R. Ober, K. F. Jensen, and M. G. Bawendi. 1997. (CdSe)ZnS core/shell quantum dots: synthesis and characterization of a size series of highly luminescent nanocrystallites. *J Phys Chem B* 101:9463–75.
13. Peng, X. G., M. C. Schlamp, A. V. Kadavanich, and A. P. Alivisatos. 1997. Epitaxial growth of highly luminescent CdSe/CdS core/shell nanocrystals with photostability and electronic accessibility. *J Am Chem Soc* 119:7019–29.
14. Hines, M. A., and P. Guyot-Sionnest. 1996. Synthesis of strongly luminescing ZnS-capped CdSe nanocrystals. *J Phys Chem B* 100:468–71.
15. Yu, W. W., E. Chang, R. Drezek, and V. L. Colvin. 2006. Mini review: water soluble quantum dots for biomedical applications. *ScienceDirect* 348:781–86.
16. Bruchez Jr., M., M. Moronne, P. Gin, S. Weiss, and A. P. Alivisatos. 1998. Semiconductor nanocrystals as fluorescent biological labels. *Science* 281:2013–16.
17. Gerion, D., F. Pinaud, S. C. Williams, W. J. Parak, D. Zanchet, S. Weiss, and A. P. Alivisatos. 2001. Synthesis and properties of biocompatible water-soluble silica-coated CdSe/ZnS semiconductor quantum dots. *J Phys Chem B* 105:8861–71.
18. Gur, L., N. A. Fromer, M. L. Geier, and A. P. Alivisatos. 2005. Air-stable all-inorganic nanocrystal solar cells processed from solution. *Science* 310:462–65.
19. O'Regan, B., and M. Gratzel. 1991. A low-cost, high efficiency solar cell based on dye-sensitized colloidal TiO2 films. *Nature* 353:737–40.
20. Wolf, E. L. 1985. *Principles of electron tunneling spectroscopy*. New York: Oxford University Press.
21. Esaki, L., and R. Tsu. 1970. *IBM J Res Develop* 14:61.
22. Dekker, A. J. 1962. *Solid state physics*. Englewood Cliffs, NJ: Prentice-Hall.
23. Pillai, S. O. 2005. *Solid state physics*. New Delhi: New Age International Publisher.
24. Dekker, C. 1999. Carbon nanotubes as molecular wires. *Physics Today* 52:22–28.
25. Li, D., Y. Wu, P. Kim, and A. Majundar. 2003. Thermal conductivity of individual silicon nanowires. *Appl Phys Lett* 83:2934–36.
26. Schwab, K., E. A. Henriksen, J. M. Worlock, and M. L. Roukes. 2000. Measurement of the quantum of thermal conductance. *Nature* 404:974–77.
27. Taylor, E. F., S. Vokos, J. O'Meara, and N. S. Thornber. 1997. Teaching Feynman's sum-over-paths quantum theory. *Comp Physics* 12:190–99.
28. Feynman, R. P. 1985. *The strange theory of light and matter*. Princeton, NJ: Princeton Science Library.
29. Monroe, C., D. M. Meekhof, B. E. King, and D. J. Wineland. 1996. A "Schroedinger Cat" superposition of an atom. *Science* 272:1131.

4

Silicon Single Crystal Is Still King

(a)

(b)

(a) A 300-mm Si wafer at Infineon. (b) Silicon boules.

Outline

Introduction

Given the importance of the orientation of silicon single crystal (SSC) for integrated circuit (IC), microelectromechanical system (MEMS), and nanoelectromechanical system (NEMS) applications, we analyze silicon crystallography and band structure in more detail in this chapter and compare them with the GaAs crystallography and band structure. This section will give the reader a better understanding of isotropic and anisotropic etchings of Si, covered in much more detail in Volume II, Chapter 4 on wet chemical etching and wet bulk micromachining—pools as tools. Subsequently, we review the growth of single-crystal Si, the doping of Si, and the oxidation of Si wafers. Although the emphasis in this book is on nonelectronic applications of miniaturized devices, such as sensing and actuating, we do briefly review active electronic devices, such as diodes [including light-emitting diodes (LEDs), photodiodes, solar cells, and Zener diodes] and two types of transistors [bipolar and metal oxide semiconductor field-effect transistor (MOSFET)]. We then analyze in some detail the scaling issues involved in the further miniaturization of Si MOSFETs. In this context, we introduce strained Si based on silicon-germanium (SiGe) layers, which expands the prospect of continued Si use in the IC industry. In closing this chapter, we review the SCS properties that conspired to make Si so important in electronic, optical, and mechanical devices that one might rightly call the second half of the 20th century the Silicon Age.

Si Crystallography

Diamond-Cubic Structure

Crystalline silicon forms a covalently bonded diamond-cubic structure, which has the same atomic arrangement as carbon in the diamond form and belongs to the more general zinc-blende classification (in which the lattice positions are not all necessarily occupied by the same atom). Silicon, with its four covalent bonds, coordinates itself tetrahedrally, and these tetrahedrons make up the diamond-cubic structure (Figure 4.1). There are only four elements that come in diamond lattices: C, Si, Ge, and α-Sn. Diamond (carbon), silicon, and germanium all have a valence of four, and all have the same crystal structure. However, diamond is an insulator, whereas the others are semiconductors as a result of the energy bandgap difference. Diamond's bandgap is 5.5 eV, silicon's is 1.1 eV, and germanium's is 0.67 eV. The large bandgap of diamond makes it an insulator. The unit cell in a diamond lattice is cubic with atoms at each corner and in the middle of each face [i.e., face-centered cubic (FCC); see Chapter 2]. In the interior there are four additional atoms located along the cube diagonals exactly one-quarter of the way down the diagonal. This structure can also be described as two interpenetrating FCC lattices, one displaced (1/4, 1/4, 1/4)* times the lattice parameter *a* with respect to the other, as shown in Figure 4.2. The lattice parameter *a* for silicon is 5.4309 Å, and silicon's diamond-cubic lattice is surprisingly wide open, with a packing density of 34% compared with 74% for a regular FCC lattice. The {111} planes present the highest packing density, and the atoms are oriented such that three bonds are below the plane. In addition to the diamond-cubic structure, silicon is known to have several stable high-pressure crystalline phases and a stress-induced metastable phase with a wurtzite-like structure [wurtzite is a sulfide mineral of zinc and iron with the composition (Zn,Fe)S], referred to as *diamond-hexagonal* silicon. The latter has been observed after ion implantation and hot indentation.

GaAs has a zinc-blende lattice (zinc blende = ZnS; ZnTe and SiC also have that same structure), which is identical to the diamond lattice, except that the lattice sites are equally partitioned between two different atoms (Ga and As). The lattice parameter *a* for GaAs is 5.65 Å.

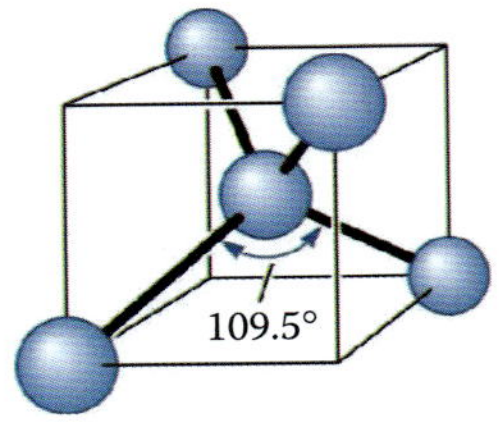

FIGURE 4.1 Covalent bonds are directional. In silicon, a tetrahedral structure is formed, with angles of 109.5° between each covalent bond.

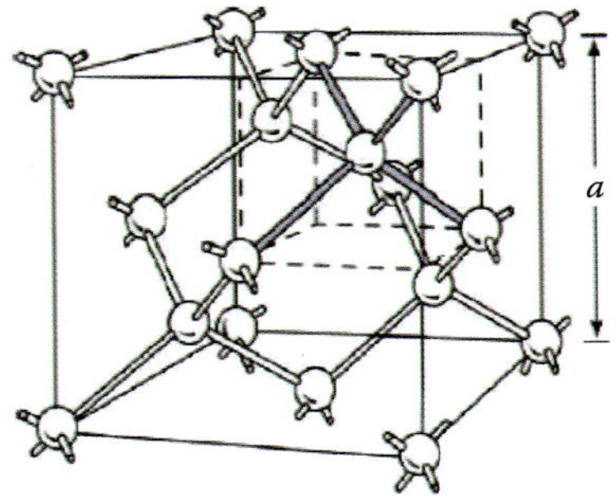

(a)

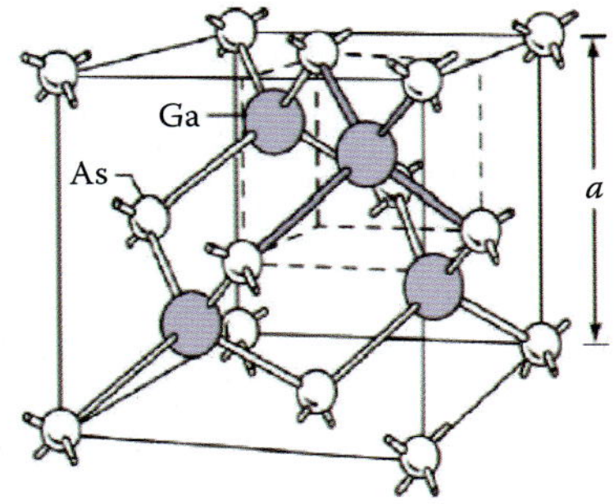

(b)

FIGURE 4.2 The "zinc-blende" lattice is a face-centered cubic (FCC) with two atoms in the base at (0, 0, 0) and (1/4, 1/4, 1/4). (a) The diamond-type lattice of Si can be constructed from two interpenetrating face-centered cubic unit cells. (b) GaAs crystal, which is identical to the diamond lattice, except that the lattice sites are equally partitioned between two different atoms (Ga and As).

* For a refresher on Miller indices to mark crystal planes with (), orientations with [], families of planes with {}, and families of orientations with < >, consult Chapter 2 on crystallography.

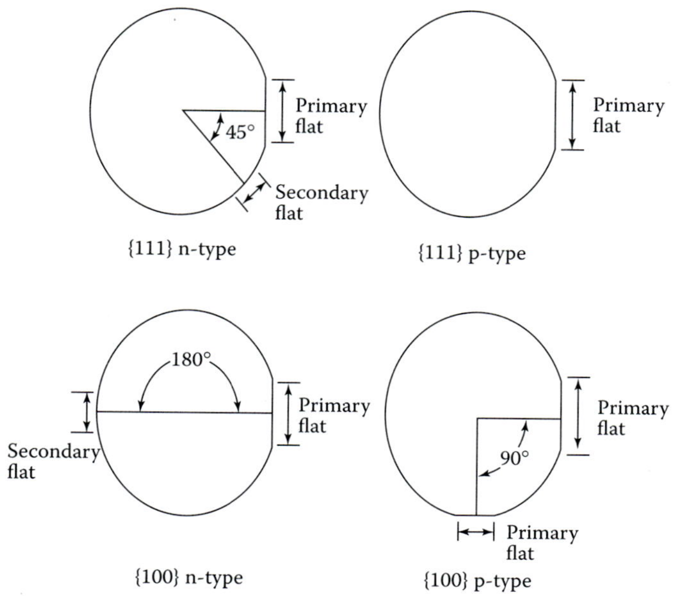

FIGURE 4.3 Primary and secondary wafer flats on silicon wafers.

When ordering silicon wafers, the crystal orientation must be specified. In IC work, [100]-oriented single-crystal Si wafers for metal oxide semiconductor (MOS) transistors and [111] for bipolar transistors are typical, but in MEMS and NEMS, other crystal orientations are used, especially [110]. The [110] wafers break much more cleanly than in other orientations. In fact, it is the only major orientation that can be cleaved with exactly perpendicular edges. The [111] wafers are used less often, as they cannot easily be etched by wet anisotropic etchants except when using special techniques such as laser-assisted etching (see Volume II, Chapter 4).[1] On a [100] wafer, the [110] direction is often made evident by a flat segment, also called an *orientation flat*. The precision on the orientation of the flat is within 3°. The flat's position on [110]-oriented wafers varies from manufacturer to manufacturer but often parallels a [111] direction. Flat areas help determine orientation, placement of slices in cassettes and in fabrication equipment (large primary flat), and help identify orientation and conductivity type (smaller secondary flat) (see also "Single-Crystal Silicon Growth"). Primary and secondary flats on (111) and (100) silicon wafers are indicated in Figure 4.3.

Crystalline silicon substrates are available as circular wafers of 100-mm (4-in) diameter and 525-μm thickness or 150-mm (6-in) diameter with a thickness of 650 μm. Larger 200-mm and 300-mm diameter wafers are currently not economically justified for MEMS or NEMS and are strictly used in the IC industry. Wafers polished on both sides, often used in MEMS, are about 100 μm thinner than substrates of standard thickness.

In Figure 4.4 we show a typical MEMS structure, a piezoresistive pressure sensor, featuring a thin suspended Si membrane, fashioned from bulk single-crystal Si in a wet anisotropic bulk micromachining etch process. Anisotropic wet etching enables the selective thinning of a silicon wafer from a starting thickness of about 425 μm to a typical final thickness of 20 μm or less for the suspended diaphragm. This machining process, which we detail in

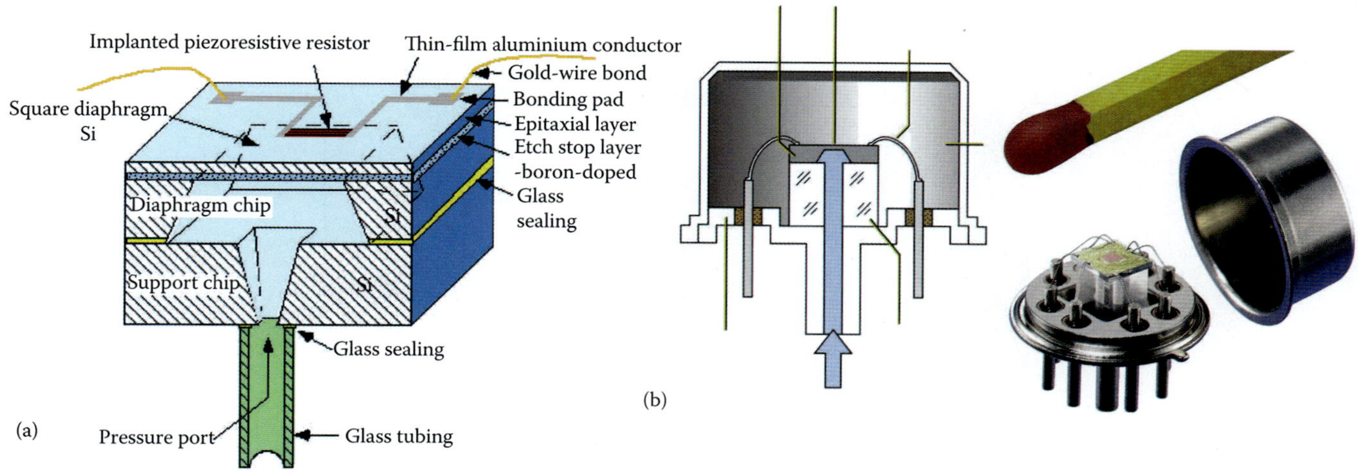

FIGURE 4.4 Piezoresistive pressure sensor. (a) Wet bulk micromachining process was used to craft this thin suspended membrane with implanted piezoresistive elements. (b) Bosch piezoresistive pressure sensor.

Volume II, Chapter 4, features very precise lateral dimensions and tight membrane thickness controls (e.g., in this case, an epitaxial Si layer that comes with a highly boron-doped etch stop layer). The thin silicon membrane suspended in the thick Si frame deforms with pressure, and the piezoresistors implanted in the membrane change resistance with the strain induced by the bending of the membrane. It was this type of Si MEMS structure that demonstrated that batch fabrication of miniature Si components did not need to be limited to integrated circuitry.

Some Important Crystallographic Planes in the Silicon Lattice

Introduction

To appreciate the different three-dimensional (3D) shapes resulting from anisotropically etched single-crystal Si (SCS), as shown in the MEMS pressure sensor in Figure 4.4, some of the more important geometric relationships between different planes within the Si lattice need further clarification. We consider only silicon wafers with a (100) plane or a (110) plane as the top surface. We will accept, for the moment, that in anisotropic alkaline etchants, the {111} family of planes has the highest atom-packing density and is nonetching compared with the other planes. Because the {111} planes are essentially not attacked by the etchant, the sidewalls of an etched pit in SCS will ultimately be bounded by this type of plane, given that the etch time is long enough for features bounded by other planes to be etched away. The types of planes introduced initially depend on the geometry and the orientation of the mask features.

In the sections below, we show how simple vector algebra (using Equation 2.14) enables one to calculate the angles between different planes or directions and thus makes it possible to predict the shapes resulting from anisotropic wet etching of single-crystal Si.

The reasons for this etching anisotropy will be revealed in Volume II, Chapter 4.

[100]-Oriented Silicon

The direction of a set of planes, we saw in Chapter 2, is the normal to that set of planes, so Si wafers with a (100) plane as the bounding surface are [100]-oriented wafers. In Figure 4.5, the unit cell of a silicon lattice is shown together with the correct orientation of a [100]-type wafer relative to this cell.[2] It can be seen that intersections of the nonetching {111} family of planes with the {100} family of planes (e.g., the wafer surface) are mutually perpendicular and lying along the <110> family of directions. Provided that a mask opening (perhaps a rectangle or a square) is accurately aligned with the primary orientation flat (the [110] direction), only {111} planes will be introduced as sidewalls from the very beginning of the

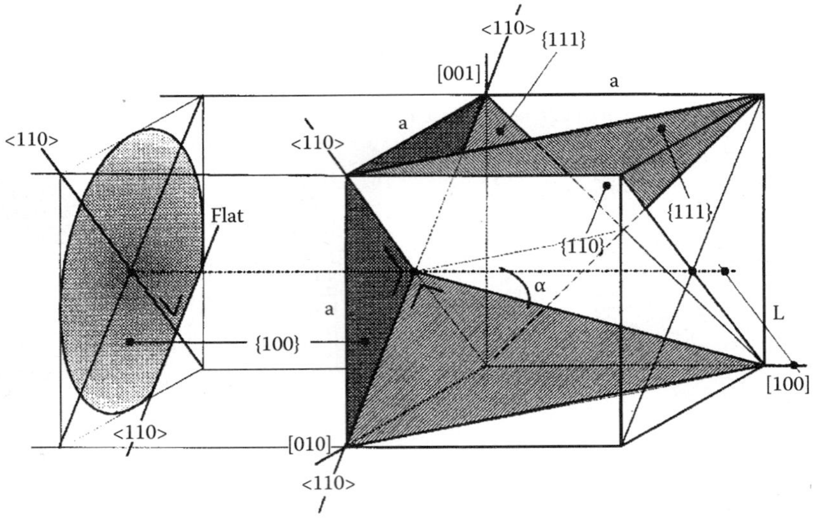

FIGURE 4.5 A (100) silicon wafer with reference to the unity cube and its relevant planes. (From Peeters, E. 1994. *Process development for 3D silicon microstructures, with application to mechanical sensor design.* PhD diss., Katholieke Universiteit Leuven, Leuven, Belgium. With permission.[2])

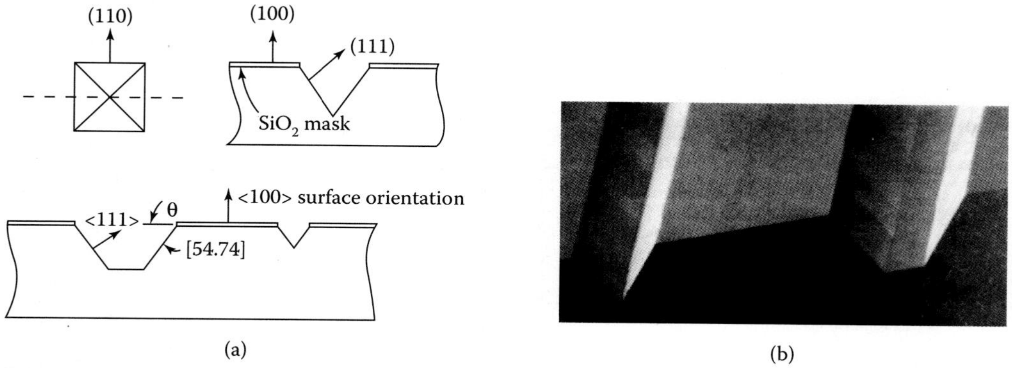

FIGURE 4.6 Anisotropically etched features in a (100) wafer with square mask (schematic) (a) and rectangular mask (scanning electron microscope micrograph of resulting V- and U-grooves) (b).

etch. Because the nonetching character of the {111} planes renders an exceptional degree of predictability to the recess features, this mask arrangement is most often used in commercial applications.

During etching, truncated pyramids (square mask) or truncated V-grooves (rectangular mask) deepen but do not widen (Figure 4.6). The edges in these structures are <110> directions, the ribs are <211> directions, the sidewalls are {111} planes, and the bottom is a (100) plane parallel with the wafer surface. After prolonged etching, the {111} family of planes is exposed down to their common intersection, and the (100) bottom plane disappears, creating a pyramidal pit (square mask) or a V-groove (rectangular mask) (Figure 4.6). As shown in Figure 4.6, no underetching of the etch mask is observed because of the perfect alignment of the concave oxide mask opening with the [110] direction. Misalignment still results in pyramidal pits, but the mask will be undercut. For a mask opening with arbitrary geometry and orientation, a circle, for example, and for sufficiently long etch times, the anisotropically etched recess in a (100) wafer is pyramidal, with a base perfectly circumscribing the circular mask opening.[2] Convex corners (>180°) in a mask opening will always be completely undercut by the etchant after sufficiently long etch times. This can be disadvantageous (e.g., when attempting to create a mesa rather than a pit), or it can be advantageous for undercutting suspended cantilevers or bridges. In the section on corner compensation in Volume II, Chapter 4, the issue of undercutting will be addressed in more detail. In corner compensation, the convex corner undercutting is compensated by clever mask layout schemes.

The slope of the {111} sidewalls in a cross-section perpendicular to the wafer surface (100) and parallel to the wafer flat is marked by the angle $\theta_{(100),(111)}$ in Figures 4.6a and 4.7. This angle, $\theta_{(100),(111)}$, was calculated in Chapter 2, using Equation 2.14, as:

$$\begin{aligned}\theta_{(100),(111)} &= \cos^{-1}\frac{\left(h_1h_2+k_1k_2+l_1l_2\right)}{\sqrt{h_1^2+k_1^2+l_1^2}\sqrt{h_2^2+k_2^2+l_2^2}}\\ &= \cos^{-1}\frac{\left(1+0+0\right)}{\sqrt{1}\sqrt{3}}\\ &= 54.74^\circ\end{aligned} \tag{2.14}$$

There are six equivalent (100) planes [the {100} family] and eight equivalent (111) planes [the {111} family] for a total of 48 combinations on how they can intersect, but these intersecting angles are only either one of two, and 54.74° is one of them. The same angle of 54.74°, derived from Equation 2.14, can also be calculated from the lattice parameter *a* and the length *L*, both marked in Figure 4.5. The off-normal angle α, i.e., the intersection of a (111) sidewall and a (110) cross-secting plane, is calculated as:

$$\begin{aligned}&\tan\alpha = \frac{L}{a}\\ &\text{with } L = a\frac{\sqrt{2}}{2} \text{ or } \alpha = \arctan\frac{\sqrt{2}}{2} = 35.26^\circ\\ &\text{and } 54.74^\circ \text{ for the complementary angle}\end{aligned} \tag{4.1}$$

The tolerance on this angle is determined by the alignment accuracy of the wafer surface with respect to the (100) plane. Wafer manufacturers typically specify this misalignment to 1° (0.5° in the best cases).

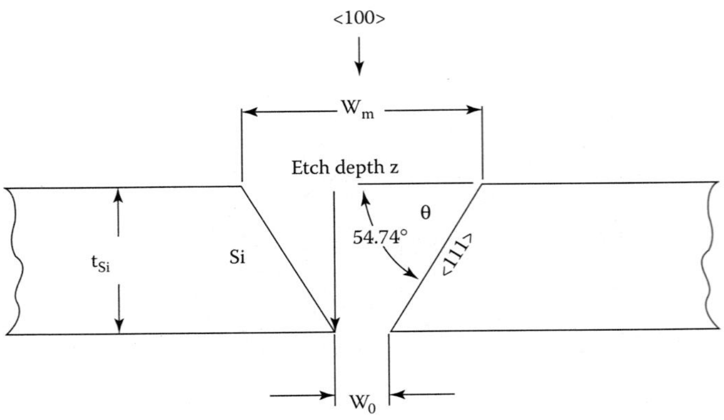

FIGURE 4.7 Relation of bottom cavity plane width W_0 with mask opening width W_m.

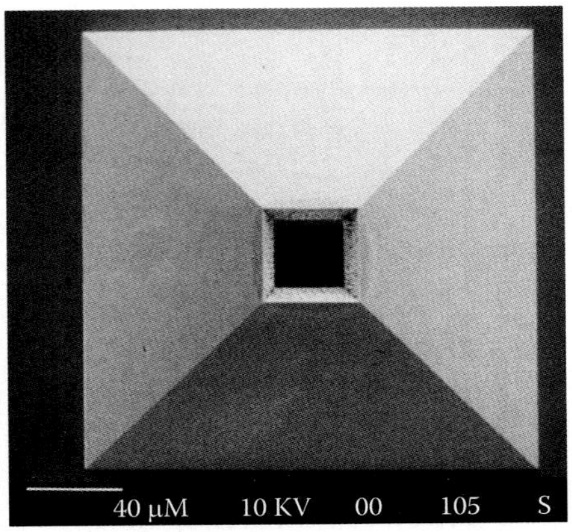

FIGURE 4.8 Orifice or via through a Si wafer made by anisotropic etching.

The 8 equivalent (111) planes [the {111} family] and 12 equivalent (110) planes [the {110} family] form 96 combinations, but they can intersect each other only at 35.26°, 90°, or 144.74°.

The width of the rectangular or square cavity bottom plane, W_0, in Figure 4.7, aligned with the <110> directions, is completely defined by the etch depth, z, the mask opening, W_m, and the above-calculated sidewall slope:

$$
\begin{aligned}
W_0 &= W_m - 2\,z \cot 54.74° \\
&= W_m - \sqrt{2}\,z
\end{aligned}
\tag{4.2}
$$

The larger the opening in the mask, the deeper the point at which the {111} sidewalls of the pit intersect. The etch stop at the {111} sidewalls' intersection occurs when the depth is about 0.7 times the mask opening. If the oxide opening is wide enough, $W_m > 849$ μm (for a typical 6 in wafer with thickness $t_{si} = z = 600$ μm), the {111} planes do not intersect within the wafer. The etched pit in this particular case extends all the way through the wafer, creating a small orifice or via (see Figure 4.8). If a high density of such vias through the Si is required, the wafer must be made very thin; otherwise, too much Si real estate is wasted.

Corners in an anisotropically etched recess are defined by the intersection of crystallographic planes, and the resulting corner radius is essentially zero. This implies that the size of a silicon diaphragm is very well defined, but it also introduces a considerable stress concentration factor. The influence of the zero corner radius on the yield load of diaphragms can be studied with finite element analysis (FEA).

One way to obtain vertical Si sidewalls instead of 54.7° sidewalls using a [100]-oriented Si wafer may be understood from Figure 4.9. It can be seen that there are {100} planes perpendicular to the wafer surface and that their intersections with the wafer surface are <100> directions. These <100> directions enclose a 45° angle with the wafer flat (i.e., the <110> directions). By aligning the mask opening with these <100> orientations, {100} facets are introduced as sidewalls. The {110} planes etch faster than the {100} planes and are not introduced. Because the bottom and sidewall planes are all from the same {100} family, lateral underetch equals the vertical etch rate, and rectangular channels, bounded by slower etching {100} planes, result (Figure 4.10).

Because the top of the etched channels are exposed to the etchant longer than the bottom, one might expect the channels in Figure 4.10 to be wider at the top than at the bottom. We use Peeters' derivation to show why the sidewalls remain vertical.[2] Assume the width of the mask opening to be W_m. At a given depth, z, into the wafer, the underlying Si is no longer masked by W_m, but rather by the intersection of the previously formed {100} facets with the bottom surface. The width of this new mask is larger than the lithography mask W_m by the amount the latter is being undercut. Let us call the new mask width W_z, the effective mask width at a depth z. The relation between W_m and W_z at any moment is given by the lateral etch rate of a {100} facet and the time that facet was exposed to the etchant at depth z:

$$
W_z = W_m + 2R_{xy}\Delta t_z \tag{4.3}
$$

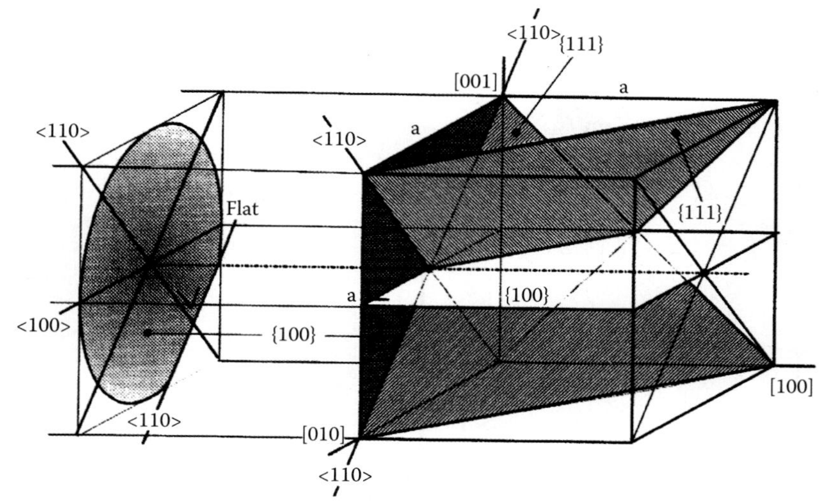

FIGURE 4.9 A (100) silicon wafer with [100] mask-aligned features introduces vertical sidewalls. (From Peeters, E. 1994. *Process development for 3D silicon microstructures, with application to mechanical sensor design.* PhD diss., Katholieke Universiteit Leuven, Leuven, Belgium. With permission.[2])

where R_{xy} = lateral underetch rate (i.e., the etch rate in the *x-y* plane)

Δt_z = etch time at depth z

The underetching, U_{xy}, of the effective mask opening W_z is given by:

$$U_{x,y} = TR_{x,y} - R_{x,y}\Delta t_z \tag{4.4}$$

where *T* is the total etch time so far. The width of the etched pit, W_{tot}, at depth *z* is further given by the sum of W_z and twice the underetching for that depth:

$$W_{tot} = W_z + 2U_{x,y} = W_m + 2TR_{x,y} \tag{4.5}$$

Or, because *T* can also be written as the measured total etch depth *z* divided by the vertical etch rate R_z, Equation 4.5 can be rewritten as:

$$W_{tot} = W_m + 2z\left(\frac{1}{R_z}\right)R_{x,y} \text{ or since } R_{xy} = R_z$$

$$W_{tot} = W_m + 2z \tag{4.6}$$

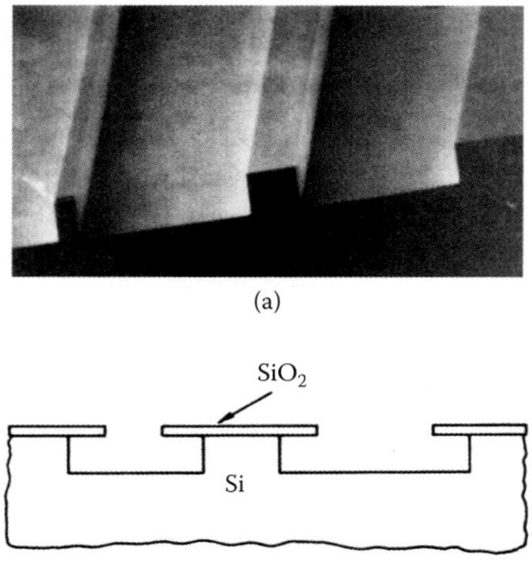

FIGURE 4.10 Vertical sidewalls in a (100) wafer obtained using anisotropic etching. (a) Anisotropically etched features in a (100) wafer with rectangular mask (scanning electron microscope micrograph). (b) Undercutting of the rectangular mask.

Therefore, the width of the etched recess is equal to the photolithographic mask width plus twice the etch depth. In other words, the walls remain vertical independent of the depth *z* because W_m does not depend on *z*.

For long etch times, {111} facets take over from the vertical {100} facets. These inward sloping {111} facets are first introduced at the corners of a rectangular mask and grow larger at the expense of the vertical sidewalls until the latter ultimately disappear altogether. Therefore, alignment of mask features with the <100> family of directions to obtain vertical sidewalls in [100]-oriented wafers is not very useful for the fabrication of diaphragms. However, it can be very effective for anticipating the undercutting of convex corners on [100] wafers. This

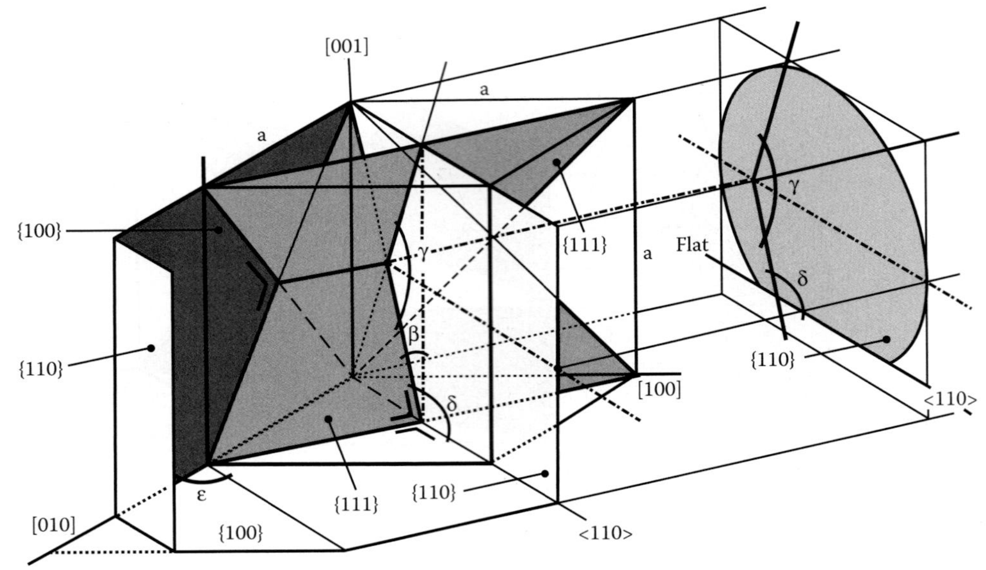

FIGURE 4.11 A (110) silicon wafer with reference to the unity cube and its relevant planes. The wafer flat is in the [110] direction. (From Peeters, E. 1994. *Process development for 3D silicon microstructures, with application to mechanical sensor design.* PhD diss., Katholieke Universiteit Leuven, Leuven, Belgium. With permission.[2])

useful aspect will be revisited when discussing corner compensation.

[110]-Oriented Silicon

Silicon wafers with a (110) plane as the bounding surface are [110]-oriented wafers (see Chapter 2). In Figure 4.11, we show a unit cell of Si properly aligned with the surface of a [110] Si wafer. This drawing will enable us to predict the shape of an anisotropically etched recess on the basis of elementary geometric crystallography. Earlier, we determined that there are 8 equivalent (111) planes [{111} family] and 12 equivalent (110) planes [{110} family] for 96 combinations that can intersect each other at only three different angles: 35.26°, 90°, and 144.74° (calculated using Equation 2.14). Four of the eight equivalent (111) planes are perpendicular to the (110) wafer surface. The remaining four are slanted at 35.26° or 144.74° with respect to the surface. Whereas the intersections of the four vertical (111) planes with the (100) wafer surface are mutually perpendicular, they enclose an angle δ or δ + γ with the wafer flat (the [110] direction). It follows that a mask opening that will not be undercut [i.e., oriented such that resulting feature sidewalls are exclusively made up by vertical and slanted {111} family planes] cannot be a rectangle aligned with the wafer flat, but must be a parallelogram with a large inside angle γ of 109.47° and a small angle of 180° – 109.46° = 70.5° (see Figure 4.12).

The angles δ and γ can also be calculated from the lattice parameter *a* as marked in Figure 4.11. We start by calculating the angle β (Figure 4.11)[2]:

$$\tan\beta = \frac{\frac{1}{2}a\frac{\sqrt{2}}{2}}{\frac{a}{2}} = \frac{\sqrt{2}}{2} \tag{4.7}$$

and from β we can deduce the other pertinent angles:

$$\gamma = 180^\circ\ -2\beta = 180^\circ\ -2\arctan\left(\frac{\sqrt{2}}{2}\right) = 109.47^\circ \tag{4.8}$$

$$\delta = 90^\circ\ -\beta = 90^\circ\ +\arctan\left(\frac{\sqrt{2}}{2}\right) = 125.26^\circ \tag{4.9}$$

$$\varphi = 270^\circ\ -\delta = 144.74^\circ \tag{4.10}$$

The large inside angle of the parallelogram γ is 109.47° (Equation 4.8), and the small angle is thus 70.53°. A groove etched in [110]-oriented wafers has the appearance of a complex polygon delineated by six (111) planes, four vertical and two slanted (Figure 4.13). The bottom of the etch pit shown in

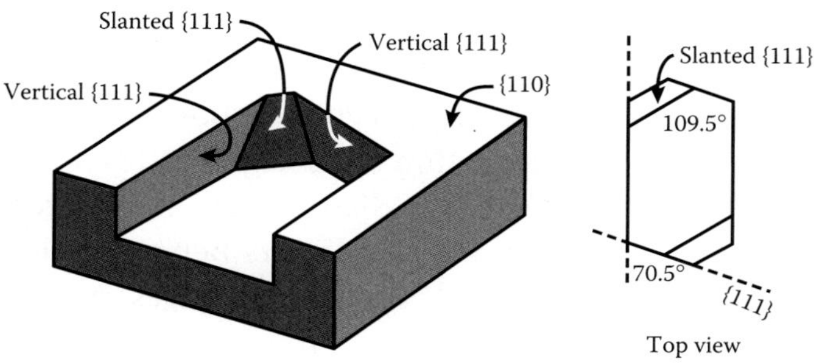

FIGURE 4.12 Illustration of anisotropic etching in [110]-oriented silicon. Etched structures are delineated by four vertical (111) planes and two slanted (111) planes from the {111} family. The vertical {111} planes intersect each other at angles of 109.5° and 70.5°.

Figure 4.13 is bounded at first by planes from the {110} and/or {100} families, depending on the etch time. At short etch times, one mainly sees a flat (110) bottom (Figure 4.12). As the {110} family of planes are etching slightly faster than the {100} family of planes, the flat (110) bottom gets progressively smaller, and a V-shaped bottom bounded by a {100} family of planes results. The angle ε as shown in Figure 4.13 equals 45°, being the angle enclosed by

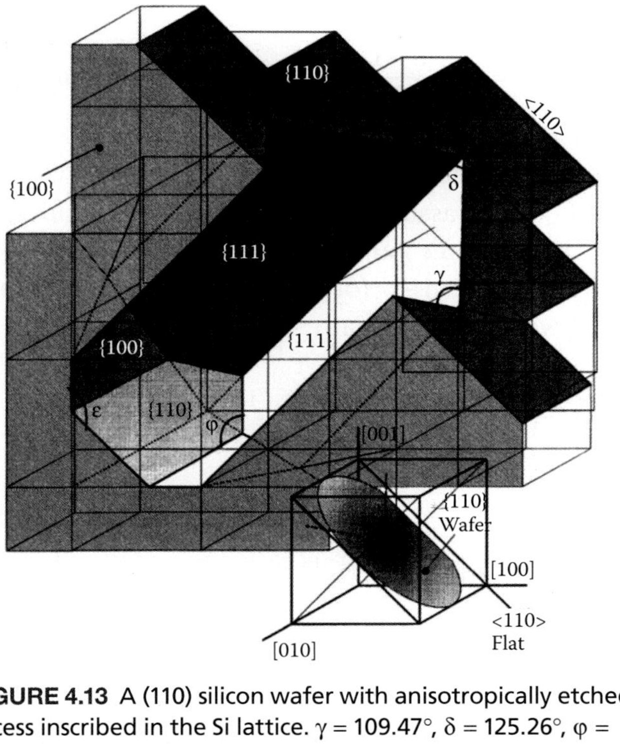

FIGURE 4.13 A (110) silicon wafer with anisotropically etched recess inscribed in the Si lattice. γ = 109.47°, δ = 125.26°, φ = 144.74°, and ε = 45°. (From Peeters, E. 1994. *Process development for 3D silicon microstructures, with application to mechanical sensor design.* PhD diss., Katholieke Universiteit Leuven, Leuven, Belgium. With permission.[2])

the intersections of {100} and {110} bottom planes. Even longer etch times result in shallow {111} planes forming at the bottom, and these eventually stop the etching process.

For [110] etching, an arbitrary window opening is circumscribed by a parallelogram with the previously derived orientation and skewness for sufficiently long etching times. Another difference between [100]- and [110]-oriented silicon wafers is that on the [110] wafers, it is possible to etch under microbridges crossing at a 90° angle a shallow V-groove [formed by (111) planes]. To undercut a bridge on a (100) plane, the bridge cannot be perpendicular to the V-groove; it must be oriented slightly off normal.[3]

Single-Crystal Structure and Conductivity

Intrinsic semiconductors are undoped semiconductors with electrical properties native to the material, whereas extrinsic semiconductors are doped semiconductors with electrical properties controlled by added impurities. At low temperature, pure silicon is an insulator because all of its valence electrons participate in bonding. Pure Si lacks electrons that can wander around freely in the crystal, and as a consequence, at absolute 0 K (T = 0 K), the conductivity (ρ) is very small. Thermal agitation at room temperature (kT ~ 0.025 eV) provides energy to electrons, and there is a finite probability that some of them will be able to break away to become conduction electrons. With intrinsic Si in thermal equilibrium, there is a balance between continuous bond breaking and bond formation. In this process, electrons and holes live fleeting existences with a recombination time on the order of microseconds.

The bandgap for a semiconductor in **k**-space (momentum space; see Chapters 2 and 3) is often represented in a simplified format as shown in Figure 4.14. In semiconductors and insulators, the Fermi level, E_F, falls within the forbidden energy gap.

The density of states (DOS) function $G(E)$ was introduced in Chapter 3 as the measure of volume that is available to hold charge carriers in a crystal. The number of states per energy interval is a complicated function, but the distribution of states near energy minima or maxima is proportional to $\sqrt{E}$.

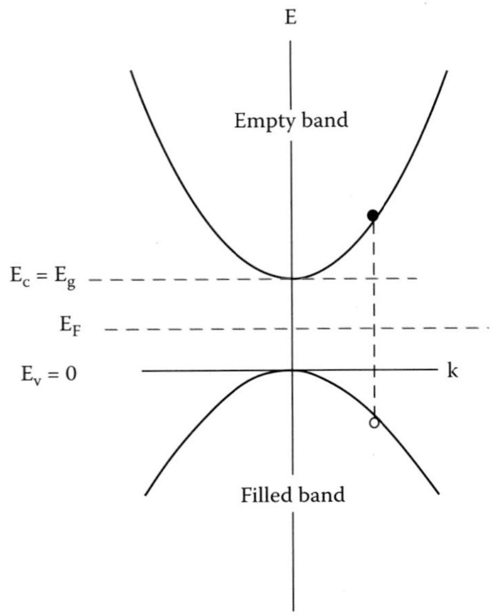

FIGURE 4.14 Energy band diagram in **k**-space for an intrinsic semiconductor. The open dot represents a hole in the valence band, and the filled dot represents an electron in the conduction band.

The DOS per unit volume for the parabolic bands in Figure 4.14 is given as:

$$G(E) = \frac{1}{2\pi^2}\left(\frac{2m_e^*}{\hbar^2}\right)^{\frac{3}{2}} E^{\frac{1}{2}} \tag{3.229}$$

With reference to Figure 4.14, where we have marked the bottom of the conduction band as E_c (= E_g) and the top of the valence band as E_v, we rewrite the DOS available for electrons at the bottom of the conduction band as:

$$G_n(E) = \frac{1}{2\pi^2}\left(\frac{2m_e^*}{\hbar^2}\right)^{\frac{3}{2}} (E - E_c)^{\frac{1}{2}} \tag{4.11}$$

with m_e^* representing the effective electron mass. The DOS available for holes at the top of the valence band is:

$$G_h(E) = \frac{1}{2\pi^2}\left(\frac{2m_h^*}{\hbar^2}\right)^{\frac{3}{2}} (E_v - E)^{\frac{1}{2}} \tag{4.12}$$

with m_h^* representing the effective hole mass. To calculate the carrier density in the states available at the bottom of the conduction band and the top of the valance band (a large 3D model of semiconductor is assumed), we use:

$$n(E, T)dE = G(E)f(E)dE \tag{3.230}$$

where we multiply the DOS with the Fermi-Dirac distribution function. This distribution takes into account that fermions (electrons, holes) cannot occupy the same quantum state. Figure 3.107 illustrates this kind of particle distribution, and we recognize that a sharp interface exists between filled and unfilled states; the function changes rapidly in a small region of a few kT around E_F. Assuming $E - E_F > 3kT$ (= 75 meV), the Fermi-Dirac equation reduces to the Boltzmann expression, and integrating from the bottom of the conduction band E_c to ∞ (or top of the conduction band) gives us the following equilibrium electron concentration $n(E, T)$:

$$n(E,T) = \int_{E_c}^{\infty} G(E)f(E)dE = \frac{1}{2\pi^2}\left(\frac{2m_e^*}{\hbar^2}\right)^{\frac{3}{2}} \times \int_{E_c}^{\infty} (E - E_c)^{1/2} e^{-(E-E_F)/kT} dE \tag{4.13}$$

and

$$n(E,T) = 2\left(\frac{m_e^* kT}{2\pi\hbar^2}\right)^{\frac{3}{2}} e^{-\left(\frac{E_C - E_F}{kT}\right)} = N_c e^{-\left(\frac{E_C - E_F}{kT}\right)} \tag{4.14}$$

where $N_c = 2\left(\frac{m_e^* kT}{2\pi\hbar^2}\right)^{\frac{3}{2}}$ is called the *effective DOS in the conduction band.*

For the concentration of holes, p, in the valence band, again assuming a Boltzmann distribution ($E - E_F > 75$ meV) and integrating $p(E, T)dE = G(E)[1 - f(E)]dE$ from $-\infty$ (or bottom of the valence band) to E_v, we similarly obtain:

$$p(E,T) = 2\left(\frac{m_h^* kT}{2\pi\hbar^2}\right)^{\frac{3}{2}} e^{-\left(\frac{E_F - E_V}{kT}\right)} = N_v e^{-\left(\frac{E_F - E_V}{kT}\right)} \tag{4.15}$$

where $N_v = 2\left(\frac{m_h^* kT}{2\pi\hbar^2}\right)^{\frac{3}{2}}$ is called the *effective DOS in the valence band.*

The values for N_c and N_v are calculated from first principles based on the detailed band structures: for Si, one obtains $N_c = 2.8 \times 10^{19}$ cm^{-3} and $N_v = 1.04 \times 10^{19}$ cm^{-3} (at 300 K). These relations for $n(E, T)$ and $p(E, T)$ are valid for any semiconductor, with or without impurities. There is a one-to-one relation between Fermi energy and local carrier density; if one adds electrons, E_F will increase, and if one removes electrons, E_F will decrease. In pure "intrinsic" semiconductors, the concentration of electron (n) and holes (p) arises only from the balance of electron and hole formation; therefore, $n = p = n_i$, where n_i is the material's intrinsic carrier concentration, and the law of mass action prescribes:

$$n_i^2 = n \times p \qquad (4.16)$$

This equation remains valid when dealing with a nonintrinsic semiconductor (doped) in thermal equilibrium. In an ideal crystal, at equilibrium, the concentration of the intrinsic carriers thus is given as a product of Equations 4.14 and 4.15, or:

$$n_i = \sqrt{N_c N_v} \exp\left(-\frac{E_g}{2kT}\right) \qquad (4.17)$$

a result that depends on the width of the bandgap ($E_g = E_c - E_v$) and temperature but is independent of the Fermi level E_F. The concentration of intrinsic free electrons/holes, n_i, generated this way can be expressed as:

$$n_i = 3.9 \times 10^{16}\, T^{\frac{3}{2}} e^{-0.56\,\text{eV}/kT}\,\text{cm}^{-3} \qquad (4.18)$$

with $n_i = 1.45 \times 10^{10}$ cm^{-3} at 300 K. This translates into a room temperature resistivity for pure Si of 2.3×10^5 Ω cm. To get a feel for this number of intrinsic carriers generated in Si, realize that at room temperature there is only one pair of thermally produced carriers in a volume of Si containing 10,000 billion silicon atoms. The intrinsic carrier density as a function of temperature is shown for intrinsic Si and GaAs in Figure 4.15 (the intrinsic carrier concentrations at room temperature are marked by bullets).*

From Equations 4.14 and 4.15, equating n and p, we derive the following expression for E_F:

$$E_F = \frac{1}{2}E_g + \frac{kT}{2}\ln\left(\frac{m_h^*}{m_e^*}\right)^{\frac{3}{2}} \qquad (4.19)$$

* The difference between calculated intrinsic carrier densities and experimental values may be off by a factor of 2.

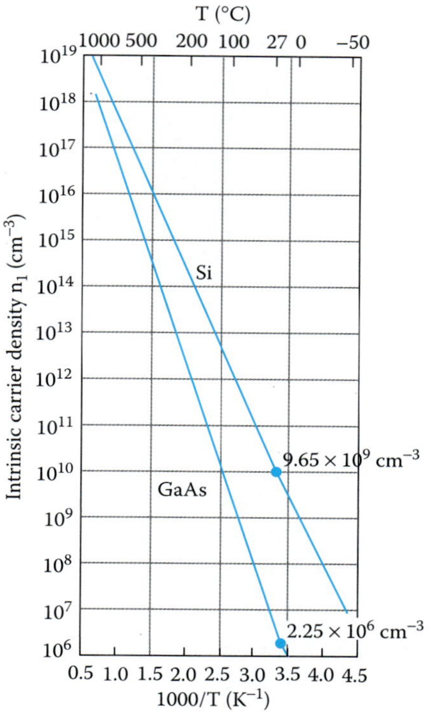

FIGURE 4.15 Intrinsic carrier concentration as a function of temperature for Si and GaAs.

In other words, the chemical potential μ, or Fermi level E_F, has some dependence on T, but if m_h^* and m_e^* are similar, that dependence becomes very small. In the latter case, the Fermi level is situated very nearly in the middle of the bandgap, as sketched in Figure 4.14. Semiconductors and insulators have qualitatively similar band structures, with the quantitative distinction that the bandgap $E_g > 3.0$ eV in insulators.

Single-Crystal Si Growth

Silicon, the second-most abundant element on the Earth after oxygen, is the principal ingredient of beach sand. To make wafers from sand, the silicon must be refined and purified. The refined silicon is melted, and trace amounts of impurities are added for doping the silicon n- or p-type and is then crystallized to form boules. Silicon boules are sliced into wafers and polished.

Silicon crystal growth comprises the primary process toward the construction of ICs and most of the existing micro- and nanomachines. In the Czochralski (CZ) crystal pulling method (Figure

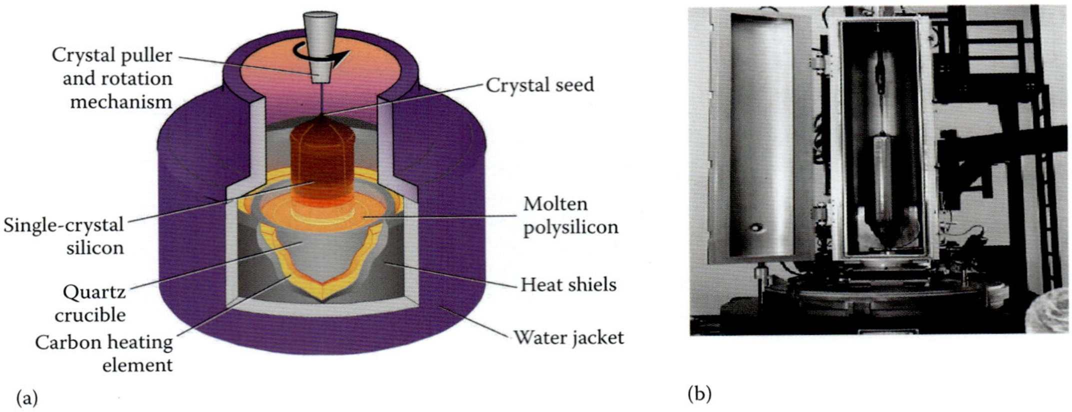

FIGURE 4.16 Schematic (a) and actual picture (b) of a Czochralski crystal puller.

4.16), a silicon crystal seed is grown into a single-crystal Si by pulling it slowly upward, at about 2 to 5 cm/h, from a molten and ultrapure (99.999999999%) polysilicon melt (Si melts at 1414°C). The molten silicon slowly rotates, nested in a silica crucible while pulling the growing crystal. Fused silica reacts with hot silicon and releases oxygen into the melt. Hence, CZ-Si has an indigenous oxygen concentration of approximately 10^{18} atoms/cm^3. The melt must not be stirred; otherwise, more oxygen is transported from the SiO_2-Si (crucible/liquid) interface to the Si (liquid)/Si (solid) interface.

Float zone crystal growth (Figure 4.17) is used to avoid all contact with crucibles to eliminate impurities such as oxygen. This method requires a slow rotating polycrystalline silicon rod, locally melted and recrystallized by a scanning RF heating coil, resulting in very pure silicon. The latter type of silicon is better suited for devices requiring a low doping level to produce, for example, low leakage diodes as required for detectors and power devices. In the float zone process, the regrowing silicon crystal rejects dopants and other impurities. Impurities tend to stay in the liquid, and refining can be accomplished with multiple melting/recrystallization passes. The latter process is called zone refining. The best purity achieved by chemical means is typically one impurity in every 1 million atoms. With zone refining this can be reduced to less than one impurity in every 10 billion atoms.

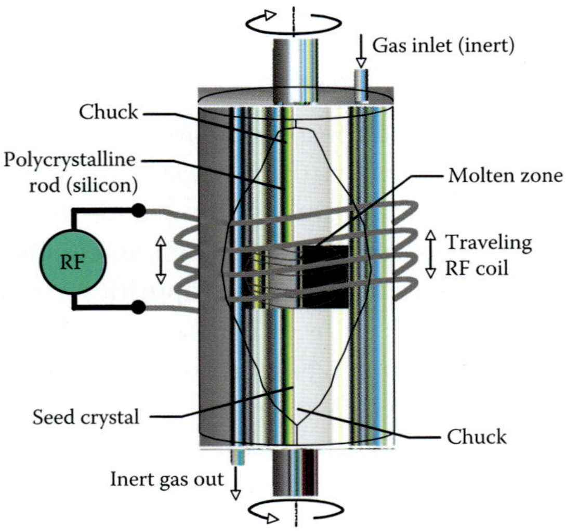

FIGURE 4.17 Float zone crystal growth.

Modern silicon boules can reach a diameter of more than 300 mm and a length of 1–2 m. They are first characterized for resistivity and crystal perfection. Resistivity is measured with a four-point probe (see Volume III, Chapter 6 on metrology), whereas x-ray or electron beam diffraction helps evaluate the crystal perfection (Chapter 2). To prepare wafers from the grown Si boule, both ends (the seed and the tail) are cut off, and the diameter of the boule is trimmed by grinding on a lathe. Based on the x-ray crystal orientation, one or more flats are grinded along the length of the ingot (Figure 4.18). The accuracy of this process is ±0.5° (in the very best case). Wafer flats (illustrated in Figure 4.3) help orientation determination and placement of slices in cassettes and fabrication equipment (large primary flat) and help identify orientation and conductivity type (smaller secondary flat). On a {100} wafer, the largest flat is oriented in the <110> direction. The drawback of these flats is that the usable area on the wafer is significantly decreased.

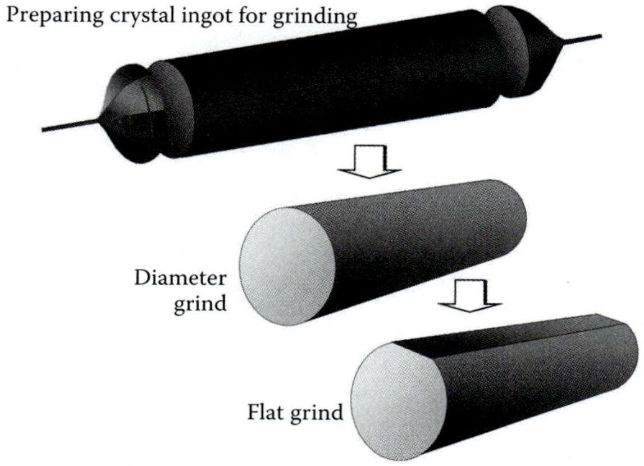

FIGURE 4.18 Ingot cap (seed and tail) cutting, and diameter and flat grind.

For some 200-mm and 300-mm diameter wafers, only a small notch is cut from the wafer to enable lithographic alignment, and no dopant type or crystal orientation information is conveyed.

The ingot/boule is then sliced into wafers, typically using inner diameter (ID) sawing, guided by argon laser markings. Mechanical grinding rounds the wafer edges. Wafer-edge rounding cannot be taken lightly because it removes microcracks at the wafer edge, makes the wafers more fracture resistant, increases the mask life, and minimizes resist beading. The edge-rounding process also makes for safer quartz boat loading by increasing the number of transfers possible of those wafers without chipping. The slices are then lapped with a mixture of hard Al_2O_3 powder in glycerin to produce a wafer with flatness uniform to within 2 μm. After a damage-removing etch, the wafers are polished, cleaned, and inspected. The slicing, lapping, edge treatment, damage removal, and polishing result in a material loss of up to 50%.

Typical specifications for 125-mm to 300-mm diameter Si wafers are summarized in Table 4.1.

Notice that the thickness control on Si wafers is limited, which has important consequences for fabricating thin Si membranes.

The Si wafer preparation process is summarized in Figure 4.19.

Doping of Si

Extrinsic versus Intrinsic Semiconductors

Intrinsic semiconductors, we saw earlier, are undoped semiconductors with electrical properties native to the material. Extrinsic semiconductors are doped semiconductors with electrical properties controlled by added impurities. Doping is the intentional addition of impurities to a material through ion implantation or diffusion. Doping by ion implantation and diffusion is different from other semiconductor processes because these processes do not create a new layer on the wafer. Instead, ion implantation and diffusion change the electrical characteristics of precise areas within an existing layer on the wafer.

On incorporation into the crystal lattice, the dopant either gives up (donor) or receives (acceptor) an electron from the crystal. In the former case, an extra electron is available (e^-), and in the latter case a hole (h^+) is created. Adding either electrons (e^-) or holes (h^+) varies the conductivity (σ) of the material. Adding more electrons makes the material more n-type, and adding more holes makes the material more p-type. Elements having one more or one less valence electron than Si (group IV) are used as substitutional donors (to make n-type) or acceptors (to make p-type), respectively. The first type of dopant is pentavalent, and the second type is trivalent. Boron (III), phosphorus (V), arsenic (V), and antimony (V) represent the more commonly

TABLE 4.1 Typical Specifications for Monocrystalline Silicon Wafers

Parameter	125 mm	150 mm	200 mm	300 mm
Diameter (mm)	125 ± 1	150 ± 1	200 ± 1	300 ± 1
Thickness (mm)	0.6–0.65	0.65–0.7	0.715–0.735	0.755–0.775
Bow (μm)	70	60	30	<30
Total thickness variation (μm)	65	50	10	<10
Surface orientation	±1°	±1°	±1°	±1°

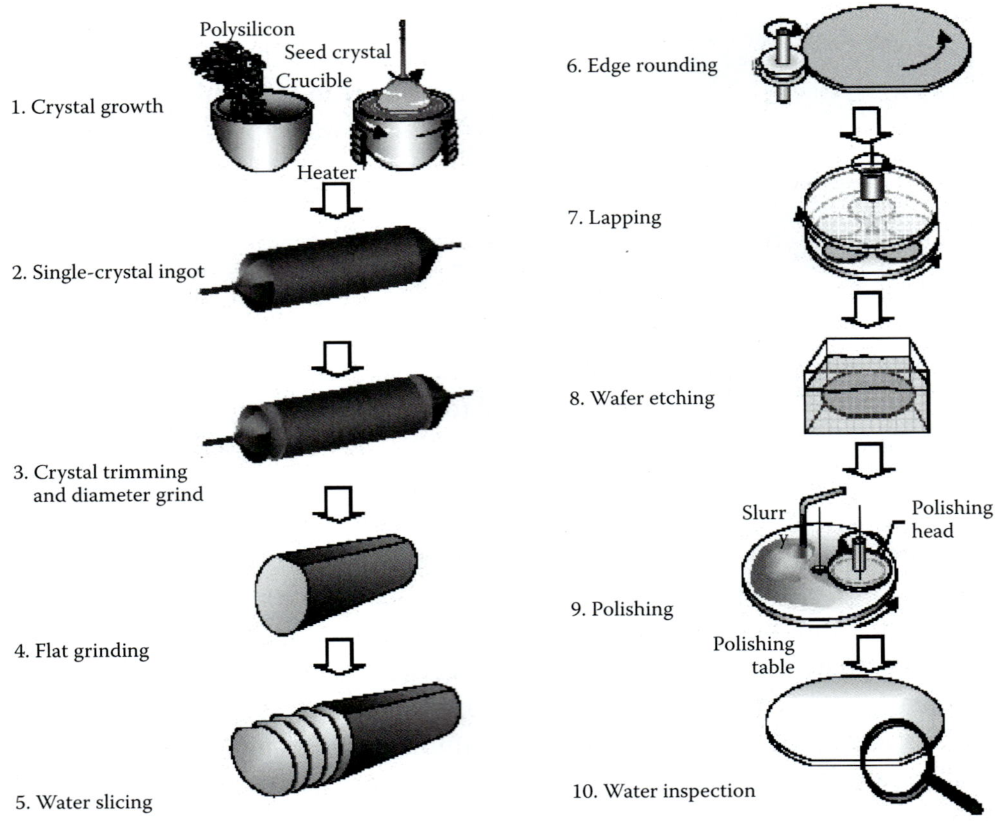

FIGURE 4.19 Silicon wafer preparation.

used dopant elements for single-crystal silicon (see Table 4.2). The two principal means of doping Si in wafer form are diffusion and ion implantation (as illustrated in Figure 4.20).

Dopants can be incorporated in the ingots during crystal growth or in sliced wafers in a planar process. Fabrication of circuit elements and micromachines also requires a method for selective n- or p-type doping of the silicon substrate. For diffusion, a uniformly doped ingot is sliced into wafers; an oxide film is grown on the wafers; and the oxide film is patterned and etched using photolithography, exposing specific sections of the silicon to be doped. The wafers are then coated with a solid doping source by spin

TABLE 4.2 Doping of Silicon with Pentavalent and Trivalent Atoms

		Concentration (atoms/cm^3)			
Dopant	Material Type	<10^{14} (very lightly doped)	10^{14}–10^{16} (lightly doped)	10^{16}–10^{19} (doped)	>10^{19} (heavily doped)
Pentavalent	n	n^{--}	n^{-}	n	n^{+}
Trivalent	p	p^{--}	p^{-}	p	p^{+}

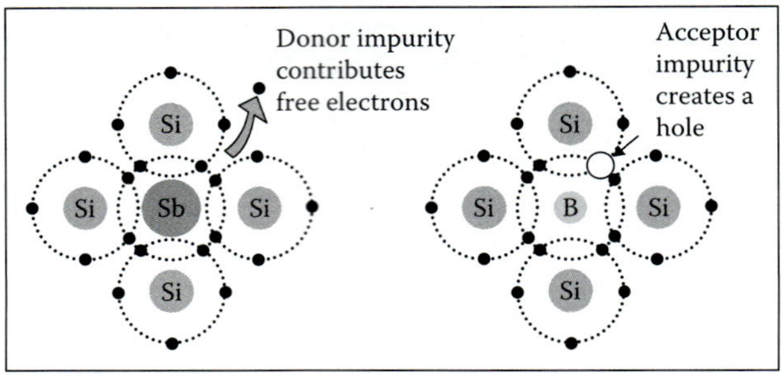

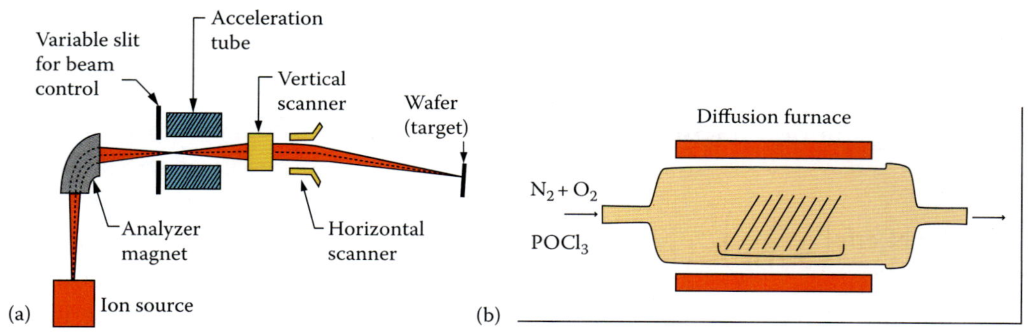

FIGURE 4.20 Ion implantation (a) and diffusion (b) to dope a silicon wafer.

coating and are heated in a furnace at 800–1250°C to drive the doping atoms into the exposed silicon areas. In ion implantation, a particle accelerator is used to accelerate doping atoms so that they can penetrate a silicon wafer to a depth of several micrometers. The process creates some lattice damage, which may be repaired by annealing the wafer at a moderate temperature for a few minutes.

Si Doping by Diffusion

Doping of silicon by diffusion is a chemical process carried out in a tube furnace to introduce a minute amount of impurities and to uniformly spread the incorporated impurities (Figure 4.20b). Dopant atoms may move into lattice vacancies, defining substitutional diffusion, or they may displace silicon atoms, moving them to interstitial positions, in an interstitial model. Finally, dopants may simply go between the Si atoms without displacing them (Figure 4.21).

Dopants are typically diffused thermally into a Si substrate at temperatures between 950°C and 1250°C. Silicon dioxide is used for a mask. Typical dopant sources used are BBr_3, $POCl_3$ (liquids), Sb_2O_3, As_2O_3, P_2O_5, B_2O_3 (solids), PH_3, AsH_3, and B_2H_6 (gases). To a first approximation, Fick's first and second laws describe the diffusion of dopants in a

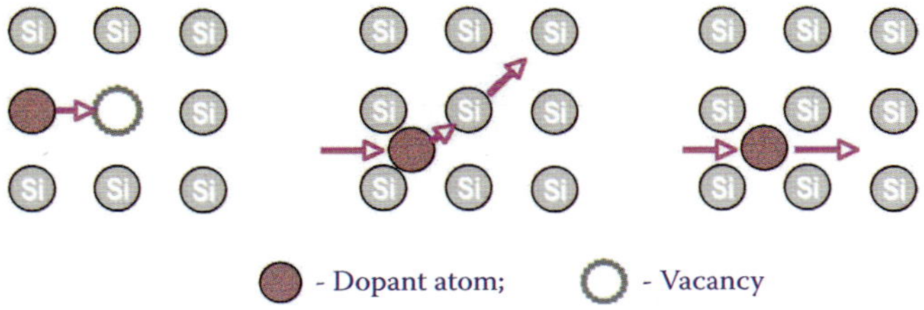

FIGURE 4.21 Diffusion mechanisms of dopant atoms in Si.

substrate. The dopant flux J is proportional to the concentration gradient as given by Fick's first law:

$$J = -D\frac{\delta N(x,t)}{\delta x} \tag{4.20}$$

where D is the diffusion coefficient (units of cm^2/s), and $\delta N(x,t)/\delta x$ is the concentration gradient of the dopant. The concentration N is a function of position x and time t, and in this expression, D is assumed constant, which is a reasonable assumption when the concentration of the dopant is low. Equation 4.20 has a negative sign, and $\delta N(x,t)/\delta x$ is negative for a decrease in the concentration with depth; consequently, the flux into the sample has a positive value.

The flux gradient ($\delta J/\delta x$) is proportional to the change of the concentration with time:

$$\frac{\delta J}{\delta x} = -\frac{\delta N(x,t)}{\delta t} = -D\frac{\delta^2 N(x,t)}{\delta x^2} \tag{4.21}$$

This is also called the *continuity equation* or Fick's second law. Solutions for Equation 4.21 have been obtained for various simple conditions, including diffusion with a constant surface concentration (N_s) and diffusion with constant total dopant amount (Q). In the first scenario, impurity atoms may be transported from a vapor source onto the semiconductor surface and diffuse into the semiconductor wafer. In the second scenario, a fixed amount of dopant is already in the semiconductor, and a heat treatment distributes it further into the material.

In gaseous doping, the initial condition at t = 0 is N(x,0) = 0, which states that the dopant concentration in the host semiconductor is initially zero. The boundary conditions are that the dopant concentration at the Si surface is constant [$N(0,t) = N_s$], whereas in the bulk, the concentration remains zero [$N(\infty,t) = 0$] during the whole doping process, resulting in the following solution for the differential Equation 4.21:

$$N(x,t) = N_s \text{erfc}\left(\frac{x}{2\sqrt{Dt}}\right) \quad (4.22)$$

where erfc = complementary error function (see Figure 4.22a and Appendix 4A)
$\sqrt{Dt}$ = diffusion length
N_s = surface concentration of the dopant (in cm^{-3})

Generally speaking, this same boundary condition also applies for the predeposition of dopant atoms onto semiconductor substrates. The temperature of the substrates during predeposition is usually kept between 800°C and 1200°C.

The total amount of dopant atoms per unit area of the semiconductor, or the dose, $Q(t)$, is given by integration of $N(x,t)$ from $x = 0$ to $x = \infty$:

$$Q(t) = \int_0^\infty N_s \text{erfc}\left(\frac{x}{2\sqrt{Dt}}\right) dx = \frac{2\sqrt{Dt}}{\sqrt{\pi}} N_s \cong 1.13 N_s \sqrt{Dt} \quad (4.23)$$

The gradient of the diffusion profile, $\delta N(x,t)/\delta x$, can be obtained by differentiating Equation 4.22:

$$\frac{\delta N(x,t)}{\delta x} = -\frac{N_s}{\sqrt{\pi Dt}} \exp\left(-\frac{x^2}{4Dt}\right) \quad (4.24)$$

Another boundary condition applies when the dopant concentration at the silicon surface is limited; in other words, the diffusing amount Q is constant and $C(\infty,t) = 0$. In this boundary condition, the surface dopant concentration decreases as the material supply is depleted. The initial condition at $t = 0$ is again $N(x,0) = 0$, and the boundary conditions are mathematically expressed as:

$$\int_0^\infty N(x,t)dx = Q \text{ and } N(\infty,t) = 0 \quad (4.25)$$

where Q is the total amount of dopant or the dose. The solution of the diffusion equation satisfying the above conditions is:

$$N(x,t) = \frac{Q}{\sqrt{\pi Dt}} \exp\left(\frac{-x^2}{4Dt}\right) \quad (4.26)$$

In this case, the resulting dopant concentration profile is Gaussian instead of an erfc function (see Figure 4.22b). At the surface $x = 0$, and Equation 4.26 at $t = 1$ becomes:

$$N_s = N_{01} = \frac{Q}{\sqrt{\pi Dt}} \quad (4.27)$$

or the surface concentration changes as $(Dt)^{-1/2}$. Therefore, the dopant surface concentration decreases

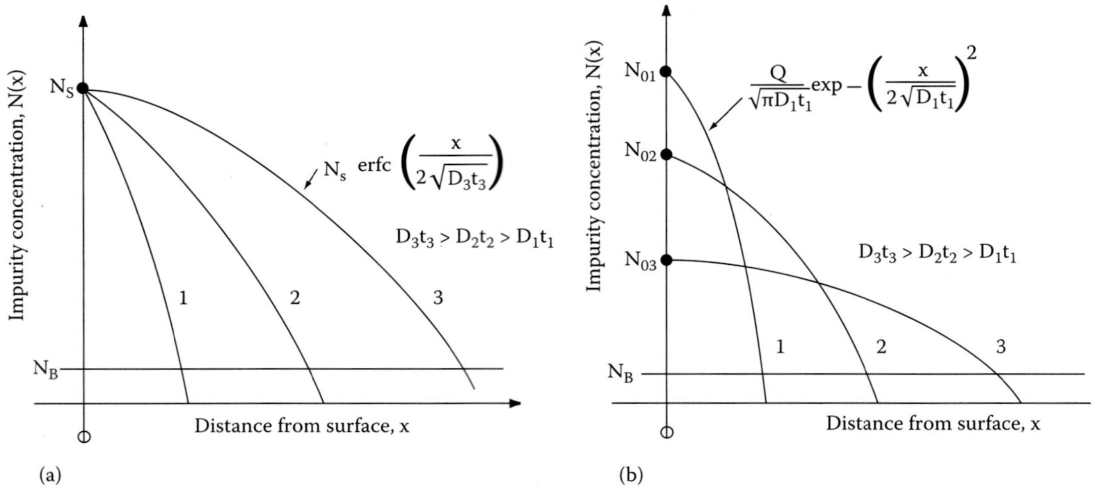

FIGURE 4.22 Diffusion profiles. (a) Normalized complementary error function (erfc) versus distance for successive diffusion times. N_s is the surface concentration, and N_B is the bulk concentration. (b) Normalized Gaussian function versus distance for successive diffusion times.

with time because the dopant moves into the semiconductor as time increases (see Figure 4.22b from N_{01} to N_{03}, with $t = 1 < t = 2 < t = 3$). The gradient of the diffusion profile is obtained by differentiating Equation 4.26:

$$\frac{\delta N(x,t)}{\delta x} = -\frac{x}{2Dt}N(x,t) \tag{4.28}$$

The gradient is zero at $x = 0$ and at $x = \infty$, and the maximum gradient is obtained at $x = \sqrt{2Dt}$. The complementary error function and the Gaussian distribution are both functions of a normalized distance, $x/2(Dt)^{1/2}$. Hence, by normalizing the dopant concentration with the surface concentration, each distribution can be represented by a single curve valid for all diffusion times, as shown in Figure 4.23.

This type of boundary condition is typical for drive-in diffusion. Drive-in diffusion is a high temperature (>800°C) operation performed on semiconductor wafers in an inert ambient. The heat treatment causes motion of dopant atoms in the direction of the concentration gradient (diffusion) and is used to drive dopant atoms deeper into the semiconductor. The dopants introduced in the substrate by a predeposition step are redistributed deeper into the substrate by such a drive-in step. A two-step diffusion sequence is commonly used to form an n-p junction; during predeposition, a diffusion layer is formed under a constant surface concentration condition, which is followed by a drive-in diffusion, or redistribution, under a constant total dopant condition. For most practical cases, the diffusion length $(Dt)^{1/2}$ for the predeposition diffusion is much smaller than that for the drive-in condition. Hence the predeposition profile can be treated as a delta function at the surface. This two-step junction formation is illustrated in Figure 4.24 with a predeposition from a constant source (erfc) and a limited source diffusion (Gaussian).

The D in the above equations is the dopant diffusivity and is given by:

$$D = D_0(T_0)\exp\left(-\frac{E_a}{kT}\right) \tag{4.29}$$

where E_a is the activation energy, whose value depends on the transport mechanism of the dopant

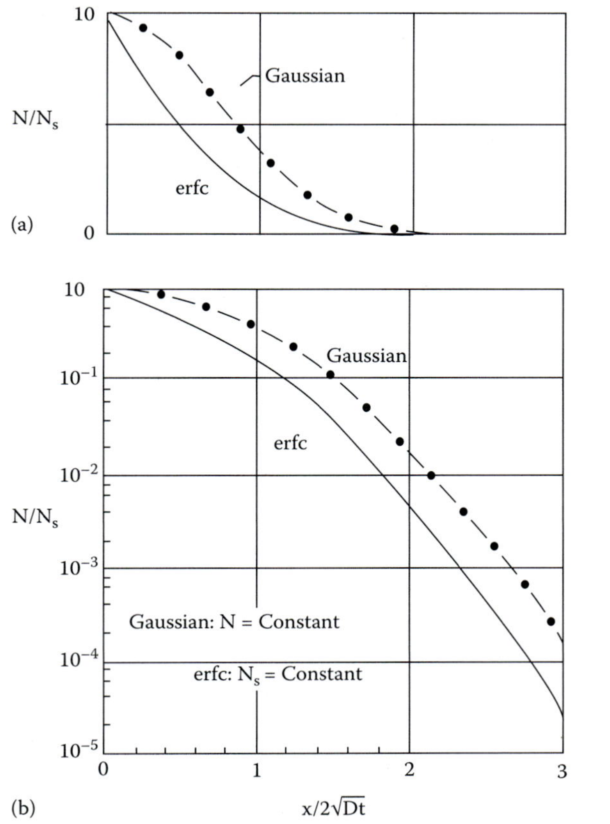

FIGURE 4.23 Normalized concentration versus normalized distance for the erfc and Gaussian functions. (a) Linear plot. (b) Semilog plot.

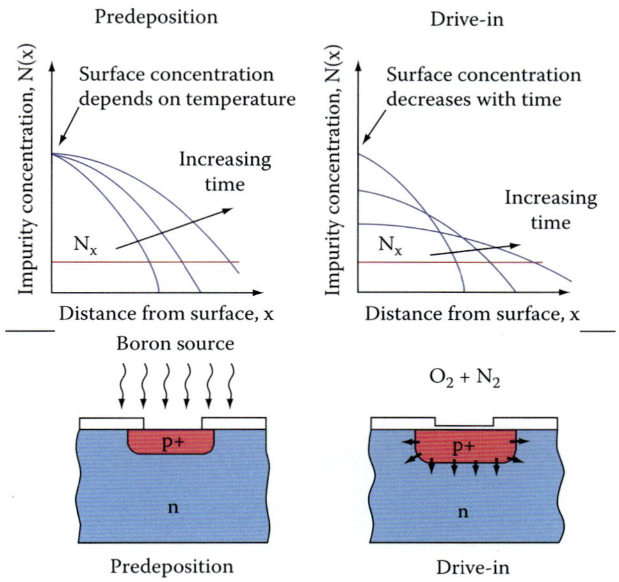

FIGURE 4.24 Two-step junction formation with a predeposition from a constant source and a limited source diffusion.

TABLE 4.3 Typical Values for D_0 and E_a for Commonly Used Dopant Atoms

Element	D_0 (cm²/s)	E_a (eV)
B	10.5	3.69
Al	8.00	3.47
Ga	3.60	3.51
In	16.5	3.90
P	10.5	3.69
As	0.32	3.56
Sb	5.60	3.95

atoms. The diffusion coefficient, D_0, is only a constant for a given reaction and concentration.

In Table 4.3, typical values for D_0 and E_a are given for commonly used dopant atoms. Dopant atoms take on several different ionization states and diffuse by several different mechanisms. The effective diffusion coefficient is the sum of the diffusivities of each species, which are in turn a function of temperature and dopant concentration. Hence, actual dopant concentration profiles deviate from the simple first-order functions given above.

The dopant diffusivities of phosphorus, arsenic, and boron in silicon at 1100°C are typically 2×10^{-13}, 1×10^{-13}, and 1×10^{-13} cm²/s, respectively. Below the intrinsic carrier concentration (~5×10^{18} cm^{-3} for Si at 1000°C), D is independent of the carrier concentration. Above the intrinsic concentration, the diffusion increases with carrier concentration and generally follows a power law as a result of the diffusion mechanism becoming point defect (vacancies and interstitials) assisted.

The point at which the diffused impurity profile intersects the background concentration is the metallurgical junction depth, x_j. At that depth, the net impurity concentration is zero. In the case of a complementary error function, the depth is defined as:

$$x_j = 2\sqrt{Dt\,\text{erfc}^{-1}\left(\frac{N_B}{N_S}\right)} \tag{4.30}$$

When dealing with a Gaussian distribution, one obtains:

$$x_j = 2\sqrt{Dt\ln\left(\frac{N_s}{N_B}\right)} \tag{4.31}$$

The latter case is illustrated in Figure 4.25, where we show the diffusion of an n-type impurity in a p-type Si background substrate.

It is important to remember that the concentration profile of the dopant in the semiconductor never changes abruptly (this would imply a zero or infinite flux), so diffusion always spreads impurities gradually around the edges of any mask. Lateral diffusion is ~85% of the junction depth. This lateral diffusion continues every time the wafer is heated, resulting in an expanding die area. Besides lateral diffusion, diffusion doping also suffers from surface contamination interference, poor doping control, dislocation generation, and difficulty in creating ultrathin junction depths. Next, we will see that for each generation of IC technology, the trend has been to reduce the junction depth, which favors ion implantation (see comparison Table 4.4).

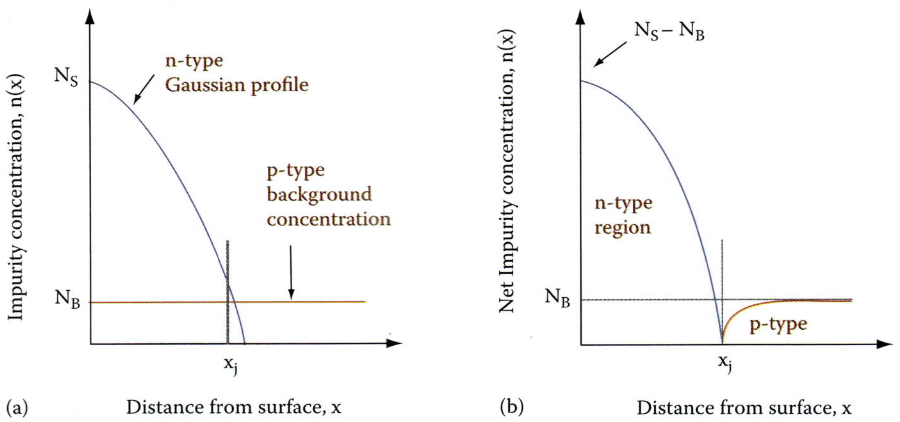

FIGURE 4.25 Diffusion of n-type impurity into a p-type Si background substrate. Determination of x_j in case of a Gaussian dopant distribution. Two representations are shown. In (a), the *y*-axis represents the impurity concentration; in (b), the *y*-axis represents the net impurity concentration (i.e., corrected for the background impurity concentration N_B).

Table 4.4 Comparison of Thermal Diffusion and Ion Implantation

Ion Implantation and Annealing	Solid-/Gas-Phase Diffusion
Advantages	
Room temperature mask	No damage created by doping
Precise dose control (10^{11}–10^{16}/cm^2)	Batch fabrication
	Cheaper
Accurate depth control	
Disadvantages	
Implant damage enhances diffusion	Usually limited to solid solubility
Dislocations caused by damage may cause junction leakage	Low surface concentration hard to reach without a long drive-in
Implant channeling may affect profile	Low dose predeposition very difficult to achieve
Very expensive	

Si Doping by Ion Implantation

Ion implantation is the introduction of selected impurities (dopants) by means of high-voltage ion bombardment to achieve desired electronic properties in defined areas. Photolithography is used to mask the regions of the device where doping is not desired. For masking materials, most films used in the IC process can be used in the ion-implantation process—photoresists, SiO_2, Si_3N_4, metal film, etc.—in contrast with diffusion, where choices are much more limited. High energy ion beam implantation is a low-temperature physical process and offers the advantage of being able to place ions at various depths in the sample, independent of the thermodynamics of diffusion and problems with solid solubility and precipitation. Ion implantation for doping was invented by Shockley in 1954. The technique is now very commonly used, with penetration depths in silicon of As, P, and B typically being 0.5, 1, and 2 μm at 1000 keV. A large range of extremely accurate doses is accessible: from 10^{11} to 10^{16}/cm^2, and even buried (or retrograde) dopant profiles are possible. Unfortunately, however, ion beams produce crystal damage because the implant energy—between 30 and 100 keV—is thousands of times the silicon bond energy. Crystal damage can reduce electrical conductivity, but fortunately most of the damage can be eliminated by annealing at 700–1000°C, which is below the diffusion temperature. The anneal temperature is high enough for Si interstitial atoms to recombine with vacant lattice sites and for the dopant atoms to move to substitutional lattice sites where they are electrically active. The combination of excellent spatial and dose control, as well as ease of manufacture, has led to widespread use of ion implantation. For the newest and future generation of complementary metal oxide silicon (CMOS) devices, which require highly doped, abrupt, and very shallow profiles (junction depths less than 30 nm!), the dopant diffusion step during annealing must be minimized.

A beam of energetic ions implants charged dopant atoms into the silicon substrate. The acceleration energy and the beam current control the implantation depth and dopant concentration. The implanted dopant atoms enter unmasked regions in the silicon lattice at high velocity. The stopping mechanism of the ions involves nuclear collisions at low energy and electronic interactions at high energy.

The acronyms associated with ion implantation include:

- Projected range, R_p: average distance traveled by ions parallel to the beam
- Projected straggle, ΔR_p: standard deviation or fluctuation in the projected range
- Lateral straggle, $\Delta R//$: standard deviation or fluctuation in the final rest position, perpendicular to the beam (also called *transverse straggle*)
- Peak concentration, N_p: concentration of implanted ions at R_p

In the absence of crystal orientation effects, the concentration profile, to a first-order approximation, is Gaussian:

$$N_i(x) = N_p \exp\left[-\frac{1}{2}\left(\frac{x - R_p}{\Delta R_p}\right)^2\right] \tag{4.32}$$

The range, R_p, is determined by the acceleration energy, the ion mass, and the stopping power of the material. Heavy ions do not travel as far into the crystal as light ions. They have a more narrow distribution (ΔR_p) than light ions. The crystalline symmetry of the silicon lattice can have a dramatic impact on the implant profile. Once an ion enters a lattice channel, the small angle scattering events on the channel walls translate to ion penetration for a long distance before electron drag or a sharp collision stops the ion. Orientation of the substrate surface away from the plane perpendicular to the beam (tilt of 3–7°) prevents channeling, which can occur along crystal planes, ensuring better reproducibility of R_p. Ion channeling leads to an exponential tail in the concentration versus depth profile. This tail results from the crystal lattice and is not observed in an amorphous material. Channeling may also be prevented by creating a damaged layer on the Si wafer surface or by depositing a surface-blocking amorphous layer such as an oxide. The three methods to limit ion channeling are compared in Figure 4.26.

The total implanted dose (ions/cm²), Q_i, can be either calculated by integrating the charges per unit time (current i) arriving at the substrate over time or also by integrating Equation 4.32 over all of x:

$$Q_i = \int i\,dt = \int_0^{+\infty} N_i(x)\,dx = \sqrt{2\pi}\Delta R_p N_p \tag{4.33}$$

Technologies studied to make the very shallow junctions needed in the latest CMOS devices are ultralow-temperature processing and rapid thermal annealing. For deep diffusions (>1 μm), implantation is used to create a dose of dopants, and thermal diffusion is used to drive in the dopant.

Diffusion is a cheaper and more simplistic method than ion implantation but can only be performed from the surface of the wafers. Dopants also diffuse unevenly and interact with each other, altering the diffusion rate. Ion implantation is more expensive and complex. Ion implantation, however, except for annealing, does not require high temperatures and allows for greater control of dopant concentration and profile. It is an anisotropic process and therefore does not spread the dopant implant as much as diffusion. This aids in the manufacture of self-aligned structures, which greatly improve the performance of MOS transistors. Ion beams can implant enough material to form new materials, for example, oxides and nitrides, some of which show improved wear and strength characteristics. Ion implanters may be purchased from Varian Semiconductor (http://www.vsea.com) and Eaton Corporation (http://www.eaton.com). This equipment is very expensive, and usually service providers are used instead.

In Table 4.4 we compare thermal diffusion and ion implantation.

Band Diagram of an Extrinsic Semiconductor

In this section we establish the band diagram and carrier density for extrinsic semiconductors. The conductivity σ (Ωm)$^{-1}$ is an intrinsic characteristic of a material and is independent of geometry. In extrinsic semiconductors, negative and positive charge carriers may contribute to the conductivity. In Chapter 3 we saw that the conductivity is

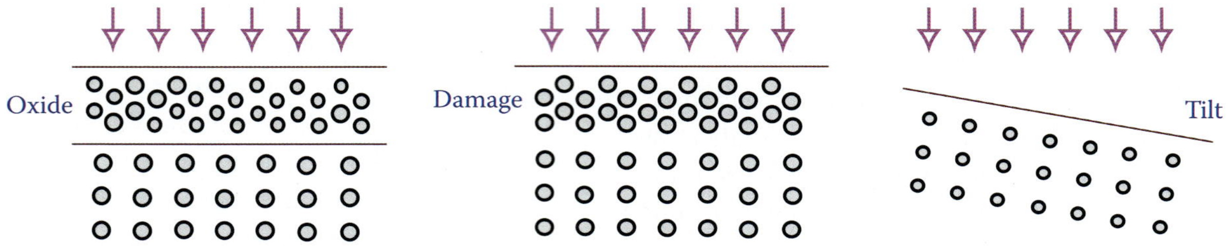

FIGURE 4.26 How to prevent ion channeling during implantation.

the product of the number of free charges and their mobility [for electrons (e^-) the mobility in $cm^2V^{-1}s^{-1}$ is symbolized as μ_e (see Equation 3.10)]:

$$\sigma = \mu_e ne \tag{3.10}$$

The positive charge carriers (h^+) are called *holes* as they correspond to an absence of electrons, and μ_h is their mobility. When both the concentration of holes (p) and the concentration of electrons (n) are considered, the conductivity σ of a solid is given as:

$$\sigma = 1/\rho = e(\mu_e n + \mu_p p) \tag{4.34}$$

where e is the electron charge, n is the electron concentration, and p is the hole concentration. In the case of a metal, we saw that resistance increases with temperature because the number of carriers remains the same but scatters more at higher temperatures. In a semiconductor, the resistance decreases as the temperature increases because more free carriers are generated (n and p increases). A semiconductor is called n-type when $e\mu_n n$ dominates the conductivity, in which case the majority of carriers are electrons, and the semiconductor is p-type if $e\mu_p p$ dominates, in which case the majority carriers are holes.

By adding a donor atom (energy E_D) to an intrinsic semiconductor at T = 0 K, an extra electron is bound to the positive charge on the donor atom. When the donor, at higher temperatures, ionizes, the neutral donor concentration N_D converts to positively charged N_D^+. The electron bound to the positive ion is in an energy state, $E_D = E_g - \Delta E$, where ΔE is the binding energy. An electron that moves onto an acceptor atom has energy E_A and coverts a neutral acceptor concentration N_A into a negatively charged N_A^-. In Figure 4.27, we show donor ions (E_D) energetically close to the bottom of the conduction band (E_C) and acceptor ions (E_A) close to the top of the valence band (E_V).

To estimate the Bohr's radius and the binding energy (ΔE) involved in the donor holding onto its electron, we use the simple Bohr model for the hydrogen atom that was introduced in Chapter 3 but adapt the pertinent parameters to describe an impurity atom in a crystal rather than a hydrogen atom in vacuum. In Chapter 3 we calculated the Bohr radius of the hydrogen atom using Equation 3.89 as:

$$a_0 = \frac{\hbar^2 4\pi\varepsilon_0}{m_e e^2} = 0.53 \text{ Å} \tag{3.89}$$

For the binding energy involved, we use Equation 3.91:

$$\Delta E = E_n = -\frac{m_e e^4}{32\hbar^2\pi^2\varepsilon_0^2 n^2} = -\frac{E_0}{n^2} \tag{3.91}$$

With the Bohr radius equal to 0.53×10^{-10} m and n = 1, we get a value of 13.6 eV = 21.8×10^{-19} J for

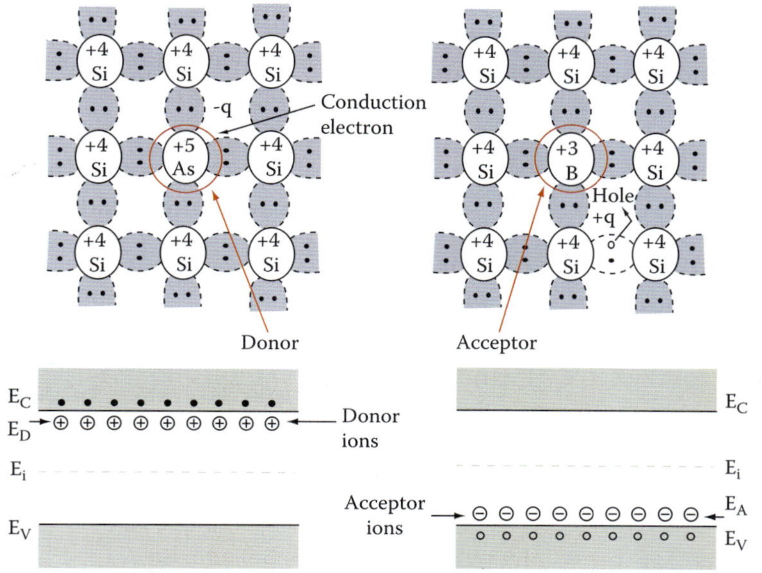

FIGURE 4.27 Donors and acceptors in the Si lattice and their position in an energy band diagram. Here E_i stands for the Fermi level of the intrinsic material.

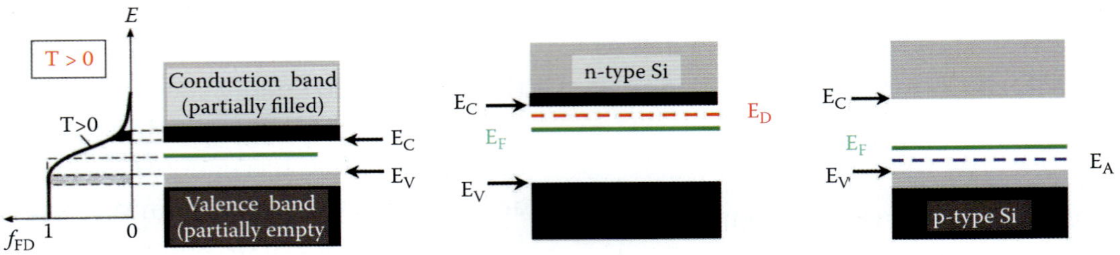

FIGURE 4.28 Band diagram for intrinsic and extrinsic Si compared. Black is filled with electrons, and gray is empty.

the energy required to remove an electron from a hydrogen atom (see also Figure 3.36). For a typical solid, we saw in Chapter 3 one must replace m_e with the electron effective mass $m_e^* \sim 0.15\ m_e$ and ε_r of 1 in the case of hydrogen with $\varepsilon_r \sim 15$ for a typical semiconductor so that a 53-Å Bohr radius is obtained from:

$$a_0 = \frac{0.53 \times \varepsilon_r}{m_e^*/m_e} = \frac{0.53 \times 15}{0.15} = 53\ \text{Å} \tag{4.35}$$

and 9 meV is calculated for the binding energy of the electron to the donor atom from:

$$\Delta E = \frac{13.6}{\varepsilon_r^2}\frac{m_e^*}{m_e}\ \text{eV} = \frac{13.6}{(15)^2} \times 0.15 = 9\ \text{meV} \tag{4.36}$$

The ionization energy of the donor tells us how far below the conduction band the donor energy level is situated ($\Delta E = E_C - E_D$). A similar calculation provides us with the distance above the valence band edge of the acceptor levels in a p-type material. This Bohr-like doping model assumes no interaction between dopant sites, i.e., a nondegenerate semiconductor.

At T = 0 K, all electrons are in the lowest available energy states, and no electrons are excited from donor states (N_D) into the conduction band (n = 0), but with a low binding energy of 9 meV at room temperature ($kT \sim 25$ meV $> \Delta E = E_C - E_D$) all of the donors will be ionized (N_D^+), and the number of free electrons in the conduction band n is given by:

$$n = N_D^+ \sim N_D \tag{4.37}$$

This is the relevant regime for most electronics; note that the density of electrons is independent of temperature. Electrons do freeze out (i.e., return to the valence band) if kT becomes too small to overcome the binding energy. From Equation 4.14 we calculate that:

$$E_F = E_c - k_B T \ln\left(\frac{N_C}{n}\right) \tag{4.38}$$

where $N_C = 2.8 \times 10^{19}$ cm^{-3} and $n \sim N_D \sim 10^{17}$ cm^{-3}; we can see that the Fermi level is about 150 meV below the conduction band (~6 kT). This situation is sketched in the middle panel of Figure 4.28, where we compare intrinsic silicon with n-type and p-type Si. Analogously, from Equation 4.15 we derive:

$$E_F = E_v - k_B T \ln\left(\frac{N_v}{p}\right) \tag{4.39}$$

Some typical values for energy levels of impurities in Si are listed in Table 4.5.

The Bohr-like doping model assumes no interaction between dopant sites; if adjacent dopant atoms are within 2 Bohr radii, then orbits overlap, and we are dealing with a degenerate semiconductor. This happens when N_D (~n) ~ N_c or when N_A (~p) ~ N_v. The degenerate semiconductor is then defined by E_F ~/> E_C or E_F ~/< E_V. In these cases, the Fermi level is situated either inside the conduction or the valence band.

TABLE 4.5 Some Typical Impurity Levels in Si

Phosphorous, P	$E_C - E_D = 44$ meV
Arsenic, As	$E_C - E_D = 49$ meV
Boron, B	$E_A - E_V = 45$ meV
Aluminum, Al	$E_A - E_V = 57$ meV
Gallium, Ga	$E_A - E_V = 65$ meV
Gold, Au	$E_A - E_V = 584$ meV
	$E_C - E_D = 774$ meV

Oxidation of Silicon

Oxidation Kinetics of Si-Silica Is One of the Keys to Silicon's Success

Silicon owes a lot of its success to its stable oxide. By contrast, germanium's oxide is soluble in water, and GaAs produces only leaky oxides. In IC design, silicon oxides, like other dielectrics, function as insulation between conducting layers, as diffusion and ion masks, as capping layers of doped oxides to prevent the loss of dopants, for gettering impurities, and for passivation to protect devices from impurities, moisture, and scratches. Doped oxides also act as a solid-state source of ions that can be diffused into a substrate at high temperatures. In micromachining, silicon oxides serve the same purposes but also very often act as sacrificial material. The stability and ease of formation of SiO_2 on a Si surface was one of the reasons that Si replaced Ge as the semiconductor of choice in the late 1960s. Silica is the distinguishing feature of silicon versus other important semiconductor materials (see Table 4.6). Silica is inert to most chemicals, protects the Si surface, and forms a reasonably good diffusion barrier (although Si_3N_4 is much better in this respect). It is thermally stable up to 1600°C (T_m is 1830°C), easy to grow, and easy to pattern and etch selectively. In wet etching, HF preferentially etches SiO_2, not Si, and in dry etching it can easily be plasma etched in CHF_3. The material also makes for an excellent dielectric (insulator) with a 9-eV bandgap, a dielectric strength of 10^7 V/cm, and a resistivity ρ of 10^{15} Ωcm.

Silicon dioxide growth involves the heating of a silicon wafer in a stream of steam or in wet or dry oxygen/nitrogen mixtures (Figure 4.29) at elevated temperatures (between 600°C and 1250°C) in a horizontal or vertical quartz furnace. Dry oxide grows more slowly, but the oxide layers are more uniform, and there are also relatively few defects at the oxide-silicon interface. Defects interfere with the proper operation of semiconductor devices, and because a dry oxide has especially low surface state charges, it forms an ideal thin dielectric for MOS transistors (30–100 Å; see below). In wet oxides, hydrogen atoms, liberated by the decomposition of the water molecules during oxidation, produce imperfections that degrade the oxide quality somewhat. A major advantage of wet oxides includes their faster growth rate, making them useful for thick layers of field oxides (>3000 Å; see below). Silicon readily oxidizes, even at room temperature, forming a thin native oxide approximately 20 Å thick. The high temperature aids diffusion of oxidant through the growing surface oxide layer to the interface to form thick oxides quickly:

$$Si + 2H_2O \text{ (gas)} \rightarrow SiO_2 + H_2 \text{ (wet oxidation)}$$
Reaction 4.1

$$Si + O_2 \text{ (gas)} \rightarrow SiO_2 \text{ (dry oxidation)}$$
Reaction 4.2

Another gas-phase oxidation method of Si is the pyrogenic method (Figure 4.29). Besides oxygen and nitrogen, the gas also contains hydrogen, which generates ultrapure steam. In the oxidizing gas stream, one also may add 2% HCl or Cl_2; this helps to remove (getter) impurities by volatizing any metal impurities ($AuCl_3$ and $CuCl_2$).

TABLE 4.6 The Question of Good-Quality Native Oxides on Semiconductors

Semiconductor (Bandgap)	Native Oxide
Si (1.1 eV)	Yes
Ge (0.7 eV)	No
GaAs (1.4 eV)	No
SiC (~2.2 eV)	No
GaN (~3 eV)	No
Diamond (5 eV)	No

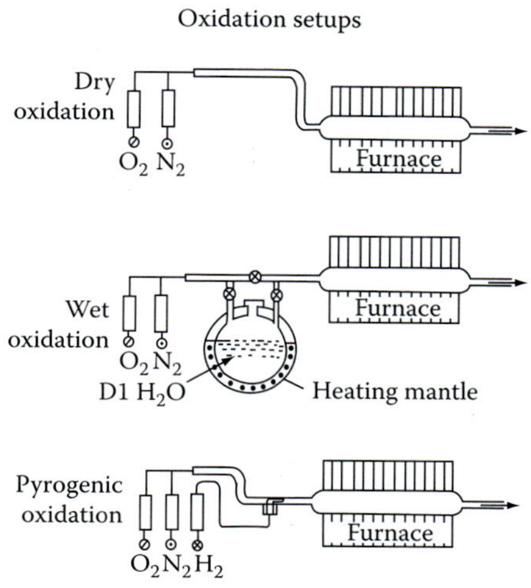

FIGURE 4.29 Oxidation setups.

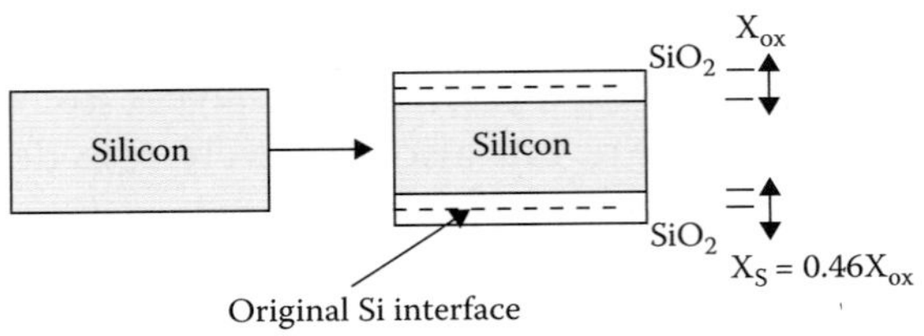

FIGURE 4.30 Silicon thickness converted, X_S. Oxidation by inward diffusion of oxidant through the growing oxide to the interface.

The ratio of silicon thickness converted (Figure 4.30), X_S, to resulting oxide thickness, X_{ox}, is proportional to their respective densities:

$$X_S = 0.46X_{ox} \tag{4.40}$$

When 10,000 Å of oxide has grown, 4,600 Å of Si will be consumed; in other words, the amount of silicon consumed is 46% of the final oxide thickness. This relationship is important for calculating step heights that form in silicon microstructures. Very good adhesion to Si leads to volume expansion upward, resulting in compressive stress in the SiO_2 film.

All gas-phase oxidation processes involve three steps in series: gas-phase transport of oxidant to the surface (F_1), diffusion through the existing oxide (F_2), and the oxidation reaction itself (F_3). The fluxes are:

$$\begin{aligned} F_1 \text{ (gas)} &= h(C^* - C_{ox}) \\ F_2 \text{ (oxide)} &= D(C_{ox} - C_i)/x_{ox} \\ F_3 \text{ (reaction)} &= k_s C_i \text{ (see Figure 4.31)} \end{aligned} \tag{4.41}$$

At equilibrium $F_1 = F_2 = F_3$, and this yields the concentration of oxidant at the interface oxide/silicon, C_i, as:

$$C_i(X_{ox}, k_s, D, P_G) \tag{4.42}$$

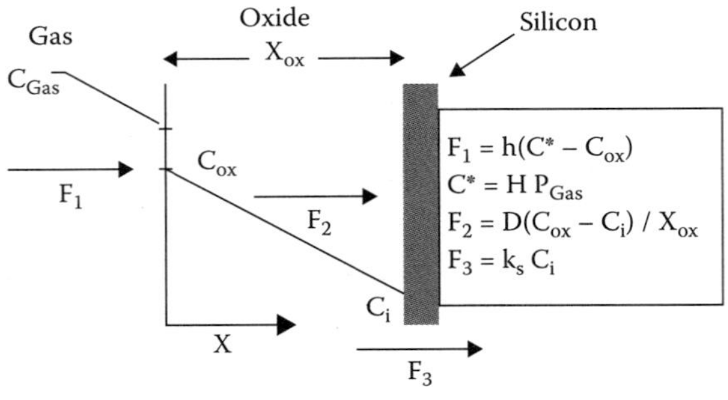

FIGURE 4.31 Fluxes in the silicon oxidation reaction.

where X_{ox} = oxide thickness
- k_S = Si oxidation rate constant, which is a function of temperature, oxidant, crystal orientation, and doping
- D = oxidant diffusivity, which is a function of temperature and oxidant
- P_G = partial pressure of the oxidant in the gas phase
- C^* = equilibrium oxidant concentration in the oxide = HP_G, with H being Henry's law constant

The oxide growth rate is given by:

$$\frac{dX_{ox}}{dt} = \frac{C_i k_s}{N} \tag{4.43}$$

where N stands for the number of molecules of oxidant per unit volume of oxide (2.2×10^{22} cm^{-3} for dry oxygen and 4.6×10^{22} cm^{-3} for wet oxidation). Solution of this differential equation in the Bruce Deal-Andy Grove model for oxidation (1965) yields:

$$X_{ox}(t) = \frac{A}{2}\left[\left(\sqrt{1+\frac{t+\tau}{A^2/4B}}\right)-1\right] \tag{4.44}$$

where A and B are rate constants:
- $A = 2D(1/k_s + 1/h)$ (units of μm)
- $B = 2DC^*/N$ (units of μm^2/h—parabolic rate constant)
- $\tau = (X_i^2 + AX_i)/B$
- h = gas-phase mass transport coefficient in terms of concentration in the solid
- C^* = equilibrium oxidant concentration in the oxide = HP_G, with H being Henry's law constant
- X_i = initial oxide thickness

The reason that wet oxide growth is much faster than dry oxidation can be found in the value for C^* in the rate constant B. In dry oxygen at 1000°C, $C^*(O_2) = 5.2 \times 10^{16}$ cm^{-3}, whereas at the same temperature in wet oxygen $C^*(H_2O) = 3.0 \times 10^{19}$ cm^{-3} (three orders of magnitude larger). This more than compensates for the fact that the diffusion coefficient of oxygen in SiO_2 is larger than that of water [$D(O_2) > D(H_2O)$]. This general relationship for silicon oxidation is illustrated in Figure 4.32. From this figure it can be seen that for thin oxides [short

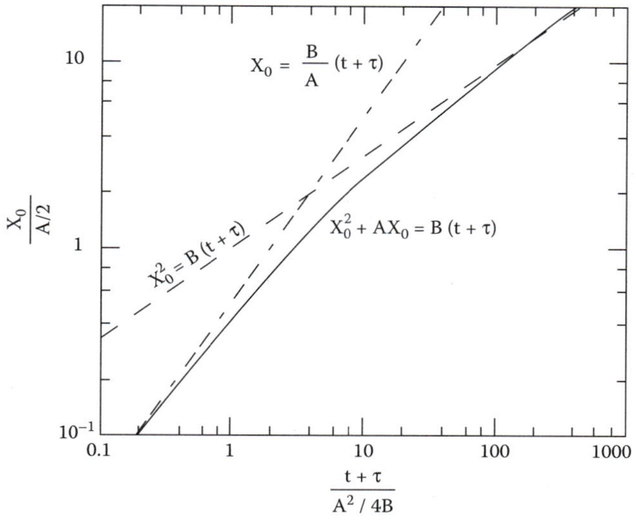

FIGURE 4.32 Silicon dioxide thickness versus growth time.

oxidation times with $(t + \tau) \ll A^2/4B]$, the process is reaction rate limited:

$$X_{ox}(t) = \frac{B}{A}(t+\tau) \tag{4.45}$$

where B is the parabolic rate constant, and B/A is the linear rate constant (in μm/h). For thick oxides (long oxidation times with $t \gg \tau$ and $t \gg A^2/4B$), we deduce from Figure 4.32 that the process is diffusion limited, and a parabolic law results:

$$X_{ox}(t) = \sqrt{B(t+\tau)} \tag{4.46}$$

where B is the parabolic rate constant. From Figure 4.32 we also conclude that oxide thickness versus growth time is initially linear when the oxidation reaction at the surface dominates but parabolic when the diffusion to the oxide/silicon interface dominates.

A, B, and τ are strong functions of temperature and the gas ambient (wet vs. dry) and weak functions of orientation and doping (both type and level). The oxidation rate is also influenced by the presence of halogen impurities or hydrogen in the oxidizing gas, the oxidizing gas pressure, and the use of a plasma or a photon flux during oxide growth.

For $X_0 < 200$ Å, the oxidation is always faster than that predicted by the Deal-Grove model, and empirical models are used to fit the data. Even the substrate cleaning method influences the oxide thickness in this regime. The Deal-Grove model is also limited to the 700–1300°C temperature range and to the 0.1–25 atm pressure range.

Short of using an ellipsometer to determine the oxide thickness (see Volume III, Chapter 6 on metrology), the silicon oxide thickness color table (Table 4.7) comes in handy. This table also lists some typical applications for oxides of various thicknesses. Charts showing the final oxide thickness as a function of temperature, oxidizing environment, and time are widely available.

TABLE 4.7 Oxide Color Table and Applications of Thermal Oxides: Observed Perpendicularly under Daylight Fluorescent Lighting

Color	Thickness (Å)				Application
Gray	Up to 100				Tunneling oxides in EEPROM (flash memories) and gate oxides (around 30 Å in 180-nm CMOS)
Tan	500				
Brown	700				Pad oxides
Blue	800				Masking oxides during implant, diffusion, etc.
Dark violet-red violet	1000	2800	4600	6500	Field oxide for isolation (e.g., LOCOS)
Light blue-metallic blue	1500	3000	4900	6800	Field oxide for isolation (e.g., LOCOS)
Green	1800	3300	5200	7200	
Yellow	2100	3700	5600	7500	
Gold with slight yellow orange	2200	4000	6000		
Orange-melon	2500	4400	6200		

Some new types of oxidation under investigation include the growth of thick oxides at high pressures and the growth of ultrathin oxides (<30 Å) in rapid thermal oxidation (RTO). High-pressure oxidation of up to 25 atm of O_2 allows for a reduced temperature in preparing thick oxide fabrication. In the case of ultrathin oxides, conventional thermal oxidation must be performed <800°C to reduce the interfacial roughness between Si and SiO_2, which leads to reduced transistor channel mobility. The solution is RTO around 1050°C; high quality gate oxides have been demonstrated this way, and present development is aimed at wet RTO with in situ steam generation (ISSG). This method is said to be yet five times faster than wet oxidation with better quality than dry oxidation.

With 20 million devices per chip today, electrically isolating them from each other—to prevent cross-talk—is becoming more of a difficult task. Historically, there have been three main choices: 1) field oxidation, 2) local oxidation of silicon (LOCOS), and 3) the latest, shallow trench isolation (STI). To isolate features using field oxides, thick field oxides (>10,000 Å) are grown and patterned over electrically inactive regions of the wafer and etched. In LOCOS, a silicon nitride mask allows for the selective thermal oxide growth in CMOS and bipolar (Bi)-CMOS processes. In STI, also known as "box isolation technique," high-aspect-ratio trenches (5 μm deep and 2 μm wide) are etched in the substrate, and then through repetitive, relatively short oxidations, the trench is refilled.

In Figure 4.33, we compare field oxidation (A) with the LOCOS process (B). It is the latter process we currently want to detail a bit further. The LOCOS process was for a time so important for the isolation of MOS devices that it became almost synonymous with the term "MOS," until recently, when LOCOS was supplanted by the "box isolation technique."

Localized Oxidation of Silicon—LOCOS

In LOCOS, as the name implies, the goal is to oxidize Si only locally, wherever a field oxide is needed.

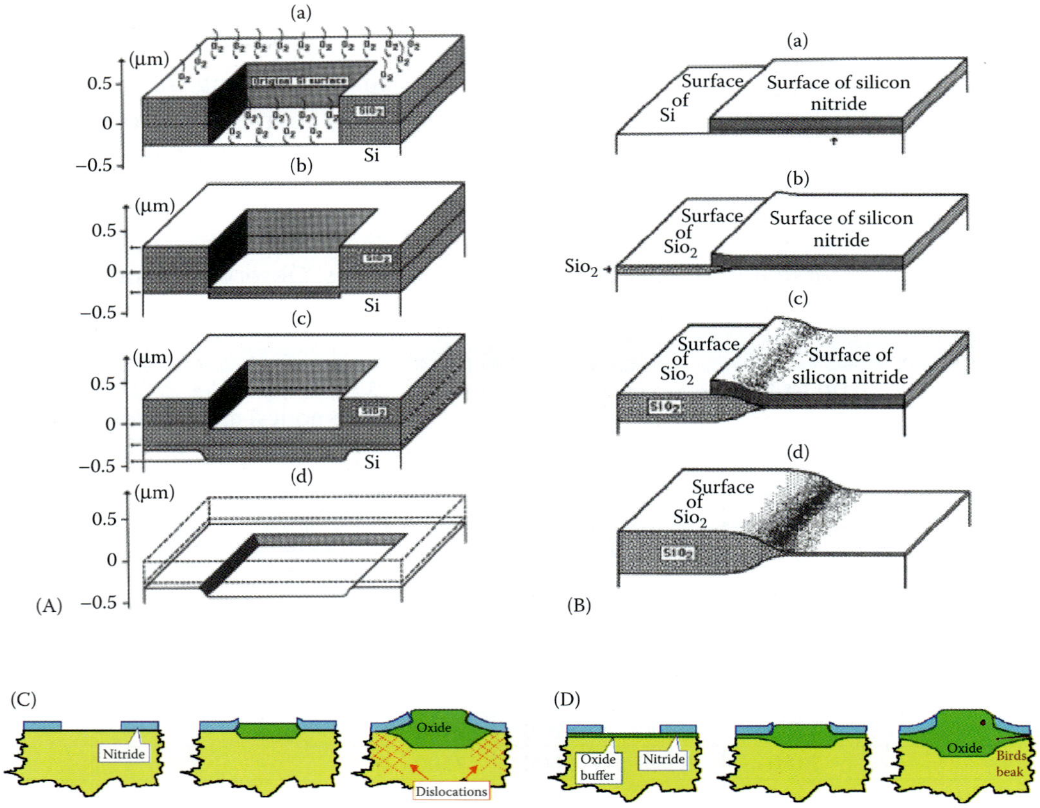

FIGURE 4.33 Device isolation through field oxide isolation (A) and local oxidation of Si (LOCOS) (B). (C) Silicon nitride is deposited directly on Si. (D) Silicon nitride is deposited on a buffer layer of oxide. See also the bird's beak.

For a local oxidation, a material that does not allow oxygen diffusion at the typical oxidation temperatures of 1000–1100°C must protect the areas of the Si not to be oxidized. The only material that is used for this purpose is silicon nitride, Si_3N_4. It can be deposited and structured easily, and it is compatible with Si IC processing. However, Si_3N_4 introduces two major problems when deposited directly on top of Si. Using a Si_3N_4 layer for protecting parts of the Si from thermal oxidation does not prevent oxygen diffusion through the oxide already formed from oxidizing the Si underneath the Si_3N_4 (lateral oxidation). Another major problem is the volume expansion that occurs in Si oxidation, which expands a given volume of Si so that the nitride mask is then pressed upward at the edges as illustrated in Figure 4.33C. With increasing oxidation time and oxide thickness, pressure under the nitride mask increases, and at some point, the critical yield strength of Si is exceeded. Plastic deformation then starts, and dislocations are generated and move into the Si. Below the edges of the local oxide a high density of dislocations may destroy the electronic function under construction. The oxidation step is not even necessary to produce the dislocations because Si_3N_4 layers are always under large stresses even at room temperature and exert shear stresses on the Si. The important message is that one cannot deposit Si_3N_4 directly on Si.

To save the LOCOS concept, one must introduce a buffer oxide. A buffer oxide between the Si_3N_4 mask and the Si substrate relieves the stress building up during oxidation. The SiO_2 buffer layer acts as a "grease" material: the hard Si_3N_4 is now pressing down on something "soft," and the stress felt by the Si does not reach the yield stress. This way one avoids dislocations, but a comparatively large lateral oxidation instead results, leading to a configuration known as "bird's beak" for the obvious reason shown in Figure 4.33D. The lateral extension of the field oxide via the bird's beak is comparable with its thickness and limits the minimum feature size. Although this was not a serious problem in the early days of IC technology, it could not be tolerated beyond the middle of the 1980s. One solution was the use of a poly-Si layer as a sacrificial layer. It was put on top of the buffer oxide below the nitride mask and structured with the same mask. It provided sacrificial Si for the "bird's beak," and the total dimension of the field oxide could be reduced. But even this is not sufficient for feature sizes around and below 1 µm.

The LOCOS process eventually became a very complicated process in its own right, and with feature sizes shrinking ever more, LOCOS reached the end of its useful lifespan in the 1990s and had to be replaced by "box isolation." The idea in box isolation is clear: etch a hole (with vertical sidewalls) in the Si wherever you want an oxide, and next simply "fill" it with oxide.

Orientation and Dopant Dependence of Oxidation Kinetics

The thermal oxidation rate, as mentioned above, is slightly influenced by the orientation of the Si substrate. The effect involves the linear oxidation rate constant used in the Deal-Grove model in the regime where the surface reaction is rate limiting. This constant was given as *B/A* (in µm/h) in Equation 4.45. The ratio of this constant, for a (111) Si plane to that for a (100) Si plane, is given by:

$$\frac{\frac{B}{A}(111)}{\frac{B}{A}(100)} = \frac{C_1(111)\exp\left(-\frac{2.0\ \text{eV}}{kT}\right)}{C_1(100)\exp\left(-\frac{2.0\ \text{eV}}{kT}\right)} = 1.68 \qquad (4.47)$$

Notice that the activation energy for both dry and wet oxidation in the surface reaction limited regime is 2 eV. Thus, a (100) surface oxidizes about 1.7 times more slowly than a (111) surface. The lower oxidation rate of (100) surfaces might be because of the fewer silicon bonds on the surface with which oxygen can react. The available bond density at a (111) plane equals 11.76×10^{14} cm^{-2} and 6.77×10^{14} cm^{-2} for the (100) plane. The linear oxidation rate for Si follows the sequence (110) > (111) > (311) > (511) > (100), corresponding to an increasing activation energy that incorporates a term for the bond density in the plane, as well as one for the bond orientation. As might be expected, steric hindrance results in higher activation energy.[4]

In anisotropic wet etching, as we will discover in Volume II, Chapter 4, the sequence of the etch rates is mostly reversed compared with that of oxidation. For example, a (100) plane etches up to 100 times faster than a (111) plane. It has been postulated that the slow etching of a (111) Si plane relative to a (100) plane may be because of its more efficient oxidation, protecting it better against etching than a (100) plane.[5] In this interpretation—etching being an electrochemical process—one assumes that anodic oxidation, just like thermal oxidation, will occur faster at a (111) Si surface. To our knowledge, no evidence to support this theory has been presented as of yet. Moreover, Seidel et al.[6] pointed out that the huge anisotropy in wet etching of Si could not be credited to the number of bonds available on different Si planes, as that number barely differs by a factor of 2 (see Volume II, Chapter 4). From the numbers observed, such an explanation could be valid for the anisotropy in thermal oxidation. There is no agreement, however, about the exact mechanism, and an understanding of the orientation dependence of the oxidation rate is still lacking. It has been attributed at various times to the number of Si-Si bonds available for reaction and the orientation of these bonds, the presence of surface steps, mechanical effects such as stress in the oxide film, and the attainment of maximum coherence across the Si/SiO_2 interface.[4] It should also be noted that the oxidation rate sequence depends on temperature and oxide thickness—one more reason to take any model with a grain of salt.[4]

In Si, n-type dopants (As and P) at high concentrations (n^{++}) lead to increases in oxidation rate, particularly at low temperatures. It is found that B/A is affected but that B remains unaffected. Ho-Plummer explained this phenomenon by postulating that point defects (vacancies) are being consumed during the volume expansion that accompanies Si oxidation.[7,8] The oxidation rate is changed much less in case of highly boron-doped (p^{++}) wafers. Whereas boron atoms are taken up by the oxide, P and As pile up at the Si/SiO_2 interface.[9]

Oxidation is also influenced by stress. Etched substrates exhibiting shaped structures oxidize at different rates than planar wafers. Oxide layers formed on silicon are under compressive stress, even in the planar case, and these stresses can be much larger on curved surfaces because the volume expansion is dimensionally confined. For example, oxidation occurring in a confined corner, in which volume expansion is more difficult, is different from oxidation of a flat surface. Retardation of oxidation is very strong for sharp corners, i.e., for r small or $1/r$ large (e.g., a factor of two retardation in some cases). Retardation is also more pronounced for lower temperatures, and there are virtually no corner effects observed for 1200°C. Interior (concave) corners show more pronounced retardation effects than exterior (convex) corners.

It is important to include stress when simulating oxide growth in the LOCOS process (see above). Si oxidation modeling using ATHENA software predicts the oxide shape much better when stress is included as a parameter (http://www.silvaco.com/tech_lib/simulationstandard/1997/aug/a3/a3.html). In Volume III, Chapter 10 we show how oxidation sharpening is used to make sharp Si needle tips for scanning tunneling microscopes (STMs) and atomic force microscopes (AFMs; Volume III, Figure 10.12).

Nonthermal Methods for Depositing SiO_2 Films

Thermal oxidation involving the consumption of a thin layer of the Si surface results in excellent adhesion and good electrical and mechanical properties. It is an excellent technique when available. However, the thermal oxidation described in the previous section is by no means the only method available to implement a SiO_2 film. Two nonthermal SiO_2 deposition methods include electrochemical oxidation of Si and chemical vapor deposition. Electrochemical oxidation of Si has been studied (see, e.g., Schmidt[10] and Lewerenz[11]) but has not led to commercial usage. Even though these anodic oxides are pinhole free, they exhibit many interface states. Madou et al.[12] investigated mechanisms for anodic oxidation of Si and the introduction of oxide "dopants" into these anodically formed oxides. Such oxide dopants can be used as a source for doping of the underlying Si at elevated temperatures. Further study material on silicon oxidation can be found in Fair[13] and Katz.[4]

Silicon dioxide is frequently used as an insulator between two layers of metallization. In such cases, some form of deposited oxide must be used rather than grown oxides. Deposited oxides can be produced by various reactions between gaseous silicon compounds and gaseous oxidizers. Deposited oxides tend to possess low densities and large numbers of defect sites, meaning they are not suitable for use as gate dielectrics for MOS transistors but are acceptable for use as insulating layers between multiple conductor layers or as protective overcoats.

For deposition of oxides at lower temperatures, we will learn in much more detail in Volume II, Chapter 7 that one may use low-pressure chemical vapor deposition (LPCVD), such as in the following reaction:

$$SiH_4 + O_2 \rightarrow SiO_2 + 2\ H_2\text{: } 450°C \qquad \text{Reaction 4.3}$$

Another advantage, besides the lower temperature, is the fact that one can dope an LPCVD silicon oxide film easily to create, for example, a phosphosilicate glass (PSG):

$$SiH_4 + 7/2\ O_2 + 2\ PH_3 \rightarrow SiO_2\text{: } P + 5\ H_2O\text{: } 700°C \qquad \text{Reaction 4.4}$$

Phosphorus-doped glass, also called *P-glass* or *PSG*, and borophosphosilicate glasses (BPSGs) soften and flow at lower temperatures, enabling the smoothing of topography. They etch much faster than SiO_2, which benefits their application as sacrificial material in surface micromachining (see Volume II, Chapter 7).

Properties of Thermally Grown SiO_2

A film of thermal SiO_2 performs excellently as a mask against diffusion of the common dopants in Si. The diffusion coefficient, *D*, of boron at 900°C in SiO_2 is 2.2×10^{-19} cm²/s compared with 4.4×10^{-16} in Si. For phosphorus, *D* equals 9.3×10^{-19} cm²/s compared with 7.7×10^{-15} in Si (from the common Si dopants, only Ga diffuses fast through the oxide with $D = 1.3 \times 10^{-13}$). For some other elements, SiO_2 forms a poor diffusion barrier. The amorphous oxide has a more open structure than crystalline quartz—only 43% of the space is occupied (see Figure 4.34). Consequently, a wide variety of impurities (especially alkali ions such as Na^+ and K^+) can readily diffuse through amorphous SiO_2. Diffusion through the open SiO_2 structure happens especially fast when the oxide is hydrated. One

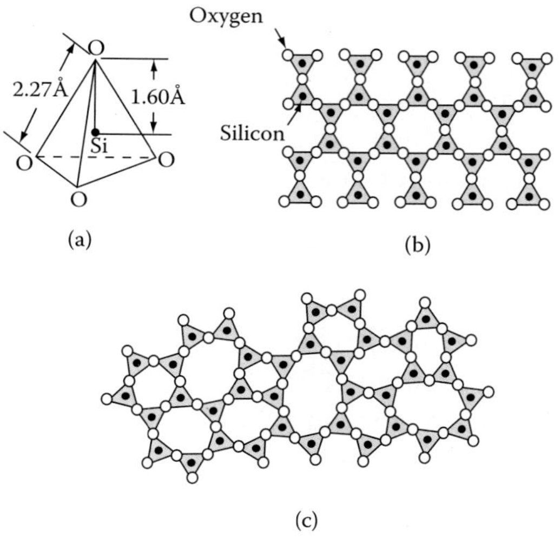

FIGURE 4.34 (a) Basic structural unit of silicon dioxide. (b) Two-dimensional representation of a quartz crystal lattice. (c) Two-dimensional representation of the amorphous structure of silicon dioxide.

of the reasons for the poor performance of silicon dioxide-based ion-sensitive field-effect transistors (ISFETs) in aqueous solutions can be traced back to the simple observation that SiO_2 in water almost behaves like a sponge for ions.[14] This is why, in such devices, SiO_2 is often topped off with the ionic barrier material Si_3N_4. The use of SiO_2 as a diffusion mask often stems from convenience; SiO_2 is easy to grow, whereas Si_3N_4 cannot be put directly onto Si without problems.

The quality of silicon dioxide depends heavily on its growth method. Dry oxidation at high temperatures (between 900 and 1150°C) in pure oxygen produces a better quality oxide than steam oxidation. Such a thermal oxide is stoichiometric, has a high density, and is basically pinhole free. Wet oxidation in steam occurs much faster but produces a lesser quality oxide, with water causing a loosening effect on the SiO_2, making it more prone to impurity diffusion. Both types of oxidation are carried out in a quartz tube. Oxide thicknesses of a few tenths of a micrometer are used most frequently, with 1–2 μm being the upper limit for conventional thermal oxides. Addition of chlorine-containing chemicals during oxidation increases the dielectric breakdown strength and the rate of oxidation while also improving the threshold voltage of many electronic devices.[15] However, too high concentrations of halogens at high temperatures can pit the silicon surface. The influence of chlorine, hydrogen, and other gases on the SiO_2 growth rate and the resulting interface and oxide quality has been a fertile research field, a short summary of which can be found in Fair.[13] It should be noted that for the IC industry, the electronic quality of the Si/SiO_2 interface is expressed through low concentrations of interface trap states, low fixed oxide surface charges, low bulk oxide trapped charges, and mobile charges (impurity ions such as Li^+, Na^+, K^+) (Figure 4.35);[16] the interface quality is of little consequence in micromachining. In micromachining, the oxide is used as a structural element, a sacrificial layer, or a dielectric in a passive device, applications where the interface quality rarely matters. One notable exception is the aforementioned micromachined ISFET, where the electronic properties of the gate oxide are key to its functioning; a perfect oxide and oxide/semiconductor interface hold as much importance in this case as it does in the IC industry.

Because of molecular volume mismatch and thermal expansion differences, a thermal silicon oxide is compressed. The compressive stress depends on the total thickness of the silicon dioxide layer and can reach hundreds of MPa. Oxide films thicker than 1 μm can cause bowing of the Si wafer. Wet oxidation relaxes the compressive stress while speeding up the SiO_2 growth.

In Table 4.8, we review some of the more significant properties of thermally grown SiO_2. The oxide is characterized by a variety of means reviewed in Volume III, Chapter 6 on metrology. Electrical tests are conducted to establish the capacitance (CV) at high and low frequencies to calculate the interface density, to measure dielectric breakdown, and for electrical stressing (with temperature ramp). Optical tests are carried out for thickness and optical constant determination using color (see Table 4.7), interferometry (needs calibrated substrate), and ellipsometry. Physical tests for thickness, surface quality, etc. involve profilometry, AFM, and cross-sections by scanning electron microscopy (SEM)/transmission electron microscopy (TEM).

Oxidation furnaces may be purchased from Thermcraft (http://www.thermcraftinc.com/profile.html) and Tystar (http://www.tystar.com) (which makes Tytan Horizontal Furnace Systems; http://www.tystar.com/furnace.htm).

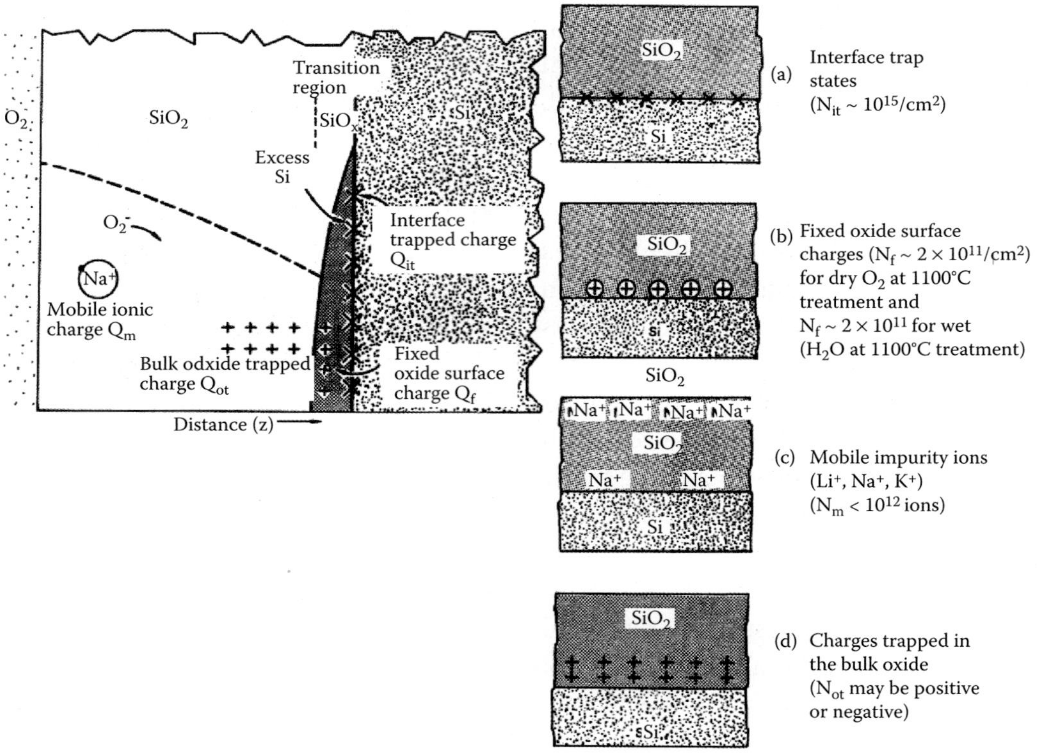

FIGURE 4.35 Location of oxide charges in thermally oxidized silicon structures as defined by Bruce Deal. Q_f and Q_{it} are much lower for (100) silicon than for (111).

Table 4.8 Properties of Thermal SiO_2

Property	Value
Density (g/cm³)	2.24–2.27 (dry) and 2.18–2.20 (wet)
Dielectric constant	3.8–3.9
Dielectric strength (V/cm)	2×10^6 (dry) and 3×10^6 (wet)
Energy gap (eV)	~8
Etch rate in buffered HF (Å/min)	1000
Infrared absorption peak (μm)	9.3
Melting point (°C)	~1700
Molecular weight	60.08
Molecules/cm³	2.3×10^{22}
Refractive index	1.46
Resistivity at 25°C (Ωcm)	3×10^{15} (dry) and 3–5 $\times 10^{15}$ (wet)
Specific heat (J/g°C)	1.0
Stress in film on Si (dyne/cm²)	2–4 $\times 10^9$ (compressive)
Thermal conductivity (W/cm°C)	0.014
Thermal linear expansion coefficient (°C⁻¹)	5×10^{-7} (0.5 ppm/°C)

The End of the Line for SiO_2?

In the following sections we learn about the gate oxide thickness crisis. The projected gate oxide thickness for 70 nm CMOS of 10–15 Å results in large tunneling currents (see Chapter 3) through the oxide. It is a challenge to find a replacement for thermal silicon dioxide with comparable dielectric properties. Oxides with high dielectric constant $\varepsilon_r >> 3.9$ are most desirable. Possible SiO_2 replacements developed at present are silicon oxynitride (SiO_xN_y), Al_2O_3, Ta_2O_5, and Hf and Zr silicates.

Si-Based Electronic Devices

Introduction

The following active electronic devices will be reviewed in this section: diodes, special-purpose diodes [including light-emitting diodes (LEDs), photodiodes, solar cells, Zener diodes, avalanche diodes, and tunneling (Esaki) diodes], and two types of transistors (bipolar and MOSFET). Si and Ge are the two most common single elements that are used to make diodes. A compound semiconductor that is commonly used is GaAs, especially in the case of LEDs because of its large direct bandgap. We analyze in some detail the scaling issues involved in the further miniaturization of Si MOSFETs, and in this context we introduce strained Si, which expands the prospect of continued Si use. Fundamental limits to MOSFET downscaling include not only electromagnetic, quantum mechanical, and thermodynamic effects but also technical and economic (practical) considerations. In the current volume Chapters 3 and 5 on quantum mechanics, and photonics, respectively and in Volume III, Chapter 3 on nanotechnology, alternative device operating principles that exploit the quantization effects, rather than being hampered by them, are investigated. These approaches include quantum wells, quantum wires, quantum dots, resonant tunneling diodes and transistors, superlattices, and molecular electronics.

Diodes

Many semiconductor devices are based on the properties of the junction between p-type and n-type semiconductors (Figure 4.36). Russell Kohl at Bell Labs

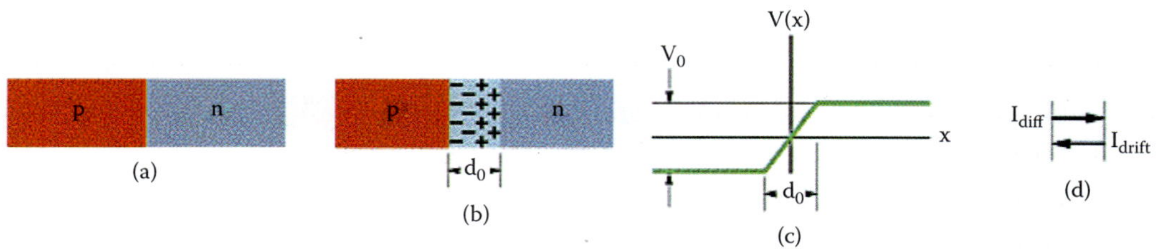

FIGURE 4.36 (a) p and n semiconductor before majority carriers cross the junction. (b) Motions of majority charge carriers across the junction uncover two space-charge layers associated with uncompensated donor ions (to the right of the plane) and acceptors ions (to the left). (c) Associated with the space charge is a contact potential difference V_0, which acts to limit the flow of majority carriers. (d) The diffusion of majority carriers across the junction produces a diffusion current, I_{diff}. The small concentration of minority carriers on either side of the junction moves in the opposite direction of the majority carriers, giving rise to a drift current, I_{drift}. In an isolated junction, the diffusion current is compensated for by the drift current, with the result that the net current through the junction is zero.

invented the solid-state p-n junction diode in 1940. This device, when exposed to light, produced a voltage of 0.5 V across the junction and represented a stepping stone to the invention of the transistor (see Chapter 1). In a diode p-n junction, electrons from the n-type diffuse into the p-type and holes diffuse the other way until the charge transfer builds up a big enough potential difference to halt the diffusion. As typical for diffusion, carriers flow from where there are many to where there are few. This motion of the majority charge carriers across the junction causes two space-charge layers to form associated with the uncompensated donor ions (to the right of the junction) and acceptors ions (to the left of the junction). The interface where the p- and n-type materials meet is called the metallurgical junction (Figure 4.36a). The two space-charge regions form a depletion zone that is free of mobile charge carriers, with the ionized donor and acceptor ions fixed in their lattice sites. This double layer of opposite net electric charges is also called a *dipole layer* or a *sandwich*. A depletion layer, d_0, typically extends a few micrometers on either side of the junction (Figure 4.36b). Associated with the space charge is a contact potential difference V_0, which acts to limit the flow of majority carriers with a self-induced electric field (10^3–10^5 V/cm), opposing further diffusion (Figure 4.36c). The diffusion of majority carriers across the junction produces the diffusion current, I_{diff}. A small concentration of minority carriers on either side of the junction moves in the opposite direction of the majority carriers and gives rise to a drift current, I_{drift} (Figure 4.36d). The electrical field is driving this current, not a concentration gradient. In a junction in equilibrium, the drift current compensates for the diffusion current, resulting in a net zero current through the junction. Even though there is no net current, electrons flow across the junction; it is just that the rates of flow in each direction are equal; such equilibrium is called *dynamic equilibrium*. In a real p-n junction, the boundaries of the depletion zones are not as sharp as shown in Figure 4.36, but the contact potential curve is smooth with no sharp corners.

The contact potential or barrier voltage, according to the Boltzmann relation, is given as:

$$V_0 = \frac{kT}{e}\ln\frac{p_p}{p_n} \qquad (4.48)$$

where p_p = hole concentration in the p-type material
p_n = hole concentration in the n-type material

The term kT/e (= V_T) is the thermal voltage (~26 mV at room temperature). With $p_p \approx N_A$ and $p_n \approx n_i^2/N_D$, where n_i is the intrinsic carrier concentration, the barrier voltage can be approximated as:

$$V_0 = \frac{kT}{e}\ln\frac{N_A N_D}{n_i^2} \qquad (4.49)$$

where N_A = concentration of immobile acceptors on the p-side of the junction
N_D = immobile donors on the n-side of the junction

For Si at room temperature, V_0 is typically around 0.67 V (with $N_A = 10^{17}$ cm^{-3} and $N_D = 10^{15}$ cm^{-3}). A typical width of the junction with these numbers is about 4 μm, and the electrical field E in this case is about 60 kV/cm.

The majority carriers in the depletion layer disappear through recombination in which electrons and hole pairs annihilate each other. When equilibrium is reached, a single Fermi level characterizes the p-n junction (Figure 4.37).

The above describes a p-n junction in an equilibrium situation. Now, we consider what happens when we apply a bias, as shown in Figure 4.38. When the p-side of a p-n junction is connected to the + terminal of a voltage source, and the n-side to the – terminal, the majority carriers are repelled by their respective terminals toward the junction. This is called *forward bias*, and with such a bias the depletion layer becomes narrower (d_F), and most electrons can go across the junction (right to left flow of electrons makes a positive current left to right because positive current flow is opposite of

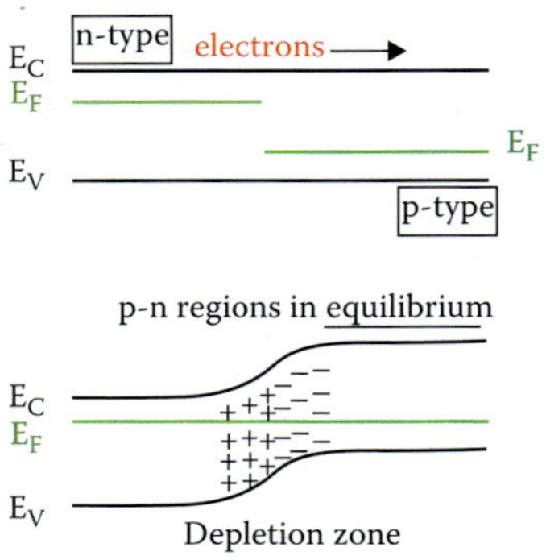

FIGURE 4.37 A single Fermi level characterizes the whole p-n junction in equilibrium.

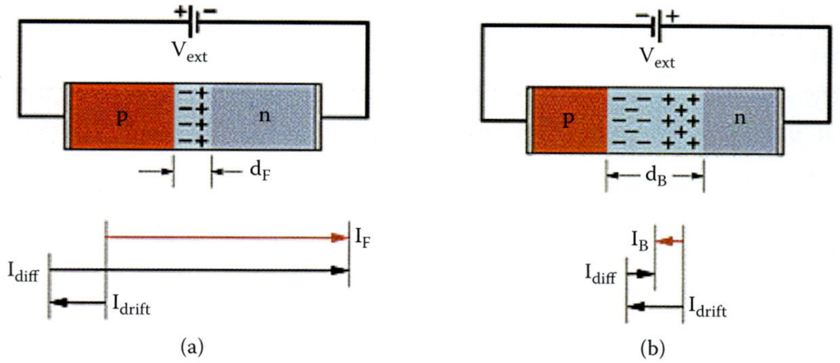

FIGURE 4.38 (a) The forward-biased connection of a p-n junction, showing the narrowed depletion (d_F) and the larger forward current I_F. (b) The reverse (back)-biased connection, showing the widening depletion zone (d_B) and the small reverse (back) current I_B.

the negative electron flow) (Figure 4.38a). With the shrinking of the depletion layer, the energy required for charge carriers to cross the depletion region decreases exponentially. Therefore, as the applied voltage increases, current starts to flow across the junction. The barrier potential of the diode is the voltage at which appreciable current starts to flow through the diode. The barrier potential varies for different materials. When connected in the opposite direction, the barrier gets higher, and practically no current flows. This is called *back bias* or *reverse bias*. The depletion layer is very wide (d_B) when the reverse bias is high (Figure 4.38b). Very few electrons get into the depletion layer from the neutral parts of the diode. Only electrons "produced" inside the depletion layer will move through it. A small leakage current, also saturation current, flows under these reverse-biased conditions.

A typical current-voltage plot (I–V) for a diode is shown in Figure 4.39. Mathematically, the forward current (I_F), with diffusion dominating, is described by an exponential as a function of the voltage. The reverse "leakage" current (I_B), where drift dominates, is almost constant over most of the reverse voltage range (negative potentials in Figure 4.39). Combined we obtain:

$$I = I_0\left(e^{\frac{eV}{\eta kT}} - 1\right) \tag{4.50}$$

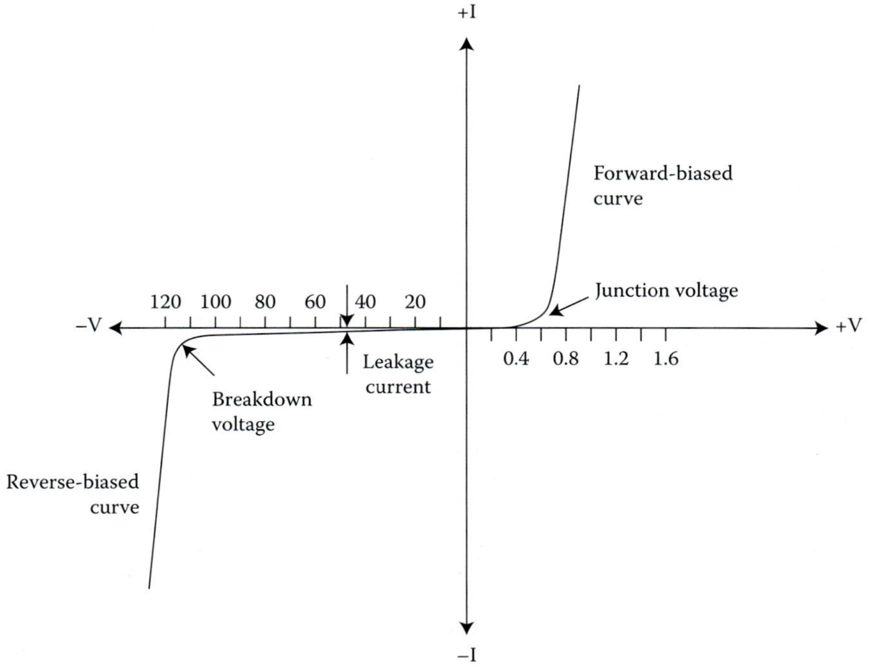

FIGURE 4.39 Typical diode current-voltage curve.

This equation is referred to as the Shockley equation. In it, η is the emission coefficient for the diode, which is determined by the way the diode is constructed and somewhat varies with diode current. For a silicon diode, η is around 2 for low currents and goes down to about 1 at higher currents. The reverse current, I_0, depends on the number of electrons per second that "appear" in the depletion layer, not on the voltage. Only the temperature and the bandgap have an influence on I_0. The simple Shockley equation does not explain the sudden increase of the leakage current at voltages above the breakdown voltage. We will explain this effect further below when discussing Zener and avalanche diodes.

One important use for diodes is to convert alternating current into direct (unidirectional) current in power supply circuits. Diodes are also useful in some logic devices, and we can make some types of digital logic circuits using only diodes and resistors.

Special-Purpose Diodes

Introduction

There are many types of specialized diodes. In this section we briefly introduce light-emitting diodes (LEDs), solar cells, Zener diodes, avalanche diodes, and tunnel or Esaki diodes. In Chapter 5 we discuss double-barrier resonant tunneling diodes (DBRTs) and compare laser diodes with quantum well lasers. The existence of the zero point energy for confined electrons in a quantum well laser (see Chapter 3) makes a quantum well laser much more efficient than a diode laser. In a quantum well, there are no allowed electron states at the very lowest energies. What that means is that the lowest conduction state in a quantum well is broad so that many electrons can be accommodated. Similarly, the top of the valence band has plenty of states available for holes. This means that it is possible for many holes and electrons to combine and produce photons with identical energy for enhanced probability of stimulated emission (lasing) (see Chapter 5). Peltier diodes are described in Volume III, Chapter 8 on actuators.

Light-Emitting Diodes

Electrons and holes can recombine in extrinsic semiconductors that have a substantial number of electrons in the conduction band and many holes in the valence band. On recombination, energy conservation requires that photons and/or phonons be produced. The only place where there are enough electrons and holes to produce bright light is at the p-n junction when a large current is flowing (i.e., in a forward-biased p-n junction). If the bandgap corresponds to the energy of a photon of visible light, then that type of radiation is produced. This is the light-emitting diode (LED), used extensively in electronic devices and cars as indicator lamps, as picture elements in color matrix displays for laptop computers, and so on. Being an indirect bandgap material, Si is currently not used for LEDs, and the most utilized material is direct bandgap GaAs (see Chapter 5). The earliest LEDs produced infrared or visible red light, but LEDs in yellow, green, and blue visible light colors are now available (Figure 4.40). These LEDs convert electrical input to light output. Greater difference in energy levels in the p- and n-sides of the diode produces greater energy change, and consequently higher-frequency, shorter-wavelength light.

A forward-biased p-n junction, showing electrons being injected into the n-type material and holes into the p-type material, is sketched in Figure 4.41. Light is emitted from the narrow depletion zone each time an electron and a hole combine across that zone. The ratio of generated photons to the electrons injected across the junction is called the quantum efficiency, which is a key measure of how well a light emitter is working. For high-performance III–V LEDs, the efficiency is around 10%.

GaAs is a direct energy gap semiconductor with an energy gap of 1.424 eV in the near-infrared portion of the electromagnetic spectrum. This semiconductor is often used in LED construction in combination with AlAs, an indirect energy gap semiconductor with an energy gap of 2.168 eV in the yellow-green portion of the electromagnetic spectrum. As the mole fraction x of $Al_xGa_{1-x}As$ is raised from zero, the energy gap of the resulting compound increases from that of GaAs to that of AlAs. When the value of x is 0.45, the semiconductor switches from a direct energy gap to an indirect energy gap. GaAs and AlAs differ in lattice constants by less than 0.2% at 250°C, which allows for

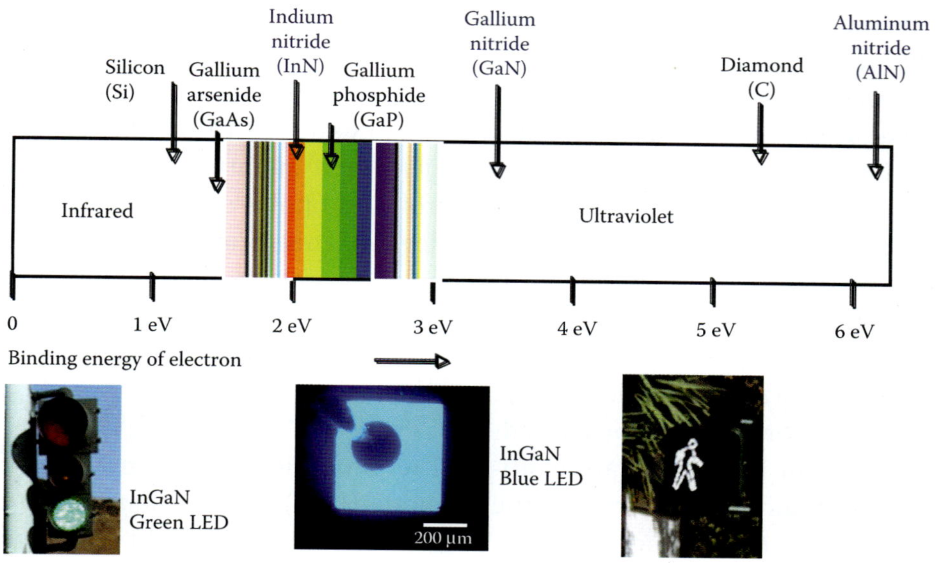

FIGURE 4.40 LEDs of all colors: the larger the difference in energy levels in the p- and n-sides of a diode, the shorter the wavelength of the light.

the growth of very high-quality AlGaAs films on GaAs substrates. Molecular beam epitaxy (MBE) is a crystal-growth technique in which material is deposited by evaporating individually controlled source materials onto a heated substrate in an ultrahigh vacuum chamber. The growth rates in MBE are very slow (0.1–2 mm/h). MBE is capable of producing high-quality AlGaAs heterostructure devices with very uniform thickness and doping characteristics. Crystal growth of LEDs by metalorganic chemical vapor deposition (MOCVD) and liquid-phase epitaxy (LPE) began in the late 1960s. By using MOCVD, it is easier to scale up to larger growth areas than with LPE, and the uniformity and surface morphology of MOCVD grown lasers are superior (MBE, LPE, and MOCVD are compared in Volume II, Chapter 7).

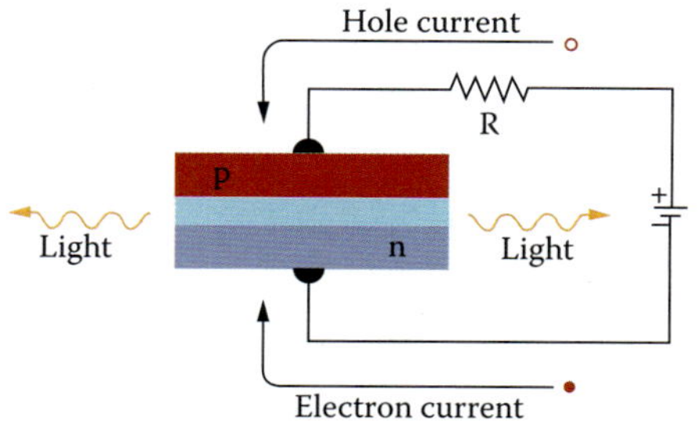

FIGURE 4.41 A light-emitting diode (LED): forward based p-n junction, showing electrons being injected into the n-type material and holes into the p-type material.

The absorption coefficient for direct and indirect bandgaps was calculated using Fermi's golden rule as $\alpha(\omega) \propto G(E) \propto (\hbar\omega - E_g)^{\frac{1}{2}}$ (Equation 3.301) and $\alpha(\omega) \propto G(E) \propto (\hbar\omega - E_g \pm \hbar\nu_p)^2$ (Equation 3.302) (with ν_p the phonon frequency and + for absorption and − for emission), respectively. Emission in an indirect bandgap semiconductor (e.g., Si) is a three-particle interaction (1 electron + 1 photon + 1 phonon); therefore, the transition is unlikely, and such materials are poor light emitters. In silicon, few excited electrons generate photons, and most recombinations result in heat rather than light; thus, the quantum efficiency is poor. In other words, silicon as an indirect bandgap is a lousy light emitter: the probability that a phonon with just the right amount of momentum will meet an electron/hole pair is not very good. The occurrence of such a photon-generating transition in a III–V material is thousands of times more likely than in silicon. As a consequence, until recently, most people in the semiconductor industry would assure you that silicon is pretty much worthless at turning electricity into light, and as we just saw, LEDs and lasers are made of the more exotic III–V semiconductors. But these materials, including gallium arsenide, gallium nitride, and indium phosphide, are anywhere from 30 to 200 times more expensive than silicon. Moreover, the manufacturing infrastructure and know-how about making ICs from silicon is useless for making chips from III–V semiconductors. To connect the manufacturing of microprocessors with that of LEDs and lasers, one would like to develop cheap, integrated optoelectronics with all components made from silicon.

In Chapter 3, we saw that one approach to getting more light out of silicon involves quantum confinement, with electron/hole pairs physically restricted to a small volume, typically less than 30 nm^3, or 300 times the size of a typical atom. This can be accomplished, for example, by embedding nanocrystals of silicon in an insulating silicon dioxide shell. Within such a nanocrystal, the energy levels of the valence and conduction bands differ significantly from those in bulk crystals, and in general, the smaller the nanocrystal, the larger the bandgap (blue shift). The latter represents an example of tuning a device's optical properties by bandgap engineering, which is achieved in this case by the fine control of the nanocrystal's growth during the manufacturing process. It turns out that quantum confinement also reduces silicon's momentum problem, thus increasing the probability that injected electrons will produce photons. In confined silicon nanostructures, one observes a blue shift for the bandgap absorption onset, and some forbidden transitions become feasible. In other words, through nanotechnology, Si might yet be made into a good candidate for LEDs and lasers after all (another approach to make Si a better photonic material is seeding the silicon with lanthanide rare earth element ions, which give off light when electrically excited).

One important reason to use LEDs is that they last longer, are brighter, and are more efficient than incandescent lights. LED lamps use as little as 10% of the electricity that an incandescent light uses and have lifespans of 100,000 h, which is more than 11 years of continuous use. A laser diode is made by putting an LED-type structure in a resonant cavity formed by polishing the parallel end faces of the semiconductor (see Chapter 5). Photons bounce between the reflective ends, colliding with atoms and stimulating the emission of additional photons that are in phase with the others in the region. In a laser, the concentrated, active region bounded by the reflective ends is known as the resonant cavity. Laser diodes are used in optical storage devices and for high-speed optical communication. As we already alluded to in Chapter 3, they are less efficient than quantum well lasers because more electrons and holes can be accommodated in the latter devices. This means that it is possible for many more holes and electrons to combine and produce photons with identical energy for enhanced probability of stimulated emission (lasing).

Photodiodes and Solar Cells

The photodiode is the inverse of the light-emitting diode (LED). Rather than giving off light it is sensitive to incoming light. Photons of energy E_g or greater create electron/hole pairs in the junction region. The electrons and holes are then swept out of the junction region by the electric field that exists before they can recombine. Photodiodes can be operated in two different modes, namely, the photovoltaic mode and the photoconductive mode. In the photovoltaic mode, like in a solar cell, the illuminated photodiode generates a voltage. However, the dependence of this voltage on the light intensity is rather nonlinear, and the dynamic range is typically small. Although the maximum sensitivity is not achieved, the area of the junction is large, and enough electron/hole pairs are produced so that significant solar energy can be taken from the device. In the photoconductive mode, one applies a reverse voltage to the diode (i.e., a back bias) and measures the resulting photocurrent. The dependence of the photocurrent on the light power can be very linear over six or more orders of magnitude of the light power. This type of diode is very sensitive and is used in CD players.

The electron/hole pairs created by light in a photodiode result in an additional component of current flowing in the diode:

$$I = I_0\left(e^{\frac{eV}{kT}} - 1\right) - I_{optical} \qquad (4.51)$$

with the first term the usual diode equation (Equation 4.50), and the second term the photocurrent, $I_{optical}$, which is proportional to the light intensity. Photodiodes are constructed so their p-n junction can be exposed to incoming photons through a clear window or lens. To increase the volume in which radiation can be absorbed, positive intrinsic negative (PIN) diodes are often used. A PIN diode has a central undoped, or intrinsic, layer, forming a p-type/intrinsic/n-type structure. An internal electric field is maintained over a wider layer with the benefit that the depletion region now exists almost completely within the intrinsic region, which is a constant width (or almost constant) regardless of the reverse bias applied to the diode. This intrinsic region can be made large, increasing the volume where electron/hole pairs can be generated. For these reasons many photodetectors are based on PIN diodes.

In solar cells, photons coming from the sun excite electrons in the valence band and raise them to the conduction band. Illumination of a p-n junction produces hole pairs that diffuse to the depletion region. The minority carriers (electrons in p-type material) are swept across the depletion region and increase the reverse current. In these cells one tries to maximize the minority carrier lifetime, τ, so that long diffusion lengths, $L = (D\tau)^{1/2}$, in excess of 100 μm, can be attained. In a p-type material, electrons are the minority carriers, and the number of electrons generated per unit area per second (generation rate *G*) in a p-type material within a diffusion length of the depletion layer is given as *GL*. The larger *G* and *L*, the more electrons reach the edge of the depletion layer, where they are collected by sweeping them across the depletion layer, resulting in a current, $I_{optical}$ (see Equation 4.51), which is given by:

$$I_{optical} = eGL \tag{4.52}$$

With a reverse bias applied to the solar cell, Equation 4.52 represents the increase in current with light intensity as illustrated in Figure 4.42b. The current is bias-independent as long as $L \gg d$, with *d* the depletion layer thickness. The depletion layer thickness is the only voltage-dependent parameter.

Under open circuit conditions, no current is measured, and *I* in Equation 4.50 is equal to zero. On illumination, an open circuit voltage, $V_{optical}$, appears across the n-p junction:

$$V_{optical} = \frac{kT}{e} \ln \frac{I_{optical}}{I_0} \tag{4.53}$$

This open circuit voltage counteracts the flow of optically generated carriers, and the junction potential (V_0; see Equation 4.48) is lowered so that an equal and opposite flow of electrons from n-type to p-type is established.

Research interest in solar cells today, with soaring oil prices, is very high. Solar energy reaching the surface of earth is about 800 W/m². The current problem with solar cells is not really the efficiency (20%) but rather the cost of production. Although it has come down by a factor of 100 in 50 years, it still has a factor of 2–5 to go to down to

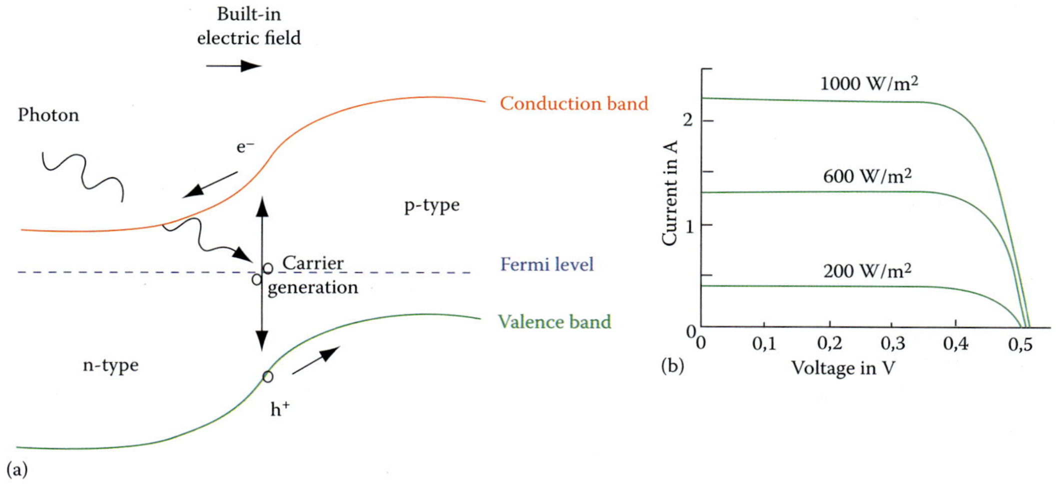

FIGURE 4.42 Illumination of a p-n junction produces hole pairs that diffuse to the depletion region. The minority carriers (electrons in p-type materials) are swept across the depletion region and increase the reverse current (a). The higher the illumination intensity the higher the reverse current (b).

be competitive with fossil plants. Another problem is that no energy can be produced at night, so energy storage is needed to provide continuous service. The highest efficiency solar cells are made of silicon single crystals. These are difficult to manufacture and are expensive. The most popular alternatives are polycrystalline and amorphous Si thin films, as well as GaAs and several of its alloys. The latter are sensitive to different parts of the solar spectrum because they have different bandgaps.

We detail progress in MEMS and NEMS applied to solar cells, including liquid junction-based solar cells (the Gräetzel cell), in Volume III, Chapter 9 on power and brains in miniature devices.

Zener, Avalanche, and Tunnel (Esaki) Diodes

In the breakdown region in the diode current-voltage curve in Figure 4.39, the diode exhibits a sudden increase in current. This is called the Zener breakdown, named for Dr. Clarence Melvin Zener of Southern Illinois University, inventor of the device. The Zener diode works at strong negative bias. If the reverse voltage is high enough to raise the top of the valence band on the p-side above the bottom of the conduction band on the n-side, then valence electrons can directly tunnel into the conduction band, and there is a sudden breakdown of the resistance (Figure 4.43). Zener diodes are used to regulate voltage: they do not allow voltage to go beyond the Zener breakdown point. They are equipped with cooling fins to keep them from melting under high voltage and high current (high power). Zener diodes

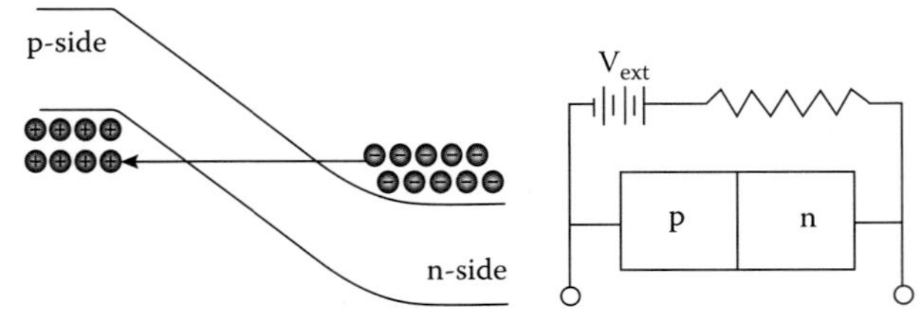

FIGURE 4.43 Band diagram for Zener diode: the top of the valence band on the p-side is above the bottom of the conduction band on the n-side, and valence electrons can directly tunnel into the conduction band, causing a sudden breakdown of the resistance. Also shown is the use of a reverse-biased p-n diode for a voltage clamping circuit. The current saturates to a value determined by the external circuit resistor, while the output voltage is clamped at the diode breakdown voltage.

are mostly used to produce an accurate reference voltage for measurement devices or analog-digital converters, and so on. Manufacturers set a precise breakdown voltage by careful doping, allowing the diode to be used as a precision voltage reference.

Some devices labeled as high-voltage Zener diodes are avalanche diodes. Avalanche diodes conduct in the reverse direction when the reverse-biased voltage exceeds the breakdown voltage. Avalanche diodes are electrically very similar to Zener diodes and are often mistakenly called Zener diodes, but they break down by the avalanche effect. This occurs when the reverse electric field across the p-n junction causes a wave of ionization, reminiscent of an avalanche, leading to a large current. In this region, the high electric field in the middle of the depletion layer accelerates electrons that are produced there (by action of heat or light) so much that they garner enough energy to "dislodge" electrons from the valence band into the conduction band. When one energetic electron can "dislodge" two or more such electrons, it starts a "chain reaction" in which these electrons can produce even more conduction electrons. Avalanche diodes are designed to break down at a well-defined reverse voltage without being destroyed. Avalanche diodes are sensitive to very low radiation because of the "multiplication" of current by the avalanche effect.

The difference between the avalanche diode (which has a reverse breakdown above about 6.2 V) and the Zener diode is that the channel length of the former exceeds the "mean free path" of the electrons, which means there are collisions between them on the way out. The only practical difference is that the two types of diodes have temperature coefficients of opposite polarities.

Yet another special-purpose diode is the tunnel diode (or Esaki diode), which, like the Zener diode, exploits tunneling across a potential barrier. The tunnel diode has the unusual characteristic that the current decreases with the applied bias voltage. Because of this negative resistance region at low voltage and its rapid response, it is capable of amplification, oscillation, and switching in electronic circuits at high frequencies. The Esaki or tunnel diode is designed from very heavily doped n and p regions. The high doping ensures a very large

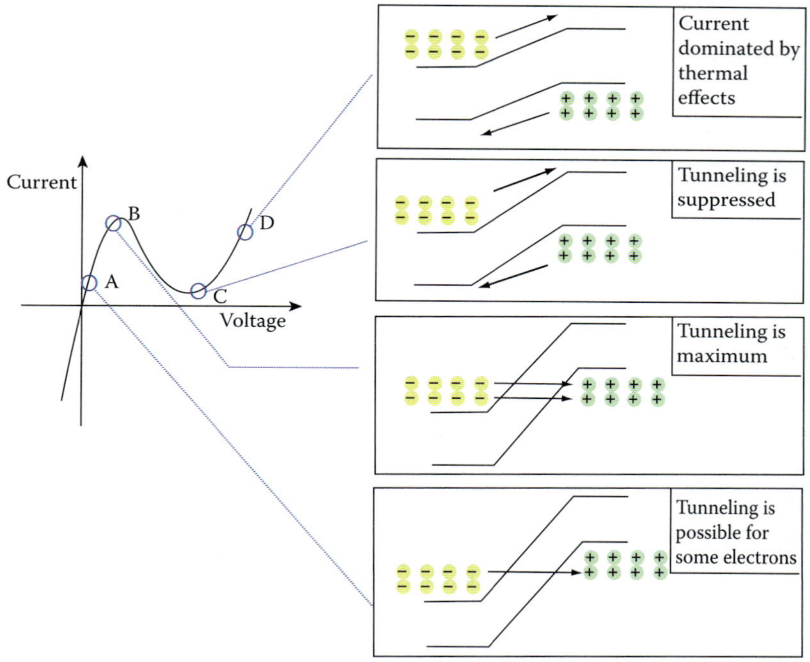

FIGURE 4.44 The tunnel or Esaki diode. For explanation, see text.

electric field in the junction so that band-to-band tunneling is possible at all biasing conditions (see Figure 4.44). The discovery of the Esaki diode was very significant as it was the first quantum mechanical electronic device where electron tunneling was clearly manifested in semiconductors, and it opened up a new field of research: tunneling in solids. Esaki described the tunnel diode in his thesis in 1957 and received the 1972 Nobel Prize for it. Shockley was so impressed by Esaki's work that at a symposium of the International Union of Pure and Applied Physics in Brussels in 1958, he announced (Figure 4.45):

> The most beautiful demonstration of the Zener effect so far achieved is presented at this symposium by L. Esaki of Tokyo. Esaki has studied p-n junctions which are very heavily doped on both sides, so that there is a built-in field of the order of 5×10^5 volt/cm. Under these conditions, current flows by the Zener mechanism even at zero voltage. Esaki finds that as forward bias is increased across the junction, this possibility of direct tunneling disappears and, as a consequence, he observes a predictable negative resistance region.

The tunnel diode couples tunneling with clever energy band alignment to create negative differential resistance as illustrated in Figure 4.44.

With reference to Figure 4.44, the operation of the Esaki diode is as follows. When applying a small forward bias, electrons on the n-side of the diode tunnel into holes on the p-side. As that bias is increased, the band alignment enables yet more electrons to find holes to tunnel into, and the current increases (A). Beyond the point (B), empty states on the p-side move higher in energy with respect to the electron energies on the n-side. Eventually, at point C, there are no empty states (holes) available

FIGURE 4.45 W. Shockley (age 48 years) and L. Esaki (age 33 years) in 1958 in Brussels for the symposium of the International Union of Pure and Applied Physics. This coincided with the World Expo held there.

for electrons to tunnel into. This causes the current to decrease to a minimum value. Eventually, the current increases again because the electrons and holes are able to overcome the reduced barrier because of their thermal distribution.

Transistors

Introduction

An integrated circuit (IC) is an electronic network fabricated in a single piece of a semiconductor material. In IC fabrication, the semiconductor surface is subjected to various processing steps in which impurities and other materials are added with specific geometrical patterns. The fabrication steps are sequenced to form 3D regions that act as transistors and interconnects that form the switching or amplification network. In this section, we detail different types of transistors.

Transistors are three-terminal devices, where the current or voltage at one terminal, the input terminal, controls the flow of current between the two remaining terminals. In Chapter 1 we saw how the name "transistor," a name coined by John R. Pierce, a scientist from Bell Laboratories, is a contraction of the terms *trans* and *resistor*. This term was adopted because it best describes the operation of the transistor—the transfer of an input signal current from a low-resistance circuit to a high-resistance circuit. Basically, the transistor is a solid-state device that amplifies by controlling the flow of current carriers through its semiconductor materials. Electronic devices made up of transistors are more interesting and have more applications than diodes.

Two types of transistors are widely used: the bipolar junction transistor (BJT), which is physically like two junction diodes back to back, and the field-effect transistor (FET), in which one uses the voltage on a control (gate) electrode to control the shape and hence the conductivity of a channel of one type of charge carrier in a semiconductor material. Only weeks after the point contact transistor was conceived by Shockley, Bardeen, and Brittain (see Chapter 1), Shockley came up with the BJT in 1948. Bardeen was the physicist; Walter Brittain, the experimentalist; and William Shockley, who became involved later in the development, was the instigator and idea man. The team won the 1956 Nobel Prize in physics for their efforts. The transistor demonstrated for the first time that amplification in solids was possible. The device was initially known simply as the junction transistor, and it did not become practical until the early 1950s. The term *bipolar* was tagged onto the name to clarify the fact that both carrier types play important roles in its operation. Historically the eldest, it was in use for many years for large integrated devices and high speed at the expense of low density and higher power consumption. The dominance of BJTs through the 1960s and 1970s was quickly overwhelmed by the simplicity of the metal–oxide–semiconductor field-effect transistor (MOSFET) (in terms of both circuit design and fabrication processes) in the 1980s. This is especially true in the very large scale integration (VLSI) era. In a MOSFET, power consumption is very low, and the current flow depends on the majority carriers only; thus, it is unipolar. The BJT's high current drive capability and superior analog performance continue to drive applications, however, and a recent trend is to combine the best of MOSFETs and bipolar devices in bipolar CMOS (BiCMOS) processes. BiCMOS is a type of IC that uses both bipolar and CMOS technologies. The logic gates are primarily made of CMOS, but their output stages use BJTs. Bipolar transistors consume more current (and therefore dissipate more power) but switch faster (see Table 4.9).

Bipolar Transistors

The BJT structure contains two p-n diodes: one between the base and the emitter, and one between the base and the collector. There are many different types of transistors, but the basic theory of their operation is all the same. The three elements of the two-junction transistor are 1) the EMITTER, which gives off, or *emits*, current carriers (electrons or holes); 2) the BASE, which controls the flow of current carriers; and 3) the COLLECTOR, which collects the current carriers (Figure 4.46). Early BJTs were fabricated using alloying—a complicated and unreliable process (Figure 4.46a). Figure 4.46a, also shows the symbol used for a BJT transistor. The "planar process," developed by Fairchild in the late 1950s, shaped the basic structure of the BJT, which is still seen even up to the present day. In the planar

TABLE 4.9 Bipolar, CMOS, and BiCMOS Compared

Bipolar	CMOS	BiCMOS
Better analog capability		High-performance analog
High power dissipation	Lower static power dissipation	Lower power dissipation than purely bipolar technology (simplifying packaging and board requirements)
Generally better noise performance and better high-frequency characteristics	Higher noise margins	
Improved I/O speed (particularly significant with the growing importance of package limitations in high-speed systems)		Flexible I/Os technology is well suited for I/O intensive applications
Low packing density	Higher packing density	
	Lower manufacturing cost per device	
Lower input impedance (high drive current)	High input impedance (low drive current)	
Low delay sensitivity to load	High delay sensitivity to load (fan-out limitations)	
Low voltage swing logic	Scaleable threshold voltage V_T	
Higher current drive per unit area, higher gain	Low output drive current (issue when driving large capacitive loads)	
High transconductance g	Low transconductance g	
Essentially unidirectional	Bidirectional capability (drain and source are interchangeable)	
Higher switching speed	A near ideal switching device	Improved speed over purely CMOS technology

process, all steps are performed from the surface of the wafer (Figure 4.46b).

A forward-biased p-n junction is comparable to a low-resistance circuit element because it passes a high current at a given voltage. In turn, a reverse-biased p-n junction is comparable to a high-resistance circuit element. By using the Ohm's law formula for power ($P = I^2R$) and assuming the current is held constant, it can be concluded that the power developed across a high resistance is greater than that developed across a low resistance. Thus, if a crystal were to contain two p-n junctions (one

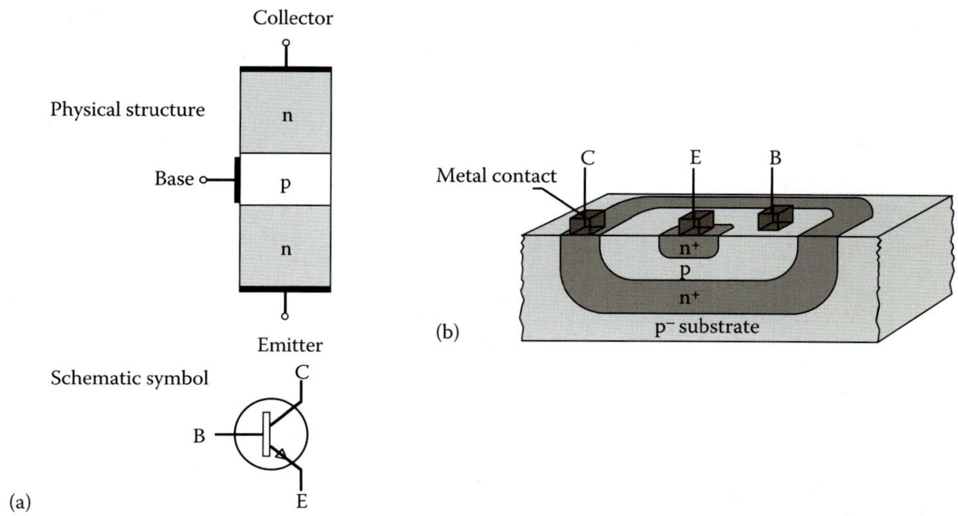

FIGURE 4.46 Bipolar npn transistor made by alloying a BJT symbol (a) and a BJT made in a planar process (b).

forward-biased and the other reverse-biased), a low-power signal could be injected into the forward-biased junction and produce a high-power signal at the reverse-biased junction. In this manner, a power gain would be obtained across the crystal. This concept is the basic theory behind transistor amplification.

MOSFETS

The most common field-effect transistor (FET) in both digital and analog circuits is the MOSFET. Nomenclature is different here than for a bipolar transistor with the base replaced by a gate (*G*), the emitter by a source (*S*), and the collector by a drain (*D*) (see Figure 4.47). The physical structure shown here consists of a p-type Si substrate, n^+ source, n^+ drain, a gate oxide (SiO_2), a polysilicon gate, a chemical vapor deposition (CVD) oxide, and a contact metal 1 (usually Al). Notice that the effective gate length is shorter than the one drawn in the mask layout ($L_{eff} < L_{drawn}$) as a result of lateral doping effects.

Silicon is the main choice of semiconductor used in MOSFET manufacture, but SiGe is also used more and more by some manufacturers (see page 269). Some other semiconductors such as GaAs are not useful in MOSFET manufacture because they do not form good insulating gate oxides. A thin oxide layer between the gate (e.g., a PolySi or so-called *poly gate*) and the semiconductor characterizes the MOSFET. This high resistivity oxide layer ($\rho > 10^{16}$ Ωcm) prevents current flow from the semiconductor to the gate. Silicon-based MOSFETs rely on the extremely high-quality interface between SiO_2, the standard gate dielectric, and silicon.

To understand a MOSFET, it is essential to recall the metal oxide semiconductor (MOS) capacitor fundamentals. In an idealized MOS capacitor, without any applied voltage, the metal work function, ϕ_m, is equal to the semiconductor work function, ϕ_s. Therefore, the Fermi level of the semiconductor, E_{FS}, is aligned with the Fermi level of the gate, E_{Fm}. As a consequence, there is no band bending in any region of the MOS capacitor. We also assume here that the oxide is free of any charges, and the semiconductor is uniformly doped. In the example MOS structure in Figure 4.48a, we use a p-type semiconductor with a negative gate bias applied. With this negative gate bias (V_{GB}), the gate charge, Q_G, is negative. Because charge neutrality must be preserved, a net positive charge, Q_C, is generated in the silicon substrate to counterbalance the negative charge on the gate. This is achieved by an accumulation of majority carrier holes under the gate (because we are dealing with a p-type semiconductor, the majority carriers are holes). The condition, where the majority carrier concentration is greater near the Si-SiO_2 interface compared with the bulk concentration, is called *accumulation*. Under this applied negative gate bias, the Fermi level of the gate is raised with respect to the Fermi level of the substrate by an amount qV_{GB}.

The energy bands in the p-type semiconductor are bent upward, bringing the valence band closer to the Fermi level, which is indicative of a higher hole concentration at the semiconductor surface right under the dielectric. Because no current flows through the device due to the presence of the insulating oxide layer, the Fermi level in the substrate

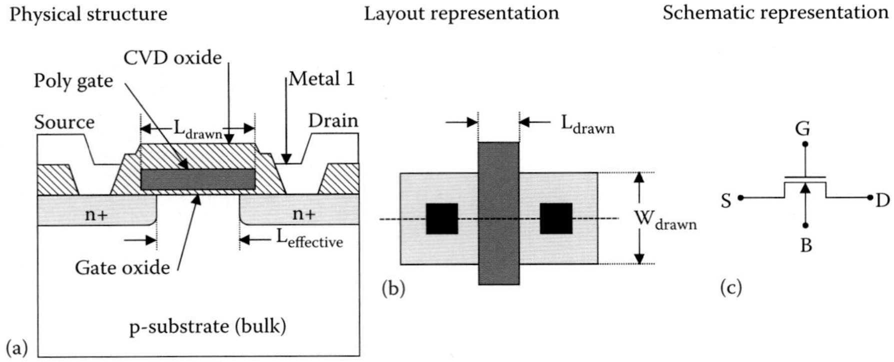

FIGURE 4.47 A MOSFET with nomenclature (a). The geometry of the areas on the Si surface is known as the layout of the chip (b). Schematic representation of a MOSFET (c).

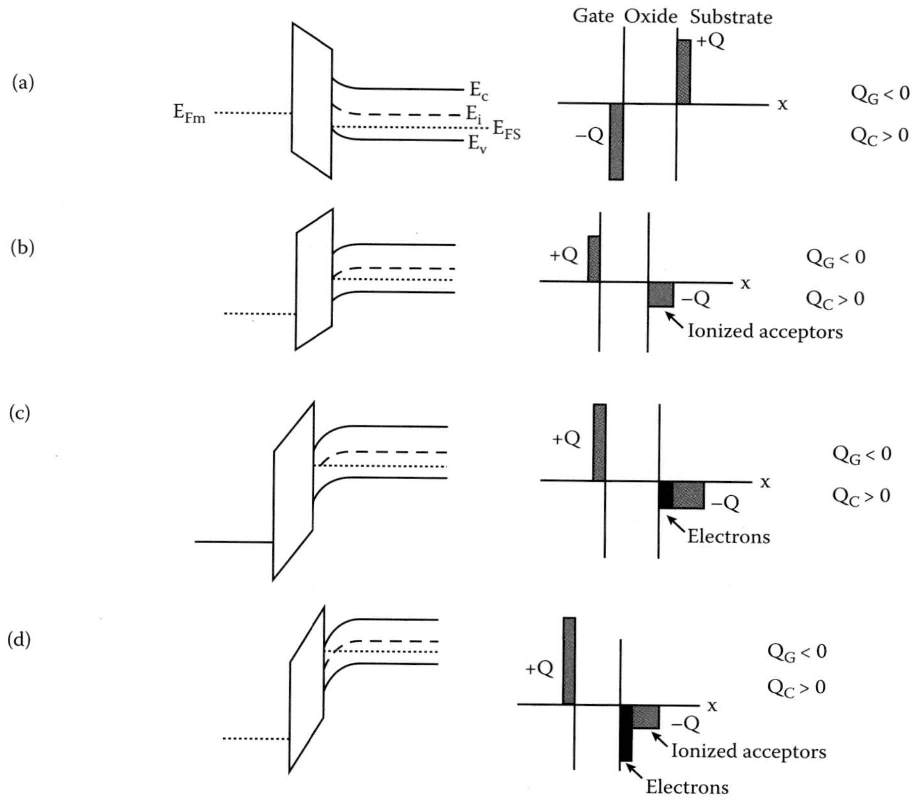

Accumulation	$\psi_s < 0$	$V_{GB} < V_{FB}$	$Q_C > 0$
Depletion	$\psi_s > 0$	$V_{GB} > V_{FB}$	$Q_C < 0$
Inversion	$\psi_s > 0$	$V_{GB} >> V_{FB}$	$Q_C < 0$

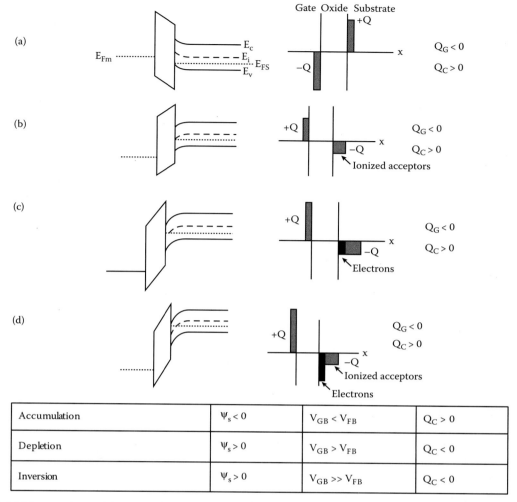

FIGURE 4.48 A MOS capacitor with a metal gate, an insulator, and a p-type semiconductor. (a) Accumulation; (b) depletion; (c) inversion; (d) strong inversion. E_i is the Fermi level for the intrinsic material, and E_{FS} is that of the doped material (because it is a p-type material the Fermi level is close to the valence band).

(the semiconductor) remains invariant even with an applied bias. The applied gate voltage (V_{GB}) divides between the gate dielectric and the semiconductor such that:

$$V_{GB} = V_{ox} + V_s \quad (4.54)$$

where V_{ox} and V_s are the voltages that drop across the oxide and the semiconductor, respectively. The bands in the oxide drop linearly over the bulk of the oxide thickness an amount equal to qV_{ox}, and the electric field in the oxide can be expressed as:

$$E_{ox} = V_{ox}/t_{ox} \quad (4.55)$$

where t_{ox} is the oxide thickness. The amount of band bending in the semiconductor is principally located at the surface and is equal to $q\psi_\sigma$, where ψ_σ is the surface potential and is negative when the band bend is upward.

When applying a positive gate bias V_{GB} to counterbalance the positive gate charge, the holes under the gate are pushed away, leaving behind ionized, negatively charged acceptor atoms (N_A^-), which create a depletion region (Figure 4.48b). The charge in the depletion region of width W and area A is exactly equal to the charge on the gate to preserve charge neutrality [$Q_G(+) = Q_c(-) = -qN_A^-AW$]. With a positive gate bias, the Fermi level of the gate is now lowered with respect to the Fermi level of the semiconductor. The bands bend downward, resulting in a positive surface potential. Under the gate, the valence band moves away from the Fermi level, which is indicative of hole depletion. When the band bending at the surface is such that the intrinsic level (E_i) coincides with the doped semiconductor Fermi level (E_{FS}), the surface region resembles an intrinsic material. The surface potential for this condition is given by:

$$\psi_s = \phi_F = \frac{1}{q}(E_i - E_F) \tag{4.56}$$

with ϕ_F, the Fermi potential, given by:

$$\phi_F = \frac{kT}{q}\ln\frac{N_A}{n_i} \tag{4.57}$$

where n_i is the charge carrier density of the intrinsic material (see Equation 4.17: $n_i \approx 1.5 \times 10^{10}$ cm^{-3} at 300 K in pure Si). Applying a larger positive gate bias, the positive charge on the gate increases further, and thermally generated electrons start collecting under the gate. With these electrons, the intrinsic surface region begins to convert into an n-type inversion layer. The negative charge in the semiconductor is now composed of both ionized acceptor atoms in the depletion layer and free electrons in the inversion layer (Figure 4.48c). In this condition, the electron concentration at the surface is still less than the hole concentration in the neutral bulk of the semiconductor, and this regime is called weak inversion. The point where $\psi_s = \phi_F$ is defined as the onset of weak inversion.

As the gate bias is further increased, the band bending also continues to increase. This causes the depletion region to become wider while the electron concentration in the inversion layer increases more and more. At the gate bias where the electron concentration in the inversion region is equal to the hole concentration in the bulk, a strong inversion layer is created (Figure 4.48d). The surface potential to achieve this strong inversion is given by:

$$\psi_s = 2\phi_F = \frac{2}{q}(E_i - E_F) \tag{4.58}$$

When the gate bias is increased beyond this value, new electrons in the inversion layer readily compensate extra positive charge on the gate and the depletion layer width does not further increase, and hence the band bending increases only very slightly. For simplicity, $2\phi_F$ is generally taken as the maximum value for ψ_s. The maximum depletion layer width, W_d, at the onset of strong inversion is then given by:

$$W_d = \sqrt{\frac{4\phi_F\varepsilon_s}{qN_A}} \tag{4.59}$$

In a real MOS device, the work function of the metal, ϕ_m, is not likely to be equal to that of the semiconductor, ϕ_s. Consider a MOS capacitor with $\phi_m < \phi_s$ at zero gate bias. In this case, electrons in the metal reside at energy levels above the conduction band of the semiconductor. Thus, electrons will flow from the metal into the semiconductor until a potential counterbalances the difference in the work functions between the two plates of the MOS capacitor. This induces a negative charge under the gate dielectric, accompanied by a downward band bending and hence a preexisting positive surface potential ψ_{ms}. With an external voltage, equal to this difference ψ_{ms}, applied to the gate, the net charge in the semiconductor disappears, and the bands return to their flat position. This applied voltage is defined as the flatband voltage, V_{FB}, and is expressed as:

$$V_{FB} = q\psi_{ms} = q(\phi_s - \phi_s) \tag{4.60}$$

The oxide and the semiconductor/oxide interface are not perfect, and another modification to the ideal picture of a MOS capacitor comes from charges that reside in the dielectric or at its interface. Such charges may originate from processing, defects in the bulk, or charges that exist at the interface between Si and SiO_2 (for an overview of the type of oxide defects, see Figure 4.35). Charges in the bulk oxide, or at the interface, induce opposite charges in the semiconductor and the gate. This again causes the bands to bend up or down, and the flatband voltage has to be adjusted to take this into account. The flatband voltage can then be expressed as:

$$V_{FB} = q\phi_{ms} - \frac{Q_o}{C_{ox}} \tag{4.61}$$

where Q_o is the oxide charge, and C_{ox} is the oxide capacitance (we also have introduced $\psi_{ms} = \phi_{ms}$). Defects located at the Si/SiO_2 interface may not be fixed in their charged state, and the amount of charges may also vary with the surface potential of the semiconductor. These defects are referred to as fast interface traps, and they impact the switching characteristics of MOSFETs.

We now put a MOS capacitor between a source (*S*) and a drain (*D*). In Figure 4.49A we show an n-channel and a p-channel device. Both n-channel and p-channel MOSFETs are extensively used. In fact, CMOS IC technology relies on the ability to use both devices on the same chip. As shown in Figure 4.49B, to create an nMOSFET and a pMOSFET on the same p-type substrate, the pMOSFET must be provided with n-type well regions. N-wells are formed by ion implantation or by diffusion. Lateral diffusion limits the proximity between structures, and ion implantation results in shallower wells compatible with today's fine-line processes. Metal 1 makes contact to the nMOSFET and metal 2 to the pMOSFET. Metals in vias connecting to transistors are often made of tungsten metal plugs, and such plugs

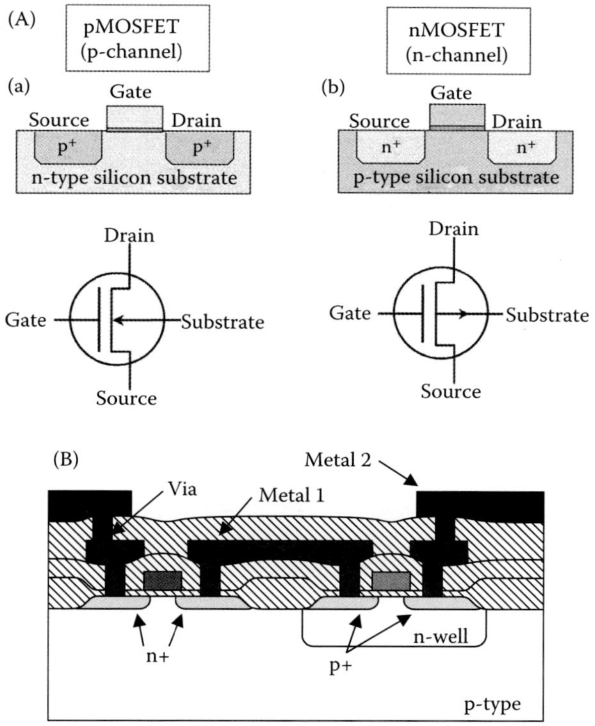

FIGURE 4.49 (A) pMOS and nMOS with symbols. (a) pMOS with p-channel: a negative voltage is applied to the gate, thus attracting holes that form a channel between the two p^+ source and drain regions. (b) nMOS with n-channel: a positive voltage is applied to the gate, thus attracting electrons that form a channel between the two n^+ source and drain regions. (B) To create an nMOSFET and a pMOSFET on the same p-type substrate, the pMOSFET must be provided with an n-well.

are also used to fill vias between successive layers of metal (Cu or Al) (see Figure 4.62 and Volume III, Figure 4.3). Tungsten is used because of the extraordinarily good conformality of CVD from WF_6. The tungsten in turn contacts the Si via an adhesion/barrier layer such as Ti/TiN. The Ti layer improves the *W* contact to the transistor and protects the underlying Si from attack by fluorine.

Depending on the type of MOSFET (i.e., p-channel or n-channel), two different situations arise. The voltage applied between the gate and the substrate (V_{GS}) generates an electric field through the oxide layer, creating an inversion channel under the gate oxide. In the case of a pMOSFET (with a p-channel) transistor, the n-type substrate is inverted to p-type in a thin layer under the gate oxide, and a p-type substrate is similarly inverted to n-type in an nMOSFET (with an n-channel) device. Because the inversion channel is of the same type as the source and drain, current can now pass from source to drain. The pMOS transistors are 2.5 times slower than nMOSFETs because of the mobility difference between electrons and holes.

By varying the potential between the gate and substrate, the inversion channel in which current flows can be altered to control the magnitude of the current, or be turned off completely. Thus, the magnitude of the drain source current, I_D, through the channel is controlled by the potential difference V_{GS} applied between the source *S* and the gate *G*. The threshold voltage, V_T, is the value of V_{GS} where the drain current just begins to flow. From Figure 4.50 typical values range from 0.3 to 0.8 V.

The drain current (I_D) versus the voltage between drain and source (V_{DS}) for different values of V_{GS} is illustrated in Figure 4.51. As shown at low V_{DS}, the drain current increases almost linearly with V_{DS}, resulting in a series of straight lines with slopes increasing with V_{GS}. At high V_{DS}, the drain current saturates and becomes independent of V_{DS}.

Using the nMOSFET in Figure 4.52, the curves in Figures 4.50 and 4.51 are explained in a bit more detail. To make an n-channel MOSFET, as shown in Figure 4.52a, we diffuse donors in the source, drain regions (forming n-regions) in an p-type substrate, and grow an oxide layer that acts as the gate oxide. With a small positive voltage on the drain and no

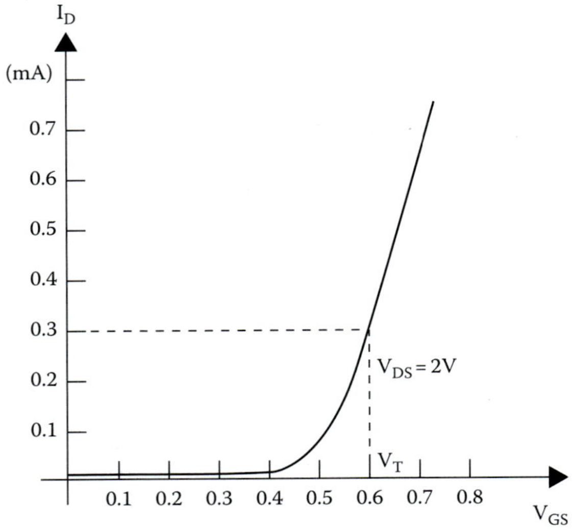

FIGURE 4.50 The value of V_{GS} where the drain current, I_D, just starts flowing is called V_T or the threshold voltage.

bias on the gate, i.e., $V_{DS} > 0$ and $V_{GS} = 0$, the drain is a reverse-biased p-n junction. Conduction band electrons in the source region encounter a potential barrier determined by the built-in potential of the source junction. As a result, no electrons enter the channel region, and hence no current flows from the source to the drain. This is referred to as the "off" state and is illustrated in Figure 4.52b.

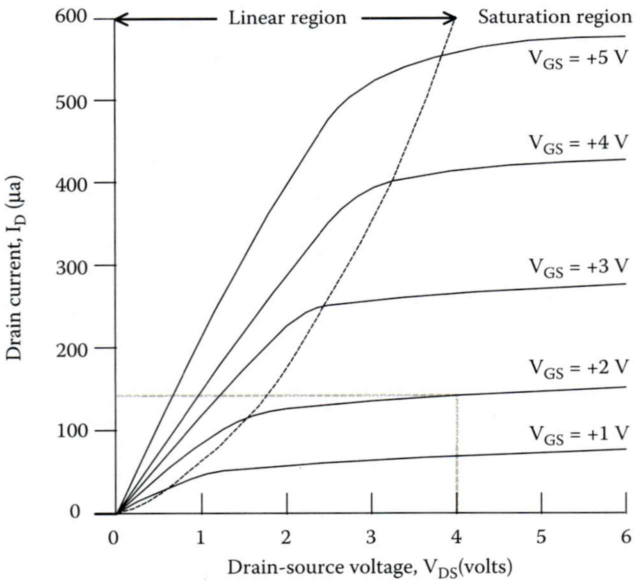

FIGURE 4.51 The drain current, I_D, versus the voltage between drain and source, V_{DS}, for different values of V_{GS}. The boundary between linear and saturation (active) modes is indicated by the upward-curving parabola.

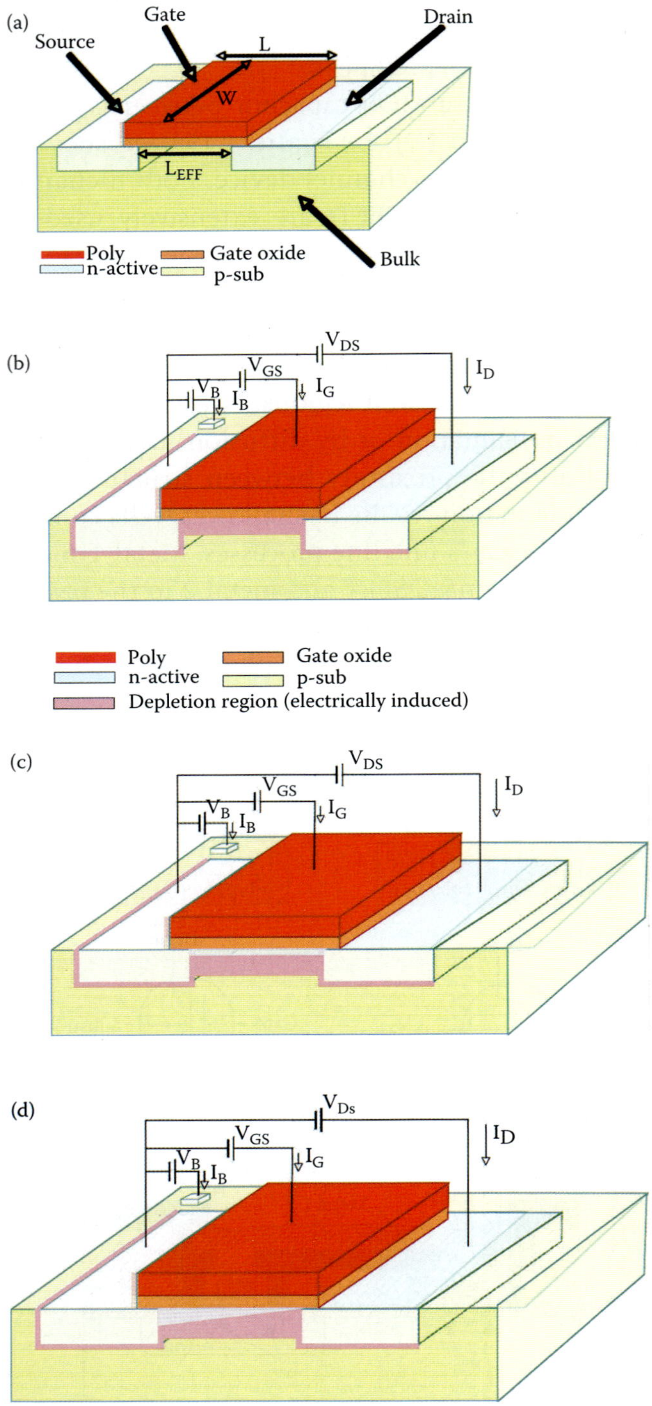

FIGURE 4.52 (a) nMOSFET. (b) Apply small V_{GS} (V_{DS} and V_B assumed to be small). Depletion region electrically induced in channel, termed "cutoff" region of operation. (c) At a critical value of V_{GS}, termed the threshold voltage (V_T), an inversion layer is created. The inversion layer supports current flow from D to S. The channel behaves as a thin-film resistor. (d) With $|V_{DS,\,max}| > |V_{GS} - V_T|$, the channel is flowing as much current as it can, so we call it pinched off (even though current continues to flow). The semiconductor bulk potential is V_B.

With a small positive bias on the gate, band bending in the channel region ($\psi_s > 0$) brings the conduction band in the channel region closer to the conduction band in the source region, reducing the height of the potential barrier to electrons. At a critical value of V_{GS}, termed the threshold voltage (V_T), an inversion layer is created (see Figure 4.52c). The inversion layer supports current flow from D to S, and electrons can now enter the channel and flow from source to drain; we say that the transistor is on. In the low drain bias regime, the drain current increases almost linearly with drain bias, and like an ideal resistor it obeys Ohm's law (see Figure 4.51).

The channel resistance is determined by the electron concentration in the channel, which is a function of the gate bias. As the gate bias is increased, the slopes of the linear portion of the I–V characteristics in Figure 4.51 gradually increase as a result of the increasing conductivity of the channel. This region, where the channel behaves like a resistor, is called the *linear region of operation*. The drain current I_D is given by:

$$I_D = \frac{1}{2}\mu_e C_{ox}\frac{W}{L}[2(V_{GS} - V_T)V_{DS} - V_{DS}^2] \qquad (4.62)$$

The dimensions W, L can be identified from Figure 4.52a, and μ_e is the effective channel mobility of the electrons, which differs from the bulk electron mobility. For small V_{DS} ($V_{DS} << V_{GS} - V_T$), the second term in parentheses can be ignored, and the expression for drain current reduces to a straight line with a slope equal to the gain factor of the device (see also below). We shall deal with the concept of effective channel mobility later. In transistors of the 1980s the V_{GS} was 5 V; in modern processes this has come down to 3.3 V, then 2.5 V, 1.8 V, 1.5 V, 1.2 V, and nowadays 1 V. The higher voltages used in the past would cause breakdown in the latest transistors. Importantly, lower V_{GS} saves power. The threshold voltage, V_T, of a MOSFET is the gate voltage, V_{GS}, required to create strong inversion (i.e., $\psi_s = 2\phi_F$; Equation 4.58) under the gate. When the semiconductor (body) is at ground potential ($V_B = 0$), the threshold voltage, V_T, equals the sum of the flatband voltage V_{FB} (Equation 4.61), the voltage across the oxide due to the depletion layer charge ($Q_d = -qN_AW_d$ with W_d given by Equation 4.59), and twice the bulk potential π_F (Equation 4.58), or:

$$V_T = q\phi_{ms} - \frac{Q_o}{C_{ox}} - \frac{Q_d}{C_{ox}} + 2\phi_F \quad \text{(nMOSFET) with}$$

$$\phi_F = \frac{kT}{q}\ln\frac{N_A}{n_i}$$

$$V_T = q\phi_{ms} - \frac{Q_o}{C_{ox}} - \frac{Q_d}{C_{ox}} - 2\phi_F \quad \text{(pMOSFET) with}$$

$$\phi_F = \frac{kT}{q}\ln\frac{N_D}{n_i}$$

and with

$$Q_d(\text{nMOSFET}) = -2\sqrt{qN_A\phi_F\varepsilon_s}$$

$$Q_d(\text{pMOSFET}) = 2\sqrt{qN_D\phi_F\varepsilon_s} \qquad (4.63)$$

Control of the threshold voltage is crucial in making MOSFETs. By using highly doped polysilicon as the gate material instead of Al, ϕ_{ms} is now just the difference in Fermi levels of the two silicon regions. Polysilicon is also more process friendly as it can withstand higher temperatures than Al. In Equation 4.63, C_{ox} (F/m^2) is given by:

$$C_{ox} = \frac{\varepsilon_{ox}}{t_{ox}} \qquad (4.64)$$

where t_{ox} is the oxide thickness. A larger C_{ox} leads to a smaller threshold. We would like a small V_T under the gate, but elsewhere we would like a large V_T to prevent channels from forming between transistors. To grow the most perfect SiO_2 layer, [100]-oriented wafers are used (less dangling bonds), slower growth rates in dry oxygen are used (leads to higher quality layer), and HCl is often added to the oxygen steam to reduce sodium incorporation in the SiO_2. In a 0.24-μm MOSFET process, t_{ox} is 5 nm (~10 atom layers), and C_{ox} is 5.6 fF/μm^2. With even thinner oxide films, say 2 nm, large tunneling currents lead to leakage and high power consumption. One solution is to look at dielectrics that feature a higher dielectric constant ε_{ox}, such as ZrO_2, HfO_2, and Al_2O_3. Finally, there is a substrate bias effect. By adding a contact to the semiconductor substrate (normally the source and body are tied together to ground), another means of controlling the threshold voltage through changing V_B results (making the transistor a four-terminal device: gate, source, drain, and substrate body):

$$\Delta V_T = \frac{\sqrt{2\varepsilon_s q N_A}}{C_{ox}}\left[(-V_B)^{\frac{1}{2}}\right](\text{nMOS})$$

$$\Delta V_T = \frac{\sqrt{2\varepsilon_s q N_D}}{C_{ox}}\left[(V_B)^{\frac{1}{2}}\right](\text{pMOS}) \tag{4.65}$$

where $\frac{\sqrt{2\varepsilon_s q N_D}}{C_{ox}}$ (for nMOS) and the same expression with N_D for pMOS are the body-effect coefficients (impact of changes in V_B) ($\varepsilon_s = 1.053 \times 10^{-10}$ F/m is the permittivity of silicon; $C_{ox} = \varepsilon_{ox}/t_{ox}$ is the gate oxide capacitance with $\varepsilon_{ox} = 3.5 \times 10^{-11}$ F/m). In the case of an n-channel device, one requires a gate voltage more positive than V_T to create an electron channel. For a p-channel, one requires a gate voltage more negative than V_T to create an electron channel.

As we saw earlier for small V_{DS}, the second term in parentheses in Equation 4.62 can be ignored, and the expression for drain current reduces to:

$$I_D = \mu_e C_{ox} \frac{W}{L}(V_{GS} - V_T)V_{DS} \tag{4.66}$$

which is a straight line with a slope equal to the gain factor of the device:

$$g = \mu_e C_{ox} \frac{W}{L} \tag{4.67}$$

and $\mu_e C_{ox}$ is the device's transconductance g'. When the drain voltage is made comparable to the gate bias voltage, the drain current increases sublinearly as predicted by Equation 4.62. This is due to a reduction of the inversion layer at the drain end of the channel. We delve into this reduction of the inversion layer length a bit deeper now.

For these larger drain biases, the drain current saturates and becomes independent of the drain bias. Naturally, this region is referred to as the saturation region. The drain current saturation is derived from Equation 4.62, which is parabolic with a maximum occurring at $|V_{DS,\,max}| = |V_{GS} - V_T|$, where $V_{DS,max}$ is also called the *pinch-off voltage*. To calculate the drain current, I_D, in this saturation region we substitute $|V_{DS,\,max}| = |V_{GS} - V_T|$ in Equation 4.62 to obtain:

$$I_D = \mu_e C_{ox} \frac{W}{L} V_{DS,max}^2 \tag{4.68}$$

When $|V_{DS,\,max}| = |V_{GS} - V_T|$, the channel is flowing as much current as it can, and the inversion layer now starts changing shape. Let us analyze why the inversion layer changes shape and why it is eventually pinched off. This is illustrated in Figures 4.52d and in more detail in Figure 4.53. The potential drop across the gate oxide determines the amount of positive charge on the gate in the nMOS structures of Figures 4.52 and 4.53. When the current flowing through the channel increases, a finite voltage drop occurs along it since it comes with its own finite resistance. This voltage opposes the applied gate bias, and as a consequence the amount of charges in the inversion layer becomes smaller at the drain end compared with the charge induced at the source end of the channel. This is illustrated in Figures 4.52d

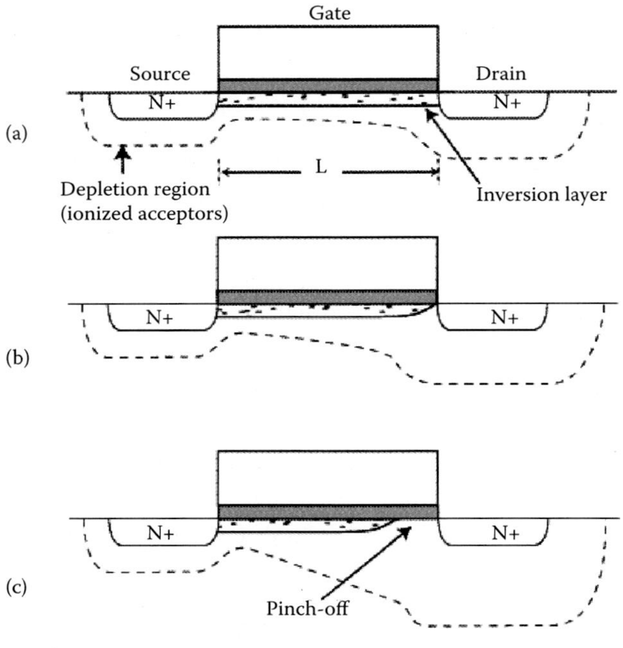

FIGURE 4.53 The pinch-off process in an nMOSFET. (a) The drain voltage is small compared with the gate bias voltage, and the charge in the inversion layer is uniform throughout the channel. (b) Increasing the drain voltage further reduces the inversion layer thickness and increases the depletion width around the reverse-biased drain junction. (c) Eventually the inversion layer completely disappears near the drain or the channel is pinched off.

and 4.53b by the smaller thickness for the channel inversion region close to the drain. Increasing the voltage on the drain also increases the depletion width around the reverse-biased drain junction, freeing more negative acceptor ions so that a fewer number of inversion layer electrons are needed to compensate the positive charges on the gate (see Figure 4.53b).

This means that the electron density in the inversion layer near the drain would decrease even if the charge density on the gate were constant. The reduction in number of carriers causes a reduction in the channel conductance as reflected in the smaller slope of I_D–V_{DS} curves as V_{DS} approaches $V_{D,sat}$ when the MOSFET enters the saturation region. Eventually, the inversion layer completely disappears near the drain. At this point, the induced charge is zero, and the conducting channel disappears or is pinched off (no channel exists in the vicinity of the drain region). In other words at a certain point there is not enough voltage difference between the gate and the semiconductor to cause a strong inversion layer at the drain end of the channel. This is shown in Figure 4.53c as a channel with zero thickness on the drain end, and this is referred to as the *channel pinch-off*.

The voltage difference over the induced channel remains fixed, and the current remains constant (or saturates). As V_{DS} is increased beyond $V_{D,sat}$, the width of the pinched-off region increases. However, the voltage that drops across the inversion layer remains constant and equal to $V_{D,sat}$. The portion of the drain bias in excess of $V_{D,sat}$ appears across the pinched-off region. In a long-channel MOSFET, the width of the pinched-off region is assumed to be small relative to the length of the channel. The applied voltage V_{DS} modulates the effective length of the conductive channel, so:

$$I_D = I'_D\,(1 + \lambda V_{DS}) \qquad (4.69)$$

where λ is the channel-length modulation (varies with the inverse of the channel length). In shorter transistors, the drain-junction depletion region presents a larger fraction of the channel, and the channel-length modulation effect is more pronounced.

In smaller devices the depletion width is similar in size to the gate length, and this affects the transistors current-voltage (I-V) curve. At short gate lengths, the electrostatic field no longer resembles that of a planar capacitor, and the FET suffers from threshold voltage shift, lack of pinch-off, increased leakage current, and increased output conductance. A linear I-V relation in a short-channel device replaces the quadratic dependence of I_D on V_{GS} in the saturation region of a long-channel device. As channel length decreases, space-charge regions at the source and drain encroach on areas normally controlled by the gate. In sufficiently small devices, the depletion regions of the two junctions can merge; this is known as punch-through. In punch-through the current is not controlled by the gate voltage but by V_{DS}^2 / L^3. Short-channel effect decreases the gate's ability to control the depletion region and allows current to flow as it normally would with a long-channel length. To defeat the short-channel effect, one must scale the device vertically, i.e., scale the lateral dimensions, but also the dimensions vertical to the wafer surface. By scaling all dimensions, the original field distribution can be recreated (see also below under "Constant Field and Constant Voltage Scaling"). However, the scaled structure has another problem now, namely, the closer proximity of source and drain-depletion region. As a consequence, the device is more susceptible to punch-through breakdown. To avoid this problem, the substrate doping is increased to reduce the depletion widths. Decreasing the junction depths also decreases the short-channel threshold voltage effects.

Complementary metal oxide semiconductor (CMOS) was first proposed in the 1960s but was not seriously considered until the severe limitations in power density and dissipation in nMOS circuits became apparent. Today CMOS is the dominant technology in IC manufacturing. CMOS uses both pMOS and nMOS transistors to form logic elements. The advantage of CMOS is that its logic elements draw significant current only during the transition from one state to another and very little current between transitions—hence power is conserved. In the case of an inverter, in either logic state one of the transistors is off. Because the transistors are in series, no current flows. A known deficiency of MOS technology is its limited load-driving capabilities (because of limited current sourcing and sinking abilities of pMOS and nMOS transistors). Bipolar transistors have higher gain, better noise characteristics, and better high frequency characteristics.

BiCMOS combines bipolar and CMOS transistors in a single IC. By retaining benefit of bipolar and CMOS, BiCMOS is able to achieve VLSI circuits with speed-power-density performance previously unattainable with either technology individually. In Table 4.9 we summarize advantages/disadvantages of each of these transistor types.

CMOS Scaling Issues

Introduction

The driver behind the fantastic pace of miniaturization in electronics technology over the past 100 years (Figure 4.54a) is economics. This progress is projected to continue with 200 million transistors per chip today and 10 billion projected by 2016 (Figure 4.54b). In 1962 the semiconductor industry was worth more than $1 billion; by 1978 it was above $10 billion; and by 1994 it was more than $100 billion. From Appendix 1B, the IC market for 2003 was $166 billion worldwide, with 2004 projected at $241 billion (data from the World Semiconductor Trade Statistics; http://www.eetimes.com/ news/semi/showArticle.jhtml). From the same appendix, we learn that the IC business is feeding an electronic equipment business, which today is in the trillion-dollar range! Traditionally, the cost/function in IC reduced by 25% to 30%/year, and this enabled the industry to grow by 15%/year. To maintain this growth, the number of functions per IC has to be continuously increased. This necessitates an increase in transistor count, an increase in clock speed (= increased operations per unit time), and a decrease of feature size (Moore's first law: Chapter 1 and Figure 4.55, this volume, and Volume II, Chapter 2). After reviewing fundamental challenges in MOSFET miniaturization, we present a short treatise on scaling of CMOS devices. The downscaling factor α has typically been $\sim\sqrt{2}$. Today, the smallest features in a CMOS transistor are approaching atomic dimensions. In this context, we cover technical and practical limits involved in downscaling Si CMOS devices before quantum mechanics takes over. In response to that, we cover suggested Si solutions. Alternatives to Si-based MOSFETs include single-electron transistors, quantum wells, wires, dots, resonant tunneling diodes and transistors, and molecular electronics.

Fundamental Barriers to MOSFET Scaling

The rate of progress in downscaling Si-based electronics, perhaps the most important development of the 20th century, is said to be running out of steam. Today the presumed limit to scaling of MOSFETs shown on the International Technology Roadmap for Semiconductors [Appendix 1A and http://public.itrs.net (from 2003 and updated in 2004)] is often projected to occur in 2018 or 2020, when half-pitch spacing of metal lines will be 18 nm and device gate

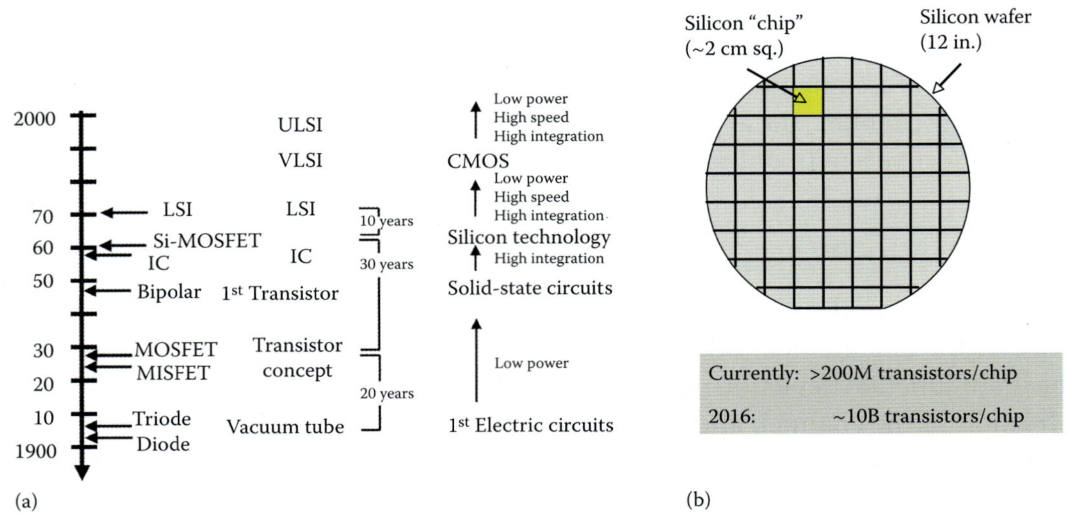

FIGURE 4.54 (a) History of electronics in the 20th century. In 100 years, the feature size was reduced by 1 million times. (b) Current and future (2016) transistors per chip.

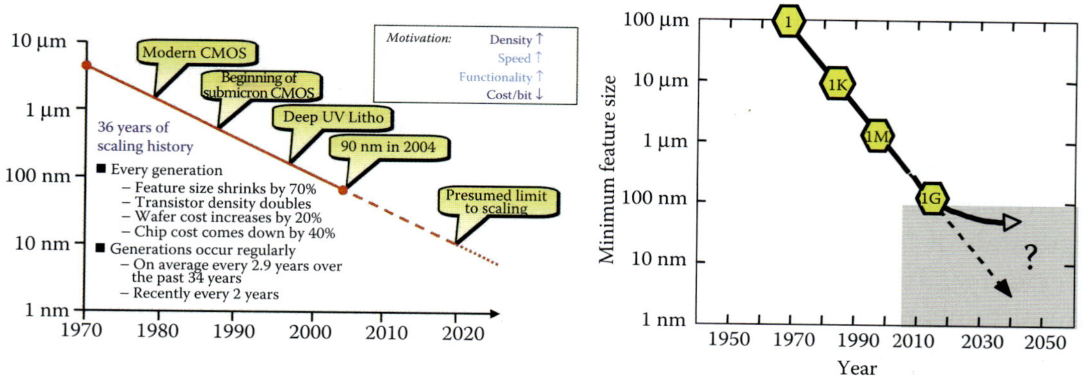

FIGURE 4.55 Two versions of Moore's first law. Scaling of minimum IC feature size.

lengths will be 7 nm (see Figure 4.55a). Technical limits are debatable and can be tackled, but fundamental limits are firm, and we discuss these ultimate limits first. Fundamental limits include electromagnetic, thermodynamic, and quantum mechanical limits.

From the electromagnetic point of view, the speed of light fundamentally limits the electron transport along IC wiring to ~60 ps/cm ($\tau > l/c$).* In today's technology, transmission times are getting longer than transistor switching times (now approaching 1 ps). In other words, as transistors are getting smaller and faster, on-chip interconnects (wires) are not getting any shorter or faster! This means that a larger percentage of chip area is devoted to metal interconnects as transistors get smaller. The capacitance of the wires is also not decreasing nearly as fast as the transistor sizes (in practice $\tau \sim RC \sim l^2$, with R and C proportional to l; see also "Constant Field and Constant Voltage Scaling").

The fundamental thermodynamic limit of any electrical device used as a switch (like MOSFETs in computers) is the energy required to produce a binary transition that can be distinguished: independent of the system architecture one must be able to distinguish between logic 0 and 1. CMOS uses an irreversible process with an inevitable intrinsic dissipation of energy. The von Neumann-Landauer[17] formula sets the theoretical limit for scaling of the energy in all irreversible technologies as (at 300 K):

$$E_{bit} > kT \ln 2 \approx 18 \text{ meV} \tag{4.70}$$

It was Landauer who realized that with the binary value of a bit unknown, erasing that bit changes the logical entropy of the system from $k\log 2$ to $k\log 1 = 0$. If the physical representation of the bit is part of a thermalized distribution (so that thermodynamics does apply), then this bit erasure is associated with the dissipation of energy in a heat current of $kT\log 2$ per bit. The practical performance of a MOSFET is limited by its thermodynamic energy efficiency, i.e., the useful work accomplished per unit energy dissipated. If a MOSFET consumes an amount of free energy E and performs N useful operations, the energy efficiency is given by:

$$\eta = E/N \tag{4.71}$$

with η in units of operations per unit energy, or operations/s/W. Using the thermodynamic energy limit per bit from Equation 4.70, this would enable an energy efficiency of $\eta = 3.5 \times 10^{20}$ operations/s/W. In the case of 90-nm VLSI technology, for minimal operations (e.g., conventional switching of a minimum-sized transistor), E is on the order of 1 fJ (femtojoule) and $\eta = 10^{15}$ operations/s/W. We see that the heat from the thermodynamic and logical entropy associated with a bit are currently five orders of magnitude lower than the electrostatic erasure energy.

From classical physics we expect a transistor gate to be a perfect barrier where no electrons flow through, but from quantum physics (Chapter 3) we know that electrons are extended wave-like objects

* It is important to note that the electric field propagates at the speed of light, but the electrons making up the electric current do not travel this fast themselves since they have mass. The individual electrons travel much more slowly. The electrons that go "into" one end of the wire are not the same electrons that "come out" of the other end.

that can "tunnel" through a thin gate, and this tunnel rate goes up exponentially as the gate gets thinner. MOSFET scaling is hampered by quantization of charge, which becomes important at a length scale of 10 nm in all materials, and energy level quantization, which becomes important in semiconductors also at a length scale of 10 nm and a factor of about 10 times lower for metals (see Chapter 3). The quantum mechanical limits on MOSFET scaling, based on the Heisenberg uncertainty principle (HUP), can be summarized as:

$$\Delta p_x \Delta x \geq \frac{h}{2\pi} = \hbar \text{ (see Equation 3.106)}$$

$$x_{min} = \frac{\hbar}{\Delta p_x} = \frac{\hbar}{\sqrt{2m_e E_{bit}}} = \frac{\hbar}{\sqrt{2m_e k_B T \ln 2}} = 1.5 \text{ nm} \tag{4.72}$$

which puts a fundamental limit on the integration density. So quantum mechanics allows for a very small switch indeed, but other factors, such as material temperature tolerance range or economics, will limit before this fundamental limit sets in. And also:

$$\Delta E \Delta t \geq \frac{h}{2\pi} = \hbar \text{ (see Equation 3.107)}$$

$$\Rightarrow \tau \sim 0.04 \text{ ps or 25 Tbits/s} \tag{4.73}$$

which sets a fundamental limit to switching speed.

Constant Field and Constant Voltage Scaling

Scaling of transistors may be achieved by reducing all dimensions (lateral and vertical) by $1/\alpha$. Suppose we scale down all transistor dimensions this way, including the gate oxide thickness (t_{ox}); then, to maintain the original electric field distribution, including the gate field, one must decrease voltages to $1/\alpha$ and increase the channel doping by α. Typically $1/\alpha = 0.7$ (30% reduction in the dimensions). The scaling variables are summarized in Table 4.10. This is called *constant field scaling* because the electric field across the gate oxide does not change when the technology is scaled.

If the power supply voltage is maintained constant, the scaling is called constant voltage. In this case, the electric field across the gate oxide increases

Table 4.10 Scaling of Variables

Scaling Variables	Before	After
Voltage	V	V/α
Gate length	L	L/α
Gate width	W	W/α
Gate oxide thickness	t_{ox}	t_{ox}/α
Junction depth	X	X/α
Channel doping	N	$N\alpha$

as the technology is scaled down. Because of gate oxide breakdown, at less than 0.8 μm only "constant field" scaling is used. As an example, let us consider the consequences of a 30% scaling effort in the constant field regime ($\alpha = 1.43$, $1/\alpha = 0.7$).

- Device/die area: $W \times L \Rightarrow (1/\alpha)^2 = 0.49$. In practice, microprocessor die size grows about 25% per technology generation! This is a result of added functionality.
- Transistor density: (unit area)/($W \times L$) $\Rightarrow \alpha^2 = 2.04$. In practice, memory density has been scaling as expected (not true for microprocessors…).
- Gate capacitance: $W \times L/t_{ox} \Rightarrow 1/\alpha = 0.7$
- Drain current: $(W/L)(V^2/t_{ox}) \Rightarrow 1/\alpha = 0.7$
- Gate delay: $(C \times V) / I \Rightarrow 1/\alpha = 0.7$ (see Figure 4.56)
- Frequency: $\alpha = 1.43$ (see Figure 4.56). In practice, microprocessor frequency has doubled every technology generation (2–3 years)!
- The processing power (number of gates/area times speed): goes up as (unit area)/($W \times L$) $\times$ frequency $\Rightarrow \alpha^3 = 2.92$
- Power: $C \times V^2 \times f \Rightarrow (1/\alpha)^2 = 0.49$
- Power density: $1/t_{ox} \times V^2 \times f \Rightarrow 1$
- Active capacitance/unit area: Power dissipation is a function of the operation frequency, the power supply voltage, and the circuit size (number of devices). If we normalize the power density to $V^2 \times f$, we obtain the active capacitance per unit area for a given circuit. This parameter can be compared with the oxide capacitance per unit area: $1/t_{ox} \Rightarrow \alpha = 1.43$. In practice, for microprocessors, the active capacitance/unit area only increases between 30% and 35%. Thus, the twofold improvement in logic density between technologies is not achieved.

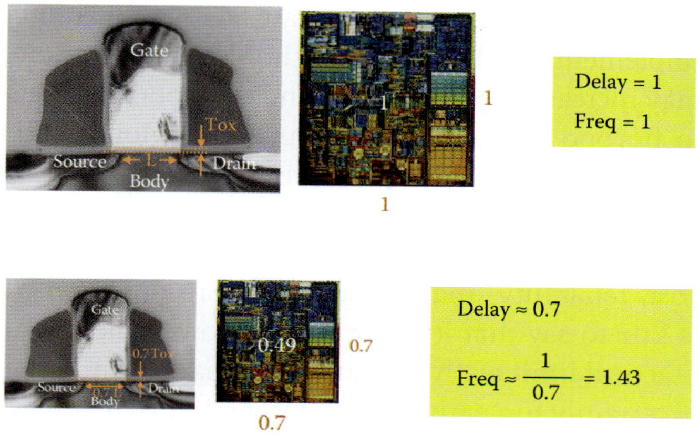

FIGURE 4.56 Scaling of delay and frequency in a MOSFET.

From the preceding, if we succeed in constant field scaling, the speed of the transistor goes up by α and the number of devices/area goes up by α^2, the processing power (number of gates/area times speed) goes up by α^3, and the heat generation/area (power density) remains unchanged (=1). Unfortunately, you cannot scale down voltage indefinitely. There are fixed voltage characteristics in MOSFETs, such as the 59-mV/decade turn-on current relation so the industry scales down voltage less aggressively than feature size and as a consequence the electric fields in the gate oxide have been increasing. If the electric field increases by E and the features are scaled down by α, then the doping goes up by $E \times \alpha$, the power density (heat) goes up by E^2, the gate leakage current increases, and transistors are more likely to fail.

As pointed out earlier, higher transistor densities are only possible if the interconnects also scale. Typical wires have a capacitance of ~0.2 fF/μm compared with 2 fF/μm for a typical gate capacitance. Reduced interconnect wire width leads to increased resistance, and denser interconnects also lead to a higher capacitance. For a wire interconnect, R and C are proportional to l, whereas the RC delay is proportional to l^2. This leads to unacceptably great delays for long wires. To account for increased parasitics and integration complexity, more interconnection layers are added with thinner and tighter layers for local interconnections and thicker and sparser layers for global interconnections and power. One also breaks long wires into N shorter segments and drives each one with an inverter or buffer in so-called *repeaters*.

In Table 4.11 constant field and constant voltage scaling of a MOSFET are compared.

Technical and Economic (Practical) Barriers to Downscaling and Possible Solutions

Introduction Besides the fundamental limits to CMOS scaling (see above), there are several technical limits to consider, and ultimately the economics might decide the ultimate fate of CMOS. However, Si devices are the smallest devices currently available

TABLE 4.11 Constant Field and Constant Voltage Scaling of a MOSFET

Parameter	Constant	Field	Constant	Voltage
Supply voltage (V)		$1/\alpha$		1
Length (L)	$1/\alpha$		$1/\alpha$	
Width (W)		$1/\alpha$		$1/\alpha$
Gate oxide thickness (t_{ox})		$1/\alpha$		$1/\alpha$
Junction depth (X_j)		$1/\alpha$		$1/\alpha$
Substrate doping (N_A)	α		α	
Electric field across gate oxide (E)	1		α	
Depletion layer thickness		$1/\alpha$		$1/\alpha$
Gate area (Die area)	$1/\alpha^2$		$1/\alpha^2$	
Gate capacitance (load) (C)		$1/\alpha$		$1/\alpha$
Drain current (I_d)		$1/\alpha$		α
Transconductance (g)	1		α	
Gate delay	$1/\alpha$		$1/\alpha^2$	
Current density		α		α^3
DC and dynamic power dissipation	$1/\alpha^2$		α	
Power density		1		α^3
Power-delay product	$1/\alpha^3$		$1/\alpha$	

in the market, and there are no other candidates smaller than Si in the near future. Introducing new materials, structures, and processes will solve problems of downsizing.

Mass Manufacturability and Economics As we scale down components, they have fewer atoms, electrons, and holes, and unless one has perfect control of the fabrication process, statistical fluctuations start coming in. Obviously MOSFET fabrication lines do not have perfect control, and statistical fluctuations get worse as the device gets smaller (the tyranny of small numbers). As a consequence, as transistors get smaller, there are greater variations in device-to-device characteristics, such as gate threshold voltage (V_T) and output current (I_D). Combined with fundamental voltage fluctuations (Johnson noise) and current fluctuations (shot noise), this leads to a greater probability of errors. As on-chip circuitry gets smaller and more complex, it is also increasingly sensitive to smaller defects. This lowers the yield and drives up costs even further. The demise of CMOS on technical grounds has been announced several times before.* However, it is more likely that one will hit an economic limit before the ultimate technical limit is reached. Because of increased sophistication of lithography tools, cost of IC fabrication lines is increasing exponentially with time (Moore's second law) and is now in the multibillion-dollar range (1967 cost of a silicon fabrication laboratory: \$2 million; 2002 cost: ~\$3 billion; projected 2015 cost: ~\$100 billion). As an example of mass manufacturability and economics, consider resolution-enhancing techniques (RET) that enable subwavelength printing of MOSFETs (see Volume II, Chapters 1 and 2). Basically, RET is a set of optical wave-front engineering techniques that allows for the pattern transfer of features smaller than the lithography wavelength. Feature size scaling is now outpacing lithography wavelength advances, forcing design compromises and complex and costly optical wave-front management techniques. It is important to recognize that at less than 90 nm there is no yield without RET, but this unfortunately vastly increases mask costs and severely constrains design flexibility. Implementing all of these technology innovations may increase manufacturing and development costs to beyond the reach of all but a few global entities. This subwavelength crisis is summarized in Figure 4.57.

A practical limit may thus be reached because of cost, reliability, and performance considerations. It is safe to say that it is not really known where that practical limit really lies, and, as remarked earlier, it is certainly too early to give up on Si. It is also clear that pushing CMOS to its ultimate limits will require revolutionary materials and device innovations that pose significant scientific and engineering barriers. These problems have caused more changes in IC technology and materials in the past 5 years than in the previous 40 years with work in SiGe, silicon on insulator (SOI), strained Si, alternative dielectrics, new metallization systems, and so on.

Carrier Mobility Limitations The effective mobility of charge carriers in the channel of a MOSFET differs from bulk mobility as a result of additional scattering mechanisms associated with the channel surface, including interface charge scattering and surface roughness scattering. These mechanisms also tend to depend on the vertical electrical field as illustrated in Figure 4.58, where E_{EFF} represents the effective electrical field. As gate length shrinks, mobility decreases, and increased doping is often used to combat this short-channel effect. Another

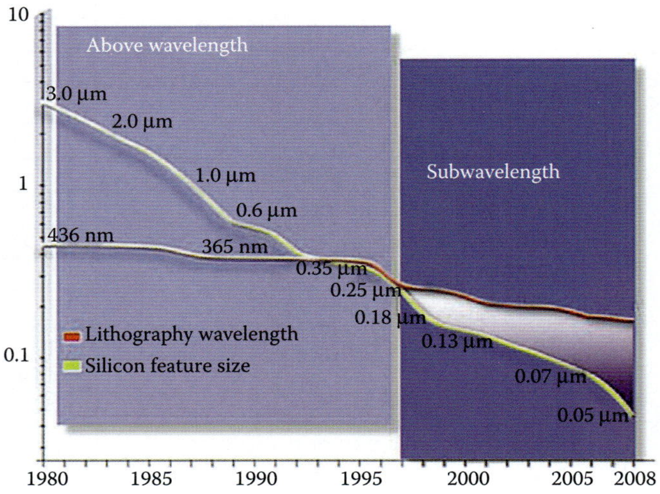

FIGURE 4.57 The subwavelength crisis.

* In the late 1980s, 0.1 µm was called the "brick wall"; today 50 nm is called the "red brick wall" and 100 nm the "fundamental limit."

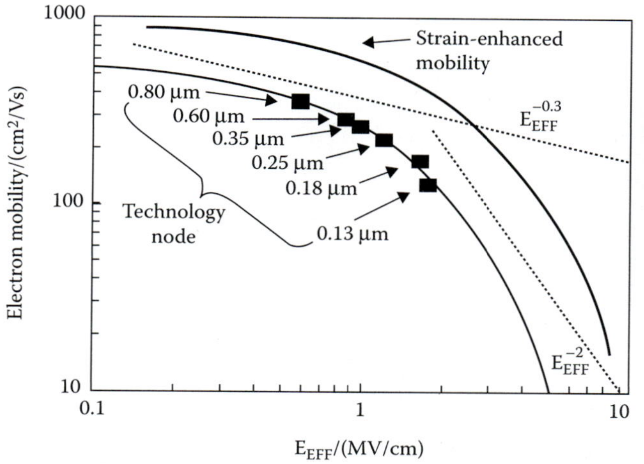

FIGURE 4.58 Mobility versus technology scaling trend for Intel process technologies. The *x*-axis is the effective electrical field E_{EFF}. (From Thompson, S.E., M. Armstrong, C. Auth, M. Alavi, M. Buehler, R. Chau, S. Cea, et al. 2004. A 90-nm logic technology featuring strained-silicon. *IEEE Trans Electron Devices* 51:1790–7.[49])

possible solution involves electron and hole mobility enhancement by using strained Si. This method increases mobility of electrons and holes in the channel of a FET by applying tensile (to increase electron mobility) or compressive strains (to increase hole mobility) to the channel. Strain alters the energy band structure and enables charge carriers to move with less resistance through silicon. The MOSFET channel is very close to the device surface, so that the strain must occur in the topmost ~0.01 μm of the silicon to realize a mobility enhancement in the channel. The effect of strain on mobility was noticed in the early 1950s, the exploitation of the phenomenon in an actual device was first carried out at MIT in the early 1990s (Gene Fitzgerald, who founded AmberWave in 1998). By August 2002, Intel used strained silicon in their HT Prescott P4 Processor. Models suggest that a strained silicon lattice can give hole mobilities up to 4× the unstrained value and electron mobilities up to 1.8× the unstrained value.[18]

One way to introduce strain in Si is to use a $Si_{1-x}Ge_x$ crystal alloy substrate. Such an alloy is possible because both materials create diamond-type lattices and their lattice constants (a) are close. Vegard's law determines the lattice constant of the alloy, which is a linear average between the lattice constants of Si (5.431 Å) and Ge (5.658 Å):

$$a_{alloy} = (1 - x) \cdot a_{Si} + x \cdot a_{Ge} \qquad (4.74)$$

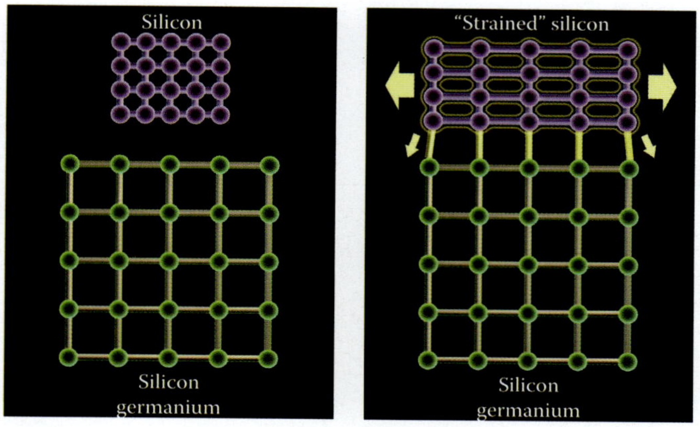

FIGURE 4.59 Formation of a strained Si lattice using an epitaxial strain-inducing template.

First, one introduces Ge in a Si lattice, and then the Si gate material is grown epitaxially using molecular beam epitaxy (MBE) on top of this $Si_{1-x}Ge_x$; the Si atoms align with the underlying lattice, causing a lattice strain in the silicon (see Figure 4.59).

The electrons flow up to 70% faster through strained silicon, resulting in a 17% increase in transistor speed and a 34% reduction in power consumption.

The SiGe devices market has grown more rapidly than was anticipated, and III–Vs producers are concerned about the switching from Si to SiGe instead of using GaAs in a number of key high-speed applications. With SiGe matching GaAs performance and beating it on cost, GaAs still edges out SiGe in terms of power-added efficiency (PAE)* (typically 40% vs. <20%). So for those applications where efficiency is less important—such as for shorter-range wireless networks like WiFi and Bluetooth—SiGe is competing hard with GaAs (see also Volume III, Chapter 5 where we compare GaAs with SiGe). An example SiGe process is the 130-nm SiGe BiCMOS technology designed for high-speed wireless and optical communications applications from Jazz Semiconductor (http://www.jazzsemi.com/process_technologies/sige.shtml). This process combines industry standard 130-nm CMOS with 200-GHz SiGe heterojunction bipolar transistors (HBTs) and npn transistors for high-performance RF and millimeter-wave ICs. This technology enables the design of the highest

* In an RF power amplifier, PAE is defined as the ratio of the difference of the output and input signal power to the DC power consumed.

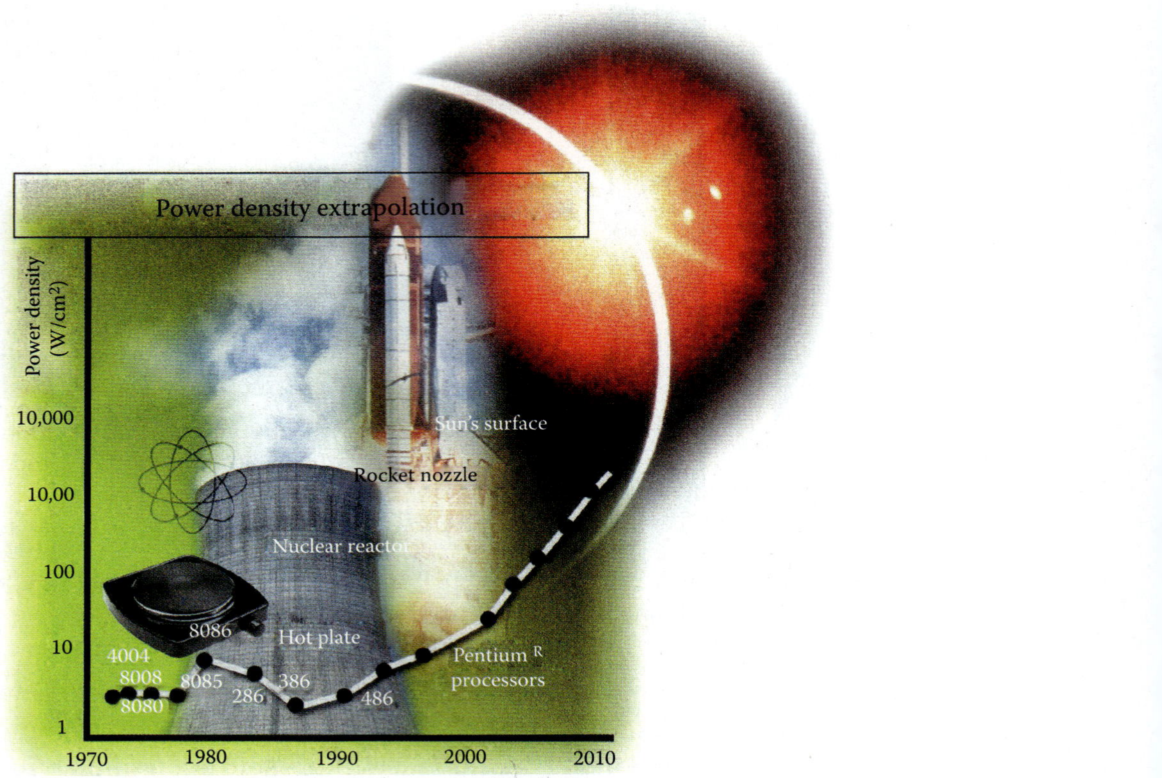

FIGURE 4.60 Power densities of various MOSFET technologies, one of the biggest challenges to further miniaturization.

performance circuits in advanced high-speed optical, wireless, and millimeter-wave applications.

Power Budget The attempts to increase the carrier mobility (see above), reduce the threshold voltage, minimize current leakage in the off state, and minimize losses in the interconnect system of MOSFETs are all somewhat interrelated. The further miniaturization of MOSFETs is limited mostly by both static (passive) and dynamic (active) power. Dynamic power is required to charge and discharge load capacitances when transistors switch and was calculated earlier as $C \times V^2 \times f$. From Figure 4.60 the latest Pentium chips dissipate about 100 W/cm²—more heat than a stove-top cooking surface.

Static power is consumed even when the chip is quiescent and is calculated from $P_{off} = I_D(off)V_{DS}$. Gate current leakage, I_D(off), is parasitic, on all the time, and uses extra power and generates extra heat. Over the years this gate leakage (10× increase per technology node) has caused the off-current [I_D(off)] to encroach on the on-current [I_D(on)] (Figure 4.61). Gate current reduces the on/off current ratio, which reduces operating margins.

Direct tunneling through the gate insulator is the dominant cause of static power dissipation. Single atom defects can cause local leakage currents 10–100 times higher than the average current, impacting reliability and generating unwanted variation between devices.

To reduce resistance and thus power consumption, both Ti and Co silicides are used as source/drain contact materials. Silicides are formed at the Si interface by an approach referred to as self-aligned-silicide (salicide), which selectively forms the silicide on the drain and source junctions as well as on the polysilicon gate. The salicide process is detailed in Volume II, Chapter 1. An important advantage of the

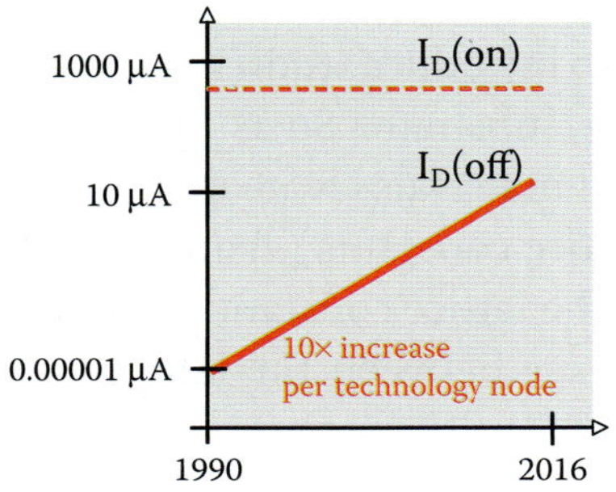

FIGURE 4.61 Evolution of the off-current [I_D(off)].

salicide process is that the entire junction area is used for contact formation, which translates into a lower overall contact resistance. The silicide contact materials on the transistors are then further contacted through chemical vapor-deposited (CVD) plugs of pure W filling the contact vias. CVD tungsten (from WF_6) exhibits extraordinarily good conformality, and the silicide adhesion/barrier layer protects the underlying Si from attack by fluorine during deposition; ensures adhesion of W to the silicon; and forms a good contact to the drain, source, and gate.

The vertical W plugs then make contact to horizontal Al (or Cu) interconnects as shown in Figure 4.62. To planarize the wafers between metallizations, IBM pioneered the damascene process, based on chemical-mechanical polishing (CMP). The damascene process is repeated many times to form the alternating layers of wires and vias that form the complete wiring system of a silicon chip. The damascene process is detailed in Volume II, Chapter 8 on single and dual damascene processes.

Solutions to reduce the excessive heat generation include using cooling fins and chip fans. Thus, it would appear that MEMS research for passive and active power management of ICs could present a tremendous new market opportunity. A wide variety of implementations of MEMS structures for cooling are feasible, but the associated costs, size, and practicality are the real issue.

Oxide Leakage and Device Isolation Today's MOSFET has evolved through many modifications to provide solutions for parasitic problems that emerged during downscaling. The gate dielectric is still mostly SiO_2-grown by thermal oxidation. As a result of continued scaling, that gate oxide has thinned down to the extent that direct tunneling has become a serious concern. As the thickness scales below 2 nm, leakage currents due to tunneling increase drastically. Despite the tremendous benefits that SiO_2 has presented toward the success of CMOS, scaling and high-power consumption are now urging process engineers to look for higher-k materials (k is the relative dielectric constant of the material, 3.9 for silicon dioxide).* High-k dielectric materials are used to minimize the tunneling current and to prevent the diffusion of dopants out from the gate. There are three major categories of high-k materials:

1. $4 < k < 10$; Si_3N_4/SiO_2 (5 ~ 6), Si_3N_4 (~7)
2. $10 < k < 20$; Ta_2O_5, Al_2O_3 (~10), TiO_2, $ZrSi_xO_y$, $HfSi_xOy$, $LaSi_xO_y$ (last three are silicates), $ZrAl_xO_y$ (aluminate), ZrN_xO_y (nitride)
3. $15 < k < 30$; ZrO_2, HfO_2, La_2O_3, Pr_2O_3, Gd_2O_3, Y_2O_3, crystal Pr_2O_3 (~30)

A prominent candidate for a gate dielectric of the future is Al_2O_3. Recent studies show that Al_2O_3 dielectrics less than 1 nm thick are superior to previously used Si_3N_4 and SiO_2 (Al_2O_3 film has been used to make 1-Gbit dynamic random access memory). As we will see in Volume II, Chapter 7, it may be deposited, for example, monolayer by monolayer with atomic layer deposition (ALD; see Volume II, Figure 7.47).

The most significant advances in reducing overall device size and packing density of transistors have come from improved isolation methods. The traditional junction isolation technique requires the p+ deep diffusion to be aligned to the n+ buried layer that is covered by a thick epitaxial layer. The area (and hence junction capacitance) is determined by alignment tolerance, area for side diffusion, and allowance for the spread of the depletion region. Modern isolation techniques such as oxide isolation and trench isolation eliminate most of these requirements. Oxide isolation processes were

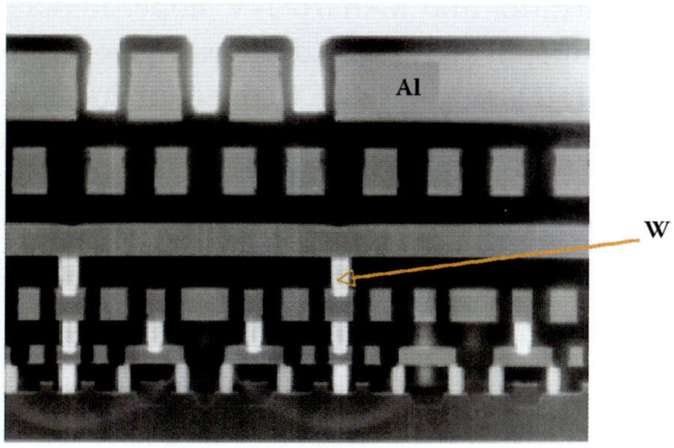

FIGURE 4.62 Multilevel aluminum metallization with tungsten plugs.

* There is no standard symbol for the dielectric constant; you may see it referred to as k, κ, ε, ε' or ε_r.

first introduced in the late 1970s. These first attempts used wet anisotropic etching (with KOH) of the [100] Si wafer with Si_3N_4 as a mask to etch V-grooves. Oxide is either deposited or grown (e.g., LOCOS) to fill the V-grooves. Because no side spread of the depletion region occurs, the packing density can be higher. These V-grooves require spacings of several tens of microns between devices, and that is not compatible with the required densities in VLSI. To further reduce the area between adjacent transistors, trench isolation is used, making use of dry etched vertical trenches. The trenches are typically 2 μm wide and 5 μm deep. The process is illustrated in Figure 4.63. After a patterning step (a), a channel stop (Chanstop) trench area is dry etched into the substrate (b); the trench is refilled with oxide (c); and chemical-mechanical polishing (CMP) removes the oxide on the nitride to get a flat surface (d).

In this same regard, silicon-on-insulator (SOI) wafers are used more and more in MOSFET construction. Differences from bulk CMOS include the fact that the transistor source, drain, and body are now surrounded by insulating SiO_2 rather than by a Si substrate (well). Advantages include reduced source and drain to substrate capacitance, lower passive current, and denser layout. Because of the high cost of SOI wafers, only some microprocessors, which command high prices and compete on speed, have embraced this technology (March 2004, Apple's Xserve G5 and AMD 90-nm processor at the end of 2004). An important SOI extension is to use

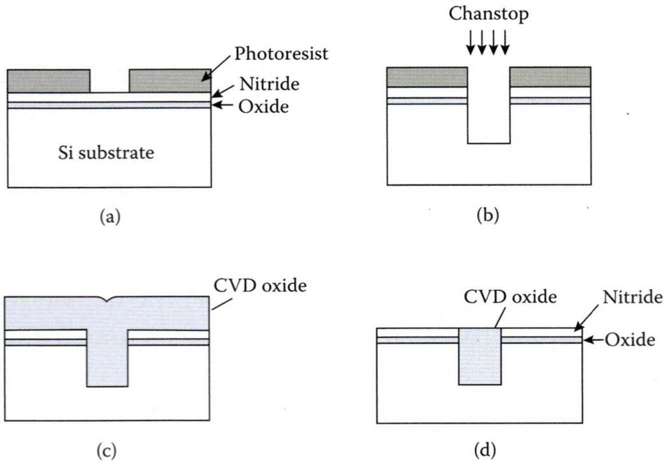

FIGURE 4.63 Trench isolation channel stops. See text for details.

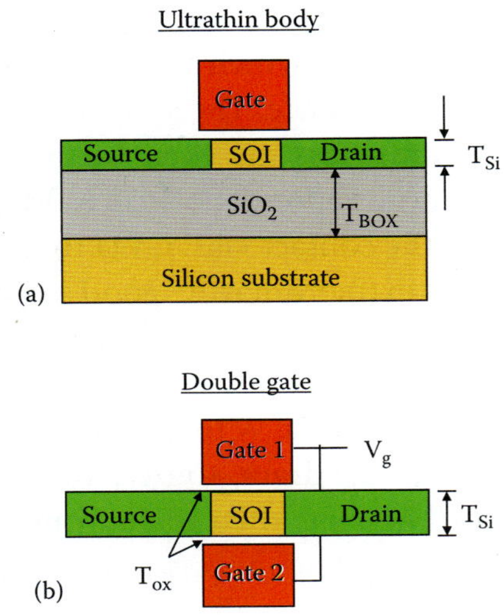

FIGURE 4.64 Ultrathin-body SOI FET (a) and a double-gate FET (b).

fully depleted SOI (Figure 4.64a). In this case, an ultrathin silicon body is used so that the entire body is depleted before the threshold voltage is reached (the body is thinner than the channel depletion width). Ultrathin-body MOSFETs feature very low leakage currents and a raised source/drain structure leading to low resistance and a ~12-nm scaling limit because of the low channel doping. The double-gate transistor (DGT) structure, proposed in the 1980s, further improves this situation. The DGT is composed of a conducting channel (usually undoped), surrounded by gate electrodes on either side (Figure 4.64b). This ensures that no part of the channel is far away from a gate electrode. Advantages of this design include the reduction of short-channel effects (e.g., a decreased V_T resulting from a reduced channel depletion charge), higher current drive capability, good electrical characteristics with a high on/off current ratio, and a sharp I–V slope. Reducing the body thickness further decreases gate current leakage, I_D(off), but has the tradeoff of increasing series resistance. This series resistance is minimized using a raised source/drain type of structure as in an ultrathin-body MOSFET or FinFET (see next section). It is projected that this technology will enable sub-10 nm scaling.

3D Transistors: Go Up, Young Man! A viable approach to stay on the Moore's curve is to deposit additional Si layers and fabricate transistors in them; in other words, start growing in the third dimension.

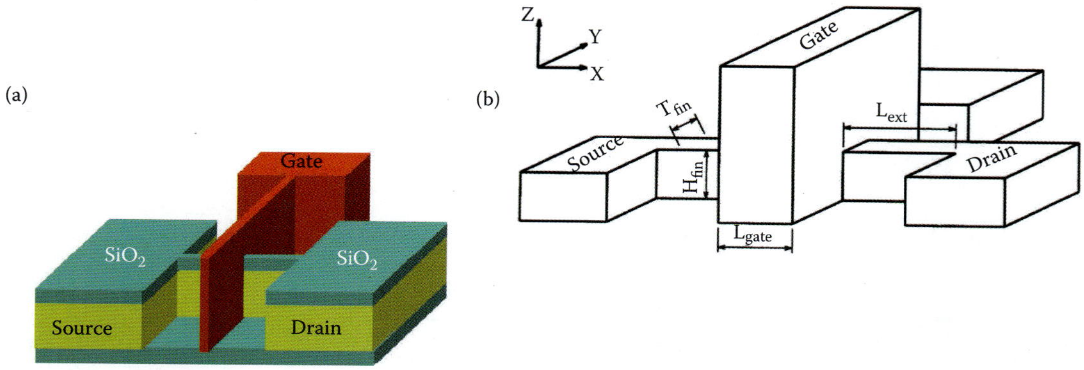

FIGURE 4.65 (a) FinFET structure. (b) Dimensioning: $H_{fin} \gg T_{fin}$, top gate oxide thickness ≫ sidewall oxide, effective channel length $L_{eff} = L_{gate} + 2 \times L_{ext}$ thickness, effective channel width $W = T_{fin} + 2 \times H_{fin}$.

Present MOSFET designs are planar, "buried" in the flat upper surface of the Si substrate. Each time you double circuit density by stacking, you can move a whole Moore's law generation without using smaller features. But interconnection is tricky, as are connections to the outside world. It is also more difficult to make high-performance transistors from deposited films, and additional processing tends to spoil previous layers. Source and drain are getting too close for reliable performance, and the gate between them is getting harder to define. 3D techniques can extend microprocessor design for several more generations, but after that, a new design paradigm will be needed to deal with increasing errors and defects.

The raised drain and source design in SOI approaches is a step in the right direction. In the FinFET approach, this concept is pushed even further; the transistor structure is put on top of the chip substrate as shown in Figure 4.65a. The term *FinFET* was coined by University of California, Berkeley researchers to describe a nonplanar, double-gate transistor built on an SOI substrate. The FinFET is a 3D device structure with a channel Si fin, sitting on the substrate, surrounded by a gate that effectively makes two to three gates in parallel. The gate wraps around the fin to form a very small channel. The Si fin has an insulator on the top, and the gate has oxide on either side (two channels). The current flows parallel to the device surface, and the channel width depends on the fin height (for dimensioning see Figure 4.65b). A lightly doped channel in a FinFET affords good gate control. Spacers are used to self-align the gate to source/drain, and the fin height can be used to increase the on-current, limited by aspect ratio. Many benefits arise from this design. Short-channel effects are reduced, and there is very low leakage current, high on/off current ratio, low voltage operation, efficient gate design (less switching power), flexibility of using multiple fins for better performance, compatibility with current manufacturing processes, and scalability to sub-10 nm.

FinFET is the leading candidate set to replace the classic MOSFETs. It is based on a 3D architecture that uses double gates. It can be manufactured with very minor modifications to the current self-aligned process flow. However, there are many issues that have to be resolved before it can be used commercially. A 25-nm transistor operating on just 0.7 V was demonstrated in December 2002 by the Taiwan Semiconductor Manufacturing Company. The "Omega FinFET" design, named after the similarity between the Greek letter omega and the shape in which the gate wraps around the source/drain structure, has a gate delay of just 0.39 ps for the n-type transistor and 0.88 ps for the p-type.

In Figure 4.66, we sketch the predicted material and device evolution for the years 2004–2020.

From the preceding, it appears that making Si-based MOSFETs smaller than 50 nm is not a problem. But the "ultimate" FET may not be made of silicon. Alternative approaches to fabricate logic switches, which we analyze in this volume Chapters 3 and 5 and in Volume III, Chapter 3 on nanotechnology, include molecular electronics, resonant tunneling diodes, and single-electron transistors.

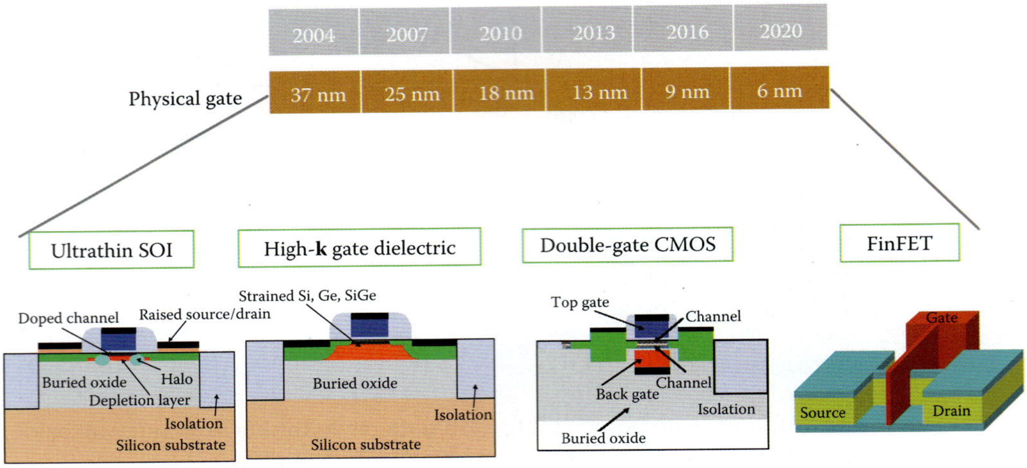

FIGURE 4.66 Evolving materials and device structures in MOSFET evolution. Halo from halo implant is a high-angle implant of well-type dopant species introduced into the device after transistor gate patterning.

Physicochemical Properties of Si

Introduction

In the previous sections, we mostly focused on Si as an electronic material. In this section we are looking at all the other properties that make Si such an attractive IC, MEMS, and NEMS material. We first consider Si as a simple substrate, where the semiconductor only plays the role as a flat convenient passive substrate. Next, we investigate the mechanical, thermal, and optical properties of Si. We finish this section with a short commentary on Si's biocompatibility.

Silicon as a Passive Substrate

Any IC, MEMS, or NEMS device needs a substrate. In some cases that substrate acts as a passive support only and does not play any other role in the operation of the device it supports. The use of Si as a passive substrate is discussed in more detail in Volume III, Chapter 1, which focuses on comparing nonlithography-based (traditional) and lithography-based (nontraditional) manufacturing methods. In Volume III, Table 1.9, we show a performance comparison of passive substrate materials in terms of cost, metallization ease, and machinability. We will see there that silicon, because of its extreme flatness, relative low cost, and well-established coating procedures, is often the preferred passive substrate—especially for thin films. The disadvantages of using Si substrates become more pronounced with increasing device size and low production volumes. An overwhelming determining factor for passive substrate choice is the final package of the device because the package often determines the volume and cost. Packaging is so important in the case of sensors that as a rule, their design should start from the package rather than from the sensor itself. For many mechanical sensor applications, single-crystal Si (SCS), based on its intrinsic mechanical stability and the feasibility of integrating sensing and electronics on the same substrate, presents an excellent substrate choice. But, as we will see in Volume III, Chapter 1, for chemical sensors, Si, with few exceptions, is merely the substrate and as such is not necessarily the most attractive option.

Silicon in Mechanical Sensors

Introduction

In mechanical sensors, the active structural elements convert a mechanical external input signal (e.g., pressure, acceleration) into an electrical signal output (voltage, current, or frequency). The transfer functions in mechanical devices describing this conversion are mechanical, electromechanical, and electrical.

In mechanical conversion, a given external load is concentrated and maximized in the active member of the sensor. Structurally active members are typically high-aspect-ratio elements such as suspended beams or membranes. Electromechanical conversion is the transformation of the mechanical quantity into an electrical quantity, such as capacitance, resistance, charge, and so on. Often, the electrical signal needs further electrical conversion into an output voltage, frequency, or current. For electrical conversion into an output voltage, a Wheatstone bridge (see Figure 4.75) may be used as in the case of a piezoresistive sensor, and a charge amplifier may be used in the case of a piezoelectric sensor. To optimize all three transfer functions, detailed electrical and mechanical modeling is required. One of the most important inputs required for mechanical models is the experimentally determined, independent elasticity constants or moduli. In what follows, we describe in more detail what makes Si such an important structural element in mechanical sensors and present its elasticity constants.

Stress-Strain Curve and Elasticity Constants

Yield, tensile strength, hardness, and creep of a material all relate to the elasticity curve (the stress-strain diagram). Stress-strain curves for several types of materials are shown in Figure 4.67. For small

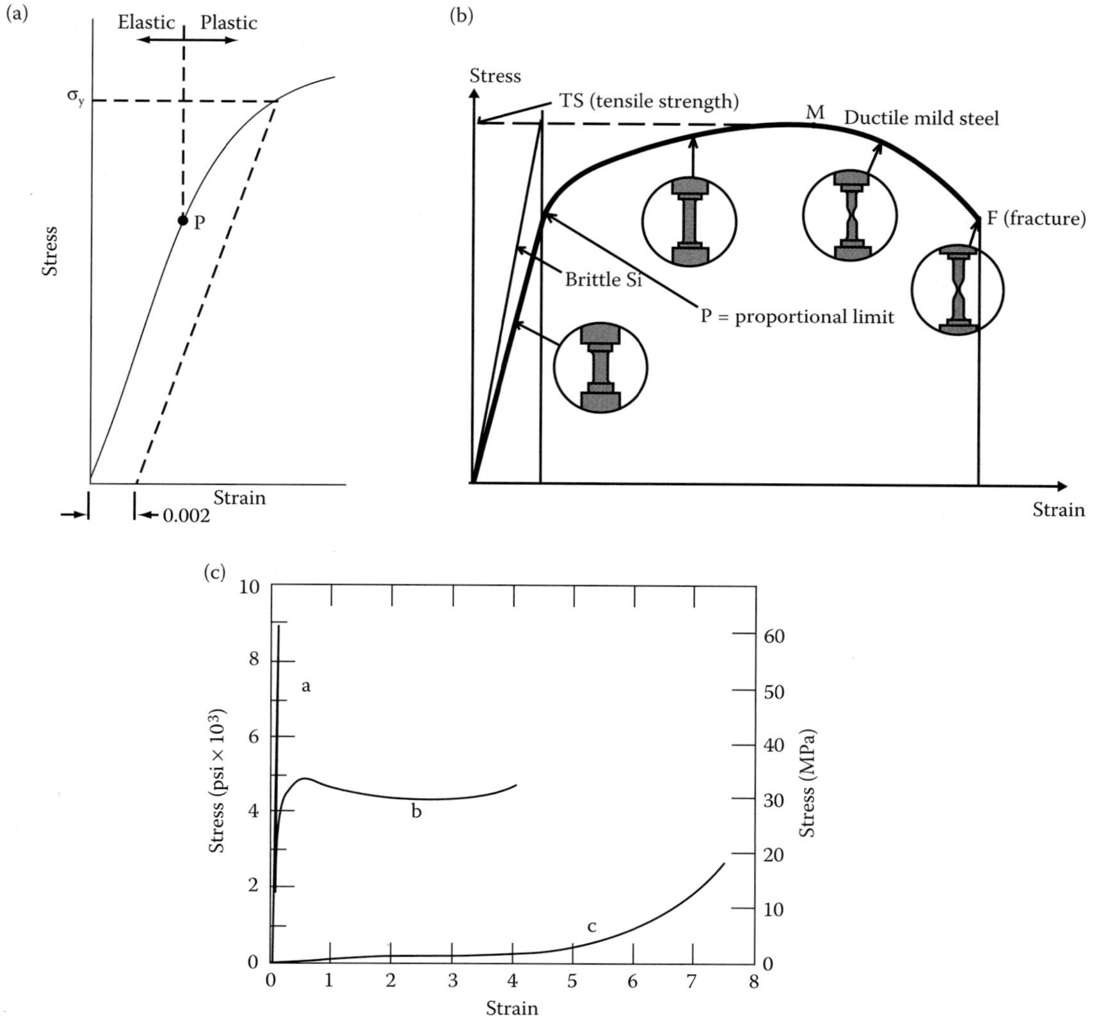

FIGURE 4.67 Typical stress-strain behaviors for different types of materials showing different degrees of elastic and plastic deformations. (a) Stress-strain curve for a typical material with the proportional limit *P*, and the yield strength σ_y, as determined using the 0.002 strain offset method. (b) Stress-strain curves for silicon and a ductile mild steel. The tensile strength (TS) of a metal is the stress at the maximum of the curve shown (*M*). Fracture is indicated by *F*. The various stages of a test metal sample are indicated as well. A high modulus material such as Si exhibits abrupt breakage; it is brittle with no plastic deformation region at all. (c) Stress-strain behavior for brittle (curve a), plastic (curve b), and highly elastic (elastomeric) (curve c) polymers.

strain values, Hooke's law (1635–1702) applies, that is, stress (force per unit area, N/m^2) and strain (displacement per unit length, dimensionless) are proportional, and the stress-strain curve is linear, with a slope corresponding to the elastic modulus *E* (Young's modulus, N/m^2). This regime, as marked in Figure 4.67a, is the elastic deformation regime (typically valid for $\varepsilon < 10^{-4}$). The magnitude of the Young's modulus ranges from 4.1×10^4 MPa (Pascal, Pa = N/m^2) for magnesium to 40.7×10^4 MPa for tungsten and 144 GPa for Invar,* whereas concrete is 45 GPa, aluminum is about 70 GPa, and elastomeric materials are as low as 10^{-3} to 10^{-2} GPa. Silicon is a stiff and brittle material with a Young's modulus of 160 GPa (Figure 4.67b). With increasing temperature, the elastic modulus diminishes.

For isotropic media, such as amorphous and polycrystalline materials, the tensile stress, σ_a, or the applied axial force per unit area and the axial or tensile strain, ε_a, are related as:

$$\sigma_a = E\varepsilon_a \tag{4.75}$$

with ε_a given by the dimensionless ratio of $(L_2 - L_1)/L_1$ ($\Delta L/L_1$), that is, the ratio of the sample's elongation to its original length. The elastic modulus may be thought of as stiffness or a material's resistance to elastic deformation. The greater the modulus, the stiffer the material. A tensile stress usually leads to a lateral strain or contraction (Poisson effect), ε_1, given by the dimensionless ratio of $(D_2 - D_1)/D_1$ ($\Delta D/D_1$), where D_1 is the original wire diameter and ΔD is the change in diameter under axial stress (see Figure 4.68). The Poisson ratio is the ratio of lateral over axial strain:

$$\nu = -\frac{\varepsilon_1}{\varepsilon_a} \tag{4.76}$$

The minus sign indicates a contraction of the material. For most materials, ν is a constant within the elastic range and fluctuates for different types of materials over a relatively narrow range. Generally, it is on the order of 0.25–0.35, and a value of 0.5 is the largest value possible. The latter is attained by materials such as rubber and indicates a material in which only the shape changes, whereas the volume remains constant. Normally, some slight volume change does accompany the deformation, and, consequently, ν is smaller than 0.5. The Poisson ratios for metals are typically around 0.33 (e.g., aluminum and cast steel are 0.34 and 0.28, respectively). For ceramics, it is around 0.25, and for polymers it is typically between 0.4 and 0.5. In extreme cases, values as low as 0.1 (certain types of concrete) and as high as 0.5 (rubber) occur.

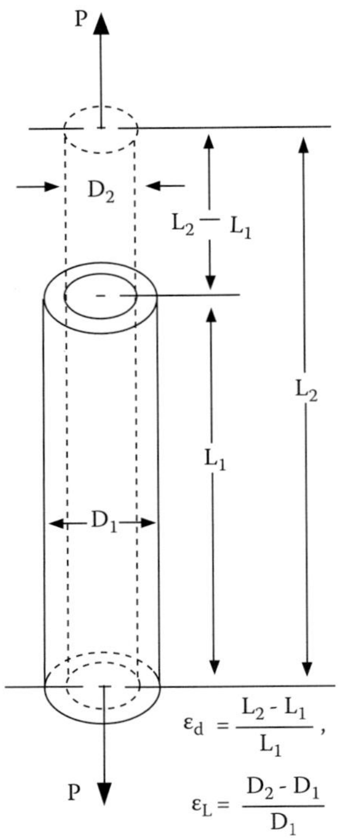

FIGURE 4.68 Metal wire under axial or normal stress; normal stress creates both elongation and lateral contraction.

For an elastic isotropic medium subjected to a triaxial state of stress, the resulting strain component in the *x*-direction, ε_x, is given by the summation of elongation and contraction:

$$\varepsilon_x = \frac{1}{E}\left[\sigma_x - \nu(\sigma_y + \sigma_z)\right] \tag{4.77}$$

and so on for the *y*- and *z*-directions (three equations in total).

For an analysis of mechanical structures, we must consider not only compressional and tensile strains but also shear strains. Whereas normal stresses create elongation plus lateral contraction with accompanying volume changes, shear stresses create shape

* Invar, an alloy formed of nickel and iron, has an extremely small coefficient of thermal expansion at room temperature. Invar is often called for in the design of machinery that must be extremely stable.

changes without volume changes (e.g., by twisting a body), that is, shear strains. The 1D shear strain, γ, is produced by the shear stress, τ, and has units of N/m^2. For small strains, Hooke's law may be applied again:

$$\gamma = \frac{\tau}{G} \tag{4.78}$$

where G is called the *elastic shear modulus* or the *modulus of rigidity*. For any 3D state of shear stress, three equations of this type will hold. Isotropic bodies are characterized by two independent elastic constants only because the shear modulus G, it can be shown,[19] relates the Young's modulus and the Poisson ratio as:

$$G = \frac{E}{2(1+\nu)} \tag{4.79}$$

Crystal materials, whose elastic properties are anisotropic, require more than two elastic constants, the number increasing with decreasing symmetry. Cubic crystals, for example, require 3 elastic constants, hexagonal crystals require 5, and materials without symmetry require 21[19,20] (see also Table 2.1). The relation between stresses and strains is more complex in this case and depends greatly on the spatial orientation of these quantities with respect to the crystallographic axes (see Chapter 2 on crystallography). Hooke's law in the most generic form is expressed in two formulas:

$$\sigma_{ij} = E_{ijkl} \cdot \varepsilon_{kl} \text{ and } \varepsilon_{ij} = S_{ijkl} \cdot \sigma_{kl} \tag{4.80}$$

where σ_{ij} and σ_{kl} = stress tensors of rank 2 expressed in N/m^2

ε_{ij} and ε_{kl} = strain tensors of rank 2 and are dimensionless

E_{ijkl} = stiffness coefficient tensor of rank 4 expressed in N/m^2 with at the most $3 \times 3 \times 3 \times 3 = 81$ elements

S_{ijkl} = compliance coefficient tensor of rank 4 expressed in m^2/N with at the most $3 \times 3 \times 3 \times 3 = 81$ elements

The first expression is analogous to Equation 4.75, and the second expression is the inverse, giving the strains in terms of stresses. The tensor representations in Equation 4.80 can also be represented as two matrices:

$$\sigma_m = \sum_{n=1}^{6} E_{mn}\varepsilon_n \text{ and } \varepsilon_m = \sum_{n=1}^{6} S_{mn}\sigma_n \tag{4.81}$$

Components of tensors E_{ijkl} and S_{ijkl} are substituted by elements of the matrices E_{mn} and S_{mn}, respectively. To abbreviate the *ij* indices to *m* and the *kl* indices to *n*, the following scheme applies:

11 → 1, 22 → 2, 33 → 3, 23 and 32 → 4, 13 and 31 → 5, 12 and 21 → 6, $E_{ijkl} \to E_{mn}$ and $S_{ijkl} \to S_{mn}$ when m and n = 1, 2, 3; $2S_{ijkl} \to S_{mn}$ when m or n = 4, 5, 6; $4S_{ijkl} \to S_{mn}$ when m and n = 4, 5, 6; $\sigma_{ij} \to \sigma_m$ when m = 1, 2, 3; and $\varepsilon_{ij} \to \varepsilon_m$ when m = 4, 5, 6.

Thus, with these reduced indices, there are six equations of the type:

$$\sigma_x = E_{11}\varepsilon_x + E_{12}\varepsilon_y + E_{13}\varepsilon_z + E_{14}\gamma_{yz} + E_{15}\gamma_{zx} + E_{16}\gamma_{xy} \tag{4.82}$$

And hence there exist 36 moduli of elasticity or E_{mn} stiffness constants. There are also six equations of the type:

$$\varepsilon_x = S_{11}\sigma_x + S_{12}\sigma_y + S_{13}\sigma_z + S_{14}\tau_{yz} + S_{15}\tau_{zx} + S_{16}\tau_{xy} \tag{4.83}$$

defining 36 S_{mn} constants, which are called the *compliance constants*. It can be shown that the matrices E_{mn} and S_{mn}, each composed of 36 coefficients, are symmetrical ($E_{mn} = E_{nm}$ and $S_{mn} = S_{nm}$); hence a material without symmetrical elements has 21 independent constants or moduli. Because of symmetry of crystals, several more of these may vanish until, for an isotropic medium, they number only two (E and ν). The stiffness coefficient and compliance coefficient matrices for cubic-lattice crystals with the vector of stress oriented along the [100] axis are given as:

$$E_{mn} = \begin{bmatrix} E_{11} & E_{12} & E_{12} & 0 & 0 & 0 \\ E_{12} & E_{11} & E_{12} & 0 & 0 & 0 \\ E_{12} & E_{12} & E_{11} & 0 & 0 & 0 \\ 0 & 0 & 0 & E_{44} & 0 & 0 \\ 0 & 0 & 0 & 0 & E_{44} & 0 \\ 0 & 0 & 0 & 0 & 0 & E_{44} \end{bmatrix} \tag{4.84}$$

$$S_{mn} = \begin{bmatrix} S_{11} & S_{12} & S_{12} & 0 & 0 & 0 \\ S_{12} & S_{11} & S_{12} & 0 & 0 & 0 \\ S_{12} & S_{12} & E_{11} & 0 & 0 & 0 \\ 0 & 0 & 0 & S_{44} & 0 & 0 \\ 0 & 0 & 0 & 0 & S_{44} & 0 \\ 0 & 0 & 0 & 0 & 0 & S_{44} \end{bmatrix} \quad (4.85)$$

In cubic crystals, the three remaining independent elastic moduli are usually chosen as E_{11}, E_{12}, and E_{44}. The S_{mn} values can be calculated simply from the E_{mn} values. Expressed in terms of the compliance constants, one can show that $1/S_{11} = E =$ Young's modulus, $-S_{12}/S_{11} = \nu =$ Poisson's ratio, and $1/S_{44} = G =$ shear modulus. In the case of an isotropic material, such as a metal wire, there is an additional relationship:

$$E_{44} = \frac{E_{11} - E_{12}}{2} \quad (4.86)$$

With this expression the number of independent stiffness constants reduces to two. The anisotropy coefficient α is defined as:

$$\alpha = \frac{2E_{44}}{E_{11} - E_{12}} \quad (4.87)$$

making $\alpha = 1$ for an isotropic crystal. For an anisotropic crystal, the degree of anisotropy is given by the deviation of α from 1. SCS has moderately anisotropic elastic properties,[21,22] with $\alpha = 1.57$, and very anisotropic crystals may have a value larger than 8. Brantley[21] gives the nonzero Si stiffness components, referred to the [100] crystal orientation, as $E_{11} = E_{22} = E_{33} = 166 \times 10^9$ N/m^2, $E_{12} = E_{13} = E_{23} = 64 \times 10^9$ N/m^2, and $E_{44} = E_{55} = E_{66} = 80 \times 10^9$ N/m^2:

$$\begin{matrix} \sigma_x \\ \sigma_y \\ \sigma_z \\ \tau_{xy} \\ \tau_{xz} \\ \tau_{yz} \end{matrix} = \left| \begin{matrix} 166(E_{11}) & 64(E_{12}) & 64(E_{12}) & 0 & 0 & 0 \\ 64(E_{12}) & 166(E_{11}) & 64(E_{12}) & 0 & 0 & 0 \\ 64(E_{12}) & 64(E_{12}) & 166(E_{11}) & 0 & 0 & 0 \\ 0 & 0 & 0 & 80(E_{44}) & 0 & 0 \\ 0 & 0 & 0 & 0 & 80(E_{44}) & 0 \\ 0 & 0 & 0 & 0 & 0 & 80(E_{44}) \end{matrix} \right| \times \begin{matrix} \varepsilon_x \\ \varepsilon_y \\ \varepsilon_z \\ \gamma_{xy} \\ \gamma_{xz} \\ \gamma_{yz} \end{matrix} \quad (4.88)$$

with σ normal stress, τ shear stress, ε normal strain, and γ shear strain. The values for E_{mn} in Equation 4.88 compare with a Young's modulus of 207 GPa for a low carbon steel. Variations on the values of the elastic constants on the order of 30%, depending on crystal orientation, doping level, and dislocation density, must be considered to have effects as well. In Volume II, Chapter 7, we learn that the Young's modulus for polycrystalline silicon is about 161 GPa. From the stiffness coefficients, the compliance coefficients of Si can be calculated as $S_{11} = 7.68 \times 10^{-12}$ m^2/N, $S_{12} = -2.14 \times 10^{-12}$ m^2/N, and $S_{44} = 12.6 \times 10^{-12}$ m^2/N.[23] A graphical representation of elastic constants for different crystallographic directions in Si and Ge is given in Worthman et al.[24] and is reproduced in Figure 4.69a–d displays E and ν for Ge and Si in planes (100) and (110) as functions of direction. Calculations show that E, G, and ν are constant for any direction in the (111) plane. In other words, a plate lying in this plane can be considered as having isotropic elastic properties. A review of independent determinations of the Si stiffness coefficients, with their respective temperature coefficients, is given in Metzger et al.[25] Some of the values from that review are reproduced in Table 4.12. Values for Young's modulus and the shear modulus of Si can also be found in Greenwood[26] and are reproduced in Table 4.13 for the three important crystal orientations.

Residual Stress in Single-Crystal Si

Most mechanical properties, such as the Young's modulus, for lightly and highly doped silicon are identical. Residual stress and associated stress gradients in highly boron-doped single-crystal Si (SCS) present some controversy. Highly boron-doped membranes, which are usually reported to be tensile, have also been reported as compressive.[27,28] From a simple atom-radius argument, a large number of substitutional boron atoms would be expected to create a net shrinkage of the lattice compared with pure silicon and that the residual stress would be tensile with a stress gradient corresponding to the doping gradient. For example, an etched cantilever would be expected to bend up out of the plane of the silicon wafer. Maseeh et al.[28] believe that the appearance of compressive behavior in heavily boron-doped single-crystal layers results from the use of an oxide etch mask. They suggest that plastic deformation of the p^+ silicon beneath the compressively stressed oxide can explain the observed behavior. Ding et al.[29] also

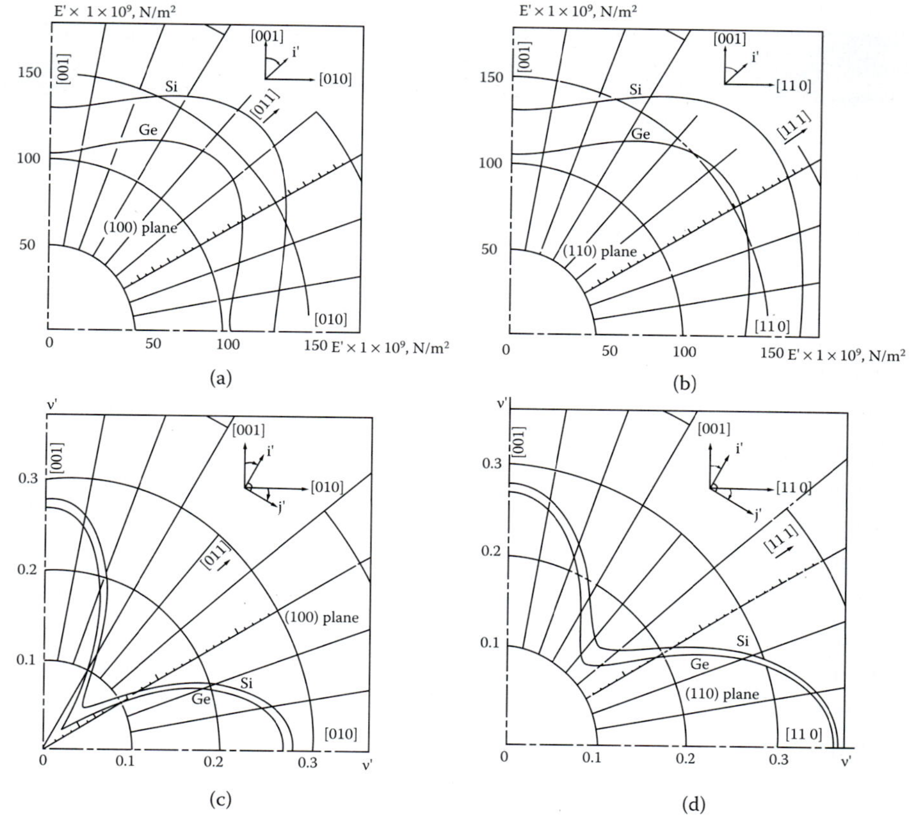

FIGURE 4.69 Elasticity constants for Si and Ge. (a) Young's modulus as a function of direction in the (100) plane. (b) Young's modulus as a function of direction in the (110) plane. (c) Poisson's ratio as a function of direction in the (100) plane. (d) Poisson's ratio as a function of direction in the (110) plane. (From Worthman, J. J., and R. A. Evans. 1965. *J Applied Physics* 36:153–56. With permission.[24])

found compressive behavior for nitride-covered p^+ Si thin membranes. These authors believe that the average stress in p^+ silicon is tensile, but great care is required to establish this fact because the combination of heavy boron doping and a high-temperature drive-in under oxidizing conditions can create an apparent reversal of both the net stress (to compressive) and of the stress gradient (opposite to the doping gradient). A proposed explanation is that, at the oxide/silicon interface, a thin compressively stressed layer is formed during the drive-in, which is not removed in buffered HF. It can be removed by reoxidation and etching in HF, or by etching in KOH.

The question of residual stress in polycrystalline Si is a heavily studied topic as we shall discover in Volume II, Chapter 7 on surface micromachining. Stress in thin polycrystalline films has dramatic effects on the reliable operation of surface micromachined mechanical elements (see also Volume II, Equation 7.29).

Table 4.12 Stiffness Coefficients and Temperature Coefficients of Stiffness for Si

Stiffness Coefficients in GPa (value in GPa = 10^9 N/m²)	Temperature Coefficient of Young's Modulus (10^{-6} K^{-1})
$E_{11} = 164.8 \pm 0.16$	−122
$E_{12} = 63.5 \pm 0.3$	−162
$E_{44} = 79.0 \pm 0.06$	−97

Source: Metzger, H., and F.R. Kessler. 1970. Der Debye-Sears Effect zur Bestimmung der Elastischen Konstanten von Silicium. *Z Naturf A* 25:904.

Yield, Tensile Strength, Hardness, and Creep

As a material is deformed beyond its elastic limit, yielding or plastic deformation (permanent, nonrecoverable deformation) occurs. The point of yielding in Figure 4.67a is the point of initial departure from linearity of the stress-strain curve and is sometimes called the *proportional* limit indicated by a letter *P*. Young's modulus of mild steel is ±30,000,000 psi, and its proportional limit (highest stress in the

TABLE 4.13 Derived Values for Young's Modulus and Shear Modulus for Si

Crystal Orientation	Young's Modulus E (GPa)	Shear Modulus G (GPa)	Temperature Coefficient (10^{-6} K^{-1})
[100]	129.5	79.0	–63
[110]	168.0	61.7	–80
[111]	186.5	57.5	–46

Source: Greenwood, J.C. 1988. Silicon in mechanical sensors. *J Phys E Sci Instrum* 21:1114.

elastic range) is approximately 30,000 psi.* Thus, the maximum elastic strain in mild steel is about 0.001 under a condition of uniaxial stress. This gives an idea as to the magnitude of the strains we are dealing with. A convention has been established wherein a straight line is constructed parallel to the elastic portion of the stress-strain curve at some specified strain offset, usually 0.002. The stress corresponding to the intersection of this line and the stress-strain curve as it bends over in the plastic region is defined as the yield strength, σ_y (see Figure 4.67a). The magnitude of the yield strength of a material is a measure of its resistance to plastic deformation. Yield strengths may range from 35 MPa (5,000 psi) for a soft and weak aluminum to more than 1,400 MPa (200,000 psi) for high-strength steels. The tensile strength is the stress at the peak of the stress-strain curve. This corresponds to the maximum stress that can be sustained by a structure in tension; if the stress is applied and maintained, fracture will result.

Crystalline silicon is a hard and brittle material, deforming elastically until it reaches its yield strength, at which point it breaks. For Si, the yield strength is 7 GPa, equivalent to a 700-kg weight suspended from a 1-mm^2 area. Both tensile strength and hardness are indicators of a metal's resistance to plastic deformation. Consequently, they are roughly proportional.[30] Material deformation occurring at elevated temperatures and static material stresses is termed *creep*. It is defined as a time-dependent and permanent deformation of materials when subjected to a constant load or stress.

Silicon exhibits no plastic deformation or creep below 500°C; therefore, Si sensors are very insensitive to fatigue failure when subjected to high cyclic loads. Silicon sensors have been cycled in excess of 100 million cycles with no observed failures. This ability to survive a very large number of duty cycles is because there is no energy-absorbing or heat-generating mechanism caused by intergranular slip or movement of dislocations in silicon at room temperature. However, single-crystal Si (SCS), as a brittle material, will yield catastrophically when stress beyond the yield limit is applied rather than deform plastically as metals do (see Figure 4.67b). At room temperature, high modulus materials, such as Si, SiO_2, and Si_3N_4, often exhibit linear-elastic behavior at lower strain and transition abruptly to brittle-fracture behavior at higher strain. Plastic deformation in metals is based on stress-induced dislocation generation in the grain boundaries and a subsequent dislocation migration that results in a macroscopic deformation from intergrain shifts in the material. No grain boundaries exist in SCS, and plastic deformation can only occur through migration of the defects originally present in the lattice or of those generated at the surface. As the number of these defects is very low in SCS, the material can be considered a perfect elastic material at normal temperatures. Perfect elasticity implies proportionality between stress and strain (i.e., load and flexure) and the absence of irreversibilities or mechanical hysteresis. The absence of plastic behavior also accounts for the extremely low mechanical losses in SCS, which enable the fabrication of resonating structures that exhibit exceptionally high Q-factors. Values of up to 10^8 in vacuum have been reported. At elevated temperatures, and with metals and polymers at ordinary temperatures, complex behavior in the stress-strain curve can occur. Considerable plasticity can be induced in SCS at elevated temperatures (>800°C), when silicon softens appreciably and the mobility of defects in the lattice is substantially increased. Huff

* In the sensor area it is still mandatory to be versatile in the different unit systems especially with regard to pressure and stress. In this book we are using mostly Pascal, Pa (=N/m^2), but in industry it is still customary to use psi when dealing with metal properties, Torr when dealing with vacuum systems, and dyne/cm^2 when dealing with surface tension.

and Schmidt[27] report a pressure switch exhibiting hysteresis based on buckling of plastically deformed silicon membranes. To eliminate plastic deformation of Si wafers, it is important that during high-temperature steps the presence of films that could stress or even warp the wafer in an asymmetric way, typically oxides or nitrides, be avoided.

Piezoresistivity in Silicon

Piezoresistance is the fractional change in bulk resistivity induced by small mechanical stresses applied to a material. Lord Kelvin discovered the effect in 1856. Most materials exhibit piezoresistivity, but the effect, Smith found in 1954, is particularly important in some semiconductors (more than an order of magnitude higher than that of metals).[31] Monocrystalline silicon has a high piezoresistivity and, combined with its excellent mechanical and electronic properties, makes a superb material for the conversion of mechanical deformation into an electrical signal. The history of silicon-based mechanical sensors started with the discovery of the piezoresistance effect in Si (and Ge) more than four decades ago.[31] The piezoresistive effect in semiconductor materials originates in the deformation of the energy bands as a result of the applied stress. The deformed bands change the effective mass and mobility of the charge carriers (electrons and holes), hence modifying the resistivity. The two main classes of piezoresistive semiconductor sensors are membrane-type structures (typically pressure and flow sensors) and cantilever beams (typically acceleration sensors) with in-diffused resistors (boron, arsenic, or phosphorus) strategically placed in zones of maximum stress (Figure 4.70).

In Chapter 2 we saw how in a 3D anisotropic crystal, the electrical field vector (**E**) is related to the current vector, **J**, by a 3 × 3 resistivity tensor (see Equation 2.4).[23] Experimentally, the nine coefficients always reduce to six, and the symmetric tensor, is given by:

$$\underset{\text{Electric Field}}{\begin{bmatrix} \mathbf{E}_1 \\ \mathbf{E}_2 \\ \mathbf{E}_3 \end{bmatrix}} = \underset{\text{Resistivity Tensor}}{\begin{bmatrix} \rho_1 & \rho_6 & \rho_5 \\ \rho_6 & \rho_2 & \rho_4 \\ \rho_5 & \rho_4 & \rho_3 \end{bmatrix}} \underset{\text{Curren}}{\begin{bmatrix} \mathbf{J}_1 \\ \mathbf{J}_2 \\ \mathbf{J}_3 \end{bmatrix}} \qquad (4.89)$$

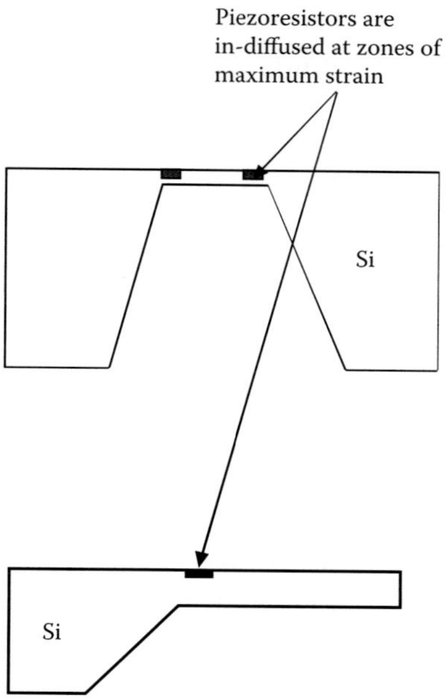

FIGURE 4.70 Si membrane and Si cantilever with in-diffused resistors.

For the cubic Si lattice, with the axes aligned with the <100> directions, ρ_1, ρ_2, and ρ_3 define the dependence of the electric field on the current along the same direction (one of the <100> directions). The cross-sensitivities, ρ_4, ρ_5, and ρ_6, relate the electric field to the current along a perpendicular direction.

The six resistivity components in Equation 4.89 depend on the normal (σ) and shear (τ) stresses in the material as defined earlier in this chapter. This can be represented as:

$$\rho = \rho_{\text{no stress}} + \Delta\rho(\sigma,\tau) \text{ or } \sigma_c \left(= \frac{1}{\rho}\right) = \sigma_{\text{no stress}} + \Delta\sigma_c(\sigma,\tau) \qquad (4.90)$$

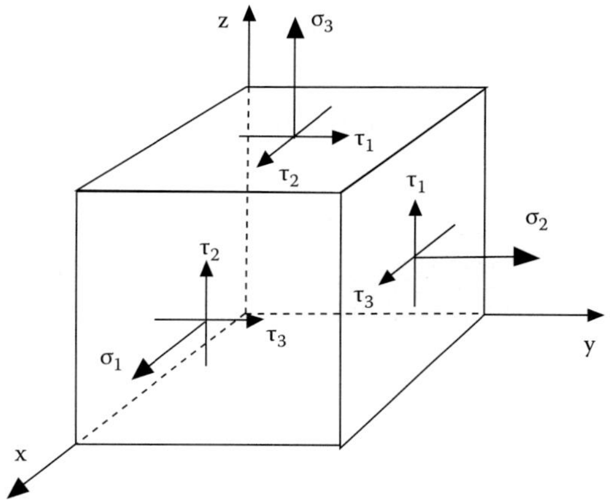

FIGURE 4.71 Piezoresistance in silicon.

where we use σ_c for conductivity to distinguish it from σ for the normal stress. For cubic crystals this can be written out as (see also Figure 4.71):

$$\begin{bmatrix} \Delta\sigma_{c,1} \\ \Delta\sigma_{c,2} \\ \Delta\sigma_{c,3} \\ \Delta\sigma_{c,4} \\ \Delta\sigma_{c,5} \\ \Delta\sigma_{c,6} \end{bmatrix} = \begin{bmatrix} \pi_{11} & \pi_{12} & \pi_{12} & 0 & 0 & 0 \\ \pi_{12} & \pi_{11} & \pi_{12} & 0 & 0 & 0 \\ \pi_{12} & \pi_{12} & \pi_{11} & 0 & 0 & 0 \\ 0 & 0 & 0 & \pi_{44} & 0 & 0 \\ 0 & 0 & 0 & 0 & \pi_{44} & 0 \\ 0 & 0 & 0 & 0 & 0 & \pi_{44} \end{bmatrix} \cdot \begin{bmatrix} \sigma_1 \\ \sigma_2 \\ \sigma_3 \\ \tau_1 \\ \tau_2 \\ \tau_3 \end{bmatrix} \quad (4.91)$$

Substituting the results of Equation 4.91 in Equation 4.89 one obtains the following three expressions for a large piece of bulk silicon:

$$\begin{aligned} E_1 &= \rho J_1 + \rho\pi_{11}\sigma_1 J_1 + \rho\pi_{12}(\sigma_2+\sigma_3)J_1 + \rho\pi_{44}(J_2\tau_3 + J_3\tau_2) \\ E_2 &= \rho J_2 + \rho\pi_{11}\sigma_2 J_2 + \rho\pi_{12}(\sigma_1+\sigma_3)J_2 + \rho\pi_{44}(J_1\tau_3 + J_3\tau_1) \\ E_3 &= \rho J_3 + \rho\pi_{11}\sigma_3 J_3 + \rho\pi_{12}(\sigma_1+\sigma_2)J_3 + \rho\pi_{44}(J_1\tau_2 + J_2\tau_1) \end{aligned} \quad (4.92)$$

For devices with finite dimensions, the influence of dimensional changes needs to be added through the Poisson ratio.

Smith[31] was the first to measure the resistivity coefficients for Si at room temperature. The coefficients are dependent on crystal orientation, temperature, and dopant concentration. Table 4.14 lists Smith's results for the three remaining independent resistivity coefficients, that is, π_{11}, π_{12}, and π_{44} of [100]-oriented Si at room temperature.[31] The piezoresistance coefficients are largest for π_{11} in n-type silicon and π_{44} in p-type silicon, about $-102.2\ 10^{-11}$ and $138\ 10^{-11}\ Pa^{-1}$, respectively. The piezoresistivity coefficients are related to the gauge factor (G_f) by the Young's modulus. The gauge factor, G_f, is the relative resistance change divided by the applied strain, or:

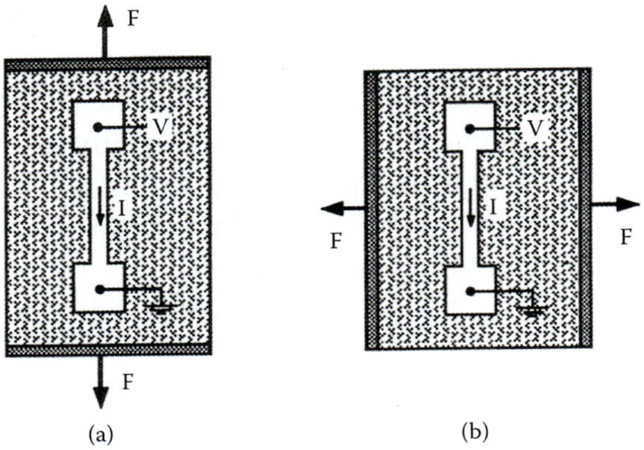

FIGURE 4.72 Longitudinal (a) and transverse (b) piezoresistors.

$$G_f = \frac{1}{\varepsilon}\frac{\Delta R}{R} \quad (4.93)$$

The gauge factor of a metal strain gauge is typically around 2; for single-crystal Si (SCS) it is 90; and for polycrystalline and amorphous Si, it is between 20 and 40.

The contribution to resistance changes from stresses that are longitudinal (σ_l) and transverse (σ_t) with respect to the current flow is given by (Figure 4.72):

$$\frac{\Delta R}{R} = \sigma_l\pi_l + \sigma_t\pi_t \quad (4.94)$$

where σ_l = longitudinal stress component, i.e., stress component parallel to the direction of the current

σ_t = transversal stress component, i.e., stress component perpendicular to the direction of the current

π_l = longitudinal piezoresistance coefficient

π_t = transversal piezoresistance coefficient

The piezoresistance coefficients π_l and π_t for (100) silicon as a function of crystal orientation are reproduced from Kanda in Figure 4.73a for

TABLE 4.14 Resistivity and Piezoresistivity Coefficients at Room Temperature, <100> Si Wafers and Doping Levels below $10^{18}\ cm^{-3}$ in $10^{-11}\ Pa^{-1}$

	ρ (Ωcm)	Direction	π_{11}	π_{12}	π_{44}	π_t	π_l
p-Si	7.8	<100>				0	0
		<110>	+6.6	−1.1	+138.1	−66.3	71.8
n-Si	11.7	<100>				+ 53.4	−102.2
		<110>	−102.2	+53.4	−13.6	−17.6	−31.2

Source: Smith, C.S. 1954. Piezoresistance effect in germanium and silicon. *Phys Rev* 94:42; and Khazan, A.D. 1994. *Transducers and their elements.* Englewood Cliffs, NJ: PTR Prentice Hall.

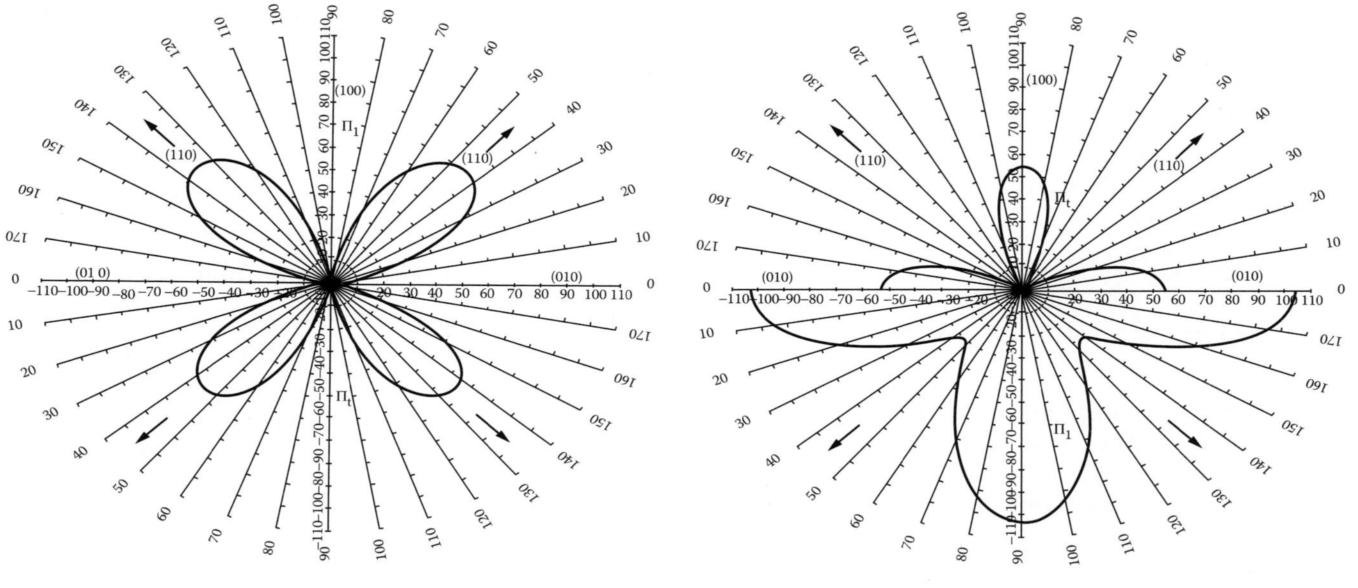

(a) (b)

FIGURE 4.73 Piezoresistance coefficients π_l and π_t for (100) silicon in the (001) plane in 10^{-12} cm^2 $dyne^{-1}$ or 10^{-11} Pa^{-1}. (a) For p-type. (b) For n-type. (From Kanda, Y. 1982. *IEEE Trans Electr Dev* 29:64. With permission.)

p-type and Figure 4.73b for n-type.[32] For lightly doped silicon (n- or p-type < 10^{18} cm^{-3}), the temperature coefficient of resistance for π_l and π_t is approximately 0.25%/°C. It decreases with dopant concentration to about 0.1%/°C at 8×10^{19} cm^{-3}. By maximizing the expression for the stress-induced resistance change in Equation 4.94, the achievable sensitivity in a piezoresistive silicon sensor is optimized. From Figure 4.73, the maximum longitudinal piezoresistive coefficient for p-type Si is $\pi_{l,[111]} = 93.5 \times 10^{-11}$ Pa^{-1} and for n-type Si it is $\pi_{l,[100]} = -102.2$ 10^{-11} Pa^{-1} (this latter value is also found in Table 4.14).

The orientation of a sensing membrane or beam (Figure 4.74) is determined by its anisotropic fabrication.

In p-type Si the piezoresistive coefficients are opposite in sign and comparable in magnitude, making them ideal for Wheatstone bridge detection (see Figure 4.75).

The surface of the silicon wafer is usually a (100) plane; the edges of the etched structures are intersections of (100) and (111) planes and are thus <110> directions. For pressure sensing, p-type piezoresistors are most commonly used. This is because the orientation of maximum piezoresistivity (<110>)

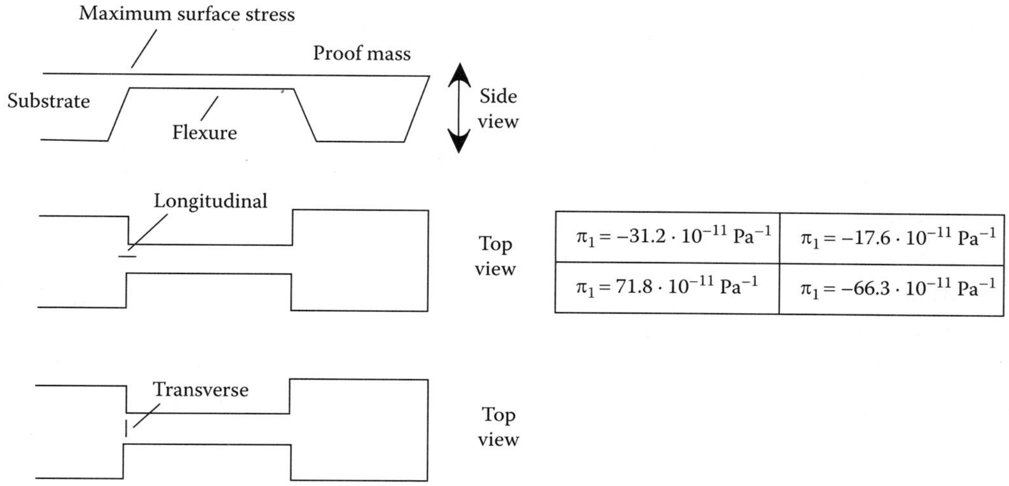

FIGURE 4.74 Bulk micromachined cantilever with longitudinal and transverse piezoresistors.

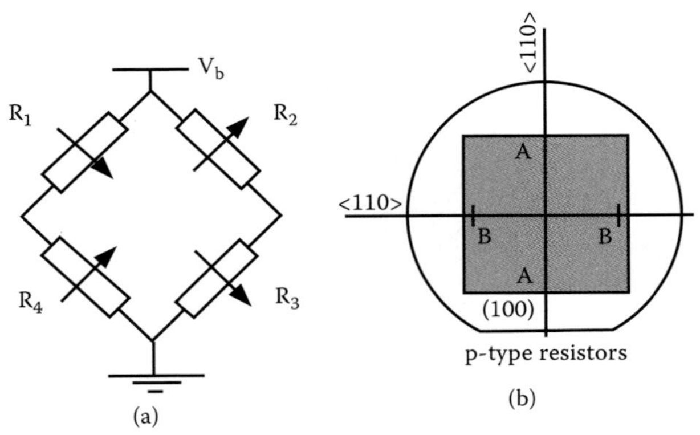

FIGURE 4.75 Measuring membrane resistance with piezoresistors. (a) Wheatstone bridge configuration of four in-diffused piezoresistors. The arrows indicate resistance changes when the membrane is bent downward (into the page). (b) Maximizing the piezoresistive effect with p-type resistors. The A resistors are stressed longitudinally and the B resistors are stressed transversally. (Based on Peeters, E. 1994. *Process development for 3D silicon microstructures, with application to mechanical sensor design.* PhD diss., Katholieke Universiteit Leuven, Leuven, Belgium. With permission;[2] and Maluf, N. 2000. *An introduction to microelectromechanical systems engineering.* Boston: Artech House.[34])

happens to coincide with the edge orientation of a conventionally etched diaphragm and because the longitudinal coefficient is roughly equal in magnitude but opposite in sign as compared with the transverse coefficient (see Figures 4.73a and 4.75b and Table 4.14).[2] Piezoresistors oriented at 45° with respect to the primary flat, that is, in the <100> direction, are insensitive to applied stress, which provides an inexpensive way to incorporate stress-independent diffused temperature sensors. Longitudinal and transverse piezoresistance coefficients are related to principal piezoresistance coefficients through coordinate transformation. With the values in Table 4.14, π_l and π_t can be calculated numerically for other primary orientations. The longitudinal piezoresistive coefficient in the <110> direction is $\pi_l = 1/2(\pi_{11} + \pi_{12} + \pi_{44})$. The corresponding transverse coefficient is $\pi_t = 1/2(\pi_{11} + \pi_{12} - \pi_{44})$. From Table 4.14, we know that for p-type resistors π_{44} is more important than the other two coefficients, and Equation 4.94 is approximated by:

$$\frac{\Delta R}{R} = \frac{\pi_{44}}{2}(\sigma_l - \sigma_t) \tag{4.95}$$

For n-type resistors, π_{44} can be neglected, and we obtain:

$$\frac{\Delta R}{R} = \frac{\pi_{11} + \pi_{12}}{2}(\sigma_l + \sigma_t) \tag{4.96}$$

Equations 4.95 and 4.96 are valid only for uniform stress fields or if the resistor dimensions are small compared with the membrane or beam size. When stresses vary over the resistors, they have to be integrated, which is conveniently done by computer simulation programs. More details on the underlying physics of piezoresistivity and its dependence on crystal orientation can be found in Kanda[32] and Middlehoek and Audet.[33]

To convert the piezoresistive effect into a measurable electrical signal, a Wheatstone bridge is often used. A balanced Wheatstone bridge configuration is constructed as in Figure 4.75a by locating four p-piezoresistors midway along the edges of a square diaphragm as in Figure 4.75b (location of maximum stress). Two resistors are oriented so that they sense stress in the direction of their current axes, and two are placed to sense stress perpendicular to their current flow. Two longitudinally stressed resistors (A) are balanced against two transversally stressed resistors (B); two of them increase in value, and the other two decrease in value on application of a stress. In this case, from Equation 4.95:

$$\frac{\Delta R}{R} \approx 70 \cdot 10^{-11}(\sigma_l - \sigma_t) \tag{4.97}$$

with σ in Pa. For a realistic stress pattern where σ_l = 10 MPa and σ_t = 50 MPa, Equation 4.97 gives us $\Delta R/R \approx 2.8\%$.[2] Thus, resistance change can be calculated as a function of the membrane or cantilever beam stress.

By varying the diameter and thickness of the silicon diaphragms, piezoresistive sensors in the range of 0–200 MPa have been made. The bridge voltages are usually between 5 and 10 V, and the sensitivity may vary from 10 mV/kPa for low-pressure to 0.001 mV/kPa for high-pressure sensors.

A schematic illustration of a pressure sensor with diffused piezoresistive sense elements is shown in Figure 4.76. In this case, p-type piezoresistors are diffused into a thin epitaxial n-type Si layer on a p-type Si substrate.[34]

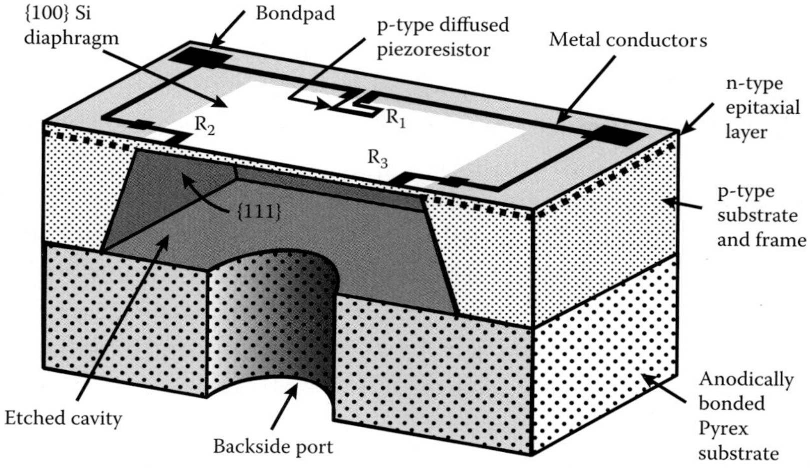

FIGURE 4.76 Schematic illustration of a pressure sensor with p-type diffused resistor in an n-type epitaxial layer. (Based on Maluf, N. 2000. *An introduction to microelectromechanical systems engineering.* Boston: Artech House.[34])

Peeters[2] shows how a more sensitive device may be based on n-type resistors when all the n-type resistors oriented along the <100> direction are subjected to a uniaxial stress pattern in the longitudinal axis, as shown in Figure 4.77. The overall maximum piezoresistivity coefficient (π_l in the <100> direction) is substantially higher for n-type silicon than it is for p-type silicon in any direction (maximum π_t and π_l in the <100> direction, Figure 4.73b). Exploitation of these high piezoresistivity coefficients is less obvious, though, because the resistor orientation for maximum sensitivity (<100>) is rotated over 45° with respect to the <110> edges of an anisotropically etched diaphragm. Also evident from Figure 4.73b is a transversally stressed resistor cannot be balanced against a longitudinally stressed resistor. Peeters has circumvented these two objections with a uniaxial, longitudinal stress pattern in the rectangular diaphragm represented in Figure 4.77. With a (100) substrate and a <100> orientation (45° to wafer flat), we obtain:

$$\frac{\Delta R}{R} \approx 53.10^{-11} \times \sigma_t - 102.10^{-11} \times \sigma_l \qquad (4.98)$$

with σ in Pa. Based on Equation 4.98, with $\sigma_t = 10$ MPa and $\sigma_l = 50$ MPa, $\Delta R/R \approx -4.6\%$. In the proposed stress pattern it is important to minimize the transverse stress by making the device truly uniaxial as the longitudinal and transverse stress components have opposite effects and can even cancel out one another. In practice, a pressure sensor with an estimated 65% gain in pressure sensitivity over the more traditional configurations could be made in the case of 80% uniaxiality.[2]

Silicon as a Mechanical MEMS Material: Summary

Here we want to refine decision criteria about the choice of an "active substrate" for mechanical micromachining applications in which the substrate is involved in determining the mechanical performance of the device crafted from/on it. Mechanical stability is crucial for mechanical sensing applications. Any sensing device must be free of drift to avoid recalibration at regular intervals. Part of the

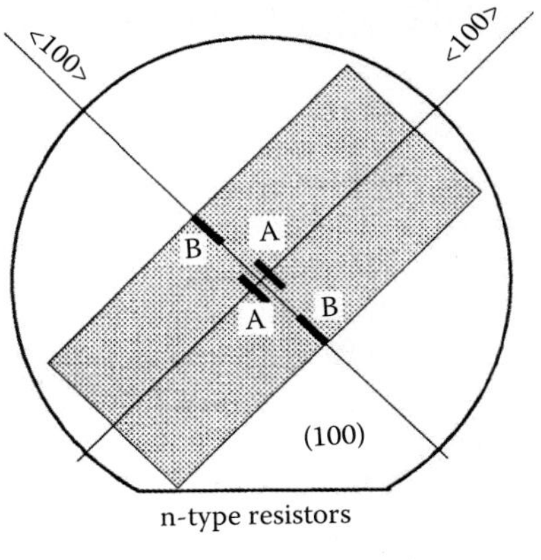

FIGURE 4.77 Higher pressure sensitivity by strategic placement of in-diffused piezoresistors proposed by Peeters.[2] The n-type resistors are stressed longitudinally with the A resistors under tensile stress and the B resistors under compressive stress. (From Peeters, E. 1994. *Process development for 3D silicon microstructures, with application to mechanical sensor design.* PhD diss., Katholieke Universiteit Leuven, Leuven, Belgium. With permission.[2])

drift in mechanical sensors may be associated with movement of crystal dislocations in the loaded mechanical part. In ductile materials, such as metals, dislocations move readily. In contrast, brittle materials, such as semiconductors, have dislocations that hardly move. Mechanical engineers often avoid using brittle materials and opt for ductile materials, even though these plastically deform, meaning that they are subject to mechanical hysteresis. The lack of plastic deformation coupled with extremely high yield strength—comparable with steel—makes Si a material superior to any metal in most applications. As a consequence, Si has been quite successful as a structural element in mechanical sensors, particularly during the past 25 years. Pressure and acceleration sensors, based on simple piezoresistive elements embedded in a Si movable mechanical member, have turned into major commercial applications. Besides the desirable mechanical and known electrical properties, this success can also be attributed to the available fabrication technology of ICs.

The significant properties that have made Si a successful material not only for electronic applications but also for mechanical applications are reviewed in Tables 4.15 and 4.16. From these tables we summarize some of the major reasons behind the success of Si as a mechanical sensor element:

- Silicon surpasses stainless steel in yield strength while displaying a density lower than that of aluminum. In fact, Si's specific strength, defined as the ratio of yield strength to density, is significantly higher than for most common engineering materials.
- The hardness of Si is slightly better than that of stainless steel; it approaches that of quartz and is higher than most common glasses.
- The Young's modulus of Si has a value approaching that of stainless steel and is well above that of quartz. From Tables 4.15 and 4.16 we also note that Si_3N_4, a coating often used on silicon, has a hardness topped only by a material such as diamond. Therefore, the combination of silicon with silicon nitride coatings can be used for highly wear-resistant components as required in micromechanisms such as micromotors (see Volume II, Chapter 7).

TABLE 4.15 Mechanical Properties of Single-Crystal Silicon (SCS) among Other Important Technological Materials

	Yield Strength (10^9 N/m^2= GPa)	Specific Strength (10^3 m^2 s^{22})	Knoop Hardness (kg/mm^2)	Young's Modulus (10^9 N/m^2= GPa)	Density (10^3 kg/m^3)	Thermal Conductivity at 300 K (W/cmK)	Thermal Expansion (10^{26}/°C)
Diamond (SC)	53	15,000	7,000	10.35	3.5	20	1.0
Si (SCS)	2.8–6.8	3,040	850–1,100	190 (111)	2.32	1.56	2.616
GaAs (SC)	2.0			0.75	5.3	0.81	6.0
Si_3N_4	14	4,510	3,486	323	3.1	0.19	2.8
SiO_2 (fibers)	8.4		820	73	2.5	0.014	0.4–0.55
SiC (6H-SiC)	21	6,560	2,480	448	3.2	5	4.2
Iron	12.6		400	196	7.8	0.803	12
Tungsten (W)	4	210	485	410	19.3	1.78	4.5
Al	0.17	75	130	70	2.7	2.36	25
AlN	16			340		1.60	4.0
Al_2O_3	15.4		2,100	275	4.0	0.5	5.4–8.7
Stainless steel	0.5–1.5		660	206–235	7.9–8.2	0.329	17.3
Quartz values // Z ⊥ Z	9		850	107	2.65	0.014	7.1 13.2 (increases with T)
Polysilicon	1.8 (annealed)			161			2.8

SC, single crystal; SCS, single-crystal silicon; T, temperature; // Z, parallel with *z*-axis; ⊥Z, perpendicular to *z*-axis.

TABLE 4.16 Silicon Single Crystal Material Characteristics

Si Parameter	Value and Comment
Atomic weight	28.1
Atoms/cm^3	5×10^{22}
Bandgap at 300 K	1.12 eV. Si has a high bandgap, making it useful electrically at high temperatures. Indirect bandgap in the near infrared. It is opaque to ultraviolet and transparent to IR.
Chemical resistance	High. Si is resistant to most acids, except combinations of HF/HNO_3 and certain bases.
Density (g/cm^3)	2.4. Si has a lower density than aluminum (2.7).
Dielectric constant	11.9 vs. 13.1 for GaAs
Dielectric strength (V/cm 10^6)	3
Dislocation density	<100 cm^2. IC-grade Si contains virtually no imperfections; thus, it is relatively insensitive to cycling and fatigue failure.
Electron mobility (cm^2/Vs)	1500
Hole mobility (cm^2/Vs)	400
Intrinsic carrier concentration (cm^{-3})	1.45×10^{10}
Intrinsic resistivity (Ωcm)	2.3×10^5 vs. 10^8 for GaAs
Knoop hardness (kg/mm^2)	850 (stainless steel is 820). Si is harder than steel and can readily be coated with silicon nitride, providing high abrasion resistance.
Lattice constant (Å)	5.43
Linear coefficient of thermal expansion at 300 K (10^{-6}/°C)	2.6. The low expansion coefficient of Si is closer to quartz than to metal, making it insensitive to thermal shock.
Melting point	1415°C. Silicon is a high-melting material, making it suitable for high-temperature applications.
Minority carrier lifetime(s)	2.5×10^{-3}
Oxide growth	Si grows a dense, strong, chemically resistant, passivating layer of SiO_2. This oxide is an excellent thermal insulator with a low expansion coefficient.
Poisson ratio	0.22
Relative permittivity	11.8
Silicon nitride	A typical coating for Si with a hardness and wear resistance only topped by diamond.
Specific heat at 300 K (J/gK)	0.713
Thermal conductivity at 300 K (W/cmK)	1.56. Si has a high thermal conductivity comparable to metals such as carbon steel (0.97) and Al (2.36).
Temperature coefficient of Young's modulus (10^{-6} K^{-1}) at 300 K	–90
Temperature coefficient of piezoresistance (10^{-6} K^{-1}) at 300 K (doping < 10^{18} cm^{-3})	–2500
Temperature coefficient of permittivity (10^{-6} K^{-1}) at 300 K	1000
Thermal diffusivity (cm^2/s)	0.9
Yield strength (GPa)	7 (steel is 2.1). IC-grade Si is stronger than steel.
Young's modulus E (GPa)	190 [111] direction. The elastic modulus is similar to that of steel (steel is 200).

Other Si Sensor Properties

Thermal Properties of Silicon

In Figure 4.78, the thermal expansion coefficient of Si, W, SiO_2, Ni-Co-Fe alloy, and Pyrex is plotted versus absolute temperature. Single-crystal Si (SCS) has a high thermal conductivity (comparable with metals such as steel and aluminum and approximately 100 times larger than that of glass) and a low thermal expansion coefficient. Because of its thermal conductivity, Si is used in some devices as an efficient heat sink. Its thermal expansion coefficient is closely matched to Pyrex glass but exhibits considerable temperature dependence. A good match in thermal expansion coefficient between the device wafer and a support substrate is often required in sensor manufacture. A poor match introduces stress, which degrades the device performance. This makes it difficult to fabricate composite structures of Pyrex and Si that are stress-free over a wide range of temperatures. Drift in silicon sensors often stems from packaging. In this respect, several types of stress relief, subassemblies for stress-free mounting of the active silicon parts, play a major role; using silicon as the support for silicon sensors is highly desirable. The latter aspect will be addressed when discussing anodic bonding of Pyrex glass to Si and fusion bonding of Si to Si in Volume III, Chapter 4 on packaging, assembly, and self-assembly.

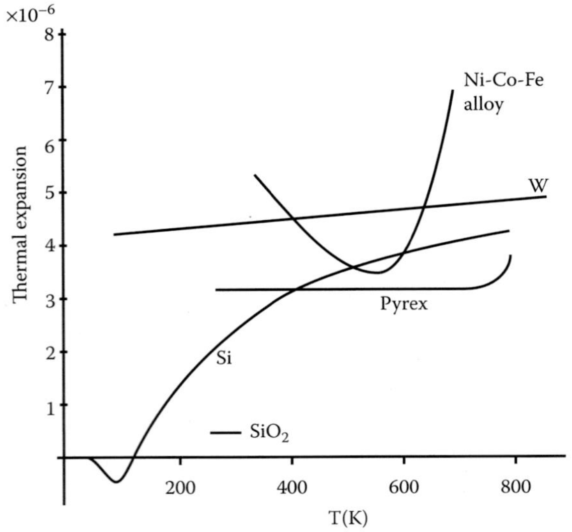

FIGURE 4.78 Thermal expansion coefficient versus absolute temperature. (From Greenwood, J.C. 1988. *J Phys E Sci Instrum* 21:1114. With permission.[26])

Although the Si bandgap is relatively narrow, by using SOI wafers, high-temperature sensors can be fashioned. For the high-temperature applications, highly doped Si, which is relatively linear in its temperature coefficient of resistance and sensitivity over a wide range, is typically used.

When fabricating thermally isolated structures on Si, such as the miniature heater element in Volume II, Figure 9.11, the large thermal conductivity of Si poses a considerable problem. For thermally isolated structures, machining in glass, quartz, or ceramics with their lower thermal conductivity represents an important alternative.

Silicon Optical Properties

Silicon features an indirect bandgap and is not an active optical material (also see Chapters 3 and 5). As a consequence, silicon-based lasers do not exist. Silicon is effective only in detecting light because the indirect bandgap makes emission of light difficult. Above 1.1 μm, silicon is transparent, but at wavelengths shorter than 0.4 μm (in the blue and ultraviolet portions of the visible spectrum), it reflects more than 60% of the incident light. One of the most established applications for silicon sensors, although not often typified as MEMS, is visual imaging with charge-coupled devices (CCDs). In a CCD imager, each element in a 2D array generates an electrical charge in proportion to the amount of light it receives. The charge, stored by CCD elements along a row, is subsequently transferred to the next element in a bucket-brigade fashion as the light input is read out line by line. The number of picture elements (pixels) on the CCD determines the resolution. In top-of-the-line CCD cameras (http://www.pctechguide.com/19digcam.htm) a 3.34-megapixel CCD is capable of delivering a maximum image size of 2048 × 1536 pixels. Si as the pixel semiconductor can be used for a wide variety of electromagnetic radiation wavelengths, from gamma rays to infrared. There is a trend now to configure pixels in clever ways to make novel optical sensors feasible.

One example of a smart pixel configuration is embodied in the retina chip shown in Figure 4.79.[35] This is an IC chip that works like the human retina, selecting only the necessary information from

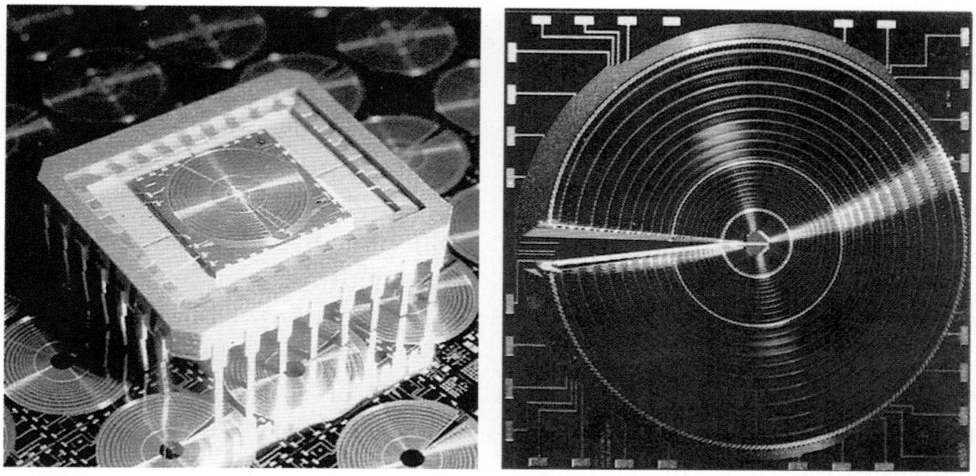

FIGURE 4.79 Photos of the retina sensor. (Courtesy of Dr. Lou Hermans, IMEC, Belgium.)

a presented image to greatly speed up image processing. The chip features 30 concentric circles with 64 pixels each. The pixels increase in size from 30 × 30 μm for the inner circle to 412 × 412 μm for the outer circle. The circle radius increases exponentially with eccentricity. The center of the chip, called the *fovea*, is filled with 104 pixels measuring 30 × 30 μm placed in an orthogonal pattern. The total chip area is 11 × 11 mm. The chip is designed for those applications where real-time coordination between the sensory perception and the system control is of the primary concern. The main application is active vision, and its potential application is expected in robot navigation, surveillance, recognition, and tracking. The system can cover a wide field of view with a relatively low number of pixels without sacrificing the overall resolution and leads to a significant reduction in required image-processing hardware and calculation time. The fast, but rather insensitive, large pixels on the rim of the retina chip pick up a sudden movement in the scenery (peripheral vision), prompting the robot equipped with this "eye" to redirect itself in the direction of the movement to better focus on the moving object with the more sensitive fovea pixels.

The area where MEMS saw one of its biggest commercial breakthroughs is in micromachined mirrors for optical switching in both fiberoptic communications and in data storage applications. Optical switches are to optical communications what transistors are to electronic signaling. What makes single-crystal Si attractive in this case is the optical quality of the Si surface. The quality of the mirror surface is of primary importance to obtain very low insertion loss even after multiple reflections. Wet bulk micromachining has an advantage here over deep reactive-ion etching (DRIE), as the latter leaves Si mirror surfaces with more losses as a result of the inevitable ripples (Volume II, Chapter 3 on dry etching). Important commercial developments in this field by companies such as Quinta Corporation, MEMX, Corning's IntelliSense, Nortel (Xros), and Lucent are summarized in Volume III, Chapter 10. We will see that, after causing a huge stock market technology bubble in 2001, this technology is slowly re-entering the market with more realistic expectations.

Biocompatibility and Si-Based MEMS

Mechanical and Chemical Components of Biocompatibility The definition and history of biocompatibility and the biocompatibility of selected materials are reviewed in Volume III, Chapter 5. Here we cover biocompatibility of Si only. Biocompatibility has both a mechanical and a chemical component. For example, it is well known that, for better biocompatibility, a smooth surface and rounded corners are preferable to a rough surface, sharp corners, and small crevices but also that for a given surface finish and shape, carbon is preferable to copper. The research community remains somewhat divided on the relative contribution of mechanical effects (e.g., surface roughness and shape) versus chemical effects (e.g., Cu vs. C) to the overall bioresponse of an implant. von Recum's work demonstrates that micromachining (microtexturing) of surfaces of different materials may help deconvolute these relative contributions.[36] Some specific examples of substrate microtexturing effects follow. An increase in biosynthetic activity and mobility of bone cells on polymer films cast from micromachined molds has been identified (this might have applications for bone regeneration and bioengineering of prosthetics). The regeneration of severed tendons only on textured surfaces has been demonstrated and might lead to the use of textured bandages to accelerate healing of tissue. Many examples in the MEMS literature also illustrate directed growth of nerve cells on chemically patterned surfaces. In a typical procedure, a hydrophobic silane pattern on a glass slide is produced by "lift-off" where the hydrophobic silane (dimethyltrichlorosilane) replaces the more familiar metallization step (see Volume II, Figure 1.37). Hydrophobic patterning is followed by the deposition of an amino silane (3-aminoethylaminopropyltrimethoxysilane) in the remaining areas, the amino group forming an attachment site for proteins.[37,38] Successful chemical patterning of glass slide surfaces is demonstrated by the growth pattern of cells. Some cells, such as fibroblasts, grow in hydrophilic areas but have difficulty adhering and spreading in hydrophobic areas. Thus, sharp edge definition is obtained along edges of hydrophilic/hydrophobic interfaces. Using this technique, neuronal growth can be controlled, and neuronal processes may be induced to follow the straight lines of the original mask pattern. It has been suggested that this type of approach could be applied to nerve repair techniques. The method and its range of applications are described in Britland et al.[39] Surface characteristics of bio-MEMS materials that should be considered are not only chemistry and surface texture but also hydrophilicity, charge, polarity and energy, heterogeneous distribution of functional groups, water absorption, and chain mobility (for a polymer).

Biocompatibility of Si in MEMS

The above background on the many aspects determining biocompatibility may explain why literature on biocompatibility of Si remains somewhat of an enigma; it is often hard to understand what is meant when a Si device is proclaimed to be biocompatible. Maluf, without providing references, claims that preliminary medical evidence indicates that silicon, even without protein adsorption control, is benign in the body.[34] In reality, it remains of considerable interest to improve the biocompatibility of silicon microdevices and to engineer surfaces with reduced protein and cell adherence for the prevention of fouling and fibrotic response—especially when dealing with active silicon devices in contact with blood (e.g., an in vivo Si-based glucose sensor).[40] Even if significant control over protein adsorption and cell adhesion is achieved, Si remains a brittle material, which may shard in the body, and, just like glass, is to be avoided in any load-bearing application in living subjects or foodstuffs (e.g., as a micromachined needle). Although the motivation for most MEMS devices (especially active ones) is usually to eliminate any adhesion between proteins and surfaces, in some cases the ability to enable protein adhesion locally and not in other areas to provide for a background signal (e.g., for cell-based assays) clearly requires two different surface chemistries. Studies at the University of Michigan and MIT, according to Issys' web site, demonstrated that boron-doped SCS has a superior biocompatibility when compared with standard SCS or polysilicon-based devices (http://www.memsissys.com/new_htm/memsmedical/memsmedicalbiocompatibility.htm). Some definitive work has been done with Si for cortical implants. Si electrode arrays were proposed as permanent cortical implants with the purpose of restoring useful vision, and their biocompatibility with neural tissue was investigated. In this context, silicon electrode arrays were left in place within a rabbit cortex for six months. Afterward, the neuron density as a function of radius from the center of the electrode shafts was used as the means of assessing damage. The researchers concluded that the electrode array showed long-term biocompatibility because normal brain tissue was observed within micrometers of the surface of the shafts.[41,42] Initial results on porous silicon from DeMonfort University (Leicester, UK) indicate that it is biocompatible with living tissue and offers the potential to bridge the gap between a mechanical device and human tissue.

Example 4.1: Design of an SOI-based high-sensitivity static piezoresistive cantilever for label-less sensing

In this example, we want to implement an integrated piezoresistor for position sensing of a static cantilever. Most cantilever design considerations involve a beam in resonance mode. Here we consider the optimal design rules for measuring bending induced in a static cantilever by differential surface stresses. This is especially important for measurements in liquids; immersion of a cantilever in water damps the resonance response to a value approximately an order of magnitude less than in air, whereas the bending response remains unaffected by the presence of water. Thundat et al. claim that static, adsorption-induced stress sensors have a sensitivity three orders of magnitude higher than dynamic based ones.[43] In practice, using optical techniques to measure deflection, surface stresses as small as ~10^{-4} Nm^{-1} have been measured.[44] Here, we want to implement a simpler piezoresistive measurement. Unfortunately this usually comes at the expense of resolution. The aim is to design the cantilever and the piezoresistor such that the same sensitivity that is obtained with the optical methods can be maintained in this case as well.

For the analysis of the degree of bending of a cantilever caused by differential stress, we examine Stoney's formula. In 1905, Stoney derived a relation relating thin-film stress to the radius of curvature of a thin substrate. In the case where the substrate is a thin cantilever, the Stoney equation is given by:

$$\frac{1}{R} = \frac{6(1-\nu)}{Et^2}\Delta\sigma \quad (4.99)$$

where *R* corresponds to the radius of curvature of the cantilever, *t* its thickness, σ the differential surface stress ($\Delta\sigma = \sigma_1 - \sigma_2$, i.e., the difference in surface stress between the top and bottom surfaces, in units of N/m), and ν and *E* are Poisson's ratio and Young's modulus for the substrate, respectively (see also Equation 6.23 in Volume III, Chapter 6, "Disk Method: Biaxial Measurements of Mechanical Properties of Thin Films" for bending of a thin disk).

From geometric considerations, the radius of curvature is related to the displacement of the free end of the cantilever, δ, and its length, *L*, as $1/R = 2\delta/L^2$. Combining this last expression with Equation 4.99, we derive the cantilever displacement as a function of the differential surface stress as:

$$\delta = \frac{3L^2(1-\nu)}{Et^2}(\sigma_1 - \sigma_2) \quad (4.100)$$

From this equation, if we assume a constant differential surface stress, maximum deflection occurs with a low *t* and *E*, a large *L*, and a low ν. To maximize sensitivity, the piezoresistor should be placed in the zone of maximal stress. The stress in the beam is zero at the centerline and increases linearly with the distance away from the centerline. The stress also increases toward the base of the cantilever so that the highest sensitivity will be achieved with a piezoresistor placed at the surface of the cantilever beam near the base. The stress at that point can be calculated to be:

$$\sigma_{max} = \frac{6L}{Wt^2}F = \frac{3Et}{2L^2}\delta \quad (4.101)$$

where *F* is the applied force (i.e., K × δ). The resulting fractional resistance change is given by the piezoresistive relation:

$$\frac{\Delta R}{R} = \pi_L\sigma_{max} = \beta\frac{6L\pi_L}{Wt^2}F = \beta\frac{3Et\pi_L}{2L^2}\delta \quad (4.102)$$

where π_L is the longitudinal piezoresistive coef) ficient of silicon at the operating temperature and at a given doping. The cantilever is oriented along the <110> crystallographic axis of silicon where the piezoresistive coefficient, π_L, is maximum. The coefficient β is a correction factor between 0 and 1 and accounts for the fact that the resistor is not limited to the surface of the cantilever but has a certain depth. Its value depends on the silicon doping depth profile and the thickness of the beam. The doping depth is determined by the implantation parameters and subsequent thermal processes. It is important that the resistor be made shallow, so that the current flows as close as possible to the surface of the cantilever. In an extreme case, a uniformly doped cantilever would exhibit no net piezoresistive response because opposite sign stresses on the top and the bottom of the cantilever would give an equal but opposite contribution to the change in resistance. Substituting Equation 4.100 in Equation 4.102, we obtain the expression for fractional resistance change in terms of surface stress:

$$\frac{\Delta R}{R} = \beta \frac{3\pi_L(1-\nu)}{t}(\sigma_1 - \sigma_2) \qquad (4.103)$$

Thinning the cantilever is the principal means to obtain the largest resistivity change. Making the overall device smaller (more specifically *W*, *L*, and *t*) is helpful for making denser arrays and is simpler to achieve with piezoresistive cantilevers, as they are easier to make smaller than the ones that are used in conjunction with an optical detection technique, principally because their surface area is not limited by the laser spot size (typically about 30 μm).

We now address the need to design in a reference cantilever. Immersion of cantilevers in a liquid results in two reported long-term drift phenomena. First, thermal effects (especially where a gold layer on one surface creates a sensitive bimetallic structure) occur slowly over a period of hours. A second slow effect, which takes up to 10 h to stabilize, seems associated with slow rearrangement of surface adsorbates.[45] Both drift phenomena imply that a differential-type measurement will be beneficial to extract the data from adjoining cantilevers and to exclude common mode drift phenomena. The latter approach was implemented by Fritz et al., who demonstrated that in doing so, a single mismatch between two DNA sequences could be detected.[46] From the above, it would also be preferable to derivatize the cantilever pairs with the required chemical coatings without using a metal deposit. This way, the temperature effect can be reduced, and the mass of the sensor will be less.

Thus, the fabrication challenge is to make an array with the thinnest possible cantilevers having an extremely shallow doped layer for the piezoresistor at its base to keep β in Equation 4.103 as close to 1 as possible. Moreover, a symmetrical Wheatstone bridge design with a built-in reference cantilever is required to reduce the described background drift and common mode vibrations. For the fabrication of large arrays of cantilever pairs, one would like to have them all as identical as possible, and, if possible, the chemical coatings should be deposited directly on the beam rather than on a deposited metal layer. Our proposed design incorporates features from various MEMS research groups around the world.

The first concerns the depth of the doped region in the piezoresistors. For very thin beams, it is difficult to confine the doped region to the surface of the beam. Because activation of dopants is achieved by annealing, dopant diffusion is unavoidable (see above section "Doping of Si"). To maximize β we rely on Harley et al., who made cantilevers under 1000 Å thick (870–900 Å) with lengths ranging from 10 to 350 μm and widths from 2 to 44 μm.[47] To reduce the depth below the capabilities of conventional implantation, this team used vapor-phase epitaxial growth to deposit the boron-doped layer. The boron atoms are incorporated into the lattice during the epitaxy, so an activating anneal is not required. Furthermore, because there is no damage-enhanced mobility from the ion-implantation step, some high-temperature steps can be tolerated. Their fabrication starts by growing and removing a 2000-Å thermal oxide from the epi Si layer of a 10-Ω cm p-type SIMOX SOI wafer, thinning the Si layer to 800 Å. A 30-s HCl clean in the epichamber removes another 100 Å before a 300-Å layer of 4×10^{19} cm^{-3} boron doped silicon is grown over the entire wafer. The boron layer defines the thickness of the resistors. The intermediate oxide layer is used as an etch stop during the process of etching the bulk silicon substrate in one of the last fabrication steps. In a first step, the cantilevers are patterned and plasma etched. Boron contacts are implanted at 1×10^{15} cm^{-2}, 30 keV, followed by a 200-Å growth of passivating thermal oxide during a 3-h anneal at 700°C. For contacting the piezoresistors, aluminum is deposited and annealed in a forming gas at 400°C for 1 h. Subsequently, a release mask is sputtered on the back of the wafer, and a Bosch DRIE (see Volume II, Chapter 3 under "Inductively Coupled Plasma") is used to release the cantilevers. The DRIE stops at the buried oxide, which is removed with a 6:1 buffered oxide etch. For implementing a reference probe, one could take a cue from Thaysen et al.,[48] who fabricated a thermally symmetrical Wheatstone bridge as shown in Figure 4.80.

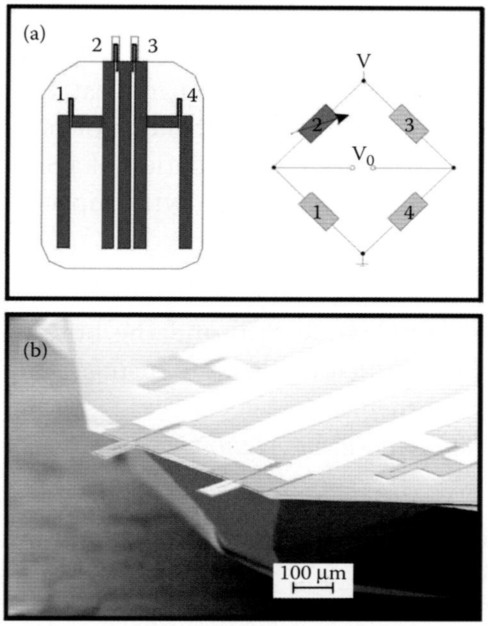

FIGURE 4.80 (a) Thermally balanced Wheatstone bridge. (b) SEM of a cantilever pair. (Sensor and reference courtesy of Dr. Anja Boisen, Technical University Denmark.)

Appendix 4A: Some Properties of Error Functions and Complementary Error Functions

$$\text{erf}(x) \equiv \frac{2}{\sqrt{\pi}} \int_0^x e^{-dy}$$

$$\text{erf}(x) \equiv 1 - \text{erf}(x)$$

$$\text{erf}(0) = 0$$

$$\text{erf}(\infty) = 1$$

$$\text{erf}(x) \cong \frac{2}{\sqrt{\pi}} x \quad \text{for } x << 1$$

$$\text{erfc}(x) \cong \frac{1}{\sqrt{\pi}} \frac{e^{-x^2}}{x} \quad \text{for } x >> 1$$

$$\frac{d}{dx} \text{erf}(x) = \frac{2}{\sqrt{\pi}} e^{-x^2}$$

$$\frac{d^2}{dx^2} \text{erf}(x) = -\frac{4}{\sqrt{\pi}} x e^{-x^2}$$

$$\int_0^x \text{erfc}(y')dy' = x\,\text{erfc}(x) + \frac{1}{\sqrt{\pi}}(1 - e^{-x^2})$$

$$\int_0^\infty \text{erfc}(x)dx = \frac{1}{\sqrt{\pi}}$$

Questions

4.1: Describe a fabrication process for a planar npn transistor.
Thanks to Andrew Hatch,UC Irvine

4.2: Consider the growth of a thermal oxide on a 4-in. <111> silicon wafer that is initially 525 μm thick and comes with a native oxide layer of 25 Å. If the oxidation process is carried out dry for 1 h at 1000°C and at 1 atm of pressure, what is the final wafer thickness? How much of this will be Si and how much will be SiO_2? What color will the resulting oxide be? Repeat these calculations for 1 hr of wet oxide growth at 1000°C. What is the ratio of the rates of wet to dry oxide growth? To solve the latter, assume that B (H_20) = 0.3 μm²/hr, B(O_2) = 0.01 μm²/hr; and B/A (H_2O) = .55 μm/hr, B/A (O_2) = 0.05 μm/hr for <111> Si: how do your results compare if you use a <100> wafer instead of a <111> wafer? (Experimental from http://www.eng.tau.ac.il/~yosish/Courses/vlsi1/I-4-1-Oxidation.pdf.)
Thanks to Andrew Hatch, UC Irvine

4.3: Great care is taken to ensure that silicon boules are single crystalline with the highest possible purity. Describe how some adverse effects might manifest themselves if the processing was not handled with enough care; i.e., what makes single crystallinity and high purity so important?
Thanks to Matt Sullivan, UC Irvine

4.4: We are trying to etch a 20-μm × 20-μm through hole in a Si wafer. What mask dimensions would you need to achieve such a through hole in a typical 6-in [100]-oriented silicon wafer (600 μm thick)? Assume that the 20-μm × 20-μm through hole is the exit hole in the bottom plane of the wafer.
Thanks to Rajtarun Madangopal, UC Irvine

4.5: How would you obtain vertical sidewalls using a [100]-oriented Si wafer?
Thanks to Rajtarun Madangopal, UC Irvine

4.6: In an experiment, boron is used to dope a silicon substrate by diffusion. The temperature for diffusion is 1100°C. The activation energy is 3.69 eV. D_0 is 10.5 cm^2/s.

(a) Determine the type of resulting semiconductor.

(b) If the surface concentration of boron is a constant: N_s = 10 /cm^3, calculate the total amount of dopant atoms per unit area of the semiconductor after diffusion 10 min later.

(c) If the diffusing amount Q is constant, which is the same as the value calculated in step (b), derive the diffusion equation and calculate the surface dopant concentration after diffusing for 1 h.

Thanks to Xi Wang, UC Irvine

4.7: A pn junction is made by doping Si at room temperature. At the *n* edge of the depletion region, N_A was measured to be 10^{11}/cm^3. Calculate the barrier voltage at the *n* edge of the p-n junction. For Si at room temperature, $N_D = 10^{15}$/cm^3 and $N_i = 10^{10}$/cm^3.
Thanks to Chris Hoo, UC Irvine

4.8: Draw the energy band diagram in **k**-space for an intrinsic semiconductor. Why can we locate the Fermi level in the middle of the band gap?
Thanks to Anna Torrents Cabestany, UC Irvine

4.9: Give an example of a (111) plane and a (110) plane intersecting at 35.26°, 90°, and 144.74°.
Thanks to Yuka Okabe, UC Irvine

4.10: Calculate the values for Young's modulus, shear modulus, and Poisson's ratio for [100]-directed Si. Given are the values of the following compliance constants: S_{11} (7.68×10^{-12} m^2/N), S_{44} (12.6×10^{-12} m^2/N), and S_{12} (-2.14×10^{-12} m^2/N).
Thanks to Jessica Lawson, UC Irvine

4.11: The identity of a semiconductor can be determined by measuring its intrinsic carrier concentrations at various temperatures. An unknown semiconductor is submerged in ice water (273 K) and boiling water (373 K). Its intrinsic carrier concentrations, n_i, were found to be 4.6×10^{18} m^{-3} and 3.1×10^{20} m^{-3} at 273 K and 373 K, respectively. What is the bandgap and identity of the semiconductor?
Thanks to William Sin, UC Irvine

4.12: At what temperature does the concentration of intrinsic carriers in intrinsic α-quartz (SiO_2) match the concentration in Si at T = 300 K? Comment on the semiconductor potential of SiO_2. How does the oxidation of a Si melt during single crystal Si growth impact the construction of ICs?
Thanks to Jungyun Kim, UC Irvine

4.13: Explain the electron and the hole flow in a semiconductor substrate using a daily life phenomena.
Thanks to Kivanc Azgin, UC Irvine

4.14: (a) Describe the physical differences between nMOS and pMOS. (b) Which is faster and why?
Thanks to John Clanton, UC Irvine

4.15: (a) Why does electron mobility decrease as gate length decreases? (b) What are some ways to counteract this problem?
Thanks to Gordon Pfeiffer, UC Irvine

4.16: Explain the relevance of plastic deformation in Si and metals. Include any role that temperature may play in plastic deformation.
Thanks to Daniel Almendarez, UC Irvine

4.17: Aluminum is to be diffused into a silicon single crystal. At what temperature will the diffusion constant (D) be 5×10^{-11} cm^2/s? (Given: E_A = 334 kJ/mole and D_0 = 8 cm^2/s.)

4.18: Calculate the number of silicon atoms per unit volume of silicon. During the growth of silicon single crystals it is often desirable to deliberately introduce atoms of other elements (known as dopants) to control and change the electrical conductivity and other electrical properties of silicon. Phosphorus (P) is one such dopant that is added to make silicon crystals n-type semiconductors. Assume that the concentration of P atoms required in a silicon crystal is 10^{17} atoms/cm^3. Compare the concentrations of atoms in silicon and the concentration of P atoms. What is the significance of these numbers from a technological viewpoint? Assume that density of silicon is 2.33 g/cm^3.

4.19: Describe briefly the crystal structure of silicon: include such details as structure type (how does it differ from FCC), lattice parameters, packing density, and highest packing density plane.

4.20: Write a paragraph on the mechanical properties of single-crystal silicon (SCS), with special emphasis on strength, plastic/elastic behavior, and hardness.

4.21: Ionization energy of dopants using the Bohr model: Consider an N-type silicon crystal, doped with phosphorus atoms. Each dopant atom replaces a silicon atom from its lattice site. It hence donates one extra electron to the system. Calculate the ionization energy of this extra electron, assuming the other 4 valence electrons, the inner electrons, and the nucleus of the P atom as a core of net charge +1. Assume the effective mass of electrons in silicon $m_e^+ = 0.2\ m_0$, and a relative permittivity $\varepsilon_r = 12$.
Thanks to Arvind Ajoy, Indian Institute of Technology Madras

4.22: In a silicon crystal, there are 8 equivalent (111) planes and 12 equivalent (110) planes that can be drawn into the structure, making for 96 intersection combinations.
(a) Sketch the 8 (111) planes, including the Miller indices for each.
(b) Sketch the orientation of Si atoms on the (111) plane.
(c) Sketch the 12 (110) planes, including the Miller indices for each.
(d) Sketch the orientation of Si atoms on the (110) plane.
(e) Perform the calculations that show that the {111} and {110} families of planes can intersect at only three defined angles, 35.26°, 90° or 144.74°.
(f) Assign one of these angles to each of the 96 combinations of planes. How many combinations yield 35.26°? 90°? 144.74°? Comment on the significance of these angles.
Thanks to Matt Sullivan, UC Irvine

4.23: Differentiate between intrinsic and extrinsic semiconductors.
Thanks to Rajtarun Madangopal, UC Irvine

4.25: What wafer orientation would you select if you had to etch under a microbridge a shallow V-groove crossing it at a 90° angle?
Thanks to Rajtarun Madangopal, UC Irvine

4.26: List three uses of wafer flats. Also justify why we need to round wafer edges.
Thanks to Rajtarun Madangopal, UC Irvine

4.27: Silicon dioxide growth involves the heating of a silicon wafer in a stream of steam or in wet or dry oxygen/nitrogen mixture at elevated temperatures.
(a) Write the oxidation reaction for both wet and dry oxidation of silicon.
(b) Calculate the silicon dioxide thickness after diffusing for 10 h for both wet and dry oxidation. [Assume that it is a long oxidation time. $C^*(O_2) = 5.2 \times 10^{16}/cm^3$, $C^*(H_2O) = 3.0 \times 10^{19}/cm^3$; $N(O_2) = 2.2 \times 10^{22}/cm^{-3}$, $N(H_2O) = 4.6 \times 10^{22}/cm^{-3}$.]
Thanks to Xi Wang, UC Irvine

4.28: There is an n-type silicon semiconductor with electron concentration of 10^{15} cm^{-3}, electron mobility of 1000 cm^2/v·s.
(a) Calculate the resistance of this semiconductor.
(b) If the temperature is 40 K, $N_c = 2.8\times10^{17}$ / cm^3, calculate the Femi level energy.
Thanks to Xi Wang, UC Irvine

4.29: In a doped piece of Si at 300 K, the mobility of minority carriers was found to be 476 cm^2/V-s with a carrier lifetime of 10^{-7} s. Determine the diffusion length of the minority carrier.
Thanks to Chris Hoo, UC Irvine

4.30: Tunneling in a reversed-biased diode is called the Zener process. What are two requirements need to be fulfilled before tunneling occurs?
Thanks to Chris Hoo, UC Irvine

4.31: Why is the <110> direction the only orientation in a [100]-silicon wafer that does cleave with exactly perpendicular edges?
Thanks to Anna Torrents Cabestany, UC Irvine

4.32: Explain why single-crystal silicon (SCS) exhibits a linear-elastic behavior at lower strain values and yields catastrophically when stressed beyond the limit? What happens to SCS at elevated temperatures?
Thanks to Jessica Lawson, UC Irvine

4.33: Describe some of the optical properties of silicon and some of the current applications in which these properties are utilized.
Thanks to Jessica Lawson, UC Irvine

4.34: Given the values below for GaAs, calculate the Fermi energy of intrinsic GaAs at room temperature (300 K). Does the difference between effective mass of electrons to holes have an impact on the Fermi energy?
Thanks to Jungyun Kim, UC Irvine

4.35: What are the three major categories of high-*k* materials, name a material from each category, and explain why engineers are searching for higher-*k* materials despite the immense benefits of SiO_2 ($k = 3.9$).
Thanks to Gordon Pfeiffer, UC Irvine

4.36: What is a tensor of second rank? Name one example. How many components does a tensor of second rank have? What about a tensor of fourth rank?
Thanks to Wei Xiong, UC Irvine

4.37: List some possible difficulties arising due to MOSFET size reduction.
Thanks to Yanan Liu, UC Irvine

4.38: For silicon dioxide growth, compare dry oxide growth and wet oxide growth.
Thanks to Yiying Yao, UC Irvine

Further Reading

Campbell, S. A. 1966. *The science and engineering of microelectronic fabrication.* Oxford, UK: Oxford University Press.

Dorf, R. C. ed. 2006. *Electronics, power electronics, optoelectronics, microwaves, electromagnetics, and radar (the electrical engineering handbook)* 3rd ed. Boca Raton, FL: CRC Taylor &Francis Group.

Glaser, A. B., and G. E. Subak-Sharpe. 1979. *Integrated circuit engineering.* Reading, MA: Addison-Wesley Publishing Company.

Hess, D. W., and K. F. Jensen. 1989. *Microelectronic processing, chemical engineering aspects, advances in chemistry series 221.* Washington, DC: American Chemical Society.

Maluf, N. 2000. *An introduction to microelectromechanical systems engineering.* Boston, Artech House.

Mayer, J. W., and S. S. Lau. 1990. *Electronic materials science: for integrated circuits in Si and GaAs.* New York: Macmillan Publishing Company.

Pillai, S. O. 2005. *Solid state physics.* 6th ed. New Delhi: New Age International (P) Limited.

Sze, S. M. 1988. *VLSI technology,* 2nd ed. New York: McGraw-Hill Book Company.

References

1. Schumacher, A., H.-J. Wagner, and M. Alavi. 1994. Mit Laser und Kalilauge. *Technische Rundschau* 86:20–23.
2. Peeters, E. 1994. *Process development for 3D silicon microstructures, with application to mechanical sensor design.* PhD diss., Katholieke Universiteit Leuven, Leuven, Belgium.
3. Elwenspoek, M., H. Gardeniers, M. de Boer, and A. Prak. 1994. Report no. 122830. University of Twente, Twente, the Netherlands.
4. Katz, L. E. 1988. *VLSI technology.* Ed. S. M. Sze. New York: McGraw-Hill.
5. Kendall, D. L. 1979. Vertical etching of silicon at very high aspect ratios. *Ann Rev Mater Sci* 9:373–403.
6. Seidel, H. 1990. *Technical digest: 1990 solid state sensor and actuator workshop.* Hilton Head Island, SC.
7. Ho, C. P., and J. D. Plummer. 1979. Si/SiO2 interface oxidation kinetics: A physical model for the influence of high substrate doping levels, I. Theory. *J Electrochem Soc* 126:1516–22.
8. Ho, C. P., and J. D. Plummer. 1979. Si/SiO2 interface oxidation kinetics: A physical model for the influence of high substrate doping levels, II. Comparison with experiment and discussion. *J Electrochem Soc* 126:1523–30.
9. Ho, C. P., J. D. Plummer, and J. D. Meindl. 1978. Thermal oxidation of heavily phosphorus-doped silicon. *J Electrochem Soc* 125:665–71.
10. Schmidt, P. F., and W. Michel. 1957. Anodic formation of oxide films on silicon. *J Electrochem Soc* 104:230–36.
11. Lewerenz, H. J. 1992. Anodic oxides on silicon. *Electrochimica Acta* 37:847–64.
12. Madou, M. J., W. P. Gomes, F. Fransen, and F. Cardon. 1982. Anodic oxidation of p-type silicon in methanol as compared to glycol. *J Electrochem Soc* 129:2749–52.
13. Fair, R. B. 1989. *Microelectronics processing: chemical engineering aspects.* Eds. Hess, D. W., and K. F. Jensen. Washington, DC: American Chemical Society.
14. Madou, M. J., and S. R. Morrison. 1989. *Chemical sensing with solid state devices.* New York: Academic Press.
15. Moreau, W. M. 1988. *Semiconductor lithography.* New York: Plenum Press.
16. Brodie, I., and J. J. Muray. 1982. *The physics of microfabrication.* New York: Plenum Press.
17. Landauer, R. 1961. Irreversibility and heat generation in the computing process. *IBM J Res Dev* 5:183–91.
18. Thompson, S. E., S. Suthram, Y. Sun, G. Sun, S. Parthasarathy, M. Chu, and T. Nishida. 2006. Future of Strained Si/Semiconductors in Nanoscale MOSFETs. *IEDM Tech Digest.*
19. Chou, P. C., and N. J. Pagano. 1967. *Elasticity: tensor, dyadic, and engineering approaches.* New York: Dover Publications.
20. Kittel, C. 1976. *Introduction to solid state physics.* New York: Wiley.

21. Brantley, W. A. 1973. Calculated elastic constants for stress problems associated with semiconductor devices. *J Appl Phys* 44:534–35.
22. Nikanorov, S. P., Y. A. Burenkov, and A. V. Stepanov. 1972. Elastic properties of silicon. *Sov Phys-Solid State* 13:2516–18.
23. Khazan, A. D. 1994. *Transducers and their elements.* Englewood Cliffs, NJ: PTR Prentice Hall.
24. Worthman, J. J., and R. A. Evans. 1965. Young's modulus, shear modulus, and Poisson's ratio in silicon and germanium. *J Appl Phys* 36:153–56.
25. Metzger, H., and F. R. Kessler. 1970. Der Debye-Sears Effect zur Bestimmung der Elastischen Konstanten von Silicium. *Z Naturf A* 25:904–06.
26. Greenwood, J. C. 1988. Silicon in mechanical sensors. *J Phys E Sci Instrum* 21:1114–28.
27. Huff, M. A., and M. A. Schmidt. 1992. *Technical digest: 1992 solid state sensor and actuator workshop.* Hilton Head Island, SC.
28. Maseeh, F., and S. D. Senturia. 1990. Plastic deformation of highly doped silicon. *Sensors Actuators* 23:861–65.
29. Ding, X., and W. Ko. 1991. Sixth International Conference on Solid-State Sensors and Actuators (Transducers '91), San Francisco, CA.
30. Callister, D. W. 1985. *Materials science and engineering,* New York: Wiley.
31. Smith, C. S. 1954. Piezoresistance effect in germanium and silicon. *Phys Rev* 94:42–49.
32. Kanda, Y. 1982. A graphical representation of the piezoresistance coefficients in silicon. *IEEE Tran Electron Devices* 29:64–70.
33. Middlehoek, S., and S. A. Audet. 1989. *Silicon sensors.* San Diego: Academic Press.
34. Maluf, N. 2000. *An introduction to microelectromechanical systems engineering.* Boston: Artech House.
35. IMEC. 1994. *Silicone detectors.* IMEC Brochure. Leuven, Belgium: Interuniversity Microelectronics Centre.
36. von Recum, A. ed. 1998. *Handbook of biomaterials evaluation: scientific, technical, and clinical testing of implant materials.* New York: Hemisphere.
37. Cooper, J. M., J. R. Barker, J. V. Magill, W. Monaghan, M. Robertson, C. D. W. Wilkinson, A. S. G. Curties, and G. R. Moores. 1993. A review of research in bioelectronics at Glasgow University. *Biosens Bioelectron* 8:R22–R30.
38. Connolly, P., G. R. Moores, W. Monaghan, J. Shen, S. Britland, and P. Clark. 1992. Microelectronic and nanoelectronic interfacing techniques for biological systems. *Sensors Actuators* 6:113–21.
39. Britland, S., G. R. Moores, P. Clark, and P. Connolly. 1990. Patterning and cell adhesion and movement on artificial substrate: a simple method. *J Anat* 170:235–36.
40. Zhang, M., T. Desai, and M. Ferrari. 1998. Proteins and cells on PEG immobilized silicon surfaces. *Biomaterials* 19:953–60.
41. Clark, L. D., D. J. Edell, V. M. McNeil, and V. V. Toi. 1992. Factors influencing the biocompatibility in insertable silicon microshafts in cerebral cortex. *IEEE Trans Biomed Eng* 39:635–43.
42. Horch, K., R. A. Normann, and S. Schmidt. 1993. Biocompatibility of silicon-based electrode arrays implanted in feline cortical tissue. *J Biomed Mater Res* 27:1393–99.
43. Thundat, T., L. A. Bottomley, S. Meller, W. H. Velander, and R. Van Tassell. 2001. *Immunoassays: methods and protocols.* Eds. Ghindilis, A. L., A. R. Pavlov, and P. B. Atanajov. Totawa, NJ: Humana Press.
44. O'Shea, S. J., M. E. Welland, T. A. Brunt, A. R. Ramadan, and T. Rayment. 1996. Atomic force microscopy stress sensors for studies in liquids. *J Vac Sci Technol* 14:1383–85.
45. Moulin, A. M. 2000. Microcantilever-based biosensors. *Ultramicroscopy* 82:23–31.
46. Fritz, J., M. K. Baller, H. P. Lang, H. Rothuizen, P. Vettinger, E. Meyer, H.-J. Güntherodt, C. Gerber, and J. K. Gimzewski. 2000. Translating biomolecular recognition into nanomechanics. *Science* 288:316–18.
47. Harley, J. A., and T. W. Kenny. 1999. High-sensitivity cantilevers under 1000 Å thick. *Appl Phys Lett* 75:289–91.
48. Thaysen, J., A. Boisen, O. Hansen, and S. Bouwstra. 2000. Atomic force microscopy probe with piezoresistive readout and a highly symmetrical Wheatstone bridge arrangement. *Sensors Actuators* 83:47–53.
49. Thompson, S. E., M. Armstrong, C. Auth, M. Alavi, M. Buehler, R. Chau, S. Cea, et al. 2004. A 90-nm logic technology featuring strained-silicon. *IEEE Trans Electron Devices* 51:1790–7.

5

Photonics

> If it only were possible to make materials in which electromagnetically waves cannot propagate at certain frequencies, all kinds of almost-magical things would happen.
>
> **Sir John Maddox, *Nature* (1990)**

Outline

Natural opals (a) and *P. nireus* butterfly (b). Periodic structure leads to striking iridescent color effects even in the absence of pigments as a result of photonic crystals, i.e., stacks of nanoscale gratings.

Introduction

Photonics involves the use of radiant energy and uses photons the same way that electronic applications use electrons. We start this chapter with a discussion of the nature of light, its near- and far-field diffraction, and how this relates to image formation and resolution limits. Next we introduce refraction, reflection, and polarization of light and launch into the Maxwell's equations for electromagnetic (EM) waves propagating in a vacuum, in matter, and at interfaces. We solve the Maxwell's equations for EM waves propagating in a vacuum, in media (without charges and with charges and currents), in right-handed and left-handed materials, and calculate their energy and momentum. From the considerations of EM waves at boundaries we derive the Fresnel equations linking reflectance and refractive index of a material and get our first exposure to bulk, surface, and localized plasmons in plasmonics. Plasmonics is of growing importance for use in submicron lithography, near-field optical microscopy, enhancement of light/matter interaction in sensors, high-density data storage, and highly integrated optic chips.

Newton's laws were found to remain the same under a Galilean transformation, but a Galilean transformation does not apply to Maxwell's equations. We will see that, in his special theory of relativity, Einstein convinced the world that whenever the speed of light is involved, the Galilean transformation has to be replaced by the Lorentz transformation.

To explain the interaction of EM radiation with materials at a level of the atomic structure of the material, we launch the Drude-Lorentz model, an upgrade of the Drude model from Chapter 3 on quantum mechanics. Using the Drude-Lorentz model, we review the distinctive optical properties of metals, insulators, and semiconductors. Besides reviewing the optoelectronic behavior of these bulk materials, we also deal with the interaction of electrons and photons with lower-dimensional materials. In the case of metals this brings us back to plasmonics, which we detail a bit further in this section. Then we review scaling of electronic and optical properties of semiconductors and compare semiconductor nanoparticles with metal nanoparticles.

We follow this treatise on optoelectronic properties of materials with a study of light scattering on nano- and microstructures (with sizes smaller, comparable, and larger than λ). We cover elastic Rayleigh and Mie scattering, as well as inelastic Raman scattering of light from these small particles. We learn how surface plasmons enhance the Raman effect in surface-enhanced Raman spectroscopy (SERS) by up to 10^{12}–10^{14} times compared with nonresonant Raman spectroscopy, and we discuss nanoscale confinement of radiation in near-field scanning optical microscopy (NSOM) for subwavelength resolution.

We proceed with a study of EM effects in periodic photonic crystals [three-dimensional (3D)], planar waveguides [two-dimensional (2D)], optical fibers [one-dimensional (1D)], and microsphere optical cavities [zero dimension (0D)]. This study includes a comparison of the behavior of photons with electrons in a vacuum and in periodic media, and we find that the analogous electron confinement structures to the aforementioned photonic structures are bulk crystals (3D), quantum wells (2D), quantum wires (1D), and quantum dots (OD) and that the tunneling of electrons through a forbidden zone has its equivalent in light leaking out of a waveguide to create an evanescent field.

Then we delve into the fascinating new topic of metamaterials, human-made structures with a negative refractive index, and explain how these could make for perfect lenses and might change the photonics field forever.

An introduction to diode lasers, quantum well lasers, and quantum cascade lasers closes the chapter.

The Nature of Light

In his theory, Maxwell described light as electric and magnetic fields, oscillating perpendicular to each other and propagating perpendicular to these oscillations as shown in Figure 5.1. Fusing two fields into one, he calculated that waves in an electromagnetic (EM) field traveled at the speed of light and conjectured that light was nothing but an EM vibration at a particular frequency. In this way Maxwell was the first physicist to show that the task of field unification is a very worthwhile exercise. Before Maxwell, there was no connection between light and electricity and magnetism: it was one of the great triumphs of science—to show that two things that appear so different are in fact described by the same simple laws!

EM waves exist in an enormous range of wavelengths/frequencies. This continuous range of wavelengths/frequencies is known as the *EM spectrum*. The entire range of the spectrum is often broken up into specific bands. The subdividing of the entire spectrum into smaller spectra is done mostly based on how each region of EM waves interacts with matter. In Figure 5.2, we show a typical EM spectrum representation. The color our eyes perceive is not a frequency measurement. The eye has three color-sensitive cell types ("cone cells") that respond to red, green, and blue, and the apparent color of light is the result of the added response from the different cone types. For example, the color "purple" is not a frequency but

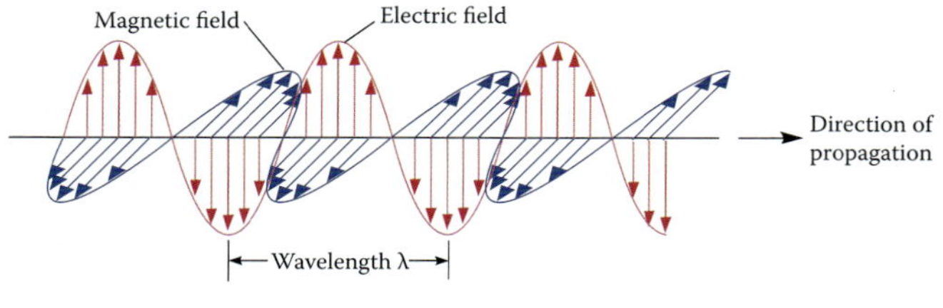

FIGURE 5.1 Maxwell's description of a light wave.

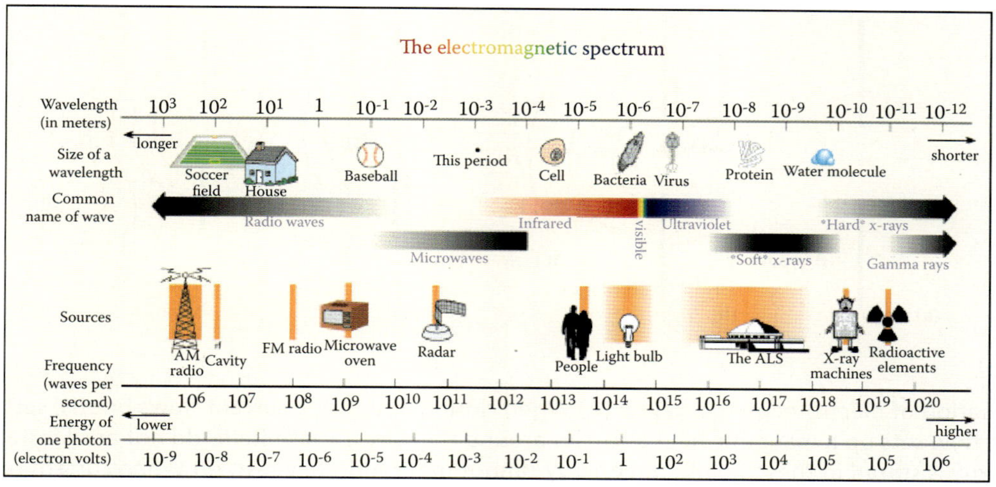

FIGURE 5.2 The electromagnetic (EM) spectrum, the source of the EM radiation, frequency, energy, wavelength, size comparison, and common name.

rather a mix of red and blue. This is great news for all types of devices: we can trick the eye by mixing three different colors (RGB) in monitors, copiers, cameras, and so on. Visible light goes from violet-blue, with a wavelength, λ, of about 400 nm and energy of about 3 eV (and since $E = h\nu = 1.24/0.4 = 3$ eV), to red, with a wavelength, λ, of about 700 nm and energy of about 1.77 eV (=1.24/0.7 = 1.77 eV). In the deep UV, and even more so at yet shorter wavelengths, materials turn transparent. Typical thermal energies at room temperature are given by kT or about 24 meV, with a radiation wavelength of 50 μm. In the microwave region, EM waves couple excellently to water molecule vibrations, making your microwave work. The radio wave regions the government assigns as "bands" to be used by ships, airplanes, police, military, amateurs, space, satellites, and radar. Cellular phones (a teenage ear infection) occupy a global system for mobile (GSM) communications band from GSM-850 MHz to GSM-1900 MHz (in the United States, Canada, and many other countries in the Americas).

The formation of EM waves is illustrated in Figure 5.3, where we drew a switch in the middle of an "antenna" wire. According to Maxwell, magnetic fields are produced in space if there is a changing electric field. In Figure 5.3A we use a battery for energy supply and in Figure 5.3B an AC source. In Figure 5.3A, when closing the switch, a current flows briefly, setting up an electric field. The current flow in turn sets up a magnetic field that circles around the wire. The + and − signs indicate the charge on each wire segment, and the black arrow is in the direction of the current. Electrical field lines are in the plane of the page and shown in red. The magnetic field lines, in accordance with the right-hand rule, are perpendicular to the plane of the page and marked by ⊗ (into the plane) and ⊙ (out of the plane). The magnetic and electrical fields do not arise instantaneously but form in time: energy is stored in these fields, and energy cannot move infinitely fast. In Figure 5.3Ba, we show a first half cycle of the AC current cycle. When the current reverses (Figure 5.3Bb), the fields reverse and we see the first disturbance moving outward. Oscillations of current and voltage at the frequency of radiation, along the simple wire antenna in Figure 5.3, cause a portion of the stored energy to be "radiated" from the antenna. These disturbances are the EM waves. The magnitudes of the electric field **E** and the magnetic field **B** decrease as $1/r$ with distance r away from the antenna. The energy density of the radiated fields falls off as $1/r^2$ because the energy carried by the EM waves is proportional to the square of the amplitude ($\mathbf{E}^2$ or $\mathbf{B}^2$). The complicated field distribution near the antenna is called *near field*. Far from the antenna, *far field*, the field lines become quite flat, and these waves are called *plane waves*. We will come back to the concept of near field and far field in the next section.

EM waves are produced whenever electric charges are accelerated, so we can produce EM waves by

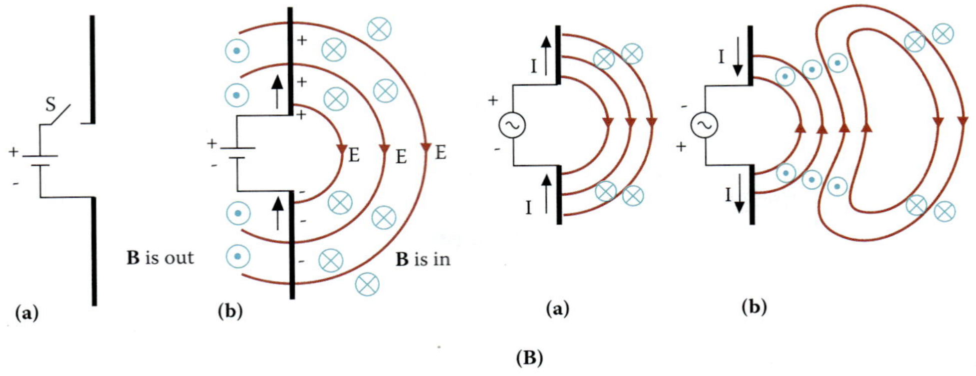

FIGURE 5.3 Formation of electromagnetic waves. (A) When closing a switch, a current flows briefly, setting up an electric field. The current flow sets up a magnetic field as circles around the wires. These fields do not arise instantaneously but form in time. Energy is stored in these fields, and energy cannot move infinitely fast. (B) When current reverses in picture (b), the fields reverse, and we see the first disturbance moving outward. These are the electromagnetic waves "...and there was light" (Genesis 1:3).

letting an alternating current flow through a wire, an antenna. The frequency of the waves created this way equals the frequency of the alternating current in the antenna. At much shorter wavelengths, the light emitted by an incandescent light bulb is caused by thermal motion that accelerates the electrons in the hot filament sufficiently to produce visible light. The inverse effect also happens: if an EM wave strikes a conductor, its oscillating electric field induces an oscillating electric current of the same frequency in the conductor. This is how the receiving antennas of radios and televisions work.

Thus, EM waves, in general, are produced by electric charges that oscillate in an antenna or perhaps in a molecule, in the case of very short wavelengths. Almost all optical effects are caused by light interacting with electric charges. An interesting exception to this is gravitational lensing. The latter occurs when the gravity of a massive celestial object, such as a galaxy, bends and focuses the light of a more distant object, resulting in a magnified, distorted, or multiple image of the original light source for a distant observer. The process was one of the predictions made in Einstein's general relativity theory.

Diffraction and Image Resolution

Far Field (Fraunhofer) and Near Field (Fresnel)

Diffraction is the apparent bending and spreading of waves when they encounter obstacles whose size is about the same order of magnitude as their wavelength. It takes place with any type of waves, including mechanical sound and water waves (Figure 5.4), as well as electromagnetic (EM) waves such as light and radio waves.

Waves either appear as plane waves, which we can envision as waves that come rolling in on a long stretch of beach, or closer to an obstruction, say, a dock, where they come as spherical waves. The latter is analogous to the circular waves expanding from a pebble dropped in a lake. Light waves, a large distance away from the obstruction (compared with the size of the obstruction), will generate an illumination pattern of light and dark. This pattern is a Fraunhofer diffraction pattern, also called *far-field diffraction*. The distances away from the obstruction must be large enough so that the light reaching the

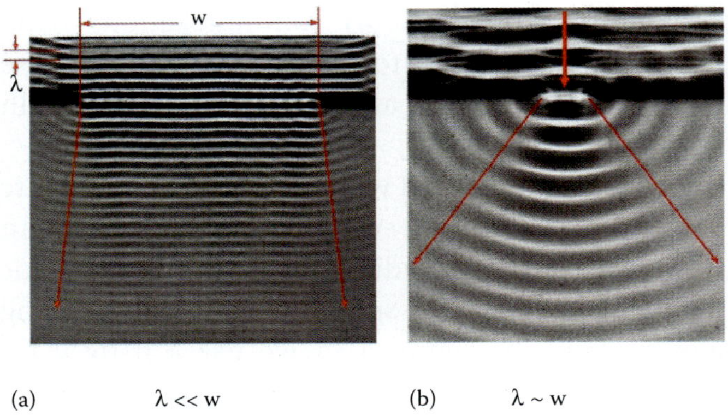

FIGURE 5.4 Diffraction in a water tank with a wavelength small compared to the aperture (a) and with a wavelength λ similar to the aperture (b).

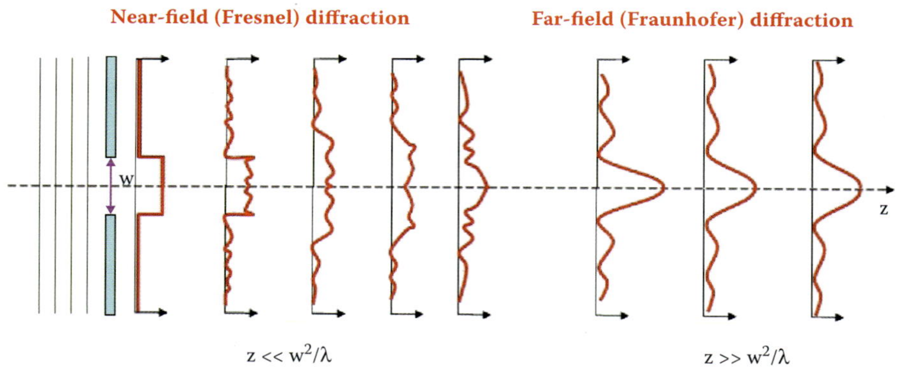

FIGURE 5.5 Fraunhofer and Fresnel diffraction regions. Aperture diameter is *w*, and *z* is the distance away from the aperture.

monitoring point, e.g., a photographic emulsion, is nearly plane waves. Spherical waves, on the other hand, produce a Fresnel diffraction pattern at a point not too far beyond the obstruction. Because these distances are close to the diffracting object, this phenomenon is also called *near-field diffraction*. Fresnel diffraction is more complex to treat mathematically than Fraunhofer diffraction. For Fraunhofer diffraction one needs distances greater than w^2/λ, where w is the size of the object (Figure 5.4). Fraunhofer diffraction works by summing the amplitudes of light waves coming from all sources. These interfere, and the intensity is determined by the square of the total amplitude.

The operational regimes for Fresnel and Fraunhofer diffraction are compared in Figure 5.5. One can calculate that Fraunhofer diffraction takes over from Fresnel diffraction if the observation point z is far enough away so that $z \gg w^2/\lambda$, with w the diameter of the aperture and λ the wavelength of the light.

What's an Image?

A major breakthrough in the understanding of image formation came in 1872, when Ernst Abbe (Figure 5.6) applied Fourier transforms to explain the phenomenon (Bragg, in 1939, applied Fourier analysis to x-ray crystallography; see Chapter 2). Image formation, according to Abbe's theory, is a two-stage, double-diffraction process: an image is the diffraction pattern of the diffraction pattern of an object.

Consider an "ideal" lens system, in which an image depicts every detail of an object as shown in Figure 5.7. In the first stage of the image formation, light rays incident on the object are scattered, and an interference pattern—a Fraunhofer diffraction pattern—is created (see also Figure 5.5).

This diffraction pattern corresponds to a forward Fourier transformation. A lens, say, a microscope objective, is essential for image formation and acts to focus light from infinity to the "focal point" at a distance from the lens known as the *focal length*, *f*. Located at the focal point is the back focal plane of the lens where the diffraction pattern can be made visible on a screen, as shown in Figure 5.8. In Figure 5.8 we show how a grating-like object is imaged and that a diffraction pattern can be seen in the back focal plane.

The second stage of image formation takes place when the scattered radiation passes beyond the back focal plane of the lens and recombines to form an

FIGURE 5.6 Ernst Abbe (1840–1905). The formula from Ernst Abbe expressing the maximum possible resolution from a light microscope on a monument erected by the University of Jena in Abbe's memory.

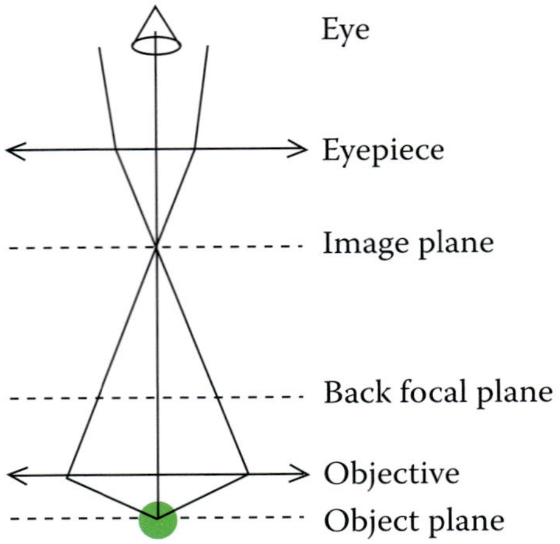

FIGURE 5.7 A lens is essential in image formation.

image. This is the back or inverse Fourier transform illustrated in Figure 5.9.

The image formed by the lens does not exactly represent the real object because not all scattered radiation enters the lens and therefore cannot all be focused on the image plane. The more of the higher orders of the fringes in Figure 5.8 can be picked up, the higher the fidelity of the image. Thus, the basic description of image formation according to Fourier optics is that a Fourier transform of the object in the back focal plane of the lens is generated, and that propagation to the image plane produces another Fourier transformation—an inverse Fourier transform—i.e., the image. Here it may also be understood why Fourier image analysis is such a powerful method for analyzing a wide variety of periodic phenomena. The Fourier transform process separates the image processing into two stages: the information contained in the diffraction pattern in the first stage reveals structural information in a straightforward manner and conveniently separates most of the signal and noise components and thus the transform may be manipulated and subsequently back-transformed in the second stage to produce a noise-filtered, better reconstructed image.

Remember from our discussion on x-ray diffraction of crystals in Chapter 2 that because there is no lens in this case, no image of the crystal lattice is generated; instead, Fresnel diffraction patterns form at finite distances and Fraunhofer diffraction patterns at infinity. Thus, x-ray diffraction methods provide a direct way to display the decomposition of x-rays in component waves (frequencies) in a forward Fourier transform. To get an image of the crystal one would need an x-ray focusing lens, which does not exist, or, as we do in practice, Fourier transform the diffraction pattern to get the real image of the crystal.

Image Resolution and Diffraction Limit

Abbe's Limit

Besides recognizing how an image is formed, Ernst Abbe was also the first to characterize the resolution of an optical system. When an object illuminated in a microscope with coherent (in-phase) light is an optical grating, the diffraction orders (N) of the grating are observed in the back focal plane of the objective, and these grating orders are then combined to form an image in the image plane (see Figures 5.7 and 5.8). The resolution of the resulting image depends on the number of grating orders that are transmitted through the optical system. Based

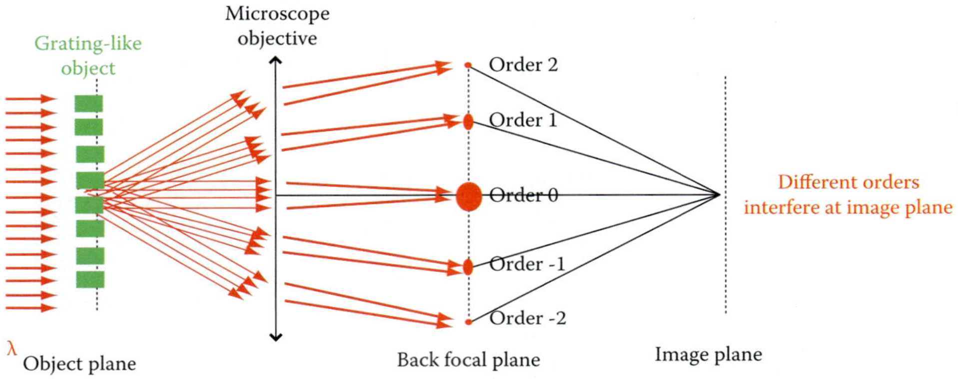

FIGURE 5.8 A grating-like object is imaged, and in the back focal plane an interference pattern—a Fraunhofer diffraction pattern—is formed that corresponds to the forward Fourier transform of the object.

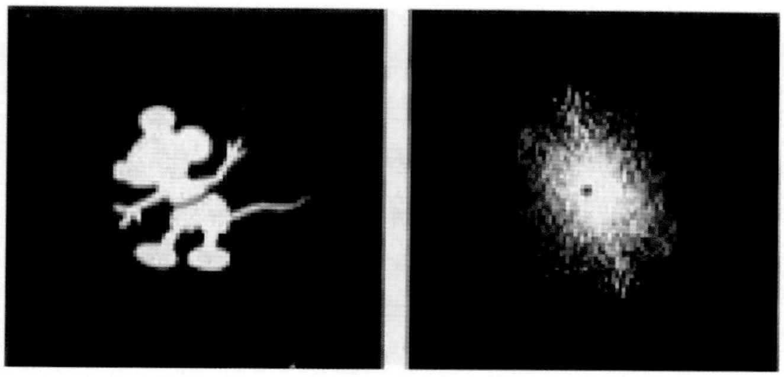

FIGURE 5.9 An inverse Fourier transform gives back the original image of Mickey Mouse on the left.

on imaging of gratings, Abbe, together with Carl Zeiss, published a paper in 1877 defining the physical laws that determined the resolving distance of an optical system. When imaging a grating with coherent light, the direction of diffracted rays is given by the grating formula:

$$d = \frac{N\lambda}{2n\sin\theta} = \frac{N\lambda}{2NA} \tag{5.1}$$

where d = grating spacing
n = refractive index in image space
λ = wavelength of the light
N = diffraction order
θ = angle of the ray of order *N* emerging from the grating
NA = numeric aperture

The grating spatial frequency ($\nu = 1/2d$) corresponding to the first-order ($N = 1$) diffracted peak is then given as:

$$\nu = \frac{1}{2d} = \frac{n\sin\theta}{\lambda} \tag{5.2}$$

For imaging with a lens, the lens needs to be able to capture the light, so it is required that $\theta = \theta_{max}$, where θ_{max} is defined by the numerical aperture NA of the lens. The latter is given by:

$$NA = n\sin\theta_{max} \tag{5.3}$$

This is illustrated in Figure 5.10, where the aim is to image a fine mask pattern onto a thin layer of photosensitive photoresist. The NA of the lens in a medium of refractive index *n* (1.0 in air) defines the angle of acceptance, $2\theta_{max}$, of the cone of diffracted light from an object that the lens can accept. It lies between 0 and 1. An NA of 0 means that the lens gathers no

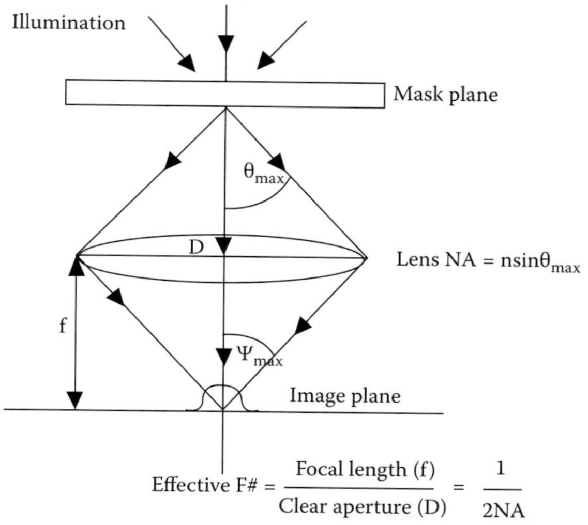

FIGURE 5.10 Relationship between the object, image, focal length, and diameter of a lens to define the numerical aperture. An important application is projection lithography, where the image on the mask is imaged onto a photoresist (see Volume II, Chapter 1).

light; an NA of 1 means that the lens gathers all the light that falls onto it. Hence the numerical aperture characterizes the ability of a lens to transmit light. It is proportional to the size of the lens, lens diameter *D*, and inversely proportional to effective *F* number (*F#*), defined as *f/D*, or the focal length *f* divided by the lens diameter *D*:

$$NA = n\sin\theta_{max} = \frac{D}{2f} = \frac{1}{2F\#} \tag{5.4}$$

The assumptions here are a large *F#*, which implies that θ is small so that $NA = \sin\theta \sim \tan\theta \sim \theta$, a situation called the *paraxial approximation*.*

At angles larger than θ_{max}, the imaging lens no longer captures light. Therefore, the highest grating spatial frequency that can be imaged by a coherent illumination system is given by:

$$\nu_{max} = \frac{1}{2d_{min}} = \frac{n\sin\theta_{max}}{\lambda} = \frac{NA}{\lambda} \tag{5.5}$$

Consequently, resolution *R* (defined here as d_{min}, i.e., slit or a line) in this case is given by:

$$R = d_{min} = \frac{\lambda}{2NA} \tag{5.6}$$

* To understand all the simplifications involved in this result consult Meinhart and Wereley (2003).[53]

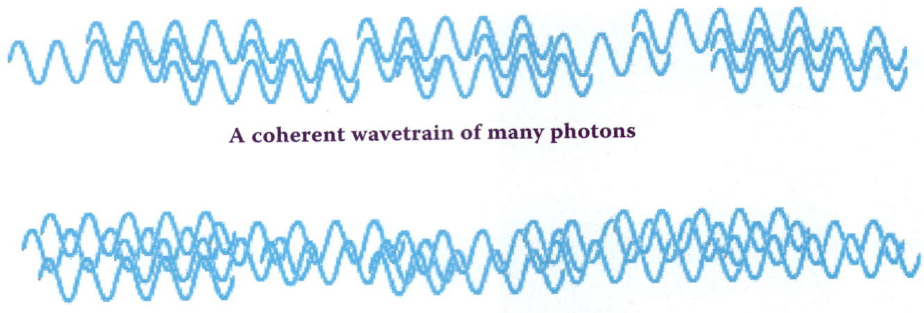

FIGURE 5.11 Spatial coherence of light.

In the case of incoherent illumination (see spatial coherence in Figure 5.11) of the same grating, each ray is diffracted by the grating and forms its own image in the wafer plane. The direction of the first diffraction peak (N = 1) for incoherent rays, incident at an angle *i*, is given by a more general grating equation:

$$d = \frac{\lambda}{2n(\sin\theta + \sin i)} = \frac{\lambda}{2NA} \tag{5.7}$$

representing the path difference for light passing through adjacent slits. For light incident normal to the grating (sin i = 0), Equation 5.7 reduces to Equation 5.1. For image formation it is now required that both *i* and $\theta = \theta_{max}$. Therefore:

$$\nu_{max} = \frac{1}{2d_{min}} = \frac{2n\sin\theta_{max}}{\lambda} = \frac{2NA}{\lambda} \tag{5.8}$$

or:

$$R = d_{min} = \frac{\lambda}{4NA} \tag{5.9}$$

so that the resolution for incoherent light is twice better (smaller features can be resolved) than the resolution of coherent light.

Rayleigh's Limit

The above imaging resolutions for coherent and incoherent light are theoretical limits. Let us calculate the practical limit to the resolution of a lens. Light passing through a circular aperture interferes with itself, creating ring-shaped diffraction patterns known as an *Airy pattern*, which blurs the image (see Figures 5.13 and 5.14). It is named for George Biddell Airy, a British astronomer, who first explained the phenomenon. In 1872, Rayleigh proposed his resolution criterion based on this Airy pattern. This criterion defines two identical incoherent light sources, separated by a distance *d*, as resolved when the maximum of the Airy disc from one point source falls on the first zero of the intensity distribution of the Airy disc from a second point source. The diffraction for a circular aperture* is given as:

$$E = \frac{2J_1\left(\frac{\pi D\sin\theta}{\lambda}\right)}{\left(\frac{\pi D\sin\theta}{\lambda}\right)} \text{ and } I = E^2 = \left[\frac{2J_1\left(\frac{\pi D\sin\theta}{\lambda}\right)}{\left(\frac{\pi D\sin\theta}{\lambda}\right)}\right]^2 \tag{5.10}$$

where E = electrical field
I = intensity
J_1 = a Bessel function of the first kind of order 1 in *x* with $x = \pi D\sin\theta/\lambda$ (see Figure 5.12)
D = diameter of the aperture
λ = wavelength of light

The expression of *I*(*r*) is called the *point spread function* (PSF), and the radius of the Airy disc is derived from it.

The PSF extends to infinity, but the "width" of the central intensity peak can be described from its first minimum where J_1 is zero, i.e., at x = 3.8 (see Figure 5.12). The radius of the PSF(*r*), at a distance *s* of the aperture, is defined by the radius of the first dark ring in the image plane, corresponds to the radius of the Airy disc, and is calculated as:

* Exercise: In Chapter 2 we calculated the diffraction of x-rays from a crystal (see Equation 2.39). Demonstrate that Equations 2.39 and 5.10 are identical expressions.

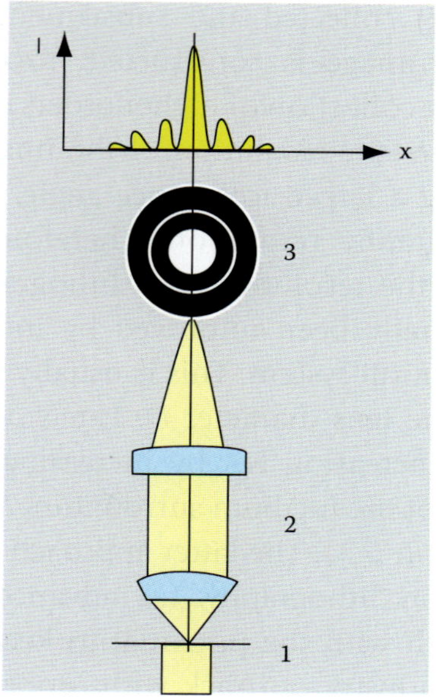

(a)

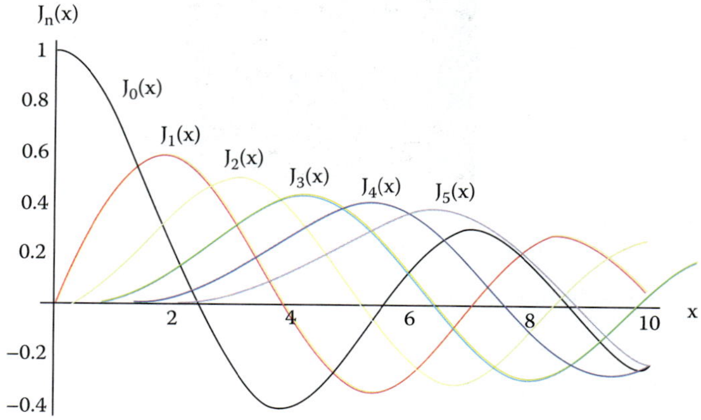

(b)

FIGURE 5.12 (a) 1) pinhole, 2) lens system, and 3) point spread function (PSF). (b) Bessel functions.

$$J_1 = 0 \text{ when } x = \frac{\pi D \sin\theta}{\lambda} = 3.8 \text{ or, for small } \theta,$$

$$D = \frac{1.22\lambda}{\theta} \text{ and since } \theta \cong \frac{r}{s} \quad (5.11)$$

$$r = \frac{1.22\lambda s}{D}$$

Fraunhofer diffraction patterns are obtained by moving the screen far enough away from the aperture. To obtain the Fraunhofer diffraction pattern at the focal point of a lens, we set the lens against the aperture and use the focal length (f) of the lens as the aperture to screen distance (i.e., replace s with f in Equation 5.11), or:

$$r = \frac{1.22\lambda f}{D} \quad (5.12)$$

This is illustrated in Figure 5.13. Or, based on Equation 5.4, for an ideal lens system of focal length f, the Rayleigh criterion yields a minimum spatial resolution as:

$$d = \frac{1.22\lambda}{2NA} \quad (5.13)$$

In this expression d is the radius of the smallest object that the lens can resolve and also the radius of the smallest spot on which a collimated beam of light can be focused. The Rayleigh criterion is illustrated in Figure 5.14.

In Figure 5.15 we show a schematic of a single lens with diameter D imaging a small circular aperture. By definition, the magnification M of a lens is given by:

$$M = \frac{S_i}{S_0} \quad (5.14)$$

where S_0 is the object distance and S_i is the image distance. Combining this expression for magnification with the Gaussian lens formula we obtain:

$$\frac{1}{S_i} + \frac{1}{S_0} = \frac{1}{f} \quad (5.15)$$

which leads to:

$$S_i = (M + 1)f \quad (5.16)$$

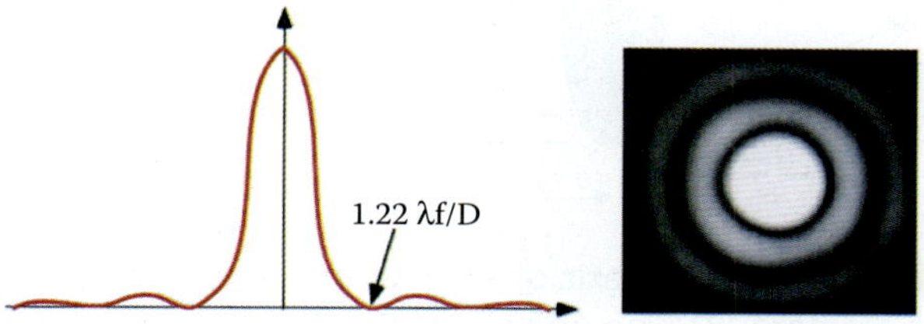

FIGURE 5.13 Image by a point source forms a circle with diameter of 2.44 $\lambda f/D$ (=2 × r) surrounded by diffraction rings (Airy pattern).

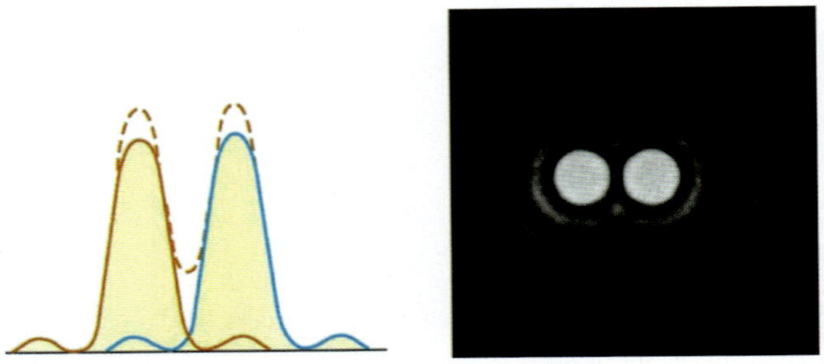

FIGURE 5.14 Rayleigh criterion: two light spots are considered resolved if the intensity minimum of one coincides with the maximum of the other $\left(d = \frac{1.22\lambda}{2NA}\right)$.

where f is the focal length of the lens. If the object distance (S_0) and the image distance (S_i) are set equal to the focal lengths, the magnification can be written in terms of the focal lengths, or:

$$M = \frac{S_i}{S_0} = \frac{f_i}{f_0} \tag{5.17}$$

In the case $S_i = S_0$ and $f_i = f_0$, the magnification M obviously equals 1. The F number ($F\#$) of such a setup is then by definition given as $F\# = f_0/D = f_i/D$ (see Equation 5.4); its numerical aperture is $NA_0 = NA_i = NA$, and these factors are linked with the magnification M via:

$$F\# = \frac{f_0}{D} = \frac{f_i}{D} = \frac{1}{2NA_0} = \frac{1}{2NA_i} = \frac{(1+M)f}{D} \tag{5.18}$$

D is the diameter of the lens or the diameter of the light beam, whichever is the smaller of the two.

The larger the NA of the projection lens in the lithography system of Figure 5.10, the greater the amount of light (containing diffraction information of the mask) collected and subsequently imaged. Because the image is constructed from diffracted light, and the collection of higher orders of diffracted light enhances the resolution of the image, a larger NA, allowing a larger acceptance angle, results in a better lithography resolution. Therefore, for a set wavelength, the resolution in a lithography system has traditionally been improved by increasing the NA of the optical system. This is usually achieved by increasing the lens diameter D. Lately, work is also directed at increasing NA by increasing the refractive index n of the medium surrounding the imaging lens (Equation 5.3). The latter approach is exploited in immersion lithography, an advanced lithography approach with the potential to keep the semiconductor industry on track with Moore's law. The physical limit to NA for lithography exposure systems using air as a medium between the lens and the wafer is 1 because the sine of any angle is always ≤1 and n = 1 for air. What happens, though, if a medium with a higher index of refraction is substituted for air, as is often done in microscopy, where oil is placed between the lens and the sample being viewed for resolution? Abbe and Zeiss developed these oil immersion systems using oils that matched the refractive index of glass. Thus, they were able to reach a maximum NA of 1.4, allowing light microscopes to resolve two points distanced only 0.2 μm apart. In Volume II, Chapter 1, we will further detail immersion lithography.

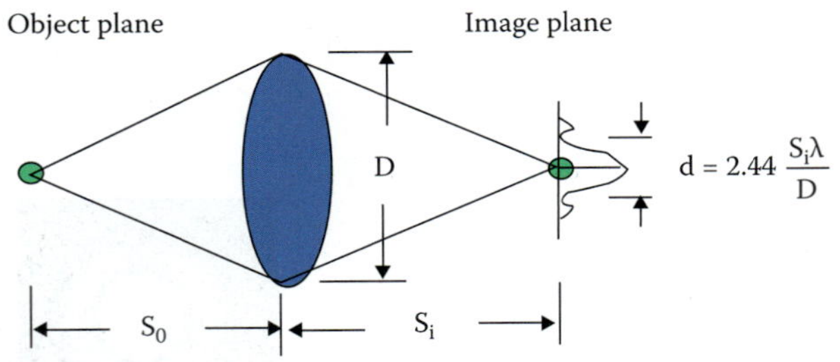

FIGURE 5.15 Schematic of a single thin lens with aperture diameter, D. The object distance is S_0, and the image distance is S_i. The point spread function through a circular aperture is the well-known Airy function of width or diameter, $d = 2.44S_i\lambda/D$ (see Equation 5.13 with $D = 2r$).

Refraction

Snellius Law

Geometric optics, also ray optics, has light propagate as "rays." These rays are bent at the interface between two dissimilar media. The "ray" is an abstract object, perpendicular to the wavefronts of the electromagnetic (EM) waves. Geometric optics provides rules for propagating these rays through an optical system, revealing how actual wavefronts propagate. Ray optics is a significant simplification of optics and fails to account for many important optical effects, such as diffraction and polarization. The latter phenomenon will only be understood properly after we introduce Maxwell's equations.

Light refracts, which means that it bends when passing from one medium to another. Thus, refraction is the change in direction of a ray of light passing from one transparent medium to another one with a different optical density. When light enters a denser medium (say, water) coming from one that is less dense (say, air), it bends toward a line normal to the boundary between the two media as illustrated in Figure 5.16. The deviation is greater for shorter wavelengths.

The ratio of the speed of the light going from one material into another material is proportional to the ratio of the sine of the angle of incidence and the sine of the angle of refraction as illustrated in Figure 5.16. This is the so-called *Snellius law* [1621; also Willebrord Snell (1580–1626)]:

$$\frac{\sin(i)}{\sin(r)} = \frac{n_1}{n_2} \quad (5.19)$$

with i being the angle of incidence and r the angle of refraction, n_1 is the index of refraction of material 1, and n_2 is the index of refraction of material 2. Two familiar examples to demonstrate refraction are a pencil in a glass of water (Figure 5.17) and a rainbow (Figure 5.18). In Figure 5.17 we show a pencil in a glass: first in an empty glass, then in one with water (positive refractive index), and then in a glass filled with a negative refractive index material.

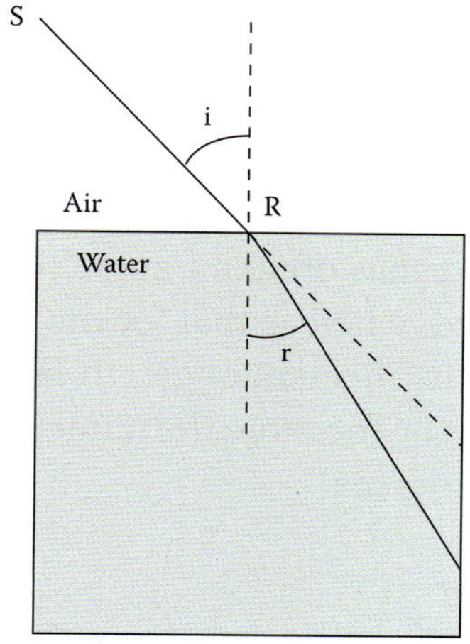

FIGURE 5.16 Snellius law illustrated: the ratio of the horizontal length, *RH*, to the diagonal, *RA*, is just the sine of the angle *RAH*, which is equal to the angle of incidence *i* measured from the line drawn perpendicular to the refracting surface at *R*. Likewise, the ratio *RH*/*RC* is just the sine of the angle of refraction *RCH*, or *r*. Thus, Snell's ratio *RC*/*RA* is the same as (sin *i*)/(sin *r*).

At the boundary between air and water, the light wave changes direction; its wavelength increases or decreases, but its frequency remains constant. We will see in the next section that the refractive index of a medium is related to the speed of light and that in a denser medium a "slow-down" of the light occurs relative to the speed of light in free space. When light slows down it appears bent. The case of negative refractive index water (Figure 5.17c) will be discussed further below under "The μ-ε Quadrant and Metamaterials." A bright vividly colored rainbow as shown in Figure 5.18 is caused by dispersion of sunlight refracted by raindrops between 0.1 and 2 mm in diameter where geometric optics applies. Complex Mie scattering effects are important for smaller drops, which do produce white fog (see further below). The light is first refracted as it enters the surface of the raindrop, reflected off the back of the drop, and again refracted as it leaves the drop. This angle is independent of the size of the drop but depends a lot on its shape and refractive index.

Definition of Refractive Index

Maxwell demonstrated that for electromagnetic (EM) waves the wave velocity v in a homogeneous, isotropic dielectric medium is determined by the magnetic permeability μ, and the dielectric constant (or permittivity), ε (for the detailed derivation see further below):

$$v = \frac{1}{\sqrt{\varepsilon\mu}} \quad (5.20)$$

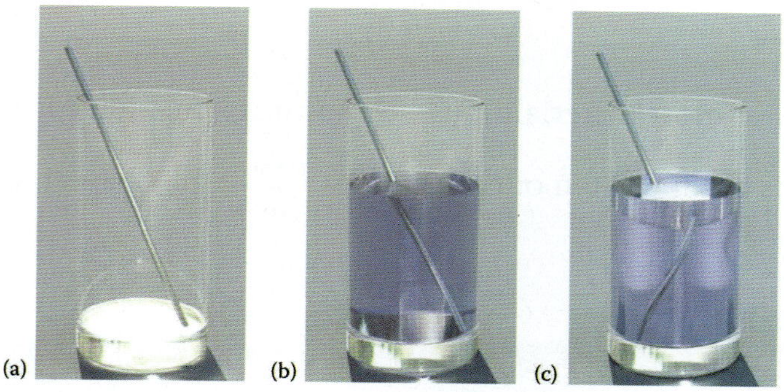

FIGURE 5.17 Refraction illustrated. (a) Empty glass: no refraction; (b) typical refraction with pencil in water with n = 1.3; (c) what would happen if the refractive index were negative with n = –1.3 (see "The μ-ε Quadrant and Metamaterials"). (From Gennady Shvets, The University of Texas at Austin.)

FIGURE 5.18 A rainbow is caused by dispersion of sunlight refracted by raindrops.

For a wave traveling in free space (where we use the permeability and dielectric constant for vacuum μ_0 and ε_0 instead of μ and ε) $v = c = 2.998 \times 10^8$ m/s. The "slow-down" factor, relative to free space, is by definition the refractive index *n*:

$$n = \frac{c}{v} \tag{5.21}$$

or also:

$$n^2 = \varepsilon\mu \tag{5.22}$$

For example, because the speed of light in air is $v = 1/\sqrt{\varepsilon_{air}\mu_{air}}$ (with the permeability and dielectric constant for air given by μ_{air} and ε_{air}, respectively), the refractive index of air is given by:

$$n = \frac{c}{v} = \frac{\sqrt{\varepsilon_{air}\mu_{air}}}{\sqrt{\varepsilon_0\mu_0}} = 1.003 \tag{5.23}$$

For most materials $\mu/\mu_0 = 1$, so the expression for *n* is often written out as $n = \frac{c}{v} = \frac{\sqrt{\varepsilon_{air}}}{\sqrt{\varepsilon_0}}$). In Chapter 3, we saw how the dielectric constant ε in the most general terms is a dielectric function dependent on both frequency (ω) and wave vector **k**: ε(ω,**k**). There is also a dispersion function for the magnetic permeability [μ(ω, **k**)]; thus, from Equation 5.23 we can expect the refractive index *n* to be frequency-dependent as well. When a material is subject to an applied electric field **E**, the internal charge distribution is distorted, which generates an electric dipole moment per unit volume called the *electric polarization* **P**. After introducing the Maxwell equations, we will link the dielectric function to this polarization **P** and to the dispersion of the refractive index *n* for metals, semiconductors, and dielectrics.

Dispersion of light is most familiar with respect to dispersion of white light on a prism. If a beam of white light enters a glass prism, what emerges from the other side is a spread-out beam of many-colored light. The various colors are refracted through different angles by the glass and are "dispersed," or spread out.

Reflectance and Total Internal Reflectance

To illustrate reflectance and total internal reflectance (TIR) we show a sequence of four images from ray optics in Figure 5.19. A light beam goes from medium 1 (say, water with refractive index n_1) into a medium 2 (say, air with refractive index n_2) at different angles θ_1. When the light coming from the water strikes the surface, part will be reflected and part will be refracted. Measured with respect to the normal line perpendicular to the surface, the reflected light comes off at an angle equal to that at which it entered, whereas that for the refracted light is larger than the incident angle. In fact, the greater the incident angle, the more the refracted light bends away from the normal.

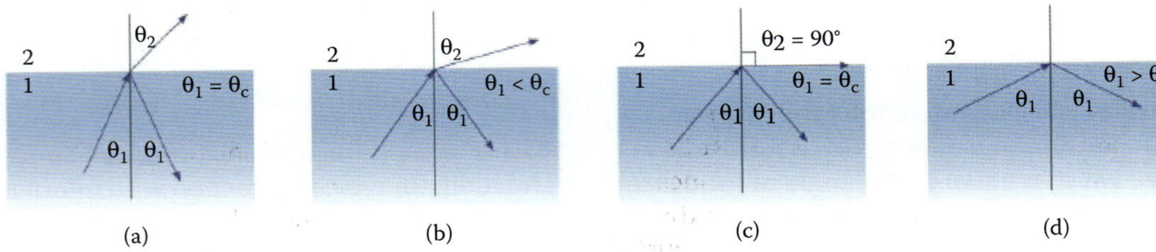

FIGURE 5.19 Illustration of reflection and total internal reflectance (TIR). Material 1 (bottom) has the higher index of refraction ($n_1 > n_2$). For explanation of (a)–(d), see text.

Thus, increasing the angle of incidence from the path shown in Figure 5.19a to the one shown in Figure 5.19b will eventually reach a point where the refracted angle is 90°, at which point the light appears to emerge along the surface between the water and air (Figure 5.19c). If the angle of incidence is increased further, the refracted light cannot leave the water (Figure 5.19d). It gets completely reflected. The interesting thing about TIR is that 100% of the light gets reflected back into the denser medium, as long as the angle at which it is incident to the surface is large enough. At a given critical angle when $\theta_1 = \theta_c$ TIR occurs (Figure 5.19c); the light cannot escape medium 1 with refractive index n_1 anymore, except for some really weak evanescent wave. Any light incident at an angle $\theta_1 > \theta_c$ is also totally reflected (Figure 5.19d). Thus, TIR is an effect that combines both refraction and reflection. The critical angle, also called the *Brewster angle*, named after the Scottish physicist Sir David Brewster (1781–1868), is given by:

$$\sin\theta_c = \frac{n_2}{n_1} \tag{5.24}$$

A common and useful application of TIR is in fiberoptics to keep light beams focused inside a fiber without significant loss (see Figure 5.20).

Further below, after launching the Maxwell equations, we calculate the reflectance R in terms of the refractive index n.

Light Polarization

Polarized light is light that has its electric vector (**E**) oriented in a predictable fashion with respect to the propagation direction. Ordinary white light is unpolarized and made up of numerous orthogonal electrical and magnetic field pairs (**E, B**) that fluctuate at all possible angles around the propagation direction. Even in short time intervals, it appears to be oriented in all directions with equal probability. Light is considered to be "linearly polarized" when it contains waves that only fluctuate in one specific plane (see Figure 5.21). Linear polarization "tries to" induce a linear charge motion, but the Lorentz force prevents this from being purely linear. For example, charge oscillation along the x-axis introduces a small E_y component (field rotation. Specifying the orientation of one of the field vectors (**E** or **B**) determines the polarization of the EM wave. The plane containing the **E**-field vector defines the polarization of the EM wave. Of course, the **B**-field vector would then always lie in a plane perpendicular to this plane containing the **E**-field vector. But according to theoretical and experimental evidence, it is the electric vector (**E**) rather than the magnetic vector (**B**) of a light wave that is responsible for all polarization effects. Therefore, the electric vector of a light wave, for all practical purposes, can be identified as the light vector in this respect.

The simplest way of producing linearly polarized light is by reflection from a dielectric surface. In the preceding section we learned that when light strikes a material such as glass, part of the light might get reflected and part of it refracted. Reflection also causes the light to become unpolarized, partially polarized, or completely polarized, depending on the angle at which the light strikes the surface, or the angle of incidence (see Figure 5.22). The light

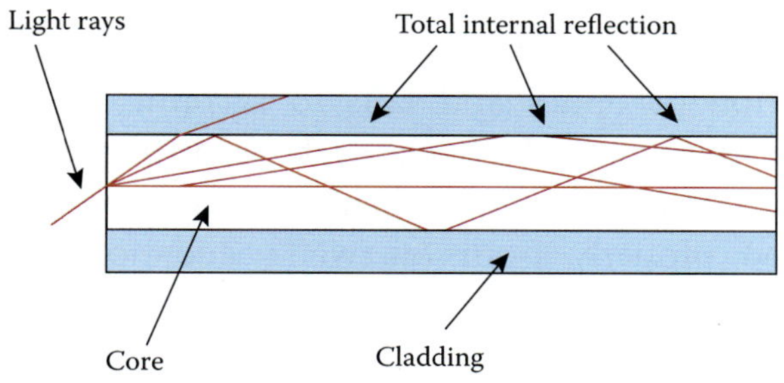

FIGURE 5.20 A fiberoptic is based on the principle of total internal reflectance.

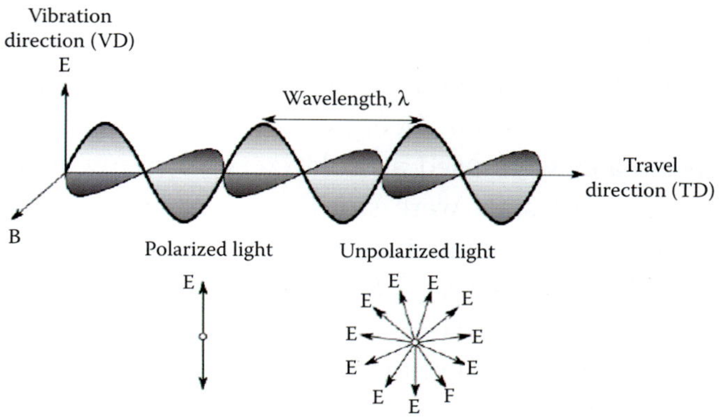

FIGURE 5.21 Unpolarized and polarized light.

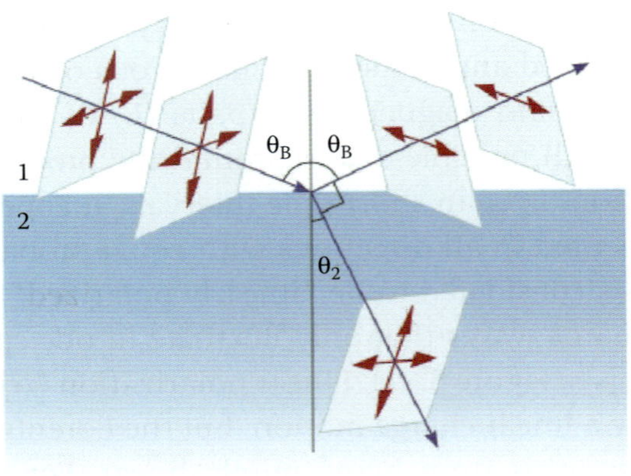

FIGURE 5.22 Reflection causes the light to become either unpolarized, partially polarized, or completely polarized, depending on the angle at which the light strikes the surface, or the angle of incidence. The angle θ_B is known as the *polarizing angle*, or the *Brewster angle*.

is unpolarized only if the incident light angle is 0°, which does not occur often. It is more common that polarization does occur, and the greater the angle of incidence, the greater the polarization. When the light strikes at an angle θ_B of 90° or perpendicular to the surface, the light is completely polarized. At this particular angle of incidence the reflectivity for light whose electric vector is in the plane of incidence becomes zero. Thus, the reflected light is linearly polarized at right angles to the plane of incidence. This angle θ_B is known as the *polarizing angle*, or the *Brewster polarizing angle*, because it was discovered by the same David Brewster of the critical angle θ_c (see previous section):

$$\tan\theta_B = \frac{n_2}{n_1} \tag{5.25}$$

Certain natural materials absorb linearly polarized light of one vibration direction much more strongly than light vibrating at right angles. Such materials are termed dichroic. Tourmaline is one of the best-known dichroic crystals, and tourmaline plates were used as polarizers for many years. Dichroic crystals were studied by Edwin Land in 1930; Land invented a polarizing material that would be called *Polaroid* and give a huge benefit to the camera industry.

Polarization of light is also possible by refraction, in which case the light is bounced off of numerous surfaces, instead of only one. Often a stack of thin films of a transparent material is used for this purpose. The beam of light that is refracted is continuously polarized because refraction causes polarization. Because of this polarization, the beam of light that is refracted is shown to increase in intensity. Polarization obviously occurs frequently: even the light from a rainbow is polarized because the angle of incidence of light off a water beam is similar to a Brewster angle.

Other natural materials exist in which the velocity of light depends on the vibration direction. These materials are called *birefringent*. We will discuss these in more detail after we have introduced the Maxwell equations.

Circular (and elliptical) polarization is possible because the propagating electric (and magnetic) fields can have two orthogonal components with independent amplitudes and phases (and the same frequency). The electric vector, at one point in time, describes a helix along the direction of wave propagation. Circular polarization may be referred to as *right* or *left*, depending on the direction in which the electric field vector rotates. The magnitude of the electric field vector is constant as it rotates. Circular polarization is a limiting case of the more general condition of elliptical polarization. A circularly polarized wave may be resolved into two linearly polarized waves, of equal amplitude, in phase quadrature (90° apart) and with their planes of polarization at right angles to each other.

Such polarization occurs in chiral* materials, for example, with helical polarizable molecules in a solution.

Maxwell's Equations

Introduction

As stated earlier, ray optics is a significant simplification of optics and fails to account for many important optical effects such as diffraction and polarization. The latter phenomena are only understood properly from Maxwell's equations. Light

* Definition of chirality by Kelvin: "I call any geometrical figure, or group of points, chiral, and say it has chirality, if its image in a plane mirror, ideally realized, cannot be brought to coincide with itself."

with its wave and particle character is not the easiest of natural phenomena to describe. Its ambivalent nature requires two complementary models for a satisfactory description of its properties. In Chapter 3, we mentioned how Dirac's quantum field theory removed the paradox of particle-wave duality: it showed that if a particle was probed in a way that was meant to demonstrate its particle-like properties, it would appear to be a particle, and if it was probed in a way that was meant to demonstrate its wave-like properties, it would appear to be a wave. Maxwell's electromagnetic (EM) wave model, the topic of this chapter, follows logically from Dirac's model and describes light as a continuous transfer of energy through space by a combination of waves of electrical and magnetic fields.

The Scottish physicist James Clerk Maxwell (1831–1879) unified in one magnificent theory—four elegant equations—all that was known of electricity and magnetism. For the first time it became apparent that EM fields could travel through space, and the theory suggested the revolutionary concept that light was nothing but an EM wave itself. Maxwell's equations are even more fundamental than Newton's three laws of motion because they agree with Einstein's relativity theory, whereas Newton's laws do not.

The refractive index of a material as introduced above comes about through the interaction of light with many, many atoms (10^{22} atoms/cm^3), and it is impossible to describe the individual response of so many electrons and atoms. So we have to make do with "effective" materials properties, such as the refractive index, an average over many atoms. This works very well for many materials because atom spacings in solids are much smaller than the wavelength of the light at hand. For visible light, e.g., green light, a volume of λ^3 corresponds to a volume of ~ (500 nm)^3 ~ (5 10^{-5} cm)^3 ~ 10^{-13} cm^3, and the number of atoms in such a volume ~$10^{-13} \times 10^{22} = 10^9$! So working with spatially averaged fields will give good results in many cases.

Before we launch into the Maxwell's equations for plane wave EM radiation, we remind the reader of some important vector relations. Because both the electric field and the magnetic field have a direction, we describe light as vector fields. Two important integral theorems we need for manipulating those equations are Gauss' divergence theorem and Stokes' theorem. We also introduce Ampère's law, Gauss' electric and magnetic laws, and Faraday's law, as they are the foundation for Maxwell's famous unifying equations.

Gauss and Stokes Theorems

The first theorem we introduce, the divergence theorem, comes from vector calculus and is also known as *Gauss'* (1777–1855) or *Green's* (1783–1841) *theorem*. The divergence of a vector field is a scalar (a number) and represents the net outward flux per unit volume (one can think of it as a "source density"). The divergence theorem relates volume and surface integrals and expresses the fact that the volume integral of the divergence of any conservative vector (or tensor) field equals the total outward flux of the vector through the surface that bounds the volume (http://mathworld.wolfram.com/ConservativeField.html). More simply stated: the sum of all sources minus the sum of all sinks gives the net flow out of a region. Mathematically:

$$\int_V \nabla \cdot \mathbf{A} dv = \oint_S \mathbf{A} \cdot d\mathbf{s} \tag{5.26}$$

where **A** is a vector field or tensor field (for example, **A** could be a stress tensor $\underset{\cdot}{\sigma}$ or an electric field vector **E**); d**s** is a differential vector element of surface area *S*, with infinitesimally small magnitude and direction normal to surface *S*; and *dv* is a differential element of volume *V* enclosed by surface *S*. We remind the reader that in vector analysis $\nabla \cdot \mathbf{A}$ is the same as div **A**. This equation applies to any volume *V* bounded by surface *S*. Effectively this comes down to saying that "field lines are conserved." We see where field lines originate on the left side of the equation or count them as they leave through the surface on the right side of the equation (Figure 5.23).

The second theorem we need to refamiliarize ourselves with is Stokes' theorem, or the curl theorem. It is named after Sir George Gabriel Stokes (1819–1903), although the first known version of the theorem is by William Thomson (Lord Kelvin). Stokes' theorem deals with the problem of a 3D

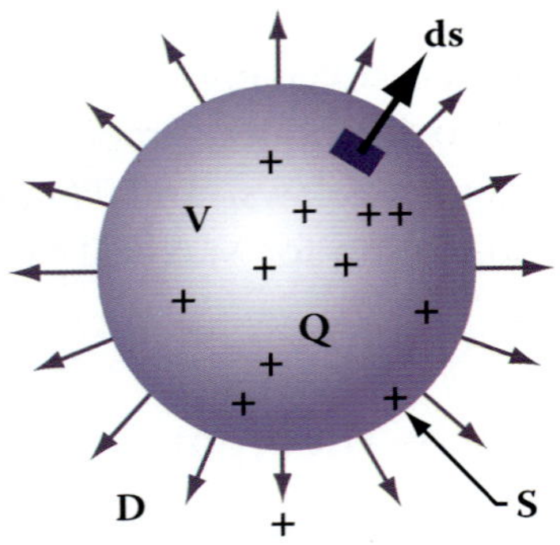

FIGURE 5.23 The divergence or Gauss' theorem illustrated for the case where **A** is equal to **D**, the electric displacement which is related to the electric field **E** via a materials-dependent constant called the *permittivity*, ε (as given by the constitutive relation, **D** = ε **E**; Equation 3.38 in Volume I). Consider a total amount of charge *Q* in a volume *V* bounded by the closed surface *S*. The area integral gives a measure of the net charge enclosed, and the divergence of the electric displacement gives the density of the sources (see Equations 5.34–5.36 below).

curve in space, and the line integral around such a curve; it relates a surface integral to a line integral:

$$\int_S (\nabla \times \mathbf{A}) \cdot \mathbf{ds} = \oint_C \mathbf{A} \cdot \mathbf{dl} \qquad (5.27)$$

where *S* is an open surface bounded by a closed curve *C*, and *d***l** is a differential vector element of path length tangential to contour *C* enclosing surface area *S*. The vector *d***s** is an outward surface normal vector of magnitude *ds*. Stokes' theorem says that if *S* is a surface in three dimensions having a closed curve *C* as its boundary, then the surface integral of the curl of a vector field over that open surface equals the closed line integral of the vector along the contour *C* bounding the open surface. Integrating the curl across a surface gives the same answer as integrating the tangential vector components along the line bounding that surface. Also, rotation in adjacent spots cancels out, and what is not canceled is present at the edge of the surface. This is a fairly remarkable theorem because *S* can be any surface bounded by the curve *C* and have just about any shape (Figure 5.24). We remind the reader that in vector analysis ∇ × **A**, rot **A**, and curl **A** are equivalent notations. Another way of saying this is that the curl of a field defines the "amount of rotation" in a field.

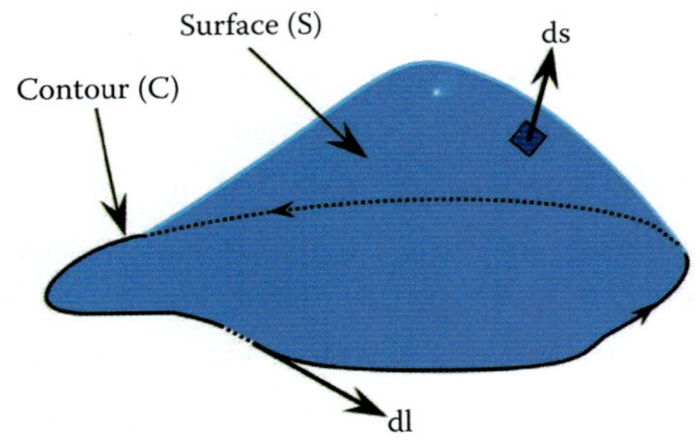

FIGURE 5.24 Illustrating Stokes' law. *S* is an open surface bounded by a closed curve *C*.

With Gauss' and Stokes' theorems Maxwell's equations can be transformed from differential format to integral, and vice versa.

Ampère's Law, Gauss' Electric and Magnetic Laws, and Faraday's Law

Ampère's Law

Ampère's law states:

> The magnetic circulation around a closed loop is proportional to the net electric current flowing through the loop, including the actual flow of electric charge as well as the displacement current, which is the time derivative of the electric flux. The positive directions of circulation, current and flux are related by the right-hand rule.

An electric current *J* running through a straight wire generates a magnetic field strength **H** (in A/m) around that wire (Figure 5.25). The integral of H along the circular path *C* is equal to the current *I*, which in turn can be obtained as the integral

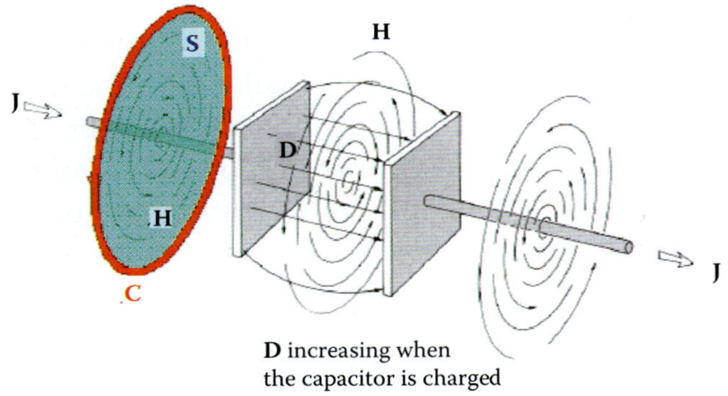

FIGURE 5.25 Ampère's law illustrated. The magnetic field is induced by electrical currents (**J**) and changes in electric field, $\frac{\partial \mathbf{D}}{\partial t}$ (the displacement current is also called *Maxwell's displacement* **D**).

of the current density **J** across the cross-section of the wire:

$$\oint_C \mathbf{H}\cdot d\mathbf{l} = I = \int_S \mathbf{J}\cdot d\mathbf{s} \tag{5.28}$$

This is the original Ampère's Law (1775–1836).

Using Stokes' theorem (Equation 5.27) we transform Equation 5.28 to:

$$\int_S (\nabla \times \mathbf{H})\cdot d\mathbf{s} = I = \int_S \mathbf{J}\cdot d\mathbf{s} \tag{5.29}$$

From the latter expression it follows that:

$$\nabla \times \mathbf{H} = \mathbf{J} \tag{5.30}$$

James Clerk Maxwell made an important correction to Ampère's law; he recognized that a changing electric field, $\frac{\partial \mathbf{D}}{\partial t}$, is equivalent to a current **J**, and he called $\frac{\partial \mathbf{D}}{\partial t}$ the displacement current. The electric flux density, also called *Maxwell's displacement* **D**, is expressed in units of As/m². The more general Ampere's law, in which a changing electric field $\left(\frac{\partial \mathbf{D}}{\partial t}\right)$ and a current (**J**) both produce a magnetic field (**H**), is then given as:

$$\nabla \times \mathbf{H} = \mathbf{J} + \frac{\partial \mathbf{D}}{\partial t} \tag{5.31}$$

The electric displacement **D**, expressed in coulombs per square meter (C/m²) is related to the electric field **E** (in units of V/m) via the permittivity, ε (ω,**k**), as:

$$\mathbf{D} = \varepsilon(\omega,\mathbf{k})\mathbf{E} \tag{3.37}$$

The permittivity ε (As/Vm) is a scalar if the medium is isotropic or a 3 × 3 matrix otherwise, and its frequency dependence defines the so-called *dielectric function*, i.e., ε(ω,**k**). To obtain $\varepsilon_r(\omega,\mathbf{k})$, the relative dielectric response of a material, one divides by ε_0 because $\varepsilon(\omega,\mathbf{k})/\varepsilon_0 = \varepsilon_r(\omega,\mathbf{k})$.

Similarly, the magnetic field or flux density **B** is linked to the magnetic field strength **H** (in A/m) within the material through a second constitutive relation:

$$\mathbf{B} = \mu\mathbf{H} \tag{5.32}$$

with $\mu = \mu_0\mu_r$ and μ_r the dimensionless relative permeability. The magnetic permeability μ is a tensor that reduces to a scalar in isotropic materials. The magnetic inductive capacity, μ, or permeability, is related to the ability of a medium, such as air, to store magnetic potential energy. The permeability of vacuum, μ_0, is 1.257×10^{-6} Hm^{-1}, and in the case of air is $\mu_{air} = 1.260 \times 10^{-6}$ Hm^{-1}. As in the case of the dielectric "constant" the response of the relative magnetic permeability μ_r of normal media to externally applied fields depends on the frequency of the field as well and on the wave vector, i.e., $\mu_r(\omega,\mathbf{k})$.

A third constitutive relation, the basis of Ohm's law (V = IR), introduced in Chapter 2, links current density **J** (A/m²) to electrical conductivity σ (in A/Vm) of the material:

$$\mathbf{J} = \sigma\mathbf{E} \tag{2.2}$$

In the most general case, the electrical conductivity σ is a second-rank tensor and is also frequency-dependent, i.e., σ(ω,**k**). For an isotropic material, σ becomes a scalar.

Gauss' Law for Electric Fields

Gauss' law for electric fields states:

> The net electric flux exiting a closed surface is proportional to the net electric charge enclosed by the surface.

Gauss' law is an extended form of Coulomb's law. Both Coulomb's and Gauss' laws relate electric fields to charges: charges make electrical fields. Consider a total amount of charge *Q* in a volume *V* bounded by a closed surface *S*. The electric charger *Q* equals the surface integral of the electric flux density over the closed surface *S*:

$$\oint_S \mathbf{D}\cdot d\mathbf{s} = Q \tag{5.33}$$

With the divergence theorem this can be rewritten as:

$$\oint_S \mathbf{D}\cdot d\mathbf{s} = \int_V \nabla\cdot\mathbf{D}dv \tag{5.34}$$

or also:

$$\int_V \nabla\cdot\mathbf{D}dv = Q = \int_V \rho_V dv \tag{5.35}$$

and finally one obtains:

$$\nabla\cdot\mathbf{D} = \rho_v \tag{5.36}$$

where ρ_V is the electric charge per unit volume (C/m^3). For linear materials, $\mathbf{D} = \varepsilon\,\mathbf{E}$ (Equation 3.38) and Equation 5.36 can be rewritten as:

$$\nabla \cdot \mathbf{E} = \frac{\rho_V}{\varepsilon} \tag{5.37}$$

From Equation 5.37 we can derive the so-called *Poisson equation* for the potential:

$$\nabla^2 V = -\frac{\rho V}{\varepsilon} \tag{5.38}$$

because $\mathbf{E} = -\nabla V$. In the case that the total charge density is zero, Equation 5.38 reduces to Laplace's Equation:

$$\nabla^2 \mathbf{V} = 0 \tag{5.39}$$

marking a constant electric field.

Gauss' Law for Magnetic Fields

Gauss' law for magnetic fields states:

> The net magnetic flux exiting a closed surface is always zero.

We know that magnetic poles always occur in pairs (north and south). Therefore, a magnetic field will be produced in space if there is a changing electric field, but this magnetic field is changing as the electric field is changing. A changing magnetic field produces an electric field that is changing—in other words, we have a self-perpetuating system. Because a single magnetic charge does not exist, the equivalent expression to Equation 5.33 for magnetism reads:

$$\oint_S \mathbf{B} \cdot \mathbf{ds} = 0 \tag{5.40}$$

This means that the magnetic field lines must form closed loops, i.e., the mathematical formulation of the fact that the field lines cannot originate from somewhere or that there are no magnetic monopoles. Using Gauss' theorem we obtain:

$$\oint_S \mathbf{B} \cdot \mathbf{ds} = \int_V \nabla \cdot \mathbf{B} dv = 0 \tag{5.41}$$

In other words:

$$\nabla \cdot \mathbf{B} = 0 \tag{5.42}$$

Faraday's Law

Faraday's law states:

> The electric circulation (or emf) around a closed loop is proportional to the negative time derivative of the magnetic flux through the loop, where the positive circulation and flux directions are related by the right-hand rule.

Faraday (1791–1867) discovered that a voltage V (or emf) is induced in a coil when a magnetic flux Φ [Webers (Wb)] in the environment changes with time, and he formulated a law stating that a changing magnetic field $\left(\frac{\partial \mathbf{B}}{\partial t}\right)$ produces an electric field (**E**):

$$\oint_C \mathbf{E} \cdot \mathbf{dl} = -\int_S \frac{\partial \mathbf{B}}{\partial t} \cdot \mathbf{ds} \tag{5.43}$$

With Equation 5.27 (Stokes) we rewrite this expression as:

$$\int_S (\nabla \times \mathbf{E}) \cdot \mathbf{ds} = -\int_S \frac{\partial \mathbf{B}}{\partial t} \cdot \mathbf{ds} \tag{5.44}$$

or:

$$\nabla \times \mathbf{E} = -\frac{\partial \mathbf{B}}{\partial t} \tag{5.45}$$

Finally, we introduce the continuity equation for the conservation of charge. Say that in a time dt, a charge dQ is lost from volume V; because of this flux of charge, a current density $\mathbf{J}$ leaves the surface element ds. We use the divergence theorem to convert the integral over the surface S into an integral over the volume V. We assume that the volume does not change, so only the charge density ρ in the volume changes. This leads to the continuity equation:

$$\frac{\partial \rho_V}{\partial t} + \nabla \cdot \mathbf{J} = 0 \tag{5.46}$$

From Equation 5.46 the current density flux flowing out of a closed surface equals the rate of decrease of the positive charge density.

Example 5.1: Place some charges on a conductor with a conductivity σ; how long does it take for the charges to spread? From the continuity relation (Equation 5.46) we have $\nabla \cdot \mathbf{J} = -\frac{\partial \rho_V}{\partial t}$. In the conductor we have Ohm's law stating $\mathbf{J} = \sigma\mathbf{E}$

(Equation 2.2), and because $\sigma\nabla \cdot \mathbf{E} = -\frac{\partial \rho_V}{\partial t}$ and using Poisson's equation $\left(\nabla \cdot \mathbf{E} = \frac{\rho}{\varepsilon}\right)$, the solution is:

$$\rho = \rho_0 e^{-\frac{t}{\tau}} \tag{5.47}$$

with the time constant $\tau = \varepsilon/\sigma$ With typical values for ε and σ one obtains a τ of 1.5 10^{-19} s. Thus, charge density decays exponentially with time. For a very good conductor, charges flow instantly to the surface to form a surface charge density and (for time-varying fields) a surface current is established. Inside a perfect conductor ($\sigma \to \infty$) and $\mathbf{E} = \mathbf{H} = 0$.

Maxwell's Elegant Summary

It was in 1864 that Maxwell summarized the above set of equations for all frequency ranges and size scales larger than atoms. He was the first to show that the four equations describing electric and magnetic fields could be rearranged to form a wave equation, with a speed of propagation equal to the speed of light. Maxwell also corrected Ampère's law by introducing an additional term for the displacement current (changing electric fields act like currents, likewise producing magnetic fields; see above). The elegant modern mathematical formulation introduced here is by Oliver Heaviside (1850–1925), who in 1884 reformulated Maxwell's equations using vector calculus. His more symmetric mathematical representation reinforces the physical symmetries between the magnetic and electrical fields.

Maxwell's equations, as remarked earlier, can be transformed from differential format to integral representation, and vice versa, by using the theorems from Gauss and Stokes. Remarkably, as we will see below, Maxwell's equations are perfectly consistent with the transformations of special relativity.

Maxwell's equations in the absence of electric charge density ρ (in units of C/m³ = 0) and current J (in units of A/m²) in a vacuum in differential and integral form are:

$$\nabla \times \mathbf{B} = \mu_0 \varepsilon_0 \frac{\partial \mathbf{E}}{\partial t} \tag{5.48a}$$

$$\oint_C \mathbf{B} \cdot d\mathbf{l} = \mu_0 \varepsilon_0 \int_S \frac{\partial \mathbf{E}}{\partial t} \cdot d\mathbf{s} \quad \text{Ampere} \tag{5.48b}$$

$$\nabla \cdot \mathbf{E} = 0 \tag{5.49a}$$

$$\int_S \mathbf{E} \cdot d\mathbf{s} = 0 \quad \text{Electrical Gauss} \tag{5.49b}$$

$$\nabla \cdot \mathbf{B} = 0 \tag{5.50a}$$

$$\oint_S \mathbf{B} \cdot d\mathbf{s} = 0 \quad \text{Magnetic Gauss} \tag{5.50b}$$

$$\nabla \times \mathbf{E} = -\frac{\partial \mathbf{B}}{\partial t} \tag{5.51a}$$

$$\oint_C \mathbf{E} \cdot d\mathbf{l} = -\int_S \frac{\partial \mathbf{B}}{\partial t} \cdot d\mathbf{s} \quad \text{Faraday} \tag{5.51b}$$

where **E** (in V/m) is the electrical field strength and **B** (in Tesla, T, or $kgA^{-1}s^{-2}$) the magnetic flux density, and μ_0 and ε_0 the permeability and electric permittivity of free space, respectively; $d\mathbf{l}$ is a line element, and $d\mathbf{s}$ a surface element. The permittivity of free space, ε_0, is 8.854 pF/m. The permeability of vacuum, μ_0, is 1.257 10^{-6} H m^{-1}. Vacuum is a linear, homogeneous, isotropic, and dispersionless medium. Except for a multiplicative scalar, electric and magnetic fields appear in these equations with remarkable symmetry, and this mathematical symmetry implies physical symmetry. We note here that the four Maxwell equations are not independent. In fact, the two divergence equations can be derived from the two curl equations.

The time-varying Maxwell's equations, in a medium and in the presence of charges and currents, in their differential and integrated forms are given as:

$$\nabla \times \mathbf{H} = \mathbf{J} + \frac{\partial \mathbf{D}}{\partial t} \tag{5.48c}$$

$$\oint_C \mathbf{H} \cdot d\mathbf{l} = \int_S \left(\mathbf{J} + \frac{\partial \mathbf{D}}{\partial t}\right) \cdot d\mathbf{s} \quad \text{Ampere} \tag{5.48d}$$

$$\nabla \cdot \mathbf{D} = \rho_V \tag{5.49c}$$

$$\int_S \mathbf{D} \cdot d\mathbf{s} = Q \quad \text{Electrical Gauss} \tag{5.49d}$$

$$\nabla \cdot \mathbf{B} = 0 \tag{5.50c}$$

$$\oint_S \mathbf{B} \cdot d\mathbf{s} = 0 \quad \text{Magnetic Gauss} \quad (5.50d)$$

$$\nabla \times \mathbf{E} = -\frac{\partial \mathbf{B}}{\partial t} \quad (5.51c)$$

$$\oint_C \mathbf{E} \cdot d\mathbf{l} = -\int_S \frac{\partial \mathbf{B}}{\partial t} \cdot d\mathbf{s} \quad \text{Faraday} \quad (5.51d)$$

where **D** [in coulombs per square meter (C/m^2)] is the electric displacement, **H** (in A/m) is the magnetic field strength, **J** is current density (A/m^2), Q (C) is the total charge, and ρ_V is the volume density of free electric charges (C/m^3). Note that no constants such as μ_0, ε_0, μ, ε, c, χ … appear when the equations are written in this way. The divergence equations basically say that flux lines start and end on charges or poles:

1. Diverging D fields relate to charges: $\nabla \cdot \mathbf{D} = \rho_V$
2. Diverging magnetic flux does not exist: $\nabla \cdot \mathbf{B} = 0$

The curl equations state that changes in magnetic fluxes give rise to electrical fields and changes in currents give rise to H fields or changing E field results in changing H field, which results in changing E field.... (Figure 5.26):

3. Rotating E fields relate to changing currents: $\nabla \times \mathbf{E} = -\frac{\partial \mathbf{B}}{\partial t}$
4. Rotating H flux lines relate to changing D, and several currents could be added (e.g., magnetization): $\nabla \times \mathbf{H} = \mathbf{J} + \frac{\partial \mathbf{D}}{\partial t}$

If we now consider solutions for time-harmonic electric and magnetic fields (e.g., sinusoidal) and replace the time derivative $\partial/\partial t$ by $-i\omega$ [because $\frac{\partial}{\partial t}(e^{-i\omega t}) = -i\omega e^{-i\omega t}$], we arrive at the following Maxwell expressions:

$$\nabla \times \mathbf{H} = \mathbf{J} - i\omega \mathbf{D} \quad (5.48e)$$

$$\oint_C \mathbf{H} \cdot d\mathbf{l} = \int_S \mathbf{J} - i\omega \mathbf{D} \quad \text{Ampere} \quad (5.48f)$$

$$\nabla \cdot \mathbf{D} = \rho_V \quad (5.49e)$$

$$\int_S \mathbf{D} \cdot d\mathbf{s} = Q \quad \text{Electrical Gauss} \quad (5.49f)$$

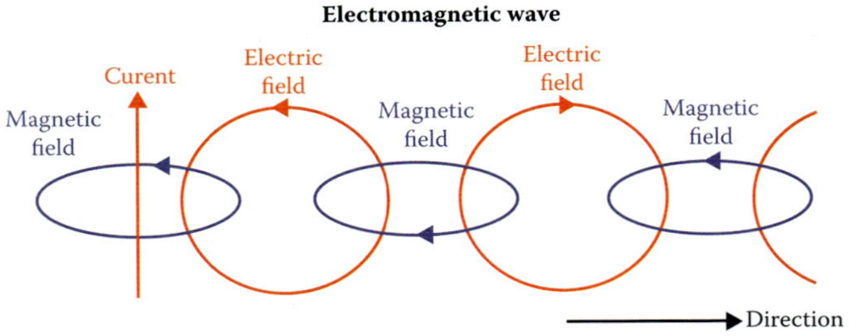

FIGURE 5.26 Changing E field results in changing H field, which results in changing E field.

$$\nabla \cdot \mathbf{B} = 0 \quad (5.50e)$$

$$\oint_S \mathbf{B} \cdot d\mathbf{s} = 0 \quad \text{Magnetic Gauss} \quad (5.50f)$$

$$\nabla \times \mathbf{E} = +i\omega \mathbf{B} \quad (5.51e)$$

$$\oint_C \mathbf{E} \cdot d\mathbf{l} = +i\omega \int_S \mathbf{B} \cdot d\mathbf{s} \quad \text{Faraday} \quad (5.51f)$$

Boltzmann was so impressed with Maxwell's equation that he used a quotation from Goethe to describe them: "*War es ein Gott der diese Zeichen schrieb*?" (Is it a God who has written these signs?). These equations, together with the continuity equation (Equation 5.46), the Lorentz force* (Equation 3.64), and the constitutive equations for ε, μ, and σ, form the foundation of all macroscopic electromagnetic phenomena.

In the case of no time variation ($\partial/\partial t = 0$), the Maxwell equations describe electrostatics and magnetostatics as summarized in Table 5.1.

In the next section we will find that Maxwell's equations allow for solutions that have the form of propagating plane waves, and that the wavelength is affected by the amount of polarization **P** of the material, leading to a refractive index n.

Plane Wave Solutions of Maxwell's Equations

Plane Wave Propagation

Earlier we saw how electromagnetic (EM) plane waves are observed when moving far enough away from the source, just like dropping a rock in a pond and looking at waves a few hundred feet away from the

* An electric charge moving in the presence of an electric and a magnetic field experiences a vector force per unit charge that is equal to the sum of the electric field and the cross product of the velocity with the magnetic field.

Table 5.1 Electrostatics (Capacitor Equations) and Magnetostatics (Inductor Equations)

Magnetostatics	
$\nabla \cdot \mathbf{D} = \rho_V$	$\nabla \cdot \mathbf{B} = 0$
$\nabla \times \mathbf{E} = 0$	$\nabla \times \mathbf{H} = \mathbf{J}$
$\oint_S \mathbf{D} \cdot ds = Q$	$\oint_S \mathbf{B} \cdot ds = 0$
$\oint_C \mathbf{E} \cdot dl = 0$	$\oint_C \mathbf{H} \cdot dl = I$
Electrical capacitance (in Farad) $C = Q/V$	Electric inductance (in Henry) $L = \Phi/I$*

*Φ is magnetic flux, and I is current.

impact point so that the wavefronts are essentially flat. When close enough to the EM source the radiated field must be considered mathematically spherical rather than planar. The first situation is called *far field*, and the second is *near field* (Figures 5.4 and 5.5). In Chapter 2, we defined a plane wave,* $e^{i(\mathbf{k}r-\omega t)}$, as a wave that is constant in a plane perpendicular to **k** (in rad/m) and is periodic parallel to it. The wave number or wave vector **k** measures the number of wavelengths in a complete cycle; thus, it has a value of $2\pi/\lambda$, and $\omega = 2\pi v = c\mathbf{k}$ (free space) is the period of the wave. In the plane wave, equation **r** is the 3D position vector. A bit simpler: a plane wave is a continuous wave (CW) whose amplitude and phase are constant in the directions transverse to the propagation direction.

We will now embark on a discussion of the interaction of light with materials. In optical experiments one measures reflectivity, transmission coefficient, luminescence, or light scattering to learn a great deal about energy band structure, impurities, phonons, defects, etc. In particular, optical measurements are used to determine the complex dielectric or complex conductivity, which is directly related to band structure.

Wave Equations for a Medium without Charges or Currents

We now derive the wave equations for plane wave propagation in a simple linear isotropic, homogeneous, nonconducting medium characterized by ε and μ without any charges or currents present (source-free region, i.e., a region where ρ_V and **J** both are zero) from the Maxwell equations. Taking the curl of Equation 5.51c leads to the homogenous vector wave equations (Equations 5.53 and 5.54) for electromagnetic (EM) radiation in such a medium we obtain:

$$\nabla \times (\nabla \times \mathbf{E}) = -\frac{\partial(\nabla \times \mathbf{B})}{\partial t} \tag{5.52}$$

L.H.S: $\nabla(\nabla \cdot \mathbf{E}) - \nabla^2\mathbf{E} = -\nabla^2\mathbf{E}$
($\nabla \cdot \mathbf{E} = \mathbf{0}$; Equation 5.49a)

R.H.S: $-\mu\varepsilon\frac{\partial^2\mathbf{E}}{\partial^2 t}$ (Equation 5.48c with
$\mathbf{J} = 0$ and $\mathbf{D} = \varepsilon\mathbf{E}$ and $\mathbf{B} = \mu\mathbf{H}$)

$$\text{or } \nabla^2\mathbf{E} = \mu\varepsilon\frac{\partial^2\mathbf{E}}{\partial^2 t} \tag{5.53}$$

and similarly

$$\nabla^2\mathbf{B} = \mu\varepsilon\frac{\partial^2\mathbf{B}}{\partial^2 t} \tag{5.54}$$

For the left-hand side (L.H.S.) of Equation 5.52 we used the "BAC-CAB" rule.† Equations 5.53 and 5.54 represent wave equations for electromagnetic (EM) waves in bulk materials. In free space, μ and ε in these equations are replaced by μ_0 and ε_0. The Laplacian ∇^2 operates on each component of **E** and **B**, so that the two vector equations (Equations 5.53 and 5.54) represent six 1D, homogeneous scalar wave equations. One of these expressions, the E_x component, in Cartesian coordinates, is written out as:

$$\frac{\partial^2 E_x}{\partial x^2} + \frac{\partial^2 E_x}{\partial y^2} + \frac{\partial^2 E_x}{\partial z^2} = \varepsilon\mu\frac{\partial^2 E_x}{\partial t^2} \tag{5.55}$$

Therefore, each component of the EM field (E_x, E_y, E_z, B_x, B_y, B_z) obeys a generic scalar differential wave equation of the type:

$$\frac{\partial^2\Psi}{\partial x^2} + \frac{\partial^2\Psi}{\partial y^2} + \frac{\partial^2\Psi}{\partial z^2} = \frac{1}{v^2}\frac{\partial^2\Psi}{\partial t^2} \tag{5.56}$$

provided that the following important identity holds:

$$v = \frac{1}{\sqrt{\varepsilon\mu}} \tag{5.20}$$

* The complex exponential representation for periodic functions is convenient for adding waves, taking derivatives of wave functions, and so on. It is equivalent to a linear combination of a sine and cosine function because $e^{i\theta} = \cos\theta + i\sin\theta$ (Euler's formula).

† We are taking advantage in this derivation of the "BAC-CAB" rule: $A \times (B \times C) = B(A \cdot C) - C(A \cdot B)$, or in the current case: $\nabla \times (\nabla \times \mathbf{E}) = \nabla(\nabla \cdot \mathbf{E}) - (\nabla \cdot \nabla)\mathbf{E} = -\nabla^2 \mathbf{E}$.

The velocity v, the propagation speed of the wave, is solely determined by the medium. For a wave traveling in free space we substitute the values for μ_0 and ε_0 in Equation 5.20 and calculate the speed for all EM waves as v = c = 2.998 10^8 m/s. At the time of Maxwell, the values for μ_0, ε_0, and c were uncertain enough that the correspondence between c and Equation 5.20 was not all that definitive. However, it was close enough that Maxwell wrote in 1864: "This velocity is so nearly that of light that it seems we have strong reason to conclude that light itself (including radiant heat and other radiations) is an electromagnetic disturbance in the form of waves propagated through the electromagnetic field according to electromagnetic laws." It is largely for this reason that this whole set of equations is named after Maxwell. When Heinrich Hertz (1857–1894) produced low-frequency radio waves (low-frequency light) and demonstrated that these waves were identical in all respects to light, save frequency, he confirmed Maxwell's hypothesis. It was also Hertz who, in 1887, discovered the photoelectric effect (see Chapter 3). Maxwell's equations show that electric and magnetic forces travel at a definite predicted speed—this is not instantaneous "action at a distance." James Clerk Maxwell's *Treatise on Electricity & Magnetism* (1873) is the last word on classical electricity and magnetism and ranks with Newton's work as one of the great accomplishments of physics. Newton's and Maxwell's equations describe all "classical physics." In Chapter 3 we introduced Bohr's correspondence principle, according to which the predictions of the quantum theory must also cover the predictions of the classical theories of physics. In other words, the above equations derived here from Maxwell's equations should also be a solution of Schrödinger's equation. We demonstrated that this is so in Chapter 3, "Solution of the Wave Equation for Free Particles."

Wave Equations for a Medium with Charges and Currents

To obtain the most general wave equations—in the presence of sources (charges, ρ) and a current (J) in a medium—we first combine the constitutive relations with the general time-varying Maxwell's equations to obtain the following two equivalent expressions (using the BAC-CAB rule again):

$$\nabla \times (\nabla \times \mathbf{E}) = \nabla(\nabla \cdot \mathbf{E}) - \nabla^2\mathbf{E}$$

$$= -\mu\frac{\partial \mathbf{J}}{\partial t} - \mu\frac{\partial^2 \mathbf{D}}{\partial t^2} = -\mu\sigma\frac{\partial \mathbf{E}}{\partial t} - \mu\varepsilon\frac{\partial^2 \mathbf{E}}{\partial t^2}$$

$$\nabla^2\mathbf{E} - \mu\sigma\frac{\partial \mathbf{E}}{\partial t} - \mu\varepsilon\frac{\partial^2 \mathbf{E}}{\partial t^2} = \nabla\left(\frac{\rho_v}{\varepsilon}\right) \tag{5.57}$$

and similarly:

$$\nabla^2\mathbf{H} - \mu\sigma\frac{\partial \mathbf{H}}{\partial t} - \mu\varepsilon\frac{\partial^2 \mathbf{H}}{\partial t^2} = 0 \tag{5.58}$$

In solving these equations, one often invokes the electric scalar and the magnetic vector potential. The magnetic vector potential **A** (in Weber per meter) is related to **B** as:

$$\mathbf{B} = \nabla \times \mathbf{A} \tag{5.59}$$

If one can find **A** for a given current distribution, then **B** can be obtained by taking the differential (curl) from **A**. This is similar to obtaining the electrical field **E** from the scalar electrical potential V ($\mathbf{E} = -\nabla V$). Substituting Equation 5.59 into the differential Faraday expression (Equation 5.51a) gives:

$$\begin{aligned} &\nabla \times \mathbf{E} = -\frac{\partial}{\partial t}(\nabla \times \mathbf{A}) \text{ or} \\ &\nabla \times (\mathbf{E} + \frac{\partial \mathbf{A}}{\partial t}) = 0 \end{aligned} \tag{5.60}$$

The electric field intensity stems from both charge accumulations (through the $-\nabla V$ term) and from the time-varying magnetic field (through the $-\frac{\partial \mathbf{A}}{\partial t}$ term), or:

$$\mathbf{E} = -\frac{\partial \mathbf{A}}{\partial t} - \nabla V \tag{5.61}$$

For the static case (with $\frac{\partial \mathbf{A}}{\partial t} = 0$), this reduces back to the familiar $\mathbf{E} = -\nabla V$, so that we can obtain **E** from V alone and **B** from Equation 5.59. With varying fields, **E** depends on both V and **A**. The inhomogeneous vector potential wave equation to be solved then becomes:

$$-\nabla^2 \mathbf{A} + \mu\sigma\frac{\partial \mathbf{A}}{\partial t} + \mu\varepsilon\frac{\partial^2 \mathbf{A}}{\partial t^2} = -\mu\sigma\nabla V \quad (5.62)$$

Using the constitutive equations and substituting Equations 5.59 and 5.61 into 5.48e, we obtain:

$$\nabla \times \nabla \times \mathbf{A} = \mu\mathbf{J} + \mu\varepsilon\frac{\partial}{\partial t}\left(-\frac{\partial \mathbf{A}}{\partial t} - \nabla V\right) \quad (5.63)$$

or also:

$$\nabla^2 \mathbf{A} - \mu\varepsilon\frac{\partial^2 \mathbf{A}}{\partial t^2} = -\mu\mathbf{J} + \nabla\left(\nabla\cdot\mathbf{A} + \mu\varepsilon\frac{\partial V}{\partial t}\right) \quad (5.64)$$

By choosing the divergence of **A** such that:

$$\nabla\cdot\mathbf{A} + \mu\varepsilon\frac{\partial V}{\partial t} = 0 \quad (5.65)$$

which is the so-called *Lorentz condition for potentials* (for static fields this condition becomes $\nabla\cdot\mathbf{A} = \mathbf{0}$). The Lorentz condition enables us to simplify Equation 5.64 to:

$$\nabla^2 \mathbf{A} - \mu\varepsilon\frac{\partial^2 \mathbf{A}}{\partial t^2} = -\mu\mathbf{J} \quad (5.66)$$

This is the nonhomogeneous wave equation for vector potential **A**; solutions are waves traveling with a velocity v equal to $\frac{1}{\sqrt{\mu\varepsilon}}$. The corresponding wave equation for the scalar potential V is obtained by substituting Equation 5.61 in Equation 5.49c, or:

$$-\nabla\cdot\varepsilon\left(\nabla V + \frac{\partial \mathbf{A}}{\partial t}\right) = \rho_v \quad (5.67)$$

With ε constant this can be written out as:

$$\nabla^2 V + \frac{\partial}{\partial t}(\nabla\cdot\mathbf{A}) = -\frac{\rho_v}{\varepsilon} \quad (5.68)$$

With the Lorentz condition (Equation 5.65) applied, this leads to:

$$\nabla^2 V - \mu\varepsilon\frac{\partial^2 V}{\partial t^2} = -\frac{\rho_v}{\varepsilon} \quad (5.69)$$

This result represents nonhomogeneous wave equation for the scalar potential *V*. Both wave equations (Equations 5.66 and 5.69) reduce to the Poisson's equations in static cases.

The above derived wave equations are the basis for modeling electromagnetic fields in microelectromechanical systems (MEMS) and nanoelectromechanical systems (NEMS).

For more details on deriving and working with Maxwell's equations, consult David K. Cheng's *Field and Wave Electromagnetics*[1] and Minoru Taya's *Electronic Composites*.[2]

Plane Wave Solutions for Media without Charges or Currents

The Overall Solutions Are Plane Waves: $\mathbf{E}(\mathbf{r},t) = \mathbf{E}_m e^{i(\mathbf{k}\cdot\mathbf{r}-\omega t)}$ *and* $\mathbf{B}(\mathbf{r},t) = \mathbf{B}_m e^{i(\mathbf{k}\cdot\mathbf{r}-\omega t)}$ The solutions of the wave equations for plane wave propagation in a simple linear isotropic, homogeneous, nonconducting medium (ρ_V and **J** both are zero) characterized by ε and μ (Equations 5.53 and 5.54) in one dimension (say with the *E* field propagating along the *z*-axis, the *B* field along *y*, and the propagation vector in the *x*-direction) are:

$$\mathbf{E}(\mathbf{r},t) = \mathbf{E}_m e^{i(k_x\cdot x-\omega t)} \text{ and } \mathbf{B} = \mathbf{B}_m e^{i(k\cdot x-\omega t)} \quad (5.70)$$

Rewriting these expressions in sines and cosines instead of the complex variable form ($e^x = \cos x + i \sin x$), one obtains as the simplest form of a light wave:

$$\mathbf{E} = \mathbf{E}_m \sin(\mathbf{k}_x x - \omega t) \text{ and } \mathbf{B} = \mathbf{B}_m \sin(\mathbf{k}_x x - \omega t) \quad (5.71)$$

Generalizing for 3D, Equation 5.70 converts to:

$$\mathbf{E}(\mathbf{r},t) = \mathbf{E}_m e^{i(\mathbf{k}\cdot\mathbf{r}-\omega t)} \text{ and } \mathbf{B}(\mathbf{r},t) = \mathbf{B}_m e^{i(\mathbf{k}\cdot\mathbf{r}-\omega t)} \quad (5.72)$$

with **r** the 3D position vector and **k** a complex propagation constant, with the real part of it corresponding to the wave vector and the imaginary part corresponding to the absorption (analogous to our treatment of x-rays in Chapter 2). The complex propagation constant **k** is obtained by substitution of this solution in the wave equations (see Equations 5.53 and 5.54).

It should be noted that plane waves are not the only solutions to Maxwell's equations; the differential wave equation allows many solutions, including spherical waves.

Constraints to the Plane Wave Solutions: **E** ⊥ **k** *and* **B** ⊥ **k** *(I)* Let us assess what we have so far: the source-free Maxwell equations (Equations 5.48a–5.51a) were manipulated in such a way as to obtain the vector wave Equations 5.53 and 5.54, and solutions for these equations are given by $\mathbf{E}(\mathbf{r},t) = \mathbf{E}_m e^{i(\mathbf{k}\cdot\mathbf{r}-\omega t)}$ and $\mathbf{B}(\mathbf{r},t) = \mathbf{B}_m e^{i(\mathbf{k}\cdot\mathbf{r}-\omega t)}$ (Equation 5.72). But we are not done yet: in addition to satisfying the vector wave equation, we must also verify these solutions satisfy the other Maxwell equations. We can write down lots of functions that satisfy the wave equations but that are not valid electromagnetic fields for the given boundary conditions (ρ_V and **J** both are zero). We first plug these proposed solutions into the divergence Equations 5.49a and 5.50a to find the following additional constraints:

$$\nabla \cdot \mathbf{E} = 0 \text{ (Equation 5.49a)} \rightarrow \mathbf{k}\cdot\mathbf{E} = 0 \qquad (5.73)$$

$$\nabla \cdot \mathbf{B} = 0 \text{ (Equation 5.50a)} \rightarrow \mathbf{k}\cdot\mathbf{B} = 0 \qquad (5.74)$$

In other words, the divergence equations require that the electric and magnetic fields are orthogonal to the propagation direction: **E** ⊥ **k** and **B**⊥ **k**. The **E** and **B** fields have no component of vibration in the direction of propagation: there are no longitudinal components, or **E** and **B** are exclusively transverse in nature.

Constraints to the Plane Wave Solutions: **E** ⊥ **B** ⊥ **k** *(II)* Let us now substitute these same solutions in the curl equations (Equations 5.48a and 5.51a) and establish how this adds the following constraints:

$$\nabla \times \mathbf{E} = -\frac{\partial \mathbf{B}}{\partial t} \text{ (Equation 5.51a)} \longrightarrow \mathbf{k}\times\mathbf{E} = \omega\mathbf{B} \qquad (5.75)$$

$$\nabla \times \mathbf{B} = \mu\varepsilon\frac{\partial \mathbf{E}}{\partial t} \text{ (Equation 5.48a)} \longrightarrow \mathbf{k}\times\mathbf{B} = \frac{\omega}{c^2}\mathbf{E} \qquad (5.76)$$

Consider $\nabla \times \mathbf{B} = \mu\varepsilon\frac{\partial \mathbf{E}}{\partial t}$ one of the curl equations: on the right-hand side, we have the time derivative of **E**, pointing in the same direction as **E** because the time derivative does not change the vector's direction. On the left, we have ∇ × **B**; this is perpendicular to **B** because the curl of any vector is orthogonal to that vector. Therefore, the first thing the curl equations tell us is that radiation's electric and magnetic fields are orthogonal to each other. Second, the curl equations also require that **E** × **B** is parallel to the propagation vector **k**. In other words, the symmetry of Maxwell's equations results in **E** ⊥ **B** ⊥ **k** as illustrated in Figure 5.1.

In electromagnetic waves, the **E** and **B** fields are obviously intimately linked. To test this further, let us try out the solutions, $\mathbf{E_x} = \mathbf{E}_m e^{i(\mathbf{k}\cdot\mathbf{x}-\omega t)}$ and $\mathbf{B}_y = \mathbf{B}_m e^{i(\mathbf{k}\cdot\mathbf{x}-\omega t)}$ [**E** along the *z*-axis, **B** along the *y*-axis, and propagation direction **k** along the *x*-axis $\left(\frac{\partial E_x}{\partial x} = 0\right)$] in Equation 5.51a, which leads to:

$$\begin{aligned} &\nabla \times \mathbf{E} = -\frac{\partial \mathbf{B}}{\partial t} \text{ or} \\ &\frac{\partial E_z}{\partial x} = -\frac{\partial \mathbf{B}_y}{\partial t} \text{ and thus } \mathbf{E}_m(ik)e^{i(\mathbf{k}\cdot\mathbf{x}-\omega t)} \\ &\quad = -\mathbf{B}_m(-i\omega)e^{i(\mathbf{k}\cdot\mathbf{x}-\omega t)} \\ &\text{or } \frac{E_m}{B_m} = \frac{\omega}{\mathbf{k}} \end{aligned} \qquad (5.77)$$

For planar waves, **k** = 2π/λ and ω = 2πv = v**k** with *v* being the speed of the wave, and because *E* and *B* are in phase we conclude that *E* and *B* are related at any point in space by the velocity of the propagating wave:

$$\frac{\mathrm{E}}{\mathbf{B}} = \frac{\omega}{\mathbf{k}} = \mathrm{v} \qquad (5.78)$$

This relates the wave vector **k**, relating to phase variations in space, to ω, the angular frequency, relating to phase variations in time. In the case of a wave in free space, v = c = 2.998 × 10⁸ m/s. Thus, boundary conditions for planar waves propagating in a medium without charges lead to Equation 5.78 expressing an important constraint for the proposed solution in a dispersion relation, i.e., a relation between the energy of the system and its corresponding momentum **k** (ω). In the case of free space, this is a linear function (ω = **k**c) with a slope *c*, the so-called *light line* (see Figure 5.27).

Summarizing, we conclude that in a source-free region for time-harmonic signals Maxwell's equations are reduced to two coupled curl equations, and these two curl equations combine to produce the wave equation for **E** or **B**. The eigenfunctions for these wave equations are plane waves described by propagation direction vectors called **k**-*vectors* that have a length of $2\pi/\lambda$ and result in wave velocity v (the **k**-vector is reciprocal to the space variation of the wavelength). The two curl equations have zero divergence, and this requires the vectors **E** and **B** to be perpendicular to the direction of propagation, whereas the **E** and **B** vectors must also be perpendicular to each other.

In the section below on electromagnetic waves at boundaries, we will learn that to solve electromagnetic problems involving adjacent regions of different constitutive parameters, it is necessary to establish the boundary conditions that the field vectors **E**, **D**, **B**, and **H** must satisfy at the interfaces between the different optical materials. These boundary conditions, we will see, are established by applying the integral form of Maxwell's equations to a small region at the interface of the contacting media.

Plane Wave Solutions for Media with Charges or Currents

We now substitute the time-harmonic electric field wave $\mathbf{E}(\mathbf{r},t) = \mathbf{E}_m e^{i(\mathbf{k}\cdot\mathbf{r}-\omega t)} = \mathbf{E}_m e^{i\mathbf{k}\cdot\mathbf{r}} e^{-i\omega\cdot t} = E_r(r)e^{-i\omega\cdot t}$ in the most general wave equation for media with charges and currents $\nabla^2\mathbf{E} - \mu\sigma\frac{\partial \mathbf{E}}{\partial t} - \mu\varepsilon\frac{\partial^2 \mathbf{E}}{\partial t^2}$ (Equation 5.57). We will assume here nonmagnetic materials with $\mu = \mu_0$ (remember that the magnetic permeability $\mu = \mu_0 \cdot \mu_r$, where μ_r is the dimensionless relative permeability, which is 1 in this case) to obtain:

$$\nabla^2\mathbf{E}(\mathbf{r}) = -i\omega\sigma\mathbf{E}(\mathbf{r}) - \varepsilon\omega^2\mathbf{E}(\mathbf{r})$$

$$\nabla^2\mathbf{E}(\mathbf{r}) = -\frac{\omega^2}{v^2}\varepsilon(\omega)\mathbf{E}(\mathbf{r}) \qquad (5.79)$$

with an effective dielectric contstant:

$$\varepsilon(\omega) = 1 + \frac{i\sigma}{\omega}$$

Substituting $E_r(\mathbf{r}) = \mathbf{E}_m e^{i\mathbf{k}\cdot\mathbf{r}}$ in Equation 5.79 leads to the following constraint expressed in a new dispersion equation:

$$\mathbf{k}^2 = \frac{\omega^2}{v^2}\varepsilon(\omega) \qquad (5.80)$$

This is a relation between the wave vector **k**, related to phase variations in space, and ω, the angular frequency, related to phase variations in time. Note that a higher temporal frequency yields a higher spatial frequency, and a large permittivity will result in a large **k** (short wavelength). The "slow-down" factor of an electromagnetic wave in a material relative to free space is by definition the refractive index $n = \frac{c}{v}$ (Equation 5.21). With $n = \frac{c}{v}$ we can rewrite Equation 5.80 in terms of the refractive index n and the velocity of light in free space c as:

$$\mathbf{k}^2 = \frac{n^2}{c^2}\omega^2 \text{ or}$$
$$|\mathbf{k}| = \frac{n}{c}\omega \qquad (5.81)$$

which in the case of free space with n = 1 simplifies back to $\omega = \mathbf{k}c$. Thus, the wave equation again imposes a constraint on the length of the **k**-vector. Below we will analyze this dispersion relation in terms of the phase velocity of a wave ($v_p = \omega/\mathbf{k} = \mathbf{E}/\mathbf{p}$) and the group velocity ($v_g = d\omega/d\mathbf{k} = d\mathbf{E}/d\mathbf{p}$) of that wave.

Characteristic Vacuum and Medium Impedance—η_0 and η

The vacuum impedance η_0 is a universal constant relating the magnitudes of the electric and magnetic fields of electromagnetic radiation traveling through free space. From Equation 5.75:

$$\eta_0 = \frac{|\mathbf{E}|}{|\mathbf{H}|} = \frac{\omega\mu_0}{|\mathbf{k}|} = \mu_0 c = \frac{|\mathbf{k}|}{\omega\varepsilon} = \frac{1}{\varepsilon_0 c} = \sqrt{\frac{\mu_0}{\varepsilon_0}} \qquad (5.82)$$

which equals 376.73 ohms. More general, for a uniform plane wave the ratio of the magnitudes of **E** and **H** is the intrinsic impedance of the medium and is designated η. Again from Equation 5.75 it follows:

$$\eta = \frac{|\mathbf{E}|}{|\mathbf{H}|} = \frac{\omega\mu}{|\mathbf{k}|} = \frac{|\mathbf{k}|}{\omega\varepsilon} = \sqrt{\frac{\mu}{\varepsilon}} \qquad (5.83)$$

Like the velocity of propagation, the intrinsic impedance is independent of the source and only determined by the properties of the medium.

Dispersion Relations in Vacuum and in Media

In Vacuum

In physics, the dispersion relation is the relation between the energy of a system and its corresponding momentum. Here we analyze the dispersion relation $\omega(\mathbf{k})$ from Equation 5.81 in terms of the phase velocity of a wave ($v_p = \omega/\mathbf{k} = E/\mathbf{p}$) and the group velocity ($v_g = d\omega/d\mathbf{k} = dE/d\mathbf{p}$).

Remember that the phase velocity of a wave ($v_p = \omega/\mathbf{k} = E/\mathbf{p}$) is not necessarily the same as the group velocity ($v_g = d\omega/d\mathbf{k} = dE/d\mathbf{p}$) of that wave. Two velocities must be defined when dealing with multiple waves, with the phase velocity corresponding to the speed of the high-frequency components and the group velocity to the speed of a localized envelope. Also, the phase velocity of the wave is the rate at which the phase of that wave propagates in space, or it is the velocity at which the phase of any one frequency component of the wave propagates and the group velocity we saw earlier also corresponds to the light "particle" speed (Chapter 3).

Because the group velocity v_g is given as $d\omega/d\mathbf{k}$, we calculate:

$$v_g = \frac{d\omega}{d\mathbf{k}} = \frac{d}{d\mathbf{k}}(\mathbf{k}v_p) = v_p + \mathbf{k}\frac{dv_p}{d\mathbf{k}} \tag{5.84}$$

In a nondispersive medium such as a vacuum $\frac{dv_P}{d\mathbf{k}} = 0$, the group and phase velocity is the same ($v_p = v_g$), or:

$$v_p = \frac{\omega}{\mathbf{k}} = v_g = \frac{d\omega}{d\mathbf{k}} = c \tag{5.85}$$

This is of course the same result we obtained earlier as a constraint when substituting a planar wave $[\mathbf{E}(\mathbf{r},t) = \mathbf{E}_m e^{i(\mathbf{k}\cdot\mathbf{r}-\omega t)}]$ in the wave equation for free space $\nabla^2\mathbf{E} = \mu\varepsilon\frac{\partial^2\mathbf{E}}{\partial^2 t}$ (Equation 5.53). Thus, the dispersion relation—the relation between the energy of the system and its corresponding momentum $\mathbf{k}(\omega)$—is a linear function of ω with a slope $c = v_g = v_p$ in the case of vacuum as shown in Figure 5.27.

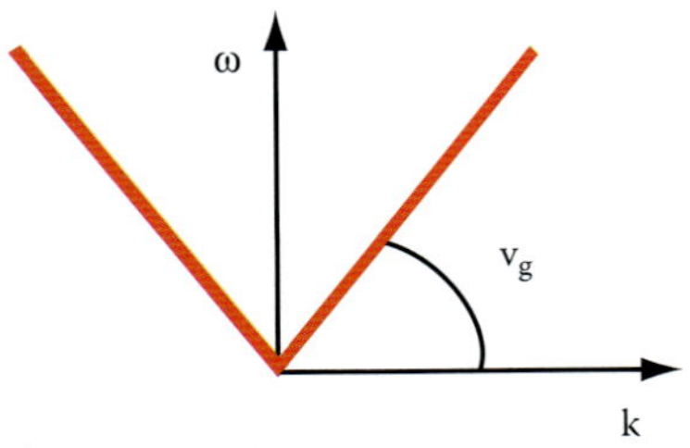

FIGURE 5.27 The dispersion relation—the relation between the energy of the system and its corresponding momentum $\mathbf{k}(\omega)$—is a linear function of ω in the case of vacuum with a slope $c = v_g = v_p$.

In a Medium

Dispersion occurs when the phase velocity v_p depends on $\mathbf{k}$ (or λ), i.e., the group velocity does not equal the phase velocity (see Equation 5.85):

$$\frac{dv_p}{d\mathbf{k}} \neq 0 \text{ or } v_g \neq v_p \tag{5.86}$$

A wave packet or group velocity with $v_g = d\omega/d\mathbf{k}$ less than the phase velocity $v_p = \omega/\mathbf{k}$ is illustrated in Figure 5.28 for a case where $v_g = v_p/2$.

We can rewrite the group velocity ($v_g = d\omega/d\mathbf{k}$) of light in a medium also in terms of refractive index n and the velocity of light c. For the group velocity we then derive for the most general case:

$$v_g = \frac{d\omega}{d\mathbf{k}} = \frac{c}{n} = \frac{c}{\sqrt{\varepsilon\mu}} \tag{5.87}$$

where we used the fact that $n(\omega) = \frac{c}{v_g}$ and $n(\omega)^2 = \varepsilon\mu$ (Equations 5.21 and 5.22), and the dispersion relation then becomes:

$$\mathbf{k}(\omega) = \frac{\omega}{v_g} = \frac{n(\omega)\omega}{c} = \frac{\omega}{c}\sqrt{\varepsilon\mu} \tag{5.88}$$

We will use this important equation to construct a μ-ε *quadrant* to classify all materials according to their magnetic permeability μ and their dielectric constant (or permittivity) ε.

From Equation 5.88 we can also calculate the group velocity v_g as a function of the refractive index:

$$v_g = \frac{c}{n + \frac{\omega dn}{d\omega}} = \frac{c/n}{1 + \frac{\omega dn}{n d\omega}} \tag{5.89}$$

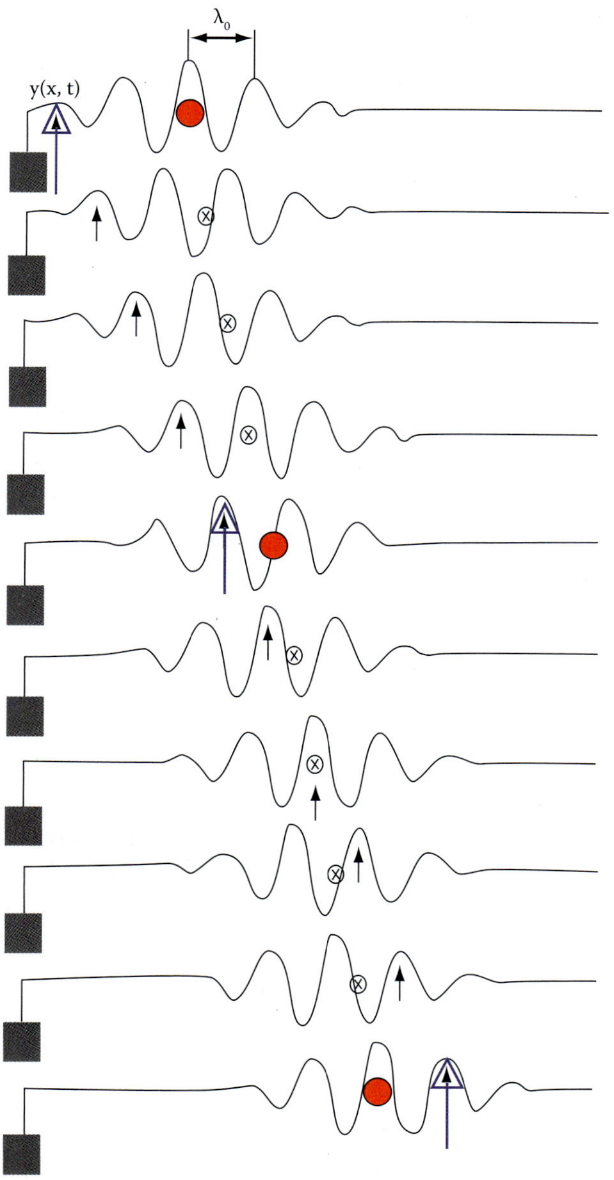

FIGURE 5.28 A wave packet or group velocity with $v_g = d\omega/d\mathbf{k}$ less than the phase velocity $v_p = \omega/\mathbf{k}$ with $v_g = v_p/2$. Phase velocity v_p is marked by blue arrows, and the group velocity v_g is marked with red dots.

and finally:

$$v_g = \frac{v_p}{1 + \frac{\omega dn}{n d\omega}} \quad (5.90)$$

The group velocity v_g equals the phase velocity when $dn/d\omega = 0$, such as in vacuum. Otherwise, because n usually increases with ω, $dn/d\omega > 0$ and v_g is less than the phase velocity v_p. This is illustrated in Figure 5.29, where we plot the refractive index as a function of ω. We see that the refractive index is positive for most frequencies ($v_g < v_p$). The exception is in regions of anomalous dispersion, regions that are not only dispersive but also absorbing, that is, near a resonance. To explain these anomalous regions we will introduce the Lorentz model further below.

We can summarize the frequency dependence of the refractive index n shown here as follows:

- Low frequencies: All transitions contribute their low-frequency response $\Rightarrow$ high n.
- Between resonances: n slowly increases = "normal dispersion."
- Close to resonances: n rapidly decreases = "anomalous dispersion."
- Above resonance, contribution vanishes $\Rightarrow n(\omega)$ mostly because of higher resonances.
- At high frequencies (~x-ray wavelengths) n approaches 1.

We more often think of the refractive index in terms of wavelength, so let us write the group velocity also in terms of the vacuum wavelength λ_0. Using the chain rule we can write:

$$\frac{dn}{d\omega} = \frac{dn}{d\lambda_0}\frac{d\lambda_0}{d\omega} \quad (5.91)$$

Now $\lambda_0 = 2\pi c/\omega$ and:

$$\frac{d\lambda_0}{d\omega} = \frac{-2\pi c}{\omega^2} = \frac{-2\pi c}{\left(\frac{2\pi c}{\lambda_0}\right)^2} = \frac{-\lambda_0^2}{2\pi c} \quad (5.92)$$

Substituting this in Equation 5.89, we obtain:

$$v_g = \frac{c}{n + \frac{\omega dn}{d\omega}} = \frac{c/n}{1 - \frac{\lambda_0 dn}{n d\lambda_0}} \quad (5.93)$$

Right- and Left-Handed Materials (Metamaterials)

From Maxwell's equations we derived that for planar waves with a positive μ and ε, the vectors **E** and **H** and **k** do constitute a right set of vectors and thus one defines right-handed materials (RHMs) as illustrated in Figure 5.30a. The direction of $\mathbf{E} \times \mathbf{H}$ is called the *beam* or *ray direction*.

In vacuum and for most materials the "right-hand rule" relates **E**, **H**, and **k** because normally

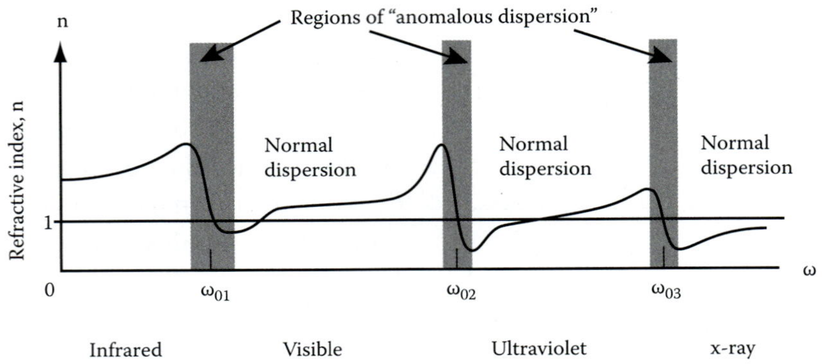

FIGURE 5.29 Refractive index n as a function of frequency ω: $n(\omega)$. The refractive index is positive, so $v_g < v_p$ for most frequencies! The exception is in regions of anomalous dispersion, absorbing regions near a resonance.

$\mu > 0$ and $\varepsilon > 0$. But in 1968, Victor G. Veselago postulated that if the permittivity ε and permeability μ of materials are negative simultaneously, there is no violation of any fundamental physical rules, and the materials will have special optical properties.[3] For a material with μ and ε negative, the vectors **E** and **H** and **k** form a left set of vectors, and one refers to them as left-handed materials (LHMs) (Figure 5.30b). For left-handed materials $n(\omega)^2 = \varepsilon\mu$ (Equation 5.22) is modified to:

$$n(\omega) = \pm\sqrt{\varepsilon\mu} \text{ which for } \varepsilon < 0 \text{ and } \mu < 0$$
$$n(\omega) = -\sqrt{\varepsilon\mu} \tag{5.94}$$

So we are forced by Maxwell's equations to choose a negative square root for the refractive index. Because no naturally occurring material or compound has ever been demonstrated with both ε and μ negative, Veselago wondered whether this apparent asymmetry in material properties was just happenstance or perhaps had a more fundamental origin. From 1996 onward, strong theoretical and experimental evidence started emerging that left-handed materials, now called *metamaterials*, could be fabricated (at least in the RF range, not yet in the optical range). In Figure 5.17 we show a picture of a drinking straw in a glass with a negative refractive index water n = −1.3 and observe an apparent bending of the straw in the opposite direction of what we are used to in the case of "normal" water. We will detail metamaterials further below. Whereas in conventional materials properties derive from their constituent atoms, in metamaterials properties derive from their constituent units. These units can be engineered as we please.

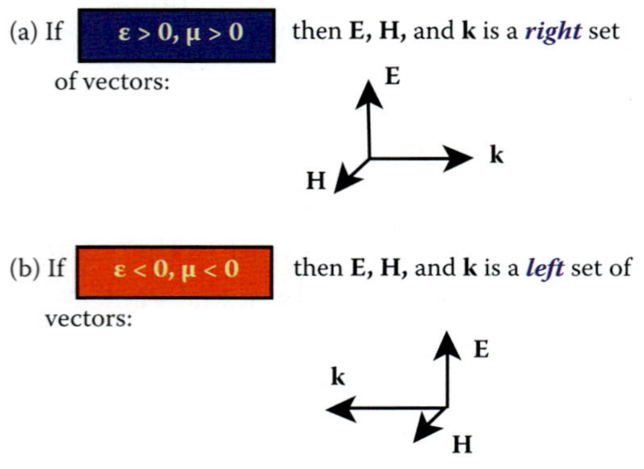

FIGURE 5.30 (a) Right-handed (RHM) and (b) left-handed materials (LHM).

Birefringence—Double Refraction

In the daily world we are most familiar with EM waves or light that propagates according to the right-hand rule (**k** is in the **E** × **H** direction) as illustrated in Figure 5.30a. Materials in which this is the case are isotropic media, and the dielectric constant ε in the constitutive relation (**D** = ε **E**; Equation 3.38) in this case is a scalar. Glass is an example of an isotropic material because it does not matter how light passes through it. The characteristics of the glass are such so that it has no different effects on the light even though the direction of the beam varies.

In nature one also has materials in which the direction of **k** is not exactly along the **E** × **H** direction; the angle between the two is then smaller than 90°. In mathematical terms, this means $\mathbf{k} \cdot (\mathbf{E} \times \mathbf{H}) > 0$. These are optically anisotropic media, such as birefringent calcite crystals ($CaCO_3$), quartz, ice, sugar, and so on. Birefringence, also called *double refraction,* was first observed and explained in the 17th century after sailors visiting Iceland brought back to Europe transparent calcite crystals (Iceland spar) that showed double images of objects that were viewed through them (Figure 5.31). The effect was first explained by the Dutch physicist Christiaan Huygens (1629–1695), who called the effect *double refraction* that involved an ordinary and an extraordinary wave. The extraordinary ray is so named because it does not exhibit ordinary Snellius law (Equation 5.19) behavior on refraction at the crystal surface. As light travels through an anisotropic material, such as calcite, the electromagnetic waves become split into two principal vibrations, which are oriented mutually perpendicular to each other and perpendicular to the direction that the waves propagate. The wave whose electric vector vibrates along the major axis is termed the ordinary or slow wave because the refractive index for this wave is greater than the refractive index for the other wave. The wave vibrating perpendicular to the slow wave is termed the *fast* or *extraordinary wave.*

The dielectric tensor in the constitutive Equation 3.38 ($\mathbf{D} = \varepsilon\mathbf{E}$) in the case of an optically anisotropic material with a single optical axis (calcite, quartz, or any tetragonal, hexagonal, or rhombohedral crystal) is a tensor of rank two given by:

$$\varepsilon = \begin{bmatrix} \varepsilon_{xx} & 0 & 0 \\ 0 & \varepsilon_{xx} & 0 \\ 0 & 0 & \varepsilon_{zz} \end{bmatrix} \quad (5.95)$$

In a crystal like this there exists a single optical axis, the ordinary optical axis, where light propagation is isotropic. In all other directions, two orthogonally polarized waves with different velocities do propagate; these are called the *extraordinary rays.*

Birefringence is widely used in optical devices, such as liquid crystal displays, light modulators, color filters, wave plates, optical axis gratings, and so on. In Volume III, Chapter 6 on metrology we encounter the phenomenon in both ellipsometry (measures the polarization of light), and in Volume II, Chapter 10 (Figure 10.8), we compare birefringence of polymer MEMS structures made with different replication techniques to evaluate their stress levels.

Birefringence should not be confused with dichroism, which itself has two related but distinctly different meanings in optics. A dichroic material is either one that causes visible light to be split up into distinct beams of different colors, or one in which light rays having different polarizations are absorbed by different amounts. Light that passes through an isotropic crystal is absorbed in the same way in all directions, but in an anisotropic crystal, the light is absorbed differently, depending on its direction of travel through the crystal. Isotropic colored crystals show no color change as they are rotated in plane polarized light, whereas anisotropic crystals show a distinct change in color when they are viewed in different directions. An anisotropic crystal may produce a number of different colors, known as *pleochroism.* When only two colors are observed, it is called *dichroism.* Such devices include mirrors and filters, usually treated with optical coatings, which are designed to reflect light over a certain range of wavelengths and transmit light that is outside that range. An example is the dichroic prism, used in some camcorders, which uses several

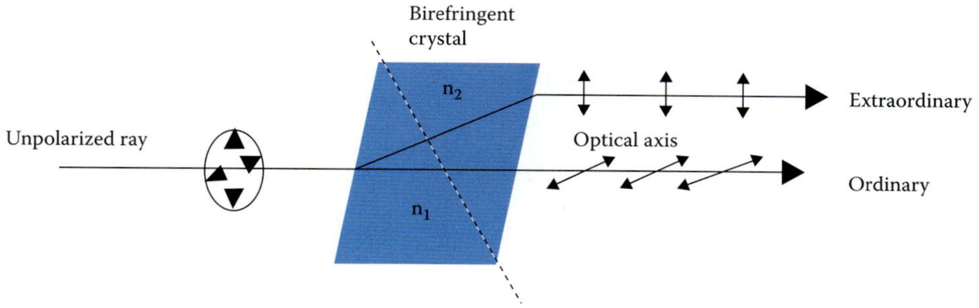

FIGURE 5.31 As light travels through an anisotropic material, such as calcite, the electromagnetic waves become split into two principal vibrations. In this double diffraction mode, an ordinary and an extraordinary wave are involved that are oriented mutually perpendicular to each other and perpendicular to the direction that the waves propagate.

coatings to split light into red, green, and blue components for recording on separate charge-coupled device (CCD) arrays. This kind of dichroic device does not usually depend on the polarization of the light. The second meaning of dichroic refers to a material in which light in different polarization states traveling through it experiences a varying absorption. The term came about because of early observations of the effect in crystals such as tourmaline. In these crystals, the strength of the dichroic effect varies strongly with the wavelength of the light, making them appear to have different colors when viewed with light having differing polarizations. This is more generally referred to as *pleochroism*, and the technique can be used in mineralogy to identify minerals. Dichroic crystals were researched by Edwin Land in 1930, who invented a polarizing material that would be called *Polaroid* and result in a huge benefit to the camera industry. The material he invented was herapathite (iodoquinine sulfate), in which the effect is not strongly dependent on wavelength, and so the term dichroic is something of a misnomer but still used.

Energy and Momentum and Electromagnetic Waves

Energy in EM—The Poynting Vector

Electromagnetic waves transfer energy from one place to another. The Poynting vector **S** describes that energy flux ($Jm^{-2}s^{-1}$). It is named after its "inventor," the English physicist John Henry Poynting (1852–1914). The vector **S** points in the direction of energy flow, and its magnitude is the power per unit area crossing a surface normal to it. The total energy, U, in an EM wave is the sum of the electrical and magnetic contributions:

$$U = \frac{1}{2}\varepsilon_0 \mathbf{E}^2 + \frac{1}{2}\frac{\mathbf{B}^2}{\mu_0} \tag{5.96}$$

From Equation 5.78, **E**/**B** = c, and because $(\varepsilon_0\mu_0)^{-1/2} = 1/c$, we can rewrite Equation 5.96 as:

$$\text{Total energy} = \varepsilon_0 \mathbf{E}^2 \tag{5.97}$$

The **E** field and the **B** field each contribute half of the total energy in an electromagnetic wave. As we just wrote out the total energy in terms of the **E** field, we may also write it out in terms of magnetic field **B** to obtain:

$$U = \frac{\mathbf{B}^2}{\mu_0} \tag{5.98}$$

Writing the total energy out in terms of both **E** and **B**, we derive:

$$U = \varepsilon_0 \mathbf{E}^2 = \varepsilon_0 \mathbf{E} c \mathbf{B} = \frac{\varepsilon_0 \mathbf{EB}}{\sqrt{\varepsilon_0 \mu_0}} = \sqrt{\frac{\varepsilon_0}{\mu_0}}\mathbf{EB} \tag{5.99}$$

Now let us get back to the Poynting vector **S**: the energy the wave transports per unit time per unit area is given by:

$$\mathbf{S} = \frac{1}{A}\frac{dU}{dt} = \frac{\mathbf{EB}}{\mu_0} \tag{5.100}$$

The direction of **S** is along **k** and perpendicular to **E** and **B** for right-handed materials; for left-handed materials, **k** is still perpendicular to **E** and **H** (we can use **B** or **H** because **B** = μ**H**) but opposite the Poynting vector as shown in Figure 5.32. Mathematically more correct, the Poynting vector is usually written out as:

$$\mathbf{S} = \frac{1}{\mu_0}(\mathbf{E} \times \mathbf{B}) \tag{5.101}$$

Because **S**, **E**, and **H** always form a right-handed set, two types of substances can be identified: right-handed substances with **S** parallel to **k**, and left-handed substances with **S** antiparallel to **k** (see also Figure 5.30). Because **k** denotes the direction of phase velocity ($v_p = \omega/\mathbf{k} = E/\mathbf{p}$) and **S** denotes the direction of group velocity ($v_g = d\omega/d\mathbf{k} = dE/d\mathbf{p}$), in right-handed materials phase velocity (along **k**)

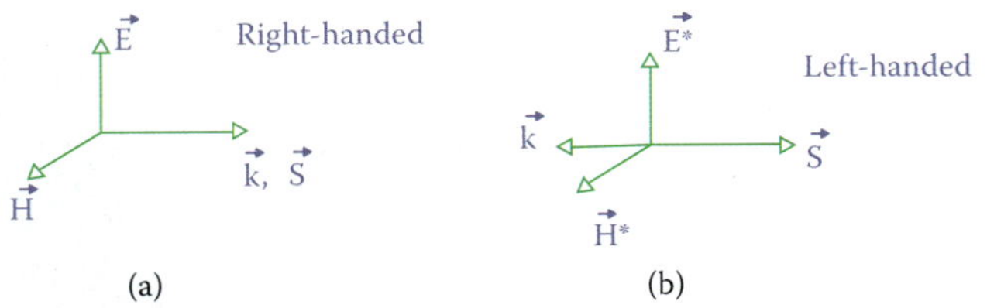

FIGURE 5.32 The phase velocity (along **k**) and group velocity (along the Poynting **E** × **H** vector) are in the same direction (a) but in left-handed materials. The group and phase velocities are in opposite directions (b) (see also Figure 5.30).

and group velocity (along the Poynting $\mathbf{E} \times \mathbf{H}$ vector) are in the same direction. In left-handed materials, group and phase velocity are in opposite directions. The left-handed substances have negative group velocity that occurs in anisotropic systems or when there is spectral dispersion. We delve into these matters deeper when we discuss metamaterials in more detail further below.

Momentum of EM: Radiation Pressure

Because magnetic waves carry energy, see above, they also deliver punch or momentum to a surface they are reflected off or absorbed in (think about a sun sail; Figure 5.33).

The force they exert on the material as a result is given by $\mathbf{F} = d\mathbf{P}/dt$. The force per unit area exerted by electromagnetic waves is called *radiation pressure*. Maxwell predicted the existence of radiation pressure and showed that if a beam of electromagnetic radiation is fully absorbed by a material, then the momentum transfer is given by:

$$\Delta \mathrm{P} = \frac{\Delta \mathrm{U}}{\mathrm{c}} \qquad (5.102)$$

with ΔU the energy absorbed by the material over a time Δt and c the speed of light. If in contrast the radiation is completely reflected, the momentum transfer becomes twice as large, or:

$$\Delta \mathrm{P} = \frac{2\Delta \mathrm{U}}{\mathrm{c}} \qquad (5.103)$$

For intermediate cases between total absorption and total reflection, one generalizes Equations 5.102 and 5.103 to read:

$$\Delta \mathrm{P} = \frac{\kappa \Delta \mathrm{U}}{\mathrm{c}} \qquad (5.104)$$

Because a real surface will always absorb and reflect some of the energy, κ is a constant between 1 and 2. From Newton's second law we can calculate the force and the pressure. The average rate of energy in the beam is related to the Poynting vector by:

$$\frac{d\mathrm{U}}{dt} = \mathrm{SA} = \frac{\mathrm{A}}{\mu_0}(\mathbf{E} \times \mathbf{B}) \qquad (5.105)$$

with A the cross-section of the object that the beam is hitting. The radiation pressure is given by:

$$\mathrm{P} = \frac{\mathrm{F}}{\mathrm{A}} = \frac{1}{\mathrm{A}}\frac{d\mathrm{P}}{dt} = \frac{1}{\mathrm{Ac}}\frac{d\mathrm{U}}{dt} = \frac{\mathrm{S}}{\mathrm{c}}$$

if light is fully absorbed and (5.106)

$$\mathrm{P} = \frac{2\mathrm{S}}{\mathrm{c}}$$

if light is fully reflected

Laser tweezers are another example dramatically illustrating the utility of radiation pressure (Figure 5.34). With optical tweezers one can manipulate nanometer and micrometer-sized dielectric particles by relying on the small forces contained in a highly focused laser beam (even a single atom can be manipulated!. Arthur Ashkin and colleagues first demonstrated laser tweezers in 1970 at Bell Labs.[4] With forces in the pN range optical tweezers can be used to manipulate and move small objects, measure small forces, and build up structures of small particles. The "Lilliputian" optical force pulls the object toward the focal volume of the beam, and

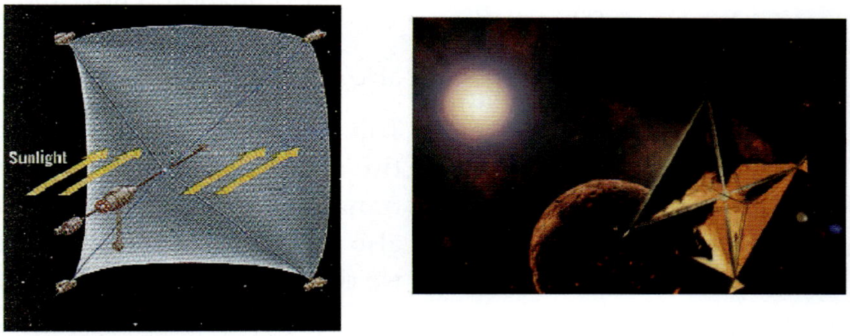

FIGURE 5.33 Sun sail. Sunlight striking the sail creates a force that pushes the spaceship away from the sun, much as the wind propels a sailboat.

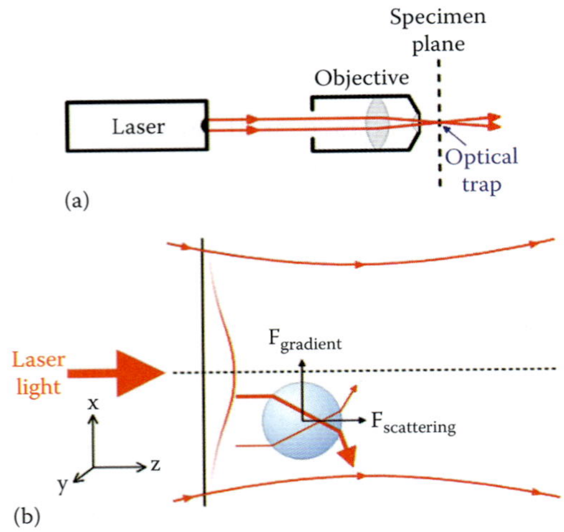

FIGURE 5.34 Laser tweezers—an example of the use of radiation pressure. (a) The beam is focused through a microscope objective. The beam waist, contains a very strong electric field gradient. Dielectric particles are attracted along the gradient to the region of strongest electric field, which is the center of the beam. (b) This spot creates an "optical trap" which is able to hold a small particle at its center. The forces felt by this particle consist of the light scattering and gradient forces due to the interaction of the particle with the light.

by moving the focal volume of the beam one can cause the object to move. In Figure 5.34a we show a laser beam focused by a microscope objective into a small spot on a specimen (e.g., a nanoparticle). This focussed spot constitutes an "optical trap" capable of holding a small particle at its center. In Figure 5.34b we illustrate how exactly such an optical trap works. Two forces are shown to be at play : 1) $F_{scattering}$ which is the scattering force and points in the direction of the incident light (z, see axis), and 2) $F_{gradient}$, which is the gradient force, due to the gradient nature of the Gaussian light intensity profile (x-y plane). As we just learned, light carries momentum and momentum is proportional to its energy in the direction of its propagation. A change in direction of the incoming light by reflection or refraction, results in a change of the momentum of the incoming photons. That momentum change, conservation of momentum dictates, causes the dielectric particles in the path of the light to undergo an equal and opposite momentum change and this is what causes a force to act on particles. When the light of a Gaussian laser beam interacts with a small particle, the light rays are bent (two example rays are illustrated in Figure 5.34b). Summing all the forces results in a resultant force for $F_{scattering}$ (z-direction) and one for $F_{gradient}$ (x-y plane). With the contribution to $F_{scattering}$ of the refracted rays larger than that of the reflected rays a restoring force is created along the z-axis, and a stable trap is established. In modern equipment the image of the trapped particle can be projected onto a quadrant photodiode to measure nm-scale displacements. A continuous wave (CW), low-power laser was used to "optically trap" individual bacteria and protozoa. Steven Chu, who was at Stanford at the time, showed that lasers could be used to manipulate molecules. He and his team famously attached polystyrene beads at the end of DNA and pulled on the beads to stretch the DNA molecule. Others have used laser tweezers to study kinesin motors (see Volume III, Figure 2.33). Laser scissors were in use earlier than laser tweezers.[5] Laser scissors are short-pulsed and high-irradiance (energy reaching the surface in a given amount of time) beams, whereas tweezers are continuous and of low irradiance. The CW lasers can cause a 1.15–1.45°C increase per 100 mW of laser power. If the heat dissipation is inefficient, the laser can produce 10 times that power. Therefore, in the case of tweezers, the target must be transparent to not store significant energy.

The Complex Refractive Index n_c *and the Complex Wave Vector* **k**

As we saw above, at optical frequencies, electric fields in homogeneous media have solutions of the form $\mathbf{E} = \mathbf{E}_m e^{i(k_x \cdot x - \omega t)}$ and $\mathbf{B} = \mathbf{B}_m e^{i(k_x \cdot x - \omega t)}$ (**E** along the z-axis, **B** along the y-axis, and **k** along the x-axis; see Equation 5.70). For nonmagnetic materials with $\mu = \mu_0$, this results in a dispersion equation $\mathbf{k}^2 = \frac{\omega^2}{v^2}\varepsilon(\omega)$ (Equation 5.80), and with $n = \frac{c}{v}$ this can also be expressed as $\mathbf{k}^2 = \frac{n^2}{c^2}\omega^2$ (Equation 5.81). This dispersion relation for the case where the refractive index n is a real number correctly describes transparent materials. To describe a material that is also absorbing and features the anomalous regions we typically observe in plots of the refractive index as a function of frequency (see, for example, Figure 5.29), we need to introduce a complex wave vector **k** and a complex refractive index n_c.

Thus, the real wave vector $\mathbf{k} = \frac{n}{c}\omega = \frac{2\pi}{\lambda / n}$ with a real refractive index n must be replaced by a complex wave vector $\mathbf{k} = \frac{n_c}{c}\omega = \frac{2\pi}{\lambda / n_c}$ with a complex refractive index n_c.

The complex refractive index n_c is given as:

$$n_c = n + i\kappa \tag{5.107}$$

with n the refractive index and κ the extinction coefficient of the material. It follows that the dispersion Equation 5.81 then is transformed as:

$$\mathbf{k}(\omega) = \pm\frac{(n + i\kappa)\omega}{c} \tag{5.108}$$

The Kramers-Krönig relation links the real and complex parts of the complex refractive index (i.e., the refractive index n and κ the extinction coefficient). If you know one, you can calculate the other one.

The macroscopic quantities n and κ in Equation 5.108 will now be derived from the microscopic functions χ and ε_r, i.e., from the complex electric susceptibility and the complex permittivity, respectively.

Substituting the complex wave vector in the proposed wave solution (Equation 5.70), we obtain:

$$\mathbf{E}_x(x,t) = \mathbf{E}_m(x,\omega)e^{\frac{\omega\kappa}{c}x}e^{i\left(\frac{\omega n}{c}x-\omega\kappa\right)} \tag{5.109}$$

damping/ absorption — propagation

which represents a traveling damped wave (Figure 5.35).

The optical intensity of this electromagnetic wave is calculated as:

$$I \propto |\mathbf{E}|^2 = I_0 e^{\frac{2\omega\kappa}{c}x} = I_0\, e^{-\alpha x} \tag{5.110}$$

with the absorption coefficient $\alpha = 2\frac{\kappa\omega}{c} = \frac{4\pi\kappa}{\lambda} = 2\mathbf{k}\kappa$. Note that $\alpha > 0$ means absorption and $\alpha < 0$ means gain (see lasers at the end of this chapter), and a propagation coefficient is defined as $\beta = \left(\frac{\omega}{c}\right)n = \mathbf{k}n$. The complex wave vector in Equation 5.108 can be written out in terms of this absorption coefficient α and the propagation coefficient β as:

$$\mathbf{k} = \mathbf{k}' + i\mathbf{k}'' = \beta - i\frac{\alpha}{2} \tag{5.111}$$

The real part of the complex wave vector ($\mathbf{k}$) is $\mathbf{k}' = \beta$ and the imaginary part $\mathbf{k}'' = \alpha/2$; in other words, the real part of the complex wave vector corresponds

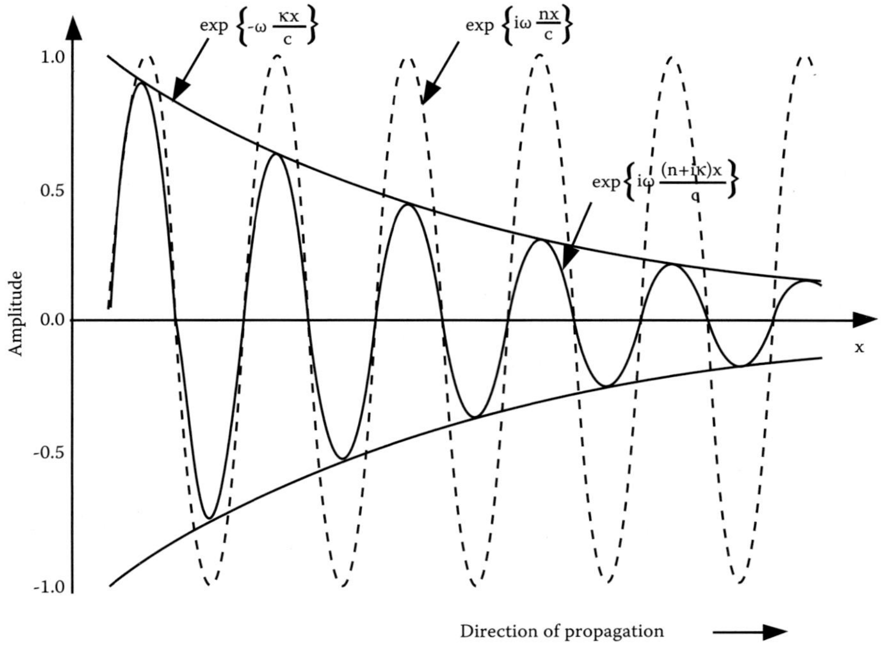

FIGURE 5.35 A traveling damped wave.

to the propagation coefficient and the imaginary part corresponds to the absorption coefficient of the material. Substituting the complex wave vector, written out in this format, into the proposed wave solution $[\mathbf{E} = \mathbf{E}_m e^{i(\mathbf{k}\cdot x-\omega t)}]$ (for simplicity we have dropped the sub x for the wave vector $\mathbf{k}$), we obtain:

$$\begin{aligned} \mathbf{E}_x(x,t) &= \mathbf{E}_m(x,\omega)e^{-k''x}e^{i(k'x-\omega t)} \\ &= \underset{\text{absorption}}{\mathbf{E}_m(x,\omega)e^{-\frac{\alpha}{2}x}}\,\underset{\text{propagation}}{e^{i(\beta x-\omega t)}} \end{aligned} \quad (5.112)$$

The complex refractive index n results from a complex electric susceptibility tensor $\chi_e(\omega)$ as:

$$\begin{aligned} n_c(\omega) &= n(\omega) + i\kappa(\omega) = \sqrt{1+\chi_e(\omega)} \\ &= \sqrt{1+\chi'(\omega)+i\chi''(\omega)} \end{aligned} \quad (5.113)$$

or also:

$$n_c(\omega) = n - i\frac{\alpha}{2k} = \sqrt{1+\chi'(\omega)+i\chi''(\omega)} \quad (5.114)$$

We still need to introduce the relationship between the complex refractive index n_c and the complex dielectric constant. To do this we take the square of Equation 5.114 and group the real parts and imaginary parts together and relate n' and n'' to ε_r' and ε_r''. From $n_c(\omega)^2 = \varepsilon_r = \varepsilon_r' + i\varepsilon_r'$, it follows that:

$$n_c(\omega) = n(\omega) + i\kappa(\omega) = \sqrt{\varepsilon_r'(\omega) + i\varepsilon_r''(\omega)}$$

$$\varepsilon_r'(\omega) = [n(\omega)]^2 - [\kappa(\omega)]^2 = 1 + \chi'(\omega) = \frac{\varepsilon'}{\varepsilon^0} \quad (5.115)$$

$$\varepsilon_r''(\omega) = 2n(\omega)\kappa(\omega) = \frac{\varepsilon''(\omega)}{\varepsilon_0(\omega)} = \chi''(\omega)$$

and:

$$\kappa(\omega) = \frac{\varepsilon_r''(\omega)}{2n} = \frac{\chi''(\omega)}{2n} \quad (5.116)$$

to finally obtain:

$$\begin{aligned} n(\omega) &= \sqrt{\frac{(\varepsilon_r')^2 + (\varepsilon_r'')^2 + \varepsilon_r'}{2}} \\ \kappa(\omega) &= \sqrt{\frac{(\varepsilon_r')^2 + (\varepsilon_r'')^2 - \varepsilon_r'}{2}} \end{aligned} \quad (5.117)$$

for transparent materials $\kappa \lll\!< 1$, and Equation 5.115 reduces to:

$$n_c(\omega) = n(\omega) = \sqrt{\varepsilon_r'(\omega)} \quad (5.118)$$

For weakly absorbing materials (dilute media) with both the real and imaginary parts of the electric susceptibility tensor $|\chi'(\omega)|$ and $||\chi''(\omega)| \lll\!< 1$, $\sqrt{1+\chi'(\omega)+i\chi''(\omega)} \approx 1 + \frac{1}{2}[\chi'(\omega)+i\chi''(\omega)]$ one obtains the following relation between n and $\chi'(\omega)$ and κ''' and $\chi'(\omega)$:

$$n(\omega) \approx 1 + \frac{1}{2}\chi'(\omega) \quad (5.119)$$

and:

$$\kappa(\omega) \approx \frac{\chi''(\omega)}{2} \quad (5.120)$$

So the absorption coefficient for small $\chi'(\omega)$ and $\chi''(\omega)$ is given by:

$$\alpha(\omega) \approx \frac{\omega}{c}\chi''(\omega) \quad (5.121)$$

The different real and imaginary partner pairs we introduced so far: $\mathbf{k}'$ and $\mathbf{k}''$, n and κ, χ' and χ'', ε' and ε'' all describe the same physics, but for some problems one set is preferable over the others. The complex wave vector pair $\mathbf{k}'$ and $\mathbf{k}''$ and the refractive index n and the extinction κ pair is mostly used when discussing wave propagation. The other pairs, χ' and χ'' and ε' and ε'', are used most often when discussing the microscopic origin of optical effects. Each imaginary component can always be calculated from the real component, and vice versa (see Kramers-Krönig relationship).

Summarizing, the polarization of matter (**P**) induced by an applied electromagnetic field (**E**) is not instantaneous, making **P** and therefore also χ, n, and **k** frequency-dependent:

$$\mathbf{P}(\mathbf{k},\omega) = \chi_e(\mathbf{k},\omega)\varepsilon_o\mathbf{E}(\mathbf{k},\omega) \Rightarrow$$

$$\begin{array}{llll} \chi'(\omega) & n(\omega) & \mathbf{k}'(\omega) = \beta = \mathbf{k}n & \text{Phase propagation} \\ \Rightarrow & \Rightarrow & & \\ \chi''(\omega) & \kappa(\omega) & \mathbf{k}''(\omega) = \frac{\alpha}{2}\mathbf{k}\kappa & \text{Absorption} \end{array}$$

Example 5.2: Magnitude of κ
With $\alpha = 1\ cm^{-1}$, then $\kappa = \alpha/2\mathbf{k}$. And **k** is about $10^7\ m^{-1}$ for visible light, so $\kappa \sim 10^{-5}$, $\ll n$. Hence, $n^2 = 1 + \chi'(\omega)$ is a good approximation, provided $\alpha < 10^4\ cm^{-1}$ (see Equation 5.115). Then we can say that refraction is solely related to χ' away from resonances.

Electromagnetic Waves at Boundaries

Reflectance R *and Transmittance* T *of Electromagnetic Radiation*

Different from mechanical waves, transferred through a medium, light does not need a medium. Electrons in materials are vibrated and emit energy in the form of photons, which propagate across the universe (photons have no mass but are pure energy). Electromagnetic (EM) waves are waves that are made up of these "photons." When these photons come in contact with physical boundaries, EM waves interact like other waves would.

Incoming radiation that interacts with an object may follow three paths: reflection, absorption, and transmission. Reflectance, *R*, is the amount of reflected radiation over the amount of incoming radiation power. Absorption, *A*, is the amount of absorbed radiation over the amount of incoming radiation power, and transmittance, *T*, is the amount of transmitted radiation over the incident power. Power continuity dictates that:

$$R + A + T = 1 \tag{5.122}$$

This expression is known as *Kirchoff's law*. For an opaque object (T = 0), what is not absorbed is reflected, or R + A = 1. For real surfaces, *A*, *T*, and *T* all vary with the incoming wavelength λ. Incoming light striking a solid object is schematically represented in Figure 5.36.

The decay of the intensity of an EM wave propagating in the *x*-direction in an absorbing material, Beer's law $[I_0 = I(x = 0)]$, is calculated in Chapter 7 as:

$$I(x) = I_0 e^{-\alpha x} \tag{5.110}$$

so that we calculate for the transmittance:

$$T = (1 - R)^2 e^{-\alpha l} \tag{5.123}$$

where *l* is the thickness of the solid piece and where we have assumed that the reflectivity of the front face and back face of that piece of material is identical.

Link between Reflectance R *and Complex Refractive Index* n_c

If we could relate the relatively easy-to-measure reflectance *R* of a material to the more difficult to access complex refractive index n_c, we then might be able to obtain both the real and imaginary parts of the complex refractive index n_c, i.e., *n* and κ, by using the Kramers-Krönig relation, which links the real and complex parts of the complex refractive index.

It was Augustin-Jean Fresnel (1788–1827) who first derived the equations that describe how light is reflected and transmitted at an interface between two media with different refractive indices. His equations give the reflection and transmission for waves parallel and perpendicular to the plane of incidence. For a dielectric medium where Snell's law (Equation 5.19) can be used to relate the incident and transmitted

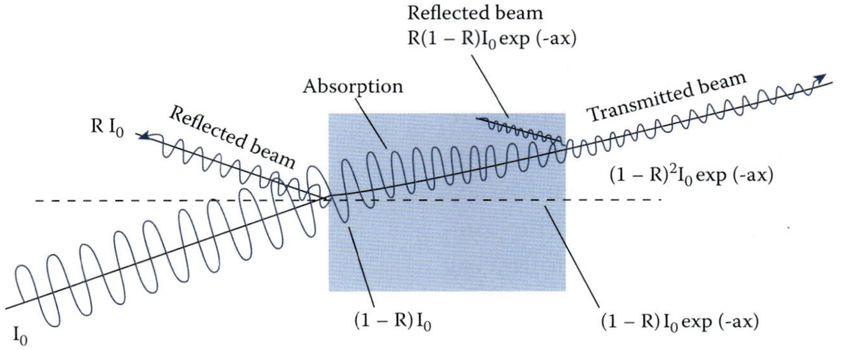

FIGURE 5.36 Radiation incident on a solid. With incoming beam of intensity I_0, externally reflected beam (RI_0), the intensity of the beam entering the solid $(1 - R_0)I_0$, the absorption of the beam inside the solid $(1 - R)I_0e^{-ax}$, that attenuated beam internally reflected $R(1 - R)I_0e^{-ax}$, and finally the transmitted beam: $(1 - R)^2I_0e^{-ax}$, where we have assumed that the reflectivity *R* is the same for the light coming in as it is for the light going out.

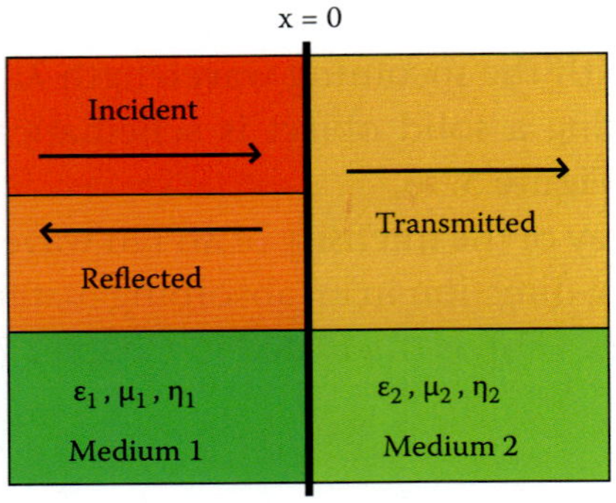

FIGURE 5.37 Wave reflection and transmission at an interface (x = 0).

angles, Fresnel's equations can be stated in terms of angles of incidence and transmission. The reflection of light that the equations predict is known as *Fresnel reflection*. In Volume III, Chapter 6 on metrology, when describing ellipsometry, we will take full advantage of these equations to relate reflectance R of a material with its complex dielectric constant n_c.

To establish the relation between R and n_c, we consider a plane wave moving through medium 1 (ε_1, μ_1, and η_1) and incident normal on a medium 2 (ε_2, μ_2, and η_2) (see Figure 5.37). We now need to express the incident (i), reflected (r), and transmitted (t) waves in each medium and then apply the boundary conditions. Because the wave is incident normal onto the surface, all **E** and **B** fields are parallel to the surface. We are using the solutions of the wave equations (Equation 5.70) [with **E** along the z-axis, **B** along the y-axis, and the propagation vector **k** along the x-axis, and where the complex propagation vector (Equation 5.108)].

Boundary conditions for EM waves at an interface are established by applying the integral form of Maxwell's equations to a small region at the interface of the contacting media. The integral equations are assumed to remain valid for regions containing discontinuous media as in the current case. We are assuming an interface between two lossless media (no charges or currents) and use the integral in Equation 5.51b to find that the following constraint is imposed on the tangential **E** field ($\mathbf{E}_t$):

$$\mathbf{E}_{1t} = \mathbf{E}_{2t} \tag{5.124}$$

This simply states that the tangential component of a field is continuous across an interface. When media 1 and 2 are dielectrics with permittivities ε_1 and ε_2, respectively, we can add that:

$$\frac{\mathbf{D}_{1t}}{\varepsilon_1} = \frac{\mathbf{D}_{2t}}{\varepsilon_2} \tag{5.125}$$

From the integral in Equation 5.48b, the tangential component of the **H** field ($\mathbf{H}_t$) also is continuous across the boundary of the two lossless media:

$$\mathbf{H}_{1t} = \mathbf{H}_{2t} \tag{5.126}$$

and for linear media we can deduce:

$$\frac{\mathbf{B}_{1t}}{\mathbf{B}_{2t}} = \frac{\mu_1}{\mu_2} \tag{5.127}$$

Similarly, the boundary conditions for the normal components of the **D** and **B** fields ($\mathbf{D}_n$ and $\mathbf{B}_n$) are obtained from the integrals in Equations 5.49b and 5.50b:

$$\mathbf{B}_{1n} = \mathbf{B}_{2n} \tag{5.128}$$

or, the normal component of a **B** field is continuous across an interface, and for linear media it holds that:

$$\mu_1\mathbf{H}_{1n} = \mu_2\mathbf{H}_{2n} \tag{5.129}$$

Finally, one also finds that:

$$\mathbf{D}_{1n} = \mathbf{D}_{2n} \tag{5.130}$$

and for linear media:

$$\varepsilon_1\mathbf{E}_{1n} = \varepsilon_2\mathbf{E}_{2n} \tag{5.131}$$

These results are summarized in Table 5.2. In the section on surface plasmonics, below, we will analyze these boundary conditions for the case of an interface between a dielectric and a conductor.

Let us denote the incident wave with $\mathbf{E}^i(x) = \mathbf{E}^i_m e^{i(k\cdot x-\omega t)}$ and $\mathbf{H}^i(x) = \frac{\mathbf{E}^i_m}{\eta_1} e^{i(k\cdot x-\omega t)}$, the reflected wave

TABLE 5.2 Boundary Conditions between Two Lossless Media

$\mathbf{E}_{1t} = \mathbf{E}_{2t} \Rightarrow \frac{\mathbf{D}_{1t}}{\varepsilon_1} = \frac{\mathbf{D}_{2t}}{\varepsilon_2}$
$\mathbf{H}_{1t} = \mathbf{H}_{2t} \Rightarrow \frac{\mathbf{B}_{1t}}{\mathbf{B}_{2t}} = \frac{\mu_1}{\mu_2}$
$\mathbf{D}_{1n} = \mathbf{D}_{2n} \Rightarrow \varepsilon_1\mathbf{E}_{1n} = \varepsilon_2\mathbf{E}_{2n}$
$\mathbf{B}_{1n} = \mathbf{B}_{2n} \Rightarrow \mu_1\mathbf{H}_{1n} = \mu_2\mathbf{H}_{2n}$

with $\mathbf{E}^r(x) = \mathbf{E}^r_m e^{i(k \cdot x - \omega t)}$ and $\mathbf{H}^r(x) = \frac{\mathbf{E}^r_m}{\eta_1} e^{i(k \cdot x - \omega t)}$, and the transmitted wave with $\mathbf{E}^t(x) = \mathbf{E}^t_m e^{i(k \cdot x - \omega t)}$ and $\mathbf{H}^t(x) = \frac{\mathbf{E}^r_m}{\eta_2} e^{i(k \cdot x - \omega t)}$. From the expression for the intrinsic impedance $\eta = \frac{|\mathbf{E}|}{|\mathbf{H}|} = \frac{\omega\mu}{|\mathbf{k}|} = \frac{|\mathbf{k}|}{\omega\varepsilon} = \sqrt{\frac{\mu}{\varepsilon}}$ (Equation 5.83), we also introduced here the fact that $\frac{\mathbf{E}^i_m}{\eta_1} = \mathbf{H}^i_m$, $\frac{\mathbf{E}^r_m}{\eta_1} = \mathbf{H}^r_m$, and $\frac{\mathbf{E}^t_m}{\eta_2} = \mathbf{H}^t_m$. Like the velocity of propagation *c*, the intrinsic impedance is independent of the source and is determined only by the properties of the medium. The two unknowns in these equations are $\mathbf{E}^r_m$ and $\mathbf{E}^t_m$, which we will now retrieve from the above-derived boundary conditions, i.e., the tangential **E** and **H** fields must be continuous at x = 0 (Equations 5.124 and 5.126). Fields parallel and perpendicular to the plane of incidence are considered *p*- and *s*-polarized, respectively. These two components are independent and can be calculated separately. Fresnel calculated the amount of light reflected and transmitted between materials of different refractive index, and for the tangential fields (||) in medium 1 he deduced:

$$\underbrace{E_{t1}(x = 0)}_{\text{Total tangential E}} = \underbrace{E^i_m(x = 0)}_{\text{Incident E}} + \underbrace{E^r_m(x = 0)}_{\text{Reflected E}}$$

$$\underbrace{H_{t1}(x = 0)}_{\text{Total tangential H}} = \underbrace{H^i_m(x = 0)}_{\text{Incident H}} \tag{5.132}$$

$$\underbrace{+H^r_m(x = 0)}_{\text{Reflected H}} = \frac{1}{\eta_1}(E^i_m - E^r_m)$$

And for medium 2 he derived:

$$E_{t2}(x = 0) = E^t(x = 0) = E^t_m$$

Therefore, at x = 0:

$$E^i_m + E^r_m = E^t_m \tag{5.133}$$

$$H_{t2} = H^t(x = 0) = \frac{1}{\eta_2} E^t_m$$

Therefore, from Equations 5.132 and 5.133, we obtain:

$$\frac{1}{\eta_1}(E^i_m - E^r_m) = \frac{1}{\eta_2} E^t_m \tag{5.134}$$

and we can solve now for the unknowns $\mathbf{E}^r_m$ and $\mathbf{E}^t_m$ in terms of the known $\mathbf{E}^i_m$:

$$E^r_m = \frac{\eta_2 - \eta_1}{\eta_2 + \eta_1} E^i_m = r(\omega) E^i_m$$

and

$$E^t_m = \frac{2\eta_2}{\eta_2 + \eta_1} E^i_m = t(\omega) E^i_m \tag{5.135}$$

where $r(\omega)$, the dimensionless reflectivity coefficient, is a complex function defined at the solid surface as the ratio of the reflected electric field to the incident electric field, or for the tangential field (||) we obtain:

$$r(\omega)_{\parallel} = \left(\frac{E^r_m}{E^i_m}\right)_{\parallel} = \left(\frac{\eta_2 - \eta_1}{\eta_2 + \eta_1}\right)_{\parallel} \tag{5.136}$$

And for $t(\omega)$, the dimensionless transmission coefficient of the tangential field (||), we get:

$$t(\omega)_{\parallel} = \left(\frac{E^t_m}{E^i_m}\right)_{\parallel} = \left(\frac{2\eta_2}{\eta_2 + \eta_1}\right)_{\parallel} \tag{5.137}$$

Equations 5.136 and 5.137 are the Fresnel expressions for the tangential fields. For the normal field (⊥) one obtains these additional Fresnel expressions:

$$r(\omega)_{\perp} = \left(\frac{E^r_m}{E^i_m}\right)_{\perp} = \left(\frac{\eta_4 - \eta_3}{\eta_4 + \eta_3}\right)_{\perp} \tag{5.138}$$

and:

$$t(\omega)_{\perp} = \left(\frac{E^r_m}{E^i_m}\right)_{\perp} = \left(\frac{2\eta_4}{\eta_4 + \eta_3}\right)_{\perp} \tag{5.139}$$

In Figure 5.38 we have sketched a planar interface, and the "incident" medium occupies the half-space $y > 0$ and has an index of refraction n_i. The

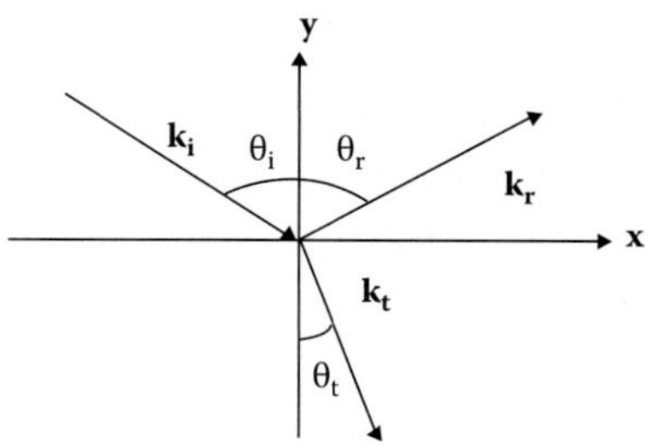

FIGURE 5.38 Light reflects and refracts according to Snell's law.

"transmission medium" occupies the half-space $y < 0$ and has an index of refraction n_t. For simplicity, we only consider the electric fields, and we obtain the four amplitude coefficients (the Fresnel equations) as:

$$r(\omega)_{\parallel} = \left(\frac{E_m^r}{E_m^i}\right)_{\parallel} = \left[\frac{\tan(\theta_i - \theta_t)}{\tan(\theta_i + \theta_t)}\right]_{\parallel} \quad (5.140)$$

$$r(\omega)_{\perp} = \left(\frac{E_m^r}{E_m^i}\right)_{\perp} = \left[-\frac{\sin(\theta_i - \theta_t)}{\sin(\theta_i + \theta_t)}\right]_{\perp} \quad (5.141)$$

$$t(\omega)_{\parallel} = \left(\frac{E_m^t}{E_m^i}\right)_{\parallel} = \left[\frac{2\sin\theta_t\cos\theta_i}{\sin(\theta_i + \theta_t)\cos(\theta_i - \theta_t)}\right]_{\parallel} \quad (5.142)$$

$$t(\omega)_{\perp} = \left(\frac{E_m^r}{E_m^i}\right)_{\perp} = \left[\frac{2\sin\theta_t\cos\theta_i}{\sin(\theta_i + \theta_t)}\right]_{\perp} \quad (5.143)$$

Note that the reflection coefficient r in Equations 5.136 and 5.138 can be positive or negative, depending on the relative sizes of the intrinsic impedances of the two media. The transmission coefficient t, on the other hand, is always positive (see Equations 5.137 and 5.139). Importantly, Equations 5.136–5.139 also hold when the media are dissipative, i.e., when η_1 and/or η_2 are complex. A complex r (or t) means that a phase shift is introduced at the interface on reflection (or transmission). Reflection and transmission are related by:

$$t(\omega) = 1 + r(\omega) \quad (5.144)$$

The same boundary conditions derived above require the components of the complex wave vector $\mathbf{k}$ (see Equation 5.111) parallel to the surface to be the same on both sides of the boundary. When a sinusoidal plane wave, with angular frequency ω and propagation wave vector $\mathbf{k}$, travels in the x-direction in a linear dielectric with a scalar permittivity and no free charge or currents present, any dissipative process leads to phase lag and a complex wave vector. Thus, a transmitted wave in a medium is attenuated because, by the dispersive relation for electromagnetic waves, the wave vector in the medium is related to the incident $\mathbf{k}$ in vacuum by $(n + i\kappa)\mathbf{k}$, or the incident wave, $\mathbf{E}^i(\mathbf{x}) = \mathbf{E}_m^i e^{i(\mathbf{k}\cdot\mathbf{x}-\omega t)}$, is modified to:

$$\begin{aligned} E_m^t &= E_m \exp[i(n + i\kappa)\mathbf{k}x - \omega t)] \\ &= \underbrace{\exp(-\kappa\mathbf{k}x)}_{\text{attenuation}}\ \underbrace{\exp[i(n\mathbf{k}x - \omega t)]}_{\text{propagation}} \end{aligned} \quad (5.145)$$

Continuity of $\mathbf{E}$ and $\mathbf{B}$ at the interface (surface of the medium) dictates that $\mathbf{E}_m^i + \mathbf{E}_m^r = \mathbf{E}_m^t$ (Equation 5.133), or:

$$\begin{aligned} kE_m^i - kE_m^r &= (n + i\kappa)kE_m^t \quad \text{or} \\ E_m^i - E_m^r &= (n + i\kappa)(E_m^i + E_m^r) \quad (5.146) \\ (n + i\kappa - 1)E_m^i &= (1 + n + i\kappa)E_m^r \end{aligned}$$

Thus, we obtain for the reflectivity at normal incidence:

$$r(\omega) = \frac{E_m^r}{E_m^i} = \left(\frac{n + i\kappa - 1}{n + i\kappa + 1}\right) = \frac{n_c - 1}{n_c + 1} \quad (5.147)$$

But the quantity measured in experiments is reflectance R, defined as the ratio of the reflected intensity to the incident intensity (see above), or:

$$R(\omega) = \frac{|E_m^r|^2}{|E_m^i|^2} = \left|\frac{n_c - 1}{n_c + 1}\right|^2 = \frac{(n-1)^2 + \kappa^2}{(n+1)^2 + \kappa^2} \quad (5.148)$$

When $R = 0$, there is no reflection, and this occurs with $n = 1$, $\kappa = 0$, and $R = 1$ when either n or κ becomes large. It is difficult to measure the phase $\theta(\omega)$ of the reflected wave, but it can be shown that it can be calculated from the measured reflectance $R(\omega)$ if this is known over all frequencies. With both $R(\omega)$ and $\theta(\omega)$ known we can use Equation 5.148 to calculate $n(\omega)$ and $\kappa(\omega)$. Above we saw that $n(\omega)$ and $\kappa(\omega)$ are related to the complex dielectric function ε_r as:

$$n_c(\omega) = n(\omega) + i\kappa(\omega) = \sqrt{\varepsilon_r'(\omega) + i\varepsilon_r''(\omega)} \quad (5.113 \text{ and } 5.115)$$

Thus, plugging the values for $n(\omega)$ and $\kappa(\omega)$ in Equation 5.113 we obtain the complex dielectric function.

The values for the refractive index n and extinction coefficient κ are not independent; they are linked by the Kramers-Krönig relations. The Kramers-Krönig relations enable us to find the real part of the response of a linear passive system if we know

Table 5.3 Optical Coefficients

Relative Dielectric Constant	$\varepsilon_r = 1 + \chi_e(\omega)$
Real part of refractive index	$n' = \sqrt{\dfrac{(\varepsilon_r')^2 + (\varepsilon_r'')^2 + \varepsilon_r'}{2}}$
Imaginary part of refractive index	$n'' = \sqrt{\dfrac{(\varepsilon_r')^2 + (\varepsilon_r'')^2 - \varepsilon_r'}{2}}$
Absorption coefficient	$\alpha = -2\dfrac{\omega}{c}n''$
Reflectivity	$r(\omega) = \dfrac{E_m^r}{E_m^i} = \left(\dfrac{n + i\kappa - 1}{n + i\kappa + 1}\right) = \dfrac{n_c - 1}{n_c + 1}$
Reflectance	$R(\omega) = \dfrac{\lvert E_m^r \rvert^2}{\lvert E_m^i \rvert^2} = \dfrac{(n-1)^2 + \kappa^2}{(n+1)^2 + \kappa^2}$

the imaginary part of the response at all frequencies, and vice versa:

$$n(\omega) - 1 = \frac{1}{\pi} P \int_{-\infty}^{\infty} \frac{\kappa(\omega')}{\omega' - \omega} d\omega'$$

$$\kappa(\omega) = \frac{-1}{\pi} P \int_{-\infty}^{\infty} \frac{n(\omega') - 1}{\omega' - \omega} d\omega' \qquad (5.149)$$

where P denotes the Cauchy principal value. Either of these equations, plus a knowledge of $R(\omega)$ at all frequencies, permits one to disentangle the separate values of $n(\omega)$ and $\kappa(\omega)$.

The previous discussion makes it clear that the refractive index of a semiconductor can also be determined from a sometimes more practical transmission measurement.

In Table 5.3 we summarize the important optical coefficient properties deduced so far. In Table 5.4 we list a number of semiconductors with their transparency range, the refractive index, and the wavelength corresponding to their bandgap.

Example 5.3*: Surface reflectance *R* of Ge. At 400 nm the complex refractive index n_c for Ge = 4.141 + i2.215. Calculate the 1) wave propagation speed *v*, 2) absorption coefficient α, and 3) reflectance *R*.

1. $v = c/n = 2.998 \times 10^8/4.141 = 7.24 \times 10^7$ m/s
2. $\alpha = 4\pi \times 2.215/400 \times 10^{-9} = 6.96 \times 10^7$ m^{-1}
3. $R = \dfrac{(4.141-1)^2 + 2.215^2}{(4.141+1)^2 + 2.215^2} = 47.1\%$

* All three examples are from Mikael Mulot's class notes.

Table 5.4 Semiconductors: Transparency Range (T), Bandgap Wavelength, and Refractive Index

Crystal	Transparency Range (μm)	λ_g (μm)	n
Ge	1.8–23	1.8	4.00
Si	1.2–15	1.1	3.42
GaAs	1.0–20	0.87	3.16
CdTe	0.9–14	0.83	2.67
CdSe	0.75–24	0.71	2.50
ZnSe	0.45–20	0.44	2.41
ZnS	0.4–14	0.33	2.20

Example 5.4: Transmittance *T* through Si. The reflectance of a 10-μm-thick Si slice at 633 nm = 35% and $\alpha = 3.8 \times 10^{-5}$ m^{-1}. Calculate the transmittance *T*.

For $\alpha l = (3.8 \times 10^{-5})(10 \times 10^{-6}) = 3.8$. As a result we obtain for $T = (1 - R)^2 e^{-\alpha l} = (1-35\%)^2 e^{-3.8} = 0.0095$.

Example 5.5: Reflectance *R* and absorption coefficient α of NaCl. The complex relative dielectric constant at 60 μm for NaCl is –16.8 + i91.4. Calculate 1) α and 2) **R**:

$$n = \frac{1}{\sqrt{2}}\sqrt{-16.8 + \sqrt{(-16.8)^2 + 91.4^2}} = 6.17$$

1. With $\kappa = \dfrac{1}{\sqrt{2}}\sqrt{+16.8 + \sqrt{(-16.8)^2 + 91.4^2}} = 7.41$

   we derive for $\alpha = \dfrac{4\pi\kappa}{\lambda} = \dfrac{4\pi \times 7.41}{60 \times 10^{-6}} = 1.55 \times 10^6\,\text{m}^{-1}$

2. $R = \dfrac{(6.17-1)^2 + 7.41^2}{(6.17+1)^2 + 7.41^2} = 76.8\%$

Bulk, Surface, and Localized Plasmons: Plasmonics

Introduction

Plasmons are electron density waves that can carry huge amounts of data and that can be squeezed into miniscule metal structures, holding the promise of a new generation of superfast computer chips, ultrasensitive molecular detectors, microscopes with improved subwavelength resolution (near-field scanning optical microscopes; NSOMs), better exploitation of nonlinear optics, sensitive optical studies of surfaces and interfaces, fabrication of an infinite variety of brightly colored metal nanoparticles, better light-emitting diodes, higher-resolution lithography, and perhaps even the possibility of altering the electromagnetic field around an object as to render it invisible (cloaking). In 2000, Harry A. Atwater from Caltech gave the name of "plasmonics" to this promising emerging field.[6]

In Chapter 3 we saw how a metal may exhibit collective longitudinal oscillations of the free electron gas density, often at optical frequencies. A plasmon is a quantum of a plasma oscillation that is excited by passing an electron through a material or by reflecting an electron or a photon off a material. It is a result of the quantization of the plasma oscillations just as photons and phonons are quantizations of light and sound waves, respectively. In other words, reflected or transmitted electrons/photons lose an amount of energy that equals an integral number of times the plasmon energy. A plasmon can also couple with a photon to create a plasma polariton. The most discussed types of polaritons are phonon-polaritons, resulting from the coupling of an infrared photon with an optical phonon; exciton-polaritons, resulting from the coupling of visible light with an exciton; and surface plasmon-polaritons, resulting from the coupling of surface plasmons with light.

Plasmons can be observed experimentally using electron energy loss spectroscopy (EELS). In EELS, electrons are accelerated to a defined energy and are then shot onto a thin foil of the material under investigation. The number and energy of the transmitted electrons are measured. These electrons are able to excite plasmons, which then appear as a more or less sharp dip in the number of transmitted electrons at an energy loss equal to the plasma frequency.

Bulk Plasmons

In Chapter 3 we explained electron oscillations in a metal using the Drude model of metals. In the Drude model the metal is treated as a 3D crystal of positively charged ions with a delocalized electron gas moving in a periodic potential of this ion grid, i.e., as a plasma. Based on that model, we calculated the complex dielectric constant of the bulk plasma as $\varepsilon(\omega,k) = \varepsilon_0\left[1 - \frac{\omega_p^2}{(\omega^2 + i\gamma\omega)}\right]$ (Equation 3.40). The plasma exhibits resonance absorption at the bulk plasmon frequency ω_p, defined as $\omega_p = \sqrt{\frac{ne^2}{m_e\varepsilon_0}}$ (Equation 3.41), and at frequencies above ω_p, the metal becomes transparent and dielectric-like, whereas below it is reflective and conductive. The bulk plasmon resonance of silver, for example, occurs at $\hbar\omega_p = 3.76$ eV. In Figure 5.27, we saw that the dispersion relation, the relation between the energy of the system and its corresponding momentum $\mathbf{k}(\omega)$, in a vacuum, is a linear function of ω with a slope $c = v_g = v_p$ (velocity of light = group velocity = phase velocity). The straight line in Figure 5.27 is referred to as the *light line*. We also established that the dispersion relation in a medium is no longer a straight line but is given by: $\mathbf{k}(\omega) = \frac{\omega}{v_g} = \frac{n(\omega)\omega}{c} = \frac{\omega}{c}\sqrt{\varepsilon\mu}$ (Equation 5.88). If we consider a highly conductive system, where the damping factor γ is very small, the real component of the dielectric constant $\varepsilon_r'(\omega,\mathbf{k}) \approx 1 - \frac{\omega_p^2}{\omega^2}$ (Equation 3.44) dominates with the imaginary component $\varepsilon_r''(\omega,\mathbf{k}) \approx \frac{\omega_p^2\gamma}{\omega^3}$ (Equation 3.45) tending to zero. In other words, Equation 3.44 is the expression for the dielectric constant when only the conduction electrons contribute, and damping is neglected. The Drude model, in this case, predicts a monotonous decrease of $\varepsilon_r(\omega,\mathbf{k})$ for decreasing frequency, and experiments confirm that $\varepsilon_r'(\omega,\mathbf{k})$ becomes zero when $\omega = \omega_p$ in such case (see Figure 3.9. When $\varepsilon_r'(\omega)$ becomes zero at $\omega = \omega_p$, the material supports free longitudinal collective modes for which all electrons oscillate in phase. The dispersion relation for bulk plasmons from Equation 5.88 is then given as:

$$\omega^2 - \omega_p^2 = c^2\mathbf{k}^2 \tag{5.150}$$

as shown in Figure 5.39, where we also mark the light line ($\omega = c\mathbf{k}$). It is clearly seen that when ω is less than ω_p there is no propagating solutions and the wave vector is imaginary.

Surface Plasmons

Surface plasmons are those plasmons that are confined to surfaces and that interact strongly with light, resulting in a polariton. They occur at the interface of a vacuum or a material with a positive

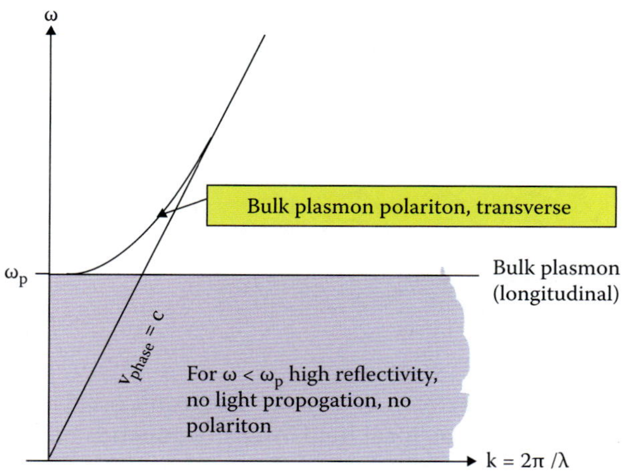

FIGURE 5.39 Dispersion of bulk plasmon. The light line is also marked ($v_{phase} = c$).

dielectric constant (a dielectric) and that of a negative dielectric constant (usually a metal or doped dielectric) (Figure 5.40).

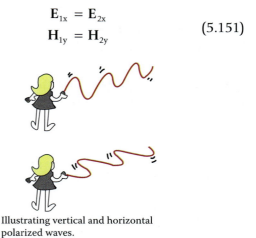

Surface plasmons are a subset of the "eigenmodes" of the free electrons in a metal or doped dielectric, with the electronic oscillations being excited parallel to the surface of the metal and a surrounding dielectric. Because all electrons may participate in the oscillations, plasmons can interact strongly with light of the right frequency. With wave vectors much smaller than the Fermi wave vectors of metals, these wave modes can be described by Maxwell's equations. Quantum mechanical considerations only are necessary for the shortest plasmon wavelengths. At low-surface plasmon wave vectors, the behavior of these surface modes can be understood intuitively. In this regime, they are essentially transverse in character, and these transverse fields generate a polarization in the dielectric, which is aligned with the stimulating field. In the metal, however, the polarization is in the opposite direction of the applied field owing to its negative dielectric constant. Now the stimulating field is creating equal and opposite electric displacements (*D*), in phase with each other across an interface. These opposing electric displacements serve to attract and confine the current to the interface, thus generating the collective electron oscillations of the surface plasmon. Figure 5.41 illustrates vertical and horizontal polarized waves. In vertical polarization (*s*-polarization from the German *senkrecht*), the **E** field is perpendicular to the plane of incidence. In horizontal polarization (*p*-polarization), the **E** field is parallel to the plane of incidence.

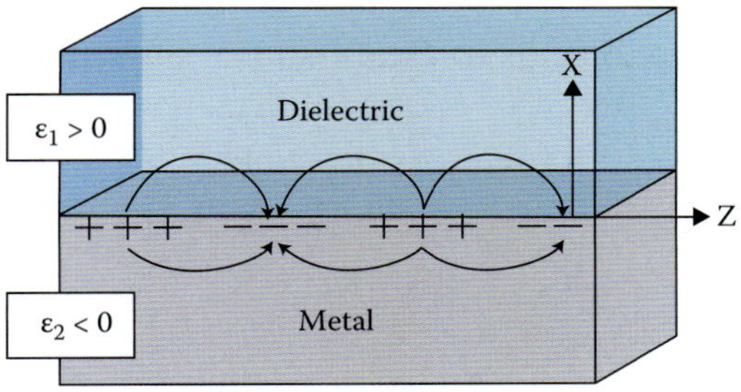

FIGURE 5.40 Surface plasmons at metal/dielectric interface.

Starting from the Maxwell's equations, we derive the characteristics for the plasmonic system shown in Figure 5.42a. Light that is *p*-polarized (see Figure 5.41) will create polarization charges at the interface as illustrated in Figure 5.42b. These charges give rise to surface plasmon modes. With an *s*-polarized incident radiation no such polarization charges are created at the interface; thus, these EM waves cannot excite surface plasmon modes.

We assume that the free current **J** is zero, that the relative permeability (μ_r) of both media to be unity, and that the wave is propagating in the *x*-direction. The EM field of surface plasmons propagates on the surface in the *x*-direction. The $\mathbf{H}_y$ is the magnetic field in the *y*-direction of the incident *p*-polarized wave. Thus, the components of the **E** and **H** fields are $\mathbf{E} = (E_x, 0, E_y)$ and $\mathbf{H} = (0, H_y, 0)$, and the interface between the metal and dielectric is at z = 0.

As we saw above, the constraints imposed on the transverse **E** and **H** field components at a boundary are that:

$$\begin{aligned} \mathbf{E}_{1x} &= \mathbf{E}_{2x} \\ \mathbf{H}_{1y} &= \mathbf{H}_{2y} \end{aligned} \tag{5.151}$$

FIGURE 5.41 The *p*-polarization: **E** field is parallel to the plane of incidence, and *s*-polarization: **E** field is perpendicular to the plane of incidence (German *senkrecht* = perpendicular).

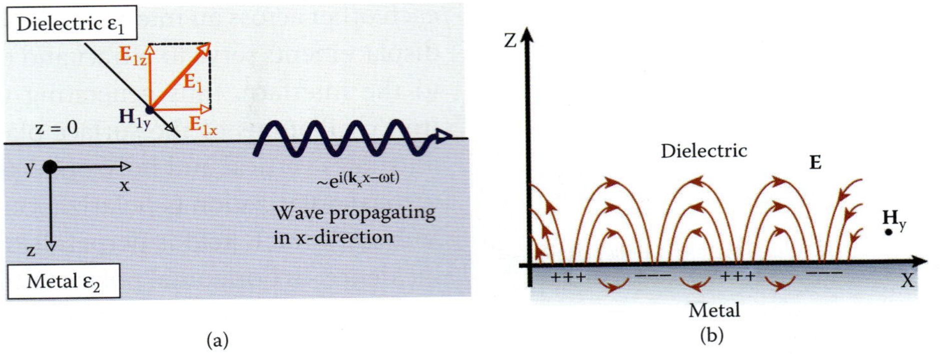

FIGURE 5.42 (a) Surface plasmons a subset of the "eigenmodes" of the free electrons in a metal. Electronic oscillations being excited parallel to the surface of the metal and a surrounding dielectric. The local field intensity depends on the wavelength. (b) Polarization charges are created at the interface between two materials.

i.e., the components of the **E** and **H** fields are continuous across the interface from metal to insulator.

It is also informative to look at the boundary condition for the normal component of the **D** field:

$$\mathbf{D}_{1z} = \mathbf{D}_{2z} \tag{5.152}$$

which can be rewritten as:

$$\begin{array}{ccc} \varepsilon_0 \mathbf{E}_{0z} + \mathbf{P}_{1z} & = & \varepsilon_0 \mathbf{E}_{0z} + \mathbf{P}_{2z} \\ \downarrow & & \downarrow \end{array} \tag{5.153}$$

Creation of the polarization charges

If one of the contacting materials is a metal, the electrons respond to the polarization by the *p*-polarized light (Figure 5.42, right), and this gives rise to surface plasmon modes. To model these waves we are looking for a localized surface mode, decaying into both materials as shown in Figure 5.43a, where the exponential dependence of the field E_z is shown. The decay in the two media is characterized by the penetration depth, i.e., the depth at which value of E_z falls to $1/e \approx 1/|\kappa_z|$ (see Equations 5.111 and 5.112: $\mathbf{k} \sim \mathbf{k}'' \sim \alpha \sim \kappa$ (because the absorption coefficient $\alpha = 2\kappa\omega/c = 2\mathbf{k}\kappa$, with κ the extinction coefficient).

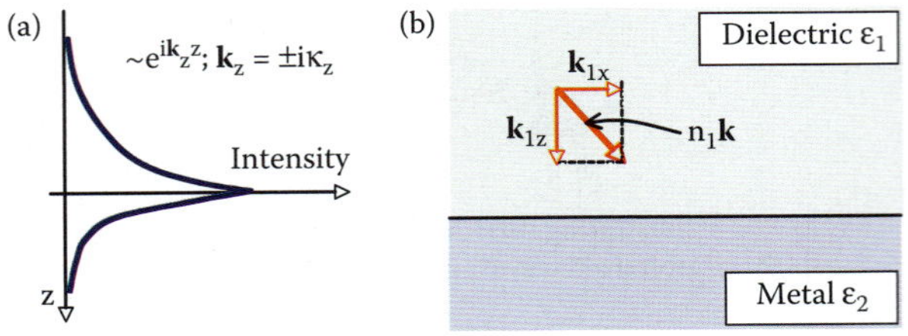

FIGURE 5.43 (a) We are looking for a localized surface mode, decaying into both materials so that $\mathbf{k}_z$ has to be imaginary; (b) n_1 is the dielectric constant of the dielectric.

Thus, a suggested solution for the surface plasmon mode of wave propagation in the dielectric may be written as:

$$\begin{aligned} \mathbf{H}_1 &= (0, \mathbf{H}_{1y}, 0) e^{i(\mathbf{k}_{1x}x + \mathbf{k}_{1z}z - \omega t)} \\ \mathbf{E}_1 &= (\mathbf{E}_{1x}, 0, \mathbf{E}_{1y}) e^{i(\mathbf{k}_{1x}x + \mathbf{k}_{1z}z - \omega t)} \end{aligned} \tag{5.154}$$

and for the metal:

$$\begin{aligned} \mathbf{H}_2 &= (0, \mathbf{H}_{2y}, 0) e^{i(\mathbf{k}_{2x}x + \mathbf{k}_{2z}z - \omega t)} \\ \mathbf{E}_2 &= (\mathbf{E}_{2x}, 0, \mathbf{E}_{2y}) e^{i(\mathbf{k}_{2x}x + \mathbf{k}_{2z}z - \omega t)} \end{aligned} \tag{5.155}$$

Substituting these suggested solutions in Maxwell's equations and applying the continuity boundary conditions, i.e., $\mathbf{k}_{1z} = \mathbf{k}_{2z}$ (because continuity holds for both **E** and **H**: Equation 5.151), we find the following additional constraint for the wave vector:

$$\frac{\varepsilon_1}{\mathbf{k}_{1z}} = \frac{\varepsilon_2}{\mathbf{k}_{2z}} \tag{5.156}$$

From the vector diagram in Figure 5.43b, we see that $n_1\mathbf{k}$ is given as:

$$(n_1\mathbf{k})^2 = \mathbf{k}_{1x}^2 + \mathbf{k}_{1z}^2 \text{ or also } \varepsilon_1(\omega, \mathbf{k}) \frac{\omega^2}{c^2} = \mathbf{k}_{1x}^2 + \mathbf{k}_{1z}^2 \tag{5.157}$$

and this enables us to establish the sign of $\mathbf{k}_z$:

$$\begin{aligned} \mathbf{k}_{1z} &= -\sqrt{(n_1\mathbf{k})^2 - \mathbf{k}_{1x}^2} \\ \mathbf{k}_{2z} &= +\sqrt{(n_1\mathbf{k})^2 - \mathbf{k}_{2x}^2} \end{aligned} \tag{5.158}$$

because $\mathbf{k}_{1z}$ and $\mathbf{k}_{2z}$ are of opposite signs, the condition in Equation 5.156 will be satisfied only if ε_1 and

ε_2 are of opposite signs as well. This is the case when one material is dielectric, $\varepsilon_1 > 0$, and the second material is metal, $\varepsilon_2 < 0$. It also follows from this equation, because $\mathbf{k}_{1z}$ and $\mathbf{k}_{2z}$ must both be imaginary, that:

$$\left(n_1\mathbf{k}\right)^2 - \mathbf{k}_{1k}^2 < 0 \quad \text{so that } \mathbf{k}_{1x} > n_1\mathbf{k} \tag{5.159}$$

and:

$$\left(n_2\mathbf{k}\right)^2 - \mathbf{k}_{2x}^2 < 0 \quad \text{so that } \mathbf{k}_{2x} > n_2\mathbf{k} \tag{5.160}$$

The condition in Equation 5.160 is always satisfied for metals because:

$$\mathbf{k}_{2x}^2 > \varepsilon_2\mathbf{k}^2 \tag{5.161}$$

with ε_2 for a metal <0. Of course, this result is to be expected from our initial assumptions. By defining the mode to have an exponential decay normal to the surface, we ensured an imaginary wave vector in this dimension. The absolute square of this positive quantity then adds to the light line wave vector to determine $\mathbf{k}_{1x}$ and $\mathbf{k}_{2x}$ (see Equations 5.159 and 5.160); hence $\mathbf{k}$ must always be greater than that of the free space field. This means that a surface plasmon has a greater momentum than a free photon at the same frequency. This makes a direct excitation (by shining light on a dielectric/metal interface) of a surface plasmon mode impossible (see inset in Figure 5.44)!

To calculate the dispersion relation for surface plasmon polaritons, we calculate $\mathbf{k}$ by substituting the expression for $\mathbf{k}_{1z}$ and $\mathbf{k}_{2z}$ from Equation 5.158 in Equation 5.156 to obtain:

$$\mathbf{k}_x = \mathbf{k}\sqrt{\frac{\varepsilon_1\varepsilon_2}{\varepsilon_1+\varepsilon_2}} = \frac{\omega}{c}\sqrt{\frac{\varepsilon_1\varepsilon_2}{\varepsilon_1+\varepsilon_2}} \tag{5.162}$$

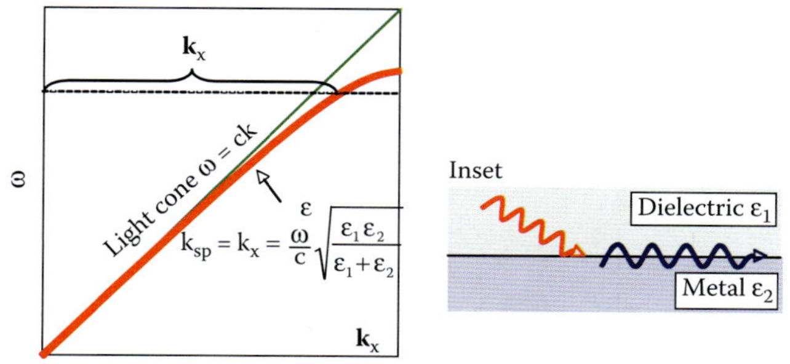

FIGURE 5.44 Dispersion of a surface plasmon. The surface plasmon mode always lies beyond the light line, i.e., it has greater momentum than a free photon of the same frequency. Inset: It is impossible to excite a plasmon mode by shining light on a dielectric/metal interface.

The surface plasmon condition is $\varepsilon_1 = -\varepsilon_2$, which occurs at high $\mathbf{k}_x$ when $\omega = \omega_{sp}$, which we calculate from:

$$\varepsilon_1 = 1 - \frac{\omega_p^2}{\omega^2} = -\varepsilon_2 \text{ so that } \omega_{sp} = \frac{\omega_p}{\sqrt{1+\varepsilon_2}} \tag{5.163}$$

In the case we deal with air as the dielectric ($\varepsilon_1 = 1$), the resonance frequency $\omega_{sp} = \omega_p/\sqrt{2}$ (see Table 5.6).

The power of plasmonics can be understood from an inspection of the dispersion relationship in Equation 5.162. From this dispersion relationship it is important to notice that the wave vector is no longer a linear function of the permittivity of a standard dielectric. Because we have the sum of dielectrics of opposite sign in the denominator, very large wave vectors are possible. Therefore, surface plasmons have the unique ability to achieve very short wavelengths ($\mathbf{k}_s = 2\pi/\lambda_{sp}$) at modest optical frequencies. The reason surface plasmon modes can achieve these anomalously high wave vectors at visible frequencies is because they are mediated by electrons rather than free space optical fields.

Because these modes are mediated by collective electron oscillations with subangstrom wavelengths, they may achieve very large optical wave vectors. From the same equation it follows that, using geometries of conductors (such as metals or doped semiconductors) with dielectrics (such as air or glass), surface plasmon modes at optical frequencies can be created with effective indices of refraction that are orders of magnitude higher than those of the constituent materials. In fact, these indices can be so high as to create x-ray wavelengths (less than 10 nm) with visible frequencies. This is of major consequence for microscopy: it is evident that deeply subwavelength focal spots cannot be formed through conventional focusing using a lens system or microscope objective. This is due primarily to the lack of high-index media at visible frequencies. But here we show that we are able to achieve a high effective refractive index with conventional optical materials. Although the permittivities of these material combinations may be modest, their geometry and interactions create an effective refractive index much larger than that available from conventional transparent media.

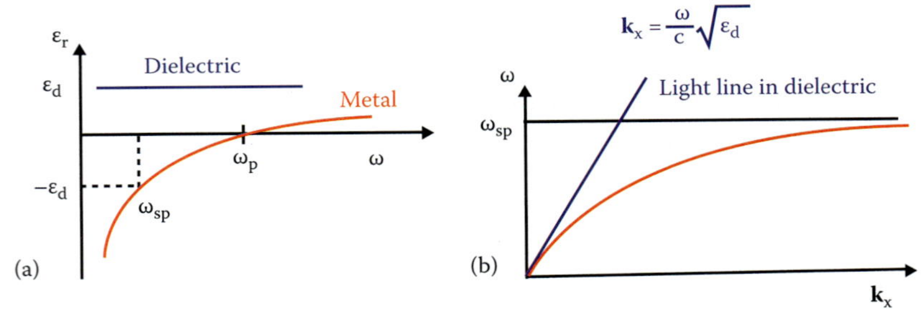

FIGURE 5.45 (a) Plot of dielectric constant for metal and dielectric. (b) Dispersion plot of surface plasmon (plot of ω, **k**).

Stimulation of surface plasmons with a purely optical field is not easy. The dispersion relation always lies at higher wave vectors than the light line ($\omega = c\mathbf{k}$) (see Figure 5.44). Because of this difference in wave vector, the plasmon field cannot efficiently couple to radiating modes. Conversely, free space optical fields cannot directly stimulate surface plasmons unless a mechanism introduces additional momentum. For example, this can be done when the light travels through a media of refractive index *n* greater than the index of refraction of the vacuum/glass. Thus, by tuning the angle of incidence of the light, the **k** vector (parallel to the surface) will match the resonant condition for SP excitation. In Figure 5.45a we show the dielectric constant of the metal (ε_m, red line) and the dielectric (blue line). At $\omega = \omega_{sp}$ we have $\varepsilon_m = -\varepsilon_d$. Figure 5.45b, is one of ω versus $\mathbf{k}_x$. At $\omega = \omega_{sp}$ (with $\varepsilon_m = -\varepsilon_d$), $\mathbf{k}_x \to \infty$, and at low frequency (where $\varepsilon_m \to \infty$) the slope of the dispersion curve approaches the light line in the dielectric, or:

$$\mathbf{k}_x \approx \frac{\omega}{c}\sqrt{\varepsilon_d} \qquad (5.164)$$

In Figure 5.45b, we see that $\mathbf{k}_x$ based on Equation 5.162 must automatically be lying below the light line based on Equation 5.164.

Special geometries are required to match the surface wave vector $\mathbf{k}_{sp}$ with that of the incoming light producing it. Surface plasmons can be excited by:

Light (photons)
- Excitation from high-index medium in the Kretschmann geometry (Figure 5.46a and b)
- Coupling using grating (Figure 5.46c)
- Coupling using subwavelength scatter points (Figure 5.46d)

Electrons

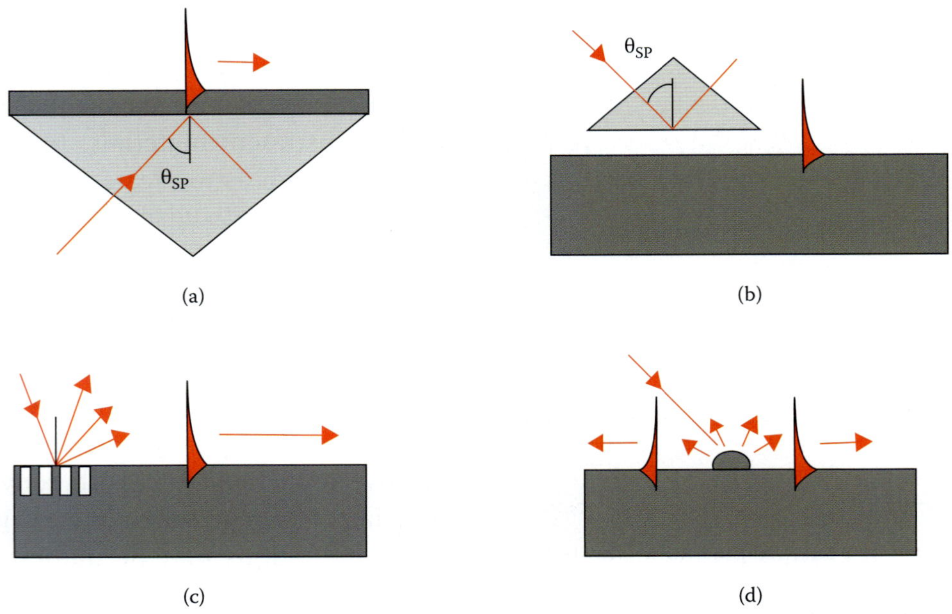

FIGURE 5.46 Excitation methods of surface plasmons. Excitation from high-index medium in Kretschmann geometry (a, b). Coupling using a grating (c). Coupling using subwavelength scatter points (d).

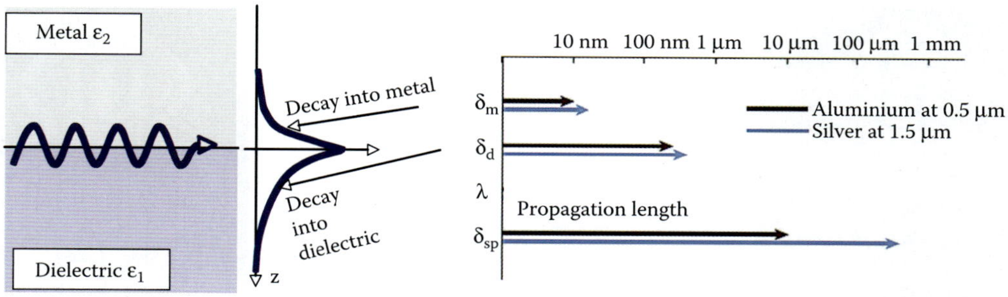

FIGURE 5.47 Decay lengths in a dielectric material (δ_d) and in a metal (δ_m), and the propagation length of the surface plasmon at resonance (δ_{sp}) for aluminum and silver.

From Equation 5.158 the penetration depth in the metal and the dielectric can be calculated:

$$
\begin{aligned}
k_{1z} &= -i\kappa_{1z} \\
k_{2z} &= +i\kappa_{2z}
\end{aligned}
\tag{5.165}
$$

($\sim e^{ik_z}$; $\kappa_z = \pm i\kappa_z$). As expected, and illustrated in Figure 5.47, the decay length in the dielectric material (δ_d) is considerably longer than in the metal (δ_m). The propagation length of the surface plasmon at resonance is the longest (δ_{sp}).

In Table 5.5 we list $\hbar\omega_{sp}$ for Au, Al, and Ag on GaN. The penetration depths into GaN are also listed.

Localized Plasmons

Whereas in semiconductor structures with reduced dimensionalities (quantum wells, quantum dots, and quantum wires) electron confinement leads to quantization of energetic states (see Chapter 3), metallic nanostructures exhibit major changes in their optical spectra because of light absorption and the creation of plasmonic waves at the surfaces of these structures. Whereas resonant coupling of light into a surface plasmon at a metal dielectric interface is difficult, coupling into surface plasmons on a metal nanoparticle is relatively easy. In the case of metal nanoparticles, surface plasmon waves are localized and are not characterized by a surface plasmon wave vector, $\mathbf{k}_{sp}$, and one refers to *localized surface plasmons*. There is no need for a special geometry to induce these localized oscillations. If the sphere is small compared to a wavelength of light, and the light has a frequency close to that of the SP, then the SP will absorb energy. The frequency of the SP depends on the dielectric function of the nanomaterial and the shape of the nanoparticle. For a gold spherical particle, the frequency is about 0.58 of the bulk plasma frequency. Thus, although the bulk plasma frequency is in the UV, the SP frequency is in the visible (close to 520 nm). Because it is possible to control the nanoparticle's surface shape, one can control the types of surface plasmons that can couple to it and propagate across it. This in turn controls the interaction of light with the surface. These effects are illustrated by the historic stained glass that adorns medieval cathedrals (Figure 5.48). In this case, the color is given by metal nanoparticles of a fixed size that interact with the optical field to give the glass its vibrant color. In modern science, these effects have been engineered for both visible light and microwave radiation. To produce optical range surface plasmon effects involves producing surfaces that have features <400 nm.

TABLE 5.5 $\hbar\omega_{sp}$ for Au, Al, and Ag on GaN and the Penetration Depths into GaN

Metal on GaN	$h\hbar sp$ (eV)	Penetration Depth (nm)
Au	2.2 (~560 nm)	33
Al	5 (~250 nm)	77
Ag	2.92 (~410 nm)	40

Plasmonic Applications I

As mentioned in the introduction of this section, plasmons have been considered a means of transmitting information on computer chips because plasmons can support much higher frequencies (into the 100-THz range, whereas conventional wires become very lossy in the tens of GHz). At first glance, guiding light signals with metal structures appear odd as metals are known for high optical losses, and it is obvious that fiberoptic networks are able to carry more information and with greater fidelity than electrons in copper or coaxial cable. Optical fibers are

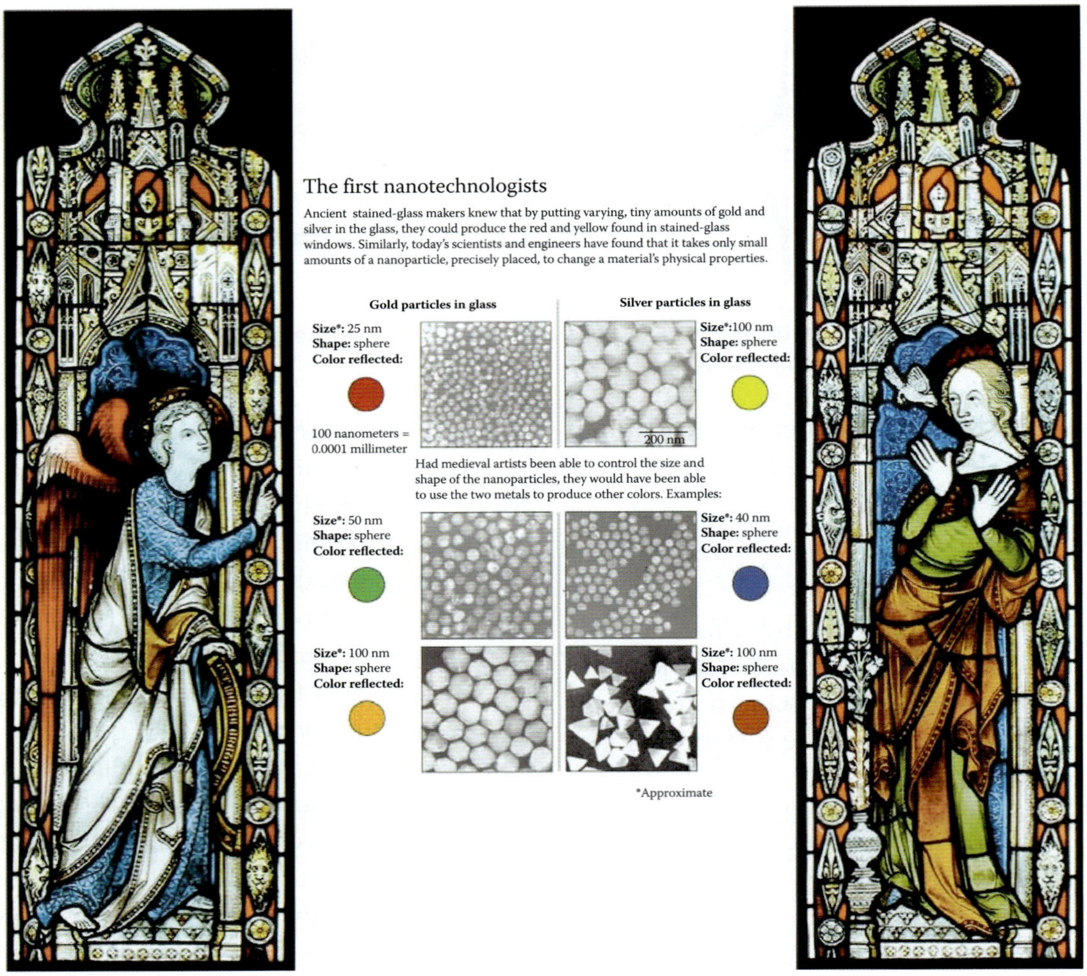

FIGURE 5.48 Medieval artisans unknowingly became nanotechnologists when they made red stained glass by mixing gold chloride into molten glass. That created tiny gold spheres, which absorbed and reflected sunlight in a way that produces a rich ruby color.

also immune to electrical interference, and they are chemically inert (glass), safe (they carry light instead of electron currents), and lightweight. They are very difficult to tap into (safe for sensitive data transfer), and finally photons travel much faster than electrons: $c = 3.0 \times 10^8$ m/s (vacuum) versus v_F (Fermi speed) = 1.57×10^6 m/s (copper wire). No wonder many scientists predicted that photonic devices would someday replace electronic ones. The reason why this has not happened is that the performance and size of photonic devices are diffraction limited so that the width of an optical fiber needs to be at least twice that of the light's wavelength. In case we use near-infrared wavelengths of about 1500 nm, we end up with photonic structures more than 15 times larger than Si-based electronic devices with features smaller than 65 nm. In other words, photonics, although high speed and high bandwidth, are harder to miniaturize than electronics because of the diffraction of light in the former (see Figure 5.49b). Moreover, in the case of a fiberoptic, one also has to deal with bending losses (see Figure 5.49a).

Plasmonics offers a way out of this dilemma as using metal to guide light avoids the optical fiber bending problem and the diffraction problem of light. Envision alternating charge densities on metal surfaces—surface plasmons—as the equivalent of an AC current traveling along an ordinary wire with the important difference that the frequency of the optical signal is more than 400,000 GHz, whereas the AC current is 60 Hz. This makes for plasmon circuits that can carry much more data than electronic ones.

Many photonic applications of surface plasmons and localized surface plasmons derive from the local

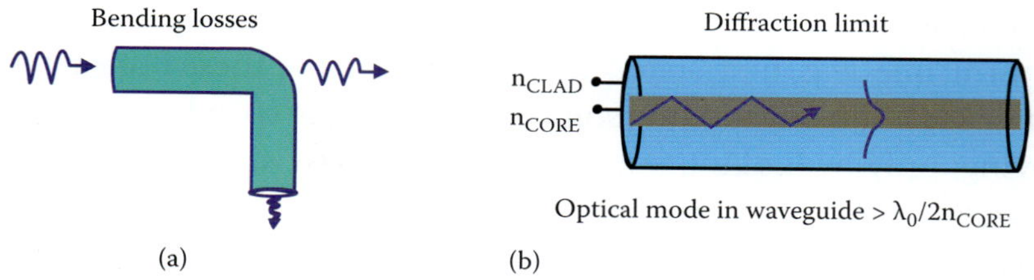

FIGURE 5.49 Problems with photonic devices. Bending (a) and diffraction losses (b).

electrical field enhancement under plasmon resonance conditions. Plasmon resonances, associated with large time-varying polarizations, can lead to extremely strong local electric fields. The local electric fields resulting from these plasmon effects are evanescent, and so they have very short ranges. The field enhancement in the small volume of the medium surrounding the metal nanoparticles or metal/dielectric interface has been used, for example, for guiding light through dimensions smaller than the wavelength of the incident light in plasmonic guiding. Plasmonic guiding in lithography is described further below and in Volume II, Chapter 2 on next-generation lithographies (NGLs) where plasmonics is proposed as a new high-resolution lithography. In Volume III, Chapter 6 on metrology and further below we learn that plasmonics also enables aperture-less near-field scanning optical microscopes (NSOMs). One other extremely interesting and important phenomenon related to plasmons is surface-enhanced Raman spectroscopy (SERS) that can be used to detect single molecules. In Raman spectroscopy the light scattered inelastically from a molecule provides information about the molecule's quantum states (see Figures 5.104 and 5.105). Discovered around 1974, it was found that a Raman signal from pyridine on a silver film is ~10^6 times larger than what one would expect based on the Raman signal for the same molecule in solution. The record enhancement inferred for SERS is now ~10^{14} (!!) if the molecules are attached to nanometer-sized metal surfaces.

Beyond Maxwell

The Lorentz Transformation

Hendrik Antoon Lorentz (1853–1928) (Figure 5.50), born in Arnhem, the Netherlands, refined Maxwell's electromagnetic theory and was the joint winner (with Pieter Zeeman) of the Nobel Prize for Physics in 1902 for his theory of electromagnetic radiation, which, confirmed by findings of Zeeman, gave rise to Albert Einstein's special theory of relativity. Lorentz proposed that light waves were caused by oscillations of an electric charge in the atom. What is remarkable is that he conceived this idea even before the existence of electrons was proved. Lorentz is also the man of the Lorentz force we introduced in Chapter 3 (Equation 3.64). He did get it a bit wrong with his proposal that the motion of a body through the ether resulted in a tiny contraction of all physical bodies in the direction of the motion—called the *Lorentz-FitzGerald contraction*, or *aether squeeze*. But the so-called *Lorentz transformation* got Einstein's attention, and he chose it over the Galilean transformation, doing away with the notion of an ether once and for all.

The speed of light c appears as a fundamental constant in Maxwell's equations, and one might ask against what relative frame of reference is this measured? Maxwell viewed light as waves moving in the medium aether. Because it was assumed that a substance filled all space and provided a medium (aether) through which light waves traveled, it was postulated that the velocity of light c should be measured with respect to the rest frame of the aether.

FIGURE 5.50 Hendrik Antoon Lorentz (1853–1928).

Relativity tells us how to relate measurements in different frames, and its principle was first stated by Galileo: no experiment inside a steadily moving ship will show that it is moving; only by looking outside can one detect motion—i.e., relative motion.

This principle is stated mathematically in the Galilean transformation for going from one reference frame to another. In the case of a reference frame moving with a speed v in the x-direction with respect to another reference frame, this yields:

$$\begin{aligned} x' &= x - vt \\ y' &= y \\ z' &= z \\ t' &= t \end{aligned} \tag{5.166}$$

Thus, in Galilean relativity one has a simple velocity addition law (Figure 5.51).

$$v = v' + v_b \tag{5.167}$$

Newton's laws were found to remain the same under the Galilean transformation. As a consequence, it is impossible to tell, by making mechanical experiments, whether a system is moving, i.e., they are invariant to a Galilean transformation. Newton articulated the idea that there is no such thing as "absolute motion" or reference frame. All "inertial" frames are equivalent, and the laws of physics are the same in all reference frames.

But there was trouble ahead: the Galilean transformation did not apply to Maxwell's equations. In other words, Maxwell's equations do not remain in the same form under the Galilean transformation (they are NOT invariant to a Galilean transformation). As a consequence, in a moving spaceship electrical and optical phenomena should be different from those in a stationary spaceship, and one should be able to use these optical phenomena to determine the absolute speed of the spaceship. Soon it was discovered that when the speed of light was involved the Galilean transformation was not applicable and that another transformation between reference frames, the Lorentz transformation, had to be invoked.

The realization that there was something special about the speed of light came from the Michelson-Morley experiment. The Michelson-Morley experiment (Figure 5.52) was yet another attempt to detect the aether. If aether existed, then the speed of light should be dependent on it. If the speed of light were constant with respect to the aether, then the measured speed of light in perpendicular directions should be different because of the movement of the earth through the aether. In 1887, A.A. Michelson and E.W. Morley refined an experiment that Michelson had performed a few years earlier. They split a beam of light in two, sending the two halves in different directions. Each ray of light was reflected back and the two beams combined again. Thus, if the earth is traveling at v with the aether, then light

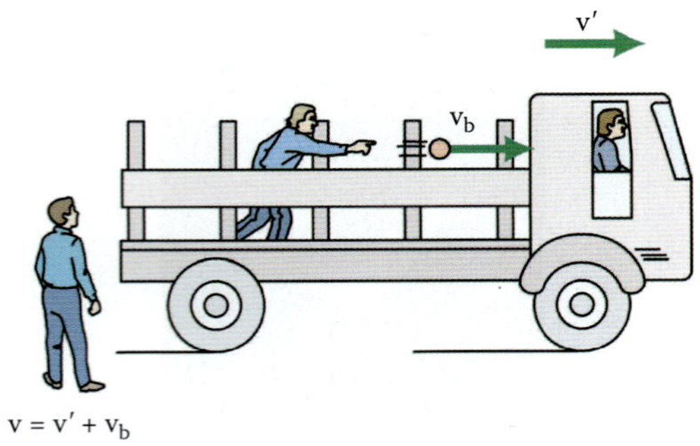

FIGURE 5.51 Illustration of Galilean addition of velocities.

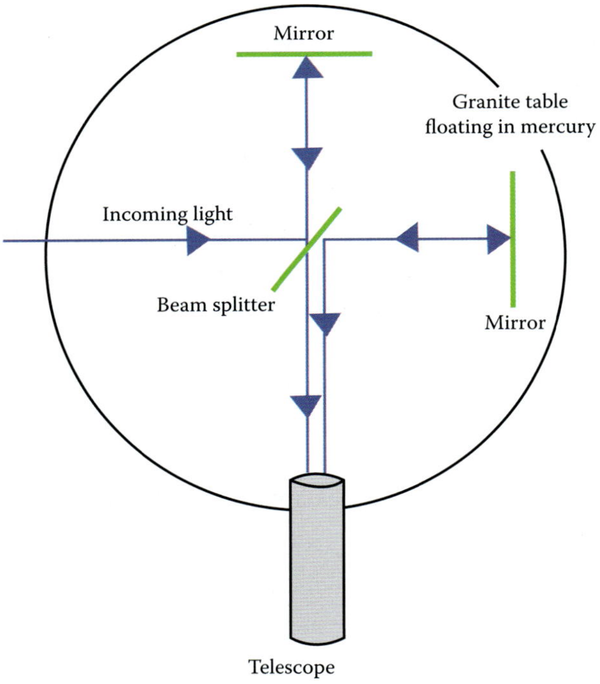

FIGURE 5.52 Michelson-Morley experiment. In 1887 Michelson and Morley set up this apparatus in which a light beam is split into two perpendicular beams and then recombined. By observing interference between the two beams, they showed that the speed of light was constant.

should have a velocity of $c + v$ (with respect to the aether). In the perpendicular direction, the velocity would just be c. By observing the interference patterns, they were able to determine the difference in the speed of light in the two perpendicular directions. Their result was astonishing: zero, or light speed, did not combine with Earth's velocity. The Michelson-Morley experiment implied that the velocity of light in vacuum, c, is independent of the frame of reference. Their result violated the Galilean law of addition of velocities (Equation 5.167).

However, the scientific community was not ready to dispense with the idea of an aether; to explain the Michelson-Morley experiment, Lorentz and the Irish physicist George Francis FitzGerald independently introduced the notion that motion through the aether resulted in a tiny contraction of all physical bodies in the direction of motion (the aether squeeze):

$$L_{parallel} = L_0\sqrt{1 - v^2/c^2} \tag{5.168}$$

In this way, the arm of the Michelson interferometer that pointed in the direction of Earth's motion ($L_{parallel}$) was supposed to be shorter than the arm at right angles. This explained the apparent absence of the relative motion of the Earth with respect to the aether, as indicated by the experiments of Michelson and Morley, and also paved the way for Einstein's special theory of relativity (see next section). In 1904, still believing in the concept of the aether, Lorentz thus replaced the Galilean transformation of reference frame with the following transformations:

$$\begin{aligned} x' &= \frac{x - vt}{\sqrt{1 - v^2/c^2}} \\ y' &= y \\ z' &= z \\ t' &= \frac{t - vx/c^2}{\sqrt{1 - v^2/c^2}} \end{aligned} \tag{5.169}$$

This transformation rendered Maxwell's equations reference frame-independent. The Lorentz transformation will result in a stationary observer recording an effect equivalent to the Lorentz contraction when observing an object in uniform motion relative to his system of coordinates. However, Einstein showed that this effect is not caused by the actual deformation of the body in question, as Lorentz had originally supposed, but by a change in the way space and time are measured. The consequences of the Lorentz transformation are mind-boggling: they imply that time is not universal, the relativity of simultaneous events, Fitzgerald-Lorentz contraction, and time dilation.

Einstein's Special and General Relativity

Special Relativity

Thus, science faced the following mutually exclusive choices: either continue to accept Newton's law and Galilean invariance or adapt Maxwell's equations and Lorentz invariance. Einstein, in 1905, introduced the theory of special relativity in which, following a suggestion made by Poincaré, he picked the Lorentz transformation instead of the Galilean transformation as the correct one. He proposed that "all physical laws should remain unchanged under Lorentz transformation."

Einstein's two postulates of special relativity are:

Postulate 1—The principle of relativity. The laws of nature are the same in all inertial frames of reference. Implications are that there is no way to detect absolute motion and that there is no preferred frame of reference (no aether).

Postulate 2—The universal speed of light. The speed of light, c, in a vacuum is the same in all inertial frames of reference. This implies that c does not depend on the source or the observer's velocity.

As a consequence of the second postulate, time appears dilated. Time-measuring instruments (clocks) must apparently slow down:

$$t' = \frac{t_0}{\sqrt{1 - v^2/c^2}} \tag{5.170}$$

and lengths seem to suffer (Lorentz) contraction for moving objects:

$$x' = \frac{x - vt}{\sqrt{1 - v^2/c^2}} \tag{5.171}$$

And finally, two events that are simultaneous in one inertial system are not, in general, simultaneous in another:

$$t_2' - t_1' = \frac{u(x_1 - x_2)/c^2}{\sqrt{1 - u^2/c^2}} \quad (5.172)$$

Einstein convinced the world that the speed of light is the same in any direction, which explains the null result of Michelson and Morley, and he also showed that there is no such thing as aether (nor any need for it): light is perfectly happy traveling in a vacuum.

Instead of the Galilean velocity addition (Equation 5.167) with $E = m_0c^2 + \frac{mv^2}{2}$, where m_0c^2 is for the stationary situation, Einstein suggested the world's most famous equation:

$$E = mc^2 \quad (5.173)$$

When applying a force to a stationary particle, we supply kinetic energy to the particle and it starts to move. From Equation 5.173, the total energy *E* and mass *m* must go up because *c* is constant, and according to the conservation of energy:

$$\frac{dE}{dt} = \mathbf{F} \cdot \mathbf{dv} \quad (5.174)$$

From Newton's second law:

$$\mathbf{F} = \frac{d(mv)}{dt} \quad (5.175)$$

and:

$$\frac{d(mc^2)}{dt} = \mathbf{v}\frac{d(mv)}{dt} \quad (5.176)$$

Multiplying both sides of Equation 5.176 with 2*m* gives us:

$$c^2 2m\frac{dm}{dt} = 2mv\frac{d(mv)}{dt} \text{ or } c^2\frac{d(m^2)}{dt} = \frac{d(m^2v^2)}{dt} \quad (5.177)$$

Integration of both sides results in:

$$m^2c^2 = m^2v^2 + C \quad (5.178)$$

The constant C can be obtained by substitution v = 0 at which $m = m_0$, or:

$$m^2c^2 = m^2v^2 + m_0^2c^2 \quad (5.179)$$

or:

$$m = \frac{m_0}{\sqrt{1 - \frac{v^2}{c^2}}} \quad (5.180)$$

Thus, Newton's laws are the low-velocity limits of Einstein's special relativity theory. The Einstein modification of p = mv is then given as:

$$p = \frac{m_0 v}{\sqrt{1 - v^2/c^2}} \quad (5.181)$$

What happens now if a constant force acts on a body? In Newtonian mechanics the body keeps picking up speed, whereas in relativity the body keeps picking up momentum, not speed. This implies that *m* must be a function of speed *v* as evident from Equation 5.180.

General Relativity

Einstein's magnum opus is the general relativity theory. Einstein recognized that his special relativity theory was violated by Newton's theory of gravity because the latter implied that gravity effects were felt instantaneously, i.e., faster than the speed of light! The general relativity theory is the geometric theory of gravitation, incorporating and extending the theory of special relativity to accelerated frames of reference and introducing the principle that gravitational and inertial forces are equivalent. From this theory, concentrations of matter or energy cause space-time to curve, and this curvature deflects the trajectories of particles, just as observed for particles in a gravitational field. Einstein was able to calculate from this new theory that gravity disturbances move at precisely the speed of light. According to Einstein, the agent of gravity then is the fabric of the cosmos, and in the presence of a massive object like the sun that fabric becomes distorted. An artistic rendition of this concept is shown in Figure 5.53.

Summarizing, all macroscopic electromagnetic phenomena can be explained in terms of Maxwell's equations, the continuity equation, and the Lorentz force equation. In Chapter 3 we saw that a description of light as purely consisting of waves fails to explain a large body of experimental facts such as the photoelectric effect and blackbody radiation. Here is where quantum mechanics comes in, with as one of its basic tenets that light and material

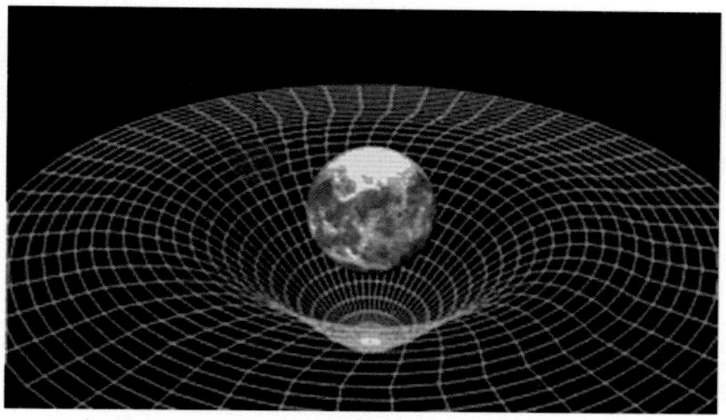

FIGURE 5.53 The fabric of the cosmos is distorted by a massive object. The Earth creates a deep hollow in the fabric of space-time.

objects both display wave and particle behavior. Special relativity becomes manifest when things are moving very fast, and general relativity comes into play when things are very massive and the warps in space and time are correspondingly large. The general relativity theory is inherently classical, and this bothers many physicists as it makes for a difficult marriage with quantum physics describing the behavior of particles and forces in the atomic and subatomic realm. A quantum theory of gravity is a holy grail for physicists. A promising approach to a quantum theory of gravity is string theory, which has been explored since the 1970s.

Two excellent books to read on the quest for the ultimate theory are Brian Greene's *The Elegant Universe*[7] and Barry Parker's *Search for a Super-Theory*.[8] A list of recommended books to read on quantum mechanics was presented in the Chapter 3 Further Reading list.

Optical Properties of Materials

Introduction

Now that we know how reflectance *R*, transmittance *T*, and absorbance *A* relate to refractive index *n* and extinction coefficient κ, as well as how these relate to electrical susceptibility $\chi(\omega)$, we must yet establish a physical model for $\chi(\omega)$. Optical properties of materials are determined by two main factors: first, how easy is it to move charges using an EM wave [$\mathbf{P} = \varepsilon_0 \chi(\omega)\mathbf{E}(\omega)$], and, second, how many charges are available (*n*)? Based on these criteria, optical behavior of materials can crudely be divided in:

- Atoms/ions, where optical transitions take place between electron orbits and lead to absorption lines, ionization energies.
- Molecules, where optical transitions again take place between electron orbits, but where one also has vibrations (in the meV range) and the optical excitation of those molecular vibrations (IR and Raman).
- Metals, with large concentration of free electrons, exhibit high reflectivity and strong absorption.
- Insulators, where electrons are strongly bound to atoms ($E_{gap} > \sim 4$ eV), resulting in a low absorption and refraction.
- Semiconductors, where electrons are "somewhat bound" ($E_{gap} < \sim 4$ eV) and are transparent in the IR and absorb in the visible and UV, and light can generate excitons (bound electron-hole pairs).

In this section we establish the optical dispersion of the dielectric constant [$\varepsilon(\omega)$], the refractive index [$(n)\omega)$], the absorption [$\alpha(\omega)$], and the reflectance [$R(\omega)$] of these types of materials based on a microscopic physical model, either the Drude model or the more sophisticated Drude-Lorentz model. First we define the polarization **P** of a material and briefly reiterate the salient points of the Drude model, encountered first in Chapter 3, and then, to accommodate both free and bound electrons (electrons bound to atom core) in a material, we introduce the Drude-Lorentz physical model for $\chi(\omega)$; this allows us to cover optical properties not only of all metals but also of insulators and semiconductors.

The Microscopic Origin of the Frequency Dependency of χ, ε, and *n* in Solids-Polarization **P**

The polarization **P** of a medium is the induced dipole moment per unit volume (C/m³) (Figure 5.54). It is not the first time we touch on polarization **P** of matter in this book. We broached the topic first when introducing the Drude model, where we discussed the polarization of the free electron gas (FEG) with respect to the immobile ions in a metal. Polarization of the medium comes about as a result of small displacements of charges in an electrical field. For

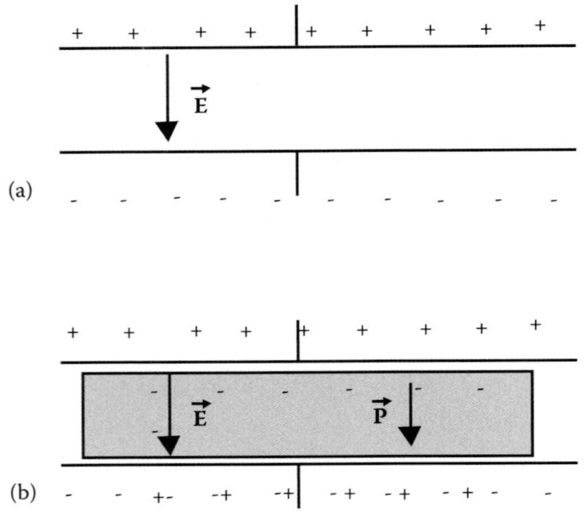

FIGURE 5.54 Polarization **P** of a medium. (a) P = 0 (no dielectric). (b) **P** ≠ 0 (dielectric). **P** points from – to +.

example, polar molecules align in a field, and nonpolar molecules have their electron cloud distorted. It is easy to see how at optical frequencies electrons might still swing along with a fast-changing electromagnetic field but that nuclei might be too heavy to follow, causing an electron cloud polarization. Different polarization mechanisms exhibit different dynamic behavior in time, and as a consequence, electromagnetic radiation absorption occurs at different frequencies. Polarization mechanisms include space-charge polarization, orientation polarization, ionic polarization, and electronic polarization (see The Drude-Lorentz Model for an example of the latter type).

As demonstrated in Chapter 3, it is the polarization $\mathbf{P}(\omega,t) = \chi_e\varepsilon_0\mathbf{E}(\omega,t)$ (Equation 3.36) of a material that is ultimately responsible for the frequency dependency of the dielectric constant ε and of the refractive index *n* because all these parameters are linked. Thus, a general and realistic relation between **P** and **E** includes the fact that the relation between **P** and **E** is dynamic:

$$\mathbf{P}(\mathbf{k},\omega) = \chi_e(\mathbf{k},\omega)\varepsilon_0\mathbf{E}(\mathbf{k},\omega) \quad (5.182)$$

The polarization of a material along the applied field direction results in an opposing response field and consequently a reduced total field. In other words, some of the external field applied over a material disappears and is replaced by polarization **P**. For a linear, homogeneous, and isotropic medium, the electric susceptibility, χ_e, is a scalar, but for an anisotropic material the electric susceptibility is a tensor, and the polarization **P** can point in a different direction from the field **E**; **D** and **E** are also not necessarily collinear. In this case, Equation 3.36 is modified to:

$$\mathbf{P}_i = \sum_j \chi_{ij}\varepsilon_0\mathbf{E}_j \quad (5.183)$$

From the expression linking the dielectric constant with the polarization **P**: $\mathbf{D} = \varepsilon_0\mathbf{E} + \mathbf{P} = (1 + \chi_e)\varepsilon_0 \mathbf{E} = \varepsilon(\omega)\mathbf{E}$ (Equation 3.37), we can establish the link between the relative permittivity $\varepsilon_r(\omega)$ (also called *dielectric constant*) and the electric susceptibility tensor as:

$$\varepsilon_r = 1 + \chi_e(\omega) \quad (5.184)$$

Because $n(\omega)^2 = \varepsilon\mu$ (Equation 5.22), the refractive index *n* can be derived from χ_e, assuming that $\mu = 1$, as:

$$n = \sqrt{1 + \chi_e(\omega)} \quad (5.185)$$

The Drude-Lorentz Model

Drude Model Recap

In the Drude physical model, we learned in Chapter 3, the only conduction mechanism is that resulting from a free electron gas (FEG). For a time harmonic electrical field $[\mathbf{E}(t) = \mathbf{E}_m e^{-\omega t}]$, we obtained the displacement *x* of the electron gas $x(t) = \frac{e\mathbf{E}(\omega)}{m_e(\omega^2 + i\gamma\omega}$ (Equation 3.34) from the equation for the momentum per electron: $m_e\frac{d^2x}{dt^2} + m_e\gamma\frac{dx}{dt} = -e\mathbf{E}(t) = -e\mathbf{E}_{(\omega)}e^{-i\omega T}$ (Equation 3.33). The macroscopic polarization **P** was then calculated as the product of this displacement *x*(*t*), and the electron density ne ($\mathbf{P} = -nex$), so that for the electric displacement we obtained $\mathbf{D} = \varepsilon(\omega,\mathbf{k})\mathbf{E} = \varepsilon_0\mathbf{E} + \mathbf{P} = \varepsilon_0\mathbf{E} - \frac{ne^2\mathbf{E}}{m_e(\omega^2 + i\gamma\omega)}$ (Equation 3.39). When electrons are free, as in a metal, the damping factor γ in this expression comes about because of electron scattering (γ = 1/τ, where τ is the electron scattering time). From the expression for the displacement field **D**, we then derived the complex dielectric constant as $\varepsilon(\omega,\mathbf{k}) = \varepsilon_0\left[1 - \frac{\omega_p^2}{(\omega^2 + i\gamma\omega)}\right]$

(Equation 3.40), where ω_p is the plasma frequency, defined as $\omega_p = \sqrt{\frac{ne^2}{m_e\varepsilon_0}}$ (Equation 3.41). If we consider a highly conductive system, where the damping factor γ tends to zero, the real component of the dielectric constant $\varepsilon_r'(\omega, \mathbf{k}) \approx 1 - \frac{\omega_p^2}{\omega^2}$ (Equation 3.44) dominates because the imaginary component $\varepsilon_r''(\omega, \mathbf{k}) \approx \frac{\omega_p^2\gamma}{\omega^3}$ (Equation 3.45) tends to zero. Based on Equation 5.118, the complex refractive index in this case can be written as $n_c(\omega) = \sqrt{\varepsilon_r'(\omega)} = \sqrt{1 - \frac{\omega_p^2}{\omega^2}}$. The Drude model predicts a monotonous decrease of $\varepsilon_r'(\omega, \mathbf{k})$ for decreasing frequency, and experiments confirm that $\varepsilon_r'(\omega, \mathbf{k})$ becomes zero when $\omega = \omega_p$. When $\varepsilon_r'(\omega)$ becomes zero at $\omega = \omega_p$, the material can support free longitudinal collective modes for which all electrons oscillate in phase. Such a collective charge oscillation is called a *plasmon*. Finally, in a metal where both bound electrons $[\varepsilon_B(\omega)]$ and conduction electrons $[\varepsilon_{r(\omega)}]$ contribute, an effective dielectric constant $\varepsilon_{Eff(\omega)}$ was defined as $\varepsilon_{Eff}(\omega) = \varepsilon_B(\omega) + \varepsilon_r(\omega)$ (Equation 3.48). As soon as there are bound charges involved, as in some metals and in all dielectrics, the Drude model fails, and the Drude-Lorentz model needs to be invoked.

The Drude-Lorentz Model

Introduction The Drude model advanced a first physical concept for $\chi(\omega)$, but as we just saw, it only accounts for the free electron contributions to the electric susceptibility. The Lorentz model takes into account the contribution stemming from bound electrons. In this physical microscopic model, atoms and molecules are replaced by a set of oscillators, resonant at some frequency:

$$\omega_0 = \sqrt{\frac{k}{m}} \tag{5.186}$$

These oscillators come with a restoring force $\mathbf{F}_r$ where:

$$\mathbf{F}_r = -k\mathbf{r} = -m\omega_0^2\mathbf{r} \tag{5.187}$$

with $\mathbf{r}$ being the oscillator's displacement. The model does not specify the nature of the oscillating dipole and makes only very basic assumptions about the restoring forces involved. For these reasons it applies to a wide variety of situations. For electronic polarization of atoms in the visible and near-infrared frequencies (10^{14}–10^{15} Hz), the mass *m* in the above equations is the mass of the electron. For molecular and lattice vibrations in the mid- and far-infrared frequencies (10^{12}–10^{13} Hz), *m* is the mass of the nucleus. Depending on the exact atom, the nucleus is three or four orders of magnitude heavier than the electron, and the interatomic force is similar to, or less than, the electron-nucleus force (*k*); thus, using $\omega_0 = \sqrt{\frac{k}{m}}$ one expects nuclear vibrations to be a factor of 50–1000 times slower than electronic resonances. As a consequence, it is infrared rather than visible light that strongly interacts with nuclear vibrations, and molecular rotations interact with even longer wavelengths in the far-infrared. Free electrons in metals and semiconductors also can be described by the Drude-Lorentz model as a special case. Because there is no restoring force ($\mathbf{F}_r$) pulling charges back to their equilibrium positions, the Drude-Lorentz model simplifies back to the Drude model.

Atomic Polarizability $\chi_{a,e}$ Imagine a nucleus and electrons connected by springs with spring constant *k* as illustrated in Figure 5.55. For electron oscillations bound to an atom, a restoring force related to the strength of the bond of the electrons to the ion cores is introduced in the electron momentum equation. The restoring force is assumed to be linear in the displacement $\mathbf{r}$, which remains true for sufficiently small amplitudes (Hooke's

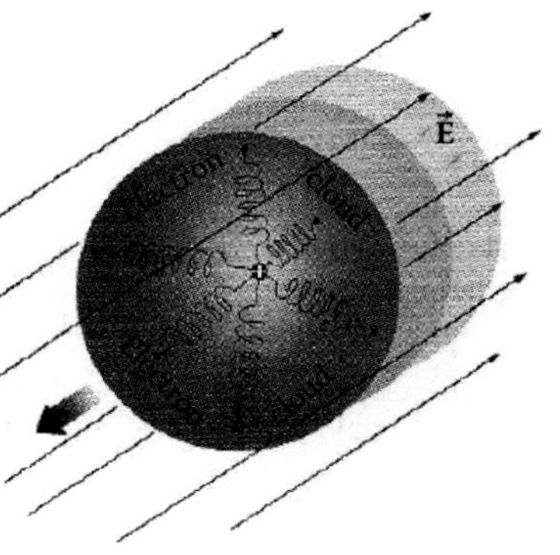

FIGURE 5.55 Oscillator model of electrons bound to an atom. The atoms and molecules are replaced by a set of harmonically bound electron oscillators, resonant at some frequency ω_0.

law). The damping force or friction loss, just as in the Drude model, is given by $F_r = m_e\gamma\frac{dr}{dt}$. This damping force is assumed to be linearly related to the charge velocity, and γ (in units of s^{-1}) is the damping coefficient. The resulting equation of motion is given as:

$$m_e\frac{d^2\mathbf{r}(\omega)}{dt^2} + m_e\gamma\frac{d\mathbf{r}(\omega)}{dt} + m_e\omega_0^2\mathbf{r}(\omega) = -e\mathbf{E}(t) = -e\mathbf{E}_m e^{-\omega t} \quad (5.188)$$

with a solution for the oscillation's amplitude given by:

$$\mathbf{r}(\omega) = -\frac{e}{m_e}\frac{\mathbf{E}(\omega)}{(\omega_0^2 - \omega^2 - i\omega\gamma)} \quad (5.189)$$

This expression is that of a driven damped oscillator where the motion matches the drive frequency. The maximum amplitude $\mathbf{r}(\omega)$ is reached when the denominator is at a minimum, i.e., when $\omega = \omega_0$ (resonance). For excitation at resonance ($\omega = \omega_0$), $\mathbf{r}(\omega)$ is entirely imaginary, and there is a 90° phase difference between $\mathbf{E}(\omega)$ and $\mathbf{r}(\omega)$. At high frequencies ($\omega >> \omega_0$) the amplitude $\mathbf{r}(\omega)$ vanishes, and at low frequencies ($\omega << \omega_0$) a finite amplitude is obtained.

For free electrons in a metal there is no restoring force ($F_r = -k\mathbf{r} = -m\omega_0^2\mathbf{r} = 0$), and this expression reduces back to the Drude AC model we derived for the case of free electrons in a metal in Chapter 3 (see Equation 3.34).

In a driven harmonic oscillator, amplitude and phase depend on frequency as sketched in Figure 5.56. At low frequency, one obtains a medium amplitude, and the displacement is in phase with the driving force. At resonance, a large amplitude is observed, and the displacement is 90° out of phase with the driving force. Finally, at high frequency, the amplitude vanishes, and displacement is in antiphase with the driving force. Note that it is the loss term $i\omega\gamma$ in Equation 5.189 that is responsible for the phase shift.

Based on Equation 5.182, we can define the polarization of a single atom as the charge (*e*) times the separation [$\mathbf{r}(\omega)$] or $\mathbf{P}_a = -e\mathbf{r}(\omega)$:

$$\mathbf{P}_a = \chi_{a,e}\varepsilon_0\mathbf{E}(\omega) \quad (5.190)$$

The atom susceptibility or atomic polarizability $\chi_{a,e}$ is a proportionality constant relating the electrical field $\mathbf{E}$ to the induced dielectric atom polarization $\mathbf{P}_a$:

$$\chi_{a,e} = \frac{\mathbf{P}_a}{\varepsilon_0\mathbf{E}(\omega)} = \frac{e^2}{m_e}\frac{1}{(\omega_0^2 - \omega^2 - i\omega\gamma)} \quad (5.191)$$

↓ (Resonance frequency) ↓ (Damping term)

This expression applies to a single atom only. The dielectric constant of a medium depends on the manner in which the atoms are assembled. Let *N* be

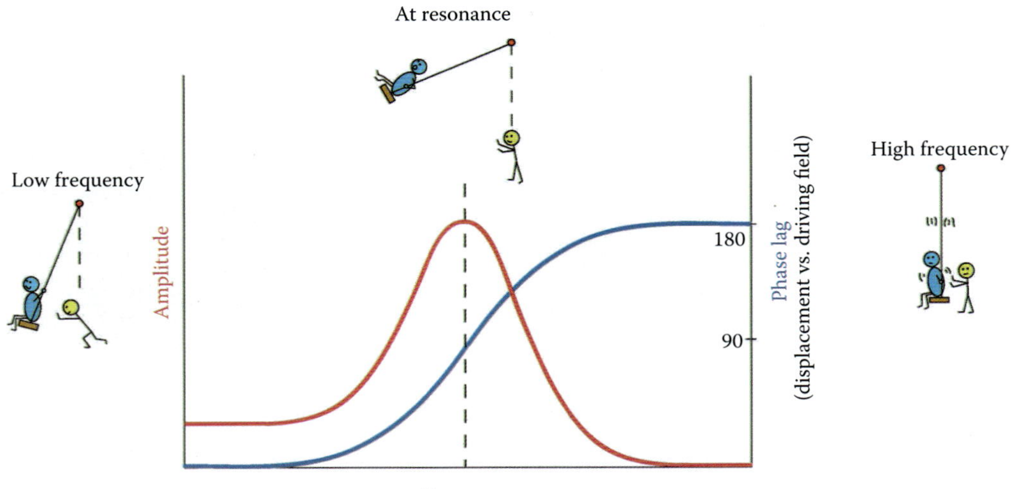

FIGURE 5.56 Amplitude and phase depend on frequency in a driven harmonic oscillator.

the number of atoms per unit volume. The polarization $\mathbf{P}$ of an assembly of atoms can be derived by summing over all the atom resonators N in a volume V and then dividing by V:

$$\mathbf{P}(\omega) = \frac{1}{V}\sum_{i=1}^{i=NV} \mathbf{P}_{a,i}(\omega) = N\langle \mathbf{P}_a(\omega)\rangle \tag{5.192}$$

where NV is the total number of atoms and $\langle \mathbf{P}_a(\omega)\rangle$ is the average dipole moment per atom. Additionally, because $\mathbf{P}(\omega) = \varepsilon_0 \chi_e \mathbf{E}(\omega)$:

$$\chi_e = \left(\frac{Ne^2}{\varepsilon_0 m_e}\right)\frac{1}{(\omega_0^2 - \omega^2 - i\omega\gamma)} \tag{5.193}$$

where the prefactor is related to the plasma frequency $\omega_p = \frac{ne^2}{m_e\varepsilon_0}$ (Equation 3.41) and the density of electrons n is replaced by the density of atoms N. The assumptions here are a linear restoring force linear in $\mathbf{r}$, a damping force linear in $d\mathbf{r}/dt$, averaged macroscopic and local fields are equal, and the dipoles are all free to point in the direction of the applied field (isotropic). Introducing the plasma frequency ω_p, we rewrite Equation 5.193 as:

$$\chi_e = \frac{\omega_p^2}{(\omega_0^2 - \omega^2 - i\omega\gamma)} \tag{5.194}$$

For the complex refractive index n, we derived earlier that $n_c^2(\omega) = 1 + \chi_e(\omega)$ (Equation 5.113), or based on Equation 5.118:

$$\varepsilon_r = \varepsilon' + i\varepsilon'' = 1 + \chi' + i\chi'' = 1 + \frac{\omega_p^2}{(\omega_0^2 - \omega^2 - i\omega\gamma)} \tag{5.195}$$

It follows that the real part of the complex dielectric constant is then given as:

$$\varepsilon' = 1 + \chi'(\omega) = 1 + \frac{\omega_p^2(\omega_0^2 - \omega^2)}{(\omega_0^2 - \omega^2)^2 + \omega^2\gamma^2} \tag{5.196}$$

and the imaginary part equals:

$$\varepsilon'' = \chi''(\omega) = \frac{\omega_p^2\gamma\omega}{(\omega_0^2 - \omega^2)^2 + \omega^2\gamma^2} \tag{5.197}$$

These real and imaginary parts of the dielectric function as a function of ω are shown in Figure 5.57a. At very high frequencies the real part of the dielectric constant approaches 1 as χ' approaches zero. If we deal with gas molecules, the dielectric

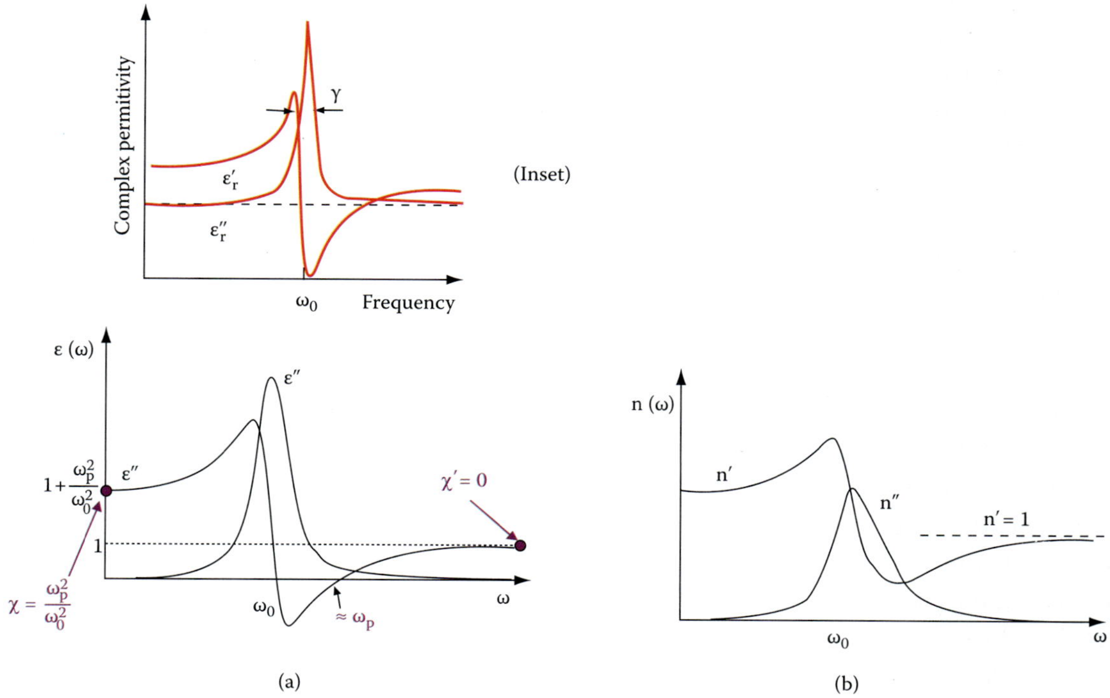

FIGURE 5.57 The real and imaginary parts of a dielectric function as a function of ω (a). The real (n) and imaginary (κ) parts of the complex refractive index n_c as a function of ω (b). (Inset) The full width at half maximum (FWHM) of the peak in ε'' is given by γ.

constant [$\varepsilon(\omega,k)$] at those frequencies approaches ε_0 [one multiplies the relative dielectric constant $\varepsilon_r(\omega,k)$ with ε_0 to obtain $\varepsilon(\omega,k)$]. In other words, the material becomes as transparent as a vacuum. With the molecule oscillators embedded in a solid, ε' approaches ε_∞ at the highest frequencies (e.g., ε_∞ = 10.9 for GaAs). At very low frequencies, we get the full contribution of χ' to the dielectric constant, i.e., $\varepsilon' = 1 + \chi'(\omega) = 1 + \frac{\omega_p^2}{\omega_0^2}$, and reach the so-called *relative static permittivity* (ε_{stat}), which is the same as the relative permittivity evaluated for a frequency of zero (e.g., ε_{stat} = 12.9 for GaAs). The real part of the dielectric constant steps from ε_{stat} to ε_∞ as the resonance is traversed from low to high frequencies. At $\omega = \omega_p$, with the damping term γ very small and $\omega_p >> \omega_0$, the real part of the dielectric constant becomes zero ($\varepsilon' = 0$). Also note that ε' drops below zero between ω_0 and ω_p. When approaching ω_0 from below, ε' starts rising to reach a maximum at $\Delta\omega = \omega - \omega_0 = -\gamma/2$ and a minimum at $\Delta\omega = +\gamma/2$. The peak (a Lorentzian line shape) in the imaginary part of the dielectric function (ε'') occurs at $\omega = \omega_0$ and has a full width at half maximum (FWHM) of γ (see Figure 5.57, inset). The quality factor (Q) of the resonance $Q = \omega_0/\gamma$ is a measure of the dissipation of the system. For an isolated molecule the damping parameter can be interpreted as the inverse of the lifetime of the excited quantum state. This is consistent with the Heisenberg uncertainty principle (HUP) ($\Delta E \Delta t \approx \frac{h}{2\pi}$ or $\Delta t \approx \frac{h}{2\pi . h\Delta v} \approx \frac{1}{2\pi\Delta v}$; Equation 3.107). If an absorption line is dampened solely by the natural lifetime of the state, one refers to natural broadening. This is also the peak that is so characteristic for bound electrons (see the example of Ag with bound d-electrons in Figure 5.64).

If one knows the ω-dependence of the dielectric function, then the ω-dependence of the refractive index can be calculated. From the relationship between the complex refractive index and the complex dielectric constant (Equations 5.115, 5.116, and 5.117), the real and imaginary parts of the complex refractive index n_c as a function of ω were plotted in Figure 5.57b. Comparing Figure 5.57a and b, one recognizes that the real part of the refractive index $n(\omega)$ approximately follows the frequency dependence of $\sqrt{\varepsilon_r'}$ (see Equation 5.118) and that the extinction coefficient κ more or less follows $\varepsilon_r''(\omega)/2n$ (see Equation 5.116). This correspondence is more and more exact if $\kappa << n$ (i.e., for weak absorption such as in gases). At high frequencies $\omega >> \omega_0$, n gradually rises, and above the highest-frequency resonance (usually in the deep UV/soft x-ray region) tends toward 1, and the group velocity v_g becomes equal to the velocity of light c; in other words, the light moves through this medium as if it were moving through a vacuum. In the region between ω_0 and ω_p, n is small over the same region where ε is negative.

At low frequencies ($\omega << \omega_0$) the refractive index n is high, which means that the group velocity $v_g < c$ because:

$$V_g = \frac{c}{n + \frac{\omega dn}{d\omega}} = \frac{c/n}{1 + \frac{\omega dn}{n d\omega}} \tag{5.89}$$

At frequencies $\omega \sim \omega_0$, the group velocity becomes very frequency-dependent as a result of the large absorption coefficient α ($\sim\kappa$ because $\alpha = 2k\kappa$).

Multiple Resonances in Rarified Media In the treatment above we focused our attention mainly on electrons moving against an immobile nucleus, and Equation 5.194 was derived for idealized rarified media where a set of atoms featured only one electronic resonance per atom. Any real optical medium does not just have a single resonant frequency but many, as a result of different oscillators, including different "atomic resonances" (atomic transitions, interband transitions), molecular, lattice vibrations (phonons, vibrational resonances), and oscillations of the free electron plasma (plasmon excitation, free electron absorption). So, Equation 5.194 must be further corrected because realistic atoms in rarified media may have many resonances, and these resonances occur as a result of the various possible motions of electrons (high ω) and of atoms (low ω). Mathematically all these resonances are accounted for by summing over all the resonances in the material, leading to an expression for the rarified medium susceptibility given by:

$$\chi_e(\omega) = \frac{N_j e^2}{\varepsilon_0 m_e} \sum_j \frac{1}{(\omega_{0,j}^2 - \omega^2 - i\omega\gamma_j)} \tag{5.198}$$

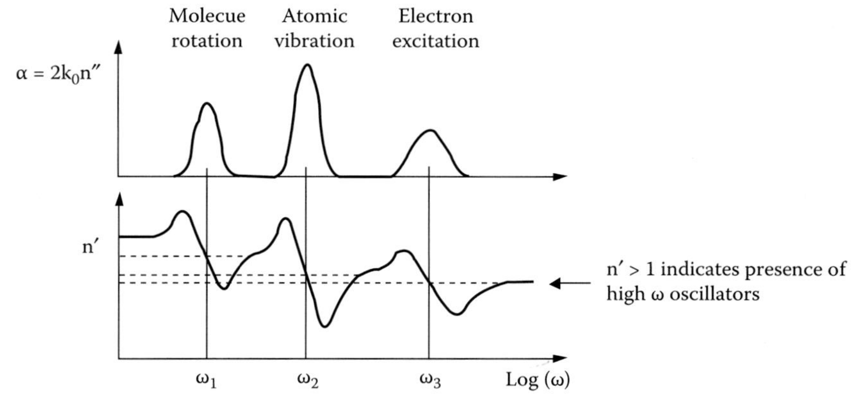

FIGURE 5.58 The real and imaginary parts of the complex refractive index n_c as a function of ω, where in the top we drew $\alpha = -2\frac{\kappa\omega}{c} = -2k\kappa$. The presence of an $n' > 1$ marks the presence of high ω oscillators.

with $\omega_{0,j}$ and γ_j the damping or decay rate of oscillator *j*. The total classical polarization **P** caused by all oscillators with resonant frequencies $\omega_{01}, \omega_{02}, \ldots \omega_{0j}$ and dissipation constants $\gamma_1, \gamma_2, \ldots \gamma_j$ is represented by:

$$\mathbf{P} = \chi_e(\omega) = \frac{N_j e^2}{m_e}\sum_j \left[\frac{1}{(\omega_{0,j}^2 - \omega^2 - i\omega\gamma_j)}\right]\mathbf{E} \quad (5.199)$$

and for the dielectric constant and refractive index we then find:

$$n_c(\omega)^2 = \varepsilon_r = 1 + \frac{N_j e^2}{\varepsilon_0 m_e}\sum_j \frac{1}{(\omega_{0,j}^2 - \omega^2 - i\omega\gamma_j)} \quad (5.200)$$

In Figure 5.58 we show an example of the dependence of the real and imaginary parts of the refractive index n_c on EM wave frequency for a rarified medium. A peak in the absorption (imaginary part) and a wiggle in the real part of refractive index signify each "resonance." The real part of the refractive index decreases step-wise as each resonance is traversed. At high frequencies $n' \to 1$ and $\varepsilon(\infty) \to 1$ (see Equation 5.118) because the frequency becomes too high for the material to respond. Typical spectral ranges for transitions are:

- Electronic: near infrared (about 1 μm) to UV (about 100 nm) (inner core electrons go up to x-ray energies)
- Vibrational: near infrared (about 1 μm) to long-wave infrared (about 50 μm)
- Rotational: far infrared (about 100–500 μm)

These three types of transitions are illustrated in Chapter 7 (Figure 7.83). The structure of a given molecule dictates the likelihood of the absorption of a photon to raise its energy state to an excited one.

In this classical picture each resonance has the same "strength," but when we treat the real situation properly with quantum theory each oscillator has a distinct transition matrix element, i.e., the oscillator strength f_j (see Chapter 3). This leads to a modification of Equation 5.200 as:

$$n_c(\omega)^2 = \varepsilon_r = 1 + \frac{N_j e^2}{\varepsilon_0 m_e}\sum_j \frac{f_j}{(\omega_{0,j}^2 - \omega^2 - i\omega\gamma_j)} \quad (5.201)$$

where f_j is the oscillator strength of each transition (~1 for an allowed transition) and $\omega_{0,j}$ is the resonance frequency of a transition between two states with an energy difference $\Delta E_j = \hbar\omega_{0,j}$ (Figure 5.59). The oscillator strength f_j in Equation 5.201 obeys the Thomas-Reich-Kuhn sum rule:

$$\sum_j f_j = Z \quad (5.202)$$

for an atom that has *Z* electrons. In other words, the total absorption integrated over all frequencies is only dependent on *Z*. However, usually one resonance dominates over all others.

The damping coefficient in Equation 5.201 corresponds to 1/lifetime (1/τ) of a state and can range from the kilohertz to the gigahertz (Figure 5.60). In an idealized case of purely radiative relaxation, the natural line width of an absorption or emission line, we explained above, is dictated by γ_j, which sets the

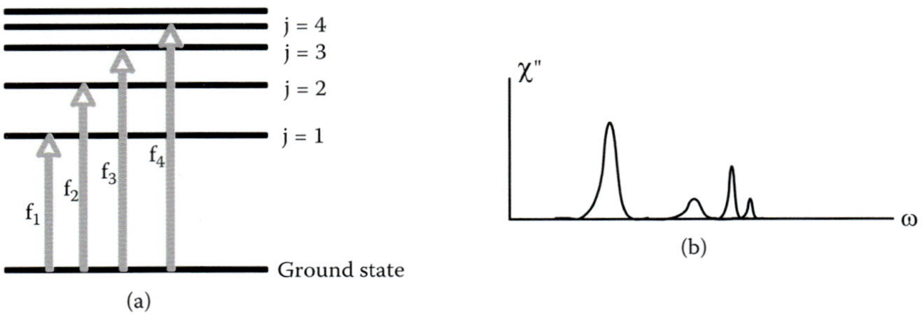

FIGURE 5.59 Real atoms have many electronic levels, so there can be many resonant transitions. Transitions between states with an energy difference $\Delta E_j = \hbar\omega_{0,j}$ (a). These transitions are reflected in absorption peaks $\alpha(\omega) \approx \frac{\omega}{c}\chi''(\omega)$, Equation 5.121 (b).

FWHM of the line (see Figure 5.57, inset). For a single simple electronic transition between two states in an atom, *fast decay* means generally high damping and leads to a broad resonance, i.e., a broad emission spectrum. Absorption associated with electronic transitions (electrons changing states) occurs in about 1 fs (10^{-15} s). The lifetime of an excited molecule depends then on how that molecule disposes of the extra energy. Because of the uncertainty principle (see Chapter 3), the more rapidly the energy is changing, the less precisely we can define the energy. Thus, long-lifetime excited states have narrow absorption peaks, and short-lifetime excited states have broad absorption peaks. In molecules, solution and solid absorption bands broaden. Electronic states in a molecule are split into many sub-bands, and in molecules each electronic state can exist in many possible vibrational and rotational states of the molecule. For an isolated molecule, the typical natural lifetime is about 10^{-8} s coming with a 5×10^{-4} cm^{-1} line width. Collisions between molecules can shorten this lifetime further, and the line width will depend on the number of collisions per second,

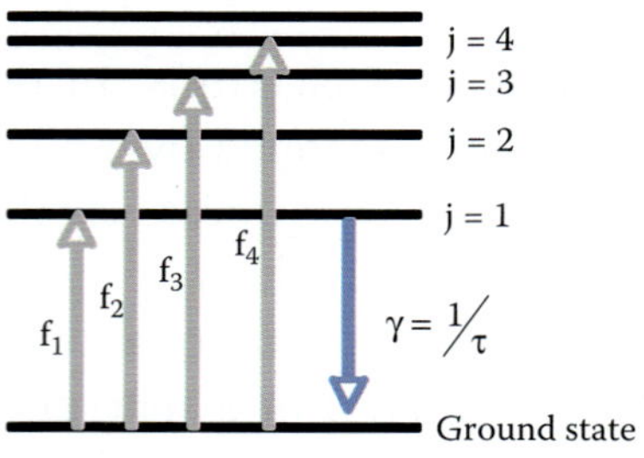

FIGURE 5.60 Idealized case—purely radiative relaxation: decay rate γ (units: s^{-1}) is related to the FWHM of an emission line: *fast decay* means a "high-damping" coefficient and a broad resonance/broad emission spectrum.

i.e., on the number density of the molecules (pressure, *P*) and the relative speed of the molecules (the square root of the temperature). A second source of line broadening in molecules is the so-called *Doppler broadening*. In the atmosphere, molecules move with velocities determined by the Maxwell-Boltzmann distribution, so molecules are in motion when they absorb, and this causes a change in the frequency of the incoming radiation as seen in the molecules' frame of reference. As a consequence, a broad emission/absorption line in a rarified medium does not necessarily mean a fast decay.

In Figure 5.61A we represent a set of calculated dispersion curves and compare a broad ($\gamma = 1$) and a narrow resonance ($\gamma = 0.3$) while keeping ω_0 and ω_p the same at 4 and 8, respectively. We compare the real and imaginary dielectric constants and the real and imaginary refractive indexes as a function of the frequency, as well as the reflectance *R* from air at normal incidence as a function of frequency. Noting that $\alpha = 2\omega\kappa/c$, we have plotted also a scaled version of α, by plotting $2\omega\kappa(\omega)$, scaled to ω_p. Note how α is skewed to higher frequencies. *R* is large in the region where *n* is small, which makes sense because in the limit of $n = 0$ then $R = 1$. In Figure 5.61B, we have colored in four distinct spectral regions, marked with *T* for transmissive ($\omega < \omega_0 - \gamma/2$), *A* for absorptive ($\omega_0 - \gamma/2 < \omega < \omega_0$), *R* for reflective ($\omega_0 + \gamma/2 < \omega < \omega_p$), and *T* for transmissive again ($\omega > \omega_p$), in other words, TART. These regions become more distinct the smaller γ and the larger ω_p. As expected for smaller damping γ, the curves are narrower and have larger maximum values. The reflectance is much higher and has sharper edges. The high *R* starts around ω_0 and falls near ω_p. This is because $\kappa >> n$ in

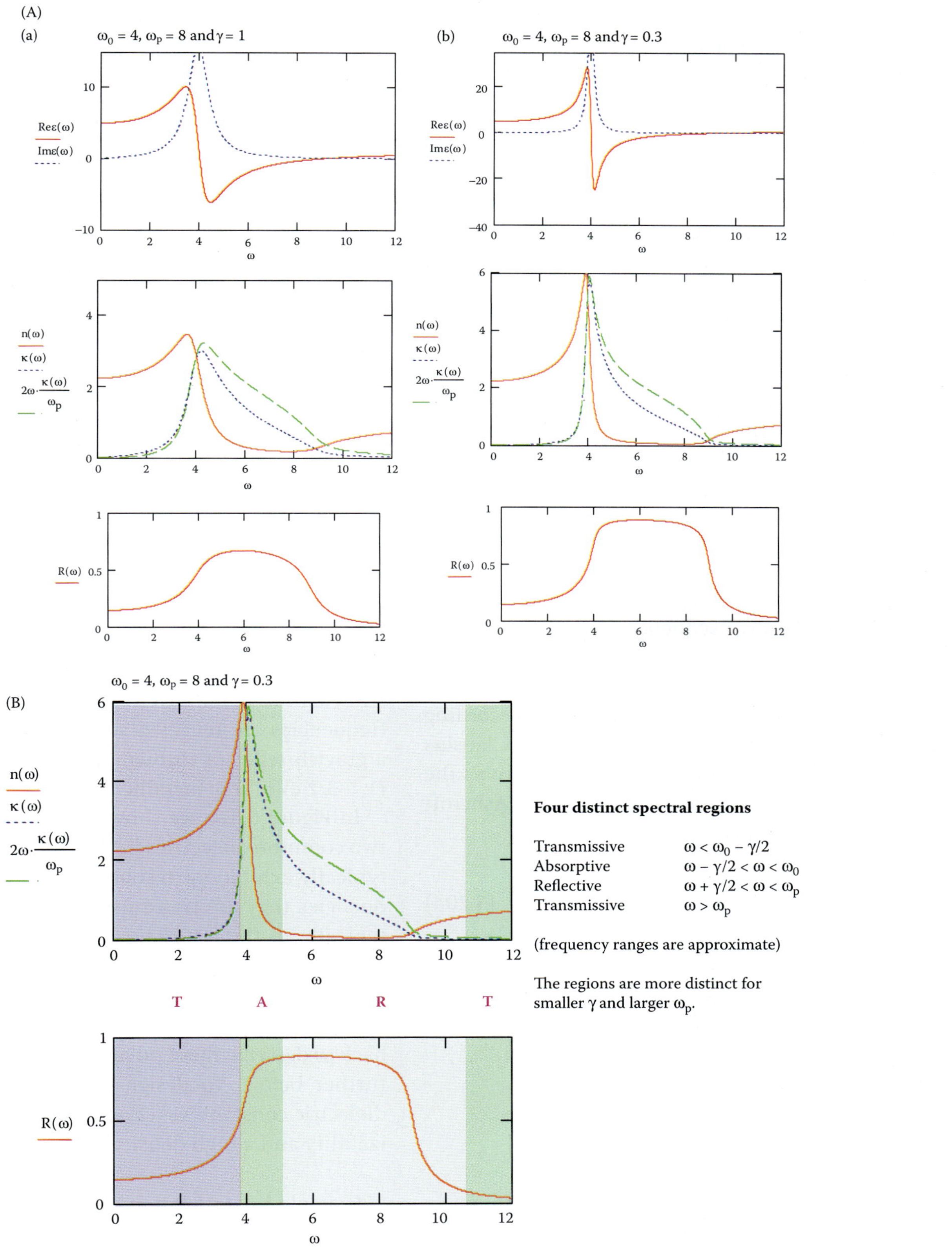

FIGURE 5.61 (A) Compare the real and imaginary parts of a dielectric function as a function of ω, the real and imaginary parts of the refractive index n as a function of ω, and the reflectivity from air at normal incidence for $\gamma = 1$ (broad resonance, a) and $\gamma = 0.3$ (narrow resonance, b). (B) We have colored in four distinct spectral regions, marked with *T* for transmissive, *A* for absorptive, *R* for reflective, and *T* for transmissive again—TART. We have plotted also a scaled version of α by plotting $2\omega\kappa$ scaled to ω_p.

the reflecting band. Above ω_0, n is well below unity. Above ω_p, n rises, but only in the high-frequency limit does n approach unity.

Local Field Corrections The calculations of the dielectric constant we have discussed until now are valid for a rarefied gas with a low density of atoms. In "dense optical media," such as solids, there are other factors that we have to consider. In a real solid, atoms may feel a field from an incident light beam, but they also experience an induced field from the surrounding atoms. In other words, the local field $\mathbf{E}_L$ is the sum of a field without matter $\mathbf{E}_0$ and an induced dipolar field $\mathbf{E}_{dipole}$ from all the other surrounding atoms, or:

$$\mathbf{E}_L = \mathbf{E}_0 + \mathbf{E}_{dipole} \tag{5.203}$$

In our discussions of dipole oscillators in solids, we should have been using $\mathbf{E}_L$ not $\mathbf{E}_0$! The degree of polarization in a solid is related to the structure of the material. As a consequence, dielectric behavior in electrostatic and alternating electric fields depends on both static and dynamical properties of the solid. The local electric field can be calculated according to the crystal structure by the method of Clausius and Mossotti. For example, for cubic structures the Clausius-Mossotti equation reads (e.g., see Ashcroft and Mermin, *Solid-State Physics*, p. 539):

$$\mathbf{E}_L = \mathbf{E}_0 + \frac{\mathbf{P}}{3\varepsilon_0}, \tag{5.204}$$

which is larger than $\mathbf{E}_0$. Similar relations can be derived for any solid. For a solid consisting of atoms of type j with a concentration N_j, we then derive for the polarization $\mathbf{P}$:

$$\mathbf{P} = \varepsilon_0 \sum_j N_j E_L \chi_{j,e} = \varepsilon_0 \sum_j N_j \left(E_0 + \frac{\mathbf{P}}{3\varepsilon_0} \right) \chi_{j,e}$$

$$= \varepsilon_0 \sum_j N_j \chi_{j,e} E_0 + \sum_j N_j \chi_{j,e} \left(\frac{\mathbf{P}}{3} \right) = (\varepsilon_r - 1)\varepsilon_0 E_0 \tag{5.205}$$

where $\chi_{j,e} = \dfrac{e^2}{\varepsilon_0 m_e} \sum_j \dfrac{f_j}{(\omega_{0,j}^2 - \omega^2 - i\omega\gamma_j)}$. But we know that $\mathbf{D} = \varepsilon_0\mathbf{E} + \mathbf{P}$, $\mathbf{D} = \varepsilon_0\varepsilon_r\mathbf{E}$, and $\mathbf{P} = \varepsilon_0(\varepsilon_r - 1)\mathbf{E}$, and this way we obtain the Clausius-Mossotti (CM) relation:

$$\frac{(\varepsilon^2 - 1)}{(\varepsilon^2 + 2)} = \frac{N\chi_{j,e}}{3} \tag{5.206}$$

This relation links the polarizability to the dielectric constant. Using $\varepsilon_r = n^2$ we find the so-called *Lorentz-Lorenz relationship*:

$$\frac{(n^2 - 1)}{(n^2 + 2)} = \frac{N\chi_{j,e}}{3} \tag{5.207}$$

And we find that the refractive index approximately follows the formula:

$$\frac{\varepsilon_r - 1}{\varepsilon_r + 2} = \frac{n_c^2 - 1}{n_c^2 + 2} = \frac{Ne^2}{3\varepsilon_0 m_e} \sum_j \frac{f_j}{(\omega_{0,j}^2 - \omega^2 - i\omega\gamma_j)} \tag{5.208}$$

We follow with some examples from the n, κ database (http://www.ioffe.ru/SVA/NSM/nk/index). In Figure 5.62a we show the refractive index n and absorption index κ versus photon energy for silicon. At visible frequencies, i.e., below the resonance, Si is "transmissive." But the high refractive index of Si still yields high reflectivity (30–50%), and all visible frequencies are reflected strongly, so Si appears as a grayish reflector throughout the visible spectrum (1.7–3.2 eV) "gray metallic."

In Figure 5.62b we show the refractive index for two common dielectrics (Si_3N_4 and SiO_2). Dielectrics typically exhibit one resonance only, and at low frequencies they are transmissive. The resonances are typically in the UV range associated with electron polarization. There is an additional resonance in the IR range for polar materials resulting from nuclear motion and orientation (see Figure 5.63). We come back to describe these polar materials in more detail further below. To design a material with a high dielectric constant, we choose ions with high polarizability and a high density of them.

Example 5.6: Estimation of $\varepsilon_r(\omega = 0)$ for Si

The reflectance R of Si increases sharply at about 3 eV, and we may take this as an estimate for ω_0. Hence $\omega_0 \approx 3 \times 1.6 \times 10^{-19} = 4.53 \times 10^{15}$ rad/s. Because $\varepsilon_r(\omega) = 1 + \dfrac{\omega_p^2}{\omega_0^2 - \omega^2 - i\gamma\omega}$, $\varepsilon_r(\omega = 0) = 1 + \dfrac{\omega_p^2}{\omega_0^2}$, so if

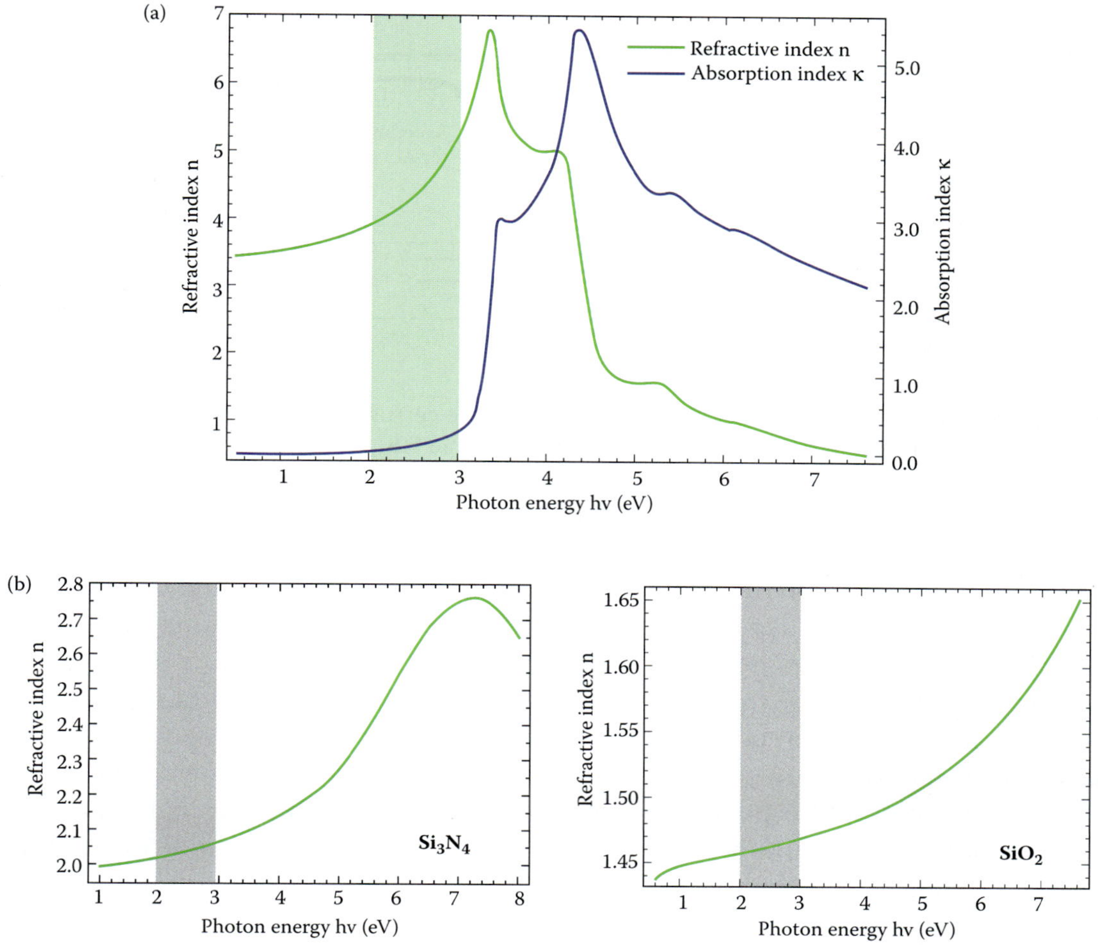

FIGURE 5.62 Examples from the *n*, κ database (http://www.ioffe.ru/SVA/NSM/nk/index.html#A3B6). (a) Refractive index *n* and absorption index κ versus photon energy for silicon. (b) The refractive index for two common dielectrics (Si_3N_4 and SiO_2). The color bands in the three figures roughly mark the visible spectrum (1.7–3.2 eV).

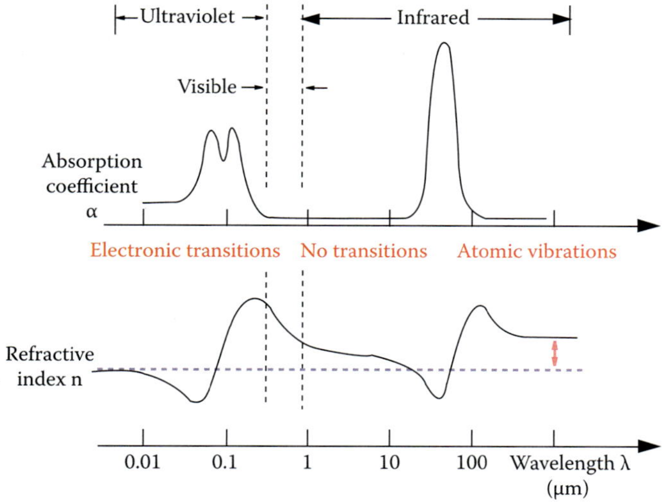

FIGURE 5.63 The absorption coefficient and the index of refraction versus wavelength for a generic polar insulator. As we derived earlier: $\alpha = 2k\kappa$.

we can determine ω_p, we can estimate $\varepsilon_r(\omega = 0)$. Now we remember that $\omega_p = \sqrt{ne^2/\varepsilon_0 m}$, and because each Si atom has four valence electrons, $n = 4N_{Si} \approx 4 \times 2\ 10^{28}\ m^{-3}$. This gives us an estimate of $\omega_p \approx 1.6 \times 10^{16}$ rad/s (corresponding to about 10.5 eV), and hence $\varepsilon_r(\omega = 0) \approx 14$. This compares to a measured value for $\varepsilon_r(\omega = 0)$ of 12, so our approximations are reasonable.

Optical Properties of Metals

Introduction

The Drude model implies that only the plasma frequency ω_p dictates the appearance of metals. This works indeed for many metals, say for aluminum, but it does not explain why copper is red, gold is yellow, and silver is colorless. In fact, the appearance

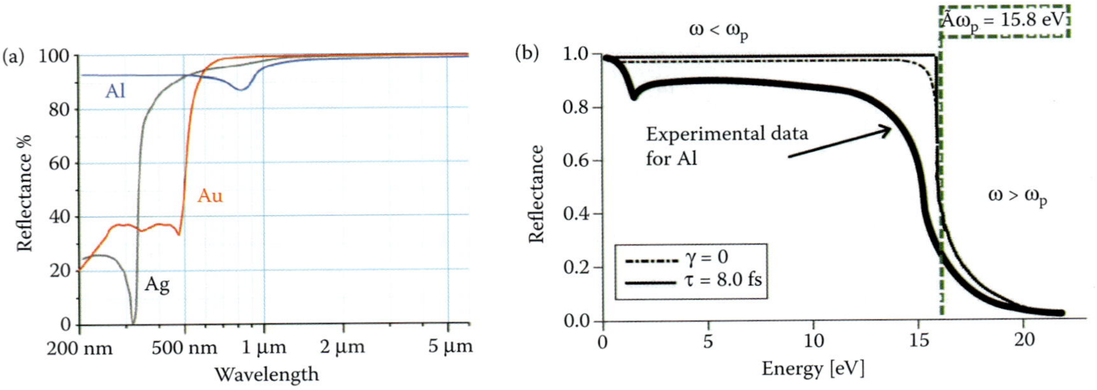

FIGURE 5.64 (a) Reflectance curves of Ag, Au, and Al. (b) Reflectance versus energy for an ideal metal with zero damping ($\gamma = 0$) and experimental data for aluminum. See also Example 5.8.

of these metals is characterized by a different edge in the reflectance spectrum (see Figure 5.64a), but the fact is that these three metals have the same number of valence electrons, so according to the Drude model they all should have been the same. Moreover, the calculated plasma frequency for all three should lie at about 9 eV, well outside the visible region, so the Drude plasma frequency cannot in itself account for the colors of Cu and Au. Colored metals have interband transitions in the visible that remove particular frequencies from the reflected spectrum. The combined effects of the free electrons and the bound d-electrons influence the reflectance properties of the metal in these cases. In Chapter 3, we pointed out that in a metal where both bound electrons $[\varepsilon_B(\omega)]$ and conduction electrons $[\varepsilon_r(\omega)]$ contribute to the dielectric constant, an effective dielectric constant $\varepsilon_{Eff}(\omega)$ must be defined as:

$$\varepsilon_{Eff}(\omega) = \varepsilon_B(\omega) + \varepsilon_r(\omega) \qquad (3.48)$$

Whereas the Drude model explains the free electron contribution $[\varepsilon_r(\omega)]$ to $\varepsilon_{Eff}(\omega)$, the bound electron effect $[\varepsilon_B(\omega)]$ can only be treated by the Lorentz model. Based on the Drude-Lorentz model, we will now review optical properties of bulk metals (3D) and metals in reduced dimensionalities in plasmonics.

Reflectance and Absorption of Electromagnetic Waves in Metals

Reflectance of Conductors Metals can be treated as plasmas because they contain equal numbers of fixed positive ions and free electrons. In Figure 5.64a, we plot the reflectance R of Al, Au, and Ag as a function of wavelength. In Figure 5.64b, we show the reflectance of an ideal metal and that of aluminum versus the energy of the incoming radiation. A characteristic feature of all metals is that they are shiny as a result of the interaction of light with the free electrons (plasma excitations) and that their reflectance is close to 100% (R = 1) below the plasma frequency ω_p but much smaller at higher ω. At low frequencies, metals with plenty of free carriers are very good reflectors. The free carriers can respond rapidly, on a time scale of an AC cycle, so as to screen the external applied AC fields. Earlier we calculated the normal incidence reflectance for any material as:

$$R(\omega) = \frac{\left|E_m^r\right|^2}{\left|E_m^i\right|^2} = \left|\frac{n_c - 1}{n_c + 1}\right|^2 = \frac{(n-1)^2 + \kappa^2}{(n+1)^2 + \kappa^2} \qquad (5.148)$$

If we consider a lightly damped (highly conductive) system, where the damping factor γ tends to zero (an ideal metal would come with zero damping $\gamma = 0$), the real components of the dielectric constant $\varepsilon_r'(\omega,\mathbf{k}) \approx 1 - \frac{\omega_p^2}{\omega^2}$ (Equation 3.44) dominate because the imaginary component $\varepsilon_r''(\omega,\mathbf{k}) \approx \frac{\omega_p^2\gamma}{\omega^3}$ (Equation 3.45) tends to zero. For the complex refractive index in this case we can write:

$$n_c(\omega) = \sqrt{\varepsilon_r'(\omega)} = \sqrt{1 - \frac{\omega_p^2}{\omega^2}} \qquad (5.209)$$

Substituting this expression in Equation 5.148 for the reflectance R, we see that when $\omega < \omega_p$, n_c is imaginary, R = 1, and the metal is shiny (say, up to the near-infrared). At higher frequencies (visible and UV), with $\omega > \omega_p$, n_c turns real, the reflectance R suddenly falls (for $\omega \rightarrow \infty$, $R \rightarrow 0$), and metals are transparent in the UV. It follows from the expression for the plasma frequency of an FEG in a metal $[\omega_p = \sqrt{\frac{ne^2}{m_e\varepsilon_0}}$ (Equation 3.41)] that the higher the conductivity (n larger), the higher the plasma frequency. And indeed, plasma frequencies ω_p are in the optical range (UV) for metals and in the terahertz to the infrared region for semiconductors and insulators (see also Table 3.2).

Without any damping ($\gamma = 1/\tau = 0$), the reflectivity R equals 1 and then drops very steeply (see rectangular box in Figure 5.64b). Introducing some damping, R becomes less than 1, and the reflectance decrease is less severe (see Figure 5.64b where we used $\tau = 8.0$ fs, a value calculated for aluminum; see also Example 5.8). When the conduction electrons can no longer move rapidly enough to screen the external applied field on the time scale of an AC cycle, only the core electrons bound to the atoms can contribute to the dielectric response; the metal behaves now as a transparent dielectric. The frequency at which the metallic-to-dielectric transition occurs is the plasmon frequency ω_p. At $\omega = \omega_p$, the real part of the dielectric constant becomes zero. Hence $n_c(\omega_p) = 0$, which means the phase velocity tends to ∞. (This occurs in the anomalous dispersion regions shown in Figure 5.29.) Another way to describe this is that the wavelength, $\lambda = 2\pi c/n_c\omega \rightarrow \infty$ as $\omega \rightarrow \omega_p$. This means that all the electrons are oscillating in phase throughout the propagation length of the material.

Absorption and Penetration Depth of Electromagnetic Radiation in Conductors Optical measurements of the complex dielectric constant $\varepsilon_r(\omega,k)$ of a plasma are equivalent to AC conductivity measurements. This can be gleaned from a comparison of the dielectric function of a free electron plasma with its AC conductivity. In Chapter 3 we calculated for the AC conductivity of a plasma:

$$\sigma(\omega) = \frac{ne^2}{m_e}\left(\frac{1}{\frac{1}{\tau} - i\omega}\right) \quad \text{(based on Equation 3.30)}$$

and for the complex dielectric constant we derived:

$$\varepsilon_r(\omega,\mathbf{k}) = \frac{\varepsilon(\omega,\mathbf{k})}{\varepsilon_0} = \left[1 - \frac{ne^2}{m_e\varepsilon_0(\omega^2 + i\gamma\omega)}\right]$$

(based on Equation 3.40)

Comparing these expressions we note that:

$$\varepsilon_r(\omega,\mathbf{k}) = 1 + \frac{i\sigma(\omega)}{\varepsilon_0\omega} \tag{5.210}$$

At very low frequencies ($\omega \ll \tau^{-1}$) we can derive a useful relationship between the conductivity of the free electron gas and the absorption coefficient (α) of the metal. The real and imaginary parts of the complex dielectric constant are $\varepsilon_r'(\omega,\mathbf{k}) = 1 - \frac{\omega_p^2\tau^2}{1+\omega^2\tau^2}$ (Equation 3.42) and $\varepsilon_r''(\omega,\mathbf{k}) = \frac{\omega_p^2\tau}{\omega(1+\omega^2\tau^2)}$ (Equation 3.43), and one appreciates that, at low frequencies, $\varepsilon_r''(\omega,\mathbf{k}) \approx \frac{\omega_p^2\tau}{\omega}$ can become much larger than $\varepsilon_r'(\omega,\mathbf{k}) \approx 1 - \omega_p^2\tau^2$. Based on Equations 5.117 for n and κ, we then derive:

$$n \sim \kappa = \left(\frac{\varepsilon_r''}{2}\right)^{\frac{1}{2}} \tag{5.211}$$

Therefore, for the absorption coefficient we calculate:

$$\alpha = \frac{2\kappa\omega}{c} = \frac{2\omega\left(\frac{\varepsilon_r''}{2}\right)^{\frac{1}{2}}}{c} \tag{5.212}$$

and with $\varepsilon_r''(\omega,\mathbf{k}) \approx \frac{\omega_p^2\tau}{\omega}$ the absorption coefficient of a metal for $\omega \ll \omega_p$ and $\omega\tau \ll 1$ is given as:

$$\alpha = \left(\frac{2\omega_p^2\tau\omega}{c}\right)^{\frac{1}{2}} \tag{5.213}$$

or, finally, recognizing that $\omega_p^2\tau = \frac{\sigma_0}{\varepsilon_0}$ and $c^2 = \frac{1}{\varepsilon_0\mu_0}$, we obtain:

$$\alpha = \sqrt{2\sigma_0\omega\mu_0} \tag{5.214}$$

Thus, the optical absorption coefficient of a metal for $\omega \ll \omega_p$ and $\omega\tau \ll 1$ is proportional to the square root of the DC conductivity and the frequency.

We calculated earlier the decay of the intensity of an EM wave, propagating in the *x*-direction in a given material, as $I(x) = I_0e^{-ax}$ (Equation 5.110). The intensity in this expression decreases to $1/e$ (or ~36%) of I_0 at a depth x for which $\alpha x = 1$. This depth is known as the *penetration depth* δ, and it is given by:

$$\delta = \left|\frac{1}{\alpha}\right| = \sqrt{\frac{1}{2\sigma_0\omega\mu_0}} \qquad (5.215)$$

Because this depth is usually very small in metals (given that σ_0 in Equation 5.215 is very large), it is also referred to as *skin depth*. Horace Lamb first described the effect in a paper in 1883. Whereas at low frequencies most of a conductor's cross-section (say, a copper wire) carries the current, the current moves to the skin of the conductor as frequency increases. This penetration depth of an electromagnetic wave in a solid is illustrated in Figure 5.65 for Al, Cu, Fe, and Pb. Skin depth is dependent on the type of metal and the frequency of the applied field. At high frequencies the skin depth is very shallow, and the fields are often considered to fall to zero within a few millimeters. The skin effect predominates at gigahertz frequencies. The skin effect has practical consequences in the design of radiofrequency (RF) and microwave circuits and to some extent in AC electrical power transmission and distribution systems.

It is important to keep in mind that Equation 5.215 is only valid when $\omega\tau \ll 1$. At higher frequencies (mid-infrared, visible, UV) this is no longer valid, and we must use the whole expression for ε_r (Equation 3.40). We also did not include possible contributions of bound electrons (see Lorentz model below).

Example 5.7: Cu ($\sigma_0 = 6.5\ \Omega^{-1}\text{cm}^{-1}$) → δ = 8.8 mm at 50 Hz, 6 μm at 100 MHz.

Example 5.8: The DC conductivity of aluminum at room temperature is $\sigma_0 = 4.1 \times 10^7\ \Omega^{-1}\text{m}^{-1}$, and the free electron density is $n = 1.81 \times 10^{29}\ \text{m}^{-3}$. What is the reflectance *R* of the surface at λ = 500 nm under normal incidence?

1. We first calculate the damping ($\gamma = 1/\tau$ from the conductivity:

$$\sigma_0 = \frac{ne^2\tau}{m_e} \text{ (Equation 3.7)} \rightarrow \tau = 8.0\ 10^{-15}\ \text{s} = 8\ \text{fs}$$

2. Next we calculate the plasma frequency ω_p:

$$\omega_p = \sqrt{\frac{ne^2}{m_e\varepsilon_0}} \text{ (Equation 3.41)} \rightarrow \omega_p = 2.4\ 10^{16}\ \text{Hz}$$

3. A 500-nm wavelength corresponds to $\omega = 2\pi c/\lambda = 3.8 \times 10^{15}$ Hz. Now we can calculate the real and imaginary parts of the dielectric function:

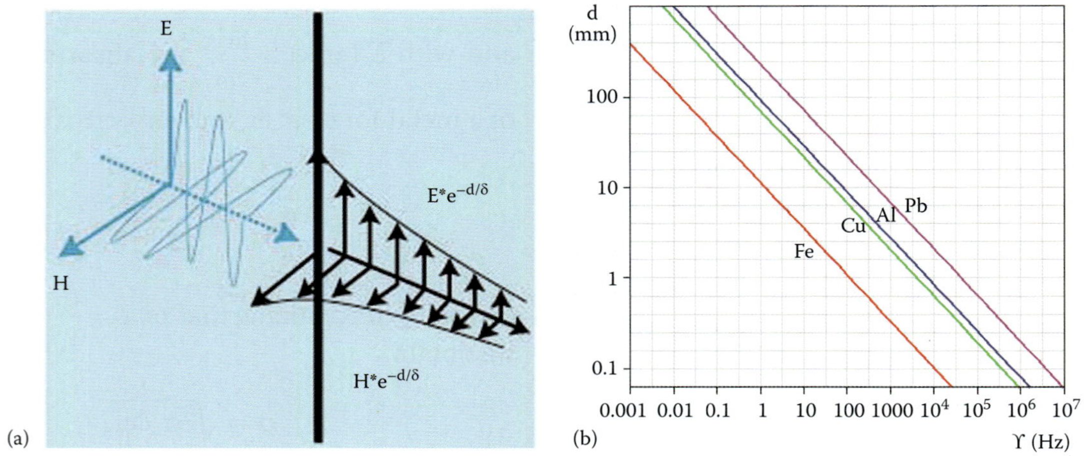

FIGURE 5.65 The depth of penetration δ of electromagnetic waves in a metal (a). Skin depth is dependent on the type of metal and the frequency of the applied field. At high frequencies the skin depth is very shallow, and the fields are often considered to be 0 in a few millimeters. Examples for Fe, Cu, Al, and Pb are shown (b).

$$\varepsilon_r'(\omega, \mathbf{k}) = 1 - \frac{\omega_p^2\tau^2}{1+\omega^2\tau^2} \text{ (Equation 3.42)} \rightarrow -39$$

$$\varepsilon_r''(\omega, \mathbf{k}) = \frac{\omega_p^2\tau}{\omega(1+\omega^2\tau^2)} \text{ (Equation 3.43)} \rightarrow 1.3$$

$$n(\omega) = \sqrt{\frac{(\varepsilon_r')^2 + (\varepsilon_r'')^2 + \varepsilon_r'}{2}}$$

(Equation 5.117) $\rightarrow n = 0.1$ and $\kappa = 6.2$

$$\kappa(\omega) = \sqrt{\frac{(\varepsilon_r')^2 + (\varepsilon_r'')^2 - \varepsilon_r'}{2}}$$

Finally, we calculate the reflectance R from:

$$R(\omega) = \frac{\left|E_m^r\right|^2}{\left|E_m^i\right|^2} = \left|\frac{n_c - 1}{n_c + 1}\right|^2 = \frac{(n-1)^2 + \kappa^2}{(n+1)^2 + \kappa^2}$$

(Equation 5.148) $\rightarrow 99\%$

Even though ω/ω_p is ~0.2, the reflectance remains very high.

To help interpret this example, in Figure 5.66 we look at some calculated plots of $\varepsilon_r(\omega)$, $n(\omega)$, $\alpha(\omega)$, and $R(\omega)$. These are plotted for $\omega_p = 10$ and for $\gamma \approx 0$ (Figure 5.66a) or $\gamma = 0.5$ (Figure 5.66b). In the limit of no damping, the refractive index $n = 0$ and $R = 1$ for $0 < \omega < \omega_p$. Above ω_p, κ is zero, and the reflectance R drops as n rises from zero toward unity. Note that even for $\varepsilon_r'' = 0$, κ and hence α are not zero. Introducing some damping causes R to be <1, and the reflectance drop at ω_p is less severe. The sharp edge in the reflectance seen at the plasma frequency is the predominant spectral feature in the optical properties of metals.

Drude-Lorentz Theory for Metals

The combined effects of the free electrons and the bound *d*-electrons determine the optical properties of some metals. In a metal where both bound electrons $[\varepsilon_B(\omega)]$ and conduction electrons $[\varepsilon_r(\omega)]$ contribute to the dielectric constant, an effective dielectric constant $\varepsilon_{Eff}(\omega)$ was earlier defined as:

$$\varepsilon_{Eff}(\omega) = \varepsilon_B(\omega) + \varepsilon_r(\omega) \quad (3.48)$$

While the Drude model explains the free electron contribution $[\varepsilon_r(\omega)]$ to $\varepsilon_{Eff}(\omega)$, the Lorentz model explains the bound electron effect $[\varepsilon_B(\omega)]$ and clarifies why copper is red, gold is yellow, and silver is colorless. All three metals have filled *d*-shells: copper has the electronic configuration [Ar].$3d^{10}.4s^1$, silver [Kr].$4d^{10}.5s^1$, and gold [Xe].$4f^{14}.5d^{10}.6s^1$. These

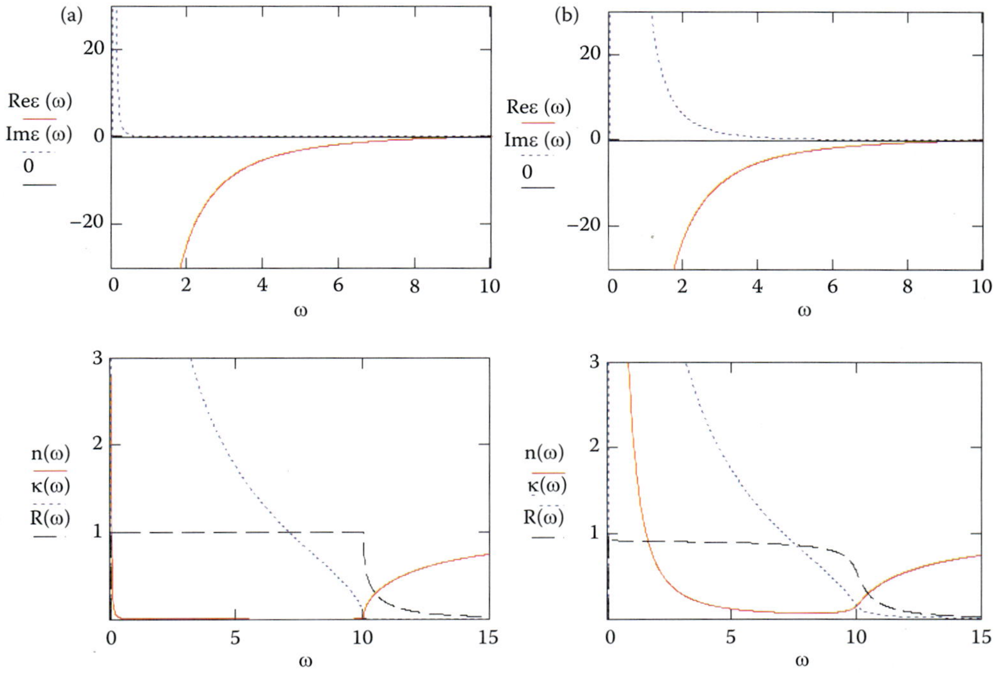

FIGURE 5.66 Calculated plots of $\varepsilon_r(\omega)$, $n(\omega)$, $\alpha(\omega)$, and $R(\omega)$. Plotted for $\omega_p = 10$ and for $\gamma \approx 0$ (a) or $\gamma = 0.5$ (b).

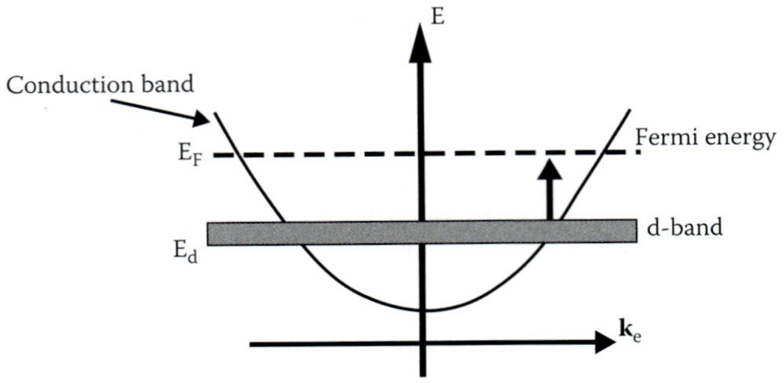

FIGURE 5.67 The combined effects of the free electrons (Drude model) and the interband transitions resulting from bound *d*-electrons (Lorentz model) influence the reflectance properties of the metal.

d-electron bands lie below the Fermi energy of the conduction band (Figure 5.67), and transitions from the *d*-band to empty states above the Fermi level occur over a fairly narrow band of energies ($\hbar\omega_0 = E_F - E_d$). These transitions are modeled as an additional Lorentz oscillator.

In Figure 5.68 we compare the optical properties of aluminum and silver in a plot of the real part and the imaginary part of the complex dielectric constant as a function of the photon energy. For most metals the only contribution to ε_{Eff} (Equation 3.48) stems from conduction electrons as illustrated by the experiments with aluminum where at $\omega = \omega_p$, $\varepsilon_{Eff}(\omega) \approx 1 - \frac{\omega_P^2}{\omega^2} \approx 0$ and $\varepsilon_B \approx 0$, with a monotonous decrease of $\varepsilon_{Eff}(\omega)$ for decreasing frequency as predicted (Figure 5.68a). In other words, for Al we observe the predicted $\varepsilon_{Eff} = 1 - \frac{\omega_P^2}{\omega^2} + i\frac{\omega_P^2}{\omega^3\tau}$ (see Equation 3.49).

For Ag, with a σ of $6.2 \times {}^7\Omega^{-1}m^{-1}$ and $\omega_p = 9.65 \times 10^{14}$ Hz (=311 nm or 4 eV), the drop in the $\varepsilon_{Eff}(\omega)$ curve toward lower frequencies is not monotonous as in the case of Al; a peak appears in the $\varepsilon''_{Eff}(\omega)$ at around 4 eV, which is reflected in the summation of the two terms [$\varepsilon_{Eff}(\omega)$; see Figure 5.68c]. Such a peak is more typical for the case of bound electrons in a dielectric. In silver, both conduction and bound electrons contribute to ε_{Eff}, and the bound electron contribution is caused by *d*-electrons (bound electrons in a *d* orbit) at 4 eV below the Fermi energy. The reflectance spectrum of silver, shown in Figure 5.68b, reveals a sharp decrease at the same 4 eV, well below the expected plasma frequency (at 9 eV). The reflectance increases again for frequencies just above 4 eV. This behavior comes about because of the *d*-band resonance at 4 eV in silver. The difference between the dielectric functions calculated from the Drude model and those observed from the experiment are a result of the d-band resonance, ε_B. The effect is to "pull" the $\varepsilon_{Eff} \to 0$ frequency in from 9 eV to about 3.9 eV. This shift in ω_p means that there is a shift in the free plasma oscillation in silver because of the *d*-electrons. This can be explained by noting that the highly polarizable *d*-electrons reduce the electric

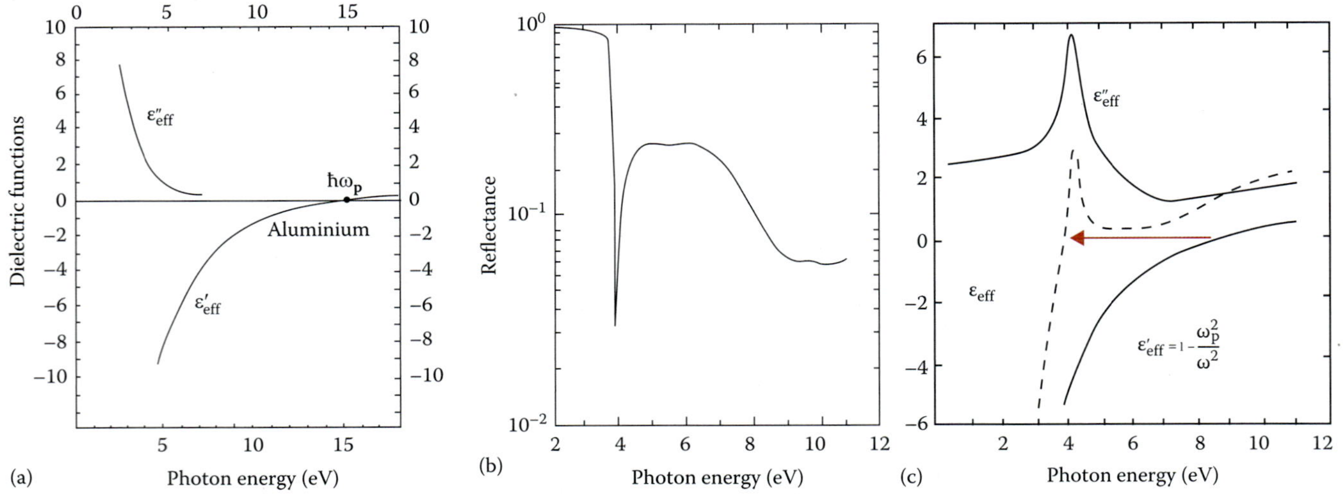

FIGURE 5.68 The real and the imaginary parts of the complex dielectric constant are plotted as a function of the photon energy for aluminum (a) and silver (c). In (b), we also show the reflectance *R* of silver versus photon energy.

field that provides the restoring force involved in these oscillations. A reduced restoring force results in a reduced oscillation frequency.

An interesting and unusual conductor in this regard is tin-doped indium oxide (ITO) with a very large EM wave penetration depth in the visible EM wave range. ITO is a semiconducting material that is transparent in the visible! The material is very useful in many low-current applications, such as liquid crystal displays. To obtain this transparency, the material combines a low electron density n with a high mobility μ_e, i.e., the electrons travel through the material with relatively few collisions. The plasma frequency of ITO depends on the Sn density and is typically around 0.7 eV ($\lambda \sim 1.7$ μm), making the material of lesser use in the near infrared ($\lambda >$ 1.7 μm) than it is in the visible. The properties of ITO, again, can be modeled by the Drude-Lorentz models.

Metal Nanoparticles

Introduction In the preceding section we used the Drude-Lorentz model to explain electrooptical properties of bulk metals; now we explore what happens with those properties when lower-dimensional metal structures are considered. We start with a comparison of metal nanoparticles with semiconductor nanoparticles, a comparison we will come back to after applying the Drude-Lorentz model to bulk and nanoparticle semiconductors (see Table 5.10). We finish this section adding some more details to plasmonics applications see the section Plasmonic Applications I, a discussion we started earlier.

Quantum Effects in Metal and Semiconductor Nanoparticles Compared In Chapter 3, we studied quantum-confined materials and found that for quantum effects to appear in an electronic conductor, the dimension along which there is quantum confinement must be smaller than the de Broglie wavelength of the electrons. The de Broglie wavelength for electrons at the Fermi energy corresponds to the size of an electron in a solid and is represented by its Fermi wavelength $\lambda_F = 2\pi/\mathbf{k}_F$ with $\mathbf{k}_F$, the Fermi wave vector, given by $(3n\pi^2)^{1/3}$ (Equation 3.238), where n is the electron density. In most metals, the electron density is very large ($n > 10^{21}\text{cm}^{-3}$) and the Fermi wavelength is of the order of a few nanometers. In semiconductors, the lower carrier density ($n > 10^{21}\text{cm}^{-3}$) gives rise to larger electron wavelengths of several tens of nanometers (see also Table 3.12). Therefore, confinement and quantization phenomena are visible in semiconductors already at dimensions of 200 nm, whereas in metals they typically are seen only at 1–2 nm. This is one important reason why quantum effects have been observed mainly in semiconductor nanostructures, which have large Fermi wavelengths, rather than in metals, which have Fermi wavelengths comparable to a few crystal lattice periods only. We can also approach the difference in quantization effects between a metal and semiconductor nanoparticle from the point of view of the separation between adjacent energy levels. The average level spacing at the Fermi level between subsequent levels can be estimated from Equation 3.237:

$$\Delta E_{n+1-n} = \frac{1}{G(E)_{3D}(E = E_F)} = \frac{2E_F}{3n_{3D}} \tag{3.237}$$

Based on this expression for a spherical Au nanoparticle with a radius R = 2 nm, an average level spacing $\Delta E = E_{n+1} - E_n$ of 2 meV only is found, whereas in Volume III, Chapter 3 we calculate the energy separation between adjacent levels for a 2-nm semiconducting CdSe particle as 0.76 eV. Obviously energy-level quantization is far more pronounced in semiconductor dots than in metal dots. In addition, the appearance of the plasmon oscillations of metal particles often obscures quantization effects in metals.

Beyond the Fermi wavelength difference, there are two additional reasons for this increased energy separation in the case of semiconductors: 1) the energy-level spacing, i.e., the energy difference $E_{n+1} - E_n$, increases for low-lying states in a 3D potential well (see Figure 3.142); and 2) the effective mass for electrons is smaller in semiconductors than in metals. With a smaller effective, we saw in Chapter 3, the "blue" shift becomes even larger (e.g., see Equation 3.251).

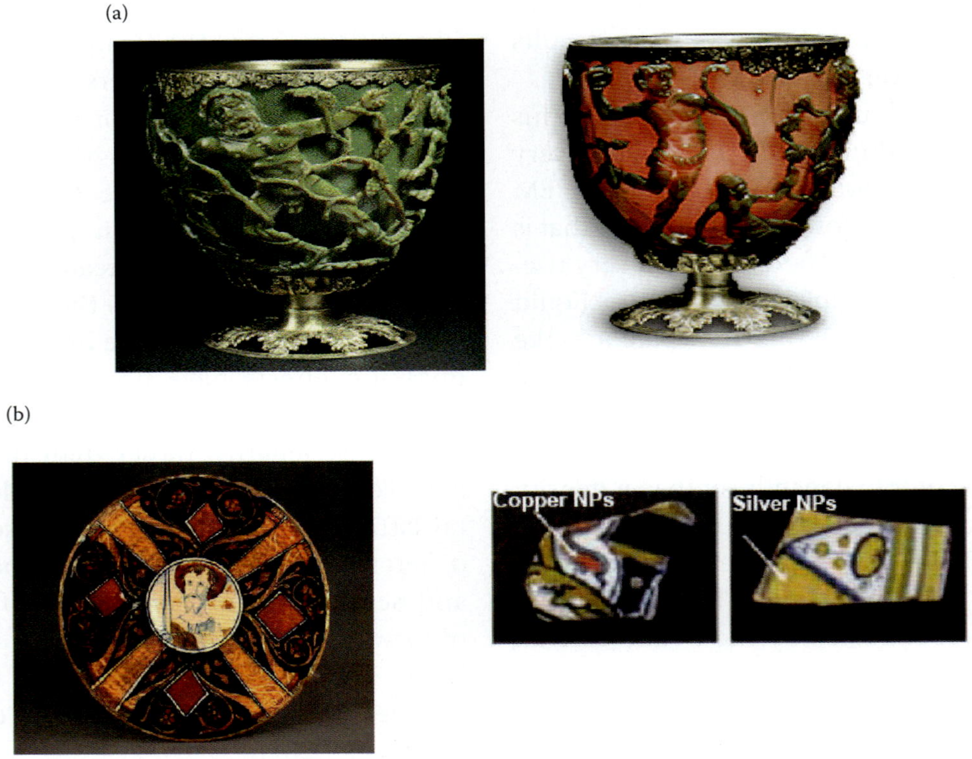

FIGURE 5.69 (a) The Lycurgus Cup (glass; British Museum, fourth century AD). When illuminated from outside, it appears green. However, when illuminated from within the cup, it glows red. The red color is caused by very small amounts of gold powder (about 40 ppm), which have an absorption peak at around 520 nm. (b) Luster decorated 16th century Renaissance pottery, Gubbio, Italy. NPs = nanoparticles.[10] (Image and information from the National Gallery of Art, Washington, DC.)

Whereas in semiconductor structures with reduced dimensionalities electron confinement leads to quantization of energetic states, metallic nanostructures exhibit major changes in their optical spectra as a result of light absorption and the creation of plasmonics waves at the surface of these structures. As mentioned above, these effects often mask the metal nanoparticle quantum effects.

Optical Properties of Metal Nanoparticles The optical properties of nanoparticles are spectacular, and they have promoted a great deal of excitement during the past few decades in metal nanoparticles. Examples of the use of metal nanoparticles for decorative purposes go as far back as the Roman time, such as those contained in the glass of the famous Lycurgus Cup (fourth century AD) pictured in Figure 5.69a. The cup can still be seen at the British Museum and possesses the unique feature of changing color depending on the light in which it is viewed. It appears green when viewed in reflected light but looks red when a light is shone from inside and is transmitted through the glass. Analysis of the glass has revealed that it contains a very small amount of tiny metal crystals containing Ag and Au in an approximate molar ratio of 14:1. It is the presence of these nanocrystals that gives the Lycurgus Cup its special color display. Other examples include decorative stained glasses in cathedrals (see Figure 5.48) and the Renaissance pottery shown in Figure 5.69b. The beautiful Maya blue, a paint often used in Mesoamerica, involves simultaneously metal nanoparticles and superlattice organization effects (see below). These examples represent applications of nanotechnology before an understanding of mechanisms or design rules had emerged. It was not until 1857 that Michael Faraday reported a systematic study of the synthesis and colors of colloidal gold: "The mere variation in the size of [gold] particles gave rise to a variety of resultant colors."[9] Before that time, colloidal gold was known and used in the Middle Ages for its purported health restorative

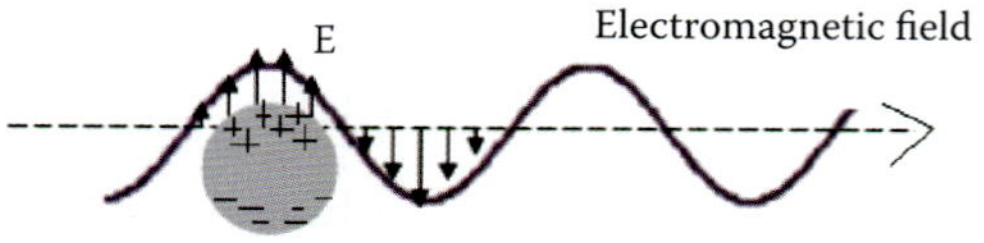

FIGURE 5.70 The interaction of an electromagnetic wave with a metal nanoparticle.

properties. Alexandria, Egypt was believed to have been the original founding place for the use of gold in medicine by a group of adepts known as *alchemists*. The alchemists developed an "elixir" made of liquid gold that purportedly had the ability to restore youth and perfect health. Paracelsus developed medicines from metallic minerals, including gold, to cure the sick.*

Earlier in this chapter we learned that 2D macroscopic metallic/dielectric interfaces require special geometries for plasmon excitation. However, in nanoparticles we saw the surface plasmon wave is localized (not traveling), and so no special excitation geometry is required (Figure 5.70). These "localized plasmons" of metallic nanoparticles may be excited directly by light absorption. For metallic nanoparticles significantly smaller than the wavelength of light, light absorption concentrates to a narrow wavelength range. Wavelengths that are absorbed by metal nanoparticles and produce such a wave are called *surface plasmon bands* or simply *plasmon bands*.

The interaction of light with metal nanoparticles is different than that for a bulk piece of metal. Bulk Au, for example, looks yellowish in reflected light, but thin films of Au look blue in transmission. The characteristic blue of gold particles (say 400–500 nm) changes to orange, via several tones of purple and red, as the particle size is reduced to ~3 nm (see Figure 5.71b). These color changes are the result of changes in the surface plasmon resonance ω_p, the frequency at which conduction electrons oscillate in response to the alternating electric field of incident electromagnetic radiation (Figure 5.71a). Only metals with free electrons (essentially Au, Ag, Cu, and the alkali metals) possess plasmon resonances in the visible spectrum, which give rise to intense colors. Gold and silver are often used in nanophotonics because of their excellent chemical stability and very good conductivity. The position and shape of the plasmon absorption band depend on the material, size, and shape of the nanoparticles, as well as on the surrounding medium, i.e., its dielectric constant.

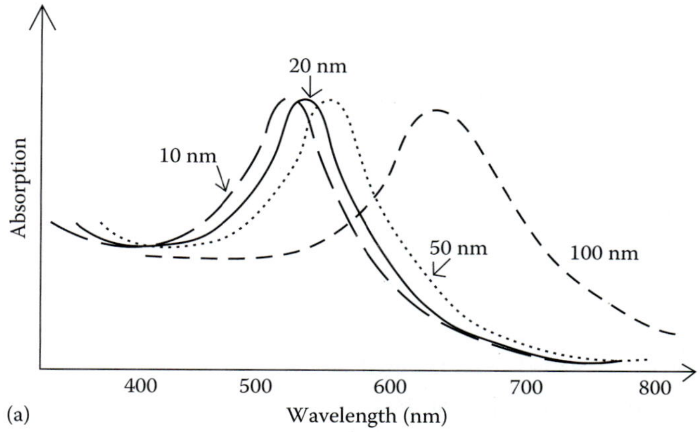

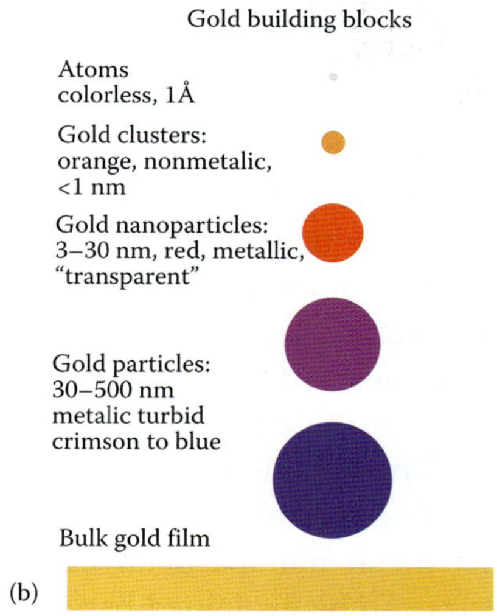

FIGURE 5.71 The surface plasmon resonance (SPR) dependence on size for gold nanoparticles (a). The properties of a material depend on the type of motion its electrons can execute, which depends on the space available for them. Thus, the properties of a material are characterized by a specific "length scale," usually in the nanometer range. If the physical size of the material is reduced below this length scale, its properties change and become sensitive to size and shape (b).

* Even today in China, remnants of the belief in the restorative properties of gold remain intact in rural villages, where peasants cook their rice with a gold coin to help replenish gold in their bodies. In India, a certain kind of fine candy is wrapped in extremely fine gold foil that is eaten along with the rest of the delicacy. It has been reported that in the early 1900s doctors would implant a $5 gold piece under the skin, such as a knee joint.

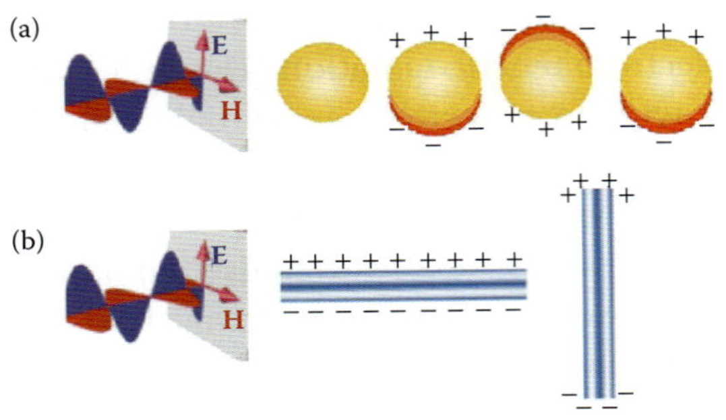

FIGURE 5.72 (a) Schematic drawing of the interaction of an electromagnetic radiation with a metal nanosphere. A dipole is induced, which oscillates in phase with the electric field of the incoming light. (b) Transverse and longitudinal oscillations of electrons in a metal nanorod.

For nonspherical particles, such as rods, the resonance wavelength depends also on the orientation of the electric field. In Figure 5.72 we illustrate the interaction of electromagnetic radiation with spherical and rod-shaped particles. Because for elongated particles the resonance depends on the particle orientation relative to the field, a collection of such particles can show transverse and longitudinal resonances. In addition, when nanoparticles are sufficiently close together, interactions between neighboring particles arise (see further below, Figure 5.76), so that the models for isolated particles do not further hold.

The electron density at the surface of a nanoparticle can induce large electric fields, and the localized surface plasmon resonance ω_p depends on a number of factors, among which particle size and shape, as well as the nature of the surrounding medium, are the most important. As described earlier, the light absorption by metallic nanoparticles is described by a size effect of the coherent oscillation of electrons, induced by an interaction with electromagnetic fields as described above.* Curiously, the major changes that metal nanoparticles exhibit in their optical spectra can be explained based simply on a further detailing of the classical Drude theory.

We use Equation 3.40 from the Drude model for the complex dielectric function and focus on the damping term γ:

$$\varepsilon_r(\omega,\mathbf{k}) = 1 - \frac{\omega_p^2}{\omega^2 + i\gamma\omega} \tag{5.216}$$

The damping term γ comes about because of scattering ($1/\gamma = \tau$). In the bulk metal, the main contributions to γ are from electron-electron scattering and electron-phonon scattering. In small particles—small compared with the wavelength of the incident light—scattering from the particle boundaries (surfaces) becomes important, and this scattering produces a damping term that is inversely proportional to the particle radius. Thus, the effect of size on plasmon resonance in metal nanospheres is contained in γ.

The real part of the dielectric constant (Equation 3.42) $\varepsilon'(\omega,\mathbf{k})$ is negative for $\omega < \omega_p$. So, for example, the calculated plasma wavelength ω_p for silver is 137 nm and for copper it is 114 nm; therefore, in both cases, $\omega < \omega_p$ and $\varepsilon_r'(\omega,\mathbf{k})$ are negative in the visible region for these particles.

Consider now a dilute dispersion of metal nanoparticles embedded in a dielectric. When shining light on such an assembly, one observes scattering of the incoming radiation, i.e., particles in the path of the EM waves continuously abstract energy from the waves and reradiate it in all directions. To classify these optical phenomena, as we will see later in this chapter, it is handy to introduce a dimensionless size parameter α, which is a measure of the size of the particles compared with the wavelength of the incoming radiation:

$$\alpha = \frac{2\pi r}{\lambda} \tag{5.217}$$

where r is the radius of the particle and λ is the wavelength of the incoming radiation. With $\alpha << 1$, one obtains Rayleigh scattering, which is inversely proportional to λ^4 (e.g., scattering of visible light by atmospheric molecules; see Equation 5.256). For $\alpha \geq 1$, one obtains Mie scattering, inversely proportional to λ (e.g., scattering of visible light by clouds). Absorption of energy by particles and molecules leads to emission at different frequencies, and the combined effects of absorption and scattering in attenuating the incoming radiant energy are quantified by the extinction, C_{ext}. For very small spherical particles with a frequency-dependent complex

* The collective dipolar oscillation at the plasma frequency is also called the *Fröhlich mode*.

dielectric function, $\varepsilon'_{m(\omega,k)} = \varepsilon' + i\varepsilon''$, embedded in a medium of dielectric constant ε_d, the Mie theory predicts[11]:

$$C_{ext} = \frac{18\pi NV\varepsilon_d^{3/2}}{\lambda} \frac{\varepsilon''}{[\varepsilon' + 2\varepsilon_d]^2 + \varepsilon''^2} \tag{5.218}$$

where λ = wavelength of the incoming light
ε_d = dielectric constant of the surrounding medium
N = number density of the metal nanospheres
V = volume of a nanoparticle
ε' and ε'' = real and imaginary parts of the metal dielectric constant ε_m

If ε'' is small or weakly dependent on ω, we expect large extinction when the resonant condition is fulfilled (Fröhlich mode):

$$\varepsilon' + 2\varepsilon_d = 0 \tag{5.219}$$

For metal nanospheres in air the plasmon resonance condition is then obtained when $\varepsilon' = -2$ (because ε_d for air = 1). Nanoparticle resonance, from the Drude model, then occurs at:

$$\varepsilon' \approx 1 - \frac{\omega_p^2}{\omega^2} \text{ or } \omega = \frac{1}{\sqrt{3}}\omega_p \approx 0.58\omega_p \tag{5.220}$$

In contrast for a thin film we found that the plasma frequency was given as $\omega = 2^{-1/2}\omega_p$ scaled relative to the bulk plasmon frequency (Equation 5.163). When the shape of the nanoparticles changes from a nanosphere to a nanorod, the surface plasmon absorption spectrum also changes as illustrated in Figure 5.73. For example, the surface plasmon absorption of gold nanorods features two bands: a strong long wavelength band in the near-infrared region resulting from the longitudinal oscillation of electrons and a weak short wavelength band in the visible region around 520 nm caused by the transverse electronic oscillation. To calculate the spectra for elongated particles (nanorods), the orientation with respect to the oscillating field must be taken into account. The absorption spectrum of the nanorods is very sensitive to the aspect ratio (R = length/width = A/B, with the length along A > B = C; see Figure 5.73b). When the aspect ratio increases, the absorption maximum of the longitudinal band is greatly red-shifted. For ellipsoidal particles under illumination, the Mie theory gives the following expression for the extinction coefficient, C_{ext}:

$$C_{ext} = \frac{2\pi V\varepsilon_d^{3/2}}{3\lambda} \sum_j \frac{(1/L_j^2)\varepsilon''}{\left[\varepsilon' + \frac{1-L_j}{L_j}\varepsilon_d\right]^2 + \varepsilon''^2} \tag{5.221}$$

(see the Gans theory for SPR of ellipsoidal nanoparticles[12,13]), where L_j is a geometry factor, with values between 0 and 1, different along each axis (j = A, B, and C):

$$L_A = \frac{1-e^2}{e^2}\left[\frac{1}{2e}\ln\left(\frac{1+e}{1-e}\right) - 1\right]$$
$$L_B = L_C = \frac{1-L_A}{2} \tag{5.222}$$

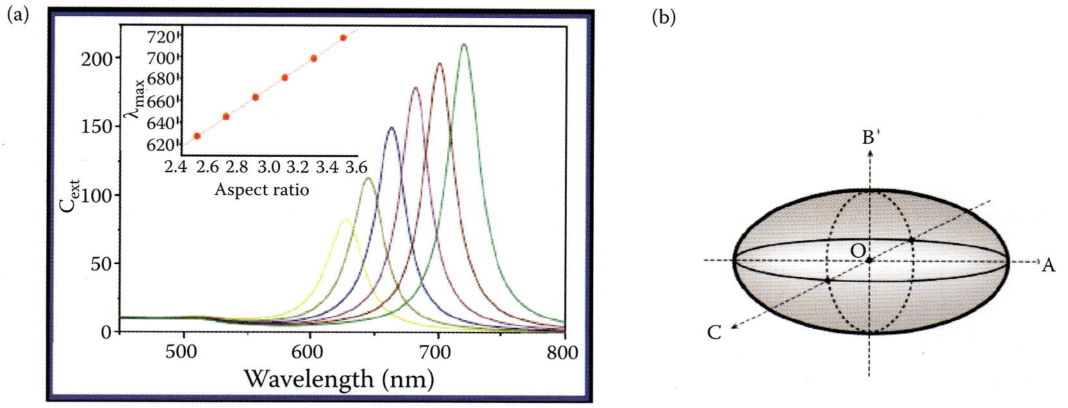

FIGURE 5.73 (a) When the aspect ratio *R* (*A*/*B*) increases, the absorption maximum of the longitudinal band is greatly red-shifted. (b) Ellipsoidal particle A > B = C.

The geometry factors L_A and L_B are connected to the aspect ratio $R = A/B$ via:

$$e = \sqrt{1-\left(\frac{B}{A}\right)^2} \tag{5.223}$$

From Equation 5.221, the resonance condition is given as:

$$\varepsilon' = \varepsilon_d\left(1-\frac{1}{L}\right) \tag{5.224}$$

so that, in air ($\varepsilon_d = 1$), the resonance frequency is $\omega_p L$.

Another type of nanoparticle that is studied heavily is a dielectric sphere coated with a nanoshell of a metal like gold. Core materials (e.g., silica) have a radius between 30 and 250 nm, and the shell thickness is in the 10–30-nm range. In 1998, Naomi Halas at Rice University demonstrated that one could create such nanostructures by growing a thin layer of gold onto a spherical silica nanoparticle.[14] The thinness of the metal shell results in a substantial redshift of the plasmon resonance. This is caused by increased electron scattering and thus an increase in the damping constant in the metal dielectric constant. Nanoshells are more sensitive than simple nanoparticles with a gain of $x = a/b$, where *a* is the radius of the particle without shell and *b* is the radius of the particle with shell (Figure 5.74).

The resonance frequency of a nanoshell is a strong function of geometry, and more resonant frequencies are possible. Without going into details, the condition for resonance in this case is given as:

$$\varepsilon'(\text{shell}) + 2\varepsilon_d = 0 \tag{5.225}$$

and the resonance frequency is:

$$\omega_{\text{nanoshell},+/-} = \frac{\omega_P}{\sqrt{2}}\left[1 \pm \frac{1}{2L+1}\sqrt{1+4L(L+1)x^{2L+1}}\right]^{\frac{1}{2}}$$

in the case of a sphere $L = 1$ and we obtain

$$\omega_{\text{nanoshell},+/-} = \frac{\omega_P}{\sqrt{2}}\left[1 \pm \frac{1}{3}\sqrt{1+8x^3}\right]^{\frac{1}{2}} \tag{5.226}$$

The plasmon energies of a nanoshell depend on the aspect ratio x (= a/b), so the plasmon energies of nanoshells can be tuned (Figure 5.74b)!

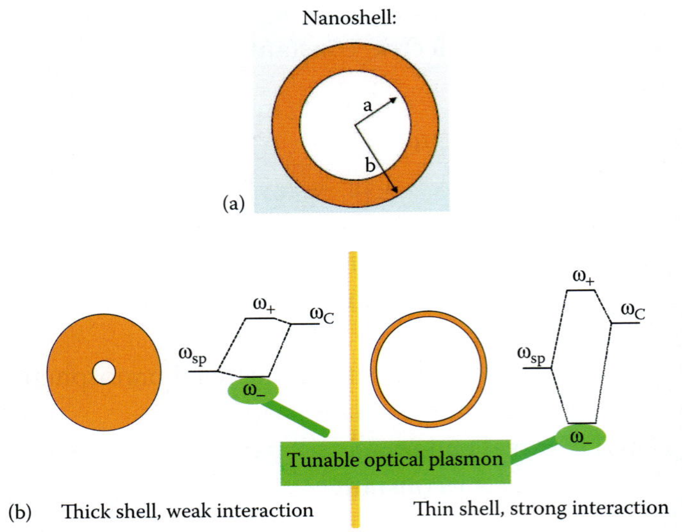

FIGURE 5.74 The resonance frequency of a nanoshell is a strong function of geometry, and more resonant frequencies are possible. The ratio of a/b = x in Equation 5.226 (a). The plasmon energies of a nanoshell depend on this aspect ratio, so the plasmon energies of nanoshells can be tuned (b).

For fabrication, a dielectric sphere is coated with a layer of amines, which binds 1–2-nm gold colloids from suspension. This is followed by a chemical treatment with $HAuCl_4$ in the presence of formaldehyde. This results in an additional layer of gold (Figure 5.75).

Younan Xia of the University of Washington (Seattle, WA) described the synthesis of gold nanocages.[15] Like nanoshells, these nanoparticles have tunable plasmon resonances because of the presence of a dielectric core. In nanocages, however, the core is empty rather than composed of silica. Gold nanocages are grown over a silver nanocube template. The monodisperse nanocubes are made by a polyol process, such as reduction of silver nitrate in ethylene glycol in the presence of poly(vinylpyrrolidone). These template particles are then exposed to a solution of gold ions, which replace the more reactive silver on the nanoparticle surface. Gold grows around the silver nanocube surface, and the silver is dissolved into solution, leaving gold nanocages.

In Table 5.6 we have compiled values for the resonance frequency and resonance condition for a variety of configurations.

Surface Plasmon Applications II

We reviewed some plasmonics applications under "Surface Plasmon Applications I." Here, having

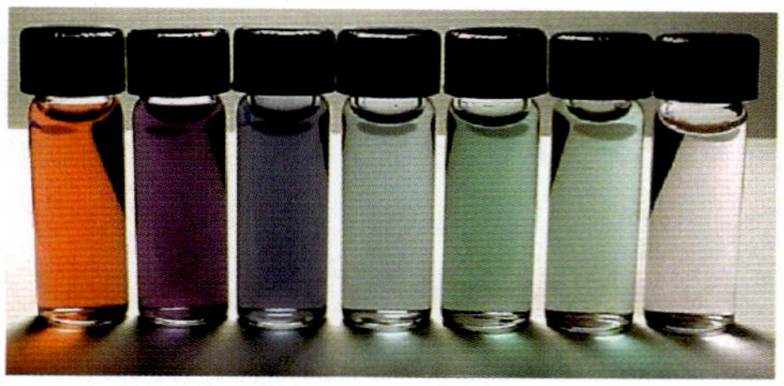

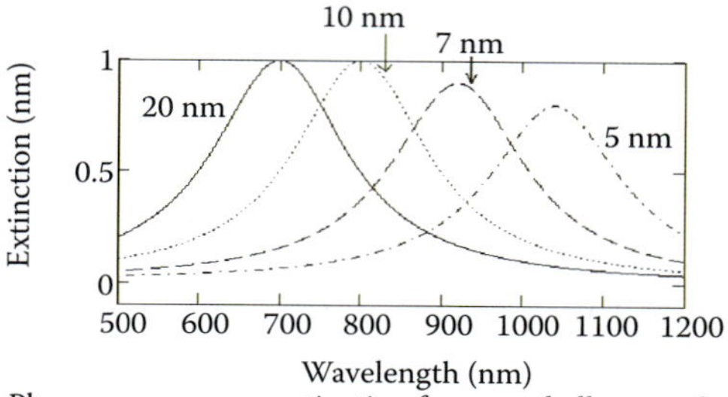

Plasmon resonance extinction for core-shell nanospheres of various shell thicknesses. Core radius is 60 nm.

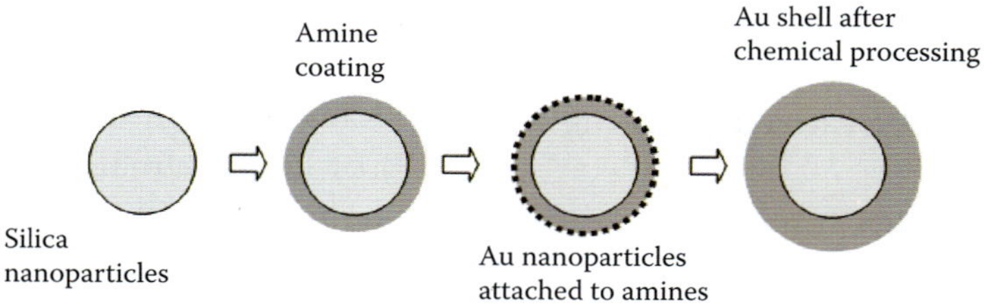

FIGURE 5.75 In general, as the shell thickness decreases, the resonance shifts to longer wavelengths. Vial on the left has solid gold colloids. Others have colloids with metal nanoshells with decreasing thickness. Vial on left absorbs IR. (From Dr. Anup Sharma.)

Table 5.6 Resonance Frequency and Resonance Condition for a Variety of Configurations

Material	Resonance Condition	Resonance Frequency
Bulk metal	$\varepsilon' = 0$	$\varepsilon' = 1 - \frac{\omega_p^2}{\omega^2} = 0,\ \omega_{\text{Bulk plasmon}} = \omega_p$
Planar surface	$\varepsilon' = -\varepsilon_d$ ($\varepsilon_d = 1$ in air)	$\varepsilon' = 1 - \frac{\omega_p^2}{\omega^2} = -\varepsilon_d$ or $\omega_{\text{Surface plasmon}} = \frac{\omega_p}{\sqrt{2}}$
Sphere	$\varepsilon' + 2\varepsilon_d = 0$ ($\varepsilon_d = 1$ in air)	$\varepsilon' \approx 1 - \frac{\omega_p^2}{\omega^2} = -2\varepsilon_d$ or $\omega_{\text{Nanosphere}} = \frac{1}{\sqrt{3}}\omega_p$
Ellipsoid	$\varepsilon' = \varepsilon_d\left(1 - \frac{1}{L}\right)$	$\varepsilon' \approx 1 - \frac{\omega_p^2}{\omega^2} = \varepsilon_d\left(1 - \frac{1}{L}\right)$ or $\omega_{\text{Ellipsoid}} = L\omega_p$
Cavity		$\omega_{\text{cavity}} = \omega_p\sqrt{\frac{L+1}{2L+1}}$ In the case of a spherical cavity: $\omega_{\text{cavity}} = \omega_p\sqrt{\frac{2}{3}}$
Nanoshell 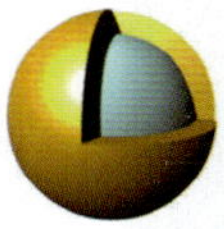	$\varepsilon'(\text{shell}) + 2\varepsilon_d = 0$	$\omega_{\text{nanoshell},+/-} = \frac{\omega_p}{\sqrt{2}}\left[1 \pm \frac{1}{2L+1}\sqrt{1 + 4L(L+1)x^{2L+1}}\right]^{\frac{1}{2}}$ In the case of a sphere: $\omega_{\text{nanoshell},+/-} = \frac{\omega_p}{\sqrt{2}}\left[1 \pm \frac{1}{3}\sqrt{1 + 8x^3}\right]^{\frac{1}{2}}$

ε' is the real part of the dielectric constant of the metal; ε_d is the dielectric constant of the contacting dielectric (= 1 in the case of air as the dielectric); *L* is the geometrical factor between 0 and 1, for a sphere L = 1 and x = a/b.

acquired some additional insights in metal nanoparticle behavior, we come back to two nanoparticle-based applications, i.e., resonant particle detection and plasmon-based circuits.

Several properties make resonant nanoparticles unique for sensor applications:

- They are very bright, making them easy to detect. The reason for this brightness is their large scattering cross-section.
- They are small enough to diffuse through biological tissues.
- They can easily be derivatized (linking to organic molecules).
- They suffer no photobleaching, making long illumination possible.
- Most of the sensor applications of localized plasmon waves are based on the EM field enhancement in the vicinity of the metal nanoparticle surface.
- The resonance can be tuned (multilayer/core-shell) for parallel detection.

Today the performance of electronic circuits is becoming rather limited when digital information needs to be sent from one point to another. With photons traveling much faster than electrons, photonics potentially offers an effective solution to this problem by implementing optical communication systems based on optical fibers and photonic circuits. Unfortunately, as we saw earlier in this chapter, traditional optical waveguides involve structures that cannot be made much smaller than $\lambda/2$ because of diffraction. Furthermore, waveguiding along bent guides is very lossy (see Figure 5.49). The latter problem is overcome in photonic bandgap structures (see below). However, the dimensions of the structures are still limited by the wavelength of light. Surface plasmon-based circuits, which merge electronics and photonics at the nanoscale by waveguiding of plasmonic excitation in closely placed metal nanoparticles, may offer a solution to this size-compatibility problem. A chain of closely spaced metallic nanoparticles allows waves of surface plasmons to propagate down the chain as illustrated in Figure 5.76. The typical size of the coupling metal nanospheres is 50 nm. The waves can be either transverse (T) or longitudinal (L) modes, and can have group velocities up to 0.1 *c* or higher. This approach might potentially be useful for propagating energy along effectively very narrow waveguides, controlling energy flow around corners, and so on. The individual particles in the chain have a specific resonance frequency, and light propagates through these "coupled resonators." The particle center frequency is tunable by material, shape, and size, and the group velocity is tunable by shape and spacing.

These surface excitations can be at any frequency between UV and IR, and so these waveguides combine the less bulky nature of metal waveguides with the high bandwidth of optical waveguides. A weakness of this technique is its high loss (6 dB/μm), and transmission has only been shown over short (~1-μm) distances. It is an active area of research for future applications.

Plasmonics has revived the general enthusiasm for photonics, and one might say that optics has evolved from lenses to fibers to metal nanoparticles

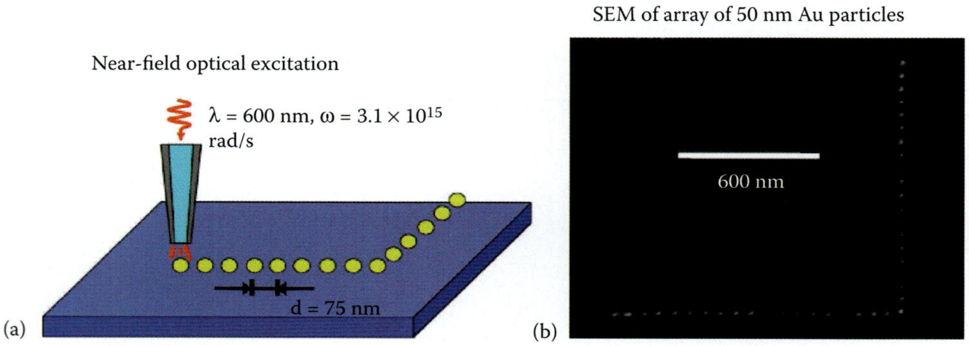

FIGURE 5.76 (a) Propagation of high-frequency EM waves along subwavelength-wide metal waveguides. Plasmonic waveguides. (b) e-beam lithography with liftoff was used to fabricate 50-nm Au nanoparticles on ITO-coated glass.[16]

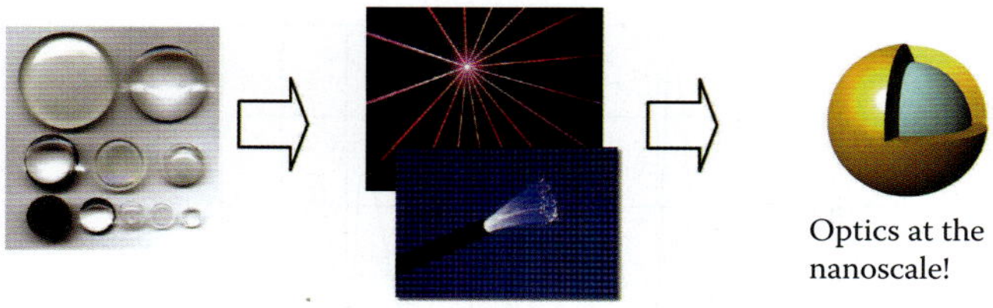

FIGURE 5.77 From lenses to fiberoptics to metal nanoparticles.

(see Figure 5.77). Once we have introduced Raman scattering later in this chapter, we will come back one more time to plasmonics applications. Specifically, we will detail its application in near-field scanning microscopy (NSOM) for subwavelength resolution and in surface enhanced Raman spectroscopy (SERS; see "Surface Plasmon Applications III").

Lorentz Model for Dielectrics

Nonpolar Dielectrics

Whereas isolated atoms have well-defined electronic states representative of the atom, as individual atoms are brought together the atoms start to interact, leading to shifts in the electronic energy levels. At the densities common in solids (~10^{22} atoms/cm^3), significant interaction occurs, causing levels to broaden out into bands. Remembering that real materials correspond to a collection of Drude-Lorentz oscillators all operating at different frequencies, the Drude-Lorentz model works surprisingly well for nonpolar dielectrics. The outer, or valence, electrons of the constituent atoms predominantly determine the characteristics of the optical properties of any solid. The tighter the valence electrons are bound, the higher the resonance frequency, so that these materials may have a transparency range that extends far into the UV. These dielectrics typically show only one resonance in the UV, and at low frequencies they are transparent (see Figure 5.62b for two common dielectrics: Si_3N_4 and SiO_2). Their bandgap, $E_g > 4$ eV, results in high resistivity and low absorption throughout the visible spectrum ($E < E_g$, $\alpha \sim 0$, and $\kappa <<< n$, so that n is real). At the higher frequencies, the refractive index n approaches 1 ($v_g = c$) as there are usually no high ω oscillators in this case, and light then passes through the material as if it were a vacuum.

Besides for their optical properties, dielectrics are also used as insulators or high-*k* dielectrics.* In Chapter 4 we emphasized the need to find loss-free high-*k* dielectrics for the next-generation CMOS. Loss is characterized through the quality factor Q. In dielectrics, a low-quality factor Q indicates that a material heavily dissipates energy from the alternating electric field by different types of adsorbing mechanisms. The quality factor (Q) of a resonance $Q = \omega_0/\gamma$ is a measure of the dissipation of the system. The loss tangent *tan*δ and quality factor Q of a material are figures of merit for a dielectric and are defined as:

$$\tan\delta = \frac{\varepsilon_r''}{\varepsilon_r'} \qquad (5.227)$$

and:

$$Q = \frac{1}{\tan\delta} \qquad (5.228)$$

Polar Dielectric Polaritons

In polar ionic materials, e.g., alkali halides such as KCl, the valence electrons are very strongly localized at the negative ion (for KCl, this would be the Cl atom), and hence the optical spectrum contains some atomic-like features, with many resonances. So, whereas the reflectance for a single transition leads to one broad reflection band, as shown in Figure 5.78a the reflectance for the polar solid KCl has many high-energy transitions (Figure 5.78b). At less than 7 eV, KCl is transmissive (KCl is transparent at visible frequencies) and displays a normal dispersion.

In Chapter 3, four types of phonons—LA, longitudinal acoustic; TA, transverse acoustic; LO, longitudinal optical; and TO, transverse optical—were illustrated in Figure 3.159. The transverse and longitudinal "optical phonons" are lattice vibrations that are called *optical phonons* because they are easily excited by light (IR radiation) (e.g., for GaAs, $\hbar\omega_{OP}$ = 36 meV). Such phonons can only arise in crystals that have more than one atom in the smallest unit cell. In a polar solid, mechanical

* High-*k* is another name and symbol used in this context for a high relative dielectric constant.

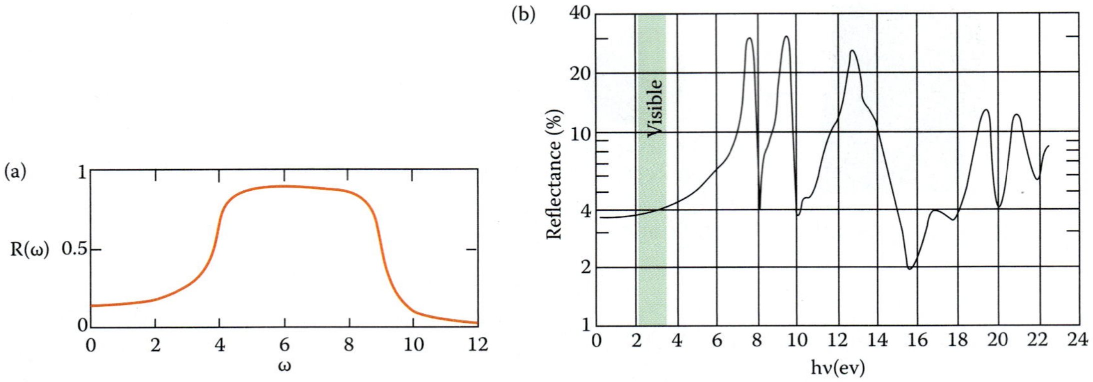

FIGURE 5.78 (a) Calculated reflectance R for a single transition leads to one broad reflection band. (b) The reflectance for the polar solid KCl has many high-energy transitions. KCl is transparent at visible frequencies.

vibrations may correspond to a mode where positive and negative ions at adjacent lattice sites swing against each other, creating a time-varying electrical dipole moment **P** that EM radiation can interact with. (EM longitudinal waves do not interact with lattice vibrations.)

The polarization $\mathbf{P}_{lattice}$ was derived in Chapter 3 as $\mathbf{P}_{lattice} = Ne(u_0 - v_0)$ (Equation 3.326), with e the effective charge on the atoms in the lattice and u_0 and v_0 the amplitudes of the excursion of atom M and atom m, respectively. The solutions for amplitudes u_0 and v_0 were calculated as $u_0 = \frac{+qE_0/M}{\omega_T^2 - \omega^2}$ and $v_0 = \frac{-qE_0/m}{\omega_T^2 - \omega^2}$ (Equation 3.324), where $\omega_T = \sqrt{\frac{2K}{\mu}}$ (Equation 3.325) is the resonant frequency of the TO mode at wave vector $\mathbf{k} = 0$, and μ is the reduced mass (see Figures 3.159 and 5.79).

The excitation near ω_T (at $\mathbf{k} = 0$) features the largest amplitude. When in this resonance condition

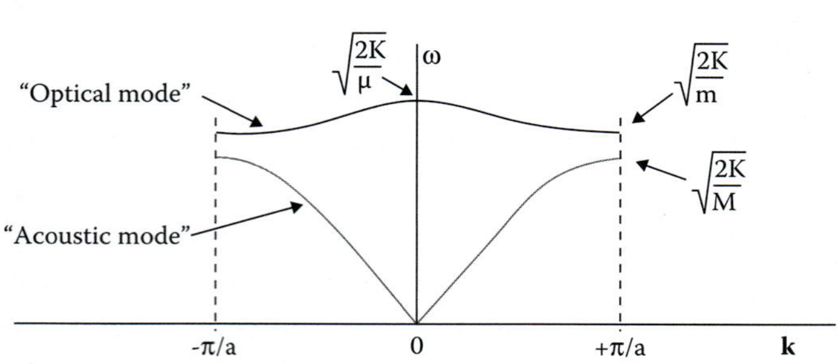

FIGURE 5.79 Two separate "phonon branches" are observed, a low-frequency or "acoustic branch" and a high-frequency "optical branch." The optical branch has a maximum at low wave vector limit: $\mathbf{k} \cong 0$ (or $\mathbf{k}a \ll 1$) and the acoustic branch at $\mathbf{k} \cong \pm\pi/a$.

the frequencies and wave vectors of optical and mechanical waves are approximately equal, then the phonon-photon coupling can change the propagation of the waves dramatically. The quantum of the coupled phonon-photon transverse wave field is called a *polariton*. In Chapter 3 we also introduced the total polarization of a material as the sum of the lattice contribution ($\mathbf{P}_{lattice}$), a bound electronic contribution ($\mathbf{P}_B$), and a free electron contribution ($\mathbf{P}_E$) as:

$$\mathbf{P} = \mathbf{P}_{lattice} + \mathbf{P}_B + \mathbf{P}_E \tag{3.327}$$

From the expression linking the dielectric constant with the polarization $\mathbf{D} = \varepsilon_0\mathbf{E} + \mathbf{P} = (1 + \chi_e)\varepsilon_0\mathbf{E} = \varepsilon(\omega,\mathbf{k})\mathbf{E}$ (Equation 3.37), we derive:

$$\varepsilon_r(\omega) = 1 + \frac{\mathbf{P}}{\varepsilon_0\mathbf{E}} = 1 + \frac{\mathbf{P}_{lattice}}{\varepsilon_0\mathbf{E}} + \frac{\mathbf{P}_B}{\varepsilon_0\mathbf{E}} + \frac{\mathbf{P}_E}{\varepsilon_0\mathbf{E}} \tag{5.229}$$

and ignoring, for the moment, the free electron contribution:

$$\varepsilon_r(\omega) = 1 + \frac{\mathbf{P}_B}{\varepsilon_0\mathbf{E}} + \frac{\mathbf{P}_{lattice}}{\varepsilon_0\mathbf{E}} = 1 + \frac{\mathbf{P}_B}{\varepsilon_0\mathbf{E}} + \frac{Ne(u_0 - v_0)}{\varepsilon_0\mathbf{E}} \tag{5.230}$$

Because $u_0 - v_0 = \frac{e\mathbf{E}_0}{\mu(\omega_T^2 - \omega^2)}$ and replacing $1 + \frac{\mathbf{P}_B}{\varepsilon_0\mathbf{E}}$ (for the bound electrons) with $\varepsilon_r = 1 + \chi_B(\omega)$ (Equation 5.115; we ignore the prime symbol here, but we are talking about the real part of the dielectric constant), we rewrite Equation 5.230 as:

$$\varepsilon_r(\omega) = 1 + \chi_B + \frac{Ne^2}{\varepsilon_0\mu(\omega_T^2 - \omega^2)} \tag{5.231}$$

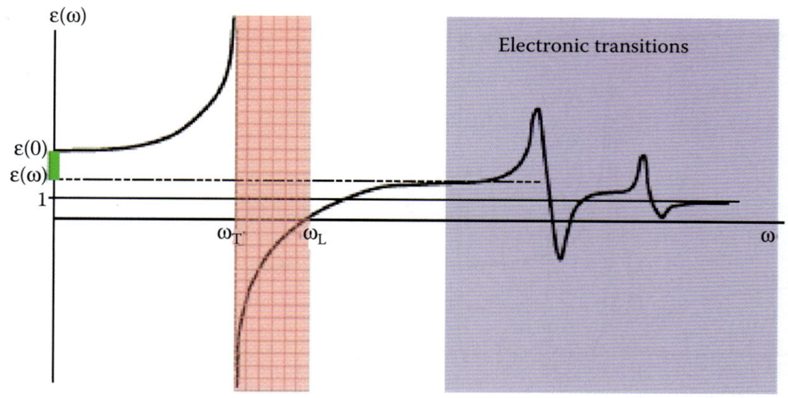

FIGURE 5.80 Dielectric constant as a function of ω for a polar material.

When considering low frequencies, this equation simplifies to:

$$\varepsilon_r(\omega) - \varepsilon_r(\infty) = \frac{Ne^2}{\varepsilon_0 \mu(\omega_T^2 - \omega^2)} \quad (5.232)$$

where $\varepsilon_r(\infty)$ is the high-frequency dielectric constant. Thus, the difference between the low- and high-frequency dielectric constant $\varepsilon_r(\omega) - \varepsilon_r(\infty) = \frac{Ne^2}{\varepsilon_0 E\mu(\omega_T^2 - \omega^2)}$ is related to charge density and mass. It is important to recognize that $\varepsilon_r(\infty)$ is not truly the value of the dielectric constant at an infinite frequency because that value is always 1. What it means is a value at frequencies well above ω_T, yet well below any electronic transitions (see Figure 5.80).

Thus, the behavior sketched of this type of resonance is similar to that of the Lorentz resonator but without the damping term (see Equation 5.195).

We can use Equation 5.232 to find a value for the low-frequency dielectric constant as $\varepsilon_r(0) = \varepsilon_r(\infty) + \frac{Ne^2}{\varepsilon_0 \mu \omega_T^2}$. The same equation can be used to find the frequency ω_L, at which $\varepsilon_r(\omega) = \varepsilon_r(\infty) + \frac{Ne^2}{\varepsilon_0 E\mu(\omega_T^2 - \omega^2)} = 0$

The latter expression with $\frac{Ne^2}{\varepsilon_0 \mu} = [\varepsilon_r(0) - \varepsilon_r(\infty)]\omega_T^2$ can be modified as:

$$\varepsilon_r(\omega) = \varepsilon_r(\infty) + \frac{[\varepsilon_r(0) - \varepsilon_r(\omega)]\omega_T^2}{(\omega_T^2 - \omega^2)} \quad (5.233)$$

This can be rewritten as:

$$\varepsilon_r(\omega) = \frac{\varepsilon_r(0)\omega_T^2 - \varepsilon_r(\infty)\omega^2}{(\omega_T^2 - \omega^2)} \quad (5.234)$$

so that we find that $\varepsilon_r(\omega) = 0$ when $\varepsilon_r(0)\omega_T^2 = \varepsilon_r(\infty)\omega^2$. The particular frequency ω where this happens is labeled ω_L, for longitudinal optical (LO) phonons. Similar to the case of plasma oscillations with a plasma oscillation frequency ω_p, ω_L is the frequency for longitudinal wavelength lattice oscillations (LO phonon frequency). We can rewrite $\varepsilon_r(0)\omega_T^2 = \varepsilon_r(\infty)\omega_L^2$ at ω_L also as:

$$\frac{\omega_T^2}{\omega_L^2} = \frac{\varepsilon_r(\infty)}{\varepsilon_r(0)} \quad (5.235)$$

This is known as the *Lyddane-Sachs-Teller relationship*, which works well for polar diatomic solids.

Example 5.9:

For GaAs: $\frac{\omega_L}{\omega_T} = 1.07$ and $\sqrt{\frac{\varepsilon_r(0)}{\varepsilon_r(\infty)}} = 1.08$

And for KBr: $\frac{\omega_L}{\omega_T} = 1.39$ and $\sqrt{\frac{\varepsilon_r(0)}{\varepsilon_r(\infty)}} = 1.38$

The resonant frequency ω_T is called the *Restrahlen frequency*, and the frequency at which the dielectric constant vanishes is ω_L [notice that ω_L is always larger than ω_T; hence $\varepsilon_r(0) > \varepsilon_r(\infty)$].

In Figure 5.81a we show the real and imaginary parts of the refractive index of AlSb, a polar material.

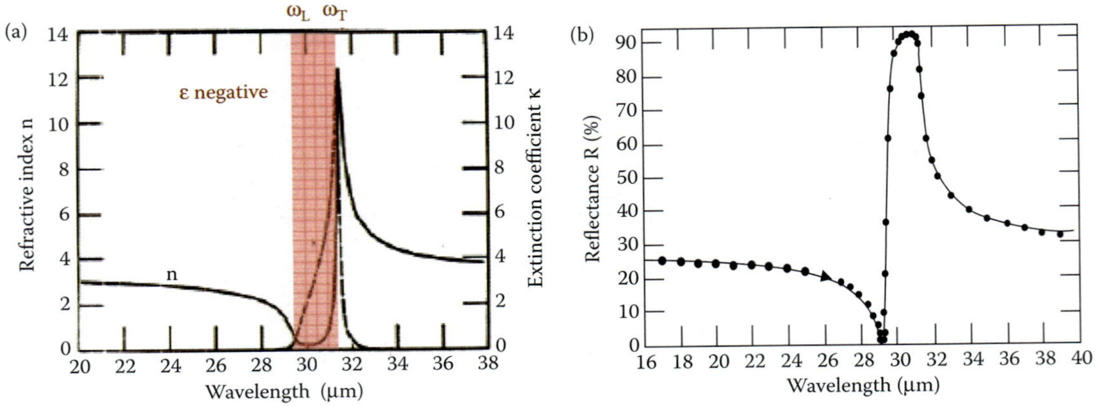

FIGURE 5.81 (a) The real (n) and imaginary (κ) parts of the refractive index of AlSb, a polar material. (b) Reflectance spectrum of AlSb with the Restrahlen band. (R, in the absence of damping, must have unity in the region $\omega_L < \omega < \omega_T$.)

Although in Equation 5.232 we have assumed that damping was not important, the damping in a real case cannot be too small as κ appears large here. Just like for electronic transitions, the reflectance R, in the absence of damping, must be one in the region $\omega_L < \omega < \omega_T$. This region is called the *Restrahlen band*, where light cannot propagate, so it must be reflected as shown in Figure 5.81b.

Summary: Generalized Dispersion Relation

Substituting the time-varying EM wave $\mathbf{E}(\mathbf{r},t) = \mathbf{E}_m e^{i(\mathbf{k}\cdot\mathbf{r}-\omega t)}$ in Maxwell's wave equation, expressed in terms of the velocity of light c and the dielectric constant $\varepsilon_r(\omega,\mathbf{k})$ $c^2\nabla^2\mathbf{E}(\mathbf{r},t) = \varepsilon_r(\omega,\mathbf{k})\frac{\partial^2\mathbf{E}(\mathbf{r},t)}{\partial^2 t}$ (because most materials have $\mu \approx 1$ and are nonmagnetic), one obtains an expression for the dispersion of EM waves in terms of the dielectric constant:

$$\varepsilon_r(\omega,\mathbf{k})\omega^2 = c^2\mathbf{k}^2 \tag{5.88}$$

where we have ignored the μ term because most materials have $\mu \approx 1$ and are nonmagnetic.

In free space $\varepsilon_r(\omega,\mathbf{k}) = 1$, and Equation 5.88 then simply reduces to $\mathbf{k} = \omega/c$, the so-called *light line.*

For the case that $\varepsilon_r(\omega,\mathbf{k})$ is real and >0, **k** is also real, and transverse EM waves propagate with a group velocity of $v_g = \frac{d\omega}{d\mathbf{k}} = \frac{c}{\sqrt{\varepsilon_r}}$ (Equation 5.87 with $\mu = 1$). With $\varepsilon_r(\omega,\mathbf{k})$ real but <0, **k** is imaginary, and the wave is damped with a characteristic skin depth δ (Equation 5.215). With $\varepsilon_r(\omega,\mathbf{k})$ complex, **k** is complex as well, and waves are damped again. At $\varepsilon_r(\omega,\mathbf{k}) = 0$, longitudinally polarized waves are possible (free electron plasmons). In longitudinal oscillations the displacement is in the same direction as the wave motion, and Drude's free electron gas (FEG), with an AC field applied, can exhibit such collective longitudinal oscillations because the displacement of all the electrons against the ion bodies of the material induces a dipole moment and an electric field opposing that displacement, as was shown in Figure 3.12.

At optical frequencies $\omega^2\tau^2 >>> 1$ (so that $1/\tau = \gamma$ tends to zero) and $\varepsilon_r(\omega) \cong \varepsilon_r'(\omega) = 1 - \frac{\omega_p^2}{\omega^2}$ (see Equation 3.42). This expression is drawn in Figure 5.82 (see also Figure 3.13) with a region (1) where the metal is transparent ($\omega > \omega_p$ so that **k** is real and there is no attenuation), and a region (2) where the waves are totally reflected ($\omega < \omega_p$ so that **k** contains an imaginary component and the wave is reflected) to give us the typical shine of a metal.

Substituting $\varepsilon_r'(\omega,\mathbf{k}) \approx 1 - \frac{\omega_p^2}{\omega^2}$ (Equation 3.44) in the dispersion expression (Equation 5.88), we find:

$$\omega^2 - \omega_p^2 = c^2\mathbf{k}^2 \tag{5.150}$$

This expression is plotted in Figure 5.83, where the reflection [corresponding to the region (2) of Figure 5.82] is shown as a forbidden frequency gap. This forbidden frequency gap falls in the region of $0 < \omega \leq \omega_p$. At those frequencies light incident on the medium does not propagate but is totally reflected, giving us shiny metals (imaginary **k**).

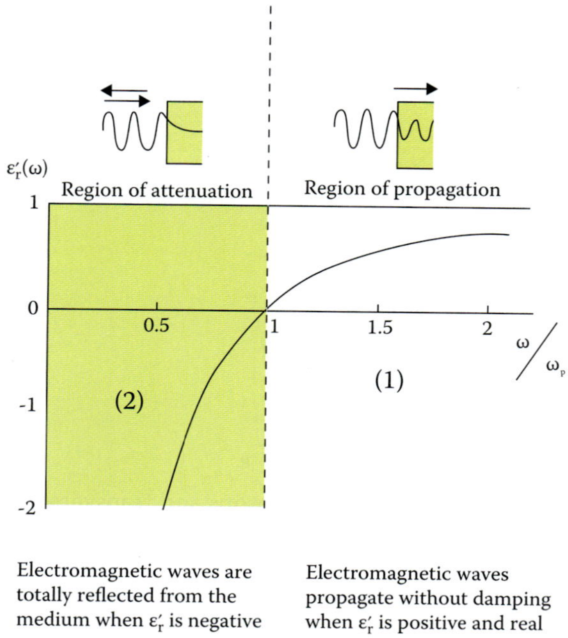

FIGURE 5.82 The dielectric function of a free electron gas versus frequency in units of the plasma frequency ω_p. Waves propagate without damping through the material when ε_r is positive and real. (Based on Kittel.)

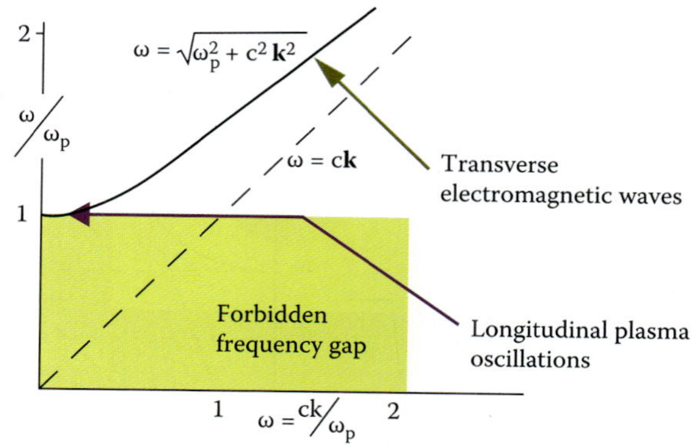

FIGURE 5.83 Dispersion relation for transverse electromagnetic waves in a plasma. The group velocity $v_g = d\omega/d\mathbf{k}$ is the slope of the dispersion curve. The straight line is for $\omega/\mathbf{k} = c$: propagation in free space. The group velocity v_g is less than the velocity of light in vacuum. The phase velocity, $v_p = \omega/\mathbf{k} > c$, is faster than the speed of light, but the latter does not correspond to the velocity of a real physical propagation of any physical quantity. A wave that propagates in the medium with the dispersion relation $\omega^2 = \omega_p^2 + c^2\mathbf{k}^2$ is also shown, as is the region of longitudinal plasma oscillation. The yellow area is the forbidden frequency gap. (Based on Kittel.)

Below the plasma frequency ($\omega < \omega_p$) there is a rapid drop in **E** field that leads to a thin penetration depth of the EM waves—the skin effect.

For $\omega > \omega_p$, the dielectric function is positive and real, and waves propagate through the transparent metal, with the dispersion relation $\omega = \sqrt{c^2\mathbf{k}^2 + \omega_p^2}$. Longitudinal plasma oscillations are shown in Figure 5.83 at $\varepsilon_r' = 0$ ($\omega = \omega_p$). For longitudinal plasma oscillations $\omega_L = \omega_P$ and from Equation 3.44, we then obtain:

$$\varepsilon_r'(\omega,\mathbf{k}) \approx 1 - \frac{\omega_p^2}{\omega^2} = 0 \tag{3.44}$$

In other words, at the plasmon frequency sustained electron oscillations are possible. For completeness sake, transverse EM waves are also shown for the vacuum situation ($\omega = \mathbf{k}c$—the light line).

In a polar dielectric, the coupling of phonons and photons in a polariton is responsible for the dispersion curves shown as solid curves in Figure 5.84. In Chapter 3, Figure 3.158b, we started by drawing the dispersion of light (ω versus **k**) in a polar solid such as GaAs assuming no interaction of light with phonons $\left(\omega = \frac{c\mathbf{k}}{\sqrt{\varepsilon(\infty)}}\right)$. However, in polar materials, light, close to the vibrational resonance frequencies ($\omega \sim 10^{14}$ rad/s), does interact with the transverse optical phonon modes, and ε_r starts to vary strongly so that ω versus **k** is not a straight line anymore as evident from Figure 5.84.

At large **k** wave vectors ($\mathbf{k} > 10^5$ cm^{-1}), we derived a positive photon-like solution with $\omega = \frac{c\mathbf{k}}{\sqrt{\varepsilon(\infty)}}$ (just like in Figure 3.158b) and a negative phonon-like mode with a singularity at $\omega = \omega_T$. With ω approaching ω_T, the index of refraction increases, and the group and phase velocities of the wave slow down (decreasing slope of the dispersion curve) to a value close to that of the lattice modes (AT mode). The wave now propagates as a strongly coupled polarization-electric field wave, with a velocity characteristic of the lattice transverse acoustic mode. At small **k** vectors ($\mathbf{k} \ll 10^4$ cm^{-1}), we obtain again two solutions: a positive phonon-like mode with $\omega = \omega_L$ (the optical phonon branch—OL), where the dielectric function vanishes, and a negative photon-like mode with $\omega = \frac{c\mathbf{k}}{\sqrt{\varepsilon(0)}}$. There is a forbidden gap established

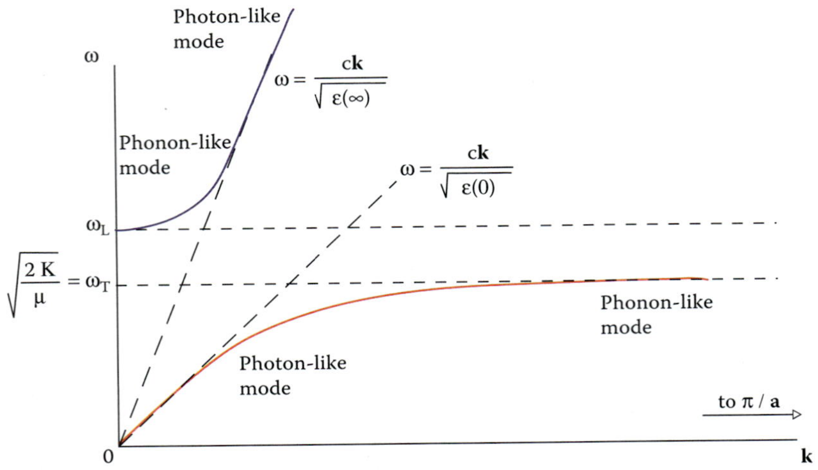

FIGURE 5.84 Dispersion relationship of light in a polar solid illustrating coupled modes of photons and transverse optical phonons. The lines labeled $\omega = \frac{ck}{\sqrt{\varepsilon(\infty)}}$ and $\omega = \frac{ck}{\sqrt{\varepsilon(0)}}$ represent electromagnetic waves in the crystal that are not coupled to the lattice vibration modes.

between the two horizontal lines at ω_T and ω_L. In this gap the EM wave attenuates as $\exp[-|\kappa|\ x]$, but the absorption is much stronger near ω_T than near ω_L as only the transverse waves couple and resonate strongly.

Lorentz Model for Semiconductors

Introduction

We saw earlier that valence electrons in insulators are relatively strongly bound to the individual atoms in the material, resulting in high resistivity and low absorption throughout the visible spectrum. In contrast for metals, where valence electrons can be considered free, we obtain a high conductivity, a negative dielectric constant at low frequencies, and high absorption and reflection. Semiconductors form an intermediate case between metals and insulators with valence electrons that are generally "somewhat" bound to the atoms in the material. Like insulators, semiconductors come with a very characteristic transmission spectrum with a transparency window for energies below the bandgap: $E < E_g$ (with $\alpha \sim 0$, $\kappa << n$, and n is real) and a fundamental absorption edge for $E \sim E_g$ (with $\alpha \neq 0$ and n is complex). The bandgap for semiconductors (<4 eV) is smaller than that of insulators (E_g > 4 eV). As a result, at room temperature these materials contain a small concentration of electrons that have broken free from the atoms, resulting in a small conductivity. The example of silicon was shown in Figure 5.62a. Optical spectroscopy is an important experimental tool for the elucidation of band structure. Even the color of a semiconductor reveals something of its band structure. Because electronic transitions from the valence to the conduction band span a fairly large range of energies, semiconductors act as a sort of long-pass filter-only reflecting light with less energy than the bandgap; this gives rise to colors as shown in Table 5.7. As we discovered already, much more is to be gained from optical absorption and reflection spectroscopy of solids than this simple table.

The band model for semiconductors was derived from quantum mechanics in Chapter 3. In this section we detail the various optical processes that can take place in a semiconductor.

Free Carrier Absorption ($E < E_g$)

EM wave absorption by free carriers in semiconductors results from the excitation of free carriers into available states higher in their respective bands (e.g.,

TABLE 5.7 Color and the Bandgap

Bandgap (eV)	Color	Examples
>3.0	White	ZnO
3.0–2.5	Yellow	CdS
2.3–2.5	Orange	GaP
1.8–2.3	Red	HgS
<1.8	Black	CdS

electrons higher into the conduction band). Free carrier absorption may be quite different for electrons than for holes; this is mainly because there are multiple valence bands available for holes, but there is only one conduction band for electrons. Free electrons located at wave vector **k** = 0, at the bottom of a single parabolic conduction band, can reach higher-energy states only if their momentum is increased. As a consequence, in n-type semiconductors, conduction band electrons may only be excited by the simultaneous absorption of a photon and the absorption or the emission of a phonon, to conserve momentum (see also direct and indirect bandgap transitions, discussed next). With p-type semiconductors, on the other hand, one can have direct (i.e., no phonons needed) transitions between the different valence bands (heavy- and light-hole bands). As a result, holes produce stronger free carrier absorption than conduction band electrons. The Drude model is sufficient to describe the free electron absorption process. The free electron density in semiconductors is often small, say $10^{14}/cm^3$, which is about 10^8 times smaller than in a metal, so the plasma frequency ω_p of semiconductors is also much smaller because $\omega_P = \sqrt{\frac{ne^2}{m_e\varepsilon_0}}$ (Equation 3.41), about 10^4 times lower than for a metal! For weakly absorbing materials such as semiconductors (relatively few conduction electrons), the electric susceptibility tensor χ' and $\chi'' \lll 1$, and one obtains for frequencies well above the plasma frequency:

$$\varepsilon_r'(\omega,\mathbf{k}) \approx 1 - \frac{\omega_P^2}{\omega^2} \tag{3.44}$$

$$\varepsilon_r''(\omega,\mathbf{k}) \approx \frac{\omega_P^2\gamma}{\omega^3} \tag{3.45}$$

and because the absorption coefficient is given as $\alpha(\omega) \approx \frac{\omega}{c}\varepsilon_r''(\omega)$ (see Equations 5.115 and 5.121), we derive:

$$\alpha(\omega) \approx \frac{\omega_P^2\gamma}{\omega^3} \approx \frac{\lambda^2\gamma}{c\lambda_p^2} \tag{5.236}$$

This describes a λ^2 dependence for free carrier absorption as a result of electrons. This type of dependence of α on λ is commonly seen in semiconductors, where dopant densities are typically 10^{16}–10^{19} cm^{-3} (ω_p in the 10^{13} rad/s range and $\lambda_p >$ ~10 μm) as compared with ~10^{22} cm^{-3} (with ω_p in the 10^{16} rad/s range and λ_p at UV wavelengths) in metals. Infrared absorption in Si is approximately proportional to λ^2. This type of absorption is also seen with metals at very high frequencies, frequencies well above the plasmon frequency. At these high frequencies ($\omega > \omega_p$) the dielectric constant n_c becomes real, and the metal acts now like a dielectric.

The form of the free-hole absorption in p-type semiconductors is more difficult to describe because it depends on the valence band structures and the hole density. In semiconductors with equal numbers of electrons and holes, the free-hole absorption dominates. As we saw above and in Chapter 3 this is because there are multiple valence bands available for holes, but there is only one conduction band for electrons. Free electrons located at wave vector **k** = 0, at the bottom of a single parabolic conduction band, may reach higher-energy states only if their momentum is increased. With p-type semiconductors, one can have such direct (i.e., no phonons needed) transitions between the different degenerate valence bands (heavy and light hole bands) because they occur at the same **k**. Because no momentum or **k** change is required, holes produce stronger free carrier absorption than conduction band electrons.

Light Absorption for Direct and Indirect Bandgap Semiconductors (E > E_g)

The bandgap for direct bandgap and indirect bandgap semiconductors in **k**-space was first shown in Figure 3.116; in the case of silicon, the valence band maximum (VBM) occurs at **k** = 0 (Γ point). However, the conduction band minimum (CBM) occurs at *L*. Thus, the VBM and the CBM occur at different points in the Brillouin zone, and this makes silicon an indirect bandgap material. GaAs is a direct bandgap semiconductor because the VBM and CBM occur at the same point in the Brillouin zone. The dominant feature of the absorption spectrum of a semiconductor is the band-to-band or interband transition: photon absorption promotes an electron from the valence into the conduction in so-called *band-to-band absorption*. In such a transition both energy and momentum must be conserved. In other words, for the energy balance we write:

$$E_i + h\nu = E_f \quad (5.237)$$

with E_i the initial and E_f the final energy and $h\nu$ the photon energy, and similarly for the momentum balance:

$$\mathbf{k}_i + \mathbf{k}_v = \mathbf{k}_f \quad (5.238)$$

with $\mathbf{k}_v$ a photon wave vector. For a bandgap E_g of 1 eV, one calculates for this photon wave vector a value of:

$$\mathbf{k}_v = \frac{\mathbf{P}}{\hbar} = \frac{E/c}{\hbar}$$

$$\approx \frac{1\,\text{ev}(1.60 \times 10^{-19}\,\text{J/ev})}{(3.00 \times 10^{8}\,\text{m/s})(1.05 \times 10^{-34}\,\text{Js})} \approx 5 \times 10^{6}\,\text{m}^{-1}$$

a quantity very small compared with a typical vertical (or direct transition) because $\mathbf{k}_i$ is of the order:

$$\mathbf{k}_i \approx \frac{2\pi}{a} \approx \frac{2\pi}{3.00 \times 10^{-10}\text{m}} \approx 2 \times 10^{10}\,\text{m}^{-1}$$

In other words, a photon in the visible range of the spectrum comes with a momentum vector much smaller than that of an electron. As a consequence, direct bandgap transitions (with an absorption coefficient α typically 10^6–10^8 m^{-1}) appear as vertical lines in an E-**k** diagram because Equation 5.238 reduces to:

$$\mathbf{k}_i = \mathbf{k}_f \text{ or } \Delta\mathbf{k} = 0 \quad (5.239)$$

For an indirect bandgap transition, momentum can only be conserved by absorption or emission of a phonon (lattice vibration), and we have:

$$\mathbf{k}_i \neq \mathbf{k}_f \text{ or } \Delta\mathbf{k} \neq 0 \quad (5.240)$$

and for the energy balance:

$$E_i + h\nu \pm h\nu_{ph} = E_f \quad (5.241)$$

with $h\nu_p$ the phonon energy, + for absorption, and – for emission. For the momentum conservation, we write:

$$\mathbf{k}_i + \mathbf{k}_{ph} = \mathbf{k}_f \quad (5.242)$$

Estimating a typical phonon energy, we obtain:

$$h\nu_p \approx 0.07\text{ eV}$$

The calculated value, although generally much less than the bandgap energy, is significant enough to change the electron's wave vector. Phonons have a momentum that is four to five orders of magnitude higher than for photons, so they can help in indirect transitions where photons cannot. Indirect absorption is a second-order process as a phonon intervention is required, so the process has a much lower transition probability (10^{-5}–10^{-6} s^{-1}) than direct absorption. The absorption α at a given frequency is an integral over all the transitions in the Brillouin zone that satisfy either Equations 5.237 and 5.238 for a direct transition or Equations 5.241 and 5.242 for an indirect transition. The need of phonons to assist in absorption and emission of light in indirect bandgap semiconductors like Si makes for weak absorption and emission of light, and most optoelectronic applications are easier implemented in direct bandgap semiconductors like GaAs. We encountered phonons first in Chapter 3, where we saw that they represent mechanical deformation waves traveling through the lattice. Compression waves do constitute a periodic deformation potential of the bandgap, and it is this periodic narrowing and widening of the bandgap that causes electron scattering and momentum transfer to photons. Electron-phonon interaction is called *Frölich interaction*.

A direct gap semiconductor features a sharp onset of absorption when the photon energy is equal to the bandgap. Indirect gap semiconductors like silicon and germanium do not have such a sharp onset of absorption as illustrated in Figure 5.85, where we compare the absorption α for Si and GaAs together with that of some other important semiconductors. Compared with atomic absorption peaks, absorption bands in crystals are broad and seem quite featureless when the photon energy is greater than the bandgap. As we discovered earlier in this chapter, despite this seemingly bland appearance, significant information on the detailed band structure of materials can be obtained from absorption and reflection spectroscopy (e.g., by taking the derivative of

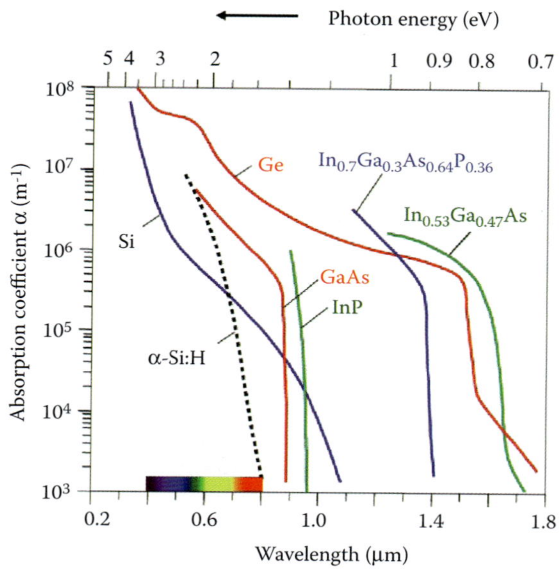

FIGURE 5.85 Absorption α of Si and GaAs and some other important semiconductors are compared.

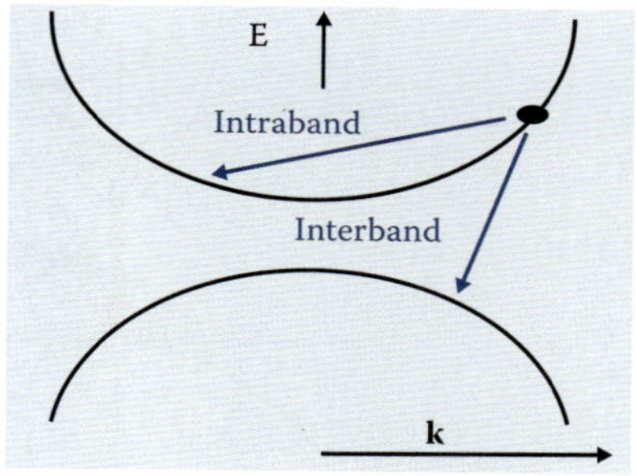

FIGURE 5.86 Interband and intraband absorption. The latter is also known as *Cerenkov radiation.*

the reflectance with respect to wavelength, temperature, electric field, pressure, etc., in so-called *modulation spectroscopy*).

A very general effect of quantum confinement in semiconductors, we saw in Chapter 3, is a widening of their optical bandgap, leading to characteristic blue shifts. Quantum confinement also changes the density of states (DOS) dramatically, and absorption and emission of light by a semiconductor depends on both the bandgap width and the availability of states. As the energy levels are packed in narrower and narrower energy ranges near the bandgap of quantum-confined structures, the optical transition strengths for absorption and emission improve the stronger the confinement. In addition to the interband transition, quantization of semiconductors also leads to new transitions between different sub-bands within the conduction band (minibands) (Figure 5.86). These transitions are called *intraband, inter-sub-band transitions*, or *Cerenkov radiation*, and they are finding more and more important applications, for example, in the quantum cascade lasers discussed at the end of this chapter. We also saw how the transition probability in indirect gap semiconductors increases with increasing confinement: confinement relaxes some selection rules, thus allowing emission (see next section) from nanoparticles that are strongly forbidden in large devices.

Emission of Photons

Luminescence comprises the generation of any light that is not induced by heat (as in incandescence) (see also Chapter 7). In photoluminescence, emission of light from a semiconductor is stimulated by using light (say, from a laser) with a photon energy $E > E_g$. In other words, the pumping wavelength is shorter than the bandgap wavelength. Hot electrons and holes are generated in this process and relax back to the bottom and the top of their bands, respectively, in a process called *thermalization*, as illustrated in Figure 5.87a for hot electrons. In this process the carriers take on a thermal distribution over the available energy levels according to statistical mechanics (Fermi-Dirac statistics because they are fermions). Subsequently, emission of light occurs: in the case of a direct bandgap, emission is strong at the bandgap energy E_g. For an indirect bandgap material, the emission (like the absorption) is a low-probability process because a transition from Γ to the *X* or *L* point requires a three-particle interaction (one photon, one electron, and one phonon) as shown in Figure 5.87b. Although the radiative lifetime is long for an indirect bandgap material, the quantum efficiency is low. Because the photon has a very small momentum **k** (~0), as a result of its long wavelength compared with that of an electron, the optically induced electronic transition basically requires that $\Delta\mathbf{k} = 0$. As a consequence, emission from an indirect bandgap material like Si is forbidden by this selection rule. Therefore, bulk Si is not a luminescent emitter, but the direct bandgap semiconductor GaAs is.

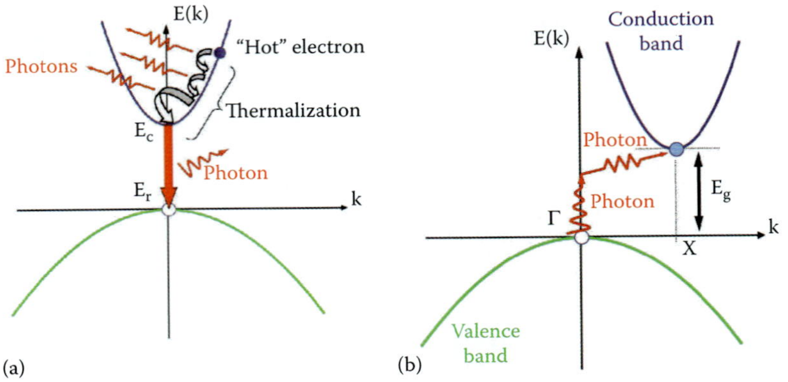

FIGURE 5.87 (a) Thermalization of hot electrons in a direct bandgap semiconductor and subsequent emission of a photon in photoluminescence. (b) For an indirect bandgap material the emission is low probability because a transition from Γ to the *X* or *L* point requires a three-particle interaction (one photon, one electron, and one phonon).

Excitons

Excitons in Large Structures We have assumed so far that when an electron-hole pair is created there are no interactions between these charge carriers in a solid. In reality, electrons and holes interact with each other through the Coulomb interaction, and an electrostatically bound electron-hole pair forms a neutral entity called an *exciton*. The latter may be compared with an electron bound to a proton to form a neutral hydrogen atom. An exciton represents an elementary excitation of a semiconductor and can be created by either a photon or an electron. Although overall neutral, an exciton carries a dipole moment and can move through the crystal and transport energy. In an exciton, the electron and hole move together as a single uncharged particle. All excitons are unstable with respect to their ultimate recombination process, in which the electron drops into the hole (Figure 5.88).

The exciton wave vector $\mathbf{k}_{ex}$ is given by:

$$\mathbf{k}_{ex} = \mathbf{k}_e - \mathbf{k}_h \quad (5.243)$$

with $\mathbf{k}_e$ the electron vector and $\mathbf{k}_h$ the hole vector. Absorption occurs only at $\mathbf{k}_{ex} = 0$, to preserve momentum and because of energy conservation:

$$h\nu = E_{ex} \quad (5.244)$$

We can treat excitons as chargeless particles capable of diffusion in a solid, but in the ultimate confinement in a molecule they can also be regarded as excited states of a single molecule. The exciton binding energy $E_n(\mathbf{k})$ is the energy required to separate the pair. In a molecule, excitons come with a large binding energy in the 1-eV range and feature a small radius of 10 Å. These are called *Frenkel excitons*, typical for organic materials. The less confined, wandering excitons in a solid are called *Mott-Wannier excitons*; they have a smaller binding energy $E_n(\mathbf{k})$ in the 5–20-meV range and a radius of about 100 Å (>>> the typical lattice constant *a*). This binding energy of Mott-Wannier excitons is of the order of kT at room temperature, and as a result exciton lines in an absorption spectrum are generally not well resolved from the band-edge absorption unless the sample is cooled. The two types of excitons are compared in Figure 5.89. Frenkel developed the theory of an exciton as a small and tightly bound entity in 1931.[17,18] E.F. Gross in 1951 discovered the Mott-Wannier excitons—large and weakly bound—in the

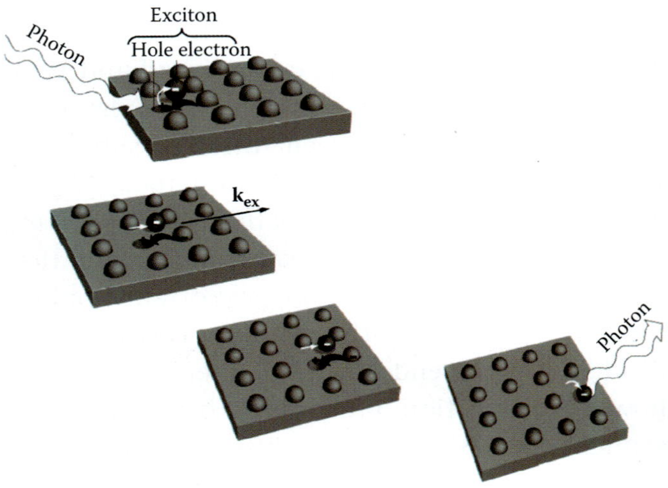

FIGURE 5.88 Life and death of an exciton.

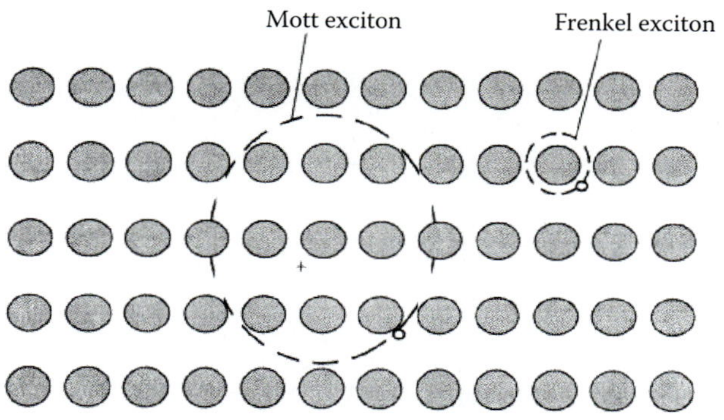

FIGURE 5.89 Frenkel excitons, typical for organic materials, and the wandering excitons in an inorganic solid, Mott-Wannier excitons, are compared.

optical spectrum semiconductor crystals of cuprous oxide.[19]

Because there is an attraction between the electron and the hole in an exciton, the photon energy required to create an exciton is lower than the bandgap energy, as illustrated in Figure 5.90. This results in absorbance and reflectance spectra revealing some structure just below the energy gap.

The exciton energy and radius can be derived from the Bohr model for hydrogen (a method we also used in Chapter 4 to calculate dopant energy levels). The Bohr model for the hydrogen atom gave us the following expression for the ground state energy (n = 1), also sometimes called the *Rydberg energy* R_y:

$$R_y = E_0 = -\frac{e^2}{8\pi\varepsilon_0 a_0} = -13.6 \text{ eV or } 21.8\ 10^{-19}\text{ J} \quad (3.91)$$

where a_0 is the Bohr radius.

The Rydberg energy is the energy required to remove an electron from the hydrogen atom. Applying the Bohr model to excitons, we modify Equation 3.91 by introducing the reduced mass $m_r^* = \frac{m_h^* m_e^*}{m_h^* + m_e^*}$, as well as ε_r, the dielectric constant of the semiconductor, to obtain:

$$R_y = 13.6\frac{m_r^*}{n^2 m_e}\left(\frac{1}{\varepsilon_r}\right)^2 \quad (5.245)$$

A bound exciton can only be found when the thermal energy $kT < R_y$. If the thermal energy is larger than this quantity, all holes and electrons behave as separate quantities. The energy levels referred to the top of the valence band are then given (for the kinetic energy term in this expression, see Chapter 3 on quantum mechanics):

$$E_{ex,n} = -\frac{R_y}{n^2} + \underbrace{\frac{\hbar^2 \mathbf{k}_{ex}^2}{2m_r^*}}_{\text{kinetic energy of exciton}} + E_g \quad (5.246)$$

The exciton ground state is found by setting n = 1; this corresponds to the ionization energy of the exciton.

In the Bohr model we derived the radii for the hydrogen atom's energy states as:

$$r_n = \frac{n^2\hbar^2 4\pi\varepsilon_0}{m_e e^2} = n^2 a_0 \quad (3.89)$$

where a_0 is the Bohr radius, and permitted orbits have radii $r_1 = a_0$, $r_2 = 4a_0$, $r_3 = 9a_0$ … for n = 1, 2, 3, ….

The Bohr radius, $a_0 = \frac{\hbar^2 4\pi\varepsilon_0}{m_e e^2}$, has a value of 0.53×10^{-10} m. Applying this to an exciton, we again introduce the reduced mass m_r^* to obtain (see Figure 5.91):

$$r_{ex,n} = 0.53\, n^2 \frac{m_e}{m_r^*}\varepsilon_r \quad (5.247)$$

Some calculated values for $E_{ex,1}$ and $r_{ex,1}$ for a variety of materials are summarized in Table 5.8. The exciton Bohr radius gives an estimate of the most probable distance of the electron from the hole. Like

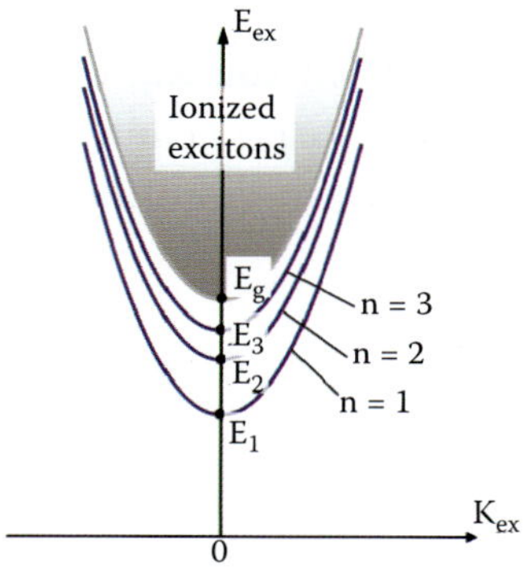

FIGURE 5.90 Exciton energy levels.

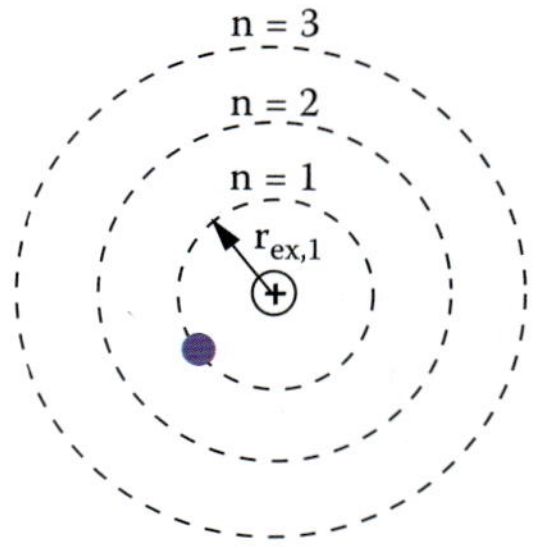

FIGURE 5.91 Bohr radii of an exciton.

Table 5.8 Values for $E_{ex,1}$ and $r_{ex,1}$ for Some Selected Bulk Materials

Semiconductor	E_g	$E_{ex,1}$ (meV)	$r_{ex,1}$ (nm)
Si	1.11	14.7	4.9
Ge	0.67	4.15	17.7
GaAs	1.42	4.2	11.3
CdSe	1.74	15	5.2
Bi	0	Small	>50
ZnO	3.4	59	3
GaN	3.4	25	11

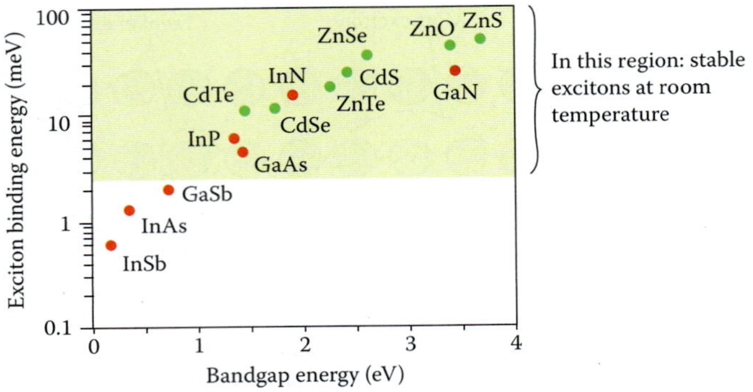

FIGURE 5.93 Stable excitons at room temperature.

the Bohr radius, the "radius" of the exciton increases with n^2. Excitonic absorption peaks appear in the absorption spectrum below the bandgap energy (Figure 5.92). Note that the absorption α decreases when $E_{ex,n}$ increases (higher n value).

Based on Equation 5.245, one can evaluate which materials will lead to stable excitons at room temperature as shown in Figure 5.93.

Excitons in Reduced Dimensionality Structures The above discussion concerned excitons in bulk semiconductors only. Let us consider what happens with their absorption spectra when we confine electron-hole pairs into semiconductor structures of the size of the Bohr radius $r_{ex,1}$ or smaller ($r_{ex,1}$ for a variety of semiconductors was introduced in Table 5.8). In Chapter 3 we similarly considered what happens with the bandgap absorption when confining the semiconductor size (answer: blue shift). Quantum confinement in semiconductor nanoparticles produces quantization of the electron and hole energy states, producing major changes in their optical spectra. The confinement of the oppositely charged carriers at reduced separations also has dramatic effects on the electron-hole Coulomb energy, and thus on exciton formation. For example, in quantum dots (QDs), excitons are generated inside the dot with a degree of confinement determined by the dot size. The physical boundary of the QD (0D) confines excitons in a well with dimensions on the order of the exciton Bohr radius (~10 nm). One expects that in these confined nanostructures the electron-hole pair interactions will be enhanced with a resulting binding energy increase. The enhanced binding between electron-hole pairs produces increased exciton energy compared with the exciton binding energies we listed for bulk semiconductors in Table 5.8. As a consequence, in confined semiconductors, just as in bulk semiconductors, the excitonic excitations are found below the conduction band edge, but in nanostructures these excitonic peaks are much more pronounced and easier to resolve, even at room temperature. In other words, the probability for exciton formation is much higher in a confined semiconductor such as in quantum wells, quantum wires, and QDs. The binding energy of a 2D exciton, for example, can be deduced from the 2D hydrogen atom problem. The exciton Rydberg energy then changes from $E_{ex,n} = -R_y/n^2$ in the 3D case to $E_{ex,n} = -R_y/(n - 1/2)^2$ in the 2D case (exciton in a well). Thus, the ionization energy for the ground state (n = 1) in the 2D case is four times larger than for the 3D case. Therefore, excitons are four times more stable in quantum wells.

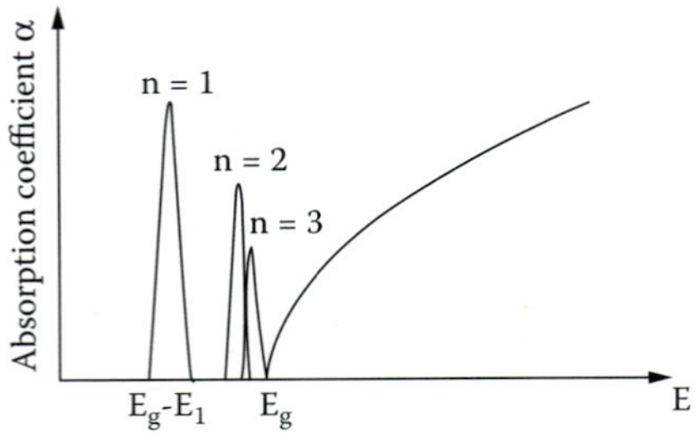

FIGURE 5.92 The hydrogen-like absorption of excitons.

Summary of Semiconductor Absorption Processes

In the preceding discussion we considered EM wave absorption by free carriers, direct and indirect band transitions, and excitons. In a real semiconductor there are also transitions of high-lying energy bands

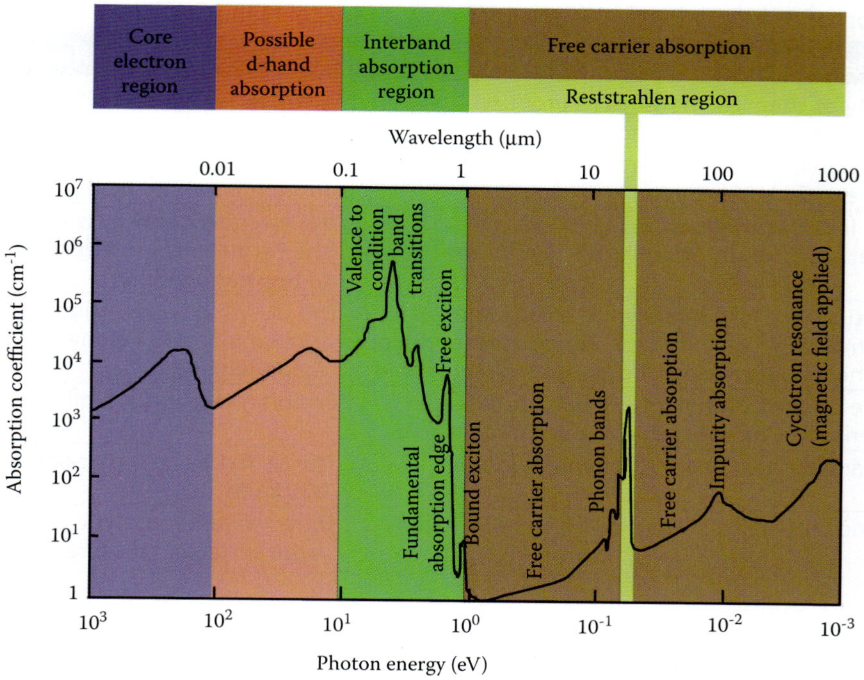

FIGURE 5.94 Different absorption processes in a semiconductor.

and from impurities (including from semiconductor dopant atoms).

Impurities create energy levels in the bandgap (= defect states). Donors and acceptors are situated in the bandgap close to the band edges. Transitions between defect states are at the origin of the impurity absorption. Deep levels are defect states close or in the middle of the bandgap (often created by transition metals such as Fe, Ni, Cu, and Au). Some deep levels generate radiation-less recombinations and should be avoided in optical components. In general, impurities in the bandgap behave like the isolated molecules in a gas as treated above. Doping of Si with donor and acceptor impurities in

Table 5.9 Fundamental Excitations and Energy Scales in Solids

Process	Energy
Resulting from "free" valence electrons	
Intraband transitions	≤ $E_{bandwidth}$ (few eV)
Single-electron oscillations	≤ 0.5 meV
Collective electron oscillations (plasmons)	≈ 2–17 eV
Resulting from bound electrons	
Inner (core and valence) electron shells	≈ 4 eV–98.5 keV
Inner-shell excitations (d-d, f-f)	≈ 100 meV–2 eV
Interband transitions (VB → CB)	≈ 10 meV–11 eV
Absorption from localized states:	
- Excitons	≈ E_{Bind} (few meV up to few eV)
- Defects (impurities, vacancies, or interstitials):	≈ 10–300 meV
- Spin-orbit splitting	≈ few eV (for d and f electrons)
- Spin waves (magnons)	≈ 10–40 meV
- Charge transfer excitations	≈ up to several eV
- Correlation effects	≈ order of eV
Resulting from coupling of photons to phonons (lattice vibrations)	
Photons: acoustic phonons (Brillouin scattering)	≈ 10–6–0.1 meV
Photons: optical phonons (polariton scattering)	≈ 40–70 meV

Table 5.10 Comparison of Semiconductor Nanoparticles with Metal Nanoparticles

Semiconductor Nanoparticles	Metal Nanoparticles
Synthesis with a variety of chemical methods to produce nanoparticles with desired size, shape, and structure	Synthesis with a sol method by using capping agents to prevent aggregation and control the size
Quantum confinement occurs when the radius of the nanoparticle is comparable to the exciton's Bohr radius	Quantum confinement occurs when the electron's motion is limited by the size of the nanoparticle
The electronic energy levels of the particle around the Fermi level are strongly affected for sizes in the 10–100-nm range	The electronic energy levels of the particle around the Fermi level are affected only for very small sizes (a few nanometers)
The UV-visible absorption is determined by the band edge transition, which is strongly size- and shape-dependent (bandgap energy increases as the size decreases) and hence highly tunable	The UV-visible absorption is determined by the surface plasmon resonance, which is size- and shape-dependent
Can exhibit band edge or defect (red-shifted) luminescence; in both cases luminescence can be very intense and can be tuned	Exhibit a very weak luminescence
The I-V behavior is characterized by the Coulomb blockade	The I-V behavior is characterized by the Coulomb blockade
Show valuable catalytic properties, owing to the large surface to mass ratio	Show valuable catalytic properties, owing to the large surface to mass ratio

semiconductors is covered in detail in Chapter 4. As for excitons, the Bohr model was used to calculate the energy levels associated with those impurities. The different absorption processes are summarized in Figure 5.94 and Table 5.9.

Comparing Semiconductor Nanoparticles (Quantum Dots) with Metal Nanoparticles

In Table 5.10 we compare semiconductor nanoparticles with metal nanoparticles (see also Figure 5.95).

Light Interaction with Small Particles

Introduction: Scattering of Light by Small Particles

When light passes through a medium with suspended particles, some of it is scattered, i.e., directed away from its original direction of travel. In the scattering process, the photon flux from the incident beam (I_0) is not lost but is redistributed over the total solid angle centered around the scattering

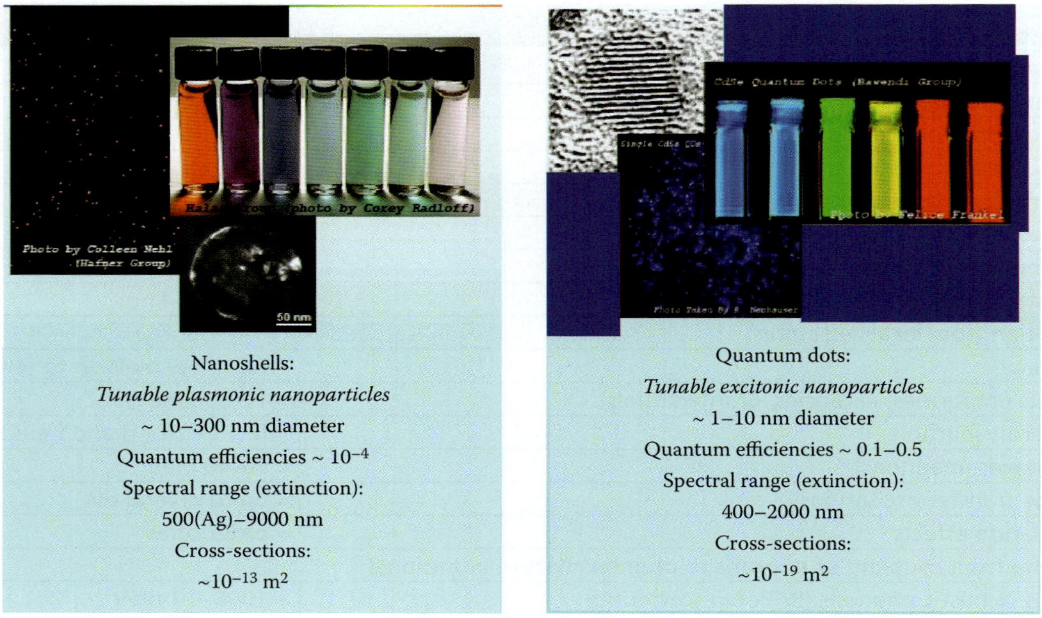

FIGURE 5.95 Plasmonic and excitonic nanoparticles are compared.

particles in the medium, and it does not change the internal energy states of the particles (elastic process) (Figure 5.96).

In contrast, absorption changes the internal energy states of the particles (inelastic process). Absorption is also very spectrally selective, whereas scattering is not. Absorption and scattering both attenuate the incoming radiant energy, and this combined effect is quantified by the so-called *extinction coefficient* (C_{ext}; see Equation 5.218 for the case of suspended spherical metal nanoparticles). To classify the interaction of light with small particles, we introduce a dimensionless size parameter α, which is a measure of the size of the particles compared with the wavelength of the incoming radiation:

$$\alpha = \frac{2\pi r}{\lambda} \tag{5.217}$$

In Table 5.11 we list the various scattering regimes of visible light on particles of various sizes. This table is very general, when applied, say, to atmospheric sciences; it illustrates that how solar radiation scatters when it impacts gases, aerosols, clouds, or the ocean surface depends on the ratio of the scatterer's size to the incoming wavelength. A large α in this application may correspond to phenomena such as sunlight on a flat ocean or sunlight on raindrops, and a small α may correspond to IR scattering of air and aerosols or microwave scattering of clouds.

When scatterers are very small compared with the wavelength of the incident radiation (radius of scatterer r is $<\lambda/10$), the scattered intensity in both forward and backward directions is equal (Figure 5.97). This is representative of a type of scattering called *Rayleigh scattering*. A strong dependence of the scattering on the frequency of the light is important to distinguish it from Mie scattering: in Rayleigh scattering, short visible waves (blue) get scattered much more than longer visible wavelengths (yellows and reds). Mie or nonmolecular scattering with $r > \lambda/10$ is much less wavelength-dependent.

Rayleigh scattering is predominant in the presence of molecules and the tiniest particles ($r < 1/10\ \lambda$). An analysis for scattering of light in this regime was first performed by Lord Rayleigh (1871), who undertook this task to explain why the sky is blue. Rayleigh scattering of light is mostly elastic, so no energy is gained or lost.

Rayleigh scattering is not always 100% elastic: in inelastic Raman scattering, 1–10% of the incoming light interacts with molecules in such a way that energy is actually gained or lost, and as a result scattered photons are shifted in frequency. Scattered light under these circumstances reveals satellite lines below and above the Rayleigh scattering peak at the incident frequency. Just like Rayleigh scattering, Raman scattering depends on the polarizability of the scattering molecules.

For particles larger than molecules and more similar in size to the wavelength of the incoming light, the angular distribution of scattered intensity becomes more complex, with more energy scattered in the forward direction than in the backward direction (see Figure 5.97). Scattering in this case is much less dependent on the wavelength of the incoming light, and this is basically why clouds are white. In this regime ($r > 1/10\ \lambda$) the scattering of light is called *Mie* or *nonmolecular scattering*. The German physicist Gustav Mie (1908) was the first to present a complete analytical solution of Maxwell's equations for the scattering of EM radiation from spherical particles (also called *Mie scattering*).[20] Mie's theory of light scattering was actually developed to explain the color of colloidal gold. In general, particles might also absorb, and Mie gave solutions for the absorption, scattering, and extinction cross-sections as a function of the scattering angle. In contrast to Rayleigh scattering, Mie solutions to scattering embrace all possible ratios of diameter to wavelength. It is so far the only "simple" way to study light scattering of any type of particle. Mie's solutions are very important in meteorological

Table 5.11 Various Types of Scattering of Visible Particles on Different Radii r

Type of Particle	Particle Diameter (μm)	Type of Scattering	Phenomena
Air molecules	0.0001–0.001	Rayleigh	Blue sky, red sunsets
Aerosols (pollutants)	0.01–1.0	Mie	Brownish smog
Cloud droplets	10–100	Geometric	White clouds

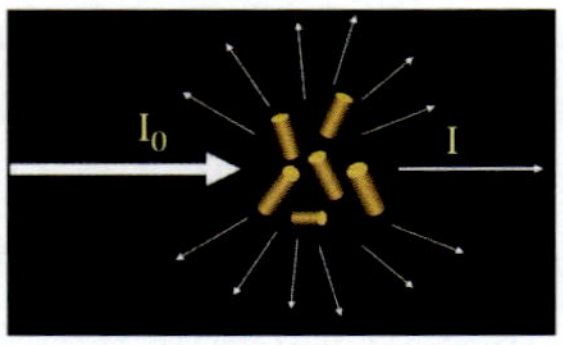

FIGURE 5.96 Incoming photon flux with intensity I_0 is scattered on particles in a medium.

optics, where diameter-to-wavelength ratios of the order of unity and larger are characteristic of many problems regarding white fog and cloud scattering, scattering of radar by raindrops (weather radar), coronas around the sun and the moon, and so on (see Figures 5.112 and 5.113). Other important Mie applications are optical particle characterization for understanding the appearance of common materials like milk, biological tissue, and latex paint and in graphic software for rendering outdoor scenes for movies.

Then finally, for particles much larger than the wavelength of the light, the ray pictures of light become appropriate: for example, when sunlight strikes raindrops between 0.1 and 2 mm in diameter, geometric optics applies, producing such effects as the rainbow illustrated in Figure 5.18.

Rayleigh Scattering (α <<< 1)—Elastic Scattering

In the Lorentz model we saw how an electric field can drive the harmonic motion of bound and free electrons in molecules (see Figure 5.98).

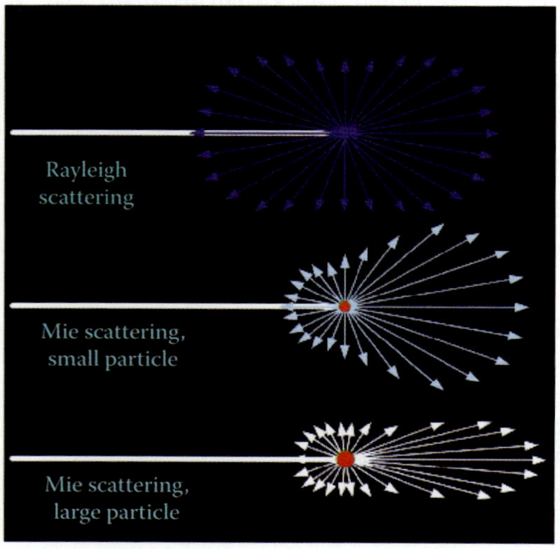

FIGURE 5.97 Rayleigh and Mie scattering compared.

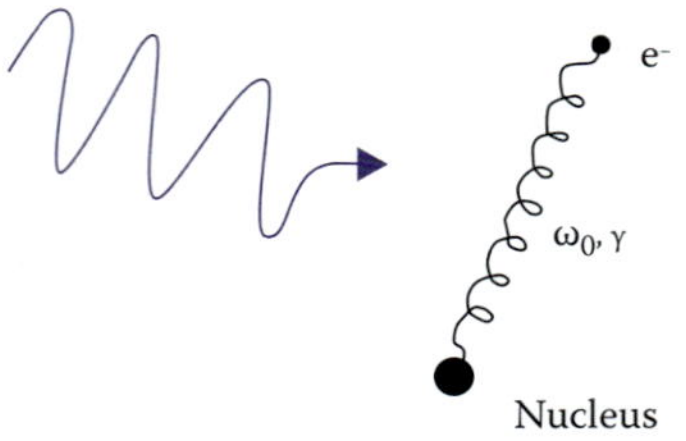

FIGURE 5.98 In the Lorentz model, an electric field drives the harmonic motion of electrons, with ω_0 the resonant frequency and γ the damping factor.

As the incident light's EM field varies, the atoms/molecules that are hit by the light are continuously redistributing their charges. When equal positive and negative charges are displaced in opposite directions, the charge distribution is called a *dipole*. The dipole moment **P** is the product of the quantity of displaced charge and the displaced distance with the vector **P** pointing in the direction of the charge displacement. The oscillating field causes the charges to oscillate at the frequency of the incoming radiation. Earlier in this chapter we explained how oscillating charges radiate EM waves (see Figure 5.3). Oscillating charges reradiate the incoming light elastically at the same frequency as the forcing field: no energy is gained or lost. However, the emitted radiation is not propagated in the same direction as the forcing field and will not necessarily have the same polarization. An isolated dipole oscillator radiates predominantly normal to the direction of the oscillation as illustrated in Figure 5.99.

A small point charge with a different refractive index n than its surrounding medium oscillates as a

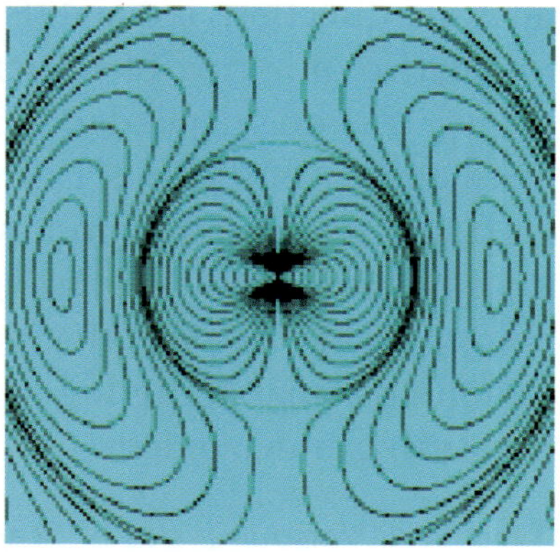

FIGURE 5.99 The electric field from an oscillating dipole (see also Figure 5.3). Formation of electromagnetic waves on a vertically oriented dipole: an oscillating dipole generates EM in a direction mainly perpendicular to the dipole axis. No waves come about along the axis of the dipole.

dipole under the influence of an incident electrical field and reradiates the incoming energy. The angle-dependent total scattered power per unit surface I_s, or the intensity of the light scattered from such an oscillating dipole at a given angle and distance *r*, is calculated as (see, e.g., the Feynman lectures):

$$I_s = \frac{P^2\omega^4}{32\pi^2\varepsilon_0 c^3 r^2}\sin^2\theta \tag{5.248}$$

where θ is the polar angle measured from the axis in a spherical coordinate system (Figure 5.100), *c* is the speed of light, and *r* is a distance large enough away to be in the far field of the dipole. Notice that in Equation 5.248 the radiation intensity is inversely proportional to the distance squared from the dipole (r^2). The magnitude of the polarization **P** induced in a molecule or small particle by an incident EM field $[E_i(\omega) = E_M e^{i\omega t}]$ was calculated earlier as $\mathbf{P} = \chi\varepsilon_0\mathbf{E}_i(\omega)$ (Equation 5.190). This radiation, with $r \lll \lambda$, constitutes Rayleigh radiation, and we detail its characteristics here.

To introduce any type of scattering properly, one also must introduce a phase function $p(\Theta)$ (in sr^{-1}) to describe the distribution of scattered radiation, be it for one particle or for a set of particles. The phase function is a dimensionless and normalized version of the scattering function, such that the integral of the phase function over 4π steradians equals 4π. The phase function for a given wavelength is a property of the medium, not of the incident radiation, provided the incident radiation is unpolarized. As the size parameter of a scatterer increases, its phase function becomes more anisotropic, with progressively more of the scattered radiance concentrated into a diffraction peak. Let the direction of light incidence in Cartesian coordinates be Ω′ and the direction of observation be Ω (Figure 5.101). The angle between these directions is Θ, the scattering angle, i.e., the angle between the incident beam and the scattered light. If Θ is <π/2, we define forward scattering, and if Θ is >π/2, we have backward scattering. Converting $\cos\Theta = \Omega'\cdot\Omega$ to polar coordinates, one derives $\cos\Theta = \cos\theta'\cos\theta + \sin\theta'\sin\theta\cos(\Phi - \varphi')$. The phase function $p(\Theta)$ is then mathematically defined as:

$$p(\Theta) = \frac{I_s(\Theta)}{\int_{4\pi} I_s(\Theta)d\omega} \tag{5.249}$$

To normalize this phase function $p(\Theta)$, we write:

$$\frac{1}{4\pi}\int_{4\pi} p(\Theta)d\omega = \int_0^{2\pi} d\phi \int_0^{\pi} p(\theta',\phi',\theta,\phi)\sin\theta d\theta = 1 \tag{5.250}$$

The angular distribution of Rayleigh scattering is governed by the classical radiation pattern for the far field of a classical dipole, i.e., a dipole radiation proportional to $\Pi\sin^2\theta$, where θ is the polar angle measured from the axis, and Π is the induced dipole moment (Equation 5.248). As illustrated in Figure 5.101, we can take the incoming radiation and break it up into two linearly polarized incident waves, one with the electric vector parallel to the scattering

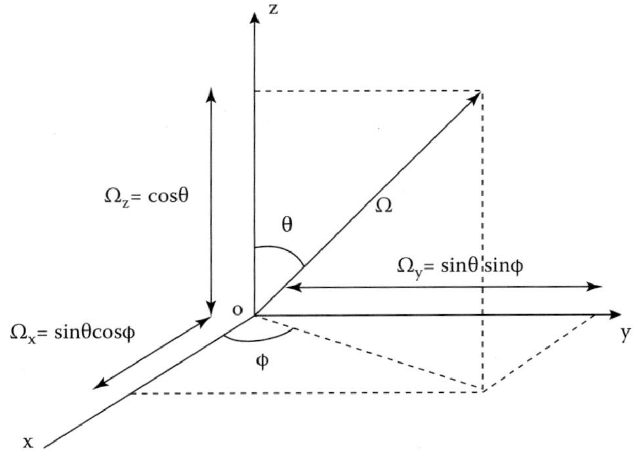

FIGURE 5.100 Relation between Cartesian and spherical coordinates. The rectangular components of the vector Ω are $\Omega_x = \sin\theta\cos\phi$, $\Omega_y = \sin\theta\sin\phi$, $\Omega_z = \cos\theta$.

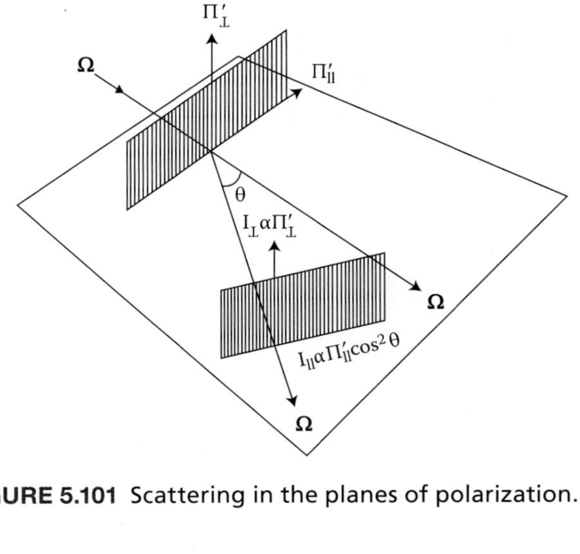

FIGURE 5.101 Scattering in the planes of polarization.

plane and the other perpendicular to the scattering plane. These waves give rise to induced dipoles. If the incident electric field lies in the scattering plane, then the scattering angle is $\pi/2 + \Theta$; if it is perpendicular to the scattering plane, the angle is $\pi/2$. Hence:

$$I_s(\Theta) \propto (I_\perp + I_\parallel) \propto \Pi_\perp \sin^2(\pi/2) + \Pi_\parallel \sin^2(\pi/2 + \Theta) \quad (5.251)$$

given that the parallel and perpendicular intensities are equal:

$$I_s(\Theta) \propto I(1 + \cos^2\Theta) \quad (5.252)$$

For Rayleigh scattering this leads to the following phase function $p(\Theta)$:

$$\frac{1}{4\pi}\int_{4\pi}(1+\cos^2\Theta)d\omega = \int_0^{2\pi}d\phi\int_0^{\pi}(1+\cos^2\theta)\sin\theta d\theta = \frac{3}{4} \quad (5.253)$$

or:

$$p(\Theta) = \frac{3}{4}(1+\cos^2\Theta) \quad (5.254)$$

Figure 5.102 illustrates the phase function for Rayleigh scattering from a dipole in both a 2D and 3D diagram. Let us look first at the 2D diagram. The phase function is symmetric in the plane normal to the incident direction, and forward scatter equals backward scatter. The maxima occur in the direction of $\theta = 0°$ and $180°$, and there is minimum radiation along the dipole axis ($\theta = 90°$). The angular distribution of the scattered intensity varies from 1.5 (6/4) in the forward and backward directions to 0.75 (3/4) at 90°, where the parallel component vanishes and the scattered radiation is plane polarized transverse to the scattering plane. In the 3D diagram we simplified things by dropping the factor ¾ in Equation 5.254 so that we obtain 2 at $\theta = 0°$ and $180°$ and 1 at 90°.

Using $\mathbf{P} = \chi\varepsilon_0\mathbf{E}_i(\omega)$ (Equation 5.190), the link between the intensity I_0 and the incoming electrical field $\left[I_0 = \frac{\varepsilon_0 c E_i(\omega)^2}{2}\right]$, and relying on the Lorentz model (Equation 5.191), we can rewrite Equation 5.248 as:

$$I = I_0\frac{e^4\omega^4}{16\pi^2 m_e^2 c^4 r^2}\left(\frac{1}{\omega_0^2 - \omega^2 - i\omega\gamma}\right)^2 \sin^2\theta \quad (5.255)$$

where I_0 is the incoming EM radiation intensity. This equation reveals that scattering is strongest near a resonance and that it is also stronger for a higher ω (or shorter λ). In the visible range, oxygen and nitrogen molecules come with $\omega_0 >>> \omega$; in

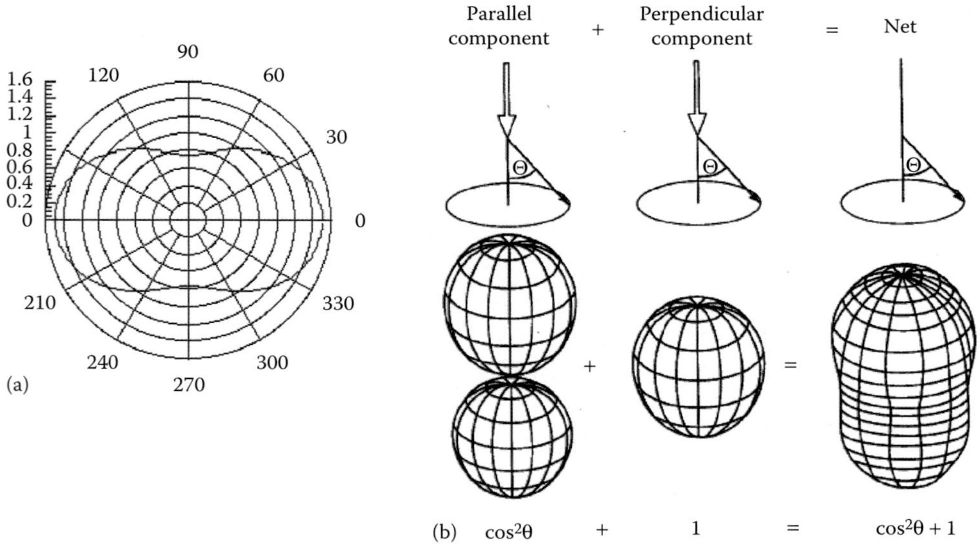

FIGURE 5.102 Rayleigh phase function $p(\theta)$: (a) 2D and (b) 3D. Light scattering from a driven dipole with $r \ll \lambda$. We can take the incoming radiation and break it up into two linearly polarized incident waves, one with the electric vector parallel to the scattering plane and the other perpendicular to the scattering plane. Of the 3D shapes, the shape on the left represents the parallel scattering component, and the sphere in the center represents the intensity from polarization out of the plane. The final peanut shape shows the total scattering pattern for a random incident polarization.

other words, the driving frequency is much less than the natural frequency, and we obtain nonresonant or Rayleigh scattering (also elastic scattering):

$$I = I_0 \frac{e^4\omega^4}{16\pi^2 m_e^2 c^4 r^2}\left(\frac{1}{\omega_0^2}\right)^2 \sin^2\theta \tag{5.256}$$

Incoming sunlight comes in a wide range of wavelengths, but the longer wavelengths hardly get scattered, whereas based on this equation, a wavelength that is two times shorter gets scattered 16 times more ($2^4 = 16$).

An example of this is observed in the blue sky, where the radiating dipoles are oxygen and nitrogen molecules. The shorter wavelengths (i.e., blue) scatter much more efficiently than longer ones and that is the reason why the sky is blue (blue light ~ 4000 Å is much larger than the scattering particles ~1 Å) [we all should be able to explain why the sky is blue (Figure 5.103)]. During sunrise and sunset, blue and yellow light is efficiently removed from the direct beam over the very long atmospheric paths at the high solar zenith angles, leading to spectacular red skies. You might wonder: because violet light has a shorter wavelength than blue, why then does the sky not appear violet? The reason is that the energy contained in the violet spectrum is much smaller than that contained in the blue spectrum and the fact that the human eye has a much lower response to violet. The Rayleigh mechanism is also of great technological importance because it governs losses in optical fibers for communication. The Rayleigh scattering results from minute local density variations that are present in the liquid glass as a result of Brownian motion and become frozen into the solid.

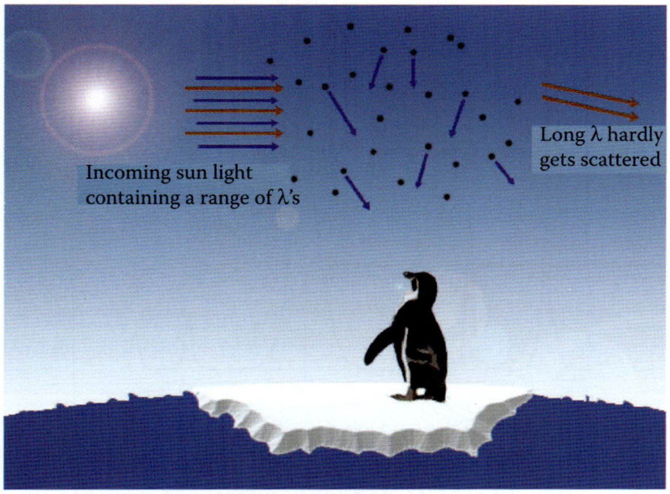

FIGURE 5.103 Why the sky is blue. (Drawing by Mary Amasia, University of California, Irvine.)

The situation just sketched for scattering molecules is equally applicable to scattering from nanoparticles (see the section on metal nanoparticles, metal optics, or plasmonics). If the scatterer is a sphere of radius *a* and a refractive index *n* and the incoming light is unpolarized, the particle's polarizability **P** can be written as a function of the refractive index *n* and diameter *d*. The intensity of the scattered light in this case is given as:

$$I = I_0 \frac{(1+\cos^2\theta)}{2r^2}\left(\frac{2\pi}{\lambda}\right)^4\left(\frac{n^2-1}{n^2+1}\right)^2\left(\frac{d}{2}\right)^6 \tag{5.257}$$

where *r* is the distance to the particle, θ is the scattering angle, *n* is the refractive index of the particle, and *d* is the diameter of the particle. Thus, the scattering intensity and energy removed from the incident beam are proportional to the sixth power of the radius (or for nearly spherical particles to the square of the volume). The scattered intensity and energy removed from the incident beam remain proportional to ω^4 or to λ^{-4}. Because the intensity of Rayleigh scattering is proportional to the sixth power of the radius but absorption, however, is proportional to the volume, scattering falls off more rapidly than absorption.

The total scattering power is given by the integral over a sphere surrounding the radiating dipole:

$$I_{total} = \int_A I_s dA = \frac{\mathbf{P}^2\omega^4}{12\pi\varepsilon_0 c^3} = \frac{4\mathbf{P}^2\pi^3 c}{3\varepsilon_0\lambda^4} I_0 \tag{5.258}$$

From Equation 5.207 (the Lorentz-Lorenz relationship), we remember that polarizability can be written in terms of refractive index *n* or also $\frac{\mathbf{P}}{E\varepsilon_0} = \frac{3}{N}\frac{n^2-1}{n^2+2} = \chi$, with the $3/(n^2+2)$ factor added to account for the local field correction, or:

$$I_{total} = \int_A I_s dA = \frac{\mathbf{P}^2\omega^4}{12\pi\varepsilon_0 c^3} = \frac{24\pi^3}{N^2\lambda^4}\left[\frac{n^2-1}{n^2+2}\right]^2 I_0 \tag{5.259}$$

Raman Scattering ($\alpha \ll 1$)—Inelastic Scattering

Raman Scattering Physics

In 1928, the Indian physicist Chandrasekhara Venkata Raman (1888–1970) from Calcutta University (Figure 5.104) discovered that the wavelength of a small fraction of radiation (1–10%) scattered by certain molecules differs from that of the incident beam. The shifts depend on the molecular structure of the molecules responsible for the scattering. Raman received the Nobel Prize in 1930. Rayleigh scattering of light (90–99%) is elastic, so no energy is gained or lost. In inelastic Raman scattering (1–10%), light interacts with molecules in such a way that energy is gained or lost, and as a result scattered photons are shifted in frequency.

Just like Rayleigh scattering, Raman scattering depends on the polarizability of the scattering molecules. IR bands, on the other hand, arise from a change in the dipole moment. In many cases, transitions that are allowed in Raman are forbidden in IR, so these techniques are often complementary. With polarizable molecules, incident light can excite vibrational modes, leading to scattered light diminished in energy by the amount of the vibrational transition energies (same as in fluorescence). Scattered light under these circumstances reveals satellite lines below the Rayleigh scattering peak at the incident frequency. These are called the *Stokes lines*. If there is enough vibrational energy pumped into the scattering molecules, it is also possible to see anti-Stokes lines (generally weaker than Stokes line) at frequencies above the incident frequency because now vibrational energy is added to the incident photon energy. Because anti-Stokes lines are generally much weaker, it is the Stokes part of the spectrum that is typically used. Summarizing, Raman spectroscopy is a method of determining modes of molecular motions, especially vibrations. It is predominantly applicable to the qualitative and quantitative analyses of covalently bonded molecules. A Raman spectrum is a plot of the intensity of Raman scattered radiation as a function of its frequency difference from the incident radiation (usually in units of wave numbers, cm^{-1}). This difference is called the *Raman shift*. A Raman spectrum of CCl_4 is illustrated in Figure 5.105.

The magnitude of the Raman shifts is independent of the wavelength of the excitation. Raman scattering results from the same quantized vibrational changes that are associated with IR absorption; therefore, Raman scattering spectra and IR absorption spectra of compounds often closely resemble each other. In Raman spectroscopy one typically uses a powerful monochromatic source (a laser), and during irradiation the spectrum of scattered light is measured at an angle (usually 90°) with a spectrometer (Figure 5.106). One important advantage of Raman over IR is that in Raman spectroscopy water does not cause any interference; thus, Raman spectra can be obtained from aqueous solutions. Also, glass or quartz cells can be used rather than NaCl or KBr disks. Raman can also involve rotational or electronic transitions of the molecules from which the scattering occurs. The molecular vibrational

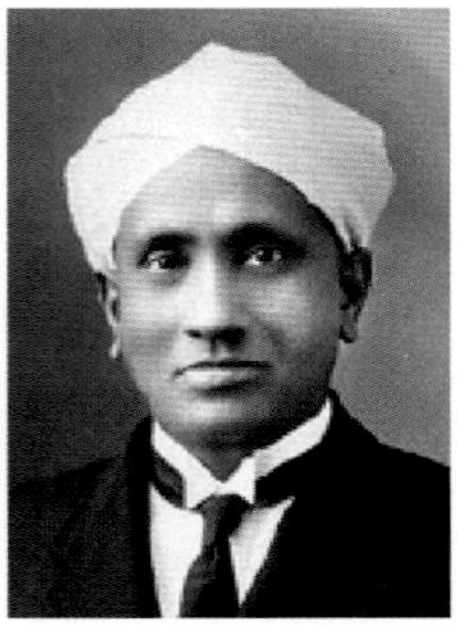

FIGURE 5.104 Indian physicist Chandrasekhara Venkata Raman (1888–1970), who received the Nobel Prize in 1930.

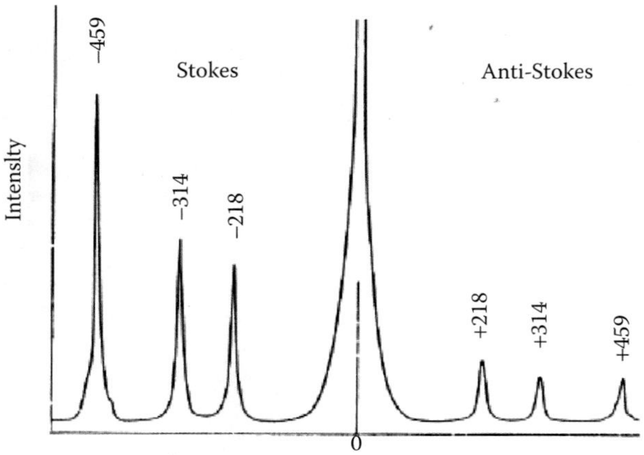

FIGURE 5.105 Raman spectrum of CCl_4. A Raman spectrum is a plot of the intensity of Raman scattered radiation as a function of its frequency difference from the incident radiation (usually in units of wave numbers, cm^{-1}). This difference is called the *Raman shift*.

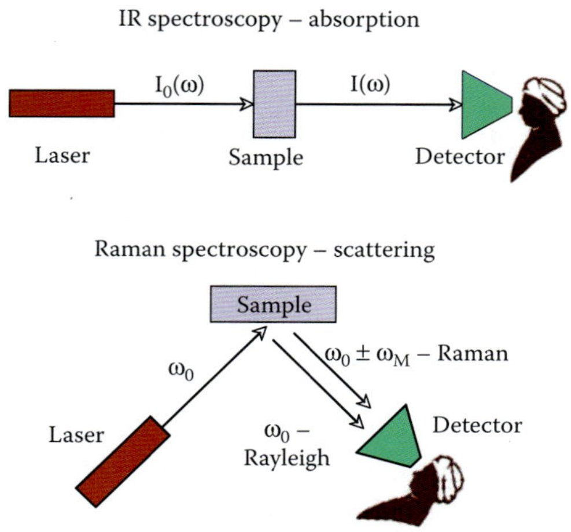

FIGURE 5.106 Raman and IR spectroscopy compared.

energies that may show up in Raman spectra were earlier calculated as $E_n = (n+\frac{1}{2})\hbar\omega$, n = 0,1,2,3,... (Equation 3.191) and the rotational energies as:

$$E_j = \frac{j(j+1)\hbar^2}{2I}, \; j = 0,1,2,... \qquad (5.260)$$

where *J* is the rotational quantum number and *I* is the moment of inertia.

Summarizing the advantages of Raman spectroscopy:

1. It provides extremely rich information on molecular vibrations and material structure.
2. There are no special requirements for sample preparations in contrast with IR spectroscopy.
3. It can also be applied noninvasively under almost any ambient condition.

And the disadvantages:

1. It is a low-sensitivity technique because of the low cross-section involved (~10^{-30} cm^2/molecule).
2. Poor reproducibility.
3. A theoretical understanding of underlying mechanisms is still incomplete.
4. A limited number of suitable substrates with Cu, Ag, and Au being the best choices (noble metals with roughness <λ of the light).

Surface-enhanced Raman scattering (SERS) strongly increases Raman signals from molecules attached to metal nanostructures. In SERS, surface plasmons of Ag or Au particles with a size of about 10–100 nm (e.g., colloidal Ag, Au) are excited by a laser. These metals have a large negative real component and a small imaginary component, e.g., the dielectric constant ε'_m of Ag is –2. The local field for a spherical noble metal nanoparticle, as we learned above (see Table 5.6), is enhanced dramatically near the SP resonance with ε'_m (= –2 for a metal like Ag) + $2\varepsilon_d$ (= 1 for air) = 0 (Equation 5.219) and is given by $E_{Local} = \frac{3\varepsilon_d}{\varepsilon'_m + 2\varepsilon_d}E_0$; so E_{local} is very large close to resonance. If the particle is elliptical, then the local field is $E_{Local} = \frac{3\varepsilon_d}{\varepsilon_d\left(1-\frac{1}{L}\right)}E_0$; In air ($\varepsilon_d$ = 1) the resonance frequency is $\omega_p L$ and the more elliptical the further the resonance is shifted to longer λs (see Figure 5.73).

The resulting electric fields cause nearby molecules to become Raman active, resulting in an amplification of their detection by up to 10^{12}–10^{14} compared with nonresonant Raman. This effect was originally observed by Fleishman but explained by Van Duyne (Northwestern University) in 1977.[21] Two signal enhancement effects play a role in SERS: EM field and chemical adsorption enhancement. If we can increase the local fields, we can obtain a larger effective scattering cross-section (higher signal), and lower detection limits result. Plasmon excitation at metal nanoparticles produces induced fields, and these fields are localized and very intense. The reason that the SERS effect is so intense is because the field enhancement occurs twice. First, the field enhancement magnifies the intensity of the incident light, which excites the Raman modes, thereby increasing the signal of the Raman scattering. The Raman signal is then further magnified by the surface just like the incident light was, resulting in a greater increase in the total output signal. At each stage, the electric field is enhanced as $\mathbf{E}^2$, for a total enhancement of $\mathbf{E}^4$. Thus, because the laser field is enhanced along with the Stokes fields, the total enhancement factor, *G*, is given by:

$$G = \mathbf{E}^2_{laser}\,\mathbf{E}^2_{Raman}$$

$$G = \left[\frac{\varepsilon_m(\omega_{laser}) - \varepsilon_d(\omega_{laser})}{\varepsilon_m(\omega_{laser}) + 2\varepsilon_d(\omega_{laser})}\right]^2 \left[\frac{\varepsilon_m(\omega'_{Stokes}) - \varepsilon_d(\omega'_{Stokes})}{\varepsilon_m(\omega'_{Stokes}) + 2\varepsilon_d(\omega'_{Stokes})}\right]^2 \qquad (5.261)$$

where $\varepsilon_m(\omega)$ is the dielectric constant of the metal, and $\varepsilon_d(\omega)$ is the dielectric constant of the medium in which the nanoparticle is embedded. The laser frequency is ω_{laser}, and ω'_{Stokes} is the Raman shifted frequency. From Equation 5.261 we want $\varepsilon_m + 2\varepsilon_d$ to be small, i.e., near the resonance, so we choose metals (with a negative ε_m). Also, because the gain depends on both E^2 laser(ω) and E^2 Raman(ω'), as the difference $\omega - \omega'$ gets larger, G will get smaller since the resonance condition is *not* met at both ω and ω'. For 30-nm gold particles, the gain is plotted for various laser and Raman wavelengths in Figure 5.107. The field enhancement of nanoparticles also increases with an increase in the radius of curvature of the nanoparticle.

If we move away from the surface of the metal nanoparticle (see Figure 5.108), the dipole decay has to be considered; for a single molecule, this gives:

$$G(d) = \left[\frac{r}{r+d}\right]^{12} G_0 \qquad (5.262)$$

with r the radius of the particle, d the distance away from the particle, and G_0 as given by Equation 5.261. And for the case of a monolayer, we obtain:

$$G(d) = \left[\frac{r}{r+d}\right]^{10} G_0 \qquad (5.263)$$

The intensity away from the nanoparticle, for a single molecule, is then written out in full as:

$$G(d) = \left[\frac{\varepsilon_m(\omega_{laser}) - \varepsilon_d(\omega_{laser})}{\varepsilon_m(\omega_{laser}) + 2\varepsilon_d(\omega_{laser})}\right]^2 \left[\frac{\varepsilon_m(\omega'_{Stokes}) - \varepsilon_d(\omega'_{Stokes})}{\varepsilon_m(\omega'_{Stokes}) + 2\varepsilon_d(\omega'_{Stokes})}\right]^2 \left[\frac{r}{r+d}\right]^{12} \qquad (5.264)$$

The decay curves for a particle and monolayer film are plotted in Figure 5.108. A typical gain G_0 for a single-metal nanoparticle is 10^6–10^7. However, single nanospheres yield relatively small SERS signals compared with a situation where many particles are close to each other (e.g., in a fractal-like aggregate); the additional average enhancement for these aggregates is ~10^5–10^6.[22] Theoretical and experimental studies indicate that the precise control of gaps between nanostructures in the sub-10-nm range (one refers to these sites as "hot junctions") is likely to be critical for the fabrication of SERS-active substrates with uniformly high Raman gain factors.

Besides the EM field, chemical adsorption also plays a role in SERS. When molecules are adsorbed to the metal surface, their electronic states can interact with the states in the metal and produce new transitions. The true nature of the chemical enhancement effect is still not fully understood.

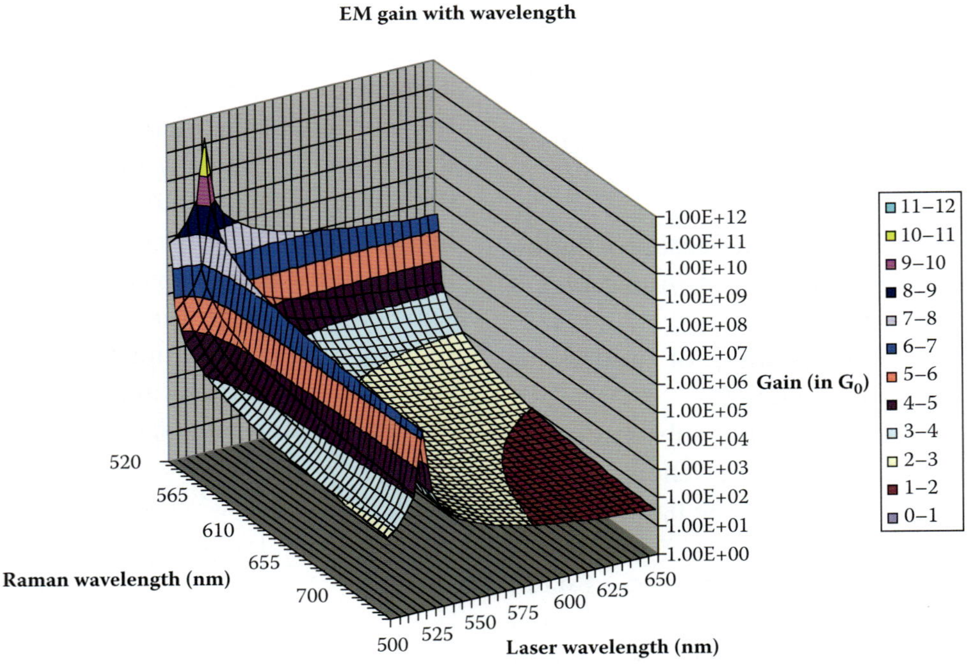

FIGURE 5.107 Gain G expressed in G_0 on 30-nm gold particles for various laser and Raman wavelengths.

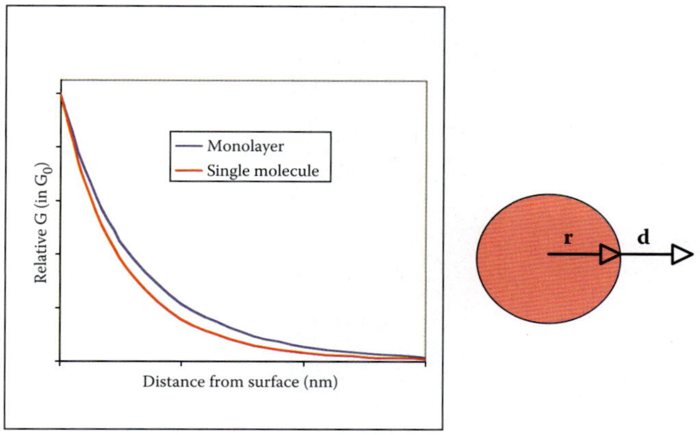

FIGURE 5.108 Decay curves $G(d)$ for a single molecule and for a monolayer (r = 30 nm).

Mie Scattering (1 < α < 10)—Elastic Scattering

For particles that are small compared with the incoming wavelength but larger than Rayleigh scatterers (i.e., $r > 1/10\ \lambda$), one has to deal with multiple waves from different molecules/atoms within the particle, and the angular distribution of elastic scattered intensity becomes more complex with more energy scattered in the forward direction. The front and the back of the particle do not experience the same incident electric field, and there is interference of the scattered waves from different parts of the particle. Forward moving waves tend to be in phase, and this gives a large resultant amplitude, whereas backward moving waves tend to be out of phase, and this results in a small resultant amplitude (Figure 5.109). Hence the scattering phase function $[p(\Theta)]$ for such a particle has a much larger forward component (forward peak) than the backward component.

We show Mie's results only pictorially here. In Figure 5.110, polar plots of the Mie-derived scattering phase function $p(\Theta)$ for selected values of α are

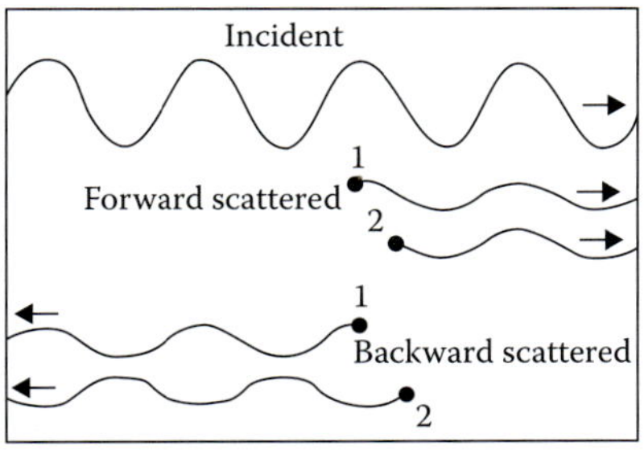

FIGURE 5.109 Incident waves, forward and backward scattering.

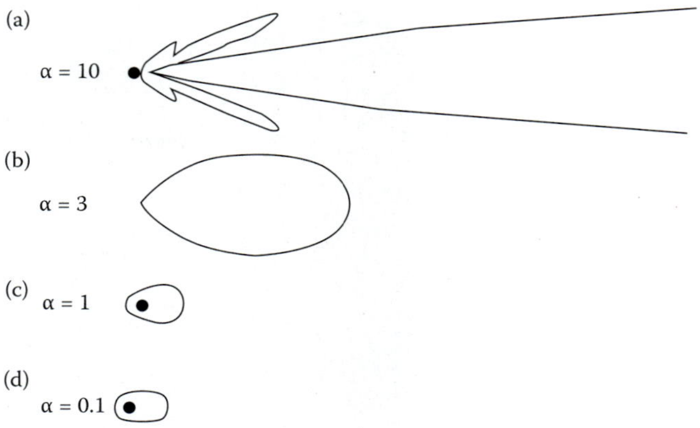

FIGURE 5.110 Polar plots of the Mie-derived scattering phase function $p(\Theta)$ for selected values of α. (a–c) Dominant forward scattering. (d) Equal forward and backward scattering (α = 0.1 = Rayleigh scattering). (b and c) Mie scattering from raindrops in clouds with r = 1 μm, and Mie scattering from an aerosol particle with r = 0.1 μm. (a) With α > 10, radiation can be described by geometric optics.

shown. At α = 0.1, forward and backward scattering are the same (Rayleigh scattering), e.g., solar light on gas (air) with molecules with a radius $r = 10^{-4}$ μm. As α increases, the forward scattering predominates more and more over backward scattering, e.g., Mie scattering from an aerosol particle with r = 0.1 μm and Mie scattering from raindrops in clouds with r = 1 μm. With α > 10, radiation can be described by geometric optics. When the size of the particle becomes larger than the wavelength, there is an extreme concentration of scattering in the forward direction with the development of maxima in different directions for different colors. The latter explain the optical phenomena of fogbows and glories (see Figure 5.113).

Typically Mie solutions are represented in terms of $K = \sigma/\pi r^2$, a dimensionless number that plays the role of the scattering area efficiency and depends on the size parameter $\alpha = 2\pi r/\lambda$ and the index of refraction n of the particles responsible for scattering. This coefficient measures the ratio of the effective cross-section of the particles to their geometric cross-section. In Figure 5.111, we plot K as a function of α.

For the lowest values of α (in the region with $\alpha \ll 1$ up to point A), the slope of the curve $(\Delta K/\Delta\alpha)$ in Figure 5.111 is positive, indicating that for a uniform set of spherical particles, scattering increases with decreasing wavelength (this typifies Rayleigh

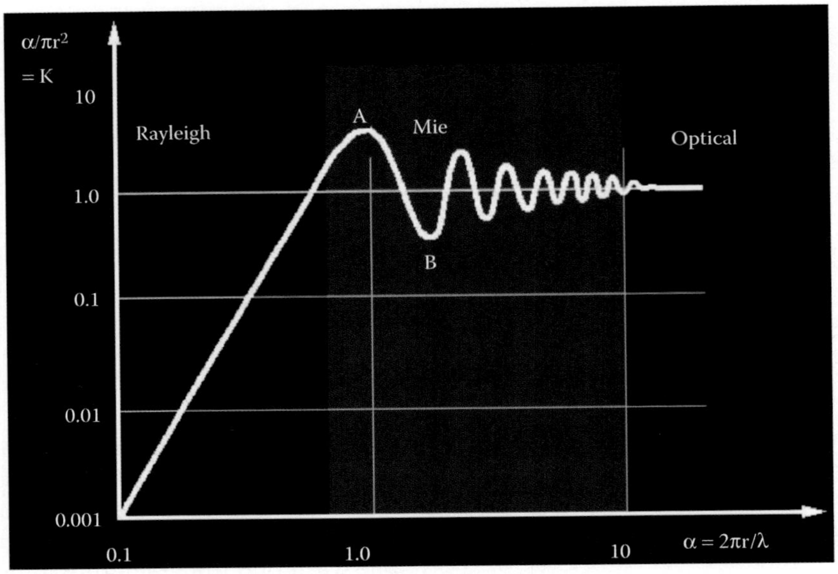

FIGURE 5.111 Scattering efficiency K as a function of size parameter α.

scattering). For a given refractive index, K is proportional to α^4. In this region, scattered radiation is evenly divided between the forward and backward hemispheres. Examples of Rayleigh scattering are scattering of solar radiation by air molecules (blue sky) and scattering of microwave radiation by raindrops. The latter is used in weather radar.

In the region between A and B the slope is negative, indicating that scattering decreases with decreasing wavelength. Examples include the scattering of sunlight by particles of haze, smoke, smog, and dust. In the region beyond point B, K exhibits oscillatory behavior. For $\alpha > 10$, K is more or less constant, and geometric optics applies.

In Figure 5.112 we show the log of the Mie phase function [log $p(\Theta)$] for Mie scattering by a cloud for red, green, and blue light. Scattering efficiencies in Mie scattering are similar for different colors, and that is why clouds are white. A large portion of the incident beam is scattered in the direction of the incident radiation (forward scattering), and a smaller portion of the incident beam is scattered in the opposite direction of the incident direction (backward scattering). Back scattering maxima occur in different directions for different colors, and this explains the beauty of fogbow and glories. A fogbow is a rare form of a rainbow. The water droplets in a fogbow are very small, so interference causes the colors to blend into pure white (Figure 5.113). The droplets then produce very weak colors. Glories are small luminous rings that surround an observer's shadow on a cloud deck or fog bank. A glory can

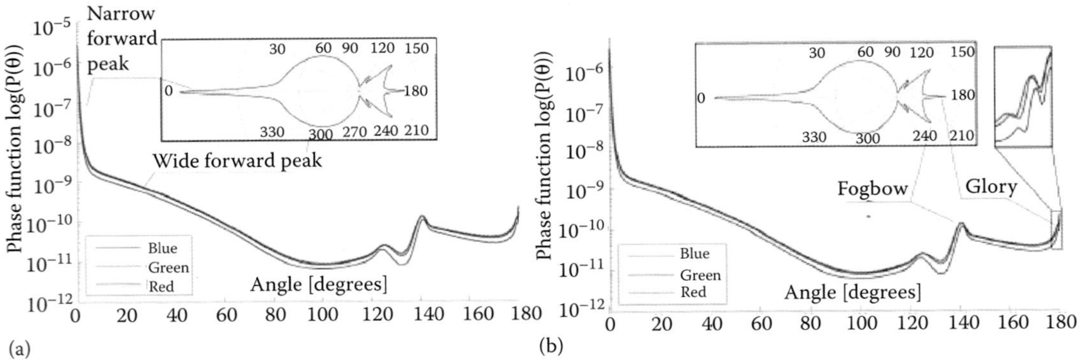

FIGURE 5.112 The log of the Mie phase function for Mie scattering of a cloud for red, green, and blue light as a function of scattering angle θ. Light is coming in from the right side. (a) Forward scattering. (b) Backward scattering. (See Figure 5.113 for an image of a fogbow and a glory.)

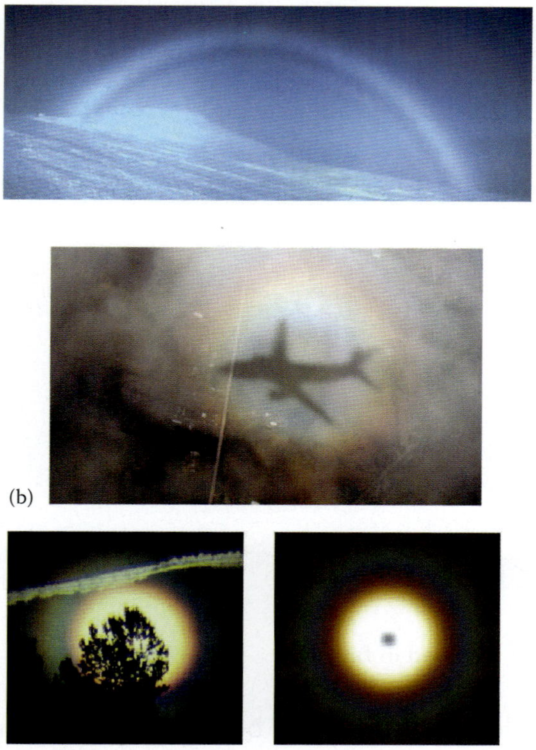

(a) (b) (c)

FIGURE 5.113 Fogbows, glories, and coronas. (a) White fogbow on Cairngorm, Scotland. (The fogbow was photographed by Simon Caldwell in February 1999.) (b) Jonathan Lansey saw this unusually bright glory while descending into Houston. (c) Corona simulation (right) matches Paul Neiman's photo of a corona with a jet contrail in mountain clouds above Nederland, Colorado (left).

often be seen surrounding an airplane's shadow. It is a pale, circular rainbow caused by the scattering and reflecting of sunlight from water droplets in clouds below the plane. A glory is not a true rainbow and is usually very small.

Finally, coronas occur when a thin cloud layer moves in front of the sun or moon. The light that shines through the cloud droplets curves (is diffracted) around the water droplets, and diffracted light casts a light larger than the original light source; several circles may appear because of the different wavelengths. The sharper they are, the more similar in size the droplets are.

Mie scattering software can be found at http://omlc.ogi.edu/software/mie.

Geometric Scattering ($\alpha > 10$)

If the scatterers are large compared with the wavelength of the incident light, then geometric optics provide a good approximation. We saw above that with $\alpha > 10$, K becomes constant (Figure 5.111). Scattering of visible radiation by cloud droplets, raindrops, and ice particles falls within this regime. Natural optical phenomena in this regime include rainbows and halos.

Plasmonic Applications 3

Introduction

We are returning to plasmonic applications for a final time in this chapter. This time we consider nanoscale confinement of EM radiation in near-field scanning optical microscopy (NSOM) for subwavelength resolution and detail NSOM applications with an emphasis on surface-enhanced Raman spectroscopy (SERS). We start by describing how NSOM manages to break through Abbe's diffraction limit. For more on NSOM see also Volume III, Chapter 6 on metrology.

Breaking the Diffraction Limit with NSOM

Earlier in this chapter, we saw how diffraction imposes a limit on the "resolution" achievable with conventional optics (see Abbe's diffraction limit, Equation 5.1). The best optical microscopes, for example, can only reach an ~200–250-nm resolution (confocal microscopy). For a long time, Abbe's law was regarded as being insurmountable: small features (or large wave numbers) of an object are lost because of the exponential evanescence of short-wavelength waves (see Figure 5.114). The evanescent near-field mode waves of the source decay exponentially away from the source, and they are never detected in conventional imaging; hence the diffraction limit on the image resolution, i.e., subwavelength features, is not resolved. One would need to amplify the near field to go beyond this diffraction limit.

For higher resolution, the textbook knowledge for a long time was that a complicated and costly electron microscope (0.1-nm resolution) or scanning tunneling microscope (STM) would be required. These techniques have a better resolution than Abbe's 200-nm limit but sacrifice many of the advantages associated with optical microscopes (i.e., nondestructiveness, low cost, high speed, reliability, versatility, accessibility, ease of use, informative contrast, spectroscopy, and real time). However, there

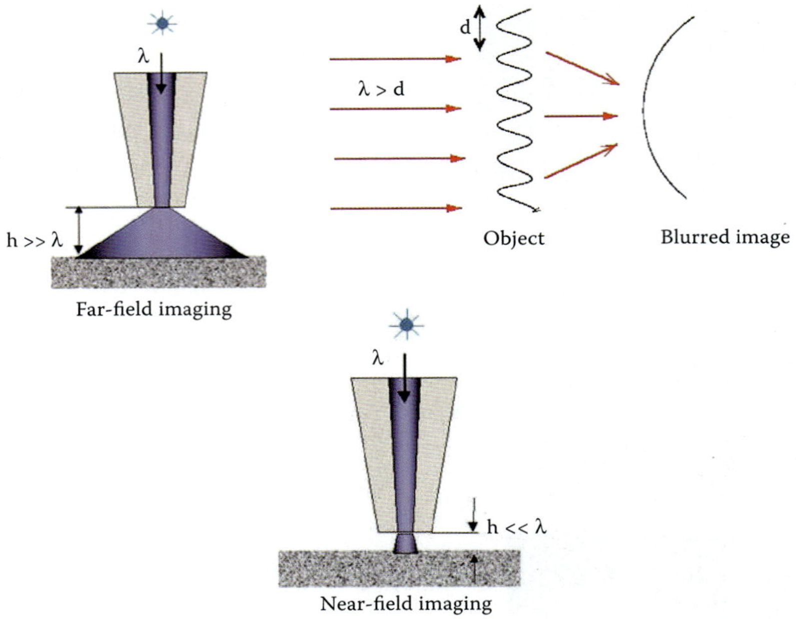

FIGURE 5.114 Near-field and far-field imaging. Small feature *d* is not resolved in far field.

is a solution, i.e., to combine a scanning proximal probe technique with optical microscopy in so-called *scanning near-field optical microscopy* (NSOM or SNOM). If we gather the light coming from the object as close as possible to the object, then the exponential evanescent waves are collected as well (see Figure 5.114). There are two fundamental differences then between near-field and far-field (conventional) optical microscopy: the size of the specimen area that is illuminated, and the separation distance between the source of radiation and the specimen. In NSOM the illuminated area is nanometric in dimensions, whereas conventional microscopes use macroscopic fields of view. In conventional far-field optical microscopy, the distance between the light source and the specimen is typically much greater than the wavelength of the incident light, whereas in NSOM, a necessary condition of the technique is that the illumination source is closer to the specimen than the wavelength of the illuminating radiation. To probe the "near field" requires one to be within 100 nm of the object. This is achieved in a variety of ways, for example, by using nanoscale apertures or by using apertureless techniques that enhance light interaction over nanoscale dimensions with the use of scattering nanoscale tips, nanospheres, and so on. Thus, NSOM is a form of lensless optics with subwavelength resolution, independent of the wavelength of the light being used. The light coming from an imaged object in the near field contains a considerable amount of nonpropagating, evanescent waves containing high lateral spatial resolution, information that decays exponentially within a distance comparable to a wavelength. It is this evanescent wave information, the information-rich light, we capture in NSOM, thus enhancing the resolution. Detecting the collected light in NSOM can be achieved with a wide variety of instruments: an Avalanche Photo Diode (APD), a Photomultiplier Tube (PMT), a CCD, or a spectrometer. The signals obtained by these detectors is then used to create an NSOM image of the surface.

For high-resolution imaging with NSOM, the tip must be positioned and held within nanometers of the sample surface during scanning. There are various feedback mechanisms that have been used for this purpose: electron tunneling, photon tunneling, impedance, and reflection measurements. The most widely adopted method is the shear-force technique in which the NSOM tip mounted on a tuning fork is dithered laterally at one of the mechanical resonances of the fork. To excite the mechanical resonance of the fork, it is rigidly mounted on a ceramic piezoelectric tube serving the purpose of a dither. The tuning fork and the tip are vibrated parallel to the sample surface. To avoid losing optical resolution, the amplitude of the vibration is kept low, usually <10 nm. As the tip approaches a sample surface, shear forces acting between tip and sample dampen the amplitude of the tip vibration. That amplitude change can be monitored by several methods and is used to generate a feedback signal to control the tip-sample gap during imaging. NSOM sorts out in two major categories, i.e., aperture near-field optics and apertureless near-field optics also called *scattering-based NSOM* (SNSOM). The optical apertures often consist of a tapered, drawn optical fiber that is coated by a metal, typically aluminum, to confine the light and prevent it from leaking out of the sides of the tip, except for a tiny hole at its very apex. Aluminum is used because of ease of application and a minimum skin depth in the visible and IR. Metal film coatings have a finite skin depth (the minimum thickness required to produce opacity) that limits the minimum probe size that can be coated with sufficient thickness to prevent penetration of light through the walls of the tip. The light emanates through the tip opening, which is in the range of ~50–100 nm. The distance to the sample is in the same range; thus, the sample senses the near-field distribution of the light. Hence, the light is confined to a dimension smaller than its wavelength, and thus near-field optics overcomes the diffraction limit. There are other aperture types, such as micromachined cantilevered AFM/NSOM tips (Si_3N_4, SiO_2) with tetrahedral tips featuring a hole at the apex. Finally, one might also use the tip of a tapered pipette filled with a light-emitting compound, which can then be excited either by light or by applying a voltage. It is also possible to use chemical luminescence. Several implementations using an aperture are illustrated in Figure 5.115.

Illumination by the tip and measuring in transmission mode is probably the easiest to interpret and gives the most signal (Figure 5.115a). Transmission NSOM allows images of the intensity of the light transmitted through the transparent sample to be obtained. The sample is illuminated via a probe aperture during the scanning, and transmitted light may be collected by an inverted optical microscope and directed to a photomultiplier tube (PMT). However, it requires a transparent sample, and so it cannot always be used. Reflection modes give less light and are more dependent on the details of the probe tip but allow one to study opaque samples. Reflection NSOM allows images of the intensity of light reflected from the sample to be taken (Figure 5.115b). The sample is illuminated via a probe aperture, and reflected light is directed by a mirror system through the objective of the inverted microscope to the PMT. In collection mode the sample is illuminated with a macroscopic light source from the top or bottom, and the probe is used to collect the light from the sample surface (Figure 5.115c). The

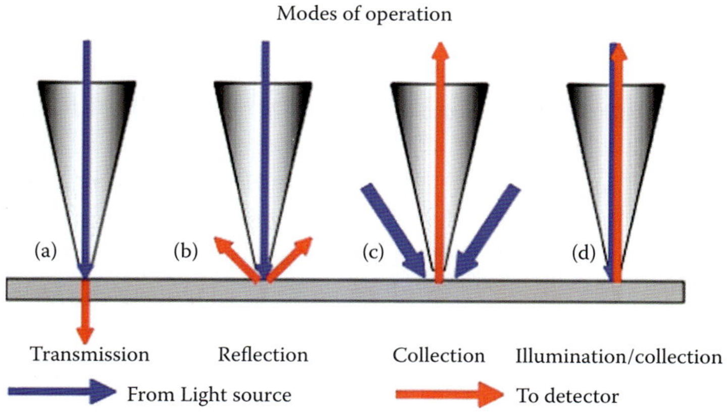

FIGURE 5.115 NSOM aperture-based proximal probes/tips. (a) Transmission mode imaging. The sample is illuminated through the probe, and the light passing through the sample is collected and detected. (b) Reflection mode imaging. The sample is illuminated through the probe, and the light reflected from the sample surface is collected and detected. (c) Collection mode imaging. The sample is illuminated with a macroscopic light source from the top or bottom, and the probe is used to collect the light from the sample surface. (d) Illumination/collection mode imaging. The probe is used for both the illumination of the sample and for the collection of the reflected signal.

illumination/collection mode provides a complement to the reflection modes, but the signal contains a large background from the light reflected without ever making it out of the probe tip (Figure 5.115d). In this hybrid mode the probe is used for both the illumination of the sample and for the collection of the reflected signal. Oscillating the tip vertically and measuring the signal at that oscillating frequency helps remove this background.

In Figure 5.116 we illustrate how an aperture-based (tapered optical fiber, coated with Al on it sides) near-field microscope can be used to probe luminescent semiconductor quantum dots (QDs). Luminescence NSOM allows images of local luminescence intensity to be gathered. The sample is illuminated via a probe aperture, and transmitted light is directed to the PMT through an inverted microscope objective and a notch filter.

Tapered coated fiber probes, as discussed above, have a resolving power of 30–50 nm at most because the amount of light that can be put through the small pinhole is limited. Apertureless near-field microscopy resolves the illumination intensity problem by shining light obliquely at a Rayleigh scatterer and periodically perturbing the interaction between the sample and its illumination. The perturbation can involve probe-induced phase shifts, or it can involve the periodic obscuration of the illuminating beam by a sharp probe. Measurement of the periodic perturbation involves lock-in detection. The perturbation is synchronous with the probe movement, and since the interaction of the sample with the probe is not the same at the low point than at the high point of the probe trajectory, lock-in detection allows the discrimination of the near-field information (which is modulated at the probe's oscillation frequency) from the far-field optical information (which varies more slowly). Apertureless SNSOM typically uses a sharp metallic subwavelength tip instead of an optical tip as described above. The sharp metal tip neither acts as excitation nor as detection light waveguide. Instead, it acts as a nanoscopic light scatterer (Rayleigh scatterer) when placed in near-field distance with respect to the sample. For "apertureless" NSOM the resolution can reach 10 nm; however, background from diffraction-limited illumination is a problem. Oblique incident outside light is scattered at the tip, and the resulting strongly enhanced local plasmonic field is used for excitation. The resulting optical response signal is collected via a lens or an objective with the collecting focus being adjusted to the tip. Exciting the surface plasmons propagating toward a tip of a tapered metal nanowire can make for a very intense local nanoscopic light source. As we learned in this chapter, plasma oscillations in a metal are vibrational modes of the electron gas density oscillating about the metallic ion cores (plasmons). Surface plasmons describe the special case in which the charges are confined to the surface of a metal. In this case, the electric field is strongest in the plane of the metallic surface. As we illustrated in Figure 5.44, surface plasmons confined to a plane do not radiate light (the surface plasmon mode always lies below the light line). However, when the local planar symmetry is disturbed, as illustrated with a couple of different configurations in Figure 5.46, plasmons can radiate. It is thus possible to design structures that take advantage of this emission of surface plasmon radiation, and this may be used to enhance radiative processes. This phenomenon leads to a giant concentration of energy on the nanoscale. The strength-of-field enhancement at the metal tip depends both on the tip material and its geometrical design. The enhancement at the tip is caused by a combination of the electrostatic lightning rod effect, which is due to the geometric configuration at the tip, and localized surface

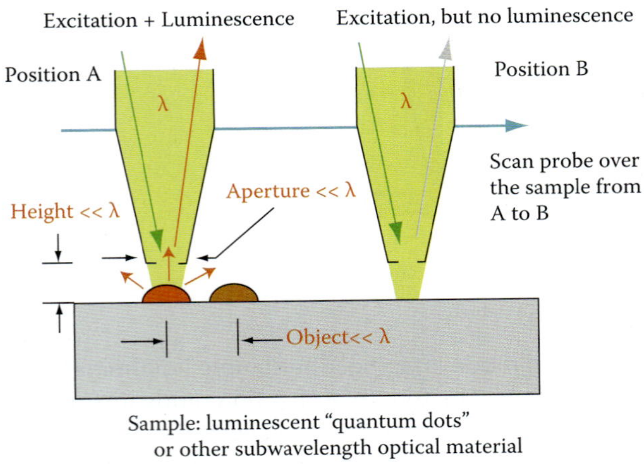

FIGURE 5.116 Use of NSOM to image a luminescent quantum dot.

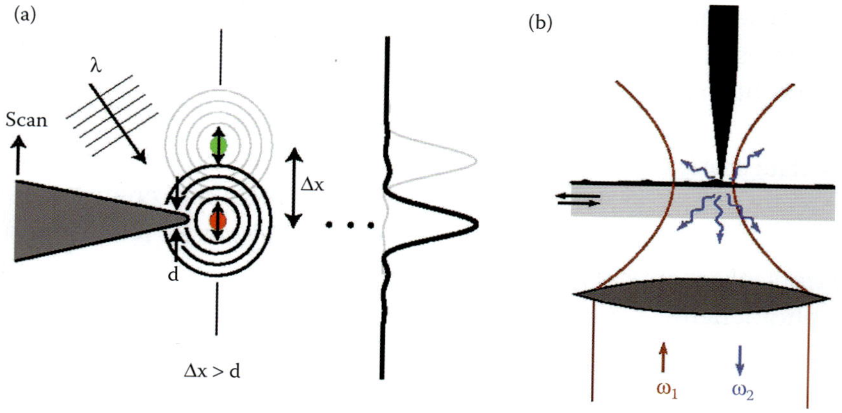

FIGURE 5.117 Local field enhancement technique. A laser-irradiated metal tip enhances the incident electric field near its apex, thereby creating a localized photon source. (a) Schematic of the method. (b) Practical implementation. A laser beam is focused on a sample surface, and a sharply pointed metal tip is positioned into the laser focus. The enhanced fields at the tip locally interact with the sample surface, thereby exciting a response that is collected by the same objective and directed on a detector. (From http://www.optics.rochester.edu/workgroups/novotny/muri03/progress04-01.htm.)

plasmon resonances that depend sensitively on the excitation wavelength (Figure 5.117).

Under "The μ-ε Quadrant and Metamaterials," we will learn that superlenses also prevent image degradation and also beat the diffraction limit established by Abbe.

Pros and Cons of NSOM

From the above, near-field scanning optical microscopy (NSOM) refers to the interaction of light with a sample close to a metal aperture or a solid metal tip, which constrains the lateral extent of the light. The light source is held in place in a manner similar to those used for other scanning proximal probe microscopes (e.g., AFM, STM; see Volume III, Chapter 6 on metrology). Advantages of the near-field interactions in NSOM include:

1. Improved spatial resolution (e.g., 10-nm resolution can be achieved with scattering-based NSOM (SNSOM)).
2. NSOM is a complementary optical tool to other scanning probe microscopes such as AFM and STM, which can only provide topographic and mechanical and electronic information, respectively.
3. A particular advantage of NSOM is to allow for a variety of powerful means of local surface analysis such as fluorescence near-field imaging and spectroscopy, which are highly sensitive and selective techniques to determine the physical and chemical properties of surfaces; specifically, surface-enhanced Raman spectroscopy (SERS) can be coupled with NSOM (see below).

Disadvantages:

1. Practically zero working distance.
2. Extremely small depth of field so that only features at the surface of specimens can be studied.
3. Extremely long scan times for high-resolution images or large specimen areas.
4. Very low transmissivity of apertures smaller than the incident light wavelength (does not apply to apertureless NSOM).
5. Fiberoptic probes are somewhat problematic for imaging soft materials because of their high spring constants.

NSOM Applications

Near-field optics technology offers an important window on the future of a wide range of vibrant new nanotechnologies. These new nanoapplications range from:

- Ultrahigh resolution optical imaging
  - Spectroscopy
  - Near-field surface-enhanced Raman spectroscopy (near-field SERS)

- Local spectroscopy of semiconductor devices
- Modification of surfaces
- Subwavelength photolithography (see Volume II, Chapter 2)
- Ultrahigh density data storage
- Laser ablation
- Deposition of various materials such as in the photodecomposition of chemical gases, i.e., near-field optics photochemical vapor deposition (NFO-PCVD)
- Near-field femtosecond studies, etc.

Several of these applications are dealt with in detail elsewhere in the book. Here we briefly describe some of them. As we saw above, measuring the Raman spectrum of scattered light provides a powerful "fingerprinting" technique for the identification of molecules (Raman spectroscopy). However, typical Raman signals are rather weak, and we saw how the EM fields carrying the fingerprinting information can be enhanced by many orders of magnitude through plasmons generated on assemblies of resonating metal nanoparticles. The combination of scanning near-field optics (NSOM), which enhances locally the Raman effect (SERS), enables localization and scanning of illumination or collection of light over a surface. This provides the means to detect trace amounts on surfaces that would otherwise be undetectable. Trace analysis down to a detection limit of about 100 molecules has been demonstrated. This is of use, for example, in environmental, biomedical, and pharmaceutical research.[23] Even single-molecule detection is possible with SERS. The important evidence of single-molecule SERS is blinking of the SERS signal and spectral diffusion of Raman shift with time. Single-molecule detection by Raman scattering is useful, especially for the detection and distinction of single DNA bases: the nucleotide bases show well-distinguished SERS spectra.[24]

In Volume II, Chapter 2, on next-generation lithographies, we will see that NSOM is used to write submicron patterns into a photoresist. Aperture NSOM in illumination mode (Figure 5.115) can also be used to write and read data on magneto-optical materials. When writing, the probe heats the medium under the probe. The domain under the write beam becomes magnetized opposite of the surrounding material. This varying domain of magnetization can then be read using the same probe but with a lower output power.

Near-field optical ultrafast laser ablation relies on extremely fast-energy deposition in the femto-/picosecond range, resulting in subsequent material ablation without thermal heat shocks or mechanical damaging (i.e., no melting, burr formation, or cracks). Near-field optics photochemical vapor deposition (NFO-PCVD) uses NFO in a vacuum as a patterning method to deposit material very locally via photodissociation.

Comparing Photons with Electrons—Photonic Crystals

Introduction

Photons and electrons exhibit wave and particle characteristics. When treated as waves, the wavelength associated with either particle is given by the de Broglie equation, i.e., $\lambda = h/mv$ (Equation 3.96), where $mv = p$ is the particle momentum and h the Planck constant. From Schrödinger's equation we established that the space propagation of electrons (Chapter 3), and, from Maxwell's equations that of photons (current chapter), can be described by plane waves as summarized in Table 5.12. Photons, we saw, satisfy Bose-Einstein statistics, whereas electrons are fermions. In Chapter 3 we also calculated the wave equations for EM waves from Schrödinger's equation because, according to Bohr's correspondence principle, the results from the older theory (Maxwell) should also follow from the newer theory (Schrödinger). The probability of finding matter at position x in quantum mechanics scales with $|\Psi_0(x,t)|^2$, and this may be compared with the probability of detecting light in Maxwell's theory scaling as $|\mathbf{E}_m(x,t)|^2$. Light waves are described by an electric (**E**) and magnetic (**B**) vector field, whereas electron wave functions are scalars. Comparing the photon and electron momentum [$\mathbf{p} = (h/2\pi)\mathbf{k}$ or $\hbar\mathbf{k}$], we find that the electron momentum is generally bigger than the photon momentum because the electron wave vector ($\mathbf{k} = m_e v/\hbar$) is bigger than the photon wave vector ($\mathbf{k} = 2\pi/\lambda$). Because the photon wave vector is

Table 5.12 Comparison of Electrons and Photons

		Electrons	Photons
Equation		Schrödinger	Maxwell's equations for light
Applicable particles distribution		Electrons are fermions and follow Fermi-Dirac statistics	Photons are bosons and follow Bose-Einstein statistics
Solution	Generic	Describes allowed energies of electrons	Describes the allowed frequencies of light
	Free space	$\psi = \psi_0\left(e^{i\mathbf{k}\cdot\mathbf{r}-\omega t} + e^{-i\mathbf{k}\cdot\mathbf{r}+\omega t}\right)$: a scalar wave function	$\mathbf{E} = \mathbf{E}_m\left(e^{i\mathbf{k}\cdot\mathbf{r}-\omega t} + e^{-i\mathbf{k}\cdot\mathbf{r}+\omega t}\right)$: a vector field described by **E** and **B**
	In a medium	Propagation of electrons affected by Coulomb potential	Propagation of light affected by the dielectric medium (refractive index)
Wavelength	Equation	$\lambda = \frac{h}{\mathbf{p}} = \frac{c}{mv}$	$\lambda = \frac{h}{\mathbf{p}} = \frac{c}{\nu}$
	Remark	Electron momentum **p** is larger so the wavelength of electrons is generally <1 nm	Photon momentum **p** of light is smaller so the wavelength of light is generally <1 μm
Momentum	Equation	$\mathbf{p} = (h/2\pi)\mathbf{k}$ or $\hbar\mathbf{k}$	$\mathbf{p} = (h/2\pi)\mathbf{k}$ or $\hbar\mathbf{k}$
	Remark	Same equation	Same equation
Energy	Equation	$E = \frac{(\hbar\mathbf{k})^2}{2m_e}$	$E = \mathbf{p}c = (h/2\pi)\mathbf{k}c = \hbar\mathbf{k}c$
	Remark	Parabolic	Linear
Speed		The velocity of an electron, *v*, can be any value smaller than *c*.	Fixed = c for electromagnetic radiation is always $3\ 10^8$ m/s
Mass		$m = 9.1 \times 10^{-31}$ kg	$m = 0$
Propagation through classically forbidden zones		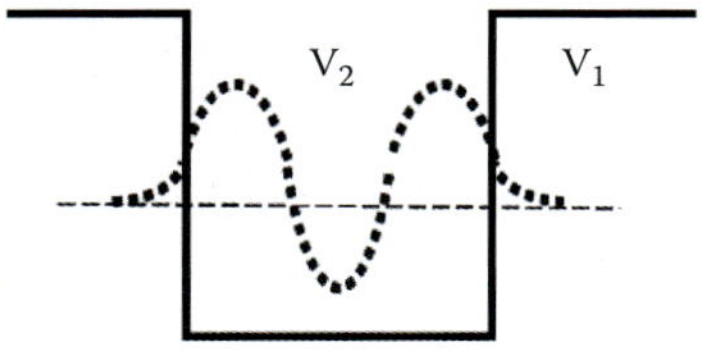 $V_1 > V_2$ Wave function for a particle in a potentail well Electron wave functions decay exponentially in forbidden zones	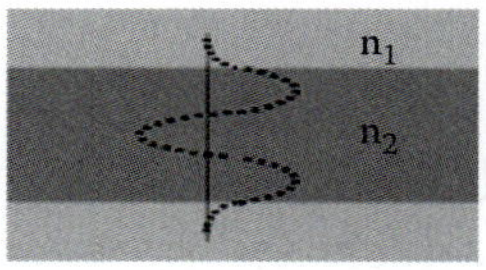 Planner waveguide, $n_1 > n_2$ Photons tunnel through classically forbidden zones. **E** and **B** fields decay exponentially and **k** vector is imaginary.
	0D (no size confinement)	Single crystal: bulk material	Photonic crystals: bulk material
Confinements	1D	 Thin film-quantum well: 1D confinement	 Planar waveguide: 1D confinement
	2D	 Quantum wire: 2D confinement	 Optical fiber: 2D confinement
	3D	 Quantum dot: 3D confinement	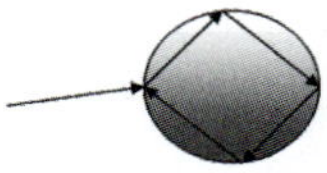 Microsphere: 3D confinement
Bandgap, forbidden energies		Electronic bandgap	Photonic bandgap
Possible states		Spin up and spin down	Polarization
Charge		$q = -1.6 \times 10^{-19}$ C	$q = 0$

small compared with the electron wave vector, the wavelength λ for an electron is of the order of the atomic spacing (<1 nm), whereas the wavelength of light is generally much larger (<1 μm). For photons, the energy $E = pc = \hbar \mathbf{k}c$, whereas for electrons, the energy $E = (h/2\pi)^2\mathbf{k}^2/2m$. Thus, whereas the energy dispersion, i.e., the variation of energy with wave vector $\mathbf{k}$ $(=2\pi n/\lambda)$, is linear for photons, for electrons it is a parabola (Figure 5.118). The energy of an electron with a 1-Å wavelength is 12,000 eV (= hard x-rays), and the energy of a photon with a 1-Å wavelength is 150 eV. Simply comparing the far superior resolution of electron microscopy with that of an optical microscopy confirms that in most instances the wavelength of an electron is far shorter than that of a photon. Moreover, electrons feature much more punch with a momentum given by $\mathbf{p} = m_e v$, where m_e is the electron's rest mass, far larger than the relativistic mass of photons of similar energies, with $m = h\nu/c^2$ (Equation 5.173). Remember also that for photons the speed is fixed, i.e., c for EM radiation is always 3×10^8 m/s, whereas the velocity of an electron, v, can be any value smaller than c.

The propagation of light is affected by the dielectric medium (refractive index), whereas the Coulomb potential affects the propagation of electrons. Solution of Schrödinger's equation in a 3D periodic Coulomb potential for an electron crystal forbids propagation of free electrons with energies within the energy bandgap. Imposing boundaries on photons by making them move in a material with a periodic dielectric constant in one, two, or three dimensions leads to photonic crystals as illustrated in Figure 5.119. A typical repeat unit for a photonic crystal is in the 100-nm range, whereas in a crystal the repeating units are the atoms (±1 nm) (see Figure 5.120). Photonic crystals generally possess photonic bandgaps (PBGs), ranges of frequency in which light cannot propagate through the structure. The forbidden range of frequencies depends on the direction of the light with respect to the photonic crystal lattice. For a sufficiently large refractive index

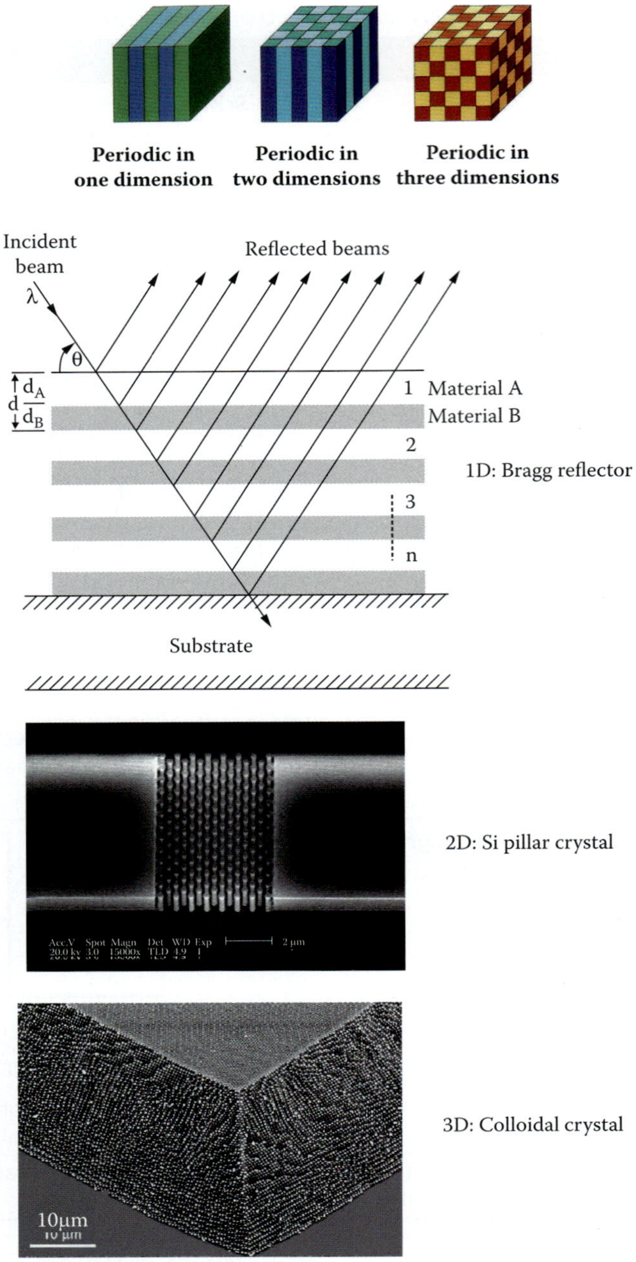

FIGURE 5.119 Photonic crystals are materials with a periodic dielectric constant in one, two, or three dimensions; typical repeat distance is 100 nm. Different colors represent different refractive indices. 1D photonic crystal: stack of alternating refractive index material in one direction; 2D photonic crystal: periodic variation of refractive index in two directions; 3D photonic crystal: periodic variation of refractive index in three directions.

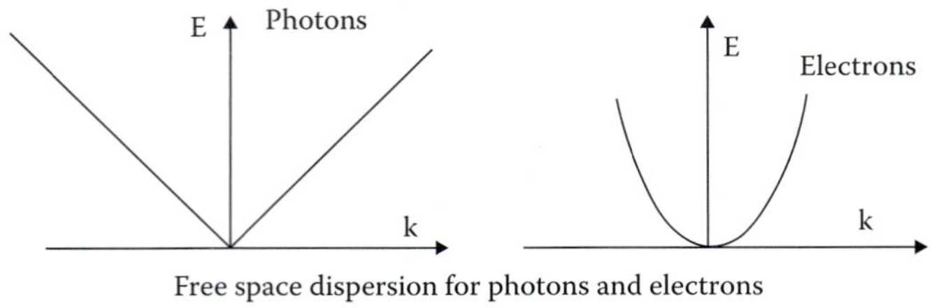

FIGURE 5.118 Dispersion relationships of photons and electrons in free space.

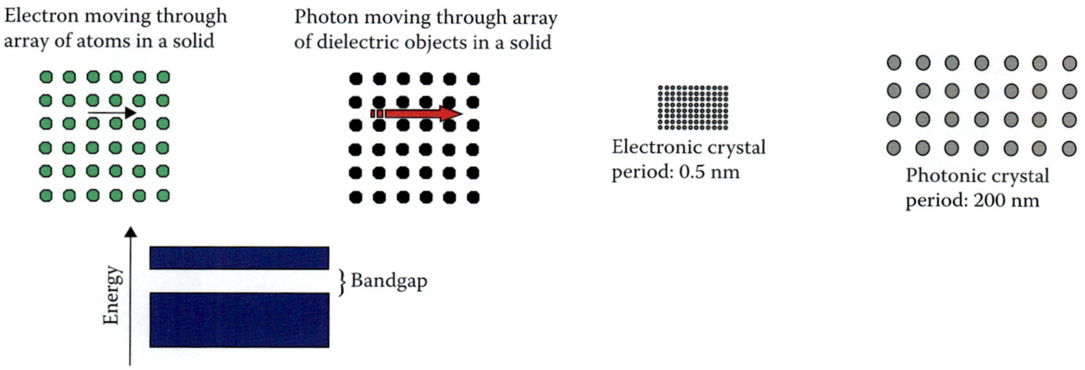

FIGURE 5.120 Similar to the periodic electron crystal lattice, one can fabricate a photonic crystal lattice. The refractive index varies with a much larger period of around 200 nm.

contrast (ratio n_1/n_2), there exists a bandgap that is omnidirectional. Each constitutive component in these periodic structures is perfectly transparent, but their periodic arraying blocks EM waves of certain frequencies. Photonic crystals may be regarded as semiconductors/insulators of light. Rather than a periodic array of atoms that scatter and modify the energy momentum of electrons, photonic crystals consist of a spatially periodic dielectric constant, with a periodicity of the order of the wavelength of light. They are mostly artificial structures that control the flow of light and are expected to lead to a new photonic era, with an impact similar to that of the electronic revolution brought on by semiconductors.

Inside a semiconductor, electrons experience a periodic potential as a result of the regularly spaced atomic cores in the crystal lattice, leading to multiple electron wave reflections and electron wave interference. The multiple reflections result in eigenmodes that are affected by the exact shape of the periodic potential $V(x)$. Felix Bloch pioneered the study of wave propagation in 3D periodic media in 1928, unknowingly extending an 1883 theorem in 1D by G. Floquet. Bloch proved that waves in such a medium could propagate without scattering, their behavior governed by a periodic envelope function multiplied by a plane wave (see Equations 3.198 and 3.199). Bloch studied the problem from the viewpoint of quantum mechanics. He postulated that electrons in a conductor scatter only from imperfections and not from the periodic ions. The same techniques can be applied to electromagnetism by casting Maxwell's equations as an eigenproblem in analogy with Schrödinger's equation.

The tunneling of electrons through a forbidden zone has its equivalent in light leaking out of a waveguide to create an evanescent field. Table 5.12 summarizes the above comparison of photons with electrons.

The Photonic Bandgap

During the past 17 years, John Maddox's lament (see quote at the beginning of this chapter) was heard loud and clear, and tremendous progress was made toward realizing photonic crystals. Stacks of dielectrics (with repetition in one direction, 1D) have been known since 1887 (Bragg reflectors). Also known and used were optically confined structures, with 1D confinement (planar waveguides), 2D confinement (fiberoptic cables), and 3D confinement (microspheres). The breakthrough in photonics in 1987 was the proposal of photonic crystals with periodicity in 2D and 3D (see Figure 5.119). This concept was introduced independently and simultaneously that year by two researchers, each following different paths. Photonic crystals were originally proposed as a means to realize two fundamental optical principles: the localization and trapping of light in a bulk material and the complete inhibition of spontaneous emission over a broad frequency range. Sajeev John, from the University of Toronto, was formulating an answer to the first question—whether localization of electrons in a disordered solid can be extended to photons in a strongly scattering medium—and he predicted that localized states of the EM field can be created in a periodic dielectric medium in three dimensions.[25] At the same time, Eli Yablonovitch, then at Bell

Communications Research in New Jersey, was trying to address the possibility of suppressing the unwanted spontaneous emission affecting semiconductor lasers, and he predicted that a periodic dielectric in thee dimensions could produce a forbidden gap in the EM spectrum.[26] Spontaneous emission of light is a natural phenomenon, which is of great practical and commercial importance. For example, in semiconductor lasers, spontaneous emission is the major sink for threshold current I_{th}, which must be surmounted to initiate lasing. In solar cells, spontaneous emission determines the maximum voltage available. Yablonovitch reasoned that if there are no optical modes available for photons to go to there cannot be spontaneous emission. From Fermi's golden rule (Equation 3.285), we know indeed that the density of final states is the density of optical modes available to the photon and an optical bandgap would indeed cut the optical modes that are available.

The origin of the photonic bandgap is best understood by looking back at the dispersion curve of a 1D photonic crystal [ω(**k**)] (i.e., a Bragg reflector). Similar dispersion curves in electronic crystals reveal the electronic bandgap. A 1D photonic crystal is simply a Bragg stack of dielectric planes, with a period of refractive index variation labeled a in Figure 5.121a. Light incident in the *z*-direction with a wave vector (**k**) is diffracted by this 1D photonic crystal into many possible directions, and the wave vectors of incoming and diffracted waves are related as:

$$\mathbf{k}_z - \mathbf{k}'_z = \pm N(2\pi/a) \tag{5.265}$$

where *N* is the multilayer number (see also Equation 2.20). Thus, the incident and diffracted *z*-components of a wave vector are shifted with respect to each other by integral multiple of $(2\pi/a)$. A special case of diffraction is the Bragg diffraction, which occurs when incident light is totally reflected by the photonic crystal (it behaves like a metal!). This occurs whenever $\mathbf{k}_z$ is a multiple of (π/a):

$$\mathbf{k}_z = \pm N(\pi/a) \tag{5.266}$$

as seen in Figure 5.120b. This wave cannot propagate, and its group velocity $(d\omega/d\mathbf{k})$ is zero. Figure

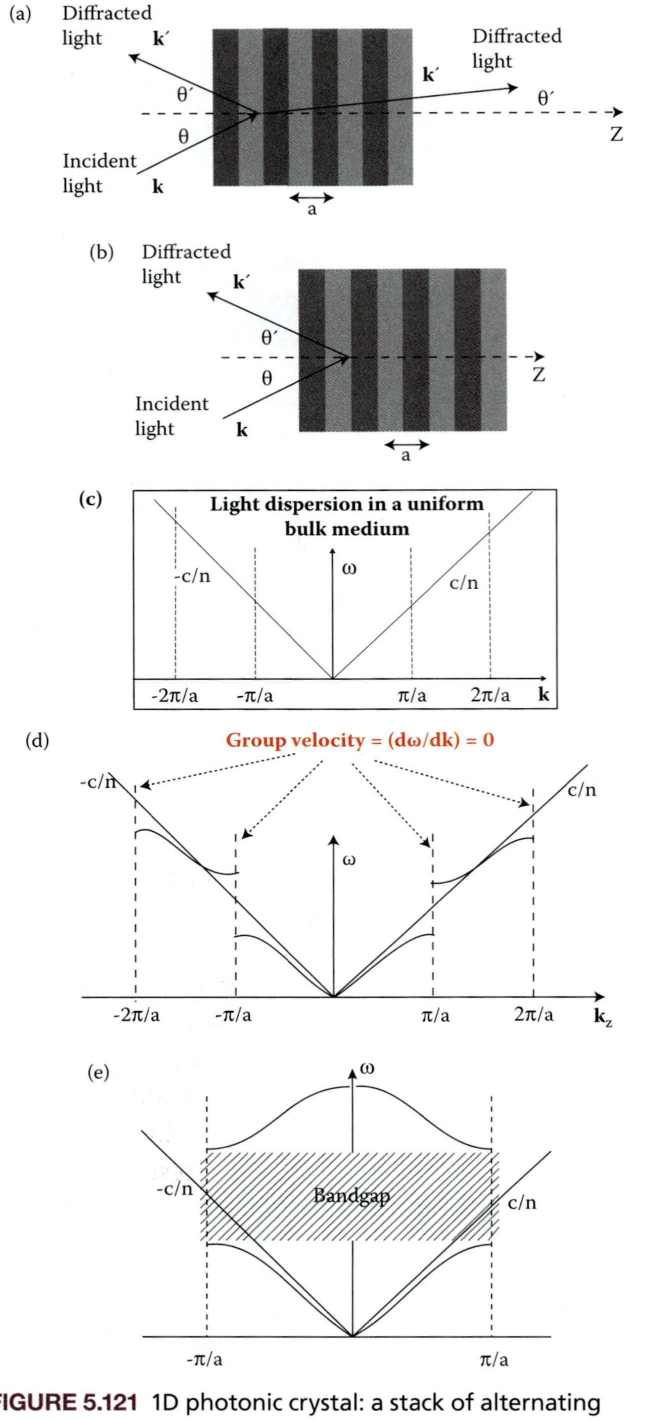

FIGURE 5.121 1D photonic crystal: a stack of alternating refractive index materials in the *z*-direction. (a) Light incident in the *z*-direction with a wave vector (**k**) is diffracted by the 1D photonic crystal into many possible directions. (b) A special case of diffraction is the Bragg diffraction, which occurs when incident light is totally reflected by the photonic crystal. (c) The dispersion character of light in a bulk medium with a uniform refractive index, *n*. (d) The dispersion curve of a 1D photonic crystal deviates from the straight-line dispersion curve of a uniform bulk medium. (e) The region of $\mathbf{k}_z$ between $-\pi/a$ and π/a is called the *first Brillouin zone*. The shaded area marks the photonic bandgap.

5.120c shows the dispersion character of light in a bulk medium with a uniform refractive index, *n*. In contrast, in Figure 5.120d, we notice that for the 1D photonic crystal, the dispersion curve deviates from the straight-line dispersion curve of such a uniform bulk medium. To ensure Bragg reflection for $\mathbf{k}_z = \pm N(\pi/a)$, the curve becomes horizontal when approaching π/a ($d\omega/d\mathbf{k} = 0$). It is necessary to plot the dispersion curve in Figure 5.120d only for values of $\mathbf{k}_z$ between $-\pi/a$ and π/a as illustrated in Figure 5.120e. All other values of $\mathbf{k}_z$ can be obtained by diffraction of waves with $\mathbf{k}_z$ between $-\pi/a$ and π/a. The region of $\mathbf{k}_z$ between $-\pi/a$ and π/a is called the *first Brillouin zone*. There is a band of frequencies that are forbidden for all possible values of $\mathbf{k}_z$. This is the bandgap for the 1D photonic crystal.

Compared with a 1D photonic crystal, it is relatively more difficult to deduce the dispersion curves for 2D and 3D photonic crystals. The software for doing this is publicly available at http://www.elec.gla.ac.uk/groups/opto/photoniccrystal/Software/SoftwareMain.htm.

Complete photonic bandgaps can only be obtained under favorable circumstances, i.e., with the right structures and sufficient (threshold) refractive index contrast. Ho et al.[27] showed theoretically that photonic "atoms" arranged in a diamond lattice structure may have a complete bandgap. Simple face-centered cubic photonic crystals, on the other hand, even with high dielectric contrast, do not yield a complete photonic bandgap.[28] The threshold refractive index contrast for complete bandgaps in 3D photonic crystals obviously depends on the exact periodic structure; e.g., for diamond it is 1.87, and for an inverse opal structure it is 2.80.

Manufacturing of 2D and 3D Photonic Crystals

In 1991, the first photonic diamond structure was fabricated by the group of Yablonovitch. This first photonic crystal was made by mechanically drilling holes, 1 mm in diameter, into a block of dielectric with a refractive index n = 3.6. The material, which became known as *Yablonovite*, prevented microwaves from propagating in any direction; in other words, it exhibited a 3D photonic bandgap.[28] The Yablonovite was the first "crystal" that showed experimentally a full 3D bandgap in the microwave range, and at these long wavelengths a mechanical fabrication method (drilling holes!) could be used. A schematic drawing of the Yablonovite is illustrated in Figure 5.122. Materials like this, in theory, guide and confine light without losses. A 3D photonic crystal provides the ultimate tool for the manipulation of light.

Several ingenious ways to produce these so-called *photonic crystals* for use in the microwave regime have been devised in addition to Yablonovitch's method. For example, in 1994 Ekmel Özbay, then at the Ames Laboratory in the United States, and his coworkers fabricated a photonic crystal with a bandgap at microwave wavelengths by stacking micromachined silicon wafers in a "woodpile" or "picket fence" structure.[29] The key to their success was the accurate alignment of successive layers—inaccurate alignment destroys the bandgap. Other structures that have bandgaps at microwave and radio frequencies are currently being used to make antennae that direct radiation away from the heads of mobile phone users.

Despite these successes in the microwave range, it took more than a decade to fabricate photonic crystals that work in the near-infrared (780–3000 nm) and visible (450–750 nm) regions of the spectrum. To block wavelengths in the infrared and visual spectrums, one needs to get a bit more creative. The main challenges are to find suitable materials and processing techniques to fabricate structures that

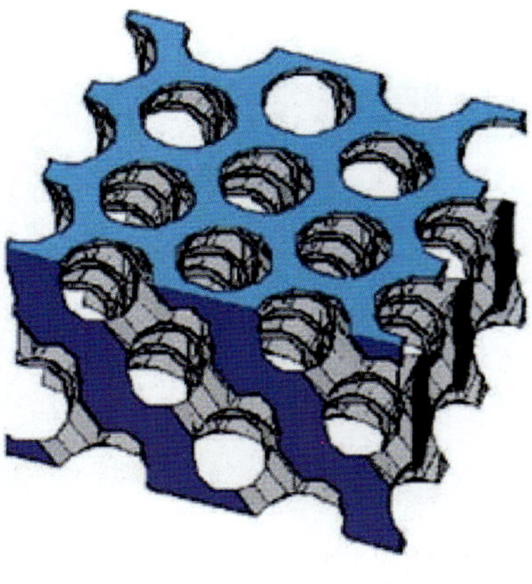

FIGURE 5.122 Yablonovite was the first crystal that showed experimentally a full 3D bandgap in the microwave range.

are about one thousandth the size of the microwave crystals. It becomes increasingly difficult to achieve the required accuracy as the dimensions of the structures are reduced and as the number of layers is increased in an attempt to make a device that operates at infrared or optical frequencies. As so often seems the case, nature is way ahead of the scientists: the iridescence of various colorful living creatures, from beetles to peacock feathers to butterflies [see chapter-opening image (b) before the Introduction of this chapter], is the result of photonic crystals. Unlike pigments, which absorb or reflect certain frequencies of light as a result of their chemical composition, the way that photonic crystals reflect light is a function of their physical structure. That is, a material containing a periodic array of holes or bumps of a certain size may reflect blue light, for example, and absorb other colors even though the crystal material itself is entirely colorless. Because a crystal array looks slightly different from different angles (unlike pigments, which are the same from any angle), photonic crystals can lead to shifting shades of iridescent color that may help some animals attract mates or establish territories. Another naturally occurring example of a photonic crystal is opal [see chapter-opening image (a) before the Introduction of this chapter]. Opals are naturally occurring 3D photonic crystals: their microstructure consists of silica spheres about 150–300 nm in diameter, which are tightly packed into repeating hexagonal or cubic arrangements. Synthetic opals use this same pattern, although they can be made from different materials. A type of photonic crystal made using the gaps between the spheres is called an *inverse opal*. Self-assembly methods enable the fabrication of large polymer opals and inverse opals at low cost. However, the contrast in the refractive index in opal is rather small, which results in a rather small bandgap. Yablonovitch pointed out that the larger the difference in refractive indices between the two repeating materials of the photonic crystal, the more extremely light is influenced.

Two-dimensional (2D) photonic crystals are relatively easy to fabricate using Si micromachining and UV and electron beam lithography. Periodic arrays of macropores in single-crystal Si were first shown by Lehmann and Föll in 1990; some examples of this work are presented in Volume II, Figure 4.75.[30] Grüning et al. realized a bandgap in the near IR using such macropores in n-type silicon etched under back side illumination.[31] If one retains a periodicity of an alternating index of refraction in the x-y plane while the z-direction remains unmodulated, a 2D photonic crystal results as shown in Figure 5.123. Grüning et al. obtained for the first time a complete 2D bandgap in the near infrared (at a wavelength of about

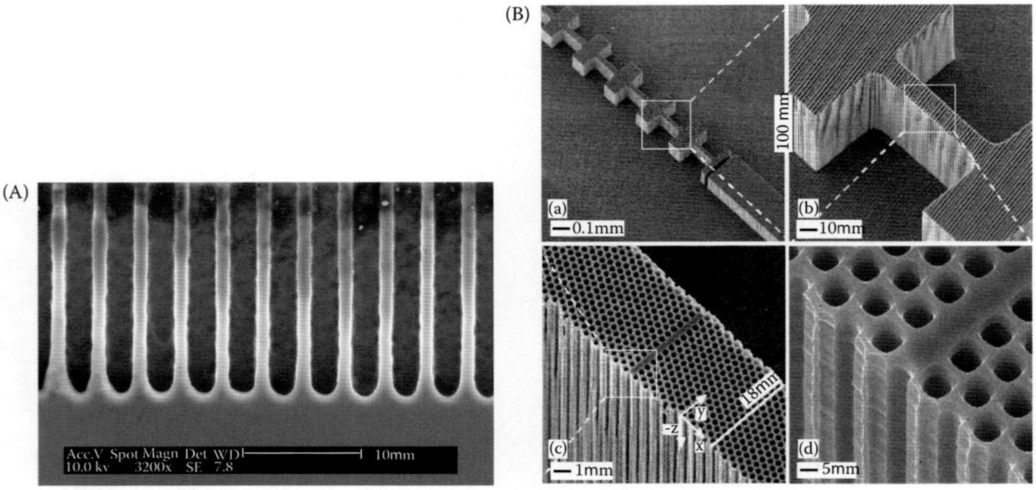

FIGURE 5.123 (A) SEM image of a slab of macroporous silicon, representing a 2D photonic crystal (a). The pore walls are about 100 μm tall and about 25 μm wide in the direction of transmission (b). Omitting some pores yielded a waveguide structure (c). The extremely smooth finish of the structure is clearly visible (d). (Courtesy of Dr. R. Wehrspohn.) (B) Macropores in n-type silicon etched under back side illumination form a 2D photonic crystal.[31]

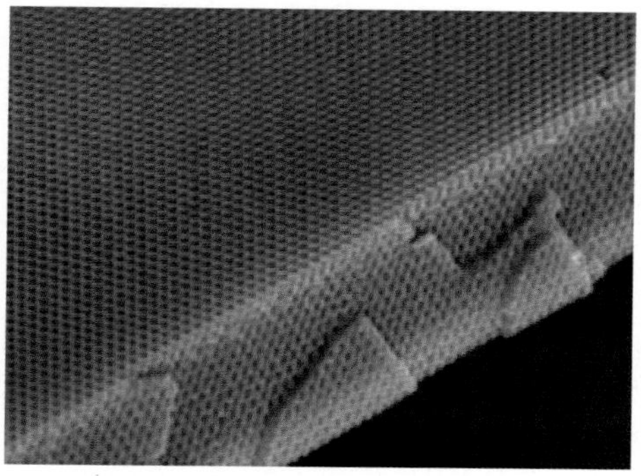

FIGURE 5.124 A 3D photonic crystal fabricated using holographic lithography.[32]

4.9 μm). A complete bandgap in this case means that a bandgap always exists independent of the polarization and propagation of the light wave.

As pointed out earlier, fabrication of 3D photonic crystals is more difficult than 2D crystals. In addition to the methods listed above, it has been achieved by sedimentation of monodisperse colloidal nanospheres of polystyrene or silica and by using holographic two-photon lithography. Two-photon lithography (see Volume II, Chapter 2) is based on the concept that electronic transitions that would normally require absorption of one UV photon for excitation can under some circumstances be accomplished by simultaneous absorption of two IR photons (two-photon overlap in time and space for exposure). When this high-intensity IR light enters a photoactive material, the excitation reaction is confined to a focal point at a precise depth in a substrate. Because two-photon absorption takes place only close to the focal point of a laser, one can fabricate structures deep within the volume of the polymer matrix. Atomic layer deposition (Volume II, Chapter 7) has also been used to build photonic crystals for coatings inside porous alumina and to fabricate inverted opals. The current experimental efforts are directed at decreasing the disorder present in the structures and increasing their scalability, and in the meantime, there is a continuous search for designing new structures with more robust bandgaps.

A 3D photonic crystal fabricated using holographic lithography is shown in Figure 5.124. In Figure 5.125 we show a self-assembled silica opal grown on a silicon substrate, and an inverse opal is made using LPCVD to fill the opal template with silicon, and wet etching yields the inverse opal silicon structure. By 1998, artificial 3D photonic crystals in the near infrared and visible were fabricated based on inverted opals, and in 1999 the first 3D photonic crystal-based optical devices (lasers, waveguides) were made. Another way of making photon crystals is by using block copolymers as described in Volume II, Chapter 2.

Defects in Photonic Crystals

As light is forbidden within the photonic bandgap, we can anticipate that we can manipulate the behavior of light using photonic crystals just as we can manipulate electrons using a semiconductor. Most proposals for devices that make use of photonic crystals do not use the properties of the crystal directly but make use of defects. Defects are made when the lattice is changed locally, and as a result, light with a frequency inside the bandgap can now propagate locally in the crystal, i.e., at the position of the defects. It remains impossible, however, to

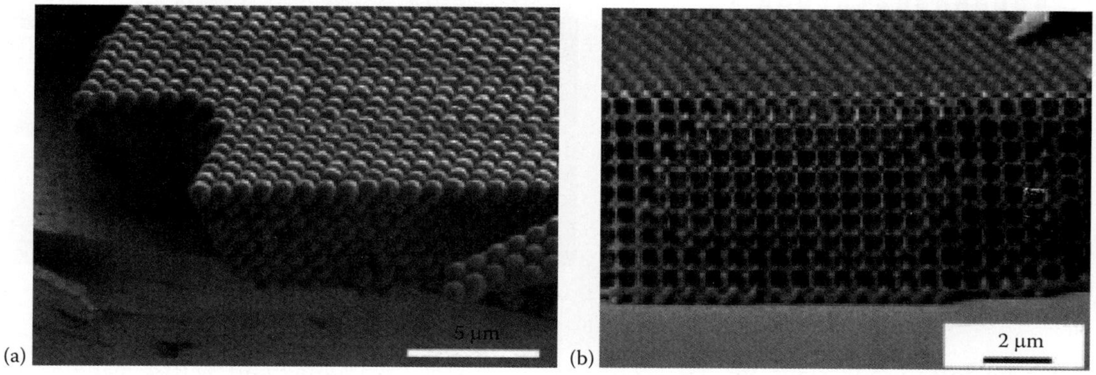

FIGURE 5.125 (a) Self-assembled silica opals grown on silicon substrate. (b) LPCVD is used to fill opal template with silicon; wet etching yields inverse opal silicon structure.[33]

propagate in the surrounding photonic crystal material. For example, if we consider the introduction of a light-emitting layer, with the same wavelength as the photonic bandgap in the PC, light emission is impossible as light is forbidden to be present in the photonic bandgap. This leads to the control of spontaneous emission as first recognized by Yablonovitch. However, if we disrupt the periodic characteristics of the photonic crystal by introducing defects, energy levels corresponding to the defects are formed in the photonic bandgap. Thus, light is now permitted at the energy levels corresponding to those defects. This in turn enables the realization of single-mode light-emitting diodes (LEDs) and zero-threshold lasers. In addition, if we introduce a line of defects, the propagation of light through the defects is possible, and as light propagation is forbidden outside the defects, a no-loss microoptical circuit can be formed. Therefore, to provide functionality to photonic crystals, the introduction of controlled defects is necessary; the importance of defects in photonic crystals is comparable to that of dopants in semiconductors.

Most of the work on introducing controlled defects is carried out in 2D photonic lattices, so-called *PC slabs*. These slabs exhibit some of the useful properties of a truly 3D photonic crystal and are far simpler to make with current IC manufacturing type of equipment. These structures can block certain wavelengths of light at any angle in the plane of the device and can even prevent light entering from certain angles in the third dimension (i.e., perpendicular to the surface). Thus, 2D photonic crystals are a good compromise for many applications and are easily incorporated within planar waveguides.

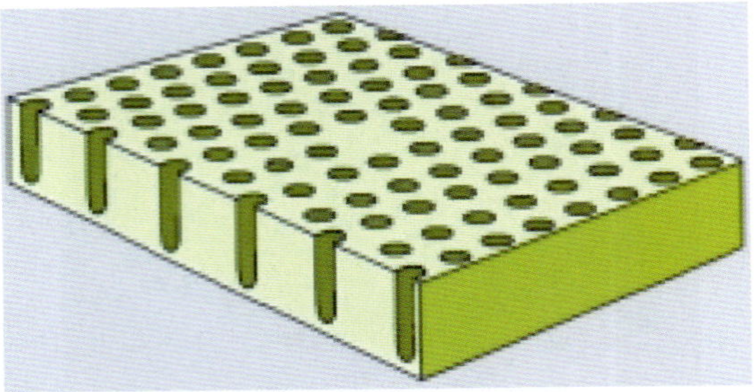

FIGURE 5.126 Point defect as high-*Q* resonator cavity in photonic crystals.

Point defects in a photonic crystal lattice (Figure 5.126) remove a single dielectric unit and produce microcavities that have size-dependent radiation modes. They are analogous to a single F-center (atom vacancy) defect in an electronic material. A microcavity resonant for wavelength λ satisfies $D = N(\lambda/2)$, where N is an integer and D is the characteristic size of the cavity. These "microcavities," just like in electronic crystals, produce defect states in the photonic bandgap. They come with a very high resonance Q (10^4–10^8).

Line defects in a photonic crystal remove lines of a dielectric material and act as waveguides confining photons (Figure 5.127). This is analogous to a line of F-centers (atom vacancies) for electronic

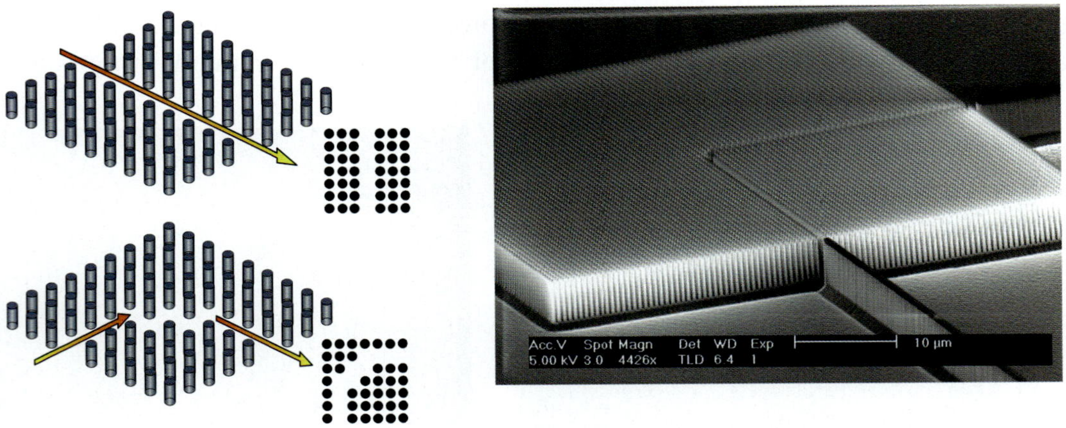

(a) (b)

FIGURE 5.127 (a) Line defects in PC slabs. (b) 2D Silicon photonic crystal waveguide bend. Zijlstra, van der Drift, De Dood, and Polman (DIMES, FOM) [Delft Institute of Microelectronics and Submicron Technology (DIMES), Foundation for Fundamental Research of Matter (FOM); http://www.dedood.demon.nl/PhotonicCrystals.html].

defects. The E field is confined to the defects and cannot propagate in the rest of the material. Unlike the conventional refractive waveguides, these photonic waveguides work by diffraction. Modes that cannot propagate through the photonic crystal lattice (i.e., those within the bandgap) can propagate easily within the waveguides, including those waveguides that feature bends at right angles. The defect mode cannot penetrate the crystal in the x-y plane because of the bandgap but extends in the z-direction. Whereas photonic crystal waveguiding is based on diffraction, conventional optical fibers and waveguides operate based on total internal reflection. Fiberoptic technology has revolutionized telecommunications owing to the speed of data transmission, which is equivalent to >3 h of television per second or 24,000 simultaneous telephone calls (0.1 kg of fiber carries the same information as 30,000 kg of copper cable). But this benefit of speed comes with a penalty: fiber optic cables are large.

For an optical fiber, the contrast in refractive index between the glass core of the fiber and the surrounding cladding material determines the maximum radius through which light can be bent without losses. For conventional glass fibers, this bending radius is a few millimeters. However, the interconnects between optical components on a dense integrated optical circuit require bending radii of 10 μm or less! This is where photonic crystal-based waveguides offer a solution. It is possible to form narrow-channel waveguides within photonic crystals by creating a row of holes in an otherwise regular pattern (see Figure 5.127). Light will now be confined within that line of defects for wavelengths that lie within the bandgap of the surrounding photonic crystal. Because the surrounding material has no available modes at this wavelength, an optical quantum well forms in the waveguide region and traps the light. Under these conditions we can introduce a pattern of sharp bends that will either cause the light to be reflected backward or directed around the sharp bends.

Thus, we can create point defects corresponding to resonant cavities, or line defects behaving like waveguides.

A great effort in studying defects in photonic crystals, as building blocks for photonic integrated circuits, has been carried out by Professor John D. Joannopoulos' group at MIT, and the interested reader can refer to their publications (e.g., *Photonic Crystals: Molding the Flow of Light*[34]).

Super Refraction or Super Prism Effect

The prism effect (see Figure 5.128) refers to separation of colors by refraction through a prism. This is related to the dispersion or variation of refractive index with wavelength, i.e., the derivative $(dn/d\lambda)$ of the refractive index with wavelength. In a normal bulk medium like glass, dispersion is small, but $dn/d\lambda$ can be made unusually large in photonic crystals. This can be understood from the dispersion curves for photonic crystals mentioned earlier (see Figure 5.121).

Photonic Crystal Optical Fibers

Photonic crystal optic fibers are a special class of 2D photonic crystals where the dimension of the medium perpendicular to the crystal plane can be hundreds of meters long. These types of fibers are known by several names: photonic crystal fiber, photonic bandgap fiber, holey fiber (photonic crystal fiber using air holes in their cross-section), microstructured fiber, and Bragg fiber (Figure 5.129).

Besides the bending losses, another major drawback associated with conventional optical fibers is that different wavelengths of light travel through the material at different speeds. Over long distances, time delays then occur between signals that are encoded at different wavelengths. This dispersion is worse if the core is large, as the light can follow more different paths or "modes" through the fiber. A pulse of light traveling through such a fiber will broaden out, thereby limiting the amount of data that can be sent through it. Adding new signals at other optical wavelengths can also increase the fiber data transfer capacity of an optical fiber, a method known as *dense wavelength division multiplexing* (DWDM). Unfortunately, the optical fiber is only transparent over a small range of wavelengths, so the number of separate conversations that can be transmitted depends on the line width of neighboring optical channels (which is currently at the subnanometer level). These problems can be solved effectively using the "holey fiber" developed by Philip Russell

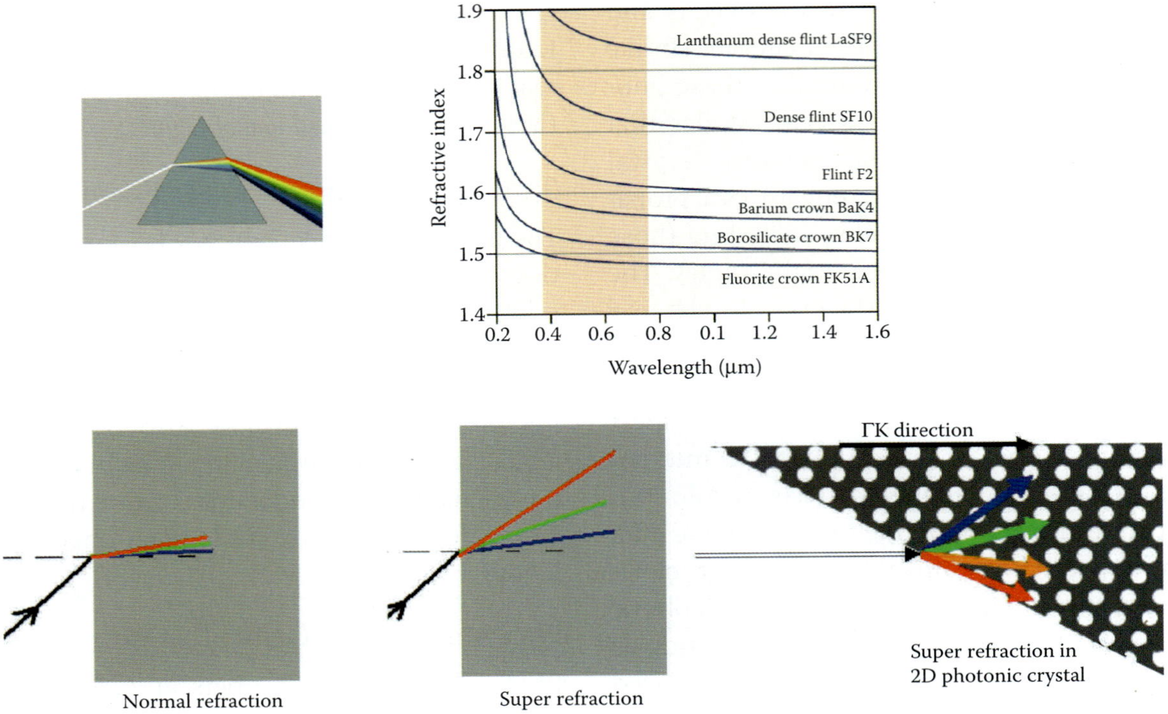

FIGURE 5.128 Super refraction or super prism effect compared with normal refraction.

and Jonathan Knight and coworkers at the University of Bath.[35] This fiber has a regular lattice of air cores running along its length and transmits a wide range of wavelengths without suffering any dispersion. One version of the photonic crystal fiber is made by packing a bundle of glass capillary tubes around a solid glass core that runs through the center of the bundle. This assembly is then heated and stretched out to create a long monolithic fiber that is only a few micrometers in diameter. The resulting fiber has the unusual property that it transmits a single mode of light, even if the diameter of the core is very large. Frequencies in the bandgap propagate within the fiber core, which is like a defect in a 2D photonic crystal. The same Bath team created an even more unusual photonic crystal fiber by removing the

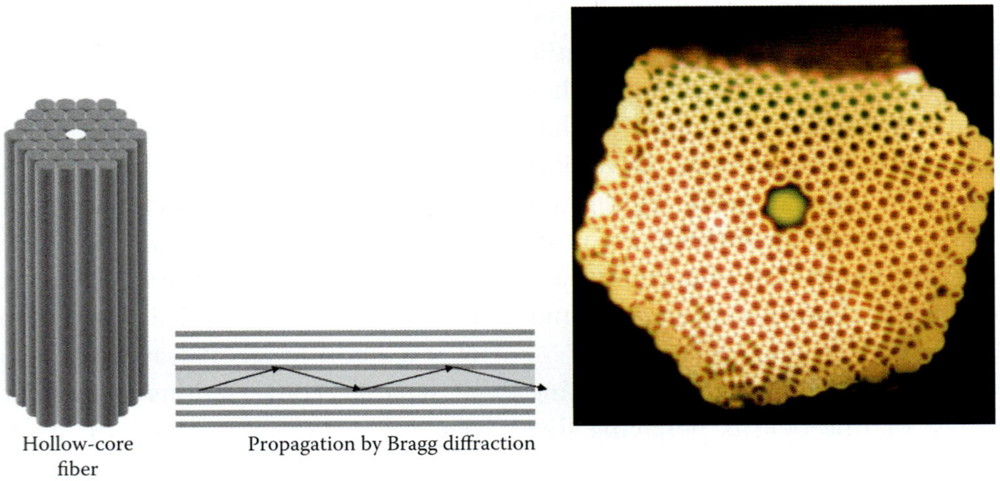

FIGURE 5.129 Photonic crystal fiber, photonic bandgap fiber, holey fiber, microstructured fiber, or Bragg fiber are all used as synonyms in this field.

central solid glass core to form a long air cavity. In this case, the light is guided along the low-refractive-index air core by the photonic bandgap confinement effect. Because the light is not guided by the glass material, very high-power laser signals could potentially be transmitted along such a fiber without damaging it. Because light travels in air, group velocity dispersion is zero for all wavelengths.

In a hollow-core fiber, even those wavelengths that experience high losses in conventional fibers can propagate. Propagation over ~1000 m has been demonstrated for wavelengths like 1.5 μm and 10 μm with an attenuation of only around ~1 dB/km (can be made lower than index guiding fiber). By filling the hollow core with gas, the fiber can be also used as a very sensitive gas sensor.

Applications

It is easy to see why photonic crystals are generating so much interest. Photonic bandgap crystals hold the promise for several fantastic integrated optics applications, including thresholdless microlasers, better signal filters, photonic computers, lossless mirrors and lossless waveguides (even in bends)/waveguide couplers, photonic bandgap optical fibers, and so on. The impact this will make can be gauged by looking at today's information technology (IT). IT derives its power from our ability to manipulate the flow of electrons and holes in semiconductors in the most intricate ways; photonic crystals promise to give us even greater flexibility because we have far more control over the properties of the larger photonic crystals than we do over the electronic properties of semiconductors. Structure possibilities are limited only by our imagination. The benefits of photonic crystals include the ultrahigh refractive indices, the ability to control spontaneous emission, to modify the state density and group velocity, and to locate light by introducing structural defects. This way, photonic crystals could eventually address many of the problems that currently limit the speed and capacity of communications networks. For example, these structures could be used to create novel light-emitting diodes (LEDs) and lasers that emit light in a very narrow wavelength range, together with highly selective optical filters that could be integrated on one and the same chip.

A short sampling of photonic crystal applications follows:

1. Low-threshold microlasers. Within a few years a number of basic photonic crystal applications are expected to start making an appearance in the marketplace. Among these will be highly efficient photonic crystal lasers and extremely bright LEDs. Point defects in photonic crystals act as high-*Q* resonator cavities and are used in low-threshold photonic crystal lasers. In photonic lasers, only the wavelength that matches the wavelength of the defect mode is amplified because it can propagate freely through the material. The intensity of the propagating light increases as it undergoes successive reflections and travels back and forth through the photonic crystal. Meanwhile, light at other wavelengths is trapped within the photonic crystal and cannot build up. This means that the laser light is emitted in a narrow wavelength range that is directly related to the diameter of the microcavity.
2. Highly efficient light-emitting diodes (LEDs). LEDs, like lasers, play a key role in optical communication systems. Encoding and sorting thousands of multiplexed wavelength channels poses quite a problem: at the transmitting end of an IT system, each wavelength channel requires a very stable light source that only emits in a very narrow range of wavelengths. Although LEDs offer high switching speeds, they emit light over a wide range of wavelengths, which makes them less suitable than lasers for dense wavelength division multiplexing (DWDM) systems. LEDs are made from photoemissive materials that emit photons once they have been excited electrically or optically. These photons are typically emitted in many different directions and also have a range of wavelengths, which is not ideal for communications applications. Photonic crystals can be used to design a mirror that reflects a selected wavelength of light from any angle with high efficiency (Figure 5.130). Moreover, they can be integrated within the photoemissive layer to create an LED that emits light at a specific wavelength and direction.

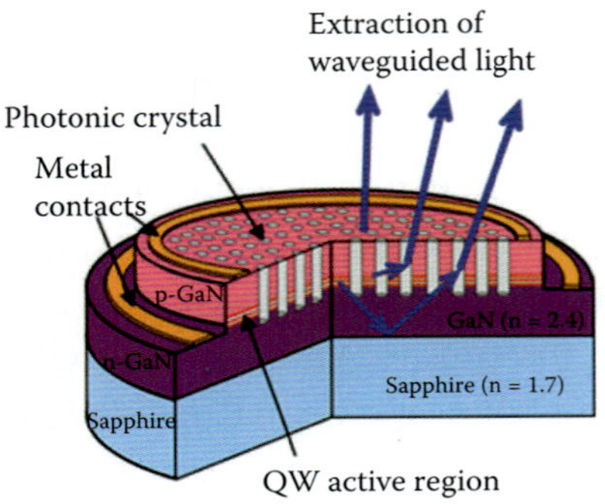

FIGURE 5.130 GaN-based diode with photonic crystal slab.

3. Filters with a very narrow range of wavelengths. Photonic crystal microcavities that are fabricated from passive materials, such as silicon dioxide and silicon nitride, are used to create filters that only transmit a very narrow range of wavelengths.
4. Micrometer-sized optical benches. To develop a micrometer-sized optical bench one first needs to develop small-scale optical "interconnects"—tiny planar waveguides that can steer light around tight corners. Although it is straightforward to route electrons around sharp bends in microchips, it is impossible to direct light in the same way using conventional glass waveguides because the losses at the bends are so large. Photonic crystal waveguides can be used to route light around micrometer-sized optical benches. Together with the very narrow range photonic crystal filters, this could lead to a viable "spectrometer on a chip." These developments are expected within the next 5 years.
5. On a 5–10-year timescale it is expected that the first photonic crystal "diodes" and "transistors" will be demonstrated. A demonstration of the first photonic crystal logic circuit could even take place in the next 10–15 years, whereas a prototype optical computer driven by photonic crystals could be available within the next 25 years.
6. Synthetic opals might find a niche in the valuable jewelry and artwork markets, and thin photonic crystal films could also be used as anticounterfeit devices on credit cards.
7. Negative refractive index. Photonic crystals can exhibit an effective negative refractive index as explained below under "The μ-ε Quadrant and Metamaterials." Whereas photonic crystals occur in nature, metamaterials do not. Metamaterials are engineered composites tailored for specific EM properties that are not found in nature and not observed in the constituent materials. Cubukcu et al.[36] have experimentally demonstrated negative refraction by a photonic crystal.

Although complex photonic crystal LED and laser structures are being developed, fundamental questions about the formation, operation, and efficiency of photonic bandgap materials remain unanswered. Real photonic crystal waveguides are far from ideal because they suffer losses as a result of diffraction and scattering and because of an optical mismatch at the surface joining the device to input and output waveguides. These loss mechanisms need to be fully understood and the devices need to be optimized for particular wavelengths before they can become commercially viable.

The μ-ε Quadrant and Metamaterials

Introduction

As we saw in Chapter 2, material properties are determined by the properties of the subunits (atoms/molecules) plus their spatial distribution. In Figure 5.131 we show a diagram of the EM properties as a function of the ratio of a, the "lattice constant" of a material/structure, and λ, the wavelength of the incoming light.

- Natural crystals (in Figure 5.131 we show the stunning violet tanzanite as an example) have lattice constants much smaller than the light wavelengths: $a \ll \lambda$. In this case, the lattice constant a is the distance between constituent atoms. Such materials are treated as homogeneous media with parameters ε and μ, which are tensors in anisotropic crystals. They have a positive refractive index, $n > 1$, and show no magnetic response at optical wavelengths ($\mu = 1$).
- In the above case, the EM wave is much longer than the typical lattice distance a ($\lambda \ggg a$), and one can apply effective medium theory; however, when a is in the same range of the

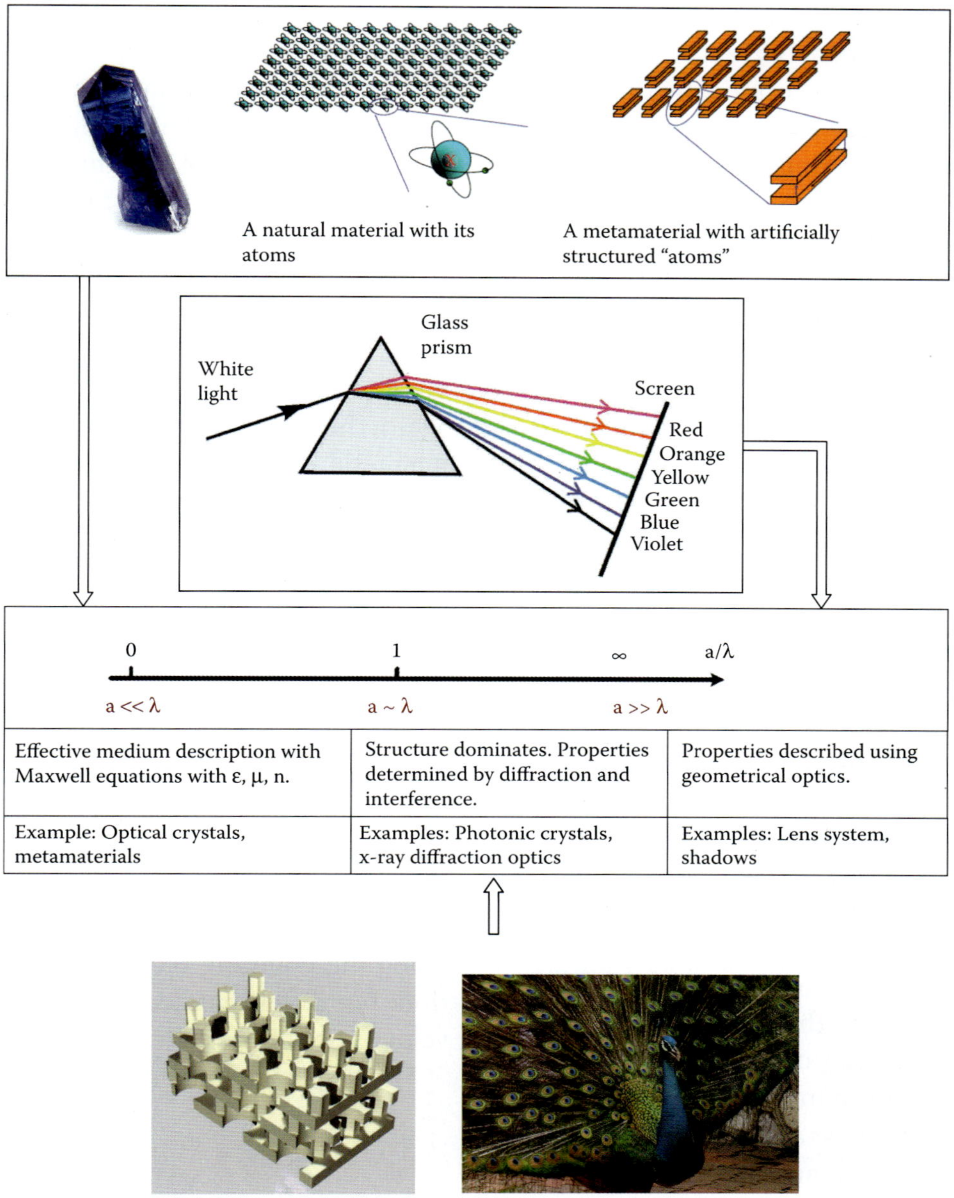

FIGURE 5.131 Electromagnetic properties as a function of the ratio a of the "lattice constant" of a material/structure and λ the wavelength of the incoming light.

wavelength of the incoming light, one defines a photonic crystal, a mesoscale material with subunits bigger than atoms but smaller than the EM wavelength. In photonic crystals *a* is the distance between repeat units with a different dielectric constant, so we can still call it a "lattice constant." Photonic crystals have properties governed by the diffraction of the periodic structures and may exhibit a bandgap for photons. They typically are not described well using effective parameters ε and μ and may be artificial or natural (in the figure we show an artificial 3D photonic crystal and peacock feathers as an example of natural photonic crystals).

- When considering macroscopic optic components with a critical dimension $a >>> \lambda$, geometric optics do apply. In this case *a* might be the repeat unit of a grating, the aperture of a lens, or the length of a side of an optical triangular prism.

Metamaterials are artificially engineered materials possessing properties (e.g., mechanical, optical, electrical) that are not encountered in naturally occurring materials. *Meta* originates from the Greek word μετα: "after/beyond." The dimensions of the unit cell are less than the wavelength of the excitation EM wave; thus, the effective media theorem can be applied. We can design and control the properties of metamaterials. Novel properties, such as negative electric permittivity, negative magnetic permeability, negative refractive index, and so on, have been explored. Photonic crystals often behave like—but are not—true metamaterials, and they do occur in nature. *Metamaterials* is a general EM term: an object that gains its (EM) material properties from its structure rather than inheriting them directly from the materials it is composed of.

The μ-ε Quadrant

We can classify all materials according to their specific conductivity σ, their magnetic permeability μ, and their dielectric constant (or permittivity) ε. Materials with $\sigma > 0$ are conductors, and those with $\sigma \approx 0$ are insulators or dielectrics. All materials have two fundamental EM parameters: permeability μ and permittivity ε, which respectively measure the capacities of a medium to form magnetic and electrical fields. The values of those parameters produce the characteristic bending of a light beam when it travels from one medium into another. In addition, because both parameters are usually positive in nature, the electric and magnetic vector field components are directed according to the "right-hand rule," which can be represented by pointing the index finger of the right hand in the direction of propagation. The thumb and middle finger are then oriented at right angles to the index finger, showing the field vector directions. Mathematically, from the boundary conditions in Maxwell's equations, we concluded earlier that for a positive μ and ε, the vectors **E** and **H** and **k** constitute a right set of vectors and define right-handed materials (RHMs) as illustrated in Figure 5.30a. Most materials do come with $\varepsilon > 1$ (e.g., $\varepsilon = 12$ for Si and 2.25 for glass), but this is not without exceptions; we know that materials with a negative ε at optical frequencies (visible, IR) include metals such as Au, Ag, and Al (see, e.g., Figure 5.68). Most materials have $\mu \approx 1$ at optical frequencies and are nonmagnetic, but if $\mu \neq 1$, one deals with a magnetic material. With $\mu > 1$ we have a paramagnetic material at hand; with $\mu < 1$ it is diamagnetic. (Diamagnetic materials have a relative magnetic permeability that is less than 1, thus a magnetic susceptibility that is less than 0, and are therefore repelled by magnetic fields.) And finally with $\mu \gg 1$ we define a ferromagnetic material, which is associated with hysteresis effects because of fields. Materials with a negative μ are again the exception; they include resonant ferromagnetic or antiferromagnetic systems at microwave frequencies.

In 1968, Victor Veselago, a Russian physicist, showed that if ε and μ are both negative, we are forced by the Maxwell's equations to choose a negative square root for the refractive index:

$$n = -\sqrt{\varepsilon\mu}, \varepsilon < 0, \mu < 0 \qquad (5.267)$$

For such a material with μ and ε negative, the vectors **E** and **H** and **k** form a left set of vectors, and one refers to them as left-handed materials (Figure 5.30b). For right-handed materials, the phase velocity ($v_p = \omega/\mathbf{k}$ along **k**) and group velocity ($v_g = d\omega/d\mathbf{k}$ along the Poynting vector **S**) are in the same direction, but in left-handed materials group and phase velocity are in opposite directions (see Figure 5.32).

The dispersion equation:

$$\mathbf{k}(\omega) = \frac{\omega}{v} = \frac{n(\omega)\omega}{c} = \frac{\omega}{c}\sqrt{\varepsilon\mu} \qquad (5.81)$$

which enables one now to construct a μ-ε plane (see Figure 5.132 for a few different representations of such a diagram). For most natural materials, ε and μ are positive, $n > 0$ (top right), and waves propagate through the medium. If either ε or μ is negative, n is imaginary, and there is no wave propagation through the material. A substance with simultaneously negative values of permittivity ε and permeability μ (bottom left quadrant) does have a negative refraction index, as $n = \sqrt{\varepsilon}\sqrt{\mu} = -\sqrt{|\varepsilon\mu|}$.

Left-Handed Materials

Left-handed materials (LHMs) have EM properties that are distinct from any known material and

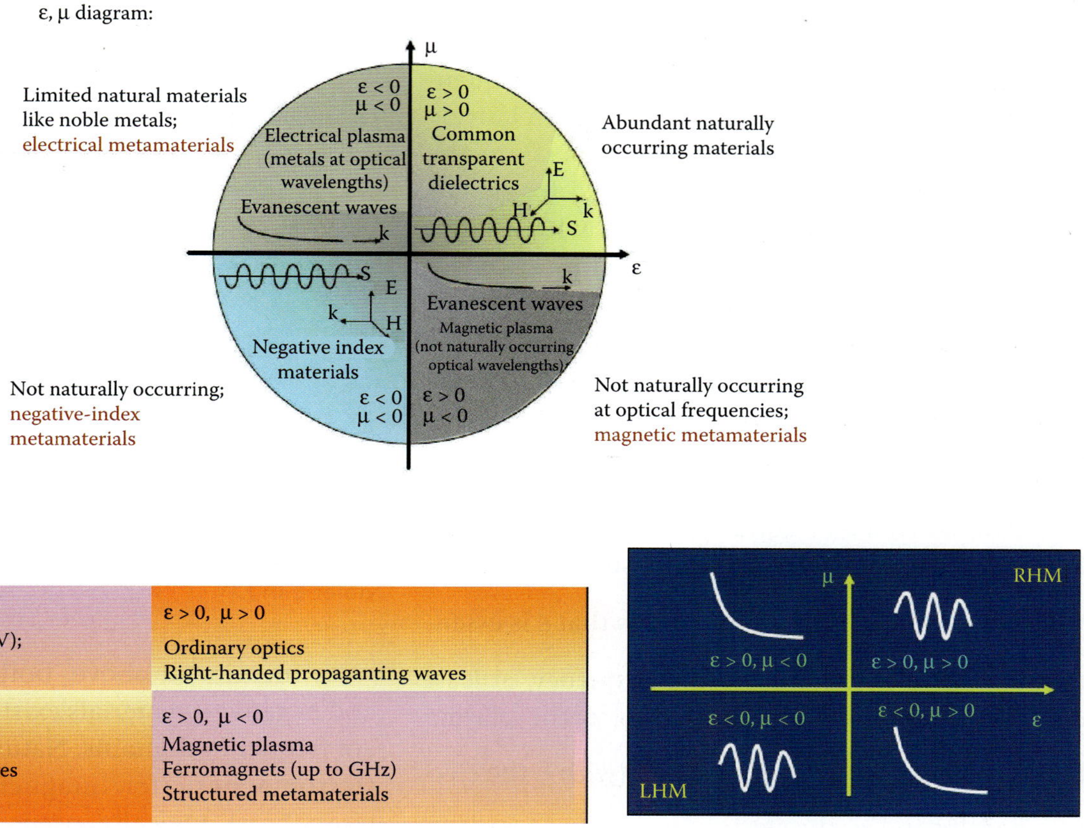

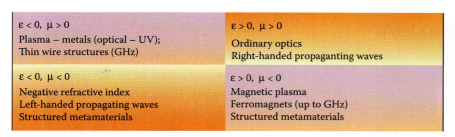

FIGURE 5.132 Various representations of ε-μ plane and typical materials.

hence are examples of metamaterials. Veselago expected that such a left-handed metamaterial would, besides a negative refractive index n, also have a reverse Doppler effect, backward Cerenkov radiation,* and make it possible to design perfect lenses and have a phase velocity directed against the flow of energy. In addition, the direction of the refracted light bends away from the normal to the interface between two media, rather than toward the normal, as in Snell's law (see metawater in Figures 5.17 and 5.133).

There are no known naturally occurring left-handed or negative-index (*n*) materials, but today experiments have shown that artificially designed structures/materials can act as negative refractive index materials in the microwave region and even in the visible.[37]

Veselago's fascinating prediction that perfect lenses could be made from metamaterials lay dormant until about 1996, when John Pendry, a physicist at Imperial College in London, showed that certain metal structures could be engineered to respond to electric fields as though the field parameters were both negative. Most electric resonances for natural materials are in the terahertz region or

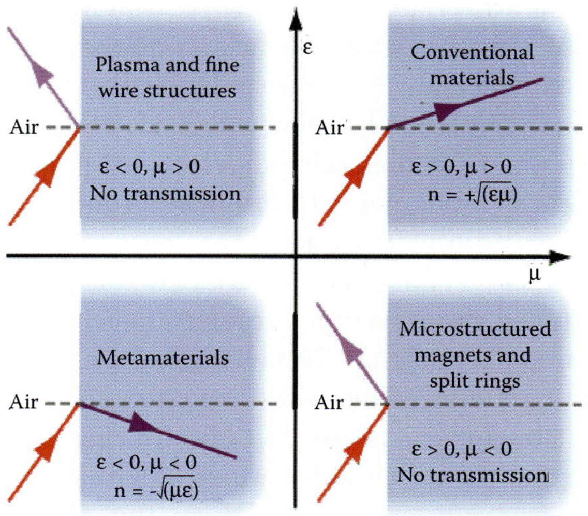

FIGURE 5.133 Refraction and the ε-μ plane.

* Cerenkov radiation (also spelled *Cherenkov*) is electromagnetic radiation emitted when a charged particle passes through an insulator at a speed greater than that of light in the medium.

higher. Magnetic natural systems typically have resonances through the gigahertz (a few magnetic systems have resonances up to terahertz frequencies, e.g., MnF_2, FeF_2). Negative ε, with relatively low loss in the visible region, we know from the Drude-Lorentz model, can be obtained for metals such as Au and Ag. Natural materials with a negative μ also exist; they include resonant ferromagnetic or antiferromagnetic systems at microwave frequencies. Unfortunately, electric and magnetic resonances do not overlap in existing materials, but, as we will see, this restriction does not exist for artificial materials!

The Drude model, we saw, describes the negative dielectric constant of conducting materials and is a basic type of material response with $\varepsilon(\omega) = 1 - \frac{\omega_p^2}{\omega(\omega + i\gamma)}$. The important point is that ε is essentially negative below the plasma frequency, at least down to frequencies comparable to γ. The plasma frequency ω_p is given by $\omega_p^2 = \frac{ne^2}{\varepsilon_0 m_e}$ (with n the electron density and m_e the electron mass), and for many metals falls in the UV region. At lower frequencies, the permittivity of most metals is imaginary. The damping factor γ is given by $\gamma = \frac{\varepsilon_0}{\sigma}$ with σ the electric conductivity. For aluminium ω_p = 15 eV and γ = 0.1 eV. On a macroscale, Pendry et al.[38,39] found a metamaterial with ε < 0 can be fashioned by a simple periodic structure composed of thin, infinite wires arranged in a simple cubic lattice (Figure 5.134). A periodic array (lattice constant a) of thin metal wires (radius r) with $r \ll a \ll \lambda$ acts as a low-frequency plasma. The effective ε is described with a modified ω_p, and the plasma frequency depends on geometry rather than on material properties. Although the concept of using wires to mimic a Drude material had been around in a practical sense since the 1940s (or earlier), Pendry reintroduced the idea in 1996 and noted the important connection with optical plasmon phenomena. With wire spacing a = 5×10^{-3} m and a radius of the wires r of 1.0×10^{-6} m, one calculates an ω_p of about 8.3 GHz (see Figure 5.134). Note that the wire model brings down the plasma frequency substantially, even to the microwave regime, but the same Drude equations apply.

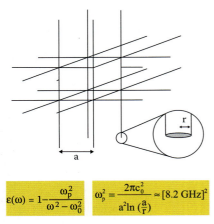

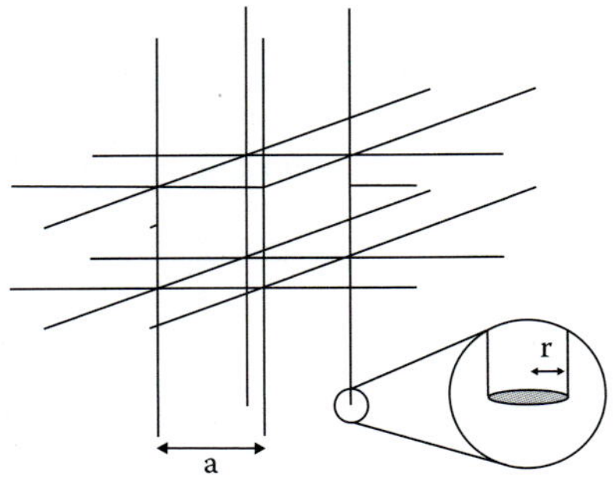

$$\varepsilon(\omega) = 1 - \frac{\omega_p^2}{\omega^2 - \omega_0^2} \qquad \omega_p^2 = \frac{2\pi c_0^2}{a^2 \ln\left(\frac{a}{r}\right)} \approx [8.2\ \text{GHz}]^2$$

FIGURE 5.134 On a macroscale, a metamaterial with ε < 0 can be fashioned by a simple periodic structure composed of thin, infinite wires arranged in a simple lattice, which mimics the response of a plasma. The plasma frequency is in the gigahertz band instead of the ultraviolet band.[39]

As mentioned above, both negative ε materials and negative μ materials exist in nature but in different frequency domains. Natural magnetic responses (ferromagnetic/ferrimagnetic/antiferromagnetic resonances) typically occur at lower than a few gigahertz, and because there are no free magnetic monopoles, there is no such thing as a magnetic plasma and conventional optics (lenses, mirrors, crystals, etc.) essentially have nothing to do with magnetic fields. Pendry et al.[40] also discussed how to obtain negative permeability in metamaterials. Magnetism in materials originates from the orbital motion of electrons and unpaired electron spins. The magnetic response of most natural materials fades away in the gigahertz region. But Pendry realized that conducting, nonmagnetic split-ring resonators (SRRs) can realize artificial magnetism even in nonmagnetic Cu. The magnetic flux-induced current loops form magnetic dipoles, and the intrinsic conductance and inductance of the split rings cause strong paramagnetic or diamagnetic activity around the resonance frequency. The magnetic response can be extended to the terahertz range or even higher frequencies with large positive or negative permeability (see Figure 5.135).

By 2001, researchers at Imperial College and Marconi Caswell (London) announced a magnetic resonance medical imaging system using a magnetic metamaterial based on Pendry's design (Wiltshire

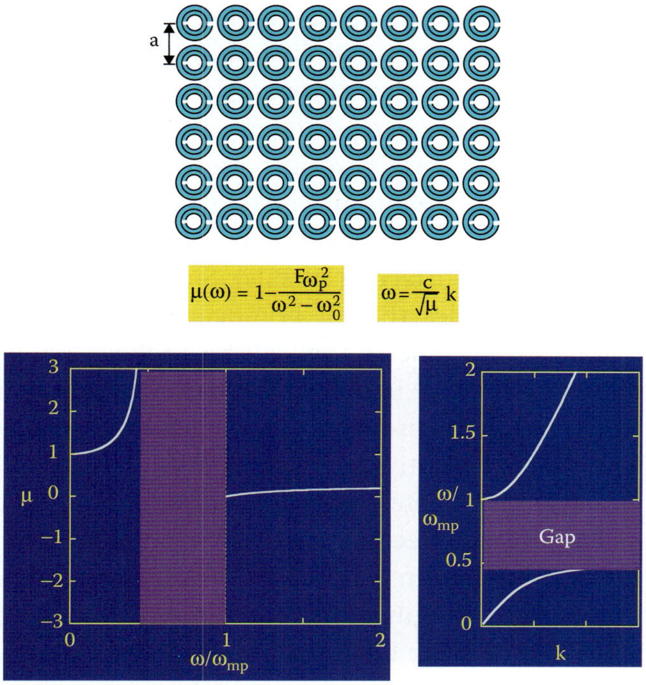

FIGURE 5.135 Pendry's split rings result in a negative μ.

et al.[41]). This group stated that "exploiting this class of materials could fundamentally change existing approaches to magnetic resonance imaging and spectroscopy."

Also in 2001, 3 years after Pendry's split ring proposal and close to 40 years after Veselago's first paper on negative refraction, physicist Richard Shelby's group at the University of California, San Diego demonstrated a left-handed composite metamaterial that exhibited a negative index of refraction for microwaves.[42] In Figure 5.136a we show two embodiments of these composite microwave metamaterial structures. The device pictured in the left photo shows sets of SRRs on common substrates alternating with separate rows of wires. The San Diego group reported that they had followed Pendry's prescriptions and succeeded in constructing a material with both a negative μ and a negative ε at microwave frequencies. The raw materials used, copper wires and copper rings, do not have unusual properties of their own and are nonmagnetic. But, when incoming microwaves fall on these alternating rows of rings and wires, a resonant reaction between the light and the whole of the ring-and-wire array sets up tiny induced currents, which contribute fields of their own. The net result is a set of fields moving to the left, even as EM energy is moving to the right. Therefore, this effective medium is a left-handed material. In Figure 5.136b a transmission measurement is shown for this metamaterial. The results show that there is a frequency region where both ε and μ are negative.

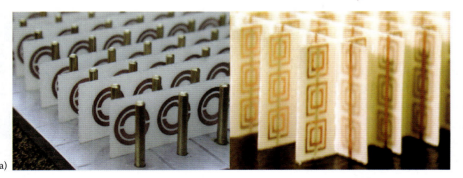

In the photo on the right in Figure 5.136a, we show a more integrated microwave metamaterial with wire and SRR on the same substrate. The latter arrangement consists of a planar pattern of copper SRRs and wires on a thin fiberglass circuit board. The SRRs and wires are arranged into a 2D structure with a repeated 5-mm lattice, with the wires located on the opposite side of the circuit board from the SRRs. This material exhibits negative refractive index at around 10.5 GHz. The San Diego team also demonstrated that microwaves were refracted to the opposite side (negative angles) of the normal as expected for a left-handed sample. From the measured angle, a refractive index of −2.7 ± 0.1 was derived.

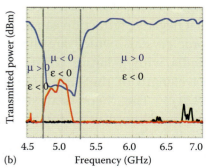

FIGURE 5.136 (a) Two embodiments of composite metamaterial structures. The structure on the left shows an SRR on a circuit board substrate and separate wires, whereas the structure on right is more integrated with wire and SRR on the same substrate.[43,44] (b) Transmission spectrum for the metamaterial on the left in (a).[44]

These metamaterials might have applications in microwave communications: a medium that focuses waves when other materials would disperse them (and vice versa) ought to be useful in improving existing delay lines, antennas, and filters.

Development of metamaterials that refract light at ever-shorter wavelengths has been fast. A negative refractive index material for the optical range, specifically for wavelengths close to 1.5 μm (200-THz frequency), was accomplished by Shalaev et al. in 2005[45] with a metal dielectric composite consisting of a 2 × 2-mm array of nanorods on a glass substrate fabricated using electron beam lithography. A "fishnet" design developed by Costas Soukoulis' group at the US Department of Energy's Ames Laboratory and produced by researchers Stefan Linden and Martin Wegener of the University of Karlsruhe was made by etching an array of holes into layers of silver and magnesium fluoride on a glass substrate.[46] This constituted the first negative refractive index metamaterial with a refractive index of −0.6 at the far red end of the visible spectrum (wavelength 780 nm). This discovery, detailed in the January 5, 2007 issue of *Science* and the January 1 issue of *Optic Letters* and noted in the journal *Nature*, marked a significant step forward from existing metamaterials that operate in the microwave or far infrared—but still invisible—regions of the spectrum.[46]

Thus, negative refraction is clearly demonstrated in Pendry-type quasihomogeneous metamaterials with unit cell dimensions less than the wavelength of the incoming radiation, consisting of interleaved arrays of wires ($\varepsilon < 0$) and SRRs ($\mu < 0$). In 2000, Masaya Notomi at NTT Basic Research Laboratories in Japan presented strong theoretical evidence that negative refraction does not imply large losses and that negative refraction can also be achieved in photonic crystals, which are inhomogeneous periodic media with a lattice constant comparable to the wavelength of the incoming radiation (see Figure 5.131).[47] The theoretical structures that Notomi imagined could in reality be composed of insulating dielectric materials, which have negligible losses even at optical wavelengths. In 2003, Cubukcu et al.[36] experimentally demonstrated negative refraction by a photonic crystal (a square array of alumina rods in air).

The physical principles that allow negative refraction in a photonic crystal arise from the dispersion characteristics of wave propagation in a periodic medium and are very different from that of the Pendry metamaterials we just studied. The dispersion surface in a photonic crystal becomes rounded near the photonic bandgap (similar to the behavior of a Fermi surface in metals). Such a dispersion surface resembles that of homogeneous dielectric materials; thus, Snell's Law can be applied, and an effective refractive index can be defined. However, the propagation direction can be negative because $d\omega/dk$ can be negative in photonic crystals. Effective refractive index states in photonic crystals have their analogy in the effective-mass states in semiconductors as illustrated in Figure 5.137. For many applications such as imaging, one requires index matching between a negative refractive index material and its surroundings so as to incur negligible losses. These requirements are more easily met with photonic crystals than with composite negative index metamaterials. Furthermore, a microwave photonic crystals can be more easily scaled to three dimensions and to optical frequencies, which is more difficult with composite negative index metamaterials.

The Ideal Lens

The performance of conventional lenses is limited by diffraction: they need a wide aperture NA for good resolution, but even so their resolution is

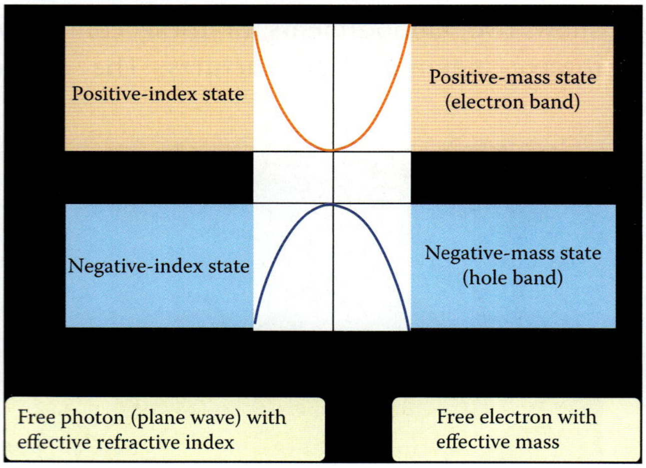

FIGURE 5.137 Effective refractive index states in photonic crystals have their analogy in the effective-mass states in semiconductors.

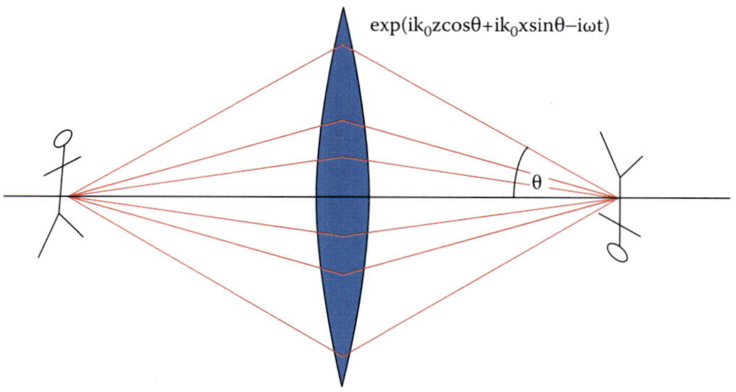

FIGURE 5.138 The resolution limit of a traditional lens.

limited by the wavelength used (Figure 5.138). The contribution to the resolution of an image from the far field is limited by the free space wavelength λ_0. Considering an object and a lens placed along the *z*-axis with rays from the object travelling in the +*z*-direction, we know that with $\theta = 90°$ (see Abbe's limit in Equation 5.1), we get a maximum value of the wave vector $\mathbf{k}_0 = \omega/c_0 = 2\pi/\lambda_0$—the shortest wavelength component of the image. Hence the resolution *R* cannot be better than:

$$R \approx \frac{2\pi}{\mathbf{k}_0} = \frac{2\pi c_0}{\omega} = \lambda_0 \tag{5.268}$$

Contributions from the near field to the image come from large values of $\mathbf{k}_x$ (or $\mathbf{k}_y$) responsible for the finest details in the source. The light emanating from the object is made up of a superposition of plane waves, and the resulting $\mathbf{k}_z$ wave vector may be written as of function of $\mathbf{k}_x$ and $\mathbf{k}_y$ as follows:

$$\mathbf{k}_z = \sqrt{\frac{\omega^2}{c_0^2} - (\mathbf{k}_x^2 + \mathbf{k}_y^2)} \tag{5.269}$$

In the far field, with energy going in the +*z*-direction, only the positive square root is taken, and all the components of the angular spectrum of the image for which $\mathbf{k}_z$ is real are transmitted through a conventional lens and refocused. However, if the second term under the square root in Equation 5.269 is larger than the first, $\mathbf{k}_z$ becomes imaginary, and the wave turns evanescent with an amplitude decaying as it propagates along the *z*-axis. What this really means is that the high-angular frequency components of the image are lost, resulting in an inability to resolve the smallest features of the object. The near-field, high-information content contributions to the image are not transmitted by the lens. For that, a lens with an NA of 1 would be required.

It is here that Pendry again had an amazing insight*: a superlens can be made from a flat slab of negative refractive index material that not only brings rays to a focus but also has the capacity to amplify the near field so that it can contribute to the image, thus removing the wavelength limitation. With the Pendry superlens with a refractive index $n = -1$ ($\varepsilon = -1$ and $\mu = -1$), transport of light in the +*z*-direction requires now that $\mathbf{k}_z$ in Equation 5.269 comes with the opposite sign (−). The evanescent wave now grows instead of decaying, and with the proper lens thickness all components of the angular spectrum are transmitted through such an ideal lens undistorted. A superlens prevents image degradation and beats the diffraction limit established by Abbe! However, the resonant nature of the amplification places severe demands on materials: they must be very low loss! The latter explains why there is so much interest in using photonic crystals rather than metamaterials for superlensing.

In Figure 5.139 we compare imaging with a traditional and a "Veselago" lens. RH and LH stand for right-handed and left-handed, respectively. A negative refractive index medium (LH) bends light to a negative angle relative to the surface normal. Light formerly diverging from a point source is set in reverse and converges back to a point. Released from the medium, the light reaches a focus for a second time. Note that the parallel component of the wave vector is always preserved in transmission but that the energy flow is opposite to the wave vector. The new Pendry lens based on negative refraction has unlimited resolution, provided that the condition $n = -1$ is met exactly. This can happen only at one frequency. Therefore, the secret of the new lens is that it can focus the near field, and to do this it must amplify the highly localized near field to reproduce the correct amplitude at the image. This

* "All this was pointed out by Veselago some time ago. The new message in this Letter is that, remarkably, the medium can also cancel the decay of evanescent waves. The challenge here is that such waves decay in amplitude, not in phase, as they propagate away from the object plane. Therefore to focus them we need to amplify them rather than to correct their phase. We shall show that evanescent waves emerge from the far side of the medium enhanced in amplitude by the transmission process." J.B. Pendry.[54]

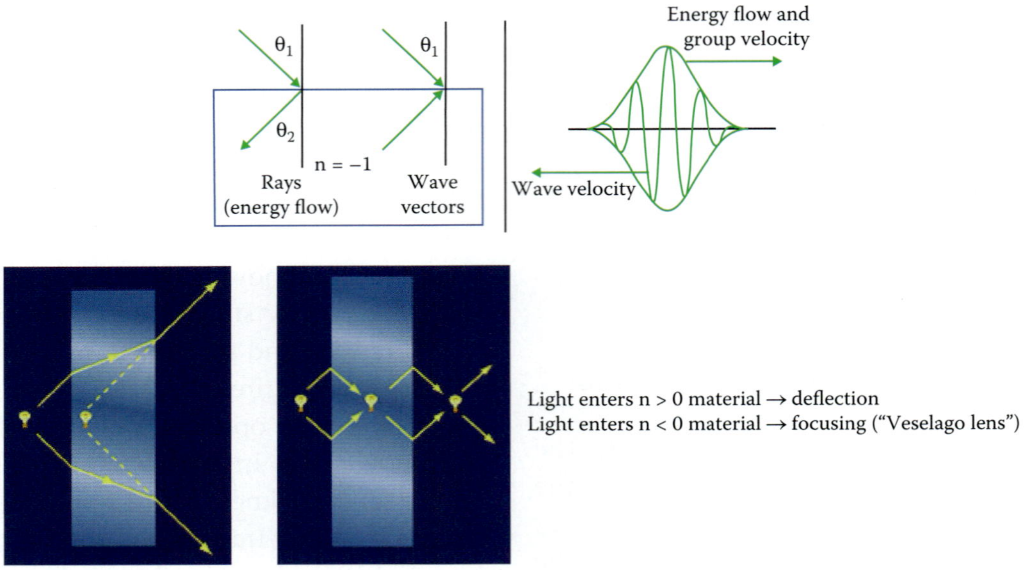

FIGURE 5.139 Comparing imaging with a traditional and a "Veselago" lens. RH and LH stand for right-handed and left-handed, respectively.

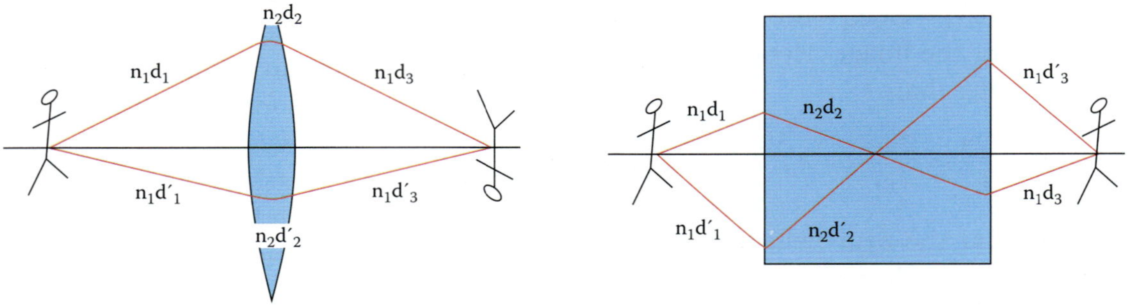

FIGURE 5.140 Fermat's principle applied to right-handed and left-handed lenses.

can be understood from Fermat's principle that light takes the shortest optical path between two points, as illustrated in Figure 5.140. For a traditional lens, the shortest optical distance between object and image is:

$$n_1d_1 + n_2d_2 + n_1d_3 = n_1d'_1 + n_2d'_2 + n_1d'_3 \quad (5.270)$$

Both paths converge at the same point because both correspond to a minimum. In the Pendry lens n_2 is negative, and the ray transverses negative optical space.

For a perfect lens, the shortest optical distance between object and image is zero:

$$n_1d_1 + n_2d_2 + n_1d_3 = 0 = n_1d'_1 + n_2d'_2 + n_1d'_3 \quad (5.271)$$

In other words, for a perfect lens the image is the object. The evanescent waves are "regrown" in a negative refractive index slab and fully recovered at the image plane as illustrated in Figure 5.141.

A poor man's near-field superlens ($\varepsilon < 0$ but $\mu = 1$) has already been demonstrated. In 2003, Zhang's group at the University of California, Berkeley

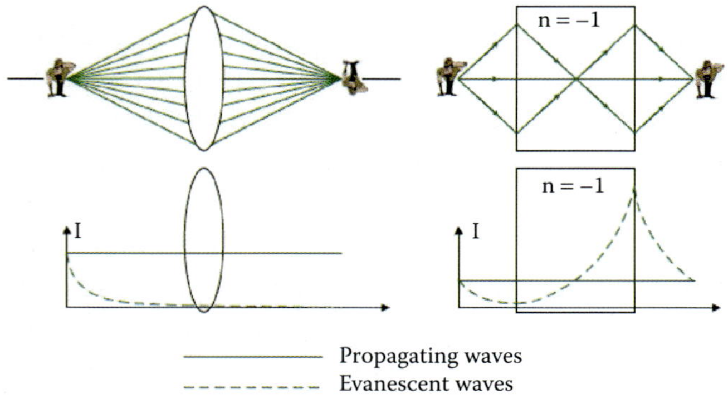

FIGURE 5.141 For a perfect lens, the image is the object. The evanescent waves are "regrown" in a negative refractive index slab and fully recovered at the image plane.

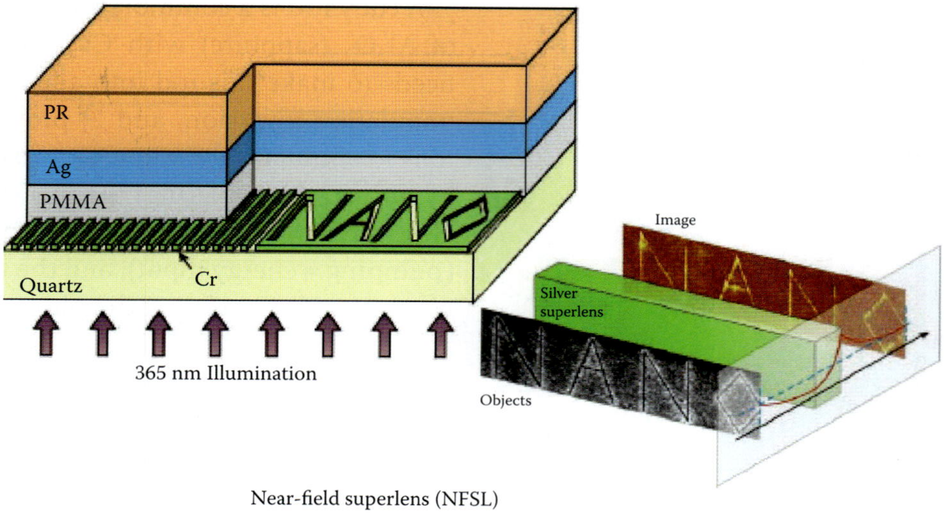

FIGURE 5.142 This poor man's superlens images 10-nm features with 365-nm light.[49]

showed that optical evanescent waves could be enhanced as they passed through a silver superlens.[48] The same group took this work one step further and imaged objects as small as 40 nm across with blue light at 365 nm with a superlens that was just 35-nm thick (see Figure 5.142).[49] In contrast, current optical microscopes can only resolve objects down to around 400 nm. With the Zhang's poor man's superlens at 365 nm, illumination features of a few tens of 10 nanometers were imaged using 365-nm light, clearly breaking Abbe's diffraction limit.

Lasers

Introduction: Masers and Lasers

LASER stands for Light Amplification by the Stimulated Emission of Radiation. The key word here is "stimulated." Most of the light emission we have mentioned so far is spontaneous; it happened as a result of randomly occurring "natural" effects. Stimulated emission refers to electron transitions that are "encouraged" by the presence of other photons. Albert Einstein put a mathematical analysis of stimulated emission forward as far back as 1916 that showed that an incident photon with $E \geq E_g$ (with E_g the semiconductor bandgap) was equally likely to cause stimulated emission of light as to be absorbed (Figure 5.143).

The emitted light has the same energy and phase as the incident light; in other words, it is coherent. Under normal circumstances, there are few excited electrons, and most are in the ground state, so we get predominantly absorption of light. If we could arrange for more excited than nonexcited electrons, then we would get mostly stimulated emission. Such a condition is called a *population inversion*. Because we get more photons out than we put in, this constitutes optical amplification, and the chain reaction of stimulated emissions is referred to as *amplification by stimulated emission of radiation*, or ASER for short. It took about 40 years after Einstein's initial theory was put forward until a technique was invented to sustain a chain reaction of stimulated emissions. The stimulated emissions system was first used in 1954 to amplify microwaves for communications by James Gordon, H.J. Zeiger, and Charles Townes in a so-called *Maser*, the microwave counterpart of a laser. It was not until 1960 when the first laser was successfully tested. The first laser used a ruby rod, giving off a beam of red laser light, the same kind one sees at a grocery checkout counter. Applications of laser technology went into full swing only in the early 1980s. Clearly, random spontaneous emission "wastes" electron transitions by giving an incoherent

FIGURE 5.143 An incident photon with $E \geq E_g$ is equally likely to cause stimulated emission of light to be absorbed.

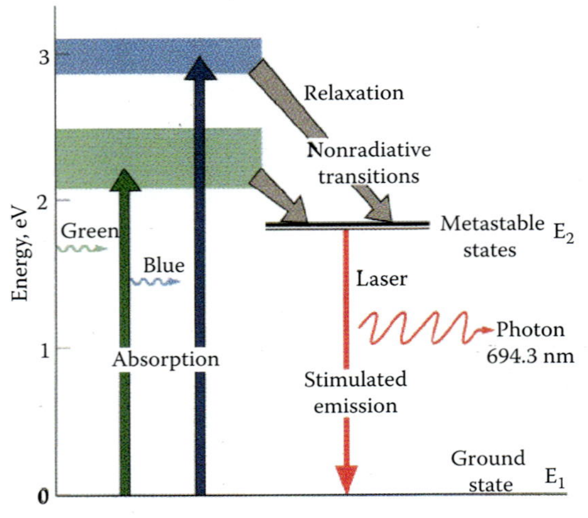

FIGURE 5.144 The energy levels of a typical laser material with metastable states at E_2.

output, so one must minimize those random transitions by using transitions for which the spontaneous emissions are of low probability, i.e., so-called *metastable states*. Therefore, the energy levels of a typical laser material look as we sketched in Figure 5.144, with E_2 representing the metastable states.

A ruby rod is a common laser material; it consists of Al_2O_3 (sapphire) with Cr^{3+} impurities. All one needs to make this rod into a laser is to 1) achieve population inversion, and 2) provide enough photons to stimulate emission. The first objective is met by filling (or populating) the metastable states with electrons generated by light from a xenon flash lamp (pumping = energy input), and the second condition is achieved by confining the photons such that they are forced to travel back and forth along the ruby rod of ruby by equipping the rod with mirrored ends. To keep the coherent emission, we must ensure that the light that completes a round trip between the mirrors returns in phase with itself. Hence the distance between the mirrors should obey $2L = N\lambda$, where N is an integer, λ is the laser wavelength, and L is the cavity length. Semiconductor lasers work in just the same way except that they achieve the population inversion electrically by using a carefully designed band structure. The ruby laser has an output at 694.3 nm. The basic lasing process is summarized in Figure 5.145.

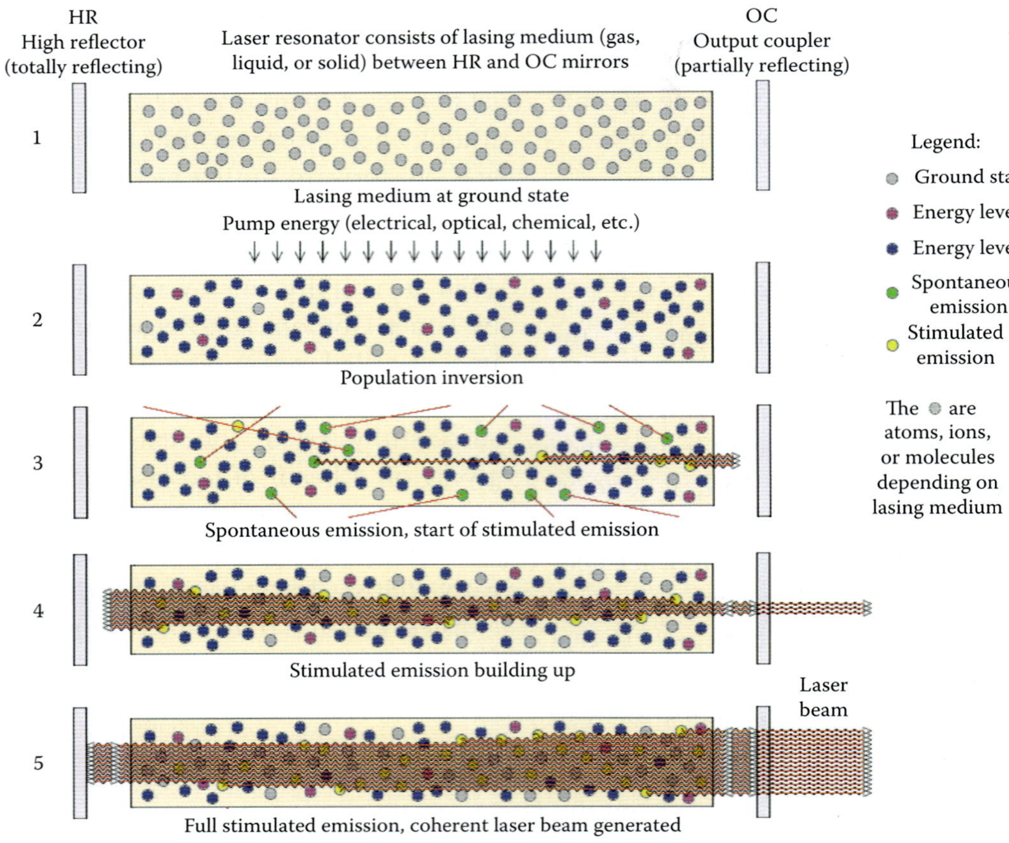

FIGURE 5.145 The basic lasing process.

From this figure, a laser comprises:

1. An active medium: the active medium may be made up by solid crystals such as ruby or neodymium-doped/yttrium aluminum garnet (Nd:YAG), liquid dyes, gases like CO_2 or helium/neon, or semiconductors such as GaAs. Active mediums contain atoms whose electrons may be excited to a metastable energy level by an energy source.
2. Excitation mechanism: excitation mechanisms pump energy into the active medium by one or more of three basic methods: optical, electrical, or chemical.
3. High-reflectance mirror: a mirror that reflects essentially 100% of the laser light.
4. Partially transmissive mirror: a mirror that reflects less than 100% of the laser light and transmits the remainder.

A laser diode is a laser where the active medium is a semiconductor, and it is similar to the light-emitting diode (LED) described in Chapter 4. Compared with other laser types, semiconductor lasers are compact, reliable, and last a very long time. The most common laser diode is formed from a p-n junction, and the excitation mechanism is the injection of an electrical current (Figure 5.146). When electron and holes find themselves present in the same region of the solid, they may recombine, resulting in spontaneous emission. This spontaneous emission is necessary to initiate the lasing action, but it is also one among several sources of inefficiency once the laser is working. It is also this spontaneous emission that makes a laser diode, in the regime below the lasing threshold, behave like an LED.

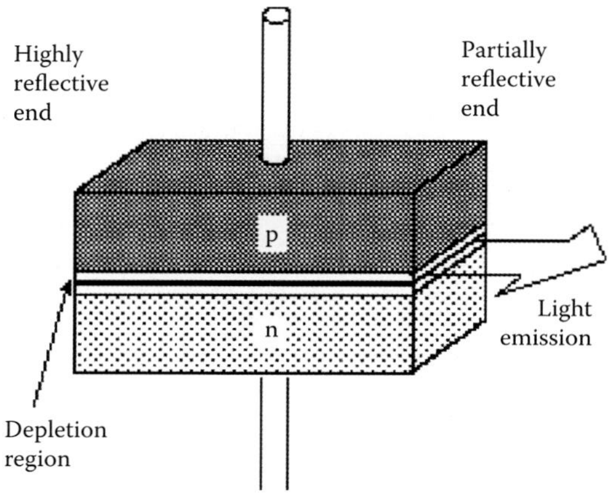

FIGURE 5.146 Simple laser diode.

In a conventional semiconductor junction diode, the energy released from the recombination of electrons and holes is carried away as phonons, i.e., lattice vibrations, rather than as the emerging photons. The difference between the photon-emitting semiconductor lasers and LEDs (on one hand) and conventional phonon-emitting (nonlight-emitting) semiconductor junction diodes (on the other hand) lies mostly in the type of semiconductor that is used in their construction. For lasers and LEDs we need a "direct bandgap" semiconductor such as gallium arsenide. Diode lasers then consist of these two basic components: an optical amplifier and a resonator. The resonator continuously recirculates light through the amplifier region and helps to focus it. The amplifier consists of the direct bandgap semiconductor junction. The two ends of the resonator structure need to be optically flat and parallel with one end mirrored and one partially reflective (see Figure 5.146). The length of the resonator is precisely related to the wavelength of the light to be emitted. The junction is forward biased, and the recombination process produces light as in an LED (incoherent). Above a certain current threshold (I_{th}) a photon moving parallel to the junction with energy equal to the recombination energy causes recombination by stimulated emission. This generates another photon of the same frequency, traveling in the same direction, with the same polarization and phase as the first photon. This means that stimulated emission causes gain in an optical wave (of the correct wavelength) in the injection region, and the gain increases as the number of electrons and holes injected across the junction increases. Finally, if there is more amplification than loss, the diode begins to "lase." Photons emitted into a mode of the waveguide will travel along the waveguide and be reflected several times from each end face before they are emitted. However, light is also lost because of absorption and by incomplete reflection from the end facets.

The spontaneous and stimulated emission processes are vastly more efficient in direct bandgap semiconductors than in indirect bandgap semiconductors; thus, silicon is not a common material for laser diodes.

Some important properties of laser diodes are determined by the geometry of the optical cavity. Generally, in the vertical direction, the light is contained in a very thin layer, and the structure supports only a single optical mode in the direction perpendicular to the layers. In the lateral direction, if the waveguide is wide compared with the wavelength of light, then the waveguide can support multiple lateral optical modes, and the laser is known as *multimode.*

Vertical cavity surface-emitting lasers (VCSELs) have the optical cavity axis along the direction of the current flow rather than perpendicular to the current flow as in conventional laser diodes (Figures 5.146 and 5.147). The active region in this case is very short compared with the lateral dimensions, so that the radiation emerges from the surface of the cavity rather than from its edge. The reflectors at the ends of the cavity are dielectric mirrors made from alternating high- and low-refractive index quarter-wave thick multilayer. Such dielectric mirrors provide a high degree of wavelength-selective reflectance at the required free surface wavelength λ if the thicknesses of alternating layers d_1 and d_2 with refractive indices n_1 and n_2 are such that $n_1d_1 + n_2d_2 = \frac{1}{2}\lambda$, which then leads to the constructive interference of all partially reflected waves at the interfaces. However, there is a disadvantage because of the high mirror reflectivities; VCSELs have lower output powers when compared with edge-emitting lasers.

The laser gain g is the amplification factor by which the radiation intensity is multiplied after traversing a certain distance L of active laser medium. If the gain g is positive, the intensity increases exponentially as a function of distance:

$$I = I_0 e^{(g \times L)} \tag{5.272}$$

with g in units of inverse distance. In other words, when the absorption coefficient $\alpha > 0$, we are dealing with absorption, and with $\alpha < 0$, we have a gain g (so $g = -\alpha$ is used for negative absorption). The increase of intensity in a laser is much greater than can be obtained by spontaneous emission mechanisms. The threshold of a laser is the state where the gain just equals the cavity losses. For electrically pumped lasers this corresponds to a certain threshold current I_{th}.

Quantum Mechanics and Lasers

Quantum Confinement

Lasers work because of the quantum properties of photons—one photon tends to cause another to be emitted—and one photon cannot be distinguished from another (they are bosons; see Chapter 3). If there are many excited atoms, the photons can "cascade," and very intense, collimated light is emitted, forming a beam of precisely the same color of light. Because photons cannot be distinguished, which atom emitted a given photon is completely uncertain, but that also means that the direction and energy of the photons can be very certain.

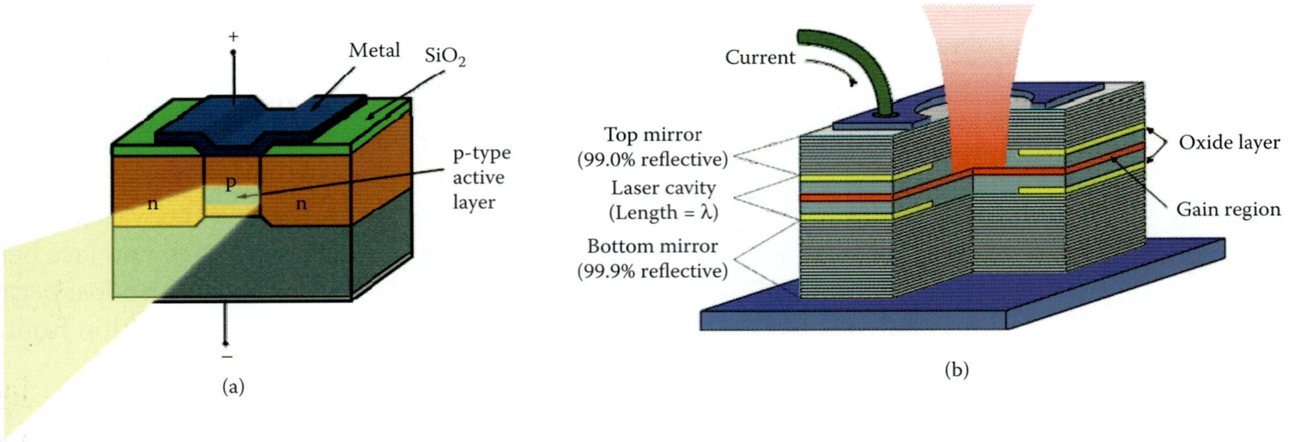

FIGURE 5.147 Comparison of an edge-emitting laser diode (a) and a vertical cavity surface-emitting laser (VCSEL) (b).

In a quantum well, we saw in Chapter 3, there are no allowed electron states at the very lowest energies (zero-point energy). What that means is that the lowest allowed conduction state in a quantum well is broad so that many electrons can be accommodated. Similarly, the top of the valence band has plenty of states available for holes. This means that it is possible for many holes and electrons to combine and produce photons with identical energy for enhanced probability of stimulated emission (lasing), making the quantum well laser much more efficient than a diode laser. The two important reasons that better lasers are enabled with lower-dimensionality semiconductors is the quantum confinement of carriers, which leads to a concentrated density of states, and a sharper energy distribution. The carrier distribution is given by n(E) = G(E)f(E) (see Equation 3.230), and bulk semiconductors have a broad carrier distribution compared with lower-dimensional semiconductor structures. Arakawa and Sakaki predicted significant increases in the gain and decreases in the threshold current I_{th} of semiconductor lasers using QDs or boxes in the active layer of the laser.[50] The gain g predicted by Asada et al.[51] for 3D, 2D, 1D, and 0D semiconductor structures is shown in Figure 5.148.

Quantum Cascade Lasers

Esaki was not only the originator of the Esaki tunnel diode, covered in Chapter 4, but also the originator

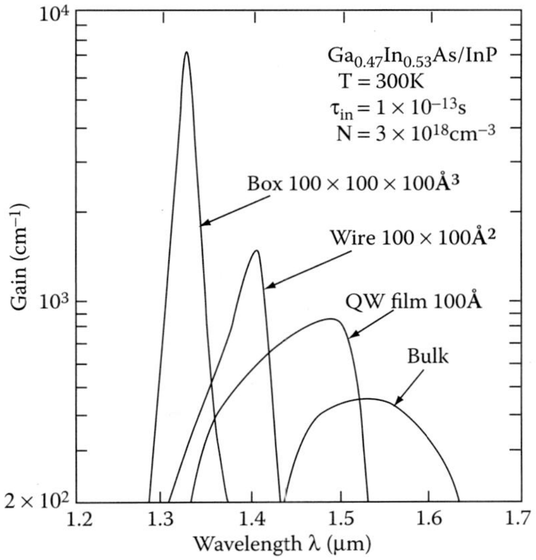

FIGURE 5.148 Gain for 3D, 2D, 1D, and 0D semiconductor structures proposed by Asada et al.[51]

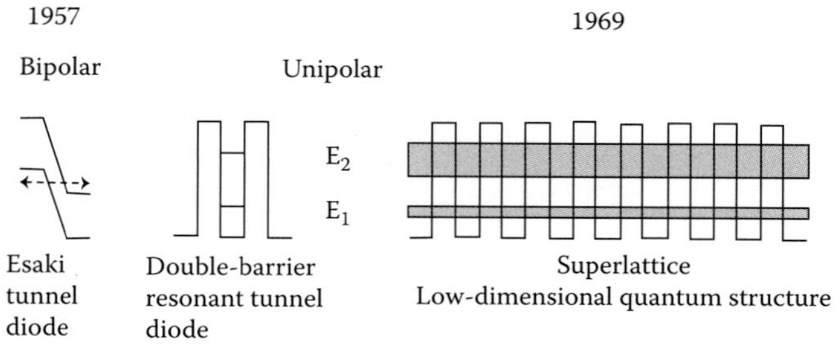

FIGURE 5.149 Esaki inventions from the 1957 Esaki tunnel diode to the double-barrier resonant diode to the superlattice. In bipolar devices, electrons and holes are involved; in unipolar devices, only one type of charged particle is involved.

of superlattices (1969) and double-barrier resonant tunneling diodes (1974). The resonant tunnel diode has even more spectacular characteristics than the Esaki tunnel diode and constitutes one of the possible contenders to replace CMOS technology. In Figure 5.149 we summarize the intellectual path followed by Esaki and colleagues in going from the first tunnel diode to the double-barrier resonant tunnel diode and finally to low-dimensional quantum structures.

Before arriving at the superlattice concept, Esaki and Tsu examined the feasibility of structural formation of potential barriers and wells that were thin enough to exhibit resonant tunneling (see Figure 5.149).

Resonant tunneling in quantum mechanics is an effect observed in quantum wells that are coupled to electron reservoirs by thin tunnel barriers and corresponds to a resonant enhancement of transmission that is seen whenever the energy in the reservoirs matches one of the quantized values in the well. With barriers that are thin enough, an electron with energy lower than that of the height of the barrier can tunnel through the barrier, as long as there is an empty state of the same energy level waiting on the other side. The well is then "in resonance." Resonant tunneling can be compared to the transmission of an EM wave through a Fabry-Pérot resonator.

The double-barrier resonant tunneling diode shown in Figure 5.150 is a variant of the tunnel diode of Chapter 4, with thin barrier layers of GaAlAs and a GaAs quantum well with sharp interfaces and layers with a width comparable with the Schrödinger wavelengths of the electrons

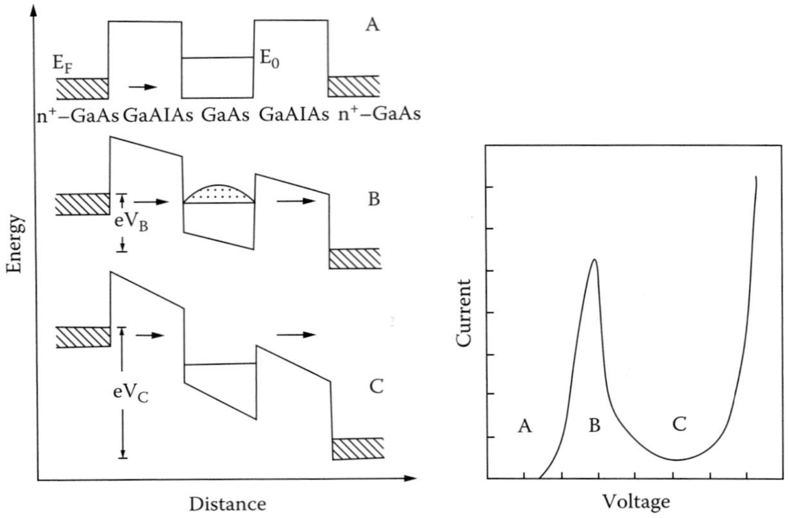

FIGURE 5.150 Semiconductor resonant tunneling diode: device schematic and current voltage plot.

permitting resonant behavior. In the device schematic we see that the contacts/leads are made of highly doped n-type GaAs (n^+ GaAs) and that two thin GaAlAs barrier layers (e.g., 20-Å thick) enclose a thin GaAs well region (e.g., 50-Å thick). The narrower bandgap material GaAs is enclosed by the considerably larger bandgap of GaAlAs to establish potential barriers at the surface of the GaAs well. Because the GaAs well is thin, the motion of electrons and holes is restricted in one dimension and is forced to occupy discrete states of energy instead of staying arbitrarily within an energy continuum. With a low bias applied, the quantized energy level in the well E_o is situated above E_F, and no current flows (A in the schematic and the current voltage plot). At a medium bias, the level in the well has lined up with the conduction band, and current flows (B in the schematic and the current voltage plot). At yet higher bias, the energy level in the well is now situated below the bottom of the conduction band, and the current decreases (C in the schematic and the current voltage plot). The current voltage plot is typical for a tunnel diode (see Figure 5.150), but whereas the Esaki diode is a bipolar device (two types of charge carriers are involved: electrons and holes) and features relatively large dimensions, the resonant tunnel diode is a unipolar device (only one type of charge carrier) with layers that have a thickness comparable with the Schrödinger wavelengths of the electrons.

A superlattice is a material with periodically alternating layers of different substances. Such structures possess periodicity both on the scale of each layer's crystal lattice and on the scale of the alternating layers. Quantum superlattices are a type of superlattice featuring arrays of periodic quantum structures: quantum wells, quantum dots, or quantum wires. Esaki and Tsu first proposed research on such artificially structured materials in 1960 to the Army Research Office (ARO). They envisioned engineered semiconductor superlattices based on a periodic structure of alternating layers of semiconductor materials with wide and narrow bandgaps using high-precision heteroepitaxial deposition methods such as molecular beam epitaxy (MBE) and metalorganic chemical vapor deposition (MOCVD) and involving III–V semiconductors. Although the proposal was favorably received by ARO, *Physical Review* rejected their first paper on the basis that it was "too speculative" and involved "no new physics." In some circles, the ARO proposal was criticized as impossible to implement in practice. The main objection being that a human-made structure with compositional variations on the order of a few nanometers would be thermodynamically unstable because of interdiffusional effects. Fortunately, it turned out that interdiffusion is negligible at the temperatures involved, and innovations and improvements in epitaxial techniques such as MBE and MOCVD made it possible to prepare high-quality heterostructures with predesigned potential profiles and impurity distributions with dimensional control close to the interatomic spacing. This great precision opened up the mesoscopic quantum regime to practical experimentation. Because a 1D potential could now be introduced along with the growth direction in the laboratory, famous examples in the history of 1D mathematical physics, including the above-mentioned resonant tunneling (1930s), Krönig-Penney bands (1931), Zener band-to-band tunneling (1934), and so on, all of which had remained textbook exercises, could for the first time be practiced in the laboratory.

The superlattice example shown in Figure 5.151A is formed by a periodic array of quantum wells produced by the growth of alternate layers of a wide-bandgap (AlGaAs) and narrower bandgap (GaAs) semiconductors in the growth direction. At 300 K, the bandgap of GaAs is 1.43 eV, whereas it is 1.79 eV for $Al_xGa_{1-x}As$ (x = 0.3). Thus, the electrons and holes in GaAs are confined in a 1D potential well of length L in the *z*-direction. Two or more quantum wells side by side give rise to a multiple quantum well (MQW) structure. This is called a *compositional superlattice.* An artificial periodic potential in a semiconductor crystal can also be fabricated through periodic n-type and p-type doped layers as shown in Figure 5.151B. Such a superlattice is called a *doping superlattice.* When the individual quantum wells are widely separated, the wave functions of the electrons and holes remain confined within the individual wells, and the array can be treated as a set of isolated quantum wells. Noninteracting MQWs like this are used to enhance/amplify absorption or emission of photons compared with what is possible with a single quantum well. In lasing, for example,

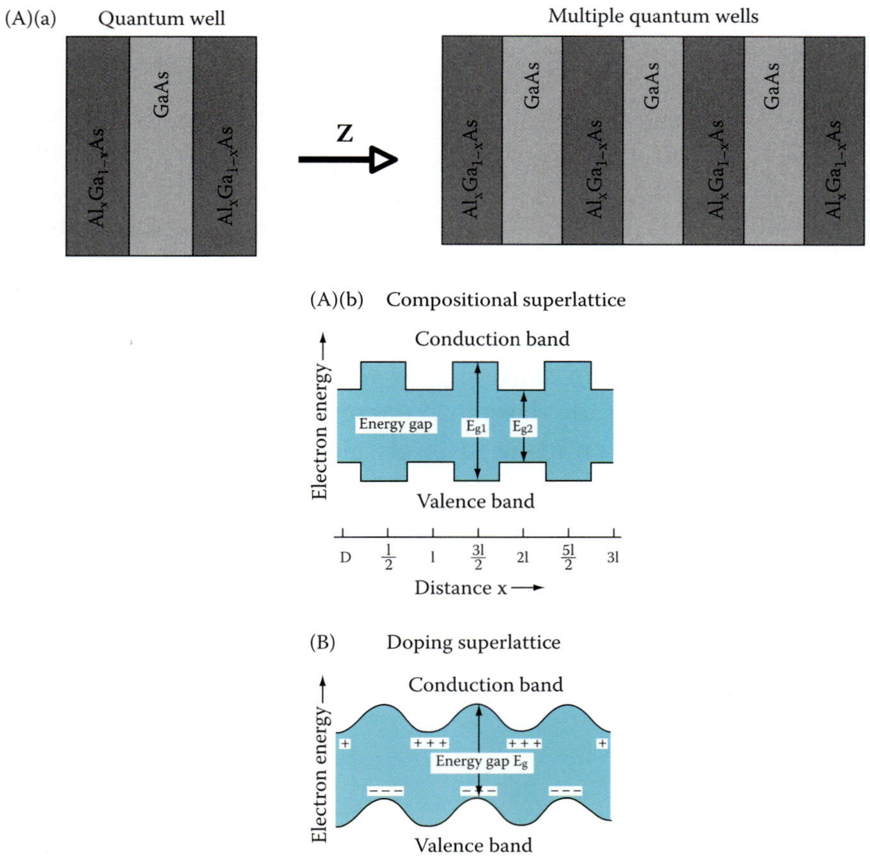

FIGURE 5.151 (A) Superlattice formed by a periodic array of quantum wells produced by the growth of alternate layers of wider bandgap (AlGaAs) and narrower bandgap (GaAs) semiconductors in the growth direction. Schematics of the arrangement (a) and the energy bands of multiple quantum wells (b). This is a compositional superlattice. (B) Artificial periodic potential in a semiconductor crystal made through periodic n-type and p-type doped layers. Such a superlattice is called a *doping superlattice.*

the stimulated emission is amplified by photons transversing through MQWs, with each well acting as an independent lasing medium.

One of the most interesting applications of superlattices is the quantum cascade laser. In the April 1994 issue of *Science*, Faist, Capasso, Sivco, Sirtori, Hutchnson, and Cho of AT&T Bell Labs reported a new type of far-infrared laser called the *quantum cascade laser*.[52] The idea of this superlattice device was first proposed by Rudolf Kazarinov (who is now with AT&T Bell Labs) and Robert Suris of A. F. Ioffe Institute in 1971 (http://www.mtmi.vu.lt/legacy/pfk/funkc_dariniai/nanostructures/superlattice.htm). The Bell Labs' work thus represents a real breakthrough that had been anticipated for almost a quarter of a century. Unlike quantum-confined lasers, which involve interband transitions between the conduction band and valence band, these quantum cascade lasers involve transitions of electrons between the various sub-bands (minibands) corresponding to different quantized levels of the conduction band. Because only electron transitions are involved, one refers to these types of lasers as unipolar lasers. There could be ~50 quantum wells in an MQW laser geometry. The barrier layers are very thin (1–3 nm), and an excited electron may emit 25–75 photons as it cascades down the ladder of subbands in the conduction band. Quantum cascade lasers have been demonstrated for wavelengths between 3 and 20 μm, which is useful for sensing atmospheric pollution. Quantum cascade lasers, operating on intraband transitions and emitting in the mid-infrared region, are used, e.g., for trace gas analysis.

In what follows we clarify how these bands within the conduction band come about. If the thickness of the wide-bandgap barrier layer in Figure 5.151 is small enough so that electrons may tunnel through, then the situation becomes similar to what happens when individual atoms are brought together in a crystal. To understand the interaction between these very closely spaced quantum wells better, we follow the thinking of Esaki and Tsu. Let us bring a set of *N* quantum wells together as shown in Figure 5.152. These wells consist again of alternate layers of GaAs (well) and AlGaAs (barriers). First, consider two wells in this array only, each with their own set of quantized levels E_n and labeled with their own quantum numbers n = 1, 2, 3, along the confinement direction (the growth direction). Reducing the separation between these two wells allows the wave functions to overlap, and the same (degenerate) energy levels E_n of the two wells now form two new states: $E_n(1)$ and $E_n(2)$ (just as we saw in Chapter 3 when bringing atoms together), where $E_n(1) = E_n - \Delta_n$ and $E_n(2) = E_n + \Delta_n$. The magnitude of the splitting $2\Delta_n$ depends on the level E_n. The higher the energy E_n, the more the wave function extends into the energy barrier between the wells, allowing for stronger interaction between the wave functions in those wells and leading to a larger splitting Δ_n. Bringing *N* wells together, instead of only two, increases the energy degeneracy for all *N* wells, producing *N* levels closely spaced in a so-called *miniband*. In the

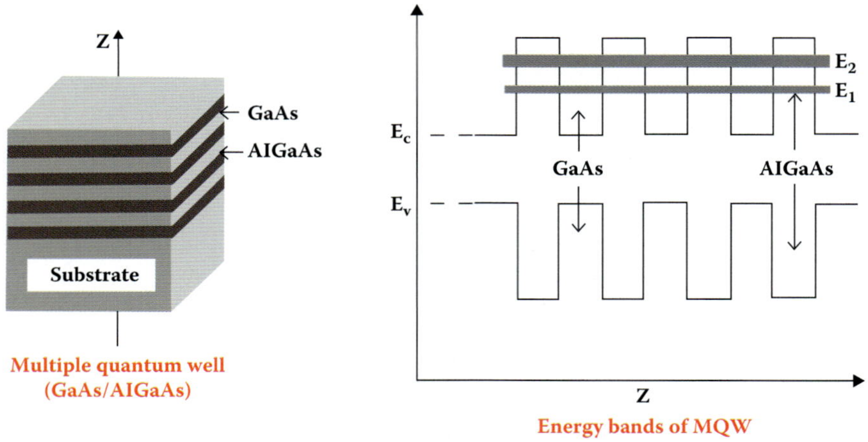

FIGURE 5.152 Schematic of formation of minibands in a superlattice consisting of alternate layers of GaAs (well) and AlGaAs (barriers).

case illustrated in Figure 5.152, individual levels E_1 and E_2 in the quantum wells are split into two minibands; the higher-energy miniband has a greater bandwidth than the lower-energy miniband for the reason explained above (larger splitting for higher-energy levels). Just as in a crystal where the periodic atomic potential leads to band formation in a superlattice, an artificial, human-made periodical potential causes the formation of minibands. Thus, the introduction of the 1D superlattice potential perturbs the band structure of the host materials, yielding a series of narrow sub-bands and forbidden gaps that arise from the subdivision of the Brillouin zone into a series of minizones. Most semiconductor devices are operated based on the interband transitions, namely, between the conduction and valence bands. These devices are usually bipolar involving a p-n junction. A new class of devices is based on the transitions between the sub-bands or minibands in the conduction or valence bands as described here. These intraband devices are unipolar and are faster than the intraband devices. Today a wide variety of engineered structures have been developed that exhibit extraordinary electronic transport and optical properties that do not exist in any natural crystals. This new degree of freedom offered in semiconductor research through advanced materials engineering has inspired many ingenious experiments, resulting in observations of not only predicted effects but also totally unknown results, and it appears that nearly half of all semiconductor physicists in the world are working now in this area. In addition to being used in quantum cascade layers, superlattice structures have been used in field effect transistors, where several quantum wells provide parallel conducting channels, increasing the device current-carrying capabilities and hence the output power. Superlattices are also used for photodetectors and for novel LEDs.

Photonic Crystal and Lasers

Earlier in this chapter we listed as an important benefit of photonic crystals the ability to control spontaneous emission and to modify the state density and group velocity of light by creating structure defects. We pointed out that these structures could be used to create novel, highly efficient lasers that emit light in a very narrow wavelength range. Point defects in photonic crystals act as high-Q resonator cavities and have been used in low-threshold photonic crystal lasers. In photonic lasers, only the wavelength that matches the wavelength of the defect mode is amplified because it can propagate freely through the material. The intensity of the propagating light increases as it undergoes successive reflections and travels back and forth through the photonic crystal. Meanwhile, light at other wavelengths is trapped within the photonic crystal and cannot build up. This means that the laser light is emitted in a narrow wavelength range that is directly related to the geometry of the microcavity.

Questions

Questions by Chuan Zhang, UC Irvine

5.1: In a traditional lithography system using i-line (365 nm) wavelength, the numerical aperture is NA = 0.8. Can we use this system to fabricate a pattern with a minimum line width of 65 nm? List several ways of increasing the resolution of the lithography system.

5.2: A beam of light with a wavelength of 500 nm goes through an aperture that has a diameter of 0.1 mm and generates an Airy disk on the image plane very far from the aperture. If the diameter of the Airy disk is 50 mm, what is the distance between the aperture and the image plane? If the distance becomes 3 times longer, then what would the size of the Airy disk be?

5.3: A beam of monochromatic red light with a frequency of 500 terahertz and a wavelength of 600 nm goes from vacuum into a piece of glass with an incidence angle of 30°. After getting into the glass, the wavelength of the light becomes 450 nm. What is the angle of refraction?

5.4: A beam of monochromatic light with a frequency of 500 terahertz and a wavelength of 600 nm transmits in water, which has a refractive index of 1.333, and then the light is totally internally reflected at the interface of water and an unknown material with an incident angle 60°. What is the minimum possible wavelength of that light when it is transmitted in the unknown media?

5.5: Make a comparison between right-handed and left-handed materials. List the two most important applications for left-handed materials.

5.6: A stationary proton with an electric charge of $1.60217653 \cdot 10^{-19}$ C and a mass of $1.6726 \cdot 10^{-27}$ kg is accelerated to 0.8c under an electric field intensity of $E_0 = 1 \cdot 10^6$ V/m; what is the weight of the particle at this speed and how long did it take to reach the final speed?

5.7: In an optical microscope, if we replace the light source (L = 550 nm) of visible light with ultraviolet (L = 275 nm), what is the percentage increase of resolution? If the numeral aperture in air is 0.8, what is the smallest distance between two parallel lines that can be resolved with the ultraviolet light source? And what if we immerse the lens system into an oil medium with a refractive factor of 1.5?

5.8: A beam of light has a wavelength of 500 nm in air. After crossing through another medium, its wavelength becomes 200 nm. Assuming the difference of permeability between this medium and air is very small, what is the relative dielectric constant for this medium?

5.9: The refractive index of a crystal is 1.7, dn/dL = -0.15 um^{-1}; what is the ratio of phase velocity to group velocity for light with a wavelength of 400 nm?

5.10: What is the optical phenomenon called *light polarization*? A beam of light becomes completely polarized after being reflected from the interface between water and an unknown medium. We know the incident angle is 60°. What is the refractive index of the unknown medium?

5.11: How does an optical fiber work? Explain why fiber losses increase if the fiber is bent? How can plasmonics avoid the bending problem?

5.12: How can surface plasmons be excited? What are some applications?

5.13: What is the difference between a direct bandgap and indirect bandgap semiconductor? Which one has a higher photon absorption and electron transition possibility? For optoelectronic applications, which kind of semiconductor is preferred?

5.14: Compare the similarities and differences between photonic crystals and metamaterials.

5.15: Why is there a resolution limit for traditional lenses? How does a superlens made from negative refractive index material exceed that limit?

5.16: What is Raman scattering? Compare Raman spectroscopy with infrared (IR) spectroscopy.

5.17: As the size of a metal particle becomes smaller and smaller, will there be any changes in its color? Explain why.

5.18: Why is the sky blue, while clouds are white?

5.19: Show that in the Maxwell equation $\Delta \times H = J_c + \frac{\partial D}{\partial t}$ all three terms have the same dimensions.

5.20: If an eye has a pupil opening (lens diameter) of 4 mm, can it resolve two point sources of 600 nm light (direct or reflected) that are 5 mm apart and 10 m away? Does the eye's resolving ability improve if the pupil enlarges to 5 mm?

References

1. Cheng, D. K. 1989. *Field and wave electromagnetics.* 2nd ed. Boston: Addison-Wesley.
2. Taya, M. 2005. *Electronic composites.* New York: Cambridge University Press.
3. Veselago, V. G. 1968. The electrodynamics of substances with simultaneously negative values of ε and μ. *Sov Phys Usp* 10:509–14.
4. Ashkin, A. 1970. Acceleration and trapping of particles by radiation pressure. *Phys Rev Lett* 24:156–59.
5. Berns, M. W., and D. E. Rounds. 1970. Cell surgery by laser. *Scientific American* 222: 98–103.
6. Atwater, H. A. 2007. The promise of plasmonics. *Scientific American* 296:56–63.
7. Greene, B. 1999. *The elegant universe.* New York: W.W. Norton.
8. Parker, B. 1987. *Search for a super-theory: from atoms to superstrings.* New York: Plenum Press.
9. Faraday, M. 1858. The Bakerian lecture: experimental relations of gold (and other metals) to light. *Philos Trans Royal Soc London* 147:145–81.
10. Padovani, S., C. Sada, P. Mazzoldi, B. Brunetti, I. Borgia, A. Sgamellotti, and A. Giulivi. Copper in glazes of renaissance luster pottery: nanoparticles, ions, and local environment. *J Appl Phys* 93:10058–63.

11. Bohren, C. F., and D. R. Huffman. 1983. *Absorption and scattering of light by small particles.* New York: John Wiley and Sons.
12. Gans, R. 1915. Fortpflanzung des Lichts durch ein inhomogenes Medium. *Ann Physik* 352:709–36.
13. Link, S., and M. El-Sayed. 1999. Size and temperature dependence of the plasmon absorption of colloidal gold nanoparticles. *J Phys Chem B* 103:4212–17.
14. Westcott, S. L., S. J. Oldenburg, T. R. Lee, and N. J. Halas. 1998. Formation and adsorption of clusters of gold nanoparticles onto functionalized silica nanoparticle surfaces. *Langmuir* 14:5396–401.
15. Chen, J., B. Wiley, Z.-Y. Li, D. Campbell, F. Saeki, H. Cang, L. Au, J. Lee, X. Li, and Y. Xia. 2005. Gold nanocages: engineering their structure for biomedical applications. *Adv Mater* 17:2255–61.
16. Maier, S. A., M. L. Brongersma, and H. A. Atwater. 2001. Electromagnetic energy transport along arrays of closely spaced metal rods as an analogue to plasmonic devices. *Appl Phys Lett* 78:16–18.
17. Frenkel, J. 1931. On the transformation of light into heat in solids. I. *Phys Rev Lett* 37:17–44.
18. Frenkel, J. 1931. On the transformation of light into heat in solids. II. *Phys Rev* 37:1276–94.
19. Gross, E. F. 1956. Optical spectrum of excitons in the crystal lattice. *Il Nuovo Cimento* 3:672–701.
20. Gustav, M. 1908. Beiträge zur Optik Trüber Medien, Speziell Kolloidaler Metallösungen. *Ann Physik* 330:377–445.
21. Jeanmaire, D. L., and R. P. V. Duyne. 1977. Surface Raman spectroelectrochemistry. Part I. Heterocyclic, aromatic, and aliphatic amines adsorbed. J Elecroanal Chem 84:1–20.
22. Shalaev, V. M. 1999. *Nonlinear optics of random media: fractal composites and metal-dielectric films.* Berlin: Springer.
23. Laserna, J. J. 1993. Combining fingerprinting capability with trace analytical detection: surface-enhanced raman spectometry. *Anal Chim Acta* 283:607–22.
24. Kneipp, K., H. Kneipp, V. B. Kartha, R. Manoharan, G. Deinum, I. Itzkan, R. R. Dasari, and M. S. Feld. 1998. Detection and identification of a single DNA base molecule using surface-enhanced raman scattering (SERS). *Phys Rev* E57:R6281–84.
25. John, S. 1987. Strong localization of photons in certain disordered dielectric superlattices. *Phys Rev Lett* 58:2486–89.
26. Yablonovitch, E. 1987. Inhibited spontaneous emission in solid-state physics and electronics. *Phys Rev Lett* 58:2059–64.
27. Ho, K. M., C. T. Chan, and C. M. Soukoulis. 1990. Existence of a photonic gap in periodic dielectric structures. *Phys Rev Lett* 65:3152–55.
28. Yablonovitch, E., and T. J. Gmitter. 1989. Photonic band structure: the face centered cubic case. *Phys Rev Lett* 63:1950.
29. Ozbay, E. 1994. Micromachined millimeter-wave photonic band-gap crystals. *Appl Phys Lett* 64:2059–61.
30. Lehmann, V., and H. Foll. 1990. Formation mechanism and properties of electrochemically etched trenches in n-type silicon. *J Electrochem Soc* 137:653–60.
31. Grüning, U., V. Lehmann, S. Ottow, and K. Busch. 1996. Macroporous silicon with a complete two-dimensional photonic band gap centered at 4.9 μm. *Appl Phys Lett* 68: 747–49.
32. Campbell, M., D. N. Sharp, M. T. Harrison, R. G. Denning, and A. J. Turberfield. 2000. Fabrication of photonic crystals for the visible spectrum by holographic lithography. *Nature* 404:53–56.
33. Vlasov, Y. A., X. Z. Bo, J. C. Sturm, and D. J. Norris. 2001. On-chip natural assembly of silicon photonic bandgap crystals. *Nature* 414:289–93.
34. Joannopoulos, J. D. 2008. *Photonic crystals: molding the flow of light.* Princeton, NJ: Princeton University Press.
35. Coen, S., A. H. Chau, R. Leonhardt, J. D. Harvey, J. C. Knight, W. J. Wadsworth, and P. S. Russell. 2001. White-light supercontinuum generation with 60-ps pump pulses in a photonic crystal fiber. *Opt Lett* 26:1356–58.
36. Cubukcu, E., K. Aydin, E. Ozbay, S. Foteinopoulou, and C. M. Soukoulis. 2003. Electromagnetic waves: negative refraction by photonic crystals. *Nature* 423:604–05.
37. Research Highlights. 2007. Invisibility cloak in sight. *Nature* 444:4–5.
38. Pendry, J. B., A. J. Holden, D. J. Robbins, and W. J. Stewart. 1998. Low frequency plasmons in thin-wire structures. *J Phys Condens Matter* 10:4785–809.
39. Pendry, J. B., A. J. Holden, W. J. Stewart, and I. Youngs. 1996. Extremely low frequency plasmons in metallic mesostructures. *Phys Rev Lett* 76:4773–76.
40. Pendry, J. B., A. J. Holden, D. J. Robbins, and W. J. Stewart. 1999. Magnetism from conductors and enhanced nonlinear phenomena. *IEEE Trans Microwave Theory Techn* 47:2075–84.
41. Wiltshire, M. C. K., J. B. Pendry, I. R. Young, D. J. Larkman, D. J. Gilderdale, J. V. Hajnal. 2001. Microstructured magnetic materials for RF flux guides in magnetic resonance imaging. *Science* 291:849–51.
42. Shelby, R. A., D. R. Smith. and S. Schultz. 2001. Experimental verification of a negative index of refraction. *Science* 292:77–79.
43. Smith, D.R., and N. Kroll. 2000. Negative refractive index in left-handed materials. *Phys Rev Lett* 85:2933.
44. Smith, D. R., W. J. Padilla, D. C. Vier, S. C. Nemat-Nasser, and S. Schultz. 2000. Composite medium with simultaneously negative permeability and permittivity. *Phys Rev Lett* 84:4184–87.
45. Shalaev, V. M., W. Cai, U. K. Chettiar, H. K. Yuan, A. K. Sarychev, V. P. Drachev, and A. V. Kildishev. 2005. Negative index of refraction in optical metamaterials. *Opt Lett* 30:3356.
46. Soukoulis, C. M., S. Linden, and M. Wegener. 2007. Negative refractive index at optical wavelengths. *Science* 315:47–49.
47. Notomi, M. 2000. Theory of light propagation in strongly modulated photonic crystals: refraction like behavior in the vicinity of the photonic band gap. *Phys Rev B* 62:10696–705.
48. Fang, N., and X. Zhang. 2003. Imaging properties of a metamaterial superlens. *Appl Phys Lett* 82:161.
49. Fang, N., H. Lee, C. Sun, and X. Zhang. 2005. Sub-diffraction-limited optical imaging with a silver superlens. *Science* 308:524–37.
50. Arakawa, Y., and H. Sakaki. 1982. Multidimensional quantum well laser and temperature dependence of its threshold current. *Appl Phys Lett* 40:939–41.
51. Asada, M., Y. Miyamoto, and Y. Suematsu. 1986. Gain and the threshold of three-dimensional quantum-box lasers. *IEEE J Quant Electron* QE-22:1915–21.
52. Faist, J., F. Capasso, D. L. Sivco, C. Sirtori, A. L. Hutchinson, and A. Y. Cho. 1994. Quantum cascade laser. *Science* 264:553–56.
53. Meinhart, C. D., and S. T. Wereley. 2003. The theory of diffraction-limited resolution in microparticle image velocimetry. *Meas Sci Technol* 14:1047.
54. Pendry, J. B. 2000. Negative refraction makes a perfect lens. *Phys Rev Lett* 85:3966.

6

Fluidics

Angry Sea at Naruto
Ando Hiroshige,
1830

Observe the motion of the surface of the water, which resembles that of hair, which has two motions, of which one is caused by the weight of the hair, the other by the direction of the curls; thus the water has eddying motions, one part of which is due to the principal current, the other to random and reverse motion.

Leonardo da Vinci, 1510

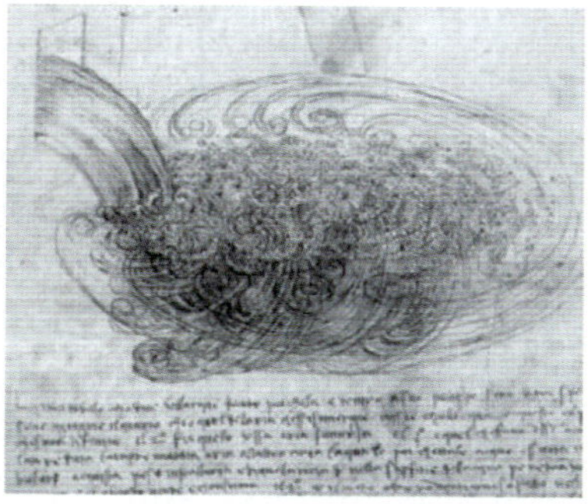

The current nano hype makes us forget that already Sir Isaac Newton described nanoscale phenomena, the so-called Newton rings (left; no, that is not spaghetti he is looking at). Newton also very clearly formulated experiments of microfluidics by describing experiments with capillary flow between parallel glass plates (below). So, neither nano nor micro are [sic] really very novel it seems!

Based on an editorial
by Andreas Manz and Harpal Minhas
for *Lab on a Chip*

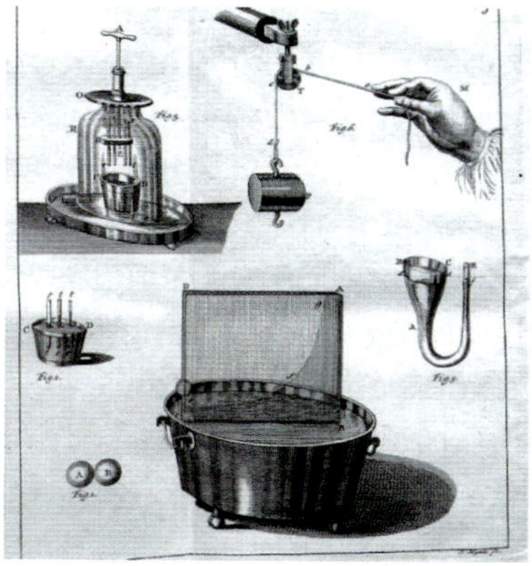

These [natural] laws may have originally been decreed by God, but it appears that he has since left the universe to evolve according to them and does not now intervene in it.

Stephen Hawking
A Brief History of Time

Outline

Introduction

Fluid mechanics (fluidics) is the study of motion of fluids (liquids and gases) and the forces associated with that motion. In Chapter 6 we introduce fluidics, compare various fluidic propulsion mechanisms, and discuss the influence of miniaturization on fluid behavior. Our treatment deals with macroscale, microscale, and nanoscale fluidic phenomena. Given the current academic and industrial interest, fluidics in miniaturized analytical equipment is detailed in this chapter as well. Example applications we discuss include electro-osmotic pumping systems, diffusive separation systems, micromixers, DNA amplifiers (PCR), compact disc diagnostic platforms, and cytometers.

Newton and da Vinci aside (see insets above), fundamental work on the physical behavior of fluids in micro- and nanomachined structures only started in the early 1980s. In the micro- and nanodomain, many factors seem to upset the old fluidics continuum order: Navier-Stokes equations (NSEs), with no-slip boundary conditions at the fluid interface, cannot always be applied, and slip flow, creep flow, intermolecular forces, etc., start suggesting molecular-based flow models instead. For gases, breakdown of macroscale physics is readily recognized, and some molecular-based modeling progress has been made. For liquids, with molecules much more closely packed, modeling has proven to be much more elusive.

Microfluidics and nanofluidics present researchers with exciting future applications. These include the characterization of individual molecules (e.g., reading the base sequence in DNA by feeding the molecule through a nanopore), studying the kinetics of very fast reactions, isolating a molecule in a nanometric space for detection purposes, further increasing the role of surfaces in catalytic reactions, manipulation of single molecules, labeling and studying of conformational changes of single molecules, applying intense electrical or magnetic fields to control chemical reactions, constructing hybrid systems by assembling biomolecules in NEMS structures, etc.

Macroscale Laws for Fluid Flow

Definition of Fluids and Types of Flow

A fluid, in our common experience, is a substance that has volume but no shape; that is, a fluid is a material that cannot resist a shear force or shear stress without moving.[1] The two most important parameters characterizing a fluid are density and viscosity. Density or specific mass, ρ, of a fluid is the mass per unit volume (in kgm^{-3}), and dynamic viscosity, η, is the fluid property that causes shear stresses when the fluid is moving; without viscosity there would be no fluid resistance. Dynamic viscosity is measured as kg/ms (SI system), that is, Pa·s or Poiseuille (Pl)*, and is given as the ratio of shear stress, τ_s, to shear rate, Γ_s:

$$\eta = \frac{\tau_s}{\Gamma_s} \tag{6.1}$$

The dynamic viscosity of air is 1.85×10^{-5} kg/ms, and for water it is 10^{-3} kg/ms. Equation 6.1 can be better understood from an inspection of Figure 6.1, in which we illustrate two large plates, of area A, separated by a fluid layer of uniform thickness, d. When one of the two plates is moved in a straight line relative to the other at an average velocity V, the shear force required for obtaining V is given by τ_{zx} (=F/A = force/area in N m^{-2}, i.e., Pa):

$$\frac{F}{A} = \tau_{zx} = \eta\frac{V}{d} \tag{6.2}$$

and V/d corresponds to the shear rate Γ_s (velocity gradient in s^{-1}). The first subscript in τ_{zx} is in the direction normal to the shearing force (z), and the second

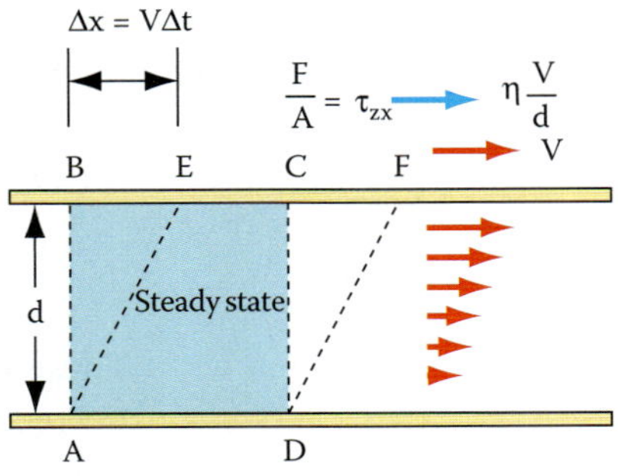

FIGURE 6.1 Schematic of shearing experiment.

* In the CGS system it is measured in poise (1 poise or P = 10^{-1} N·sm^{-2}) and 1 Pl = 10 P.

subscript is in the direction of the force (x). Equation 6.2 can be generalized as:

$$\tau_{zx} = \eta \frac{dV}{dz} \tag{6.3}$$

This relationship is known as Newton's Law of viscosity, and it relates the velocity gradient in the fluid to the shear stress. Thus, viscosity refers to friction between layers. The normal component of stress (F_n) for a fluid at rest is called the pressure, and the tangential component (F_t) is the shear stress (Figure 6.2). A solid can resist applied shear by deforming, and the stress is proportional to the strain, but a fluid deforms continuously under applied shear, and the stress is proportional to strain rate.

The kinematic viscosity, ν in m^2/s, or also myria (=10,000) Stokes (ma St, where the Stokes is the CGS unit, 1 $m^2/s = 10^4$ St), is the ratio of the absolute viscosity to density given by:

$$\nu = \frac{\eta}{\rho} \tag{6.4}$$

Its value is 1.43×10^{-5} m^2/s for air and 10^{-6} m^2/s for water. For Newtonian fluids such as water and air, the shearing stress and the velocity are linearly related so that the viscosity is only a function of the nature of the fluid.[2] For non-Newtonian fluids such as milk and blood,* the viscosity is also a function of the velocity gradient (see Figure 6.3). Most often, the viscosity in these cases diminishes when the velocity gradient increases, and these types of materials are called *pseudoplastics*. A Bingham plastic is a viscoelastic material that behaves as a rigid body at low stresses but flows as a viscous fluid at high stress. Once a critical shear stress (or "yield stress") is exceeded, the material flows as a Newtonian fluid.

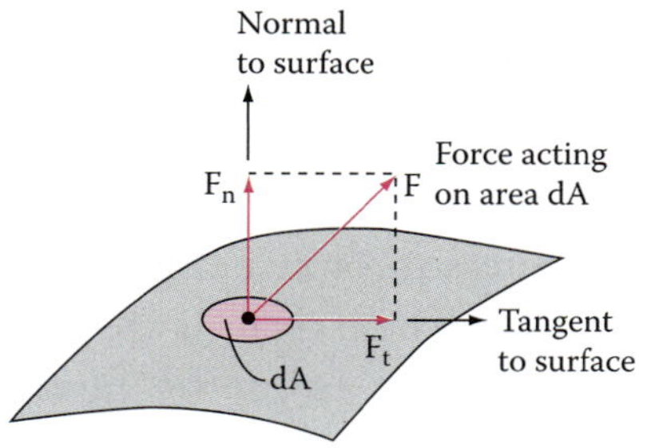

FIGURE 6.2 Pressure and shear stress in a fluid.

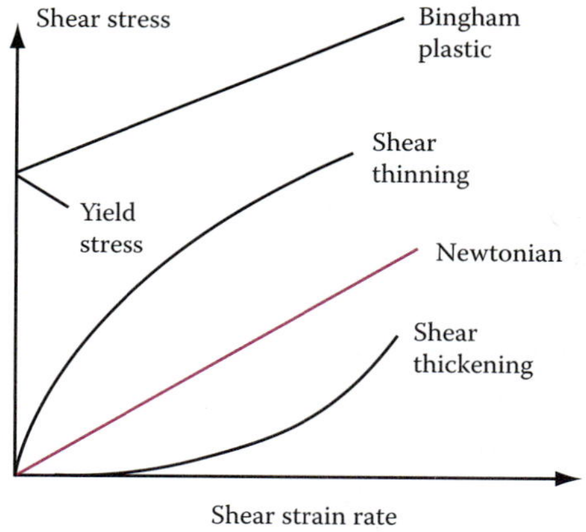

FIGURE 6.3 Newtonian and non-Newtonian fluids compared. The viscosity usually diminishes when the velocity gradient increases, and these types of materials are called *pseudoplastics.*

Non-Newtonian flow behavior of fluids can offer real advantages. Paint, for example, should be easy to spread, so it must have a low viscosity at the high shear caused by the paintbrush. Many cleaning fluids and furniture waxes have similar properties. The causes of non-Newtonian flow depend on the colloid chemistry of the particular product.

The fundamental laws governing fluid motion are the so-called Navier-Stokes equations (NSEs) (see Navier and Stokes in Figure 6.4). The NSEs are 3D vector equations and represent the differential forms of three equations: 1) the conservation of mass, 2) the conservation of linear momentum, and 3) the conservation of energy (this third equation

FIGURE 6.4 (a) French mathematician L. M. H. Navier (1758–1836) and (b) English mathematician Sir G. G. Stokes (1819–1903) formulated the Navier-Stokes equations by including viscous effects in the equations of motion.

* Although serum is a Newtonian fluid, the presence of red blood cells makes whole blood non-Newtonian. The same is true for mayonnaise; the oil in it is Newtonian, but the emulsion with yolk makes it non-Newtonian.

that makes up the complete set of NSEs will not be separately derived here). The Navier-Stokes equations are applicable to the description of the motion of a fluid volume at an arbitrary location in a flow field at any instant of time. One often classifies flows as a means to simplify the assumptions governing the partial differential Navier-Stokes equations. Lower dimensional flows reduce the complexity of analytical and computational solutions. We start by introducing the NSEs and show how they simplify depending on the type of flow. Next we derive two of the actual NSEs.

1. The Conservation of Mass Equation. The continuity equation in differential form is given as:

$$\frac{\partial \rho}{\partial t} + \nabla \cdot (\rho \mathbf{V}) = 0 \tag{6.5}$$

Mass is neither created nor destroyed; there are no sources or sinks, or the net inflow rate, i.e., (inflow – outflow), must equal the accumulation rate. For one direction (e.g., the x-direction with velocity component u) this yields:

$$\frac{\partial \rho}{\partial t} + \frac{\partial (\rho u)}{\partial x} = 0 \tag{6.6}$$

A flow is classified as incompressible if the density remains constant. For an incompressible fluid $\left(\rho = \text{constant or } \frac{\partial \rho}{\partial t} = 0\right)$, this results in:

$$\nabla \cdot \underset{\sim}{u} = 0 \text{ (for 1D) and } \nabla \cdot \mathbf{V} = 0 \text{ (for 3D)} \tag{6.7}$$

Note that the above is true even for unsteady flows. Liquid flows typically are incompressible, but gas flows are compressible, especially at high speeds. The Mach number, Ma = **V**/c, is the ratio of flow velocity to the speed of sound (c) and may be regarded as the ratio of inertial forces to elastic ones. The Mach number is a good indicator of whether compressibility effects are important (Figure 6.5).

2. The Conservation of Momentum. In vector notation, the conservation of momentum is given as:

$$\underset{a}{\rho \frac{\partial \mathbf{V}}{\partial t}} + \underset{b}{\rho (\mathbf{V} \cdot \nabla) \mathbf{V}} = \underset{c}{-\nabla p} + \underset{d}{\eta \nabla^2 \mathbf{V}} + \underset{e}{\rho g} \tag{6.8}$$

FIGURE 6.5 Ma = V/c. With Ma < 0.3 (~100 m/s), incompressible; with Ma < 1, subsonic; with Ma = 1, sonic; with Ma > 1, supersonic; and with Ma >> 1, hypersonic.

where g is the gravitational acceleration, p is the driving pressure, and t is time. The gravitational acceleration is related to the gravitational potential Ω as $\mathbf{g} = -\nabla\Omega$. Each term in Equation 6.8 has the dimension of $\{m/L^2t^2\}$.

The term a is the local acceleration of the fluid element. For a steady flow this term is zero. Steady flow implies no change at any point with time; transient terms in the NSE are then also zero:

$$\frac{\partial \mathbf{V}}{\partial t} = \frac{\partial \rho}{\partial t} = 0 \tag{6.9}$$

Transients usually describe a starting or developing flow (Figure 6.6). A change in coordinate system (e.g., Cartesian, cylindrical, spherical) may facilitate the mathematical treatment. For example, from Figure 6.6, for a fully developed pipe flow, velocity $\mathbf{V}(r)$ is a function of radius r, and pressure $p(z)$ is a function of distance z along the axis.

The next term in Equation 6.8 is the term b, which represents the convective acceleration of the fluid particle, and it predicts how the flow differs from one location to the next at the same instant of time. Uniform flow has no convective acceleration. The c term is the pressure acceleration caused by the "pumping" action. The d term is the viscous deceleration term resulting from the fluid's frictional resistance. The e term is the acceleration caused by gravity (weight). Regions where frictional effects are dominant are called viscous regions. They usually are close to solid surfaces (Figure 6.7). Regions where frictional forces are small compared with inertial or pressure forces are called inviscid. Such fluids are

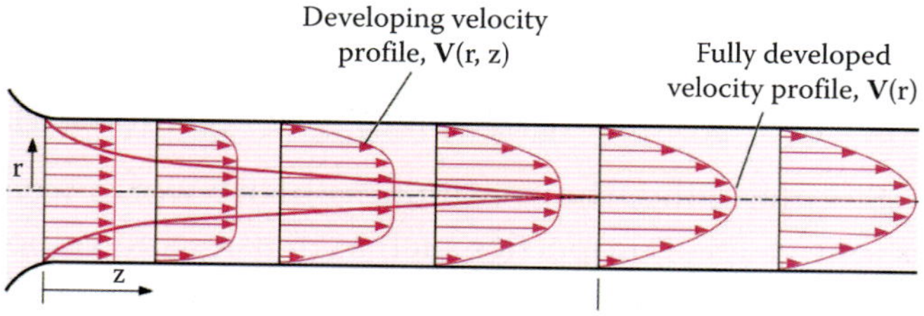

FIGURE 6.6 From a developing velocity profile to a fully developed velocity profile.

considered to be ideal, and in that case the *d* term in Equation 6.8 can be neglected:

$$\rho\frac{\partial \mathbf{V}}{\partial t}+\rho(\mathbf{V}\cdot\nabla)\mathbf{V}=-\nabla \mathbf{p}+\rho \mathbf{g} \qquad (6.10)$$

This expression is known as the Euler equation of motion. Even though this relation is simpler than the Navier-Stokes, an analytical solution is still not generally possible because the advective accelerations make the equations nonlinear.

Internal flows (e.g., inside a pipe), as illustrated in Figure 6.6, are dominated by the influence of viscosity throughout the flow field, whereas in external flows, viscous effects are limited to the boundary layer and wake as illustrated in Figure 6.8.

In the case of very slow fluid motion, the total acceleration term (a + b) as well as the acceleration caused by gravity (*e*) are zero. Equation 6.8 then reduces to the equation for Stokes flow or creep flow:

$$\nabla p=\eta\nabla^2\mathbf{V} \qquad (6.11)$$

This equation applies when dealing with very viscous fluids, capillary flow, and certain molten metals. Flows are also distinguished as being laminar, turbulent, or transitional, as shown in Figure 6.9.

An important concept in fluidics concerns the idea of streamlines. A streamline is a path traced out by a

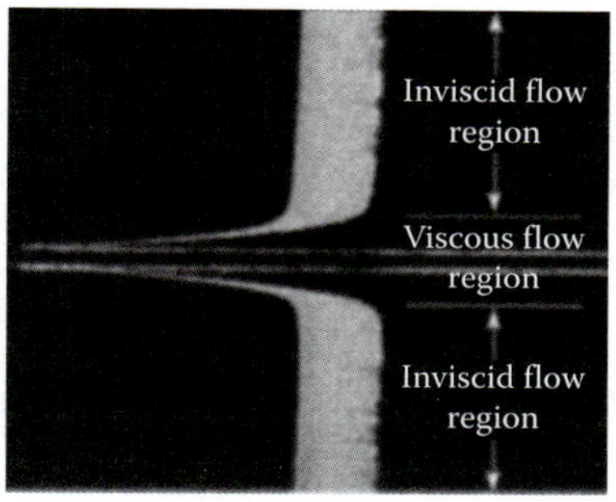

FIGURE 6.7 Inviscid and viscous regions in a liquid flow.

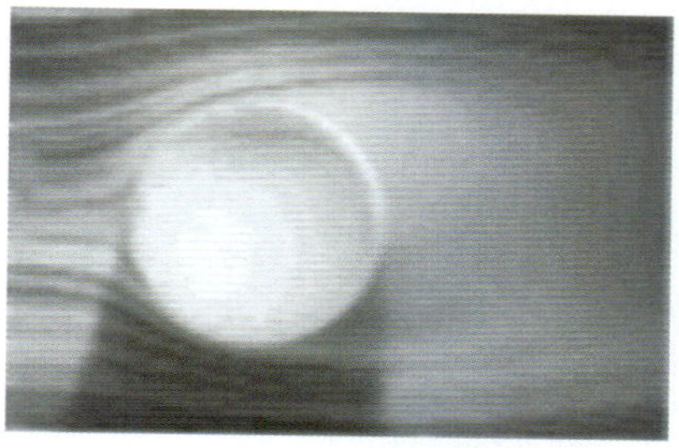

FIGURE 6.8 External flow. Viscous effects are limited to the boundary layer and wake around the object in the flow.

massless particle as it moves with the flow. Velocity is tangent to a streamline at every point.

In the laminar case, the fluid motion is highly ordered with smooth streamlines. In the case of turbulence, one has a highly disordered fluid motion characterized by velocity fluctuations and eddies. In the transitional case, one obtains a flow that contains both laminar and turbulent regions. We will see below that it is the Reynolds number, $Re = \rho VL/\eta$, that is the key parameter in determining whether a flow is laminar or turbulent. A rotational flow is one where streamlines close in on themselves while in an irrotational flow the streamlines do not close in on themselves.

Boundary Conditions

Continuum Hypothesis

In a continuum all quantities are defined everywhere and vary continuously from point to point. To understand this better, consider an "average" density, ρ, averaged over a fluid "particle" with volume L^3 as illustrated in Figure 6.10. For the "average density," it

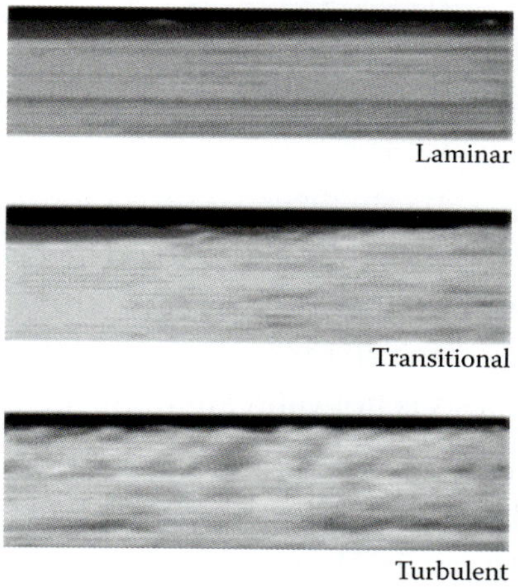

FIGURE 6.9 Laminar, transitional, and turbulent streamlines.

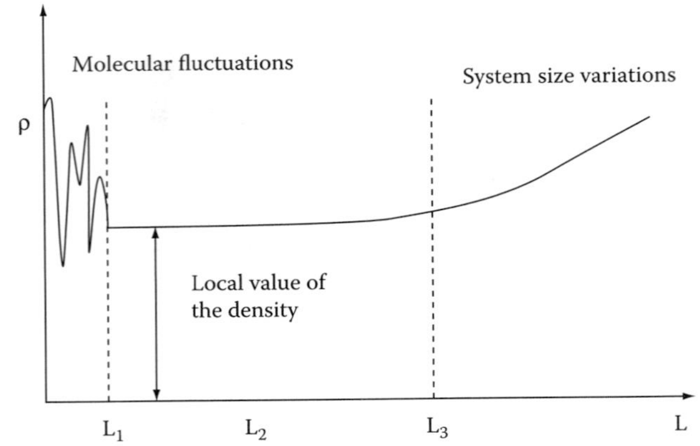

FIGURE 6.10 Schematic representation of the variation of instantaneous density of the fluid particle with respect to its volume.

holds that $\rho = N \cdot m/L^3$, with *N* the number of molecules and m the mass of each molecule. The continuum assumption fails when the sampling volume becomes too small. For the smallest particle size (L_1), molecular fluctuations may cause density variations. Also, for large volume elements, of the scale of the system itself (L_3), density can evolve because of spatial variations caused by changes in, for example, pressure (e.g., shocks) or temperature. It is only in the intermediate, "mesoscopic" size regime (L_2), the plateau in Figure 6.10, where it is possible to establish equations of fluid flow assuming a continuous medium. The continuum hypothesis thus assumes that in the L_2 regime (with $L_1 \ll L_2 \ll L_3$), the fluid can be treated as a continuum (i.e., not made up of individual molecules) with the properties of the system in the plateau region. Liquid particles in the plateau regime typically will have many, many molecules, even in a sample volume as small as a 1-μm cube. Liquid particles in a typical microchannel (e.g., more than a couple of micrometers wide) therefore still obey the Navier-Stokes equations. It is only when considering nanochannels that the situation needs to be re-evaluated. In the case of gases, it is easier to go beyond the NSE behavior, and even with sample volumes of 1 μm³, deviations become apparent. For example, the continuum hypothesis is not valid anymore for:

1. Rarefied gases: $L_2 \sim \lambda$ (mean free path), as discussed in Volume III, Chapter 7, "Breakdown of Continuum Theory in Electrostatics" (see page 380).
2. Knudsen diffusion: diffusion within very small pores where *d*, the pore diameter, $< \lambda$ and $L_3 \sim d$ (i.e., no plateau), a situation we encountered when discussing evaporation from a Knudsen cell in Volume II, Chapter 7 (see Equation 7.5).

Summarizing, the continuum theory works well as long as local properties such as density and velocity can be defined as averages over elements large compared with the microscopic structure of the fluid but small enough compared with the scale of macroscopic phenomena. Importantly, we will learn below that the flow also should not be too far from thermodynamic equilibrium.

No-Slip Condition

A fluid in direct contact with a solid "sticks" to the surface as a result of viscous effects as illustrated in Figure 6.11. This sticking is responsible for the generation of wall shear stress τ_w, surface drag $D = \int \tau_w dA$, and the development of the boundary layer. The fluid property responsible for this no-slip condition is the viscosity, η. According to the no-slip condition, for a fluid in contact with a solid wall, the velocity of the fluid must equal that of the wall, as illustrated $v_{fluid} = v_{wall}$ in Figure 6.11.

This no-slip condition at the fluid-solid interface in momentum space has an equivalent in energy space with the analogous requirement for a no-temperature jump at the fluid-solid interface. Together these restrictions ensure that there are no finite discontinuities that would involve infinite velocity/temperature gradients. The latter, in continuum modeling, would lead to infinite viscous stress/heat flux. In other words, the fluid velocity must be zero relative to the solid, and the fluid temperature must equal that of the surface. This ultimately requires thermodynamic equilibrium between fluid and solid.[3] In the section on microscale laws for fluid flow, where we cannot

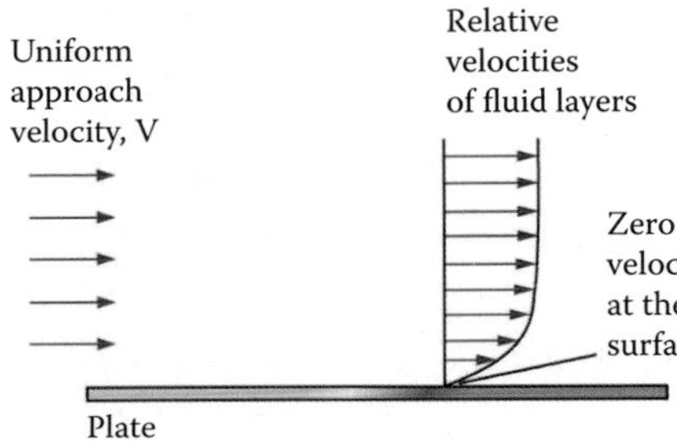

FIGURE 6.11 No-slip condition: zero velocity at the wall.

necessarily assume that the no-slip and the no-temperature jumps at the fluid-solid interface apply, we investigate what happens when microscopic effects start playing a role in determining fluid behavior.

Navier-Stokes Equations Derivation

Continuity Equation

Abbreviated, the derivation of the NSEs follows. The continuity equation can be derived with reference to Figure 6.12, where we apply the conservation of mass to a small control volume $\Delta V = \Delta x \Delta y \Delta z$, through which there is fluid flow. The rate of accumulation of mass in ΔV is:

$$\frac{\partial \rho}{\partial t}\Delta V \qquad (6.12)$$

The rate of inflow across a control surface plus the rate of change of mass inside the control volume ΔV equals the rate of outflow across the control surface:

Inflow through $S = \rho u \Delta y \Delta z + \rho v \Delta x \Delta z + \rho w \Delta x \Delta y$

Rate of outflow through S

$$= \left[\rho u + \frac{\partial(\rho u)}{\partial x}\Delta x\right]\Delta y \Delta z + \left[\rho v + \frac{\partial(\rho v)}{\partial y}\Delta y\right] \Delta x \Delta z + \left[\rho v + \frac{\partial(\rho w)}{\partial z}\Delta z\right]\Delta x \Delta y \qquad (6.13)$$

and, because mass is neither created nor destroyed, we apply Equation 6.12 and divide by $\Delta V = \Delta x \Delta y \Delta z$, resulting in the earlier introduced continuity Equations 6.5 and 6.7:

$$\frac{\partial \rho}{\partial t} + \nabla \cdot (\rho \mathbf{V}) = 0 \text{ (compressible)} \qquad (6.5)$$

$$\nabla \cdot \mathbf{V} = 0 \text{ (incompressible)} \qquad (6.7)$$

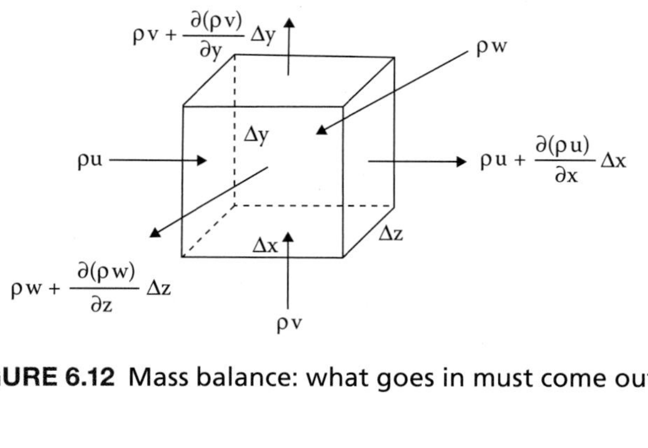

FIGURE 6.12 Mass balance: what goes in must come out.

Momentum Equation

Now we will set up the momentum equation. Even in a steady flow, a fluid particle can change momentum by moving to a position where its velocity is different; this acceleration requires that a force be applied to the fluid particle. This situation is equivalent to that of a change in temperature of a fluid particle over time Δt, during which it also has moved over a distance (Δx, Δy, Δz). This heat transfer problem may be expressed as:

$$dT = \frac{\partial T}{\partial x}\Delta x + \frac{\partial T}{\partial y}\Delta y + \frac{\partial T}{\partial z}\Delta z + \frac{\partial T}{\partial t}\Delta t \Rightarrow$$
$$\frac{dT}{dt} = \frac{\partial T}{\partial x}u + \frac{\partial T}{\partial y}v + \frac{\partial T}{\partial z}w + \frac{\partial T}{\partial t} \qquad (6.14)$$

This can be written as a total substantive derivative (also called the Stokes derivative) or:

$$\frac{DT}{Dt} = \frac{\partial T}{\partial x}u + \frac{\partial T}{\partial y}v + \frac{\partial T}{\partial z}w + \frac{\partial T}{\partial t} = \frac{\partial T}{\partial t} + \mathbf{V}\cdot\nabla T \qquad (6.15)$$

The operator in both the heat transfer example and in convective fluid transport is the same, i.e.:

$$\frac{D}{Dt} = \frac{\partial}{\partial t} + \mathbf{V}\cdot\nabla \qquad (6.16)$$

This equation sums the local rate of change in $\frac{\partial}{\partial t}$ and the convective rate of change in $\mathbf{V}\cdot\nabla$. Momentum is a vector quantity, with the total rate of change of momentum for an incompressible fluid (per unit volume) given as:

$$\frac{D}{Dt}(\rho \mathbf{V}) = \rho(\mathbf{V}\cdot\nabla)\mathbf{V} + \rho\frac{\partial \mathbf{V}}{\partial t} \qquad (6.17)$$

We now apply Newton's Second Law to Equation 6.17: for a small fluid element ($F = ma = \rho \Delta x \Delta y \Delta z a$) the rate of change of momentum equals the sum of all applied forces on that fluid element. For the sum of the forces in the *x*-direction, we may write that $\sum F_x = \rho \Delta x \Delta y \Delta z \frac{Du}{Dt}$ (*u* being the *x*-vector of speed **V**). We first sum up all the body forces in the *x*-direction: the *x*-component of the gravity force acting on the fluid element is $\Delta F_g = \rho \Delta x \Delta y \Delta z g_x$

(weight), and the force caused by the pressure gradient in the x-direction is $\Delta F_p = -\frac{\partial p}{\partial x}\Delta x\Delta y\Delta z$. Next, we consider the surface forces. The latter are exerted on the fluid body as it interacts with its surroundings, and both normal stresses and shear stresses are involved as shown in Figure 6.13. The forces in the x-direction caused by viscous stresses are given as $\Delta F_s = \frac{\partial\sigma_{xx}}{\partial_x}\Delta x\Delta y\Delta z + \frac{\partial\tau_{yx}}{\partial_y} + \Delta x\Delta y\Delta z + \frac{\partial\tau_{zx}}{\partial z}\Delta x\Delta y\Delta z$. Thus, for an intermediate solution we find for the total change of momentum in the x-direction:

$$\begin{aligned}\sum F_x &= \Delta F_g + \Delta F_s + \Delta F_p\\ &= \rho\Delta x\Delta y\Delta z g_x + \frac{\partial\sigma_{xx}}{\partial x}\Delta x\Delta y\Delta z\\ &\quad + \frac{\partial\tau_{yx}}{\partial y}\Delta x\Delta y\Delta z + \frac{\partial\tau_{zx}}{\partial z}\Delta x\Delta y\Delta z - \frac{\partial p}{\partial x}\Delta x\Delta y\Delta z\\ &= \rho\Delta x\Delta y\Delta z\frac{Du}{Dt}\end{aligned}$$

and after dividing by $\Delta x\Delta y\Delta z$ we obtain

$$\rho\frac{Du}{Dt} = -\frac{\partial p}{\partial x} + \rho g_x + \left(\frac{\partial\sigma_{xx}}{\partial x} + \frac{\partial\tau_{yx}}{\partial y} + \frac{\partial\tau_{zx}}{\partial z}\right) \quad (6.18)$$

In Chapter 4 we saw that the stress gradients in Equation 6.18 can be rewritten as:

$$\begin{aligned}\left(\frac{\partial\sigma_{xx}}{\partial x} + \frac{\partial\tau_{yx}}{\partial y} + \frac{\partial\tau_{zx}}{\partial z}\right) &= 2\eta\frac{\partial^2 u}{\partial x^2} + \eta\frac{\partial}{\partial y}\left(\frac{\partial u}{\partial y} + \frac{\partial v}{\partial x}\right)\\ &\quad + \eta\frac{\partial}{\partial z}\left(\frac{\partial u}{\partial z} + \frac{\partial w}{\partial x}\right)\end{aligned} \quad (6.19)$$

where viscosity, η, for the first time enters our derivation. This expression can be simplified by

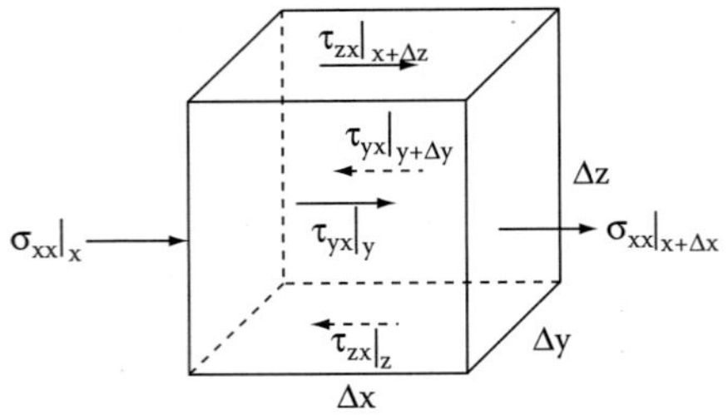

FIGURE 6.13 Surface forces exerted on an infinitesimal control volume $\Delta x\Delta y\Delta z$ include normal stresses and shear stresses.

using the continuum assumption (Equation 6.7, for 1D):

$$\begin{aligned}&= \eta\frac{\partial}{\partial x}\left(\frac{\partial u}{\partial x} + \frac{\partial v}{\partial y} + \frac{\partial w}{\partial z}\right) + \eta\left(\frac{\partial^2 u}{\partial x^2} + \frac{\partial^2 u}{\partial y^2} + \frac{\partial^2 u}{\partial z^2}\right)\\ &\qquad\qquad\Downarrow \qquad\qquad\qquad\qquad\qquad \Downarrow\\ &= 0,\ \text{since } \nabla\cdot u = 0 \qquad\qquad \eta\cdot\nabla^2\cdot u\end{aligned} \quad (6.20)$$

Thus, we obtain the equation for the x-momentum as:

$$\rho\frac{Du}{Dt} = -\frac{\partial p}{\partial x} + \eta\nabla^2 u + \rho g_x \quad (6.21)$$

and similarly for the y- and z-directions. We can write all three momentum equations together, using vector notation, and using Equation 6.17, we obtain Equation 6.8, i.e., the irrotational NSE for constant density and viscosity:

$$\rho\frac{\partial\mathbf{V}}{\partial t} + \rho(\mathbf{V}\cdot\nabla)\mathbf{V} = -\nabla p + \eta\nabla^2\mathbf{V} + \rho\mathbf{g} \quad (6.8)$$

The gravitational acceleration **g** is the negative of the gradient of the potential energy ($-\nabla\Omega$). We also introduce the reduced NSEs here; the latter formulation will enable us to discuss "similar" situations (see the concept of similarity in Volume III, Chapter 7 on scaling). The nondimensional NSEs will come in handy, for example, when we compare the efficiency of different separation techniques further below. We first define nondimensional variables, according to Table 6.1.

To obtain the dimensionless velocity, we write: $\mathbf{V}^* = \mathbf{V}/V$, for dimensionless length: $\mathbf{x}^* = \mathbf{x}/L$, for dimensionless gravity: $\mathbf{g}^* = \mathbf{g}/g$, and for dimensionless time we get: $t^* = ft$, with f the characteristic frequency. For pressure: $p^* = p/p_0 - p_\infty$, and the nabla or del operator as $\nabla^* = L\nabla$. Substituting these parameters in the NSE results in:

$$\begin{aligned}&\rho V f\frac{\partial\mathbf{V}^*}{\partial t^*} + \frac{\rho V^2}{L}(\mathbf{V}^*\cdot\nabla^*)\mathbf{V}^*\\ &= -\frac{P_0 - P_\infty}{L}\nabla^* p^* + \frac{\eta V}{L^2}\nabla^{*2}\mathbf{V}^* + \rho g\mathbf{g}^*\end{aligned} \quad (6.22)$$

To nondimensionalize Equation 6.22, we multiply all terms by $L/(\rho V^2)$, which has primary dimensions $\{m^{-1}L^2t^2\}$, so that the dimensions cancel. After rearrangement:

Table 6.1 Scaling Parameters Used to Nondimensionalize the Continuity and Momentum Equations, along with Their Primary Dimensions

Scaling Parameter	Description	Primary Dimensions
L	Characteristic length	{L}
V	Characteristic speed	$\{Lt^{-1}\}$
f	Characteristic frequency	$\{t^{-1}\}$
$p_0 - p_\infty$	Reference pressure difference	$\{mL^{-1}t^{-2}\}$
g	Gravitational acceleration	$\{Lt^{-2}\}$

$$\underbrace{\left[\frac{fL}{V}\right]}_{\text{Strouhal}}\frac{\partial \mathbf{V}^*}{\partial t^*}+(\mathbf{V}^*\cdot\nabla^*)\mathbf{V}^* = -\underbrace{\left[\frac{p_0-p_\infty}{\rho V^2}\right]}_{\text{Euler}}\nabla^* p^* + \underbrace{\left[\frac{\eta}{\rho VL}\right]}_{\text{Inverse Reynolds}}\nabla^{*2}\mathbf{V}^* + \underbrace{\left[\frac{gL}{V^2}\right]}_{\text{Inverse Froude}}\mathbf{g}^*$$

which can be written as:

$$[\mathrm{St}]\frac{\partial \mathbf{V}^*}{\partial t^*}+(\mathbf{V}^*\cdot\nabla^*)\mathbf{V}^* = -[\mathrm{Eu}]\nabla^* p^* + \left[\frac{1}{\mathrm{Re}}\right]\nabla^{*2}\mathbf{V}^* + \left[\frac{1}{\mathrm{Fr}^2}\right]\mathbf{g}^* \tag{6.23}$$

In some textbooks $\mathbf{g}^*$ is replaced by $-\mathbf{k}$ (vector along the z-axis); the x- and y-axes are assumed to be in a horizontal plane relative to the earth's surface, so one can write $\mathbf{g} = -9.81\ \mathbf{k}(\mathrm{m/s^2})$.

Writing the NSEs in dimensionless form not only reduces the number of free parameters, it also shows the appropriate limiting form of the equations if the dimensionless numbers approach extreme values. To normalize Equation 6.23, we must choose scaling parameters L, V, etc., that are appropriate for the flow being analyzed, such that all nondimensional variables are of order of magnitude unity, i.e., their minimum and maximum values are close to 1.0 (e.g., $t^* \sim 1$, $V^* \sim 1$, $p^* \sim 1$, etc.). If the NSEs are properly normalized, one can compare the relative importance of the terms in the equation by comparing the relative magnitudes of the nondimensional parameters Strouhal (St), Euler (Eu), Froude (Fr), and Reynolds (Re).

With Re tending to zero, we obtain Stokes or creep flow (with St ~ 1 and Fr ~ 1) (see Equation 6.11):

$$\underbrace{[\mathrm{Eu}]\nabla^* p^*}_{\text{Pressure forces}} = \underbrace{\left[\frac{1}{\mathrm{Re}}\right]\nabla^{*2}\mathbf{V}^*}_{\text{Viscous forces}} \tag{6.24}$$

and because $p^* \sim 1$ and $\nabla^* \sim 1$:

$$\mathrm{Eu} = \frac{p_0 - p_\infty}{\rho V^2} \sim \frac{1}{\mathrm{Re}} = \frac{\eta}{\rho VL}$$

and:

$$p_0 - p_\infty \sim \frac{\eta V}{L} \tag{6.25}$$

This result is very different from inertia-dominated flow (see below), where $p_0 - p_\infty \sim \rho V^2$. Equations 6.11 and 6.25 are linear equations that are much easier to solve for simple geometries than the full NSEs.

Example 6.1: Falling Sphere (I): Viscid Regime

With a small sphere, of radius r, falling through a viscous fluid the inertial forces are small compared with viscous forces (Re < 1). In this case, fluid motion is smooth and transfer is molecular. That is, momentum transfers from molecule to molecule through the fluid, and fluid flows in layers. The analytical solution of the NSEs for creeping flow around the falling sphere gives a drag force that, as predicted by Equation 6.25, is a linear function of velocity V and viscosity η:

$$F_D = 6\pi\eta Vr \tag{6.26}$$

Equation 6.26 is the Stokes' Law (1851, George Gabriel Stokes), where V is the terminal velocity. When an object starts to fall (free fall), it accelerates downward until it reaches a terminal, maximal velocity V. At this velocity, the forces acting on the body, i.e., drag forces, gravity, and buoyancy, have reached equilibrium, and acceleration stops due to gravity working on the body. In other words:

$$\sum F = 0 = \underbrace{F_D}_{\text{Friction}} + \underbrace{F_g}_{\text{Gravity}} + \underbrace{F_b}_{\text{Buoyancy}} \text{ or}$$

$$0 = 6\pi\eta Vr + \frac{8\pi r^3}{6}(\rho_p - \rho_f)g \text{ or} \tag{6.27}$$

$$V = \frac{2(\rho_p - \rho_f)gr^2}{9\eta}$$

The terminal velocity thus depends on the density (ρ_f) and viscosity (η) of the fluid, as well as the shape, size (L), and density of the object (ρ_p).

Viscid (Stokes), Inviscid (Euler), and Froude Flow

Einstein, in his 1905 paper on Brownian motion, established the following unexpected relation:

$$D = \mu k_B T \qquad (6.28)$$

linking *D*, the diffusion constant, and μ, the mobility of the particles, where k_B is Boltzmann's constant, and *T* is the absolute temperature. The mobility μ is the ratio of the particle's terminal drift velocity to an applied force, $\mu = V/F_D$. Substituting this in Equation 6.26 from Example 6.1 one obtains the Stokes-Einstein formula for the molecular diffusion coefficient—another example of Stokes flow:

$$D = \frac{k_B T}{6\pi\eta r} \qquad (6.29)$$

where k_B = Boltzmann's constant ($1.38 \times 10\ \mathrm{m}^{-23}\ \mathrm{J\ K}^{-1}$)
T = the absolute temperature (K)
η = the absolute viscosity (kg/m·s), and
r = the hydrodynamic radius

It is this very relation that allowed Perrin, using colloid particles as molecules, to measure Boltzmann's constant (Figure 6.14).

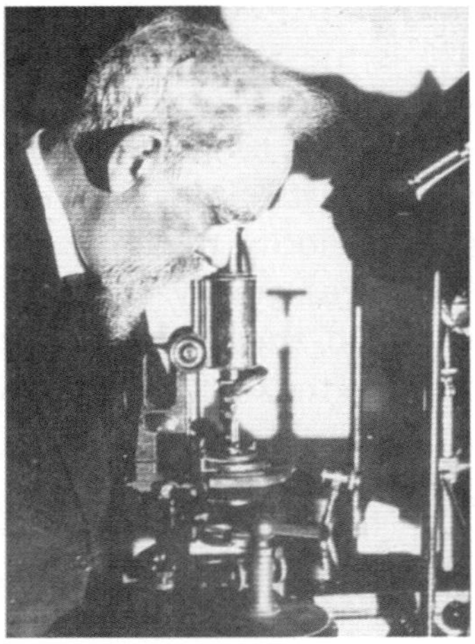

FIGURE 6.14 J. Perrin (1870–1942). Nobel Prize for Physics in 1926. "For his work on the discontinuous structure of matter, and especially for his discovery of sedimentation equilibrium." "I do not believe that it is possible to study Brownian motion with such precision," Einstein wrote to Perrin in 1910.

In Volume III, Chapter 7 on scaling, we derived a drag coefficient C_D for flow around a submerged object as:

$$C_D = \frac{F_D}{\frac{1}{2}\rho V^2 A} \qquad \text{(Volume III, Equation 7.27)}$$

where ρ = the density of the fluid
V = the free stream velocity (far from the object)
A = the characteristic area of the body: the projected area; the projected area is the area "seen" by the fluid area; for a spherical particle with diameter *L*, the area $A = \pi L^2/4$ (see Figure 6.15)*
F_D = the drag force

Keep in mind that this drag coefficient C_D is analogous to the friction factor in pipe flow (see below).

As derived in Volume III, Chapter 7 on scaling, we may express the drag coefficient C_D and friction force F_D also in terms of the Reynolds number Re:

$$C_D = F_D/\rho l^2 v^2 = f\,(\mathrm{Re}) \qquad \text{(Volume III, Equation 7.8)}$$

A plot of drag coefficient versus Reynolds number is shown in Figure 6.16. For a given Reynolds number, the best design for a race car, submarine, or airplane is the one that minimizes the drag coefficient C_D (see also Volume III, Figure 7.8). In the Stokes (viscid)

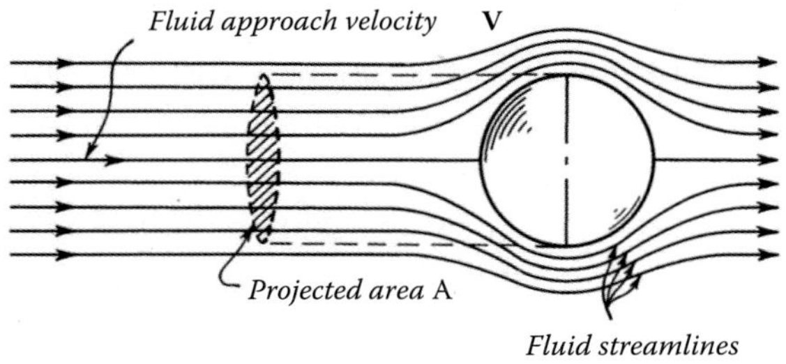

FIGURE 6.15 The projected area is the area "seen" by the fluid area. For a spherical particle $A = \pi L^2/4$ in a fluid stream.

* A = frontal area (thick body, cylinder, cars, projectiles)
= plan area (wide/flat body, wings, hydrofoils)
= wetted area (surface ships, barges)
This is a simplification, as both frictional and pressure drag forces should be considered, and factor A has a different meaning in both. For a more rigorous treatment, see Granger (1995).[6]

regime with values of Re < 1 (see Figure 6.16) one obtains:

$$C_D = \frac{24}{Re} \tag{6.30}$$

In inviscid flow, the net viscous forces are negligible compared with pressure and/or inertia forces:

$$[St]\frac{\partial \mathbf{V}^*}{\partial t^*} + (\mathbf{V}^* \cdot \nabla^*)\mathbf{V}^* =$$

$$-[Eu]\nabla^* p^* + \left[\frac{1}{Fr^2}\right]\mathbf{g}^* \left[\frac{1}{R}\right]\nabla^{*2}\,\mathbf{V}^* \tag{6.31}$$

With a very large Reynolds number Re (= ρ**V**L/η) the inertial effects dominate over the viscous effects and we obtain Euler's expression (Equation 6.10). Elimination of the viscous term changes the partial differential equations from mixed elliptic-hyperbolic to hyperbolic. This affects the type of analytical and computational tools used to solve the equations. This result is very different from viscous flow, with the drag force, $F_D = C_D\,\rho V^2 A/2$, now proportional to ρV^2. The Euler equation often is used in aerodynamics (Figure 6.17).

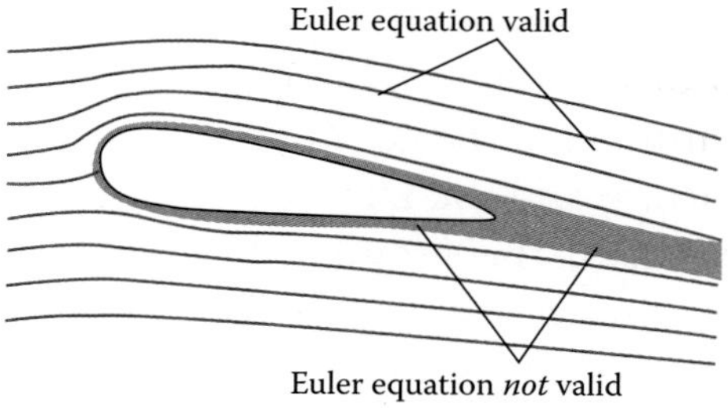

FIGURE 6.17 Regions of validity of the Euler equation in aerodynamics.

Example 6.2: Falling Sphere (II): Inviscid Flow

At high Reynolds numbers (Re > 1000), with negligible viscous forces and inertia-dominated flow, Newton's resistance law predicts that the force on the falling sphere will be proportional to the fluid pushed away and the relative velocity between the sphere and the fluid. In this case the flow is large, fluid motion is chaotic, and transfer is with blocks of molecules moving in all directions, causing eddy currents. For Re ⋙ 1, the flow pattern is very complicated; thus, very little theory exists to predict the relationship between C_D and Re. As the flow increases, wake drag becomes an important factor. The streamline pattern becomes mixed at the rear of the sphere and at very high Reynolds numbers completely separates in the wake. From experimental results shown in Figure 6.16, for 1000 < Re < 10^5, C_D is about 0.44. In the transition regime, the drag coefficient is:

$$C_D = \frac{24}{Re}\left(1 + \frac{Re^{2/3}}{6}\right) \tag{6.32}$$

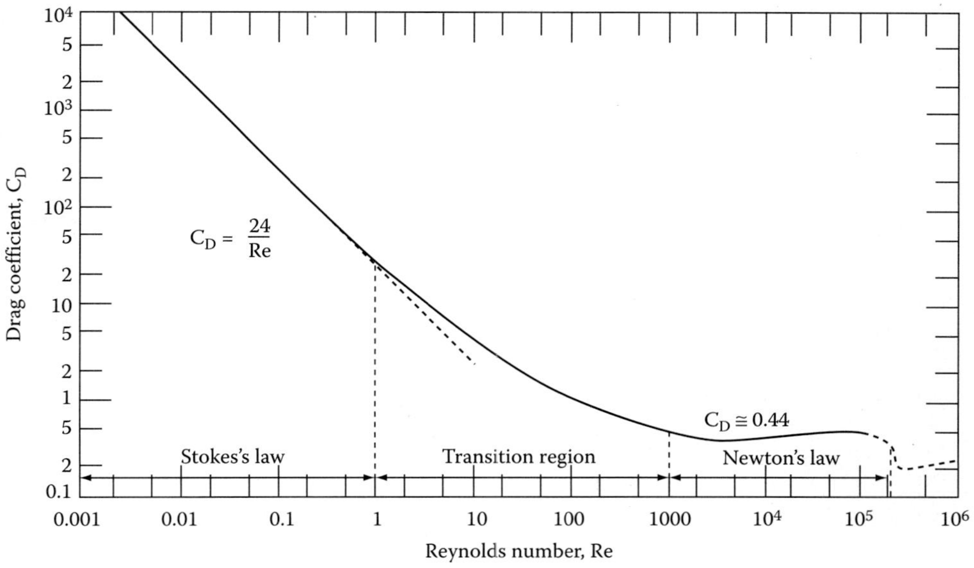

FIGURE 6.16 Drag coefficient C_D of a sphere versus Reynolds number Re.

The Froude number in Equation 6.31 represents the ratio of inertial to gravitational forces. If the Froude number is large, gravitational forces are not important and the $1/Fr^2$ term can be neglected. In many cases, the Froude number is large and gravity is unimportant; exceptions are in modeling ships and waves. Waves involve vertical displacements and therefore, gravity is involved (see also Volume III, Chapter 7, Equations 7.24–7.26) (Figure 6.18).

For a more detailed derivation of the NSEs, including the conservation of energy equation, refer to White,[4] Allen,[5,6] and Denn.[1] The NSEs cannot be solved unless some terms drop out and the equations simplify. There are about 80 known exact solutions to the NSEs. Various computational fluid dynamic (CFD) codes for solving dynamic behavior of fluids in microsystems built on these equations are commercially available (see Volume III, Chapter 6).

Navier-Stokes Applications

Introduction

Here, we present five applications of the NSEs. As warm-up applications we derive the expression for hydrostatics and calculate the Bernoulli expression from the Euler equation for flow. We then move on to pressure-driven flow between two large plates, derive the Couette flow, and finally introduce Hagen-Poiseuille's law for pipe flow.

Hydrostatics

An extreme simplification of the NSEs involves hydrostatics where the velocity **V** = 0 (Figure 6.19).

FIGURE 6.18 In modeling waves the Froude number is important.

With gravity in the z-direction, only the following NSE term survives:

$$\rho g = \frac{\partial p}{\partial z} \tag{6.33}$$

More generally, this can be interpreted as saying that pressure varies only in the direction of net acceleration. With a constant density, Equation 6.33 can be integrated to Figure out how pressure varies with depth, or:

$$\int dp = p(z) = \int \rho g dz = \rho g \int dz = \rho g z + p_0 \tag{6.34}$$

Because the fluid is not moving, η does not play a role.

Bernoulli's Equation

Remember that there is no flow across a streamline because at every point the streamline is a tangent to the velocity. Therefore, a streamline can be thought of as if it were a solid boundary. The superposition of irrotational, inviscid flow solutions gives us tools to construct velocity distributions but does not yet tell us anything about the pressure distribution in the flow. For that we need to derive the Bernoulli equation connecting pressure, velocity, and elevation. To derive that equation, we start from the

FIGURE 6.19 The Hoover Dam: for an incompressible fluid at rest, pressure increases linearly with depth. As a consequence, large forces can develop on plane and curved submerged surfaces. The water behind the Hoover Dam, on the Colorado River, is approximately 715 ft. deep, and at this depth the pressure is 310 psi. To withstand the large pressure forces on the face of the dam, its thickness varies from 45 ft. at the top to 660 ft. at the base.

NSEs for inviscid flow, i.e., the dimensional Euler expression:

$$\frac{\partial \mathbf{V}}{\partial t} + (\mathbf{V} \cdot \nabla)\mathbf{V} = -\frac{\nabla p}{\rho} + \mathbf{g} \quad (6.10)$$

and we write the gravity vector as:

$$\mathbf{g} = -g\nabla z \quad (6.35)$$

with z-positive upward. We also recall the following vector calculus identity:

$$(\mathbf{V} \cdot \nabla)\mathbf{V} = \frac{1}{2}\nabla(\mathbf{V} \cdot \mathbf{V}) - \mathbf{V} \times (\nabla \times \mathbf{V}) \quad (6.36)$$

With these two last equations we can rewrite the Euler equation as:

$$\begin{aligned} &-\rho g \nabla z - \nabla p = \frac{\rho}{2}\nabla(\mathbf{V} \cdot \mathbf{V}) - \rho \mathbf{V} \times (\nabla \times \mathbf{V}) \\ &\Rightarrow \frac{\nabla p}{\rho} + \frac{1}{2}\nabla(\mathbf{V}^2) + g\nabla z = \mathbf{V} \times (\nabla \times \mathbf{V}) \end{aligned} \quad (6.37)$$

Now we take the dot product with differential length $d\mathbf{s}$ along a streamline:

$$\frac{\nabla p}{\rho} \cdot d\mathbf{s} + \frac{1}{2}\nabla(\mathbf{V}^2) \cdot d\mathbf{s} + g\nabla z \cdot d\mathbf{s} = \mathbf{V} \times (\nabla \times \mathbf{V}) \cdot d\mathbf{s} \quad (6.38)$$

Because $d\mathbf{s}$ and $\mathbf{V}$ are parallel:

$$\begin{aligned} &\mathbf{V} \times (\nabla \times \mathbf{V}) \text{ is perpendicular to } \mathbf{V} \\ &\mathbf{V} \times (\nabla \times \mathbf{V}) \cdot d\mathbf{s} = 0 \end{aligned} \quad (6.39)$$

Thus, the term on the right in Equation 6.38 = 0. The $d\mathbf{s}$ term can be expanded as:

$$d\mathbf{s} = dx\,\mathbf{i} + dy\,\mathbf{j} + dz\,\mathbf{k}$$

$$\text{So: } \nabla p \cdot d\mathbf{s} = \frac{\partial p}{\partial x}dx + \frac{\partial p}{\partial y}dy + \frac{\partial p}{\partial z}dz = dp \quad (6.40)$$

Putting this last equation in the expression for flow results in:

$$\frac{dp}{\rho} + \frac{1}{2}d(\mathbf{V}^2) + g\,dz = 0 \quad (6.41)$$

Integration along a streamline then results in:

$$\int \frac{dp}{\rho} + \frac{1}{2}d(\mathbf{V}^2) + g\,dz = \text{constant} \quad (6.42)$$

For an incompressible fluid this can be rewritten as:

$$\frac{p}{\rho} + \frac{V^2}{2} + gz = \text{constant} \quad (6.43)$$

What happens if we impose the additional restriction that the flow be irrotational? In the above derivation we relied on the fact that the restriction in Equation 6.39 holds, but now we also add to this that $\nabla \times \mathbf{V} = \mathbf{0}$, so that the direction of $d\mathbf{s}$ is irrelevant. What this means is that in an irrotational flow Bernoulli's equation can be applied at any two points in the flow, not just at two points on a streamline! A more popular formulation of Bernoulli's equation is the following (where $\gamma = g \times \rho$):

$$\begin{aligned} &\frac{p}{\gamma} + z + \frac{\mathbf{V}^2}{2g} = \text{constant} \\ &\frac{p}{\gamma} + z = \text{piezometric head} \\ &\frac{\mathbf{V}^2}{2g} = \text{velocity (dynamic) head} \\ &\frac{p_1}{\gamma} + z_1 + \frac{\mathbf{V}_1^2}{2g} = \frac{p_2}{\gamma} + z_2 + \frac{\mathbf{V}_2^2}{2g} \end{aligned} \quad (6.44)$$

Example 6.3

Given: The velocity in an outlet pipe from a reservoir is 6 m/s and h = 15 m (see Figure 6.20).
Find: Pressure at point A.
Solution: Bernoulli's equation

$$\begin{aligned} &\frac{p_1}{\gamma} + z_1 + \frac{V_1^2}{2g} = \frac{p_A}{\gamma} + z_A + \frac{V_A^2}{2g} \\ &\frac{0}{\gamma} + h + \frac{0}{2g} = \frac{p_A}{\gamma} + 0 + \frac{V_A^2}{2g} \\ &p_A = \gamma\left(h - \frac{V_A^2}{2g}\right) = 9810\left(15 - \frac{18}{9.81}\right) \\ &p_A = 129.2 \text{ kPa} \end{aligned}$$

Unidirectional Flow between Two Infinite Plates

In Figure 6.21 we consider the unidirectional flow of a fluid between two infinite parallel plates. The plates are stationary, and the steady, incompressible flow is driven by a pressure gradient in the x-direction. We first decide what velocity components are nonzero

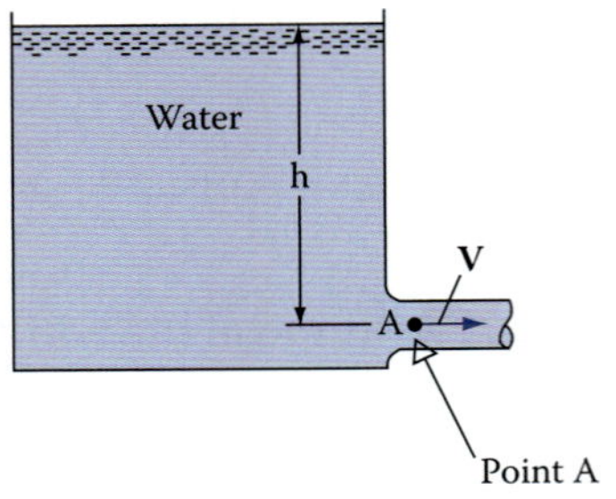

FIGURE 6.20 Velocity in outlet pipe from reservoir is 6 m/s and h = 15 m.

in this situation: w and v are = 0 and u is nonzero [$u = u(y)$].

This situation again simplifies the NSEs considerably:

$$0 = -\frac{\partial p}{\partial x} + \eta \frac{\partial^2 u}{\partial y^2}$$
$$0 = -\frac{\partial p}{\partial y} - \rho g$$
$$0 = -\frac{\partial p}{\partial z} \tag{6.45}$$

After integrating the first expression twice and applying the no-slip conditions at the plates (this sets u = 0 at z = +h and at $-h$), we obtain:

$$u = \frac{1}{2\eta}\left(\frac{\partial p}{\partial x}\right)(y^2 - h^2) \tag{6.46}$$

This solution is parabolic, as illustrated in Figure 6.22. To obtain the volumetric flow rate, we integrate:

$$Q = \int_{-h}^{h} u dy = \int_{-h}^{h} \frac{1}{2\eta}\left(\frac{\partial p}{\partial x}\right)(y^2 - h^2) dy$$
$$Q = -\frac{2h^3}{3\eta}\left(\frac{\partial p}{\partial x}\right) \tag{6.47}$$

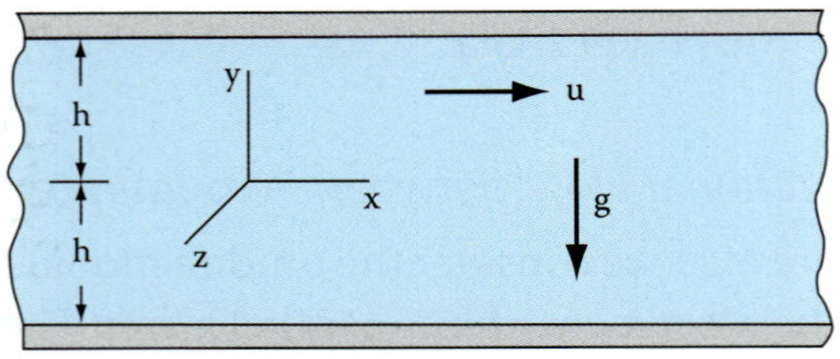

FIGURE 6.21 Unidirectional flow of a fluid between two large parallel plates. The plates are stationary, and the steady, incompressible flow is driven by a pressure gradient.

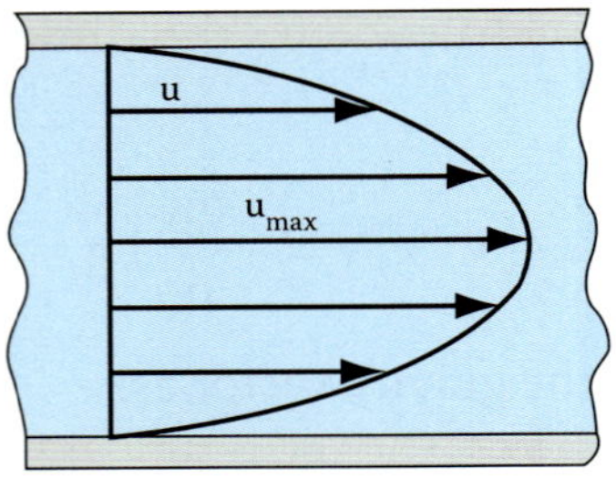

FIGURE 6.22 Parabolic velocity profile for flow of a fluid between two large parallel plates. Adjacent horizontal layers flow at different speeds, sliding over one another.

Fully Developed Couette Shear Flow

To illustrate Couette flow the same arrangement shown in Figure 6.22 is used; however, this time one plate is moving at a velocity U with respect to the other fixed plate (Figure 6.23). We simplify the Navier-Stokes equations with the same assumptions used above, except that here the no-slip boundary condition at the upper boundary is u(b) = U. This results in:

$$u = U\frac{y}{b} + \frac{1}{2\eta}\left(\frac{\partial p}{\partial x}\right)(y^2 - by) \tag{6.48}$$

Making this expression dimensionless leads to:

$$\frac{u}{U} = \frac{y}{b} + \frac{b^2}{2\eta U}\left(\frac{\partial p}{\partial x}\right)\left(\frac{y}{b}\right)\left(1 - \frac{y}{b}\right) \tag{6.49}$$

We can define $p = -\frac{b^2}{2\mu U}\left(\frac{\partial p}{\partial x}\right)$, which determines the effect of the pressure gradient as illustrated in

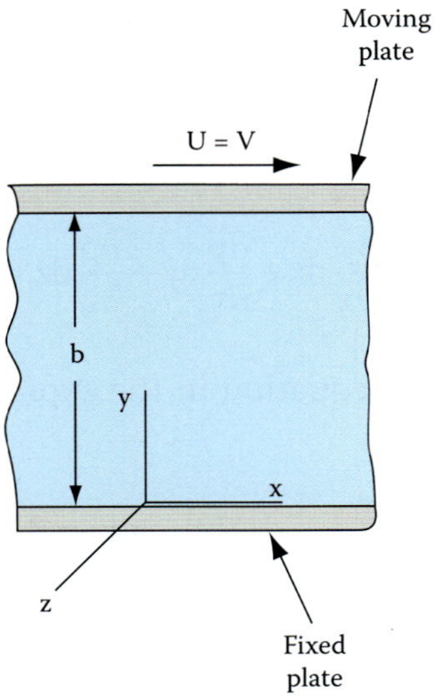

FIGURE 6.23 Couette flow between two infinitely large parallel plates.

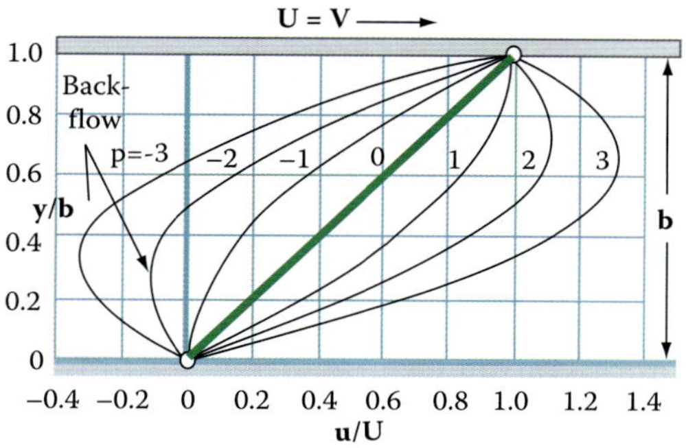

FIGURE 6.24 The dimensionless velocity as a function of reduced height and *p* value. Green slanted line is without pressure gradient.

Figure 6.23. In this case there is no pressure gradient $\left[\left(\frac{\partial p}{\partial x}\right)=0\right]$, and Equation 6.49 reduces to:

$$u = U\frac{y}{b} \tag{6.50}$$

The latter result corresponds with the slanted green line in Figure 6.24, where we show the dimensionless velocity as a function of reduced height and p value.

Hagen-Poiseuille's Law for Steady Duct Pipe Flow

Let us now consider pressure-driven flow in a pipe as shown in Figure 6.25. We assume steady and laminar flow, with flow only in the *z*-direction, so that $V_z = f(r)$ and $v_r = 0$ and $v_\theta = 0$ (we are working with cylindrical coordinates here):

$$\begin{aligned} 0 &= -\rho g \sin\theta - \frac{\partial p}{\partial r} \\ 0 &= -\rho g \cos\theta - \frac{1}{r}\frac{\partial p}{\partial \theta} \\ 0 &= -\frac{\partial p}{\partial z} + \eta\left[\frac{1}{r}\frac{\partial}{\partial r}\left(r\frac{\partial v_z}{\partial r}\right)\right] \end{aligned} \tag{6.51}$$

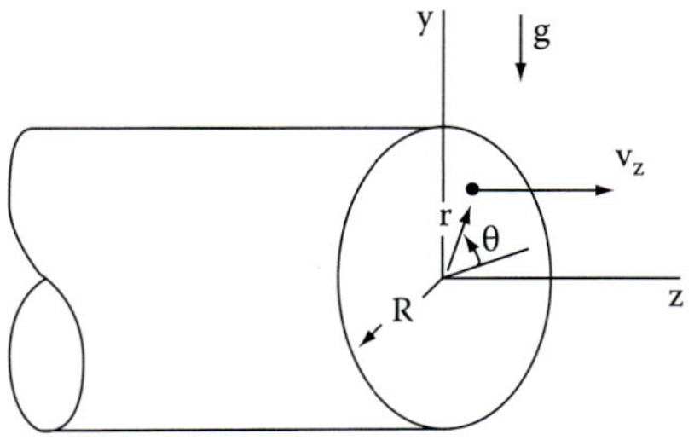

FIGURE 6.25 Pipe flow and cylindrical coordinates.

Solving these equations with the no-slip condition applied at r = R (the walls of the pipe), we obtain a parabolic velocity profile:

$$v_z = \frac{1}{4\eta}\left(\frac{\partial p}{\partial z}\right)\left(r^2 - R^2\right) \tag{6.52}$$

Experimentally, fluid viscosity may be determined based on this Hagen-Poiseuille law for laminar flow of a fluid through a capillary. Either the pressure difference over the capillary is measured for a known flow rate or the flow rate may be measured for a fixed pressure drop. The average velocity, *V*, often is used for convenience instead of flow rate, *Q*; the two are related through $V = Q/\pi R^2$. With reference to Figure 6.26, we calculate the volumetric flow rate *Q* as:

$$\begin{aligned} Q &= 2\pi\int_0^R v_z r\,dr \\ Q &= -\frac{\pi R^4}{8\eta}\left(\frac{\partial p}{\partial z}\right) \end{aligned} \tag{6.53}$$

The pressure gradient in macroscopic laminar flow in the capillary is given as $\Delta p/l$, where l is the length of the capillary. Substituting *Q* for average velocity and using $\Delta p/l$, we rewrite Equation 6.53 as the Hagen-Poiseuile equation or:

$$V = -\frac{R^2\Delta p}{8\eta l} \tag{6.54}$$

This clearly predicts the pressure drop over narrow capillaries to be very high and scaling as the inverse of R^2. The fact that the volumetric flow rate *Q* reduces as the fourth power of the radius (Equation 6.53) can be understood as follows. A gas or a liquid is at rest at the capillary wall and the fluid velocity, which reduces as the square of the channel radius (from Equation 6.54), is maximal at the center of the tube. In a wide tube, larger velocities can thus be achieved. This, coupled with the trivial factor of the larger diameter, results in a *Q* proportional to R^4.

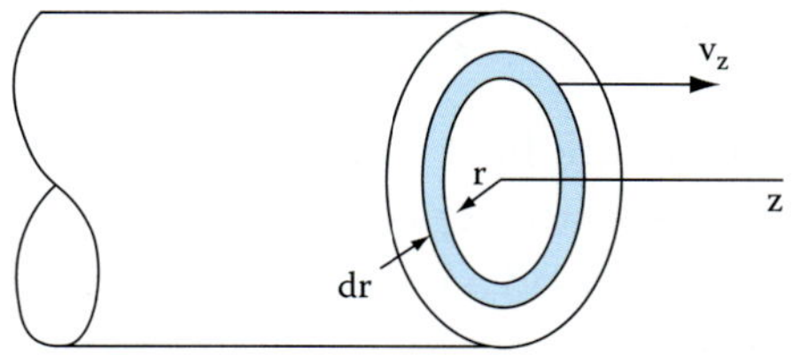

FIGURE 6.26 Volumetric flow rate in pipe flow.

What these scaling laws mean is that a 10 times reduction in the radius of the tube will mean a 10^4 reduction in volumetric flow and a 10^2 increase in pressure. Consequently, distributed micropumping, with each pump responsible for only a fraction of the total fluid movement, is expected to be important in microsystems. The required high pressures associated with mechanical pumping are a good reason to look at alternative propulsion techniques. Piezoelectric, electro-osmotic, electrowetting, and electrohydrodynamic pumping are all surface forces and scale more favorably than pressure pumping, which is a volume force, so they represent some considerable advantages for moving fluids in narrow capillaries.

Heat Conduction

Introduction

Many microdevices are actuated by thermal means (see Volume III, Chapter 8 on actuators). We are interested in learning about scaling laws in heat conduction, such as the total amount of heat required to start a thermal actuator, its heating and cooling rate, the thermal stresses or distortions expected in thermal sensors and actuators, and possible damage to delicate MEMS components. The three fundamental modes of heat transmission are conduction, convection, and radiation. They are governed by, respectively, Fourier's Law, Newton's Law of Cooling, and Stefan-Boltzmann's Radiation Law.

Fourier's Law of Heat Conduction

The total amount of heat, Q, flowing through a rectangular slab of material held at temperatures T_h and T_c (Figure 6.27) is proportional to the surface area A of the slab, the temperature difference between the two surfaces, and the time t during which heat flows, and it is inversely proportional to the distance the heat must travel ($\sim L^{-1}$), or:

$$Q = -\kappa \frac{A(T_h - T_c)t}{L} \tag{6.55}$$

where κ is the bulk heat conductivity in W/m K, and L is the plate surface separation. The value of κ is a measure of how good a heat conductor a given material is—for solids, κ generally increases with temperature. However, for most engineering materials at

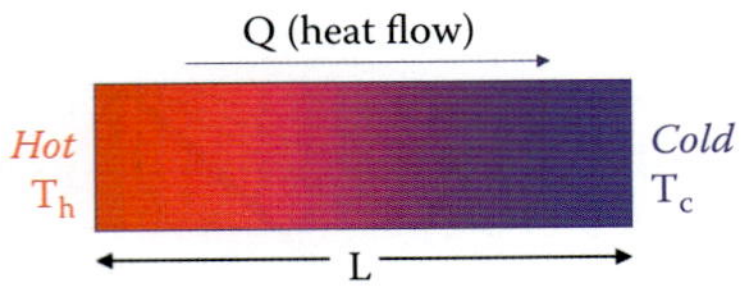

FIGURE 6.27 Heat conduction in a solid held at two different temperatures.

common operating temperatures, κ can be regarded as constant. Its value is 2.6 10^{-2} W/mK for air, 385 W/mK for copper, 50 W/mK for steel, 0.01 W/mK for Styrofoam, and 0.59 W/mK for water. The second law of thermodynamics requires the minus sign in Equation 6.55: thermal energy transfer from a thermal gradient must be from a warmer to a colder region. Heat transfer by conduction, from Equation 6.55, scales as l^{-1}, or a 10 times reduction in size of the device results in the same reduction in heat transfer.

Heat conduction or heat flux, q, is the heat flow per unit area and time, that is, $q = Q/At$ or, according to Fourier's Law in 1D Cartesian coordinates:

$$q = -\frac{\kappa}{L}(T_h - T_c) = -\kappa \frac{dT}{dx} \tag{6.56}$$

Conservation of energy for a small control volume within a larger structure requires that the stored energy in that volume (E_{st}) can be written as the sum of energy in (E_{in}) minus energy out (E_{out}) and the energy generated or lost in it ($E_{\text{internal conversion}}$):

$$E_{st} = E_{in} - E_{out} + E_{\text{internal conversion}}$$

$$\rho c_p \frac{\partial T}{\partial t} = \frac{\partial}{\partial x}\left(\kappa \frac{\partial T}{\partial x}\right) + q_m \tag{6.57}$$

where c_p is the heat capacity under constant pressure, ρ is the mass density, and q_m is the rate of internal energy conversion ("heat generation or source" or "heat sink"). The energy storage term (E_{st}) represents the rate of change of thermal energy stored in the matter in the absence of a phase change. The internal energy conversion term ($E_{\text{internal conversion}}$) in the case it acts as a source might be the result of chemical, electrical, or nuclear energy. This heat diffusion equation assumes there is no heat loss through the system boundary. The solution to this differential equation provides the temperature distribution in a stationary medium.

More generally, Fourier's Law of Heat Conduction may be written in vector notation as:

$$\mathbf{q} = -\kappa \nabla T \quad (6.58)$$

This equation determines the heat flux vector **q** for a given temperature profile *T* and thermal conductivity κ (a 3 × 3 tensor).

Just as we introduced kinematic viscosity in Equation 6.4, it is useful to introduce thermal diffusivity λ_d, in m²/s, which is related to thermal conductivity via:

$$\lambda_d = \frac{\kappa}{\rho c_p} \quad (6.59)$$

where ρ and c_p are, respectively, the mass density and the specific heat under constant pressure. Thermal diffusivity is a measure of how fast heat conducts. Its value is 2.24×10^{-5} m²/s for air and 10^{-7} m²/s for water.

We can now apply Fourier's Law to the generalized version of Equation 6.57, assuming κ is constant, and introduce the thermal diffusivity to yield the generalized heat diffusion equation as:

$$\frac{1}{\lambda_d}\frac{\partial T}{\partial t} = \nabla^2 T + \frac{q_m}{\kappa} \quad (6.60)$$

A common example for heat generation is the resistance heating of an electrical conductor. From Equation 6.60, scaling of the rate of heat conduction is measured by λ_d, and from Equation 6.59, we observe that $\lambda_d \sim \rho^{-1}$ and, since density ~ volume ~ l^3, this leads to the simple conclusion that a 10 times reduction in size results in 1000 times faster heat exchange.

Thermal diffusivity and kinematic viscosity (defined as the ratio of viscosity and density) have the same dimensions, and their ratio defines the important dimensionless Prandtl number:

$$\text{Pr} = \frac{\nu}{\lambda_d} \quad (6.61)$$

The interpretation of this number is the ratio of the fluid velocity boundary layer thickness to the fluid temperature boundary layer. This fluid property enables one to choose an optimum thermal transfer medium. For water, Pr is 10, and for air it is 0.71. For the most efficient heat transfer between two components at different temperatures, while minimizing viscous energy losses, one chooses media with a Pr <<< 1. This can be accomplished, for example, by using liquid metals (Pr = 10^{-3} to 10^{-1}). On the other hand, for lubrication of parts at different temperatures and keeping low thermal transfer between them, one chooses fluidic media with a Pr >>> 1 such as silicone oils (Pr = 10 to 10^7).

Convection: Newton's Cooling Law

Fourier's Law, as explained above, describes heat transport in solids through conduction. Convection heat transfer is the mode of energy transfer between a solid surface and an adjacent liquid or gas that is in motion, and it involves the combined effects of conduction and fluid motion. In Figure 6.28 we show velocity and temperature variation of air flowing over a hot substrate and illustrate the difference between forced and natural convection.

In Newton's Cooling Law, the heat flux between a solid surface at temperatures T_s and the bulk temperature of a fluid, T_∞, is proportional to the temperature difference, the exposed area *A*, and the convective heat transfer coefficient *h*:

$$q = hA(T_s - T_\infty) \quad (6.62)$$

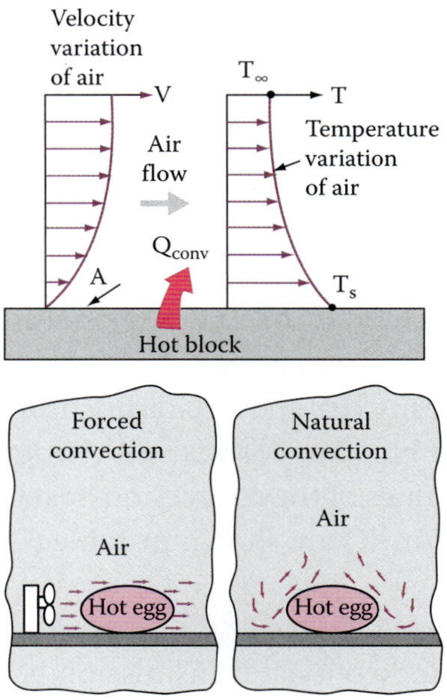

FIGURE 6.28 Velocity and temperature variation of air flowing over a hot substrate, and the difference between forced and natural convection.

The units of h are W/m²K. For most experiments, h is related to the fluid velocity u, or $h \sim u$, which in turn is proportional to l^{-1}. As in the case of conductive heat transfer, 10 times reduction in size will give 10 times reduction in heat transfer. It is important to keep in mind that the energy exchange at the solid/fluid interface is by conduction and that this energy is then convected away by fluid flow.

Convective heat transfer also can be discussed in terms of the Nusselt number (William Nusselt, 1882–1957), the dimensionless heat transfer coefficient:

$$Nu = hL/\kappa \tag{6.63}$$

with L (in W/mK) the characteristic length and κ the thermal conductivity of the solid, so this number may be viewed as the ratio of convection to conduction for a layer of fluid. With Nu = 1, we have pure conduction, and higher values mean that heat transfer is enhanced by convection. In a boundary layer situation, L is the thickness of the boundary layer. To determine a numerical value for Nu in a forced convection situation, one uses the relationship $Nu = \alpha (Re)^{\beta} (Pr)^{\gamma}$, and for a free convection at low velocity $Nu = \alpha(Re)^{\beta}(Pr)^{\gamma}(Gr)^{\delta}$. The parameters α, β, γ, and δ in this last expression are determined by the experimental setup. From the dimensionless numbers—the Reynolds number (Re), the Prandtl number (Pr), and the Grashof number (Gr)—Re has the dominant effect on h. The Grashof number Gr is defined as:

$$Gr = \frac{L^3 \rho^2 g \beta \Delta T}{\eta} \tag{6.64}$$

with L the characteristic length (m), ρ the density (kg/m³), g the acceleration resulting from gravity (9.080665 m/s²), β the coefficient of volumetric expansion (K⁻¹), ΔT the temperature difference between the wall and the surrounding bulk (°C), and η is the viscosity (Pa·s). The Grashof number is a ratio between the buoyancy forces and viscous forces.

Typical values of the convection heat transfer coefficient h (W/m²K) are shown in Table 6.2.

Radiation

The third mode of heat transmission is the result of electromagnetic wave propagation, which occurs both in fluidic media and in an absolute vacuum (Figure 6.29).

Table 6.2 Typical Values of the Convection Heat Transfer Coefficient h [W/m²K]

Free convection	**h**
Gases	2–25
Liquids	50–1,000
Forced convection	
Gases	25–250
Liquids	50–20,000
Convection with phase change	
Boiling or condensation	2,500–100,000

Radiation is the energy emitted by matter that is at a finite temperature. Radiant heat transfer is proportional to the fourth power of the absolute temperature, whereas we learned that conduction and convection are linearly proportional to temperature differences. The fundamental Stefan-Boltzmann's law $E = \sigma T^4$ (Chapter 3):

$$E = 5.67 \times 10^{-8} T^4 \tag{3.71}$$

where T is the absolute temperature, and the constant is independent of surface, medium, and temperature. The ideal emitter, or blackbody, is one that gives off radiant energy according to Equation 3.71. All other surfaces emit somewhat less than this amount, and the thermal emission from many surfaces (gray bodies) can be represented by:

$$E = \varepsilon\, 5.67 \times 10^{-8} T^4 \tag{6.65}$$

where ε, the emissivity of the surface, ranges from zero to one.

Example 6.4: Thermal Problem in Miniaturization

Let us apply the insights gathered so far to a simple practical problem, i.e., the calculation of

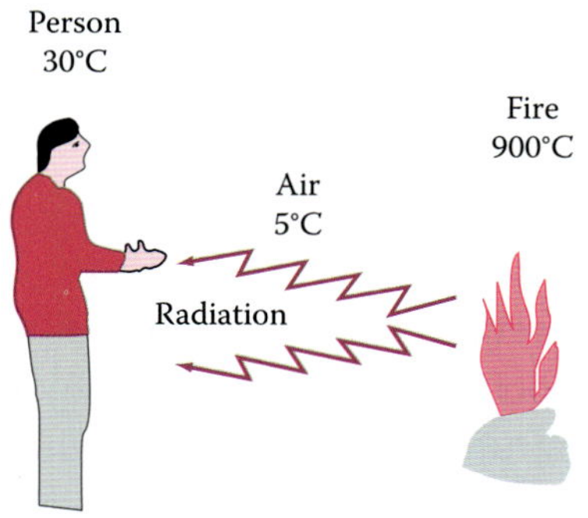

FIGURE 6.29 Example of radiation heat.

the thermal time constant associated with the sudden transfer of an object from one temperature to another. The heat flow into a material of volume V_0, at a uniform temperature T_0, through boundary surface A, when the volume is suddenly immersed in an environment of temperature T_∞, is equated to the rate of increase of the internal energy of the material in volume V_0, or:

$$q = hA\left(T_0 - T_\infty\right) = -\rho c_p V_0 \frac{dT}{dt} \tag{6.66}$$

Solving this differential equation gives us the temperature of the solid as a function of time:

$$\frac{T - T_\infty}{T_0 - T_\infty} = e^{-\left(\frac{hA}{\rho c_p V_0}\right)t} \tag{6.67}$$

The term on the left represents the dimensionless temperature, and it can be verified that the temperature in the solid decays exponentially with time and that the shape of the curve is determined by the time constant τ_c given by:

$$\tau_c = \frac{\rho c_p V_0}{hA} \tag{6.68}$$

To develop a feeling for the above expression, we compare how fast a Ni rod of 0.5-in. diameter and one with a 50-μm radius come to equilibrium with their surroundings. For a cylinder, with $V_0 = \pi r^2 l$ and $A = 2\pi r l$, the above equation transforms to:

$$\tau_c = \frac{\rho c_p r}{2h} \tag{6.69}$$

Assuming $h = 10$ W/m²K, $\rho = 8900$ kg m^{-3}, and $c_p = 0.44$ J kg^{-1}/K gives a $\tau_c = 1.24$ s for the bigger cylinder. If we assume equilibration with the surrounding temperature, T_∞, in 4 τ_c, equilibration is reached in 5 s. For the 50-μm Ni rod, equilibrium is reached after 0.039 s.

Biot's Number

When a solid body is placed in a surrounding of a different temperature, the body heats up or cools down, and internal temperature gradients that cause thermal stresses are set up. The dimensionless number that characterizes the magnitude of the established thermal gradient is the Biot number (Bi), the ratio of the average surface heat/transfer coefficient for convection from the entire surface (h), to the conductivity of the solid (κ), for a characteristic dimension L, or:

$$\mathrm{Bi} = \frac{hL}{\kappa} \tag{6.70}$$

As L becomes smaller, the Biot number goes down. For Bi ≪ 1, internal temperature gradients become small, and the body can be treated as having a uniform temperature. Consequently, with small devices there is less worry about thermal stresses induced by thermal gradients, and systems with larger internal heat generation capacity by volume can be built.[7] A representative application of the described effect is the micromachined planar Taguchi gas sensor shown in Volume II, Figure 9.11. Not only is the power required to operate at 350°C reduced by more than a factor of 4 (225 mW instead of 1 W), it also can be heated very quickly to high temperatures without breaking.[8]

With small things, we can heat and cool down many times a second (~20 Hz in the example of the 50-μm Ni rod) while thermal stress is minimized. This finds its application in fast thermal detectors such as thermocouples, high-performance heat sinks for high-density microelectronic devices, and fast actuators such as thermopneumatic valves in μ-PCR chambers, fast gas sensors, etc.

Reynolds Number, Friction Factor, and Hydrodynamic and Thermal Boundary Layers

Introduction

Three important concepts need some further dusting off before we comfortably can analyze fluidics in the microdomain and nanodomain, i.e., the Reynolds number, the friction factor, and the hydrodynamic and thermal boundary layers. The hydrodynamic boundary layer is compared with the electrical double layer and Nernst diffusion layer in Chapter 7.

Reynolds Number

The Reynolds number introduced in Equation 6.23 characterizes fluid flow as laminar or turbulent and includes fluid properties ρ and η, the characteristic length L, and average velocity V:

$$\mathrm{Re} = \frac{\rho V L}{\eta} \tag{6.71}$$

Characteristic length, as the term implies, is the length that best represents the body under consideration. In a capillary, for example, the diameter d, where it is called the hydraulic diameter (in m), is far more important in determining the nature of the flow than the total length of the capillary, L_t. The Reynolds number Re characterizes a liquid flow within a conduit of characteristic length L or around an object of characteristic length L. It may be viewed as the ratio of shear stress resulting from turbulence (i.e., inertial forces or forces set up by acceleration or deceleration of the fluid) to shear stress caused by viscosity. Similarly, it can be viewed as the ratio of convective forces over diffusional forces. For Re <<< 1, viscous and diffusional forces dominate, whereas for Re >>> 1, convective and inertial forces do. Similarity of flow will always be maintained if the Reynolds number has an identical value for the bodies being considered.[6] In laminar flow the resistance, caused by intermolecular viscous forces, is proportional to velocity, with turbulence the resistance, generated along the channel perimeter and related to channel shape, particle size, and concentration is proportional to the square of the velocity. Laminar flow is characterized by smooth flow of fluid in layers that do not mix: such flow also is called streamline flow. Turbulent flow is characterized by eddies and swirls that mix layers of fluid together. At sufficiently high velocity, a fluid flow can change from streamline to turbulent flow. The onset of turbulence can be found from the Reynolds number. If Re = 2000 or below, the flow is streamline; if 2000 < Re < 3000, the flow is unstable; and if Re = 3000 or above, the flow is turbulent.

Example 6.5:

Equation 6.71 may be used to calculate the Reynolds number for the flow of air ($\eta = 1 \times 10^{-6}$ m^2s^{-1}) and water ($\eta = 15 \times 10^{-6}$ m^2s^{-1}) in a 50-µm-diameter capillary. Assuming a velocity V = 500×10^{-6} ms^{-1} (10 times the diameter of the pipe per second), we obtain 0.0016 for air and 0.025 for water. Both are considerably less than 1, indicating that if fluids behave as a continuum, as is assumed in classical fluids, microsystems will operate in a viscous-dominated Stokes regime.

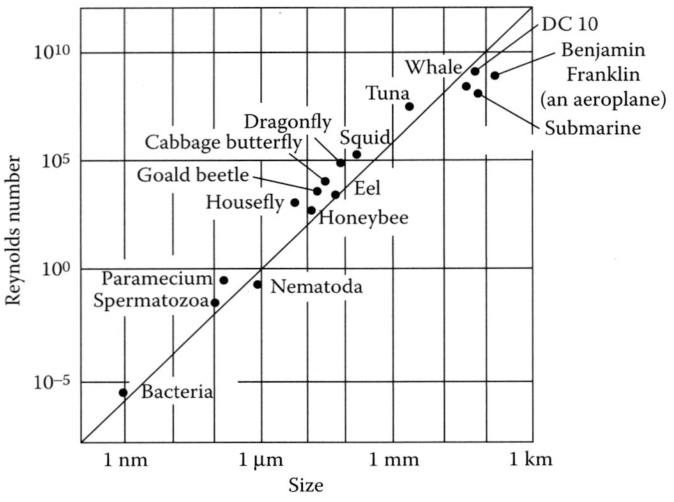

FIGURE 6.30 Reynolds number for mobile machines in their fluidic environment as a function of their size.

A tiny rotor blade in an air-filled, 10-µm-wide capillary will only be able to rotate in milliseconds rather than microseconds. The consideration of moving small things in water is even more daunting, because water is like syrup to the submerged micromachine. Fluid systems in the microdomain are damped heavily and exhibit slow response times. In Figure 6.30, the Reynolds number for mobile machines in their fluidic environment is plotted as a function of size (length), and Table 6.3 lists some additional example Reynolds numbers. Bacteria must be seen as truly autonomous micromachines. Swimming in an aqueous environment, such small vessels are up against the tide with an Re of 10^{-5} (see Table 6.3). Mechanical engineers have avoided designing such machines, but that is exactly how nature has done it, and bacteria are the most abundant life form on Earth.

Friction Factor

We touched upon frictional resistance when discussing a falling sphere and when deriving pipe flow. When a fluid flows through a fluidic channel it experiences frictional resistance, and as a consequence, useful mechanical energy is degraded into unusable heat energy. Knowledge of frictional characteristics is vital for the design of any fluidic system. For a tube of circular cross-section, friction loss or head loss can be determined in terms of shear stress at the wall, τ:

$$\tau = \frac{\Delta pL}{41} \quad (6.72)$$

Table 6.3 Example Reynolds Numbers and Hydrodynamic Layer Thickness

Description	Re	
Earth's tectonic plates (100 km at 10^{-8} m/s, kinematic viscosity is 10^{20} m²/s)	10^{-23}	
Glacier	10^{-11}	
Bacteria in water (1 μm at 10 μm/s and kinematic viscosity ν in water = 10^{-6} ma St)	10^{-5}	
Sperm cells in semen	10^{-3}	
A small marble falling in honey	10^{-2}	
Fish in a tropical aquarium	10^{2}	
Human swimmer	10^{5}	
Bird	10^{6}	
Car (1.5 m long at 90 km/h, kinematic viscosity ν in air = 15 × 10^{-6} ma St)	2.5 ×10^{6}	Size of δ, hydrodynamic boundary layer (mm)
		~0.9
Airplane taking off at 300 km/h, 9-m wingspan	50 × 10^{6}	1.2
Pleasure boat (10-m long at 21.6 km/h or 1,166 knots, and kinematic viscosity υ in water = 10^{-6} ma St)	60 × 10^{6}	1.3
Large fish in the ocean	10^{8}	

where l is the length of the tube in meters and L is the internal diameter of the pipe.

With reference to Figure 6.31, Equation 6.72 is easily derived from the fact that the force balance is given by:

$$p\frac{\pi L^2}{4} = (p - \Delta p)\frac{\pi L^2}{4} + \tau\pi L l \quad \text{or}$$

$$\tau = \frac{\Delta p L}{4l}$$

The shear stress, τ, at the wall is a function of five independent parameters: τ = f(η, ρ, V, L, k). The surface roughness of the pipe is given as k and can be described as a length. With five independent variables, the solution is very complex. However, the complexity can be reduced by use of only two nondimensional variables. The nondimensional shear stress, called the friction coefficient or Fanning friction factor, C_D, can be tabulated as a function of the Reynolds number and surface roughness k only. The Fanning friction factor C_D is defined as $2\tau/\rho V^2$ and

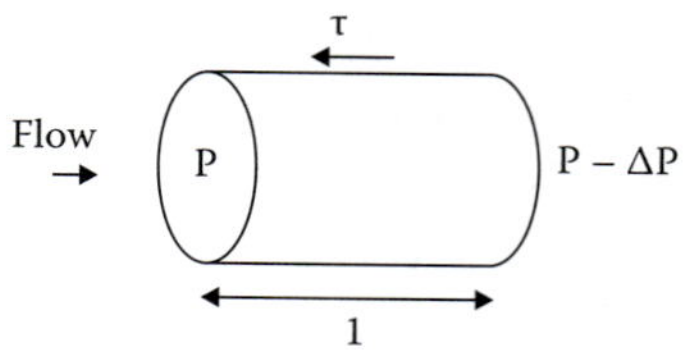

FIGURE 6.31 Friction in pipe flow ($\tau = F_D/A$).

it is used mostly for heat transfer problems in pipe flow. In other words (see also Volume III, Equation 7.27 with $\tau = F_D/A$):

$$C_D = \frac{F_D}{\frac{1}{2}\rho V^2 A} = \frac{\tau}{\frac{\rho V^2}{2}} = f\left(\frac{\rho V L}{\eta}, \frac{k}{L}\right) \tag{6.73}$$

with k/L the relative surface roughness. Civil engineers replace the Fanning friction factor frequently with the Moody friction factor, C_M, related to C_D as:

$$C_M = 4C_D = \frac{8\tau}{\rho V^2} \tag{6.74}$$

Yet differently, Coulson and Richardson define the friction factor as $\tau/\rho V^2$.[9] The friction factor for pipe flow of all incompressible Newtonian fluids is a unique function of the Reynolds number. This is a very important insight. Only two groups of variables need to be studied experimentally to obtain a relation that is universally valid for a wide class of fluids, geometries, and flow parameters. For a smooth pipe, Equation 6.73 reduces to C_D = f(Re) only. In laminar flow, we calculate from Equation 6.54 that the pressure drop, Δp, is given as:

$$\Delta p = -\frac{Q8\eta l}{\pi R^2}\left(\text{with } L = 2R \text{ and } V = \frac{Q}{\pi R^2}\right)$$
$$= -\frac{32\eta l V}{L^2} \tag{6.75}$$

Q is the volumetric flow rate, and from the balance of forces (see Equation 6.72):

$$\tau = \frac{\Delta p L}{4l} = \frac{8\eta V}{L} \tag{6.76}$$

For macroscopic laminar flow in long tubes, the friction factor, C_D, thus correlates to the Reynolds number as:

$$C_D = \frac{\tau}{\frac{\rho V^2}{2}} = f\left(\frac{\rho V L}{\eta}\right) = 16\frac{\eta}{\rho V L} = \frac{16}{R_e} \tag{6.77}$$

where C_D is a constant (for Re < 2200), also called the *Poiseuille number*, that depends on the cross-sectional shape but typically has a value of 16 for tube flow (see linear part of curve in Figure 6.32). Compare this with a Fanning friction factor for a sphere of 24 (see Figure 6.16). In case the Moody friction factor, C_M, for pipe flow is plotted, the slope is 64 instead.

The friction factor can be regarded as the ratio of the net imposed external force to the inertial force. Figure 6.32 represents Re versus friction factor C_D for different pipe sizes, relative pipe roughness, densities, and a broad range of viscosities.[1] Data from all pipe sizes and the entire viscosity range overlap and cover more than four decades in Re. At Reynolds numbers below 2200, the viscous shear effects are so large they balance the driving effect of the pressure, and a laminar, creeping, or Stokes flow condition results. For this type of flow, acceleration is negligible. Stokes flow is obtained either by a slow flow, that is, V is small, L is small, or η is large. It is in this regime that Equation 6.54, the Hagen-Poiseuille equation, is valid. At large enough Reynolds numbers, above 4000, a laminar flow becomes unstable and eventually turns turbulent. The region between 2100 and 4000 is a transition region, where transition depends on the size and frequency of disturbances inherent to the type of flow, the roughness of the boundaries, temperature, boundary flexibility, etc. For turbulent flow, the friction factor/Reynolds number dependence is weaker, but no simple theory exists and we must rely on experimental results. An empirical equation for turbulent flow is $f = 0.079\ Re^{-1/4}$.[1,5] The onset of turbulent flow, as is obvious from Figure 6.32, is related to the relative roughness of the pipe wall. As illustrated in Figure 6.33, the relative roughness is represented by the dimensionless group k/L. The larger that ratio, the earlier turbulence starts. For micromachined tubes, one might expect a large k/L and turbulence to kick in quickly. But some simple calculations show that the Reynolds numbers are so low that turbulence is not likely to be observed in microsystems even with a large k/L.

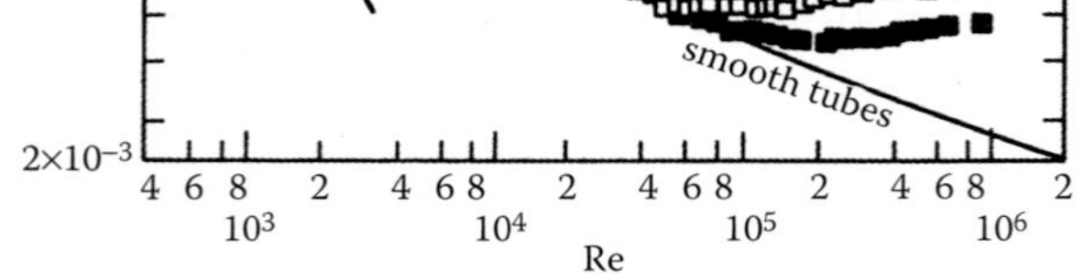

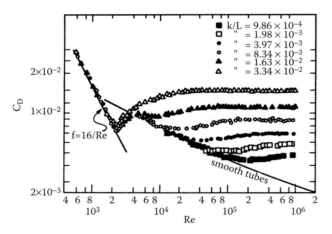

FIGURE 6.32 The Reynolds number, Re, is plotted versus friction factor, C_D, for different pipe sizes, densities, and a broad range of viscosities. Data from all pipe sizes and the entire viscosity range overlap. Different relative roughness factors k/L (caused by sand-roughened pipes) cause different onsets for turbulence. (From Denn, M. M. 1980. *Process fluid mechanics.* Englewood Cliffs, NJ: Prentice Hall. With permission.)

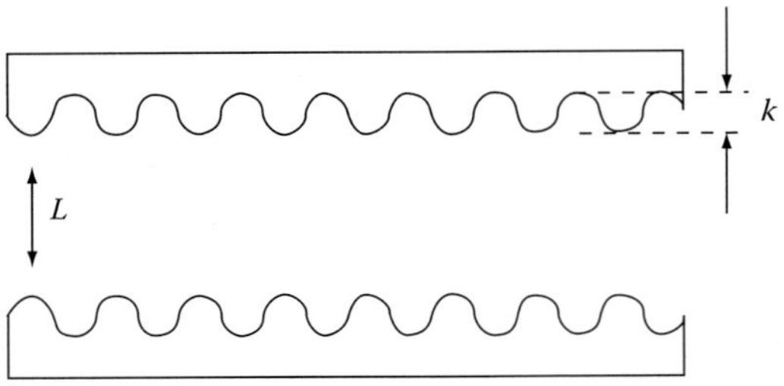

FIGURE 6.33 Schematic of a pipe wall of uniform, regular roughness.

Boundary Layers

The primary resistance to heat transfer and mixing by convection is controlled within a thin layer of stagnant fluid adjacent to the immersed body. In this hydrodynamic boundary layer, the flow velocity, V, varies from zero at the surface, that is, the no-slip condition, to the value in the bulk of the fluid, that is, V_∞. A quantity δ, the upper limit of the boundary layer, corresponds to the layer thickness when V has reached 0.99 V_∞. For laminar flow around an object of length L, δ is given by:

$$\delta \approx \frac{L}{\sqrt{\frac{VL}{\nu}}} \approx \frac{L}{\sqrt{Re}} \tag{6.78}$$

As the Reynolds number gets larger, the boundary layer gets thinner. From Figure 6.34 we can deduce how Equation 6.78 comes about. From the continuity equation it follows that:

$$\frac{\partial v}{\partial y} = -\frac{\partial u}{\partial x} \tag{6.79}$$

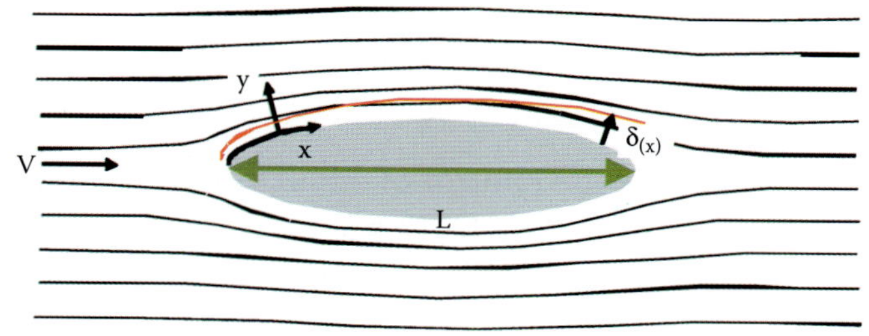

FIGURE 6.34 Boundary layer dimensions.

The scaling of the term, $\partial u/\partial x$, goes as V/L. If the length scale in the direction normal to the body is denoted as δ, i.e., the boundary layer thickness, and the normal velocity scale is denoted as U, it follows from the estimate $\partial v/\partial y \sim U/\delta$ that:

$$U \sim \frac{\delta}{L} V \quad (6.80)$$

The stream-wise equation of momentum balance, with p the pressure, and the scaling of each term annotated, is then given as:

$$u\frac{\partial u}{\partial x} + v\frac{\partial u}{\partial y} = -\frac{1}{\rho}\frac{\partial p}{\partial x} + \nu\left(\frac{\partial^2 u}{\partial x^2} + \frac{\partial^2 u}{\partial y^2}\right)$$

$$\frac{V^2}{L}, \quad \frac{V^2}{L}, \quad \frac{V^2}{L}, \quad \nu\frac{V}{L^2}, \quad \nu\frac{V}{\delta^2}$$

multiply by L/V^2 to obtain:

$$1, \quad 1, \quad 1, \quad \frac{\nu}{VL}, \left(\frac{L}{\delta}\right)^2 \frac{\nu}{VL} \quad \text{or also,}$$

$$1, \quad 1, \quad 1, \quad \frac{1}{Re}, \left(\frac{L}{\delta}\right)^2 \frac{1}{Re} \quad (6.81)$$

The case of interest here is that of a large Reynolds number (but not so large as to cause turbulence), so that $1/Re \ll 1$. The viscous terms in the Navier-Stokes equation in general are small compared with the other terms when the Reynolds number is large. However, this cannot be true everywhere; there must be at least some small region where viscosity is powerful enough to bring the tangential velocity to zero at the boundary. The small zone where viscosity remains important, even at larger Re, is in the boundary layer, δ. This means that in Equation 6.81 the last term remains, and we obtain Equation 6.78:

$$u\frac{\partial u}{\partial x} + v\frac{\partial u}{\partial y} = -\frac{1}{\rho}\frac{\partial p_d}{\partial x} + \nu\left(\frac{\partial^2 u}{\partial x^2} + \frac{\partial^2 u}{\partial y^2}\right) \quad \text{or}$$

$$\left(\frac{L}{\delta}\right)^2 \frac{\nu}{VL} \sim 1 \text{ or } \delta \sim \frac{L}{\sqrt{Re}}$$

Example 6.6

1. For a car being driven at 90 km/h (with $\nu = \pm 15 \times 10^{-6}$ ma St), the thickness of the hydrodynamic boundary layer is about 1 mm (see Table 6.3 for other examples).
2. In Volume II, Chapter 7, in the section on chemical vapor deposition (CVD), we calculated the average boundary layer thickness over a substrate on which a thin film is deposited, using the boundary layer model from Prandtl (Volume II, Equation 7.13) and obtained:

$$\delta = \frac{2L}{3\sqrt{Re}} \quad \text{(Volume II, Equation 7.13)}$$

When a fluid at one temperature flows along a solid surface that is at another temperature, a thermal boundary layer develops in addition to the hydrodynamic boundary layer. The thickness of the thermal boundary of a solid object in a fluid medium depends on the velocity of the fluid over the solid surface and the properties of the fluid medium. The equation to estimate the thermal boundary layer thickness, δ_T, is given by:

$$\delta_T \approx \frac{L}{\sqrt{\frac{VL}{\lambda_d}}} \approx \frac{L}{\sqrt{Pe}} \quad (6.82)$$

where λ_d is the mean free path. This equation is equivalent to Equation 6.78 for the estimation of hydrodynamic boundary layer thickness, where the Péclet number ($Pe = VL/\lambda_d$) replaces the Reynolds number. The Péclet number represents the ratio of convection over diffusion forces. The higher the Péclet number, the more the influence of flow dominates over molecular diffusion. Temperature and velocity profiles are identical when the Prandtl number equals 1, as then Pe and Re are identical. This is approximately the case for most gases ($0.6 < Pr < 1.0$). The Prandtl number for liquids, however, varies widely, ranging from very large values for viscous oils to very small values (on the order of 0.01) for liquid metals. One can deduce that $\delta/\delta_T \sim Pr^{1/2}$. In air ($Pr = 0.71$) the hydrodynamic boundary layer is thinner than the thermal boundary layer ($\delta = 0.84\ \delta_T$). In water ($Pr = 10$) the situation is reversed; the hydrodynamic boundary layer is thicker than the thermal one ($\delta = 3.12\ \delta_T$).

In Chapter 7 we learn that the relation between δ_d, the thickness of a steady-state diffusion layer, and δ_h, the thickness of the stagnant hydrodynamic layer on an electrode in a stirred solution, is given by $\delta_d = (D/\nu_k)^{1/2}\delta_h$. For $D = 10^{-5}$ cm^2 s^{-1} and $\nu_k = 10^{-2}$ cm^2 s^{-1}, this leads to $\delta_d = 0.1\ \delta_h$ (see note on page 581 "Nonlinear Diffusion Effects on Ultramicroelectrodes"). Obviously, the thickness of the diffusion boundary layer is considerably smaller than that of the hydrodynamic boundary layer. To prevent random convective motion from affecting transport to and from the sensing electrode, one wants the diffusion layer thickness to be smaller than the hydrodynamic boundary layer and δ_h to be regular. The latter explains the use of rotating electrodes, stirring bars, etc. The diffusion layer is in turn considerably thicker than the electrical double layer at an electrode surface.

Breakdown of Continuum Theory in Fluidics

Introduction

Microfluidics is the control of small volumes of liquids (fL–μL, 10^{-15}–10^{-6} L). In a short period of time, analytical chemists moved from flasks, to test tubes, to microfluidic chips and learned about some of the amazing physics in microfluidics (Table 6.4). Whenever an application demands reproducible handling of small amounts of liquid, microfluidics comes in handy to reduce liquid consumption, for a faster response, to afford disposable hardware, efficient automation (e.g., several functions on the same chip and parallel systems), reduction of power budget, portable devices, etc. In the micro- and nanodomain, Navier-Stokes equations (NSEs), with no-slip boundary conditions at the fluid interface, often seem to fail, and slip flow, creep flow, rarefaction, viscous dissipation, compressibility, intermolecular forces, etc., start demanding the introduction of molecular-based flow models. The breakdown of macroscale physics in gases readily is recognized, and some progress in molecular-based modeling has been made. For liquids, with molecules much more closely packed, evidence of deviation from macroscale fluidics laws is more recent, and modeling has proven to be much more elusive. For gases, breakdown of continuum theory starts being noticed

Table 6.4 Some Physics in Microfluidics

Channel of 100 μm × 100 μm × 1 cm	**Volume = 0.1 μL**		
Typical protein sample: concentration 0.1 mg/mL; average molecular weight = 50 kDa = 50,000 g/mol	Concentration = 2 μM	Number of molecules in microchannel, 2 pmol = 1.2 ×10^{11} molecules	
Channel of 100 nm × 100 nm × 1 μm	**Volume = 10 aL**		
Protein solution at 2 μM in nanochannel	Number of moles in nanochannel = 2 10^{-22}	Number of molecules in nanochannel = 12	
Pressure vs. flow rate	A few inches of water gives 100s of μm/s in a typical channel (l = 1 cm, r = 50 μm)		
Reynolds number ($Re = \rho VL/\eta$)	Small, Re < 100, no turbulence, viscous and diffusional forces dominate		
Péclet number ($Pe = VL/D$)	Large, 10 < Pe < 10^6; slow diffusion		
Diffusion coefficients	Small molecule	D = 500 μm^2/s	25 μm in 1 s
	Protein	D = 70 μm^2/s	8 μm in 1 s
	0.5-μm sphere	D = 0.5 μm^2/s	0.7 μm in 1 s
Surface/volume area	Huge: at small size scales the ratio of molecules in solution to molecules adhering to surfaces goes down and evaporation of droplets goes up		
Slip condition	Slip may occur at the fluid/solid boundary		

typically around 1 μm, whereas for liquids this happens around 10 nm.

Once we understand the behavior of fluids at the micro- and nanoscale better, we might be able to exploit many new phenomena. For example, viscous flow in nanochannels could be used to orient and assemble nanowires, separation without membranes is already possible in so-called diffusion-based *H*-separators, and electro-osmotic pumping results in attractive plug flow profiles for better analytical separation (see below).

The No-Slip and Slip Condition

Introduction

It is at an interface between solid and fluid where macroscopic meets microscopic and where microfluidics turns complex: near the contact, the physical properties of the fluid are not the same as in the bulk, and the influence of nanoscale interactions at the interface extends to macroscopic distances (Figure 6.35; see also van der Waals forces in Chapter 7). The no-slip condition means that at a boundary, fluids have zero velocity relative to the boundary. The fluid velocity is equal to that of the solid boundary. Conceptually, one can think of theoutermost molecule of a fluid sticking to the surfaces past which it flows. Stokes, in "On the effect of the internal friction of fluids on the motion of pendulums" (1851),[10] showed that such no-slip condition led to remarkable agreement with a wide range of experiments, including the capillary tube experiments of Poiseuille (1840)[11] and Hagen (1839).[125] However, the no-slip assumption at the interface is an idealization, assuming moderately strong attractive forces between fluid and wall. At the micro- and nanoscale, new phenomena arise because surface forces, and all types of microeffects, ignored at the macroscale, become important. These include friction, surface tension, air bubbles, liquid evaporation, porosity, osmotic effects, van der Waals forces, electrostatic forces, etc. In the case of slip, a slip length, L_s, must be introduced. Slip length is defined as the distance behind the interface at which the liquid velocity extrapolates to zero (see Figure 6.36a and b). In the case of electro-osmosis, with liquids moving in microchannels because of an electrical field, we will see that the electro-osmosis flow velocity V_{EOF} may be regarded as the slip velocity. Phenomena that can be explained assuming slip include a nonconstant pressure gradient in a long microduct and a flow rate higher than predicted from nonslip boundary conditions, electric currents needed to move micromotors that are a lot higher than expected, and micromachined accelerometers operating at atmospheric pressure that are overdamped.

Slip in Gases

Gas flow in devices with dimensions that are on the order of the mean free path of the gas show significant slip. When the mean free path of a gas, λ, approaches the separation, *L*, of the container plates (conditions easily met in a rarefied gas), the laws of high-pressure behavior begin to err. As the

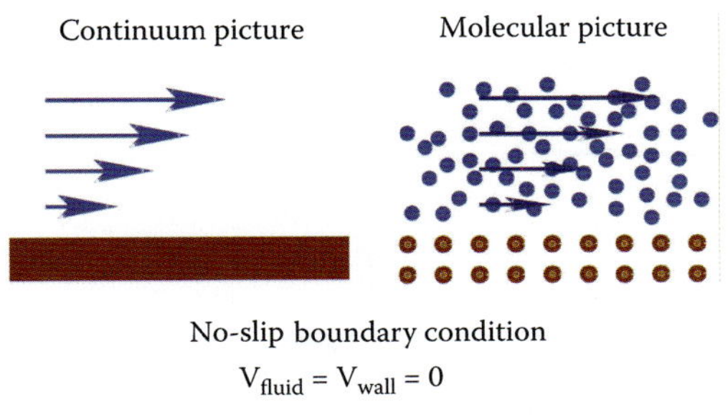

FIGURE 6.35 From continuum theory to the molecular picture at the interface between a solid and a fluid.

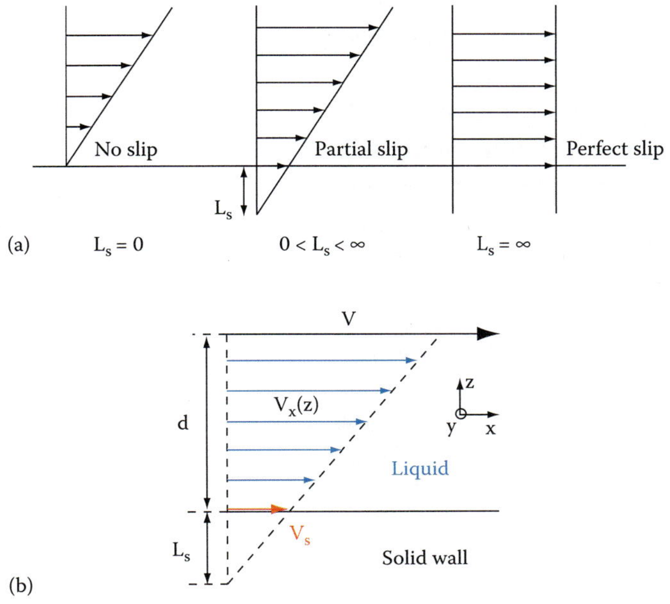

FIGURE 6.36 (a) To slip or not to slip. (b) Determination of slip length L_s.

density is lowered, the gas seems to lose its grip, so to speak, on solid surfaces. In gas flow it begins to slip over the surface, and in the conduction of heat a discontinuity of temperature develops at the boundary of the gas (see below). The same phenomena can be seen at high pressure with fluidic vessels of microdimensions. At the standard atmospheric conditions, the mean free path of air, λ, is about 0.06 μm, and with micrometer-sized channels in air, the no-slip condition does not apply; the friction factor starts to decrease with diameter reduction. The Knudsen number is a dimensionless number given by the ratio of the mean free path of the molecules to the characteristic dimension *L*:

$$\mathrm{Kn} = \frac{\lambda}{L} \tag{6.83}$$

which measures the deviation of the state of a gas from continuum behavior. As the value of the Knudsen number increases, noncontinuum effects become more important. If the characteristic dimension of the system, L in this case, is much larger than the mean free path λ, continuum equations apply (Kn << 1). A typical way to determine the conditions of flow is as follows:

For Kn < 0.001, continuum flow regime
For 0.001 < Kn < 0.1, slip flow regime
For 0.1 < Kn < 10, transition regime
For 10 < Kn, free molecular flow regime

These different Knudsen number regimes are summarized in Figure 6.37.

For tube flow (Equation 6.77), the corrected Navier-Stokes result for the friction coefficient is written as:

$$C_D = \left(\frac{16}{\mathrm{Re}}\right)\left(\frac{1}{1+8\frac{\lambda}{L}}\right) \tag{6.84}$$

For λ of a size approaching *L*, C_D is approximately 1/9 the value expected from n analysis assuming that the velocity at the wall equals 0. Experiments of flow resistance in 0.5-μm channels by Pfahler et al.[12] show a higher flow rate. This phenomenon is important for gas damping of oscillating microstructures, as they almost always operate in the size regime where λ approaches *L*.[13] Liu et al.[14] measured the pressure distribution of gaseous flow in microchannels by lining the channels with a series of pressure sensors. This group confirmed the Navier-Stokes equations with slip boundary conditions (first used by Arkilic et al.[15]).

Example 6.7

The macroexpression for the pressure drop through a tube is given in Equation 6.54 with the pressure gradient ~1/R². Deviations in gas flow start showing up already in channels with a hydraulic diameter of 100 μm. This is the case, for example, in a gas chromatography column of that diameter and a length of 1 m (making the length-to-diameter ratio 10,000). Because the channel is very narrow, the viscous effect becomes dominant over the inertial effect. For a long and narrow channel, the pressure and density changes between inlet and outlet are very large. This causes the flow to accelerate near the outlet and leads to a higher mass flow rate than predicted based on incompressible flow. As a consequence, when modeling this so-called compressible creep flow, one needs to include the compressibility effect by incorporating a large change in pressure and density, even though the gas is traveling at a very low subsonic speed.[16]

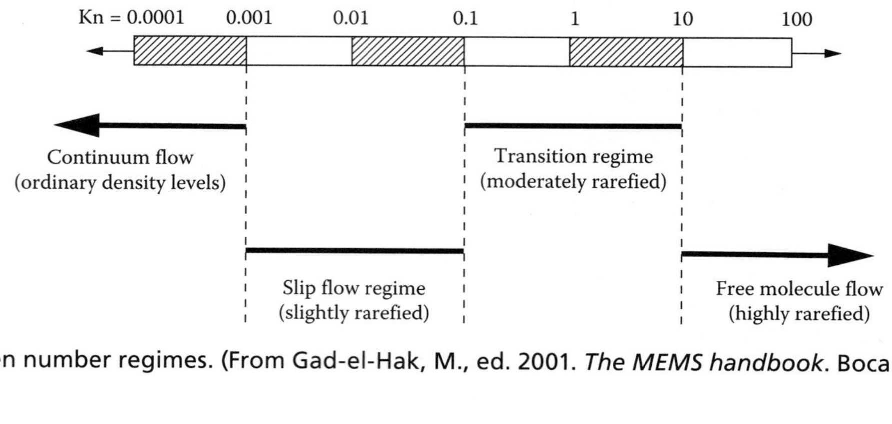

FIGURE 6.37 Knudsen number regimes. (From Gad-el-Hak, M., ed. 2001. *The MEMS handbook*. Boca Raton, FL: CRC Press.)

Corresponding to Equation 6.84, there also is a corrected friction factor for the gas velocity at the surface of a small aerosol particle in the presence of slip (the so-called Cunningham slip correction factor), given as:

$$C_D = \frac{24}{Re\left(1+\frac{2.52\lambda}{L}\right)} \tag{6.85}$$

This constitutes the corrected Equation 6.30, derived from continuum theory and valid for particles larger than 0.1 μm. For particles less than 0.1 μm, the primary transport mechanism is through Brownian motion and diffusion, with a diffusion coefficient *D* calculated from Equation 6.29, the Stokes-Einstein expression.

Let us analyze also how slipping of gas molecules at a solid surface influences the effectiveness of heat transport. As fluidic channels become smaller, the effect of boundary layers becomes dominant, and classical theories break down. In the case of a gas, the effectiveness of heat transport from one plate to another is controlled by the mean free path λ of the gas molecules, that is, the average distance before the next collision. The mean free path λ from the kinetic theory of gases is derived as:

$$\lambda = \frac{1}{\sqrt{2}\pi\rho a^2} = \frac{kT}{\sqrt{2}\pi p a^2} \tag{6.86}$$

where ρ = the density (the number of molecules per unit volume)
T = the temperature
k = the Boltzmann constant
p = the pressure
a = the size of the molecule

In Chapter 3 we calculated the bulk heat conductivity κ of a mole of solid (Equation 3.59); for a gas the same equation holds, except for the additional term for the gas density (ρ):

$$\kappa = \frac{1}{3}\rho c_v v\lambda \qquad \text{(3.59 adapted for gases)}$$

and with the mean velocity *v* given as $v = \sqrt{\frac{3kT}{m}}$ (3.20)

where c_v = the constant volume-specific heat capacity
λ = mean free path of the molecules
v = mean velocity, and
m = mass of the gas molecule

From Equation 6.86 we recognize that $\lambda \sim 1/\rho$, so that an increase in density decreases the mean free path with the product ρλ in Equation 3.59 remaining constant. Consequently, κ is not a function of density. We also confirm, from another inspection of Equation 3.59, that κ will increase with increasing temperature. What happens, though, when λ approaches *L*? As the density ρ decreases, no corresponding increase in λ occurs because λ is limited by *L*. In other words, boundary conditions start limiting the average value of λ within the gas, and the thermal conductivity goes down; the linear relation between heat flux and temperature gradient and the no-jump temperature condition at the solid-fluid interface are no longer accurate. A qualitative corrective factor to the heat flux in Equation 6.56, considering the effect on both plates, sometimes is used for small gaps:

$$q = -\frac{\kappa}{L+2g}(T_h - T_c) = -\kappa\frac{dT}{dx} \tag{6.87}$$

where, for conduction in air, oxygen, nitrogen, carbon dioxide, methane, and helium, experimental *g* values satisfy $2.4\lambda < g < 2.9\lambda$. The macroscopic prediction for the heat flux exceeds the microscale prediction by less than 5% when $L > 12\lambda$. For hydrogen, however, the experimental value for *g* is 11.7λ, and the condition $L < 47\lambda$ must be satisfied.[17] Equation 3.59 also determines the boundary for microscale conduction in solids and liquids.

For solid-state thermal conduction in metals Equation 3.59 holds, with λ the electron mean free path, C_v the electron-specific heat, and v the Fermi velocity (see also Chapter 3). In dielectrics, fluids, and semiconductors, λ is the phonon mean free path, C_v the phonon-specific heat, and *v* the average speed of sound. Experimental *g* values for solids are not yet available. However, calculations indicate that macroscale exceeds microscale results by less than 5% for $L > 7\lambda$. Consequently, the boundary between the microscale and macroscale regimes equals L = 7λ for the conduction across a layer, where *L* is the layer thickness and λ the mean free path of the dominant carrier of heat. Given the dimensions of a phonon at least 10 times smaller than λ of a gas at atmospheric pressure, we might not observe deviations from the Navier-Stokes equations in liquids until the dimensions of the containing vessels are in the 100-Å range.

Maps for heat transfer regimes in microstructures, immediately showing whether a certain heat transport in a given device can be analyzed with macroscale theory, have been developed.[17]

Slip in Liquids

Deviations from the Navier-Stokes equations in liquid are more difficult to demonstrate than in a gas. Liquids do not have a well-advanced molecular-based theory as dilute gases; for example, there is no Knudsen number for liquids, and the concept of a mean free path in a liquid is very different than for a gas. Gravesen et al.[13] in 1993 reviewed some of the literature describing deviations from Navier-Stokes equations in liquid flow in narrow channels but could not find enough consistency in the data to conclude that there was a breakdown of the macrotheory. More clarity has emerged since 1993. In controlled experiments, generally with dimensions of microns or smaller, apparent violations of the no-slip boundary condition clearly have been demonstrated in liquids. We usually assume that viscosity in liquids is independent of the dimensions of the flow channel. However, experiments by Harley et al.[18] suggest that in microchannels liquid flow takes on values as much as an order of magnitude greater than that expected from a constant viscosity. Pfahler et al.[12] similarly measured a flow significantly higher than expected from macrotheory when using isopropanol and silicone oil. Israelachvili[19] showed that the apparent viscosity of a liquid is the same as that in the bulk once the film is thicker than a dozen molecular layers (~5 nm); below that thickness the viscosity grows by several orders of magnitude. Tretheway et al.,[20] using a 30 × 300-μm channel made hydrophobic with a 2.3-nm-thick monolayer of octadecyltrichlorosilane, measured fluid slip velocities within 450 nm of the wall of about 10% of that in the center instead of zero. This slip condition has the effect of increasing the total volume flow. In the case of a hydrophilic surface, velocity profiles remained consistent with Stokes flow and the well-accepted no-slip boundary (see Figure 6.38). Slip of liquids was also measured with hexadecane flowing over an atomically smooth surface, such as mica, that again was made nonwettable by deposition of a hydrophobic monolayer.[21]

From the above, we conclude that slip causes a "shift" in the velocity profile. It also has been established that slip increases with the contact angle,[22] decreases with increasing surface roughness,[22] and increases with viscosity and driving rate.[23]

In flow channels approaching molecular diameters, "molecular" effects turn yet more dramatic. When the channel is less than 10 molecular diameters, of the order of a few nanometers, the fluid loses its liquid-like behavior and assumes some solid-like characteristics.[24] The structure of the molecules changes from a random order to a discrete number of ordered layers, with an "apparent" viscosity of the fluid 10^5 times the "regular" viscosity. The molecular time scale for molecules in these nanotubes slows down by a factor of 10^{10}.[25] In Figure 6.39, the friction force between two parallel plates, separated by a few

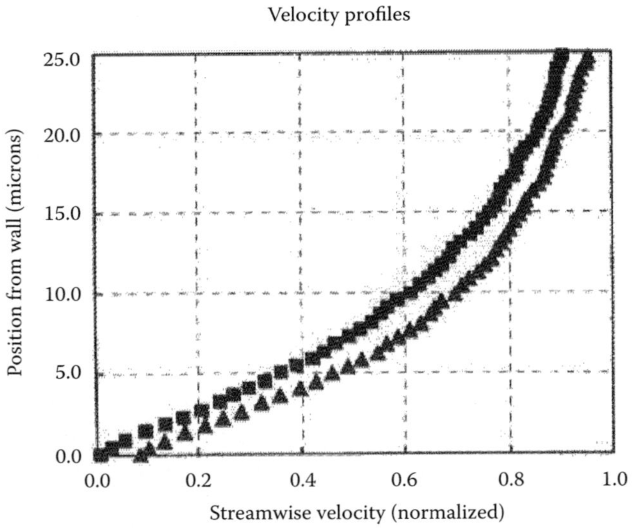

FIGURE 6.38 Measurement of slip length. Square symbols: hydrophilic coating ($L_s = 0$), triangle symbols: hydrophobic coating ($L_s = \sim 1$ μm). The velocity at 450 nm from the wall was measured at 10% from that in the middle of the flow (instead of 0). (Tretheway, D. C., and C. D. Meinhart. 2002. Apparent fluid slip at hydrophobic microchannel walls. *Phys Fluids* 14: L9.)

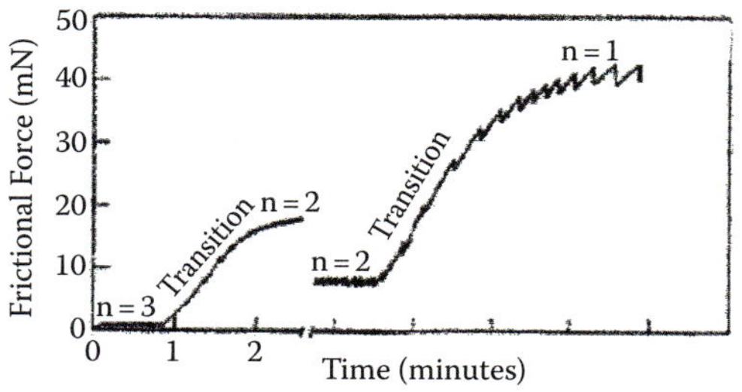

FIGURE 6.39 Friction forces between two parallel plates. (Ho, C. M. 2001. 14th IEEE International Conference MEMS, Interlaken, Switzerland, 375.)

molecular layers of OMCTS (octamethylcyclotetrasiloxane), sliding with respect to each other while the distance between them decreases (horizontal axis), is shown. Two very striking features are apparent. First, the frictional forces increase in a stepwise fashion as the distance between the plates is reduced and settles on a stable value corresponding to an integer number of molecular layers (1, 2, or 3). Second, a saw-tooth behavior is observed, corresponding to a stick-stick motion of the plates when the molecular layers are rearranging themselves while moving with respect to each other. When working with macromolecules such as DNA and proteins in nanochannels, effects as described here are to be expected.

Instead of continuum theory, molecular dynamics methods are used to study interfaces at the molecular level. In molecular dynamics one integrates Newton's equations of motion for each individual particle. A commonly used force between the interacting molecules is the Lennard-Jones force introduced in Chapter 7 (Equation 7.22).

Forces at Interfaces

Introduction

Molecules condense when attractive intermolecular bond energies are comparable with or greater than their thermal (i.e., kinetic) energy. These condensed materials have a surface: a transition region between condensed phase [S(olid) or L(iquid)] and a gas phase. An interface is a region between two condensed phases. At interfaces, atoms and molecules possess energies and reactivities significantly different from those of the same species in the bulk. That imbalance leads to surface free energy. The imbalance of forces at a gas/liquid interface usually is termed the interfacial tension (in units of N/m or also J/m^2), whereas for interfaces involving solids, it is called "surface energy." We start by illustrating the surface forces at the liquid/gas interface, i.e., surface tension. Next we study the solid/liquid interface to elucidate beading of droplets and capillary action.

Surface Tension

Many of the properties of a liquid, including surface tension, capillary action, and beading, are caused by cohesion and/or adhesion of molecules. In liquids, the rearrangement after creating a new interface is very fast, and the surface tension is an equilibrium value. Liquids and gases are separated by a meniscus and differ only in density, not structure. Molecules at the surface of a liquid experience forces around and below, whereas molecules in the middle are attracted from all sides, so the net attraction on molecules at the top is downward (Figure 6.40). Increasing the surface area results in more molecules that are exposed and a higher overall energy, so the surface area tends toward the smallest possible area to reduce the overall energy of the system. A liquid thus arranges itself to have less surface area. Water forms spherical drops because a sphere exhibits less surface to volume.

The energy required to break through the surface of a substance or to spread a drop of the substance out into a film is the surface tension. Fluids behave as though they have a surface composed of an elastic skin that is always in tension. Surface tension, with reference to Figure 6.41, is the amount of energy

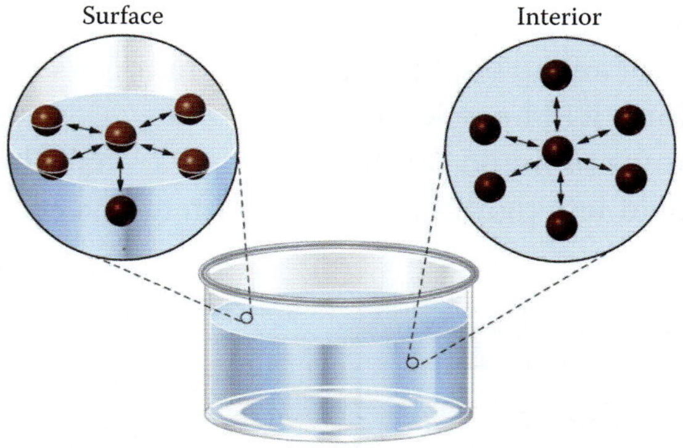

FIGURE 6.40 The molecular origin of surface tension. Molecules at the surface of the liquid experience forces around and below, whereas molecules in the middle are attracted from all sides, so the net attraction on molecules at the top is downward.

required to stretch or increase the surface area of a liquid by a unit area. The force exerted by a planar film supported on a rectangular frame with one movable bar of length *l*, as shown in Figure 6.41, is given by:

$$F_s = 2\gamma l \quad (6.88)$$

The factor of two is introduced because the liquid film has two surfaces. The work required to create new surface area is then:

$$W = 2\gamma l d \text{ and } 2\gamma = \frac{dW}{dA} \quad (6.89)$$

An interfacial energy γ is associated with any interface between two phases (J m^{-2}) (also called a surface tension: Nm^{-1}). An increase in area increases the system's free energy, which is not thermodynamically favorable. Water has the highest surface tension of all common liquids.

Drops and Bubbles

The forces on a liquid sphere (say, a drop of water) of radius *r* can be calculated as follows. Balance of forces in Figure 6.42 requires that $\sum F = 0$. So we can write:

$$F_{in} - F_{out} - F_{surface} = 0$$

$$p_{in}\pi r^2 - p_{out}\pi r^2 - \gamma 2\pi r = 0$$

$$(p_{in} - p_{out})\pi r^2 = \gamma 2\pi r$$

$$p_{in} - p_{out} = \frac{2\gamma}{r} \text{ or also } \Delta p = \frac{2\gamma}{r} \quad (6.90)$$

In this example there is only one liquid/gas surface. For a soap bubble, with two liquid/gas surfaces, the force will be twice as great.

Atoms of liquid on the surface of a small droplet are held less tightly compared with atoms on a flat

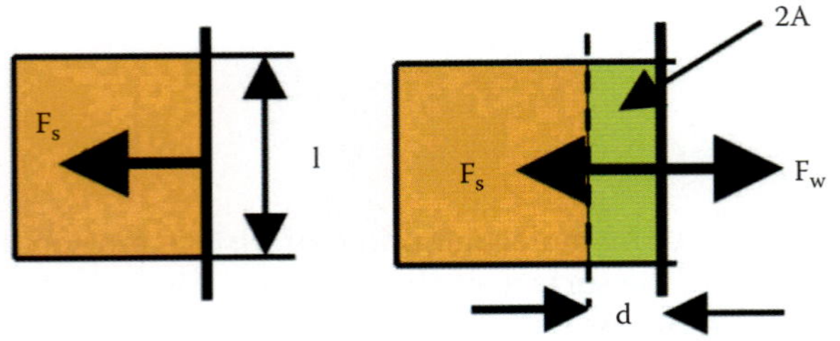

FIGURE 6.41 The work involved in expanding a liquid film by 2A (=2dl).

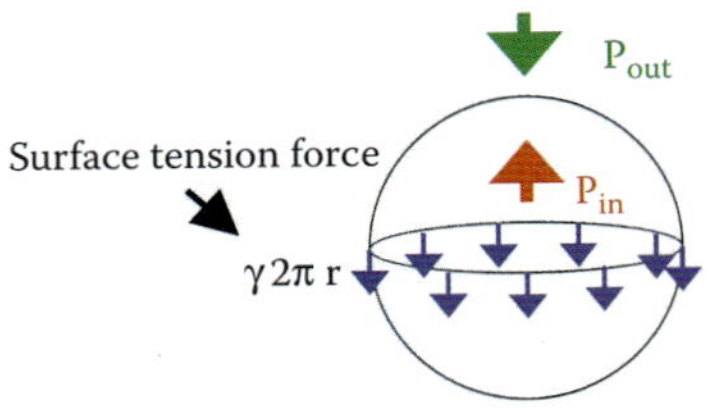

FIGURE 6.42 Pressure difference across a curved surface.

liquid surface. High curvature effectively reduces the coordination number of the surface atoms, making them easier to evaporate. This is manifested as an increase in the vapor pressure *p* of a liquid droplet compared with that of a bulk liquid p_0 and is described in the Kelvin equation. In the case of a positive curvature, the liquid in drops has a higher vapor pressure than in the bulk. In the case of negative curvature, it has a lower vapor pressure than in the bulk (e.g., liquid in pores). The Kelvin equation for a droplet is given as:

$$\ln\frac{p}{p_0} = \frac{2\gamma\bar{V}}{RTr} \quad (6.91)$$

In this equation:

r = the drop radius
γ = the surface tension
$\bar{V}$ = the molar volume of the liquid
R = the gas constant
T = the absolute temperature

From Equation 6.91 the vapor pressure *p* relative to the bulk p_0 can be calculated as:

$$p = p_0 \exp\left(\frac{2\gamma\bar{V}}{RTr}\right) \quad (6.92)$$

Evaporation of a droplet of diameter *d* at time *t* is given as:

$$d^2 = d_0^2 - \beta t \quad (6.93)$$

where d_0 is the initial diameter, and β is a constant independent of drop size. The time it takes for the drop to disappear is then:

$$t = \frac{d_0^2}{\beta} \quad (6.94)$$

The latter scales as l^2, and clearly miniaturization favors faster evaporation. Fast evaporation of small

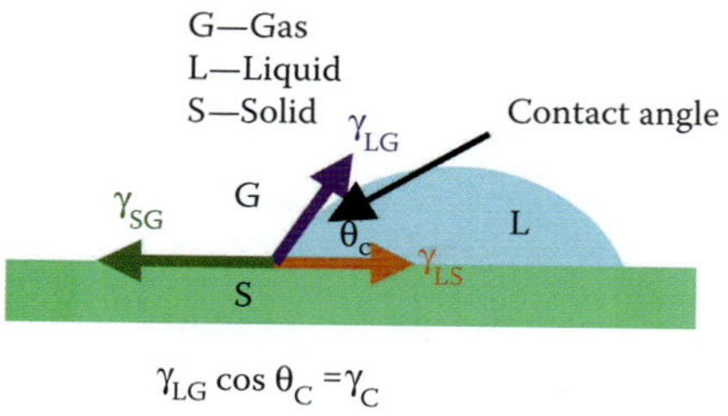

FIGURE 6.43 The horizontal force balance between these three interfacial tensions represents Young's equation.

liquid droplets often poses a considerable problem in MEMS. A droplet ejected from a small nozzle might evaporate before it reaches the substrate, which makes storing small amounts of chemicals on a liquid platform very cumbersome for the same reason.

Wetting and Contact Angle–Beading on Solid Surfaces

We now introduce a solid surface in a liquid and define adhesion forces and wetting of a solid surface. From the previous section we know that increasing the surface area of any liquid requires work, and this effort per unit surface area is the liquid's surface tension (with a unit of N/m and symbol γ). When liquid of a sessile drop comes in contact with both a solid and a gas, as shown in Figure 6.43, a contact angle θ_c of between 0 and 180° is set up. Three interfacial tensions come into play: γ_{SG}, the solid/gas interface tension; γ_{SL}, the solid/liquid interface tension; and γ_{LG}, the liquid/gas interface tension. The horizontal force balance between these three interfacial tensions represents Young's equation:

$$\begin{aligned}
&\sum \gamma = 0 \\
&\gamma_{LS} + \gamma_{LG}\cos\theta_c - \gamma_{SG} = 0 \\
&\gamma_{LS} + \gamma_{LG} 2\pi r \cos\theta_c - \gamma_{SG} = 0 \\
&\frac{\gamma_{LS} - \gamma_{LG}}{2\pi r} + \gamma_{LG}\cos\theta_c = 0 \\
&\frac{-\gamma_{LS} + \gamma_{LG}}{2\pi r} = \gamma_{LG}\cos\theta_c = \gamma_c
\end{aligned} \tag{6.95}$$

where θ_c is the contact angle. The critical surface tension γ_c is an intrinsic characteristic of the surface. Liquids with $\gamma < \gamma_c$ completely wet the surface ($\theta = 0°$). Liquids with $\gamma > 90°$ are said to not wet the surface ($\gamma_{LS} > \gamma_{SG}$).

The surface of plant leafs, especially of the lotus flower (Nelumbo nucifera), the symbol of purity in some Asian religions, can show extreme hydrophobicity or superhydrophobicity to water. This makes the deposition of water drops on their surface almost impossible (large water contact angles $\theta > 150°$) (Figure 6.44). Such hydrophobic surfaces also show a remarkable self-cleaning effect. Even after emerging from mud, the lotus leaves do not retain dirt when they unfold. The prospect of self-cleaning surfaces is obviously of significant commercial interest.

Two botanists, Barthlott and Neinhuis, from the University of Bonn, discovered the reason for this self-cleansing effect.[27] The general opinion used to be that the smoother a surface, the less dirt and water adhere to it. However, when inspected with an SEM, lotus leaves show a combination of nano- and microstructures that make for a rough surface at the micro- and nanoscale. The explanation for the resulting effect lies in two physical characteristics of the

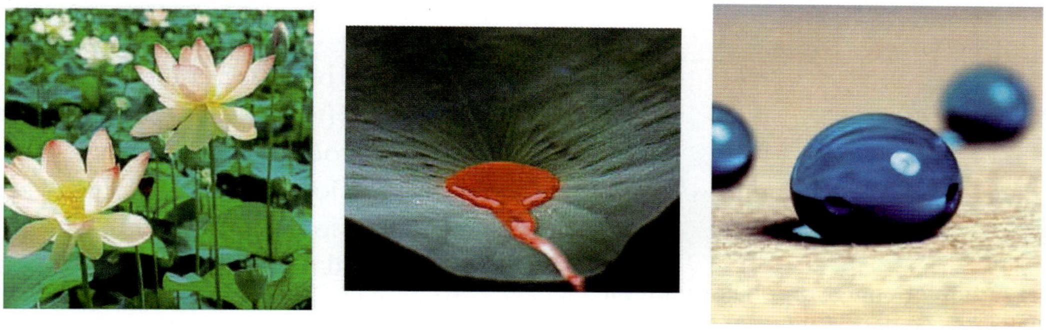

FIGURE 6.44 Extreme hydrophobicity of lotus plant leafs. The far right photo represents a commercial application of the Lotus Effect; i.e., a BASF aerosol spray for paper, wood, leather, etc., makes those materials superhydrophobic.

lotus leaf, i.e., the properties of these microstructures repel water, and the nanostructures found on top of the microstructures are made of waxy materials that are poorly wettable. The combination of the chemistry, the nanostructures, and the non-adherence of dirt and water to the micro-rough surface is what Barthlott and Neinhuis named the Lotus Effect. The roughness of a surface improves the wettability for hydrophilic surfaces ($\theta < 90°$). A drop on such a surface will appear to sink into the hydrophilic surface. For hydrophobic surfaces ($\theta > 90°$) the wettability decreases as it gets energetically too expensive to wet a rough hydrophobic surface. The result is an increased water-repellency. Energetically the best configuration for the drop is then on top of the roughness spikes like "a fakir on a bed of nails" (Figure 6.45).

A droplet on an inclined superhydrophobic surface rolls off instead of sliding. When the droplet rolls over a contamination, the particle is picked up from the surface, and the energy of absorption into the rolling droplet is higher than the surface stiction force between the dirt particle and the surface. Usually the force needed to remove a particle is very low because of the minimized contact area between the particle and the surface. As a result, the droplet cleans the leaf by rolling off the surface (Figure 6.46). See also Volume III, Chapter 3, page 144 "Superhydrophobicity and the Lotus Effect".

Capillary Action

The beading, illustrated in Figure 6.44, occurs when a polar substance is placed on a nonpolar surface and cohesive forces compete with adhesive forces. There is also a competition between adhesive and cohesive forces in capillary action. If adhesive forces are greater than cohesive forces, the liquid surface is attracted to its container more than the bulk molecules. Therefore, the meniscus is U-shaped (e.g., water in glass) (Figure 6.47Aa left). If cohesive forces are greater than adhesive forces, the meniscus is curved downward (Figure 6.47Aa right). In capillary action, liquids rise spontaneously in a narrow tube (Figure 6.47Ab).

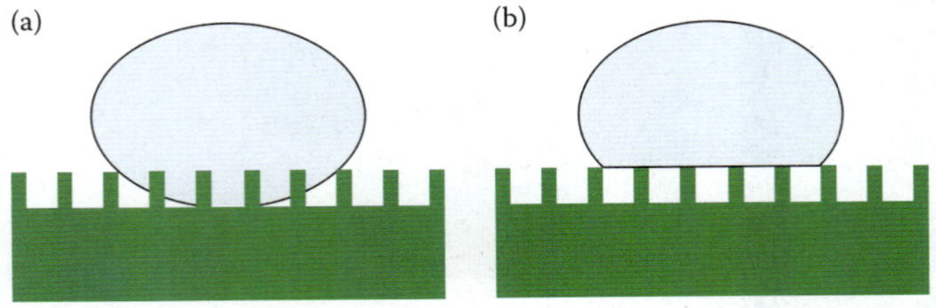

FIGURE 6.45 (a) A droplet on a hydrophilic rough surface seems to sink into the gaps. (b) A droplet on a rough hydrophobic surface sitting on the spikes. (From http://lotus-shower.isunet.edu/the_lotus_effect.htm.)

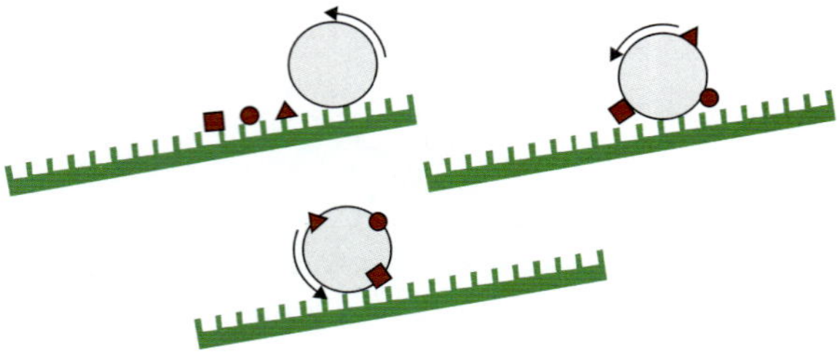

FIGURE 6.46 Self-cleaning behavior of a superhydrophobic surface by droplets rolling off instead of sliding off. (From http://lotus-shower.isunet.edu/the_lotus_effect.htm.)

We are all familiar with water being sucked up by a hydrophilic capillary (e.g., in a glass capillary) as a result of positive capillarity. Less evident from daily experience is that water may have to be pushed up in the case of a hydrophobic capillary as a result of negative capillarity (e.g., in a Teflon tube). Figure 6.47B illustrates the two types of capillarity schematically. In the positive capillarity case, liquid in a capillary rises as to seek a lower interfacial energy. The liquid will rise until it lifts a column of liquid balancing the energy reduction gained by rising. The energy reduction by rising over a distance l in the capillary is given by $(\gamma_{SG} - \gamma_{SL})\, 2\pi rl$. In this expression, r is the capillary radius, and the work performed in raising the liquid column is given by $\Delta p \pi r^2 l$, with Δp the pressure difference between the top and bottom of the raised column. By equating these two energies, we can deduce the pressure drop across a liquid/gas interface in a small capillary as:

$$\Delta p = \frac{(\gamma_{SG} - \gamma_{SL})}{r} = \frac{2\sigma_L \cos\theta_c}{r} \tag{6.96}$$

where it has been assumed, as is generally the case, that $\gamma_{LG} = \sigma_L$ (the surface tension of the liquid). The pressure produced by wetting, calculated from this expression for typical values of the gas/liquid surface tension ($\gamma_{LG} = \sigma_L = 0.073$ N/m at room temperature and 0.058 N/m at 100°C) and a capillary radius of 10 μm, is in the range of 0.01 MPa. This is roughly the same as achieved with piezoelectric pumps, which are much larger in size.

Capillary forces scale with l^1, and as we will see in Volume III, Chapter 7 on scaling, the smaller the exponent of l, which is 1 in this case, the more favorable the scaling. The Kelvin equation for a liquid in a cylinder (say, in a pore of a solid) as shown in Figure 6.47 is, in analogy with Equation 6.91 for a droplet, given as:

$$\ln\frac{p}{p_0} = \frac{\gamma \bar{V}}{RTr} \tag{6.97}$$

In this equation:

- r = the cylinder radius
- γ = the surface tension
- $\bar{V}$ = the molar volume of the liquid
- R = the gas constant
- T = the absolute temperature

The difference is a factor of 2 because for a liquid column there is only 1 interface.

An air bubble can completely block a microchannel, and Colgate et al.[28] report that the pressure difference, Δp, needed to move a gas bubble in a straight channel is empirically given by:

$$\Delta p = \frac{2\gamma_f}{r} \tag{6.98}$$

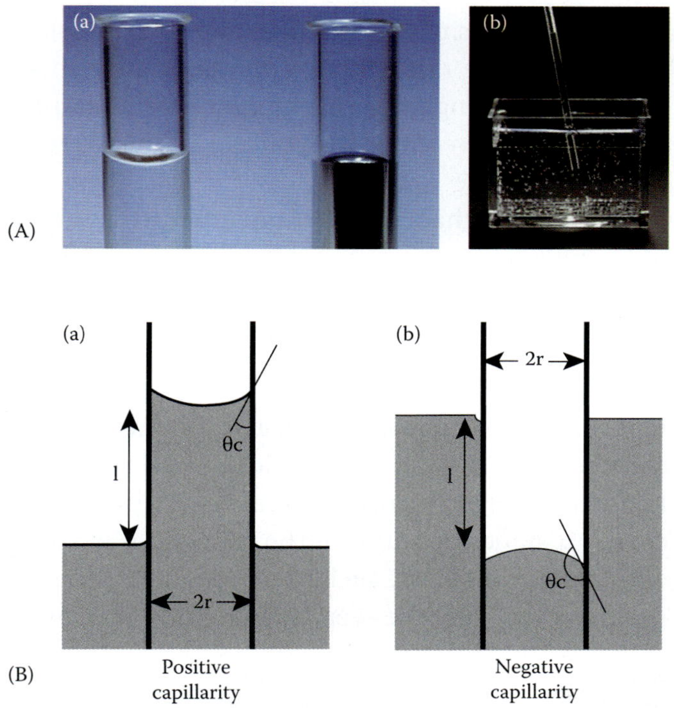

FIGURE 6.47 (A) (a) Positive (left) and negative (right) capillarity; (b) Liquid rises in a capillary. (B) Capillary phenomena: positive (a) and negative (b) capillarity.

Mathematically this equation, not surprisingly, has the same form as the expression derived in Equation 6.90 or a liquid sphere with γ_f now a frictional surface tension. The pressure differences in Equations 6.90, 6.96, and 6.98 all increase with decreasing channel dimension or bubble size.

Mixing, Stirring, and Diffusion in Low Reynolds Number Fluids

The Péclet Number

Stokes or low Reynolds number flow, we saw above, occurs when Re << 1; this might be the case because ρ, *V*, or *L* are very small, e.g., in microorganisms, MEMS, particles, bubbles, etc., or it might happen because η is very large, e.g., lava flow, honey, etc. In these regimes viscous forces dominate, and inertial strategies do not help for moving microsystems in a medium (swimming) nor are they any help in pushing microvolumes of fluid (mixing). The question we address in this section is how to mix reactants in a small microreactor in the absence of turbulence? The primary resistance to mixing by convection, we saw earlier, is controlled by a thin layer of stagnant fluid adjacent to a solid surface. In this hydrodynamic boundary layer, the flow velocity, *V*, varies from zero at the surface, that is, the no-slip condition, to the value in the bulk of the fluid, that is, V_∞. Laminar flow around an object of length *L* and boundary layer δ is given by:

$$\delta \sim \frac{L}{\sqrt{Re}} \tag{6.78}$$

As the Reynolds number becomes smaller, the boundary layer gets larger. So one appreciates that mixing in the microdomain (< 1 mm) is difficult because flows are predominantly laminar, and mechanical stirring is difficult in restricted fluidic channels. Stated differently: with very low Re numbers, inertial convective mass transport effects (like turbulence) are not feasible. Or also: at the microfluidic level, mixing is like trying to stir syrup into honey, and two liquids, traveling side-by-side through a narrow channel, only become mixed after several centimeters because mass transport in the microdomain is traditionally limited to simple diffusion (Figure 6.48). To decide which transport

FIGURE 6.48 Laminar flows do not mix. Taking advantage of low Reynolds number (Re) and high Péclet number (Pe).

type dominates, diffusion or convection, one must inspect the Péclet number.

The Péclet number represents the ratio of mass transport by convection to mass transport by diffusion (see also Table 6.4):

$$Pe = \frac{VL}{D} \tag{6.99}$$

The higher the Péclet number, the more the influence of flow dominates over molecular diffusion (see also Equation 6.82). In liquids, the diffusion coefficient of a small molecule typically is about 10^{-5} cm²/s. With a velocity of 1 mm/s, in a channel of 100-μm height, the Péclet number is on the order of 100. This elevated value suggests that the diffusion forces are acting more slowly than the hydrodynamic transport phenomena: for mixing by diffusional forces one must have Pe < 1.

Definitions of Mixing, Stirring, and Chaotic Advection

Mixing means to homogenize a solution and involves convection and diffusion. In mixing, concentration gradients are smoothened until they disappear (Figure 6.49).

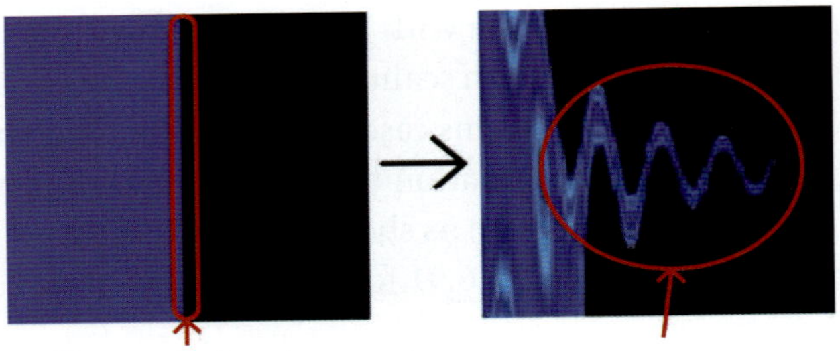

FIGURE 6.50 Stirring is a mechanical process that stretches and folds fluid elements, increasing regions with concentration gradients.

Stirring is convection meant to reduce the mixing time. It involves a mechanical process that stretches and folds fluid elements, increasing regions with concentration gradients (Figure 6.50).

At low Reynolds numbers, stirring is like kneading dough for making bread, with stretching of fluid elements to increase the diffusional interface and folding to decrease distance over which species have to diffuse. Creating chaotic pathlines for dispersing fluid species effectively in smooth and regular flow fields is called "chaotic advection."[30] Chaotic advection results in rapid distortion and elongation of the fluid/fluid interface, increasing the interfacial area across which diffusion occurs, which increases the mean values of the gradients driving diffusion, leading to more rapid mixing. Designing for chaos refers to devising mixer configurations that produce chaotic advection. The concept is illustrated in Figure 6.51.

An example chaotic advection structure is shown in Figure 6.52.

Implementing Micromixers

Introduction

From Table 6.4 we gleaned that, assuming a diffusion constant D of 10^{-5} cm²s⁻¹, a molecule will take 500 s to diffuse over a distance of 1 mm but only 0.5 ms to cross a distance of 1 μm. In the case one is working with a macromolecule such as myosin, the diffusion coefficient, D, becomes very small (1.2×10^{-11} cm²/s

FIGURE 6.49 Mixing homogenizes concentration gradients (t_a is some intermediate time).

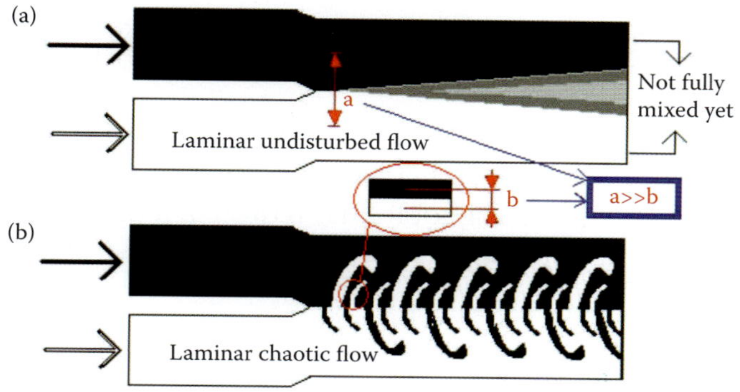

FIGURE 6.51 The concept of chaotic advection. (a) Two flows come together in a small conduit and only limited mixing takes place. (b) Laminar chaotic advection reduces the time it takes to mix in a microchamber.

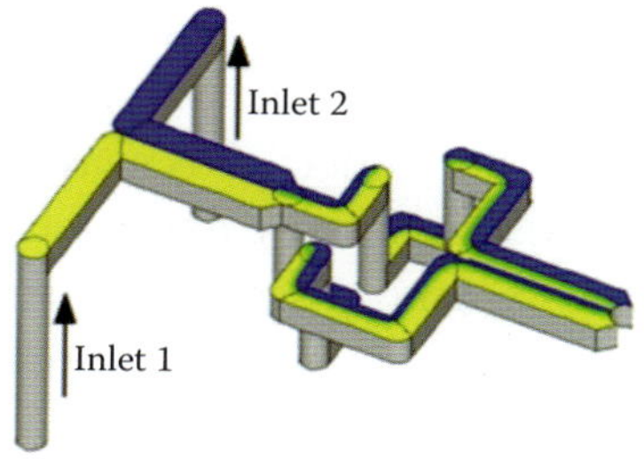

FIGURE 6.52 Mixer configuration for chaotic advection in the microdomain. Two fluids entering the inlet ports laminate at the first horizontal junction, producing two side-by-side fluid streams. Successive vertical separation and horizontal reuniting of fluid streams increase the number of laminates with each stage and, thus, the contact area between the two fluids.

compared with 3×10^{-6} cm²/s for fluorescein), and one might need to look for alternatives for inducing fast mixing because pure diffusional mixing is now too slow. This is an important, current technological challenge, for example, in the design and manufacture of fast DNA and protein arrays for molecular diagnostics: assays in arrays need to be married with a small, simple fluidic manifold, whereas reaction rates in the microfluidic chambers holding the arrays must remain high, despite low Reynolds numbers and high Péclet numbers.

Passive Micromixers

Several passive micromixer designs aim at reducing the diffusion path length. We start this analysis from the Einstein-Smoluchowski equation:

$$x^2 = 2Dt \qquad (6.100)$$

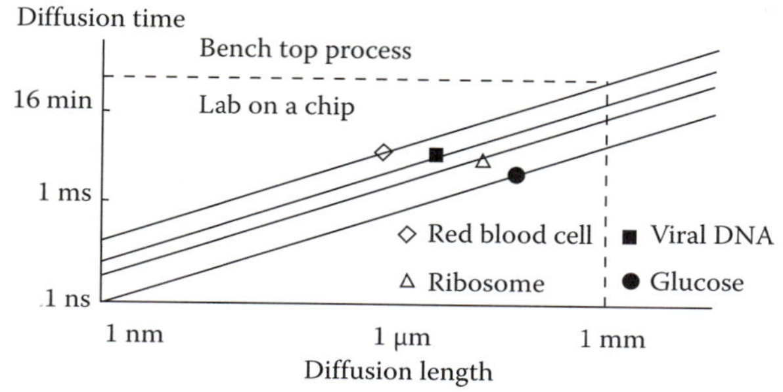

FIGURE 6.53 Diffusion times and lengths in the lab and on a chip.

where D is the diffusion coefficient, x the distance over which a species has diffused, and t is the average time between collisions. From this equation, the diffusional mixing time is of the order x^2/D. Miniaturization, as illustrated in Figures 6.53, can thus be expected to lead to smaller diffusive mixing times (scales with l^2). To help things along, in passive mixer designs the interfacial area is increased to reduce diffusion length. This may be accomplished using sinusoidal, square-wave, or zigzag channels. As illustrated in Figure 6.54, the smaller the distance between regions of different concentration, the larger the diffusive flux. Multi-laminated flow down-scales the diffusion length.

Mixing as illustrated in Figure 6.55 can also be improved on by using a fractal approach, a design akin to how nature uses passive mixing (Figure 6.55). In nature, only the smallest animals rely on diffusion for transport; animals made up of more than a few cells cannot rely on diffusion anymore to move materials within themselves. They augment transport with hearts, blood vessels, pumped lungs, digestive tubes, etc. These distribution networks

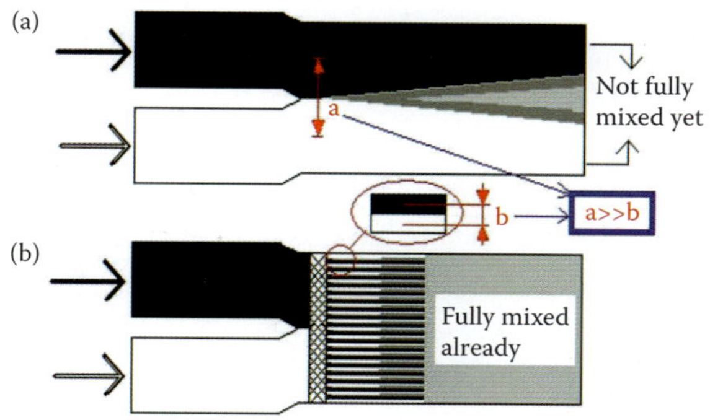

FIGURE 6.54 (a) Two flows come together in a small conduit and only limited mixing takes place. (b) The smaller the distance between regions of different concentration, the larger the diffusive flux. Multi-laminated flow down-scales the diffusion length.

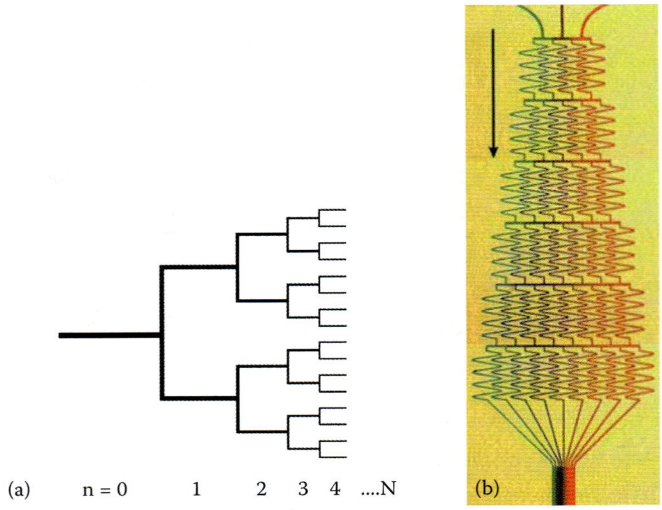

FIGURE 6.55 (a) The branching structures found in mammalian circulatory and respiratory systems have, through natural selection, evolved over many millennia to their current state of minimizing the cost function associated with the amount of biological work required to operate and maintain the system. (b) Biomimetic microfluidic device used for studying concentration gradients in a network of branching serpentines, in which three dyes are injected at the top before combining in a single channel at the bottom. (Whitesides, G. M. 2003. The "right" size in nanobiotechnology. *Nat Biotech* 21:1161–65.)

typically constitute fractals. Fractals are an optimal geometry for minimizing the work lost as a result of the transfer network while maximizing the effective surface area (see Volume III, Chapter 9).

In Figure 6.56 we demonstrate a T-shaped diffuser/mixer, where two fluids A and B enter a T-shaped intersection. Beyond the T, the width of the collection channel is reduced. In the narrowed rectangular cross-section of the fluidic structure, with width w and height b, the diffusive mixing time is proportional to:

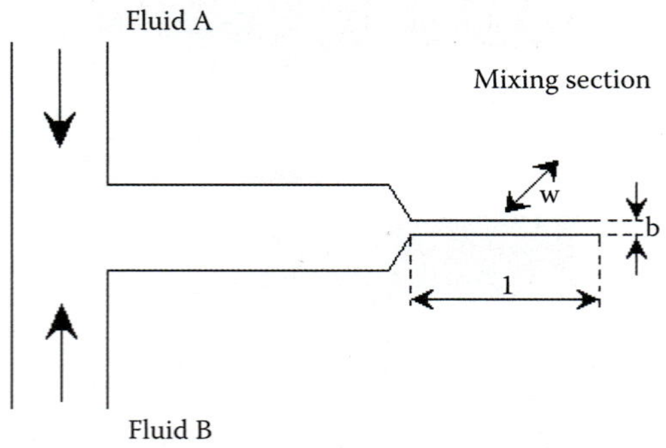

FIGURE 6.56 T-shaped diffuser/mixer, forcing fluids to mix by diffusion in a constriction of the microchannel. The length of the constriction is *l*, the width is *w*, and the height is *b* (unchanged from the wide channel section).

$$\tau \sim \frac{w^2}{D} \tag{6.101}$$

The necessary mixing length of the mixer is estimated by assuming that the two fluids at a velocity of Q/wb, with Q the total flux, require a length L_m to achieve mixing, or:

$$L_m \sim \frac{Qw}{bD} \tag{6.102}$$

At a fixed height b, narrowing the channel width w can reduce the length L_m to achieve mixing in a shorter time.

Two adjoining laminar flows A and B in a T-diffuser/mixer may be used as a filter as we illustrate in Figure 6.57. As the two parallel laminar flows contact, diffusion may extract one compound from one stream into the other. This way, components with higher diffusivity can be extracted from a sample stream (say, A) into a second diluent flow (say, B). This simple system does not require a membrane and is somewhat akin to field flow fractionation. Example applications include competitive immunoassays, determination of diffusion coefficients, and reaction kinetics parameters.

Diffusive extraction is also the underlying principle of the H-filter, illustrated in Figure 6.58. This continuous filter consists of two T-diffuser/mixers: one for mixing and one for separation. In the H-filter a proper design of a microfluidic channel allows for controlled extraction of analytes with high diffusion coefficients from components of the sample with lower diffusion

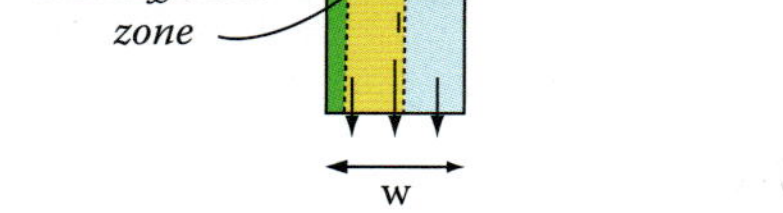

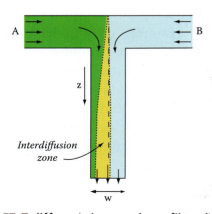

FIGURE 6.57 T-diffuser/mixer used as a filter. (Ismagilov, R. F., A. D. Stroock, P. J. A. Kenis, and G. Whitesides. 2000. Experimental and theoretical scaling laws for transverse diffusive broadening in two-phase laminar flows in microchannels. *Appl Phys Lett* 76:2376–78.)

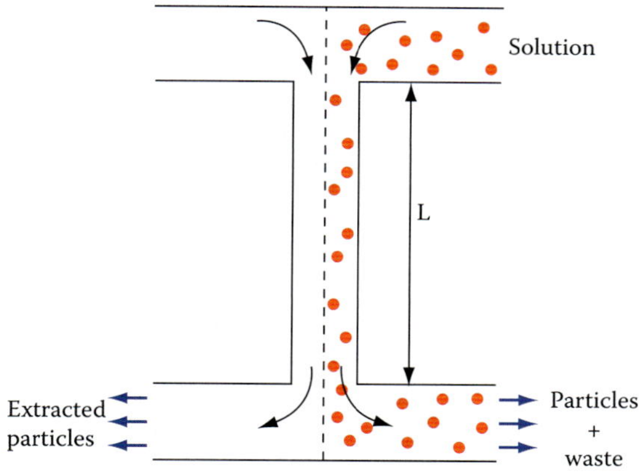

FIGURE 6.58 The H-filter is based on separation by diffusion between laminar flows. (Brody, J. P., P. Yager, R. E. Goldstein, and R. H. Austin. 1996. Biotechnology at low Reynolds numbers. *Biophys J* 71:3430–41.)

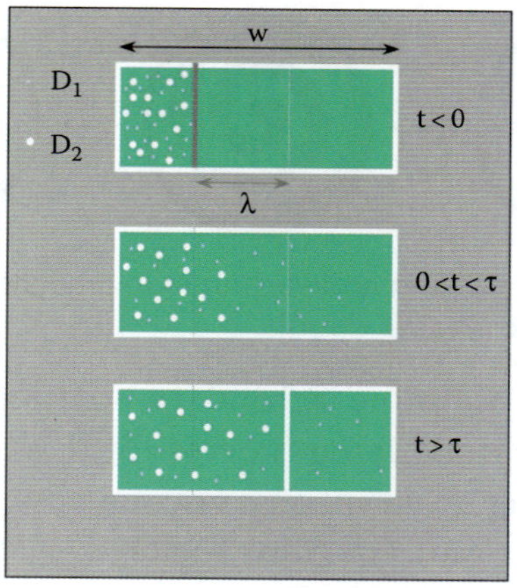

FIGURE 6.59 Design parameters of the H-filter. The fixed residence time in the transverse channel provides a "diffusion window" for the fastest components in one stream to diffuse to the other parallel stream and thus be separated.

coefficients. In the center of the device shown in Figure 6.58, streams move in parallel, and diffusion causes equilibration of small molecules across the channel, whereas larger particles do not equilibrate during the transit time of the device. The fixed residence time in the transverse channel provides a "diffusion window" for the fastest components in one stream to diffuse to the other parallel stream and thus be separated. Simplifying, the design rules of such filter reillustrated in Figure 6.59, the width of the conduit, w, accomodates two flows parallel to each other and the design obeys the following relation $w^2 \ll D_1\tau$ and λ the overlap region (with fast and slow diffusing particles mixed the smaller particles come with the largest diffusion coefficient $D_1 > D_2$) obeys: $\lambda^2 \gg D_2\tau$. Typical values are $\lambda = 10$ μm, $w = 40$ μm, and $t = 5$ s.

Applications of the H-filter include cell lysis, nonmotile sperm separation, extracting a wide range of analytes from the mucins in saliva, etc.

For a final example of a passive mixer in low Reynolds number fluids, we return to chaotic advection mixing. A first example of this type of mixing was illustrated in Figure 6.52. As a second example we use the herringbone micromixer developed by Stroock et al. (2002),[34] shown in Figure 6.60.

In this mixing approach, staggered herringbone microgrooves in the bottom of a flow channel in the

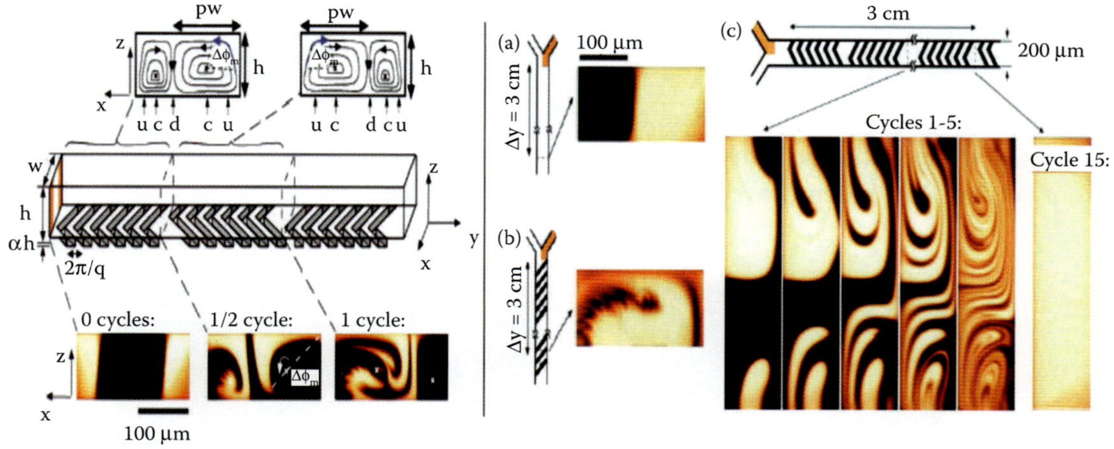

FIGURE 6.60 Micromixing by generating chaotic advection: the staggered herringbone mixer. By patterning grooves in the bottom of a microchannel, secondary flows are promoted that stretch and fold fluid elements, increasing areas with concentration gradients and so increase diffusive mass transport and mixing. In (a) no grooves are patterned, and parallel-flowing separated fluids undergo almost no mixing. By patterning grooves in several configurations, different secondary flows are promoted with smaller (b) or larger (c) mixing effects. (Stroock, A. D., S. K. W. Dertinger, A. Ajdari, I. Mezi, H. A. Stone, and G. M. Whitesides. 2002. Chaotic mixer for microchannels. *Science* 295:647–651.)

FIGURE 6.61 Mixing and stirring in low Reynolds number fluids is like kneading dough.

y-direction promote secondary rotating velocities (v_x, v_z). The underlying process for this type of efficient mixing, even in very viscous flow, is depicted in Figure 6.61 and is called the "baker's transformation. The main property of this transformation is that fluid pair elements will separate exponentially in "chaotic advection."

As illustrated in Figure 6.62, the width of the fluid folds (Δx) decreases exponentially, and the length of the interface (l) grows exponentially:

$$\Delta x(n) = \Delta x_0 (2)^{-n}$$

and

$$l(n) = l_0 (2)^n \tag{6.103}$$

The mixing length L_m was estimated as[34]:

$$L_m \approx w \log Pe \tag{6.104}$$

where w is the channel width, and Pe, the Péclet number, is calculated from the width and average flow velocity. This estimation, confirmed by experiments, leads to mixing lengths several tens of times larger than the width w for Péclet numbers on the order of 10^5.

Active Mixers

One also can use active perturbation to create time-dependent flows so that chaotic advection can occur. Different sources can be employed: pressure,

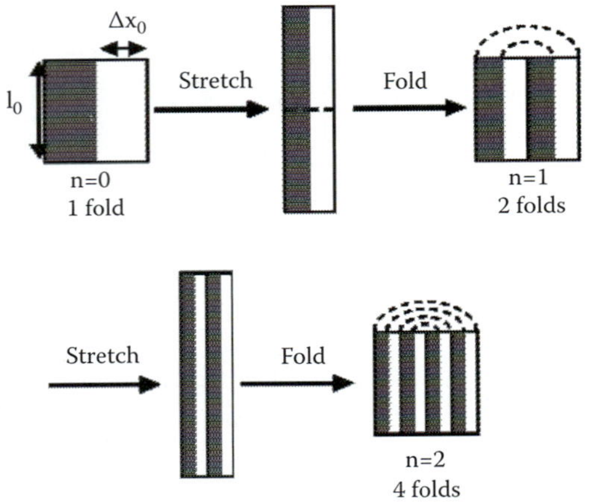

FIGURE 6.62 The baker transformation. Chaotic advection.

ultrasonic/piezoelectric, electrohydrodynamic, electro-osmosis, and mechanical stirring.

The mixing in the narrow section of the T-diffuser/mixer in Figure 6.56, for example, can be dramatically improved through hydrodynamic focusing, the technique used in cell and particle flow sorting. Two side flows "squeeze" or "hydrodynamically focus" an inlet flow. By using hydrodynamic focusing, the diameter of a central liquid jet can be reduced to just a few nanometers, which allows for mixing within this kind of beam in the tens of microseconds. This work, by Knight et al.,[35] is illustrated in Figure 6.63. The diameter of the jet (W_f) in the collection channel (W_c) is controlled by the ratio of the flow rates coming out of the three channels that bring the fluids together at the intersection. Knight et al. showed that with flow ratios of >100, W_f can be made to be around 30 nm.

Applications of hydrodynamic focus-based mixing include the study of fast reaction kinetics (like protein folding), which depend on fast mixing of reactants.

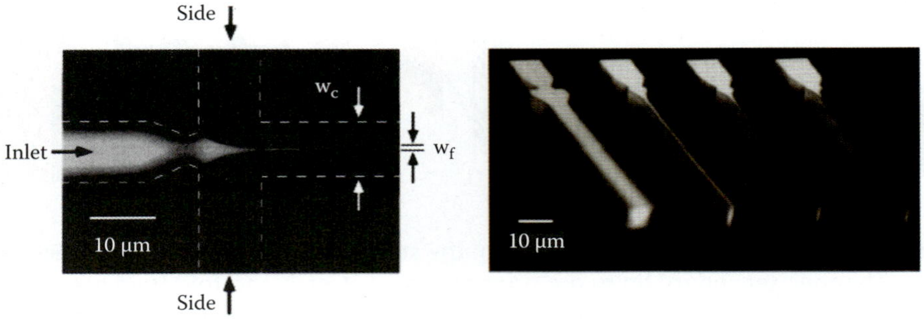

FIGURE 6.63 Geometry of a hydrodynamic focusing structure allowing for the reduction of the size of the jet and thereby enabling effective mixing transverse to the jet. (Knight, J. B., A. Vishwanath, J. P. Brody, and R. H. Austin. 1998. Hydrodynamic focusing on a silicon chip: mixing nanoliters in microseconds. *Phys Rev Lett* 80:3863–66.)

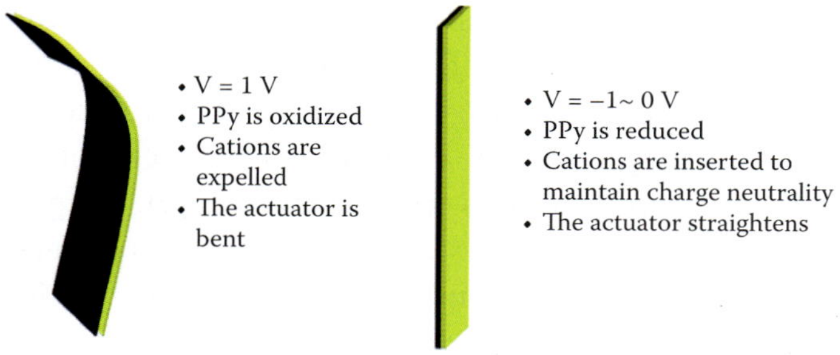

FIGURE 6.64 Principle of PPy/Au actuator. Mixing is achieved by activation of PPy-based actuators in microchannels.

Another active mixing approach pursued by the Madou group at the University of California, Irvine, involves the use of gold/polypyrrole actuators. Small rectangular gold/polypyrrole bilayer structures make for little flaps that can be made to move by biasing the polypyrrole with respect to a counter electrode in the same solution (see Figure 6.64). These actuators have been used to promote flows around a sensor in close proximity to the flapping bilayers. In doing so, convective mass transport is achieved, and the response in the sensor is increased over diffusive limits. Arrays of these mixing flaps have been fabricated as well with individual actuators ranging in dimensions from several millimeters to a few micros (see Casadevall i Solvas, X. 2009. *Active micromixing and in-vivo protection of sensors with gold/polypyrrole actuators*. PhD diss., University of California, Irvine, Irvine, CA; and Casadevall i Solvas, X., R. A. Lambert, L. Kulinsky, R. H. Rangel, and M. J. Madou. 2009. Au/PPy actuators for active micromixing and mass transport enhancement. *Micro and Nanosystems* 1:2–11).

The idea of the flexible flaps presented above is based on biomimetics. In 1977 Purcell, discussing swimming of microorganisms in fluids at very low Reynolds numbers,[36] presented the physical requirements that would allow microsystems to move (or propel fluids) in the microdomain (Figure 6.65). A simple single-hinged, rigid system (Figure 6.65a) would permanently move back and forth from its initial to its final position without accomplishing any real progress. To circumvent this limitation, multiple-hinged systems (Figure 6.65b) and flexible propelling structures (Figure 6.65c) can be used. Both configurations can be implemented easily with

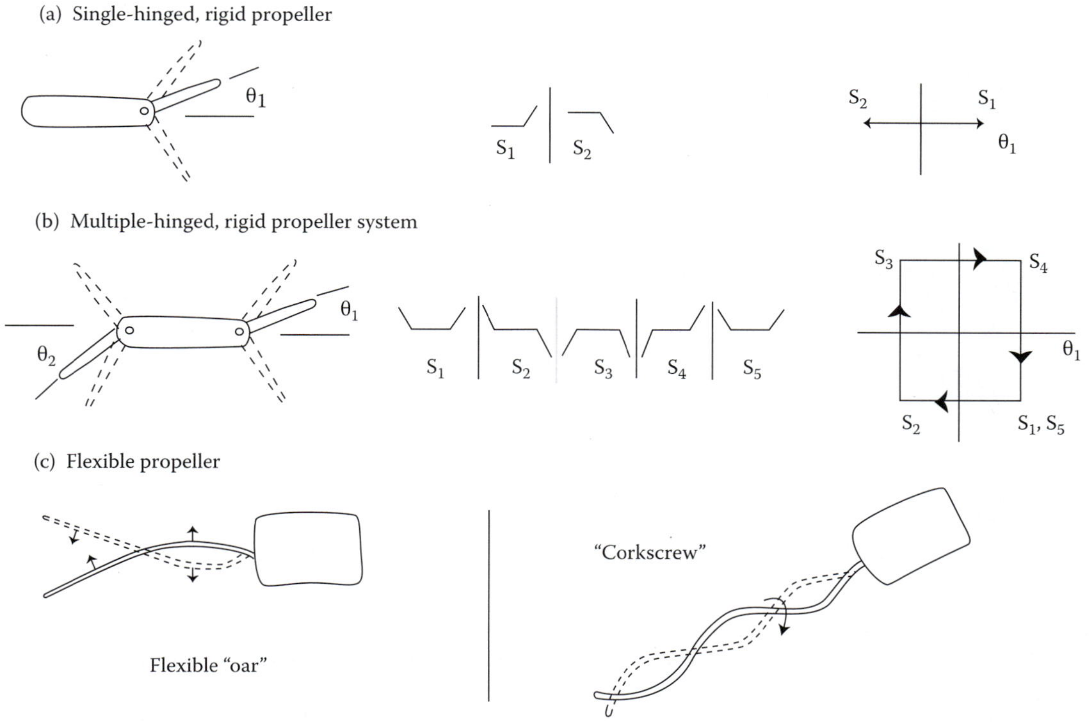

FIGURE 6.65 Swimming of microorganisms in fluids at very low Reynolds numbers. A simple single-hinged, rigid system (a) would permanently move back and forth from its initial to its final position without accomplishing any real progress. To circumvent this limitation, multiple-hinged systems (b) and flexible propelling structures (c) can be used. (Purcell, E. M. 1977. Life at low Reynolds number. *Am J Phys* 45:3–11.)

the flexible gold/polypyrrole actuators shown in Figure 6.64. In microvolumes of fluid, because of the flexible nature of these actuators and the possibility to actuate several of them simultaneously, these devices could be used to promote flows of sufficient complexity for chaotic advection to develop, and so enhance mixing in a targeted, isolated fashion, minimizing device and reagent volumes.

For a good review and an excellent comparison of active and passive micromixing, see Nguyen and Zhigang.[37] The summarizing comparison between active and passive mixing in Figure 6.66 is borrowed from this review.

Chemical Reactions in Microchambers—Microreactors

Introduction

The ultimate objective of passive and active micromixers in many instances is to speed up a chemical reaction by enhancing the mass transport to reaction sites in a chemical microreactor. As pointed out earlier, this finds a key application in the development of faster DNA and protein arrays for molecular diagnostics. These array assays need to be accommodated in a small reaction chamber, whereas reaction rates must remain high, despite low Reynolds numbers and high Péclet numbers. Because of these chemical reactions, the earlier described equations for fluid motion now need to be expanded with terms that take into account the generation (source) or consumption (sink) of reagents dissolved in the fluid. The proper design of chemical microreactors leads to faster and more sensitive assays.

Damköhler Numbers

Diffusive transport is captured by Fick's second law (see Equation 4.21 in Chapter 4), stating that the flux gradient $\left[-\frac{\delta F(x,t)}{\delta x}\right]$ is proportional to the change of the concentration with time. For a diffusion coefficient D, which is independent of position, this equation can be written for 3D using the ∇ operator as:

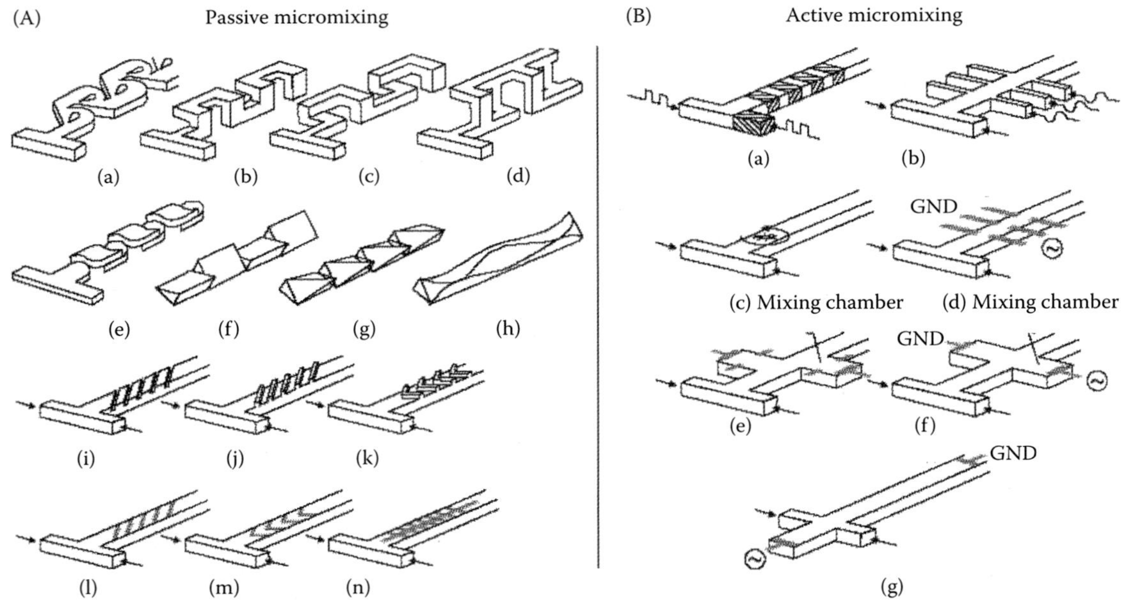

FIGURE 6.66 Comparing passive (A) and active (B) micromixing. Passive micromixers: modified Tesla structure (A) (a), C-shape (A) (b), L-shape (A) (c), connected out-of-plane L-shapes (A) (d), twisted microchannel (A) (e), other designs of twisted channel (A) (f–h), slanted ribs (A) (i), slanted grooves (A) (j), staggered-herringbone grooves (A) (k), and patterns for surface modification in a micromixer with electrokinetic flows (A) (l–n). Active micromixers: serial segmentation (B) (a), pressure disturbance along the mixing channel (B) (b), integrated microstirrer in the mixing channel (B) (c), electrohydrodynamic disturbance (B) (d), dielectrophoretic disturbance (B) (e), electrokinetic disturbance in the mixing chamber (B) (f), and electrokinetic disturbance in the mixing channel (B) (g). (Nguyen, N. T., and W. Zhigang. 2005. Micromixers—a review. *J Micromech Microeng* 15:R1–R16.)

$$\frac{\partial C}{\partial t} = D\nabla^2 C \quad (6.105)$$

For steady-state conditions the concentrations are constant in time, and there is no change in the mass stored with time, or $\nabla^2 C = 0$ (Laplace's equation). If, besides diffusion, we also consider convective or advective transport (mass flux due to the velocity V of the flow), we need to add a term:

$$\frac{\partial C}{\partial t} = D\nabla^2 C - \nabla CV \quad (6.106)$$

Expanding the advective term we obtain:

$$\frac{\partial C}{\partial t} = D\nabla^2 C - V\nabla C - C\nabla V \quad (6.107)$$

For transport in a steady-state flow field $\nabla V = 0$, so that:

$$\frac{\partial C}{\partial t} = D\nabla^2 C - V\nabla C \quad (6.108)$$

In case the velocity of the fluid is zero, we recover the diffusion equation. Finally, when reactions take place, we add a term for production (source) or removal (sink) of chemicals. This results in:

$$\frac{\partial C}{\partial t} = D\nabla^2 C - V\nabla C + r\,(= kC) \quad (6.109)$$

where r is the net mass produced or consumed per unit volume per unit time (moles/L^3t). For each species in the system an Equation 6.109 is required. The general form of the r term is $r = kC$, that is, r is the product of concentration and a reaction rate coefficient, k.

Depending on the type of reaction, r takes on different forms; in a first order–type reaction (e.g., encountered in radioactive decay), $r = -\lambda C$, where λ is a decay constant. In the case of second order kinetics, with reagents A and B, $r = -kC_A C_B$.

One can distinguish between two extreme cases depending on the Damköhler numbers. The Damköhler numbers (Da) are dimensionless numbers used to relate chemical reaction time scales to other phenomena occurring in a system. They are named after the German chemist Gerhard Damköhler, 1908–1944. There are several Damköhler numbers, and their definition varies according to the system under consideration. A global definition for the Damköhler number is:

$$Da = \frac{\tau_C}{\tau_M} \quad (6.110)$$

where τ_C is the characteristic time of the chemical reaction, and τ_M is the mixing time that can be controlled by hydrodynamics or by diffusion. So a small Damköhler number refers to a very slow chemical reaction compared to all other processes. Introducing a dimensionless concentration C^* (dividing C by an arbitrary concentration C_r); a dimensionless time t^* (dividing t by an arbitrary time t_r); and dimensionless lengths x^*, y^*, and z^* (by dividing by the characteristic length L), we nondimensionalize Equation 6.109 as:

$$\left[\frac{L^2}{Dt_r}\right]\frac{\partial C^*}{\partial t^*} = \nabla^2 C^* - \left[\frac{VL}{D}\right]\nabla C^* + \left[\frac{kL^2}{D}\right]C^*$$

Fourier Péclet Damköhler

$$\text{or } Fo\frac{\partial C^*}{\partial t^*} = \nabla^2 C^* - Pe\nabla C^* + PeDa_I C^*$$

$$\text{where } Da_I = \frac{kL}{V}$$

$$\text{and also } Fo\,\frac{\partial C^*}{\partial t^*} = \nabla^2 C^* - Pe\Delta C^* + Da_{II} C^*$$

$$\text{where } Da_{II} = \frac{kL^2}{D} \quad (6.111)$$

The Fourier number Fo (L^2/Dt_r) is a dimensionless quantity that can be used to characterize a diffusion length for a particular time. For example, if $D = 10^{-9}$ m^2/s and $t_r = 1$ year, then the diffusion length is about 0.18 m. The first Damköhler number ($Da_I = kL/V$) measures the tendency for reaction versus advective transport. The second Damköhler number ($Da_{II} = kL^2/D$) measures the tendency of reaction relative to the tendency for diffusive transport. With a small Da_I or Da_{II}, convection and/or diffusion is much faster than chemical reaction, and the concentration in the reactor is uniform. The system is reaction rate-controlled, and convection and/or diffusion have no influence. With a large Da_{II}, the chemical reaction is very fast so that the concentration, say on an electrode, might fall to zero, and the system is transport-limited.

Since the Biot number (see above) will tend to decrease for microdevices, it is expected that micro-chemical reactors will have nearly uniform temperatures. Thus, the wall effects will be very important from the point of view of keeping a high-temperature reaction going. The tendency will therefore be for the presence of small Damköhler numbers (weak reaction) due to heat losses through the reactor walls that reduce the reaction rate and consequently increase the chemical time, while the small dimensions reduce the diffusion time. For a general chemical reaction A → B of n-th order, the Damköhler number is defined as:

$$Da = kC_0^{n-1}t \tag{6.112}$$

where k = kinetics reaction rate constant
C_0 = initial concentration
n = reaction order
t = time

Electrochemical Reactions

Here we set up an expression for the current on an electrochemical sensor with transport of electroactive species to and from the electrode by diffusion, migration, and convection. From Fick's first law (Equation 4.20) we derive for the flux or current resulting from electroactive species *i*:

$$\underset{\text{Diffusion}}{J_i(x) = -D_i\frac{\delta C_i(x)}{\delta(x)}} \; \underset{\text{Migration}}{- z_i\mu_i FC_i\frac{\delta\phi}{\delta x}} \; \underset{\text{Convection}}{+ C_iV(x)} \tag{6.113}$$

where J_i = flux of species *i* in mol/cm² s
z_i = charge number of species *i* in eq/mol
μ_i = mobility of species *i* in cm²mol/Js
C_i = concentration of species *i* in mol/cm³
ϕ = electrostatic potential in *V*
F = the Faraday constant
D_i = diffusion coefficient of species *i* in cm²/s
V = fluid velocity in cm/s

In this expression for an electrochemical reaction, the new term we consider is migration transport. For migration of species, say negatively charged DNA molecules to a positively charged electrode, the solution must have a low conductivity as otherwise the field does not penetrate the liquid (see Chapter 7). As we learn in Chapter 7, to obtain reliable analytical data on amperometric electrochemical sensors, the migration term needs to be minimized in favor of the diffusion and convection terms. This, we will learn, is accomplished by the addition of an indifferent electrolyte suppressing the migration of the electro-active species of interest.

In Table 6.5 we summarize the important transport phenomena determining the response of an electrochemical sensor.

Electrochemicals are studied in detail in Chapter 7 where they are also compared with optical sensors.

Fluid Propulsion

Introduction

Many options present themselves for fluid propulsion, including mechanical means, dc and ac electrokinetics, acoustic pumping, electrowetting, centrifugal, electrohydrodynamic and magnetohydrodynamic, etc. Acoustic, electrohydrodynamic, and magnetohydrodynamic pumping techniques are reviewed in Volume III, Chapter 8 on actuators, and in Volume III, Chapter 1 we compare and summarize the most important new pumping methods and their MEMS integration (see Volume III, Table 1.8). In the current section we detail the theory behind dc and ac electrokinetics, electrowetting, and centrifugal fluidics.

Electrokinetic Effects—Direct Current (DC)

Introduction to Electrophoresis and Electro-Osmosis

An electrical field works on both charged particles in a fluid and on the fluid itself when the fluid is placed in a narrow capillary with electrically charged walls. The force on the particles in the fluid leads to electrophoresis; the force on the fluid in a narrow column leads to electro-osmosis. Both phenomena are of great importance in miniaturization science: electrophoresis for separation of compounds and electro-osmosis for movement of fluids. Mathematically, it does not make a difference whether the fluid passes a charged wall or a charged particle moves through a fluid.

Table 6.5 Transport Phenomena on an Electrochemical Sensor

Transport	Migration	Diffusion	Convection
Response to	Potential gradient	Concentration gradient	Pressure gradient
Acts on	Ions only	All solutes	Solutes + solvent
Flux density (from Equation 6.113) J =	$-zF\mu C\frac{\partial \phi}{\partial x}$ ↖ mobility ↘	$-D\frac{\partial C}{\partial x}$ ↗ diffusivity ↙	CV
Relationship between μ and D	Einstein derived a simple relation between D and μ as: $D = \mu RT$		
Combining migration with diffusion flux	$J = -D\left[\frac{\partial C}{\partial x} + \frac{zF}{RT}C\frac{\partial \phi}{\partial x}\right]$ (Nernst-Planck flux equation using the Einstein relation between D and μ.)		
Conservation law	$\frac{\partial C}{\partial t} = -\frac{\partial J}{\partial x}$ for planar transport; $\frac{\partial C}{\partial t} = -\frac{\partial J}{\partial r} - \frac{2J}{r}$ for spherical transport		

Electrophoresis

Electrophoresis is the migration of ions in a separation medium under the influence of an electric field. An ion with charge q (Coulombs) in an electric field E (Vm^{-1}) feels an acceleration force qE (Newtons). A steady-state speed is reached when the accelerating force (qE) equals the frictional force generated by the separation medium, or:

$$qE = fV_E \tag{6.114}$$

where f = the friction coefficient or friction factor
V_E = the electrophoretic steady-state speed of the ion

The expression for the electrophoresis velocity V_E of a charged species in a separation medium can then be written as:

$$V_E = \mu_E E \tag{6.115}$$

with μ_E ($m^2/V/s$) the electrophoretic mobility (= q/f) and E the applied electrical field. The friction coefficient for a spherical particle of radius r moving through a fluid of viscosity η, we learned from Equation 6.26, is given by 6πηr, and its electrophoretic mobility consequently equals:

$$\mu_E = \frac{q}{6\pi\eta r} \tag{6.116}$$

The frictional coefficient f depends on the size of the molecule in free solution. For a solid sphere f is given by 6πηr, so that $f \sim MW^{1/3}$ because r ~ MW (where MW is molecular weight). Other models are more appropriate for differently shaped molecules, as summarized in Table 6.6.

According to the free-draining coil model, a randomly coiled polymer chain in motion hardly interacts with the solvent within it at all. DNA is a good example of a free-draining coil with each unit of the polymer contributing equally to the friction created by its movement through a fluid. Because the charge q is proportional to N (number base pairs), and $f \sim (MW)^1 \sim N$, we conclude that:

$$\mu_E = \frac{q}{f} = \text{constant} \tag{6.117}$$

As a consequence, DNA strands of varying size cannot be separated by electrophoresis in solution.

TABLE 6.6 Friction Factor f for Variously Shaped Molecules

Model	Proportionality
Sphere	$f \sim (MW)^{1/3}$
Random coil	$f \sim (MW)^{1/2}$
Long rod	$f \sim (MW)^{4/5}$
Free-draining coil	$f \sim (MW)^{1}$
MW = molecular weight.	

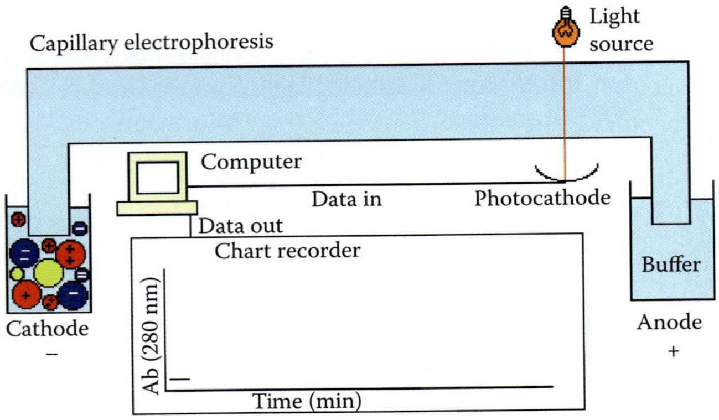

FIGURE 6.68 Capillary electrophoresis schematic. A sample may be injected in the capillary through pressure application (e.g., a syringe) on the sample vial or suction at the outlet of the capillary. A more elegant solution for sample introduction is through electro-osmotic pumping, discussed below. The chart recorder in this case plots absorbance (Ab at 280 nm) versus time.

This is why we need to perform electrophoresis in a porous or tortuous material such as a gel. In this case, the solute moves snake-like through the polymer pores. In this model, the frictional term is $f \sim N^2$. This leads to:

$$\mu_E = \frac{q}{f} \approx \frac{N}{N^2} \approx \frac{1}{N} \tag{6.118}$$

which is the type of behavior normally observed in DNA electrophoresis.

The difference in ion mobility gives rise to a separation in bands of the various solutes, and the narrower the bands, the better the separation. Traditionally, electrophoresis was carried out on columns packed with gels (e.g., a polyacrylamide gel) or on gel-covered plates (Figure 6.67).

Electrophoresis in narrow capillaries, as shown in Figure 6.68, was pioneered by researchers such as Virtanen[38] and Jorgenson,[39] who recognized important advantages of this approach. Performing electrophoresis in small-diameter capillaries allows the use of very high electric fields because the small capillaries efficiently dissipate the heat that is produced. Increasing the electric fields produces very efficient separations and reduces separation times. Capillaries are typically of 50-μm inner diameter and 0.5 to 1 m in length. The applied potential is 20 to 30 kV. Because of electro-osmotic flow (see next section), all sample components migrate toward the negative electrode. A small volume of sample (10 nL) is injected at the positive end of the capillary, and the separated components are detected near the negative end of the capillary. CE detection schemes include absorbance (see the example in Figure 6.68), fluorescence, electrochemistry, and mass spectrometry. The capillary also can be filled with a gel, which eliminates the electro-osmotic flow. Separation is accomplished as in conventional gel electrophoresis, but the capillary allows higher resolution, greater sensitivity, and on-line detection.

To cool the capillary during electrophoresis, a Peltier device, forced air convection, or a flowing liquid bath may be used.

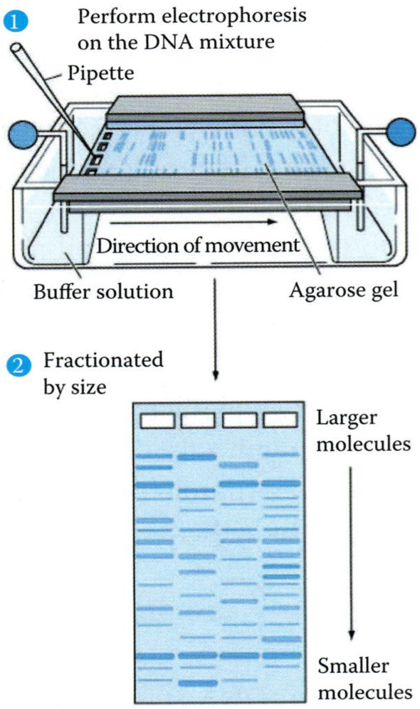

FIGURE 6.67 Gel electrophoresis.

Electro-Osmosis

Capillary electrophoresis to some degree always is accompanied by electro-osmosis. The inside wall

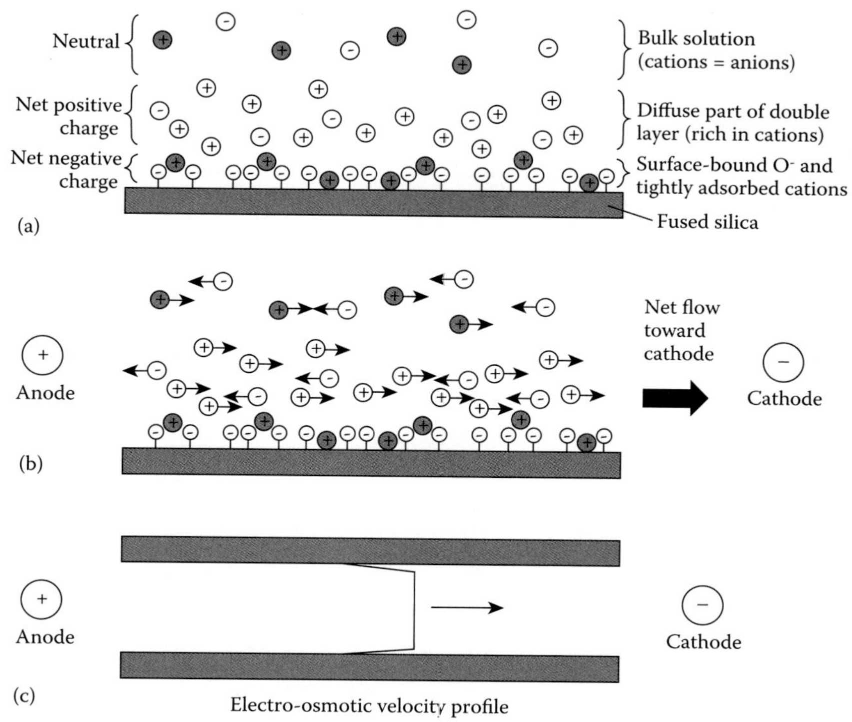

FIGURE 6.69 A schematic diagram showing the origin of electro-osmotic flow. (a) Electric double layer is created by negatively charged silica surface and excess cations in the diffuse part of the double layer in the solution near the wall. The wall is negative, and the diffuse part of the double layer is positive. (b), Predominance of cations in diffuse part of the double layer produces net electro-osmotic flow toward the cathode when an external field is applied. (c) Electro-osmotic velocity profile is uniform over more than 99.9% of the cross-section of the capillary. Capillary tubing is required to maintain constant temperature in the liquid. Temperature variation in larger diameter tubing causes bands to broaden.

of a fused silica capillary is covered with silanol groups (Si-OH). Silanol groups have a pK_a near 2 and consequently carry a negative charge ($Si\text{-}O^-$) at pH > 2. Positive compensating charges (cations) are either in an immobile layer immediately adjacent to the glass surface, the Stern layer, or in the diffuse part of the double layer, referred to as the Gouy-Chapman layer. The Gouy-Chapman layer stretches a bit further out into the solution, ~10 nm further away when the ionic strength of the medium in the capillary is 1 mM.* In a tangential electric field, excess cations in the mobile part of the diffuse double layer are attracted to the cathode, and this imparts a pumping action onto the whole fluid column, which moves toward the cathode (Figure 6.69).[40] This pumping describes electro-osmotic or also electroendosmotic flow.

* For a binary electrolyte in aqueous solution, the double-layer thickness ranges from 3 to 300 nm for electrolyte concentrations of 10^{-2} to 10^{-6} M, respectively.

The electro-osmotic velocity, V_{EOF}, is given by:

$$V_{EOF} = \mu_{EOF} E \tag{6.119}$$

where μ_{EOF} = the electro-osmotic mobility
E = the applied electrical field

The electro-osmotic mobility (q/f) is given by:

$$\mu_{EOF} = \frac{\varepsilon \varsigma}{4\pi\eta} \tag{6.120}$$

where ε = the dielectric constant
ζ = electrokinetic or zeta potential
η = is the fluid viscosity

The structure of the electrical double layer (EDL) is detailed in Chapter 7. Here we list some of the most salient features of the EDL model. A thin layer of liquid along the capillary wall never moves; liquid movement begins only at a shear plane in the diffuse double layer at a distance, x_0, away from the wall. The electrokinetic potential ζ (from which all electrokinetic effects can be derived) represents the value of the electrostatic potential, ψ, at x_0. Its value at any point is determined from the profile of the electrostatic potential as a function of distance from the capillary wall into the moving fluid (see Figure 7.25). The thickness δ of the diffuse layer itself is defined as the distance from the immobile Stern layer to a point at which the electrostatic potential has dropped to 37% of the zeta potential. The amount and type of ionic species in the solution can greatly influence the zeta potential; for example, the larger the concentration of added indifferent electrolyte in the moving solution, the smaller the ζ (see Figure 7.25). The zeta potential of an aqueous solution in contact with glass can have a magnitude as high as 100 mV.[3] Besides pH and ionic strength, surface impurities also might affect the electrokinetic potential, and all conspire to make electro-osmosis a rather difficult pumping mechanism to control. Despite these barriers, several new miniaturized bioanalytical products based on electrokinetic effects are now on the market (e.g., visit www.nanogen.com and www.calipertech.com). Most product applications concern genomic analysis; medical and environmental diagnostics have not been able to garner the same amount of funding today.

Generalizing, Equations 6.115 and 6.119 may be expressed as:

$$V = \mu_a E \quad \text{or} \quad V = L_d / t \cdot V / L_t \qquad (6.121)$$

where V = the velocity of ions or fluid column
- μ_a = the apparent mobility (i.e., the vector sum of electro-osmotic and electrophoretic mobilities)
- E = the electric field strength
- L_d = the length of the column from injection to detector
- t = the time required for solute to migrate from the injection point to the detector
- V = the applied voltage between the two ends of a capillary with total length L_t

As pointed out earlier, mathematically, electro-osmosis and electrophoresis are indistinguishable: either the charge moves through the medium or the medium moves through the charged capillary. In the case of electro-osmosis, ζ is the electrokinetic potential at the capillary wall; in the case of electrophoresis, it is the electrokinetic potential at the charged particle. In absolute terms, electro-osmotic mobility generally is larger than electrophoretic mobility. Under typical experimental conditions, say for DNA separations (300 V/cm), the electro-osmotic mobility of DNA molecules within fused silica capillaries is 1.25 to 3 times larger than their electrophoretic mobility.

In some cases electro-osmosis is undesirable; for instance, in gel-filled capillaries the gel may ooze out of the capillary or positively charged proteins may adsorb onto the capillary walls. Electro-osmosis may be suppressed by chemical modification of the capillary surface with a neutral EOF-suppressing polymer (e.g., by coating with polyacrylamide[41] or methyl cellulose). By the adjustment of the buffer pH and/or buffer concentration, or by the application of a radial electric field, the EOF also may be modulated by the user.[40] The latter may change both the magnitude and the polarity of the zeta potential.

In the section below on scaling in analytical separation techniques, we will further detail advantages and disadvantages of miniaturizing electrophoresis and electro-osmosis and separation equipment in general.

Flow Profiles

The flow through a separation media such as a capillary may be a consequence of an applied pressure (i.e., hydrodynamic) or of an applied field (i.e., electro-osmotic). In a pressure-driven system, an external pump provides the pumping force. As shown in Figure 6.70a, a cross-section of a hydrodynamic velocity flow profile driven by a pressure difference between the ends of a column is parabolic (see Equation 6.46 and Figure 6.22). It is the fastest at the center and slows to zero at the walls. When injecting a solute, the parabolic flow profile makes for a less attractive analytical tool, as it causes band broadening (see also the section below on scaling in separation techniques). Electro-osmosis, in which the driving force is generated very close to the capillary wall, gives a uniform flow over more than 99.9% of the cross-section of a capillary column, as shown in Figure 6.70b. The speed of the flow falls off immediately adjacent to the capillary wall. To illustrate

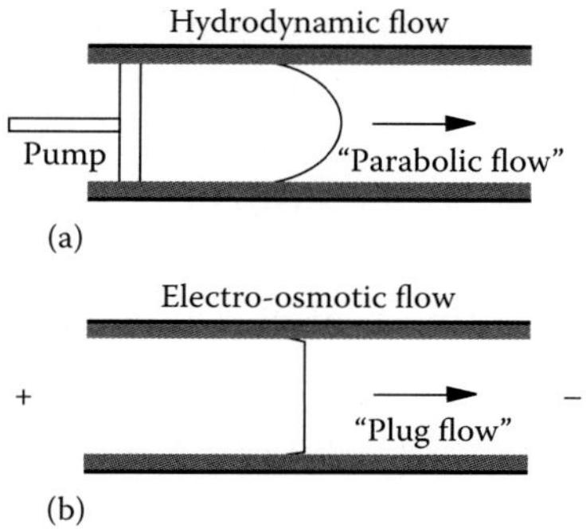

FIGURE 6.70 A hydrodynamic velocity profile (a) compared with an electro-osmotic velocity profile (b).

this with a practical example, in a typical DNA electrophoresis experiment, the double layer δ is around 30 Å thick, and x_0 (see Figure 7.25) is even thinner. Compared with the inner diameter of a capillary of 50 μm, the region of velocity variation is negligible. Flat flow profiles are expected when the capillary radius is greater than about 7 times the double-layer thickness. The rapid flow of buffer and analyte with a flat velocity profile is attractive, as it can be used to pull all analytes, regardless of their mobility, past a detector without contributing significantly to band broadening.

In hydrodynamic injection of a sample in a capillary, the injected volume is derived from the volumetric flow rate, *Q*, in m^3s^{-1}. The flow rate *Q*, of an incompressible fluid with viscosity η (N/s/m²) in a capillary with inner diameter *d* (m) and total length of the capillary L_t (m), resulting from an applied pressure Δp (Nm^{-2}), is calculated from Hagen-Poiseuille's Law (Equation 6.54) as:

$$Q = \frac{\Delta p \pi d^4}{128 \eta L_t} \qquad (6.54)$$

This equation applies for laminar flow of a fluid through a capillary. To obtain the volume, *V*, the volumetric flow rate *Q* is multiplied by *t*, where *t* is the injection time:

$$V = Q \cdot t \qquad (6.122)$$

Conventional pumps must generate extremely high pressures when narrow capillary diameters are used, and they are not well suited for the delivery of small samples. This is where electro-osmotic pumping comes in.

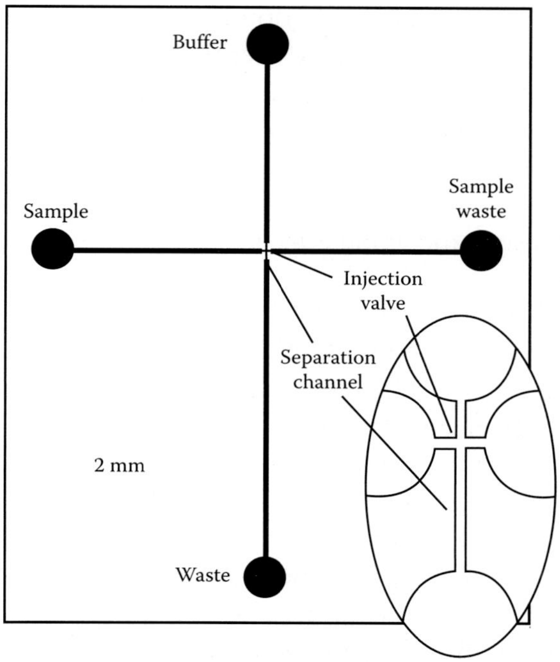

FIGURE 6.71 Microchip used for high-speed electrophoretic separations. Inset: Enlargement of the injection valve and separation channel. (Based on Jacobson, S. C., C. T. Culbertson, and J. M. Ramsey. 1998. *Solid-state sensor and actuator workshop*. Hilton Head Island, SC: Transducers Research Foundation.)

Electrokinetic Injection

Electrokinetic injection is popular in micromachined fluidics (see Figure 6.71), but it is more generally applied in capillary gel electrophoresis, in which the liquid is too viscous to allow for hydrodynamic injection. In electrokinetic injection, a capillary is dipped in sample solution, and a voltage is applied between the ends of the column. The moles of each ion, mol_i, absorbed into the capillary in *t* seconds are given by:

$$mol_i = \mu_a \left(E \frac{\kappa_b}{\kappa_s} \right) t \pi r^2 C_i \qquad (6.123)$$

where μ_a = the apparent mobility of the analyte
E = the applied electric field
r = the radius of the capillary
C_i = the sample concentration ($molm^{-3}$)
κ_b/κ_s = the ratio of the conductivities of the buffer and the sample

Electrokinetic injection leads to sampling bias, because a disproportionately large quantity of the species with higher electrophoretic mobility migrates into the tube and can cause problems for quantitative analysis. Moreover, electro-osmotic flow rates tend to change over time because the zeta potential changes over time, thus leading to changes in μ_a in Equation 6.123. Below, we will see that certain types of electro-osmotic injection methods in miniaturized flow channels avoid this type of sample biasing and aging effects.

Micromachined Capillary Electrophoresis Equipment

Planar Fluidic Networks In free-flow, traditional capillary electrophoresis work by Jorgenson et al.,[42] electro-osmotic flow rates of 1.7 mm s^{-1} were reported, certainly fast enough for many analytical purposes. The technology for generating this flow rate does not involve moving parts and is easily implemented in micromachined channels. Using micromachined channels in a substrate instead of glass capillaries offers the potential for mass production, the choice of many materials (e.g., different plastics besides glass), and the integration of many processing steps on one chip. One needs only one electrode in the reservoirs at each end of the different flow channels, and these electrodes may be thin films deposited on the substrate or needle electrodes inserted into the reservoirs from above (bed-of-nails approach). Harrison et al.[43] achieved electro-osmotic pumping with flow rates up to 1 cm s^{-1} in 20-μm capillaries micromachined in glass. Compared with conventional capillary electrophoresis, where separations usually are in the order of 10–20 min, microfabricated electrokinetic devices are capable of producing high quality separation in seconds or even submilliseconds as first demonstrated by Jacobsen et al.[44,45] In Figure 6.71 we show a schematic of the microchip used by Jacobson et al.[46] for submillisecond electrophoresis using a field strength of 53 kV cm^{-1} and a separation length of 200 μm. Two dyes, rhodamine B (RB) and dichlorofluorescein (DCF), were resolved in 0.8 ms using this system.[47]

The earliest patents describing electrophoresis and electro-osmosis in a micromachined channel with side branches in a planar substrate are by Pace (filing date June 1988)[48] and by Soane et al.[49] (priority filing date Feb 28, 1990). The latter is the more generic patent of the two, as it covers all insulating substrates for making fluidic channels, whereas the former patent is confined to Si. Si, besides being expensive, is a semiconductor and needs to be coated with an insulator before it can be used in an electrokinetic application. When using SiO_2 as the insulator, the oxide tends to break down and age rapidly in contact with water, making Si far from ideal as a substrate for an electrokinetic device. As far back as 1991, Manz et al.[50] demonstrated the injection, mixing, separation, and reaction of fluids in a manifold of micromachined flow channels without the use of mechanical valves.

Manz et al.[51] described a set of reservoirs connected by an H-like pattern of capillary channels etched into a glass substrate. This setup is schematically reproduced in Figure 6.72. Application of a voltage between reservoir 1 (sample) and reservoir 4 (injection waste) results in flow of sample across the center of the H. Subsequent application of a potential between reservoir 2 (carrier) and reservoir 5 (waste) directs the geometrically confined sample plug from the center bar of the H into the elongated leg of the H, that is, into the separation channel and toward the detector. By applying a potential between reservoirs 1 and 4 for an extended period of time, the composition of the fluid plug in the center of the H will be identical to that of the sample, thus avoiding the bias errors described under electrokinetic injection above. In addition, because the injected volume is geometrically defined, the amount of injected sample is reproducible and independent of changes in the electro-osmotic flow over time. The key to the tightest possible valving of liquids at intersecting capillaries in a manifold, as shown in Figure 6.72, is the suppression of convection and diffusion effects. Harrison et al.[43] demonstrated that these effects can be well controlled by appropriate and simultaneous application of voltages to the intersecting channels. Fast separations in micromachined fluidics are possible by confining the smallest possible sample plug at an intersection.[52,53] Ramsey achieved this by controlling the voltages at all four terminals of a simple cross structure, as shown in Figure 6.73, and pinching the plug at the intersection. In this way, the injected sample plug is of the order of the channel width. In the pinched injection mode demonstrated in Figure 6.73, analyte is pumped electrophoretically and electro-osmotically from reservoir 1 to 2 (left to right) with mobile phase from reservoir 3 (top) and

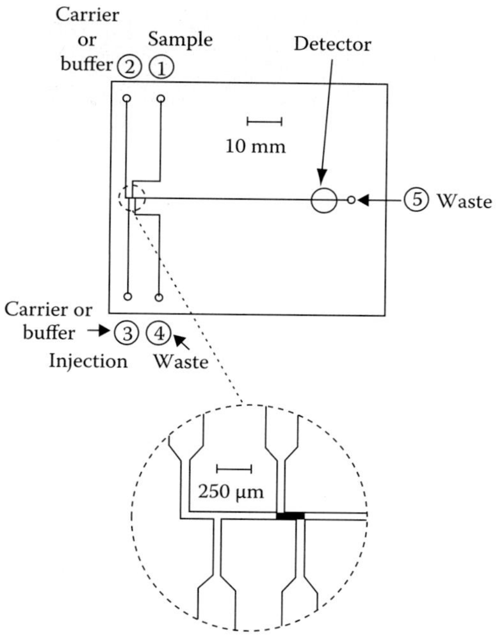

FIGURE 6.72 Schematic drawing of an electrokinetic injection process using a branched microflow system. Layout of the glass chip with integrated sample injector. Channel cross-sections: 50 × 12 (thin channels) and 250 × 12 μm (broad channels), respectively. After application of a high voltage between reservoirs 1 (sample) and 4 (injection waste), the geometrically defined injection volume (double-"T" injection, circled area) is filled by electrophoretic migration of the sample ions. After loading is completed, application of a high potential between reservoirs 2 (buffer) and 5 (waste) drives the injected sample plug into the separation channel and causes electrophoretic separation of the sample components. The detection is based on laser-induced fluorescence (LIF). The T-intersection where the selected sample zones are withdrawn is enlarged in the circled image. (Based on Effenhauser, C. S., A. Manz, and H. M. Widmer. 1995. Manipulation of sample fractions on a capillary electrophoresis chip. *Anal Chem* 67:2284–87.)

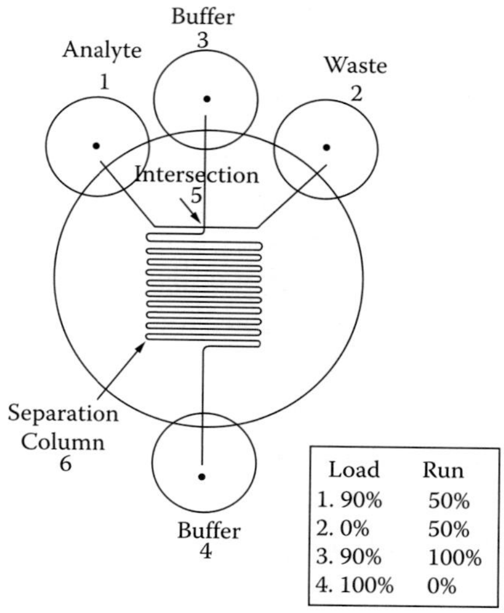

FIGURE 6.73 Pinched injection. Load and run modes (see text). (Based on Ramsey, J. M. 2000. Lockheed Martin Energy Research Corporation. US Patent 6,010,607.)

reservoir 4 (bottom) traveling to reservoir 2 (right). The appropriate voltages, as indicated in Figure 6.73, are applied to reservoirs 1, 3, 2, and 4, whereas reservoir 2 is grounded, thereby spatially constricting the analyte in the injection cross or intersection 5. This plug can be injected into the separation column 6 by applying a bias between electrodes 3 and 4 (4 is grounded now) while the potential of reservoirs 1 and 2 is kept at half the value of reservoir 3.

The above plumbing examples are simple; more complex manifolds exist that enable serial and more complex analytical procedures. These procedures include flow injection analysis, metering, mixing, reaction, dilution, aliquoting, pre- and postcolumn derivatization, addition of masking agents, and preconcentration (perhaps using isoelectric focusing techniques; see below). An interesting manifold configuration by Burggraf et al.[55] involves synchronized cyclic capillary electrophoresis (SCCE). In this technique a sample, through a sequence of voltage applications, is cycled through a loop made up of intersecting capillaries; after each pass the sample gets separated better into its constituent compounds.[55] A single electrokinetic channel can only be optimized to a point; once an optimum critical channel length and a minimal dead volume detector and injector have been designed (see scaling in analytical separations), other means to improve throughput must be implemented. For example, to obtain high throughput with integrated fluidics, one important need is for serial-to-parallel converters.[56]

Optimal Size of Fluidic Channels—Scaling With 1-mm-wide conduits and flow velocities not greater than a typical centimeter-per-second range, the Reynolds numbers for flow are so low that all flow is laminar. Mixing then occurs solely by diffusion, and the time required for molecules to travel a distance x is given by $\tau = x^2/2D$ (see Equation 6.100). For homogeneous flow to be established, bringing two streams of different compositions together at a Y-junction requires that the molecules diffuse over

500 μm (one-half of the channel width). For a molecule such as fluorescein with a D of 3×10^{-6} cm²/s, this process will take 400 s, and with a flow velocity of 1 cm/s, the channel must be 4 m long before the mix has homogenized. Reducing the channel width to 70 μm and using the same flow rate, mixing is established in 2 s over a distance of 2 mm only. This suggests an upper limit of about 100 μm for the channel width, based on the need for fast mixing by diffusion. At very small dimensions, another limitation lurks. Assume that one wants to detect an analyte at 1 nM concentration in a cubic micrometer volume. The number of molecules in that small volume is only 0.6 (see also Volume III, Table 7.5). In other words, about half of the time nothing there will be detectable. Single-molecule detection is possible, for example, with fluorescence burst counting as demonstrated by Haab et al.[57] and Wahl et al.,[58] but it is far from common; 1000 molecules is a much more comfortable number with which to work. To have about 1000 molecules available in a 1-nM solution (a quantity readily detected), a volume of 12 μm³ is needed. So, for optimal detection chamber dimensions should not be much smaller than 10 μm. With a channel cross-section of 2×10^{-3} mm² and a flow velocity between mm/s and cm/s the flow rate is in the nL/s range. Often, flow control must be within 1% or 1 pL/s. This tight tolerance is impossible to reach with today's micropumps, which cannot provide accurate control in the nL/s range. With electrokinetics flow control better than 1% is easily achieved.[19]

Isoelectric Focusing Isoelectric focusing (IOF) is the migration of charged particles (electrophoresis) under pH gradients to a location in the buffer, where they have zero net charge (isoelectric point). The pH gradient is set up between a cathode and an anode with the cathode at the higher pH.[59] This ingenuous process simultaneously concentrates and separates. Individual charged species, such as proteins, are rendered immobile as electrophoresis pushes them into the pH gradient until they arrive at the pH that renders them neutral, that is, at their isoelectric point (Ip) or isoelectric pH (IpH). Small amphoteric molecules (called *ampholytes*), which comprise a multitude of varying Ips, form the pH gradient. The ampholytes in an agarose or polyacrylamide gel carry a net charge, and a linear pH gradient is built up when an electric field is applied in the so-called prefocusing step (see Figure 6.74).

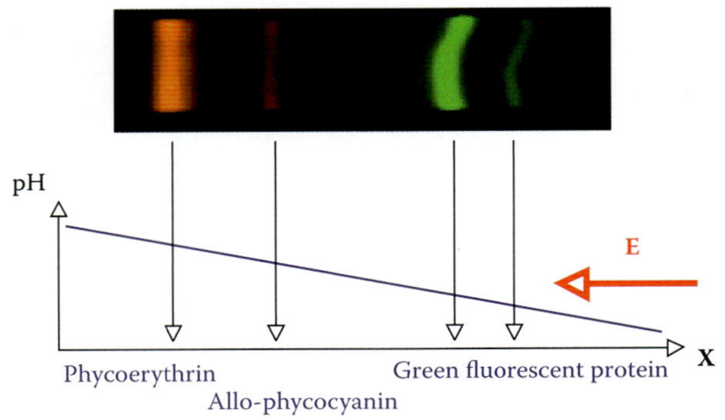

FIGURE 6.74 Isoelectric focusing (IOF).

In this brief treatment of the subject, we are particularly interested in the advantages and disadvantages of downscaling a static IOF system compared with downscaling a traditional electrophoresis system.

The focusing effect of the electrical force is counteracted by diffusion, which is directly proportional to the analyte (e.g., proteins) concentration gradient in the zone. At steady state, a balance between electrophoresis and sample diffusion holds and can be described by:

$$C\mu E = D\frac{dC}{dx} \tag{6.124}$$

where C = the sample concentration
dC/dx = the concentration gradient
E = the applied field strength
D = the diffusion coefficient
μ = the electrophoresis mobility

At steady state, this differential equation yields a Gaussian $[C(x) = C_0 \exp(-x^2/2\sigma^2)]$ for the concentration distribution, resulting in individual bands with bandwidth σ:

$$\sigma = \pm\sqrt{\frac{D\dfrac{dx}{dpH}}{E\left(\dfrac{d\mu}{dpH}\right)}} \tag{6.125}$$

As the values of E and $d(\text{pH})/dx$ increase, the distribution width decreases. At a separation of 3σ, bands are considered resolvable, and the minimum pH separation between adjacent zones is described as:

$$\Delta pH_{min} = \Delta L_{min}\frac{dpH}{dx} = 3\sigma\frac{dpH}{dx} \quad (6.126)$$

The resolving power is deduced by setting ΔpH equal to ΔpI and replacing σ with Equation 6.125:

$$\Delta pI_{min} = 3\sqrt{\frac{DdpH}{\left(\frac{d\mu}{dpH}\right)V}} \quad (6.127)$$

To arrive at Equation 6.127, we also substituted the electric field E by V/L, with V the applied potential and L (= dx) the length of the separation medium. From this equation, the minimum resolvable isoelectric point difference between adjacent focused bands is independent of L and dependent only on individual protein characteristics, the total pH difference, and the applied potential.[60] With pH and voltage gradient held constant, a sharper pH gradient (i.e., a shorter column) thus is expected to lead to a better separation. Exceptionally short columns might enable shorter analysis time and imaging of the entire separation with an inexpensive line array charge-coupled device (CCD).[60] In principle IOF downscales better than capillary electrophoresis, where the capillary length determines the resolution (see further below under scaling in analytical equipment). The maximum allowable temperatures dictate the minimum IOF separation length. Higher temperatures decrease the viscosity in an exponential fashion, thereby increasing the diffusivity of the sample. As the separation channel is shortened below 1 mm, Herr et al.[60] calculate a 23% increase in diffusivity compared with macroscale IOF (the macroscale device in the comparison is 20 cm long, and both have a 50-μm diameter channel). As a compromise, this research group chose an 8-mm-long channel. In Figure 6.75 capillary electrophoresis and capillary isoelectric focusing are compared.

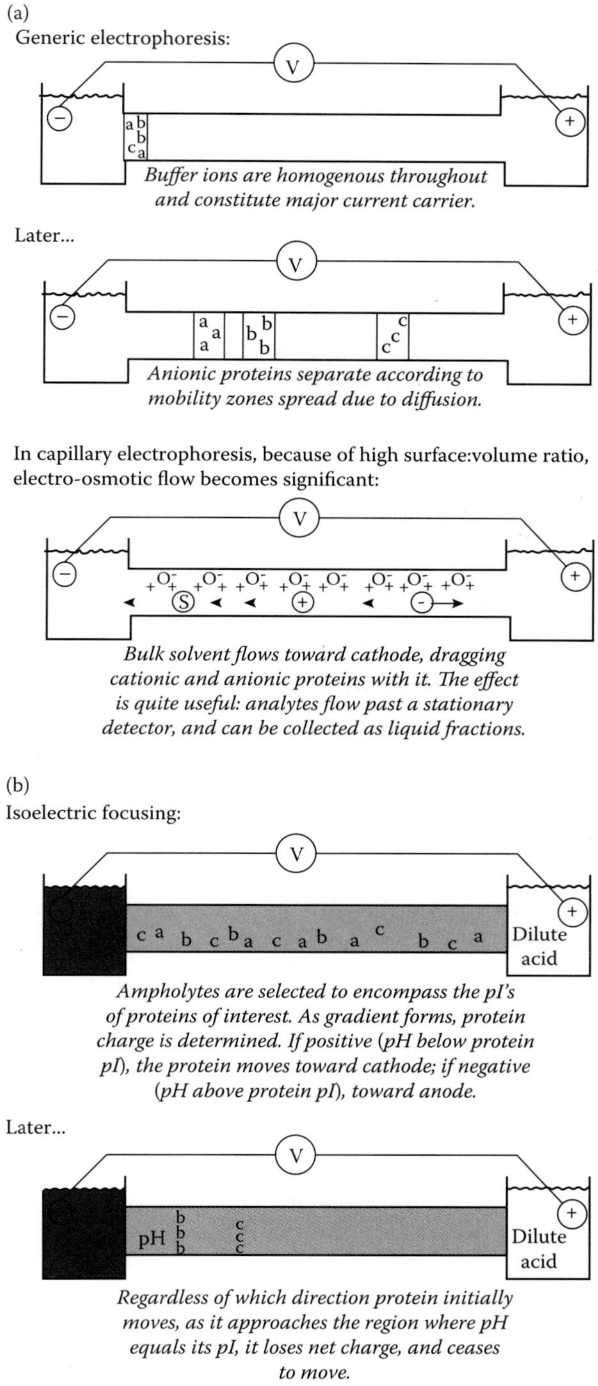
(a)
Generic electrophoresis:
Buffer ions are homogenous throughout and constitute major current carrier.
Later...
Anionic proteins separate according to mobility zones spread due to diffusion.
In capillary electrophoresis, because of high surface:volume ratio, electro-osmotic flow becomes significant:
Bulk solvent flows toward cathode, dragging cationic and anionic proteins with it. The effect is quite useful: analytes flow past a stationary detector, and can be collected as liquid fractions.
(b)
Isoelectric focusing:
Dilute acid
Ampholytes are selected to encompass the pI's of proteins of interest. As gradient forms, protein charge is determined. If positive (pH below protein pI), the protein moves toward cathode; if negative (pH above protein pI), toward anode.
Later...
pH
Dilute acid
Regardless of which direction protein initially moves, as it approaches the region where pH equals its pI, it loses net charge, and ceases to move.

FIGURE 6.75 (a) Generic electrophoresis. (b) Isoelectric focusing (IOF).

Open Electrophoresis Electrokinetics can be exploited to guide and manipulate charged particles (DNA, RNA, PCR amplicons, polynucleotides, proteins, enzymes, antibodies, nanobeads, and even micron scale semiconductor devices) on a planar chip equipped with an array of closely spaced planar metal film electrodes.[61] The ability to produce well-defined electric fields with such electrode arrays (see, e.g., the Nanogen DNA array; Volume II, Chapter 8, Example 8.2) enables one to electrophoretically transport those charged particles to or from any microlocation on the surface of the device submerged in the analyte solution. This so-called open electrophoresis technique does not use fluidic conduits but closely spaced, individually addressable electrodes. We explore this topic further below when using the Nanogen DNA array for dielectrophoresis (DEP) and in Volume III, Chapter 4 on packaging, assembly, and self-assembly, where we use the same chip for electric field-driven assembly.

Surface Electrophoresis As a further extension of the open electrophoresis concept, consider Wickramasinghe's use of AFM tips for molecular manipulation.[62] Applying a field between an AFM tip and a conductive surface, DNA electrophoresis over the shaft of the AFM tip was demonstrated at unprecedented speeds (10^4 times faster than CE) (Figure 6.76). We saw earlier that in free solution electrophoresis, DNA behaves as a free-draining coil, so that both charge and hydrodynamic drag scale linearly with molecular size (Equation 6.117). This is why DNA sequencing is not normally possible in free solution. Traditional DNA sequencing is done in a gel or polymer matrix, which presents a series of obstacles or obstructions, providing friction that depends on molecular size. Small DNA molecules migrate more easily through the pores of the gel, and thus have greater electrophoretic mobility than large DNA molecules. In the AFM case, electrophoretic separation of molecules such as DNA takes place in the nanometer-thick water films covering the shaft of the AFM tip (Figure 6.76A). In this layer, the Debye screening length is greater than the water film thickness, and surface charges are not screened totally by the water molecules. This might explain these encouraging results.

Electrokinetic Effects—Alternating Currents (AC)

Introduction

The AC electrokinetic effects we discuss here are *dielectrophoresis* (DEP), *electrorotation* (ROT), and *traveling wave dielectrophoresis* (TWD). Pohl introduced the term dielectrophoresis in 1951 to describe the motion of particles caused by dielectric polarization effects in nonuniform electric fields.[63,64] The effect was first recorded more than 2500 years ago, when it was discovered that rubbed amber attracts bits of fluff and other matter. *Electrorotation* is the rotation of particles in rotating electric fields, and traveling wave dielectrophoresis describes the translational movement of particles exposed to traveling electric fields.[65]

In electrophoresis, motion of a particle is determined by a net intrinsic electrical charge carried by that particle. In dielectrophoresis, on the other hand, motion is determined by the magnitude and polarity of charges induced in the particle by an applied field. In electrophoresis, direct current (DC) or low-frequency electric fields, usually homogeneous, are applied; in dielectrophoresis, on the other hand, alternating current (AC) fields of a very wide range of frequencies (in principle, there is no upper limit) are used, and the field must be inhomogeneous.[66]

The dielectrophoresis effect can be understood with reference to Figure 6.77. A particle placed in an electric field becomes electrically polarized as a result of partial charge separation, which leads to

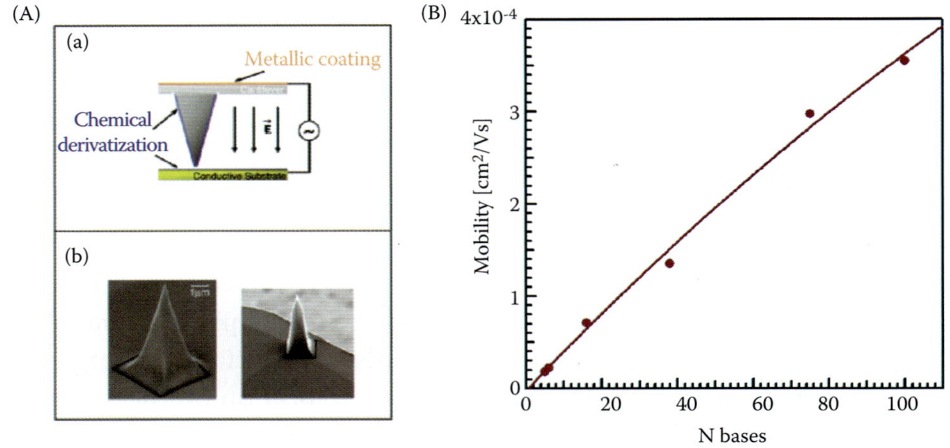

FIGURE 6.76 DNA electrophoresis. (A) (a) AFM setup and (b) AFM tips. (B) DNA mobility as a function of base pair number.

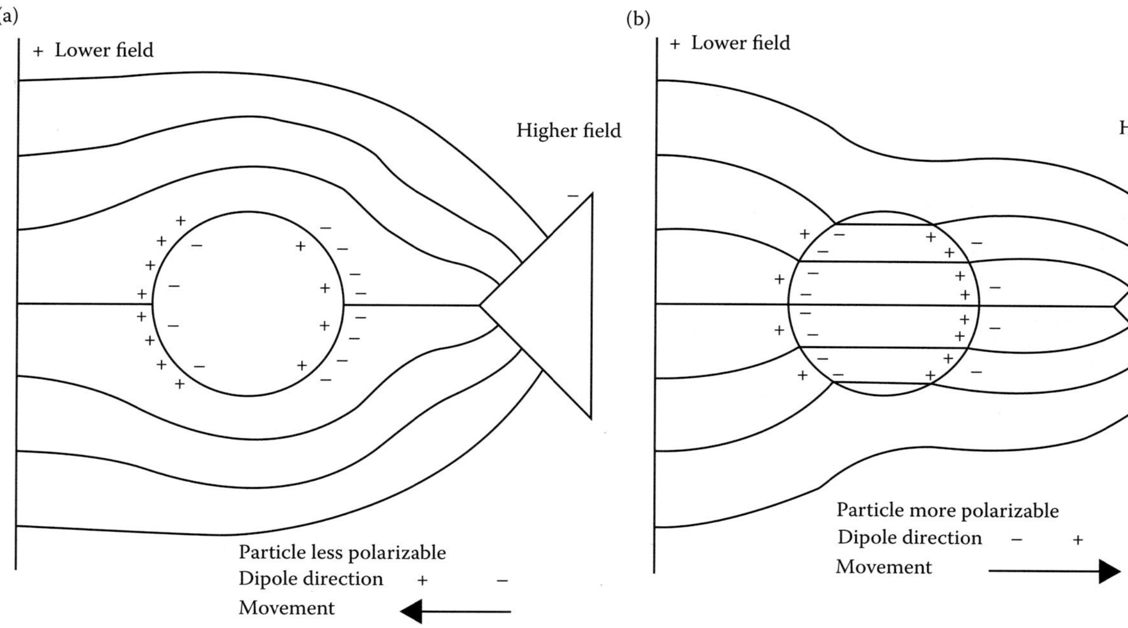

FIGURE 6.77 Negative (a) and positive (b) dielectrophoresis.

an induced dipole moment. The dipole moment is a consequence of the generation of equal and opposite charges ($+q$ and $-q$) at the boundary of the particle. This induced surface charge is only about 0.1% of the net surface charge normally carried by cells and microorganisms and is generated within about a microsecond. In a nonuniform electric field, the field on one side of the particle will be stronger than that on the other side, and a net dielectric force, F_{DEP}, is exerted. The magnitude of the induced dipole depends on the polarizability of the particle with respect to that of the medium. If a suspended particle has polarizability higher than the medium, the dielectrophoresis force will push the particle toward regions of higher electric field in so-called "positive dielectrophoresis." In negative dielectrophoresis, the medium has a higher polarizability than the suspended particles, and the particles are driven toward regions of low field strength. Dielectrophoresis causes a motion of neutral matter by polarization of the neutral matter in a nonuniform electric field. The method is particularly suited for the separation, aggregation, selective trapping, manipulation, and identification of cells and microorganisms.[66]

Dielectrophoresis Force

Spherical particles of radius r and complex permittivity ε_p, suspended in a fluid of absolute complex dielectric permittivity ε_m, are subject to a dielectric force given by:

$$\mathbf{F}_{DEP} = 2\pi r^3 \varepsilon_m \alpha_r \nabla \mathbf{E}^2 \tag{6.128}$$

where $\nabla \mathbf{E}^2$ = the gradient of the square of the electric field (rms, or root mean square value) quantifying the nonuniformity of the electric field

∇ = the del vector operator

α_r = the real component of the Clausius-Mossotti factor, that is, the effective polarizability of the particle with respect to its suspending medium; the latter term may thus be written as:

$$\alpha_r = \mathrm{Re}\left(\frac{\varepsilon_p^* - \varepsilon_m^*}{\varepsilon_p^* + 2\varepsilon_m^*}\right) \tag{6.129}$$

In the above expression:

Re = "the real part of"

ε_p^* and ε_m^* = the complex dielectric properties of the particle and its medium, with $\varepsilon^* = \varepsilon - j\sigma/\omega$, in which $j = \sqrt{-1}$

ε = permittivity

σ = conductivity

ω = angular frequency of the applied field ($\omega = 2\pi f$)

For a vacuum, which has no mobile charges, ε is 1, and for a conductor, it is infinite. Because $\varepsilon_p^* - \varepsilon_m^*$ can be positive or negative, depending on the relative magnitudes of ε_p^* and ε_m^*, controlled movement from and to areas of high electric field strengths is possible. In other words, a material with a higher dielectric constant will experience a force tending to move it into a stronger electric field, displacing a material with a lower dielectric constant in the process. A positive value for α_r in Equation 6.129 implies a positive DEP force; a negative value produces a negative DEP. Because the field $\boldsymbol{E}$ in Equation 6.128 appears as ∇E^2, reversing the bias does not reverse the DEP force. Ac voltages in a frequency range from 500 Hz to 50 MHz have been used. Based on Equation 6.129, one recognizes that the factor α_r is frequency-dependent. At low frequencies polarizability mainly is determined by the conductivity, and at high frequencies it is determined by the permittivity. Theoretically, the value of α_r may be between +1.0 and −0.5. Key in dielectrophoresis is the presence of a material of substantially different dielectric constant from its surroundings. In an aqueous conductive solution, ions shield the dielectric material from the external applied field. Varying the field in time reduces the shielding effect of the ions somewhat. If the mobile dielectric materials are embedded in a sufficiently low resistance medium, then the dielectrophoresis effect can be much more readily observed. For a rigorous derivation of Equation 6.128, we refer to Huang et al.[67] and Wang et al.[68]

Using DEP, individual biological cells have been moved and positioned by altering the ac frequency so as to switch from a "capture" mode using positive DEP to "release" mode using negative DEP.[69] Particles can also be separated using DEP. This is illustrated in Figures 6.78 and 6.79. Figure 6.78a reveals how cell type A switches from negative to positive DEP at a changeover frequency of about 10^5 Hz. From Figure 6.78b we learn that there is a frequency range above 10^5 Hz where cell type B displays negative dielectrophoresis, whereas A exhibits positive dielectrophoresis. This means that one cell type will move to higher electric field regions, while the other cell type will do the opposite, effectively separating the two types of cells. This effect is exploited in Figure 6.79, where we separate *Listeria* from whole blood. An array of 100-planar Pt electrodes [the Nanogen chip (see Volume II, Chapter 8, Example 8.2)] is used in this experiment. In the first panel, blood contaminated with *Listeria* is placed over the chip. In the second panel, a 10-kHz, 10-Vpp ac signal is applied, driving the red blood cells to low electric field regions (i.e., between the Pt electrodes) and the *Listeria* to high electric field regions (i.e., on the Pt electrodes).

In the third panel we wash off the red blood cells, leaving the *Listeria* highly enriched on the Pt electrodes where they stick better than the red blood cells on the insulating regions (Si_3N_4) between the Pt electrodes. In the fourth panel only *Listeria* is left on the chip. Finally, in the fifth panel, we release the *Listeria* by removing the bias and washing again. In Table 6.7 we summarize some pros and cons of using dielectrophoresis for particle separation.

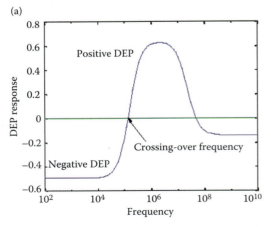

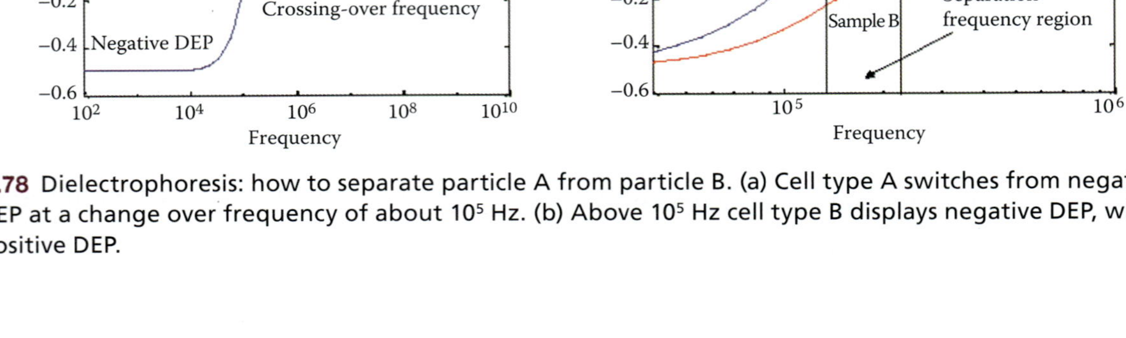

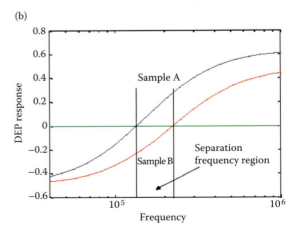

FIGURE 6.78 Dielectrophoresis: how to separate particle A from particle B. (a) Cell type A switches from negative to positive DEP at a change over frequency of about 10^5 Hz. (b) Above 10^5 Hz cell type B displays negative DEP, whereas A exhibits positive DEP.

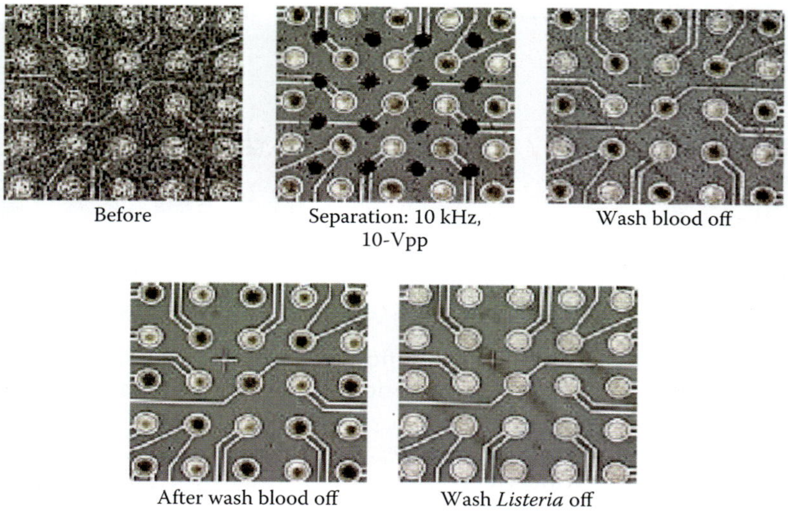

FIGURE 6.79 Dielectrophoretic separation of *Listeria* from whole blood.

One problem associated with traditional methods of DEP separation highlighted in Table 6.7 is that the DEP force, which is proportional to ∇E^2, rapidly decays as the distance from the planar electrodes increases. As a consequence, a common problem to most current dielectrophoresis devices using planar electrodes is the incapability to achieve high throughputs and high efficiencies. With a given flow channel cross-section, many targeted particles might flow over the planar electrodes without experiencing a force, its mean distance to the electrode surface being too long, making it necessary to reflow the same sample several times. This has prevented DEP from being used in high-volume applications. To correct this problem, the author's University of California, Irvine, team is using 3D electrodes to reduce the mean distance of any targeted particle contained in a channel or chamber to the closest electrode surface, as illustrated in Figure 6.80. Numerical simulations demonstrating such advantage were obtained by Park,[70] and the experimental verification was carried out by Martinez-Duarte.[71]

The 3D electrodes are implemented using the C-MEMS process detailed in Volume III, Chapter 5 (page 266 "Microfabrication of Carbon: C-MEMS Process"). In 3D C-MEMS dielectrophoresis, carbon 3D structures up to several hundreds of micrometers high are implemented as electrodes in the flow channel. An example of 65-μm-high posts is shown in Figure 6.81.

The difference in trapping efficiency of yeast cells, when only flowing the sample once, between 2D and 3D electrodes increases as flow rate increases, with the 2D efficiency always being lower as shown in Figure 6.82.

Microfabricated 3D electrodes also have been implemented by Wang[72] and Voldman,[73] who used electroplated gold electrodes for particle separation, applying the lateral field established by the electrodes that form the channel side walls and for particle trapping for cytometry purposes, respectively. More

Table 6.7 Pros and Cons for Particle Separation by Dielectrophoresis

Pros	Cons
Intrinsic dielectric properties are taken advantage of	Speed: clinical sample requires ml volume (1)
No immunolabeling involved	Throughput: can only separate two types of cells (2)
No narrow channels to clog	Selectivity/sensitivity
Simple and low cost	
Easy to scale down or scale up	
Can move even semiconductor components	

1) This problem may be solved using 3D volumetric DEP methods (described next).
2) This problem could be solved by using an array of signal generators.

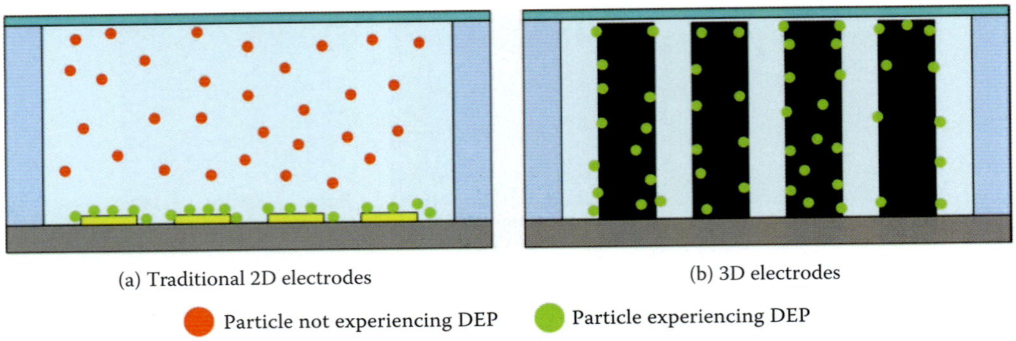

FIGURE 6.80 (a) In traditional 2D DEP, many particles might flow by without experiencing any DEP force. (b) In 3D DEP, the mean distance from the particle to the electrode surface decreases; thus, the number of particles experiencing DEP force greatly increases.

recently, Iliescu[74] has used complex microfabrication techniques to obtain 3D doped silicon electrodes in a multistep process. In a nontraditional way, Lapizco-Encinas et al.[75] have induced DEP on bacteria and proteins using insulator volumetric microelectrodes, in a technique that has become known as insulator DEP. The main concept behind this approach is the use of two metal electrodes, excited by thousands of volts, to generate a uniform electric field between them, which is then disrupted and turned nonuniform by the insulator structures in between. Another truly 3D DEP approach is that of Fatoyinbo et al.,[76] who use a drilled conductor-insulator-conductor laminate to obtain wells with electrodes all along the top and bottom surfaces of the laminate.

The final goal is to improve throughput and efficiency in DEP devices for a commercial product featuring high selectivity at high flow rates. Immediate applications might include cell enrichment for cell therapy, sample purification, drug screening, water analysis, and many other applications may become possible in the future. Throughput refers to inducing DEP on as many particles as possible without consideration of selectivity between targeted and nontargeted particles. This is also known as *maximal retention efficiency*.[77] Selectivity refers to how well one can induce DEP to a targeted particle without affecting nontargeted ones. It is also referred to as *maximal purification efficiency*.[77]

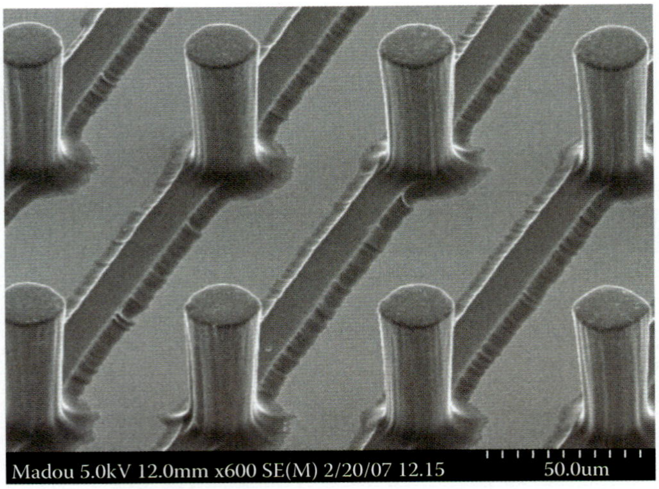

FIGURE 6.81 Carbon posts (65 μm high) for use as electrodes in C-MEMS dielectrophoresis (C-MEMS DEP).

Traveling Wave Dielectrophoresis

If, instead of using a stationary electric field, as is used in DEP, a particle is subjected to a moving electric field, the particle can be moved by an effect known as traveling wave dielectrophoresis, or TWD. To produce these traveling waves, electrode

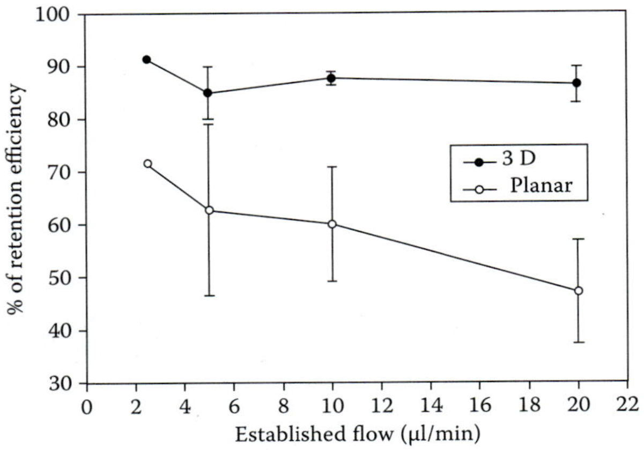

FIGURE 6.82 Higher retention efficiency at all flow rates are obtained with 3D DEP compared with 2D DEP. The difference between them sharply increases with flow rate. (Martinez-Duarte, R., H. A. Rouabah, N. G. Green, M. Madou, and H. Morgan. 2007. *Proceedings of the Eleventh International Conference in Miniaturized Systems for Chemistry and Life Sciences: microTAS 2007*. Paris.)

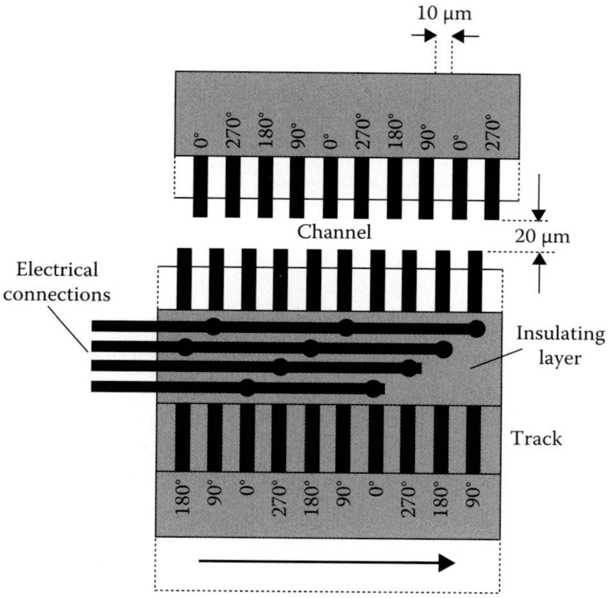

FIGURE 6.83 TWD setup. Electrode structures for producing traveling electric fields along channels or tracks. When addressed with sinusoidal voltages of the phase sequences shown, a traveling field is established by the propagation direction indicated by the arrow. The characteristic dimensions shown are suitable for manipulating yeast, erythrocytes, and parasites such as *Cryptosporidium*, for example. Correspondingly larger or smaller dimensions are required for larger bioparticles (e.g., white blood cells, plant cells, pollen) or smaller ones (e.g., bacteria, viruses), respectively. (Burt, J. P. H., R. Pethig, and M. S. Talary. 1998. Microelectrode devices for manipulating and analysing bioparticles. *Trans Inst MC* 19:1–9. With permission.)

geometries with tracks of electrodes in a channel, as shown in Figure 6.83, may be used. In this arrangement, electrodes are fabricated on a bottom-insulating substrate, and a second insulator layer is used to define a channel along the length of the electrode track. Another set of four electrodes is then deposited on this second insulator layer, and vias are used to link every fourth electrode together and to provide external connections. The four electrodes are addressed with sinusoidal voltages of 90° phase separation (0°, 90°, 180°, and 270°). The used phase quadrature voltages are equal in amplitude (around 1 to 5 V peak-peak) and of a frequency range from about 1 kHz to 10 MHz. TWD provides conveyor tracks over which particles, suspended by negative dielectrophoresis, are moved. By changing the magnitude and frequency of the energizing voltages, the speed and direction of particles in the channel can be controlled.[65]

TWD, in combination with DEP, enables selective and nonselective trapping of particles. The selective trap only holds desirable subpopulations, whereas a nonselective trap retains all particles from a fluid flow. Particles moving in a channel by TWD may be captured nonselectively and held immobile by suddenly changing the frequency and phases of the applied voltages so as to encourage positive dielectrophoresis. By returning to a TWD-type of field, all the particles, or a subset of them, can be made to resume their movement along the channel. In the case of selective trapping, the TWD remains in effect, and trapping is accomplished by the choice of the electric field frequency. The frequency may be chosen such that some particles are levitated by negative dielectrophoresis and thus keep on moving under the influence of the TWD, whereas others are captured by positive dielectrophoresis and are immobilized at the electrode surfaces.[65]

Particle Levitation

The gravitational settling force pulls down a spherical particle with radius r and mass density γ_1, suspended in a fluid with mass density γ_2, and is given by:

$$\mathbf{F}_g = \frac{4}{3}\pi r^3(\gamma_2 - \gamma_1)\mathbf{g} \tag{6.130}$$

where **g** is the gravitational acceleration vector. A particle will hover above the electrode surface when the opposing gravitational and negative dielectrophoretic forces balance:

$$\mathbf{F}_{DEP} + \mathbf{F}_g = 0 \tag{6.131}$$

Or, from Equations 6.128 and 6.130:

$$2(\gamma_2 - \gamma_1)\mathbf{g} = -3\,\varepsilon_m \alpha_r\, \nabla \mathbf{E}^2 \tag{6.132}$$

From Equation 6.132, levitation is independent of particle size, as both DEP and gravitational force scale as l^3. Levitation height only depends on the intrinsic properties of the particle, such as permittivity, conductivity, and mass density. By working at very high frequencies, where the polarizability factor primarily depends on the permittivity rather than the conductivity, levitation may also be made independent of the conductivity of the solution. A quantitative estimate of the levitation height requires knowledge of how the factor ∇E^2 varies as a function of height h above the electrode plane. Rousselet et al.[79] derived the following relationship between ∇E^2 and h:

$$\nabla \mathrm{E}^2 = \mathrm{AV}^2 \exp(-\mathrm{hk}) \tag{6.133}$$

where V = applied voltage
A = constant
k = another constant inversely proportional to the width of the electrodes[79]

Combining this expression with Equation 6.132, the expression for stable levitation height is derived as:

$$\mathrm{h} = \frac{1}{\mathrm{k}} \ln\left[\frac{-3\,\varepsilon_{\mathrm{m}}\alpha_{\mathrm{r}}\,\mathrm{AV}^2}{2(\gamma_2-\gamma_1)\mathrm{g}}\right] \tag{6.134}$$

This type of dielectrophoretic levitation was demonstrated, for example, with latex beads and interdigitated electrodes of widths and separations ranging from 5 to 40 μm, with ac signals of 0–10 V (rms) in the frequency range 1 kHz to 10 MHz. Maximum levitation was observed at 1 MHz. At this frequency it was confirmed experimentally that levitation height was independent of particle size and solution conductivity.[79] It also was shown that red blood cells could be separated from latex beads easily. Whereas the erythrocytes were attracted to the electrodes by positive dielectrophoresis, the latex particles levitated and could be separated by flowing liquid over the surface. By de-energizing the electric field, the red blood cells also could subsequently be removed by flowing more liquid.[79]

Particles possessing different dielectric properties levitate at different heights in a flat field-flow-fractionation (FFF) chamber through the balance of DEP and gravitational forces.[68] The different particles are levitated into streamlines with different velocities, as shown in the parabolic flow profile illustrated in Figure 6.84, and can be separated this way. Even continuous separation of particles is possible.[80]

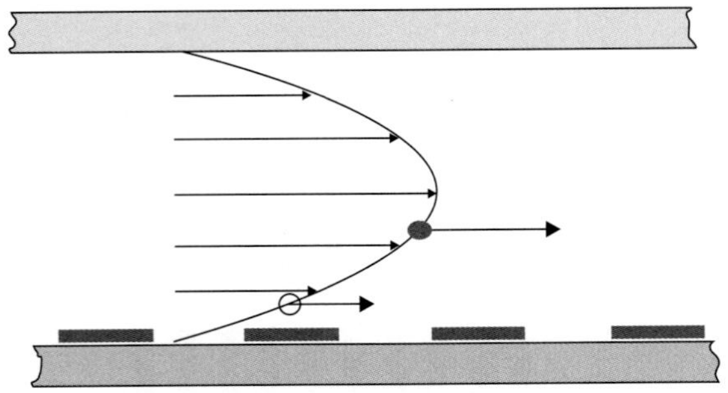

FIGURE 6.84 Field-flow fractionation (FFF). Outline of the combined DEP-FFF separation procedure. Particles with different dielectric properties or density are levitated by negative dielectrophoresis to different planes of the parabolic velocity profile of the liquid flowing through the chamber. (Rousselet, J., G. H. Markx, and R. Pethig. 1998. Separation of erythrocytes and latex beads by dielectrophoretic levitation and hyperlayer field-flow fractionation. *Colloids Surfaces* A 140:209–16. With permission.)

Electrorotation

Electrorotation (ROT) is the imaginary component of the ac electrokinetic effect, whereas its real component is the above-introduced dielectrophoresis (DEP). The rotating electric field usually is generated using four electrodes positioned at right angles to one another and energized with phase-quadrature signals of frequencies between 50 Hz and 100 MHz. The ROT torque acting on a particle positioned in the center of this electrode assembly is given by:

$$\mathrm{T}_{\mathrm{ROT}} = -4\pi\,\mathrm{r}^3\mathrm{E}_{\mathrm{m}}\alpha_{\mathrm{i}}\,\mathrm{E}^2 \tag{6.135}$$

where α_i is the "imaginary part" of the Clausius-Mossotti factor. As with impedance spectroscopy, there is a direct relationship between the real and imaginary components of the ac electrokinetic effect, so that the electrorotation frequency spectrum can be calculated from the measured dielectrophoresis response, and vice versa. In Figure 6.85 the frequency variations of α_r and α_i for viable and nonviable yeast cells are shown. The polarity and peak values of α_i are determined by the rate of change of α_r. The frequency dependence of magnitude and polarity of α_r and α_i is the basis of the three ac electrokinetic applications discussed here.

Rotation analysis chambers in a biofactory-on-a-chip are used to monitor dielectric properties of particles.[65] Electrorotation has been shown to be a sensitive method for monitoring the physiological viability of cells.[82]

Scaling Considerations

Depending on particle size, different forces compete with the DEP force. For example, a typical biological cell with a radius of 1 μm and a density of

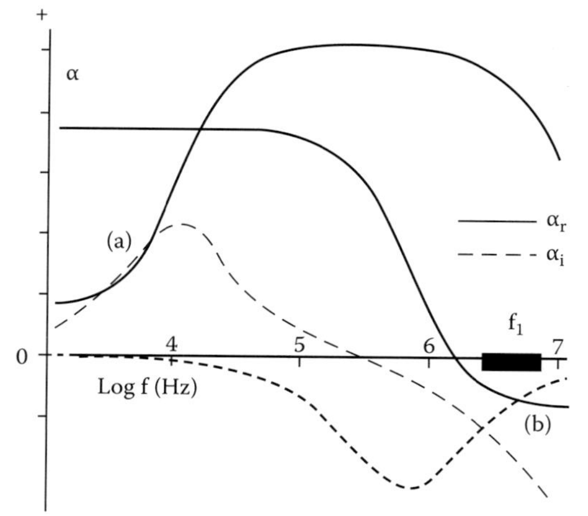

FIGURE 6.85 Alpha plots. The frequency variations of the real and imaginary polarizability parameters α_r and α_i of Equations 6.128 and 6.135 for viable (a) and nonviable (b) yeast cells in a suspending medium of conductivity 0.8 mS/m. Only the nonviable cells will exhibit traveling field dielectrophoresis (in the frequency range f_1) and thus the viable cells will remain immobilized at the electrodes. (Burt, J. P. H., R. Pethig, and M. S. Talary. 1998. Microelectrode devices for manipulating and analysing bioparticles. *Trans Inst MC* 19:1–9; and Talary, M. S., J. P. H. Burt, J. A. Tame, and R. Pethig. 1996. Electromanipulation and separation of cells using travelling electric fields. *J Phys D Appl Phys* 29:2198–203. With permission).

1.05 kg m^{-3} suspended in an aqueous solution at room temperature feels both the gravitational and the Brownian force of around 2×10^{-15} N each. For the DEP force to dominate, it should be at least 10 times larger than these competing forces; to affect this, ∇E^2 must be around 9×10^{12} V^2m^{-3} (with an α of 0.5).[66] Both the DEP and the gravitational effect scale as l^3, and, with increasing particle size, no scaling advantage of one over the other emerges. With submicrometer particles, on the other hand, the randomizing Brownian effect, being proportional to l^{-1}, becomes the dominating force competing with DEP. With particles below 0.1 μm, for example, values for ∇E^2 greater than 10^{17} V^2m^{-3} are required to make the DEP force 10 times larger than the Brownian. This requires field strengths in excess of 10^6 Vm^{-1}, and at a suspending medium conductivity higher than 0.1 Sm^{-1}, these high fields cause local heating, leading to hydrodynamic effects. The latter interfere with the DEP effect of submicron particles significantly.[66] Moreover, at these fields cell damage might become an issue. Müller et al.[83] used submicron-scale electrodes to achieve DEP levitation and trapping of particles as small as 14 nm, but, as predicted, hydrodynamic effects interfered significantly.

Downscaling of the electrodes is very favorable in dielectrophoresis. By using thin film metal electrodes rather than wire or pin electrodes, one approaches the size of the typical particles to be separated. In Figure 6.86, the expression for ∇E^2 for the specific represented electrode configuration is given. Maintaining the applied voltage at 1 V rms and reducing the radius of curvature 100-fold produces a 1000-fold increase in ∇E^2. Lower voltages can thus be used to produce the same DEP force on a particle in the same relative position. Despite the small cross-sectional area of the electrodes, the volume of liquid they energize is very large by comparison, so that heat produced by the electrical current is efficiently dissipated, and, moreover, surface electrochemical processes are reduced.[66]

Applications

In a 1983 patent, Batchelder[84] describes the use of dielectric forces to selectively position, transport,

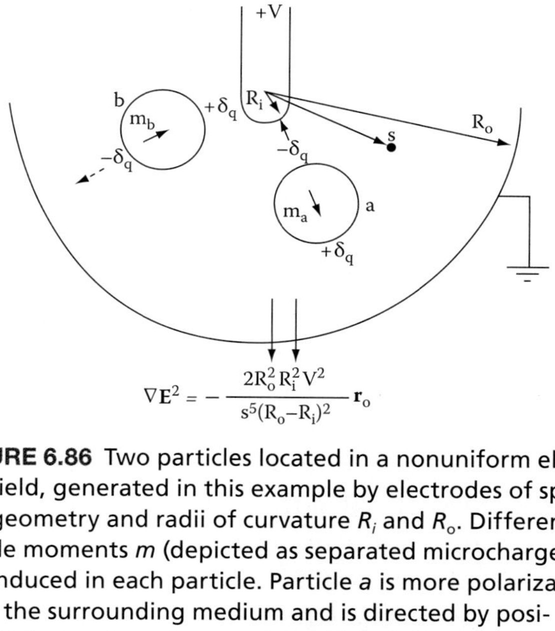

$$\nabla \mathbf{E}^2 = -\frac{2R_o^2 R_i^2 V^2}{s^5(R_o - R_i)^2}\mathbf{r}_o$$

FIGURE 6.86 Two particles located in a nonuniform electric field, generated in this example by electrodes of spherical geometry and radii of curvature R_i and R_o. Different dipole moments *m* (depicted as separated microcharges *q*) are induced in each particle. Particle *a* is more polarizable than the surrounding medium and is directed by positive dielectrophoresis toward the high field at the inner electrode, whereas particle *b* is weakly polarizable and moves toward the low-field region at the outer electrode under the influence of negative dielectrophoresis. The force causing this motion is proportional to ∇E^2, described by the equation in the figure, where *s* is the distance of the particle from the inner electrode, *V* is the applied (rms) voltage, and r_0 is a unit-radius vector. (Pethig, R., and G. H. Markx. 1997. Applications of dielectrophoresis in biotechnology. *TIBTECH* 15:426–32. With permission.)

mix, and react one or more chemicals within a reaction chamber. The Batchelder patent was ahead of its time in terms of vision for microfluidics and constituted a laboratory-on-a-chip (LOC) long before academics introduced the buzzword. The emphasis in this patent is on the dielectrophoresis mechanism, and little is said about micromachining. Although metal electrodes are patterned with lithography techniques, the reaction chambers are made with more traditional manufacturing methods.

Once microelectrode technology and MEMS made their entry, it became apparent that dielectrophoresis would be a promising tool for the selective manipulation and separation of cells, bacteria, viruses, and perhaps even biomacromolecules such as DNA and proteins. A limitation may be cell damage, electroporation, and electrofusion, which start at field strengths of $1\text{–}3 \times 10^5$ Vm^{-1}. However, in dielectrophoresis, the field is rarely higher than 10^5 Vm^{-1}. In practice, no irreversible cell damage has been reported in DEP experiments.[66] Another possible limitation is that of particles being smaller than 10 nm; these might be too difficult to manipulate because of the hydrodynamic effects interfering with the DEP force.

Batch and continuous separation of particles has been achieved through positive and negative DEP in FFF chambers. A typical DEP-FFF separation chamber is shown in Figure 6.87 and consists of two glass plates spaced a small distance (typically <300 μm) apart. The inner surfaces of the glass plates are provided with interdigitated electrodes.[80,85] Working in the frequency range marked f_1 in Figure 6.85 separates viable and nonviable yeast cells. The nonviable cells levitate in this frequency range and are transported with the fluid flow, whereas the viable ones remain immobilized at the electrodes.[81] As just one more example of the use of this type of separation, we refer to the removal of human breast cancer cells from hematopoietic $CD34^+$ stem cells. The $CD34^+$ stem cells are levitated higher and carried faster by the fluid flow and exit the separation chamber earlier than the cancer cells. Using on-line flow cytometry, efficient separation of the cell mixture is observed in less than 12 min, and $CD34^+$ stem cell fractions with a purity >99.2% are obtained.[85]

Electrorotation is used, for example, to determine the dielectric properties of human leukocyte

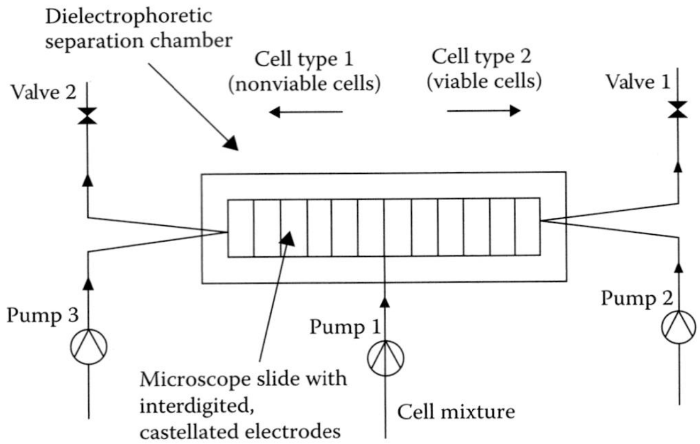

FIGURE 6.87 Outline of a continuous dielectrophoretic separation system. Valves, frequency generators, and pumps are all under computer control. (Markx, G. H., and R. Pethig. 1995. Dielectrophoretic separation of cells: continuous separation. *Biotech Bioeng* 45:337–43. With permission.)

subpopulations.[86] Pethig et al.[65] are experimenting with combinations of DEP, ROT, and TWD elements on a single substrate to make a so-called biofactory-on-a-chip.

Electrowetting

Electrical control of interfacial tension between a liquid and a solid provides another means of fluid pumping without moving mechanical parts. Altering the apparent surface tension at the solid/liquid interface becomes possible by applying a potential over the interface. This phenomenon is known as electrowetting and is described by the Lippmann equation:

$$\gamma_{SL} = \gamma_{SL}^{max} - \frac{\varepsilon_r \varepsilon_0}{2\delta}(V - V_{pzc})^2 \quad (6.136)$$

where V_{pzc} represents the potential of zero charge for the solid/liquid (electrolyte) interface, and γ_{SL}^{max} is the surface tension corresponding to $V = V_{pzc}$. The thickness of the diffusion layer in the solution is given by δ, and ε_r and ε_0 are the relative permittivity of the solution and the permittivity of vacuum, respectively. A voltage applied across the interface will alter the surface tension and cause the liquid to move within the capillaries. The flow velocities achieved with electrocapillary pressure are several centimeters per second or nearly two orders of magnitude higher than the velocities achieved by electro-osmotic pumping.[87]

Surface tension becomes important only when dimensions drop well below 1 mm. As in the case of electro-osmotic pumping, subtle uncontrollable changes at the solid/electrolyte interface make this actuation principle difficult to control and electrolyte specific.

Light has been used to control the wettability of the surface of certain photoresponsive materials. The exact mechanisms of the surface wettability changes of these materials on illumination still are being debated; however, it is generally accepted that the effect is caused by a photoelectrochemical reaction at the surface.[88] With ZnO, UV irradiation is responsible for generating electron-hole pairs on its surface. These holes react with the lattice oxygen to form surface oxygen vacancies in a process described by Zhang et al.[89] in the following equations:

$$ZnO + 2h\nu \rightarrow 2h^{+} + 2e^{-}$$

$$O^{2-} + h^{+} \rightarrow O^{-}$$

$$O^{-} + h^{-} \rightarrow \frac{1}{2}O_2 + V_0 \text{ (oxygen vacancy)} \quad \text{Reaction 6.1}$$

In ambient air, water and oxygen from the surroundings compete with lattice oxygen to dissociatively absorb on those vacancies. In solution the defect sites, however, kinetically favor the absorption of hydroxyl groups over oxygen,[90] and with the surface attracting more hydroxyl (–OH) groups, the surface becomes more hydrophilic.

This switch from a hydrophobic to a hydrophilic surface has been extensively documented. Sun et al.,[91] for example, showed changes of contact angle from 109° before to 5° after UV exposure for a droplet of water on a ZnO surface.

Other wide bandgap semiconductor oxides, such as Ta_2O_5, TiO_2, indium–tin oxide, $InTaO_4$, and In_2O_3, also exhibit optowetting. These materials reverse their characteristic and tend to go back to their hydrophobic state if the UV-exposed sample is placed in a dark room; over time the hydroxyl groups absorbed on the defect sites are replaced by oxygen atoms. This is because the surface is thermodynamically unstable after the hydroxyl absorption, and oxygen is more strongly bonded on the defect site than a hydroxyl group.

Because there is some degree of reversibility, these semiconductors show promise as a wettability switch for future microfluidic applications.

Centrifugal Fluidic Platform—CD Fluidics

Introduction

An attractive approach to fluid propulsion is centrifugation in fluidic channels and reservoirs crafted in a CD-like plastic substrate as shown in Figure 6.88. The centrifugal force caused by the rotation of the CD results in the release and flow of reagents and analytes. No external pumps and valves are needed—fluidic flow can be controlled by the spinning rate. The centrifugal force generated in such an instrument overcomes (depending on the rotation speed) the capillary forces in the fluidic network and moves fluid in a valveless manner from reservoir chamber

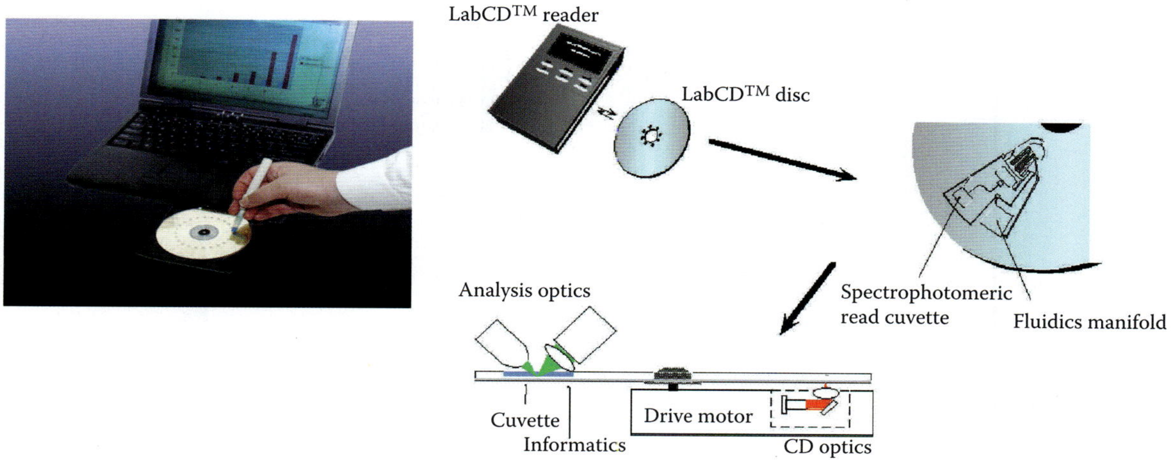

FIGURE 6.88 Centrifugal platform on a CD. The name LabCD™ was introduced by Gamera, the first fluidic CD company (now defunct).

to reaction chamber and eventually to a waste site. A whole range of fluidic functions, including valving, mixing, metering, sample splitting, and sample separation, has been implemented this way.[92] Such analytical functions make for a laboratory-on-a-disc with applications in diagnostics and drug discovery. Analytical measurements may be fluorescent or absorption based, and informatics embedded on the same disc could provide test-specific information (merging of fluidics with informatics).

Prototyping methods of this microfluidic platform are covered in Volume III, Chapter 5 on selected new materials and processes for MEMS and NEMS. In Volume II, Chapter 10 on replication techniques and LIGA, we cover manufacturing of the fluidic CD. The aspect of the CD platform we are interested in here is the physics behind its operation and the comparison of CD fluidics to mechanical, osmotic, and acoustic pumping.

Centrifugal Platform Operation

In a centrifugal platform, the centrifugal force provides the pumping force, whereas hydrophilic and hydrophobic valves are used to inhibit flow. In Figure 6.89a, we show a schematic of a centrifugal platform together with relevant parameters; Figure 6.89b and c demonstrates two types of hydrophobic valves, and Figure 6.89d is a hydrophilic or capillary valve. For a hydrophobic CD surface, to make a valve, a flow restriction can be made in the flow channel by reducing its radius from r_1 to r_2. A positive pressure is then needed to push the fluid across this restriction. From Equation 6.96, the pressure p is determined by r, the radius of the capillary, and the required pressure Δp to overcome a sudden narrowing in capillary tubing is calculated from Equation 6.96 as:

$$\Delta p = 2\sigma_L \cos\theta_c \left(\frac{1}{r_1} - \frac{1}{r_2}\right) \qquad (6.137)$$

Equation 6.137 is independent of the length of the restriction and flow rate. Once the capillary downstream is wetted, the pressure needed to continue flow is given by Equation 6.54 (Hagen-Poiseuille's Law). These hydrophobic restrictions can be used as passive valves in fluidic networks: by setting the pumping pressure below the pressure associated with the restriction, no flow will take place unless the pumping pressure is increased. Mixers, diluters, sample splitters, and sample integrators have been built using passive hydrophobic valving in combination with an external mechanical pump.[93] Of course, the same principles can be applied in combination with centrifugal, electrokinetic, or acoustic pumping. As shown in Figure 6.89c, one can also create a hydrophobic valve by locally changing the hydrophobicity of the surface, i.e., by locally increasing θ_c. In the case of a contact angle less than 90° (a hydrophilic surface), to make a valve the capillary is widened, which leads to the need of increased pressure as more energy is required now to wet a larger surface area (Figure 6.89d).

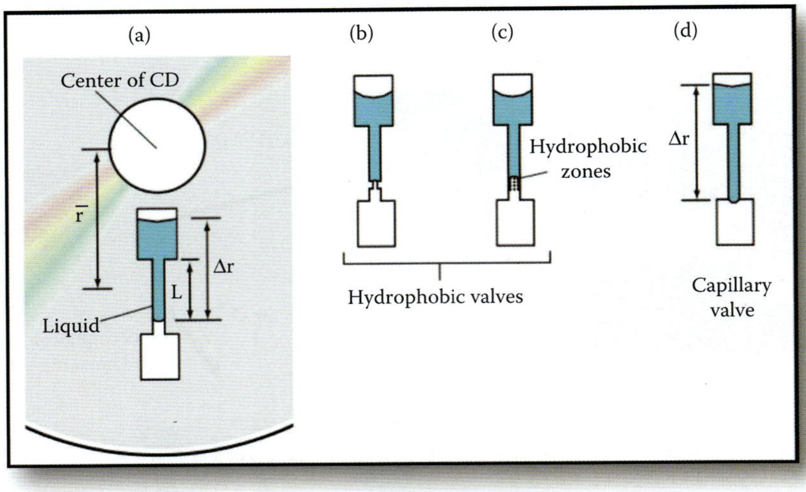

FIGURE 6.89 Centrifugal platform on a CD. (a) CD platform, (b and c) hydrophobic valves, and (d) hydrophilic capillary valve.

Assuming a capillary valve, when spinning the CD, the two opposing forces at work can be described as follows (see Figures 6.89a and 6.90).

The pumping force (P_c) caused by spinning of the disc is given by:

$$\frac{dP_c}{dr} = \rho\omega^2 r \qquad (6.138)$$

where ρ = the density of the liquid
ω = the angular velocity of the CD platform
r = the distance between a liquid element and the center of the CD

Integration of Equation 6.138 from $r = R_1$ to $r = R_2$ gives:

$$\Delta P_c = \rho\omega^2(R_2 - R_1)\frac{R_2 + R_1}{2} = \rho\omega^2 \Delta R \cdot \bar{R} \qquad (6.139)$$

The capillary force (P_s) caused by interface tension is given by:

$$\Delta P_s = \frac{\gamma \cos\theta_c \cdot C}{A} \qquad (6.140)$$

where γ = the interface tension
θ_c = the contact angle (in a practical example θ_c may be 61.5°)
A = the cross-section area of the capillary
C = the associated contact line length

The burst frequency is the frequency at which the fluidics are released from their reservoirs. The liquid will be released when ΔP_c is greater than ΔP_s; consequently, the burst frequency calculated from Equations 6.139 and 6.140 is given by:

$$f_b \geq \left(\frac{\gamma \cos\theta_c \cdot C}{\pi^2 \rho \cdot \bar{R} \cdot \Delta R \cdot 4A}\right)^{\frac{1}{2}} = \left(\frac{\gamma \cos\theta_c}{\pi^2 \rho \cdot \bar{R} \cdot \Delta R \cdot d_H}\right)^{\frac{1}{2}} \qquad (6.141)$$

Here d_H is the so-called hydrodynamic diameter of the channel.

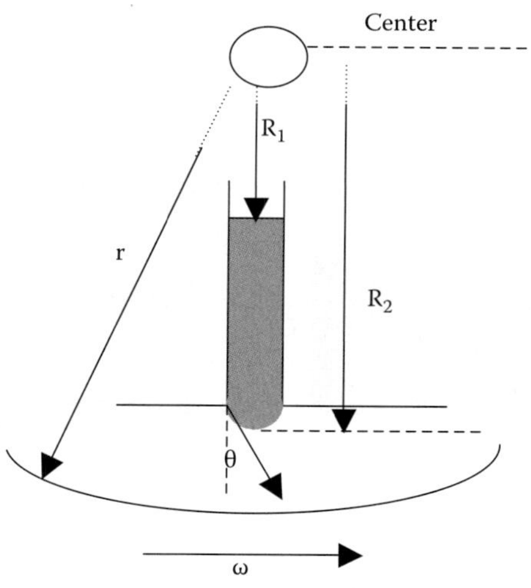

FIGURE 6.90 Schematic illustration of fluid propulsion in centrifugal microfluidics.

Example 6.8: Two-Point Calibration Disc

As an early first example of the use for the CD fluidic platform, the author and his team, by appropriately choosing the channel diameter and reservoir dimensions and locations, designed an automated two-point optrode calibration system, capable of five sequential fluid movement steps as shown in Figure 6.91a and b. The liquid flows in the order of calibrant 1, wash 1, calibrant 2, wash 2, and sample to an optrode chamber by increasing the rotating speed (see Figure 6.91a for the placement of these fluidic chambers, top figure is an actual photograph of the CD and the bottom is a schematic). Figure 6.91b shows the calculated burst frequencies of the reservoirs in the two-point calibration platform. The same optrode chamber is used for measuring calibrants and sample to eliminate artificial system errors common with devices using separate chambers for measuring sample and calibrants.

To demonstrate the functionality of this two-point calibration system, we deposited a potassium ion selective membrane in the optrode chamber using a microdrop delivery setup. The membrane was composed of poly(vinyl chloride) (PVC), valinomycin (ionophore), dioctyl sebacate (plasticizer), potassium tetrakis(4-chlorophenyl) borate, and chromoionophore for detection at a 640-nm wavelength. The slope (sensitivity) of a dose-response curve (absorption of the 640-nm wavelength vs. concentration) was determined by the measurements of calibrant 1 and calibrant 2.[94] The ion concentration of unknown samples was then deduced from this calibration curve.

Example 6.9: The CD Platform as a Microscope

Horacio Kido and Jim Zoval from the author's research team recognized that the optical

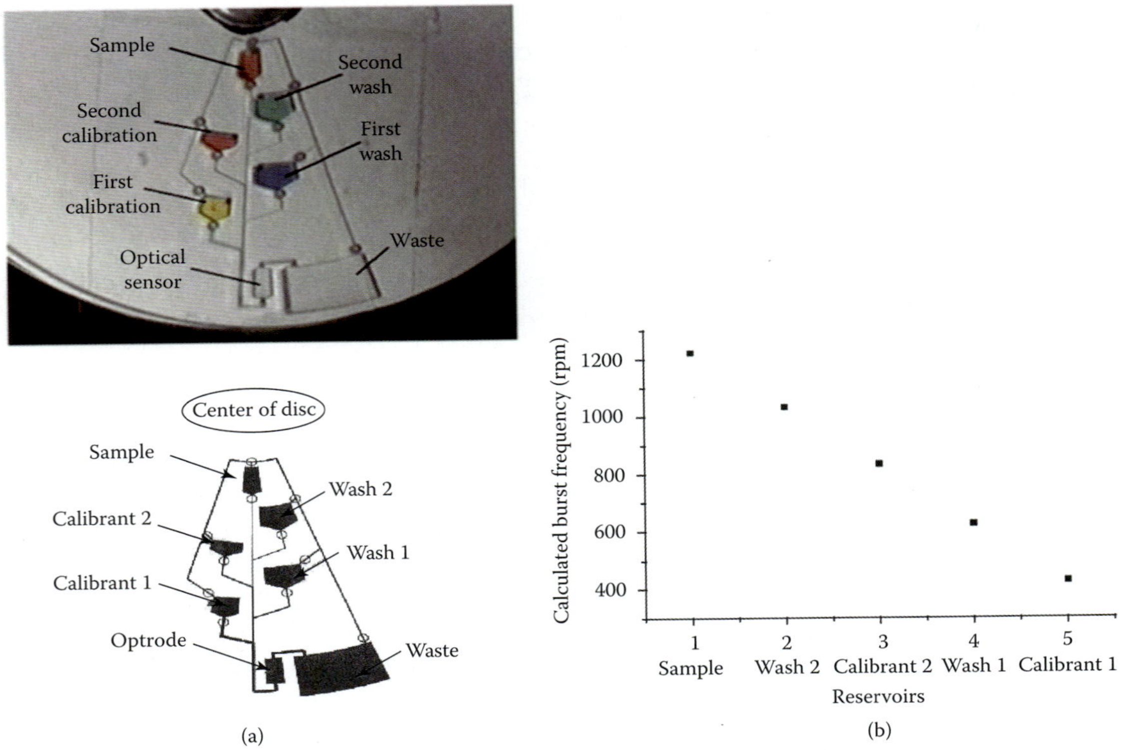

FIGURE 6.91 Two-point calibration of an optrode on a CD. (a) Functional diagram and photo of a two-point calibration unit on a CD. (b) A plot of calculated burst frequencies.

disc drive is a sophisticated laser scanning microscope designed to characterize and identify micrometer-sized features at a rate of about a megahertz (Figure 6.92). By taking the voltages from the photodetector and sending them to a computer using a fast A/D converter, an image of what is on the disc can then be reconstituted using simple graphics software. It does not matter whether the image is a pit in the CD (see Figure 6.92) or objects placed on the disc (see Figure 6.93).

Example 6.10: Using the Coriolis Force on the CD

On a spinning CD, the Coriolis force can be used to switch fluid flow between two reservoirs depending on the direction of the CD's rotation. The Coriolis force, we remember, is a fictitious force exerted on a body when it moves in a rotating reference frame. It is called a fictitious force because it is a by-product of a coordinate transformation. To understand the Coriolis force

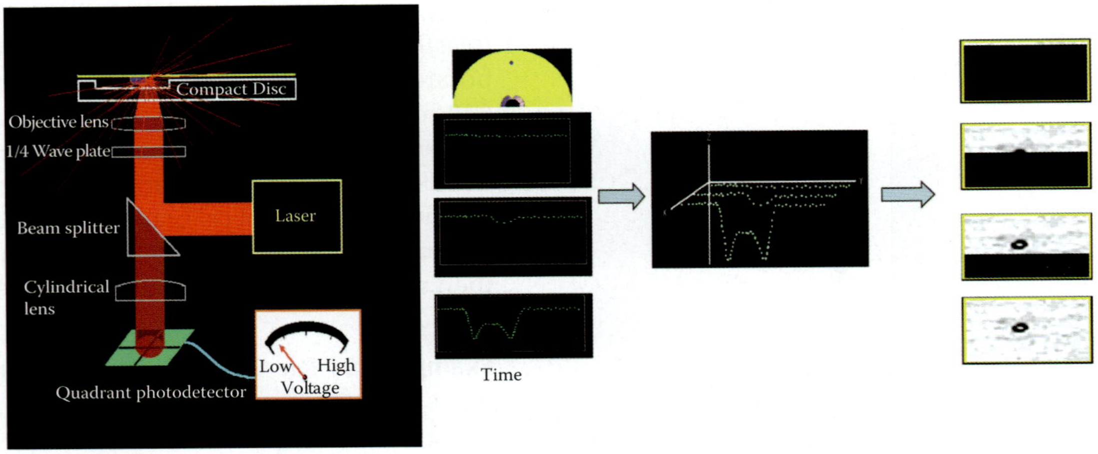

FIGURE 6.92 Using the CD as a microscope.

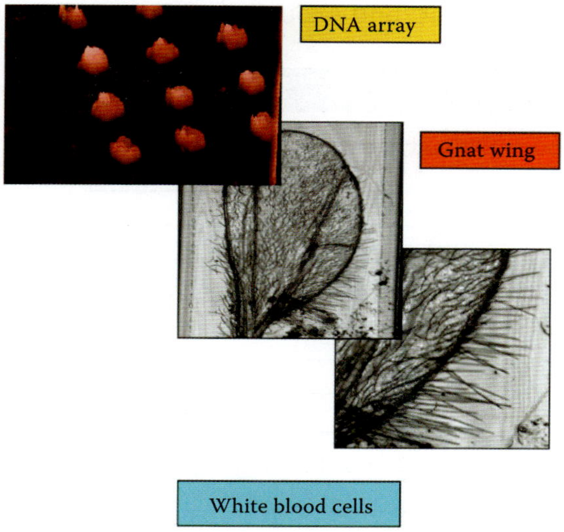

FIGURE 6.93 Images from the CD microscope.

on the CD, it is convenient to rewrite the Navier-Stokes equation (Equation 6.8) in a rotating reference frame. This results in an expression for the Coriolis force given as:

$$F_{Coriolis} = -2\rho \cdot \mathbf{u} \times \mathbf{u} \tag{6.142}$$

The formula implies that the Coriolis force is perpendicular both to the direction of the velocity of the moving mass and to the frame's rotation axis. In this expression, ρ is the fluid density, **ω** is the angular velocity vector, and **u** is the flow velocity vector. The Coriolis force is transverse to the CD radius and acts to transport the fluid stream left or right depending on the direction of the CD rotation (see Figure 6.94a). The Coriolis force thus can be used to switch fluid flow between a left reservoir and a right reservoir depending on the direction of the CD rotation (Figure 6.94b).

Some other applications of the CD, including cultivation of *C. elegans* worms, can be found in the References.[95,96] A recent review article on CD fluidics can be found in Madou et al.[97]

In Table 6.8 we compare four popular fluidic platforms: mechanical, osmotic, centrifugal, and acoustic. This table should be read together with the more detailed Table 1.8 from Chapter 1 in Volume III. For more background on the physics behind acoustic fluidics see Volume III, Chapter 8 on actuators.

Scaling in Analytical Separation Equipment

Introduction

When incorporating sensors in microfluidics one finds that most sensing techniques scale poorly in the microdomain. As we analyze in Chapter 7, different detection schemes exhibit different signal scaling factors. For example, in an amperometric technique, in which an analytical current is the signal, the output exhibits a (l^2) scaling (i.e., proportional to the area of the electrode); an optical absorption technique scales as (l^3), and potentiometric techniques are, in principle, size-independent. In general we find that most detection methods are less sensitive

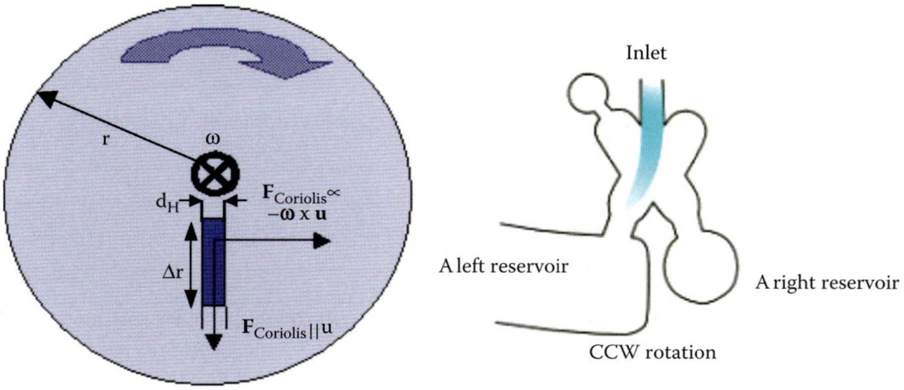

(a) (b)

FIGURE 6.94 Coriolis force for flow switching. (a) Forces shown on a disc rotating at the angular velocity vector **ω** pointing into the paper plane. When this disc spins in the counterclockwise direction, the Coriolis force switches the fluid in the channel to the left, and when this disc spins in the clockwise direction, the fluid moves to the right. (b) When the CD rotates counterclockwise (CCW), the Coriolis switching valve diverts fluid to the waste reservoir (left reservoir), and when the CD rotates clockwise, fluid is diverted to the extraction chamber (right reservoir).

Table 6.8 Comparison of CD Fluidics with Mechanical, Osmotic, and Acoustic Pumping (see also Table 1.8 , Chapter 1 in Volume III)

Mechanical (e.g., with Johnson and Johnson's blister pouch HIV test; see Figure 1.12 in Volume III, Chapter 1)
– Scales as l^3
+ No contact of sample with electrodes
+ Generic (all types of sample can be used)
+ Innovation in the blister pouch
+ Solves liquid and vapor valving
– Difficult to further miniaturize
– Difficult to multiplex
Electrokinetic (electro-osmosis) (e.g., with the microchip in Figure 6.71 used for high-speed electrophoretic separations)
+ Scales as l^2
– High-voltage source is not convenient
– Many parameters influence propulsion force
– Not generic (only limited types of samples can be used)
– Mixing difficult to implement
– Sample contact with capillary walls modifies the electro-osmotic force (EOF)
+ First products on the market
/ Solves liquid valving but not for vapors!
+ Better for high-throughput screening and smaller samples
Centrifugal (e.g., with the two-point calibration CD in Figure 6.91)
+ Scales a bit better than l^3
+ Compatible with a wide range of samples
+ Mixing easy to implement
+ Sample preparation easier than with an electrokinetic set-up
+ Simple and inexpensive CD player for drive
+ No fluid contact
+ Established
+ Generic
/ Solves liquid valving elegantly but not vapor valving
+ Most functions demonstrated
+ Cell work easier
+ Better for diagnostics
Acoustic (e.g., with White's flexural plate wave device; see Volume III, Chapter 8)
+ Scales as l^2
+ No fluidic contact
– Research and development phase
+ Generic
– Does not solve valving yet
– ZnO technology to implement the acoustic transducer still difficult to reproduce
– Easy to further miniaturize

when used in microchambers in microfluidic platforms. In this chapter, we saw already that all flow in small tubes is laminar (making mixing a challenge), that evaporation from small droplets is very fast, and that surface tension in the microdomain becomes very appreciable. These represent mostly negative aspects associated with scaling down any analytical equipment involving fluid handling. Positive attributes of miniaturizing such equipment are the savings on expensive reagents, short analysis times, and fast heating/cooling (e.g., for PCR), which all dictate small devices.

We now look into one class of analytical equipment involving fluid handling, namely, separation equipment, with some more detail. Electrophoresis, as described earlier in this chapter, is only one of many separation techniques used in analytical chemistry; other methods include gas chromatography (GC) and many varieties of liquid chromatography. Miniaturization in electrophoresis and chromatography has been a study topic for many years. The goal is not only to separate smaller amounts of chemical compounds, but theory also predicts that a reduction in dimensions of the separation column should result in an enhancement of analytical performance, such as a shorter separation time and more efficient heat dissipation.[98,99] To put recent advances in miniaturization of separation devices in context, we need to reconsider the fundamentals of separation methodology. The following theoretical analysis will shed light on scaling laws applicable to analytical separation systems. We closely follow the derivations presented by Manz et al.[98] One principal goal of separation techniques is to create the narrowest possible separation bands, w, to increase the separation efficiency. Several mechanisms contribute to band broadening, each with its own standard deviation σ, and we need to understand each of the contributing mechanisms to appreciate the effect of miniaturization on separation performance.

Commonly Used Terms in Separation Chemistry

Analytical separation of compounds often is affected by forcing the sample compounds, suspended in a carrier medium or mobile phase, through a selectively absorbing medium or stationary phase immobilized in some sort of flow channel. The mobile phase in chromatography may be a liquid or a gas. The stationary phase (the one that stays in place inside the flow channel or column) is most commonly a viscous liquid coated on the inner side of a capillary tube or on the surface of solid particles packed in the column. Alternatively, the solid particles in the column may form the stationary phase themselves. To force the sample and carrier phase (i.e., the eluent), through the column, either a pressure or electrical gradient, respectively, defining chromatography techniques and electrophoresis techniques, can be employed. The fluid emerging from the column is called the *eluate*, and the process of passing liquid or gas through a chromatography column is called *elution*. The partitioning of solutes between the mobile and stationary phases gives rise to separation of the various components in the sample. Columns are either packed or open tubular. A packed column is filled with particles as the stationary phase or the particles are coated with the stationary phase. An open columnar column is a narrow capillary in which the inside wall itself forms the stationary phase or a stationary phase is coated on that wall. The center of an open tubular column is hollow. Based on the mechanisms of interactions of the solute with the stationary phase, different types of chromatography are distinguished: adsorption, partition, ion exchange, molecular exclusion (also called *gel filtration*), and affinity chromatography.[100]

An electrochemical or optical detector, in the case of liquid chromatography, or a thermal conductivity detector (TCD) or flame ionization detector (FID), in the case of gas chromatography, may detect solutes eluted from a column. A chromatograph is a graph showing the response of such a detector as a function of elution time (Figure 6.95). Ideally, a solute applied as an infinitely narrow band at the inlet of a separation column emerges with a Gaussian shape at the outlet. In less ideal circumstances, the band becomes asymmetric. Different compounds are retained for different amounts of time, called *retention times* (t_r), onto the immobilized medium through which they are forced. Retention time t_r is the time needed after injection of the mixture for each component to reach the detector. The retention time for a solute often is corrected for the time it takes the unretained mobile phase (e.g., air) to travel through the column (t_m). This so-called adjusted retention time t'_r may be written as:

$$t'_r = t_r - t_m \tag{6.142}$$

For any two components 1 and 2, the relative retention, α, is given as:

$$\alpha = \frac{t'_{r2}}{t'_{r1}} \tag{6.143}$$

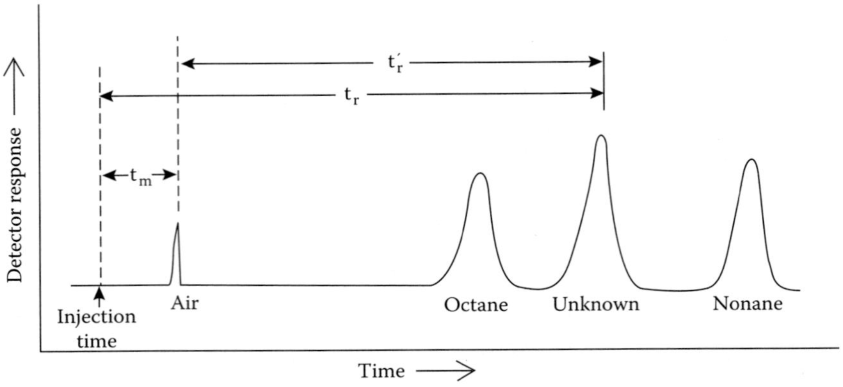

FIGURE 6.95 Schematic gas chromatogram showing measurement of retention times.

in which $t'_{r2} > t'_{r1}$ so that α is always larger than 1. The larger the α, the better the separation between compounds 1 and 2. The relative retention is independent of flow rate and is used to identify compounds, even when the flow rate has changed between experiments. For each compound in the column, the capacity factor, k', is defined as:

$$k' = \frac{t_r - t_m}{t_m} \quad (6.144)$$

A larger capacity factor signifies a compound better retained onto the column. The capacity factor in Equation 6.144 corresponds to the ratio of time the solute spends in the stationary phase over the time it spends in the mobile phase. The latter ratio in turn is based on the partition coefficient K, an equilibrium constant defined by the ratio of the concentration C_s of solute in the stationary phase over C_m, the concentration of the solute in the mobile phase:

$$k' = \frac{C_s V_s}{C_m V_m} = K\frac{V_s}{V_m} \quad (6.145)$$

where V_s is the volume of the stationary phase and V_m the volume of the mobile phase. From Equations 6.143–6.145, we can conclude that:

$$\alpha = \frac{t'_{r2}}{t'_{r1}} = \frac{k'_2}{k'_1} = \frac{K_2}{K_1} \quad (6.146)$$

The relative retention of two solutes is proportional to the ratio of their partition coefficients.

Two factors contribute to the efficiency of separation of components on a column. One is the difference in elution times between peaks; the other is the width of the peaks. Different compounds in an applied sample on a column separate into bands, and the further apart they are, the better the separation performance of the setup. The resolution of two peaks is defined as:

$$\text{resolution} = \frac{\Delta t_r}{w_{av}} = \frac{\Delta V_r}{w_{av}} \quad (6.147)$$

in which Δt_r or Δw_r is the separation between peaks in units of time (t) or volume (V), and w_{av} is the average width of the two peaks in corresponding units. A band of solute in a column invariably spreads as it travels through a separation medium and emerges at the detector with a standard deviation σ; the narrower the bands (w), the better the resolution.

Band Broadening in the Column

Individual solute bands in a column generally assume symmetric concentration profiles that can be described in terms of a Gaussian distribution curve and its standard deviation, σ_x. The Gaussian curve is given by:

$$C = \frac{m}{\sqrt{4\pi Dt}} e^{-\frac{x^2}{4Dt}} \quad (6.148)$$

where D = diffusion coefficient (m^2/s) (10^4 faster in gases than in liquids)
m = moles per unit cross-sectional area
C = concentration (mol m^{-3})
t = time
x = distance along the column (total length L_t) from the center of the band (at the band center, x = 0)

The standard deviation of this distribution curve, σ_x, is used to describe the bandwidth length and is given by:

$$\sigma_x = \sqrt{2Dt} \quad (6.149)$$

i.e., the width at time t of a band progressing in the column.

If a solute moves a distance x at the linear flow rate u_x (m s^{-1}), the time it has been on the column is $t = x\,u_x^{-1}$ (when $x = L_t$, then $t = t_r$). Or we can write that:

$$\sigma_x^2 = 2Dt = 2D\frac{x}{u_x} = \left(\frac{2D}{u_x}\right)x = Hx \quad (6.150)$$

and:

$$H = \frac{\sigma_x^2}{x} \quad (6.151)$$

In this equation, plate height H (also called the *height equivalent to a theoretical late*) is the proportionality constant between the variance (σ^2) of the band and the distance it has traveled (x). Thus, plate height is simply a quantity relating the width of a band to the distance traveled through the column. The number of theoretical plates, N, in the entire column is the total length L_t divided by the plate height:

$$N = \frac{L_t}{H} = \frac{Lx}{\sigma_x^2} = \frac{L_t^2}{\sigma_x^2} = \frac{16L_t^2}{w^2} \quad (6.152)$$

because $x = L$ and $\sigma = w/4$. The above analysis can be made more generic; the bandwidth parameter, σ, for a flow channel can be expressed not only in terms of length in m (σ_x, see above) but also in terms of time in s (σ_t), and volume in m^3 (σ_v). Rewriting Equation 6.152 with L and w (or σ) in units of time instead of length leads to:

$$N = \frac{16t_r^2}{w^2} = \left(\frac{t_r^2}{\sigma^2}\right) \quad (6.153)$$

where t_r is the retention time of the peak, and w is the width at the base of the peak in units of time.

The smaller the H and thus the larger the N, the narrower the bands and the better the separation. Plate heights are ~0.1 to 1 mm in gas chromatography (GC), ~10 μm in high-performance liquid chromatography (HPLC), and <1 μm in capillary electrophoresis (CE).[101] In HPLC the grains size of the matrix is reduced: this leads to an increased number of plates N, but at the cost of a much increased pressure drop, Δp, across the length of the column.

The resolution between two compounds in a chromatography technique (Equation 6.147) now can be expressed in terms of number of theoretical plates in the column, N:

$$\text{resolution} = \frac{\sqrt{N}}{4}\left(\frac{\alpha - 1}{\alpha}\right)\left(\frac{k_2'}{1 + k_{av}'}\right) \quad (6.154)$$

In this equation, α is the relative retention, k'_2 is the capacity factor for the strongest retained component of two compounds, and k'_{av} is the average capacity factor for the two components. Resolution, like band spreading, scales as $N^{1/2}$. Because the number of theoretical plates is proportional to column length, doubling the column length increases the resolution by $\sqrt{2}$. Resolution also increases with increasing α and k'_2. Increasing the capacity factor only works up to a certain level, as retention times might become too long and the peaks too broad.

Golay[102] calculated the height equivalent of a theoretical plate, H, as a function of flow rate (u) as:

$$H = \underset{A}{\frac{2D_m}{u_x}} + \underset{B}{\frac{1 + 6k' + 11k'^2}{96(1+k')^2}\left(\frac{d^2}{D_m}\right)u_x} + \underset{C}{\frac{2k'}{3(1+k')^2}\left(\frac{d_f^2}{D_s}\right)u_x} \quad (6.155)$$

In this equation, D_m and D_s (m^2s^{-1}) correspond, respectively, to the diffusion coefficients of the sample molecules in the mobile and stationary phase; u_x (ms^{-1}) is the linear flow rate of the mobile phase, k' is the capacity factor of the specific compound, d is the diameter of the capillary; and d_f is the thickness of the stationary phase layer. The A term in Equation 6.155 ($2D_m/u_x$) is the result of longitudinal diffusion (spreading out along the length of the column) and contributes an amount H_D to the expression for H. This type of diffusion is called Taylor-Aris dispersion. This diffusional term can be derived from Equations 6.150 and 6.151 as:

$$H_D = \frac{\sigma^2}{L_t} = \frac{2D_m}{u_x} \quad (6.156)$$

An expression given by Taylor[103] approximates D_m in Equation 6.156 as:

$$D_m \sim Pe^2 D \qquad (6.157)$$

where the Pe number is given by:

$$Pe = \frac{ud}{D_m} \qquad (6.158)$$

Formula 6.157 works at high Péclet numbers; for microsystems Pe is typically between 0.1 and 100. For Péclet numbers in the upper range, diffusion along the flow can be increased by several orders of magnitude compared with molecular diffusion, a feature very useful in micromixers.

The faster the linear flow, the less time is spent in the column, and the less diffusional broadening occurs. Whereas the *A* term in the Golay equation is associated with longitudinal diffusion, *B* is the result of mass transfer in the mobile phase and contributes H_m to the expression of *H*. The *C* term is associated with mass transfer to the stationary phase, and it contributes a term H_s to *H* as a result of the finite time it takes for a solute to equilibrate between the mobile phase and the stationary one. Decreasing the thickness of the stationary phase d_f and increasing the temperature both reduce H_s. The thickness uniformity of coatings in capillaries is one of the most difficult parameters to control in micromachined columns. As a result, the current commercial micro gas chromatographs have miniaturized detectors and injectors but no micromachined columns (see Figure 1.22). Glass capillary columns with inner diameters as small as 100 μm continue to outperform micromachined columns because of the difficulty in maintaining coating uniformity in micromachined structures. Equation 6.155 was derived for round fluid conduits; Wong et al.[104] have derived a modified Golay equation for rectangular channels with different aspect ratios. These authors calculated that the difference between slip and no-slip flow in those channels becomes appreciable (as high as 12%) if the column size is reduced to 10 μm.[104]

For capillary electrophoresis in an open tube, the expression for the number of theoretical plates simplifies as both the *B* term and the mass transfer *C* term disappear, reducing plate height to submicron values. Moreover, the high surface-to-volume ratio improves heat dissipation, enabling the application of higher voltages and thus faster separations. The method provides unprecedented resolution as compared to a packed wider column. In the case of capillary electrophoresis, miniaturized planar columns can easily outcompete traditional glass capillaries.

Other Band-Broadening Effects

In practice, bands broaden not only because of diffusional terms (σ_{diff}), which cause broadening in the column itself, but also because of injector, connector, and detector finite volumes. Total band broadening is calculated from a summing of the squares of the standard deviations, that is, the variances (variance is additive but standard deviation is not):

$$\sigma^2 = \sigma^2_{diff} + \sigma^2_{inj} + \sigma^2_{det} \qquad (6.159)$$

These are all important parameters to calculate when designing a separation system. Injection and detection variances in traditional separation systems usually can be made smaller than a quarter of the diffusional variance. To contribute less than 10% of the total band broadening, injection and detection volume must be less than $\sigma_v/2$ of the column, and the detector response time must be faster than $\sigma_t/2$. If in injection the band is applied as a sharp-edged zone of width Δt (in units of time), the contribution to the variance of the final bandwidth is:

$$\sigma^2_{inj} = \frac{(\Delta t)^2}{12} \qquad (6.160)$$

A similar equation holds for broadening in the detector that requires a time *t* to pass through. Systems must be designed such that the dead volume can be kept minimal and the capillary heated and cooled very quickly and uniformly with a minimal power budget; that is, the heat capacity of the system must be minimized.

To retain a small footprint while making a separation channel longer, the channel can be made to turn back and forth or made into a spiral. Unfortunately, distance and field strength vary laterally across the width of the channel in those turns; that is, a "racetrack effect" is generated. Racetrack effects cancel the benefit of additional length. Molho et al.,[105] using Memcad 4.6, designed and fabricated "compensated" turns in PMMA, greatly reducing dispersion compared with an uncompensated corner. Using turns with large radii of curvature and/or very narrow channel widths, Culbertson et al.[106] fabricated glass microchips with less dispersion in the turns. These authors also built a separation channel in glass in the shape of a spiral 25 cm long and fitting on a 5 × 5-cm piece of real estate. Negligible band broadening from the turns and separation efficiencies of more than 1,000,000 theoretical plates in less than 46 seconds with an applied potential of 1100 V/cm for dichlorofluorescein dye were observed.[106]

Scaling

Miniaturization of the detector in a miniaturized analysis system, we alluded to earlier, is quite a challenge, especially with detector chamber sizes in the order of picoliters. By fixing the number of theoretical plates, *N*; the retention time, *t*; and the heating power per length (for capillary electrophoresis), the resulting flow channel dimensions and operating conditions for different separation techniques are compared. In Table 6.9 we reproduce calculations by Manz et al.[98,99] comparing capillary electrophoresis (CE), liquid chromatography (LC), and supercritical fluid chromatography (SFC). Listed are flow channel lengths, diameter, operating voltage or pressure, minimum sample bandwidths (expressed in σ_x, σ_t, and σ_v), peak capacity, length/diameter ratio of eluting peaks (σ_x/d), and the required detector and injector specifications.

Still following Manz et al.'s analysis,[99] we now introduce a set of dimensionless or reduced parameters to make comparisons of separation systems easier. By introducing reduced parameters, including the Reynolds number, Péclet number (flow rate), Fourier number (elution time), and Bodenstein

TABLE 6.9 Calculated Parameter Sets for a Given Separation Performance

Parameter		Electro-Osmotic Chromatography			Liquid Chromatography (LC)		Supercritical Fluid Chromatography (SFC)	
Number of theoretical plates	N	100k	1M	10M	100k	1M	100k	1M
Analysis time	1 (k′ = 5) (min)	1	1	1	1	1	1	1
Heating power	P/L (W/m)	1.1	1.1	1.1				
Capillary inner diameter	d (μm)	24	7.6	2.4	2.8	0.9	6.9	2.2
Capillary length	L (cm)	6.5	21	65	8.1	26	20	64
Pressure drop	p (atm)				26	2600	1.4	140
Voltage	U (kV)	5.8	58	580				
Peak capacity	n	180	570	>2000	220	700	220	700
Signal bandwidth	σ_x (mm)	0.21	0.21	0.21	0.56	0.56	1.4	1.4
	σ_t (ms)	42	13	4.2	70	22	70	22
	σ_v (pl)	94	9.4	0.94	3.3	0.33	52	5.2
Detection volume	V (pl)	47	4.7	0.47	0.8	0.08	1.2	12
Response time	t (ms)	21	6.5	2.1	16	5	16	5
Injection pulse	p × t (s × atm)				1.5	49	0.075	2.4
	U × t (s × kV)	0.41	1.3	4.1				
Stop time	t (s)	3.3	3.3	3.3	5.1	5.1	5.1	5.1

Note: Obtained with capillary electro-osmotic (EC), liquid (LC), and supercritical fluid chromatography (SFC). Assumed constants are diffusion coefficients 1.6×10^{-9} m²/s (LC, EC) and 10^{-8} m²/s (SFC); viscosities of the mobile phase 10^{-3} Ns/m² (LC, EC) and 5×10^{-5} Ns/m² (SFC); electrical conductivity of the mobile phase 0.3 S/m (EC); electrical permittivity × zeta potential, 5.6×10^{-11} N/V (EC); and heating power 1.1 W/m (EC).

Source: After Manz, A., E. Verpoorte, C. S. Effenhauser, N. Burggraf, D. E. Raymond, D. J. Harrison, and H. M. Widmer. 1993. Miniaturization of separation techniques using planar chip technology. HRC *J High Resol Chromatogr* 16:433–36; and Manz, A., N. Graber, and H. M. Widmer. 1990. Miniaturized total chemical analysis systems: a novel concept for chemical sensing. *Sensors Actuators B* B1:244–48.

number (pressure drop), systems are easier to compare because all reduced parameters remain constant regardless of the size of the system.[107] First, we define some of these reduced parameters in terms of those variables that can be assumed constant over the entire range of interest, including inner capillary diameter (d), mobile phase viscosity (η_m), average diffusion coefficient (D_m) of the sample in the mobile phase, and a Poiseuille number, C, of 32 for a circular cross-section. The diameter, d, is also the characteristic length here. With those constants, other quantities can be grouped into dimensionless forms—for example, volume, V; column length, L; linear flow rate, u; retention time, t; and pressure drop, Δp. A reduced volume parameter, for example, is obtained by division through d^3. Thus, one obtains:

$$w = \frac{V}{d^3} \text{ and } s_v = \frac{\sigma_v}{d^3} \tag{6.161}$$

Similarly, time-related parameters such as the migration time, t_0; the retention, t, for a compound with a capacity factor k'; and the time bandwidth, σ_t, can be reduced to their dimensionless terms, known as Fourier numbers, in the following fashion:

$$\tau_0 = \frac{t_0 D_m}{d^2}, \tau = \frac{tD_m}{d^2}(k'+1), \text{ and } s_t = \frac{\sigma_t D_m}{d^2} \tag{6.162}$$

The first expression describes the ratio of the time required for a molecule to migrate in a field-driven flow through the capillary from end to end to the time required to diffuse from wall to wall. The second term defines the analogous expression for chromatography. Terms with a length dimension are reduced by dividing by d:

$$\lambda = \frac{1}{d}, h = \frac{H}{d}, \text{ and } s_x = \frac{\sigma_x}{d} \tag{6.163}$$

where λ is the so-called reduced length. The linear flow rate of a nonretained component or the mobile phase is reduced to the Péclet number, Pe, given earlier as:

$$Pe = \frac{ud}{D_m} \tag{6.158}$$

which represents the average linear flow rate divided by the absolute value of the average rate of diffusion orthogonal to the direction of flow, D_m/d. The applied pressure can be reduced to the so-called Bodenstein number as:

$$\Pi = \frac{\Delta pd^2}{\eta_m D_m \Phi} = \frac{uL}{D_m} \tag{6.164}$$

The Bodenstein number thus represents the average linear flow rate divided by the absolute value of the average rate of longitudinal diffusion, D_m/L. In this expression, Φ is the Poiseuille number (32 for a circular cross-section). The applied voltage can be reduced to the "electrical" Bodenstein number, Ψ:

$$\Psi = \frac{U\varepsilon\xi}{\eta_m D_m} \tag{6.165}$$

with ζ the zeta potential. Both Bodenstein numbers, describing the reduced pressure and voltage, can be rewritten as the so-called flow equations:

$$\Psi = \lambda Pe \tag{6.166}$$

in electro-osmotic and electrophoretic flow, and:

$$\Pi = \lambda Pe \tag{6.167}$$

for pressure-driven chromatography. Also, the Fourier numbers can be further simplified as:

$$\tau_0 = \frac{\lambda}{Pe} \text{ and } \tau = \frac{\lambda}{Pe}(k'+1) \tag{6.168}$$

The equations for reduced plate height, h, and the Péclet number, Pe, lead to the following simplified Golay equation for band broadening in pressure-driven flow, neglecting the third term (C) in Equation 6.155:

$$h = \frac{2}{Pe} + \frac{1+6k'-11k'^2}{96(1+k')^2}Pe \tag{6.169}$$

for a field-driven flow we simply obtain:

$$h = \frac{2}{Pe} \tag{6.170}$$

The reduced bandwidth is then given by:

$$s_x^2 = \lambda h \tag{6.171}$$

and the plate number by:

$$N = \frac{\lambda}{h} \text{ and } N = \lambda\frac{Pe}{2} \tag{6.172}$$

Table 6.10 Reduced Parameter Set for the Example Separation Systems in Table 6.9

Parameter	CE	Capillary LC	Capillary SFC
Number of theoretical plates, N	100,000	100,000	100,000
Analysis time, t (k′ = 5) (min)	1	1	1
Capillary inner diameter, d_c (μm)	24	2.8	6.9
Capillary length, L (cm)	6.5	8.1	20
Reduced length, λ	2700	29,000	29,000
Péclet number (reduced flow rate), Pe	100	14	14
Fourier number (reduced retention time), τ	28	2100	2100
Bodenstein number (reduced pressure decrease), Π	—	400,000	400,000
Electric Bodenstein number (reduced voltage decrease), ψ	260,000	—	—

CE = capillary electrophoresis; LC = liquid chromatography; SFC = supercritical flow chromatography.
Source: Manz, A., E. Verpoorte, C. S. Effenhauser, N. Burggraf, D. E. Raymond, D. J. Harrison, and H. M. Widmer. 1993. Miniaturization of separation techniques using planar chip technology. HRC *J High Resol Chromatogr* 16:433–36; and Manz, A., N. Graber, and H. M. Widmer. 1990. Miniaturized total chemical analysis systems: a novel concept for chemical sensing. *Sensors Actuators B* B1:244–48.

It can be recognized here how the number of theoretical plates is in a way representative of the Péclet number of the system. From Equations 6.172 and 6.163, the smaller D_m, the higher the number of plates, and the larger the velocity and the length of the column, the more the separation is "efficient."

With these reduced parameters, the examples of Table 6.9 now are compared again in Table 6.10. We conclude that the values for the reduced variables for LC and SFC are identical, regardless of differences in capillary diameter, lengths, diffusion coefficients, and viscosities. This can be used conveniently to deduce the influence of changing the capillary diameter on retention time, pressure, and signal bandwidth for a given number of theoretical plates and a single set of reduced parameters, as illustrated in Table 6.11.

The scaling for diffusion-controlled separations, in which the time scale is proportional to d_2, is summarized in Table 6.12. All reduced parameters remain constant regardless of the size of the system; that is, hydrodynamic diffusion, heat diffusion, and molecular diffusion effects behave in the miniaturized system as in the original large system. Reducing the characteristic length d (i.e., the tube diameter) by a factor of 10 makes for a 100-times faster analysis. The pressure required is 100 times higher, and, more importantly, the voltage requirements remain unchanged in electrophoresis/electro-osmosis systems. Miniaturization in diffusion-controlled

Table 6.11 Calculated Parameter Set for an Open Tubular Column LC System

	Diameter d (μm)					Reduced Parameter
	1	2	5	10	20	
Length, L (m)	0.45	0.9	2.3	4.5	9	λ = 450,000
Time, t (min)	0.12	0.5	3	12	50	t = 11,800
Pressure, Δp (atm)	8700	2200	350	87	22	Π = 17,000,000
Peak, σ_x (μm)	450	890	2200	4500	8900	$s_x = 447$
Peak, σ_t (ms)	7.4	30	190	740	3000	$s_t = 12$
Peak, σ_v (pl)	0.35	2.8	44	350	2800	$s_v = 354$

Note: One million theoretical plates at zero retention (Péclet number Pe = 38). Assume diffusion coefficient is $D_m = 10^{-3} m^2 s$.
Source: Manz, A., E. Verpoorte, C. S. Effenhauser, N. Burggraf, D. E. Raymond, D. J. Harrison, and H. M. Widmer. 1993. Miniaturization of separation techniques using planar chip technology. HRC *J High Resol Chromatogr* 16:433–36; and Manz, A., N. Graber, and H. M. Widmer. 1990. Miniaturized total chemical analysis systems: a novel concept for chemical sensing. Sensors Actuators B B1:244–48.

Table 6.12 Proportionality Factors for Some Mechanical Parameters in Relation to the Characteristic Length, d, in a Diffusion-Controlled System

	Diffusion-Controlled System
Space, d	d
Time, t	d^2
Linear flow rate, u	L/d
Volume flow rate, F	d
Pressure drop (laminar flow) Δp	L/d^2
Voltage (electro-osmotic flow), U	Constant
Electric field, U/L	L/d
Reynolds number, Re	Constant
Péclet number, reduced flow rate, Pe	Constant
Fourier number, reduced elution time, τ	Constant
Bodenstein number, reduced pressure, Π	Constant
Reduced voltage, ψ	Constant

Source: After Manz, A., E. Verpoorte, C. S. Effenhauser, N. Burggraf, D. E. Raymond, D. J. Harrison, and H. M. Widmer. 1993. Miniaturization of separation techniques using planar chip technology. HRC J High Resol Chromatogr 16:433–36; and Manz, A., N. Graber, and H. M. Widmer. 1990. Miniaturized total chemical analysis systems: a novel concept for chemical sensing. Sensors Actuators B B1:244–48.

systems leads to a higher rate of separation while maintaining separation efficiency.

Separation efficiency in capillary electrophoresis, in terms of theoretical plates per second, can be estimated from Equations 6.170 and 6.172:

$$N = \frac{\Psi}{2} \propto U \tag{6.173}$$

where Ψ is the reduced voltage so that *N* is proportional to U, the applied voltage; the higher the applied voltage, the higher the number of theoretical plates. The voltage cannot be increased too much though, because heat evolution quickly becomes the limiting factor. In standard capillary electrophoresis, the maximum allowable heat generation is about 1 W/m. Because of the higher surface-to-volume ratio in micromachined channels, heat dissipation is faster, which permits higher electric fields than in standard capillary electrophoresis. For now, we will admit 1 W/m as an upper limit even for micromachined capillaries. Keeping the power per unit length constant means:

$$\frac{P}{L} = \frac{UI}{L} = \text{const} \tag{6.174}$$

where *I* is the current through the capillary. The resulting upper limit for the voltage is then determined by the geometry of the capillary in terms of the reduced characteristic length ($\lambda = 1/d$):

$$U_{max} \propto \lambda \tag{6.175}$$

The plate number *N* can thus reach values up to N_{max}, where:

$$N_{max} \propto \lambda \tag{6.176}$$

The minimum migration time is given by:

$$t_0 \propto \frac{L^2}{U} \tag{6.177}$$

and:

$$t_{min} \propto L^2 d \tag{6.178}$$

The maximum number of plates obtainable per second is then given by:

$$\frac{N_{max}}{t_{min}} \propto \frac{1}{L^2 d^2} \tag{6.179}$$

clearly demonstrating the benefits of reducing the inner diameter of the capillary for rapid separation in capillary electrophoresis.

The reduced height, *h*, of a chromatographic system according to Equation 6.169 shows a minimum, h_{min}, at an optimum reduced flow rate, Pe_{opt} (Péclet number). To reduce losses in pressure drop and to minimize analysis time, one must operate close to this optimum. Analogous to Equation 6.172, we can write the maximum number of theoretical plates as:

$$N_{max} = \frac{\lambda}{h_{min}} \quad (6.180)$$

where h_{min} is constant for optimum Péclet number, Pe_{opt}, at a fixed capacity factor k'. This makes Equation 6.180 for pressure-driven chromatography equivalent to Equation 6.176 for electrophoresis. Equations 6.178 and 6.179 are also equally valid for capillary electrophoresis and capillary chromatography.

The reduced parameter analysis shows that a downscale of 1/10 of the original size (diameter of a tube) reduces the related time variables (analysis time, required response time of a detector) to 1/100. The pressure requirements increase by a factor of 100, but the voltage requirements (for electrophoresis/electro-osmosis) remain constant. The main advantage of capillary electrophoresis is thus higher speed of separation with a comparable efficiency.[108]

Example 6.11: Droplet Microfluidics

In a new branch of microfluidics, a rich variety of droplets are made by emulsification in microfluidic systems. Droplet-based microfluidics allows for independent control of each droplet, thus generating microreactors that can be individually transported, mixed, and analyzed.[109] The generation of droplets involves the disruption of interfacial forces between two streams of immiscible fluids either via flow-induced shear forces or other surface forces (e.g., electrowetting, dielectrophoresis). An often-used breakup method of streams of immiscible fluids is in the confined geometry of a microfluidic T-junction (see Figure 6.56). For a detailed theoretical treatment of droplet formation in a T-junction, we refer to the work by De Menech et al.[110] The droplets thus formed can be filled with reagents to perform bioassays in discretized femtoliter to nanoliter volumes separated by immiscible fluids as opposed to more conventional microfluidics, where reactions are not confined either spatially or temporally within the microchannels. Rapid mixing is possible because there is relative fluid motion between the immiscible fluids at the droplet interfaces that create recirculation of flows inside the droplets. The sizes of the droplets are extremely monodisperse (within 2%) and can be controlled over a wide range (typically 1–100 μm in diameter).[111–115] Because of high surface area to volume ratios at the microscale, heat and mass transfer times and diffusion distances are shorter, facilitating faster reaction times. Because multiple identical microreactor units can be formed in a short time, parallel processing and experimentation can easily be achieved, allowing large data sets to be taken efficiently and enabling high throughput screening and syntheses. Droplet microfluidics also offers greater potential for increased throughput and scalability than continuous single-phase flow microfluidics. In the past 5 years, several groups have used digital microfluidics to form irregular particles,[116] double emulsions,[117] hollow microcapsules,[118] and microbubbles.[74,119–123]

As a curious return to fluidic logic (see the fluidic logic chips from 1965 in Volume III, Figure 1.2), MIT's Neil Gershenfeld et al.[124] are building rudimentary bubble-based logic devices. They constructed an entire family of chips modeled on the architecture of existing digital circuits that mimic the functions of electronic components: logic gates (AND/OR/NOT), memory, and even a microfluidic ring oscillator constructed from three AND gates (see Figure 6.96). These digital microfluidic devices show the nonlinearity, gain, bistability, synchronization, cascadability, feedback, and programmability required for scalable universal computation. Instead of using high and low voltages to represent a bit of information, the presence or absence of a bubble is used. This means that a bubble can not only carry some kind of payload—say a therapeutic drug or chemical reagents (see above)—but

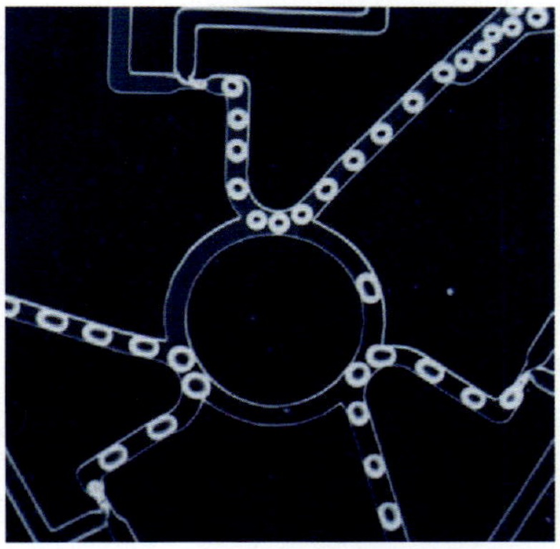

FIGURE 6.96 A bubble microfluidic ring oscillator constructed from three AND gates.

it also can carry information at the same time. The bubble chips are 100 times slower than an average microprocessor, but among other uses they could enhance the memory capabilities of microfluidic chips: thousands of reagents could be stored on a single chip—much like conventional data storage technology—using counters to dispense exact amounts and built-in logic circuits to deliver the reagents to targeted spots. They could also be used to sort biological cells in assays and to make programmable print heads for inkjet printers, for example. With increasing complexity in large-scale microfluidic processors, bubble logic provides an on-chip process control mechanism integrating chemistry and computation.

Acknowledgments

Special thanks to Dr. Suman Chakraborty (Indian Institute of Technology Kharagpur), Dr. Nahui Kim (University of California, Irvine), and Marleen Madou.

Questions

Questions by Omid Rohani, UC Irvine

6.1: We partially fill an open vessel with a liquid and rotate it at a constant angular velocity ω about a vertical axis. After some time the liquid also rotates at the same angular velocity as the vessel and the surface of the liquid develops into a steady shape. What is the equation that describes the free surface of the liquid?

6.2: Consider the flow of blood through a micro or nano channel. List 4 assumptions that introduce potential errors when modelling blood flows through such a micro channel using the Navier-Stokes equations.

6.3: What conclusions can one draw regarding liquid flows in micro and nano channels based on the currently available data in the literature? What do you know about turbulence in the macro scale and in the micro and nano scale? How can one enhance good mixing of two chemicals in a microreactor?

6.4: A lawn sprinkler consists of two arms equipped with nozzles at right angles which rotate about a pivot in a horizontal plane. Assuming the flow rate through each nozzle equals Q and the nozzle's exit area is A, find the velocity of the water leaving the nozzle, in two cases:

(a) There is no friction at the pivot.
(b) There is a constant frictional torque T_f at the pivot.

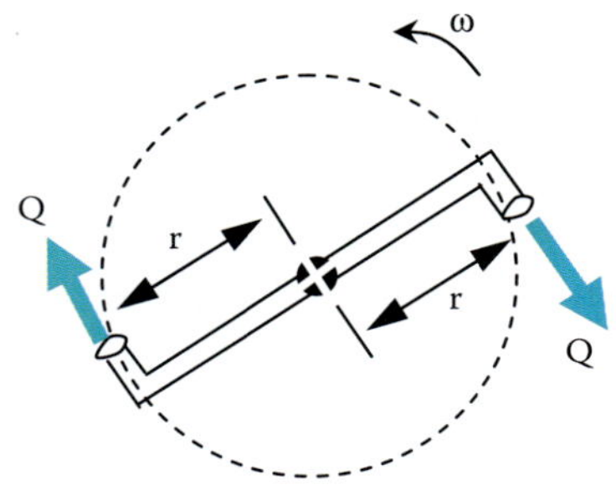

6.5: Derive an equation for the flow rate Q as a function of U, b, $\frac{dp}{dx}$, η for fully developed Couette shear flow, where U is the velocity of a moving plate and b is the height of the channel between this moving plate and a fixed bottom plate. We will assume that the flow is in the x-direction and b is in the y-direction. The viscosity is represented by η and $\frac{dp}{dx}$ is the pressure gradient in the x-direction.

6.6: What are the advantages of microfluidics? Give an example application.

6.7: Compare the breakdown of continuum theory for gasses and liquids.

6.8: What techniques can be implemented for mixing two reagents in the microscale (in low Reynolds number liquids)?

6.9: Given is a mixture of two proteins of similar molecular mass. Protein A comes with a MW = 12,400 and a pI = 10.7 and protein B has a MW = 12,800 and a pI = 7.5 (pI is the isoelectric point). You are told that capillary zone electrophoresis, CZE, is the best way to separate these two proteins.

(a) Do you agree with the suggestion of using CZE or would you employ another method? Briefly support your decision and your choice of separation method.
(b) Explain the principle behind capillary electrophoresis and draw a schematic illustrating its operation.
(c) What will be the elution order if you perform the separation using a buffer with a pH of 5.0? Why?

6.10: When do the macroscopic laws of gas flow begin to fail and when does the no-slip condition cease to apply?

6.11: Explain some of the changes in the behavior of a liquid when the channel diameter decreases to below ten molecular diameters (the channel diameter of the order of a few nanometers).

6.12: What is electrophoresis? Explain why DNA strands of varying size cannot be separated by electrophoresis in a free solution.

6.13: Describe how electro-osmotic phenomena can cause a pumping effect in a fused silica micro capillary.

6.14: Using the Hagen-Poiseuille law for flow rate, explain why conventional pumping is not useful for the delivery of small samples? What pumping method do you suggest for delivery of small samples? How does this method change the flow profile?

6.15: In a centrifugal fluidic platform (CD-fluidics), what constitutes the pumping force, and how are valves for liquids implemented?

6.16: A simple and accurate viscometer can be made from a length of capillary tubing. If the flow rate and pressure drop are measured, and the tube geometry is known, the viscosity can be computed. A test of a certain liquid in a capillary viscometer gave the following data: flow rate: 880 mm^3/s; tube diameter: 500 μm; tube length: 1 m; pressure drop: 1.0 MPa. Determine the viscosity of the liquid.

6.17: The accepted transition Reynolds number for flow in a circular pipe is $Re \approx 2300$. For flow through a 6 cm diameter pipe, at what velocity will this occur at 20°C for (a) airflow and (b) water flow?

6.18: List five electrokinetic effects taken advantage of in the MEMS field and list one application for each approach.

6.19: Why are more theoretical plates achievable in capillary electrophoresis than in the following separation techniques?
(a) Slab electrophoresis
(b) Liquid chromatography

6.20: Compare/contrast how separation is achieved in chromatography and electrophoresis.

6.21: List 6 different methods to propel fluids through micro channels.

6.22: Why is there a lower limit to the size of an uncharged particle one can move with dielectrophoresis (~14 nm) and not on the size of a charged particle one can move in electrophoresis? What types of particles (size and charge) can one move with electro-osmosis?

6.23: If a 100 bp long DNA is to be sequenced in a microfabricated channel, calculate the length of a channel needed to get single base pair separation. What voltage (*E* field) would you pick to perform the separation in a polyacrylamide gel. Make the necessary assumptions! The width and height of the channel is 20 μm × 20 μm.
Thanks to Dr. Rashid Bashir, Purdue University.

6.24: Demonstrate that $\delta/\delta_T \sim Pr^{1/2}$.

Further Reading

Abraham, S., E. H. Jeong, T. Arakawa, S. Shoji, K. C. Kim, I. Kim, and J. S. Go. 2006. Microfluidics assisted synthesis of well-defined spherical polymeric microcapsules and their utilization as potential encapsulants. *Lab Chip* 6:752–56.

Anna, S. L., N. Bontoux, and H. A. Stone, 2003. Formation of dispersions using "flow focusing" in microchannels. *Appl Phys Lett* 82:364–66.

Belder, D. 2005. Microfluidics with droplets. *Angewandte Chemie* 44:3521–22.

Cubaud, T., M. Tatineni, X. Zhong, and C. M. Ho. 2005. Bubble dispenser in microfluidic devices. *Phys Rev E* 72:4.

Cubaud, T., and C. M. Ho. 2004. Transport of bubbles in square microchannels. *Phys Fluids* 16:4575–85.

Fair, R.B., 2007. Digital microfluidics: is a true lab-on-a-chip possible? *Microfluidics Nanofluidics* 3:245–81.

Forsberg, F., J. B. Liu, D. A. Merton, N. M. Rawool, and B. B. Goldberg. 1995. Parenchymal enhancement and tumor visualization using a new sonographic contrast agent. *J Ultrasound Med* 14:949–57.

Garstecki, P., I. Gitlin, W. DiNuzio, and G. M. Whitesides. 2004. Formation of monodisperse bubbles in a microfluidic flow-focusing device. *Appl Phys Lett* 85:2649–51.

Gerdts, C. J., V. Tereshko, M. K. Yadav, I. Dementieva, F. Collart, A. Joachimiak, R. C. Stevens, P. Kuhn, A. Kossiakoff, and R. F. Ismagilov. Time-controlled microfluidic seeding in nL-volume droplets to separate nucleation and growth stages of protein crystallization. *Angewandte Chemie* 45:8156–60.

Goldberg, B. B., J. S. Raichlen, and F. Forsberg. 2001. *Ultrasound contrast agents*. London: Dunitz Ltd.

Gordillo, J. M., Z. Cheng, A. M. Ganan-Calvo, M. Márquez, and D. A. Weitz. 2004. A new device for the generation of microbubbles. *Phys Fluids* 16:2828–34.

Gunther A., and K. F. Jensen. 2007. Multiphase microfluidics: from flow characteristics to chemical and materials synthesis. *Lab Chip* 7:935.

Hettiarachchi, K., E. Talu, M. L. Longo, P. A. Dayton, and A. P. Lee. 2007. On-chip generation of microbubbles as a practical technology for manufacturing contrast agents for ultrasonic imaging. *Lab Chip* 7:463–68.

Jensen, K., and A. P. Lee. 2004. The science and applications of droplets in microfluidic devices. *Lab Chip* 4:31N–32N.

Kobayashi, I., K. Uemura, and M. Nakajima. 2007. Formulation of monodisperse emulsions using submicron-channel arrays. *Colloids Surfaces A Physicochem Eng Aspects* 296:285–89.

Liu, K., H. J. Ding, J. Liu, Y. Chen, and X. Z. Zhao. 2006. Shape-controlled production of biodegradable calcium alginate gel microparticles using a novel microfluidic device. *Langmuir* 22:9453–57.

Nisisako, T., S. Okushima, and T. Torii. 2005. Controlled formulation of monodisperse double emulsions in a multiple-phase microfluidic system. *Soft Matter* 1:23–27.

Nisisako, T., and T. Torii. 2007. Formation of biphasic Janus droplets in a microfabricated channel for the synthesis of shape-controlled polymer microparticles. *Adv Mater* 19:1489–93.

Sakai, S., I. Hashimoto, and K. Kawakami. 2007. Agarose-gelatin conjugate for adherent cell-enclosing capsules. *Biotech Lett* 29:731–35.

Sakai, S., I. Hashimoto, and K. Kawakami. 2006. Usefulness of flow focusing technology for producing subsieve-size cell enclosing capsules: application for agarose capsules production. *Biochem Eng J* 30:218–21.

Song, H., D. L. Chen, and R. F. Ismagilov. 2006. Reactions in droplets in microfluidic channels. *Angewandte Chemie* 45:7336–56.

Tan, Y. C., V. Cristini, and A. P. Lee. 2006. Monodispersed microfluidic droplet generation by shear focusing microfluidic device. *Sensors Actuators B Chem* 114:350–56.

Teh, S.-Y., R. Lin, L. Hung, and A. Lee. 2008. Droplet microfluidics. *Lab Chip* 8:198–220.

Utada, A. S., E. Lorenceau, D. R. Link, P. D. Kaplan, H. A. Stone, and D. A. Weitz. 2005. Monodisperse double emulsions generated from a microcapillary device. *Science* 308:537–41.

Wei, K., and S. Kaul. 1997. Recent advances in myocardial contrast echocardiography. *Curr Opin Cardiol* 12:539–46.

Woodward, A., T. Cosgrove, J. Espidel, P. Jenkins, and N. Shaw. 2007. Monodisperse emulsions from a microfluidic device, characterised by diffusion NMR. *Soft Matter* 3:627–33.

Xu, J. H., S. W. Li, Y. J. Wang, and G. S. Luo. 2006. Controllable gas-liquid phase flow patterns and monodisperse microbubbles in a microfluidic T-junction device. *Appl Phys Lett* 88:3.

Yobas, L., S. Martens, W. L. Ong, and N. Ranganathan. 2006. High-performance flow-focusing geometry for spontaneous generation of monodispersed droplets. *Lab Chip* 6:1073–79.

Zheng, B., C. J. Gerdts, and R. F. Ismagilov. 2005. Using nanoliter plugs in microfluidics to facilitate and understand protein crystallization. *Curr Opin Struct Biol* 15:548–55.

Zheng, B., L. S. Roach, and R. F. Ismagilov. 2003. Screening of protein crystallization conditions on a microfluidic chip using nanoliter-size droplets. *J Am Chem Soc* 125: 11170–71.

Zhou, C. F., P. T. Yue, and J. J. Feng. 2006. Formation of simple and compound drops in microfluidic devices. *Phys Fluids* 18:14.

References

1. Denn, M. M. 1980. *Process fluid mechanics.* Englewood Cliffs, NJ: Prentice Hall.
2. Potter, M. C., and J. F. Foss. 1982. *Fluid mechanics.* Okemos, MI: Great Lakes Press.
3. Barron, A. E., and Blanch, H. 1995. DNA separations by slab gel and capillary electrophoresis: Theory and practice. *Separation and Purification Methods* 24:1–118.
4. White, F. M. 1994. *Fluid mechanics.* New York: McGraw-Hill.
5. Allen, T. J., and R. L. Ditsworth. 1972. *Fluid mechanics.* New York: McGraw-Hill.
6. Granger, R. A. 1995. *Fluid mechanics.* New York: Dover Publications.
7. Ozisik, M. N. 1977. *Basic heat transfer.* New York: McGraw-Hill.
8. Corcoran, P., H. V. Shurmer, and J. W. Gardner. 1993. Integrated tin oxide sensors of low power consumption for use in gas and odour sensing. *Sensors Actuators B* B15–16:32–37.
9. Coulson, J. M., and J. F. Richardson. 1999. *Chemical engineering.* New York: Pergamon.
10. Stokes, G. G. S. 1851. On the effect of the internal friction of fluids on the motion of pendulums. *Trans Cambridge Phil Soc* 9:8–106.
11. Poiseuille, J. L. M. 1840. Experimental investigations upon the flow of liquids in tubes of very small diameter. *Rheological Memoirs* 1.
12. Pfahler, J., J. Harley, and H. Bau. 1991. *ASME 1991: micromechanical sensors, actuators, and systems.* Atlanta: ASME, 49–60.
13. Gravesen, P., J. Branebjerg, and O. S. Jensen. 1993. Microfluidics: a review. *J Micromech Microeng* 3:168–82.
14. Liu, J., and Y.-C. Tai. 1995. *Proceedings: IEEE micro electro mechanical systems (MEMS '95).* Amsterdam: IEEE, 209–15.
15. Arkilic, E. B., K. S. Breuer, and M. A. Schmit. 1994. *Applications of microfabrication to fluid mechanics.* Chicago: ASME, 57–66.
16. Wong, C. C., D. R. Adkins, G. C. Frye-Mason, M. L. Hudson, R. Kottenstette, C. M. Matzke, J. N. Shadid, and A. G. Salinger. 1999. *Proceedings: microfluidic devices and systems II.* Santa Clara, CA: SPIE, 120–29.
17. Flik, M. I., B. I. Choi, and K. E. Goodson. 1991. *ASME 1991: micromechanical sensors, actuators, and systems.* Atlanta: ASME, 31–46.

18. Harley, J., H. Bau, J. N. Zemel, and V. Dominko. 1989. *Proceedings: IEEE micro electro mechanical systems (MEMS '89)*. Salt Lake City, UT: IEEE, 25–28.
19. Israelachvili, J. N. 1986. Measurement of the viscosity of liquids in very thin films. *Colloid Interface Sci* 110:263–71.
20. Tretheway, D. C., and C. D. Meinhart. 2002. Apparent fluid slip at hydrophobic microchannel walls. *Phys Fluids* 14:L9.
21. Pit, R., H. Hervet, and L. Léger. 2000. Direct experimental evidence of slip in hexadecane: solid interfaces. *Phys Rev Lett* 85:980–83.
22. Zhu, Y., and S. Granick. 2001. Rate-dependent slip of Newtonian liquid at smooth surfaces. *Phys Rev Lett* 87:096105.
23. Craig, V. S., J. Neto, and C. Williams. 2001. Shear-dependent boundary slip in an aqueous Newtonian liquid. *Phys Rev Lett* 87:054504.
24. Israelachvili, J. N. 1992. *Intermolecular and surface forces: with applications to colloidal and biological systems (colloid science)*. New York: Academic Press.
25. Gee, M. L., P. M. McGuiggan, J. N. Israelachvili, and A. M. Homola. 1990. Liquid to solidlike transitions of molecularly thin films under shear. *J Chem Phys* 93:1895–906.
26. Ho, C. M. 2001. 14th IEEE International Conference MEMS, Interlaken, Switzerland, 375.
27. Barthlott, W., and C. Neinhuis. 1997. The purity of sacred lotus or escape from contamination in biological surfaces. *Planta* 202:1–8.
28. Colgate, E., and H. Matsumoto. 1990. An investigation of electro-wetting-based microactuation. *J Vac Sci Technol* A8:3625–33.
29. Kenis, P. J. A., R. F. Ismagilov, and G. M. Whitesides. 1999. Microfabrication inside capillaries using multiphase laminar flow patterning. *Science* 285:83–85.
30. Aref, H. 1984. Stirring by chaotic advection. *J Fluid Mech* 143:1–21.
31. Whitesides, G. M. 2003. The "right" size in nanobiotechnology. *Nat Biotech* 21:1161–65.
32. Ismagilov, R. F., A. D. Stroock, P. J. A. Kenis, and G. Whitesides. 2000. Experimental and theoretical scaling laws for transverse diffusive broadening in two-phase laminar flows in microchannels. *Appl Phys Lett* 76:2376–78.
33. Brody, J. P., P. Yager, R. E. Goldstein, and R. H. Austin. 1996. Biotechnology at low Reynolds numbers. *Biophys J* 71:3430–41.
34. Stroock, A. D., S. K. W. Dertinger, A. Ajdari, I. Mezi, H. A. Stone, and G. M. Whitesides. 2002. Chaotic mixer for microchannels. *Science* 295:647–651.
35. Knight, J. B., A. Vishwanath, J. P. Brody, and R. H. Austin. 1998. Hydrodynamic focusing on a silicon chip: mixing nanoliters in microseconds. *Phys Rev Lett* 80:3863–66.
36. Purcell, E. M. 1977. Life at low Reynolds number. *Am J Phys* 45:3–11.
37. Nguyen, N. T., and W. Zhigang. 2005. Micromixers—a review. *J Micromech Microeng* 15:R1–R16.
38. Virtanen, R. 1974. Zone electrophoresis in a narrow-bore tube employing potentiometric detection. *Acta Polytech Scand* 123:1–67.
39. Jorgenson, J. W., and K. D. Lukacs. 1981. High resolution separations based on electrophoresis and electroosmosis. *J Chromatogr* 218:209–14.
40. Barron, A. E., and H. Blanch. 1995. DNA separations by slab gel and capillary electrophoresis: theory and practice. *Separation Purification Methods* 24:1–118.
41. Cobb, K. A., V. Dolnik, and M. Novotny. 1990. Electrophoretic separations of proteins in capillaries with hydrolytically stable surface structures. *Anal Chem* 62:2478–83.
42. Jorgenson, J. W., and E. J. Guthrie. 1983. Liquid chromatography in open-tubular columns. *J. Chromatog* 255: 335–48.
43. Harrison, D. J., Z. Fan, K. Fluri, and K. Seiler. 1994. *Technical digest: 1994 solid-state sensor and actuator workshop*. Hilton Head Island, SC.
44. Jacobson, S. C., R. Hergenröder, A. W. Moore, and J. M. Ramsey. 1994. *Technical digest: 1994 solid-state sensor and actuator workshop*. Hilton Head Island, SC.
45. Jacobson, S. C., J. Kutter, C. Culbertson, and J. Ramsey. 1998. *Micro total analysis systems '98*. Banff, Canada, 315.
46. Jacobson, S. C., and J. M. Ramsey. 1998. Microchip structures for submillisecond electrophoresis. Microchip structures for submillisecond electrophoresis. *Anal Chem* 70: 3476–80.
47. Jacobson, S. C., C. T. Culbertson, and J. M. Ramsey. 1998. *c-state sensor and actuator workshop*. Hilton Head Island, SC.
48. Pace, S. 1990. E. I. Du Pont De Nemours & Co. US Patent 4,908,112.
49. Soane, D. S., and Z. Soane. 1998. Soane Biosciences . US Patent 5,750,015.
50. Manz, A., J. C. Fettinger, E. Verpoorte, H. Ludi, H. M. Widmer, and D. J. Harrison. 1991. Micromachining of monocrystalline silicon and glass for chemical analysis systems: a look into next century's technology or just a fashionable crazy? *Trends Anal Chem* 10:144.
51. Manz, A., D. J. Harrison, E. M. J. Verpoort, J. C. Fettinger, A. Paulus, H. Ludi, and H. M. Widmer. 1992. Planar chips technology for miniaturization and integration of separation techniques into monitoring systems: capillary electrophoresis on a chip. *J Chromatogr* 593:253–58.
52. Ramsey, J. M. 1999. Lockheed Martin Energy Systems. US Patent 6,001,229.
53. Ramsey, J. M. 2000. Lockheed Martin Energy Research Corporation. US Patent 6,010,607.
54. Effenhauser, C. S., A. Manz, and H. M. Widmer. 1995. Manipulation of sample fractions on a capillary electrophoresis chip. *Anal Chem* 67:2284–87.
55. Burggraf, N., A. Manz, E. Verpoorte, C. S. Effenhauser, H. M. Widmer, and N. F. de Rooij. 1994. A novel approach to ion separation in solution: synchronization cyclic capillary electrophoresis (SCCE). *Sensors Actuators B* 20:103.
56. Bousse, L., A. R. Kopf-Sill, and J. W. Parce. 1998. *Systems and technologies for clinical diagnostics and drug discovery.* Ed. G. E. Cohn. San Jose, CA: SPIE.
57. Haab, B. B., and R. A. Mathies. 1998. *Systems and technologies for clinical diagnostics and drug discovery.* Ed. G. E. Cohn. San Jose, CA: SPIE.
58. Wahl, M., R. Erdmann, K. Lauritsen, and H.-J. Rahn. 1998. *Systems and technologies for clinical diagnostics and drug discovery.* Ed. G. E. Cohn. San Jose, CA: SPIE.
59. Righetti, P. G. 1990. *Immobilized pH gradients: theory and modelling.* Amsterdam: Elsevier.
60. Herr A. E., J. I. Molho, J. G. Santiago, T. W. Kenny, D. A. Borkholder, G. J. Kintz, P. Belgrader, and M. A. Northrup. 2000. *Technical digest: 2000 solid-state sensor and actuator workshop.* Hilton Head Island.
61. Heller, M. J., A. H. Forster, and E. Tu. 2000. Active microelectronic chip devices which utilize controlled electrophoretic fields for multiplex DNA hybridization and other genomic applications. *Electrophoresis* 21: 157–64.

62. Martin, Y., and H. K. Wickramasinghe. 1987. Magnetic imaging by "force microscopy" with 1000 Å resolution. *Appl Phys Lett* 50:1455–57.
63. Pohl, H. A. 1951. The motion and precipitation of suspensoids in divergent electric fields. *J Appl Phys* 22:869–71.
64. Pohl, H. A. 1978. *Dielectrophoresis*. Cambridge, UK: Cambridge University Press.
65. Pethig, R., J. P. H. Burt, A. Parton, N. Rizvi, M. S. Talary, and J. A. Tame. 1998. Development of biofactory-on-a-chip technology using excimer laser micromachining. *J Micromech Microeng* 8:57–63.
66. Pethig, R., and G. H. Markx. 1997. Applications of dielectrophoresis in biotechnology. *TIBTECH* 15:426–32.
67. Huang, Y., X. B. Wang, F. F. Becker, and P. R. C. Gascoyne. 1997. Introducing dielectrophoresis as a new force field for field-flow-fractionation. *Biophys J* 73:1118–29.
68. Wang, X.-B., J. Vykoukal, F. F. Becker, and P. R. C. Gascoyne. 1998. Separation of polystyrene microbeads using dielectrophoretic/gravitational field-flow-fractionation. *Biophys J* 74:2689–701.
69. Suehiro, J., and R. Pethig. 1998. The dielectrophoretic movement and positioning of a biological cell using a three-dimensional grid electrode system. *J Phys D Appl Phys* 31:3298–305.
70. Park, B. Y., and M. J. Madou. 2005. 3-D electrode designs for flow-through dielectrophoretic systems. *Electrophoresis* 26:3745–57.
71. Martinez-Duarte, R., H. A. Rouabah, N. G. Green, M. Madou, and H. Morgan. 2007. *Proceedings of the Eleventh International Conference in Miniaturized Systems for Chemistry and Life Sciences: microTAS 2007*. Paris.
72. Wang, L., L. Flanagan, and A. P. Lee. 2007. Side-wall vertical electrodes for lateral field microfluidic applications. *J Microelectromech Systems* 16:454–61.
73. Voldman, J., M. L. Gray, M. Toner, and M. A. Schmidt. 2002. A microfabrication-based dynamic array cytometer. *Anal Chem* 74:3984–90.
74. Iliescu, C., L. Yu, G. Xu, and F. E. H. Tay. 2006. A dielectrophoretic chip with a 3-D electric field gradient. *J Microelectromech Systems* 15:1506–13.
75. Lapizco-Encinas, B. H., B. A. Simmons, E. B. Cummings, and Y. Fintschenko. 2004. Dielectrophoretic concentration and separation of live and dead bacteria in an array of insulators. *Anal Chem* 76:1571–79.
76. Fatoyinbo, H. O., D. Kamchis, R. Whattingham, S. L. Ogin, and M. P. Hughes. 2005. A high-throughput 3-D composite dielectrophoretic separator. *IEEE Trans Biomed Eng* 52:1347–49.
77. Becker, F. F., X. B. Wang, Y. Huang, R. Pethig, J. Vykoukal, and P. R. C. Gascoyne. 1994. The removal of human leukaemia cells from blood using interdigitated microelectrodes. *J Phys D Appl Phys* 27:2659–62.
78. Burt, J. P. H., R. Pethig, and M. S. Talary. 1998. Microelectrode devices for manipulating and analysing bioparticles. *Trans Inst MC* 19:1–9.
79. Rousselet, J., G. H. Markx, and R. Pethig. 1998. Separation of erythrocytes and latex beads by dielectrophoretic levitation and hyperlayer field-flow fractionation. *Colloids Surfaces A* 140:209–16.
80. Markx, G. H., and R. Pethig. 1995. Dielectrophoretic separation of cells: continuous separation. *Biotech Bioeng* 45:337–43.
81. Talary, M. S., J. P. H. Burt, J. A. Tame, and R. Pethig. 1996. Electromanipulation and separation of cells using travelling electric fields. *J Phys D Appl Phys* 29:2198–203.
82. Goater, A. D., J. P. H. Burt, and R. Pethig. 1997. A combined travelling wave dielectrophoresis and electrorotation device applied to the concentration and viability determination of cryptosporidium. *J Phys D Appl Phys* 30:L65–69.
83. Müller, T., S. Fiedler, T. Schnelle, K. Ludwig, H. Jung, and G. Fuhr. 1996. *Biotech Tech* 10:221–26.
84. Batchelder, J. S. 1983. Method and apparatus for dielectrophoretic manipulation of chemical species. US Patent 4,390,403.
85. Huang, Y., J. Yang, X.-B. Wang, F. F. Becker, and P. R. C. Gascoyne. 1999. The removal of human breast cancer cells from hematopoietic CD34+ stem cells by dielectrophoretic field-flow-fractionation. *J Hematother Stem Cell Res* 8:481–90.
86. Yang, J., Y. Huang, X. Wang, X.-B. Wang, F. F. Becker, and P. R. C. Gascoyne. 1999. Dielectric properties of human leukocyte subpopulations determined by electrorotation as a cell separation criterion. *Biophys J* 76:3307–14.
87. Prins, M. W. J., W. J. J. Welters, and J. W. Weekamp. 2001. Fluid control in multichannel structures by electrocapillary pressure. *Science* 291:277–80.
88. Asakuma, N., T. Fukui, M. Toki, K. Awazu, and H. Imai. 2003. Photoinduced hydroxylation at ZnO surface. *Thin Solids Films* 445:284–87.
89. Zhang, Z., H. Chen, J. Zhong, G. Saraf, and Y. Lu. 2007. Fast and reversible wettability transitions on ZnO nanostructures. *J Electr Mater* 36:895–99.
90. Feng, X., L. Feng, M. Jin, J. Zhai, L. Jiang, and D. Zhu. 2004. Reversible super-hydrophobicity to super-hydrophilicity transition of aligned ZnO nanorod films. *J Am Chem Soc* 126:62–63.
91. Sun, R. D., A. Nakajima, A. Fujishima, T. Watanabe, and K. Hashimoto. 2001. Photoinduced surface wettability conversion of ZnO and TiO_2. Thin films. *J Phys Chem B* 105:1984–90.
92. Madou, M. J., and G. J. Kellogg. 1998. *Systems and technologies for clinical diagnostics and drug discovery*. Eds. G. E. Cohn and A. Katzir. San Jose, CA: SPIE.
93. McNeely, M. R., M. K. Spute, N. A. Tusneem, and A. R. Oliphant. 1999. *Microfluidic devices and systems II*. Eds. C. H. Ahn and A. B. Frazier. Santa Clara, CA: SPIE.
94. Madou, M. J., Y. Lu, S. Lai, Y.-J. Juang, L. J. Lee, and S. Daunert. 2000. *Technical digest: 2000 solid sensor and actuator workshop*. Hilton Head Island, SC.
95. Kim, N., C. M. Dempsey, C. J. Kuan, J. V. Zoval, E. O'Rourke, G. Ruvkun, M. J. Madou, and J. Y. Sze. 2007. Gravity force transduced by the MEC-4/MEC-10 DEG/ENaC channel modulates DAF-16/FoxO activity in Caenorhabditis elegans. *Genetics* 177:835–45.
96. Kim, N., C. M. Dempsey, J. V. Zoval, J. Y. Sze, and M. J. Madou. 2007. Automated microfluidic compact disc (CD) cultivation system of Caenorhabditis elegans. *Sensors Actuators B Chemical* 122:511–18.
97. Madou, M. J., J. Zoval, G. Jia, H. Kido, J. Kim, and N. Kim. 2006. Lab on a CD. *Ann Rev Biomed Eng* 8:601–28.
98. Manz, A., E. Verpoorte, C. S. Effenhauser, N. Burggraf, D. E. Raymond, D. J. Harrison, and H. M. Widmer. 1993. Miniaturization of separation techniques using planar chip technology. HRC *J High Resol Chromatogr* 16:433–36.
99. Manz, A., N. Graber, and H. M. Widmer. 1990. Miniaturized total chemical analysis systems: a novel concept for chemical sensing. *Sensors Actuators B* B1:244–48.
100. Nölting, B. 2005. *Methods in modern biophysics*. Berlin: Springer.

101. Harris, D. C. 1999. *Quantitative chemical analysis*. New York: W. H. Freeman.
102. Golay, M. J. E. 1958. *Gas chromatography*. New York: Academic Press.
103. Taylor, G. I. 1953. Dispersion of soluble matter in solvent flowing slowly through a tube. *Proc Roy Soc A* 219:186–203.
104. Wong, C. C., D. R. Adkins, G. C. Frye-Mason, M. L. Hudson, R. Kottenstette, C. M. Matzke, J. N. Shadid, and A. G. Salinger. 1999. *Microfluidic devices and systems II*. Santa Clara, CA: SPIE.
105. Molho, J. I., A. E. Herr, B. P. Mosier, J. G. Santiago, T. W. Kenny, R. A. Brennen, and G. Gordon. 2000. A low dispersion turn for miniaturized electrophoresis. *Technical digest: 2000 solid-state sensor and actuator workshop,* Hilton Head, SC.
106. Culbertson, C. T., S. C. Jacobson, and J. M. Ramsey. 2000. *Technical digest: 2000 solid-state sensor and actuator workshop.* Hilton Head, SC.
107. Manz, A., and W. Simon. 1987. Potentiometric detector for fast high-performance open-tubular column liquid chromatography. *Anal Chem* 59:74–79.
108. Manz, A., J. C. Fettinger, E. Verpoorte, and H. Ludi. 1991. Micromachining of monocrystalline silicon and glass for chemical analysis systems: a look into the next century's technology or just a fashionable craze. *Trends Anal Chem* 10:144–49.
109. Fair, R. B. 2007. Digital microfluidics: is a true lab-on-a-chip possible? *Microfluid Nanofluid* 3:245–81.
110. De Menech, M., P. Garstecki, F. Jousse, and H. A. Stone. 2008. Transition from squeezing to dripping in a microfluidic T-shaped junction. *J Fluid Mech* 595:141–61.
111. Anna, S. L., N. Bontoux, and H. A. Stone. 2003. Formation of dispersions using "flow focusing" in microchannels. *Appl Phys* Lett 82:364–66.
112. Zhou, C. F., P. T. Yue, and J. J. Feng. 2006. Formation of simple and compound drops in microfluidic devices. *Physi Fluids* 18:14.
113. Tan, Y. C., V. Cristini, and A. P. Lee. 2006. Monodispersed microfluidic droplet generation by shear focusing microfluidic device. *Sensors Actuators B Chem* 114:350–56.
114. Woodward, A., T. Cosgrove, J. Espidel, P. Jenkins, and N. Shaw. 2007. Monodisperse emulsions from a microfluidic device, characterised by diffusion NMR. *Soft Matter* 3:627–33.
115. Yobas, L., S. Martens, W. L. Ong, and N. Ranganathan. 2006. High-performance flow-focusing geometry for spontaneous generation of monodispersed droplets. *Lab Chip* 6:1073–79.
116. Nisisako, T., and T. Torii. 2007. Formation of biphasic Janus droplets in a microfabricated channel for the synthesis of shape-controlled polymer microparticles. *Adv Mater* 19:1489–93.
117. Nisisako, T., S. Okushima, and T. Torii. 2005. Controlled formulation of monodisperse double emulsions in a multiple-phase microfluidic system. *Soft Matter* 1:23–27.
118. Utada, A. S., E. Lorenceau, D. R. Link, P. D. Kaplan, H. A. Stone, and D. A. Weitz. 2005. Monodisperse double emulsions generated from a microcapillary device. *Science* 308:537–41.
119. Gordillo, J. M., Z. D. Cheng, A. M. Ganan-Calvo, M. Marquez, and D. A. Weitz. 2004. A new device for the generation of microbubbles. *Phys Fluids* 16:2828–34.
120. Garstecki, P., I. Gitlin, W. DiLuzio, G. M. Whitesides, E. Kumacheva, and H. A. Stone. 2004. Formation of monodisperse bubbles in a microfluidic flow-focusing device. *Appl Physi Lett* 85:2649–51.
121. Cubaud, T., and C. M. Ho. 2004. Transport of bubbles in square microchannels. *Phys Fluids* 16:4575–85.
122. Cubaud, T., M. Tatineni, X. L. Zhong, and C. M. Ho. 2005. Bubble dispenser in microfluidic devices. *Phys Rev* E72:4.
123. Hettiarachchi, K., E. Talu, M. L. Longo, P. A. Dayton, and A. P. Lee. 2007. On-chip generation of microbubbles as a practical technology for manufacturing contrast agents for ultrasonic imaging. *Lab Chip* 7:463–68.
124. Prakash, M., and N. Gershenfeld. 2007. Microfluidic bubble logic. *Science* 315:832–35.
125. Hagen, G. H. L. 1839. III Über die Bewgung des Wassers in engen cylindrischen Röhren, *Poggendorfs Annalen der Physik und Chemie* 2(46):423–42.

7

Electrochemical and Optical Analytical Techniques

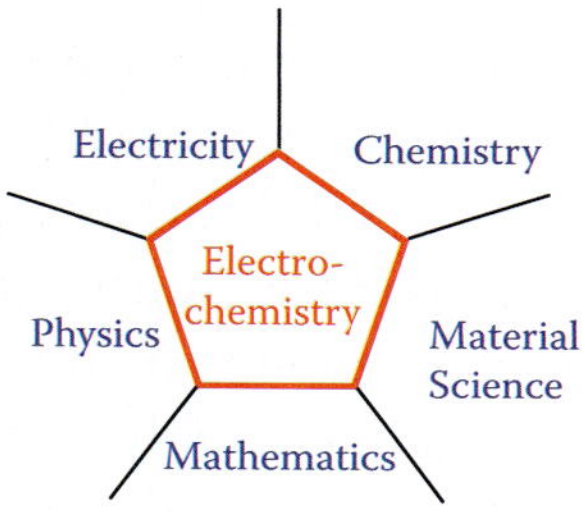

The History of Electrochemistry (calendar from http://www.basinc.com).

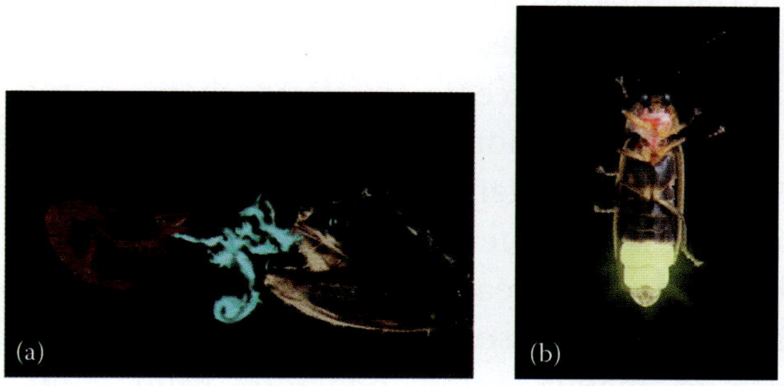

(a) Shrimp vomiting a bioluminescent substance toward a predator. (b) Firefly exhibiting bioluminescence.

Introduction

In all electrochemical and many optical sensors, molecules interact with a solid surface and generate an analytical signal that can be used to identify and sometimes quantify an unknown analyte. This chapter begins with a detailed study of the nature of the attractive and repulsive forces at an interface between a liquid and a solid (mostly in the form of small particles and bulk electrodes). Forces at the liquid/solid interface are the same as those that control the interaction between liquid molecules but are modulated by the presence of the solid. Intermolecular nonbonding forces between molecules, ions, or molecules and ions are much weaker than intramolecular bonding forces, such as covalent bonds within a molecule. Nevertheless, intermolecular forces sum to an impressive total binding energy, often several hundreds of kilojoules per mole, when many are present within a macromolecule or in a solid. The intermolecular forces we will study include the classical (coulomb) electrostatic forces, van der Waals interactions, and the quantum mechanical Casimir force. These forces, when exerted between individual molecules, extend typically over a few nanometers only, but when molecules come together into condensed phases, the force range at the surface is extended to several tens of nanometers. Intermolecular forces explain the electrical double layer (EDL) at the interface of liquid/solid surfaces, surface tension, viscosity, vapor pressure, boiling points, and many other important phenomena, including DNA hybridization and self-assembly. Hydrophilic and hydrophobic interactions, also based on intermolecular forces, are treated in Chapter 6 on fluidics.

Electrochemistry is the study of reactions taking place at the interface of a solid conductor and an electrolyte. Using sensor examples, we introduce some of the most important concepts in electrochemistry, i.e., the EDL, potentiometry, voltammetry, two- and three-electrode systems, Marcus theory of electron transfer, reaction rate- and diffusion rate-controlled electrochemical reactions, and ultramicroelectrodes (UMEs). We also learn how different materials, such as metals, semiconductors, solid electrolytes, and mixed conductors, exposed to a solution offer the possibility for building different types of electrochemical sensors. Besides insights in electrochemical sensor construction, this section prepares the reader for an appreciation of the role of electrochemistry in the manufacture of microelectromechanical systems (MEMS) and nanoelectromechanical systems (NEMS), i.e., in electrochemical machining as covered in Volume II, Chapter 4 (electrochemical etching, a subtractive technique) and Volume II, Chapter 8 (electrochemical deposition, a forming technique). We conclude the electrochemistry section with more electrochemical sensor examples, including ion-sensitive field-effect transistors (ISFETs), potentiometric and amperometric immunosensors, and glucose sensors.

Many researchers use MEMS and NEMS to miniaturize optical components or whole instruments for absorption, luminescence, or phosphorescence measurements. The latter techniques, like the electrochemical methods reviewed, constitute a branch of analytical chemistry, a branch that is concerned with the production, measurement, and interpretation of electromagnetic data arising from either emission or absorption of radiant energy by matter. We compare the sensitivity of the various optical sensing techniques and analyze how amenable they are to miniaturization (scaling laws).

We end this chapter by comparing the merits and problems associated with electrochemical and optical measuring techniques.

Intermolecular Forces

Introduction

For the nonchemist, a few simple chemistry concepts are recalled here. A covalent bond, holding a molecule together, is a bonding or intramolecular force, whereas the attraction between molecules is a nonbonding or intermolecular force. Ionic chemical bonding results from the maximization of the electrostatic forces between large numbers of cations and anions in ionic solids. Intermolecular forces between molecules, between ions in solution, or between molecules and ions are much weaker than intramolecular forces (e.g., leading to a binding energy of only 16 kJ/mol vs. 431 kJ/mol for HCl) (see Figure 7.1).

Covalent bonds (300–400 kJ/mol) determine molecular architecture, whereas noncovalent bonds, typically 10–100 times weaker, determine molecular conformations. Being individually weak, noncovalent interactions nevertheless sum to an impressive total, often several hundreds of kilojoules per mole, when many are present within a given

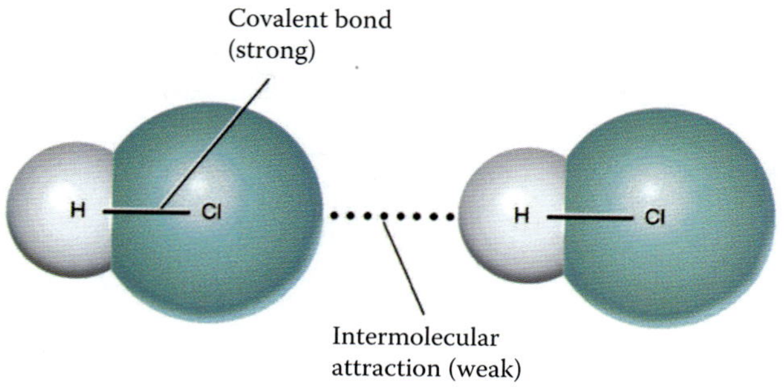

FIGURE 7.1 In a HCl molecule, there is a bonding or intramolecular bond that is covalent and strong and a nonbonding or intermolecular hydrogen bond that is weak.

macromolecule, between macromolecules in a solution, or between the building blocks making up a solid. The intermolecular energy term (used in the analytical calculations involving energy) usually includes the classical (coulomb) electrostatic and van der Waals interactions. The nonbonding forces we will analyze below are:

- Charge-charge interactions (Coulomb's law in a dielectric medium)
- Permanent and induced dipole interactions (van der Waals interactions):
  - Charge-dipole interactions
  - Dipole-dipole interactions
  - Charge-induced dipole interactions
  - Dipole-induced dipole interactions
  - Dispersion (induced dipole-induced dipole) interactions
- Hydrogen bonds: donor group-acceptor group interactions (a very important case of dipole-dipole interactions)
- Casimir force: a quantum mechanical effect (a special type of van der Waals interaction, i.e., the retarded van der Waals force also known as the *Casimir-Polder force*)

Hydrophilic versus hydrophobic interactions, a very important group of van der Waals interactions, were treated in Chapter 6 on fluidics.

In Table 7.1 we compare bonding and nonbonding forces.

Properties controlled by intermolecular forces include vapor pressure, boiling point, molecular crystals, protein folding, colloidal stability, rheological behavior, self-assembly, surface tension, viscosity, and so on. To understand how important intermolecular forces are, the two following quotes help:

> The very existence of condensed phases of matter is conclusive evidence of attractive forces between molecules, for in the absence of attractive forces, the molecules in a glass of water would have no reason to stay confined to the glass...
>
> **Anthony J. Stone**
> *The Theory of Intermolecular Forces* (1996)

> The most important forces dictating the properties of biomolecules are the non-bonded [intermolecular] interactions...
>
> **Alexander D. MacKerell, Jr.**
> *Journal of Computational Chemistry* (2004)

In other words, in the absence of ionic bonding and interatomic or intermolecular forces, all matter would exist in the gas form, even at 0 K. Thus, intermolecular forces play a crucial role in many of the properties of liquids, gases, and solids. When a substance melts or boils, the intermolecular forces are broken (not the covalent bonds), and when a substance condenses, intermolecular forces are formed.

Intermolecular forces at a solid/liquid interface are the same forces that control the interaction between the constituent molecules of the solution modulated by the presence of foreign molecules at the solid/liquid interface. In this section, we establish the operational range of these forces and find that forces between individual molecules typically only extend over a few nanometers. However, when molecules come together into condensed phases, the force range at the surface is extended to several tens of nanometers.

A couple of examples of intermolecular forces at work follow. When intermolecular forces are weak, little energy is required to overcome them. Consequently, boiling points are low for such compounds. The vapor pressure of a liquid depends on intermolecular forces such that, when the intermolecular forces in a liquid are strong, the vapor pressure is low. The boiling point is related to vapor pressure and is lowest for liquids with the weakest intermolecular forces. Surface tension increases with increasing intermolecular forces, and because surface tension

Table 7.1 Comparison of Bonding (Intramolecular), Ionic, and Nonbonding (Intermolecular) Interactions

Force	Model	Basis of Attraction	Energy (KJ/mol)	Example
Bonding				
Ionic		Cation–anion	400–4000	NaCl
Covalent	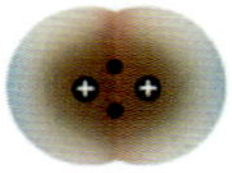	Nuclei-shared e^- pair	150–1100	H—H
Metallic	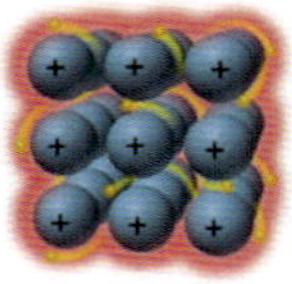	Cations–delocalized electrons	75–1000	Fe
Nonbonding (intermolecular)				
Ion-dipole		Ion charge–dipole charge	40–600	Na^+····OH_2
H-bond	δ^- δ^+ δ^- —A—H·······:B—	Polar bond to H-dipole charge (high EN of N,O,F)	10–40	:Ö(H)—H····:Ö(H)—H
Diploe-bond		Dipole charges	5–25	I—CI····I—CI
Ion-induced dipole		Ion charge-polarizable e^- cloud	3–15	Fe^{2+}····O_2
Dipole-induced dipole	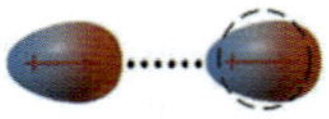	Dipole charge-polarizable e^- cloud	2–10	H—CI····CI—CI
Dipersion (London)	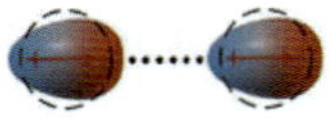	Polarizable e^- cloud	0.05–40	F—F····F—F

is the energy needed to reduce the surface area of a liquid, it is necessary to pull molecules apart against the intermolecular forces of attraction (cohesion) to increase surface area. Viscosity increases with increasing intermolecular forces because increasing these forces increases the resistance to flow. Other factors, such as the possibility of molecules tangling together, can also affect the viscosity. Liquids with long molecules that tangle together are expected to have higher viscosities. Intermolecular forces explain all the data of the liquids listed in Table 7.2, and we will learn in the pages that follow and elsewhere in the book that they explain many other important phenomena, from DNA hybridization to self-assembly.

Many observed phenomena in nature have their basis in intermolecular forces! For example, in Volume III, Chapter 7 on scaling laws, we will learn that the gecko's ability to walk on ceilings is due to van der Waals attractive forces between the spatulae (1 billion of them) on the gecko's feet and the wall. This example shows how short-range forces can have long-range (macroscopic) effects. Intermolecular forces exert their influence especially in the micro- and nanoworld. In the nanoworld, gravitational forces are negligible, and forces such as van der Waals and the quantum mechanical Casimir force take over. At distances less than a micrometer, the Casimir force (a special type of van der Waals force) becomes the dominant force between two neutral objects in a vacuum. In Volume III, Chapter 8 on actuators, we calculate that two plates with an area of 1 cm^2 separated by a distance of 1 μm have an attractive Casimir force of about 10^{-7} N. At separations of 10 nm, the Casimir effect produces the equivalent of 1 atm. Obviously,

TABLE 7.2 Properties of Some Liquids at 20°C

Substance	Molecular Weight (amu)	Vapor Pressure (mmHg)	Surface Tension (J/m²)	Viscosity (kg/m · s)
Water, H_2O	18	1.8×10^1	7.3×10^{-2}	1.0×10^{-3}
Carbon dioxide, CO_2	44	4.3×10^4	1.2×10^{-3}	7.1×10^{-5}
Pentane, C_5H_{12}	72	4.4×10^2	1.6×10^{-2}	2.4×10^{-4}
Glycerol, $C_3H_8O_3$	92	1.6×10^{-4}	6.3×10^{-2}	1.5×10^0
Chloroform, $CHCl_3$	119	1.7×10^2	2.7×10^{-2}	5.8×10^{-4}
Carbon tetrachloride, CCl_4	154	8.7×10^1	2.7×10^{-2}	9.7×10^{-4}
Bromoform, $CHBr_3$	253	3.9×10^0	4.2×10^{-2}	2.0×10^{-3}

Intermolecular forces explain the data of the liquids in this table.

this is a force to reckon with in MEMS and NEMS devices.

Interparticle Interaction Energy (*U*) and Interaction Forces (*F*)

The interaction energy between two particles (atoms or molecules) is defined as the pair potential (energy), $U(r)$, which depends on the intermolecular separation, r. The force, $F(r)$, between the two particles is obtained as the negative of the gradient of the pair potential:

$$F(r) = -\frac{dU(r)}{dr} \tag{7.1}$$

At very large separations (as $r \rightarrow \infty$), the interparticle interaction is negligible ($U \rightarrow 0$, $F \rightarrow 0$). At shorter separations, the particles attract ($F < 0$, particles are pulled together). A dimer is more stable than the two isolated particles, hence $U(r) < 0$. At very short separations ($r \rightarrow 0$), the particles repel each other ($F > 0$, particles are pushed apart). At $r = r_e$, $U(r)$ is at its minimum (U_{min}), and the interparticle force, $F = 0$. This is summarized in Figure 7.2.

The diagram in Figure 7.2 shows the total potential energy between two particles. A power law with r raised to the exponent $-n$ generally models the intermolecular potential energy function:

$$U(r) = Cr^{-n} = \frac{C}{r^n} \tag{7.2}$$

where C is a constant. The general expression for $F(r)$ is then:

$$F(r) = -\frac{nC}{r^{n+1}} \tag{7.3}$$

For the case of attractive interactions, $U(r) < 0$, whereas for repulsive interactions, $U(r) > 0$. To understand the exact shape of the curve in Figure 7.2, we must introduce the exact equations for repulsive and attractive potentials first. The most important repulsive term corresponds to the limit of compressibility of matter. At short internuclear separations, there are both electrostatic repulsions between the atomic nuclei and between the electrons (both valence and core) on neighboring atoms. The latter is based on the Pauli exclusion principle. The Pauli exclusion principle is quantum mechanical in origin and only allows for two electrons with opposite spin per energy level. The importance of the Pauli principle was exemplified in Chapter 3, where we saw that without the Pauli principle all material would collapse! The total short-range repulsive interaction is generally modeled by steep $1/r^{12}$ dependence on interatomic distance (Figure 7.2, left

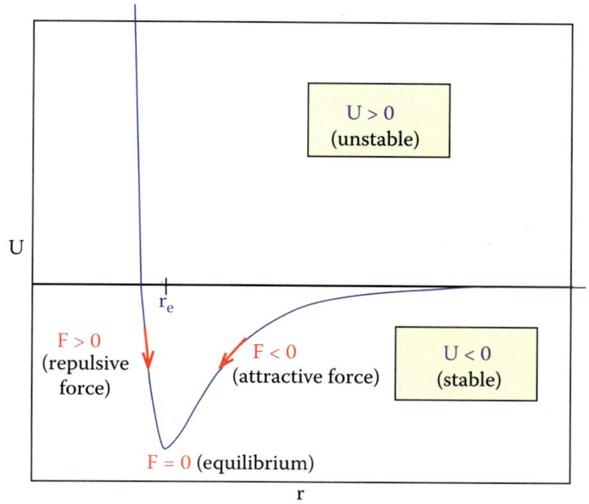

FIGURE 7.2 Interparticle interaction energy (*U*) and forces (*F*) as a function of interparticle distance (*r*).

side of the curve). Forces that scale with an exponent of n ≥ 3 involve short-ranged forces, whereas values of $n < 3$ mark long-range forces. To appreciate where this distinction comes from, we need to briefly introduce the scaling of forces between sets of molecules.

Consider a region of space (solid, liquid, or gas) with a molecular density ρ. The summation of all interaction energies, U_{Total}, of a molecule with all other molecules in a spherical region of size L in that space is then given as:

$$U_{Total} = \int_{\sigma}^{L} U(r)\rho 4\pi r^2 dr = \int_{\sigma}^{L} \frac{\rho C 4\pi r^2}{r^n} dr \qquad (7.4)$$

or:

$$U_{Total} = \frac{-4\pi C\rho}{(n-3)\sigma^{n-3}}\left[1-\left(\frac{\sigma}{L}\right)^{n-3}\right]$$

where σ is the molecular diameter. This important equation reveals that molecular forces contribute to interaction energies for n ≥ 3. At distances $L >> \sigma$, intermolecular forces become inefficiently weak and do not contribute to U_{Total}; however, with $n < 3$, the size of the system L becomes important. The latter is the case, for example, for gravity (distant planets and stars) and Coulomb interactions where n = 1, i.e., both are very long-ranged forces that fall off very slowly. For n ≥ 3, the intermolecular potential becomes important and is the reason why bulk properties of a material are size-independent (water boils at the same temperature in a cup as it does in a bucket) unless we go to submicrometer size distances, where properties of material change as intermolecular forces start taking over. The Pauli-based repulsive interaction, for example, is very short ranged (n = 12).

Types of Interparticle Interaction Forces

Coulomb Interactions

The Coulomb electrostatic interaction potential between two charges (Figure 7.3) is represented by the following equation:

$$U(r) = \frac{Q_1 Q_2}{4\pi\varepsilon_0 \varepsilon r} \qquad (7.5)$$

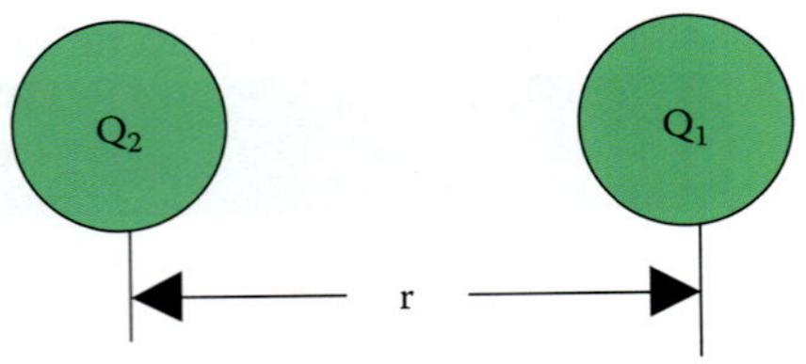

FIGURE 7.3 Coulomb interaction between two point charges or monopoles.

(Equation 7.2 with n = 1 and $C = \frac{Q_1 Q_2}{4\pi\varepsilon_0\varepsilon}$). In Equation 7.5, $Q_1 = z_1 e$, where e is the magnitude of the electron charge (1.602×10^{-19} C) and z_1 is an integer value. In the denominator, ε_0 is the permittivity of free space (8.854×10^{-12} J^{-1} C^2 m^{-1}), and ε is the relative permittivity of the medium between the ions (ε = 1 in a vacuum or ε > 1 in a gas or liquid). In the case of repelling like charges $U(r)$ is positive, and in the case of attracting opposite charges $U(r)$ is negative. This Coulomb interaction potential is additive when charges are grouped in crystals. With ions in solution, the Coulomb interaction is modulated by the dielectric constant of the medium. Highly polar solvents have high dielectric constants. Water, for example, has a dielectric constant ε = 78 (at interfaces this number is much lower as movement of the water dipoles is more restricted). This leads to a large reduction in the Coulombic potential between ions in polar solvents and can lead to the solvation or hydration of a salt in a solution (Figure 7.4). The hydration of cations and anions is responsible for the high solubility of many ionic compounds in water.

When molecules are in close contact, $U(r)$ is typically ~10^{-18} J, corresponding to about 200–300 kT (where k is the Boltzmann constant and T the temperature in Kelvin) at room temperature. To contrast

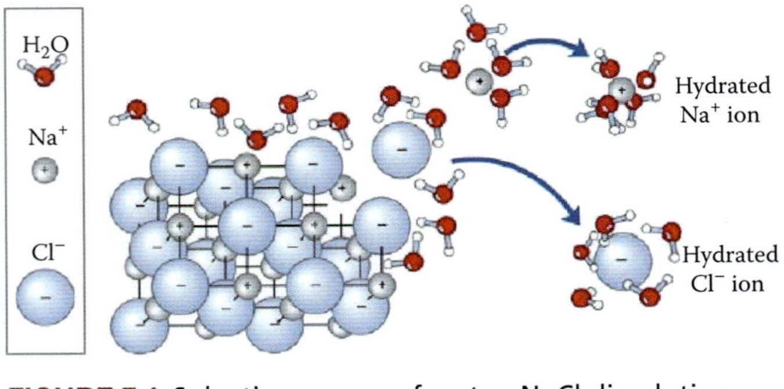

FIGURE 7.4 Solvation power of water. NaCl dissolution.

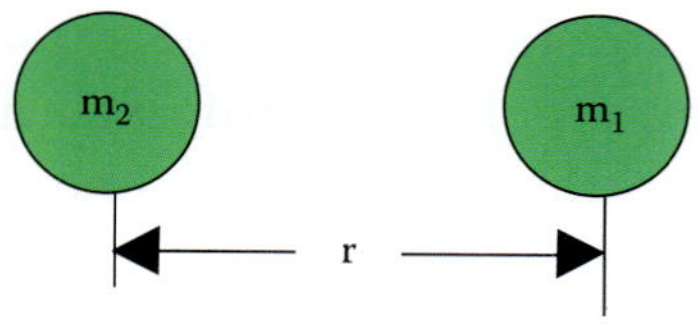

FIGURE 7.5 Gravitational interaction between two molecules.

this Coulomb interaction with the gravity interaction of molecules (Figure 7.5), consider that the gravitational interaction energy is given by:

$$U(r) = \frac{gm_1m_2}{r} \tag{7.6}$$

This is equivalent to Equation 7.2 with n = 1 and $C = gm_1m_2$. In this expression the gravitational constant $g = 6.67 \times 10^{-11}$ Nm^{2}kg^{-1}, and for molecules in contact this leads to a negligible gravitational interaction energy $U(r)$ of about 10^{-52} J compared with 10^{-18} J for the Coulomb interaction energy.

The Coulomb interaction is an electrostatic interaction, i.e., such an interaction may occur between charged atomic or molecular species (ions = monopoles) or between asymmetric charge distributions in neutral molecules (e.g., dipoles, quadrupoles, etc.). A dipole is an asymmetric charge distribution in a molecule, where there is no net charge; however, one end of the molecule is negative (partial charge = −q) relative to the other (partial charge = +q). Molecules may possess higher-order electric multipoles, arising from their nonspherical charge distributions (Figure 7.6). Each type of multipole has an associated multipole moment: the monopole moment is the charge of the atom/molecule; the dipole moment is a vector whose magnitude is the product of the charge (q) and distance (λ) between the charge centers ($\mu = q\lambda$). Higher-order multipole moments have tensor properties. A polar molecule is one that possesses a permanent dipole moment: there is an asymmetric charge distribution, with one end of the molecule relatively negative ($-q$) with respect to the other ($+q$).

The interaction energy between a stationary point charge Q_2 and a permanent fixed dipole μ_1 separated by a distance r is given as (Figure 7.7):

$$U(r) = -\frac{Q_2\mu_1\cos\theta}{4\pi\varepsilon_0\varepsilon r^2} \tag{7.7}$$

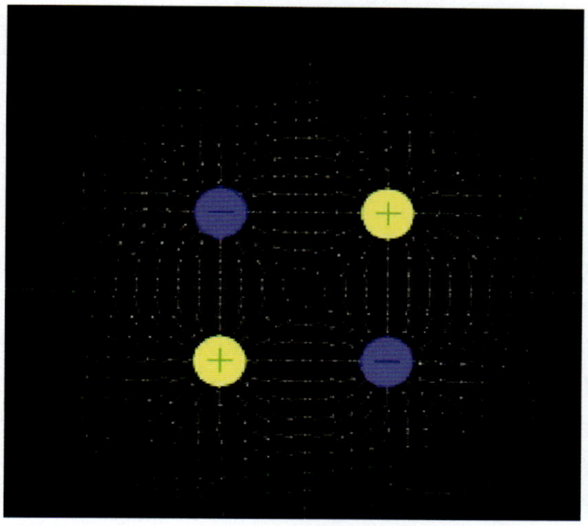

FIGURE 7.6 Electric quadrupole. Four charges, two positive and two negative. The arrows depict the electric fields, and the lines are the equipotential surfaces.

with $Q_2 = z_2e$ in units of C and μ_1 ($q_1\lambda$) in units of Cm [often expressed in Debye (D), where D = 3.336 10^{-30} Cm]. The interaction energy now depends on the orientation of the dipole relative to the charge (θ) and the sign of the charge. In the case of a permanent dipole that can freely rotate, one obtains a new expression:

$$U(r) = -\frac{Q_2^2\mu_1^2}{6(4\pi\varepsilon_0\varepsilon)^2 kTr^4} \tag{7.8}$$

with k being the Boltzmann constant. Notice that temperature dependence now appears in the denominator. One example of this type of interaction is the condensation of water (μ = 1.65 D) caused by the presence of ions in water. During a thunderstorm, ions are created that nucleate raindrops in thunderclouds (ionic nucleation).

When an electric charge (i.e., an ion = electric monopole) or asymmetric charge distribution (e.g.,

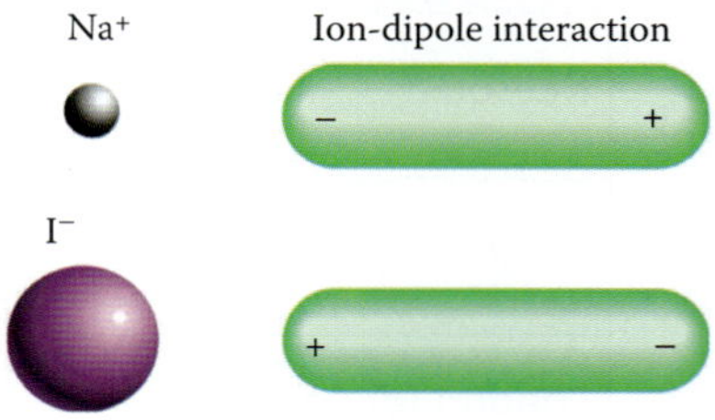

FIGURE 7.7 Ion-dipole interaction.

dipole, quadrupole, etc.) gives rise to an electric field (E) that causes the polarization of neighboring nonpolar atoms or molecules, an induced dipole is created. The magnitude of the induced dipole is:

$$\mu = \alpha E \tag{7.9}$$

where α is the polarizability of the neutral atom or molecule. Polarizability is the ease with which the electron distribution in the atom or molecule can be distorted. Polarizability increases as the number of electrons increases and as the electron cloud becomes more diffuse (larger molecules). An ion-induced dipole interaction energy is given as:

$$U(r) = -\frac{Q_1^2\alpha_2}{2(4\pi\varepsilon_0\varepsilon)^2 r^4} \tag{7.10}$$

Van der Waals Forces

To explain deviations from ideal gas behavior (equation of state by Boyle and Gay-Lussac: $PV = RT$), van der Waals, in 1873, proposed a modification to the equation of state of a gas as follows:

$$\left(P + \frac{a}{V^2}\right)(V - b) = RT \tag{7.11}$$

This new expression accounted for the fact that molecules have a finite volume (b) and that there is an attractive force between these molecules (a). This relation correctly predicts gas behavior over a wider pressure range.

The attractive intermolecular forces between gas molecules are now known as *van der Waals forces.* The three types of interactions that constitute van der Waals forces are summarized in Table 7.3. The exponent $n = 6$ indicates that these types of van der Waals forces are short range. The attractive interactions between dipoles, quadrupoles, and between permanent multipoles are sometimes named after Keesom. Another source of attraction is induction (i.e., polarization), which is the interaction between a permanent dipole or multipole on one molecule with an induced dipole or multipole. This interaction is named after Debye. The third type of attraction, London forces, is named after the German-American physicist Fritz London, who himself called it *dispersion forces*. This is the only attraction experienced by noble gas atoms, but it is operative between any pair of molecules, regardless of their symmetry. London dispersion forces are weak intermolecular forces that arise from the attractive forces between transient dipoles (or multipoles) in molecules without permanent dipole (or multipole) moments. We will see that the London component is often the most dominant interaction from the three interactive forces.

Dipole-Dipole Interactions—Keesom The expression for the interaction energy of two fixed dipoles, μ_1 and μ_2, that are not coplanar (i.e., there is an angle ϕ between them) is calculated as:

$$U(r) = -\frac{\mu_1^2\mu_2^2 f(\theta_1,\theta_2,\phi)}{4\pi\varepsilon_0\varepsilon r^3} \tag{7.12}$$

where r is the distance between the centers of mass of the two molecules. The expression for two freely rotating dipoles, the so-called *Keesom equation*, is given as:

$$U(r) = -\frac{\mu_1^2\mu_2^2}{3(4\pi\varepsilon_0\varepsilon)^2 kTr^6} \tag{7.13}$$

Table 7.3 Types of van der Waals Forces

Interaction Component	Origin of Interactions	Equation
Keesom	Dipole-dipole	$U(r) = -\frac{\mu_1^2\mu_2^2}{3(4\pi\varepsilon_0\varepsilon)^2 kTr^6}$
Debye (induction)	Dipole-induced dipole	$U(r) = -\frac{\mu_1^2\alpha_2}{(4\pi\varepsilon_0\varepsilon)^2 r^6}$
London (dispersion)	Induced dipole-induced dipole	$U(r) = -\frac{3\alpha^2 h\nu}{4(4\pi\varepsilon_0\varepsilon)^2 r^6}$

Here a temperature dependence of the interaction energy appears. Dipole-dipole interactions are generally weak. These dipole-dipole interactions become significant when two interacting dipoles approach each other closely, e.g., in $O^{\delta-}$—$H^{\delta+}$, $N^{\delta-}$—$H^{\delta+}$, $F^{\delta-}$—$H^{\delta+}$, etc. Such strong dipole-dipole interactions are the underlying reason for the well-known hydrogen bonds (see below). The electrostatic interaction between higher-order multipoles can be treated in an analogous fashion to that described here for dipole-dipole interactions. The distance dependence of the interaction energy between an *n*-pole and an *m*-pole (assuming no molecular rotation) is given as:

$$U(r) \propto \pm \frac{1}{r^{(n+m-1)}} \tag{7.14}$$

where *n* and *m* are the orders of the multipoles: n, m = 1 (monopole), 2 (dipole), 3 (quadrupole), 4 (octopole), etc. (Table 7.4). It is noted that the higher the order of the multipole, the shorter ranged the interaction is, i.e., the more rapidly it falls off with distance.

Dipole-Induced Dipole Interactions—Debye An induced dipole interacts in an attractive fashion with a fixed, permanent dipole (μ_1). The inductive interaction energy between the fixed permanent dipole (μ_1) associated with molecule 1 and the induced dipole on molecule 2 (which has no fixed dipole but has certain polarizability α_2 and is a distance *r* away from molecule 1) is given by:

$$U(r) = -\frac{\mu_1^2 \alpha_2 \left(1 + 3\cos^2\theta\right)}{2(4\pi\varepsilon_0\varepsilon)^2 r^6} \tag{7.15}$$

Table 7.4 Interaction Energies between an n-Pole and an m-Pole

Interaction Energies for...	
Monopole-monopole (coulomb)	$U(r) \propto r^{-1}$
Monopole-dipole	$U(r) \propto r^{-2}$
Dipole-dipole	$U(r) \propto r^{-3}$
Dipole-quadrupole	$U(r) \propto r^{-4}$
Quadrupole-quadrupole	$U(r) \propto r^{-5}$

In the case where the permanent dipole is free to rotate, the so-called *Debye interaction* energy results:

$$U(r) = -\frac{\mu_1^2 \alpha_2}{(4\pi\varepsilon_0\varepsilon)^2 r^6} \tag{7.16}$$

As the direction of the induced dipole follows that of the permanent dipole, the effect survives even if the permanent dipole is subjected to thermal reorientations. Debye interactions are thus independent of temperature. Dipoles (and higher multipoles) may also be induced by neighboring molecules featuring permanent charges (monopoles) or by high-order multipoles. Similar expressions can be derived for the interactions between these permanent and induced multipoles.

Induced Dipole-Induced Dipole—Dispersive or London Forces The intermolecular forces between nonpolar molecules and closed shell atoms (e.g., rare gas atoms such as He, Ne, and Ar) are the "London" or dispersion forces, attractive forces that arise as a result of temporary dipoles, which are induced in atoms or molecules (Figure 7.8). Keesom and Debye interactions require the presence of at least one permanent dipole, but London interactions do not. Hence, London interactions exist between all molecules. It is also generally the dominant contribution to van der Waals forces. The London interaction component was known for a long time, but the theory only evolved after the development of quantum mechanics. Nonpolar molecules exhibit London forces because electron density moves about a molecule probabilistically. There is a high chance that the electron density will not be evenly distributed throughout a nonpolar molecule. When an uneven distribution occurs, a temporary multipole is created, which may interact with other nearby multipoles. London forces are also present in polar molecules, but they are usually only a small part of the total interaction force. This phenomenon is the only attractive intermolecular force between neutral atoms (e.g., helium) and is the major attractive force between nonpolar molecules (e.g., nitrogen or methane). Without London forces, there would be no attractive force between noble gas atoms, and they could not then be obtained in a liquid form.

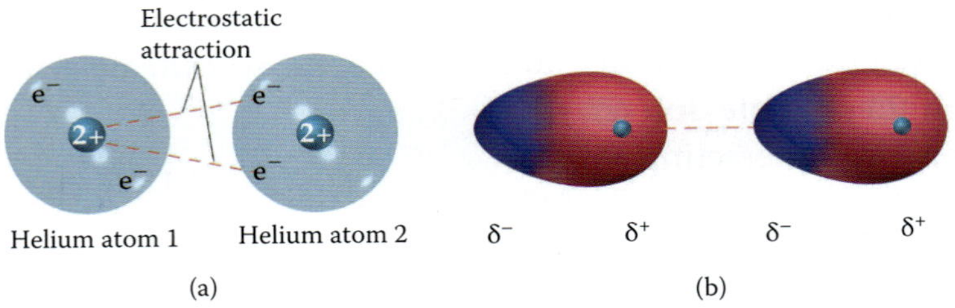

FIGURE 7.8 Induced dipole-induced dipole interaction. Dispersion or "London" forces in helium atoms (a) before the dipole and (b) with the dipole.

A general expansion for the attractive energy resulting from these dispersion forces can be written as:

$$U(r) = -\left\{\frac{C_6}{r^6} + \frac{C_8}{r^8} + \frac{C_{10}}{r^{10}} + \cdots + \frac{C_n}{r^n}\right\} \tag{7.17}$$

where the C_n terms are constants and r is the interatomic distance. The first term represents the instantaneous dipole-dipole interaction and is the dominant variable; therefore, the higher terms (which contribute less than 10% of the total dispersion energy) are often omitted when calculating dispersion energies. A reasonable approximation to the dispersion energy is found in the London formula (derived by London in 1930, based on a model by Drude), in which the dispersion energy between two identical atoms/molecules is given as:

$$U(r) = -\frac{C_6}{r^6} = -\frac{3\alpha^2 h\nu}{4(4\pi\varepsilon_0\varepsilon)^2 r^6} \tag{7.18}$$

where α is the polarizability and $h\nu$ is the ionization energy of the atom/molecule. The polarizability α is squared in Equation 7.18 because the magnitude of the instantaneous dipole depends on the polarizability of the first atom/molecule. The strength of the induced dipole (in the second atom/molecule) also depends on α. When molecules are in close contact, $U(r)$ is typically ~10^{-21}–10^{-20} J/mol, corresponding to about 0.2–2 kT at room temperature, which is a comparable magnitude to the thermal energy. London forces become stronger as the atom (or molecule) in question becomes larger. This is because of the increased polarizability of molecules with larger, more dispersed electron clouds. London forces provide molecular level reasoning that at room temperature small molecules such as Ar, He, and CH_4 are gases, whereas larger molecules, like hexane and decane, are liquids, and yet even larger molecules are solids. Interactions involving charged species (e.g., ions) tend to be stronger than polar-polar interactions.

All intermolecular forces are anisotropic (except those between two noble gas atoms), which means that they depend on the relative orientation of the molecules. The induction and dispersion interactions are always attractive, but the electrostatic interaction changes sign on rotation of the molecules (can be attractive or repulsive, depending on the mutual orientation of the multipoles). When molecules are in thermal motion, as they are in the gas and liquid phase, the electrostatic forces are diminished to a large extent because the molecules thermally rotate and thus probe both repulsive and attractive parts of the potential. The thermal averaging effect is much less pronounced for induction and dispersion forces.

Further below we will see that the London-van der Waals forces are related to the quantum mechanical Casimir effect introduced in Chapter 3 (Figure 3.69). The latter is sometimes referred to as the *retarded* van der Waals force.

Example of Dipole-Dipole Interaction: Hydrogen Bonding Earlier we saw that weak dipole-dipole interactions can become significant when two interacting dipoles approach each other closely, e.g., in $O^{\delta-}$—$H^{\delta+}$, $N^{\delta-}$—$H^{\delta+}$, $F^{\delta-}$—$H^{\delta+}$, and so on. Such strong dipole-dipole interactions are the underlying reason for hydrogen bonds. Hydrogen bonding is an intermolecular linkage of the form X—H—Y, where a hydrogen atom is bound to one electronegative atom (e.g., X=N, O, F) and interacts with a second electronegative atom (Y) that has an available pair of free electrons. For example, X—H is the hydrogen bond donor (e.g., O—H, peptide N—H), and Y is the hydrogen bond acceptor (e.g., :O, :N, peptide C=O). The donor is, in this case, an acid (proton donor), whereas the acceptor is a base (proton acceptor). A hydrogen bond is stronger than most other nonbonded interactions. Typical hydrogen bond strength is 10–40 kJ mol^{-1} (about 10% of a covalent bond). A rare strong H-bond may be 60 kJ mol^{-1} when the donor and acceptor are very close for significant orbital overlap.

Hydrogen bonding is especially strong when H is attached to F, O, or N because these three elements all have high electronegativity and are small enough to get close to. Most biological H-bonds are ~2.8 Å (e.g., O—H...O is 2.7 Å and O—H...N is 2.9 Å). Water forms hydrogen bonds: the O—H bonds act as H-bond donors, and the O lone pairs act as H-bond acceptors (Figure 7.9). The proton in the water molecule is unshielded and makes an electropositive end to the bond, which explains the partial ionic character of the hydrogen bond. Most of the unique properties of water are the result of this hydrogen bonding and its polar nature. Water, we saw above, is a good solvent for ionic species because it has a high dielectric constant. Each molecule of water in the solid state (ice, see Figure 7.10) is engaged in 4.0 hydrogen bonds (accepting two and donating two), and each molecule of water in the liquid state is engaged in approximately 3.4 hydrogen bonds that are transitory in nature. With fewer hydrogen bonds, the oxygen atoms may on average be closer to each other (denser). About 15% of H-bonds break when ice melts. As a consequence, water is denser than ice.

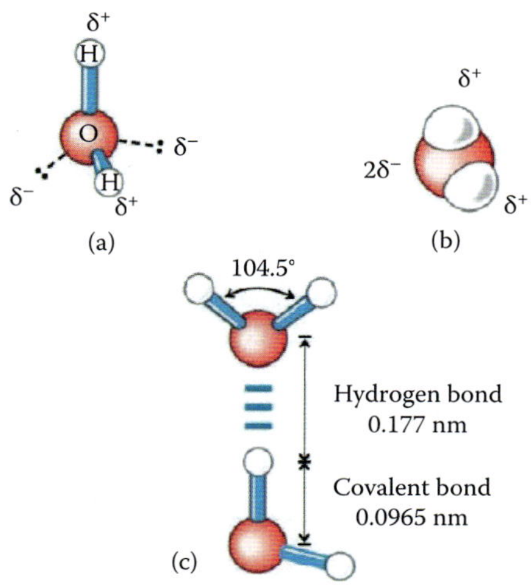

FIGURE 7.9 Hydrogen bonding in water. The structure of a water molecule (a) makes it a perfect hydrogen bond donor (b) and acceptor (c).

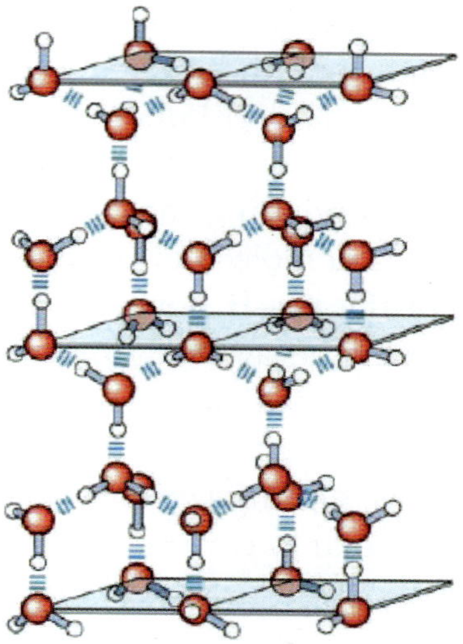

FIGURE 7.10 In ice, water molecules form an H-bonded network that fully satisfies the H-bonding potential (4 per H_2O).

The hydrogen bond endows specificity to an interaction as particular atoms are required as donors and acceptors. It also imposes geometrical constraints; hydrogen bonds, like covalent bonds, are highly directional. The X—H—Y bond is optimally a straight line (see Figure 7.9c), where the donor H bond tends to point directly at the acceptor electron pair.

The major component (around 80%) of a hydrogen bond is electrostatic, whereas other contributions are inductive and dispersion forces. It is the latter forces that impose the directionality in the direction of the acceptor lone pair. Hydrogen bonding plays a very important role in biology, being responsible for the structural organization of proteins and other biomolecules. Actually, the existence of life on Earth is a direct consequence of the hydrogen bonding present in water and the wide temperature range over which water remains liquid (0–100°C). Also, the tetrahedral three-dimensional (3D) structure of ice and the local structure of liquid water is a consequence of hydrogen bonding. We will come back to hydrogen bonding in discussions on protein folding, DNA, and self-assembly

(Volume III, Chapters 2–4). In the gas phase, the optimal H-bond strength between water molecules is ~23 kJ mol^{-1}. The H-bonding in condensed phases of water is cooperative (nonadditive), and the strength of H-bonding increases with increasing number of water molecules, as this increases the polarization of the O—H bonds. This is observed as an increase in the average dipole moment (μ) per water molecule, which increases from 1.85 D (isolated H_2O) to 2.4–2.6 D (liquid H_2O at 0°C).

In Table 7.5, we summarize what we have learned about intermolecular forces thus far. We describe the type of interaction, give a model and an example, and list the distance dependence of the energy.

The Lennard-Jones Potential

Having introduced the equations for repulsive and attractive intermolecular forces we are now ready to combine them to describe the exact shape of the curve in Figure 7.2 for the total potential energy of interaction between two neutral atoms or polar molecules as a function of the distance between them. In combining forces, one typically groups the attractive forces between polar and nonpolar molecules in the so-called *van der Waals force*. This refers to all electrostatic induction and dispersion interactions between polar and nonpolar molecules, varying as r^{-6}. From Table 7.3, van der Waals forces include Keesom (dipole-dipole), Debye (dipole-induced dipole), and London dispersive interactions (induced dipole-induced dipole). These short-range forces (n = 6) can all be expressed as:

$$U(r) = -\frac{C}{r^6} \tag{7.19}$$

The most important component of the van der Waals force, as seen earlier, is the London term.

Table 7.5 Intermolecular Energy Dependence on Distance for Various Types of Interactions

Type of Interaction	Model	Example	Dependence of Energy on Distance
(a) Charge-charge Longest-range force; nondirectional	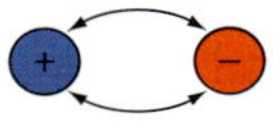	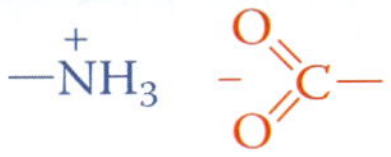	1/r
(b) Charge-dipole (fixed) Depends on orientation of dipole	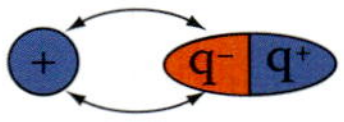		$1/r^2$
(c) Dipole-dipole Depends on mutual orientation of dipole	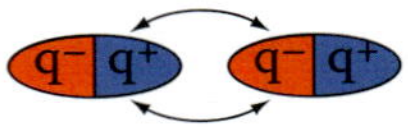	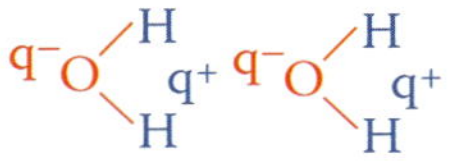	$1/r^3$
(d) Charge-induced dipole Depends on polarizability of molecule in which dipole is induced	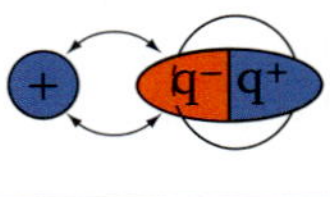	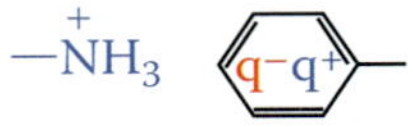	$1/r^4$
(e) Dipole-induced dipole Depends on polarizability of molecule in which dipole is induced	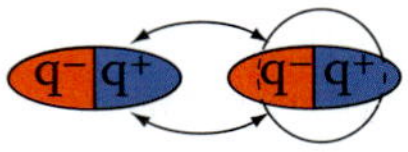	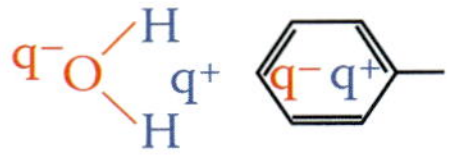	$1/r^6$
(f) Dispersion-London forces Involves mutual synchronization of fluctuating charges		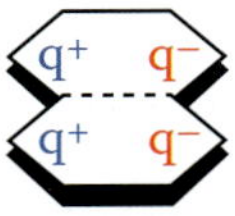	$1/r^6$
(g) Fermi repulsion Occurs when outer electron orbitals overlap	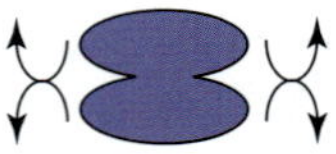	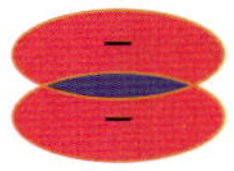	$1/r^{12}$
(h) Hydrogen bond Charge attraction + partial covalent bond	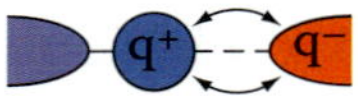		Length of bond fixed

Regardless of other interactions found within a complex system of atoms and molecules, there will almost always be a contribution from van der Waals forces. Taking into account both repulsive and attractive interactions, a good approximation to the total potential energy of interaction between two neutral atoms or polar molecules is given by the Lennard-Jones potential:

$$U(r) = +\frac{B}{r^{12}} - \frac{C}{r^6} \tag{7.20}$$

The C/r^6 term is the attractive van der Waals term, also known as the *dispersive term* (Equation 7.19). The term B/r^{12} describes the hard-core repulsion, also known as *Fermi repulsion*, and stems from the Pauli exclusion principle. Atoms are treated here as hard spheres with diameter σ [in the extreme; $U(r) = \left(\frac{\sigma}{r}\right)^{\infty}$]. This leads to a potential expression of the following form:

$$U(r) = 4\varepsilon\left\{\left(\frac{\sigma}{r}\right)^{12} - \left(\frac{\sigma}{r}\right)^{6}\right\} \tag{7.21}$$

as illustrated in Figure 7.11. The force between the two molecules is derived from the Lennard-Jones potential as:

$$F(r) = -\frac{dU}{dr} = 24\varepsilon\left\{2\left(\frac{\sigma^{12}}{r^{13}}\right) - \left(\frac{\sigma^{6}}{r^{7}}\right)\right\} \tag{7.22}$$

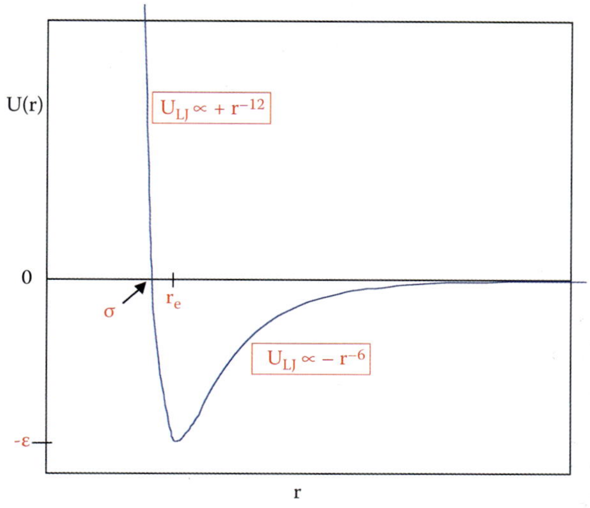

FIGURE 7.11 The Lennard-Jones potential.

In Equation 7.21, ε is the depth of the energy well and σ is the separation at which U(r) = 0. The equilibrium separation, i.e., the position of the m inimum in $U(r)$, occurs at $r_e = 2^{1/6} \times \sigma$ [$U(r) = -\varepsilon$]. The *van der Waals' radii* (σ) of atoms, which is the size of an atom or the preferred distance atoms will pack to if no other significant interactions occur, may be calculated from this equation. The increase in ε when going from lighter to heavier rare gases reflects the increase in the attractive dispersion forces because of the higher polarizability that occurs as a result of the lower effective nuclear charge experienced by the outer electrons in the heavier rare gas atoms. As molecules are larger and (usually) more polarizable than rare gas atoms, typical binding energies caused by dispersion forces are of the order $\varepsilon \approx 10$ kJ mol^{-1} per intermolecular interaction. The interactive attraction forces are the derivative of the van der Waals term, and these contributions (polar or nonpolar) all decrease as $1/r^7$. In all cases, substituting numerical values in Equation 7.22, these forces do not exceed a range of a few nanometers.

Example 7.1:

The magnitude of the dispersion energy for two methane molecules separated by 0.3 nm is approximately −4.7 kJ mol^{-1}, i.e., $\varepsilon \approx 4.7$ kJ mol^{-1}.

Calculating Interaction Forces between Surfaces and Particles

Introduction

Thus far, we have mainly discussed intermolecular forces at work between various types of molecules. Now we turn our attention to the detailed nature of the attractive and repulsive forces at an interface between a liquid and a solid. Based on this knowledge, we will be ready to introduce the electrode/liquid interface and the electrochemical principles that enable electrochemical sensors.

Van der Waals Forces on Solid Surfaces

Van der Waals forces for molecules, we saw, vary as $1/r^7$ (Equation 7.22), with r being the distance between them. In 1937, Hamaker calculated the

Table 7.6 Hamaker Constants H for Various Materials in 10^{-20} J

Material	Vacuum	Medium Water	n-Dodecane
6H-SiC	25	13	1
tetra-ZrO_2	20	8.8	6.8
β-Si_3N_4	18	7.0	5.0
α-Al_2O_3	15	5.2	3.6
Y_2O_3	13	4.0	2.6
MgO	12	3.2	1.9
$MgAl_2O_4$	13	3.5	2.1
SiO_2	6.5	0.83	0.15

Source: Bergström, L. 1997. *Adv Colloid Interface Sci*, 70:125.[64]

total van der Waals interaction force between two spherical particles (an agglomeration of molecules/atoms) by adding contributions for each atom of the interacting particles and found*:

$$F(r) = -\frac{H\sigma}{12L_0^2} \quad (7.23)$$

where σ = particle radius

L_0 = equilibrium separation distance of r, the distance where repulsive quantum forces equal attractive van der Waals forces (for liquids, 0.16 nm; for solids, 0.3 nm)

H = Hamaker constant

A typical value of the Hamaker constant *H* for SiO_2 in water is 0.83×10^{-20} J. For the Hamaker constant of some selected materials in different media, see Table 7.6.

The Hamaker constant *H* is defined as:

$$H = \pi^2 C\rho_1\rho_2 \quad (7.24)$$

where ρ_1 and ρ_2 are the atom densities in the two spherical particles and *C* is the coefficient in the interaction potential energy equation.

Based on Equation 7.23, the energy of adhesion between two spheres is:

$$U(r) = -\int F(r)dr = \int -\frac{H\sigma}{12L_0^2}dr = -\frac{H\sigma}{12L_0} \quad (7.25)$$

where $r = L_0$ is the equilibrium distance.

Example 7.2:

Considering only van der Waals attraction, what is the force and energy of adhesion between two particles of radius 1 μm with $H = 10^{-19}$ J? (L_0 = 3 Å).

The force of adhesion is calculated as:

$$F(r) = -\frac{H\sigma}{12L_0^2} = -\left(\frac{10^{-19}\,J \times 10^{-6}m}{12 \times (3 \times 10^{-10})^2 m^2}\right) = -9.26 \times 10^{-8}N$$

and the energy of adhesion is given by:

$$U(r) = -\int F(r)dr = \int -\frac{H\sigma}{12L_0^2}dr = -\frac{H\sigma}{12L_0}$$

$$= \left(\frac{10^{-19}\,J \times 10^{-6}m}{12 \times (3 \times 10^{-10})m}\right) = -28 \times 10^{-18}\,J = -6800\,kT$$

The theoretical energy of adhesion between two flat plates (van der Waals approach) is given as:

$$U(r) = -\frac{H}{12\pi L_0^2} = -2\gamma \quad \text{or}$$
$$\frac{H}{24\pi L_0^2} = \gamma \quad (7.26)$$

where γ is the surface energy (for solids) or the surface tension for liquids (when considering the interface between two liquids), and L_0 is the equilibrium distance between the two planes [the surface tension $\gamma = (\partial G/\partial A)$ is the change of Gibbs free energy per area change, i.e., a measure of the energy required to produce a unit area of new surface]. Energy of 2γ is necessary to separate two planes bound by van der Waals forces. The strength of the van der Waals interaction is essentially a function of the surface area of contact. The larger the surface area, the stronger the interaction will be. For solids, the surface energy depends on the exact crystal structure of the surface, and as a consequence the surface tension might be anisotropic. Surface energy of solids is not only a material property, it also is significantly influenced by the environment and may change dramatically based on the history

* This is a simplified equation in which we assume that the distance between particles is significantly smaller than the particle radius (σ << *L*).

Table 7.7 Surface Energies of Different Materials

Material	Surface Energy (mJ/m²)
Teflon	20
Paraffin wax	26
Polypropylene	28
Polyethylene	36
Lead iodide	130
Silica	462
Iron	1360
Mica	4500

of the surface. Factors influencing surface energy include organic contamination, roughness/edges/high-energy sites, wetting agents, humidity, and the crystallography. The contact angle of water on silica changes from approximately 10° to 45° with heat treatment, and the contact angle of water on TiO_2-silicone thin film changes from 12° to ~3° on irradiation with UV. The surface energies of some different materials are listed in Table 7.7. The surface of an object determines its optical appearance, stickiness, wetting behavior, frictional behavior, and chemical and electrochemical reactivity.

From Equation 7.26, the van der Waals force between two planes is:

$$F(r) = \frac{H}{6\pi L_0^3} \tag{7.27}$$

The van der Waals interactions potentials between surfaces, molecules, and particles are summarized in Table 7.8.

Perhaps it is somewhat troubling for the observant reader that the potentials for sphere and plate interactions all have *n* values less than 3(n = 1); did not we say earlier that intermolecular forces exist only for n ≥ 3? To explain this, we consider the sphere-plane interaction sketched in Figure 7.12. Integrating the interaction of every element on the sphere with the plate, we obtain:

$$U(r) = -\int_0^\infty \frac{\pi\rho C}{6(L+z)^3} 2\pi\sigma\rho z dz \quad \text{or}$$
$$U(r) = -\frac{\pi^2\rho^2 C\sigma}{6} = -\frac{H\sigma}{6L_0} \tag{7.28}$$

We recognize that the bulk material contribution to the interaction terms causes *n* to collapse to 1,

Table 7.8 van der Waals Interaction Potentials for Various Configurations

Configuration	Schematic	Expression of the Interaction Energy U(r)
Molecule-molecule		$U(r) = -\frac{C}{r^6}$
Molecule-plane		$U(r) = -\frac{\pi C\rho}{6L_0^3}$
Sphere-plane	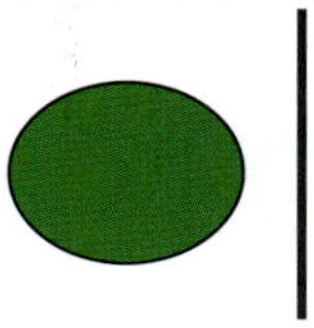	$U(r) = -\frac{H\sigma}{6L_0}$
Sphere-plane	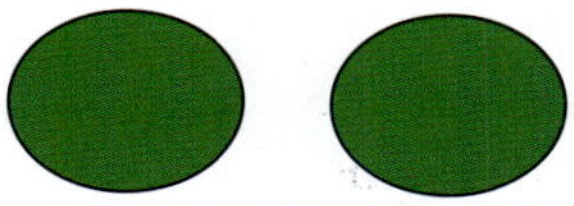	$U(r) = -\frac{H\sigma}{12L_0}$
Plane-plane		$U(r) = -\frac{H}{12\pi L_0^2} = -2\gamma$

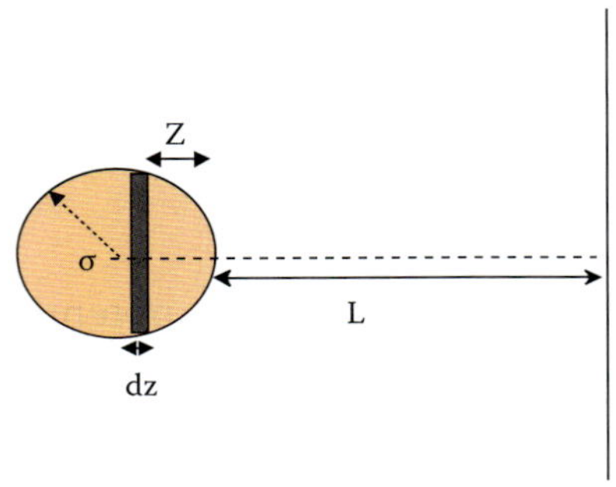

FIGURE 7.12 The van der Waals interaction energy, $U(r)$ for a sphere/plane system.

FIGURE 7.13 Solid/liquid interface as a fractal.

as a result of the integration of a large number of individual sphere elements with the plate. This also explains the longer range of van der Waals forces, tens of nanometers instead of a few nanometers, when considering solid planes and spheres instead of molecules.

In large objects with small surface area A to volume V ratio (A/V), the physical and chemical properties are primarily defined by the bulk (inside), whereas in small objects with a large A/V ratio, the properties are strongly influenced by the surface.

Two important references to consult on intermolecular forces are Rigby et al.[1] and Stone.[2]

Electrostatic Forces at the Solid/Liquid Interface

Introduction

Most chemical sensors eventually involve a solid in direct contact with an analyte in its environment. Therefore, the better one can control the sensor surface, the better one can control that interface and thus say something more definite about the analyte. Here we will further elaborate on solid/liquid interfaces before we discuss any types of electrochemical sensors in detail. Liquid/solid interfaces are very complex, often involving fractals* (e.g., beach, trees, snowflakes; see Figure 7.13) rather than atomically sharp transitions. This implies, as we will see, that perfect sensor selectivity will be hard to achieve (too many different types of binding sites with different binding constants/selectivities).

There is a tendency for charged species to be attracted to or repelled from the solid/solution interface. This gives rise to a separation of charge, and the layer of solution with a different composition from the bulk solution is known as the *electrical double layer* (EDL), which is sketched for a metal in an aqueous solution in Figure 7.14.

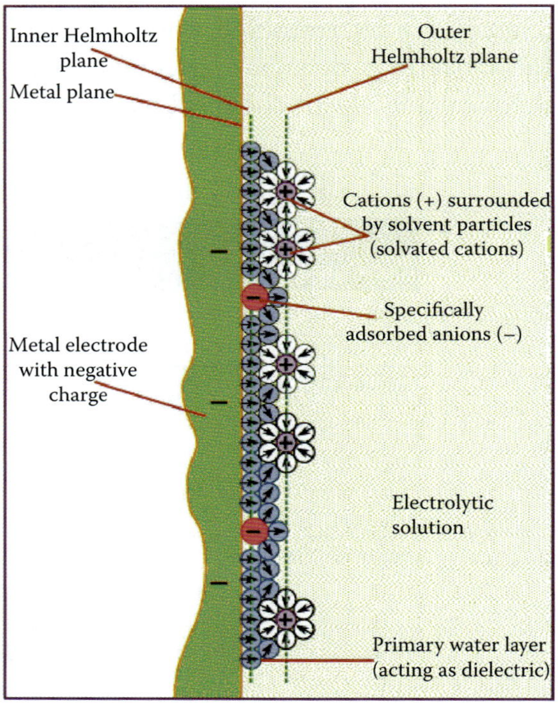

FIGURE 7.14 The electrical double layer (EDL) for a metal in an aqueous solution. The metal/electrolyte interface, showing fully and partially solvated ions. Completely solvated ions are located in the outer Helmholtz plane (OHP). Ions with weakly bound solvation shells (mostly anions) may lose part of their solvation shell and form a chemical bond with the surface (so-called *specific adsorption*) and are located in the inner Helmholtz plane (IHP).

* A fractal, we will see in Volume III, Chapters 7 and 9, maximizes surface area while minimizing the amount of work involved in reaching all points on that surface.

There are a number of theoretical descriptions of the structure of the double layer, including 1) the Helmholtz model (1879), 2) the Gouy-Chapman model (1910–1913), 3) the Stern model (1924), and 4) the Grahame model (1947).

The Electrical Double Layer Models

Helmholtz Model Helmholtz proposed that a layer of oppositely charged ions in the solution balances the metal surface charge. When a metal is in contact with an ionic solution, such as water (polar) with dissolved electrolytes, the metal surface has a high density of electrons, giving it a locally negative charge. The solvated positive ions (cations), such as H^+, Ca^{++}, and Mg^{++}, align themselves along the metal surface, producing a rigid electric double layer—no thermal motion of ions is considered. In the Helmholtz model, the potential dies off both sharply and linearly as we move away from the electrode, i.e., no interactions occur further away from the first layer of adsorbed ions (Figure 7.15). The electrolyte has no effect in this model.

The relation between the stored charge density ρ in the electrical double layer and the potential drop *V* across its plates is derived from the Poisson equation (slightly modified Equation 5.37):

$$\nabla \cdot E = \frac{\rho(r)}{\varepsilon_0 \varepsilon_r}$$

where E = electrical field

ε_o = permittivity of free space

ε_r = relative permittivity (which is dimensionless); relative permittivity effectively indicates the extent to which a dielectric has greater permittivity than a vacuum (ε_r is 1 for vacuum and very close to 1 for air but 78–80 for water at room temperature)

r = distance

ρ(r) = charge density

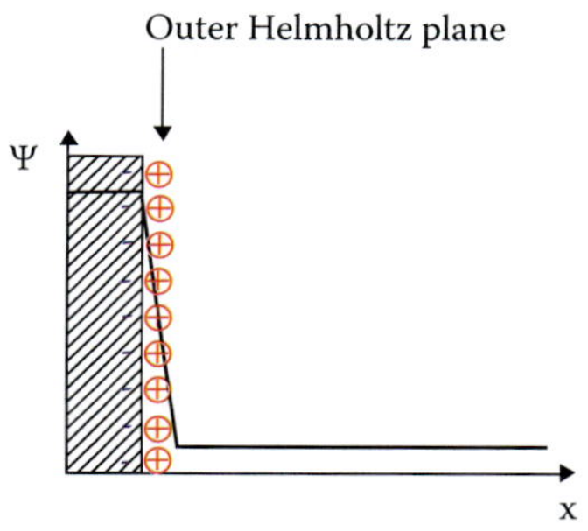

FIGURE 7.15 The Helmholtz model showing the outer Helmholtz plane (OHP).

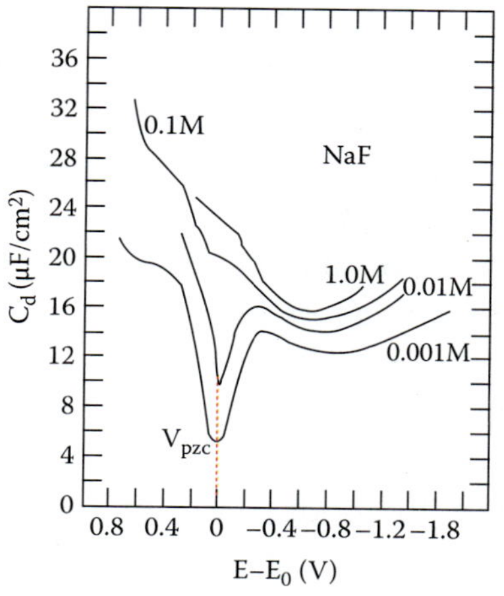

FIGURE 7.16 Differential capacitance as a function of applied voltage for a Hg electrode in NaF solutions at various concentrations.

The electrical field *E* is the differential of the electrical potential ψ [remember E = –∇V, where ∇ is the gradient operator (also called the *del vector operator*)]; thus, Equation 5.37 may also be rewritten as (replacing the potential *V* with ψ*):

$$\nabla^2 \cdot \Psi(r) = -\frac{\rho(r)}{\varepsilon_0 \varepsilon_r}$$

In the Helmholtz model E = ψ/d, or:

$$\rho(r) = \frac{\varepsilon_0 \varepsilon_r V}{d} \tag{7.29}$$

with *d* the distance between the plates. The differential capacitance, C_d, in this model is given as:

$$C_d = \frac{\partial \rho(r)}{\partial V} = \frac{\varepsilon_0 \varepsilon_r}{d} \tag{7.30}$$

The weakness of this model is that it predicts a constant differential capacitance, which contradicts experimental evidence gleaned from Figure 7.16, where we show the differential capacitance C_d as

* Note that in this electrochemistry section we use *E*, *V*, ϕ, and ψ interchangeably to symbolize a voltage or potential. Use of these different symbols arose in different application fields, and touching on all these application fields in this section we decided to retain the symbol most familiar to the practitioners in each of these fields.

a function of voltage and observe a strong voltage and electrolyte concentration dependence. At low electrolyte concentrations, the dip or U-shape in the C_d versus V curve occurs near the point of zero charge (pzc), V_{pzc}. The pzc of a solid surface (of an electrode or of a dispersion of colloidal particles) is the voltage or pH (pH_{pzc}) at which there are no net charges left at the solid's surface. The terms isoelectric point (IEP) and pzc are often used interchangeably.

The Helmholtz model only approximates the real picture illustrated in Figure 7.16 for very high electrolyte concentration and at very low temperatures where Brownian motion and diffusion are negligible; for most cases, however, we clearly need a more sophisticated model.

Gouy-Chapman In the Gouy-Chapman (G-C) model, because of the thermal motion of the ions in the solution, the population of positive charges (i.e., cations like H^+, Ca^{++}, and Mg^{++}) decreases exponentially with increasing distance from the metal surface (balance between Coulomb attraction and Brownian motion) (see Figure 7.17). The double layer is not compact anymore but of variable thickness with ions that are free to move. The model assumes a Poisson-Boltzmann (PB) distribution of the ions away from the surface and pictures ions as point charges that do not interact with each other. This model explains the effects of an applied potential and the nature of the electrolyte on the differential capacitance, C_d.

To introduce the notion of thermal motion of ions in an aqueous solution, a Maxwell—Boltzmann

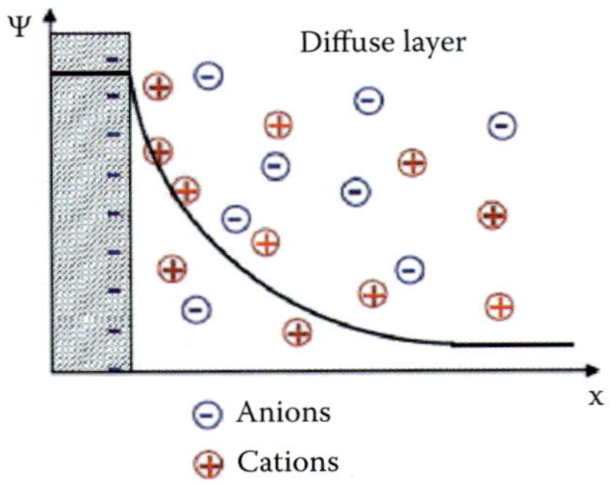

FIGURE 7.17 The Gouy-Chapman model. Thermal motion of the ions causes ions in the outer Helmholtz plane to diffuse away from the surface.

distribution (see Chapter 3, Equation 3.12) is introduced in the Poisson's equation to obtain the Poisson-Boltzmann distribution.

The number of ions at a point x close to the surface, where the field is $\psi(x)$, relative to a point far from the surface, where the potential is zero [$\psi(\infty) = 0$], in a Boltzmann distribution, is given by:

$$\frac{n_i(x)}{n_i(\infty)} = \exp\left(\frac{-w_i}{kT}\right) = \exp\left(\frac{-z_i e\psi(x)}{kT}\right) \quad (7.31)$$

where $n_i(x)$ is the number of ions at x, and w_i is the work required to bring an ion from an infinite distance away from the surface to the point x, i.e., $z_i e\psi(x)$ with z_i the charge on ion i. Combining the Boltzmann equation with the Poisson equation results in the Poisson-Boltzmann expression, whose solution is the potential profile $\psi(x)$:

$$\frac{d^2\psi}{dx^2} = -\frac{1}{\varepsilon_r\varepsilon_0}\sum_i z_i e n_i(\infty)\exp\left(\frac{-z_i e\psi}{kT}\right) \quad (7.32)$$

When dealing with a single symmetrical electrolyte ($|+z| = |-z|$), the PB equation has only two terms: one for the $+z$ and one for the $-z$ ions:

$$\frac{d^2\psi}{dx^2} = -\frac{zen(\infty)}{\varepsilon_r\varepsilon_0}\left[\exp\left(\frac{-ze\psi}{kT}\right) - \exp\left(\frac{ze\psi}{kT}\right)\right] \equiv \frac{2zen(\infty)}{\varepsilon_r\varepsilon_0}\sinh\left(\frac{ze\psi}{kT}\right) \quad (7.33)$$

where z = charge of the symmetrical electrolyte
n = electrolyte concentration (number of ions/m^3)
ε_r = dielectric constant of medium
ε_0 = permittivity of vacuum (F/m)
k = Boltzmann constant (J/K)
T = temperature (K)
e = elementary charge (C)

The two exponential terms in Equation 7.33 combine to form a hyperbolic sine function, and although this differential equation is readily solved, it is helpful to make a further assumption, i.e., that the potential is sufficiently low so that the sinh function may be replaced by its argument, i.e., for low potentials

$|z\psi| \leq 25$ mV, sinh y ≈ y (this is accurate whenever the argument < 1), and it follows that:

$$\frac{d^2\psi}{dx^2} = \frac{2z^2e^2n(\infty)}{\varepsilon_r\varepsilon_0 kT}\psi = \kappa^2\psi;$$

$$\kappa = \text{"Debye parameter"} = \sqrt{\frac{2z^2e^2n(\infty)}{\varepsilon_r\varepsilon_0 kT}} \quad (7.34)$$

This is called the *Debye-Hückel approximation*. The resulting differential equation may be solved using the following boundary conditions: at x = 0 (the surface), $\psi = \psi_0$ and at x = ∞ (in the bulk of the solution), ψ = 0. This yields an exponential function for ψ(*x*), with a decay constant made up of a cluster of constants captured by κ and referred to as the *Debye parameter*. It describes the decay length [(κ^{-1}) or the Debye length in units of length]:

$$\psi = \psi_0 e^{-\kappa x} \quad (7.35)$$

The potential profile ψ(*x*) is illustrated in Figure 7.18.

Once the potential profile is known, it may be substituted into the Boltzmann equation (7.31) to yield the ion concentration profiles:

$$C_i(x) = C \exp\left[-\frac{z_i e}{kT}\psi_0 \exp(-\kappa x)\right] \quad (7.36)$$

The concentration profiles for both cations and anions in the case of C = 10^{-2} M NaCl and ψ_0 = 25 mV are shown in Figure 7.19. The location of x = κ^{-1} is marked, and this is regarded as the "thickness of the double layer."

The effect of the electrolyte concentration on the potential ψ as a function of distance from the surface for two different concentrations is illustrated in Figure 7.20.

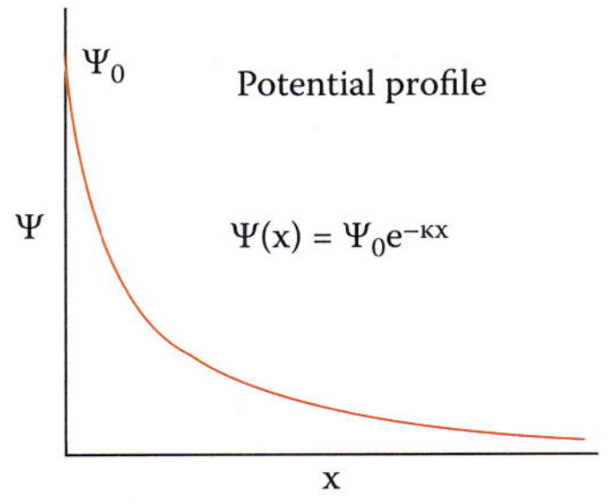

FIGURE 7.18 Potential profile in the double layer.

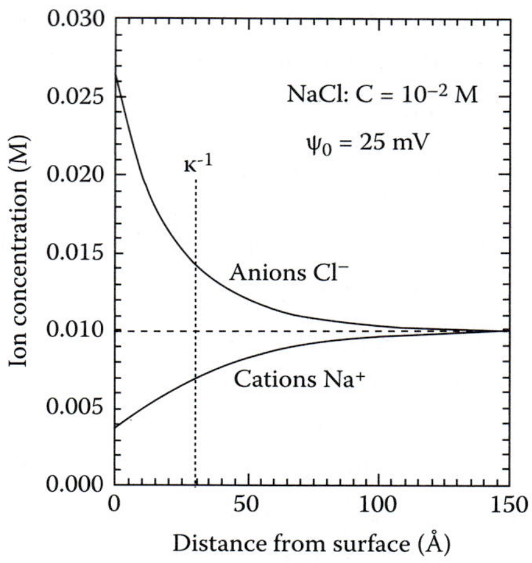

FIGURE 7.19 Concentration profile for cations and anions with ψ_0 = 25 mV and C = 10^{-2} M.

The range of values for the Debye length in water at 25°C can be evaluated from $\kappa^{-1} \approx \frac{0.30}{z\sqrt{C(M)}}$ nm [where *C*(*M*) is the concentration in moles], which give values of $0 < \kappa^{-1} < 1000$ nm. The lower the electrical conductivity of a solution, the higher the Debye-Hückel length, although it is hard to make x = κ^{-1} much longer than a few tens of nanometers (value at 1 μM of salt in water). This is because higher-resistivity water samples are hard to prepare.

The Debye-Hückel length for a multivalent electrolyte is given by:

$$\kappa = \sqrt{\frac{e^2}{\varepsilon_r\varepsilon_0 kT}\sum_{i=1}^{n} C_i z_i^2} \quad (7.37)$$

where z_i = electrolyte valence
C_i = ion concentration (number of ions/m³)

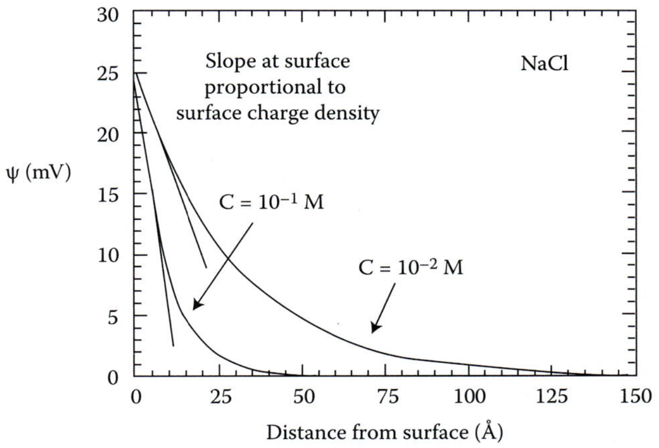

FIGURE 7.20 Potential as a function of distance from the surface for two different concentrations with ψ_0 = 25 mV.

ε_r = dielectric constant of the medium
ε_0 = permittivity of vacuum (F/m)
k = Boltzmann constant (J/K)
T = temperature (K)
e = elementary charge (in C)

Higher-valency ions are more effective in screening surface charge; therefore, a 3-3 electrolyte has a shorter Debye length at the same concentration than a 1-1 electrolyte.

Whereas the Helmholtz model represents the solid/electrolyte interface well at the high concentration limit, the Gouy-Chapman (G-C) model best represents the high dilution limit. Because ions are regarded solely as point charges, the G-C model permits them to pack without limit into the layer adjoining the surface. Because ions are characterized only by their charge (i.e., they have no chemistry), they cannot specifically adsorb to surfaces in excess of the charge required for neutralization, nor can they adsorb to surfaces of the same charge. In the G-C model, the differential capacitance C_d (Equation 7.30) varies with potential in a U-shape as illustrated in Figure 7.21. Here, we see that the differential capacitance versus voltage takes on at least one aspect found in the experimental data presented in Figure 7.16, i.e., a U-shaped minimum at the *pzc*, V_{pzc}. The G-C model still leads to an unlimited increase in differential capacitance C_d with applied bias, although again in violation of experimental evidence. A yet more sophisticated model is needed.

Stern Model The Stern model combines the concepts of the Helmholtz and G-C models. Based on minimizing the free energy at the interface, one obtains a double layer with the positive ions of a finite size closest to the metal surface constrained into a rigid outer Helmholtz plane (OHP), whereas outside this plane, the positive ions are dispersed as in the G-C model (see Figure 7.17). The monolayer of ions in the OHP (also the "Stern layer") directly adjacent to the surface generally effects partial neutralization of the surface charge. The remainder of the surface charge is neutralized in a G-C diffuse layer adjacent to the Stern layer. The Stern layer is assumed to reside between the surface and a plane (the Stern plane), passing through the centers of the first layer of hydrated ions, a distance δ from the surface (OHP). The distance δ is generally 2–3 Å, and the potential at this location is called the *Stern potential* or ψ_H. The Stern potential is the effective surface potential. The differential capacitance C_d (see Equation 7.30) is now the sum of two series capacitors, i.e., C_H, the voltage-independent capacitance of the charge at the OHP, and C_D, a truly diffuse charge, or:

$$\frac{1}{C_d} = \frac{1}{C_H} + \frac{1}{C_D} \tag{7.38}$$

With large ion concentrations or at large polarizations, C_D is large, and the voltage-independent C_H dominates the differential capacitance. Typical double-layer capacitance (C_H) for a metal is 10–40 μF/cm². At low concentrations, C_D varies with potential in a U-shape and so does C_d, as illustrated in Figure 7.22. Here we see that the differential capacitance versus voltage starts to look more like the experimental data presented in Figure 7.16.

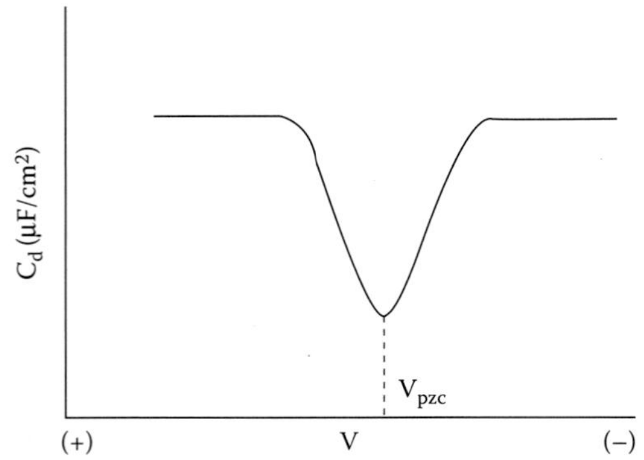

FIGURE 7.21 Variation of the differential capacitance C_d with voltage *V*, showing the minimum at the point of zero charge, V_{pzc}.

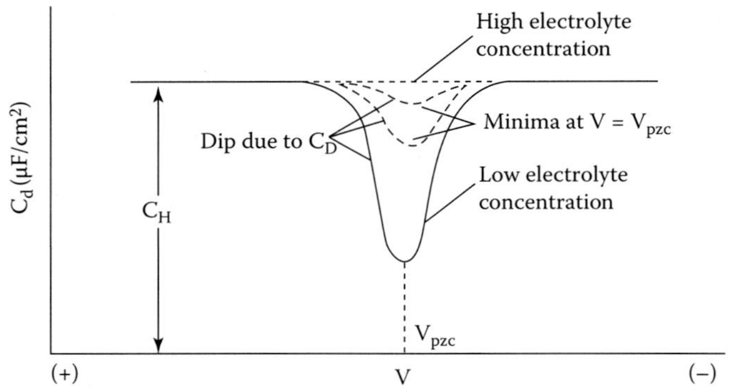

FIGURE 7.22 Differential capacitance versus applied voltage according to the Stern-Gouy-Chapman models.

Grahame Model The Grahame model further refines the Stern model. It assumes that specifically adsorbed ions are desolvated and may approach the electrode surface closer than the outer Helmholtz plane (OHP), and thus feel a greater potential, than the solvated ions. In this model, an inner Helmholtz plane (IHP) passes through the center of specifically adsorbed ions, and an OHP passes through solvated and nonspecifically adsorbed ions (see Figure 7.14). In the case of a metal immersed in pure water, the electrons within the metal cause the polar water molecules to adsorb to the surface and orient themselves so as to create two thin planes of positive and negative charge. Most materials spontaneously develop a thin native oxide coat (e.g., SiO_2 on Si) and surface charges when exposed to air, or even more so when submerged in an electrolyte. In the case of glass or Si (with its native SiO_2 film), the Si-OH (silanol) groups are partially ionized, and depending on the pH, the surface becomes more or less negatively charged. The adsorbed charges and ionized charges are located in the IHP, and the compensating solvated cations are in an opposing rigid OHP and the diffuse G-C layer. Both the IHP and OHP layers are very narrow (in the range of atomic dimensions). The diffusion layer is larger and extends from the OHP to the bulk (~100 Å in a solution >10^{-2} M). With the Grahame model, shifts in the potential dip toward more positive or more negative values than the pzc can be explained.

In Figure 7.23, we illustrate three example mechanisms that generate surface charges at the solid/liquid interface: ion adsorption (e.g., Ca^{2+} onto a lipid layer), ion dissociation of surface groups (e.g., solid-COOH → solid $-COO^- + H^+$), and isomorphous substitution. When a surface is fully ionized, a charge occupies about 0.5 nm^2.

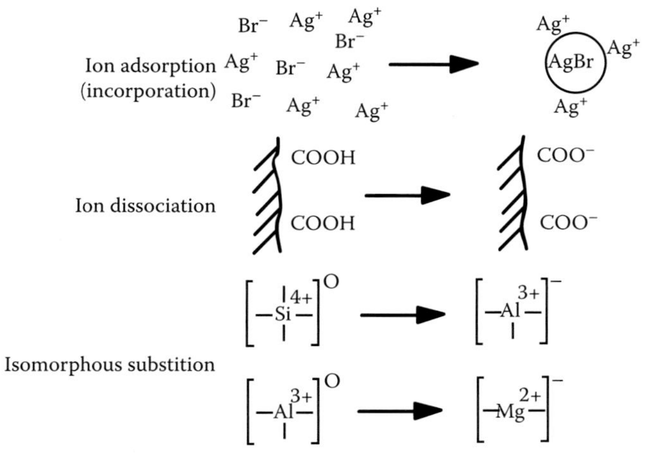

FIGURE 7.23 Examples of surface charges.

The Zeta Potential and Electrokinetic Phenomena When a liquid moves over a solid surface, a thin layer of liquid along that solid surface stays motionless. Liquid movement begins only at a shear plane in the diffuse double layer at a distance, x_0, away from the wall. The shear plane (slipping plane) is an imaginary surface separating the thin layer of liquid bound to the solid surface that shows elastic behavior different from the rest of the liquid showing normal viscous behavior. Adsorbed species are rigidly held inside the shear plane while outside of it there is free movement. The shear plane can therefore roughly be associated with the outer Helmholtz plane (OHP). The value of the electric potential at the shear plane with respect to the bulk solution is called the *electrokinetic* or *zeta potential* (ζ). The electrokinetic potential ζ—from which all electrokinetic effects (Figure 7.24) can be derived—represents the value of the electrostatic potential, ψ, at x_0. Its value at any point is determined from the profile of the electrostatic potential as a function of distance from the wall into the moving fluid. The thickness δ of the diffuse layer is defined as the distance from the immobile OHP or Stern layer to a point at which the electrostatic potential has dropped to 37% of the zeta potential. The amount and type of ionic species in the solution greatly influence the zeta potential; for example, the larger the concentration of added indifferent electrolyte in the moving solution, the smaller the ζ (see Figure 7.25). The electrokinetic effects that are controlled by the zeta potential are electrophoresis, electro-osmosis, streaming potential, and sedimentation potential (see Figure 7.24). As listed in Figure 7.24, the movement of charged particles in an electric field is termed *electrophoresis*. When the charged solid surface is fixed and the electric field causes a movement of the liquid, this is called *electro-osmosis*. Forcing a liquid through a capillary or porous plug induces a difference of electric potentials called *streaming potential*, and forced movement of charged solid particles in a liquid,

- Electrophoresis: Movement of particles (e.g., ions or colloidal particles) in a stationary fluid by an applied electric field.
- Electro-osmosis: Movement of liquid past charged solid surfaces (or perhaps through a membrane) under the influence of an applied electric field. The effects are often studied in a fine capillary to maximize the solid surface area to liquid volume.
- Streaming potential: Creation of an electric field as a liquid moves past a stationary charged surface.
- Sedimentation potential: Creation of an electric field when a charged particle moves relative to stationary fluid due to the gravitational force.

Phenomenon	Electric Field	Solid Phase	Liquid Phase	Property
Electrophoresis	Applied	Mobile	Fixed	Mobility
Electro-osmosis	Applied	Fixed	Mobile	Pressure/ velocity
Streaming potential	Generated	Fixed	Mobile	Potential
Sedimentation potential	Generated	Mobile	Fixed	Potential

FIGURE 7.24 Electrokinetic effects controlled by the zeta potential.

e.g., as a result of gravitation, induces a difference of electric potentials, i.e., the *sedimentation potential*. To a certain approximation, all four of the electrokinetic effects are different manifestations of the same phenomenon. As a rough approximation, the electrophoretic mobility (the ratio of the velocity of particles to the field strength), induced pressure difference in electro-osmosis, streaming potential, and sedimentation potential are all proportional to the zeta potential (ζ). The size of particles that can be calculated from these experiments corresponds to the particle dimensions plus the double-layer thickness.

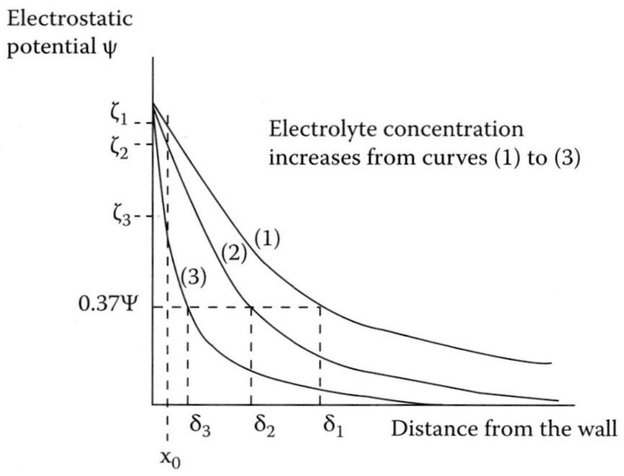

FIGURE 7.25 The electrokinetic potential, ζ, is the electrostatic potential, ψ, at a distance, x_0, from the surface, that is, the shear plane where the fluid movement begins. The concentration and type of ions in the solution influence the position of the curves and thus the electrokinetic potential. Ions and impurities at the surface influence these curves as well.

Marian Smoluchowski (1921) was the first to properly derive an equation to calculate the zeta potential from the electrokinetic mobility, via:

$$v = \frac{\varepsilon_r \varepsilon_0 \zeta}{\eta} E \quad \text{or} \quad \mu_E = \frac{\varepsilon_r \varepsilon_0 \zeta}{\eta} \tag{7.39}$$

where v = velocity

ε_r = media dielectric constant

ε_0 = permittivity of free space

ζ = zeta potential

E = electric field

η = medium viscosity

μ_E = electrophoretic mobility

In this simplified derivation, the electrostatic driving force is opposed by the frictional force only, and it is assumed that $\kappa a \gg 1$, where κ is the Debye parameter and a is the particle radius. In other words, the electrical double layer thickness is assumed to be much smaller than the particle radius. Only when there is a low zeta potential, large colloidal particles, and high ionic strengths does this equation produce good results. For example, with values of zeta potentials with practical meaning, i.e., $\zeta <$ 120 mV, the error is negligible when $\kappa a > 100$. The zeta potential of an aqueous solution in contact with glass can have a magnitude as high as 100 mV.[3]

Many commercial zetameters have the above Smoluchowski equation built into the software, usually together with the electric permitivity and viscosity data for water at 25°C.[4]

In the past, electro-osmosis and electrophoresis were the most popular methods to determine zeta potentials. Recently, electro-osmosis also has been studied for implementation as the pumping mechanism in commercial fully automatic instruments. However, pH, ionic strength, and surface impurities all affect the electrokinetic potential and conspire to make electro-osmosis a rather difficult pumping mechanism to control (see Chapter 6).

Electrophoresis and electro-osmosis potential were also covered in Chapter 6 on fluidics (both AC and DC electrokinetic phenomena are discussed). Here we dwell a bit further on the stability of colloidal dispersions as they constitute one of the key avenues toward making nanoparticles.

Colloidal Dispersions Colloidal dispersions are among the most important types of systems encountered both in nature and in the manufacture of a wide variety of products, and their electrokinetic properties are among the most important in effecting their characterization. Colloids also form one of the most important gateways toward bottom-up nanotechnology (i.e., nanochemistry). (See Volume III, Chapter 3 on the comparison of top-down and bottom-up approaches in nanotechnology.)

Colloids are mixtures in which one material is dispersed in another, with the dispersed material (the "dispersoid") divided into particles that range in size from approximately 1 nm to 10 μm. Colloids have many properties distinct from those of solutions of ordinary-sized molecules or from those of mixtures of macroscopic phases (e.g., slurries of visible particles). The dispersed material may either be dissolved (i.e., in the case of solutions of macromolecules), giving a lyophilic colloid, or they may constitute a separate phase, giving a lyophobic colloid. We limit the present discussion to lyophobic colloids, which may be further classified according to the types of phases that compose them, as shown in Table 7.9. We further limit our study to "sols," i.e., solid particles dispersed in liquids.

One of the features distinguishing the behavior of lyophobic colloids is their inherent instability. The stability ("shelf life") of a lyophobic colloid is kinetic in nature rather than thermodynamic, i.e., it depends on how slowly the inexorable processes of self-aggregation take place. Aggregation is often accompanied by sedimentation, which ultimately leads to macroscopic phase separation. The aggregation of colloidal particles is a spontaneous process because of the integrated effect of the attractive van der Waals interactions between the molecules in the different particles. As we saw above, these van der Waals forces consist chiefly of London or "dispersion" forces. Although these interactions are extremely short ranged, with a potential that varies as $-C/r^6$ (where C is a constant dependent on molecular properties, and r is the distance between the molecules; see Equation 7.18), the integrated effect between all the molecules of one colloid particle with those of a nearby colloid particle is much longer ranged, with a potential that varies as $-H\sigma/12L_0$ (Equation 7.25). In this expression, H is the Hamaker constant and L_0 is the equilibrium distance between repulsive and attractive forces between the particle surfaces. The constant H depends on C and the molecular density ρ in the particles (see Equation 7.24). These

Table 7.9 Lyophobic Colloids

Dispersed Phase	Dispersion Medium	Common Name	Examples
Liquid	Gas	Aerosol	Fog, fine sprays
Solid	Gas	Aerosol	Smoke, viruses, tear "gas"
Gas	Liquid	Microfoam	Shaving cream, beer foam
Liquid	Liquid	Emulsion	Milk, mayonnaise
Solid	Liquid	Sol	Paint, ink, clay
Gas	Solid	Solid microfoam	Foam plastics, marshmallow
Liquid	Solid	Solid emulsion	Butter, cheese, opal, pearl
Solid	Solid	Micro(nano) composite	Injection-molded objects

integrated attractive forces between colloid particles are referred to as *long-range van der Waals forces* (see Figure 7.26a). They are generally important when ordinary colloid particles are within 100 nm of each other and become large as the particle surfaces approach one another. The result is that colloid particles aggregate quickly as external effects, such as Brownian motion and/or bulk fluid motion (e.g., by stirring), bring them within range. In aqueous media, one effect that can slow down the aggregation process is the presence of electric charge on the surfaces of the particles as shown in Figure 7.26b. Solid surfaces in contact with water generally acquire a charge by one or a combination of the mechanisms we illustrated earlier in Figure 7.23 (e.g., oxide particles acquire charge via the pH-dependent ionization of their superficial –OH groups). These charged solid surfaces attract a cloud of ions of opposite charge (counterions) to their vicinity. As the particles approach, these clouds of like charge begin to overlap, setting up repulsion.

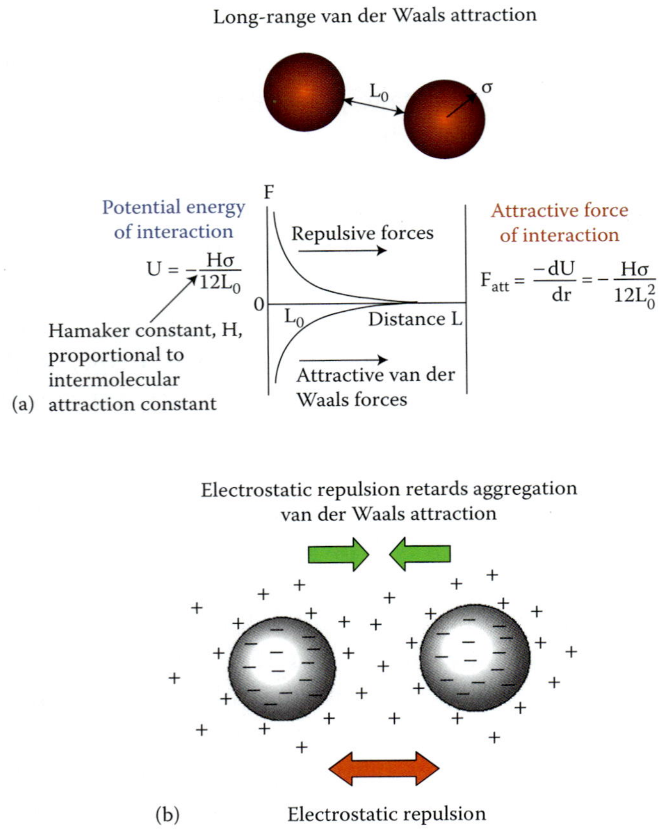

FIGURE 7.26 (a) The aggregation of colloidal particles is a spontaneous process because of the integrated effect of the attractive van der Waals interactions between the molecules in the different particles. These consist chiefly of London or "dispersion" forces, that is, forces caused by synchronous oscillations of the electron clouds of nearby molecules causing strong attraction between their temporal dipoles. (b) In aqueous media, one effect that may slow down the aggregation process is the presence of electric charge on the surfaces of the particles. Such charged surfaces attract a cloud of ions of opposite charge (counterions) to their vicinity. As the particles approach, these clouds of like charge begin to overlap, setting up repulsion.

When this electrostatic repulsion is put together with the long-range van der Waals attraction, a total interaction curve of the type shown in Figure 7.27 results. This adding of these forces was first carried out by Derjaguin and Landau in Russia and Verwey and Overbeek in the Netherlands in the 1940s in the so-called the *DLVO theory*. The curve that results when combining these forces has an energy barrier, and approaching particles must possess sufficient kinetic energy to overcome that barrier if they are to aggregate. If the barrier is large relative to *kT*, the probability of aggregating (the sticking probability) during a collision is low. In other words, the rate of aggregation of colloidal particles depends on the height of the DLVO barrier. For a given colloid (given Hamaker constant *H* and given particle radius σ), the electrostatic repulsion (barrier height) depends on the zeta potential and the double-layer thickness, which in turn depends on the ion concentration and valency of the electrolyte.

The stability of hydrophobic colloids thus depends on the zeta potential: when the absolute value of the zeta potential is above 50 mV, the dispersions are very stable because of mutual electrostatic repulsion. When the zeta potential is close to zero, the coagulation (formation of larger assemblies of particles) is very fast, causing fast sedimentation. For a 1-1 electrolyte and a given colloid, the effect of increasing the salt concentration on the barrier is shown in Figure 7.28. Small changes in concentration *C* have a large effect on the barrier height, and it is possible to identify, for any given case, a particular concentration (called the CCC, or *critical coagulation concentration*) at or above which rapid aggregation occurs. For a monovalent electrolyte, the CCC is generally between 50 and 200 mM. For counterions of higher

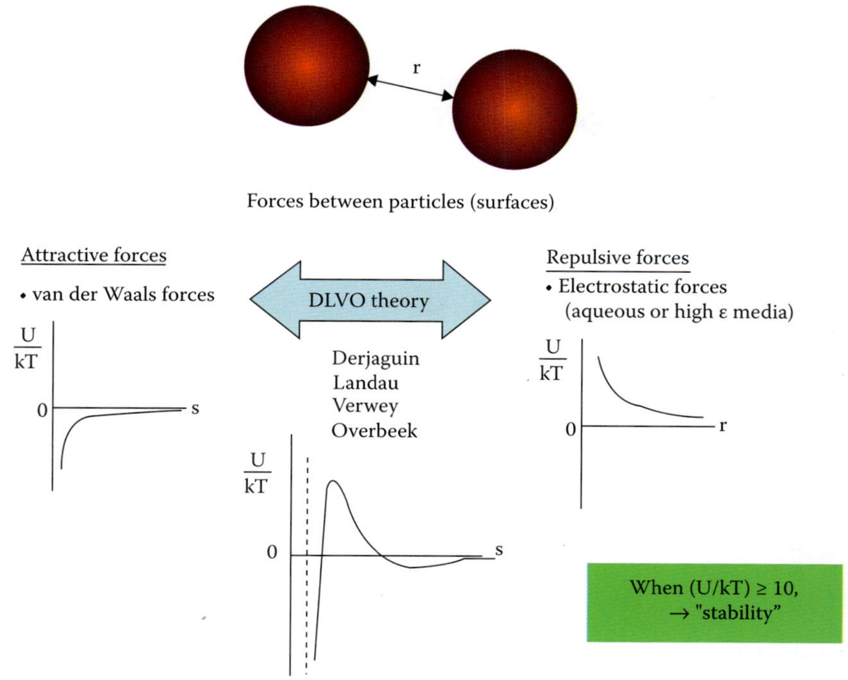

FIGURE 7.27 When electrostatic repulsion is put together with long-range van der Waals attraction, a total interaction curve of the type shown here results.

valence, the CCC decreases drastically, in accord with the Schulze-Hardy rule:

$$\mathrm{CCC} \propto \frac{1}{Z_i^6} \tag{7.40}$$

where Z_i is the valence of the counterions.

Casimir Forces

An important prediction of quantum theory is the existence of irreducible fluctuations of electromagnetic fields even in a vacuum and at 0 K, i.e., the existence of a so-called *zero point energy* (ZPE). The

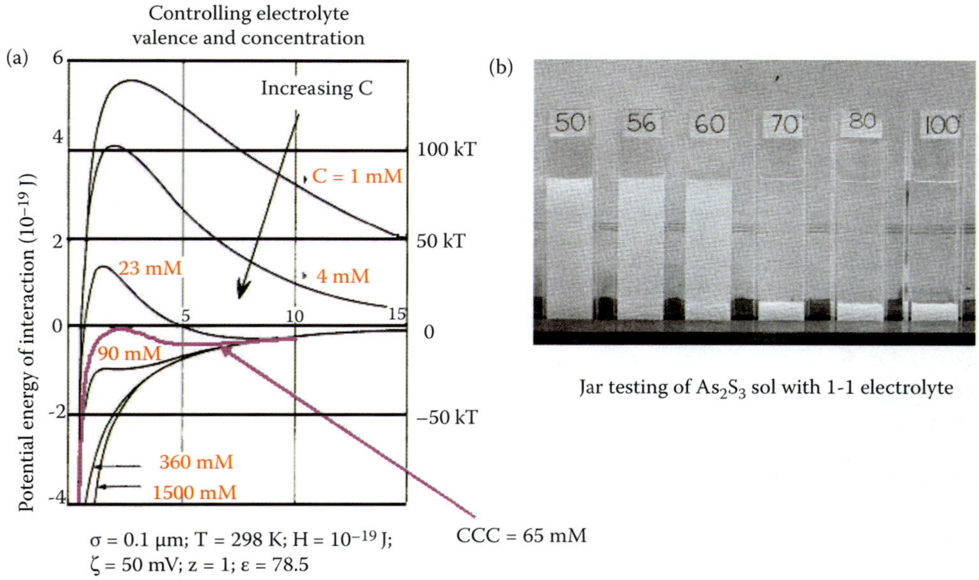

FIGURE 7.28 (a) For a 1-1 electrolyte and a given colloid, the effect of increasing the salt concentration on the barrier is shown. Small changes in *C* have a large effect on the barrier, and it is possible to identify, for any given case, a particular concentration (called the *CCC*, or *critical coagulation concentration*) at or above which rapid aggregation will occur. This can be determined experimentally by (b) "jar testing." For a monovalent electrolyte, the CCC is generally between 50 and 200 mM. For counterions of higher valence, the CCC drops drastically, in accord with the Schulze-Hardy rule, as shown.

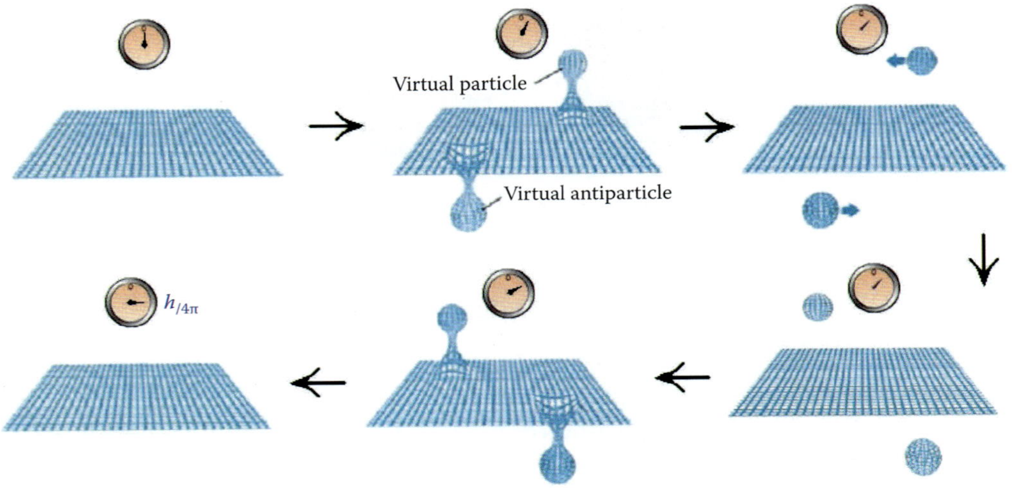

FIGURE 7.29 Virtual particles pop out of a vacuum and wander around for an undefined time and then pop back, thus giving the vacuum an average zero point energy. (From http://www.casimir.rl.ac.uk/zero_point_energy.htm.)

latter can be described as a turmoil of virtual particles that come in and out of existence, violating the energy-momentum conservation of the system for very short periods of time, as described by Heisenberg's uncertainty principle (Equations 3.106 and 3.107) (Figure 7.29). In Chapter 3, we saw how this zero point energy can be measured through the Casimir force, a small attractive force between two close parallel, uncharged conductive plates in a vacuum (Figures 3.69 and 7.30). This force, inversely proportional to the fourth power of the distance between the plates (see Volume III, Chapter 8 on actuators), comes about because the fluctuating virtual particles exert a "radiation pressure" on the plates, which on average is greater outside the plates than between them, as shown in Figure 7.30. The Dutch theoretical physicist Hendrik Casimir predicted this startling phenomenon in 1948 while he was working at Philips Research Laboratories in Eindhoven on, of all things, a problem concerning colloids!

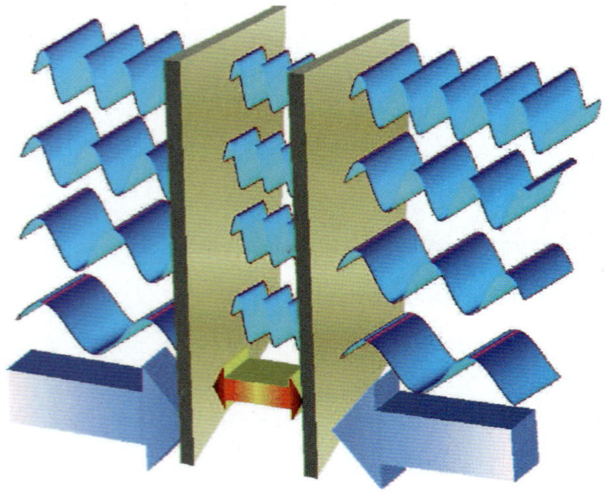

FIGURE 7.30 The Casimir force. Fluctuating virtual particles exert a "radiation pressure" on the plates that on average is greater outside the plates than between them. (Based on http://www.casimir.rl.ac.uk/zero_point_energy.htm.)

At the time of his discovery, Casimir was studying the properties of "colloidal solutions." The properties of such solutions, we know now, are determined by van der Waals forces. One of Casimir's colleagues, Theo Overbeek (of the DLVO theory fame), realized that the theory that was used at the time to explain van der Waals forces did not properly explain some of the experimental evidence on colloids. The prevailing theory Overbeek found lacking was the one by Fritz London, who, in 1930, described the interaction potential between two atoms as a result of fluctuations of the atomic electric dipole moments as $U \sim 1/r^6$ (see Equation 7.18). Overbeek asked Hendrick Casimir to investigate this problem, and by 1948, Casimir and Polder[5] had extended the London force to larger separations and found that retardation effects weaken this force, and that the attractive atom-surface force at large distances comes with a different power law than in the simpler short-range case. If the distance between fluctuating charges is large enough, it takes a finite amount of time for the electromagnetic field from one charge to reach the other (i.e., the speed of light *c* is finite). By the time the second charge responds to the field, the momentary charge configuration at the first position will have changed, and the charge fluctuations will fall out of step with each other. The strength of interaction is therefore always weakened, and its distance dependence is changed; for example, $1/r^6$ energy between small particles becomes $1/r^7$. This retarded van der Waals force is also known as the *Casimir-Polder force*.

Casimir noticed that this result could be interpreted in terms of vacuum fluctuations. He asked himself what would happen if there were two mirrors—rather than two molecules—facing each other in a vacuum. It was this work that led to his famous prediction of an attractive force between reflecting plates. In the late 1950s and early 1960s, Evgenii Lifshitz took into account the dielectric properties of the media and thermal fluctuations and developed a general theory for van der Waals type forces, of which the Casimir-Polder force is a limiting case.[6]

Although experimental evidence of an attractive force between neutral atoms and between neutral atoms and surfaces has existed for a long time, it was not until 1993 that Ed Hinds' group at Yale was able to measure that atom-surface force at larger distances or in the retarded or Casimir-Polder regime.[7]

Although the Casimir force is too small to be observed for plates that are several meters apart, it can be measured if the plates are within micrometers of each other. For example, we calculate in Volume III, Chapter 8 on actuators that two plates with an area of 1 cm^2 separated by a distance of 1 μm have an attractive Casimir force of about 10^{-7} N, roughly the weight of a water droplet that is half a millimeter in diameter. Although this force might appear small, at distances below 1 μm the Casimir force becomes the strongest force between two neutral objects. Indeed, at separations of 10 nm—about 100 times the typical size of an atom—the Casimir effect produces the equivalent of 1 atm. Although we do not deal directly with such small distances in everyday life, they are important in nano- and microscale structures (NEMS and MEMS). With the decrease in device dimensions, additional forces have to be considered that are normally neglected in macrosystems. For example, one of the principal causes of malfunctioning in surface micromachined MEMS devices, we discuss in Volume II, Chapter 7, is stiction, i.e., the collapse of movable elements onto nearby surfaces, resulting in their permanent adhesion. This can occur during fabrication, especially because of capillary forces, or during operation. The Casimir effect is potentially an important underlying mechanism causing this phenomenon. The problem is technologically important because it adversely affects the production yield on batch-fabricated devices.

An overview of recent measurements of the Casimir force is presented in Volume III, Table 8.21.

Electrochemistry

Introduction

Electrochemistry is the study of reactions taking place at the interface of a solid conductor and an electrolyte (an ionic conductor with ions as the mobile charge-carrying species). Using sensor examples, we introduce some of the most important concepts in electrochemistry, i.e., potentiometry, voltammetry, two- and three-electrode systems, Marcus theory of electron transfer, reaction rate- and diffusion rate-controlled electrochemical reactions, and ultramicroelectrodes (UMEs). In this same section, we also learn how different materials such as metals, semiconductors, solid electrolytes, and mixed conductors exposed to a solution offer the possibility for building different types of electrochemical sensors. As summarized in Table 7.10, different types of mobile charge carriers characterize these different materials. Metal electrodes (e.g., Pt) feature electrons as the mobile charge carriers, whereas semiconductors (e.g., n-Si) have electrons and/or holes to play that role. Solid electrolytes (e.g., LaF_3) possess mobile ions. Insulators (e.g., SiO_2), however, have no free charge carriers, whereas mixed conductors (e.g., IrO_x) have both ions and electrons.

We conclude this section by providing more electrochemical sensor examples, including ion-sensitive field-effect transistors (ISFETs), potentiometric

Table 7.10 Charge Carriers for Different Electrode Materials and Different Types of Sensors Based on Their Interface with a Solution

Material (Examples)	Charge	Type of Electrochemical Sensor
Metals (Pt, Au, etc.)	Electrons	Electrodes of the first kind sensor for redox species with the highest electron exchange current $i_{0,e}$
Semiconductors (GaAs, Si, etc.)	Electrons and holes	pH sensor for the least conductive semiconductors (wide bandgap), redox couple sensitive for the more conductive semiconductors (narrow bandgap)
Insulators (SiO_2, Al_2O_3, etc.)	No mobile charges	pH-sensitive surface in an ISFET
Ionic conductor (LaF_3, ZrO_2(s), $RbAg_4I_5$, sea water, etc.)	Ions as mobile carriers	For detection of ionic species with the highest ion exchange current $i_{0,i}$
Mixed conductor (Pd, plasma, IrOx, etc.)	H_2 in Pd [$H^+ + e^-$], IrO_x[$H^+ + e^-$],	Sensor application depends on the relative magnitude of electron exchange and ion exchange currents $i_{0,e}$ and $i_{0,i}$

and amperometric immunosensors, and glucose sensors.

Interface Processes

In electrochemistry, one studies reactions in which charged particles (ions or electrons) cross the interface between two phases of matter. Typically this involves a solid phase (the electrode) and a conductive solution (or electrolyte). These interface processes are represented as chemical reactions and are known generally as *electrode reactions*. Electrode reactions take place within the double layer and produce a slight unbalance of the electric charges in both the electrode and the solution. External manipulation of the interfacial potential difference affords an important way to exert control over an electrode reaction. The interfacial potential difference at a solid/electrolyte interface is limited to only a few volts at most. This does not seem like very much until one realizes that this potential spans only within a layer of water molecules and ions that attach themselves to the electrode surface—normally only a few atomic diameters thick. Thus, a very small voltage produces a very large potential gradient. For example, a potential difference of 1 V across a typical 10^{-8}-cm interfacial boundary layer amounts to a potential gradient of 100 million V/cm—a very significant electrical field and hence an important potential influence on the charge transfer reaction.

Charge transport in electrodes occurs via motion of electrons (or holes), and charge transport in the electrolyte occurs via motion of ions (positive and negative). With reference to Figure 7.31, it is easily recognized that for current to flow across the solid/liquid interface (in this case across a metal/electrolyte solution interface) an electrochemical reaction must occur. A process at the solid/electrolyte interface where charges (e.g., electrons) are transferred across this interface is called a *Faradaic process*. For example, an *oxidant* O (say, Ag^+) gets reduced in a *cathodic* reaction (reaction on the *cathode*) to become a *reductant* R (say, $Ag^+ + e^- \rightarrow Ag$):

$$O + ne^- \rightarrow R \qquad \text{Reaction 7.1}$$

An oxidant O is defined as an oxidizing agent, i.e., a reactant that oxidizes another reactant and which is itself reduced. A reductant R is a reducing agent, i.e., a reactant that reduces another reactant and which is itself oxidized. For a complete circuit, a *counter electrode* (CE) must also be present in the *electrochemical cell* of Figure 7.31, so that an *anodic reaction* (on the *anode*) may occur (e.g., $Cd{-}2e^- \rightarrow Cd^{2+}$):

$$R' - ne^- \rightarrow O' \qquad \text{Reaction 7.2}$$

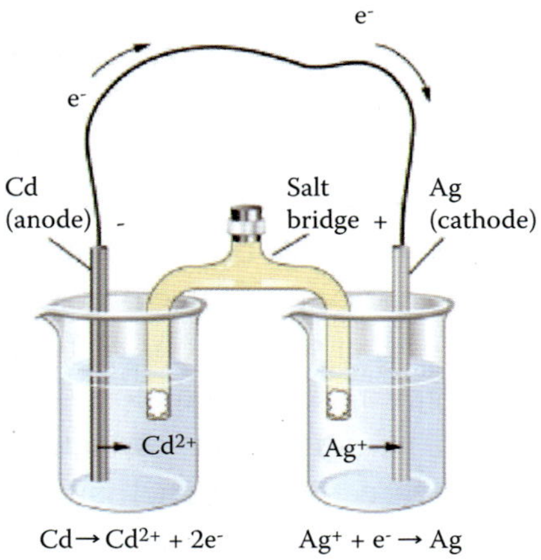

FIGURE 7.31 An electrochemical cell with an anode and a cathode and a salt bridge connecting the two half-cells.

Redox reactions (Reactions 7.1 and 7.2) are electron transfer processes where an oxidizing agent receives electrons, becoming a reducing agent; and a reducing agent gives off electrons, becoming an oxidizing agent. Oxidizing (O and O′) and reducing agents (R and R′) form so-called *redox couples* (ox + ne$^-$ ⇔ red, e.g., $Ag^+ + e^- \rightarrow Ag$). The *anode* of the electrochemical cell in Figure 7.31 is the electrode where the oxidation occurs, and the *cathode* is the electrode where the reduction takes place.

An *electrochemical cell* typically consists of two electronic conductors (also called *electrodes*) connected through an external wire and two *half-cells* filled with an ionic conductor (called an *electrolyte*) and a redox couple. A *salt bridge* is an ionic solution connecting the two half-cells (half-reactions) that completes the circuit. No further electrode reaction would take place if electrical neutrality were not maintained via this salt bridge. The salt bridge (also *Luggin capillary*) maintains electrical neutrality in the two half-cells by migration of ions through the porous material and also prevents mixing of the half-cell solutions.

If the anode and cathode solutions were allowed to mix, the chemical driving force would lead to the transfer of electrons directly between reactants rather than through the external circuit (wire, load). A salt bridge is often made by equipping a U-shaped glass tube with Vycor frits at both ends and filling it with a viscous, aqueous ionic solution that allows charge transport without permitting the two solutions to mix with each other. The solution may be an agar gel loaded with KCl or KNO_3. The K^+, Cl^-, and NO_3^- ions have similar mobilities and migrate in opposite directions with similar velocities. This minimizes the appearance of *junction potentials* in the bridge. Junction potentials occur at the interface of solutions with different ion concentrations as, for example, between the electrolyte solutions of a reference electrode and the sample. These potential differences are caused by the different mobilities of ions at the interface of the two solutions (Figure 7.32). There is no need for concern about junction potentials as long as they are kept small and constant during an analysis.

The salt bridge is one part of an electrochemical sensor that, we will see further below, is particularly cumbersome to miniaturize (see ISFET section).

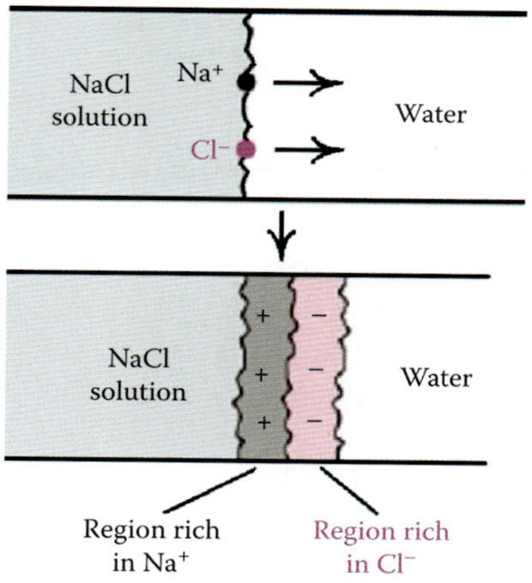

FIGURE 7.32 Ions diffuse from a region of high ionic strength to a region of lower ionic strength and will migrate (diffuse) at different rates, depending on their relative charge and size (*z*/*r*). This creates regions with higher net negative charge and regions of higher net positive charge at the interface (or junction). The difference in local charge represents some work (Δ*G* free energy), or chemical potential, called a *junction potential*.

At each electrode, an electrochemical reaction occurs. This reaction is called a *half-cell reaction* (because there are two electrodes in a typical cell at which reactions occur). There are two fundamental types of half-cell reactions: an oxidation reaction (in Figure 7.31, the oxidation reaction is $Cd \rightarrow Cd^{2+} + 2e^-$) and a reduction reaction (in Figure 7.31, the reduction reaction is $Ag^+ + e^- \rightarrow Ag$). In writing the overall reaction the charge must be conserved so that the number of electrons lost in oxidation equals the number of electrons gained in reduction:

$$Cd(s) + 2Ag^+(aq) \rightarrow 2Ag(s) + Cd^{2+}(aq)$$

Reaction 7.3

Reactions occur pairwise; one cannot have oxidation without reduction. Because all redox reactions occur pairwise, i.e., reduction and oxidation always occur at the same time, we cannot measure the cell potential for just one half-cell reaction. This means that we must establish a relative scale for cell potentials.

The standard hydrogen electrode (SHE) serves as the ultimate reference electrode (REF E) for calculating the potential of a single electrode. Hydrogen gas is bubbled at 1 atm into a 1 M hydrochloric acid solution at 25°C (298 K) with a platinum electrode (Pt black) providing the reaction surface (Figure 7.33).

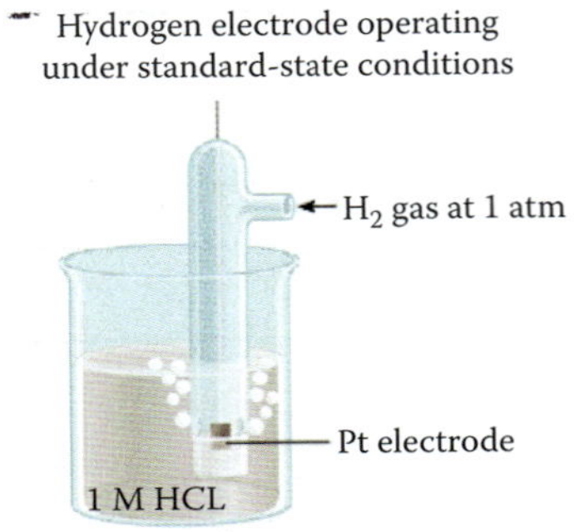

FIGURE 7.33 The standard hydrogen electrode (SHE).

Under these standard-state conditions, the potential for the reduction of H^+ is taken to be exactly zero:

$$2\ H^+\ (1\ M) + 2\ e^- \rightarrow H_2\ (1\ atm)\ E^0 = 0\ V$$

Reaction 7.4

The voltage associated with a reduction reaction at an electrode with all solutes present at 1 M and all gases at 1 atm is called the *standard reduction potential*. Now recall that the cell potential is the difference between two electrode potentials (the anode and the cathode):

$$E^0_{cell} = E^0_{red}(\text{cathode}) - E^0_{red}(\text{anode}) \qquad (7.41)$$

Table 7.11 lists a number of standard reduction potentials, all measured relative to the SHE. The redox reactions are spontaneous, in the forward direction as written in this table, if the standard EMF of the cell is positive [E^0_{red}(cathode) > E^0_{red}(anode)]. Comprehensive listings of E^0_{red} values for most half-cell reactions are available in the *Lange's Handbook of Chemistry* and in the *CRC Handbook of Chemistry and Physics*.

Because half-cell potentials are measured relative to SHE, they reflect the spontaneity of redox reactions relative to SHE, with more positive potentials resulting in more potent oxidants (oxidants that want to be reduced) and more negative potentials resulting in more potent reductants (reductants do not want to be reduced; they spontaneously oxidize). Using a table of standard reduction potentials, any species on the left of a given half-reaction will react spontaneously with any species appearing on the right of any half-reaction that appears below it, where reduction potentials are listed from highest and most positive to lowest and most negative (see Figure 7.34).

TABLE 7.11 Number of Standard Reduction Potentials All Measured Relative to the Standard Hydrogen Electrode (SHE)

Standard Potential (V)	Reduction Half-Reaction
+2.87	$F_2(g) + 2e^- \rightarrow 2F^-(aq)$
+1.51	$MnO_4^-(aq) + 8H^+(aq) + 5e^- \rightarrow Mn^{2+}(aq) + 4H_2O(l)$
+1.36	$Cl_2(g) + 2e^- \rightarrow 2Cl^-(aq)$
+1.33	$Cr_2O_7^{2-}(aq) + 14H^+(aq) + 6e^- \rightarrow 2Cr^{3+}(aq) + 7H_2O(l)$
+1.23	$O_2(g) + 4H^+(aq) + 4e^- \rightarrow 2H_2O(l)$
+1.06	$Br_2(l) + 2e^- \rightarrow 2Br^-(aq)$
+0.96	$NO_3^-(aq) + 4H^+(aq) + 3e^- \rightarrow NO(g) + H_2O(l)$
+0.80	$Ag^+(aq) + e^- \rightarrow Ag(s)$
+0.77	$Fe^{3+}(aq) + e^- \rightarrow Fe^{2+}(aq)$
+0.68	$O_2(g) + 2H^+(aq) + 2e^- \rightarrow H_2O_2(aq)$
+0.59	$MnO_4^-(aq) + 2H_2O(l) + 3e^- \rightarrow MnO_2(s) + 4OH^-(aq)$
+0.54	$I_2(s) + 2e^- \rightarrow 2I^-(aq)$
+0.40	$O_2(g) + 2H_2O(l) + 4e^- \rightarrow 4OH^-(aq)$
+0.34	$Cu^{2+}(aq) + 2e^- \rightarrow Cu(s)$
0	**$2H^+(aq) + 2e^- \rightarrow H_2(g)$**
−0.28	$Ni^{2+}(aq) + 2e^- \rightarrow Ni(s)$
−0.44	$Fe^{2+}(aq) + 2e^- \rightarrow Fe(s)$
−0.76	$Zn^{2+}(aq) + 2e^- \rightarrow Zn(s)$
−0.83	$2H_2O(l) + 2e^- \rightarrow H_2(g) + 2OH^-(aq)$
−1.66	$Al^{3+}(aq) + 3e^- \rightarrow Al(s)$
−2.71	$Na^+(aq) + e^- \rightarrow Na(s)$
−3.05	$Li^+(aq) + e^- \rightarrow Li(s)$

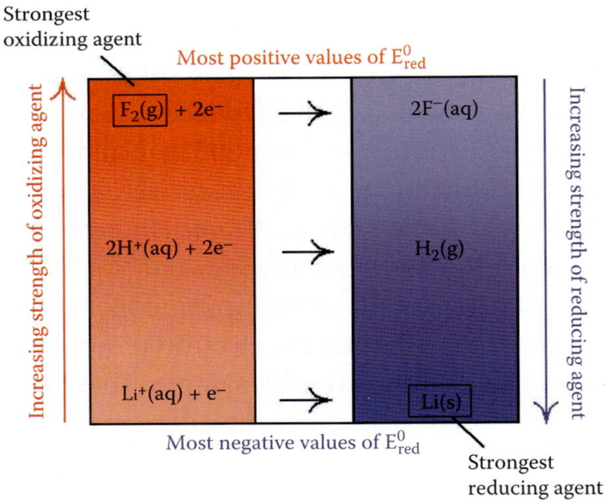

FIGURE 7.34 In a table of standard reduction potentials, any species on the left of a given half-reaction will react spontaneously with any species appearing on the right of any half-reaction that appears below it when reduction potentials are listed from highest and most positive to lowest and most negative.

Half-cell reactions are reversible, i.e., depending on the experimental conditions, any half reaction can be either an anode or a cathode reaction. Oxidation potentials can be obtained from reduction potentials by changing the sign, so Equation 7.41 can be rewritten as:

$$E^0_{cell} = E^0_{red}(\text{cathode}) + E^0_{ox}(\text{anode}) \qquad (7.42)$$

The standard reduction potential is an intensive property, and changing a stoichiometry coefficient of a reaction does not change the reduction potential [e.g., both half-cells $Zn^{2+}(aq, 1\ M) + 2e^- \rightarrow Zn(s)$ and $2Zn^{2+}(aq, 1\ M) + 4e^- \rightarrow 2Zn(s)$ come with a standard reduction potential of $E^0 = -0.76$ V]. This intensive property is also not affected by either the size of the electrode or by the volume of electrolyte, but the concentration of Zn ions must be 1 M.

The potential difference at any concentration is called the *electromotive force* (EMF = cell voltage = cell potential = E_{cell}). This electromotive force depends on:

- The reactions that occur on the electrodes
- The concentrations of reactants and products (see Nernst equation further below)
- The temperature (assumed to be 25°C unless otherwise stated)

A shorthand electrochemical cell notation has been developed for convenience:

- The anode is always on the left and the cathode on the right.
- All redox forms of reagents present should be listed with phase and concentration in parentheses, $ZnSO_4$ (*aq*, 1 M).
- Single lines represent a change of phase: single vertical line "|" (e.g., *s* to *l* or *l* to *g*).
- A comma should be used to separate two components in the same phase.
- Double lines represent a salt bridge or membrane: double vertical line "||".

Example 7.3:

- $Fe(s)\ |\ Fe^{+2}$ (1M) || H^+ (1M) | $H_2(g)$ | Pt (1M)
- Cd^{+2} (1M) | Cd(s) || Ag^+ (1M) | AgCl(s) | Ag(s)

Exercise 7.1: A spontaneous redox reaction produces energy that can do electrical work as in the Daniell cell (1836) with a zinc electrode (in 1 M Zn^{2+}) and a copper electrode (in 1 M Cu^{2+}) as shown in Figure 7.35. Calculate the cell EMF (a), and represent this cell in shorthand notation (b).

From Table 7.11: (a) $Cu^{2+}(aq) + 2e^- \rightarrow Cu(s)\ E^0_{red} = +0.34$, and from the same table $Zn^{2+}(aq) + 2e^- \rightarrow Zn(s)\ E^0_{red} = -0.76$ V; the latter reaction will spontaneously run to the left or $Zn(s) - 2e^- \rightarrow Zn^{2+}(aq)\ E^0_{ox} = +0.76$ V, so the overall reaction (using Equation 7.42):

$$Cu^{2+}(aq) + 2e^- \rightarrow Cu(s) \quad E^0_{red} = +0.34$$

$$Zn(s) - 2e^- \rightarrow Zn^{2+}(aq) \quad E^0_{ox} = +0.76\ V$$

$$Zn(s) + Cu^{2+}(aq) \rightarrow Zn^{2+}(aq) + Cu(s)\ EMF = +1.1\ V$$

(b) The Daniell cell in shorthand notation:

$$Zn(s)\ |\ Zn^{2+}\ (1M)\ ||\ Cu^{2+}\ (1M)\ |\ Cu(s)$$

With a measured EMF of = 1.10 V at 25°C.

Exercise 7.2: Is the following redox reaction spontaneous?

$$Mg^{2+} + 2Ag \rightarrow Mg + 2Ag^+$$

Given that:

$$Ag^+ + e^- = Ag \quad E^0_{red} = +0.80\ V$$

$$Mg^{2+} + 2e^- = Mg \quad E^0_{red} = -2.37\ V$$

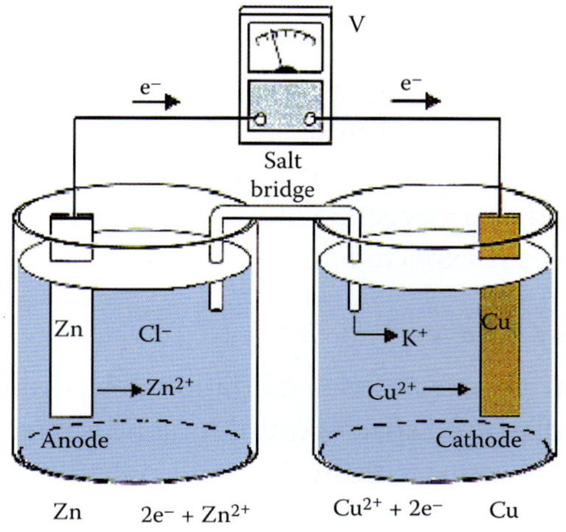

FIGURE 7.35 The Daniell cell.

Summing the two half-cells to obtain the above total cell reaction, we obtain (using Equation 7.42):

$$2Ag - 2e^- = 2Ag^+ \quad E^0_{ox} = -0.80 \text{ V}$$

$$Mg^{2+} + 2e^- = Mg \quad E^0_{red} = -2.37 \text{ V}$$

$$Mg^{2+} + 2Ag \rightarrow Mg + 2Ag^+ \quad EMF = -3.17 \text{ V}$$

This is not a spontaneous reaction because the EMF of the cell is negative.

The above examples all concern Faradaic processes in which electrons cross the solid/electrolyte interface. In a *non-Faradaic* process, there are no charge transfer reactions across the interface because they are thermodynamically and kinetically unfavorable. Under these conditions, processes such as adsorption and desorption can still take place and modify the structure of the electrode/solution interface. During the charging of the double layer, a transient external capacitive current flows:

$$C_d = \frac{dQ}{dV} = \frac{idt}{dV} \tag{7.43}$$

Further below, we will see that the value of C_d can be obtained from a cyclic voltammogram by measuring the current, *i*, at a given voltage sweep rate, *dV/dt*.

Faradaic and non-Faradaic processes occur simultaneously during electrode reactions. The electrochemical stability window of an electrochemical cell is defined as the potential range over which an electrode in a solvent with an inert/indifferent electrolyte (salt) can be polarized without substantial Faradaic currents. These stability ranges depend on the nature of the solvent, the salt, the electrode, and the presence of contaminants. Nonaqueous electrochemistry often provides stability windows as wide as 4–5 V, whereas in aqueous electrochemistry the hydrolysis of water limits the range typically to 1.5 V. Thermodynamically, the voltage difference between hydrogen and oxygen evolution at ideally reversible electrodes is expected at 1.23 V.

Faraday's Law

Electron transfer causes oxidation (electrons jump from a dissolved species onto the solid) or reduction (electrons transfer from the solid onto a species in solution) to occur at the interface. This process is governed by Faraday's law, which dictates the amount of chemical produced (in moles) by the amount of charge passed through the interface:

$$\text{moles} = \frac{Q}{nF} = \frac{i \cdot t}{nF} \tag{7.44}$$

where i = current passed (in amperes, A)
t = time of current flow (in seconds, s)
n = number of electrons involved in the reaction
F = Faraday (= 96,500 C mol^{-1})
Q = charge (in coulombs, C)

The current, *i*, multiplied by the time the current flows, *t*, gives us the total number of coulombs ($Q = i \cdot t$). The charge on 1 mol of electrons is called the *Faraday* (*F*), i.e., 1.6×10^{-19} C multiplied by the number of Avogadro (N_A) (eN_A) 6.02×10^{23} or 96,500 C/mol.

Thermodynamic Interpretation of the Electromotive Force

Starting with some standard thermodynamic relations, we find from dU = dQ + dW (U = internal energy, Q = heat, and W = work) and dQ = TdS that $dU = TdS - PdV + dW_{electrical}$ (*S* is entropy, –PdV is mechanical work, and $+dW_{electrical}$ is electrical work). Because $dG_T = dH_T - TdS$ (*H* is enthalpy) and $dG_T = dU_{T,P} + PdV - TdS$ (*U* is internal energy, *P* pressure, and *V* volume); therefore, $dG_T = TdS - PdV + dW_{electrical} + PdV - TdS$, and we arrive at the important relationship between Gibbs free energy and work done in an electrical system (most electrochemical cells work at constant *P* and *T*), i.e., the so-called *Gibbs function*:

$$dG_{T,P} = dW_{electrical} \tag{7.45}$$

The Gibbs function is at the heart of electrochemistry because it identifies the amount of work we can electrically extract from a system. Now the maximum electrical work done by an electrochemical cell equals the product of the charge flowing and the potential difference across which it flows or $W_{electrical} = EQ$. The charge, $Q = n \cdot N_A \cdot e$ where *n* is the number of moles of electrons transferred per mole of reaction, N_A is Avogadro's number (6.02×10^{23}), and *e* is the charge on an electron (-1.6×10^{-19} C). As we just calculated above, $N_A \cdot e = F$ (1 Faraday is 96,485 C/mol), and we obtain $W_{electrical} = nFE$. By convention, we identify work, which is negative with work being done by the system on the surroundings. Negative free energy change is identified as a spontaneous process. Therefore, we must choose a convention for the sign of the cell potential consistent with these equations, the convention regarding reaction spontaneity, and the sign for ΔG. If we choose to associate a positive potential with a spontaneous process, then we need a negative sign in the equation relating ΔG to E:

$$\Delta G_{T,P} = -nFE \tag{7.46}$$

Note how a measurement of a cell potential directly calculates the Gibbs free energy change for the process. With ΔG negative and E positive, the reaction is spontaneous, and one defines a voltaic (galvanic) electrochemical cell as a cell that generates an electrical current (e.g., battery or fuel cell). With ΔG positive and E negative, the reaction is nonspontaneous, as encountered in an electrolytic electrochemical cell, and it occurs as a result of passage of current from an external power source (e.g., electrolysis of water and the charging of a battery). A galvanic cell is one that is allowed to operate in its natural direction. An electrolytic cell is one operated in reverse. A galvanic cell produces a potential to drive an external load. An electrolytic cell is one that consumes external electricity. Note that the electrode polarity changes between galvanic and electrolytic. The nomenclature of the electrode also changes. However, the definition for reduction and oxidation is constant: Oxidation Is Loss of electrons and Reduction Is Gain of electrons, or "OIL RIG."

All thermodynamic measurements are based on differences between states; there is no absolute value for any property (with the exception of entropy). To quantify thermodynamics, we choose by convention a reference state called the *standard ambient temperature,* pressure, and concentration state [temperature = 298 K (25°C), pressure = 1 bar (10^5 Pa), concentration = 1 molal (mol of solute/kg of solvent)]. The use of atmospheres as a unit of pressure is widespread, and it is so close to the value of 1 bar (1.0134 bar = 1 atm) that the reference state often used is that of 1 atm. The differences are seldom significant. Concentrations are more reliably measured in molality (amount/mass of solvent) rather than molarity (amount/volume of solvent). However, for the following reasons: 1) volume is easier to measure than mass, and 2) density of water (the most common solvent) is close to 1, the most commonly used reference state for concentration is that of 1 M (mol/L). Reference states are indicated by the superscript °, e.g., C° or P°.

The propensity for a given material to contribute to a reaction is measured by its activity, a. This describes how "active" this substance is in the reaction compared with how it would behave if it were present in its standard state. Activity scales with concentration or partial pressure:

$$a \propto C/C^\circ \text{ or } a \propto P/P^\circ \tag{7.47}$$

Generally speaking, a substance's activity increases as the amount of that substance available for reaction increases. However, because of intermolecular interactions (which increase as a substance's concentration or partial pressure increases), there are deviations from a direct correspondence as pressure and concentration increase. These differences can become very significant at very high concentrations. We then define activity as:

$$a = \gamma C/C^\circ \text{ or } a = \gamma P/P^\circ \tag{7.48}$$

Activity coefficients γ are close to 1 for dilute solutions and for low partial pressures, but they change with concentration, temperature, other species, and so on. Generally, we ignore activity coefficients for educational simplicity, but careful work requires its consideration. In the case of pressure, the coefficient used is often called *fugacity,* but it is the same thing and behaves in the same way. Activity a is unitless, and because activity coefficients are close to 1 for dilute solutions, the reference states for partial pressure and concentration have a numerical value of 1. Therefore, we often approximate activity by concentration (M) or partial pressure (atm). The reference state for a solid material is itself, and the same is true for a pure liquid. A solvent is usually present in huge excess (compared with the solutes); therefore, it is mostly like a pure liquid. In all these cases, a = 1. In other words, if we increase the amount of a pure solid, a pure liquid, or a solvent, the reaction goes on for a longer time (but it does not go any faster).

If you increase the amount of these materials in a system, it will not increase the reaction rate, but it may increase the duration of a reaction. The point is that the "activity" of the system does not change with amount of these materials, so they are always referenced to themselves.

The Nernst Equation

We now want to know what happens with the cell potential if the concentrations and pressures are not in the standard state. We first derive how the chemical potential, μ, changes with activity by integrating the expressions for the dependence of amount of material on the Gibbs function, leading to the following relationship:

$$\mu = \mu^0 + RT \ln a \tag{7.49}$$

When analyzing a chemical reaction such as aA + bB → cC + dD mathematically, we form the following reaction quotient:

$$Q = \frac{a_C^c a_D^d}{a_A^a a_B^b} \tag{7.50}$$

where products appear in the numerator and reactants in the denominator. It requires the activity of each reaction participant, where each term is raised to the power of its stoichiometric coefficient. Terms in the reaction quotient that arise from the participation of pure solids, pure liquids, or solvents are left out as they all equal to 1. If we ignore activity coefficients, then activities for solutes and gases are numerically equal to their concentration or partial pressures (because the activity is the ratio of these terms with the number 1). One often writes the reaction coefficient Q with concentration and pressure shown explicitly, but remember that we are really dealing with activities. The reaction quotient is unitless because it has only activity in its expression, and all activities are unitless. The Gibbs free energy has the same activity dependence as the chemical potential:

$$G = G^0 + RT \ln a \tag{7.51}$$

When we apply this to a reaction, the reaction quotient comes into play, which gives us:

$$\Delta G = \Delta G^0 + RT \ln Q \tag{7.52}$$

Q contains all the activities of both reactants and products, all raised to their proper stoichiometric coefficients. When all participants have unit activity (a = 1), then Q = 1 and ln Q = 0, or:

$$\Delta G = \Delta G^0 \tag{7.53}$$

The reaction proceeds and Q changes until finally ΔG = 0. At that point, the reaction stops as equilibrium is reached with the reaction quotient $Q = Q^*$. The reaction does not stop at Q^* but simply has reached the situation where the forward and reverse reaction rates are equal. Chemical activity is still ongoing with just as much vigor as before, but it is happening equally in both directions—products are reacting to turn into reactants just as fast as reactants are reacting to turn into products. This is called a *dynamic equilibrium*, rather than a *static equilibrium*.

$$\begin{aligned} 0 &= \Delta G^0 + RT \ln Q^* \\ \Delta G^0 &= -RT \ln Q^* \\ \text{and } Q^* &= K_{eq} \end{aligned} \tag{7.54}$$

This special Q^* is renamed K_{eq}, the equilibrium constant. From the expression for the Gibbs free energy as a function of activity (Equation 7.51), we can derive an expression in terms of the cell potential by using the relation between the free energy and the cell potential *E* (Equation 7.46):

$$\Delta G = \Delta G^0 + RT \ln Q = -nFE = -nFE^0 + RT \ln Q \tag{7.55}$$

where *n* is the number of electrons transferred in the process. By rearranging, we obtain the very important Nernst equation:

$$E = E^0 - \frac{RT}{nF}\ln Q = E^0 - \frac{0.0592}{n}\log Q \tag{7.56}$$

This expression relates the dependence of the cell potential *E* on the reaction quotient Q. The logarithmic term tells us that if Q changes by a factor of 10, this will change the cell potential by 59.2 mV, or about 60 mV for a one-electron process (n = 1). This "rule of thumb" relationship will show up often in sensors, batteries, fuel cells, and so on.

The above equations allow us to relate measured cell potentials to standard Gibbs free energies of reaction. These, in turn, are related to a reaction's equilibrium constant (see Equation 7.54).

Example 7.4:

Consider the Daniell cell of Figure 7.35 again. What is the potential of the cell if $[Cu^{2+}] = 0.01$ M and $[Zn^{2+}] = 1.00$ M? (We will assume that the activity coefficients are = 1.)

$$E = 1.10 - \frac{0.0592}{n}\log\frac{1.00}{0.01} = 1.10 - 0.030\ \log(100)$$

$$= 1.041\,V$$

Note that the cell potential is decreased by about 60 mV from the standard cell potential. This accounts for a decrease in Cu^{2+} concentration of two orders of magnitude, where our rule of thumb predicted 120 mV; however, because the reaction involves two electrons (n = 2), we obtain a 60-mV change in potential only.

Example 7.5:

The Nernst equation demonstrates that potential depends on concentration. A cell made of the same materials, but with different concentrations, will also produce a potential difference, e.g.:

$$Cu \mid Cu^{2+}\ (0.001M) \parallel Cu^{2+}\ (1.00M) \mid Cu$$

What is standard cell potential E^0 for this cell? What is the cell potential E? What is n, the number of electrons transferred? Which electrode, anode, or cathode will be in the numerator?

$$E = E^0 - \frac{0.0592}{n}\log\left(\frac{[Cu^{2+}]_{anode}}{[Cu^{2+}]_{cathode}}\right)$$

$$= 0 - \frac{0.0592}{2}\log\frac{0.001}{1.00} = +0.089\ V$$

The standard cell potential in a concentration cell clearly must be 0 for the two electrodes, because at standard conditions both are identical. The number of electrons transferred is two. Remember that we are calculating the cell potential for the cell as written in the definition. The right side is the cathode and thus refers to a reduction process. The other is for oxidation. If we write those two reactions, we see that the aqueous solution from the cathode appears as a reactant. Hence the cathode aqueous solution is in the denominator, and conversely the anode side is in the numerator.

Example 7.6:

Consider the cell:

$$Pt \mid I^-\ (1.00M),\ I_2\ (1.00M) \parallel Fe^{2+}\ (1.00M),\ Fe^{3+}\ (1.00M) \mid Pt$$

with a standard cell potential given by = 0.771 V – 0.536 V = +0.235 V.

$$\Delta G^0 = -\,nFE^0 = -\,2(96{,}485\ C/mol)(0.235\ J)$$

$$= -45{,}348\ J/mol$$

This is the free energy change leading to the equilibrium constant, K_{eq}, for the reaction as follows:

$$\ln K_{eq} = -\frac{\Delta G^0}{RT} = -\frac{(-45{,}348\ J/mol)}{(8.314\ J/Kmol)(298\ K)} = 18.3034$$

$$K_{eq} = e^{18.3034} = 8{,}8910^7$$

Exercise 7.3: A metal in contact with a sparingly soluble salt of the metal, we will see below, is called an *electrode of the second kind*. Another common name for such an electrode is an anion electrode. Examples include reference electrodes (REF Es), such as Ag/AgCl(s) and $Hg/Hg_2Cl_2(s)/Cl^-$ [saturated calomel electrode (SCE)]. These reference electrodes are often more convenient to use than the above-reviewed SHE. Consider the Ag/AgCl electrode shown in Figure 7.36. This REF E is made from an Ag wire coated with AgCl(s) immersed in an NaCl or KCl solution.

The reactions involved are $AgCl \leftrightarrow Ag^+ + Cl^-$ and $Ag^+ + e^- \leftrightarrow Ag(s)$ [$E^0_{red} = +0.80$ V]. Derive the Nernst equation for this electrode.

$$Ag^+ + e^- \leftrightarrow Ag(s) \quad E^0_{red} = +0.80\ V$$

Chemical reaction;

$$\frac{AgCl \leftrightarrow Ag^+ + Cl^-\ \text{no electrons involved}}{AgCl + e^- \leftrightarrow Ag(s) + Cl^-}$$

$$E = E^0 - \frac{0.0592}{1}\log\frac{a_{Ag}a_{Cl^-}}{a_{AgCl}}\ \text{but}\ a_{Ag}\ \text{and}\ a_{AgCl} = 1$$

$$\text{or}\ E = 0.80 - 0.0592\log a_{Cl^-}$$

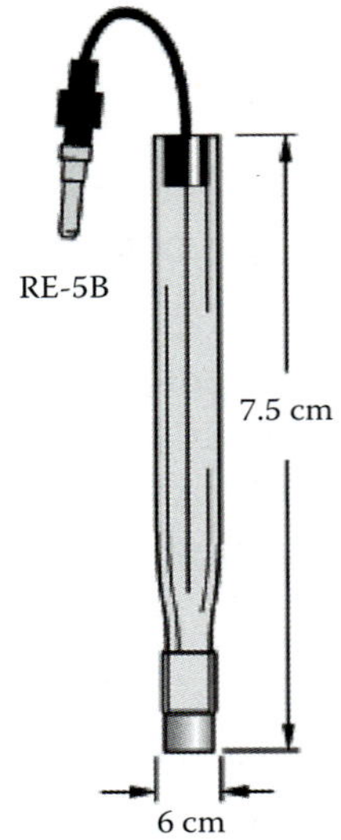

FIGURE 7.36 A typical commercial Ag/AgCl reference electrode. (From http://www.bioanalytical.com.)

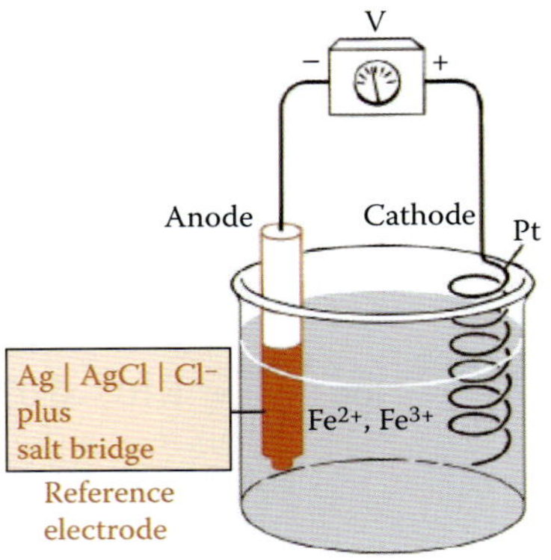

FIGURE 7.37 Potentiometric sensor in a two-electrode cell. The entire left half-cell is made into a self-contained reference electrode containing appropriate solutions and a salt bridge. The chloride concentration in the reference electrode is held constant by the fixed solubility of KCl.

From this expression, Cl^- (an anion) activity determines the potential of this electrode, and the electrode can be used as a Cl^- sensor (60-mV voltage change per decade of Cl^- concentration change). The purpose of a reference electrode is to provide a stable potential, independent of solution composition, against which other potentials can be reliably measured. Filling a glass envelope, which is equipped with a small pinhole to make ionic contact with the outside solution, with a saturated Cl^- solution that surrounds a Ag/AgCl wire produces a reliable reference electrode. The constant a_{Cl^-} in the above equation provides for a constant E ($E = 0.22$ V vs. SHE at 25°C). Only when enough solvent or ions from the outside solution enter the glass envelope does the reference potential start drifting. It is obvious that the larger the REF E electrolyte reservoir, the longer the reference electrode will last without need to refill the reservoir with fresh saturated Cl^- solution. This explains why miniaturization of a reference electrode is such a challenge; smaller means less electrolyte capacity and thus a shorter lifetime.

Two-Electrode and Three-Electrode Cell Configurations

There are two common electrode configurations used in electrochemistry: two-electrode and three-electrode cells. In a two-electrode configuration, an electrochemical cell is equipped with one working electrode (WE) and one reference electrode (REF E) (Figure 7.37). The electrode at which the reaction of interest occurs is called the *working electrode*. An *amperometric measurement* with such a cell occurs when current is passed between the WE and REF E by applying a potential between the two. The kinetics of the electrochemical reaction are driven by the applied potential, and we measure the diffusion-controlled current flowing across the electrode/solution interface. This current is directly proportional to the bulk concentration of the analyte present in the solution (see Equation 7.124). Passage of a significant current through a reference electrode shifts the electrode away from equilibrium and may even destroy its materials (e.g., in the case of Ag/AgCl, the AgCl could all be reduced to Ag). Moreover, a significant resistance decrease (iR) results for less conductive media (in a salt water solution, this is not a problem since R is typically 100 Ω, so with $i < 10$ μA, $iR < 1$ mV) or when the WE is very large. As a consequence, a two-electrode cell should only be used when the iR drop is less than 1 mV or when using *micro-* or *ultramicroelectrodes*. In *potentiometric measurements*, only very small currents are drawn, and a two-electrode system can always be used confidently in this case (as long as a high-impedance voltmeter is used so that a measurement of the potential difference will be accurate). In potentiometry, the electron transfer reaction is kinetically facile, i.e., the equilibrium electron exchange current $i_{0,e}$ (on an electron conductor) or ion exchange current $i_{0,i}$ (on an ion conductor) is large. The cell potential responds to changes in the activity of the analyte species present in the solution in a well-defined manner described by the Nernst equation. The cell potential varies in a linear manner with the logarithm of the analyte activity.

Microelectrochemical sensors are often based on a two-electrode system with a WE and a pseudoreference electrode because a real reference electrode is too complicated and expensive to implement. This is possible because the small working electrode draws so little current that a larger pseudoreference electrode (i.e., a Ag/AgCl film) is not shifted from equilibrium enough to cause large errors.

In a three-electrode electrochemical cell (Figure 7.38) one deals with a WE, a REF E, and a counter electrode (CE) or auxiliary electrode. A Luggin capillary is often used to position the sensing point of a reference electrode at a desired point in a cell close to the working electrode.

With three-electrode systems, a *potentiostat* is used. A potentiostat is an instrument that controls the voltage difference between a WE and a REF E (Figure 7.39). The potentiostat implements this control by injecting current into the cell through the auxiliary electrode or CE. In almost all applications, the potentiostat measures the current flow between

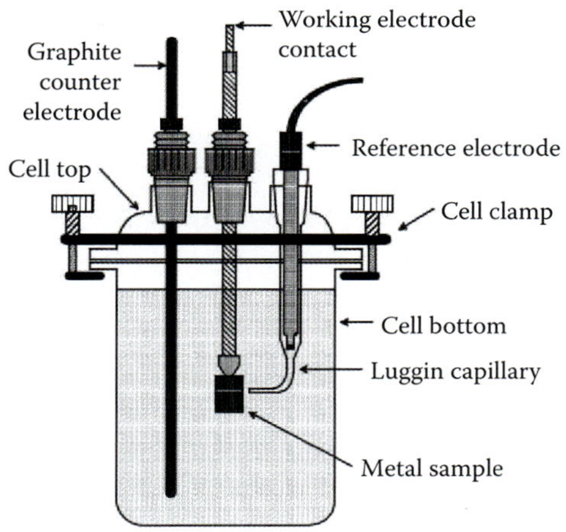

FIGURE 7.38 In this three-electrode cell, the counter electrode (CE) is a graphite rod, the working electrode (WE) is located in the center, and the reference electrode (REF E) is brought close to the working electrode surface via a Luggin capillary.

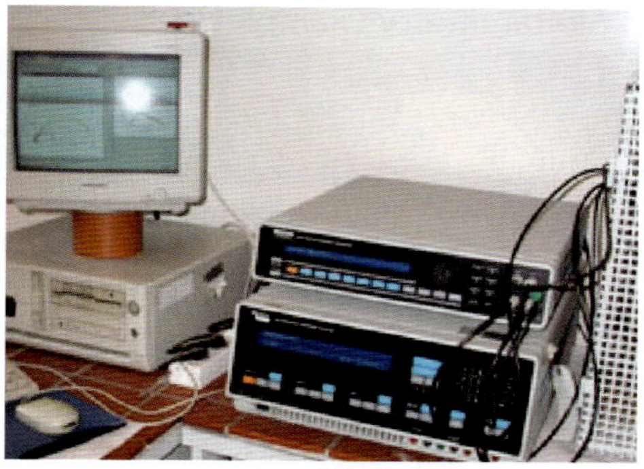

FIGURE 7.39 Solartron 1286 potentiostat and 1250 frequency response analyzer. (See also http://www.bioanalytical.com.)

the working and auxiliary electrodes. A high impedance is placed in front of the reference electrode to limit the current so that the reference potential remains constant. A cell like this avoids the internal polarization of the reference electrode and its potential degradation. The system also compensates for a major portion of the cell *iR* drop.

A potentiostat can be converted to work in *galvanostat* mode. Whereas in potentiostat mode one controls the potential and measures the ensuing current, in galvanostat mode one controls the current and measures the ensuing voltage. *Voltammetry* is an electroanalytical method in which the potential of an indicator electrode is varied in a definite manner with time, and the current flowing through the indicator electrode is measured. Voltammetric techniques are classified according to the type of voltage perturbation applied to the working or indicator electrodes, i.e., the way in which the voltage signal input varies with time. The form of the input $V(t)$ function determines the form of the resulting current response (see some examples in Table 7.12). The current/potential response curve is called a *voltammogram*. The voltammetry method relies on the fact that the current measured reflects rate-determining diffusion of the analyte species from the bulk solution to the surface of the indicator electrode, where it is readily oxidized or reduced. The solution contains supporting electrolyte, i.e., an excess of nonreactive electrolyte (e.g., KCl or NaCl), to conduct current besides the redox couple. This ensures that the redox species reaches the electrode by diffusion but not by migration. Under such conditions of diffusion control, the measured current is linearly proportional

Table 7.12 Voltammetry Examples

Name of Techniques	Potential Excitation Signal		Mass Transfer	Measurement	Analytical Relation to Bulk Concentration	Typical Display
Polarography (DC or normal)	Slow linear scan (or constant E)	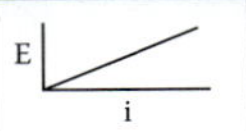	Diffusion	i vs. E	$i_d \propto C$	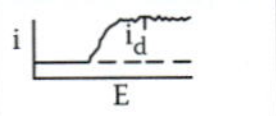
AC Polarograph	Slow linear scan + low amplitude sine wave	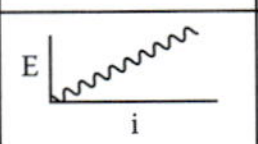	Diffusion	i_{ac} vs. E	$i_a \propto C$	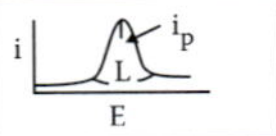
Pulse polarography	Square voltage pulses of increasing amplitude	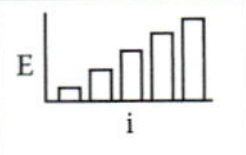	Diffusion	i vs. E	$i_d \propto C$	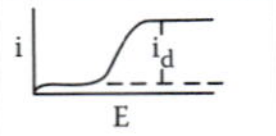
Differential pulse polarography	Square voltage pulses of constant amplitude + linear ramp	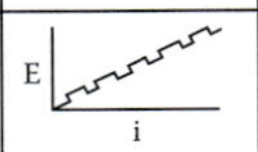	Diffusion	Δi vs. E	$i_p \propto C$	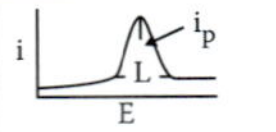
Single-sweep voltammetry	Linear scan E	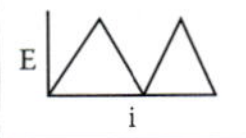	Diffusion	i vs. E	$i_p \propto C$	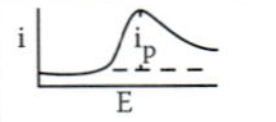
Cyclic voltammetry	Triangular scan E	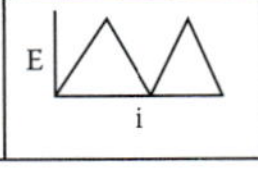	Diffusion	i vs. E	$i_p \propto C$	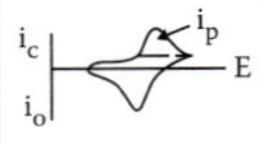

to the bulk concentration of the analyte species (see Equation 7.124).

A most important example of voltammetry is cyclic voltammetry, an electrochemical method in which information about an analyte is obtained by measuring current (i) as a function of applied potential using a triangular waveform for excitation (Figure 7.40). We will detail the features of a cyclic voltammogram after we elucidate the difference between kinetic and diffusion-controlled electrochemical reactions.

In Figure 7.41 we compare two- and three-electrode systems.

Types of Electrodes

Introduction

We distinguish electrodes based on the nature and number of phases between which electron transfer occurs and/or based on the nature of the current-carrying species within them. Based on the nature and number of phases, one typically recognizes electrodes of the first kind and electrodes of the second kind. Based on the charge carrier in the solid, we can also distinguish between metals, semiconductors, insulators, solid electrolytes, and mixed conductors. As summarized in Table 7.10, different types of mobile charge carriers characterize the different types of electrode materials. Metal electrodes (e.g., Pt) feature electrons as the mobile charge carriers, whereas semiconductors (e.g., n-Si) have electrons and/or holes that play that role. Solid electrolytes

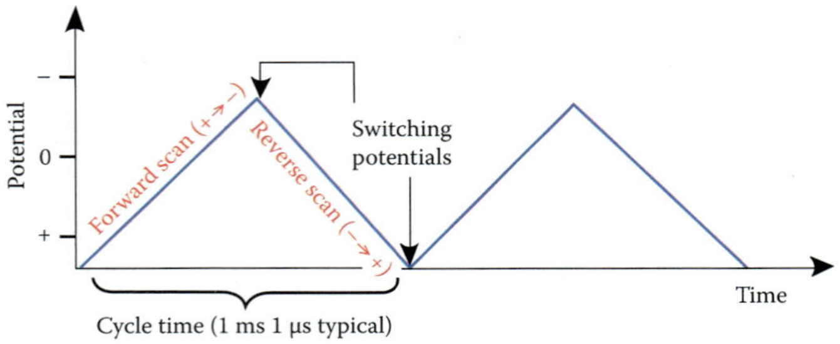

FIGURE 7.40 Triangular waveform used to register a cyclic voltammogram.

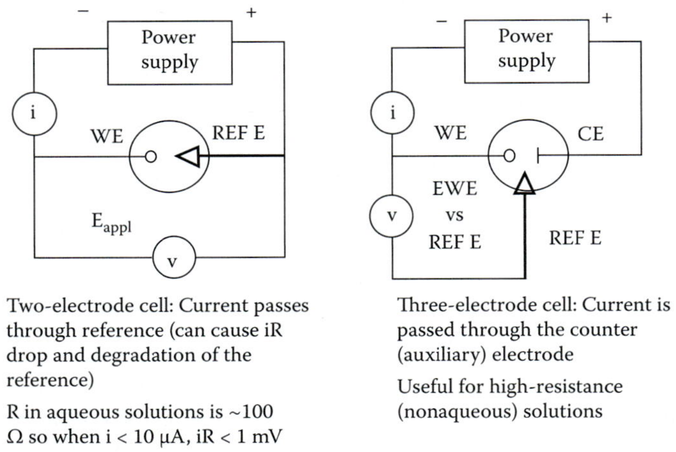

FIGURE 7.41 Two-electrode versus three-electrode system.

(e.g., LaF_3) have mobile ions, whereas insulators (e.g., SiO_2) possess no free charge carriers. Mixed conductors (e.g., IrO_x) have both ions and electrons for conduction.

We treat ion-selective electrodes (ISEs), also called *membrane electrodes*, as a separate category of electrodes. Membrane electrodes include pH glass membrane electrodes, crystalline solid electrolyte electrodes, polymeric electrodes, and membrane-based gas-sensing probes.

Electrodes of the First and Second Kind

Electrodes of the First Kind A metal in contact with its cations or a metal in contact with a redox couple in solution is called an *electrode of the first kind*. Examples include $Cu^{2+}/Cu(s)$, $Zn^{2+}/Zn(s)$, the standard hydrogen electrode (SHE) (i.e., Pt black in contact with H_2/H^+), Ag^+/Ag [nonaqueous reference electrode (REF E)], and $Cl^-/Cl_2(g)/Pt$. A working electrode (WE) (e.g., Pt) that changes potential against a suitable reference electrode and senses the redox couple concentration present in solution is an indicator electrode of the first kind. The electrode response of this type with respect to a REF E is given by the Nernst equation:

$$E = E^0 + \frac{0.0592}{n}\log a_{M^{n+}} \tag{7.57}$$

Often these sensors respond to cations, so they also may be called *cation electrodes*. With an electrode of the first kind, the fastest electron-exchange reaction, i.e., the reaction with the largest exchange current density $i_{0,e}$ (the current at zero external current and with no net reaction), determines the potential of the electrode. Often there are different redox species involved in establishing the equilibrium potential, in which case we speak about a mixed potential: zero external current but net reaction (e.g., corrosion). The inert species in solution (e.g., 1 M NaCl), also called *indifferent electrolyte*, do not, at zero or low bias, exchange electrons with the platinum electrode. Instead, they provide solution conductivity. All redox couples have a known redox potential as measured against a standard reference electrode (e.g., SHE). Platinum is the most common metal indicator electrode to measure these redox potentials. Pt is mostly inert, does not participate in reactions, and allows electron transfer to and from solution. Gold is also an inert metal indicator electrode used for the same purpose. Carbon electrodes are often used because many redox reactions are very fast on a carbon surface (see also C-MEMS in Volume III, Chapter 5). Note that not all metals are electrodes of the first kind; metals such as Fe, Al, and W are not electrodes of the first kind because they have relatively thick surface oxide coatings that are pH-sensitive.

Electrodes of the Second Kind Metals in contact with their sparingly soluble salt are *electrodes of the second kind*, also called *anion electrodes*. Examples include Ag/AgCl(s) and $Hg/Hg_2Cl_2(s)/Cl^-$ (saturated calomel electrode; SCE). As seen from the Nernst expression for an electrode of the second kind (see also Exercise 7.3 on a Ag/AgCl electrode):

$$E = E^0 - \frac{0.0592}{n}\log a_{A^{n-}} \tag{7.58}$$

In this case, anion activity determines potential. These electrodes make great reference electrodes because of the low solubility of their salt (potential very stable). As said before, a reference electrode is an electrode that does not change its potential with solution composition. By making the anion concentration around the sensor element constant, one obtains such a situation. Fixing the anion concentration is often accomplished by mounting the electrode in a glass envelope and putting a saturated solution of the anion in contact with the sparingly soluble salt of the metal wire, which itself is encapsulated in a very slowly leaking glass envelope. This is illustrated for a Ag/AgCl REF E in Figure 7.42.

The purpose of a reference electrode is to provide a stable potential against which other potentials can be reliably measured. Criteria for the selection of a good reference electrode include stability (time, temperature) and reproducibility. Importantly, the potential should not be altered by passage of a small current, i.e., the electrode should be perfectly nonpolarizable (Figure 7.43). The definition for an ideal nonpolarizable electrode is that there is no charge transferred across the electrode/electrolyte interface (in the case of a Ag/AgCl electrode, all the charge transfer is between the Ag and the

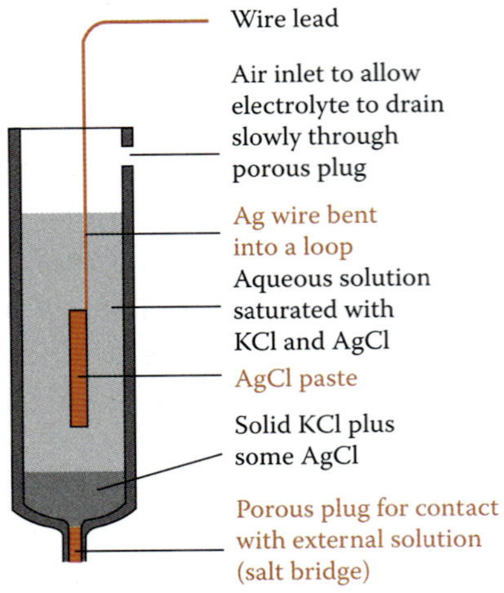

FIGURE 7.42 Construction of a Ag/AgCl reference electrode.

AgCl instead), and ΔV remains = 0 for a wide range of currents. An ideal polarizable electrode (IPE) is one where the charge transfer resistance is very large, so that Δi = 0 for a large range of voltages. The behavior of such an interface when the potential across the electrode is changed resembles that of a plain capacitor. An IPE follows the standard capacitor equation. The reference electrode should also be easily constructed and convenient for use. The standard hydrogen electrode is difficult to use because Pt black is easily poisoned by organics, sulfide, cyanide, and so on. Also, hydrogen is explosive, and sulfuric and hydrochloric acids are very strong acids; however, SHEs feature one of most reproducible potentials (±1 mV). More convenient reference electrodes include the Ag/AgCl and the saturated calomel electrode (Hg/Hg_2Cl_2) (which has problems with mercury toxicity though).

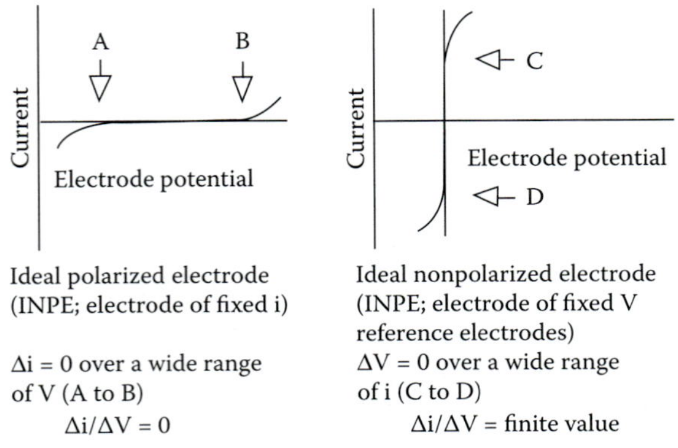

FIGURE 7.43 Concept of polarization. A good reference electrode is ideally nonpolarizable.

When the above reference electrodes are used in nonaqueous media, there might be significant junction potentials. Nonaqueous media can also cause greater noise, and electrolyte precipitation may clog the electrode frit. Finally, the chemistry may be water-sensitive.

Electrochemical reference electrodes are very difficult to miniaturize, and when miniaturized they do not last very long because the internal electrolyte quickly mixes with the external solution. The development of miniature chemical and biological sensors has been hampered by this problem. Because of these problems, many researchers have chosen to implement pseudo-reference electrodes in miniature sensors. These pseudo-reference electrodes, for example, a Ag/AgCl wire without the saturated anion solution, only mimick the behavior of a real thermodynamically justified electrode. These problems will be illustrated below when introducing the ISFET.

Earlier, we pointed out that some metals with a relatively thick surface oxide coating may act as electrodes of the second kind, with the oxide coat setting up a pH response. This may be interesting as an alternative for pH glass electrodes. Glass pH electrodes have their shortcomings, such as extreme fragility, relatively slow response times, short lifetime, and very high impedance. Moreover, glass electrodes cannot be used in strong alkaline, HF, or organic solutions and are very difficult to miniaturize. As a consequence, various metal oxides have been tested as alternate, fast, robust, long-lifetime, low-impedance pH sensors that are easier to miniaturize. Most studies have focused on Sb/Sb_2O_3 (widely used for acid-based titrations or in solutions containing HF; however, voltage drift with this electrode is considerable), Bi/Bi_2O_3 (used in alkaline media), Pt/PtO_2, Ir/IrO_x, Ru/RuO_2, Os/OsO_2, Rh/RhO_2, Ti/TiO_2, and Sn/SnO_2. These oxides all represent mixed conductors with both electrons and ions (protons) exchanging rapidly at the interface. Depending on the relative magnitude of $i_{0,e}$ versus $i_{0,i}$, the electrode acts as a redox probe, a pH sensor, or both. For most mixed conductors $i_{0,e} \gg i_{0,i}$, i.e., one expects severe redox interference when used as pH sensors. Of the metal oxides listed, IrO_x holds the most promise as a very fast, stable, low-impedance pH sensor with surprisingly small redox interference.[8] As an example, we introduce some of our own work on Ir/IrO_x as a solid-state ion-selective electrode (ISE). Whereas this oxide previously had been prepared by electrochemical growth (potential cycling), anodic electrodeposition, AC reactive sputter coating (Ir target and oxygen gas), and thermal oxidation, we showed for the first time that an electrode prepared by melt-oxidation in a lithium carbonate (Li_2CO_3) exhibits a very reproducible, fast pH response (faster than a few tens of seconds) with low redox interference. The Nernstian slope (58 mV/pH) is close to ideal. Other good characteristics of this electrode include stability, even in solutions containing F^-. The ease of miniaturization using integrated circuit (IC) techniques is an additional benefit. The world's smallest CO_2 sensor, based on thin-film IrO_x pH sensing, is shown in Volume II, Figures 4.92–4.97. An example IrO_x-based pH probe (A) and some experimental results (B) are shown in Figure 7.44. From the SEM picture of an electrode cross-section (C), the IrO_x coating on the Ir wire appears as a 20-μm-thick, nonporous film with a columnar grain structure. In Figure 7.50 we use a Ir/IrO_x wire as described here to make a CO_2 gas sensor.

SensIrOx, a private enterprise, demonstrated effective pH measurement in extreme temperature and pressure conditions using IrO_x probes as shown in Figure 7.44 (http://www.sensirox.com). For example, in food production or biological process monitoring, IrO_x probes provide a safe, nonglass solution. They are also well-suited for measurements in high salt solutions, viscous sludges and slurries, or for wide-ranging pH. A pH sensor based on IrO_x, suitable for measuring total acidity number in nonaqueous applications (e.g., engine oil), was also developed by SensIrOx.

Ion-Selective Electrodes

Introduction Ion-selective electrodes (ISEs) are membrane electrodes that exchange ions at a membrane/solution interface. Ideally, these electrodes respond to one ion only, and their equilibrium potential does not involve redox reactions. ISEs use a membrane that exchanges ions fast, where the electric potential across the membrane/liquid interface depends on the concentration of the ion with the fastest (largest) ion exchange current, $i_{0,i}$. The difference in voltage across the membrane is measured between two REF Es: an internal and an external one, as sketched for the case of F^- sensing with a LaF_3 ISE (developed in 1967 by Ross) in Figure 7.45. The ion from solution that exchanges sites within the single-crystal LaF_3 is F^-, which is mobile in the LaF_3 crystal lattice, even at room temperature.

The electrical potential difference across the membrane boundaries of an ISE can be described by:

$$\Delta E = \frac{0.0592}{n} \log \frac{[C^+]_{outer}}{[C^+]_{inner}} \tag{7.59}$$

where n is the charge on the analyte ion (1 in the case of F^-), $[C^+]_{outer}$ is the ion concentration in the

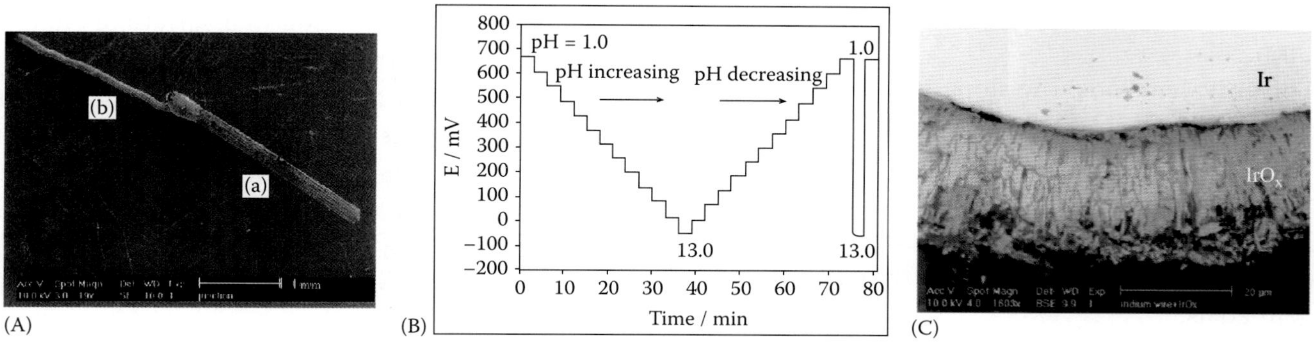

FIGURE 7.44 (A) Solid-state pH sensor: Ir wire with IrO_x coat (a) and Au wire contact (b). (B) A pH titration curve with IrO_x, no hysteresis. (C) SEM of an electrode cross-section. Note this same type of pH sensor was used to construct the CO_2 gas sensor in Figure 7.50.

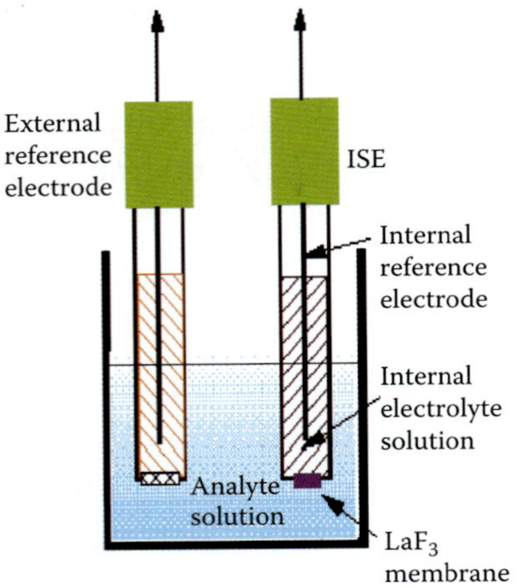

FIGURE 7.45 A F^- ion-selective electrode based on a single crystal of the solid electrolyte LaF_3 as the membrane.

unknown solution, and $[C^+]_{inner}$ is its concentration in the inner solution. Because the $[C^+]_{inner}$ is constant, the above equation can be simplified to:

$$\Delta E = \frac{0.0592}{n}\log\left[C_{outer}\right] \tag{7.60}$$

There is no change in the inner reference solution because there is no contact between the inner reference electrolyte and the unknown sample solution. Contact with the sample solution is established with the outer membrane surface and the external reference electrode, which is in ionic contact via a Luggin capillary. The potential of the whole electrochemical cell (at 25°C) is then given as:

$$E_{cell} = E_{const} + \frac{0.0592}{n}\log a_{M^{n+}} \tag{7.61}$$

where n is the valence of the ion that can be – (for anions) or + (for cations) and a is the activity of the ion.

For univalent ions, the slope is 59.16 mV/decade, and for divalent ions, the slope is 29.58 mV/decade. From the slope, it is possible to tell whether the electrode is behaving properly.

The selectivity coefficient describes the relative response of any electrode to interfering species and is defined as:

$$k_{A,X} = \frac{\text{response to interfering X}}{\text{response to principal ion A}} \tag{7.62}$$

The smaller the value of k, the lesser the response to X. For example, for F^- sensing with a LaF_3 sensor in the presence of Cl^-, the selectivity coefficient is $k_{F^-,Cl^-} = 10^{-4}$; in other words, Cl^- ions must be present in a concentration 10,000 times higher than F^- to set up a similar signal strength. At pH > 8, OH^- anions become a major interference for F^- measurement. At low pH, F^- exists only as HF (which this sensor will not pick up). Therefore, the sample must be buffered at a pH between 5 and 7 for reliable F^- analysis.

Some requirements that all membranes in ion-selective electrodes must meet include minimal solubility, a finite electrical conductivity, and selectivity for one ion only. The solubility of an ISE electrode in analyte solutions typically approaches zero. In general, the finite conduction of a membrane takes the form of migration of singly charged ions within the membrane (e.g., F^- in LaF_3 and Na^+ and/or Li^+ in a glass electrode). Finally, a membrane or some species contained within the membrane matrix must be capable of selectively exchanging ions with analyte ions.

The types of membrane electrodes include glass electrodes, crystalline solid-state electrodes, polymer electrodes, and gas-sensing probes.

The selectivity of a membrane electrode is linked with the uniqueness of the binding site for the intended ion from solution. Ideally, there would be only one type of binding site in the membrane with exclusive selectivity for the intended ion. In practice, surfaces are very complex, often involving fractals as encountered on the beach, trees, snowflakes, and so on, rather than atomically sharp transitions (see Figure 7.13). This implies that a binding site at the surface might find itself in a variety of energetic configurations with different associated binding constants/selectivities for the analyte. As a consequence, absolute selectivity with one sensor material cannot be achieved. In Volume III, Chapter 10 on MEMS and NEMS, we illustrate how using an array of sensors in an electronic tongue or in an electronic nose might present a possible alternative to more selective sensing.

Glass Electrodes A traditional pH (pondus Hydrogenii; literally hydrogen exponent)

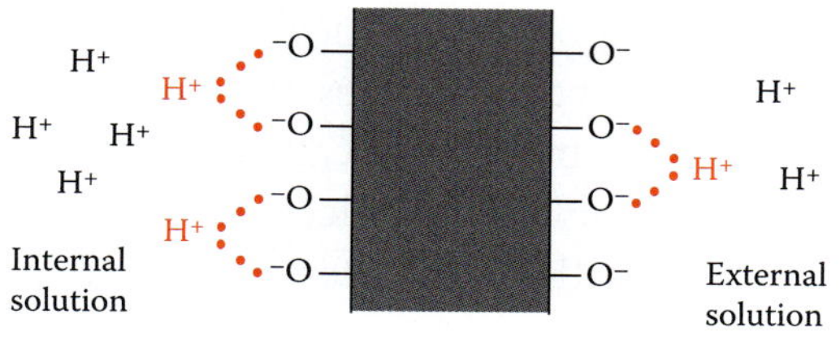

FIGURE 7.46 The Donnan potential on a thin pH-sensitive glass membrane is established on both sides of the membrane.

measurement with a glass electrode qualifies as a potentiometric ion-selective electrode (ISE) (e.g., a thin glass layer with a composition of 22% Na_2O, 6% CaO, and 72% SiO_2), and it is the most common solid-state ISE. The H^+ ion is the only one that significantly exchanges with the outer hydrated glass layer (see Figure 7.46). The proton exchange at the glass/solution interface results in a charge buildup, establishing the so-called *Donnan potential*. The Donnan potential is also established at the glass membrane/internal solution interface and is kept constant by the fixed pH of the internal electrolyte (inner reference), whereas the Donnan potential on the other side of the membrane is determined by the analyte solution pH (Figure 7.46).

Na^+ or Li^+ ions that are present in the glass carry the electrical current by migrating from one side to the other. A combination electrode used for pH measurements utilizes a combination of a silver reference electrode and a glass electrode as shown in Figure 7.47.

The pH response of a glass electrode, based on Equation 7.61 and the fact that pH = –log [H^+], is given by:

$$E_{cell} = E_{const} = 0.0592\ \text{pH} \qquad (7.63)$$

A glass electrode is calibrated with several buffers close to the pH of the unknown solution. It is also important that the glass electrode be kept hydrated when not in use. If not, the hydrated gel layer of the glass dries out, and the electrode has to be reconditioned/rehydrated for several hours before use.

The selectivity of a glass electrode strongly depends on the glass composition, with a Li_2O-containing glass having a selectivity coefficient k_{H^+,Na^+} of ≈ 10^{-11} – 10^{-13} and a Na_2O-containing glass having a k_{H^+,Na^+} of ≈ 10^{-10}. An "alkaline error" can result when cations other than H^+ are present in large excess in the solution. These cations can exchange for H^+ in the outer glass gel layer. However, the glass can be carefully selected to reduce the effect of interfering ions. On the other hand, by adding alumina to the glass composition, selectivity can be changed for the determination of other univalent ions such as Na^+.

The resistance (impedance) of a glass pH sensor is very high (100 MΩ to more than 1000 MΩ), which requires the input amplifier of the pH meter to be very high; the input impedance of the meter must be at least 100 times greater than that of the sensor (high-input impedance amplifier with field effect transistor input stage). However, very high

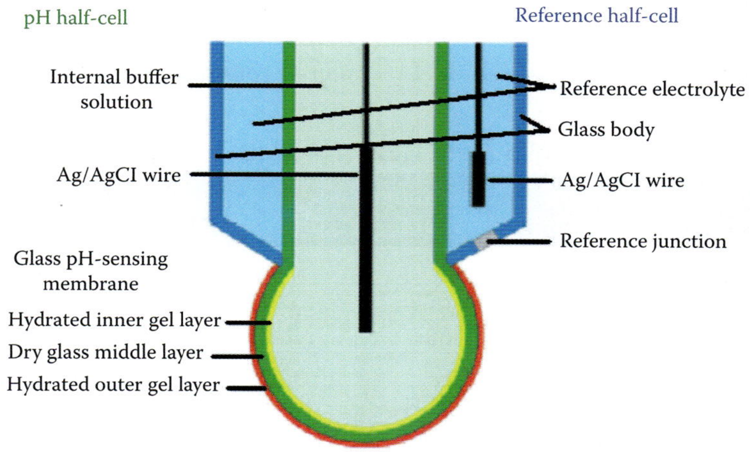

FIGURE 7.47 A combination pH sensor.

impedance can make the measurement noisy, and the smaller the sensor, the bigger this problem becomes. We will come back to this aspect of miniaturized electrochemical sensors when covering ISFETs. We saw earlier how glass pH electrodes have many shortcomings, such as extreme fragility, relatively slow response time, short lifetime, and very high impedance. We have also shown how IrO_x holds promise as a very good alternative, as a stable, low-impedance pH sensor with small redox interference (although hydrogen is detrimental).

Crystalline Solid-State ISEs Solid-state ISEs are based on inorganic crystals such as single-crystal LaF_3, as we explored earlier for F^- sensing. Doping of the LaF_3 with EuF_2 creates anion vacancies in the crystal, allowing F^- anions to jump back and forth between the solution and the vacancies in the solid. This leads to a dynamic equilibrium that establishes a Donnan potential at the crystal/solution interface (F^- has a larger ion exchange current $i_{0,i}$ than any other ion at the LaF_3/electrolyte interface). The F^- sensor cell in Figure 7.45 can be annotated as:

Ag(s) | AgCl(s) | NaCl, NaF(aq) | LaF_3 crystal | ext. soln || saturated KCl and AgCl(aq) | AgCl(s) | Ag(s)

The response of this electrode is given by:

$$E_{cell} = E_{const} - 0.0592 \log[F^-] = E_{const} + 0.0592 \, pF^- \quad (7.64)$$

The solid-state membrane may be a single crystal, polycrystalline, or a mixed pellet of an ionic salt. Mixed pellets of Ag_2S (with a solubility constant k_{sp} of 10^{-31}) and AgX respond to X^-, where X = Cl, Br, I. The charge transporting species in Ag_2S electrodes is Ag^+. Ion-selective electrodes for Cu^{2+}, Cd^{2+}, and Pb^{2+} can be made by mixing CuS, CdS, or PbS with Ag_2S. All mixed pellet Ag_2S electrodes also respond to Ag^+, so this will be an interferent. Some example solid-state ISEs for anion sensing are shown in Table 7.13.

Polymer-Based ISEs Liquid- and polymer-based ISEs use a mobile carrier (also ion exchanger, chelator, or ionophore = ion carrying, from the Greek phoros, meaning "bearing") in an organic solvent or a polymer matrix, respectively, to render that liquid or polymer ion-selective. We will restrict ourselves here to the discussion of polymer-based ISEs because they are easier to miniaturize than liquid-based ISEs. The ionophore is typically a hydrophobic species that is a chelator of the species of interest. An ion-selective polymeric membrane, say for a potassium ion sensor, is made by mixing a K^+ ionophore such as 1% valinomycin (a natural-occurring antibiotic) with 33% polyvinyl chloride (PVC) and 66% plasticizer (to make the rigid plastic more flexible: plasticized PVC acts like a sponge). The valinomycin (Figure 7.48) in the membrane selectively exchanges potassium ions from the solution (with a selectivity for K^+ over Na^+ of $k_{K^+,Na^+} = -4.3$) setting up a Donnan potential at the interface. The atomic radius of K^+ is 1.33 Å and that of Na^+ is 0.95 Å. With only this difference to work with, valinomycin manages to select for the K^+ ion over the Na^+ ion by a factor of nearly 5000.

No significant amount of counterions from solution may enter the membrane phase, or selectivity will be lost. To achieve this so-called *Donnan exclusion of counterions*, counterions must also be present

Table 7.13 Examples of Solid-State ISEs for Anions

Ion	Concentration Range (M)	Membrane Material	pH Range	Interfering Species
F^-	$10^{-6} - 1$	LaF_3	5–8	OH^- (0.1 M)
Cl^-	$10^{-4} - 1$	AgCl	2–11	CN^-, S^{2-}, I^-, $S_2O_3^2$, Br^-
Br^-	$10^{-5} - 1$	AgBr	2–12	CN^-, S^{2-}, I^- S^{2-}
I^-	$10^{-6} - 1$	AgI	3–12	
SCN^-	$10^{-5} - 1$	AgSCN	2–12	S^{2-}, I^-, CN^-, Br^-, $S_2O_3^2$
CN^-	$10^{-6} - 10^{-2}$	AgI	11–13	S^{2-}, I^-
S^{2-}	$10^{-5} - 1$	Ag_2S	13–14	

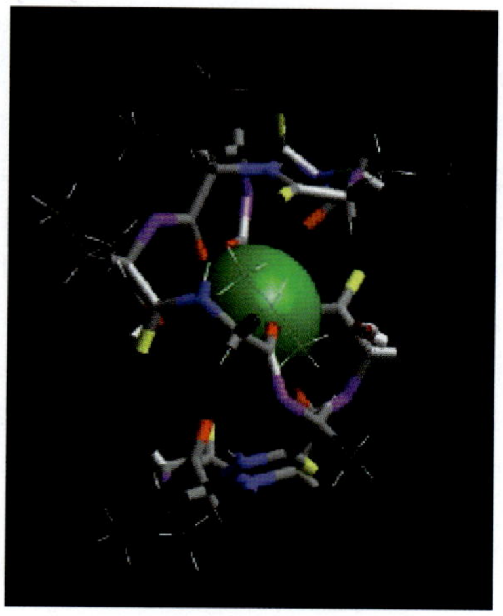

FIGURE 7.48 Valinomycin ionophore for K^+ ion.

in the membrane itself. To accomplish this, lipophilic counterions are added to the membrane cocktail [e.g., from sodium tetrakis[3,5-*bis*(trifluoromethyl) phenyl]borate or NaTFPB]. Addition of lipophilic salts to the membrane improves selectivity and also reduces the impedance of the probe. In a 3D ISE embodiment, as shown in Figure 7.49a the membrane potential is measured between an internal and an external reference electrode. Equation 7.61 gives the Nernstian response of polymeric membrane electrodes.

Ionophores for anions are harder to find than ionophores for cations because anions come in all types of shapes (e.g., SCN^- is linear but SO_4^{2-} is stetrahedral), whereas cations are mostly spherical (see Figure 7.48). Anion exchangers are typically R_4N^+ salts, and they respond according to the hydrophobicity of the anions; this is the Hofmeister series:

$$ClO_4^- > SCN^- > I^- > NO_3^- > Br^- > Cl^-$$

By using a combination of thin- and thick-film deposition techniques, one can fabricate planar ISEs on an inexpensive substrate as shown in Figure 7.49b. Thin-film silver electrodes are vapor deposited through a stencil mask on a polyimide substrate. The top layer of the Ag is converted to AgCl by dipping in a $FeCl_3$ solution. Silk screening of hydrogels with the appropriate electrolytes completes the internal and external reference electrodes. The internal reference side is now covered with the ISE membrane using a drop delivery technique, whereas the external reference side is covered with a dielectric membrane (also using drop delivery) with a small laser-cut hole for Luggin capillary. By simply changing the ionophore in the membrane, different ISEs can be fabricated. Although we show only one ISE in this figure, many planar sensors can be made in parallel (batch fabrication!). By switching from a 3D to 2D architecture, mass production of minute, polymeric ISEs (e.g., 2 × 3 mm per sensor) becomes possible, resulting in inexpensive (disposable), reproducible biosensors for ions such as Ca^{2+}, Na^+, K^+, NO_3^-, Cl^-, and so on. For ease of manufacturing, planar ISEs often do not use an internal reference solution. This makes the electrodes easier to handle and store, but voltage drift tends to be larger because there is no pinning of the Donnan potential on the internal reference side with a fixed reference ion concentration. The manufacturing approach sketched here is contrasted further below (under ISFETs) with an approach where the electronics are integrated. In Volume III, Chapter 1 we elaborate on the MEMS batch and continuous mode manufacture of arrays

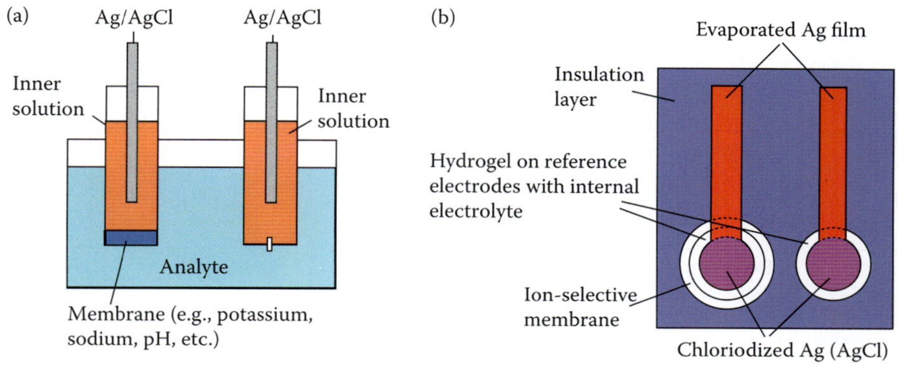

FIGURE 7.49 (a) A 3D measuring setup for polymeric ISEs. (b) A miniaturized 2D polymeric ISE.

of ISEs. In Table 7.14 we list membrane compositions and selectivities of important ISEs.

Membrane-Based Gas-Sensing Probes Gases that react with water, which causes freeing or absorbing of protons in the electrolyte, can be detected by a pH-sensitive element, e.g., a glass or an IrO_x electrode. A gas-permeable membrane (microporous), such as a 100-μm-thick Gore-Tex® film, separates the unknown sample from a pH probe and a reference electrode immersed together in a small internal solution (no ions can cross the gas-permeable membrane). The pH of the internal solution provides one with a measure of the gas content of the analyte solution. Gases that can be quantified this way include CO_2, NH_3, and H_2S. A direct proportionality exists between the concentration of the neutral gas and the measured pH. For example, in the case of the Severinghaus-type CO_2 sensor, depicted in Figure 7.50, the indicator electrode is only H^+-sensitive, and the potential of the reference is constant because of the constant Cl^- concentration in the internal electrolyte. Therefore, the potential of the pH sensor in the internal reservoir is given as:

$$E = E^0 + 0.059 \log a_{H^+} \tag{7.65}$$

The CO_2 gas penetrates through the gas-permeable membrane and reacts with the inner electrolyte,

Table 7.14 Membrane Compositions and Selectivities of Important Ion-Selective Electrodes

Analyte	Membrane Type	Membrane composition	Linear Range (–log a_i)	log $K^{pot}_{i,j}$
H^+	Glass	72.2% SiO_2, 6.4% CaO, 21.4% Na_2O (mol%)	2–12	Na^+:–11 K^+:–11
H^+	Polymer	Tri-*n*-dodecylamine, PVC, *bis*-2-ethylhexylsebacate, tetraphenylborate	3–12	Na^+: –10.4 K^+: –9.8 Ca^{2+}: –11
Na^+	Glass	11% Na_2O,18% Al_2O_3, 71% SiO_2	1–6	K^+: –2 Ag^+: +2.6 NH_4: –4.2 H^+: 1–2.5
Na^+	Polymer	ETH 227, PVC, 2-nitrophenyloctyl ether, tetraphenylborate	1–3	Li^+: +0.3 K^+: –1.7 Ca^{2+}: –0.4 Mg^{2+}: –2.4
K^+	Polymer	Valinomycin, PVC, dioctyladipate	0–5	Na^+: –4.2 NH_4^+: –1.9 Ca^{2+}: –3.3 Mg^{2+}: –4.4
Li^+	Polymer	ETH 1810, PVC, 2-nitrophenyloctyl ether, tetraphenylborate	<2–4	Na^+: –2.2 K^+:–2.3 NH_4^+: –2.3 Ca^{2+}–27 Mg^{2+}: –4 H^+: +1
Li^+	Polymer	Dodecylmethyl-14-crown-4, PVC, 2-nitrophenyloctyl ether, tetraphenylborate	0–5	k^+: –1.8 Na^+:–2.1 $NH_4{}^+$: –2.8 Ca^{2+}: –4.2 Mg^{2+}: –4.6 H^+: –3.2
Ca^{2+}	Polymer	ETH 1001, PVC, 2-nitrophenyloctyl ether, tetraphenylborate	2–7	Na^+:–5.5 K^+:–5.4 Mg^{2+}: –4.9
Ca^{2+}	Polymer	Ca-di-(*n*-decyl) phosphate, PVC, di-(*n*-octylphenyl) phosphonate	0–5	Na^+:–4.4 K^+:–4.5 Mg^{2+}: –4.9
Mg^{2+}	Polymer	ETH 5282, PVC, 2-nitrophenyloctyl ether, KTpClPB	<2–5	Na^+:–3.8 K^+:–1.4 Ca^{2+}: –2.4
Ag^+	Solid state	Ag_2S	1–7	Cu^{2+}: –6 Pb^{2+}: –6 to –9 H^+: –5 Hg^{2+}: –2

which contains a high fixed amount of Cl^- and bicarbonate ($NaHCO_3$) in a hydrogel matrix:

$$CO_2 + H_2O \rightleftharpoons H^+ + HCO_3^- \quad (7.66)$$

and:

$$a_{H^+} = \frac{Kp_{CO_2} a_{H_2O}}{a_{HCO_3^-}} \quad (7.67)$$

Because the activities of water and HCO_3^- are constant, one obtains:

$$E = E^{0'} - 0.059\,pH \text{ and also } E = E^{0''} - 0.059\,pCO_2 \quad (7.68)$$

The Severinghaus sensor shown in Figure 7.50a was fabricated by the author and his team using the same Ir/IrO_x wire illustrated in Figure 7.44.[8] For the reference electrode, a Ag/AgCl wire was used. Figure 7.50a is a schematic of the CO_2 sensor, whereas Figure 7.50b is a photo of the PVC tube-based sensor.

The world's smallest MEMS-fabricated electrochemical CO_2 sensor, incorporating thin-film Ir/IrO_x and Ag/AgCl electrodes, mounted in a catheter, together with an oxygen and a pH sensor, also made by this author and his team, is shown in Volume II, Figures 4.92–4.97. These miniaturized sensors are used clinically to determine blood CO_2 concentration.

Mass and Charge Transport in Solutions

As we saw in Chapter 6 on fluidics there are basically three modes of mass transport in solution: diffusion, migration, and convection (natural and mechanical). Diffusion from a region of high concentration to a region of low concentration occurs whenever the concentration of an ion or molecule at the surface of an electrode is different from that in the bulk of the solution. This will occur whenever a chemical change takes place at a surface. Convection takes place when mechanical means are used to carry reactants toward the electrode and to remove products from the electrode. The most common means of convection is to stir the solution using a stirring bar. Other methods include rotating the electrode and incorporating the electrode into a flow cell. Migration happens when charged particles in solution are attracted or repelled from an electrode that has a surface charge. The flux of material to and from the electrode surface is a complex function of all three modes of mass transport, and for most analytical techniques, one wants diffusion to dominate over migration. The reason, as we will find out further below, is that at the limit in which diffusion is the only significant means for the mass transport, the diffusion-limited current, i_l, in a voltammetric cell is simply related to the bulk concentration c_{bulk} $(x = \infty)$ by:

$$i_l = nFAD_0\left[\frac{c_{bulk}(x=\infty)}{\delta}\right] \quad (7.124)$$

where n = number of electrons transferred in the redox reaction
F = Faraday's constant
A = area of the electrode

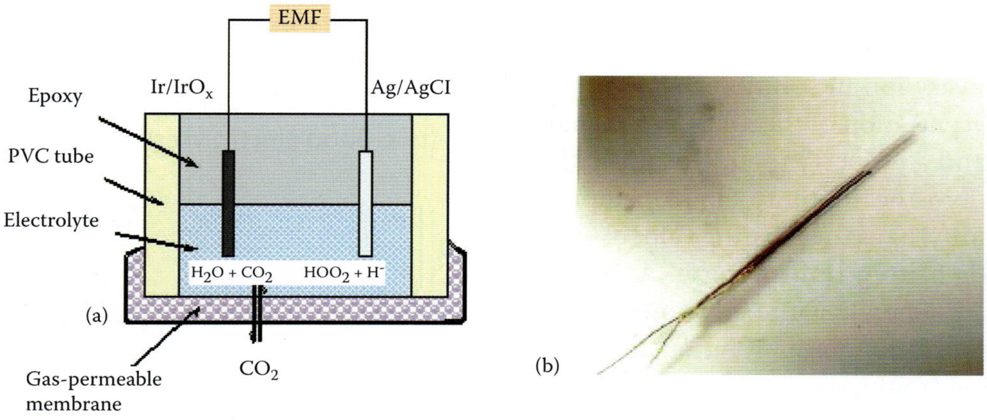

FIGURE 7.50 (a) Schematic of Severinghaus-type CO_2 sensor. (b) Photo of the sensor in a PVC tube. Notice the two wires: an Ir/IO_x wire and an Ag/AgCl wire. (By Dr. Marc Madou.)

D_0 = diffusion coefficient for the reactant or product
c_{bulk} $(x = \infty)$ = concentration of the analyte in the bulk solution
δ = thickness of the diffusion layer

For the above equation to be valid, migration and convection must not interfere with the formation of the diffusion layer around the electrode surface. Adding a high concentration of an inert supporting electrolyte to the analytical solution can eliminate migration. The large excess of inert ions ensures that few reactant and product ions will move as a result of migration. Although convection may be easily eliminated by not physically agitating the solution, in some situations it is desirable either to stir the solution or to flow the solution through an electrochemical flow cell. Fortunately, the dynamics of a fluid moving past an electrode results in a small diffusion layer, typically of 0.001–0.01-cm thickness; in this stagnant layer the rate of mass transport by convection drops to zero. Below our calculations for δ, the thickness of the steady-state diffusion layer, will reveal it to be only a fraction of the thickness of the stagnant hydrodynamic boundary layer, δ_h (see footnote on page 582).

The conductance *S* of an electrolyte, in units of Siemens (S), is given by:

$$S = \frac{1}{R} = \frac{\rho A}{l} \tag{7.69}$$

where R = resistance of the electrolyte
A = active area of contact
l = length (thickness) of electrolyte layer
ρ = conductivity–intrinsic property of the electrolyte (in S/cm)

The conductivity ρ of the electrolyte has a contribution from every ion in solution that is proportional to the concentration of the ions (*c*), the charge of the ions (*z*), and a property that determines their migration-velocity—the mobility, μ. In Chapter 6 we defined mobility, μ, as the limiting velocity of the ion in an electric field of unit strength. As an electric field (F = eEz, where *e* is the electronic charge) forces ions to move through the solution, an opposing force resulting from frictional drag is generated. The latter is represented by the Stokes equation:

$$F_D = 6\pi\eta V r \qquad \text{(see Equation 6.26)}$$

where η = viscosity of the solution
r = Stokes radius of the ion
V = ion velocity

When the opposing frictional force exactly counterbalances the force exerted on the ion by the electrical field, the ion attains a steady-state velocity called the *terminal velocity*. This terminal velocity is termed the mobility (μ) when the electric field strength is unity:

$$\mu = \frac{ze}{6\pi\eta r} \tag{7.70}$$

With this expression for mobility μ, we can write the following expression for conductivity ρ:

$$\rho = F\sum_i z_i \mu_1 c_i \tag{7.71}$$

This is often expressed as molar ionic conductivity Λ, which is given as:

$$\Lambda = \frac{\rho}{c} \tag{7.72}$$

where *c* is in moles and Λ is in units of S m²/mol. Some examples of molar ionic conductivity for ions in water at 25°C and at infinite dilution are given in Table 7.15.

At infinite dilution, no interionic interactions take place; thus, the molar conductivity of a salt C_cA_a can simply be calculated from its constituent ions as:

$$\Lambda_{C_cA_a} = c\Lambda_c + a\Lambda_a \tag{7.73}$$

Example 7.7:

Calculate the molar conductivity of KCl and $BaCl_2$ (two good electrolytes):

- $\Lambda_{KCl} = \Lambda_{K^+} + \Lambda_{Cl^-} = (74 + 77) \times 10^{-4}$ S m²/mol
  $= 151 \times 10^{-4}$ S m²/cm
- $\Lambda_{BaCl_2} = \Lambda_{Ba^{+2}} + 2\Lambda_{Cl^-} = [127 + (2 \times 77)] \times 10^{-4}$ S m²/mol $= 281 \times 10^{-4}$ S m²/mol

A useful measure of how much of the current is carried by the cations versus the anions of a salt

Table 7.15 Molar Ionic Conductivity

Ion	Λ(10^{-4} S m²/mol) at Infinite Dilution (25°C)
H^+	350
OH^-	200
Ba^{2+}	127
K^+	74
Cl^-	77
HCO_3^-	45
TBA^+	24

is the transference or transport number. The fractions of current carried by the positive and negative ions are given by their transport numbers t_+ and t_-, respectively:

$$t_+ = c_+\Lambda_+ / \Lambda_{C_cA_a} \tag{7.74}$$

$$t_- = a_-\Lambda_- / \Lambda_{C_cA_a} \tag{7.75}$$

where $c_+\Lambda_+ + a_-\Lambda_- = \Lambda_{C_cA_a}$. Because each mole of current passed through the cell corresponds to 1 mole of electrochemical change at each electrode, the amount of ions transported in the electrolyte also equals 1 mole. Thus:

$$t_+ + t_- = 1 \tag{7.76}$$

Example 7.8:

Calculate the transference numbers for the cations in KCl and $BaCl_2$ (two good electrolytes):

- t_+ (KCl) = 74/151 = 0.49
- t_+ ($BaCl_2$) = 127/281 = 0.45

Solvent and concentration affect ionization and therefore both ionic conductivity and transference numbers.

For multiple species, Equation 7.76 is generalized to:

$$\sum_i t_i = 1 \tag{7.77}$$

This equation is valid when more than one type of ion is in the solution. Now the transport numbers of ions are determined by the conductivity (ρ) of the electrolyte (Equation 7.71). The transport numbers may be represented as the ratio of the contribution to conductivity made by a particular ion to the total conductivity of the solution, or based on Equation 7.71:

$$t_i = \frac{\rho_i}{\rho} = \frac{z_i\mu_i c_i}{\sum_i z_i\mu_i c_i} \tag{7.78}$$

This equation gives us a clue how to eliminate migration for an analyte ion of interest, say, redox species *i*. Current flow resulting from electrostatic attraction (or repulsion) of analyte ion *i* by the electrode is reduced to a negligible level by the presence of a high concentration of supporting or indifferent electrolyte. By increasing the concentration for all the other ions but ion *i*, the transport number for ion *i* is minimized. The only way for an analyte ion, present in much lower concentration than the supporting electrolyte, to reach the electrode is by diffusion. Typically, a supporting electrolyte concentration 50–100 times greater than the analyte concentration will reduce electrostatic transport of the analyte to a negligible level and minimize the iR drop in the solution as well. Under the latter circumstances, qualitative and quantitative analysis of an ion analyte, using Equation 7.124 (see further below), is possible.

Energetics of Metals, Semiconductors, Membranes, and Mixed Conductors in Contact with an Electrolyte

Introduction

In Chapters 3 and 4, we introduced the band theory of solids and learned how to draw energy diagrams for solid-state junctions, such as metal/semiconductor and *p*-type and *n*-type semiconductor combinations. In this section, we derive the energy diagrams for solid/liquid junctions that explain the operation of these materials as sensor electrodes, in fuel cells, in batteries, as liquid junction solar cells, ISFETs, during corrosion processes (including the anisotropic and isotropic etching of semiconductors), and so on.

When electrons or ions pass through a solid/electrolyte interface, a reversible electrode potential E_{rev} is set up. This equilibrium is a dynamic equilibrium, meaning that the rates of charge carriers passing through the interface in either direction are equal. We denote the forward electron/ion transfer and the backward electron/ion transfer rates as $\vec{i}$ and $\overleftarrow{i}$, respectively. At equilibrium, the exchange current density is the same in both directions, or $\vec{i} = \overleftarrow{i} = i_0$. Thermodynamic equilibrium between two phases is reached when the electrochemical potentials on either side of the interface become equal. To refresh the reader's understanding of this process, we briefly reconsider here the relation between chemical potential μ, electrochemical potential $\bar{\mu}_e$, and Fermi level E_F. Every substance has a unique propensity to contribute to a system's energy. We call this property the chemical potential μ. When the substance is a charged particle (such as an electron or an ion with charge z) we must include the response of the particle to an electrical field in addition to its chemical potential. This sum is called the *electrochemical potential*, which in the case of electrons is given by $\bar{\mu}_e = \mu + zF\phi$ with F the Faraday constant and ϕ the potential drop. It can be demonstrated that the Fermi level, E_F, in a given material is equivalent to the electrochemical potential of the electrons, $\bar{\mu}_e$, in that material expressed per particle, or:

$$E_F = \frac{\bar{\mu}_e}{N_A} \tag{7.79}$$

As explained in Chapters 3 and 4, when bringing a solid in contact with another solid, the Fermi levels line up, and a potential drop ϕ is established across the interface. The potential drop is always the largest in the phase with the highest resistance (Ohm's law). For example, for a metal/semiconductor contact the whole potential drop is located in the semiconductor space charge (there are many more charge carriers in a metal than in a semiconductor), but for a metal/electrolyte contact the potential drop is all in the electrolyte (there are many more charge carriers in the metal than in the electrolyte).

Metal/Electrolyte Interface

Introduction In this section we detail the electron exchange current $i_{0,e}$ at a metal/electrolyte contact, as well as the current i as a function of the applied overvoltage η. We first derive these properties of the metal/liquid junction phenomenologically before we approach the same problem semiquantitatively, introducing the energetics of the solid/liquid interface. For simplicity in equations we will abbreviate $i_{0,e}$ as i_0.

Phenomenological Description of the Exchange Current $i_{0,e}$ and the Relationship between Current and Metal Electrode Overpotential We will first assume that the reactions at the metal electrode involve the two partners of the same redox system (say, Fe^{3+} and Fe^{2+}), so that at equilibrium no macroscopic changes take place at the electrode. The forward current $\vec{i}$ ($Fe^{3+} + e^- \rightarrow Fe^{2+}$) and the backward current $\overleftarrow{i}$ ($Fe^{2+} - e^- \rightarrow Fe^{3+}$) cancel each other out, i.e., the exchange current density is the same in both directions, or $|\vec{i}| = |\overleftarrow{i}| = i_0$ and the total current $i = \vec{i} - \overleftarrow{i} = 0$. This situation is called a *dynamic equilibrium at a potential E^0*. When no thermodynamic equilibrium is established for zero total current ($i = 0$ but $i \neq i_0$), $\vec{i}$ and $\overleftarrow{i}$ belong to different redox couples, and macroscopic changes occur at $i = 0$. In this case the potential at $i = 0$ is called a *mixed potential, E^{mix}* (e.g., in iron corrosion two redox couples, one involving oxygen and one iron, are responsible for this mixed potential).

Overpotentials occur when an external current is superimposed on the electrochemical cell. The electrode then assumes a voltage E_i different than the equilibrium value E^0 to an extent governed by the current i. The difference between E_i and the equilibrium voltage is known as the *overpotential*, generally denoted by η:

$$\eta = E_i - E^0 = \phi - \phi^0_{eq} \tag{7.80}$$

The rate of electron transfer between the metal and the redox couple, v, can be simply described as in homogeneous chemical kinetics:

$$v = k_{ox} c^s_{red} - k_{red} c^s_{ox} \tag{7.81}$$

From the absolute rate theory it follows that:

$$k_{ox} = A \exp\left[-\frac{\Delta G^*_{ox}(\phi)}{kT}\right] \qquad k_{red} = A \exp\left[-\frac{\Delta G^*_{red}(\phi)}{kT}\right] \tag{7.82}$$

where ΔG^*_{ox} and ΔG^*_{red} are activation energies.

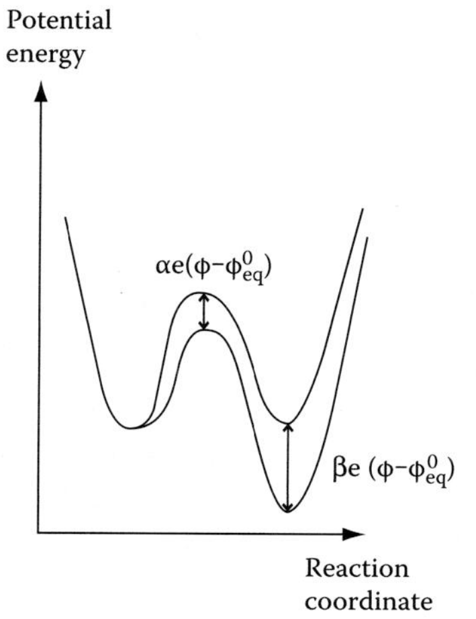

FIGURE 7.51 Reaction coordinate in the metal redox system.

The activation terms are dependent on the electrode potential but not the pre-exponential factor *A*, and with reference to Figure 7.51 we can write:

$$\Delta G^*_{ox}(\phi) = \Delta G^*_{ox}(\phi^0_{eq}) - \alpha e(\phi - \phi^0_{eq})$$

$$\Delta G^*_{red}(\phi) = \Delta G^*_{red}(\phi^0_{eq}) + \beta e(\phi - \phi^0_{eq})$$

$$\text{with } \alpha = -\frac{1}{e}\frac{\partial \Delta G^*_{ox}(\phi)}{\partial \phi}\Big|_{\phi^0_{eq}}$$

$$\text{and } \beta = \frac{1}{e}\frac{\partial \Delta G^*_{red}(\phi)}{\partial \phi}\Big|_{\phi^0_{eq}} \quad (7.83)$$

The Gibbs energy of activation and reaction are correlated by:

$$\Delta G^*_{ox}(\phi) - \Delta G^*_{red}(\phi) = G_{ox} - G_{red}$$

and

$$\Delta G^*_{ox}(\phi^0_{eq}) = \Delta G^*_{red}(\phi^0_{eq}) = \Delta G^*_{eq} \quad (7.84)$$

Assuming a transfer reaction in which the electrostatic energy of the ionic species is not significantly affected by the electrode potential (referred to as an *outer sphere electrode reaction*), it follows that:

$$\Delta G^*_{ox}(\phi) - \Delta G^*_{red}(\phi) = -e(\phi - \phi^0_{eq}) \quad (7.85)$$

Based on Equations 7.83 and 7.85, we obtain:

$$\alpha + \beta = 1 \quad (7.86)$$

We further develop Equation 7.81 to obtain:

$$i = ek_0 c^s_{red} \exp\left[\frac{\alpha e(\phi - \phi^0_{eq})}{kT}\right] - ek_0 c^s_{ox} \exp\left[-\frac{(1-\alpha)e(\phi - \phi^0_{eq})}{kT}\right] \quad (7.87)$$

$$\text{where } k_0 = A\exp\left(-\frac{\Delta G^*_{eq}}{kT}\right)$$

To connect voltages to concentrations we recall the Nernst expression (Equation 7.56), which we rewrite here as:

$$\phi_{eq} = \phi^0_{eq} + \frac{kT}{e}\ln\left(\frac{c^s_{ox}}{c^s_{red}}\right) \quad (7.88)$$

Using the expression for the overpotential η (Equation 7.80), we can finally derive the so-called *Butler-Volmer equation* for charge transfer reactions at a metal/electrolyte interface as:

$$i = i_0\left[\exp\left(\frac{\alpha e\eta}{kT}\right) - \exp\left(-\frac{(1-\alpha)e\eta}{kT}\right)\right] \quad (7.89)$$

This relationship is illustrated in Figure 7.52a for three different values of α. The transfer coefficient

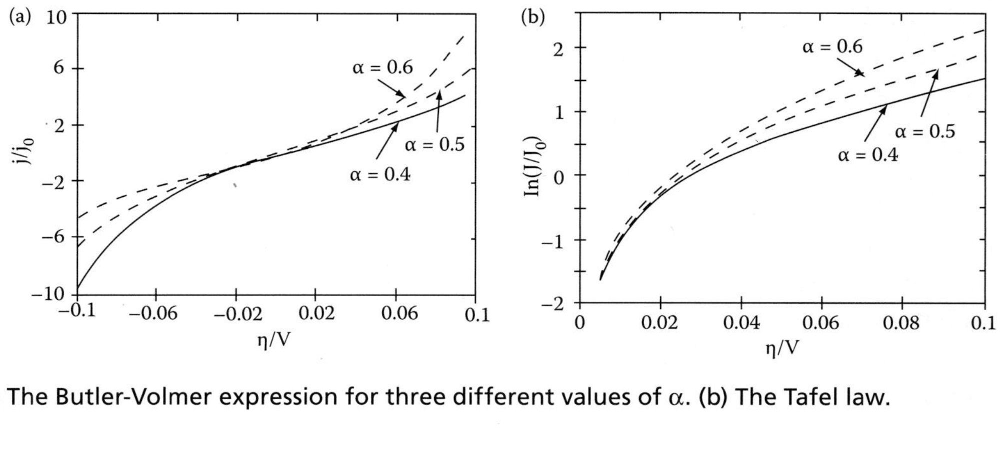

FIGURE 7.52 (a) The Butler-Volmer expression for three different values of α. (b) The Tafel law.

α determines the symmetry (or lack thereof) of the current potential curve. Notice that the Butler-Volmer analysis is based on surface concentrations. The exchange current density i_0 in Equation 7.89 is given by:

$$i_0 = ek_0\left(c_{red}^{s}\right)^{1-\alpha}\left(c_{ox}^{s}\right)^{\alpha} \quad (7.90)$$

For low overpotentials (η) the system behaves as an ohmic resistance:

$$i = i_0 \frac{e\eta}{kT} \quad (7.91)$$

It follows that the charge transfer resistance, R_{ct}, is given by:

$$R_{ct} = \frac{N_A kF}{N_A e i_0} = \frac{RF}{F i_0} \quad (7.92)$$

For large overpotentials the Butler-Volmer equation becomes logarithmic (Tafel regime):

$$\eta = a + \log i \quad (7.93)$$

The Tafel regime is illustrated in Figure 7.52b.

Semiquantitative Description of the Exchange Current $i_{0,e}$ and the Relationship between Current and Metal Electrode Overpotential

Equilibrium Current Exchange Density $i_{0,e}$ To derive the electron transfer rate at the metal/redox electrolyte interface semiquantitatively we must establish the energy density of empty and filled states that electrons can tunnel to and from on both sides of the interface. The question, as in the case of electron transfer between two solids, is whether there will be electron transfer from the metal electrode to the redox molecules in solution (or vice versa). This, we find, depends on the alignment of the density of states of the redox molecules relative to the Fermi level of the metal electrode. In Chapter 3, we learned how to calculate the density of states in solids, and here we do the same for redox couples in solution. In molecules, the lowest empty states are the lowest unoccupied molecular orbitals (LUMOs), and the highest filled states are the highest occupied molecular orbitals (HOMOs). The Fermi level in a solid metal is like the HOMO in a molecule (Figure 7.53).

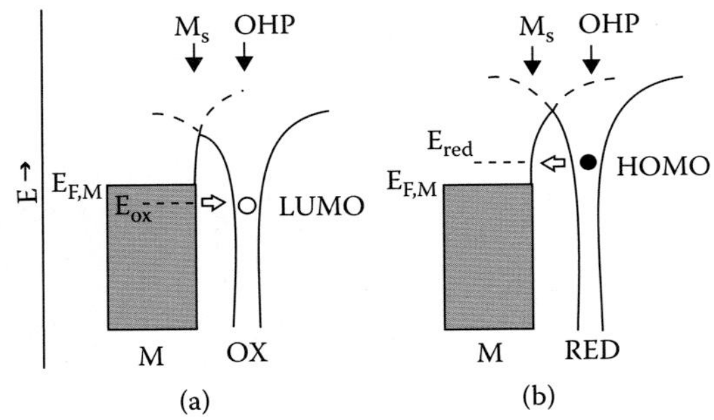

FIGURE 7.53 Solid metal/redox molecule interface energy levels. Fermi level of a solid metal and (a) energy levels of an oxidized (LUMO) and (b) a reduced (HOMO) redox species.

We can understand the concept of density of states of a redox couple in a solution in the following way. When a molecule is in the oxidized state, it has empty levels centered at E_{ox}. Once an electron occupies the oxidized state, the molecule relaxes to a lower energy level, marked as E_{red} in Figure 7.54. The relaxation includes the reorientation of the surrounding solvent molecules and changes in the chemical bonds of the reacting molecule. For example, consider the simple redox couple, Fe^{3+}/Fe^{2+}. The solvent polarization for Fe^{3+} is different from that of Fe^{2+}. Following the electron transfer $Fe^{3+} \rightarrow Fe^{2+}$, the dipoles of the solvent molecules reorient. The extra negative charge on the central ion causes the negative poles of the solvent dipoles to rotate away from the central ion, making the potential at the central ion higher and more attractive to electrons. This explains why E_{red} is always below E_{ox}. The shift in energy level

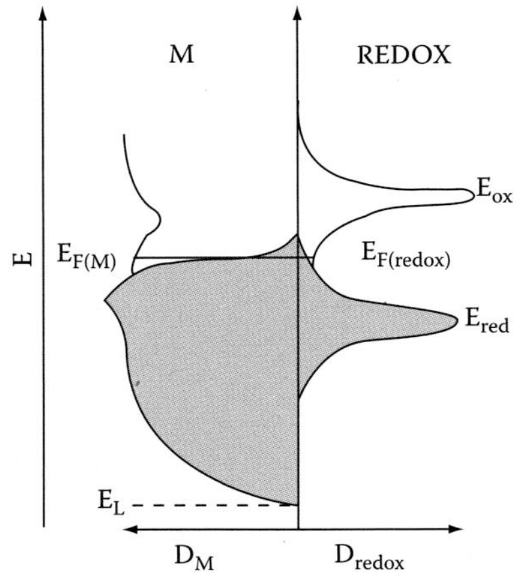

FIGURE 7.54 Time-fluctuating energy levels of the redox system can be regarded as bands of energy levels.

between oxidized (unoccupied) and reduced (occupied) levels is called the *Frank-Condon shift*. The Frank-Condon principle states that during electron transfer from one level to another, atoms and ions are effectively frozen in their positions; thus, they can move or relax only before or after the electron transfer. The amount of Frank-Condon shift, or $E_{ox} - E_{red}$, is given by:

$$E_{ox} - E_{red} = 2\lambda \tag{7.94}$$

where λ is called the *reorganization energy*. The reorganization energy contains contributions from all involved bonds and solvent polarization and is normally on the order of a few tenths of electron volts. In some cases, such as ions or metal ion complexes, the reorganization caused by bond length and angle changes is small, and the main contribution results from solvent polarization.

According to Morrison,[9] the time-fluctuating energy levels of the redox system can be regarded as bands of energy levels and are described by the densities of the oxidized and reduced states. The state density of electrons in the reduced $[D_{red}(E)]$ and oxidized $[D_{ox}(E)]$ states is given by the product of the probability density distribution, $W(E)$, and the concentration, c, or:

$$D_{red}(E) = c_{red}W_{red}(E) \text{ and } D_{ox}(E) = c_{ox}W_{ox}(E) \tag{7.95}$$

The total density of states for a redox couple is defined as:

$$D_{redox}(E) = D_{red}(E) + D_{ox}(E) \tag{7.96}$$

Similarly, for the density of states of electrons in the metal:

$$D_M(E) = D_{M(e)}(E) + D_{M(h)}(E) \tag{7.97}$$

where *e* means a filled state and h an empty one. The probability density distribution curves in solution, $W(E)$, take the form of:

$$W_{red}(E) = W_{red}(E_{red})\sqrt{\frac{c_{ox}}{c_{red}}}\exp\left(-\frac{\lambda_{red}}{4kT}\right)\exp\left\{-\frac{(1-\beta)\left[E - E_{F(redox)}\right]}{kT}\right\} \tag{7.98}$$

and:

$$W_{ox}(E) = W_{ox}(E_{ox})\sqrt{\frac{c_{red}}{c_{ox}}}\exp\left(-\frac{\lambda_{ox}}{4kT}\right)\exp\left\{-\frac{\beta\left[E - E_{F(redox)}\right]}{kT}\right\} \tag{7.99}$$

where β is the same symmetry factor we encountered in the Butler-Volmer expression; if $\lambda_{ox} = \lambda_{red}$, then $\beta = 0.5$. Equations 7.98 and 7.99 describe Gaussian distributions of energy states around E_{red} and E_{ox}, respectively, and these two curves, for $c_{ox} = c_{red}$, intersect at E^0. This intersection corresponds to the Fermi level of the redox species, $E_{F(redox)}$:

$$E_{F(redox)} = E^0 = (1/2)\,(E_{ox} + E_{red}) \tag{7.100}$$

Based on Equation 7.96 it then also holds that:

$$\frac{1}{2}D_{redox}\left[E_{F(redox)}\right] = D_{red}\left[E_{F(redox)}\right] = D_{ox}\left[E_{F(redox)}\right] \tag{7.101}$$

Consequently, we derive from Equations 7.96, 7.98, 7.99, and 7.101:

$$D_{redox}(E) = \frac{1}{2}D_{redox}\left[E_{F(redox)}\right]\left\{\exp\left\{\frac{\beta\left[E - E_{F(redox)}\right]}{kT}\right\} + \exp\left\{-\frac{(1-\beta)\left[E - E_{F(redox)}\right]}{kT}\right\}\right\} \tag{7.102}$$

This expression is sketched in Figure 7.55.

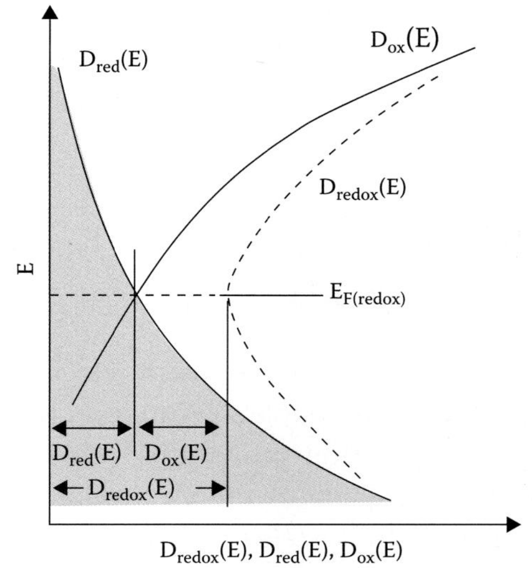

FIGURE 7.55 The state density of electrons in the reduced $D_{red}(E)$ and oxidized $D_{ox}(E)$ and the total density of states for a redox couple $D_{redox}(E) = D_{red}(E) + D_{ox}(E)$.

Because we derived expressions for the density of state functions on both sides of the interface, we are now in a position to calculate the equilibrium current exchange density i_0 across the metal/electrolyte interface. Electron transfer takes place by isoenergetic tunneling across the 0.3–0.5-nm compact layer at the metal electrode interface (OHP, see Figure 7.14). The rate of the reaction will depend on the density of states in the metal and in the redox couple. Using Fermi's golden rule (FGR) approximation (see Chapter 3, Equation 3.285), we calculate the probability of transitions of electrons between the eigenstates associated with the metal and the redox system. In this model one assumes that electron exchange occurs between levels of equal energies in both phases (the metal and the redox species in this case), and an electronic coupling strength, H, between metal electrode and the redox state modulates that transition probability. We can express the forward (cathodic) and backward (anodic) currents as a function of the energy level as:

$$\vec{i}(E) = ek_c D_{M(e)}(E) D_{ox}(E) \tag{7.103}$$

and:

$$\overleftarrow{i}(E) = ek_a D_{M(h)}(E) D_{red}(E) \tag{7.104}$$

where k_c and k_a are electron tunnel constants. Because we are dealing with electrons we also have to introduce the Fermi-Dirac distribution functions:

$$\vec{i}(E) = ek_c(E) D_M(E) f\left[E - E_{F(M)}\right] D_{redox}(E)(1-f)\left[E - E_{F(redox)}\right] \tag{7.105}$$

and:

$$\overleftarrow{i}(E) = ek_a(E) D_M(E)\left\{(1-f)\left[E - E_{F(M)}\right]\right\} D_{redox}(E) f\left[E - E_{F(redox)}\right] \tag{7.106}$$

With the Fermi level of the metal electrode being level with the redox couple's Fermi level [$E_{F(M)} = E^0 = E_{F(redox)}$], the oxidation probability (electron transfer from the molecule to the electrode) equals the reduction probability (electron from the electrode to the molecule), i.e., the exchange current density is the same in both directions, or $|\vec{i}| = |\overleftarrow{i}| = i_0$ and the total current $i = \vec{i} - \overleftarrow{i} = 0$. At equilibrium it also holds that the tunneling constant $k_t = k_a = k_c$, i.e., principle of microreversibility:

$$\begin{aligned} i_0 &= \vec{i}_0(E) = ek_t(E) D_M(E) f\left[E - E_{F(M)}\right] D_{redox}(E)(1-f)\left[E - E_{F(redox)}\right] \\ &= ek_t(E) D_M(E)\left\{(1-f)\left[E - E_{F(M)}\right]\right\} D_{redox}(E) f\left[E - E_{F(redox)}\right] = \overleftarrow{i}_0(E) \end{aligned} \tag{7.107}$$

The overall electron exchange current is obtained by integration of the microscopic current elements over the entire energy range:

$$i_0 = \int_{-\infty}^{\infty} ek_t(E) D_M(E) f\left[E - E_{F(M)}\right] D_{redox}(E)(1-f)\left[E - E_{F(redox)}\right] dE \tag{7.108}$$

Assuming $D_M(E)$ and k_t are energy-independent, Equation 7.108 can be somewhat simplified, and with further development this leads to:

$$i_0 = ek_t D_M\left[E_{F(M)}\right] \frac{1}{2} D_{redox}\left(E_{F(redox)}\right) B(\beta) \tag{7.109}$$

$$B(\beta) = \int_{-\infty}^{\infty} \exp\left\{\frac{\beta\left[E - E_{F(redox)}\right]}{kT}\right\} f\left[E - E_{F(M)}\right] dE \approx \frac{\pi kT}{\sin\left[(1-\beta)\pi\right]} \tag{7.110}$$

Ninety percent of the integrand falls in a zone of 0.25 eV around the Fermi level, and we finally obtain:

$$i_0 \approx \frac{e\pi kT}{2\sin\left[(1-\beta)\pi\right]} k_t\left[E_{F(M)}\right] D_M\left[E_{F(M)}\right] D_{redox}\left[E_{F(redox)}\right] \tag{7.111}$$

This expression is plotted in Figure 7.56. Figure 7.56a repeats the time-fluctuating energy levels of the redox system and the energy bands in the metal. The relation $k_t(E)$ is illustrated in Figure 7.56b, and $i_0(E)$ is shown in Figure 7.56c.

Out of Equilibrium: Current under Polarization Applying an overvoltage shifts the redox states of the molecule in solution relative to the Fermi level of the metal electrode. This triggers electron transfer between the electrode and the molecule in an energy

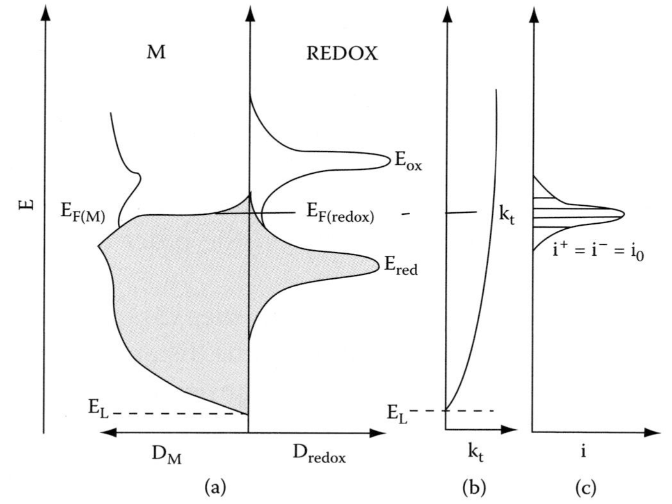

FIGURE 7.56 Using Fermi's golden rule, we can calculate the current exchange density i_0 across the metal/electrolyte interface in the presence of a redox couple. (a) Time-fluctuating energy levels of the redox system can be regarded as bands of energy levels. (b) The relation of tunneling constant k_t and energy E. (c) Electron exchange current density i_0 versus energy. The lowest energy level is E_L.

range close to the Fermi level. Out of equilibrium, the Fermi levels are offset by the overvoltage η or the total current $i(\eta) = \overleftarrow{i}(\eta) - \overrightarrow{i}(\eta) \neq 0$. Based on Equation 7.80 we rewrite the overvoltage here as:

$$\eta = E_{F(M)} - E_{F(redox)} \tag{7.112}$$

With this simple expression we can rework Equations 7.105 and 7.106 as:

$$\overrightarrow{i}(\eta) = e\int_{-\infty}^{\infty} k_t(E,\eta) D_M(E) f\left[E - E_{F(redox)} + e\eta\right] D_{redox}(E)(1-f)\left[E - E_{F(redox)}\right] dE \tag{7.113}$$

$$\overleftarrow{i}(\eta) = e\int_{-\infty}^{\infty} k_t(E,\eta) D_M(E)\left\{(1-f)\left[E - E_{F(redox)}\right] D_{redox}(E) f\left[E - E_{F(redox)}\right] dE \tag{7.114}$$

Using Equation 7.102, we find another Butler-Volmer-type expression (Equation 7.89: remember $\alpha + \beta = 1$) for the current i versus η:

$$i = i_0 \frac{k_t\left[E_{F(M)}\right]}{k_t\left[E_{F(redox)}\right]}\left[\exp\left(\frac{(1-\beta)e\eta}{kT}\right) - \exp\left(\frac{-\beta e\eta}{kT}\right)\right] \tag{7.115}$$

This expression explicitly contains the electron tunneling rate constant k_t, for which there are semi-quantitative data (see section below). In Figure 7.57, we illustrate forward and backward currents across the metal/electrolyte interface.

Notice that all the potential changes fall within the less conductive medium, i.e., the electrolyte in this case.

Electron Tunneling Transfer Rate k_t According to the Marcus theory (1992 Nobel Prize),[10] the electron transfer rate k_t in Equation 7.115 depends on the activation energy for electron transfer, ΔG^*_{ACT}, and the electronic coupling, H_{DA}, between the two electron-exchanging species (say, a redox enzyme and a metal electrode). The electronic coupling is often very difficult to calculate and is widely treated as a fitting parameter to experimental data. With reference to Figure 7.58, at the interception point of the potential energy curves, the acceptor energy level (A) is equal to the donor level (D). The energy

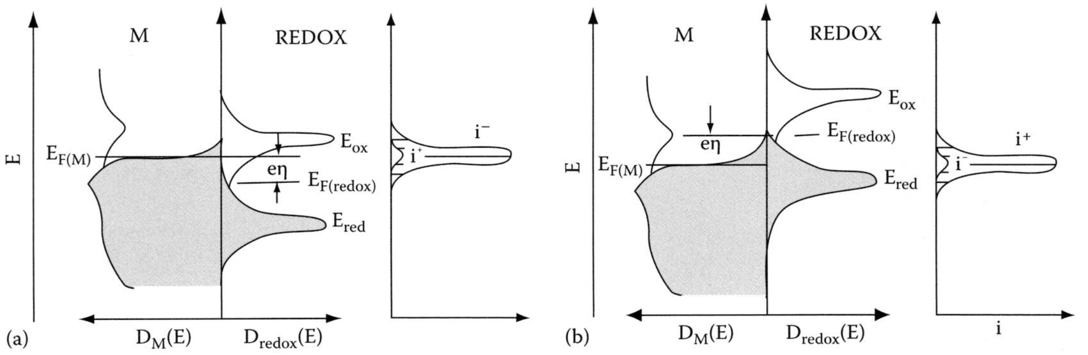

FIGURE 7.57 Applying a voltage across the metal/electrolyte interface that is different from the equilibrium potential leads to cathodic (a) or anodic (b) currents that are different from the equilibrium exchange current i_0.

barrier (activation energy) or charge transfer at this point is given as:

$$\Delta G^{*}_{ACT} = \frac{\lambda}{4}\left(1+\frac{\Delta G^{0}}{\lambda}\right)^{2} \quad (7.116)$$

where ΔG^0 is the free energy difference between initial and final states. In Figure 7.58 we made $\Delta G^0 = 0$ for simplicity so that both curves have a minimum at the same energy level. If thermal fluctuations change the distance between *D* and *A* (see reaction coordinate in Figure 7.58), then the electron transfer between donor and acceptor will be modified with a probability factor of:

$$k_t = \frac{1}{\sqrt{4\pi\lambda k_B T}}\exp\left[-\frac{(\Delta G^0+\lambda)^2}{4k_B T\lambda}\right] \quad (7.117)$$

In the case of strong coupling between the donor and acceptor orbitals, i.e., a large H_{DA}, electron transfer takes place as soon as the donor and acceptor levels line up. In this case, the electron transfer rate k_t is simply given as:

$$k_t = \nu \exp\left[-\frac{(\Delta G^0+\lambda)^2}{4k_B T\lambda}\right] \quad (7.118)$$

where ν is the frequency of the nuclear coordinate, which is typically between 10^{12} and 10^{14} s^{-1}. This extreme situation represents the so-called *adiabatic electron transfer*. On the other hand, in the weak coupling limit or the nonadiabatic case, the electron transfer rate, obtained by applying Fermi's golden rule (Equation 3.285), is given by:

$$k_t = \frac{2\pi}{\hbar}H_{DA}^2\frac{1}{\sqrt{4\pi\lambda k_B T}}\exp\left[-\frac{\Delta G^{*}}{k_B T}\right]$$
$$\text{where } \Delta G^{*} = \frac{\lambda}{4}\left(1+\frac{\Delta G^{0}}{\lambda}\right)^{2} \quad (7.119)$$

The above considerations include only one nuclear coordinate, but the result is essentially the same when multiple nuclear coordinates are considered except that the reorganization energy, λ, contains contributions from all involved bonds and solvent polarization. As mentioned before, the reorganization energy λ is normally on the order of a few tenths of electron volts.

Because H_{DA} in general diminishes exponentially with distance (like any orbital), the distance dependence of the rate of electron transfer thus can be described by:

$$k_t = k_0 e^{-\beta d} e^{-\frac{(\Delta G^0+\lambda)^2}{4k_B T\lambda}} \quad (7.120)$$

in which β (the attenuation factor) is a constant scaling the distance dependence and k_0 is the rate at close contact (~10^{13} s^{-1}). The medium between donor and acceptor strongly influences the rate of the electron transfer process through solvation and the value of β. Many researchers have focused on determining the electron transfer rate k_t as a function of the donor-acceptor separation that they systematically controlled experimentally. If donor and acceptor are separated by a vacuum, the interaction (through space interaction) is much less (β ≈ 2.8–3.8 $Å^{-1}$) than when a protein, DNA (β ≈ 1.4 $Å^{-1}$), or a solvent (through solvent coupling), like benzene, is the medium (β ≈ 1 $Å^{-1}$). In contrast, for highly conjugated organic bridges, β is much smaller, lying in the range of 0.2–0.6 $Å^{-1}$. With DNA, proteins immobilized on electrodes, and saturated organic molecules, β is found between these two limits. But if the energy level of the bridge becomes too low, it can work as a trap for electrons, and the transfer stops (now the electrons localize on the bridge).

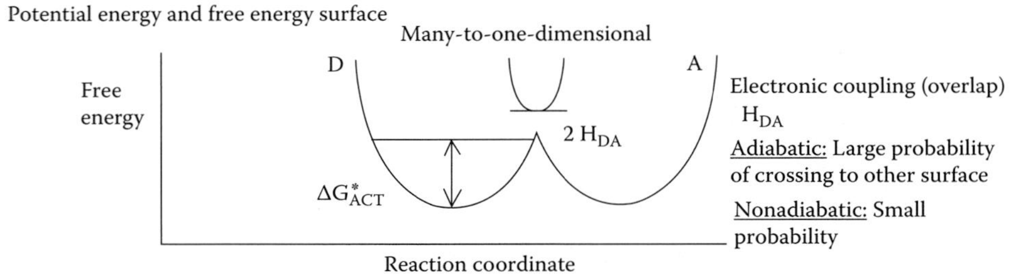

FIGURE 7.58 Potential energy and free energy surface, where H_{DA} is the electronic coupling and the electron transfer (ET) is thermal instead of optical.

Equation 7.120 is of utmost importance when attaching large molecules with reaction sites or redox centers to electrode surfaces. If the redox center is too far removed from the metal surface, no electron transfer can occur. Direct electron transfer between enzymes and metal electrodes, for example, is normally very slow as a consequence of steric insulation of the redox center in the protein. To illustrate this point, the "depth" of the redox center in glucose oxidase enzyme (GOx) is 8.7 Å, whereas for horseradish peroxidase (HRP) it is 4.1 Å. This is too deep below the protein surface for efficient electron transport to the underlying metal electrode. One solution to electronically connect to such a deep-lying redox center [flavin adenine dinucleotide ($FAD/FADH_2$) in the case of GOx (Figure 7.59)] is to rely on an electron acceptor molecule (mediator) to establish electrical communication with the electrode. Enzyme mediators are used in current (second-generation) commercial glucose sensors. Before introducing a second and even a third generation of these sensors, we consider how a first generation of glucose sensors operated.

When an enzyme reacts with its substrate S, product *P* is generated. In the case of GOx, this reaction is given as:

$$\text{Glucose (S)} + O_2 \xrightarrow{\text{GOx}} \text{glucono-}\delta\text{-lactone} + H_2O_2\ \text{(P)}$$
Reaction 7.5

As illustrated in Figure 7.60a, in this process the redox center in the enzyme is reduced from FAD to $FADH_2$, and it is the oxygen present in the sample that regenerates the enzyme by acting as an electron acceptor. Oxygen is called a *cosubstrate of the enzyme* in this case. To make a sensor based on this reaction, one might follow the anodic oxidation of H_2O_2 or the cathodic reduction of O_2. A first generation of glucose sensor was made this way.

In a second generation of sensors, the rationale was to monitor the redox center of the enzyme more efficiently. In the case of GOx, that means monitoring the $FAD/FADH_2$ redox couple directly. But, as we just saw, that redox center is located too far from the surface of the enzyme to be electrically accessible. The solution is to rely on a fast mediator redox system (MED_{red}/MED_{ox}) to establish electrical communication with the electrode (see Figure 7.60b). These mediators are chosen to regenerate FAD from $FADH_2$ much faster than oxygen does; in other words, a mediator redox couple where MED_{ox} is a much better electron acceptor than oxygen. The ferrocene/ferrocinium redox couple self-exchanges electrons very rapidly and is used as a diffusing mediator to shuttle electrons between the protein redox center and the electrode. Mediators can also be provided at much larger concentrations than

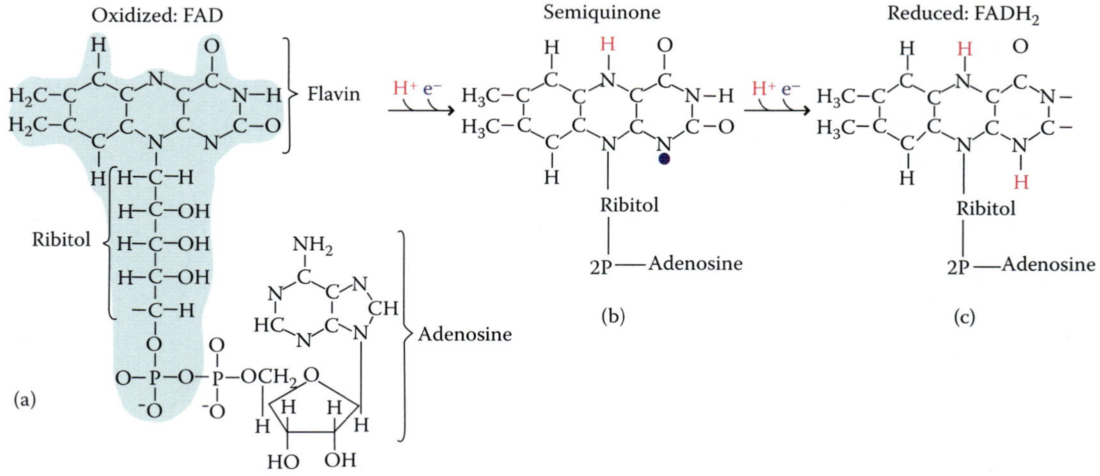

FIGURE 7.59 The redox center, flavin adenine dinucleotide ($FAD/FADH_2$). In the case of glucose oxidase (GOx) this center is buried 8.7 Å deep within the enzyme. The reduced form of the redox center or hydroquinone form is $FADH_2$ (a), and the oxidized form or the quinone form is FAD (c). Flavin adenine dinucleotide (FAD) can accept one or two hydrogen atoms. The addition of one electron together with a proton (i.e., a hydrogen atom) generates a semiquinone intermediate (b). The semiquinone is a free radical because it contains an unpaired electron (denoted by a blue dot), which is delocalized by resonance to all the flavin ring atoms. The addition of a second electron and proton (i.e., a second hydrogen atom) generates the reduced form, $FADH_2$.

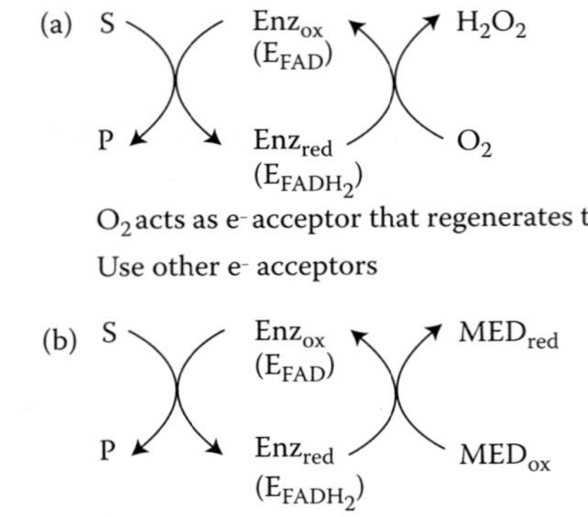

FIGURE 7.60 Glucose oxidase (GOx) reaction with its substrate (*S*) glucose in the presence of oxygen (a) and a mediator (b).

oxygen, which also makes electron transfer faster. Besides a higher current, another added benefit of these mediated enzyme-based sensors is that they are independent of the oxygen content in the sample, a factor that sometimes limits Reaction 7.5. During the regeneration of FAD, ferrocene (MED_{red}) is produced and can be measured by anodic oxidation. Alternatively, the decrease in ferrocinium (MED_{ox}) can be monitored by cathodic reduction. Medisense/Abbott uses a ferrocene derivative (carboxy ferrocene) in their commercial glucose sensor for diabetes management. This sensor is also used as an example of an amperometric device at the end of this section.

Third-generation glucose sensors were pioneered by Adam Heller and involve wiring the enzyme redox center directly to the electrode.[11,12] The $FADH_2$ redox centers of GOx can be electrically "wired" to an electrode by precipitation on the surface of thin films of the electrostatic adduct of GOx, which is a polyanion at neutral pH, and polycationic copolymers of vinyl ferrocene. This approach has the benefit of immobilizing the mediator molecules. The best approach, also invented by Heller, is currently the use of an electroactive enzyme-containing hydrogel. This is made by cross-linking nonprecipitating electrostatic adducts of polyanionic GOx and an excess of a polycationic osmium complex redox center containing polyvinyl pyridine-based redox polymer.[13,14] This cocktail creates a 3D matrix that is reversibly electroreduced and electrooxidized, extending electrochemistry from the 2D surface of the metal electrode to the 3D volume of the electron-conducting hydrogel. These wiring gels are highly permeable to facilitate the transport of reactants and products. In the volume of the hydrogel, electrons and holes are transported through collisions between Os^{2+} and Os^{3+} redox centers, which are tethered to the polyvinyl pyridine backbone, but with polymer segments that remain mobile within the polymer. A more recent optimized version of such a polymer, designed by Fei Mao,[15] is shown in Figure 7.61. In this example, the enzyme is wired using a polymer chain that folds along the enzyme structure.

The state-of-the-art glucose sensors are reviewed as the biggest bio-MEMS commercial success in Volume III, Chapter 10.

Semiconductor/Electrolyte and Insulator/Electrolyte Interface The semiconductor/electrolyte interface behaves significantly different than the metal/electrolyte interface. First, because there are many more free charge carriers in the solution than in the semiconductor, all potential changes in the electrical double layer occur on the semiconductor side, not in the electrolyte. Second, because of the

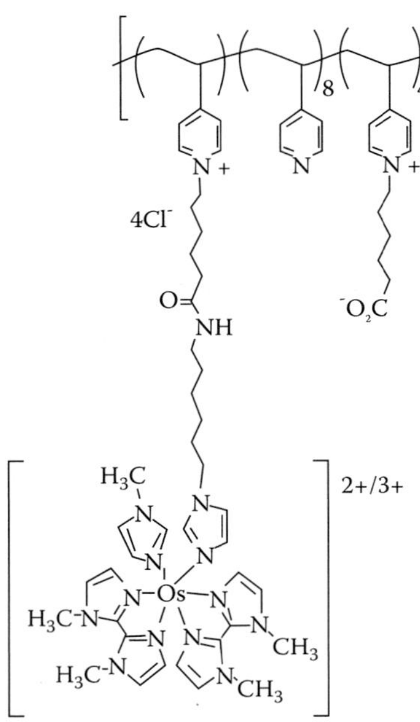

FIGURE 7.61 Os^{2+} and Os^{3+} redox centers are tethered to the polyvinyl pyridine backbone, but polymer segments remain mobile within the polymer. A more recently optimized version of such a polymer, designed by Fei Mao. (Mao, F., N. Mano, and A. Heller. 2003. *J Am Chem Soc* 125:4951.[15])

bandgap in the semiconductor, many electron transitions between redox levels and semiconductor are forbidden. In the case of an insulator/electrolyte interface, all electron transitions are forbidden. In Figure 7.62, we show what happens when we bring the same redox couple in contact with an n- and p-type semiconductor of the same material (e.g., *n*- and *p*-type Si). As in the case of the metal/electrolyte interface, equilibrium is reached when Fermi levels have lined up across the interface, which leads to a space-charge layer in the semiconductor. The reductant of a redox couple (filled electron levels) may inject electrons in the empty conduction band of the semiconductor if those energy levels overlap. Similarly, the oxidant may inject holes in the valence band if there is energy overlap in those bands.

A redox couple with energy levels that overlap only with the semiconductor bandgap does not lead to any current. Mathematically, this is expressed in the overlap integrals of Equations 7.113 and 7.114. In the case of semiconductors, there is a small range, known as the *forbidden bandgap*, where the overlap integral vanishes. In the case of an insulator that gap is large, and the overlap integral is always zero. To illustrate the previous points, consider Figure 7.63, where we show an n-type semiconductor/electrolyte contact for three different redox systems. For redox system 1, with its standard redox potential (E^0_{redox}) close to the conduction band edge, a high rate of electron exchange can be expected, and an accumulation layer forms in the semiconductor. For redox system 2, a depletion layer is formed in the semiconductor, and the electron exchange is expected to be low as the redox system significantly overlaps with the forbidden bandgap. In the case of redox system 3, the barrier height in the semiconductor is even higher (close to E_g, the bandgap), but (E^0_{redox}) is now close to the valence band edge and electron exchange is possible again. In the latter case, empty levels (holes) must be available in the valence band for electrons of reductant molecules to jump into. These holes can be generated, for example, by shining light on the semiconductor or by a strong oxidant. These holes react with the semiconductor itself and, in the presence of the appropriate electrolyte solution, may etch/corrode the semiconductor material. In Volume II, Chapter 4, in the section on wet bulk micromachining, we will expand on the concepts developed here to explain the difference between isotropic and anisotropic etching of silicon. We will also learn that current/overvoltage curves (i-η) on semiconductor/electrolyte contacts take on distinct diode characteristics. In liquid-junction solar cells, appropriate reductants compete effectively with the oxidation/corrosion of the semiconductor because they react more avidly with the holes in the valence band than the semiconductor material itself.[16]

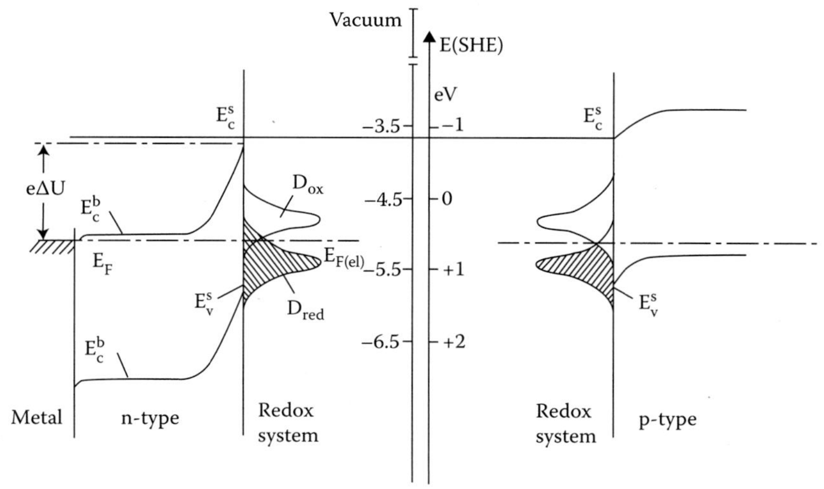

FIGURE 7.62 Energy diagram of the same redox couple in contact with *n*- and *p*-type Si (two energy scales are shown: electron in vacuum at infinity and SHE).

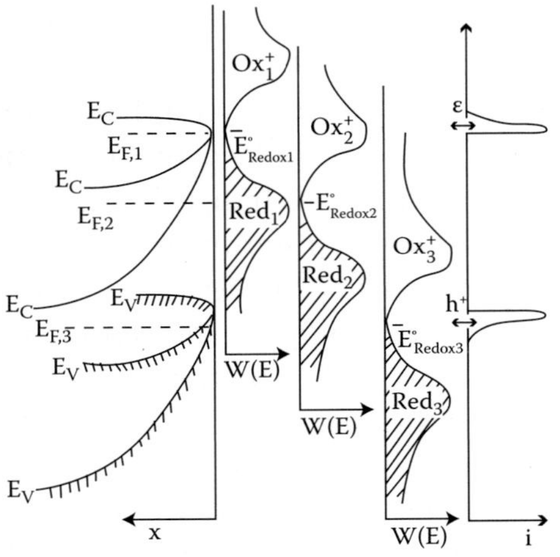

FIGURE 7.63 Energy diagram of three different redox couples in contact with n-type silicon.

For a certain value of the applied potential, there is equality between the number of electrons removed from and supplied to the semiconductor electrode. In this situation there will be no space-charge region in the semiconductor, and the potential is called the *flat band potential* E_{fb} (= E_F in the semiconductor). This potential can be established from impedance measurements on the semiconductor/electrolyte interface. From the flat band potential, the band edges of the semiconductor can be calculated if the dopant concentration is known (see Equations 4.38 and 4.39) (for details, see Madou and Morisson[15]). In Figure 7.64, we show a number of common semiconductor band edges and redox couples all referenced to the same reference electrode (REF E) on the right and with respect to an electron at infinity in vacuum on the left. Using this figure one can, in principle, predict the expected electron exchange for any semiconductor redox couple system by evaluating the overlap integral of energy levels on both sides of the interface (again using Equations 7.113 and 7.114). For example, in liquid-junction solar cells, for the reductant to compete effectively for holes in the valence band and avoid corrosion of the solar cell, one chooses a redox system where the maximum density of states of the reductant (the peak of the Gaussian) is at the same level as the valence band edge of the semiconductor.[15]

In a flat band situation, a potential drop remains on the solution side, and this remaining potential drop can be changed by ion adsorption on the semiconductor surface. Semiconductors in general develop a thin oxide coat on their surface. The hydroxyl groups at the surface of these oxide films will be more or less ionized, depending on the pH of the solution. This makes for a flat band potential,

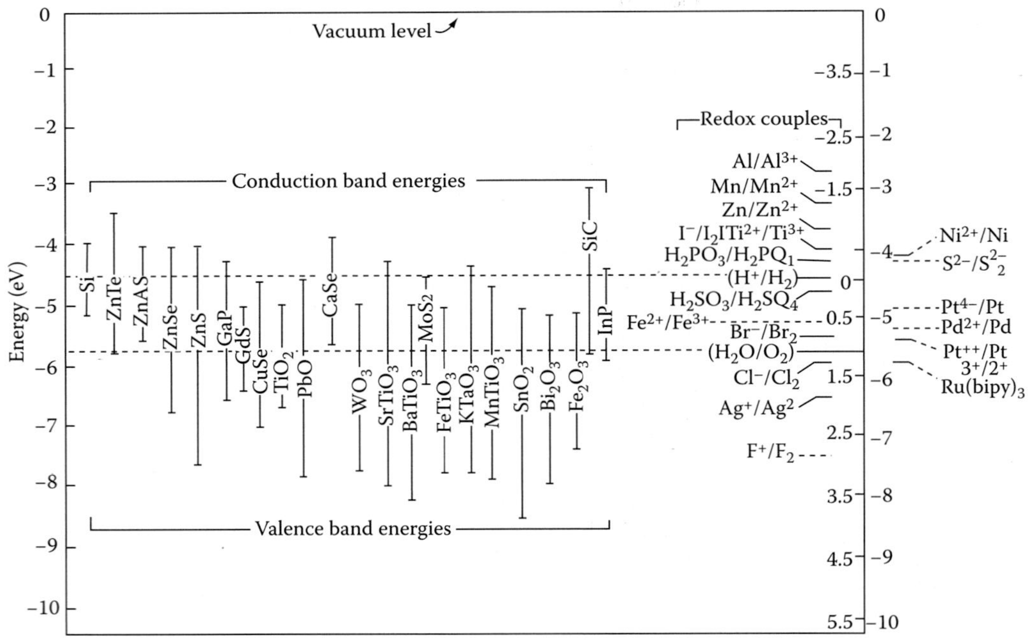

FIGURE 7.64 Relative energy levels of some common semiconductor and redox systems. The scale on the left is the energy with respect to an electron at infinity in vacuum. The scale on the right is the standard hydrogen electrode (SHE).

E_{fb}, that is a function of the pH of the solution (typically about 59 mV/pH unit change). If the redox system is not pH-dependent, this means that we can shift the semiconductor band edges with respect to the redox levels in solution. This is exploited in semiconductor electrochemistry to probe the density of energy levels of redox couples by changing the pH in small steps and observing the measured currents across the interface.[9]

With semiconductors, considerable electron transfer can only occur if the redox potential is located close to the band edges. In such a case, the electron equilibrium exchange current, $i_{0,e}$ is high but lower than for a metal. Any ion exchange processes, e.g., proton exchange, at the semiconductor surface would have difficulties competing with the electron exchange for establishing the open circuit voltage ($i_{0,e} >> i_{0,i}$). In principle, the pH dependence of the flat band potential of a semiconductor could be used as a good pH sensor, but only if the electron exchange current does not overwhelm the ion exchange current. In other words, in the presence of redox couples, a wide bandgap semiconductor or, better yet, an insulator, would make the better solid-state pH sensor. However, the pH sensitivity of an insulator normally cannot be measured because a thick insulator has too high of an impedance. Further below, we will see that the solution is to use an ion-sensitive field-effect transistor (ISFET), where the dielectric is so thin that a substantial electric field across it can influence the drain-source current of the transistor. These devices enable one to measure potentials on insulator/electrolyte contacts, ensuring that no electron exchange current ($i_{0,e}$) will interfere with the ion exchange current ($i_{0,i}$).

Membrane/Electrolyte and Mixed Conductor/Electrolyte Interfaces At an ionically conductive membrane/electrolyte interface, ideally only the fast exchange of one type of ion with the solution determines the open circuit potential. Because ions do not follow Fermi-Dirac statistics, the theory of ion transfer at the solid/liquid differs from that of electron exchange. Despite this obvious difference and the fact that ions must leave behind their hydration shell if they want to cross the phase boundary, the calculations for the equilibrium ion exchange current, $i_{0,i}$, and the polarization curves (i-η) lead to results very similar to those obtained for electron transfer at the metal/electrolyte interface. The ionic species that carries the greatest number of electric charges across the phase boundary is potential determining (largest $i_{0,i}$). For a "good" ion-selective membrane, the ion exchange current $i_{0,i}$ should be 10^{-5} A cm^{-2}. For membranes with an ion exchange current $i_{0,i} < 10^{-7}$ A cm^{-2}, mixed potentials are common. A Butler-Volmer-type expression for reactions under kinetic control and a diffusional expression similar to the ones derived for metals can be also derived for the membrane electrolyte interface but will not be developed here. For a more detailed treatise on ion exchange at solid/electrolyte interfaces consult Madou and Morrison[16] and Camman.[17]

We touched on mixed conductors, with ions and electrons as the charge transport species, when introducing IrO_x as a pH probe on page 557 "Electrodes of the Second Kind." Depending on the relative magnitude of $i_{0,e}$ versus $i_{0,i}$, a mixed conductor acts as a redox probe, a pH sensor, or both. For most mixed conductors $i_{0,e} >>> i_{0,i}$, one expects severe redox interference when used as pH sensors. As mentioned before, for reasons not well understood, from the noble metal mixed conductor oxide electrodes, IrO_x holds the most promise as a very fast, stable, low impedance pH sensor with surprisingly small redox interference.[8]

Cyclic Voltammetry and Ultramicroelectrodes

Cyclic Voltammetry

Earlier, we mentioned that the most important example of voltammetry is cyclic voltammetry, an electrochemical method in which information about an analyte is obtained by measuring current (i) as a function of an applied potential using a triangular waveform for excitation (Figure 7.40). We also explained the need of adding a high concentration of an inert supporting electrolyte to the solution (the large excess of inert ions ensures that few reactant and product ions will move as a result of migration). A typical, potentiostat-controlled, three-electrode system to carry out these experiments was shown in Figure 7.38.

Species in the electrolyte must be transported to and from the electrode before electrode reactions can occur. As species are being consumed or generated at the electrode surface via electrochemical reactions, the concentration of these species at the electrode surface will become smaller or larger, respectively, than in the bulk of the electrolyte. Suppose we are dealing with a redox species of concentration c_{bulk} in the bulk of the electrolyte (x = ∞) being consumed at the electrode. The concentration gradient at the electrode is then given as:

$$\frac{dc}{dx} = \frac{c_{bulk}(x = \infty) - c_{surface}(x = 0)}{\delta} \quad (7.121)$$

where $c_{surface}$ (x = 0) is the concentration of the redox species at the electrode surface, and δ represents the boundary layer thickness. From the thermodynamic relationship between the potential imposed on the metal electrode by the potentiostat and the concentration of the reacting redox species at its surface (i.e., the Nernst equation, Equation 7.56), we conclude that the concentration difference between surface and bulk leads to an overpotential:

$$\eta_c = E - E^0 = \frac{RT}{nF} \ln \frac{c_{surface}(x = 0)}{c_{bulk}(x = \infty)} \quad (7.122)$$

i.e., the expression for the concentration polarization, η_c, where n is the number of electrons involved in the reaction. For a planar electrode, which is much larger than the diffusion layer thickness, the measured current is given by Faraday's law (Equation 7.44):

$$i = nFAD_0 \left[\frac{c_{bulk}(x = \infty) - c_{surface}(x = 0)}{\delta} \right] \quad (7.123)$$

with D_0 (cm^2 s^{-1}) representing the diffusion coefficient of the electroactive species and n the number of electrons transferred. At a certain overpotential, η_c, all of the reacting redox species arriving at the electrode are immediately consumed, and from that overpotential onward, the concentration of the electroactive ion at the surface decreases to 0, that is, $c_{surface}(x = 0) = 0$, and we reach the limiting current density i_l:

$$i_l = nFAD_0 \left[\frac{c_{bulk}(x = \infty)}{\delta} \right] \quad (7.124)$$

The largest current (the diffusion-limited current) is established when every arriving species at the electrode is consumed immediately so that the surface concentration of it is zero, as illustrated in Figure 7.65. In the case shown we consider a cathodic reduction reaction (say, of $Fe^{3+} + e^- \rightarrow Fe^{2+}$, at a potential $E \ll E^0$).

Equation 7.124 shows that the limiting current is proportional to the bulk concentration of the ionic species. On this basis, classical amperometric (current-based) sensors are used as analytical devices (such as the amperometric glucose sensors we introduced earlier). We can rewrite $c_{surface}(x = 0)/c_{bulk}(x = \infty)$ in Equation 7.122 in terms of i and i_l to obtain:

$$i = i_l \left(1 - e^{\frac{nF\eta_c}{RT}} \right) \quad (7.125)$$

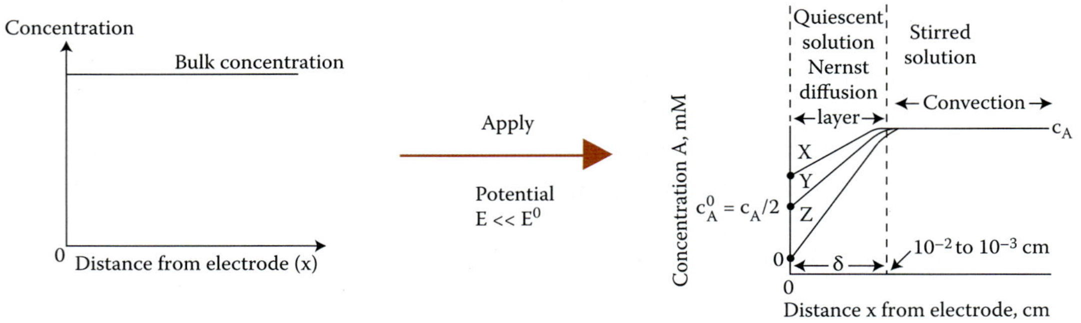

FIGURE 7.65 Applying a potential at a metal electrode that is sufficient to reduce a species in solution ($E \ll E^0$) sets up a concentration gradient that evolves over time (*X*, *Y*, *Z*). The concentration of c_A at various times after application of a voltage sufficient to consume c_A. At *Y*, the concentration of c_A has fallen to half of that in the bulk.

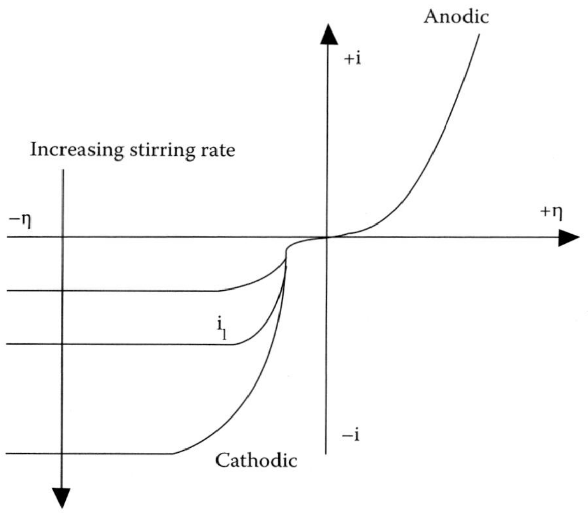

FIGURE 7.66 A diffusion-limited current is indicated for different stirring rates. The limiting current appears as a horizontal straight line limiting the current that can be achieved at a large value of the overpotential, whereas activation-controlled currents keep on increasing. This situation could come about when the concentration of oxidant is much lower than that of reductant.

This expression is illustrated for a cathodic diffusion-limited reaction in Figure 7.66. The cathodic current quickly becomes diffusion-limited, whereas the activation-controlled anodic current keeps on increasing. This situation could come about when there is very little oxidant (say, 10^{-4} M) but an abundance of reductant (say, 1 M), so that a diffusion plateau is not obtained in the anodic branch of the i-η curve.

The concentration overpotential and the activation overpotential have different origins. In the case of activation overpotential, the slope of i versus η increases with η, whereas in the transport-limiting case, the slope of i versus η decreases with increasing η and effectively becomes zero for sufficiently large values of the overpotential.

As shown in Figure 7.66, stirring of the electrolyte strongly affects the limiting current. Higher stirring rates promote convective transport of the ions toward the electrode and result in a smaller boundary layer δ.

For reversible systems, the i-E response shows an anodic and a cathodic plateau as illustrated in Figure 7.67a (assume that O and R are at the same concentrations). This can be ascertained by writing the surface concentration values for oxidant and reductant $c_{o,surface}(x = 0)$ and $c_{r,surface}(x = 0)$ in the Nernst equation in terms of i and i_l to obtain:

$$E = E^0 + \frac{RT}{nF}\ln\frac{c_{o,surface}(x=0)}{c_{r,surface}(x=0)}$$

$$= E^0 - \frac{RT}{nF}\ln\frac{D_0/\delta_0}{D_R/\delta_R} + \frac{RT}{nF}\ln\frac{i_l - i}{i}$$

or:

$$E = E_{\frac{1}{2}} + \frac{RT}{nF}\ln\frac{i_l - i}{i} \qquad (7.126)$$

The value of $E_{1/2}$ is the half-wave potential or the potential where the current is one-half the limiting current ($i_l/2$), and it is characteristic for the specific redox couple O/R only (qualitative analysis), whereas the height of the current plateau is proportional to the analyte concentration (quantitative analysis). In most cases, $E_{1/2}$ is the same as the reaction's standard state potential. A voltammetric wave and the half-wave potential $E_{1/2}$ are illustrated in Figure 7.67a. From Equation 7.126, a plot of E

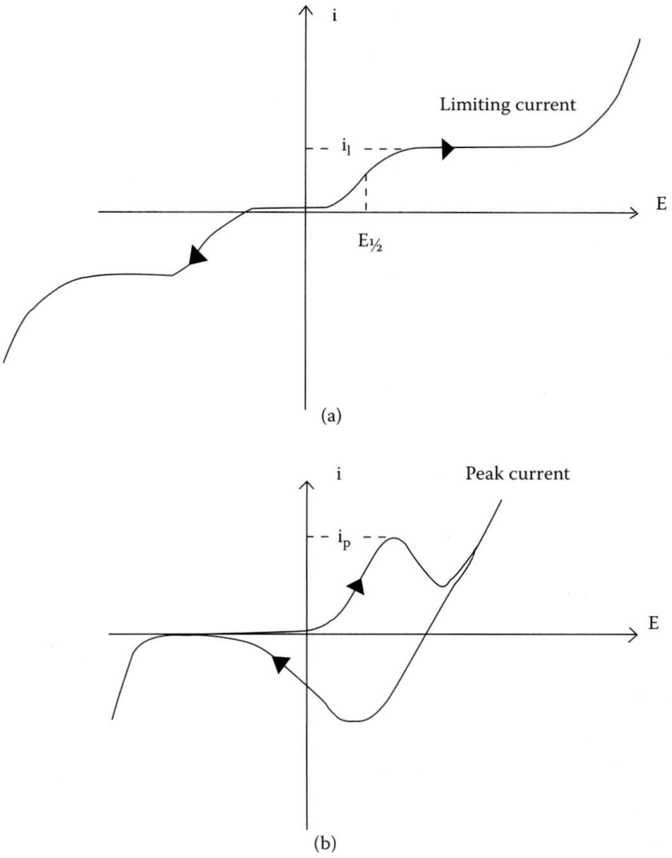

FIGURE 7.67 Cyclic voltammograms with slow (a) and fast (b) sweep rates.

versus ln $(i_l - i)/i$ yields a straight line with a slope of RT/nF (or $0.059/nV$ in a log plot) at room temperature and is used to determine the reversibility of a reaction.

The voltammetric wave in Figure 7.67a applies only when sweeping the voltage ($v = dE/dt$, where v is the voltage sweep rate) relatively slow. At higher sweep rates, cyclic voltammograms are obtained, as illustrated in Figures 7.67b and 7.68. When changes in the potential are made slowly, at such a rate that the rates of diffusion of oxidized and reduced species to and from the electrode can always keep the system in equilibrium with the bulk of the solution at all times (the steady-state value for δ is achieved in a short period compared with the measurement), the cyclic voltammogram exhibits a plateau corresponding to i_l as shown in Figure 7.67a. Most amperometric sensors work within this diffusion plateau, where operation is more forgiving for small potential drifts because the current is independent of the applied potential in that plateau. This also explains why commercial amperometric sensors mostly come without real reference electrodes (less expensive two-electrode system rather than the more complicated and more expensive three-electrode system): the correct knowledge of the voltage is not that important in a current plateau. The time necessary for diffusion to and from the electrode is crudely given by $t \approx \delta^2/2D$ (for a more correct expression, see Equation 7.128), where δ is the thickness of the diffusion layer. The diffusion layer δ will grow with $t^{1/2}$ so that a diffusion layer thickness of 10^{-4} cm is built up in 1 ms and of 10^{-2} cm in 10 s (using a typical value for the diffusion coefficent D of 5×10^{-6} cm^{2}s^{-1}). In the absence of convection δ keeps on increasing, but with convection present it finally approaches a steady-state value.

At fast sweep rates, the potential rapidly moves into the diffusion-limited regime. As a consequence, the concentration gradient is very large, and the initial current is very high. The rate of potential change is too rapid for diffusional processes to maintain equilibrium with the bulk of the solution. As a result, when sweeping the potential the current will not be maintained. Instead, it will peak and then decay as illustrated in Figure 7.67b. The peak value of the current, i_p, for a reversible reaction is given by the Randles-Sevcik equation:

$$i_p = 2.69\ 10^5 n^{\frac{3}{2}} A (Dv)^{\frac{1}{2}} c(x = \infty) \qquad (7.127)$$

where $c(x = \infty)$ is the bulk concentration, A is the electrode area, and v is the sweep rate (dE/dt). Comparing this expression for the peak current with the expression for the limiting current i_l in Equation 7.123, we see that the main difference is the appearance of the square-root sweep rate dependency.

A more detailed fast-sweep cyclic voltammogram is shown in Figure 7.68. In the forward direction, voltage window A to B, no electron transfer occurs. Between B and C, the voltage is sufficient for a reduction reaction to take place, and concentrations of reactants and products are given by the Nernst equation at the surface. At potential D, a maximum current is reached as diffusion limitation sets in. Because the solution

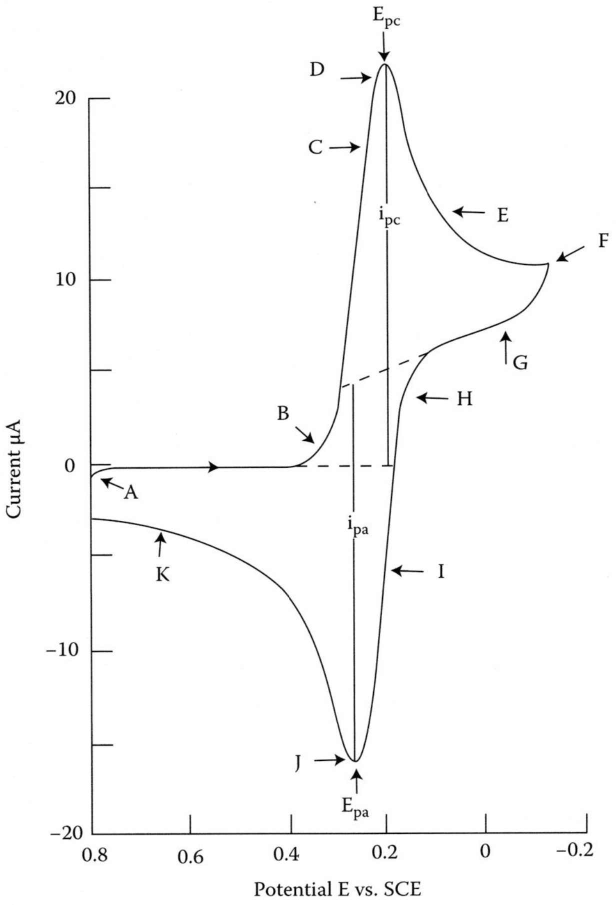

FIGURE 7.68 Fast-sweep cyclic voltammogram (*i-E*); see text.

is not stirred, δ increases with time, resulting in a decrease in current (D to E). At potential F we start the reverse scan, and we observe less current as the potential increases (F to G) until reduction no longer occurs. The reverse reaction (oxidation) then takes off at H to I (if we are dealing with a reversible reaction) and reaches a peak at J. The current decreases again because of the widening of the diffusion layer over time (J to K). A cyclic voltammogram provides a wealth of information. The cathodic (i_{pc}) and anodic peak (i_{pa}) currents at the cathodic (E_{pc}) and anodic peak (E_{pa}) potentials are all important parameters. For a reversible process, $E_{1/2} = E^0$ is the midpoint between E_{pc} and E_{pa}; also, $\Delta(E_{pa} - E_{pc}) = 0.0592/n$, independently of the sweep rate v (where n = number of electrons in reaction). For such a Nernstian process, the peak currents are proportional to the square root of the sweep rate (diffusional control) (see Equation 7.127) and $i_{pc}/i_{pa} = 1$. For a quasireversible reaction, $\Delta(E_{pa} - E_{pc}) > 0.0592/n$, which increases with increasing v. For an irreversible reaction, there is no return wave or the two waves do not overlap.

Higher sensitivities can be reached by increasing the sweep rate v (i.e., 10^{-7}–10^{-8} M; i.e., 10–100 times higher than with the slow sweep technique). This gain in sensitivity is limited by the capacitive, non-Faradaic current, because the capacitive current is linearly proportional to v [$i_C = C_H v$; see Equation 7.43, where $v = dV/dt$ (V was used instead of E in this equation)] and increases faster than the Faradaic peak current. In the next section, we will see that the use of ultrasmall electrodes enables one to measure at higher sweep rates (as high as 20,000 Vs^{-1}) because the Faradaic current on those types of electrodes is relatively larger, allowing for more sensitive measurements.

Nonlinear Diffusion Effects on Ultramicroelectrodes

Electrochemical reactions are influenced by the electrode size relative to the thickness of the diffusion layer, δ. With microstructures of the dimensions of the diffusion layer, macroscale electrochemistry theory breaks down, and we can exploit some unexpected beneficial effects afforded through miniaturization. An ultramicroelectrode (UME) is an electrode having one of its dimensions (called the *critical dimension*) equal to or smaller than 20 μm.

For a small, spatially isolated electrode, with its size reduced to a range comparable with the thickness of the diffusion layer, nonlinear diffusion caused by curvature effects must be taken into account. In Figure 7.69a we illustrate various types of UMEs, including disc shaped (most common are planar or recessed), spherical (drop), band, ring, array, and so on. The diffusional fields for various types of microelectrode shapes and a semi-infinite planar electrode are shown in Figure 7.69b. The analysis shows that as the curvature effects become more pronounced, more diffusion of ions from all

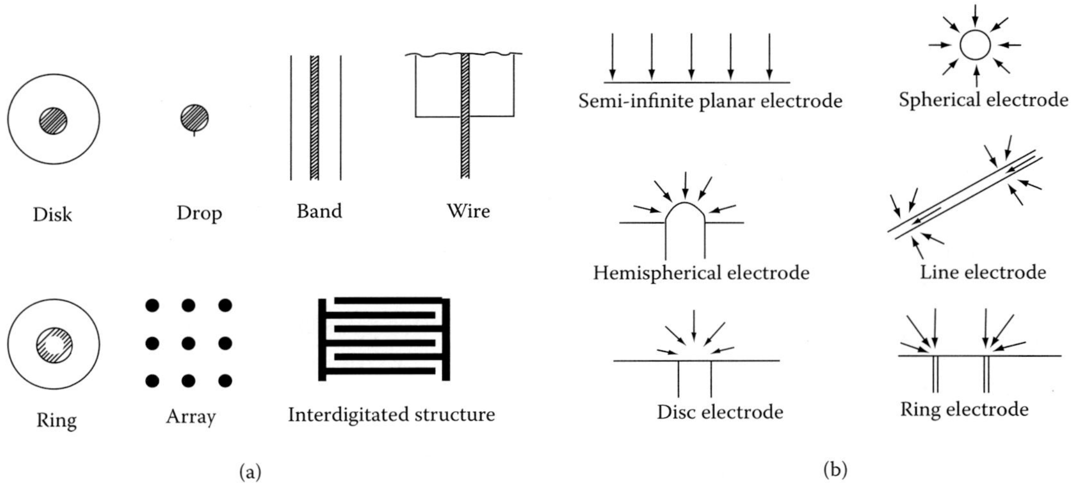

FIGURE 7.69 (a) Examples of types of ultramicroelectrodes (UMEs). (b) Diffusional fields around a semi-infinite planar electrode and various UMEs.

directions takes place, thus increasing the supply of reactants to the electrode.

The diffusion layer thickness arising from linear diffusion is time-dependent and given by [16,18]:

$$\delta = (\pi D_0 t)^{\frac{1}{2}} \tag{7.128}$$

δ is essentially the distance moved by a molecule during the experiment. Substituting Equation 7.128 into Equation 7.124, we obtain the *Cottrell equation for planar diffusion*:

$$i_1 = nFAc(x=\infty)\left(\frac{D_0}{\pi t}\right)^{\frac{1}{2}} \tag{7.129}$$

This equation represents the current-versus-time response on an electrode after application of a potential step sufficiently large to cause the surface concentration of electroactive species to reach zero. This equation, at short times after the potential step application, is appropriate regardless of electrode geometry and rate of solution stirring as long as the diffusion layer thickness is much less than the hydrodynamic boundary layer thickness.* Nonlinear diffusion at the edges of microelectrodes results in deviation from the simple Cottrell equation at longer times. The total current time relation with the correction term is given as:

$$i_1 = \frac{nFAD_0^{\frac{1}{2}}c(x=\infty)}{\pi^{\frac{1}{2}}t^{\frac{1}{2}}} + \frac{nFAD_0 c(x=\infty)}{r} \tag{7.130}$$

At short times, when the diffusion layer is thin compared with radius *r*, Equation 7.129 holds. At longer times and for small electrodes, Equation 7.130 predicts that the correction term can become significant. In practice, a study on ultrasmall electrodes often starts with a linear regression of the measured current versus $t^{-1/2}$ after application of a potential step. The intercept gives the steady-state term. This term, the diffusion-limited current, $i_{1,m}$, at sufficiently long times on UMEs for some important electrode shapes, is given by[18]:

$$\begin{aligned} i_{1,m} &= \pi r n F D_0 c_{bulk}(x=\infty) && \text{DISC} \\ i_{1,m} &= 2\pi r n F D_0 c_{bulk}(x=\infty) && \text{HEMISPERE} \\ i_{1,m} &= 4\pi r n F D_0 c_{bulk}(x=\infty) && \text{SPHERE} \end{aligned} \tag{7.131}$$

In the correction term of Equation 7.130, the electrode surface area *A* is divided by the radius *r*. Hence the principal location of charge transfer appears to be on the outer edge of the electrode. This constitutes a very favorable scaling. Phenomena scaling with the linear dimension to the power of one becomes important in the microdomain. Surface tension exemplifies another scaling law with the power of one as a very dominant force in the microdomain. In the current case, $i_{1,m}$ is proportional to the electrode radius ($\sim r^1$), whereas the background current i_c (associated with the charging current of the Helmholtz capacitance) is proportional to the area ($\sim r^2$); the ratio of the Faradaic current (the charging current) to background currents should increase with decreasing electrode radius ($1/r$). For electrodeposition, this means that smaller features will be plated faster; for sensor applications, this translates into a higher S/N ratio or improved sensitivity. In the previous section, we saw that scanning the potential faster ($v = dE/dt$) leads to higher sensitivities (see Equation 7.127), limited by the even faster increase in capacitive current. We recognize here that the capacitive contribution of a microelectrode is relatively small, so faster scan rates are possible. The latter makes amperometric sensing with microelectrodes possible at unprecedented sensitivities.

With a single microelectrode, the analytical current remains small, and the analytical gain is relative. However, using an array of microelectrodes (see example in Figure 7.69) with all electrodes connected in parallel and separated by an insulating layer provides an elegant solution to this problem.

* In diffusion the parameter *D* is analogous to the kinematic viscosity v_k in convection. These parameters are important in determining the thickness of the diffusion layer and the hydrodynamic boundary layer, respectively. For aqueous solutions, the diffusion coefficient is typically 1000 times smaller than the kinematic viscosity ($D = 10^{-5}$ $cm^2\,s^{-1}$ and $v_k = 10^{-2}\,cm^2\,s^{-1}$). It can be shown that the relation between δ, the thickness of a steady-state diffusion layer, and δ_h, the thickness of the stagnant layer on an electrode in a stirred solution, is $\delta = (D/v_k)^{1/2}\delta_h$. For $D = 10^{-5}\,cm^2\,s^{-1}$ and $v_k = 10^{-2}\,cm^2\,s^{-1}$, $\delta = 0.1\,\delta_h$. Thus, the thickness of the diffusion boundary layer, δ, is considerably smaller than that of the hydrodynamic boundary layer. In an unstirred solution, δ is not well defined, and all types of disturbances can affect the transport. To prevent random convective motion from affecting transport to and from the sensing electrode, we want the diffusion layer thickness to be smaller than the hydrodynamic boundary layer and δ_h to be regular.

In arrays of ultrasmall electrodes, the electrodes behave as an equivalent number of individual microelectrodes when the spacing between them is sufficiently large. This configuration enables analytical currents in a higher range compared with the single-electrode case. This gain is temporary because the diffusion layers around the individual microelectrodes expand across the insulating surface at a rate given by Equation 7.128. The time at which the overlap occurs is both a function of electrode size and spacing in the array, resulting in a decrease in the steady-state current. In an analytical experiment, short compared with the time frame of overlap, more sensitive measurements at reasonably high currents become possible. Another gain, besides higher currents and improved S/N ratio through higher mass transport, results from using an array because the signal can be averaged over many electrodes in parallel (an $\sqrt{n}$ improvement in S/N, with n the number of electrodes). An example of an array of UMEs based on carbon nanotubes is shown in Figure 7.70.[19]

Advantages of using UMEs include:

- High mass transfer rates at ultrasmall electrodes make it possible to experiment with shorter time scales (faster kinetics).
- An array of closely spaced ultrasmall electrodes can collect electrogenerated species from adjacent electrodes with very high efficiency.
- At high scan rates, greater than 500 V/s, the data obtained from conventional electrodes are useless because they are impossible to interpret as a result of growing *iR* drop and increasing influence of charging current. With UMEs, because of reduction in size and hence the current, the *RC* and *iR* effect are greatly reduced so that UMEs can be used for fast-scan voltammetry up to 10^5–10^6 V/s.
- Electrochemical measurements in high-resistivity media, including possibly in air, become feasible.[20]
- Microelectrodes can be used as probes for local environment and processes, in which the small physical size of the electrode is important.
- The steady-state diffusion layer thickness is equal to the electrode dimension. Hence, any process with a larger characteristic dimension, such as convection, will not influence the steady-state current. This means that these electrodes can be used independent of the flow characteristics of the medium in which they are used.
- Another important analytical advantage is that greatly reduced background signals afford a greater S/N.

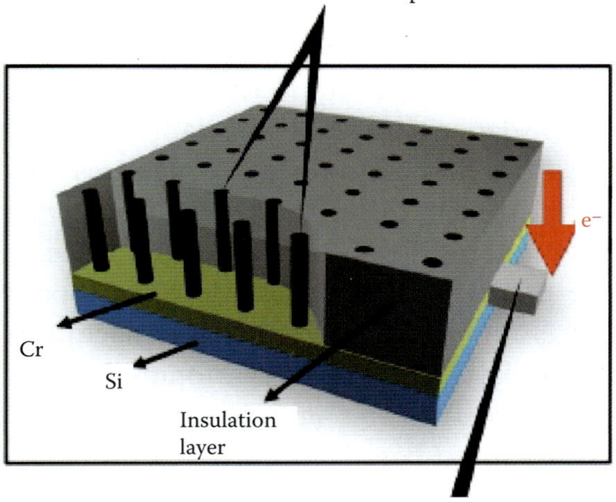

FIGURE 7.70 Arrays of CNT ultramicroelectrodes. (Lin, Y.H, W. Yantasee, F. Lu, J. Wang, M. Musameh, Y. Tu, and Z.F. Ren. 2004. *Dekker encyclopedia of nanoscience and nanotechnology*. Eds. J.A. Schwartz, C. Contescu, and K. Putyera. New York: Taylor & Francis.[19])

For example, microarray electrochemical detectors were applied in liquid chromatography for the detection of carbamate pesticides in river water.[21] The detection limits, in the subnanogram range (50–430 pg), represent a 60-fold improvement over other reported liquid chromatography detectors such as fluorescence detection and electrochemical detection with a single macroelectrode. In this early work, the electrochemical detector array consisted of a mixture of graphite and Kel-F, the so-called *Kelgraf electrode* with a microarray-like structure. Platinum microelectrodes have been microfabricated as detector electrodes for capillary electrophoresis.[22]

Micromachining small reaction chambers in which L, the outer boundary of the electrochemical cell, is made smaller than the diffusion layer thickness, δ, also increases the sensitivity of an electrochemical sensor. In Equation 7.124, δ is now replaced by a smaller L, resulting in a higher diffusion-limited current i_l. The latter finds its application in the design of thin-layer, channel-type flow cells.

When considering a small metal disc embedded in a photoresist, nonlinear diffusion can be neglected for large electrodes (r large or $i_{l,m}$ small). If r becomes small (i.e., of the dimension of and smaller than δ), the contribution of $i_{l,m}$ to the total current becomes larger than i_l; that is, an enhanced mass transport to small electrodes occurs compared with large area electrodes. As pointed out above, we can expect that nonlinear behavior increases the plating rate of small features over large features.

But another factor influences the deposition rate on a small electrode, i.e., the level or position of the conductor surface with respect to surrounding insulating surfaces. The correction term in Equation 7.130 was derived for an inlaid microdisc electrode for which the electrochemically active substrate coincides with the dielectric insulation level (Figure 7.71a). If the conductor disc is recessed within an insulating medium (Figure 7.71b), the diffusion inside the hole must be considered as well, leading to a slight modification of the correction term $i_{l,m}$:

$$i_{l,m} = \frac{nFAD_0 c(x=\infty)}{r+L} \qquad (7.132)$$

in which L stands for the depth of the recession. Mass transport is only enhanced compared with

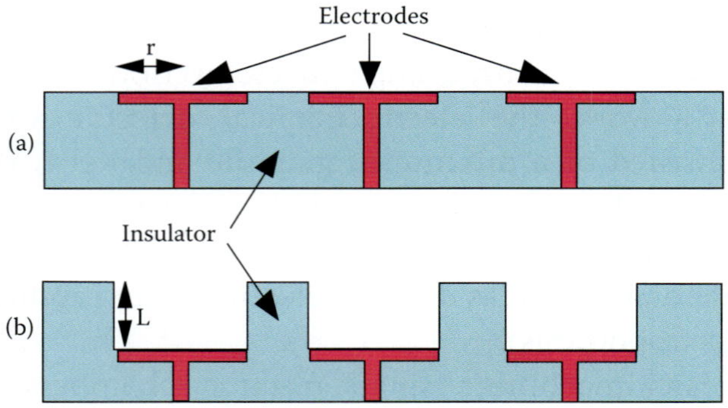

FIGURE 7.71 Nonlinear diffusion on inlaid and recessed electrodes. (a) Inlaid microelectrodes. (b) Recessed microelectrodes.

large area electrodes for small values of L. Van der Putten et al.[23] found that the metal deposition rate at smaller recessed areas (with $r \leq L$) was greater than on larger areas because overfilling was obtained faster for smaller recessed features. When plating features in the LIGA process (see Volume II, Chapter 10), we may deal with situations where a metal layer is deeply recessed ($r \ll L$) in a layer of insulating photoresist. As very high-aspect-ratio features (L/r large) can be made with LIGA, the thickness of the diffusion layers is increased artificially by the microstructured photoresist layer because it is more difficult for convective flow to reach the bottom of the deep gaps.

A microelectrode of the same size or smaller than the diffusion layer thickness could be expected to plate faster than a larger electrode because of the extra increment of current resulting from the nonlinear diffusion contribution (Equation 7.130). On the other hand, as derived from Equation 7.132, describing the current to a metal electrode recessed in a resist layer, the nonlinear diffusion contribution increases with decreasing radius r but decreases with increasing resist thickness, L. High-aspect-ratio features consequently will plate slower than low-aspect-ratio features. Moreover, the consumption of hydrogen ions in the high-aspect-ratio features during electroplating causes the pH to increase locally. Because no intense agitation is possible in these crevices, an isolating layer of metal hydroxide might form, preventing further metal deposition. This all contributes to make the deposition rate, important for economical production, much slower than the rates expected from the linear diffusion model for the current density in large, low-aspect-ratio structures.

The treatise of electroplating in micropatterned resists is important for both micromachining and the IC industry. We cover the topic of electroplating in microstructures in Volume II, Chapter 8 on electrochemical-based forming. Additional reading on this topic is suggested (e.g., Masuko et al.,[24] Dukovic,[25] and Leyendecker et al.[26]).

Scaling in Electrochemistry/Summary

Potentiometric devices such as ISEs or ISFETs (see examples in the following section) measure a voltage and are scaling invariant (voltage is an intrinsic parameter). Amperometric devices, on the other hand, measure a current and are scaling-sensitive (current is an extrinsic parameter). Miniaturization efforts have gone predominantly into the miniaturization of potentiometric devices, although more benefits can be derived from miniaturizing amperometric ones.[16] Scaling of amperometric detector electrodes was addressed in the preceding section. Here we summarize several benefits that derive from miniaturization of amperometric detectors, including higher sensitivity and the possibility of measuring in solutions with higher resistivity. Closely spaced ultramicroelectrodes (UMEs) are highly efficient in collecting electrogenerated species, and the high mass transfer rate makes it possible to experiment with shorter time scales. Nonlinear diffusion effects in amperometric microelectrodes lead to improved sensitivities for sensors with electrode dimensions of a size comparable to the diffusion layer thickness. To compensate for the decrease of the overall absolute current level, an array of microelectrodes is used. Using an array of microelectrodes in a thin, chamber-like configuration could further enhance sensitivity.

Other electrochemical devices such as batteries, as we discuss in Volume III, Chapter 9 on power in miniature devices, scale disadvantageously as the battery capacity is based on its volume.

Example Electroanalytical Devices

Potentiometric Sensors

Ion-Sensitive Field-Effect Transistors, Extended Gate Field-Effect Transistors, and Isn't FET The cable (signal line) connecting an ion-sensitive electrode (ISE) to a high-impedance voltmeter is typically about 1 m long. Because the electrical impedance of a sensor goes up as its size goes down, noise for MEMS- and NEMS-based ISEs with long signal lines is often intolerably high (this is a special problem in hospital settings that constitute an electronically very noisy environment). Integrating the ISE membrane directly onto the gate of a metal oxide semiconductor field-effect transistor (MOSFET) reduces the signal line to practically zero. This arrangement is known as an *ion-sensitive field-effect transistor* (ISFET). Pieter Bergveld invented the ISFET in 1970 at the University of Twente, Enschede, the Netherlands.[27] The operation of ISFETs, also called *CHEMFETS* (chemically sensitive FETs), and the technical and fundamental problems associated with their construction and use, were treated in detail by Madou et al.[16] and will only be summarized here. The structure of the ISFET is essentially a MOS transistor rendered ion-sensitive by eliminating its metal gate electrode to expose the gate insulator either directly to a solution or by putting an ISE membrane on top of the dielectric to interface with the solution (Figure 7.72). In the MOSFET, the metallic gate electrode controls the drain-source current, I_{ds}, through the external voltage, V_{gs}. The metallic gate electrode is insulated from both drain and source by a thin silicon dioxide layer and controls the magnitude of drain-source current electrostatically (see Chapter 4). The only difference between the electrical circuits of a MOSFET and an ISFET is the replacement of the metal gate electrode of the MOS transistor by a series combination of a reference electrode, electrolyte, and a chemically sensitive gate insulator or gate with an ion-selective membrane on top (Figure 7.73). As we saw earlier, insulators do not support any electron exchange at the interface with the solution, but hydroxyl groups at the surface ionize and are charged differently depending on the value of the pH of the surrounding solution. This electrostatic charge influences the drain-source current of the transistor.

The pH sensitivity of an insulator normally cannot be measured because a thick insulator has too high an impedance. A voltmeter, measuring the voltage between an insulator and a reference electrode, would go in overload. Typical dielectric gate materials used as pH sensors are very thin films of Al_2O_3, Ta_2O_5, and Si_3N_4 (even this dielectric has plenty of –OH surface groups in an aqueous medium). With these pH-sensitive thin-film dielectric gate materials immersed in a solution, the hydrogen ion concentration in solution influences the drain-source current, I_{ds}. The reason is that the dielectric is so thin that a substantial electric field across it can influence the magnitude of the drain-source current just as a gate voltage on the metal electrode in a MOSFET would. In an actual ISFET, the drain current value is kept

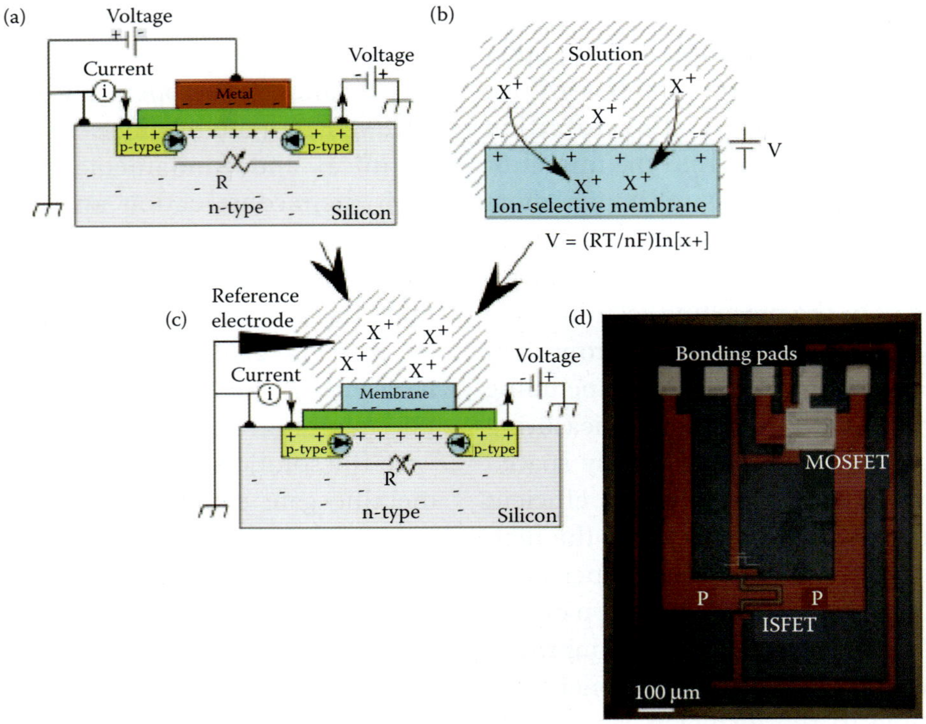

FIGURE 7.72 (a) Classical MOSFET. (b) Ion-selective membrane (ISE) for cation X^+ produces a Nernstian response. (c) Metal gate is replaced by the ion-selective membrane. (d) ISFET device (reference electrode not shown). A differential measurement between the ISFET and an on-board MOSFET helps reduce drift.

constant by means of electronic feedback, and if the pH value of the solution changes, the potential of the gate changes, and, therefore, also the drain current.

Besides using gate dielectrics directly, any of the membranes discussed above may be deposited on top of the gate dielectric.

Three important lessons were learned from working with ISFETs.

1. NEED FOR A REFERENCE ELECTRODE. In the early ISFET days, it was erroneously believed that an ISFET could work without the reference electrode (REF E). Early devices had been found to operate without them. However, it was later discovered that this was an effect caused by leaky SiO_2 gate oxides, and that provided a current return path, which, for reliable operation, should be provided via the REF E.[16] Although a FET device relates a signal through a dielectric via an electrical field, the countercharge of the charge built up in front of the dielectric needs to arrive somehow to the other plate of the capacitor, and this is only possible if a current return path is provided.

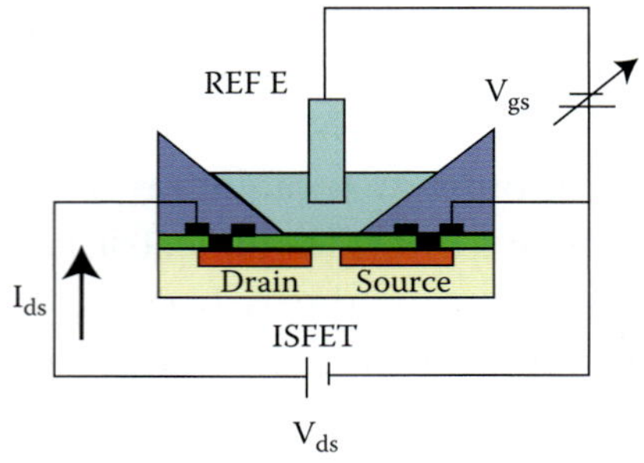

FIGURE 7.73 ISFET operation. Ions do not have to pass through the ion-sensitive membrane. The membrane's control of the drain-source current is solely based on electrostatic effects.

2. DEBYE SHIELDING. If a charged group is attached to a long chain (spacer arm) extending from the gate insulator, say, at a distance d into the solution, that charge may be neutralized by counterions in solution, diminishing the surface potential. The higher the ionic strength of the solution, the more the charge will be shielded. This is illustrated in Figure 7.74, where we consider a film with 4′-nitrobenzo-18-crown-6 ether groups on spacer arms. Calculated curves

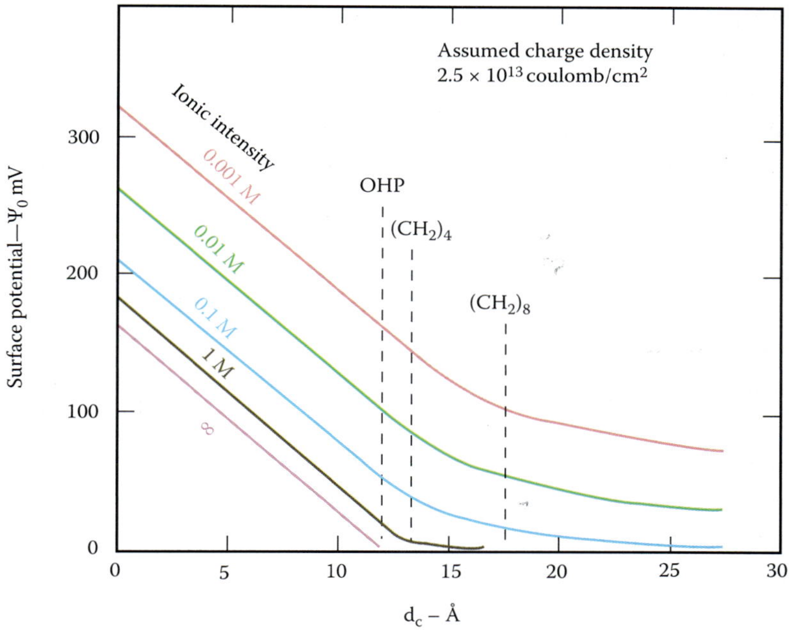

FIGURE 7.74 Calculated surface potential versus chain length d_c. Charges on spacer arms are shielded by ions from solution. This might make the signal of a large charged molecule ionic strength-dependent. (Matsuo, T., and H. Nakajima. 1984. Characteristics of reference electrodes using a polymer gate, ISFET. *Sensors Actuators* 5:293.[28])

of surface potential versus spacer chain length (d_c) by Matsuo and Nakajima[28] are shown. It can be seen that at the highest concentration of the indifferent electrolyte in the solution (∞), the surface potential becomes zero when charged centers extend all the way to the OHP or beyond. This observation is extremely important for the construction of derivatized FET devices. Obviously, sensitivity will be lost if the charge groups are big; sensitivity will depend on the ionic strength of the analyte solution. Attempts to build immunologically sensitive FET devices are hampered by this effect. Moreover, smaller charged molecules can go between the large forest of proteins on the surface and set up a mixed potential at the FET surface. Furthermore, the nonavailability of perfectly polarizable electrodes, nonselective protein interactions with the electrode, and the influence of pH and ionic strength (through Debye shielding) all make the prospect of developing selective and sensitive direct potentiometric or capacitive immunosensors rather dim.

3. IC COMPATIBILITY AND DRIFT. After more than 37 years, ISFETs are still plagued with at least four critical problems. 1) Putting a very high-impedance FET in a conductive solution requires tremendous care in packaging the device; 2) implementing a true reference electrode with a Luggin capillary and a sufficient amount of REF E to survive for an extended period, is very difficult to implement in a CMOS-compatible technology; 3) even a modest-size REF E renders the ISFET large, and the hope for IC miniaturization benefits are lost; and 4) the membranes in an ISFET have no internal electrolyte to fix the Donnan potential on the inside of the membrane. As a consequence, the drift of an ISFET is typically large compared with that of an ISE. A differential measurement between the ISFET and an on-board MOSFET as illustrated in Figure 7.72d does not help reduce this drift source. The on-board reference MOSFET only reduces solid-state-based drift sources.

Most sensor experts have come to appreciate that for chemical sensors, one should only implement on-board electronics if it is absolutely essential for the integrity of the signal. The extended gate field effect transistor (EGFET) device was conceived to avoid some of the problems associated with the leakage problems of an ISFET.[29] By depositing the ISE membranes a short distance away on conducting lines extending from the FET gates, encapsulation is made easier, and the sensors are light-insensitive. The EGFET seemed like an elegant solution at the time (Figure 7.75), but i-STAT (now Abbott), a commercial entity selling blood electrolyte analysis equipment, eliminated all on-board electronics and abandoned the EGFET approach after extensive research.

For multiuse electrochemical reference electrodes, thick hydrogel films with a large capacity to store reference solution are preferred for long-term stability, and for electrochemical sensor electrodes, thick ion-selective membranes are better at preventing pinholes. Consequently, optimized chemical sensor materials end up much thicker than the common IC electronics layers and are more easily prepared with thick-film technology. An additional reason for the slow commercialization of silicon-based electrochemical sensors is that the currently available electrochemical and chemical sensing technologies are multiuse, proven, and cost-effective. Usually, it is simpler and less expensive to build electrochemistry on a nonsilicon substrate (see Volume III, Figure 1.19).

NOTE: During a 37-year period, we went from a long-signal line (e.g., 1 m long) of a large-sized ISE to a zero signal line in an ISFET. In an EGFET, the signal line was extended back to about 100 μm, but with the latest hybrid ISEs, built on plastic substrates, the signal lengths are again from a couple of millimeters to 1 cm long. In other words, we went from ISEs to ISFETs to something that is not even a FET anymore (Isn't FET!). Hopefully, this lesson will be taken at heart in the development of nano-FETs and nanowires.

Potentiometric Immunosensors

From Immunoassays to Immunosensors Immunoassays, introduced in the late 1950s, constitute a major class of medical diagnostic tests that, along with DNA probes, have become an established tool

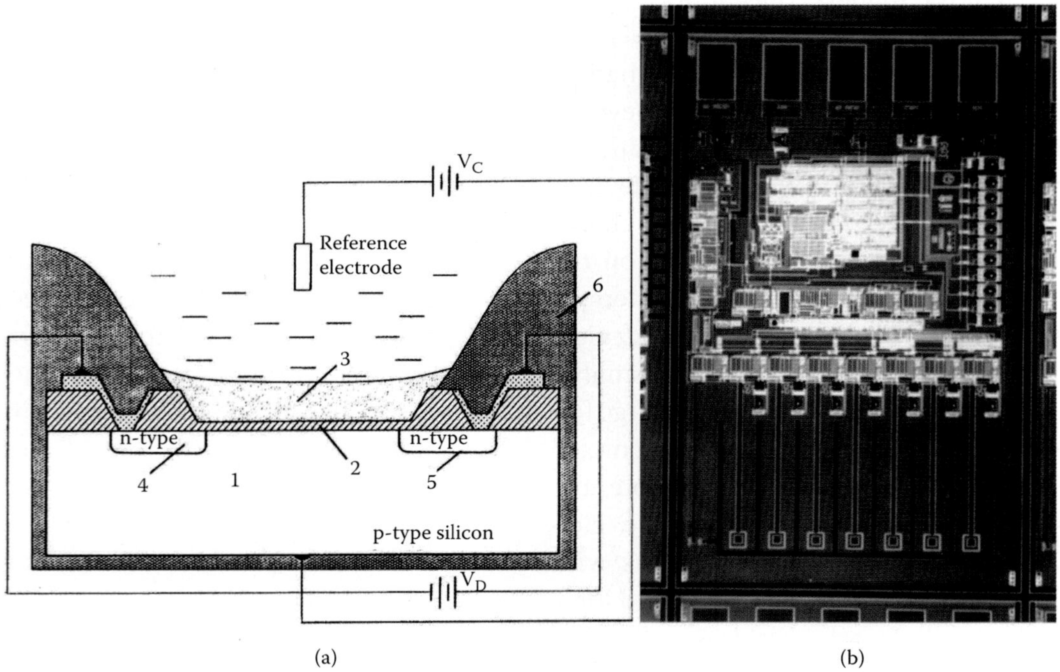

FIGURE 7.75 Integration of chemical-sensitive material and FETs. (a) Chemistry layers on the gate of the FET: ion-sensitive field effect transistor (ISFET). Si substrate (1), insulator (2), chemically sensitive membrane (3), source (4), drain (5), and insulating encapsulant (6). (b) Chemistry layers extended away from the FET: extended gate field effect transistor (EGFET). The seven small squares at the bottom of the photo are the extended gates. The rectangular pads on top are the contact pads.

in analyzing clinically important analytes. Immunoassays are used to detect hormones, vitamins, metabolites, diagnostic markers for therapeutic drug monitoring, and diagnostic procedures for detecting infections. Clinical analytes, as we can learn from Volume III, Figure 10.49, are often present in extremely low concentrations (e.g., 0.1 ng/mL).

The immunoassay principle is based on the use of antibodies (Ab) and a signal-generating irreversible chemical reaction of an antibody with an antigen (Ag, the analyte). The irreversibility of the reaction stems from the large association constants (*Ka*) involved in the antigen-antibody binding; typical values range between 10^5 and 10^9 M^{-1}. The antigen-antibody association constants are composed of large forward (k_1) and small reverse (k_{-1}) rates. These kinetic parameters make antibodies very selective for the analyte of interest but also make the binding irreversible. In fact, "reversible" immunosensors have been developed with lower *Ka*'s to allow for repeated measurements. The ability of natural cells to monitor biochemical events is mimicked in immunoassays. Therefore, these assays constitute the ultimate in selectivity and sensitivity. The utility of antibodies for detecting and isolating antigens can hardly be overstated, given the wide spectrum of extremely powerful antibody applications developed over the years. The use of antibodies is further enhanced by their relative stability in various chemical modification reactions that alter the antibody structure without destroying their capacity to bind to antigens. The two major configurations for immunoassays are 1) solid-phase/heterogeneous assays (often in microtiter plate wells) that require immobilization and separation steps, and 2) solution-phase/homogeneous assays that only involve mixing and measuring.

The binding event for an immunoassay may be sensed directly or via a label; the latter is the more common approach. For example, in an enzyme-linked immunosorbent assay (ELISA), developed in 1971, immunochemical affinity is relied on for selectivity and chemical amplification of an enzyme label for sensitivity. In an ELISA, the enzyme label on the bound Ab-Ag, in the presence of its substrate, generates a reaction product that is measurable (e.g., with GOx as the enzyme and glucose as its substrate, hydrogen peroxide is produced, which may be detected electrochemically). In a typical procedure, titer wells are coated with antibody and known quantities of antigen-containing patient sample, and labeled antigens (e.g., with GOx enzyme) are incubated in the titer wells until the antigen-antibody reaction is complete. Washing removes the unbound antigens, and adding substrate (glucose in the example case) to the bound antigen leads to an electrochemical active reaction product (H_2O_2 in this case). A crucial step is the removal of unbound and nonselectively bound antigens. This is achieved by binding the antibodies to a solid surface, such as a microtiter well surface. Antigens bound to the fixed antibodies remain stuck to the solid surface, whereas washing and decanting removes unbound and nonselective bound antigens. One appreciates that one can either detect the antigen or the antibody in an immunoassay.

The assay illustrated in Figure 7.76 is a competitive electrochemical assay, with native antigen

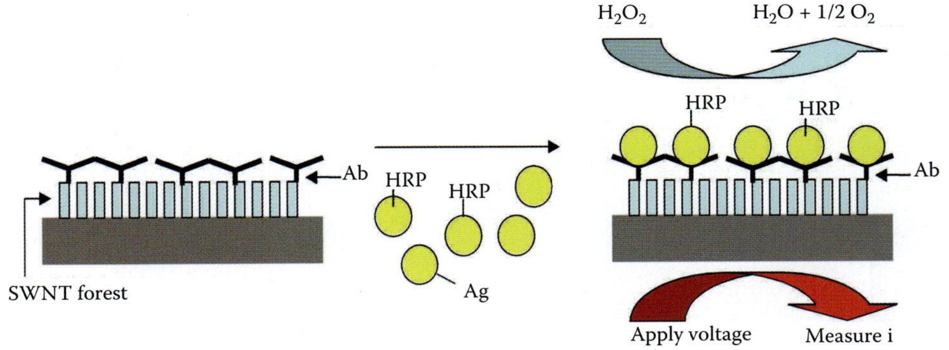

FIGURE 7.76 Enzyme-linked immunosorbent assay (ELISA). Antibody is bound to a forest of single-wall nanotubes. The current is inversely proportional to the amount of nonlabeled antigen (Ag; i.e., native antigen); antigen was prebound on antibody (Ab). Applying a voltage enables one to measure a current proportional to the amount of oxygen.

competing with labeled antigen for reaction sites on a solid surface. The solid surface in the example consists of a forest of single-wall carbon nanotubes (SWNTs) with antibody bound to it. An electrode of SWNTs has the potential of leading to higher sensitivity. The label on the antigen we use in this example of an ELISA is the enzyme horseradish peroxidase (HRP), which converts peroxide to oxygen and water. After competitive binding of the native and labeled antigen on the solid surface and washing off of unbound antigen, the substrate of the HRP enzyme, hydrogen peroxide, is added. The HRP converts hydrogen peroxide to oxygen, which diffuses to the SWNT electrode surface. Current *i* measures the amount of oxygen at the electrode with an applied potential E. The more native species (analyte in the patient's blood), the more effective the competition with labeled species and the lower the resulting signal (less oxygen means less native antigen). Frustratingly, this means that the less native species (analyte), the higher the signal, and vice versa.

In Figure 7.77, an alternative noncompetitive or sandwich assay is illustrated we again use an electrochemical antigen-measuring system, for example). The SWNT forest is coated first with a suitable antibody (Ab1), and a patient sample containing the antigen (Ag; analyte) is then added and incubated until the antigen-antibody reaction is complete. A washing step removes unbound antigen, and a second antibody (Ab2) labeled with enzyme is added and incubated until the antigen binds with this second antibody. A new wash removes the unbound labeled antibody, whereas adding substrate for the enzyme (hydrogen peroxide) leads again to a measurable reaction product.

Antibodies/antigens have also been chemically tagged with fluorescent, magnetic, radioactive (e.g., ^{125}I,^{14}C,^{3}H), and other assorted compounds, called *reporters* or *labels*, as a way of facilitating antigen or antibody detection or isolation under a variety of experimental conditions. Different types of immunoassays include Western blotting, agglutination, precipitation, radioimmunoassay (RIA), enzyme-linked immunosorbent assay (ELISA), bioluminescent immunoassay, homogeneous immunoassay, magnetic particle immunoassay, immunofluorescence, and flow cytometry assay. Typical sensitivities in microgram per milliliter for various immunoassays are listed in Table 7.16.

A major problem introduced through the use of most labeling techniques is that a separation step is the only possible way to differentiate between free and bound labeled species because both contribute to the signal. An assay involving such a separation step is called a *heterogeneous assay*. These assays are simpler to develop but somewhat more complicated to perform. A homogeneous assay, in contrast, requires no separation because the signal-producing reaction is affected in some way by the binding of the antigen to the antibody without interference from the unbound species.

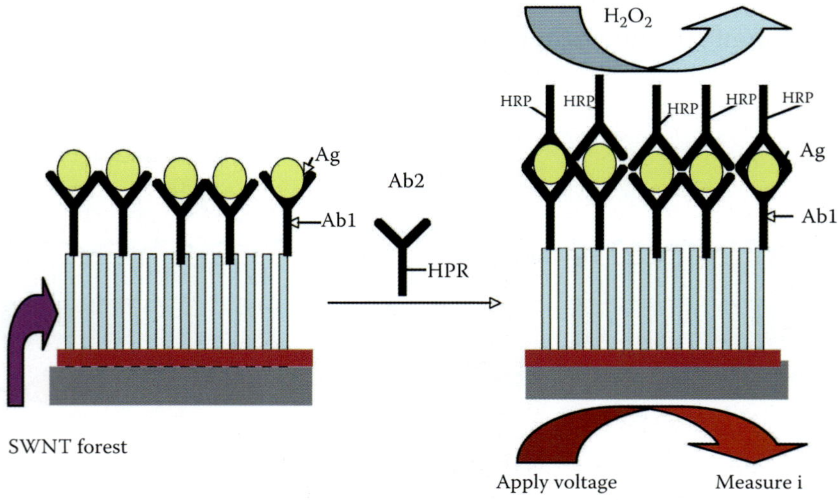

FIGURE 7.77 Noncompetitive assay or sandwich assay.

Table 7.16 Sensitivity of Various Immunoassays

Assay	Sensitivity* (μg antibody/mL)
Precipitation reaction in fluids	20–200
Precipitation reaction in gels	
Mancini radial immunodiffusion	10–50
Ouchterlony double immunodiffusion	20–200
Immunoelectrophoresis	20–200
Rocket electrophoresis	2
Agglutination reactions	
Direct	0.3
Passive agglutination	0.006–0.06
Agglutination inhibition	0.006–0.06
Radioimmunoassay	0.0006–0.006
Enzyme-linked immunosorbent assay (ELISA)	<0.0001–0.01
ELISA using chemiluminescence	<0.0001–0.01†
Immunofluorescence	1.0
Flow cytometry	0.06–0.006

*The sensitivity depends on the affinity of the antibody and the epitope density and distribution.
†Note that the sensitivity of the chemiluminescence-based ELISA assays can be made to match that of a radioimmunoassay.

Ideally, the assay and its accompanying instrumentation should not require a label, and detection thereof, to ascertain the binding of any type of biological affinity pair. It should also be inexpensive, easy to use, and compact. The latter assays are called *direct binding* or *label-less assays*. They continue to be a topic of intense research.

An immunosensor is a sensor monitoring either an antigen or an antibody based on the selective binding of antigen-antibody pairs. The interest in immunosensing is to adapt the above-described assays (Figures 7.76 and 7.77) into a sensor format and to make measurements by simply bringing the sample in contact with the sensor. Our interest here is to explore how micromachining could help this cause. It is especially challenging to develop a homogeneous immunosensor where a washing step is not required. Micromachining might result in a microinstrument, perhaps with some microfluidics flushing the unbound species from the sensor surface. Better yet, micromachining might open the road to label-less detection, as demonstrated with the static cantilever approach for DNA SNP detection in Example 4.1.

Several difficulties still must be overcome in immunosensors, such as assays involving too many steps to qualify as sensors, the deposition of uniform protein coatings, unselective protein adsorption, distinguishing between bound and unbound antigen, and the inherent irreversibility of immunosensors.

Micromachined Potentiometric Immunosensors

We highlight a few developments here to illustrate how new detection schemes and micromachining techniques affect the commercialization of immunosensing. For a more detailed review on immunosensors with commercial potential, we refer to Buerk,[30] Madou et al.,[31] and the March/April 2000 issue of *Fresenius' Journal of Analytical Chemistry*.[32]

Early on, immunosensor work was geared toward direct detection of antigens or antibodies with potentiometric techniques, that is, by simply measuring a voltage. Antigen-antibody binding on the surface of a potentiometric-sensing electrode leads to a new charge distribution. In principle, this redistribution of charges should be measurable. The effect of direct binding on the potential of a sensing electrode was observed on polymeric membrane-coated electrodes, as well as on metal electrodes. However, the signal was found to be small (1–5 mV), making the precision of the measurement poor.[33] Also, ISFETs (see above) have been attempted as one special embodiment of potentiometric devices, and several "immunoFETs" were made in laboratories around the world.[34] Unfortunately, as Janata et al.[35,36] concluded, the potential changes on potentiometric devices are caused by a mixed potential generated from the antigen-antibody binding, as well as other charge transfer reactions occurring at the electrode surface/electrolyte interface. To make a potentiometric sensor work, Janata et al. pointed out that a substrate surface with an infinitely high-charge transfer resistance is needed, as only then will protein charges exclusively account for the potential generated. Such a perfect polarizable surface has yet to be identified (see Figure 7.43). Not even densely packed Langmuir-Blodgett films qualify because they do not block all current across the interface.[37] In Figure 7.78A, an "immunoFET" is schematically shown with a membrane (perhaps a Langmuir-Blodgett film) and attached antibody molecules.

An alternative, direct-measuring immunosensor is the capacitive affinity sensor (Figure 7.78B). This immunosensor is based on the capacitance change associated with the antigen-antibody binding between an insulated pair of interdigitated electrodes on a dielectric surface. The capacitance change arises because the antibody, a large molecule with a small dielectric constant, displaces water with a high dielectric constant. Again, as with the potentiometric sensor, unless a perfect polarizable interface (i.e., a perfect stacking of only the antigen-antibody complexes at the capacitor surface) is realized, capacitive changes cannot unequivocally be associated with the reaction of interest. Along this line, Interuniversity Microelectronics Centre (IMEC, in Leuven, Belgium) is developing capacitive biosensors that allow for the detection of affinity binding of biomolecules (e.g., antigens,

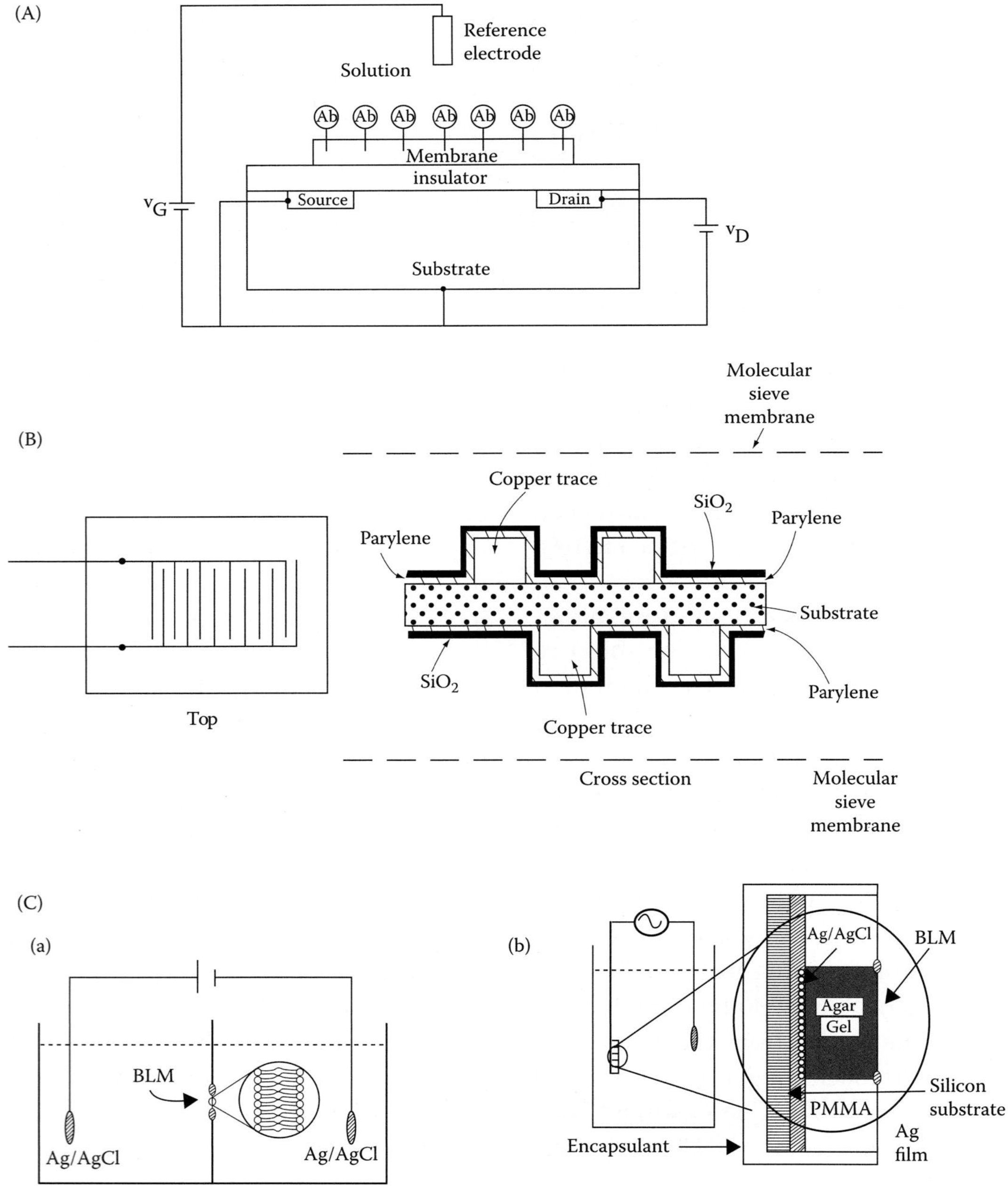

FIGURE 7.78 Schematic diagram of an immunoFET (A). Examples of micromachined, direct immunosensors (B). (C) Lipid membrane suspended in a pinhole (a) and more robust Langmuir-Blodgett lipid layer deposited on a hydrogel (b). See text.

DNA). The presence of DNA sequences or antigen-antibody binding can be detected by looking for impedance changes between closely spaced pairs of electrodes, as sketched in Figure 7.79 (http://www.wtec.org/loyola/mcc/mems_eu/Pages/Chapter-5.html). Nanoscaled interdigitated electrode arrays, as shown here, are made with deep UV lithography. Electrode widths and spacings from 500 nm to 250 nm are produced on a dielectric substrate (0.5 mm × 1 mm). In Figure 7.78Ca, a lipid membrane is suspended in a pinhole between two water reservoirs. Upon exposure to certain analytes, a current across the membrane is observed. This very brittle configuration is rendered a little bit more robust by depositing a thin Langmuir-Blodgett lipid film on a hydrogel as shown in Figure 7.78Cb.

The nonexistence of perfectly polarizable electrodes, nonselective protein interactions with the electrode, and the influence of pH and ionic strength (through Debye shielding) make the prospect of developing selective and sensitive direct potentiometric or capacitive immunosensors rather dim.

Amperometric Sensors

Glucose Sensor The amperometric glucose sensor we are using as a commercial example is a second-generation (see page 571 "Electron Tunneling Transfer Rate k_t," for a discussion of three generations of glucose sensors), disposable glucose sensor strip for diabetes management that relies on an electron acceptor (mediator) to establish electrical communication between the redox center in the glucose oxidase (GOx) enzyme and the electrode. This MediSense/Abbott sensor, the Precision QID, is shown in Figure 7.80. The sensor consists basically of a two-electrode system with counter and reference electrodes combined in a Ag/AgCl electrode. Blood is sufficiently conductive, and the sensor works in the diffusion-limited regime; thus, there is no need for a real reference electrode that is expensive to implement. The working electrode is made of carbon. A second, dummy working electrode has a mediator, but no enzyme for background correction; the mediator used is carboxy-ferrocene. A differential current measurement between the two working electrodes makes for more precise measurement of the additional ferrocene generated in regeneration of the FAD (see Figure 7.60, bottom). All films are silk-screened onto a PVC substrate. Single sensors are less than $0.50 each.

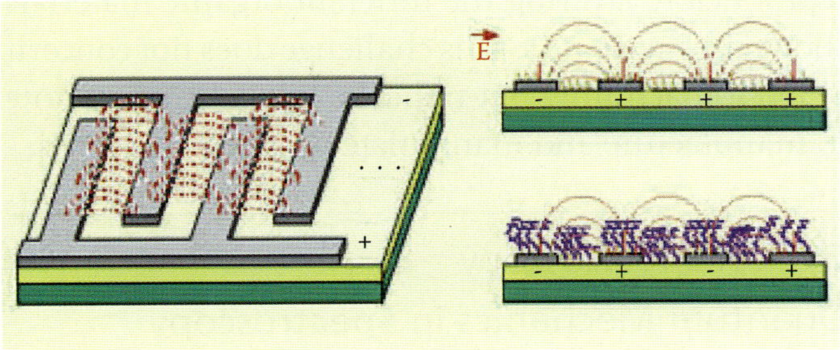

FIGURE 7.79 IMEC capacitive biosensor for DNA and immunosensing. (From IMEC, Leuven, Belgium.)

Amperometric Immunosensor A one-step amperometric immunosensor developed by this author and his coworkers is shown in Figure 7.81. The approach combines the fundamentals of enzyme immunoassay, electrochemistry, and gel filtration.[38] All the components necessary for the assay are incorporated into the sensor; only the sample has to be added for analysis. The base of the sensor consists of working and reference electrodes deposited onto an appropriate substrate. To make the components as cheaply as possible, one may use thick-film technology on a

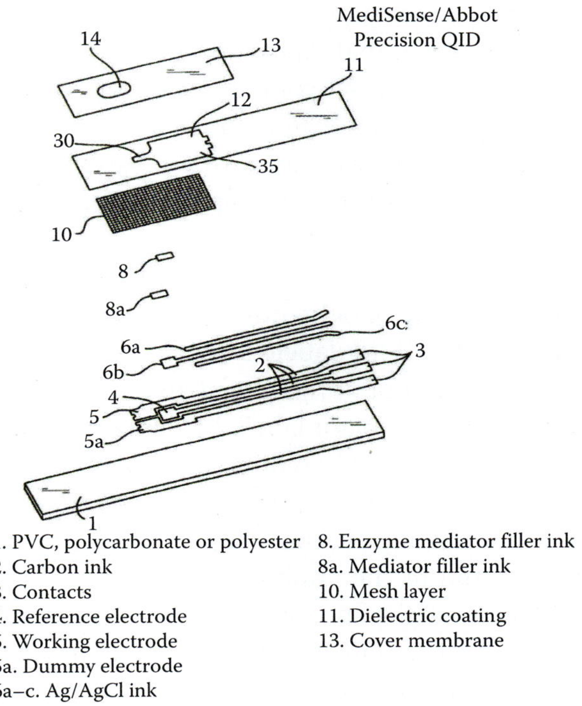

FIGURE 7.80 Example of a disposable amperometric glucose sensor. MediSense/Abbott Precison QID.

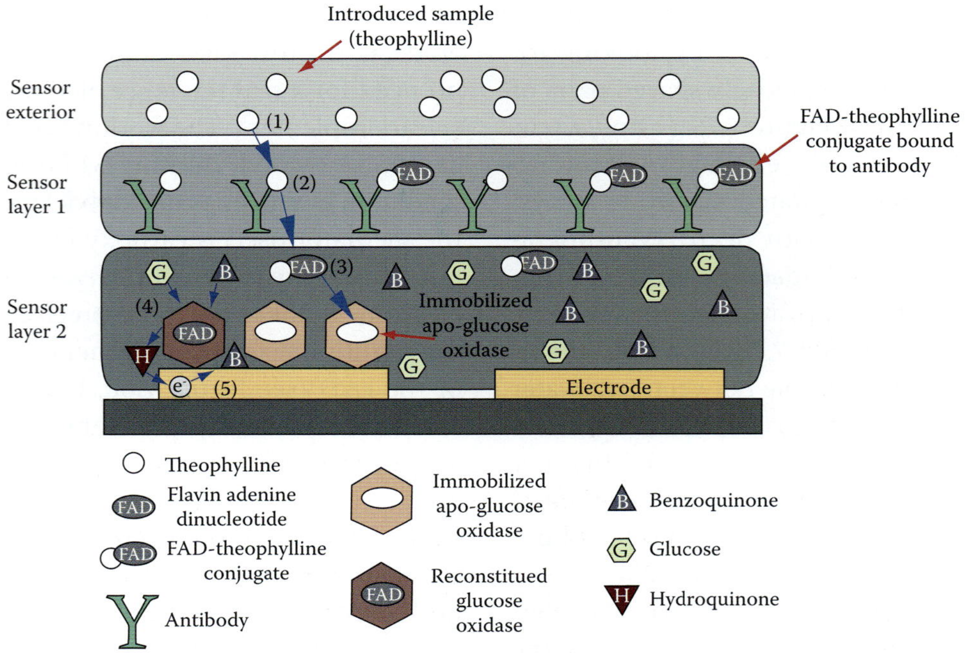

FIGURE 7.81 Layer configuration of immunosensor using theophylline as example reaction scheme. Theophylline sample diffuses into first layer (1). Free theophylline displaces prebound FAD-theophylline conjugate (2). FAD-theophylline diffuses into the second layer and activates apo-glucose oxidase (3). Benzoquinone is consumed and hydroquinone is produced (4). Hydroquinone is oxidized at the metal electrode, leading to an increase in current (5).

plastic substrate. Layered onto the electrode surface is a mixture of an inactive apoenzyme (apo-GOx), an enzyme substrate (glucose in case the enzyme is GOx), and a mediator—reagents necessary to generate an electrical signal. Above the apoenzyme matrix, there is a wettable gel containing the antibody specific for the analyte of interest, topped with a layer of analyte labeled with a prosthetic group or cofactor capable of activating the apoenzyme. A drop of sample placed on the surface of the sensor immediately hydrates the sensor with a fixed volume dictated by the total volume of hydrogel in the device. Sample analyte and cofactor-labeled analyte compete for a limiting amount of immobilized antibody in layer 1. The amount of cofactor-labeled analyte that diffuses to the apoenzyme layer is directly proportional to the concentration of sample analyte. If the sample does not contain antigen, the labeled antigen remains firmly bound to the antibody layer and will not be able to activate the apoenzyme. Activation of apoenzyme by cofactor-labeled analyte results in catalytic breakdown of the enzyme substrate. Reduction or oxidation of one of the reaction products results in a current proportional to the amount of apoenzyme activated, and thus to the concentration of analyte in the sample. The working electrode detects a product of the enzyme reaction with the substrate in an amperometric mode. Using theophylline for analyte and apo-GOx as an inactive enzyme that FAD renders active demonstrates this simple, easy to use, and quantitative immunodiagnostic system. The principle of the new assay is generic and may be applicable to the development of sensors for a variety of analytes with biomedical interest. The detection limit of the FAD/apo-GOx system is well below 10^{-10} M, adequate for the analysis of a majority of analytes. This detection limit can be reduced further by optimization of the system. A further challenge is to find ways of reproducibly layering the different organic materials shown in Figure 7.81. This challenge does not concern Si micromachining but rather involves learning how to manufacture and manipulate hydrogel materials.

Optical Spectroscopy

Quantum Mechanics in Spectroscopy

Optical spectroscopy is the study of the interaction between matter and light. By measuring and

TABLE 7.17 The Energy of the Electromagnetic Radiation Determines the Type of Interaction with Matter

EM Radiation	Interaction Type	Energy/Mole	Energy/Photon
γ-rays	Nuclear transitions	10^7 kJ/mole	10^{-14} J/photon
x-rays	Core e^-	10^6 kJ/mole	10^{-16} J/photon
Ultraviolet	Valence e^-	10^3 kJ/mole	10^{-18} J/photon
Visible	Valence e^-	10^2 kJ/mole	10^{-19} J/photon
Infrared	Molecular vibrations	10^1 kJ/mole	10^{-20} J/photon
Microwave	Molecular rotations	10^{-2} kJ/mole	10^{-22} J/photon
Radio	Nuclear spin	10^{-4} kJ/mole	10^{-24} J/photon

quantifying that interaction, it becomes an analytical technique that can be used to identify (wavelength) and quantify (intensity) the constituents in a sample. When electromagnetic energy is applied to matter, it can be absorbed, emitted, transmitted, or induce chemical changes (reactions). The precise mode of interaction depends on the energy of the electromagnetic radiation involved, as illustrated in Figure 5.2 and Table 7.17. From quantum mechanics (Chapter 3), we know that particles are restricted to fixed energy levels: molecules can only rotate and vibrate at certain frequencies, and electrons in atoms can only move at certain velocities. Quantum numbers define the electronic, vibrational, and rotational states of a species. These quantum numbers are obtained from the Schrödinger equation, the solution to which is a wavefunction (ψ) describing these states. As a consequence, molecules absorb energy as quanta (photons with energy $E = h\nu$) based on criteria specific to each molecular structure. This is what makes the absorption process selective to specific wavelengths of the electromagnetic spectrum ($E = hc/\lambda$, with c the velocity of light at 2.9972×10^{10} cm/s). Molecular rotation involves an entire molecule moving, and the energy change, ΔE, involved is very small (~0.1 kcal/mol). Bond vibrations involve a few nuclei moving, and ΔE is therefore larger (~1–10 kcal/mol), whereas electronic states involve an electron moving out of its bonding orbital, with ΔE the largest among the three (~100 kcal/mol). It is normally the electron spin angular momentum, the orbital angular momentum, and the symmetry properties of the two wavefunctions (initial and final state) that determine whether an energy transition is allowed or forbidden. The feasibility of a transition is given by selection rules in terms of the quantum numbers of the states involved.

UV/VIS Absorption Spectroscopy

Absorption is the most widely used spectroscopic technique because most materials absorb electromagnetic radiation. Absorption of a photon increases the energy of a molecule from a ground state to an excited state (Figure 7.82a). When a molecule emits light, the energy level of the molecule is decreased (Figure 7.82b). The total energy of a molecule is the sum of electronic, vibrational, rotational, translational, and spin orientation energies, some of which are sketched in Figure 7.83. The mechanism of absorption of radiation energy is different in each case; however, the fundamental process is the absorption of a certain amount of energy. Valence electrons are excited in UV/VIS (PHz) spectroscopy, whereas more energetic x-rays excite core electrons in x-ray spectroscopy, and gamma rays (EHz) induce nuclear transitions in Mössbauer spectroscopy. At lower energies, radiofrequencies (MHz) involve the nuclear spin in nuclear magnetic resonance (NMR); microwaves (GHz) involve electron spin and are used in electron spin resonance; and in microwave spectroscopy, molecular rotations are probed. Infrared (THz) excites molecular vibrations and is relied on in IR and Raman spectroscopy.

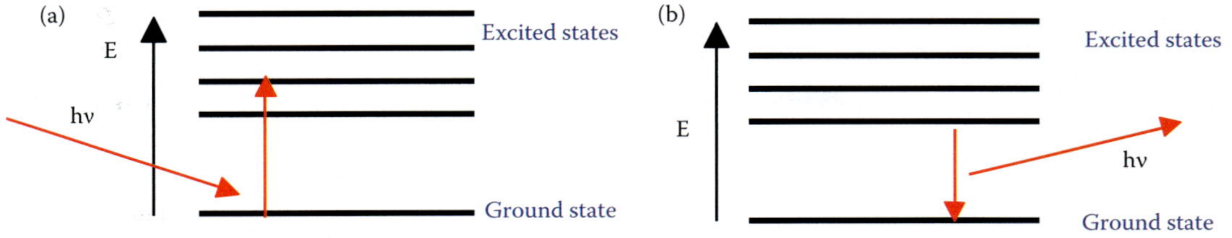

FIGURE 7.82 Absorption of a photon (a) and emission of a photon (b).

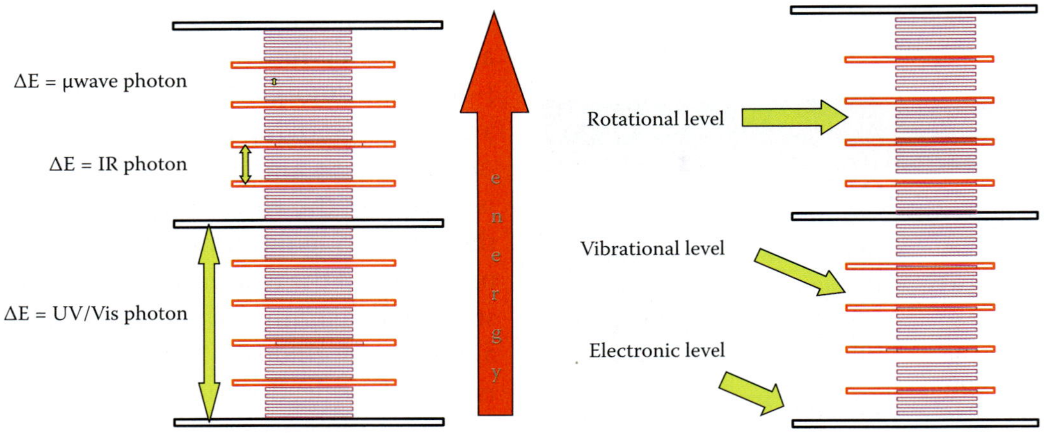

FIGURE 7.83 The total energy of a molecule is the sum of electronic, vibrational, rotational, translational, and spin orientation energies, some of which are sketched here.

Let us focus for a moment on the absorbance, *A*, of visible or UV light (from 160–780 nm) by a solution containing absorbing molecules. Groups of atoms that are responsible for color or pigmentation are called *chromophores*. They come with characteristic optical absorptions and usually contain alternating single and double bonds (such as in aromatic rings). The transmittance of light, *T*, through the solution is given as $T = P/P_0$, with P_0 the incoming light intensity (i.e., *P* at x = 0) and *P* the light leaving the absorbing medium (e.g., at x = b) (see Figure 7.84). Transmittance *T* has values between 0 and 1 and is independent of P_0. The absorbance may also be expressed as $A = \log(P_0/P) = -\log T$. Absorbance or optical density is given by Beer's law (also Beer-Lambert law):

$$A = \varepsilon bc \tag{7.133}$$

where ε is the molar absorptivity or absorption coefficient (wavelength-dependent but also a function of solvent, ionic strength, temperature, and pressure): absorbance of a solution when b = 1 cm and c = 1 M. The molar absorptivity or molar extinction coefficient has units of $M^{-1}\ cm^{-1}$ if *b* is in centimeters and *c* is M. The extinction coefficient of a chemical species at a given wavelength is a measure of how strongly the species absorbs and refers to a single wavelength, usually the absorption maximum. Additionally, *c* is the molar concentration, and *b* is the distance into the absorbing medium (optical path length typically 1 cm).

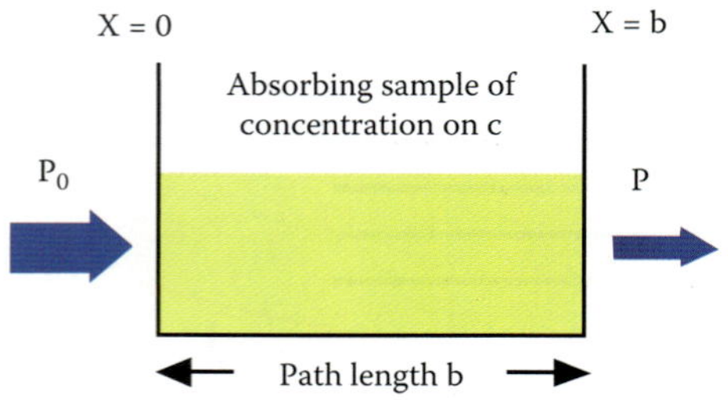

FIGURE 7.84 Absorbance of visible light in a solution of concentration c.

Equation 7.133 can be derived as follows. In a solution element *dx*, the intensity of the incoming light, *P*, is decreased by *dP* because of absorption. With *dP* given by:

$$dP = -Pc\beta dx \tag{7.134}$$

where c = molar concentration
P = light intensity
β = probability of a photon being absorbed

We can rewrite Equation 7.134 as $dP/P = -c\beta dx$ or:

$$\int_{P_0}^{P} \frac{dP}{P} = -c\beta \int_0^b dx \quad \text{or} \quad \ln P - \ln P_0 = -c\beta x \tag{7.135}$$

where P_0 is the value of *P* at x = 0. In other words:

$$\ln \frac{P_0}{P} = c\beta b = 2.303 \log \frac{P_0}{P} \tag{7.136}$$

The latter expression can be rewritten in the familiar format:

$$A = \log \frac{P_0}{P} = \frac{\beta}{2.303} cb = \varepsilon cb \tag{7.137}$$

The lowest energy transition (and the most often observed in UV/VIS spectroscopy) is typically that of an electron going from the highest occupied molecular orbit (HOMO) to the lowest occupied molecular orbit (LUMO). For any bond (pair of electrons) in a molecule, the molecular orbitals are a mixture of the two contributing atomic orbitals; for every bonding orbital "created" from this mixing (σ, π), there is a corresponding antibonding orbital of symmetrically higher energy (σ*, π*) (see Chapter 3 and Figure 7.85). The lowest energy-occupying orbitals are typically the σ orbitals; likewise, the corresponding antibonding σ* orbitals are of the highest energy. The π orbitals are of somewhat higher energy than σ orbitals, and their complementary antibonding orbitals are somewhat lower in energy than σ*. Unshared electron pairs lie at the energy of the original atomic orbital, shown as n in Figure 7.85. Most often this energy is higher than π or σ (because no bond is formed, there is no lowering of the energy for *n*).

Electronic absorption spectroscopy of molecules usually measures $\pi \rightarrow \pi^*$ and $n \rightarrow \pi^*$ transitions. Even if the incoming light has the correct energy (i.e., the frequency is exactly right) for an optical transition, not much may happen if a selection rule is broken, i.e., the specific transition is forbidden. However, selection rules are rarely absolute, and in practice even forbidden transitions can still occur (especially in solution), but such transitions will usually have low intensities. For example, $n \rightarrow \pi^*$ electronic transitions are "forbidden" (so they are inefficient), whereas $\pi \rightarrow \pi^*$ electronic transitions are "allowed" (so they are very efficient). The *n* orbitals do not overlap well with the π* orbital, so the probability of this excitation is small.

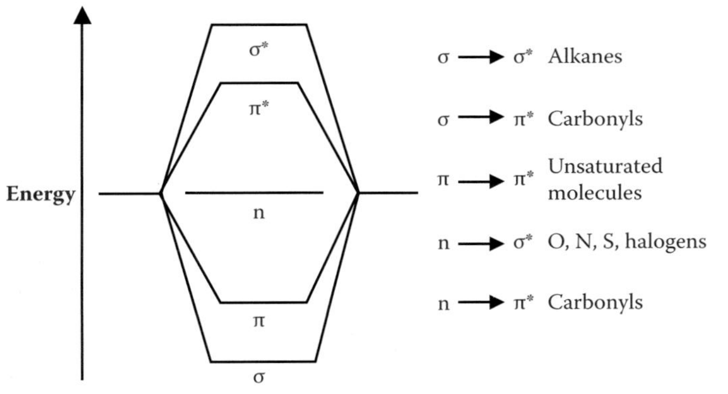

FIGURE 7.85 Electronic transitions in UV spectroscopy with example molecules.

Molar absorptivities, ε, are very large for strongly absorbing chromophores, with $\varepsilon > 10{,}000\ M^{-1}\ cm^{-1}$; a weak chromophore is one with $\varepsilon = 10\text{–}100\ M^{-1}\ cm^{-1}$. The magnitude of ε reflects both the size of the chromophore and the probability that light of a given wavelength will be absorbed when it strikes the chromophore. A general equation stating this relationship may be written as follows: $\varepsilon = 0.87 \times 10^{20}\ Ta$, where *T* is the transition probability (0–1) and *a* is the chromophore area in square centimeters. Chromophores are typically about 10 Å long; therefore, a transition of T = 1 will have an ε of 10^5.

Absorption measurements are widely applicable but offer only medium sensitivity in the 10^{-4}–10^{-5} M range. Optimization of the measurement procedure can extend the range to 10^{-6} M or even 10^{-7} M. Selectivity is only moderate, and separation before measurement is often required.

Standard absorption spectrometers are shown schematically in Figure 7.86. Light sources are tungsten-halogen (visible) and hydrogen or deuterium discharge (UV) lamps; a typical monochromator is a Czerny-Turner, and detectors are photomultiplier tubes (PMT) or diode arrays (which are less sensitive than PMTs but allow for faster spectral scanning).

Light intensity passing through the sample is measured directly in the path of the incoming light. In the system depicted in Figure 7.86a (a so-called

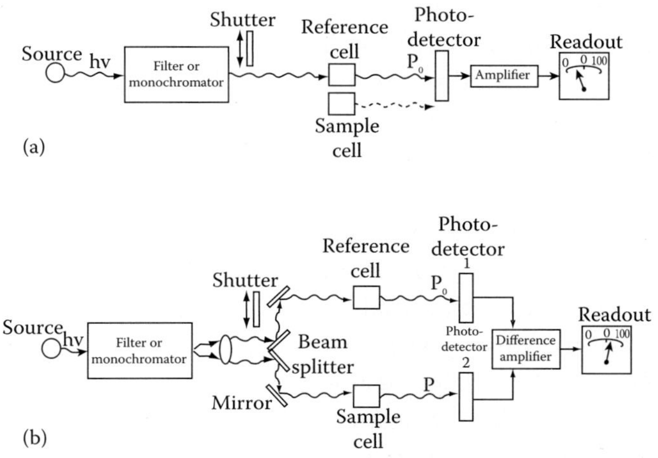

FIGURE 7.86 Two types of optical absorption spectrometers. In the system depicted in (a), the user has to switch sample and reference cell. The system in (b), using a beam splitter, accomplishes the referencing automatically.

single-beam instrument), the user has to switch between sample and reference cell manually. Spectra of absorbing species and solvent are additive. Therefore, one takes the spectrum of a blank solution and subtracts this spectrum from that of the sample solution. The measurement setup in Figure 7.86b is a double-beam system, where a beam splitter accomplishes this referencing automatically. After passing through a monochromator, a light beam splits into two beams; one goes to the sample and the other one to the reference. The beams pass to each photodetector at alternate times by virtue of a chopper. Double-beam instruments compensate for all but very short-term electrical fluctuations because measuring P_0 and P is made almost simultaneously. When many wavelengths must be measured, double beam is a time saver. A single beam is simpler and easier to maintain and is used for quantitative analysis at a single wavelength because it is cheaper.

A fiberoptic-based absorption measurement instrument is very useful when the sample cannot be placed in a cuvette, for example, in a flowing solution or in a 96-well plate reader.

IR and Raman Spectroscopy

The above analysis focused on UV and visible light absorption spectroscopy. A few words about IR absorption and Raman spectroscopy follow. As discussed in Chapter 5, IR absorption and Raman are complementary techniques based on vibrational transitions. They are not used much for quantitative work (their detection limits are not very good) and are most useful for the identification of molecules. As such, they are more often used in studies of structure and conformation of complex molecules. In Raman spectroscopy one measures the frequency (wavelength) change of the incident radiation through inelastic scattering with molecules in a wide range of wavelengths (not only those corresponding to vibrational levels). Scattered light under these circumstances reveals satellite lines below and above the Rayleigh scattering peak at the incident frequency. Raman scattering depends on the polarizability of the scattering molecules (see Chapter 5).

In IR absorption spectroscopy one measures the absorption of the IR radiation at any specific wavelength that matches (in energy) a vibrational transition of the molecule. If the selection rules for optical transitions were all the same for IR and Raman, the peak intensity may be different, but the energy shifts observed in Raman spectra should be identical to the energies of its IR absorption. However, IR absorption selection rules require that a vibrational mode of the molecule have a change in dipole moment or charge distribution associated with it. For example, homonuclear molecules, N_2, Cl_2, and H_2, are not IR active, but they are Raman active because the polarizability of the bond varies periodically in phase with the stretching vibrations. Thus, IR and Raman differ with regard to which vibrational energy transitions are active as a result of selection rules. Some important advantages of Raman over IR are as follows. First, sample preparation for Raman is easy: water is an excellent solvent for Raman, but it is a serious interference for IR absorption. Also, Raman spectra can be measured with normal transparent sample holders like glass or quartz, whereas IR measurement requires KBr or other atmospherically unstable window materials. Moreover, because Raman can also involve rotational or electronic transitions of the molecules from which the scattering occurs, a much wider range of energy sources can be used (to interact with the electronic, vibrational, and rotational energies of molecules as calculated in Chapter 3). Raman intensity increases with the fourth power of the frequency of the source (see Chapter 5, Equation 5.261), and a high Raman peak intensity can be obtained by using a short wavelength laser source.

Photodecomposition of samples may be avoided by using red or near-infrared irradiation sources, such as a diode laser (782 or 830 nm) or a neodymium-doped/yttrium aluminium garnet (Nd-YAG) laser (1064 nm). Higher-temperature studies are possible in Raman spectrometry because IR radiation at higher temperature does not affect the measurement; this would interfere with IR spectroscopy, but Raman spectrometry measures the shift (energy difference). On the other hand, IR is better than Raman for measuring molecules with a highly asymmetrical charge distribution and has a lower detection limit than Raman [except when comparing with surface-enhanced Raman spectroscopy (SERS); see Chapter 5]. Also, Raman must use much more intense sources than IR to get a measurable signal, and IR instruments are also much cheaper.

Many variations of powerful Raman spectroscopy have been developed, such as surface-enhanced Raman spectroscopy (SERS), where surface plasmons of Ag or Au surfaces or nanoparticles are excited by a laser. The resulting electric fields cause nearby molecules to become Raman active, resulting in an amplification of their detection by a factor up to 10^{11}! This effect was originally observed by Fleischmann[39] and was explained by Van Dyune in 1977.[40] Surface plasmon resonance (SPR) and SERS were discussed in Chapter 5.

Scaling in Absorption Spectroscopy and Cavity Ring-Down Spectroscopy

From Beer's law, optical absorption of any type of wavelength through a material scales poorly in the microdomain because it depends on the path length b (see Equation 7.133). Therefore, miniaturization of the absorption path length always decreases the sensitivity! This is unfortunate because besides being the most generic optical spectroscopic method (all molecules absorb electromagnetic radiation at some frequency), the intrinsic molecular cross-sections are typically largest for absorption (compared with any other optical technique, such as fluorescence).

To miniaturize an absorption spectrometer without losing sensitivity, the optical absorption path can be folded and put on a small footprint by bouncing the light from a set of mirrors. Although an improvement, the reflectivity of the mirrors limits the maximum number of bounces typically to about 100, thereby limiting the ultimate total optical path length, and thus the sensitivity of the miniaturized instrument. The best alternative absorption method, originally devised for use in the visible, UV, and infrared spectra, was developed during the 1980s by O'Keefe[41] at Stanford University and later by Saykally[42] at the University of California, Berkeley and Meijer[43] at the University of Nijmegen. This new method is the so-called *cavity ring-down spectroscopy* (CRDS), a highly sensitive method for absorption spectroscopy using pulsed lasers. It can be viewed as a multipass technique affording 10–1000 times longer path lengths than traditional multipass reflection cells. The problem in conventional multipass approaches lies in the requirement for the light beam to traverse distinct paths for each pass. In CRDS one does not take measure of the absorbed signal strength for a given distinct sample path, but instead measures the time rate of absorption of a sample located within a closed optical cavity.

In CRDS, the time required for a laser pulse to decay inside an optical cavity, also termed an *optical resonator*, formed from two highly reflective dielectric plano-concave mirrors, is measured.

A schematic is shown in Figure 7.87. In this setup, pulsed laser radiation enters the cavity through the first mirror, M1, and makes a number of round trips (n) between the mirrors.

The mirrors are ultrasmooth surfaces with ~0.05 nm rms surface roughness and may feature a reflectivity R = 0.99999 in the visible part of the spectrum. The intensity of the laser pulse trapped between the two mirrors decays exponentially with time at a rate determined by the round-trip loss experienced by the laser pulse (Figure 7.88).

The round-trip time is controlled by the cavity length L, and at each bounce on the second mirror M2, some light leaks from the cavity and is detected at a photomultiplier (PMT). For a 1-m cavity, the round-trip time is 6.7 ns. The intensity of the light pulse in the cavity decays exponentially

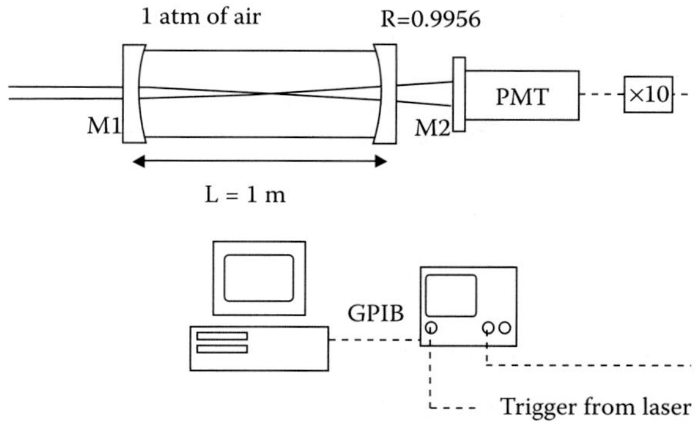

FIGURE 7.87 CRDS, a typical experimental setup. Pulsed narrow band radiation from a tunable (dye) laser system is introduced in the ring-down cavity formed by two plano-concave mirrors (M1 and M2). The mirrors are coated for optimum reflectivity in the desired wavelength range. To ensure that all transverse modes are detected with equal probability, the photomultiplier tube (PMT) that is used to measure the ring-down transients is placed directly behind the output mirror (M2).

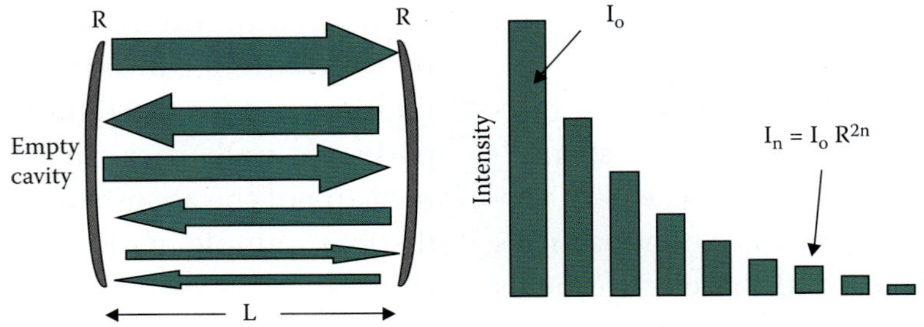

FIGURE 7.88 Decay of light intensity of the laser pulse as a function of the number of round-trips (*n*).

and is characterized by a ring-down lifetime, τ, or the time it takes the intensity to decrease to 1/*e* of its initial value. For mirror reflectivities of 99.99%, the ring-down time is on the order of 30 μs, during which time the pulse has made 5000 round trips for a total path length of 10 km. The decay curve is fit to a single exponential function to determine the ring-down time constant. In the case of an empty absorption cell (i.e., ε = 0), the decay time or ring-down time is approximately given by:

$$\tau_0 = \frac{L}{c\,|\ln(R)|} \cong \frac{L}{c(1-R)} \tag{7.138}$$

where L = absorption cell length (longest distance between the mirrors)
R = reflectivity of the optical cavity mirrors
c = speed of light

With the absorption cell filled, one obtains:

$$\tau = \frac{L}{c(1-R+\varepsilon L)} \tag{7.139}$$

with ε equal to the Beer's law absorption coefficient of a sample in the cavity.

Two intensity decay curves for an empty and filled optical resonance cavity are compared in Figure 7.89.

From Equations 7.138 and 7.139 we can derive the absorption coefficient as:

$$\varepsilon = \frac{1}{c}\left(\frac{1}{\tau} - \frac{1}{\tau_0}\right) \tag{7.140}$$

Typically the ring-down time can be determined with a relative accuracy of 10^{-3}. With a mirror reflectivity loss of (1 − R) = 10^{-4}, this implies that line-integrated absorption coefficients of better than 10^{-7} (absorption coefficients less than 10^{-9} cm^{-1} in a 1-m cavity) can be measured. In comparison, conventional absorption experiments can only measure absorption coefficients of 10^{-6} cm^{-1}. The effective sample path lengths in CRDS may be in the tens of kilometers, and the method is in principle intensity-independent. Clever setups that exploit variations in CRDS (e.g., measuring monolayers instead of bulk gas adsorption) are being investigated; see, for example, Pipino et al.[44,45] CRDS has been performed in a large spectral region, from the UV (200 nm) to the infrared (10 mm), almost without gaps. A large variety of molecules have been studied, including radicals, ions, clusters, and even molecules in the solid phase. It is a type of spectroscopy that can be applied in many environments, e.g., open air, cells, jets, supersonic expansions, discharges, and flames. Thus far, no MEMS research group has tackled a micromachined CRDS. This is not too surprising given the difficulty of microfabricating highly reflective dielectric plano-concave mirrors. Lithography-based MEMS are typically very good at making projected shapes, but they are not so good at making truly 3D surfaces as required here.

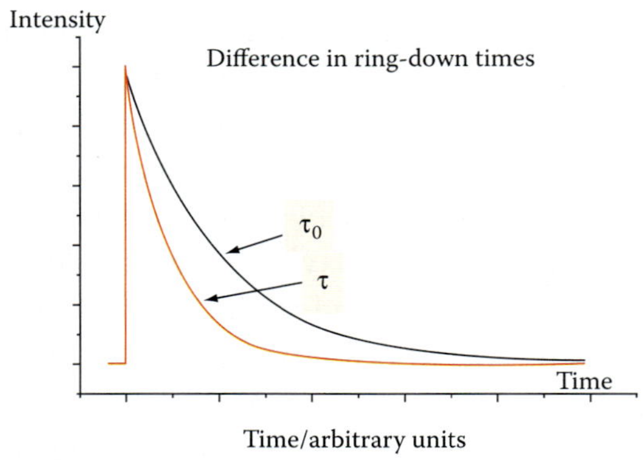

FIGURE 7.89 Difference in ring-down times for filled and empty optical resonance cavity (τ_0 is empty and τ is filled).

Luminescence

Introduction

Luminescence is the emission of light by an atom or molecule. In the case of photoluminescence, the first step in the process is the absorption of a photon, but in chemiluminescence, the light emission is caused by a chemical reaction in which chemical energy is converted into light energy. Bioluminescence is a type of chemiluminescence where the production and emission of light in a living organism are the results of a chemical reaction. We also make a distinction between stimulated emission and spontaneous emission. In stimulated emission, an excited species interacts with the incoming oscillating electric field and gives up its energy to the incident radiation. Stimulated emission, as we saw in Chapter 5, is an essential part of laser action. In spontaneous emission, the excited species gives up its energy of its own accord. Fluorescence, phosphorescence, and chemiluminescence are all spontaneous emission processes.

Photoluminescence

Fluorescence and Jablonski Diagrams Photoluminescent processes are divided into fluorescence and phosphorescence. The difference between fluorescence and phosphorescence is best illustrated in an electronic state diagram or Jablonski diagram as shown in Figure 7.90. Electronic transitions responsible for absorption spectra give rise to singlet and triplet states, and hence create fluorescence and phosphorescence, respectively. In the singlet state, electrons spin in the ground and excited states are paired, creating a net spin of zero. In the triplet state, the electron spins in the ground and excited states are not paired, so there is a net spin (Figure 7.91). Fluorescence does not involve change in spin (the transition is spin allowed) and is short lived (10^{-9}–10^{-6} s).

On the other hand, phosphorescence involves a change in electron spin and is long lived (10^{-3}–100 s) (Figure 7.92). Although it is forbidden, it still happens but with a low probability. The lifetime of an excited molecule depends on how the molecule disposes of the extra energy. Because of the uncertainty principle, the more rapidly the energy is changing, the less precisely we can define the energy (see Chapter 3). Therefore, long-lived excited states have narrower absorption peaks, and short-lived excited states have broader absorption peaks.

The blue lines in Figure 7.90 are radiative processes: absorption (1), fluorescence (2), and phosphorescence (3). The red lines are nonradiative processes: vibrational relaxation or deactivation (4), intersystem crossing to triplet state from a singlet state (5), and internal conversion (6). When a fluorophore, a small molecule, or a part of a larger molecule that can be excited by light to emit fluorescence absorbs a photon from an incandescent lamp or a laser, an excited electronic singlet state might result. This excited state exists for a finite time (typically 10^{-10}–10^{-9} s), during which the fluorophore undergoes conformational changes and is subject to interactions with its molecular environment. These processes have two important consequences. First,

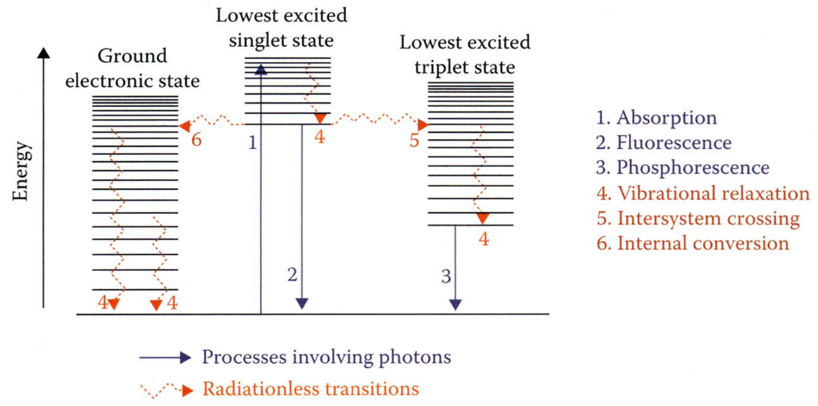

FIGURE 7.90 Electronic state diagram or Jablonski diagram.

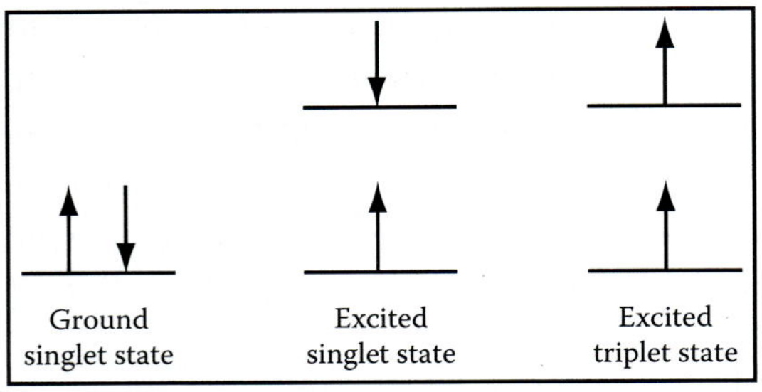

FIGURE 7.91 Singlet ground state and excited singlet and triplet states.

the energy of the excited singlet state is partially dissipated, yielding a more relaxed singlet state from which fluorescence originates. Second, not all the molecules initially excited by absorption return to the ground state by fluorescence emission (i.e., there is a certain limited yield to the process). Other processes such as collisional quenching, Förster resonance energy transfer (FRET) (see 7.147), and intersystem crossing (Figure 7.90, red line 5) may also depopulate the excited singlet state.

George Stokes, professor of mathematics and physics at Cambridge University, interpreted the light-emitting phenomena described here. He formulated the law that fluorescent light is of longer wavelength than the exciting light in 1852 (Stokes law or the Stokes shift), coining the term "fluorescence" instead of the earlier-used term "internal dispersion" in 1853. Stokes law is illustrated in Figure 7.93.

A photon that is absorbed by a molecule does not merely kick that molecule into a higher energetic state, but the molecule also acquires some vibrational energy. It was Franck-Condon who recognized that the photon absorption process is instantaneous, with the rearrangement of almost inertia-free electrons (10^{-16} s) being extremely fast (10^{-16} s) and nuclear motions much slower (10^{-13} s). Nuclear motions are negligible during the time required for an electronic excitation. Because the nucleus does not move during the excitation, the most probable electronic transitions are vertical transitions. In Figure 7.93, the energy diagrams show the potential energy of a molecule as a function of the distance between the nuclei, and according to the Franck-Condon principle, excitation is represented by a vertical arrow. This vertical arrow hits the upper curve not at its lowest point (corresponding to a nonvibrating state), but somewhere higher. After absorbing a photon, the molecule finds itself in a nonequilibrium state and begins to vibrate, lowering the potential energy of the molecule in a radiation-free manner. When the molecule fluoresces, the light originates from near the bottom of the upper potential curve and follows a vertical arrow down until it strikes the lower potential curve. The emission arrow does not

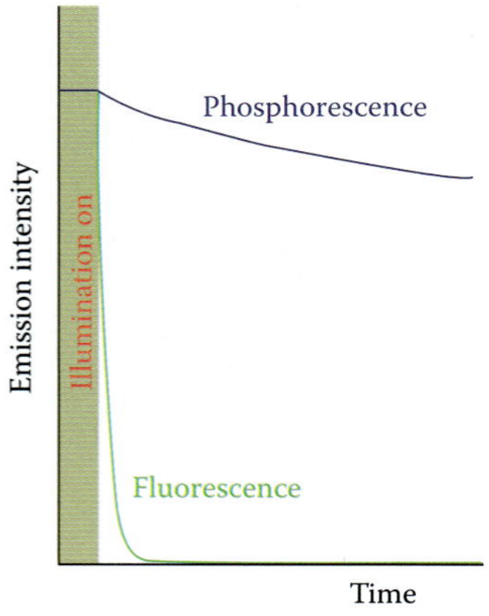

FIGURE 7.92 Decay of phosphorescence and fluorescence.

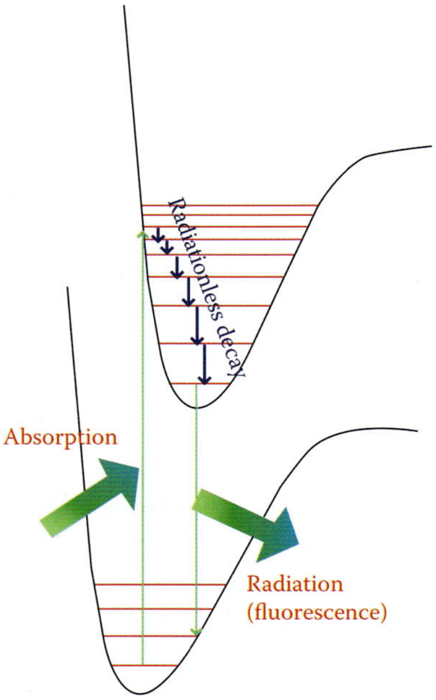

FIGURE 7.93 Stokes law illustrated: fluorescence is at a longer wavelength than absorption. These energy diagrams show the potential energy of a molecule as a function of the distance between the nuclei and, according to the Franck-Condon principle, excitation is represented by a vertical arrow.

strike the lower curve at its deepest point, so again, some excitation energy gets converted into vibrational energy. Because of energy dissipation during the excited state lifetime, the energy of the emitted photon is lower and therefore is of longer wavelength than the excitation photon. This difference in energy or wavelength is called the *Stokes shift*. The Stokes shift is fundamental to the sensitivity of fluorescence techniques because it allows emission photons to be detected against a low background, isolated from excitation photons. There is approximate mirror symmetry between the fluorescence and absorption bands when the shapes of the potential curves in the ground state and the excited state are similar.

The fluorescence quantum yield, which is the ratio of the number of fluorescence photons emitted to the number of photons absorbed, is a measure of the relative extent to which these processes occur. Mathematically, the fluorescence quantum yield is expressed as:

$$\Phi = \frac{\text{Number of emitted photons}}{\text{Number of absorbed photons}} \quad (7.141)$$

For a highly fluorescent molecule, such as fluorescein, $\phi = 1$, and for a nonfluorescent molecule, $\phi = 0$. The quantum yield can also be defined in terms of various rate constants in the Jablonski diagram. This is shown for a simple system in Figure 7.94, where k_r is the radiative pathway and k_{nr} is the nonradiative pathway:

$$\phi = \frac{k_r}{k_r + k_{nr}} \leq 1 \quad (7.142)$$

The fluorescence lifetime, τ, is the time delay between the absorbance and the emission and is given by:

$$\tau = \frac{1}{k_r + k_{nr}} \quad (7.143)$$

Whereas emission of fluorescence occurs in all directions, only the photons emitted at a 90° angle to the incident radiation are monitored to avoid interference from transmitted photons. This is shown schematically in Figure 7.95. An excitation monochromator selects λ_{ex}, the λ of light that the molecule absorbs, and the emission monochromator selects λ_{em}, one of the λ of light emitted by the molecule. Fluorescence detection sensitivity may be severely compromised by background signals, which may originate from endogenous components used in the sample or sample holder (referred to as *autofluorescence*, perhaps of the plastic substrate) or from unbound or nonspecifically fluorescent bound probes (referred to as *reagent background*).

There are many types of fluorescent measuring techniques used in biotechnology. Fluorescence microscopes resolve fluorescence as a function of spatial coordinates in two or three dimensions for microscopic objects (<0.1-mm diameter). Flow cytometers measure fluorescence per cell in a flowing stream, allowing subpopulations within a large sample to be identified, quantitated, and sorted. Flow cytometry involves the use of a beam of laser light projected

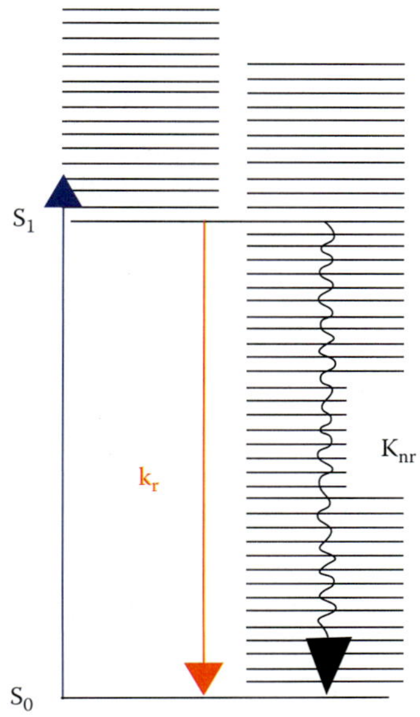

FIGURE 7.94 Simplified Jablonski diagram with rate constants for radiative (k_r) and nonradiative processes (k_{nr}).

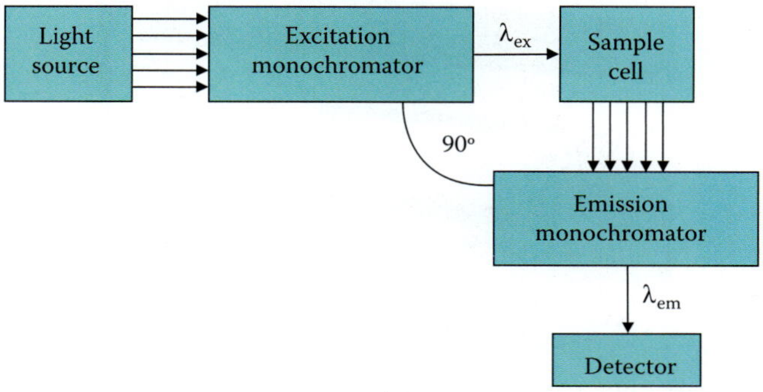

FIGURE 7.95 Simplified fluorescence measurement setup.

through a liquid stream that contains cells or other particles, which, when struck by the focused light, give off signals that are picked up by detectors. These signals are then converted for computer storage and data analysis, and can provide information about various cellular properties. Spectrofluorometers and microplate readers (Figure 7.96) measure the average properties of bulk samples (microliter to milliliter).

Fluorescence scanners resolve fluorescence as a function of spatial coordinates in two dimensions for macroscopic objects such as electrophoresis gels, blots, and chromatograms. In contemporary instruments, the excitation source is usually the 488-nm spectral line of the argon-ion laser. All these mentioned fluorescence instruments are attractive candidates for miniaturization.

Photoluminescence in Analytical Chemistry and Its Scaling Luminescence intensity *F* of an analytical sample is given by:

$$F = kP_0\varepsilon bc \tag{7.144}$$

where k = constant

P_0 = incident radiant power (W/m^2, also called *intensity*)

c = concentration of the analyte

ε = absorptivity of the sample

b = path length of the sample holder

Fluorescence intensity is quantitatively dependent on the same parameters as absorbance, but a major difference is that luminescence is also dependent on the number of incident photons (P_0). The expression in Equation 7.144 can be derived as follows.

FIGURE 7.96 A typical microplate (96 wells) and microplate reader: the MRX II microplate reader from Dynex.

The fluorescence emission intensity *F* is given by $F = K'(P_0 - P)$, with P_0 the power of the incident radiation and *P* the power of the radiation after traversing through a path length *b* of the medium. The constant *K′* depends on the quantum efficiency of the fluorescence and the light collection efficiency. In the case of absorption, one measures transmission (*T*) and determines *A* as a ratio of P_0 and *P*, whereas in the case of *F*, the absolute number of photons is counted. This situation provides fluorescence spectroscopy with a much lower detection limit, and is pictured in Figure 7.97.

We can write $F = K'(P_0 - P)$ as:

$$F = K'\left(P_0 - P_0 10^{-\varepsilon bc}\right) \text{ or } F = K'P_0\left(1 - 10^{-\varepsilon bc}\right) \tag{7.145}$$

Expanding the above equation as a Taylor's series, one obtains:

$$F = K'P_0\left(2.303\,\varepsilon bc - \frac{(2.303\varepsilon bc)^2}{2!} + \frac{(2.303\varepsilon bc)^3}{3!} + \right) \tag{7.146}$$

If $\varepsilon bc \le 0.05$, then $10^{-\varepsilon bc} \sim 1–2.3\varepsilon bc$, in which case Equation 7.145 reduces to Equation 7.144 because $F = K'\ P_0\ 2.3\varepsilon bc$ or $F = kP_0\varepsilon bc$.

According to Equation 7.144, fluorescence scales with path length in the same way absorption does. Thus, in theory, both scale poorly into the microdomain, but in the case of fluorescence, we can turn up the number of incoming photons to increase the signal. Therefore, lasers are typically used for excitation because of their higher photon flux, but one needs to be careful about sample decomposition (photobleaching). The lower limit of detection with fluorescence is frequently 10 times better than with absorption measurements, and it is also more specific and less susceptible to interferences because fewer materials absorb and also emit light. Furthermore, fluorescent molecules have different lifetimes, so that the emission from a shorter-lifetime fluorescent interferent might be left to decay before the species of interest is measured (letting the background die out). The concentration detection range is very wide with a reach of three to six decades. With fluorescence, 2–10% accuracy is common because many issues affect the quantum efficiency, ϕ.

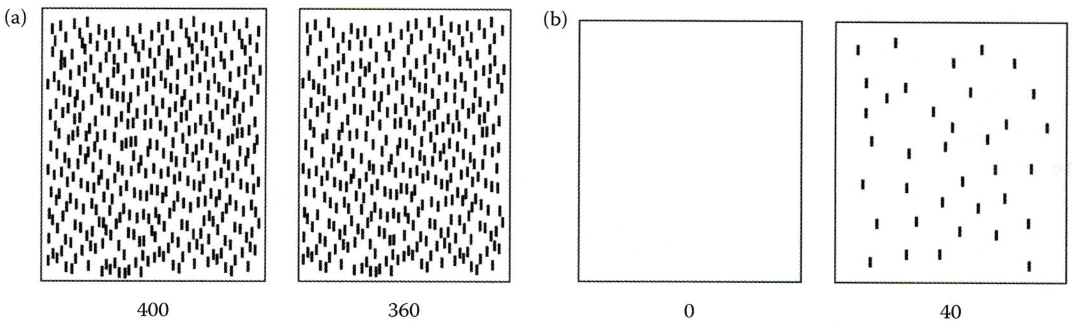

FIGURE 7.97 In absorbance spectroscopy the number of incoming and outgoing photons at the same wavelength are compared (a); in fluorescence the number of emitted photons are counted (b).

Fluorophores, Quantum Dots, and Fluorescent Proteins

Fluorophores A fluorophore is a small molecule, or a part of a larger molecule, that can be excited by light to emit fluorescence by absorbing a photon from an incandescent lamp or a laser. William Perkin, an English chemist, synthesized a coal-tar dye, aniline purple (the first synthetic dye, also known as *mauve* and *mauveine*), in 1856. His breakthrough attracted the attention of numerous synthetic chemists, and a variety of dyes were synthesized soon afterward. Perkin is acknowledged as the founder of the synthetic dye industry. Adolph von Baeyer, a German chemist, synthesized the well-known fluorescent dye fluorescein in 1871 (see Figure 7.98), and Paul Ehrlich, a German bacteriologist, used the fluorescent dye uranin (sodium salt fluorescein) to track the pathway of secretion of aqueous humor in the eye in 1882. The latter represents the first case of the use of in vivo fluorescence in animal physiology. A fluorescent probe is a fluorophore designed to bind to a specific region of a biological structure. Fluorescent probes can be used to tag proteins, nucleotides, lipids, oligosaccharides, and other biological molecules. These fluorescence-tagged molecules may then be used, for example, to localize and monitor interesting biological processes in living cells.

The molar extinction coefficient ε (see Equation 7.133) is a direct measure of a dye's ability to absorb light. This capability is clearly important in determining the amount of light a molecule can generate via fluorescence emission. Most fluorophores in common use have molar extinction coefficients at their wavelength of maximal absorption, ranging between 5,000 and 200,000 cm^{-1} M^{-1}. The Stokes shift is the difference in wavelength between the fluorescence excitation maximum and the fluorescence emission maximum. Interestingly, dyes with large Stokes shifts tend to have relatively small extinction coefficients, whereas dyes with small Stokes shifts tend to have relatively large extinction coefficients. The fluorescence quantum yield is a measure of the efficiency with which the excited molecule is able to convert absorbed light into emitted light. It is defined as the fraction of absorbed photons that are converted into fluorescence emission (see Equation 7.141). Typical quantum yields for commonly used fluorophores range from 0.05–1.0. The wavelength of absorption of a fluorophore is related to the size of the chromophores, with smaller chromophores leading to higher energy (shorter wavelength). As we just learned, the intensity of a fluorescence measurement is related to the probability of the event and the wavelength of the absorbed light.

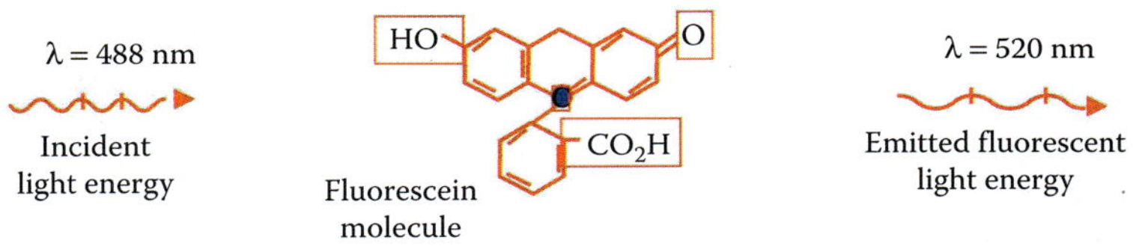

FIGURE 7.98 Fluorescein molecule.

Fluorescence is favored in molecules with structural rigidity. The quantum yields for fluorene and biphenyl, for example (see Figure 7.99), are 1 and 0.2, respectively. The increased rigidity of fluorene stabilizes the π-π^* singlet state, leading to higher quantum yield. Chelation also can lead to increased fluorescence. For polyatomic molecules in solution, the discrete electronic transitions are replaced by rather broad energy spectra called the *fluorescence excitation spectrum* and *fluorescence emission spectrum*, respectively. Quantum yield of fluorescence of most molecules decreases with increasing temperature as a result of collisional deactivation of the singlet state. Solvents containing heavy atoms, such as those containing halogens, also decrease fluorescence. Heavy atoms promote intersystem crossing to the triplet state. This decreases fluorescence quantum yield but increases phosphorescence quantum yield (see below). Unless the fluorophore is irreversibly destroyed in the excited state (an important phenomenon known as *photobleaching*), the same fluorophore can be repeatedly excited and detected. Photobleaching is an irreversible photochemical process that causes a molecule to lose its fluorescent property. Countermeasures for photobleaching include the use of shorter illumination, reducing the excitation intensity, and the use of "antifade" reagents. The latter sometimes act by reducing oxygen concentration to prevent formation of singlet oxygen (see below, Reaction 7.6). However, photobleaching has also led to the development of a useful analytical method called *Fluorescence Recovery After Photobleaching* (FRAP). The FRAP method is used to measure the ability of a molecule to move around in a cell over time. A fluorophore is covalently attached to the migrant molecule, and using an epifluorescence microscope, light is focused onto a small subset of the fluorescent molecules. After initial imaging of the amount of fluorescence, a very intense light pulse is flashed onto these same molecules, causing photobleaching. When these molecules are monitored again, the bleached area will now appear dark. If the surrounding fluorescent molecules that were not photobleached are able to diffuse, they will mix with the photobleached molecules, and the dark area will slowly become bright as fluorescent molecules move into it.

Early investigations showed that many substances autofluoresce when irradiated with UV light. In the 1930s, Haitinger and others developed the technique of secondary fluorescence using fluorochrome stains to stain specific tissue components, bacteria, or other pathogens that do not autofluoresce (using acridine orange). In the early 1950s, Coons and Kaplan developed immunofluorescence using fluorescein-tagged antibodies to localize antigens in tissues. They labeled antipneumococcal antibodies with anthracine, allowing them to detect both the organism and the antibody in tissue using UV-excited, blue fluorescence. Two commonly used fluorescent dyes that are covalently bound to antibodies are fluorescein, which emits an intense green fluorescence when excited with blue light (Figure 7.98), and rhodamine, which emits a deep red fluorescence when excited with green-yellow light. Some typical fluorescent dyes are shown in Figure 7.100.

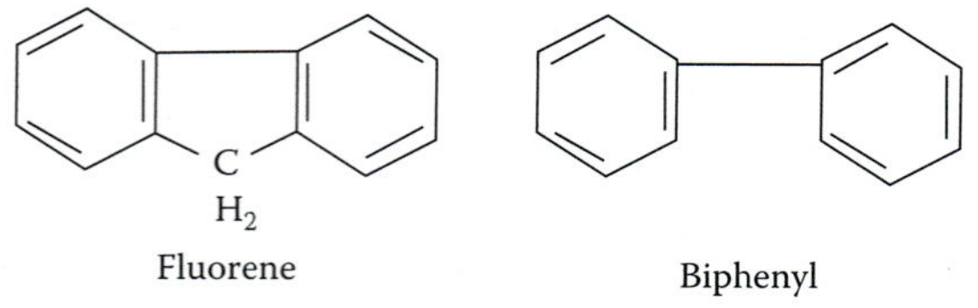

FIGURE 7.99 Fluorene and biphenyl. The fluorescence quantum yields for fluorine and biphenyl are 1 and 0.2, respectively. Fluorene has the higher rigidity and thus the higher quantum yield.

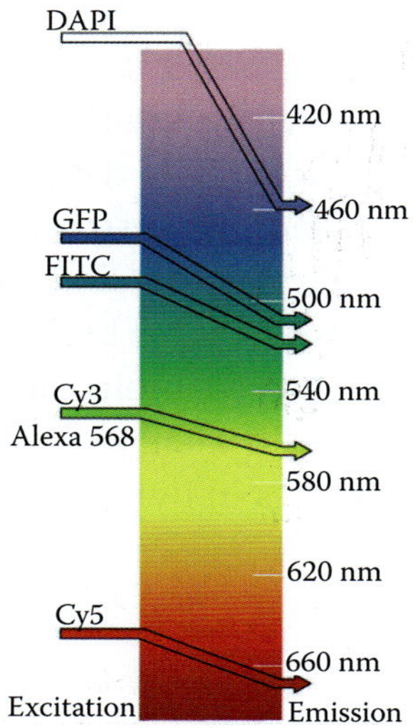

FIGURE 7.100 Typical fluorescent dyes. CY3, indotrimethinecyanines; CY5, indopentamethinecyanines; DAPI, 4′,6-diamidino-2-phenylindole; FITC, fluorescein isothiocyanate; GFP, green fluorescent protein.

Because of their particular structural nature, some fluorescent probes can bind noncovalently to special biomolecules in cells in either a specific or a nonspecific fashion. The binding can dramatically affect the fluorescence quantum yield and wavelength, which is the basis of DNA and RNA probes, protein probes, lipid probes, etc. In Table 7.18, we show some of the fluorophores that are commonly used for lighting up specific cell organelles.

Today there are fluorescent probes for almost any analyte. The binding of hydrogen or metal ions changes the electronic structure of a number of fluorophores, thus affecting the absorption properties and/or the fluorescence properties of the probes. For example, various fluorescein-labeled proteins have quantum yields in the range of 0.2–0.7 at pH 8, but drop off rapidly with decreasing pH, making a fluorescent pH probe. Certain lipophilic ester derivatives of fluorescent probes are cell membrane permeable. Cellular esterases convert these compounds into the free-acid forms of the probes, which are trapped inside the cell and are able to chelate calcium, making for good intercellular calcium probes. Fluorogenic enzyme substrates are fluorescent probes that are converted by specific enzymes into products that have either increased fluorescence or shifted spectra. These are useful probes for monitoring enzyme activities.

In fluorescence quenching, excited molecules relax to ground states via nonradiative pathways instead of fluorescence emission. Nonradiative processes include vibrational relaxation, collision with quenching molecules, and intersystem crossing. Molecular oxygen quenches by increasing the probability of intersystem crossing:

$$M^*(\text{singlet}) + O_2(\text{triplet}) \rightarrow M^*(\text{triplet}) + O_2(\text{singlet})$$

Reaction 7.6

Other typical quenchers are aliphatic and aromatic amines. Polar solvents, such as water, quench fluorescence by orienting around the excited state dipoles. Emission transfer to nonfluorescent molecules, formation of a nonfluorescent complex, aggregation forming a nonfluorescent aggregate, and emission transfer to a fluorescent molecule with an excitation wavelength equal to the emission wavelength of the first fluorescent molecule are all mechanisms that lead to quenching of the fluorescent light signal.

Förster resonance energy transfer (FRET), named after the German scientist Theodor Förster, is a radiationless energy transfer mechanism, where dipole-dipole resonance energy transfers between two chromophores. A donor chromophore in its excited state can transfer energy by a nonradiative long-range dipole-dipole coupling mechanism to an acceptor chromophore in close proximity (typically <100 Å). When both molecules are fluorescent, the term *fluorescence resonance energy transfer* is often used, although the energy is not transferred by fluorescence (there is no emission of a photon during this transfer). The phenomenon is easily detected because instead of the donor fluorescence, the emission of the acceptor is observed. The absorption spectrum of the acceptor must overlap with the fluorescence emission spectrum of the donor as shown in Figure 7.101. Donor and acceptor

Table 7.18 Specific Organelle Probes

Probe	Site	Excitation	Emission
BODIPY	Golgi	505	511
NBD	Golgi	488	525
DPH	Lipid	350	420
TMA-DPH	Lipid	350	420
Rhodamine 123	Mitochondria	488	525
DiO	Lipid	488	500
dil-Cn-(5)	Lipid	550	565
diO-Cn-(3)	Lipid	488	500

BODIPY, borate-dipyrromethene complexes; NBD, nitrobenzoxadiazole; DPH, diphenylhexatriene; TMA, trimethylammonium.

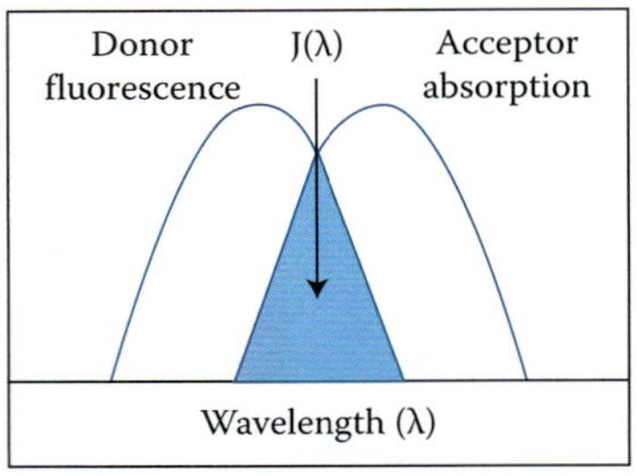

FIGURE 7.101 For fluorescence resonance energy transfer, the absorption spectrum of the acceptor must overlap the fluorescence emission spectrum of the donor. The larger the overlap J(λ), the stronger the FRET.

transition dipole orientations must be approximately parallel. As the intensity of the fluorescence of the acceptor fluorophore increases, the intensity of the fluorescence of the donor fluorophore decreases. Because the interaction between the two fluorophores is a typical dipole-dipole interaction, FRET is dependent on the inverse sixth power of the intermolecular separation (see Equations 7.13 and 7.147), making it useful over distances comparable with the dimensions of biological molecules. Summarizing, the FRET efficiency is determined by three parameters:

1. The distance between the donor and the acceptor
2. The spectral overlap of the donor emission spectrum and the acceptor absorption spectrum
3. The relative orientation of the donor emission dipole moment and the acceptor absorption dipole moment

The FRET efficiency *E* is defined by the Förster equation:

$$E = \frac{1}{1+\left(\frac{r}{R_0}\right)^6} \tag{7.147}$$

From this expression, the efficiency *E* depends on the donor-to-acceptor separation distance, *r*, with an inverse sixth order law as a result of the dipole-dipole coupling mechanism, with R_0 being the Förster distance of this pair of donor and acceptor at which the FRET efficiency is 50%. The donor-to-acceptor separation distance *r* can be from 0–100 Å.

FRET is an important technique for investigating a wide variety of biological phenomena that produce changes in molecular proximity.

A nice illustration of the FRET principle is found in the use of molecular beacons shown in Figure 7.102. The molecular beacon is a single-stranded nucleic acid molecule with a stem-and-loop structure. The loop portion of the molecular beacon "hairpin" is complementary to a target DNA sequence. The hairpin stem is formed by two sequences that are complementary to each other. In hairpin formation, the quencher, attached to one end of the probe, quenches the fluorophore attached to the other end of the probe, and no fluorescence is emitted. On hybridization of the molecular beacon to its target, the stem dissociates, causing the quencher and the fluorophore to move away from each other, resulting in the restoration of fluorescence.

The rate of fluorescence emission depends on the amount of time that the molecule remains within the excited state (say, excited state with lifetime τ_f). Optical saturation occurs when the rate of excitation exceeds the reciprocal of τ_f. Molecules that remain in the excited beam for extended periods have higher probability of interstate crossings and thus phosphorescence (see below). Usually, increasing dye concentration can be the most effective means of increasing signal when energy is not the limiting factor (i.e., using laser-based confocal systems) and until the dye concentration reaches the limit of the inner filter effect.

Quantum Dots Quantum dots, also known as *nanocrystal semiconductors*, are crystals composed of materials from periodic groups II–VI, III–V, or IV–VI

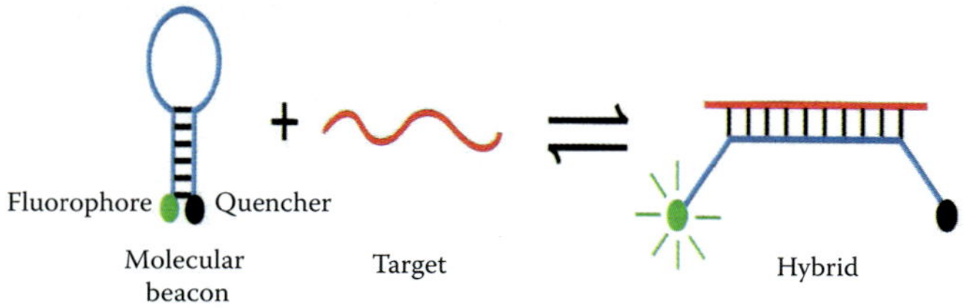

FIGURE 7.102 Schematic presentation of a molecular beacon. In the hairpin loop structure, the quencher (black circle) forms a nonfluorescent complex with the fluorophore (green circle). On hybridization of the molecular beacon to a complementary sequence, the fluorophore and quencher are separated, restoring the fluorescence.

(e.g., CdS, CdSe, PbS, PbSe, PbTd, CuCl). These nanocrystals are used as crystalline fluorophores and can be conjugated to biomolecules as an alternative to the above-reviewed organic dyes. As we saw in Chapter 3, quantum dots (QDs) form a unique class of semiconductors because of their small size, ranging from 2–10 nm (10–50 atoms) in diameter. At these sizes, material properties become size-dependent. In Figure 7.103, we illustrate how smaller dot size leads to emitted light at shorter wavelengths.

With QDs, by adjusting the size of the dot or its composition, the emission wavelength is tunable from blue to the near infrared. CdS and ZnSe dots of decreasing size emit from the blue light region to the near-UV light, different-sized CdSe dots emit light across the visible spectrum; and InP, and InAs QDs emit in the far-red and near-infrared. Elongated QDs (quantum rods) show linearly polarized emission, whereas the fluorescence emission from spherical CdSe dots is circularly polarized or nonpolarized.[46] QDs have a broad absorption spectrum (any light source "bluer" than the dot of interest works) and yield narrow emission. For example, the emission spectra of single ZnS-capped CdSe QDs are as narrow as 13 nm (full-width at half-maximum; FWHM) at room temperature.[47] Symmetric, narrow emission spectra minimize overlap of adjacent colors. Some of the manufacturing techniques of QDs, such as wet chemistry and template chemistry (zeolite, alumina templates), are detailed in Volume III, Chapter 3. See also earlier in this chapter where the Derjaguin, Landau, Verwey, and Overbeek theory (DLVO) was introduced to explain colloid stability. We will learn how crystal size can be controlled by the growth conditions. Before 1993, QDs were often prepared in aqueous solution with added stabilizing agents (e.g., thioglycerol or polyphosphate) to avoid colloid precipitation (see colloid stability, page 519). This procedure yielded low-quality QDs with poor fluorescence efficiencies and large size variations [relative standard deviation (RSD) > 15%]. In 1993, Bawendi and coworkers[48] synthesized better luminescent CdSe QDs using a high-temperature organometallic procedure instead. The resulting nanocrystals had nearly perfect crystal structures and exhibited much narrower size variations RSD < 5%), but the fluorescence quantum yields were still relatively low (~10%). At that time, even if one could make QDs in a narrow size range, they still came with two major problems: poor fluorescence and hydrophobicity, with the latter preventing their use in biological experiments. Today, these problems have been overcome. The addition of semiconductor caps such as ZnS or CdS were found to dramatically increase the fluorescence quantum yield to 45% or higher. A quantum dot often used in this regard is shown in Figure 7.104, a configuration with a CdSe core and a ZnSe shell. The core determines the nanocrystal color, and the shell, of a higher bandgap material (ZnS), dramatically enhances not only the brightness but also the chemical stability. The ZnS protects

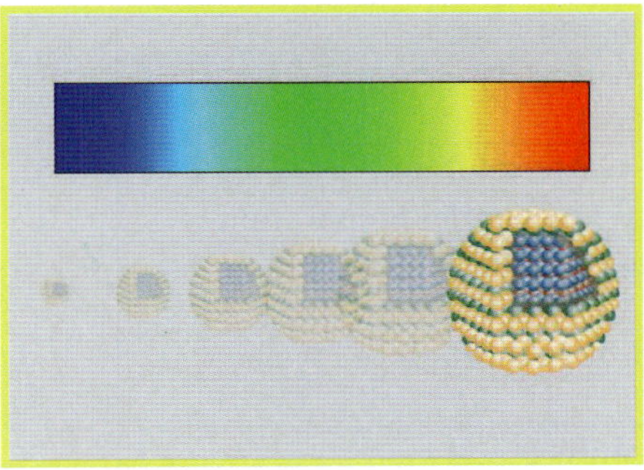

FIGURE 7.103 Quantum dots. Quantum dots change color with size because additional energy is required to "confine" the semiconductor excitation to a smaller volume.

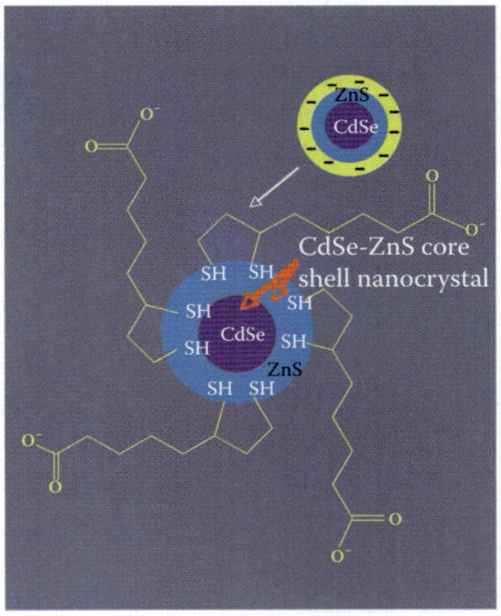

FIGURE 7.104 Nanocrystal with CdSe core and a ZnSe shell. The surface is derived with dihydrogenlipoic acid (DHLA) groups that terminate with negatively charged carboxyl groups. The addition of this coat makes the particles water soluble. To conjugate quantum dots to biomolecules, avidin or protein-G with a positively charged tail may be conjugated to the negatively charged DHLA coat.

the core from oxidation and prevents the CdSe from leeching into the surrounding solution.

Compared with traditional organic dyes such as rhodamine 6G and fluorescein, CdSe nanocrystals show similar or lower quantum yields at room temperature. The lower quantum yields are compensated by their larger absorption cross-sections and much reduced photobleaching rates. Bawendi and coworkers[48] estimated that the molar extinction coefficients of CdSe QDs are about 10^5–10^6 M^{-1} cm^{-1}, depending on the particle size and the excitation wavelength. Chan and Nie[47] estimated that single ZnS-capped CdSe QDs are ~20 times brighter and ~100–200 times more stable than single rhodamine 6G molecules (see also comparison of dyes and QDs in Table 7.19). CdSe cores with ZnS shells are the best QDs for bioapplications with an easily reproducible chemistry and a wide range of emissions.

QDs are generally negatively charged, owing to the colloid stabilizer molecules adsorbed on their surface.[49] These stablizers include hydrophobic tri-*n*-octylphosphine oxide (TOPO), which strongly coordinates to the surface metal atoms, whereas tri-*n*-octylphosphine (TOP) and tributylphosphine (TBP) are weakly bound. These surface stabilizers prevent flocculation in organic solvents and must be replaced with surface stabilizers for use of the same particles in aqueous environments. A few different methods for making nanocrystals water soluble are available. Methods include derivatizing the surface with mercaptoacetic acid or dithiothreitol, encapsulating in phospholipid micelles, derivatizing the surface with dihydrogenlipoic acid (DHLA; groups that terminate with negatively charged carboxyl groups, as shown in Figure 7.104), and coating them with amine-modified polyacrylic acid. In yet another approach, Alivisatos et al.[50] use a silica/siloxane coating for creating water-soluble, ZnS-capped CdSe QDs. During their process, 3-(mercaptopropyl) trimethoxysilane (MPS) is directly adsorbed onto the nanocrystals, and the TOPO molecules are displaced. A silica/siloxane shell is formed on the surface by the introduction of a base and hydrolysis of the silanol groups. The addition of these coats makes the particles water soluble, and therefore more useful for biological applications.

The water-soluble QDs are now ready to be conjugated to all sorts of biomolecules. The surface area of a single QD is large enough to bind multiple biomolecules. Two to five protein molecules and 50 or more small molecules (such as oligonucleotides or peptides) may be conjugated to a single 4-nm QD. To attach QDs to biomolecules in a simple

Table 7.19 Comparison of Organic Dyes and Quantum Dots

Organic Dye	Quantum Dot
Broad absorption and emission profiles spectrum	Broad absorption but narrow emission. Sharper spectrum (FWHM ~25–40 nm)
Low photobleaching threshold. Fades quickly ~100 ps	High resistance to photobleaching and photo and chemical degradation ~5–40 ns
Unstable	Stable output over time (100 times more stable)
One dye excited at a time	Multicolor imaging
Low quantum yield (usually smaller than quantum dots)	High quantum yield. Fluorescent yield is 20 times larger
	Can be made water soluble
	Requires much less sample preparation
Large Stokes shifts come with low molar extinction coefficients	Excited at one wavelength for multiple colors
Can diffuse through cell walls	QDs cannot diffuse through cell walls
Smaller	Bulkier
	Large Stokes shifts
	Wide range (UV-IR)
	High molar extinction coefficient (~10–100 times organic)

electrostatic fashion, avidin or protein-G with a positively charged tail may be conjugated to the negatively charged DHLA coat of a QD. Other scenarios for QD to biomolecule conjugation are illustrated in Figure 7.105. In general, one deals with reactive functional groups, including primary amines, carboxylic acids, alcohols, and thiols. Bioconjugated QDs are finding more and more applications, including DNA hybridization, multiplexed optical encoding, and high-throughput analysis of genes and proteins, immunoassays, and in vivo cellular imaging. Some important advantages of QDs for the latter application are the extremely high photostability of QDs, which allows real-time monitoring or tracking of intracellular processes over long periods (minutes to hours); the ability to use multicolor nanocrystals to simultaneously image multiple targets; and the fact that nanocrystals are believed to be less toxic to the cell than organic dyes. This will allow, for example, viral particles to be followed in vivo, drug molecules to be analyzed in biological systems, and tumor cells to be tracked in real time.

Micro- or nanobeads filled with QDs, as shown in Figure 7.105e, constitute one of the most fascinating new applications of QDs. In this configuration, polystyrene or latex beads are filled with QDs of various combinations of color and intensity (by using different concentrations of the same QD). Specific capturing molecules such as peptides, proteins, and oligonucleotides are then covalently linked to the beads and thus are encoded by the beads' spectroscopic signature, like a barcode on a product. The use of six colors and 10 intensity levels could, in theory, encode 1 million protein or nucleic acid sequences. A single light source is sufficient for reading all the QD-encoded beads. To determine whether an unknown analyte is captured, conventional assay methodologies may then be applied.[51] Because large numbers of barcode variations are possible, each perhaps specific to a different DNA sequence, the application of QD-encoded beads to DNA microarray assays is obvious. Traditional DNA microarrays are composed of short DNA oligomers attached to an inert substrate (glass slide) and typically contain a grid of 10^5–10^6 features (spots), each with a different DNA molecule (a DNA probe) at a known location in the grid. Fluorescently labeled target DNA or RNA hybridizes to these complementary probes. The hybridized array is scanned with a laser to check for a fluorescent signal for each spot, and because we know where each DNA probe is located, we know which ones hybridize successfully. In the case of QD-encoded beads,

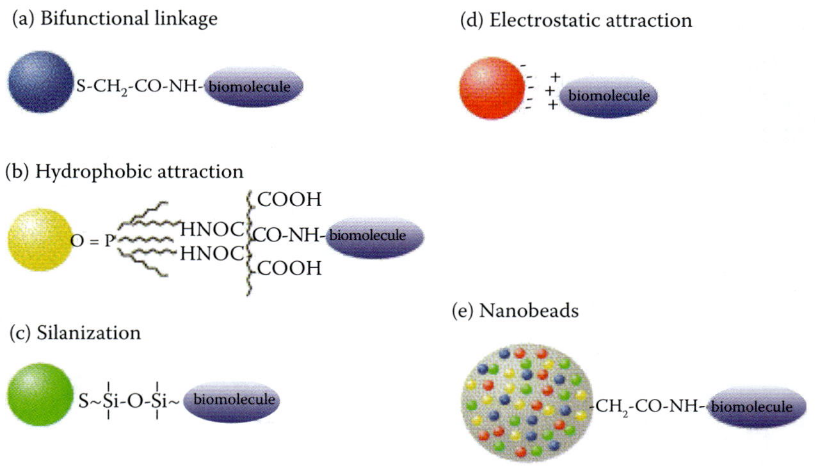

FIGURE 7.105 Quantum dot-biomolecule conjugation schemes. (a) Use of a bifunctional ligand such as mercaptoacetic acid for linking QDs to biomolecules. (b) Tri-*n*-octylphosphine oxide (TOPO)-capped QDs bound to a modified acrylic acid polymer by hydrophobic forces. (c) QD solubilization and bioconjugation using a mercaptosilane compound. (d) Positively charged biomolecules are linked to negatively charged QDs by electrostatic attraction. (e) Incorporation of QDs in microbeads and nanobeads. (From Chan, W. C. W., D. J. Maxwell, X. Gao, R. E. Bailey, M. Han, and S. Nie. 2002. Luminescent quantum dots for multiplexed biological detection. *Curr Opin Biotechnol* 13:40.[46])

multiplexed assays are simply created by mixing the appropriate beads in a solution rather than placing them on a solid surface. We do not need a grid location anymore because each particle comes with its own barcode. One can envision these 3D liquid arrays (Figure 7.106) as the successors of the current generation of gene chips (e.g., see the Nanogen DNA chip in Volume II, Figure 8.51). These new liquid "bar-coding" technologies have the potential to offer significant advantages over current planar chip devices, such as faster binding kinetics (3D vs. 2D), decreased hybridization time, more massive parallel processing, and ease of adding new targets. These novel multiplexed multicolor assays are being developed for massive parallel biosensing and analytical detection.

An entirely different bar-coding principle was developed by Natan and coworkers[52] and uses metallic nanobars. These authors report the development of metallic nanobarcodes for multiplexed bioanalysis. Fluorescence is still used for target detection, but the barcodes are read by measuring the reflectivity differences between different metals. See also the BeadArray™ from Illumina further below in Figure 7.120.

It is still unclear how similar technologies, such as encoded nanobars and QD-tagged microbeads and encodable radiofrequency devices, will compete in genomic, proteomic, and clinical diagnostic applications.

From the above, it appears that QDs could eventually constitute a big improvement over fluorescent dyes. Unlike fluorescent dyes, which tend to decompose and lose their ability to fluoresce, QDs maintain their integrity, withstanding many more cycles of excitation and light emission (they do not bleach easily!). Some of the disadvantages associated with QDs are the possibility of their irreversible aggregation and their relatively bulky size (they cannot diffuse through a cell membrane) compared with most organic fluorophores. In Table 7.19, we compare organic dyes with QDs.

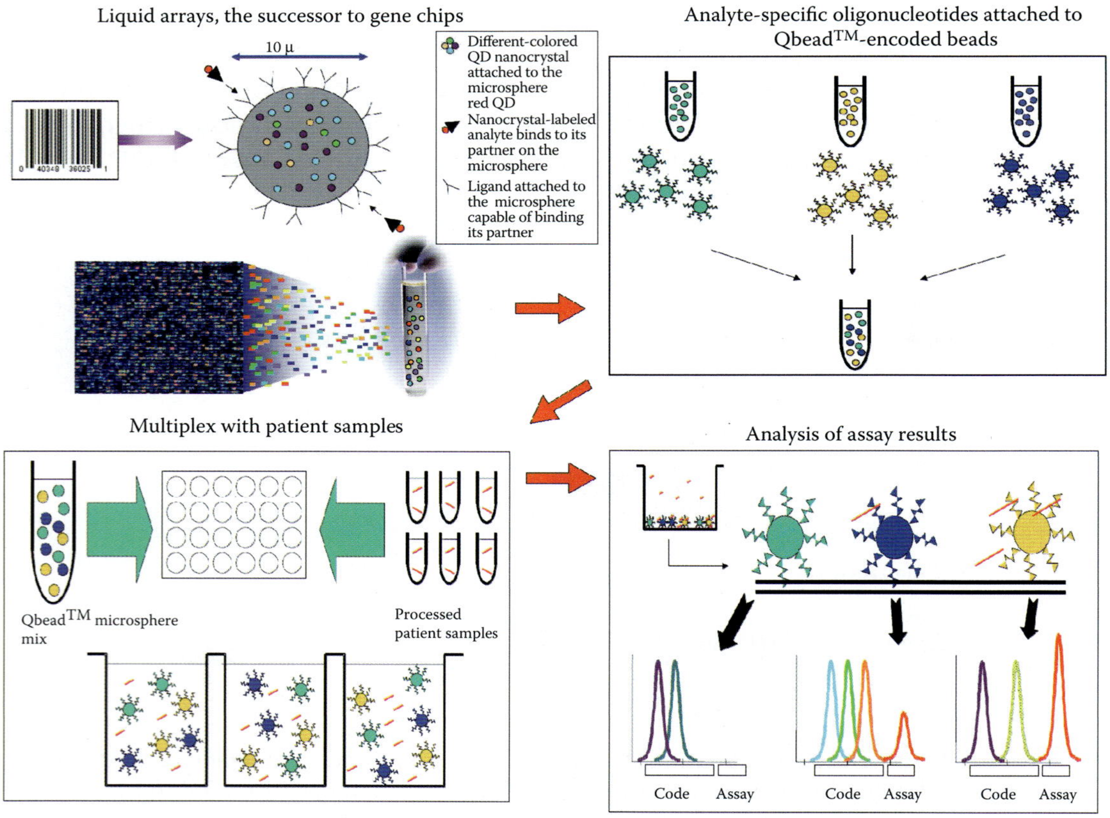

FIGURE 7.106 Spectral barcodes. How to use liquid arrays. The Qbead™ technology is based on the use of QD particles to create spectral barcodes that enable high levels of multiplexing for genetic analysis. As few as 10 assays per well or potentially as many as millions of assays per well may be encoded. Quantum Dot Corporation now known as Invitrogen.

FIGURE 7.107 Dr. Osamu Shimomura.

Fluorescent Proteins As an example of a fluorescent protein, we will focus on green fluorescent protein (GFP). GFP is an intensely fluorescent probe, with a quantum efficiency of approximately 80% and a molar extinction coefficient ε of 2.2 × 10^4 cm^{-1} M^{-1}. The GFP from the molecular system of the bioluminescent jellyfish *Aequorea victoria* is making an impact on biotechnology today. It was Osamu Shimomura who first noticed GFP in 1962 (Figure 7.107) (http://en.wikipedia.org/wiki/Green_fluorescent_protein). At first, a mere footnote in a scientific paper about a small, bioluminescent jellyfish called *Aequorea aequorea*, or *A. victoria* as the natives from the Bay of Puget Sound call it, the study of this jellyfish's glow became Shimomura's life's work.

The gene encoding GFP was first isolated from the photocytes in the umbrella of the jellyfish *A. victoria*, which was then cloned and introduced into cells of other species. The structure of the GFP molecule, along with a picture of the *A. victoria* jellyfish, is shown in Figure 7.108. This extremely stable protein is made up of 238 amino acids. Its fluorescent properties are unaffected by prolonged treatment with 6 M guanidine HCl, 8 M urea, or 2-day treatment with various proteases such as trypsin, chymotrypsin, or papain. The excitation maxima of GFP are at 395 nm (A state) and 470 nm (B state), and its peak emission is at 509 nm.

The structure of the GFP protein is a marvel of molecular engineering. Looking like a glowing green lantern, it consists of an outer coat, which is a symmetrical large barrel made up of 11 strands of β-pleated sheets. The barrel even has a top and bottom lid made up by shorter protein helices. A much longer helix is centered inside this barrel, and the chromophore itself is nestled in the middle segment of this helix. The cylinder has a diameter of 30 Å and a length of 40 Å. The GFP molecule emits green light when activated by blue or UV light. In vivo, in the jellyfish, there is a radiationless energy transfer from aequorin (AEQ), a bioluminescent protein (see below) present in the same jellyfish that emits at 469 nm to excite GFP, which in turn then emits at 509 nm. This is possible because of the overlap between the emission spectrum of AEQ and the excitation spectrum of GFP [see bioluminescence resonance energy transfer (BRET) in Figure 7.116].

The mechanism of luminescence emission in GFP involves the formation of an internal chromophore, p-hydroxybenzylidene-imidazolone, by an unusual cyclization of three neighboring amino acid residues (Ser-Tyr-Gly at the 65–67 positions of the primary

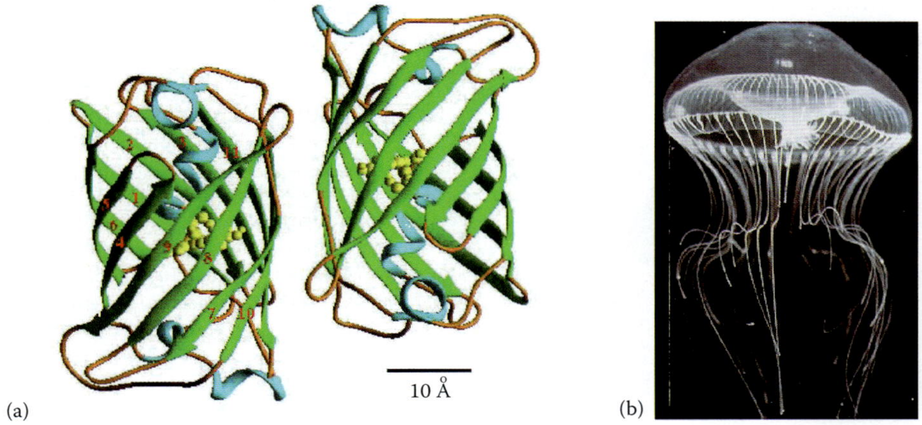

FIGURE 7.108 (a) The overall shape of the GFP protein. Eleven strands of sheet (green) form the walls of the cylinder; short segments of helices (blue) cap the top and bottom of the β barrel (can). (b) The *Aequorea victoria* jellyfish.

sequence). While the wild-type GFP emits in the green region, by introducing amino acid changes in the region close to the chromophore, all types of GFP color variants can be made. This enables the production of GFP variants (mutants) with tailor-made optical properties. Miyawaki et al.[53] have combined the protein calmodulin and the peptide M13 (binds to calmodulin on exposure to Ca^{2+}) with GFP and produced an optical sensing probe for Ca^{2+} with a detection limit in the 100 to 10 billionth of a mol per liter. The sensitivity range depends on the calmodulin mutant used. The spatial resolution of this fluorescent probe is good enough to allow one to compare concentrations of Ca^{2+} between different compartments of one cell.

In 1994, Chalfie et al.[54] coupled the gene for GFP to the gene of another protein of interest. This technique is known as *gene fusion*. The resulting gene fusion, or protein chimera, is then transferred to an organism (e.g., bacterial, yeast, or mammalian cell) so that the latter may produce a protein of interest (protein chimera) that combines the characteristics of both partner protein molecules.[54] After transfer of this target gene into the organism, one can measure the success of the gene transfer simply by shining UV light on the transformed cells. If the cells produce green light, it indicates that the transfer has occurred. GFP has been used to produce animals [e.g., fruit flies, zebrafish, mice, a dog (http://www.ekac.org/transgenic.html), and a bunny (http://www.ekac.org/gfpbunny.html)], as well as plants that light up brightly (see Figure 7.109). This fluorescence can be generated either in the whole animal or plant, or just in some desired part of the organism, such as the cells in a tumor.

FIGURE 7.109 Alba, the fluorescent bunny. (From http://www.ekac.org/gfpbunny.html.)

Further developments have made it possible for the GFP to blink intermitently.[55] This was the first demonstration of light switching of molecules induced at room temperature and of monitoring this switching on a molecule-by-molecule basis.[56] Applications in optoelectronic devices and computing are obvious future targets for GFP. Another protein that is often mentioned in the context of molecular computing and optical switching, bacteriorhodopsin,* requires the measurement of absorbance of a large number of molecules; thus, it appears less efficient for these applications. For more GFP literature, see http://www.imb-jena.de/www_sbx/gottfried/gfp/gfp_literature.html. Books on GFP are listed at http://pantheon.cis.yale.edu/~wfm5/gfp_gateway.html.

Chemiluminescence and Bioluminescence

Chemiluminescence Emission of light by matter (luminescence) has always been known to humans: lightning, the aurora borealis, and light emission by bacteria in the sea or by decaying organic matter are common natural phenomena. Scientific investigation of the luminescence phenomena began with a discovery by Vincenzo Casciarolo, a Bolognian shoemaker and alchemist who in 1603 prepared, serendipitously, an artificial phosphor known as the *Bolognian stone* (or Bolognian phosphor), which glows after exposure to light. Chemi- and bioluminescence measurements with luminometers have become extremely popular in recent years. A luminometer consists of a sample chamber, a detector [usually a photomultiplier tube (PMT)], a signal processing method, and a signal output display. In chemiluminescence, the light emission (sometimes called *cold light*†) is caused by a chemical reaction in which chemical energy is converted to light energy.

* Bacteriorhodopsin (BR) is a photon-driven ion pump. BR is a seven-helical transmembrane protein with a retinal cofactor. It is found in the bacterium *Halobacterium salinarium*, where it converts light to a proton gradient, which in turn is used by a second membrane protein called *ATP synthase* to generate chemical energy in the form of ATP. ATP is then used by the cell to drive a multitude of vital processes. Bacteriorhodopsin movies can be found at http://www.utexas.edu/depts/pharmacology/gonzales/brhod.html.

† To create light, another form of energy must be supplied. The two common ways are incandescence and luminescence. Incandescence refers to the emission of light from heat energy, e.g., the tungsten filament of an ordinary incandescent light bulb glows "white hot."

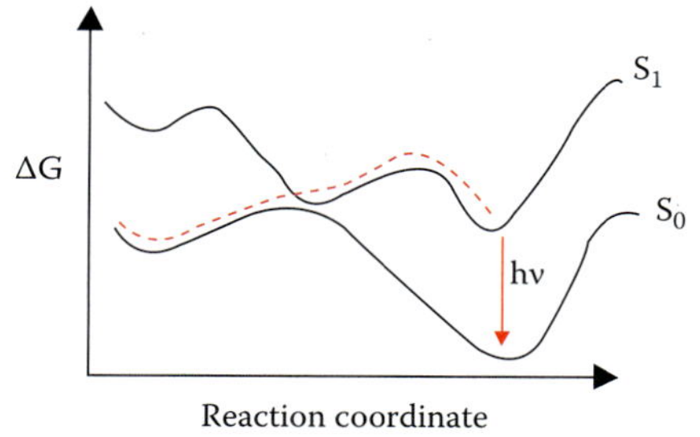

FIGURE 7.110 Chemiluminescence occurs when the ground state of one molecule crosses with the excited state of a rearranged or fragmented molecule.

Chemiluminescence occurs when the ground state of one molecule crosses with the excited state of a rearranged or fragmented molecule, as sketched in Figure 7.110.

These reactions can occur in the gas and liquid states and can last from less than a second to more than a day. Reagents containing the group —CO—NH—NHR react with oxygen, hydrogen peroxide, and many other oxidizing compounds to yield chemiluminescent products. A typical example is that of luminol, given as:

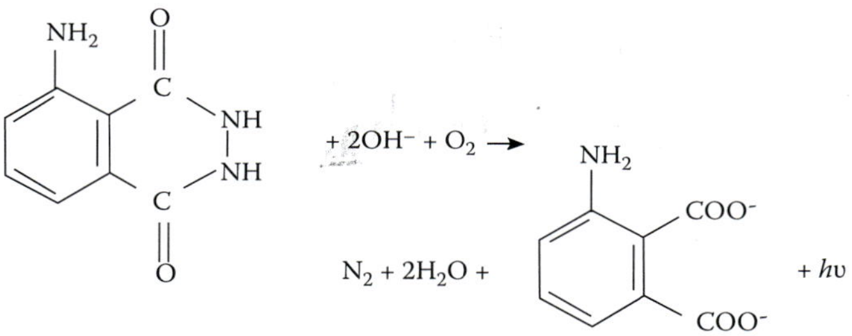

Luminol 3-aminophthalate ion

Reaction 7.7

When luminol is placed in an alkaline solution with hydrogen peroxide and a catalyst such as iron, manganese, copper, nickel, or cobalt, the luminol is oxidized to 3-aminophthalate ion in the excited state, causing fluorescence. The luminol creates light via oxidation because two oxygen atoms easily replace the two nitrogen atoms. As this reaction occurs, nitrogen gas is discharged, leaving the luminol in an excited state and releasing additional energy as light. Amino acids, fructose, glycerols, thiols, and serum albumin can also react with luminol to produce an intense light. No excitation source is needed to produce the glow, and a PMT is used to measure and detect low amounts of light in a simple luminometer.

The most important and well-known use of luminol is in the field of forensic science. Blood, which is slightly alkaline, contains cells, water, enzymes, proteins, and hemoglobin. Hemoglobin reacts with the luminol, allowing for the detection of very small amounts of blood. Once an area is suspected to have blood (even if the area has been cleaned) luminol can be applied. The lights are turned off, and after a few seconds a glow appears (Figure 7.111). Many other organic compounds can be oxidized using enzymes such as uricase and oxygen to yield hydrogen peroxide:

$$\text{uric acid} + O_2 \xrightarrow{\text{uricase}} \text{allantoin} + H_2O_2 \qquad \text{Reaction 7.8}$$

The hydrogen peroxide can then be detected with luminol (Reaction 7.7). Examples of compounds that are often detected in this manner are glucose, cholestereol, choline, uric acid, amino acids, aldehydes, and lactate. Biosensors with enzymes immobilized on optical fibers have been developed for the in situ detection of many compounds using this principle.

Another well-known chemiluminescence system is Cyalume from American Cyanamid. In this case a simplified format of the reaction is given as:

$$\text{phenyl oxalate ester} + H_2O_2 \rightarrow \text{intermediate}$$
$$(\text{I}) + \text{products I} + \text{flurophor}(\text{F}) \rightarrow \text{F}^* + \text{products}$$
$$\text{F}^* \rightarrow \text{F} + \text{hv}$$

Reaction 7.9

The same reaction is reproduced in Figure 7.112. Here, the phenyloxalate ester is shown, and the reactive intermediate (*I*) is dioxetane. The excited molecule can also transfer energy to another molecule, which then emits light. Different fluorophores (*F*)

FIGURE 7.111 Blood detection using luminol.

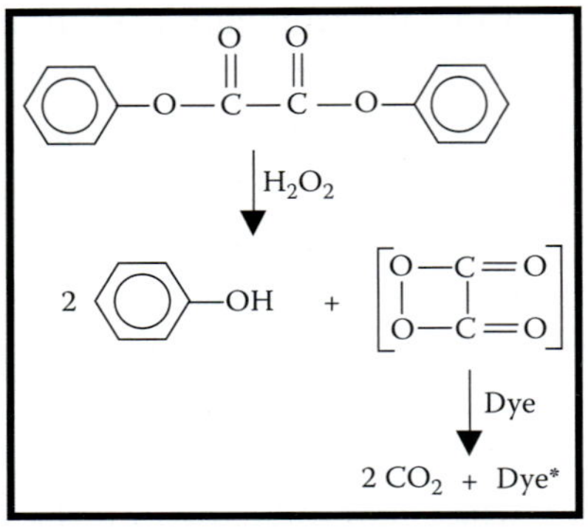

FIGURE 7.112 Oxalate ester in contact with H_2O_2 makes the reactive intermediate dioxetane (*I*). This intermediate reacts with a variety of dyes to give different colors.

reacting with intermediate *I* give different colors used in "glow-in- the-dark" light sticks (Figure 7.113). For red light, rhodamine B is used; the common green comes from 9,10-*bis*(phenylethynyl) anthracene; 9,10-diphenylanthracene results in blue.

A "glow-in-the-dark" light stick is a plastic tube with a glass vial inside. To activate such a light stick, one bends the plastic stick, which breaks the glass vial containing the hydrogen peroxide. This allows the hydrogen peroxide that was inside the glass to mix with the chemicals in the plastic tube. Once these substances contact each other, a reaction sets in that might last as long as 12 hours. When mixed with hydrogen peroxide and a dye, phenyl oxalate gives a quantum yield of 5%, which is not as efficient as a firefly (see below) but still very useful.

One of the widest analytical implementations of chemiluminescence is in the detection of NO or O_3.

FIGURE 7.113 Oxetane intermediate reacts with a variety of dyes to give different colors in light sticks. Plastic tubes have a glass vial inside that breaks on bending and lets the reagents react.

Together they react to produce electronically excited NO_2^*, which emits light throughout the red and near-infrared spectral regions (600–2800 nm):

$$NO + O_3 \rightarrow NO_2^* + O_2$$
$$NO_2^* \rightarrow NO_2 + h\nu \qquad \text{Reaction 7.10}$$

Using an excess of either component, one can detect trace amounts of the other. For example, ozone from an electrogenerator can be drawn together with an atmospheric sample into a reactor, where chemiluminescence is monitored by a photomultiplier in a luminometer. Linearity is obtained for NO concentrations from 1 ppb to 1%. This is the method of choice for measuring NO concentrations from ground level to altitudes up to 20 km. Atmospheric ozone is detected by chemiluminescence, resulting from the reaction of ozone with rhodamine B adsorbed on a silica gel. This reaction has a linear response in the range of 1–400 ppb. Ozone is also detected by the chemiluminescence from its reaction with ethylene.

Bioluminescence Bioluminescence is a type of chemiluminescence where the production and emission of *cold light* in a living organism is the result of a chemical reaction. Bioluminescence is caused by enzymes and other proteins present in the organism. Because it is a biochemically driven reaction/phenomenon, there is no need for excitation sources. These proteins can be used as markers or reporters of other biochemical processes in biomedical research. It is found in many marine animals, both invertebrate (e.g., some cnidarians, crustaceans, squid) and vertebrate (some fishes), in some terrestrial animals (e.g., fireflies, some centipedes), and in some fungi and bacteria (see Table 7.20).

The two insets in the heading of this chapter—a shrimp vomiting a bioluminescent substance to keep a predator at bay and a firefly—are examples of bioluminescence. The anglerfish lives more than a mile deep in the ocean. On her forehead, the female has a "fishing rod" tipped with a lighted "artificial worm" that she dangles over her mouth to attract her next meal. Deep-sea predators, like the anglerfish, need their own light to attract both prey and

Table 7.20 Organisms in Nature That Are Bioluminescent

Group	Features of Luminous Displays
Bacteria	Organisms glow constantly; system is autoinduced
Fungi	Mushrooms and mycelia produce constant dim glow
Dinoflagellates	Flagellated algae flash when disturbed
Coelenterates	Jellyfish, sea pansies, and comb jellies emit flashes
Annelids	Marine worms and earthworms exude luminescence
Mollusks	Squid and clams exude luminous clouds and have photophores
Crustacea	Shrimp, copepods, ostracodes; exude luminescence; also have photophores
Insects	Fireflies (beetles) emit flashes; flies (Diptera) glow
Echinoderms	Brittle stars emit trains of rapid flashes
Fish	Many bony and cartilaginous fish are luminous; some use symbiotic bacteria; others are self-luminous; some have photophores

mates. Fireflies also use their flashes to attract mates. The pattern differs from species to species. In one species, the females sometimes mimic the pattern used by females of another species. When the males of the second species respond to these "femmes fatales," they are eaten! The organism itself, or certain symbiotic bacteria living in the organism, can control the light emission. In a firefly, about 100% of the ATP energy is converted into light. By comparison, a normal electric light bulb gives off only 10% of its energy as light, whereas the rest is wasted as heat. Molecular details vary from organism to organism, but each involves a luciferin, a light-emitting substrate; a luciferase, an enzyme that catalyzes the reaction; ATP, which is the source of the energy; and molecular oxygen, O_2:

$$\text{luciferin} + \text{luciferase} + \text{ATP} \rightarrow$$
$$\text{luciferyl adenylate-luciferase} + \text{pyrophosphate}$$
$$\text{luciferyl adenylate-luciferase} + O_2 \rightarrow$$
$$\text{oxyluciferin} + \text{luciferase} + \text{AMP} + h\nu$$

Reaction 7.11

For a simplified reaction scheme, see Figure 7.114. Firefly luciferase was originally prepared by extraction of firefly tails and purified by crystallization. Luciferases and luciferins from different bioluminescent organisms might differ greatly. Today genetic cloning of various luciferases allows for its recombinant production. Genetic engineering of the luciferases and luciferins can extend both the performance and capabilities of bioluminescence.

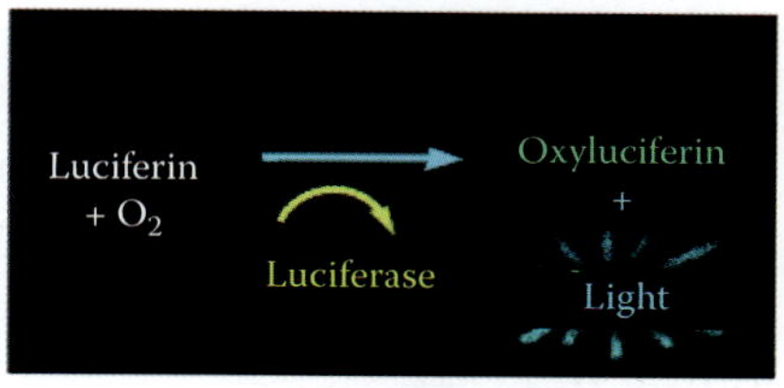

FIGURE 7.114 Luciferin is the substrate and luciferase the enzyme that catalyzes the reaction; ATP (not shown) is the source of energy. Molecular oxygen O_2 is also needed.

The more ATP available, the brighter the light. In fact, firefly luciferin and luciferase are commercially available for measuring the amount of ATP in biological materials (e.g., Millipore's Milliflex Rapid). A well-designed luminometer can detect as little as 0.6 pg of ATP or 0.1 fg of luciferase with six decades of linearity.

Sometimes the luciferin and luciferase (as well as a cofactor such as oxygen) are bound together in a single unit called a *photoprotein*. This molecule can be triggered to produce light when a particular type of ion is added to the system (frequently calcium ions) (see Figure 7.115). This mechanism can be used for a very sensitive Ca^{2+} detector. An example of such a photoprotein is aequorin.

Aequorin, like green fluorescent protein (GFP), is a photoprotein native to the jellyfish *A. victoria* (see Figure 7.108). Aequorin is composed of two distinct units, apoaequorin and an organic chromophore known as *coelenterazine*, a molecule belonging to the luciferin family. In the presence of molecular oxygen, the two components of aequorin reconstitute spontaneously, forming the functional protein.

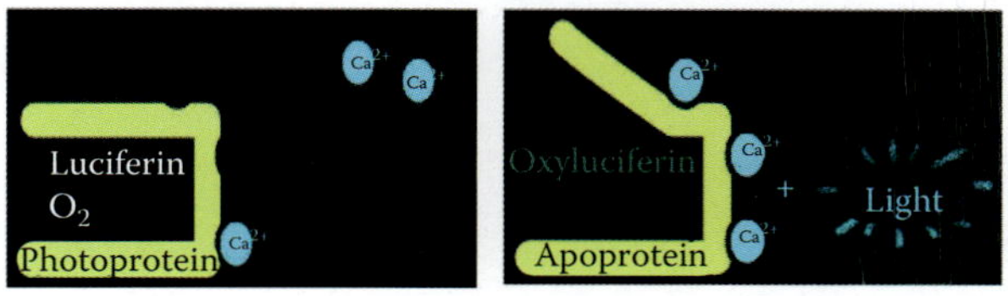

FIGURE 7.115 Luciferin, luciferase, and oxygen are bound together in a photoprotein.

Aequorin emits a flash of blue light (469 nm) that lasts 3–10 s on exposure to calcium ions, which produces a direct, ultrasensitive signal that can be measured with a luminometer:

$$\text{Aequorin} + 2\text{Ca}^{2+} \rightarrow$$
$$\text{blue fluorescent protein: } 2\text{Ca}^{2+} + \text{CO}_2 + h\nu$$

Reaction 7.12

When conjugated to other proteins, such as antibodies or streptavidin, or to small molecules such as biotin, the aequorin conjugates can be used as detection labels in bioluminescence-based assays (BIAs). Adding a calcium salt solution and measuring the amount of light produced during the bioluminescent reaction determines the amount of label present. Because of this calcium-dependent light emission, the aequorin complex has been used as an intracellular Ca^{2+} indicator.

The jellyfish *A. victoria* contains two photoactive proteins. In bioluminescence resonance energy transfer (BRET), a radiationless energy transfer takes place from aequorin, which emits at 469 nm (blue), to the GFP (see Figure 7.116), which then emits at 509 nm, i.e., the green light of the jellyfish (Figure 7.108).

Aequorin can be microinjected into cells or introduced into cells by complementary DNA (cDNA) transfection (see Volume III, Chapter 2 on nature as an engineering guide). An additional tool is provided by the targeting of aequorin to specific compartments of the cell (see Brini et al., 1999)[65] by introducing in the cDNA codes for targeting the protein to specific localities. Similar to GFP, AEQ-cDNA attached to a promoter provides a useful reporter gene.

Scaling in Chemiluminescence and Bioluminescence Earlier, we saw that fluorescence and absorption both scale with path length b, so that both techniques scale badly when miniaturizing the sample path. In the case of fluorescence, the situation is a bit better than for absorption because one can at least turn up the number of incoming photons to increase the signal intensity (see Equations 7.137 and 7.144). In chemiluminescence there is no path length, b, dependence, and intensity, I_{CL}, is only a function of the chemiluminescence efficiency, Φ, and the rate of production of the activated complex (intermediate), c^*:

$$I_{CL} = \phi \frac{dc^*}{dt} \tag{7.148}$$

Luminometers are considerably simpler than fluorescence instrumentation; they need no excitation source—the chemical reaction provides the input energy—and because emitters are usually single wavelength, they also need no wavelength selection device (filter or spectrophotometer). Essentially one only needs a light detector, a sample holder, and a mixing system (e.g., to add the Ca^{2+}; see Reaction 7.12). The latter could be further simplified through the use of a microfluidic system (e.g., see the CD-based fluidic system described in Chapter 6 and Volume III, Chapter 5). Spectral overlap between acceptor and donor excitation is of no concern, and photobleaching by an excitation source is avoided. Chemiluminescent

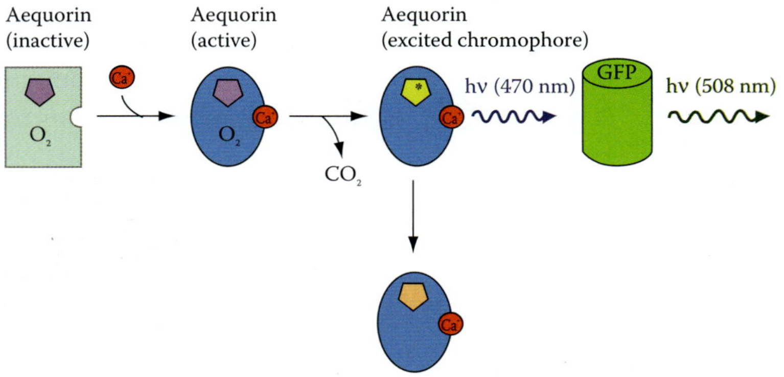

FIGURE 7.116 Bioluminescence resonance energy transfer (BRET) in *Aequorea victoria* from aequorin to GFP.

reactions can be carried out in front of the detector for higher photon collection. Perhaps the most important advantages of chemiluminescence are its high specificity—very few reactions result in light emission—and its very high sensitivity because there is no stray light. Chemiluminescence may be up to 1,000 times more sensitive than fluorometry and 100,000 times more sensitive than absorption spectroscopy. In some instances single molecules have been detected. Bioluminescence is just an extremely efficient subclass of chemiluminescence, and it provides better reliability because the enzyme is naturally evolved to protect the photon emitter. Disadvantages of chemiluminescence include the limited number of donor acceptor pairs available, i.e., Renilla luciferase, bacterial luciferase, firefly luciferase, and aequorin, and the fact that the emission spectra of bioluminescent proteins are broad.

Phosphorescence

Phosphorescence is the emission of photons with a change in spin multiplicity. Because the process violates the spin selection rule, it should not be allowed, but selection rules are broken, and phosphorescence is commonly observed. More specifically, phosphorescence, as we saw above, occurs when the molecule returns to the electronic ground state from the excited triplet state by emission of a photon (i.e., a radiative transition between states of different multiplicities) (see Figures 7.90 and 7.91). Because the transition is spin forbidden and is often slow (i.e., long lifetime, with τ in the 10^{-4}–10-s range), it is also referred to as *delayed luminescence* or *afterglow*. It occurs at much longer wavelength compared with fluorescence. Phosphorescence is susceptible to O_2 and collisions with solvent molecules. Triplet states are rapidly deactivated under these conditions. Because of the long radiative lifetime of phosphorescent transitions, internal conversion and other radiationless transfers of energy compete so effectively with phosphorescence. For many molecules it is only observed at low temperatures or in rigid frozen media (e.g., solvent mixtures at 77 K) and in the absence of oxygen. As illustrated in Figure 7.117, photon absorption occurs between electronic levels with the same spin multiplicity.

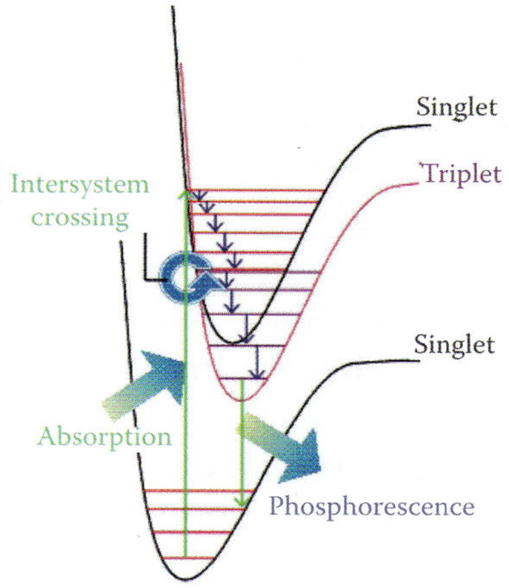

FIGURE 7.117 The phosphorescence process.

The radiationless transition between states of different multiplicity, shown here, is known as *intersystem crossing* (ISC). This may be followed by phosphorescence (weak emission from a long-lived state) to the ground electronic state. Fluorescence is measured much more often than phosphorescence because the lifetime of fluorescence (10^{-8}–10^{-4} s) is shorter than the lifetime of phosphorescence (10^{-4}–10^{2} s), allowing other processes to occur before a molecule has a chance to phosphoresce, thus making fluorescence more likely than phosphorescence.

Example Optoanalytical Devices

Introduction

To be able to compare electrochemical with optical sensors, we explore some examples of optical sensors here, just as we did for electrochemical sensors earlier in this chapter. Common types of optical sensors include fiberoptic, planar waveguide, evanescent wave, interferometric, and surface plasmon resonance (SPR). Under fiberoptic sensors, one further distinguishes between extrinsic and intrinsic sensors and direct and indirect ones. In an extrinsic optic fiber sensor, the fiber is only used as a transmission channel to take light to and from the sensing elements. In an intrinsic sensor, the fiber itself acts as a sensing element; a portion of the optical fiber cladding is removed and replaced with a chemically selective layer. The sensor is then placed directly into the media to be analyzed. Interaction of the analyte with the chemically selective layer creates a change in absorbance, reflectance, fluorescence, or light polarization. A direct fiberoptic sensor measures the intrinsic optical properties of the analyte, and an indirect sensor measures the absorbance or fluorescence of an immobilized indicator dye or label. Planar waveguides are media in which the optical propagation is confined to a plane instead of a cylindrical fiber. Typical materials used for waveguides include plastic $\lambda > 450$ nm, glass $\lambda > 350$ nm, fused silica $\lambda < 350$ nm, and germanium crystals $\lambda > 1{,}000$ nm. The essential feature in interferometric sensing is the modulation of interference behavior by analyte binding. For example, the illumination of transparent thin-film coatings on reflective surfaces produces the so-called *Fabry-Pérot fringes*—alternating high- and low-intensity features—in which the resulting reflection spectrum and binding of an analyte to such film will shift these interference fringes. Evanescent wave sensors use the interaction with the electromagnetic field that extends a very short distance away from the surface of the light-guiding medium. The interaction of the evanescent wave with the sensing layer is used to get information on the analyte binding. It can involve IR (vibrational), UV-visible (electronic) absorption band, or Raman spectroscopic transitions. Surface plasmon resonance operates on the same principle as evanescent wave sensing, except that the planar waveguide is replaced by a metal/dielectric interface. Surface plasmons are electromagnetic waves that propagate between a metal and a dielectric material.

We encountered several of these types of optical sensors already in Chapter 5, where we introduced evanescent waveguide-type and SPR sensors. Additional optical sensors we review here are a fiberoptic-based continuous immunosensor, an interferometric immunosensor, a fiberoptic-based nanobead array for detection and analysis of DNA sequences, and PEBBLEs (**p**robes **e**ncapsulated **b**y **b**iologically **l**ocalized **e**mbedding) nanosensors. PEBBLEs is a general term that describes a family of matrices and nanofabrication techniques used to miniaturize optode technologies, thus filling a niche that lies between micro-optodes and free molecular probes.

Continuous Immunosensors—Optodes

A fiberoptical chemical sensor (FOCS) is often called an *optode* (also optrode), a word derived from optical electrode. An optode measures changes in the optical properties of a coating or membrane caused by the analyte. The sensors generally have a fiber for the light source and a return fiber to a spectrometer. If the sensor uses a phase change, it has a reference fiber and a sensing fiber. If the analyte is present, the optical parameter of the sensing fiber is changed. This creates a phase difference between the light traveling through the two fibers. If intensity is measured, instead of phase difference, the optode has an important advantage over electrochemical sensors in that no reference is needed.

The sensor consists of three main parts: a light source, optode, and detector. The main part of the sensor, the so-called optode, contains an appropriate indicator that changes its optical properties in dependence on the analyte. In basic fiberoptic sensor configurations, the active sensing region of the optical fiber may be located either on the end facet of the fiber (end-fiber configuration), on the side of the fiber (side-fiber configuration), or as an interrupting porous section inserted in the fiber (interrupted-fiber configuration). The end-fiber configuration is commonly applied for absorbance, luminescence, or fluorescence sensors, using the fiber as a light guide to and from the sample. The side-fiber configuration is usually applied for sensing schemes based on evanescent field measurements using the fiber as a waveguide and as the active transducer in sections with removed cladding. The interrupted-fiber configuration incorporates the active sensing region directly in the fiber by introducing a porous section. Thus, direct absorbance measurements are facilitated along with a large surface area. The fiberoptic immunosensor sensor shown in Figure 7.118 is an end-fiber configuration.

A key issue with immunosensors is their seemingly inherent irreversibility. Barnard and Walt.[57] introduced an ingenious way to make an immunoassay continuous by using the controlled release of labeled antibodies to maintain sensitivity. They used fluorescein-labeled antibody (F-Ab) and Texas Red-labeled immunoglobulin G antigen (TR-Ag) in a competitive immunoassay based on Förster

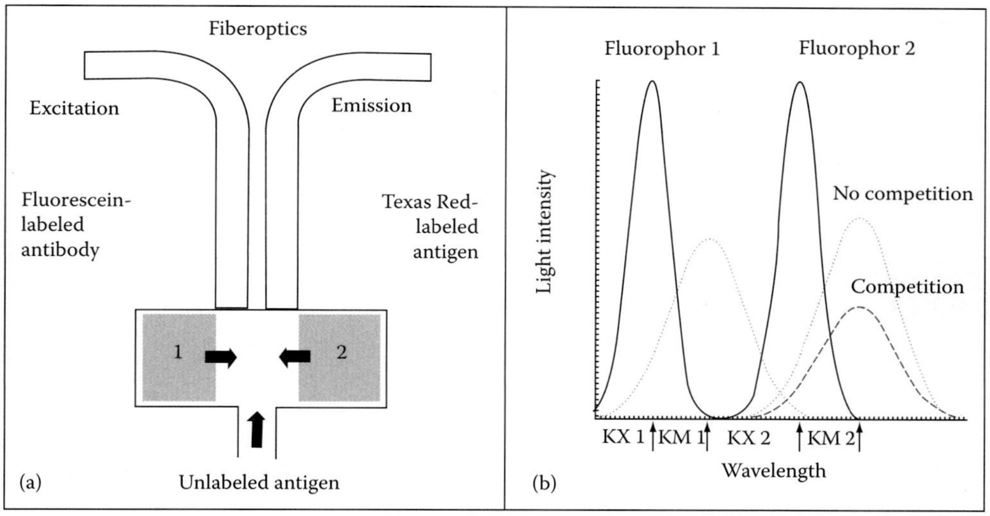

FIGURE 7.118 A new approach to immunosensing. A continuous immunomeasurement is enabled by implementing a polymer-release mechanism. Labeled antibody and antigen are released from polymer reservoirs, and fluorescence is measured by fiberoptics (a). The fluorescence emission from the second fluorophore (Texas Red) is reduced when unlabeled antigen competes with the reaction (b). (From Barnard, S. M., and D. R. Walt. 1991. Chemical sensors based on controlled release polymer systems. *Science* 251:927. With permission.[57])

resonance energy transfer (FRET) (see Figure 7.101). When the antibody (Ab) and antigen (Ag) bind, the two fluorophor labels are very close to one another, allowing nonradiative transfer of energy. When the fluorescein molecule is excited by light, it donates energy to the Texas Red fluorophor, enhancing its fluorescent light emission. To circumvent the irreversibility of the immunoassay, these researchers used a controlled-release delivery system capable of sustaining a constant release of fresh immunochemicals as shown in Figure 7.118. Specifically, they used ethylene-vinyl acetate as the controlled-release delivery system of reagents into the reaction chamber, i.e., the sensing region of the optical fiber. Unlabeled Ag diffused into the reaction chamber and competed with the Ag-TR for the available binding sites on the F-Ab, and bound species diffused out of the reaction chamber while a constant release of reagents maintained a constant concentration of fresh reagents in the reaction chamber. The fluorescence changes induced by the competition reaction were used to calculate the amount of unlabeled antigen in the sample solution. In cases where response time is not very critical but continuous monitoring is of major importance, this technique seems promising. A micromachined version of the sensor, shown in Figure 7.118, seems imminently feasible. This could involve micromachining a reaction chamber, release polymer reservoirs, and a waste dump, all connected with flow channels, and a window in the reaction chamber for the fiberoptics. Plastic injection molding or photoformed glass would be good approaches here.

Interferometric Immunosensor

The interferometric immunosensor the author and collaborators developed at SRI International is based on Langmuir-Blodgett technology.[58] As early as 1936, Irving Langmuir and Katherine Blodgett developed, in a most ingenious way, a method to measure film thickness comprising a small fraction of a wavelength of light. It consisted of building multilayers of barium stearate on a slide using a series of steps in a staircase fashion, as shown in Figure 7.119a. The method they used is known today as the Langmuir-Blodgett deposition method (see Volume II, Figure 8.13 in for a typical setup). Since then, much progress has been made in automating this deposition technology. The difference in thickness of two adjacent steps, as shown in Figure 7.119a, corresponds to a double layer of stearate (approximately 48 Å) and forms an interference-based color gauge that tracks thickness increases. When a film of unknown thickness is coated on the

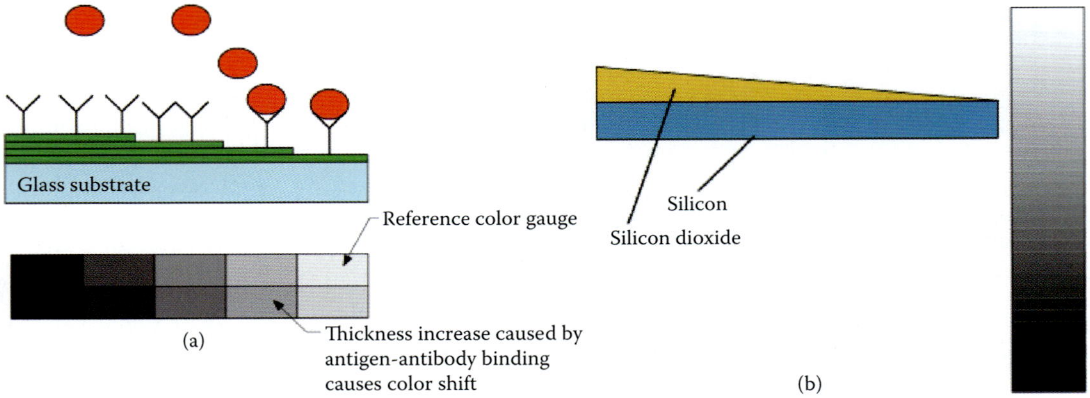

FIGURE 7.119 Optical immunosensors. (a) Staircase made by Langmuir-Blodgett technology (additive technique) or etching (substractive technique); (b) wedge made by pulling sample out of an etchant (substractive technique).

steps and is compared with an uncoated slide with steps, the eye can measure differences of thickness by means of color comparison down to about 10 Å. Greater sensitivity is obtained by illuminating films with sodium light rather than white light. Differences of thickness equal to 5 Å have been made plainly visible to the eye this way. This means that adsorbed layers of molecules or atoms having an optical thickness of 5 Å can be made visible to the eye without the aid of optical apparatus and can be measured with a probable error of about 2.5 Å. Some preliminary experiments were carried out with this simple step technique in the Madou laboratory while he was at SRI International laboratory at SRI International. When albumin was adsorbed on one-half of the staircase slide, the slide then was exposed to antialbumin. A thickness increase was observed via a color shift. What does this mean from the biosensor point of view? For many years, ellipsometers have been used to detect the deposition of antibody onto antigen preadsorbed on a flat reflecting surface. In this way, any molecule with a molecular weight greater than 60,000 (e.g., viruses, antibodies, proteins) could be detected. The ellipsometer is a very involved and expensive instrument (see Volume III, Chapter 6), whereas the system described above can perform the same task in a simple and inexpensive way. However, as pointed out earlier, the Langmuir-Blodgett deposition is a difficult manufacturing proposition. Fortunately, steps can also be built using etching techniques and using inorganic materials with optimum refractive indices for maximum interference effect.[58] Later, we found yet a better way to make interference slides: a substrate pulled at a uniform rate from an etchant solution forms a continuous optical wedge, making for a more elegant and faster manufacturing option than either Langmuir-Blodgett or staircase etching (Figure 7.119b). We believe the wedge approach is an excellent illustration of how micromachining could benefit immunosensor development by making inexpensive and disposable optical transducers possible. Whereas the wedge approach is more appropriate for the detection of larger molecules, the detection of smaller molecules is accomplished more easily with the electrochemical approach.

Bead Arrays for Detection and Analysis of DNA Sequences

Conventional DNA microarrays are manufactured by spotting or synthesizing probes at known locations onto 2D substrates, as we review in Volume II, Chapter 8 on chemical-, photochemical-, and electrochemical-based forming. Above, we discussed 3D liquid arrays (Figure 7.106) as an alternative for these planar gene chips with the idea of conjugating biomolecules to beads that are filled with various combinations of quantum dot (QD) types at various concentrations that can be used as "barcodes." Here we consider yet another option for DNA assaying, one that is already commercialized by Illumina (a San Diego biotech company) and is vying against Affymetrix, the industry pioneer, and other rivals, including Agilent Technologies, GE Healthcare, NimbleGen Systems, ParAllele BioScience, and Perlegen Sciences.

Illumina's BeadArray™ technology is based on the random self-assembly of a bead pool onto a patterned substrate.[59] The patterned substrate may be an etched flat surface (glass or Si) or a bundle of fiberoptic fibers with etched cavities to accommodate one bead per fiber. Illumina's fiber bundle-based chip-making process is illustrated in Figure 7.120.

This elegant design starts by making a pool of hundreds of thousands of tiny glass beads (3-μm diameter), each batch coated with one (among 1500) type of oligonucleotide, DNA fragments that will seek and latch onto a specific sequence of a DNA sample. Each bead has only one type of oligonucleotide attached to it but consists of hundreds of thousands of copies. Next, the batches of beads are mixed to form a mixed "pool" (see Figure 7.120, panel 1). When a bundle of 50,000 optical fibers is dipped into this pool (Figure 7.120, panel 2), a single bead adheres to the end of each fiber (Figure 7.120, panel 3). The beads occupy the wells in a random distribution, and each bead is represented by, on average, about 30 instances within the array. An array of 8 × 12, or 96 fiber bundles (i.e., 96 independent arrays) with 50,000 light-conducting strands per bundle, is constructed to be compatible with standard microtiter plates used to carry out sample preparation and enzymatic steps before exposure of the beads in the array to the sample (Figure 7.120, panel 4). As a part of the array manufacturing process at the company, a decoding process is implemented to map the precise location of a specific bead type on the array.[59]

The decoding strategy involves sequential hybridization of differentially labeled probes to an address site on the array.[59] The differential labeling uses three states: carboxyfluorescein (FAM), labeled green; cyanine 3 (Cy3), labeled red; and not labeled. During any given cycle of the process, a bead is green, red, or blank. Dye-labeled oligonucleotides hybridize to the address sequence of each probe type in the hybridization or "decoding" process (Figure 7.121). Labeled decode oligonucleotides are hybridized to the arrays at high concentrations, which allows for rapid (several minutes) hybridizations, followed by washing to remove nonspecific signal and background. If one assigns a number to each state—0 to blank, 1 to green, and 2 to red—then each cycle of the process generates a trinary digit. If we look at one hypothetical probe, then the first round may be red, which generates 2 for the first digit. The second round is green and generates 1, so the number is now 21. The third round is red, so the number is now 212. Each round just adds a new digit to the

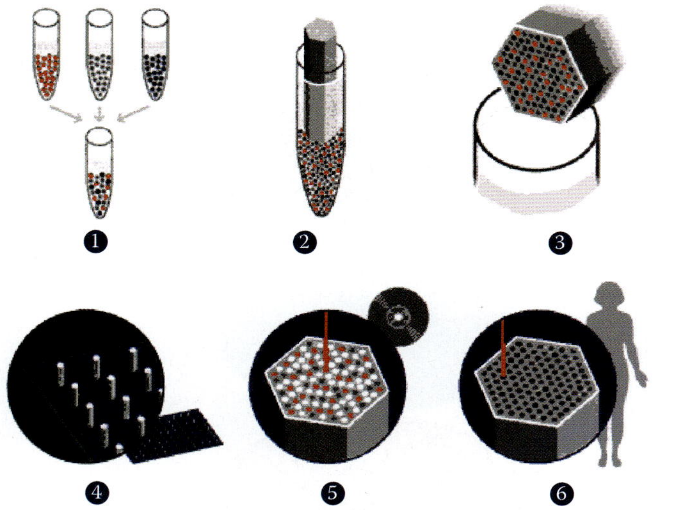

FIGURE 7.120 Illumina's BeadArray technology is based on the random self-assembly of a bead pool on fiber bundles; see text. (From Gunderson, K. L., S. Kruglyak, M. S. Graige, F. Garcia, B. G. Kermani, C. Zhao, D. Che, T. Dickinson, E. Wickham, J. Bierle, et al. 2004. Decoding randomly ordered DNA arrays. *Genome Res* 14:870.[59])

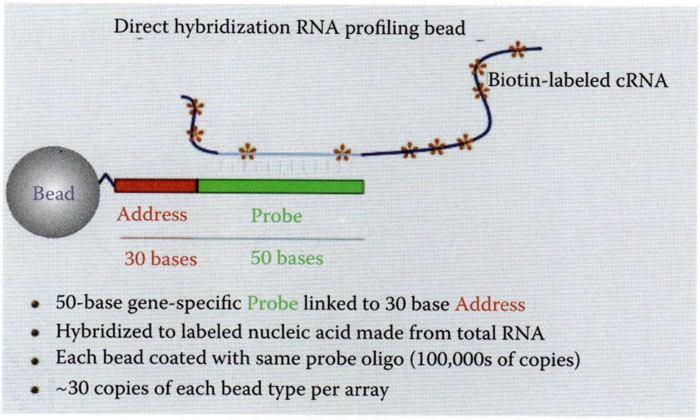

FIGURE 7.121 Gene-specific probes are attached to beads, which are then assembled into arrays. For simplicity, the diagram only shows one oligomer attached to the bead; actual beads have hundreds of thousands of copies attached per bead. The 50-base gene-specific probe is attached to a 30-base address. (From Gunderson, K. L., S. Kruglyak, M. S. Graige, F. Garcia, B. G. Kermani, C. Zhao, D. Che, T. Dickinson, E. Wickham, J. Bierle, et al. 2004. Decoding randomly ordered DNA arrays. *Genome Res* 14:870.[59])

number. This continues until there are sufficient digits to uniquely identify each probe. One digit can uniquely identify three probes (0, 1, or 2). Two digits can uniquely identify nine probes (00, 01, 02, 10, 11, 12, 21, 22, or 23). Three digits can identify 27 probes (we will leave anything further as an exercise for the reader!). Rehybridization continues until there are sufficient data to unambiguously determine the identity of each bead. The physical location of each bead type is thus mapped by using a decoding algorithm[59] that tracks oligonucleotides successfully hybridizing to a specific address. With this compiled information, the physical location of each bead is mapped, and this unique map of all the oligo on the chip is stored on a CD-ROM that is shipped to the customer with the chip. The decoding process also serves as a quality control step for each feature of each array. The address sequences are designed to have no significant homology to genomic sequences within the species queried. Customers put all this to work by dipping the array into a specially treated DNA sample. As the sample wets the beads, each oligo looks for the unique DNA sequence it is built to detect. If the sequence is present, the oligo chemically binds to it and then lights up when a finely focused laser is beamed through the array; if the sequence is absent, the oligo stays dark. The resultant array is compared with the unique map on the CD-ROM to determine which oligonucleotide-bead combination is in each well. The decoded arrays may be used for a number of applications, including gene expression analysis and genotyping.

A high degree of miniaturization can be achieved by exploiting the intrinsic size of the beads and patterned substrate. For example, the density of a randomly assembled 300-nm diameter bead array is more than 40,000 times higher than a typical spotted microarray.[60]

PEBBLE Sensors

With most electrochemical electrodes and optodes, when it comes to penetrating individual live cells, even the introduction of a submicrometer sensor tip causes biological damage. Individual molecular probes (free sensing dyes), on the other hand, are physically small enough to enter biological cells without damage. Unfortunately, they often suffer from chemical interference between the probe and the cellular components. To solve this problem, Kopelman[61] developed PEBBLE (**p**robes **e**ncapsulated **b**y **b**iologically **l**ocalized **e**mbedding) nanosensors. These are nanoscale spherical devices consisting of sensor molecules entrapped in a chemically inert matrix, as shown in Figure 7.122a. PEBBLE is a general term that describes a family of matrices and nanofabrication techniques used to

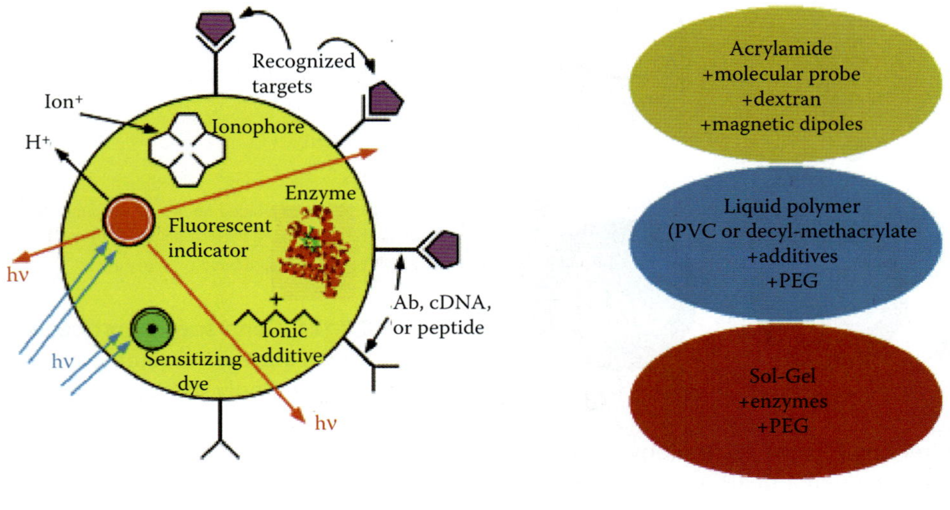

(a) (b)

FIGURE 7.122 (a) PEBBLE (**p**robes **e**ncapsulated **b**y **b**iologically **l**ocalized **e**mbedding) nanosensors. (b) Three different matrices are commonly used depending on the sensor materials. (http://www.umich.edu/~koplab/index.htm.)

miniaturize optode technologies, thus filling a niche that lies between micro-optodes and free molecular probes (naked indicator dye molecules). PEBBLEs are designed specifically for minimally invasive analyte monitoring in viable, single cells with applications for real-time analysis of drug, toxin, and environmental effects on cell function. The main classes of PEBBLE nanosensors are based on matrices of polyacrylamide hydrogel, sol gel silica, and cross-linked decyl methacrylate (Figure 7.122b).[62] These matrices have been used to fabricate PEBBLE nanosensors for H^+, Ca^{2+}, K^+, Na^+, Mg^{2+}, Zn^{2+}, Cl^-, O_2, NO, glucose, etc. that range from 30–600 nm in size.

The protective coating of a PEBBLE eliminates interferences, such as protein binding and/or membrane/organelle sequestration, which might alter dye response. Conversely, the nanosensor matrix also provides protection for the cellular contents, enabling dyes that might usually be toxic to cells to be used for intracellular sensing. In addition, the inclusion of reference dyes allows quantitative, ratiometric fluorescence techniques to be used. A number of delivery methods of PEBBLE nanosensors to single cells have been tried, including gene gun, picoinjection, liposomal delivery, and sequestration (phagocytosis and pinocytosis) into macrophages. The gene gun can best be thought of as a shotgun for nanoparticles. The gene gun can be used to deliver one to thousands of PEBBLEs per cell into a large number of cells very quickly (http://www.umich.edu/~koplab/research2/analytical/NanoScaleAnalysis.html). Cell viability is excellent: 98% viability compared with control cells. Commercially available liposomes can also be used to deliver PEBBLEs to cells. In this case, liposomes are prepared in a solution of PEBBLEs and then placed in the cell culture, where they fuse with the cell membranes and empty their contents (the PEBBLE-containing solution) into the cell. Dourado and Kopelman[63] detailed the specific advantages of these types of nanoscale dimension optical sensors. They argue that the absolute detection limit decreases with the PEBBLE radius r as r^3 (good!), and the response time is reduced as r^2 (good!). The signal-to-noise (S/N) ratio, on the other hand, decreases with r (bad!), but at least not with r^3 (luckily!) under standard working conditions. Other features that improve as sensors get smaller include required sample volume, invasiveness, spatial resolution, dissipation of heat in sensor and/or sample, toxicity, and materials cost. Features that typically worsen include fluorophore leaching and photodamage to sensor and/or sample.

Comparison of Optical versus Electrochemical Sensors

In this chapter, we gave examples of both optical and electrochemical immunosensors. In Table 7.21, pros and cons of both approaches are summarized. Several features make electrochemical techniques attractive in immunosensing (especially of small molecules). Because accuracy is achieved

Table 7.21 Comparison of Optical and Electrochemical Immunosensors

	Optical	Electrochemical
Instrument cost	Often expensive	Always inexpensive
Sensor cost	Fair	Low
Turbid solutions	Problematic	Usually no problem
Dynamic range	Limited	Very wide
Selectivity	Good	Fair
Sensitivity	Very good	Good
Simplicity of method	Often simple	Simple
Color	Sometimes problematic	No problem
Species size	Better for larger molecules	Better for smaller molecules
Electromagnetic interference (EMI)	No	Yes
Resistance to radiation and corrosion	Yes	No
Cross-talk	No	Yes
Ambient light	Problem (must be modulated)	No problem
Response curve	Sigmoidal	Nernstian (potentiometric) or linear

independent of color and turbidity of the solution, rapid measurements are often obtained with less tedious sample preparation. In principle, analysis of a whole range of analytes in matrices as complex as whole blood can be done. Electrochemical probes are easy to use as analytical tools for measuring ions and neutral molecules, typically in a range of 1 M to 1 μM. Examples of even wider dynamic ranges also exist. Coupled with chemical amplification (enzyme or liposome), lower detection limits of picomoles per liter have been attained. In addition, the sensor is inexpensive. Using optical techniques, one partner of the immunochemical complementary pair is typically immobilized onto a surface with some optical characteristics modified when binding takes place. In optical sensing, the signal is not subject to electrical interference, no reference electrode is required, and simultaneous measurement of two or more analytes is possible by measuring at two or more wavelengths. Whereas classical optical assays rely mostly on absorption spectroscopy, luminescence in the form of fluorescence or phosphorescence (depending on the excited state being single or triplet) is more important for immunosensors. Electrochemical detection tends to be better for small molecules, whereas optical techniques are better suited for large molecule detection.

From a more global perspective, comparing electrochemical with optical methods, we start by pointing out again that electrochemical methods are not affected by color and turbidity of the solution, are very low cost, and have the potential to do elemental speciation (e.g., Fe^{2+} vs. Fe^{3+}).

Acknowledgments

Special thanks to Dr. Mike Tierney, Dr. Sapna Deo, Dr. Jim Zoval, Robin Gorkin, and Jonathan Siegrist.

Questions

7.1: Show how the Butler-Volmer expression leads to an ohmic resistance at low over-potentials.
Thanks to Dhondup Pemba, UC Irvine

7.2: What is a Jablonski diagram?
Thanks to Elisa Morganti, UC Irvine

7.3: Calculate the decay time or ring-down time in cavity ring-down spectroscopy (CRDS).
Thanks to Arezoo Motavalizadeh Ardekani, UC Irvine

7.4: How does the electric double layer (EDL) arise and what are the four theoretical models that attempt to describe the structure of the EDL?
Thanks to Derek So, UC Irvine

7.5: Are ISFETs potentiometric or amperometric devices? Explain the difference between the two types of devices. What does MOSFET stand for and what are the differences between the electrical circuits of an ISFET and a MOSFET?
Thanks to Han Wu, UC Irvine

7.6: In infrared absorption spectroscopy one measures the absorption of the infrared radiation at specific wavelengths that match the energy of vibrational transitions of a molecule. Raman spectroscopy is a complementary technique also based on vibrational transitions. In what ways does Raman spectroscopy differ from infrared (IR) spectroscopy and what advantages does it offer over IR spectroscopy?
Thanks to Kiavash Koko, UC Irvine

7.7: Explain how the Fermi level of a metal electrode regulates electron transfer from the metal electrode to the redox molecules in solution
Thanks to Neil Olson, UC Irvine

7.8: What are two typical optical absorption spectrometers? Explain how they work. Compare these two types.
Thanks to Omid Rohani, UC Irvine

7.9: How do absorbance spectroscopy and fluorescence differ experimentally, and consequently what benefits does fluorescence have to offer?
Thanks to Christopher Clawson, UC Irvine

7.10: Match the following terms with the appropriate definitions:

(1) Fluorophore	(a) Crystals composed of materials from periodic groups II-VI, III-V, or IV-VI and can be conjugated to biomolecules

(2) Stokes shift	(b) Method used to measure the ability of a molecule to move around in a cell over time
(3) FRAP	(c) Intense fluorescent probe used in many biological studies
(4) Quantum dots	(d) Difference in wavelength between fluorescence excitation maximum and the fluorescence emission maximum
(5) Green Fluorescent Protein	(e) Molecule that can be excited by light to emit fluorescence

Thanks to Youssef Farhat, UC Irvine

7.11: Given the concentration of Cu^{2+} in a first solution, design an amperometric sensor to measure the concentration of any other unknown Cu^{2+} solution.
Thanks to Chuan Zhang, UC Irvine

7.12: What is the most commonly used type of spectroscopic technique? How does one representative technique of this type work? Does it scale well in the micro domain compared to other spectroscopic techniques?
Thanks to Irena Aleksieva, UC Irvine

7.13: What are the advantages of using ultra-micro electrodes (UMEs)?
Thanks to Megan Jacks, UC Irvine

7.14: The Ag/AgCl reference electrode is an electrode of the first or second kind? How is the potential of this electrode fixed constant?
Thanks to Mary Amasia, UC Irvine

7.15: Derive absorbance using $dP = -Pc\beta dx$, where c = molar concentration, P = light intensity, β = probability of a photon being absorbed.
Thanks to Kim Bui, UC Irvine

7.16: What does it mean that the standard reduction potential is an intensive property?
Thanks to Kevin Ko, UC Irvine

7.17: Why is the miniaturization of a reference electrode such a challenge?
Thanks to Michelle Tu, UC Irvine

7.18: Given the following galvanic cell and half reactions:
$Au^{3+} + 3e^- \rightarrow Au$ $E^0 = 1.50$
$Tl^+ + e^- \rightarrow Tl$ $E^0 = -0.34$
Determine the following:
(a) Overall cell reaction and E^0_{cell}
(b) $\Delta G°$ for the cell reaction at 25°C
(c) Calculate E_{cell} at 25°C when $[Au^{3+}] = 1.0 \times 10^{-2}$ M and $[Tl^+] = 1.0 \times 10^{-4}$
Thanks to Andrew Sumarto, UC Irvine

7.19: There are two carbon electrodes arranged in an electrochemical cell. On one hydrogen is evolving and on the other Ag is being deposited. 1 h after passing current, 95 cm^3 H_2 (g) (19°C, 99.19 KPa) is collected on the hydrogen electrode, while 0.8368g Ag is deposited on the Ag electrode. Use the given information to calculate the current through the circuit.
Thanks to Xi Wang, UC Irvine

7.20: Match each term with the proper definition:
Thanks to Muhammad Uneib, UC Irvine

7.21: What's the ratio of the van der Waals forces between two atoms at a distance of 1 nm and the same atoms at a distance of 3 nm?
Thanks to Elaheh Shekaramiz, UC Irvine

7.22: Under what conditions do oxide semiconductors or nonoxide semiconductors with a native oxide fail to function as pH sensors?
Thanks to Geethika Alapati, UC Irvine

7.23: Explain the differences between electrodes of the first kind and electrodes of the second kind.
Thanks to Nelson Jen, UC Irvine

7.24: Draw the equivalent electrical circuit of a metal/electrolyte interface with the electrode at a potential so that a redox reaction occurs.

7.25: Calculate the potential of a battery with a Zn bar in a 0.5 M Zn^{2+} solution and Cu bar in a 2 M Cu^{2+} solution.

7.26: Show in a cyclic voltammogram the transition from kinetic control to diffusion control and why it really happens.

7.27: Derive how the capacitive charging of a metal electrode depends on the potential sweep rate.

7.28: Calculate the potential difference between two Pt electrodes immersed in two different solutions connected via a salt bridge. The first solution has 1 M KCl and 10^{-1} M Fe^{2+} and the second has 1 M KCl and 10^{-5} M Fe^{3+}.

7.29: What are the requirements of a good reference electrode? What is a double junction reference electrode, and in what cases is it useful?

7.30: Define sensitivity, detection limit, and selectivity in an analytical method.

7.31: What is the effect of the following parameters on the quantum yield of fluorescence?
(a) Raising the temperature.
(b) Adding a heavy atom to the solvent.
(c) Lowering solvent viscosity.
(d) What is the physical process that makes phosphorescence rare?

7.32: In the last few years, applications of quantum dots have become increasingly popular among analytical chemists. Quantum dots are now commercially available for a variety of applications, and many examples of their uses can be found in the literature. Briefly describe what quantum dots are and describe their applications.

7.33: Briefly state the difference between fluorescence and phosphorescence.

7.34: In potentiometry:
(a) What is a "junction potential"?
(b) What does the acronym ISE stand for?
(c) What are the components of ISEs?
(d) Give an example of an ISE.

7.35: How does Raman spectroscopy work?
Thanks to Elisa Morganti, UC Irvine

7.36: In a fluorescence measurement setup, which emitted photons are monitored? Why?
Thanks to Omid Rohani, UC Irvine

7.37: Define photobleaching. Identify an analytical technique that uses photobleaching and describe what kind of information it can provide about the membrane of a cell.
Thanks to Kiavash Koko, UC Irvine

7.38: What problem of conventional multipass approaches does cavity ring-down spectroscopy (CRDS) overcome, and how does it do this?
Thanks to Christopher Clawson, UC Irvine

7.39: What are fluorescence, chemiluminescence, and bioluminescence spectroscopy? Give an example of each technique. Why are chemiluminescence and bioluminescence spectroscopy preferred over fluorescence spectroscopy?
Thanks to Irena Aleksieva, UC Irvine

References

1. Rigby, M., E. B. Smith, W. A. Wakeham, and G. C. Maitland. 1986. *The forces between molecules.* Oxford, UK: Oxford University Press.
2. Stone, A. J. 1997. *The theory of intermolecular forces.* Oxford, UK: Oxford University Press.
3. Barron, A. E., and H. Blanch. 1995. DNA separations by slab gel and capillary electrophoresis: theory and practice. *Separ Purif Meth* 24:1–118.
4. Ohshima, H., T. W. Healy, and L. R. White. 1982. Accurate analytic expressions for the surface charge density/surface potential relationship and double-layer potential distribution for a spherical colloidal particle. *J Colloid Interface Sci* 90:17–26.
5. Casimir, H. B. G., and D. Polder. 1948. The influence of retardation on the London-van der Waals forces. *Phys Rev* 73:360–72.
6. Dzyaloshinskii, I. E., E. M. Lifshitz, and L. P. Pitaevskii. 1961. General theory of van der Waals' forces. *Soviet Physics Uspekhi* 4:153–76.
7. Sukenik, C. I., M. G. Boshier, D. Cho, V. Sandoghdar, and E. A. Hinds. 1993. Measurement of the Casimir-Polder force. *Phys Rev Lett* 70:560.
8. Yao, S., M. Wang, and M. Madou. 2001. A pH electrode based on melt-oxidized iridium oxide. *J Electrochem Soc* 148:H29–H36.
9. Morrison, S. R. 1980. *Electrochemistry at semiconductor and oxidixed metal electrodes.* New York and London: Plenum Press.
10. Marcus, R. A., and N. Sutin. 1985. Electron transfers in chemistry and biology. *Biochim Biophys Acta* 811:265–322.
11. Degani, Y., and A. Heller. 1989. Electrical communication between redox centers of glucose oxidase and electrodes via electrostatically and covalently bound redox polymers. *J Am Chem Soc* 111:2357–58.
12. Heller, A. 1990. Electrical wiring of redox enzymes. *Acc Chem Res* 23:128–34.
13. Gregg, B. A., and A. Heller. 1991. Redox polymer films containing enzymes. 2. Glucose oxidase containing enzyme elctrodes. *J Phys Chem* 95:5976–80.
14. Heller, A. 1992. Electrical connection of enzyme redox centers to electrodes. *J Phys Chem* 96:3579–87.
15. Mao, F., N. Mano, and A. Heller. 2003. Long tethers binding redox centers to polymer backbones enhance electron transport in enzyme "wiring" hydrogeis. *J Am Chem Soc* 125:4951–57.
16. Madou, M. J., and S. R. Morrison. 1989. *Chemical sensing with solid state devices.* New York: Academic Press.

17. Camman, K. 1979. *Working with ion-selective electrodes.* New York: Springer-Verlag.
18. Fleischmann, M., S. Pons, D. R. Rolison, and P. P. Schmidt. 1987. *Utah conference on ultramicroelectrodes.*
19. Lin, Y. H., W. Yantasee, F. Lu, J. Wang, M. Musameh, Y. Tu, and Z. F. Ren. 2004. *Dekker encyclopedia of nanoscience and nanotechnology.* Eds. J. A. Schwartz, C. Contescu, and K. Putyera. New York: Taylor & Francis.
20. Madou, M. J., and S. R. Morrison. 1991. *Sixth International Conference on Solid-State Sensors and Actuators* (Transducers '91). San Francisco.
21. Anderson, J. L., K. K. Whiten, J. D. Brewster, T.-Y. Ou, and W. K. Nonidez. 1985. Microarray electrochemical flow detectors at high applied potentials and liquid chromatography with electrochemical detection of carbamate pesticides in river water. *Anal Chem* 57:1366–73.
22. Reay, R. J., R. Dadoo, C. W. Storment, R. N. Zare, and G. T. A. Kovacs. 1994. *Technical digest: 1994 solid-state sensor and actuator workshop.* Hilton Head Island, SC.
23. van der Putten, A. M. T., and J. W. G. de Bakker. 1993. Geometrical effects in the electroless metallization of fine metal patterns. *J Electrochem Soc* 140:2221–28.
24. Masuko, N., T. Osaka, and Y. Ito, eds. 1996. *Electrochemical technology: innovation and new developments.* Tokyo and Amsterdam: Kodansha and Gordon and Breach Science Publishers.
25. Dukovic, J. O. 1993. Feature-scale simulation of resist patterned electrodeposition. *IBM J Res Dev* 37:125–40.
26. Leyendecker, K., W. Bacher, W. Stark, and A. Thommes. 1994. New microelectrodes for the investigation of the electroforming of LIGA microstructures. *Electrochim Acta* 39:1139–43.
27. Bergveld, P. 1970. Development of an ion-sensitive solid-state device for neurophysiological measurements. *IEEE Trans Biomed Eng* 17:70–71.
28. Matsuo, T., and H. Nakajima. 1984. Characteristics of reference electrodes using a polymer gate, ISFET. *Sensors Actuators* 5:293.
29. Lauks, I., J. Van der Spiegel, W. Sansen, and M. Steyaert. 1985. *Third International Conference on Solid-State Sensors and Actuators (Transducers '85).* Philadelphia.
30. Buerk, D. G. 1993. *Biosensors theory and applications.* Lancaster, PA: Technomic Publishing.
31. Madou, M. J., and J. Joseph. 1993. Immunosensors with commercial potential. *Immunomethods* 3:134–52.
32. Fresenius, W., and I. Luderwald. 2000. *Fresenius' journal of analytical chemistry.* New York: Springer.
33. Aizawa, M., S. Kato, and S. Suzuki. 1977. Immunoresponsive membrane. I. Membrane potential change associated with an immunochemical reaction between membrane-bound antigen and free anti-body. *J Membr Sci* 2:125–32.
34. Janata, J. 1975. An immunoelectrode. *J Am Chem Soc* 97:2915–16.
35. Janata, J., and G. F. Blackburn. 1984. Immunochemical potentiometric sensors. *Ann N Y Acad Sci* 428:286–92.
36. Collins, S., and J. Janata. 1982. A critical evaluation of the mechanism of potential response of antigen polymer membranes to corresponding antiserum. *Anal Chim Acta* 136:93–99.
37. Anzai, J.-I., and T. Osa. 1990. Langmuir-Blodgett membranes in chemical sensor applications. *Selective Electrode Rev* 12:3–34.
38. Joseph, J., A. Jina, and M. Madou. 1991. *Annual meeting of the AIChE.* Anaheim, CA.
39. Fleischmann, M., P. J. Hendra, and A. J. McQuillan. 1974. Raman spectra of pyridine adsorbed at a silver electrode. *Chem Phys Lett* 26:163–66.
40. Jeanmaire, D. L., and R. P. V. Duyne. 1977. Surface raman spectroelectrochemistry. Part I. Heterocyclic, aromatic and aliphatic amines adsorbed on the anodize silver electrode. *J Electroanal Chem* 84:1–20.
41. O'Keefe, A., and D. A. G. Deacon. 1988. Cavity ring-down optical spectrometer for absorption measurements using pulsed laser sources. *Rev Sci Instrum* 59:2544–51.
42. Paul, J. B., and R. J. Saykally. 1995. Cavity ringdown laser absorption spectroscopy of the jet-cooled aluminum dimer. *Chem Phys Lett* 242:395–400.
43. Engeln, R., G. Berden, R. Peeters, and G. Meijer. 1998. Cavity enhanced absorption and cavity enhanced magnetic rotation spectroscopy. *Rev Sci Instrum* 69:3763–69.
44. Pipino, A. C. R. 1999. Ultra-sensitive surface spectroscopy with a miniature optical resonator. *Phys Rev Lett* 83: 3093.
45. Pipino, A. C. R. 2000. Monolithic, folded resonator for evansecent wave cavity ring-down spectroscopy. *Appl Opt* March:20.
46. Chan, W. C. W., D. J. Maxwell, X. Gao, R. E. Bailey, M. Han, and S. Nie. 2002. Luminescent quantum dots for multiplexed biological detection. *Curr Opin Biotechnol* 13:40–46.
47. Chan, W. C. W., and N. Sie. 1998. Quantum dot bioconjugates for ultrasensitive detection. *Science* 281:2016–18.
48. Murray, C. B., D. J. Norris, and M. G. Bawendi. 1993. Synthesis and characterization of nearly monodisperse CdE(E=S,Se,Te) semiconductor nanocrystallites. *J Am Chem Soc* 115:8706–15.
49. O'Neil, M., J. Marohn, and G. McLendon. 1990. Dynamics of electron-hole pair recombination in semiconductor clusters. *J Phys Chem* 94:4356–63.
50. Gerion, D., F. Pinaud, S. C. Williams, W. J. Parak, D. Zanchet, S. Weiss, and A. P. Alivisatos. 2001. Synthesis and properties of biocompatible water-soluble silica-coated CdSe/ZnS semiconductor quantum dots. *J Phys Chem B* 105:8861–71.
51. Han, M., X. Gao, J. Z. Su, and S. M. Nie. 2001. Quantum-dot-tagged microbeads for mltiplexed optical coding of biomolecules. *Nat Biotechnol* 19:631–35.
52. Nicewarner-Pena, S. R., R. G. Freeman, B. D. Reiss, L. He, D. J. Pena, I. D. Walton, R. Cromer, C. D. Keating, and M. J. Natan. 2001. Submicrometer metallic barcodes. *Science* 294:137–41.
53. Miyawaki, A., J. Liopis, R. Heim, J. M. McCaffery, J. A. Adams, M. Ikura, and R. Y. Tsien. 1997. Fluorescent indicators for Ca2+ based on green fluorescent proteins and calmodulin. *Nature* 388:882–87.
54. Chalfie, M., Y. Tu, G. Euskirchen, W. W. Ward, and D. C. Prasher. 1994. Green fluorescent protein as a marker for gene expression. *Science* 263:802–05.
55. Dickson, R. M., A. B. Cubitt, R. Y. Tsien, and W. E. Moerner. 1997. On/off blinking and switching behavior of single green fluorescent proteins. *Nature* 388:355.
56. Garcia-Parajo, M. F., G. M. J. Segers-Nolton, J. A. Veerman, J. Greve, and N. F. van Hulst. 2000. Real-time light-driven dynamics of the fluorescence emission in single green fluorescent protein. *Biophys J* 78:384A.
57. Barnard, S. M., and D. R. Walt. 1991. Chemical sensors based on controlled release polymer systems. *Science* 251:927–29.
58. Joseph, J., and K. Itoh. 1992. Etching interference slides. US Patent 5,169,599.

59. Gunderson, K. L., S. Kruglyak, M. S. Graige, F. Garcia, B. G. Kermani, C. Zhao, D. Che, T. Dickinson, E. Wickham, J. Bierle, et al. 2004. Decoding randomly ordered DNA arrays. *Genome Res* 14:870–77.
60. Michael, K. L., L. C. Taylor, S. L. Schultz, and D. R. Walt. 1998. Randomly ordered addressable high-density optical sensor arrays. *Anal Chem* 70:1242–48.
61. Xu, H., J. W. Aylott, and R. Kopelman. 2002. Fluorescent nano-PEBBLE sensors for the real-time measurement of glucose inside living cells. *Analyst* 127:1471–77.
62. Barker, S. L. R., H. A. Clark, and R. Kopelman. 2002. *Biomedical diagnostics science and technology.* Eds. W. T. Law, N. Akmall, and A. M. Usmani. New York: Marcel Dekker.
63. Kopelman, R., and S. Dourado. 1996. *Chemical, biochemical, and environmental fiber sensors VIII*. Ed. R. A. Lieberman. Bellingham, WA: SPIE-International Society for Optical Engineering.
64. Bergström, L. 1997. *Adv Colloid Interface Sci* 70:125.
65. Brini, M., P. Pinton, T. Pozzan, and R. Rizzuto. 1999. Targeted recombinant Aequorins: Tools for monitoring [Ca21] in the various compartments of a living cell. *Microscopy Research and Technique* 46:380–89.

Index

A

B

C

F

G

H

I

K

L

M

N

O

Q

R

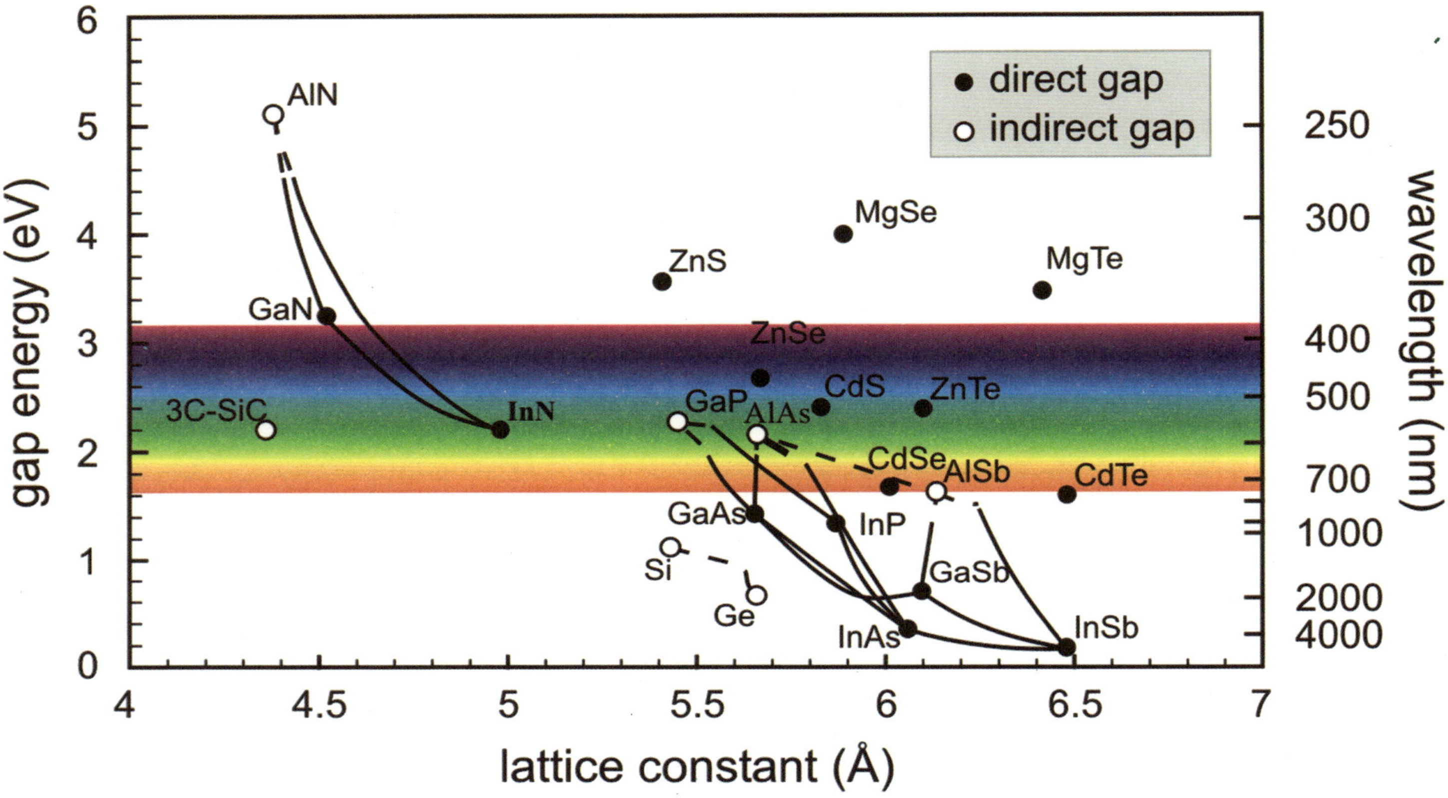

direct gap
indirect gap
gap energy (eV)
wavelength (nm)
lattice constant (Å)
0
1
2
3
4
5
6
4
4.5
5
5.5
6
6.5
7
250
300
400
500
700
1000
2000
4000
AlN
GaN
InN
3C-SiC
ZnS
MgSe
MgTe
ZnSe
CdS
ZnTe
GaP
AlAs
CdSe
AlSb
CdTe
GaAs
InP
Si
Ge
GaSb
InAs
InSb